Spanish Relativity Meeting 2010

(ERE 2010)

Gravity as a Crossroad in Physics

Journal of Physics: Conference Series Volume 314

Granada, Spain
6-10 September 2010

ISBN: 978-1-61839-286-2
ISSN: 1742-6588

Printed from e-media with permission by:

Curran Associates, Inc.
57 Morehouse Lane
Red Hook, NY 12571

Some format issues inherent in the e-media version may also appear in this print version.

Printed by Curran Associates, Inc. (2011)

Institute of Physics
Dirac House, Temple Back
Bristol BS1 6BE UK

Phone: 44 1 17 929 7481
Fax: 44 1 17 920 0979

techtracking@iop.org

Additional copies of this publication are available from:

Curran Associates, Inc.
57 Morehouse Lane
Red Hook, NY 12571 USA
Phone: 845-758-0400
Fax: 845-758-2634
Email: curran@proceedings.com
Web: www.proceedings.com

TABLE OF CONTENTS

PLENART SECTION

SESSION DEDICATED TO THE MEMORY OF S BRIAN EDGAR

SESSION IA

SESSION IB

SESSION IIIA

SESSION IIIB

SESSION IIIC

SESSION IVB

SESSION IVC

SESSION VA

SESSION VB

SESSION VC

POSTERS

PREFACE

The 2010 edition of the Spanish Relativity Meeting (ERE2010) took place in Granada from 6–10 September 2010, and was hosted by the Instituto de Astrofísica de Andalucía (IAA - CSIC). This event represented the 34th edition of Encuentros Relativistas Españoles (ERE), an international conference devoted to relativity and gravitation and organized every year by one of the Spanish groups working in this area. The particular scientific flavour of the 2010 edition was captured by the subtitle of the conference, 'Gravity as a Crossroad in Physics'.

Our underlying rationale was to present gravitational physics as a scientific 'locus' for the interaction between (separate) communities in physics. It is a remarkable property of gravity that its specific problems provide a framework that calls for the interchange of ideas, concepts and methodologies from very different communities. In this edition we aimed to reflect this interdisciplinary perspective in the scientific programme. Each day during the week was devoted to a particular 'dialogue' between two communities who share some of their ultimate goals, but differ in their conceptual background, methodology or technical approach. These 'dialogues' were envisaged as opportunities to compare alternative viewpoints, maintaining a focus on their complementary nature. This led to the organization of the week as follows:

Day 1: Fundamental vs Effective Approaches in Theoretical Gravity
This day compared approaches to gravity that differ conceptually in their understanding of the nature of the basic physical degrees of freedom of the theory, namely confronting viewpoints supporting the fundamental status of such degrees of freedom with other research programs favouring some emergence mechanism. Gravitational analogues were also discussed on this day.

Day 2: Geometric vs Quantum Field/String Theory Approaches to Quantum Gravity
This day was focused on quantum gravity. A particular emphasis was placed on the comparison between geometric approaches to the quantization of general relativity (e.g. loop quantum gravity in the context of the canonical program) and approaches leaning on or evolving from a (quantum) field theory treatment of gravity (e.g. string/M-theory).

Day 3: Theoretical Cosmology vs Physical Cosmology
This day addressed the current challenges in cosmology from a double perspective. On the one hand, offering an analysis of the large scale picture of the universe emerging from the accumulated body of observational data and, on the other hand, assessing the theoretical attempts to explain such a picture putting a special emphasis on the role of gravity.

Day 4: Relativity vs Astrophysics
This day was focused on astrophysical problems where general relativity plays a fundamental role. Challenges and difficulties encountered by relativists modelling specific astrophysical scenarios were disucssed as well as the problems found by

astrophysicists needing general relativity as a key conceptual ingredient. Particular emphasis was placed on gravitational waves and compact objects.

Day 5: Mathematical Relativity vs Numerical Relativity
This day discussed fundamental problems in general relativity, and more generally in gravity physics, where a close collaboration between relativists in the geometry/analysis community on the one hand, and relativists in the numerical community on the other hand, can prove to be particularly successful and insightful.

The contributions in this volume have been organized in two blocks, corresponding to plenary and parallel sessions during the conference. In both cases we have kept the chronological order of the presented talks. The only exception to this rule is the parallel session dedicated to the memory of the late S Brian Edgar, labeled as IV.A during the conference, which we have placed immediately after the plenary session contributions.

The result of the 'dialogue experience' at the conference was extremely satisfactory and gratifying. Scientific sessions were thrilled by tantalizing and inspiring discussions, often continued in long walks around the Cármenes of the old city. In this spirit, we wish to thank all of the participants of the ERE meeting for their enthusiasm and especially the contributors to these proceedings for their synthesis effort.

Granada, 25 July 2011

Víctor Aldaya, Carlos Barceló and José Luis Jaramillo

Dates

Issue 1 (2011)

Journal of Physics: Conference Series **314** (2011) 012001 doi:10.1088/1742-6596/314/1/012001

What can Classical Gravity tell us about Quantum Structure of Spacetime?

T. Padmanabhan

IUCAA, Pune University Campus, Ganeshkhind, Pune 411007, INDIA.

E-mail: paddy@iucaa.ernet.in

Abstract. Several features of classical gravity, combined with the existence of Davies-Unruh temperature of horizons, support the following paradigm: Gravitational field equations in a wide class of theories, including Einstein's theory, should be viewed as describing the thermodynamic limit of the statistical mechanics of (as yet unknown) atoms of spacetime. I present the conceptual evidence for this emergent paradigm and discuss several facets of this approach.

1. Introduction

Classical gravity has no obvious $\hbar$ dependence while quantum theory has and hence — if we take the question in the title literally — classical gravity cannot tell us anything about the quantum structure of spacetime! But there is one effect, viz., the thermodynamics of spacetime horizons [1] which brings together the principles of quantum theory and gravity. This will be the main anchor in the attempt to learn about the quantum structure of spacetime using 'classical' gravity.

The entire paradigm [2] can be summarized in one sentence: Gravity is an emergent phenomenon like gas dynamics or elasticity with the gravitational field equations having the same status as, say, the equations of fluid dynamics/elasticity. Historically, this paradigm originated with Sakharov [3] and were implemented in different ways by Jacobson [4], Volovik [5], Bei-Lok Hu [6] and others. I will now elaborate on this theme drawing mainly from the work I was involved in.

2. Can such a top-down view help?

This programme is a "top-down" approach to quantum spacetime in the sense of zooming in from the top to smaller and smaller spatial scales. (Some people use the word "top-down" to mean exactly the opposite; I will use "top-down" the way I have defined it.) One may wonder whether such an approach will be really useful because one is attempting to determine the features of the microscopic theory from knowing its properties at the macroscopic scales. To reassure you, let me describe at least two other examples in which the deeper theory leaves a signature on the 'top layer'.

(i) *Boltzmann's conjecture of atoms*: Classical thermodynamics of a gas/fluid uses variables like density, pressure etc. in the continuum description. But the fact that such a fluid can store and exchange heat energy *cannot* be understood within the continuum theory. Boltzmann had the insight to suggest that thermal phenomena *demand* the existence of microscopic degrees of

Journal of Physics: Conference Series **314** (2011) 012001 doi:10.1088/1742-6596/314/1/012001

freedom in matter. In fact, the law of equipartition, expressed as $E/[(1/2)k_BT] = N$ relates two thermodynamic variables E, T (which are well-defined for a continuum fluid) to N, the number density of microscopic degrees of freedom, which cannot be interpreted in the continuum limit. The Avogadro's number, closely related to N, was determined even before we understood what exactly it counts and without any direct evidence for the molecular structure of matter. This is splendid example of our being able to say something about microscopic structure from the features of macroscopic theory.

(ii) *Equality of inertial and gravitational masses*: A more dramatic example is provided by Einstein's use of principle of equivalence which, of course, was known for centuries. Einstein realized that $m_i = m_g$ is not a trivial algebraic accident which should be taken for granted, as others before him have done, but requires an explanation. This led him to the description of gravity in terms of the geometry of spacetime. The relation $m_i = m_g$ was a signature of the deeper theory discernible in the approximate [top-layer] description.

These two examples show that for a top down approach to be useful, you need to ask the right questions! One way is to pick up features of the theory that are usually taken for granted ('algebraic accidents') — or not even noticed — and demand deeper explanations for them. This is the procedure I will follow in this programme to probe the quantum structure of spacetime from known aspects of classical gravity.

3. The conceptual road map

In this approach, it is necessary to make a clear distinction between quantum description of spacetime structure and a theory of quantum gravity.

Classical field equations of gravity *also* describe the classical dynamics of the spacetime because of the geometrical interpretation. In the emergent gravity paradigm, these field equations have a status similar to the equations of fluid mechanics or elasticity. So, if this paradigm is correct, one should *not* expect quantizing a classical theory of gravity to lead us to the quantum structure of spacetime any more than quantizing the equations of elasticity or hydrodynamics will lead us to atomic structure of matter! Quantizing the elastic vibrations of a solid will lead only to phonon physics [6] just as quantizing a classical theory of gravity will lead to graviton physics. The latter could be as different from a description of quantum structure of spacetime just as phonon physics is from that of the atom.

Combining the principles of GR and quantum theory is probably not just a technical problem that could be solved by sufficiently powerful mathematics. It is a more of a conceptual issue and the decades of failure of sophisticated mathematics in delivering quantum gravity indicates that we should try a different approach. This is very much in tune with item (ii) mentioned in Sec. 2. Einstein did not create a sophisticated mathematical model for m_i and m_g and try to interpret $m_i = m_g$. He used thought experiments to arrive at a conceptual basis in which $m_i = m_g$ can be embedded naturally so that $m_i = m_g$ will cease to be an algebraic accident. Once this is done, physics itself led him to the maths that was needed.

Of course, the key issue is what could play the role of a guiding principle similar to principle of equivalence in the present context. *For this, my bet will be on the thermodynamics of horizons.*[2, 7] A successful model will have the connection between horizon thermodynamics and gravitational dynamics *at its foundation* rather than this feature appearing as a result derived in the context of certain specific solutions to the field equations. There are three technical points closely related to this conjecture which needs to be recognized if this approach has to succeed:

- One must concentrate on the general context of observer dependent, local, thermodynamics associated with the local horizons, going beyond the *black hole* thermodynamics. Black hole horizons in the classical theory are far too special, on-shell, constructs to provide a sufficiently general back-drop to understand the quantum structure of spacetime. The preoccupation with the black hole horizons loses sight of the conceptual fact that all horizons are endowed with temperature as perceived by the appropriate class of observers. Observer dependence [8] of thermal phenomena is a feature and not a bug!

- One should also think beyond Einstein's theory and use the structure of, say, Lanczos-Lovelock models of gravity [9] in exploring the microstructure of spacetime. Previous work (starting from [10]) has shown that the interpretation of gravity as an emergent phenomenon transcends Einstein's theory and remains applicable to (at least) all Lanczos-Lovelock models. Exploiting this connection will allow us to discriminate between results of general validity from those which are special to Einstein's theory in $D = 4$. Irrespective of whether Lanczos-Lovelock models are relevant to real world, they provide a good test-bed to see which concepts/results are robust and general.
- A corollary is that one should not think of entropy of horizons as being proportional to their area. This result, which is true in Einstein's theory, fails for all higher order Lanczos-Lovelock models [11]. But all the general thermodynamic features still continue to remain valid. Because area brings in several other closely related geometrical notions, restricting oneself to Einstein's theory leads to an incorrect view of what entropy and quantum microstructure of spacetime could be.

4. Thermodynamics of horizons

1. Temperature of horizons

One can associate a temperature with any null surface that can act as horizon for a class of observers, in any spacetime (including flat spacetime). This temperature is determined by the behaviour of the metric close to the horizon and has *nothing to do with the field equations* (if any) which are obeyed by the metric.

The simplest situation is that of Rindler observers in flat spacetime with acceleration κ who will attribute a temperature $k_BT = (\hbar/c)(\kappa/2\pi)$ to the Rindler horizon — which is just a $X = T$ surface in the flat spacetime having no special significance to the inertial observers. While these results are usually proved for an eternally accelerating observer, they also hold in the appropriately approximate sense for an observer with variable acceleration [12]. In general, this result can be used to compare the vacuum state in a freely falling frame with that in a locally accelerated frame in an approximate manner.

An immediate consequence, not often emphasized, is that *all* thermodynamic variables must become observer dependent. If one considers a "normal" gaseous system with "normal" thermodynamic variables (T, S, F etc.....) as a highly excited state of the inertial vacuum, it is obvious that a Rindler observer will attribute to this gas different thermodynamic variables compared to what an inertial observer will attribute. Thus thermal effects in the accelerated frame brings in [8, 13] a new level of observer dependence even to *normal* thermodynamics.

More generally, whenever a bifurcation horizon divides the spacetime into two causally disconnected regions R and L, the global vacuum state $|\text{vac}\rangle$ of a quantum field theory, described by a functional $\langle \text{vac}|\phi_L, \phi_R\rangle$ in terms of the field modes in R and L, can be expressed in the form $\langle \phi_L|e^{-(\pi/\kappa)H_R}|\phi_R\rangle$ where H_R is the Hamiltonian describing the dynamics in one of the wedges [14] and κ is the suitably defined acceleration. Tracing out the modes ϕ_L beyond the horizon will lead to a thermal density matrix for the observables in the right wedge $\rho(\phi'_R, \phi_R) \propto \langle \phi'_R|e^{-(2\pi/\kappa)H_R}|\phi_R\rangle$ corresponding to the horizon temperature $T = \kappa/2\pi$. This result only depends on the near horizon geometry having the approximate form of a Rindler metric and is independent of the field equations of the theory.

2. Entropy of horizons

It seems reasonable to assume that *all* horizons should also possess entropy which is again, by necessity, observer dependent. One would have expected that if integrating out certain field modes leads to a thermal density matrix ρ, then the entropy of the system should be related to lack of information about the *same* field modes and should be given by $S = -\mathrm{Tr}\ \rho \ln \rho$. This entropy (called entanglement entropy) (i) is proportional to area of the horizon and (ii) is divergent without a cut-off [15]. Such a divergence makes the result meaningless and thus we fail to attribute a unique entropy to horizon using just QFT in a background metric. That is, while the temperature of the horizon can be obtained through the study of QFT in an external geometry, one cannot understand the entropy of the horizon by the same procedure.

This is because, in sharp contrast to temperature, the *entropy* associated with a horizon in the theory depends on the field equations of the theory, which we will now briefly review. Given the principle of equivalence (interpreted as gravity being spacetime geometry) and principle of general covariance, one could still construct a wide class of theories of gravity. For example, if we take the action functional

$$A = \int d^D x \sqrt{-g} \left[L(R^{ab}_{cd}, g^{ab}) + L_{\mathrm{matt}}(g^{ab}, q_A) \right] \tag{1}$$

(where L_{matt} is the matter Lagrangian for some matter variables denoted symbolically as q_A) and vary the metric with appropriate boundary conditions, we will get the field equations (see e.g., chapter 15 of [16]):

$$\mathcal{G}_{ab} = P_a^{\ cde} R_{bcde} - 2\nabla^c \nabla^d P_{acdb} - \frac{1}{2} L g_{ab} \equiv \mathcal{R}_{ab} - \frac{1}{2} L g_{ab} = \frac{1}{2} T_{ab} \tag{2}$$

where $P^{abcd} \equiv (\partial L / \partial R_{abcd})$. A nice subclass of theories in which the field equations remain second order in the metric is obtained if we choose L such that $\nabla_a P^{abcd} = 0$. The most general scalar functionals $L(R^{ab}_{cd}, g^{ij})$ satisfying this condition are specific polynomials in curvature tensor which lead to the Lanczos-Lovelock models [9] with the field equations:

$$P^{de}_{ac} R^{bc}_{de} - \frac{1}{2} L \delta^b_a \equiv \mathcal{R}^b_a - \frac{1}{2} L \delta^b_a = \frac{1}{2} T^b_a; \quad \mathcal{R}^b_a \equiv P^{de}_{ac} R^{bc}_{de} \tag{3}$$

and the structure of the theory is essentially determined by the tensor P^{ab}_{cd} which has the algebraic symmetries of curvature tensor and is divergence-free in all indices.

These field equations, in general, have black hole solutions with horizons in asymptotically flat spacetime. Studying the physical processes occurring in such spacetimes, one can obtain an expression for the entropy of the horizon (called Wald entropy [11]) in such theories. The resulting form of the entropy turns out to be closely related to the Noether current J^a which is conserved due to the diffeomorphism invariance of these theories under $x^a \to x^a + q^a(x)$. Computing the Noether current for Lanczos-Lovelock models, one can show that the entropy is given by:

$$S_{\mathrm{Noether}} = \beta \int d^{D-1} \Sigma_a\, J^a \equiv \frac{\beta}{2} \int d^{D-2} \Sigma_{ab}\, J^{ab} = \frac{2\pi}{\kappa} \oint_{\mathcal{H}} P^{abcd} \epsilon_{ab} \epsilon_{cd} d\sigma \tag{4}$$

where $\beta^{-1} = \kappa/2\pi$ is the horizon temperature and $J^a \equiv \nabla_b J^{ab}$ with $J^{ab} = -J^{ba}$ ensuring $\nabla_i J^i = 0$. In the final expression the $(D-2)$-dimensional integral is on a space-like bifurcation surface with ϵ_{ab} denoting the bivector normal to the bifurcation surface. Thus horizon entropy is given by an integral over the horizon surface of the P^{abcd}, which we may call the *entropy tensor* of the theory.

In the simplest context of $L \propto R$, leading to $P^{ab}_{cd} \propto (\delta^a_c \delta^b_d - \delta^a_d \delta^b_c)$, we have $\mathcal{R}^a_b \propto R^a_b, \mathcal{G}^a_b \propto G^a_b$ and one recovers Einstein's equations and the entropy will be proportional to the area of the horizon. But in general, the entropy of the horizon is *not* proportional to the area and depends on the theory. This feature makes it impossible to identify the entanglement entropy as the entropy of the Lanczos-Lovelock models. (For the possible approach to tackle this issue within the emergent paradigm, which I will not discuss here, see ref. [17])

This dichotomous situation as regards temperature versus entropy is the first indication that the thermodynamics of the horizon, probed by QFT in a external gravitational field, is just the tip of an iceberg. As we will see the emergent paradigm provides a better understanding of these features.

5. Thermodynamic interpretation of field equations and action functionals

I stressed in Sec. 2 that for the top-down approach to be of some use, we need to identify the 'algebraic accidents' in the top level description which are usually taken for granted without a demand for explanation. I will briefly summarize four such issues in classical gravity, which can give us a clue about the microscopic theory.

(1) Gravitational field equations become a thermodynamic identity:

It can be shown that [18] the field equations in any Lanczos-Lovelock model, when evaluated on a static solution of the theory which has a horizon, can be expressed in the form of a thermodynamic identity $TdS = dE_g + PdV$. Here S is the correct Wald entropy of the horizon in the theory, E_g is a geometric expression involving an integral of the scalar curvature of the sub-manifold of the horizon and PdV represents the work function of the matter source. The differentials dS, dE_g etc. should be thought of as indicating the difference in S, E_g etc between two solutions in which the location of the horizon is infinitesimally displaced.

This equality between field equations on the horizon and the thermodynamic identity — originally obtained [19] for spherical horizons in Einstein's theory where $S = (1/4)4\pi a^2, T = \kappa/2\pi, E_g = a/2$ and P is the pressure of the source — has now been demonstrated for an impressively wide class of models [20] like stationary axisymmetric horizons and evolving spherically symmetric horizons in Einstein gravity, static spherically symmetric horizons and dynamical apparent horizons in Lanczos-Lovelock gravity, generic, static horizon in Lanczos-Lovelock gravity, three dimensional BTZ black hole horizons, FRW cosmological models in various gravity theories and even in the case Horava-Lifshitz gravity.

Incidentally, while Davies-Unruh temperature scales as $\hbar$ the entropy scales as $1/\hbar$ (coming from inverse Planck area), thereby making TdS independent of $\hbar$! This is reminiscent of the fact that in normal thermodynamics $T \propto 1/k_B, S \propto k_B$ making TdS independent of k_B. In both cases, the effects due to discrete microstructure (indicated by non-zero $\hbar$ or k_B) disappears in the continuum limit thermodynamics. Thermal phenomena requires microstructure but thermodynamical laws are independent of it!

There are significant differences between this identity $TdS = dE_g + PdV$ to which field equations reduce to and the so called Clausius relation $TdS = dE_m$ (used, for example, by Jacobson [4]) which are not often appreciated.

- In addition to the obvious existence of the work term PdV it should be stressed that E_m used in the Clausius relation is related to *matter stress tensor* while E_g in the first law is a purely geometrical construct built out of the metric. The origin of these differences can be traced to two different kinds of virtual displacements of the horizons considered in these two approaches to define the infinitesimal differences [21].
- More importantly, *while $TdS = dE_g + PdV$ holds in widely different contexts,* it has been found to be *impossible* to generalize $TdS = dE_m$ beyond Einstein's theory without introducing additional assumptions (like dissipation), the physical meaning of which remains unclear.

(2) Gravitational field equations and local entropy balance:

The above result has a simple generalisation in terms of local Rindler horizons. Let us choose any event $\mathcal{P}$ and introduce a local inertial frame (LIF) around it with Riemann normal coordinates $X^a = (T, \mathbf{X})$. Let k^a be a future directed null vector at $\mathcal{P}$ and we align the coordinates of LIF such that it lies in the $X - T$ plane at $\mathcal{P}$ and transform from the LIF to a local Rindler frame (LRF) with acceleration κ along the X axis. Let ξ^a be the approximate Killing vector corresponding to translation in the Rindler time. Usually, we shall do all the computation on a time-like surface (called a "stretched horizon") with $N^2 = -\xi_a\xi^a =$ constant, infinitesimally away from the Rindler horizon $\mathcal{H}$ which corresponds to $N^2 = 0$. An infinitesimal displacement of a local patch of the stretched horizon in the direction of its normal r_a, by an infinitesimal proper distance ϵ, will change the proper volume by $dV_{prop} = \epsilon\sqrt{\sigma}d^{D-2}x$ where σ_{ab} is the metric in the transverse space. The flux of energy through the surface will be $T^a_b\xi^b r_a$ and the corresponding entropy flux can be obtained by multiplying the energy flux by $\beta_{\rm loc} = N\beta$. Hence the 'loss' of matter entropy to the outside observer because the virtual displacement of the horizon has engulfed some matter is $\delta S_m = \beta_{\rm loc}\delta E = \beta_{\rm loc}T^{aj}\xi_a r_j dV_{prop}$. Recalling from Eq. (4) that $\beta_{loc}J^a$ gives the gravitational entropy current, the change in the gravitational entropy is given by $\delta S_{\rm grav} \equiv \beta_{loc} r_a J^a dV_{prop}$ where J^a is the Noether current corresponding to the local Killing vector ξ^a given by $J^a = 2\mathcal{G}^a_b\xi^b + L\xi^a$. As the stretched horizon approaches the true horizon, $Nr^a \rightarrow \xi^a$ and $\beta\xi^a\xi_a L \rightarrow 0$. Hence we get, in this limit: $\delta S_{\rm grav} \equiv \beta\xi_a J^a dV_{prop} = \beta 2\mathcal{G}^{aj}\xi_a\xi_j dV_{prop}$. Comparing $\delta S_{\rm grav}$ and δS_m we see that the field equations $2\mathcal{G}^a_b = T^a_b$ can be interpreted as the entropy balance condition $\delta S_{grav} = \delta S_{matt}$ thereby providing direct thermodynamic interpretation of field equations as local entropy balance in local Rindler frame.

In the conventional approach, there is no explanation for the fact that the entropy of the horizon is related to a Noether current arising from the diffeomorphism $x^a \rightarrow x^a + \xi^a(x)$. In the emergent paradigm, the spacetime is analogous to a solid made of atoms and $x^a \rightarrow x^a + \xi^a(x)$ is analogous to the deformation of an elastic solid. When such a deformation leads to changes in accessible information — like when one considers the virtual displacements of horizons — it is understandable that the gravitational entropy changes (See Sec. 7).

(3) Holography of gravitational action functionals:

If the gravitational dynamics and horizon thermodynamics are so closely related, with field equations becoming thermodynamic identities on the horizon, then the action functionals of the theory (from which we obtain the field equations) must contain information about this connection. This clue comes in the form of another unexplained algebraic accident related to the structure of the action functional.

Gravity is the only theory known to us for which the natural action functional preserving symmetries of the theory contain second derivatives of the dynamical variables but still leads to second order differential equations. (As we see from Eq. (3), this feature is shared by all the Lanczos-Lovelock action functionals.) It can be shown that this result will be true for actions that can be separated into a surface term and a bulk term with the surface term being an integral over $\partial_a(q^A\pi^a_A)$ where q^A are the dynamical variables and π^a_A are the canonical momentum. All Lanczos-Lovelock action functional have this form [10, 22]. This structure allows one to interpret all these action functionals, including Einstein-Hilbert action, as providing the momentum space description (see p. 292 of [16]) of the theory.

In the conventional approach this is hardly taken note of as something worthy of explanation and, in fact, there is no explanation possible within the standard framework. In contrast, this result has direct links with the thermodynamic interpretation. To see this, note that: (a) The field equations can be obtained from varying the bulk term after ignoring (or by canceling with a counter-term) the surface term. (b) But if we evaluate the surface term on the horizon of any solution to the field equations of the theory, one obtains the entropy of the horizon! How does the surface term, which was discarded before the field equations were obtained know about the entropy associated with a solution to those field equations?!

The answer lies in the fact mentioned before, viz. that the surface and bulk term of the Lagrangian are related in a specific manner

$$\sqrt{-g}L_{\rm sur} = -\partial_a \left(g_{ij} \frac{\delta\sqrt{-g}L_{\rm bulk}}{\delta(\partial_a g_{ij})} \right) \tag{5}$$

thereby duplicating the information about the horizon entropy [22]. This duplication of information allows one (i) to construct [24] a suitable variational principle based purely on the surface term and (ii) to obtain the full action [7] from the surface term alone using the entropic interpretation. (Incidentally, in the Riemann normal coordinates around any event $\mathcal{P}$ the gravitational action reduces to a pure surface term, again showing that the dynamical content is actually stored on the boundary rather than in the bulk.)

4. Gravitational action as free energy of spacetime

The reason gravitational actions have a surface and bulk terms is because they are related to the entropy and energy of a static spacetimes with horizons, adding up to make the action the free energy of the spacetime [23]. This is most easily seem for any Lanczos-Lovelock model by writing the time component of the Noether current in the form:

$$L = \frac{1}{\sqrt{-g}}\partial_\alpha \left(\sqrt{-g}\, J^{0\alpha}\right) - 2\mathcal{G}^0_0 \tag{6}$$

Only spatial derivatives contribute in the first term on the right hand side when the spacetime is static. Integrating this to obtain the action it is is easy to see (using Eq. (4)) that the first term gives the entropy and the second term can be interpreted as energy.

Finally, I stress that the real importance of these results arise from the fact that they hold for all Lanczos-Lovelock models in an identical manner.

6. The Avogadro number of the spacetime

The results described in the previous sections suggest that there is a deep connection between horizon thermodynamics and the gravitational dynamics. Because the spacetime can be heated up just like a body of gas, the Boltzmann paradigm ("If you can heat it, it has microstructure") suggests a reinterpretation of gravity as an emergent phenomenon. This motivates the study of the microscopic degrees of freedom of the spacetime exactly the way people studied gas dynamics *before* they understood the atomic structure of matter.

There exists, fortunately, an acid test of this paradigm. Boltzmann's conjecture led to the equipartition law $\Delta E = (1/2)k_B T \Delta N$ relating the number density ΔN of microscopic degrees of freedom required to store an energy ΔE at temperature T and to the determination of Avogadro number of a gas. If our ideas are correct, we should be able to relate the E and T of a given spacetime to determine the number density of microscopic degrees of freedom of the spacetime. Remarkably enough, this can be done directly from the field equations [25]. In a hot spacetime, Einstein's equations *imply* the equipartition law

$$E = \frac{1}{2}k_B \int_{\partial\mathcal{V}} \frac{\sqrt{\sigma}\, d^2x}{L_P^2} \left\{ \frac{Na^\mu n_\mu}{2\pi} \right\} \equiv \frac{1}{2}k_B \int_{\partial\mathcal{V}} dn\, T_{\rm loc} \tag{7}$$

(where $T_{\rm loc} = (Na^\mu n_\mu/2\pi)$ is the local acceleration temperature and $\Delta n = \sqrt{\sigma}\, d^2x/L_P^2$) thereby allowing us to read off the number density of microscopic degrees of freedom. We again see that gravity is holographic in the sense that the number density Δn scales as the proper area $\sqrt{\sigma}\, d^2x$ of the boundary of the region rather than the volume. (In the case of a gas, we would have got an integral over the volume of the form $dV(dn/dV)$ rather than the area integral.) We also notice that, in Einstein's theory, the number density (dn/dA) is a constant with every Planck area contributing a single degree of freedom.

The true value of this result again rests on the fact that it holds true for all Lanczos-Lovelock models! For an Lanczos-Lovelock model with an entropy tensor P^{ab}_{cd} one gets the result

$$E = \frac{1}{2}k_B \int_{\partial\mathcal{V}} dn T_{loc}; \qquad \frac{dn}{dA} = \frac{dn}{\sqrt{\sigma}d^{D-2}x} = 32\pi P^{ab}_{cd}\epsilon_{ab}\epsilon^{cd} \tag{8}$$

where ϵ_{ab} is the binormal on the codimension-2 cross-section. All these gravitational theories are holographic and the density of microscopic degrees of freedom encodes information about the theory through the entropy tensor. *I consider these results as the most direct evidence for the emergent paradigm of gravity.*

The holographic nature of gravity which I have alluded to several times shows that area elements play a significant role in the microscopic description of the theory. This is directly related to the fact that the basic unit of the theory is the Planck *area* $\mathcal{A}_P \equiv (G\hbar/c^3)$. (Only by taking a square root, rather artificially, one obtains the Planck *length*; also see, [26]) Classical gravity, in fact, should be described using $\mathcal{A}_P$ rather than using G with Newton's law of gravity written in the form $F = (\mathcal{A}_P c^3/\hbar)(m_1 m_2/r^2)$. This has the crucial consequence that one cannot really take $\hbar \to 0$ limit at fixed $\mathcal{A}_P$ and call it classical gravity. Gravity is intrinsically quantum mechanical at all scales [27] because of the microstructure of spacetime just as normal matter is intrinsically quantum mechanical due to the atomic structure. One cannot study classical elasticity, say, by taking the $\hbar \to 0$ limit in a crystal lattice. This is because such a limit (which can give elasticity from lattice dynamics) will also make all the electrons in the atom collapse!

7. Entropy density of spacetime an its extremum

So far we have been faithfully following the 'top-down' philosophy of starting from known results in classical gravity and obtaining consequences which suggests an alternative paradigm. For example, the results in the last section were obtained by starting from the field equations of the theory, rewriting them in the form of law of equipartition and thus determining the density of microscopic degrees of freedom.

Ultimately, however, we have to start from a microscopic theory and obtain the classical results as a consequence. As a first step in this direction, I will now describe a rather attractive procedure of inverting the arguments of the last section and obtaining the field equations from the density of microscopic degrees of freedom.

We know that the thermodynamical behaviour of any system can be described by an extremum principle for a suitable potential (entropy, free energy ...) treated as a functional of appropriate variables (volume, temperature ,....). If our ideas related to gravitational theories are correct, it must be possible to obtain the field equations by extremising a suitably defined entropy functional. The fact that null surfaces block information suggests that the entropy functional should be closely related to null surfaces in the spacetime. This expectation turns out to be correct. The Lanczos-Lovelock field equations can indeed be obtained [28] by extremising an entropy functional associated with the null vectors in the spacetime.

The mathematics involves associating with every null vector in the spacetime an entropy functional S_{grav} and demanding $\delta[S_{grav} + S_{matter}] = 0$ for *all* null vectors in the spacetime where S_{matter} is the relevant matter entropy. Recall that 'how gravity tell matter to move' can be determined by demanding the validity of special relativistic laws for all locally inertial observers. Similarly, 'how matter curves spacetime' can be determined by demanding that the entropy maximization should hold for *all* local Rindler observers. The entropy functional we will use can be interpreted as the entropy of the null surface (which these observers perceive as horizon) plus matter entropy. The form of S_{matter} relevant for this purpose turns out to be

$$S_{\rm matt} = \int_{\mathcal{V}} d^kx\sqrt{-g}\,T_{ab}n^a n^b \tag{9}$$

where n^a is a null vector field and T_{ab} is the matter energy-momentum tensor. This expression can be interpreted as a limiting case of the entropy flux associated with the energy flux into the local Rindler horizon as viewed by the Rindler observers. The integration could be over this spacetime volume ($k = D$) or even over a nontrivial sub-manifold ($k < D$), say, a set of null surfaces. (This does not affect the variational principle or the resulting equations but the interpretation could be that of minimizing the rate of entropy production.) The simplest choice for S_{grav} is a quadratic expression in the derivatives of the null vector:

$$S_{grav} = -4\int_{\mathcal{V}} d^kx\sqrt{-g}\,P_{ab}^{cd}\nabla_c n^a\nabla_d n^b \tag{10}$$

where P_{ab}^{cd} is a tensor having the symmetries of curvature tensor and is divergence-free in all its indices. So the P^{abcd} in Eq. (10) can be expressed as $P^{abcd} = \partial L/\partial R_{abcd}$ where L is the Lanczos-Lovelock Lagrangian and R_{abcd} is the curvature tensor [2]. This choice will also ensure that the equations resulting from the entropy extremisation do not contain any derivative of the metric which is of higher order than second. (More general possibilities exist which I will not discuss here.) Hence the expression for the total entropy becomes:

$$S[n^a] = -\int_{\mathcal{V}} d^kx\sqrt{-g}\left(4P_{ab}^{cd}\nabla_c n^a\nabla_d n^b - T_{ab}n^a n^b\right), \tag{11}$$

We now vary the vector field n^a in Eq. (11) after adding a Lagrange multiplier function $\lambda(x)$ for imposing the condition $n_a\delta n^a = 0$. Straight forward algebra (see e.g., Section 7.1 of ref. [2]) now shows that the condition $\delta[S_{grav} + S_{matter}] = 0$ leads to the equations

$$(2\mathcal{R}^a_b - T^a_b + \lambda\delta^a_b)\,n_a = 0\,, \tag{12}$$

with $\mathcal{R}^a_b \equiv P_{bi}^{jk}R^{ai}_{jk}$ which we demand should hold for all null vector fields n^a. Using the generalized Bianchi identity and the condition $\nabla_a T^a_b = 0$ we obtain [2, 28] from Eq. (12) the equations

$$\mathcal{G}^a_b = \mathcal{R}^a_b - \frac{1}{2}\delta^a_b L = \frac{1}{2}T^a_b + \Lambda\delta^a_b \tag{13}$$

where Λ is a constant. These are precisely the field equations for gravity in a theory with Lanczos-Lovelock Lagrangian L.

In addition to providing a purely thermodynamic extremum principle for the field equations of gravity, the above approach also has the following attractive features.

- The extremum value of the entropy functional, when computed on-shell for a solution, leads to the Wald entropy. This is a non-trivial consistency check on the approach because it was not designed to reproduce the Wald entropy. It also shows that when the field equations hold, the total entropy of a region $\mathcal{V}$ resides on its boundary $\partial\mathcal{V}$ which is yet another illustration of the holographic nature of gravity.
- In the semi-classical limit, one can show [29] that the gravitational (Wald) entropy is quantized with $S_{\rm grav}$ [on-shell] $= 2\pi n$. In the lowest order Lanczos-Lovelock theory, the entropy is proportional to area and this result leads to area quantization. More generally, it is the gravitational entropy that is quantized. The law of equipartition for the surface degrees of freedom is closely related to this entropy quantization because both arise from the existence of discrete structures on the surfaces in question.

- The entropy functional in Eq. (11) is invariant under the shift $T_{ab} \to T_{ab} + \rho_0 g_{ab}$ which shifts the zero of the energy density. This symmetry allows any low energy cosmological constant, appearing as a parameter in the variational principle, to be gauged away thereby alleviating the cosmological constant problem to a great extent [30]. I will not discuss this issue here.

There is another way of interpreting Eq. (12) which is more in tune with the emergent perspective of gravity [31]. Note that, while Eq. (12) holds for any vector field once the normalization condition is imposed through the Lagrange multiplier, the entropy was originally attributed to *null* vectors and hence it is natural to study Eq. (12) when $n^a = \ell^a$, the null normal of a null surface $\mathcal{S}$ in the spacetime and project Eq. (12) onto the null surface. If $\boldsymbol{\ell}$ is the normal to $\mathcal{S}$, then such a projection leads to the equations:

$$R_{mn}\ell^m q^n_a = 8\pi T_{mn}\ell^m q^n_a; \quad R_{mn}\ell^m \ell^n = 8\pi T_{mn}\ell^m \ell^n \tag{14}$$

where $q_{ab} = g_{ab} + \ell_a k_b + \ell_b k_a$ with k^a being another auxiliary null vector satisfying $\boldsymbol{\ell} \cdot \boldsymbol{k} = -1$. The q_{ab} with $q_{ab}\ell^b = 0 = q_{ab}k^b$ acts as a projector to $\mathcal{S}$ (see ref. [31] for details). It is possible to rewrite the first equation in Eq. (14) in the form of a Navier-Stokes equation thereby providing a hydrodynamic analogy for gravity. This equation is known in the literature and is called Damour-Navier-Stokes (DNS) equation [32] and is usually derived by rewriting the field equations. Our analysis [31] provides an entropy extremisation principle for the DNS equation which makes the hydrodynamic analogy natural and direct.

It may also be noted that the gravitational entropy density — which is the integrand $\Im_{grav} \propto (-P^{cd}_{ab}\nabla_c \ell^a \nabla_d \ell^b)$ in Eq. (10) — obeys the relation:

$$\frac{\partial \Im_{\text{grav}}}{\partial(\nabla_c \ell^a)} \propto (-P^{cd}_{ab}\nabla_d \ell^b) \propto (\nabla_a \ell^c - \delta^c_a \nabla_i \ell^i) \tag{15}$$

where the second relation is for Einstein's theory. This term is analogous to the more familiar object $t^c_a = K^c_a - \delta^c_a K$ (where K_{ab} is the extrinsic curvature) that arises in the (1+3) separation of Einstein's equations. (More precisely, the projection to 3-space leads to t^c_a.) This combination can be interpreted as a surface energy momentum tensor in the context of membrane paradigm [33] because t_{ab} couples to δh^{ab} on the boundary surface when we vary the gravitational action (see, e.g., eq.(12.109) of [16]). Equation (15) shows that the entropy density of spacetime is directly related to t^c_a and its counterpart in the case of null surface. This term also has the interpretation as the canonical momentum conjugate to the spatial metric in (1+3) context and Eq. (15) shows that the entropy density leads to a similar structure.

Further, the *functional* derivative of the gravitational entropy in Eq. (10) has the form, in any Lanczos-Lovelock model:

$$\frac{\delta S_{\text{grav}}}{\delta \ell^a} \propto \mathcal{R}_{ab}\ell^b \propto J_a \tag{16}$$

Previous work [2, 25, 8] has shown that the current $J_a = 2\mathcal{R}_{ab}\ell^b$ plays a crucial role in interpreting gravitational field equations as entropy balance equations. In the context of local Rindler frames, when ℓ^a arises as a limit of the time-like Killing vector in the local Rindler frame, J_a can be interpreted as the Noether (entropy) current associated with the null surface. In that case, the generalization of the two projected equations in Eq. (14) to Lanczos-Lovelock model will read as

$$J_a \ell^a = \frac{1}{2} T_{ab}\ell^a \ell^b; \quad J_a q^a_m = \frac{1}{2} T_{ab}\ell^a q^b_m \tag{17}$$

which relate the gravitational entropy density and flux to matter energy density and momentum flux. (The second equation in the above set becomes the DNS equation in the context of Einstein's theory.) All these results, including the DNS equation, will have direct generalization to Lanczos-Lovelock models which can be structured using the above concepts. We again see that all these ideas find a natural home in the emergent paradigm.

Acknowledgements
I thank the organizers of ERE2010 meeting for excellent hospitality.

References
[1] Bekenstein J D 1972 *Nuovo Cim. Lett.* **4** 737–740; Hawking S W 1975 *Commun. Math. Phys.* **43** 199–220.; Davies P C W (1975) *J. Phys.* A **8** 609–616; Unruh W G (1976) *Phys. Rev.* D **14** 870.
[2] Padmanabhan T., *Rep. Prog. Phys.*, **73** (2010) 046901, [arXiv:0911.5004].
[3] Sakharov A D 1968 *Sov. Phys. Dokl.* **12** 1040.
[4] Jacobson T 1995 *Phys. Rev. Lett.* **75** 1260.
[5] Volovik G E 2003 *The universe in a helium droplet*, (Oxford University Press).
[6] B. L. Hu, (2010) *Gravity and Nonequilibrium Thermodynamics of Classical Matter* [arXiv:1010.5837]; *General Relativity as Geometro-Hydrodynamics* [arXiv:gr-qc/9607070].
[7] T. Padmanabhan, *Phys. Reports*, **406**, 49 (2005) [gr-qc/0311036].
[8] Padmanabhan T 2009 *A Dialogue on the Nature of Gravity* [arXiv:0910.0839]
[9] Lanczos C 1932 *Z. Phys.* **73** 147; Lovelock D 1971 *J. Math. Phys.* **12** 498.
[10] T. Padmanabhan, *Dark Energy: Mystery of the Millennium*, Albert Einstein Century International Conference, Paris, July 2005, *AIP Conference Proceedings* **861**, 858, [astro-ph/0603114]
[11] Wald R. M., *Phys. Rev. D* (1993), **48** 3427, [gr-qc/9307038]; Iyer V. and R. M. Wald, (1995), *Phys. Rev. D* **52** 4430, [gr-qc/9503052].
[12] D. Kothawala, T. Padmanabhan , *Phys.Letts.*, **B 690** , 201-206 (2010) [arXiv:0911.1017]; A. Raval, B.L. Hu, Don Koks, *Phys.Rev*, **D55**, 4795 (1997) [gr-qc/9606074].
[13] Marolf D, Minic D and Ross S 2004 *Phys.Rev.* **D69** 064006.
[14] Lee T D 1986 *Nucl. Phys.* B **264** 437; Unruh W G and Weiss N 1984 *Phys. Rev.* D **29** 1656.
[15] See, for e.g., L. Bombelli et al., *Phys.Rev.*, **D34**, 373 (1986); M. Srednicki, *Phys.Rev.*, **D71**, 66 (1993); T. Nishioka *et al. J.Phys.*, **A42**, 504008 (2009) [arXiv:0905.0932].
[16] T.Padmanabhan (2010) *Gravitation: Foundations and Frontiers*, Cambridge University Press, UK.
[17] Padmanabhan T. *Phys. Rev. D* **82**, 124025 (2010). [arXiv:1007.5066]
[18] Kothawala D and Padmanabhan T 2009 *Phys. Rev.* D **79** 104020 [arXiv:0904.0215].
[19] Padmanabhan T 2002 *Class. Quan. Grav.* **19** 5387 [gr-qc/0204019].
[20] There is large literature on this subject; see e.g., ref.110-124 of [2].
[21] D. Kothawala, *The thermodynamic structure of Einstein tensor*, [arXiv:1010.2207].
[22] A. Mukhopadhyay, T. Padmanabhan, *Phys.Rev.*, **D 74**, 124023 (2006) [hep-th/0608120].
[23] Sanved Kolekar, T. Padmanabhan , *Phys.Rev.*, **D 82**, 024036 (2010) [arXiv:1005.0619].
[24] T. Padmanabhan , *Gen.Rel.Grav.*, **38**, 1547-1552 (2006); *Int.J.Mod.Phys.*, **D 15**, 2029 (2006) [gr-qc/0609012]; *Adv. Sci. Lett.* , **2**, , 174 (2009) [arXiv:0807.2356].
[25] This was done in 2004 in the form of a relation $E = 2TS$ in Einstein's theory and is now generalized to all Lanczos-Lovelock models. T. Padmanabhan, *Class.Quan.Grav.*, **21**, 4485 (2004) [gr-qc/0308070]; *Mod. Phys. Lett.* **A 25**, 1129 (2010), [arXiv:0912.3165]; *Phys. Rev.*, **D 81**, 124040 (2010), [1003.5665].
[26] J.Makela, *On distance and area,* [arXiv:1011.2052].
[27] T. Padmanabhan , *Mod.Phys.Letts. A* , **17**, 1147 (2002). [hep-th/0205278]; *Gen.Rel.Grav.*, **34**, 2029-2035 (2002) [gr-qc/0205090].
[28] Padmanabhan T., (2008), *Gen.Rel.Grav.*, **40**, 529-564 [arXiv:0705.2533]; Padmanabhan T. and Paranjape A., (2007), *Phys.Rev.* D, **75**, 064004 [gr-qc/0701003].
[29] D. Kothawala, T. Padmanabhan, S. Sarkar, *Phys.Rev.*, **D78**, 104018 (2008) [arXiv:0807.1481]
[30] T. Padmanabhan, *Gen.Rel.Grav.*, **40**, 529 (2008) [arXiv:0705.2533]; *Adv. Sci. Lett.*, **2,** 174 (2009) [arXiv:0807.2356].
[31] Padmanabhan T., *Entropy density of spacetime and the Navier-Stokes fluid dynamics of null surfaces*, [arXiv:1012.0119].
[32] T. Damour, (1979), Thèse de doctorat d'État, Université Paris (available at http://www.ihes.fr/~damour/Articles/).
[33] R.H. Price and K.S. Thorne, (1986), *Phys. Rev.* **D 33**, 915.

Journal of Physics: Conference Series **314** (2011) 012002 doi:10.1088/1742-6596/314/1/012002

Status of Hořava gravity: A personal perspective

Matt Visser

School of Mathematics, Statistics, and Operations Research
Victoria University of Wellington
Wellington, New Zealand

E-mail: matt.visser@msor.vuw.ac.nz

Abstract. Hořava gravity is a relatively recent (Jan 2009) idea in theoretical physics for trying to develop a quantum field theory of gravity. It is not a string theory, nor loop quantum gravity, but is instead a traditional quantum field theory that breaks Lorentz invariance at ultra-high (presumably trans-Planckian) energies, while retaining approximate Lorentz invariance at low and medium (sub-Planckian) energies. The challenge is to keep the Lorentz symmetry breaking controlled and small — small enough to be compatible with experiment. I will give a very general overview of what is going on in this field, paying particular attention to the disturbing role of the scalar graviton.

1. Introduction

Is Lorentz symmetry truly fundamental? Or is it just an "accidental" low-momentum emergent symmetry? Opinions on this issue have undergone a radical mutation over the last few years. Historically, (for good experimental and theoretical reasons), Lorentz symmetry was considered absolutely fundamental — not to be trifled with — but for a number of independent reasons the modern viewpoint is more nuanced.

- First, there is now a well developed phenomenological framework that permits precision tests of Lorentz invariance. (See for instance [1, 2, 3, 4].) While all observational data to date are compatible with exact Lorentz invariance, the existence of this theoretical framework shows that at least some (limited) types of Lorentz symmetry breaking are compatible with current observations.
- Second, there are very many and varied theoretical scenarios that suggest (but do not unavoidably guarantee) that at sufficiently high energies (typically trans-Planckian energies) Lorentz symmetry might eventually break down. (See for instance [5, 6].)

Faced with this theoretical situation it makes very good sense to ask:

- What if anything are the benefits of Lorentz symmetry breaking?
- What can we do with it?
- Why should we care?
- Where are the bodies buried?

In this regard, one of the more significant developments of the last decade was the appearance in early 2009 of a paper by Hořava [7] which laid out a strategy for making a virtue of Lorentz symmetry breaking by *using* it as a key ingredient in developing a superficially plausible quantum field theoretic candidate for a theory of quantum gravity. Hořava's model is not a string theory, nor loop quantum gravity, but is instead a traditional (point-particle) quantum field theory that breaks Lorentz invariance at ultra-high (presumably trans-Planckian) energies. Since the mere existence of any such field-theoretic model (independent of its actual phenomenological viability) flew in the face of much of the "perceived wisdom" regarding the features one might reasonably expect of any realistic theory of "quantum gravity", there was very rapid and extremely extensive interest in Hořava's proposal.

As of March 2011, Hořava's original paper [7] has been cited over 425 times, with some 65% of the citing articles being published. If we add the closely related articles [8, 9] this rises to some 450 articles, with again some 65% of the citing articles being published. This is indicative of a very highly active research field. On the other hand, there is at least one note of caution that should be raised: The unpublished 35% of citing articles cannot all be attributed simply to time-lag, nor to non-journal conference-proceedings articles. The two main reasons underlying this state of affairs seem to be:

- There has been somewhat of a tendency to charge full steam ahead with applications, (typically cosmological), without first fully understanding the foundations of the model.
- For that matter, there is still considerable disagreement as to what precise version of Hořava's model is "best".

Consequently, when reading the relevant literature, a certain amount of caution is advisable. In the current article I will give a rather personal interpretation of the current state of affairs in Hořava gravity, with some indications of possible future directions.

2. Basic framework: Field theory without gravity

My interpretation of the central idea of Hořava gravity is this: One abandons ultra-high-energy Lorentz invariance as fundamental, recognizing that phenomenologically one need "merely" attempt to recover an approximate low-energy Lorentz invariance. Typical dispersion relations are then of the form

$$\omega = \sqrt{m^2 + k^2 + \frac{k^4}{K^2} + \dots} \tag{1}$$

This point of view is nicely compatible with the "analogue spacetime" programme [10, 11], which is one reason I personally was very quick to get involved. More precisely, in condensed matter language, a so-called "critical" Lifshitz point in $(d+1)$ dimensions corresponds to a dispersion relation which satisfies

$$\omega \to k^d \qquad \text{as} \qquad k \to \infty. \tag{2}$$

To recover Lorentz invariance, at "low" momentum (but still allowing $k \gg m$) the dispersion relation should satisfy

$$\omega \to \sqrt{m^2 + k^2} \qquad \text{as} \qquad k \to 0. \tag{3}$$

Note that every quantum field theory (QFT) regulator currently known to mankind either breaks Lorentz invariance explicitly (e.g. lattice QFT), or does something worse, often something outright unphysical. For example: Pauli–Villars violates unitarity; Lorentz-invariant higher-derivatives violate unitarity; dimensional regularization is at best a purely formal trick with no direct physical interpretation (and which requires a Zen-like approach to gamma matrix algebra). The standard viewpoint is this: If the main goal is efficient computation in a corner of parameter space that we experimentally know to be Lorentz invariant to a high level of precision, then by all means, go ahead and develop a Lorentz-invariant perturbation theory with an unphysical regulator — hopefully the unphysical aspects of the computation can first be isolated, and then banished by renormalization. This is exactly what is done, (very efficiently and very effectively), in the "standard model of particle physics". (See for instance [12].)

Journal of Physics: Conference Series **314** (2011) 012002 doi:10.1088/1742-6596/314/1/012002

Hořava's approach can be interpreted by taking on a non-standard viewpoint: If one has reason to suspect that Lorentz invariance might ultimately break down at ultra-high (presumably trans–Planckian) energies, then a different strategy suggests itself. Maybe one could *use* the Lorentz symmetry breaking as part of the QFT regularization procedure? Could we then at least keep intermediate parts of the QFT calculation "physical"? (Note that "physical" does not necessarily mean "realistic", it just means we are not violating fundamental tenets of quantum physics at intermediate stages of the calculation.)

Consider for example a specific "physical" but Lorentz-violating regulator based on the dispersion relation

$$\omega^2 = m^2 + k^2 + \frac{k^4}{K_4^2} + \frac{k^6}{K_6^4}. \tag{4}$$

We call this a "trans–Bogoliubov" dispersion relation [11], based on analogy with the standard condensed-matter Bogoliubov dispersion relation:

$$\omega^2 = k^2 + \frac{k^4}{K^2}. \tag{5}$$

The corresponding QFT propagator (momentum-space Green function) is

$$G(\omega, k) = \frac{1}{\omega^2 - [m^2 + k^2 + k^4/K_4^2 + k^6/K_6^4]}. \tag{6}$$

Note the rapid fall-off as spatial momentum $k \to \infty$. This improves the behaviour of the integrals encountered in Feynman diagram calculations (QFT perturbation theory). Specifically: In any (3+1) dimensional scalar QFT, with arbitrary polynomial self-interaction, this is enough (after Wick rotation and normal ordering), to keep all individual Feynman diagrams *finite*. For details see [13], the original Hořava article [7], and related work by Anselmi [14, 15]. More generally, with language again borrowed from condensed matter, for a Lifshitz point of order z in $(d+1)$ dimensions we have the dispersion relation:

$$\omega^2 = m^2 + k^2 + \sum_{n=2}^{z} g_n \frac{k^{2n}}{K^{2n-2}}. \tag{7}$$

The equivalent QFT propagator [Green function] is now

$$G(\omega, k) = \frac{1}{\omega^2 - [m^2 + k^2 + \sum_{n=2}^{z} g_n\, k^{2n}/K^{2n-2}]}. \tag{8}$$

Key theoretical results are that:

- In a $(d+1)$ dimensional scalar QFT with $z = d$, and arbitrary polynomial self-interaction, this is enough (after Wick rotation and normal ordering) to keep all Feynman diagrams individually *finite* [7, 13].
- Gravity is a little trickier, but you can at least argue for power-counting renormalizability of the resulting QFT [7, 16].

These results are unexpected, seriously unexpected. And yes, there are still significant technical difficulties, (of which more anon).

Journal of Physics: Conference Series **314** (2011) 012002 doi:10.1088/1742-6596/314/1/012002

3. Basic framework: Gravity

Consider the standard ADM decomposition for $(d+1)$ dimensional gravity:

$$\mathcal{S} = \int \sqrt{-g_d}\; N \;\left\{ \mathrm{tr}[K^2] - \mathrm{tr}[K]^2 + {}^{(d)}R \right\}\; \mathrm{d}^d x\, \mathrm{d}t. \tag{9}$$

Here one has split spacetime into space+time, while ${}^{(d)}R$ is intrinsic curvature of space, and K is the extrinsic curvature of space in spacetime. On the one hand this ADM decomposition leads to "canonically quantized gravity" (with all its problems), and on the other hand it is classically very useful for numerical relativity. The key to Hořava gravity is now to develop a non-standard extension of the ADM formalism:

- Choose a "preferred foliation".
- Decompose $\mathcal{L}$ = (kinetic term) – (potential term).
- Add extra "kinetic" and "potential" terms, beyond what you expect from Einstein–Hilbert.
- You cannot now reassemble $\mathcal{L}$ into a simple ${}^{(d+1)}R$.
- You have now implicitly reintroduced the aether, more on this below.

Specifically:

- Consider the "kinetic energy"

  $$\mathcal{T}(K) = g_K \left\{ (K^{ij}K_{ij} - K^2) + \xi K^2 \right\} = g_K \left\{ K^{ij}K_{ij} - \lambda K^2 \right\}, \tag{10}$$

  noting that standard general relativity would enforce $\xi \to 0$, (that is $\lambda \to 1$). Take the kinetic action to be

  $$\mathcal{S}_K = \int \mathcal{T}(K)\, \mathrm{d}V_{d+1} = \int \mathcal{T}(K)\sqrt{g}\, N\, \mathrm{d}^d x\, \mathrm{d}t. \tag{11}$$

  This contains only two time derivatives (hiding in K) — this is good, and is necessary (though not by itself sufficient) to suppress ghosts.
- There is also a hidden "scalar graviton" when $\xi \neq 0$ — this is potentially very bad, and I will have much more to say on this below.
- Now consider the most general "potential energy" in $(d+1)$ dimensions:

  $$\mathcal{S}_\mathcal{V} = \int \mathcal{V}(g, N)\; \sqrt{g}\, N\, \mathrm{d}^d x\, \mathrm{d}t, \tag{12}$$

  where $\mathcal{V}(g, N)$ is some scalar built out of the spatial metric and the lapse function, and their spatial derivatives.

4. Simplifying ansätze

It is roughly at this stage that Hořava makes his two great simplifications:

- Detailed balance.
- Projectability.

Even after almost two years it is still somewhat unclear whether these are *only* "simplifying ansatzë" or whether they are in some sense fundamental to Hořava's model. In particular, Thomas Sotiriou, Silke Weinfurtner, and I have argued that "detailed balance" is not fundamental [17, 18], and we have been carefully thinking about the issue of "projectability".

Journal of Physics: Conference Series **314** (2011) 012002 doi:10.1088/1742-6596/314/1/012002

Detailed balance: What is Hořava's detailed balance condition? It is the assertion that the potential $\mathcal{V}(g, N)$ is a perfect square. That is, there is a "pre-potential" $W(g, N)$ such that:

$$\mathcal{V}(g, N) = \left(g^{ij} \frac{\delta W}{\delta g_{jk}} g^{kl} \frac{\delta W}{\delta g_{li}} \right). \tag{13}$$

This simplifies some features of Hořava's model, it makes other features much worse. In particular, if you assume Hořava's detailed balance, and try to recover the Einstein–Hilbert action in the low-energy regime, then:

- You are forced to accept a non-zero cosmological constant of the wrong sign.
- You are forced to accept intrinsic parity violation in the purely gravitational sector.

Now the second item I could live with, albeit uncomfortably, (observationally there is not the slightest hint of parity violation in gravitational physics), but dealing with the the first item will require some mutilation of detailed balance anyway, so we might as well go the whole way and discard detailed balance entirely.

Projectability: What is Hořava's projectability condition?

$$N(x, t) \to N(t) \quad \text{(whence we can subsequently set } N(t) \to 1\text{)}. \tag{14}$$

In standard general relativity the "projectability condition" can always be enforced locally as a gauge choice. Furthermore for "physically interesting" solutions of general relativity it seems that this can always be done (more or less) globally. For instance:

- For the Schwarzschild spacetime this "projectability condition" holds globally in Painlevé–Gullstrand coordinates.
- For the Reissner–Nordström spacetime this "projectability condition" holds for $r \geq Q^2/2m$ in Painlevé–Gullstrand coordinates, (which covers the physically interesting region of the spacetime down to some point deep inside the inner horizon).
- For the Kerr spacetime this condition holds globally (for the physically interesting chronology respecting $r > 0$ region) in Doran coordinates. (So that the geometry is "projectable" down to some point deep inside the inner horizon).
- The FLRW cosmologies also automatically satisfy this "projectability condition".

For this purely pragmatic reason we decided to put "projectability" off to one side for a while, and first deal with "detailed balance" [17, 18] (see [19] and [20] for more recent discussions).

Pragmatic projectable parity preserving model: Abandoning detailed balance, but retaining projectability and parity invariance, it follows that $\mathcal{V}(g)$ must be built out of scalar curvature invariants — and so it is calculable in terms of the Riemann tensor and its derivatives. This tells us the marginal operators in the potential energy must be constructible from objects of the form

$$\Big\{ (\text{Riemann})^d, [(\nabla\text{Riemann})]^2 (\text{Riemann})^{d-3}, \text{etc...} \Big\}. \tag{15}$$

In $d + 1$ dimensions this is a long but finite list. Note that all of these theories should be well-behaved as QFTs (at least in terms of being power-counting renormalizable). All of these theories should have (in condensed matter language) "$z = d$ Lifshitz points". In the specific case $d = 3$ we have the short and rather specific list:

$$\Big\{ (\text{Riemann})^3, [\nabla(\text{Riemann})]^2, (\text{Riemann})\nabla^2(\text{Riemann}), \nabla^4(\text{Riemann}) \Big\}. \tag{16}$$

But in 3 dimensions the Weyl tensor automatically vanishes, so we can always decompose the Riemann tensor into the Ricci tensor, Ricci scalar, plus the metric. Thus we need only consider the much simplified list:

$$\Big\{(\text{Ricci})^3, [\nabla(\text{Ricci})]^2, (\text{Ricci})\nabla^2(\text{Ricci}), \nabla^4(\text{Ricci})\Big\}. \tag{17}$$

Once you look at all the different ways the indices can be wired up this is still relatively messy. In (3+1) dimensions there are only five independent marginal terms (renormalizable by power counting):

$$R^3, \quad RR^i{}_j R^j{}_i, \quad R^i{}_j R^j{}_k R^k{}_i; \quad R\,\nabla^2 R, \quad \nabla_i R_{jk}\,\nabla^i R^{jk}. \tag{18}$$

Now add all possible lower-dimension terms (relevant operators, super-renormalizable by power-counting):

$$1; \qquad R; \qquad R^2; \; R^{ij}R_{ij}. \tag{19}$$

If one absolutely insists on destroying parity invariance "by hand" one could add the relevant (super-renormalizable) operator:

$$R^{ij}C_{ij}, \tag{20}$$

where C_{ij} is the Cotton tensor (which is parity odd since it depends on the 3-dimensional Levi–Civita tensor for its definition). Suppressing parity violation, this results in a potential $\mathcal{V}(g)$ with nine terms and nine independent coupling constants. The Einstein–Hilbert piece of the action is

$$\mathcal{S}_{\text{EH}} = \zeta^2 \int \Big\{(K^{ij}K_{ij} - K^2) + R - g_0\,\zeta^2\Big\}\sqrt{g}\,N\,\mathrm{d}^3x\,\mathrm{d}t. \tag{21}$$

The "extra" Lorentz-violating terms become:

$$\begin{aligned}\mathcal{S}_{\text{LV}} \;=\; \zeta^2 \int \Big\{&\xi\,K^2 - g_2\,\zeta^{-2}\,R^2 - g_3\,\zeta^{-2}\,R_{ij}R^{ij} - g_4\,\zeta^{-4}\,R^3 - g_5\,\zeta^{-4}\,R(R_{ij}R^{ij}) \\ &- g_6\,\zeta^{-4}\,R^i{}_j R^j{}_k R^k{}_i - g_7\,\zeta^{-4}\,R\nabla^2 R - g_8\,\zeta^{-4}\,\nabla_i R_{jk}\,\nabla^i R^{jk}\Big\}\sqrt{g}\,N\,\mathrm{d}^d x\,\mathrm{d}t. \end{aligned} \tag{22}$$

From the normalization of the Einstein–Hilbert term:

$$(16\pi G_{\text{Newton}})^{-1} = \zeta^2; \qquad \Lambda = \frac{g_0\,\zeta^2}{2}; \tag{23}$$

so that ζ is identified as the Planck scale. The cosmological constant is determined by the free parameter g_0, and observationally $g_0 \sim 10^{-123}$, (renormalized after including vacuum energy contributions). In particular, the way we have set this up we are free to choose the Newton constant and cosmological constant independently (and so to be compatible with observation). Key features of the model are:

- The Lorentz violating term in the kinetic energy leads to an extra scalar mode for the graviton, with fractional $O(\xi) = O(\lambda - 1)$ effects at all momenta. Phenomenologically, this behaviour is potentially dangerous and should be carefully investigated.
- The various Lorentz-violating terms in the potential become comparable to the spatial curvature term in the Einstein–Hilbert action for physical momenta of order

$$\zeta_{\{2,3\}} = \frac{\zeta}{\sqrt{|g_{\{2,3\}}|}}; \qquad \zeta_{\{4,5,6,7,8\}} = \frac{\zeta}{\sqrt[4]{|g_{\{4,5,6,7,8\}}|}}. \tag{24}$$

- The Planck scale ζ is divorced from the various Lorentz-breaking scales $\zeta_{\{2,3,4,5,6,7,8\}}$.

- One can drive the Lorentz breaking scale arbitrarily high by suitable adjustment of the dimensionless couplings $g_{\{2,3\}}$ and $g_{\{4,5,6,7,8\}}$.

Based on his intuition coming from "analogue spacetimes", Grisha Volovik has for many years been asserting that the Lorentz-breaking scale should be much higher than the Planck scale [21, 22, 23, 24]. This model very naturally implements that idea in a concrete and explicit manner.

5. Potential problems

Where are the bodies buried?

- *Projectability:* Among other things, this yields a spatially integrated Hamiltonian constraint rather than a super-Hamiltonian constraint. This modifies the equations of motion away from those of standard general relativity. Some authors have made a virtue out of necessity by using the Hamiltonian/super-Hamiltonian distinction to implement "dark dust" at the level of the equations of motion [25].
- *Prior structure:* Is the preferred foliation "prior structure"? Or is it dynamical? Can it be made dynamical by giving up projectability? Can it be viewed as a variant of Einstein–aether theory? (See [26, 27, 28, 29].) Note that I particularly wish to avoid using the phrase "diffeomorphism invariance" since many authors fail to distinguish between "active" and "passive" diffeomorphisms. Invariance under active diffeomorphisms is equivalent to the absence of prior structure, and Hořava's model (because of the preferred foliation) is not diffeomorphism invariant in this active sense *as long as the preferred foliation is non-dynamical.* On the other hand, invariance under passive diffeomorphisms is just coordinate re-parameterization invariance, and *any* theory can be rendered coordinate re-parameterization invariant by introducing enough prior structure. In particular, Hořava's models are *always* diffeomorphism invariant in this passive sense.
- *Scalar graviton:* As long as $\xi \neq 0$ there is a spin-0 scalar graviton, in addition to the spin-2 tensor graviton. This is disturbing for a number of reasons, both theoretical and phenomenological.
- *Hierarchy problem*? Even if the theory is finite, it may still require fine tuning in order to be compatible with observation. Finite does not necessarily mean realistic. It is worth repeating the well-known fact that even finite QFTs still need to be renormalized: One still needs to go through the process of rewriting the bare parameters in terms of the renormalized parameters that are actually available for experimental observation.
- *Beta functions*? Can one go beyond power-counting renormalizability? (See for instance [30, 31], though note that those authors used detailed balance as an essential part of their formalism.)What is the renormalization group (RG) flow? Explicit calculations have so far been extremely limited. (See for instance [32].)

All of these issues have the potential for raising violent conflicts with empirical reality.
In particular note that:

- In the projectable model the unwanted scalar mode has negative kinetic energy all the way from the conformal point ($\xi = 2/3$, $\lambda = 1/3$) to the GR point ($\xi = 0$, $\lambda = 1$). The scalar mode is elliptic (unstable) in this entire region. The often conjectured RG flow from the conformal point $\lambda = 1/3$ in the UV to the GR point $\lambda = 1$ in the IR is not viable.
- From a theoretical perspective the scalar mode also leads to undesirable behaviour such as exhibiting the potential for strong coupling at low energies [33, 34, 35], instabilities, and over-constrained evolution. (See for instance [19, 20] for overviews).

Journal of Physics: Conference Series **314** (2011) 012002 doi:10.1088/1742-6596/314/1/012002

Conformal point	General relativity
$\lambda = 1/3$	$\lambda = 1$
$\dots\dots\dots\dots\star \longrightarrow\longrightarrow\longrightarrow$	$\longrightarrow\longrightarrow\longrightarrow\longrightarrow\longrightarrow\star\dots\dots\dots\dots$
$\xi = 2/3$	$\xi = 0$

Table 1. Conjectured RG flow in Hořava's original model: From conformal invariance in the UV to general relativity in the IR. Note that the RG flow is confined to a region where the scalar mode is a ghost (negative kinetic energy).

- In addition, the spin-0 scalar graviton is potentially dangerous for a number of purely phenomenological reasons:
  - Binary pulsar? (The scalar mode is expected to provide an extra energy-loss mechanism. But how significant is this energy loss?)
  - PPN physics? (In principle the scalar mode might affect the PPN parameters and modify solar system physics. Could it be detected via precise solar system tests?)
  - Eötvös experiments? (Scalar modes generically affect the universality of free-fall. Can this effect in Hořava-like models be suitably suppressed to remain compatible with observations?)

  I emphasize that if one wishes to attack any of these phenomenological issues one will need to make a very precise commitment as to which particular version of Hořava's model one is dealing with, and also carefully check it for unrelated problems. (See for instance [34] for a recent analysis, though that article is more focussed on the non-projectable version of Hořava's model.)

Considerable careful thought on these issues is still needed.

6. Potential solutions

Several options for improving the behaviour of Hořava-like models have been explored.

Healthy extensions: One way of dealing with at least some of the issues arising from the scalar mode is through the so-called "healthy extension" (also called the "consistent extension") of Hořava's model [36, 37, 38]. The central idea of this "consistent extension" is to make the "preferred foliation" dynamical. Effectively one is then dealing with a stably causal spacetime with a preferred dynamical "cosmic time".

- In the "healthy extension" models one enforces $\lambda > 1$ ($\xi < 0$) to avoid negative kinetic energies for the scalar mode.
- One then needs to go beyond projectability to make the spin zero mode hyperbolic, thus greatly increasing the number of terms appearing in the potential energy part of the action, and greatly increasing the overall complexity of the model. (Thus one is considering a version of the general Hořava model, with neither detailed balance nor projectability.)
- To obtain desirable low-energy limit one has to *postulate* that the parameter ξ has RG flow down to $\xi = 0$ from below. (That is $\xi \to 0^-$; equivalently $\lambda \to 1^+$.)
- Because the preferred foliation is now dynamical, the healthy extension models are related to so-called Einstein-aether models. (See in particular [29], and [26, 27, 28] for more background.)

- Somewhat unusually, when linearized around flat space the scalar mode (though not the spin-2 graviton) acquires a rational polynomial dispersion relation

$$\omega^2 = \frac{p(k^2)}{q(k^2)}, \tag{25}$$

where in 3 space dimensions degree$[p]$ = degree$[q] + 3$.
- Recently it has been noted that this picture qualitatively continues to hold in (2+1) dimensions, though then degree$[p]$ = degree$[q] + 2$. Furthermore the scalar mode is in (2+1) dimensions the *only* local degree of freedom [39]. This makes (2+1) Hořava gravity a useful toy model to work with, insofar as it captures the essential physics of the scalar mode, but without the complications attendant on the presence of the spin 2 graviton.
- Strong coupling may still be an issue [20, 40, 41, 42].

Conformal point	General relativity
$\lambda = 1/3$	$\lambda = 1$
$\cdots\cdots\cdots\star\cdots\cdots\cdots$	$\cdots\cdots\star \longleftarrow\longleftarrow\longleftarrow\longleftarrow$
$\xi = 2/3$	$\xi = 0$

Table 2. Conjectured RG flow in "healthy extension"/"consistent extension" models. The RG flow is now postulated to come from a region where both the scalar mode and spin 2 graviton have positive kinetic energy.

Analogue spacetimes: Another option is to search for modifications of Hořava's original models based on hints coming from the "analogue spacetime" programme. A model based on a Bose liquid defined on a lattice has been proposed by Xu and Hořava [43]. Unfortunately, while this particular model has interesting stability properties, it typically violates Lorentz invariance at low energy, which makes realistic model building implausible. On the other hand, the general idea of being guided by the analogue spacetime programme may prove useful in slightly different contexts.

Extra symmetries: One can try to tame the scalar mode by adding an extra gauge symmetry. See for instance [44]. In that particular model the extra symmetry does *not* seem to be related to diffeomorphism invariance in any way, and its physical significance is thereby somewhat obscured.

Other: Other more radical proposals are to somehow merge Hořava-like models with causal dynamical triangulations (CDTs) in the UV. See particularly [45], with additional background discussion in [9, 44].

Summary: Overall, the choices seem to be:
- Approximately decouple the scalar mode?
- Kill the scalar mode with more symmetry?
- Kill the scalar mode with less symmetry? Impose even more drastic a priori restrictions on the metric? (Even more restrictions than projectability.)

For various reasons, none of these approaches is as yet fully satisfying.

7. Discussion

The (generalized) Hořava-like models naturally provide a class of tempting models, of varying degrees of complexity and physical viability, that are rather interesting to work on. My personal view is that considerably more work needs to go into understanding the fundamentals of these models, specifically the role played by the scalar mode. Without a deeper understanding of the fundamental framework one is operating in, detailed phenomenological studies (and in particular specific applications to cosmology and astrophysics) are simply premature. Specifically, one needs more than hand-waving "of course it runs to general relativity in the IR" arguments. There may be subtle (or even not so subtle) qualitative deviations from general relativity due to the preferred foliation, and really pinning that issue down would be a good idea before investing more time on detailed applications. Overall, this is still a very active field. While the initial feeding frenzy has somewhat subsided, considerable ongoing interest remains. There are also some very real physics challenges remaining.

Acknowledgments

This research was supported by the Marsden Fund administered by the Royal Society of New Zealand. I also wish to thank the CSIC-IAA (Granada) for hospitality, and to thank Thomas Sotiriou and Silke Weinfurtner for their comments.

References

[1] D. Mattingly, "Modern tests of Lorentz invariance", Living Rev. Rel. **8** (2005) 5. [gr-qc/0502097].
[2] D. Mattingly, "Have we tested Lorentz invariance enough?", PoS **QG-Ph** (2007) 026 . [arXiv:0802.1561 [gr-qc]].
[3] T. Jacobson, S. Liberati, D. Mattingly, "TeV astrophysics constraints on Planck scale Lorentz violation", Phys. Rev. **D66** (2002) 081302. [hep-ph/0112207].
[4] T. Jacobson, S. Liberati, D. Mattingly, "Lorentz violation at high energy: Concepts, phenomena and astrophysical constraints", Annals Phys. **321**, 150-196 (2006). [astro-ph/0505267].
[5] V. A. Kostelecky, S. Samuel, "Spontaneous breaking of Lorentz symmetry in string theory", Phys. Rev. **D39** (1989) 683.
[6] D. Colladay, V. A. Kostelecky, "Lorentz violating extension of the standard model", Phys. Rev. **D58** (1998) 116002. [hep-ph/9809521].
[7] P. Hořava, "Quantum gravity at a Lifshitz point", Phys. Rev. **D79** (2009) 084008. [arXiv:0901.3775 [hep-th]].
[8] P. Hořava, "Membranes at quantum criticality", JHEP **0903** (2009) 020. [arXiv:0812.4287 [hep-th]].
[9] P. Hořava, "Spectral dimension of the universe in quantum gravity at a Lifshitz point", Phys. Rev. Lett. **102** (2009) 161301. [arXiv:0902.3657 [hep-th]].
[10] M. Visser, "Acoustic black holes: Horizons, ergospheres, and Hawking radiation", Class. Quant. Grav. **15** (1998) 1767-1791. [gr-qc/9712010].
[11] C. Barceló, S. Liberati, M. Visser, "Analogue gravity", Living Rev. Rel. **8** (2005) 12. [gr-qc/0505065].
[12] S. W. Weinberg, *The quantum theory of fields*, volumes I–III, (Cambridge University Press, England, 2000).
[13] M. Visser, "Lorentz symmetry breaking as a quantum field theory regulator", Phys. Rev. **D80** (2009) 025011. [arXiv:0902.0590 [hep-th]].
[14] D. Anselmi, M. Halat, "Renormalization of Lorentz violating theories", Phys. Rev. **D76** (2007) 125011. [arXiv:0707.2480 [hep-th]].
[15] D. Anselmi, "Weighted scale invariant quantum field theories", JHEP **0802**, 051 (2008). [arXiv:0801.1216 [hep-th]].
[16] M. Visser, "Power-counting renormalizability of generalized Hořava gravity", [arXiv:0912.4757 [hep-th]].
[17] T. P. Sotiriou, M. Visser, S. Weinfurtner, "Phenomenologically viable Lorentz-violating quantum gravity", Phys. Rev. Lett. **102** (2009) 251601. [arXiv:0904.4464 [hep-th]].
[18] T. P. Sotiriou, M. Visser, S. Weinfurtner, "Quantum gravity without Lorentz invariance", JHEP **0910** (2009) 033. [arXiv:0905.2798 [hep-th]].
[19] S. Weinfurtner, T. P. Sotiriou, M. Visser, "Projectable Hořava–Lifshitz gravity in a nutshell", J. Phys. Conf. Ser. **222** (2010) 012054. [arXiv:1002.0308 [gr-qc]].
[20] T. P. Sotiriou, "Hořava–Lifshitz gravity: a status report", J. Phys. Conf. Ser. **283** (2011) 012034. [arXiv:1010.3218 [hep-th]].
[21] G. E. Volovik, "Fermi-point scenario of emergent gravity", PoS **QG-Ph** (2007) 043 [arXiv:0709.1258 [gr-qc]].

[22] G. E. Volovik, "Emergent physics: Fermi point scenario", Phil. Trans. Roy. Soc. Lond. **A366** (2008) 2935-2951. [arXiv:0801.0724 [gr-qc]].
[23] G. E. Volovik, "$z = 3$ Lifshitz-Hořava model and Fermi-point scenario of emergent gravity", JETP Lett. **89** (2009) 525-528. [arXiv:0904.4113 [gr-qc]].
[24] G. E. Volovik, "The superfluid universe", [arXiv:1004.0597 [cond-mat.other]].
[25] S. Mukohyama, "Dark matter as integration constant in Hořava–Lifshitz gravity", Phys. Rev. **D80** (2009) 064005. [arXiv:0905.3563 [hep-th]].
[26] T. Jacobson, D. Mattingly, "Gravity with a dynamical preferred frame", Phys. Rev. **D64** (2001) 024028. [gr-qc/0007031].
[27] T. Jacobson, D. Mattingly, "Einstein-aether waves", Phys. Rev. **D70** (2004) 024003. [gr-qc/0402005].
[28] C. Eling, T. Jacobson, D. Mattingly, "Einstein-aether theory", [gr-qc/0410001].
[29] T. Jacobson, "Extended Hořava gravity and Einstein-aether theory", Phys. Rev. **D81** (2010) 101502. [arXiv:1001.4823 [hep-th]].
[30] D. Orlando and S. Reffert, "On the renormalizability of Hořava–Lifshitz-type gravities," Class. Quant. Grav. 26, 155021 (2009) arXiv:0905.0301 [hep-th].
[31] D.Orlando and S.Reffert, "On the perturbative expansion around a Lifshitz point," Phys. Lett. B 683, 62 (2010) [arXiv:0908.4429 [hep-th]].
[32] G. Giribet, D. L. Nacir, F. D. Mazzitelli, "Counterterms in semiclassical Hořava–Lifshitz gravity", JHEP **1009** (2010) 009. [arXiv:1006.2870 [hep-th]].
[33] C. Charmousis, G. Niz, A. Padilla, P. M. Saffin, "Strong coupling in Hořava gravity", JHEP **0908** (2009) 070. [arXiv:0905.2579 [hep-th]].
[34] I. Kimpton, A. Padilla, "Lessons from the decoupling limit of Hořava gravity", JHEP **1007** (2010) 014. [arXiv:1003.5666 [hep-th]].
[35] A. Padilla, "The good, the bad and the ugly .... of Hořava gravity", J. Phys. Conf. Ser. **259** (2010) 012033. [arXiv:1009.4074 [hep-th]].
[36] D. Blas, O. Pujolas, S. Sibiryakov, "On the extra mode and inconsistency of Hořava gravity", JHEP **0910** (2009) 029. [arXiv:0906.3046 [hep-th]].
[37] D. Blas, O. Pujolas, S. Sibiryakov, "Consistent extension of Hořava gravity", Phys. Rev. Lett. **104** (2010) 181302. [arXiv:0909.3525 [hep-th]].
[38] D. Blas, O. Pujolas, S. Sibiryakov, "Models of non-relativistic quantum gravity: The good, the bad and the healthy", [arXiv:1007.3503 [hep-th]].
[39] T. P. Sotiriou, M. Visser, S. Weinfurtner, "Lower-dimensional Hořava–Lifshitz gravity", [arXiv:1103.3013 [hep-th]].
[40] A. Papazoglou, T. P. Sotiriou, "Strong coupling in extended Hořava–Lifshitz gravity", Phys. Lett. **B685** (2010) 197-200. [arXiv:0911.1299 [hep-th]].
[41] D. Blas, O. Pujolas, S. Sibiryakov, "Comment on 'Strong coupling in extended Hořava–Lifshitz gravity'", Phys. Lett. **B688** (2010) 350-355. [arXiv:0912.0550 [hep-th]].
[42] A. Wang and Q. Wu, "Stability of spin-0 graviton and strong coupling in Hořava–Lifshitz theory of gravity," Phys. Rev. D **83** (2011) 044025 [arXiv:1009.0268 [hep-th]].
[43] C. Xu, P. Hořava, "Emergent gravity at a Lifshitz point from a Bose liquid on the lattice", Phys. Rev. **D81** (2010) 104033. [arXiv:1003.0009 [hep-th]].
[44] P. Hořava, C. M. Melby-Thompson, "General covariance in quantum gravity at a Lifshitz point", Phys. Rev. **D82** (2010) 064027. [arXiv:1007.2410 [hep-th]].
[45] J. Ambjorn, A. Gorlich, S. Jordan, J. Jurkiewicz, R. Loll, "CDT meets Hořava–Lifshitz gravity", Phys. Lett. **B690** (2010) 413-419. [arXiv:1002.3298 [hep-th]].

Journal of Physics: Conference Series **314** (2011) 012003 doi:10.1088/1742-6596/314/1/012003

Black hole entropy: lessons from loop quantum gravity

J. Fernando Barbero G.

Instituto de Estructura de la Materia, CSIC, Serrano 123, 28006 Madrid, Spain

E-mail: fbarbero@iem.cfmac.csic.es

Abstract. The purpose of this contribution is is to discuss black hole entropy in the loop quantum gravity framework. Special attention is paid to the description of the microscopic degrees of freedom responsible for the entropy, the statement of the combinatorial problems that must be solved in order to count them, and the behaviour of the entropy as a function of the horizon area. In particular I will review the derivation of the Bekenstein-Hawking law and its logarithmic corrections. I end with a comparison between the results derived within loop quantum gravity and the ones obtained by other approaches.

Any fundamental approach to quantum gravity should be able to explain the origin of the black hole (BH) microscopic degrees of freedom and describe the appearance of objects that behave like macroscopic black holes. I will argue in this contribution that the existing model in loop quantum gravity (LQG) not only gives a plausible explanation for the behaviour of black hole entropy as a function of the area but also makes some intriguing predictions.

The study of black hole entropy is a natural testing ground for LQG. A rough time line including some of the key steps in the development of the subject is the following:

- Smolin ('95): modeling black holes as inner boundaries in LQG, horizons and Chern-Simons theories, quantum group methods,... [1]
- Krasnov ('96-'98): role of Chern-Simons models, use of spin networks, distinguishability of punctures on the horizon, statistical mechanics issues,... [2]
- Rovelli ('96): introduction of combinatorial methods for black holes in LQG, distinguishability of punctures at the horizon, derivation of the Bekenstein-Hawking law,... [3]
- Ashtekar, Baez, Corichi and Krasnov ('98-'00): black holes as isolated horizons (IH's), Hamiltonian formulation for the sector of general relativity consisting on spacetimes with IH's as inner boundaries, role of the Immirzi parameter γ, derivation of the Bekenstein-Hawking law,...[4, 5]
- Domagała, Lewandowski and Meissner ('04): reformulation of the combinatorial problems involved in the definition of the BH entropy, precise counting and determination of γ,...[6]
- Corichi, Díaz Polo, F-Borja ('07): Entropy quantization of the entropy as a function of the area. [7, 8, 9].
- Agulló, Barbero, Díaz Polo, F-Borja, Villaseñor ('08): Detailed structure of the entropy as a function of the area [8, 9].
- Engle, Noui and Perez ('09): $SU(2)$ formulation from first principles [10].

It must be said, however, that other significant contributions have appeared in the literature (see, for example, [11, 12, 13]) that suggest and explore alternative models for black holes within the spirit of LQG. I think that it is fair to say that circa 1998 there was a reasonable description of BH's within LQG although the subject has recently seen a revival with the appearance of new results of methods that I will briefly mention and describe in this paper.

From a fundamental point of view a key development in the field was the introduction of isolated horizons (IH's) as a means to describing black holes with LQG inspired methods. In order to arrive at the concept of an isolated horizon one starts by first introducing non-expanding horizons Δ, that is, null, 3-dim submanifolds of $(\mathcal{M}, g_{ab})$ such that

- Δ has the topology of $S^2 \times \mathbb{R}$.
- The expansion of any null normal vanishes $q^{ab}\nabla_a \ell_b = 0$ on Δ.
- The Einstein field equations hold on Δ with T_{ab} satisfying the (mild) condition that $-T^a_{\ b}\ell^b$ is a future directed, causal vector. (This is true for minimally coupled matter fields.)

This definition, in particular the second condition, guarantees that the area of Δ is constant, there is no matter flux through it, and the horizon geometry $(q_{ab}, \mathcal{D})$ is time-independent. Isolated horizons are defined as non-expanding horizons with an –essentially unique– null normal ℓ that is a symmetry of the horizon geometry ($\mathcal{L}_\ell q_{ab} = 0\,, [\mathcal{L}_\ell, \mathcal{D}] = 0$). It is important to point out here that the IH concept is significantly weaker than that of a Killing horizon, in fact the IH definition strives to capture the minimum structure necessary to keep intact (some of) the laws of black hole thermodynamics in order to define a subsector of GR with as many degrees of freedom as possible (in fact an infinite number of them in the subsector defined by space-times admitting IH's as inner boundaries). It is also important to point out that, unlike event horizons, isolated horizons are defined locally.

In order to model black holes in LQG (at least in a semiclassical regime) one considers a reduction of general relativity consisting of spacetimes with isolated horizons as inner boundaries. The key reason why this is useful is because this system admits a Hamiltonian description that can be taken as the starting point for quantization. This idea is similar in spirit, but different, to the study of the quantization of mini and midisuperspace models for which a subset of the gravitational field configurations is selected by imposing symmetry restrictions on the metrics defining the configuration space.

Isolated horizons come in different flavors that can be classified according to the symmetries of the metric on the horizon as

- Type I: maximal symmetry corresponding to spherical geometry.
- Type II: 2-dim symmetry group. Axi-symmetric.
- Type III: 1-dim symmetry group of diffeomorphisms generated by the null normal ℓ.

Up to date only type I and II have been considered in the literature because they describe relatively simple situations. Here I will only consider the sector of GR consisting of space times with an inner boundary which is a type I isolated horizon[1] of fixed classical area a_κ. Instead of using the standard metric formulation it proves useful to use real Ashtekar variables (A^i_a, E^a_i) ($SU(2)$ connection and triad) to describe the classical dynamics of the system, this is so because we will use LQG methods to quantize the system. It is also convenient to partially gauge fix the internal $SU(2)$ to $U(1)$ on the horizon by projecting in the internal direction given by the (internal) vector r^i, (where $r^i E^a_i = \sqrt{|\det q|}r^a$ and r^a is the unit normal to the 2-sphere S) and consider the intrinsic geometry of Δ given by the pull-back of $A^i r_i =: W$ to S. I want to mention here that there are recent proposals [10, 14], that have expanded the somewhat heuristic ideas of [11], to avoid this partial gauge fixing and work within a purely $SU(2)$ model. As many

[1] Type II isolated horizons are used to model rotating black holes.

qualitative results coincide for all the different proposals appearing in the literature, I will not discuss particular details here but only mention those issues for which the difference is significant (regarding, for example, the predicted logarithmic corrections).

The fact that the inner space-time boundary is an isolated horizon has two important consequences: The first is the appearance of a boundary condition of the form

$$F = -\frac{2\pi}{a_\kappa} 8\pi G\gamma \Sigma^i r_i \tag{1}$$

where Σ^i is the pull-back to S of $\eta_{abc}E^a_i\eta^{ij}$ and $F := dW$. The IH boundary conditions imply the reducibility of the pulled-back connection. The second is the important fact that the symplectic form picks a $U(1)$ Chern-Simons surface term of the form

$$\frac{1}{2\pi}\frac{a_\kappa}{4\pi\gamma\ell_P^2}\oint_S d_1W \wedge d_2W\,,$$

where here a_κ (arbitrary at this stage) is the classical area of the horizon. Of course, in addition of these new features introduced by the implementation of the isolated horizon conditions and the presence of an inner space-time boundary, one has to take into account the standard constraints appearing in the Ashtekar formulation of general relativity in order to fully define the dynamics of the system.

Once I have briefly introduced the classical framework used to describe black holes in loop quantum gravity I will discuss quantization. The first step in this process is the construction of a suitable Hilbert space. The presence of an inner boundary in space-time and, most crucially, the appearance of a surface term in the symplectic structure, will have interesting consequences. The most important one is the possibility of finding a Hilbert space for the model that can be written as the tensor product of a certain bulk Hilbert space $\mathcal{H}_V$ –very similar to the ones customarily used in LQG– and a surface Hilbert space $\mathcal{H}_S$. These will not be independent but will be related by quantizing the boundary conditions (1) and using them to select a subspace that will define the kinematical Hilbert space for the system. The dynamics will be encoded in the quantum version of these constraints as in the usual canonical quantization of GR in terms of connection variables.

The details of the choice for $\mathcal{H}_V$ and $\mathcal{H}_S$ are dictated, precisely, by the need to implement the quantum boundary condition (1). It will be important now to use the fact that, in the bulk, the gravitational variable will be taken to be a $SU(2)$ connection, whereas we have a $U(1)$ connection in the inner boundary. The volume Hilbert space $\mathcal{H}_V$ will be a subspace of the LQG Hilbert space $L^2(\bar{\mathcal{A}}, \mu_{AL})$ (where $\bar{\mathcal{A}}$ denotes the space of generalized $SU(2)$ connections, μ_{AL} is the Ashtekar-Lewandowski measure and we mode out $SU(2)$ gauge transformations that reduce the identity on the horizon). A suitable basis for this Hilbert space is defined by using spin networks (that may end at the inner horizon S). Remember that spin networks are graphs with edges (consisting of analytic curves that meet at vertices) labeled by spins j_e and vertices labeled by intertwiners i_v. Those spin networks with edges piercing the horizon transversally (at the so called punctures labeled by an index I) play an special role because the quantum numbers of the edges at the intersections appear in a crucial way in the implementation of the quantum boundary conditions. Important quantum numbers are the $j_I \in \mathbb{Z}/2$ labels at the punctures and the $m_I = -j_I, \ldots, j_I$ spin components defined with the help of the (internal) r_i vector. The final description of the volume Hilbert space admits a convenient representation [5, 6] as an orthogonal sum. Consider a finite set of points on S, $P = \{P_1, P_2, \ldots, P_N\}$ labeled by pairs of numbers (j_I, m_I), $I = 1, \ldots, N$ such that $j_I \in \mathbb{N}/2$ and $m_I \in \{-j_I, -j_I + 1, \ldots, j_I\}$, we have then

$$\mathcal{H}_V = \bigoplus_{(P,j,m)} \mathcal{H}_V^{P,j,m}\,,$$

where the sum is extended to all the possible subsets P (including $\emptyset$) and (j,m)-labelings. In the previous expression $\mathcal{H}^{\emptyset}$ is spanned by the spin networks that do not pierce the horizon, whereas each of the Hilbert spaces $\mathcal{H}_V^{P,j,m}$ is spanned by spin networks such that the edges piercing the horizon carry the labels (j_I, m_I).

I will briefly consider now the construction of the surface Hilbert space $\mathcal{H}_S$. The appearance of a $U(1)$ Chern-Simons surface term in the symplectic structure forces us to impose a "pre-quantization condition" on the classical horizon area $a_\kappa = 4\pi\gamma\ell_P^2\kappa$ ($\kappa \in \mathbb{N}$) for consistency because we will take as $\mathcal{H}_S$ the Hilbert space of the $U(1)$ Chern-Simon theory defined on a punctured sphere. As in the case of the volume Hilbert space, $\mathcal{H}_S$ also admits a representation as an orthogonal sum. If we choose a finite sequence of points on S, $\vec{P} = (P_1, P_2, \ldots, P_N)$ labeled by numbers $b_i \in \mathbb{Z}_\kappa$, $i = 1, \ldots, N$ satisfying $b_1 + \cdots + b_N = 0$ the surface Hilbert space can be written in the form [6]

$$\mathcal{H}_S = \bigoplus_{(\vec{P},b)} \mathcal{H}_S^{\vec{P},b},$$

where the sum is extended to all the possible sequences $\vec{P}$ and b-labelings (we introduce a 1-dim Hilbert space $\mathcal{H}_S^{\vec{\emptyset}}$ for the empty sequence).

Once this is done we are ready to describe the Kinematical Hilbert space $\mathcal{H}_{kin}$ for the system. The main idea is to implement the quantum boundary condition (1) in $\mathcal{H}_V \otimes \mathcal{H}_S$ in order to take into account that S is a section of an isolated horizon Δ. In practice it is convenient [5] to enforce an integrated version of (1) acting on state vectors $\Psi \in \mathcal{H}_V \otimes \mathcal{H}_S$

$$(\mathbb{I} \otimes \underbrace{e^{i\int_R \hat{F}}}_{\text{holonomies}})\Psi = (\underbrace{e^{-\frac{2\pi i}{a_\kappa} 8\pi G\gamma \int_R (\hat{\Sigma}\cdot r)}}_{\text{fluxes}} \otimes \mathbb{I})\Psi\,. \tag{2}$$

The solutions to this equation can be written as orthogonal sums of subspaces of the type $\mathcal{H}_V^{P,j,m} \otimes \mathcal{H}_S^{\vec{P},b}$ such that the labels defining them satisfy the following conditions:

(i) The points in $P = \{P_1, \ldots, P_N\}$ coincide with the elements in the sequence $(P_1, \ldots, P_N)$.

(ii) $b_i = -2m_i \,(\mathrm{mod}\kappa)$ for $i = 1, \ldots, N$.

We finally find that the black hole kinematical Hilbert space can be written in the following form [5, 6]

$$\mathcal{H}_{kin} = \bigoplus_{\vec{P},j,m} \mathcal{H}_V^{P,j,m} \otimes \mathcal{H}_S^{\vec{P},b(m)}\,.$$

A pictorial representation of the elements appearing in the previous construction is shown in Fig. 2

Several comments are in order now:

(i) The key fact that explains why the previous construction can be made to work is that the eigenvalues of the l.h.s. and r.h.s. of the quantum boundary conditions (2) can be matched precisely if a_κ is the pre-quantized value of the area introduced above. This means that there is a non trivial space of solutions to the quantum matching conditions and a large enough kinematical Hilbert space.

(ii) Classically the bulk fields determine the surface fields by continuity but quantum mechanically something qualitatively different happens: distributional objects must be introduced so there is no obvious way to extend them to the inner horizon by looking at their values in the bulk. This is a reflection of the well known difference between the quantum and classical configuration spaces and explains the appearance of the *independent surface d.o.f.* that account for the BH entropy. This is a genuine quantum effect.

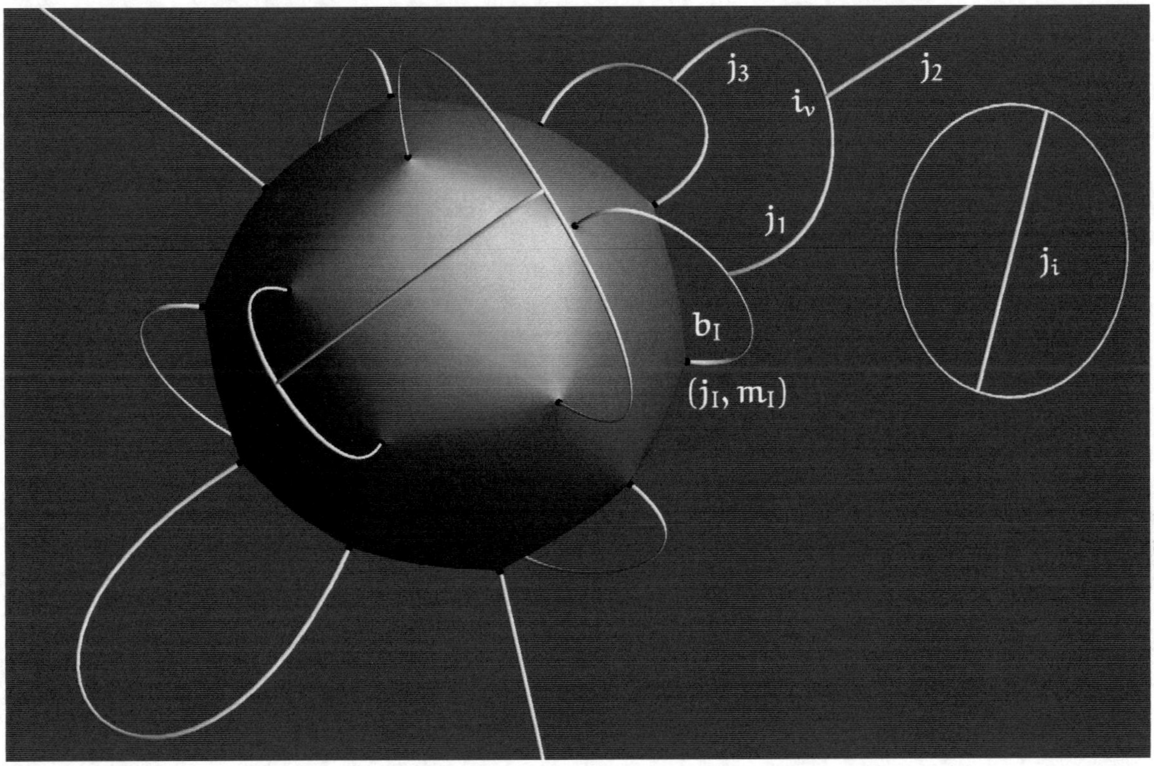

Figure 1. Schematic representation of the Hilbert space for black holes in LQG. The surface represents the inner horizon. The kinematical Hilbert space for the model is built as a direct sum of spaces of the form $\mathcal{H}_V^{P,j,m} \otimes \mathcal{H}_S^{\vec{P},b(m)}$. Each of them is labeled by a certain spin network and a set of quantum numbers on the horizon satisfying the restrictions explained in the main text.

(iii) It is important to notice that there are two different concepts of area in the construction: the classical pre-quantized area a_κ –fixed for the isolated horizon– and the area eigenvalue of the bulk area operator associated to the 2-sphere S given by

$$Area_S = 8\pi\gamma\ell_P^2 \sum_{I=\text{punct.}} \sqrt{j_I(j_I+1)}\,.$$

The actual implementation of the dynamics of general relativity requires us to quantize the constraints and solve them in the kinematical Hilbert space introduced above. This program is certainly not easy, and cannot be carried to completion at the present moment –the difficulties of appropriately dealing with the scalar constraint immediately come to mind. However, it is actually possible to avoid many issues associated with this problem because the functional differentiability of the Hamiltonian constraint in the classical theory forces the lapse to be zero on S. This means that it is not necessary to take it into account[2]. The other constraints (the Gauss law and the diffeomorphism constraint) act as expected and are dealt with in the customary way: the Gauss law translates into the requirement that the physical states must be $SU(2)$ invariant (a fact that can be incorporated in the choice of spin network basis). Diffeomorphisms, on their

[2] Although one one has to assume the existence of *some* solutions to Hamiltonian constraint in $\mathcal{H}_V$ for every choice of (j, m), a condition that is accepted to be mild.

Journal of Physics: Conference Series **314** (2011) 012003 doi:10.1088/1742-6596/314/1/012003

hand, act non-trivially on the surface and volume states but in such a way that the full physical state must be invariant. The only thing that matters, in practice, is the number of edges that pierce the horizon (on the so called punctures) and their quantum nmbers but not their location. We can then formally write

$$\mathcal{H}_{phys} = \bigoplus_{j,m} \mathcal{H}^{b(m),j,m}$$

where the labels are required to satisfy the conditions written above. An interesting comment to make at this point is that the area assigned to S by the bulk geometry is controlled by the j's, whereas the intrinsic horizon degrees of freedom are represented by the b's.

We are ready now to define the black hole entropy [5]. To this end we consider the Hilbert space of "black hole states"

$$|(m_1, j_1, \ldots, m_N, j_N), \text{the rest of the graph}\rangle_V \otimes |(b_1, \ldots, b_N)\rangle_S, \qquad N \in \mathbb{N}_0\,,$$

satisfying the conditions $b_i \equiv -2m_i \pmod{\kappa}$, for $i = 1, \ldots, N$. We introduce now an area interval $[a_\kappa - \delta, a_\kappa + \delta]$ with a δ of the order of ℓ_P^2. The entropy is defined by by tracing out the bulk degrees of freedom to get a density matrix describing a maximal entropy mixture of surface states with area eigenvalues in the interval $[a_\kappa - \delta, a_\kappa + \delta]$. This amounts to counting the number of allowed lists $(b_1, \ldots, b_N)$ of non-zero elements of $\mathbb{Z}_\kappa$ satisfying $b_1 + \cdots + b_N = 0$, and such that $b_i = -2m_i \pmod{\kappa}$ for some *permissible* list of spin components $\vec{m}$. Here *permissible* means that there exists a list of non-vanishing spin labels $(j_1, \ldots, j_N)$ such that each m_I is a spin component of j_I and

$$a_\kappa - \delta \leq 8\pi\gamma\ell_P^2 \sum_{I=1}^{N} \sqrt{j_I(j_I+1)} \leq a_\kappa + \delta\,.$$

The problem of computing the entropy has been reduced in this way to a very concrete combinatorial problem whose solution allows, in principle, to compute the black hole entropy as a function of the horizon area. However, this combinatorial problem can be cast is a simpler form that is specially useful to understand the detailed behaviour of the entropy. This was done by Domagała-Lewandowski [6]. According to them

The entropy S of a quantum horizon of the classical area a according to Quantum Geometry and the Ashtekar-Baez-Corichi-Krasnov framework is

$$S(a) = \log \mathfrak{n}(a),$$

where $\mathfrak{n}(a)$ is 1 plus the number of all finite sequences $\vec{m} := (m_1, \ldots, m_N)$ of non-zero elements of $\frac{1}{2}\mathbb{Z}$, such that the following equality (the projection constraint) and inequality are satisfied:

$$\sum_{I=1}^{N} m_I = 0, \quad \sum_{I=1}^{N} \sqrt{|m_I|(|m_I|+1)} \leq \frac{a}{8\pi\gamma\ell_P^2}\,.$$

The condition $\sum_I m_I = 0$ is known as the projection constraint. By relying on this recipe it is not difficult to obtain a plot of the entropy versus the area (see Fig. 2)

The entropy is plotted for each area eigenvalue for the smallest black holes. For some of them it is zero because there are no solutions to the projection constraint (this means that not all the eigenvalues of the area are actually represented). We can see the expected linear growth as a function of the area. In order to reproduce the Bekenstein-Hawking law exactly, the Immirzi

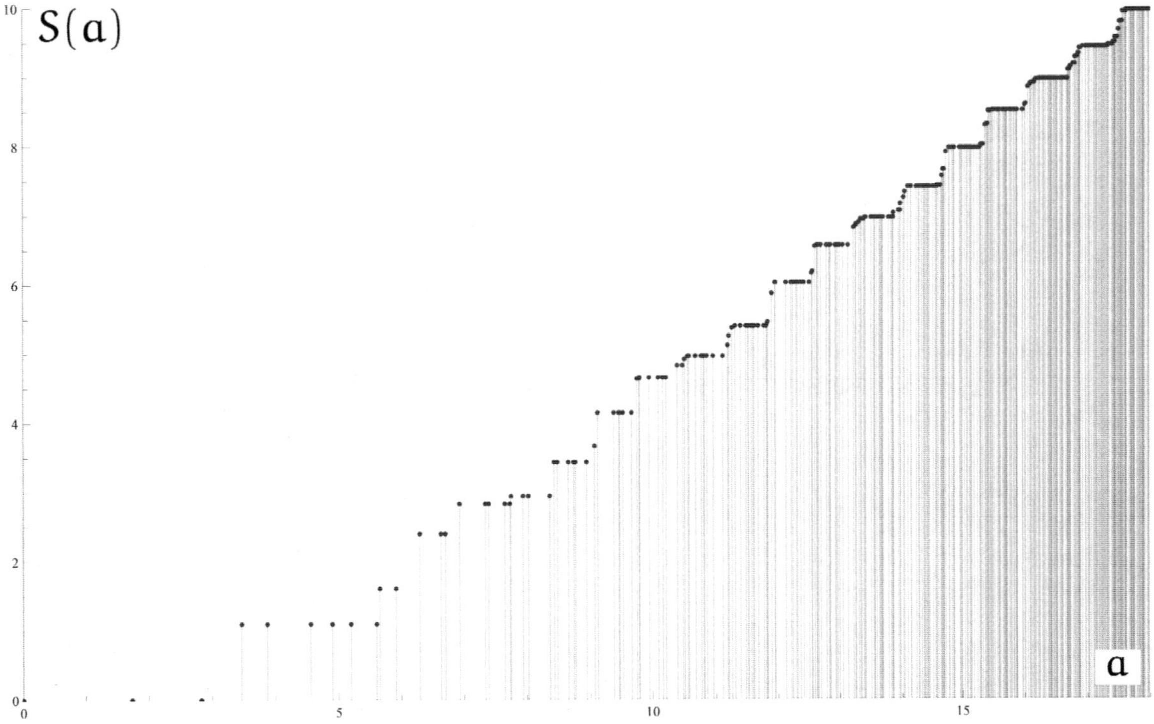

Figure 2. Black hole entropy computed according to the Domagała-Lewandowski prescription (area measured in units of $4\pi\gamma\ell_P^2$).

parameter γ must be fixed. This choice is universal i.e. it works for other types of black holes modeled as type II isolated horizons (rotating and/or distorted), dilatonic black holes, black holes in gravity models with non minimally coupled matter,etc. The exact value of γ obtained in the Domagała-Lewandowski approach differs from the one obtained by using other proposals (such as the $SU(2)$ model of [10]). In addition to the sought for linear growth, the plot shows the existence of an unexpected behaviour detected first by the authors of [7]: the entropy of the smallest black holes is effectively quantized. This is a genuinely new phenomenon made even more remarkable by the fact that the step size remains constant while the separation between consecutive area eigenvalues shrinks very quickly. A number of questions come to mind at this moment. What does this behaviour mean? Is it possible to explain it in fundamental terms? Does it persist for macroscopic black holes?

The nature of the combinatorial problem defining the entropy is such that it is actually possible to develop efficient counting methods to deal with it. First it is possible to characterize the degeneracies in the area eigenvalues. This is done by using diophantine equations. It is also possible to exactly count the number of black hole configurations contributing to the entropy for each value of the black hole area (this requires the solution of the projection constraint and the condition involving an inequality). Finally, it is possible to code this information in generating functions and use them to study the detailed behaviour of the entropy.

Generating functions are very useful devices in combinatorics. They can be efficiently used to encode and manipulate the information relevant to the solution of a given combinatorial problem. In the case of black holes in LQG, the relevant generating functions have been described in [15]. If one follows, for example, the Domagała-Lewandowski recipe, the black hole generating function

Journal of Physics: Conference Series **314** (2011) 012003 doi:10.1088/1742-6596/314/1/012003

is given by [16, 9]

$$G(z,x_1,x_2,\ldots)=\left(1-\sum_{i=1}^{\infty}\sum_{\alpha=1}^{\infty}(z^{k_\alpha^i}+z^{-k_\alpha^i})x_i^{y_\alpha^i}\right)^{-1}.$$

In the previous formula the variables x_i are associated to square-free integers p_i, z is an extra variable needed to account for the projection constraint and the numbers (k_α^i, y_α^i) are obtained from the solutions to the Pell equation $x^2 - p_i y^2 = 1$ associated with each squarefree integer p_i. The coefficient of the term $z^0 x_1^{q_1}\cdots x_i^{q_i}\cdots$ gives the number of sequences $\vec{m}$ satisfying the projection constraint and such that $a=\sum_i q_i\sqrt{p_i}$ (in units such that $4\pi\gamma\ell_P^2 = 1$). Despite the apparent infinite number of terms, for a given value of a only a finite number of variables and terms is needed. This generating function solves the problem of counting the number of sequences $\vec{m}$ such that

$$2\sum_I\sqrt{|m_I|(|m_I|+1)}=a\,,\quad \sum_I m_I=0$$

for a given value of the area a. In order to take into account the inequality in the definition of the entropy one has to introduce a Laplace-Fourier transform representation leading to the formula [17, 15]

$$e^{S_{\leq}(a)}=\frac{1}{(2\pi)^2 i}\int_0^{2\pi}\int_{x_0-i\infty}^{x_0+i\infty}s^{-1}\Big(1-2\sum_{k=1}^{\infty}e^{-s\sqrt{k(k+2)}}\cos\omega k\Big)^{-1}e^{as}\mathrm{d}s\,\mathrm{d}\omega$$

This is an *exact* expression for the entropy. It allows, in particular, to write down the equation that γ must satisfy in order to obtain the 1/4 factor of the Bekenstein-Hawking law. The position of the poles of the integrand for $\omega = 0$ determines the growth of the entropy as a function of the area (in particular the one with the largest real part s_0). It is straightforward to justify that the entropy growths linearly and, hence satisfies the Bekenstein-Hawking law. Also, the previous expression provides a simple way to find the value of the Immirzi parameter $\pi\gamma = s_0$. It is important to mention at this point that the previous formula provides also a (γ-independent!) logarithmic correction to the linear growth of the entropy. $-(1/2)\log(a/\ell_P^2)$. This is particularly relevant because, while there are a number of models (connected and not connected to LQG) for the black hole entropy that conform to the BH law, in many cases they predict different logarithmic corrections. This fact makes it possible, in principle, to choose among them on physical grounds. For example, the $SU(2)$ proposals [11, 10] predict a logarithmic correction given by $-(3/2)\log(a/\ell_P^2)$.

By carefully modifying the black hole generating functions it is possible to isolate parts of the black hole degeneracy spectrum and build it by adding "steps". For example, introducing a "step counter" (a function of the labels of the black hole configurations) it is possible to partition the space of black hole configurations in such a way that the contributions to the single steps seen in the black hole "staircase" are singled out. From the point of view of the generating functions this can be achieved easily by introducing an extra variable. For the peak counter $P := 3K + 2N$ introduced[3] in [18] the generation function is

$$\left(1-\sum_{i=1}^{\infty}\sum_{\alpha=1}^{\infty}\nu^{3k_\alpha^i+2}(z^{k_\alpha^i}+z^{-k_\alpha^i})x_i^{y_\alpha^i}\right)^{-1}.$$

This counter plays a fundamental role in the analysis of the detailed behavior of the entropy (see [9]).

[3] Here K is twice the total spin on the horizon and N the number of punctures.

It is natural to confront the results presented here with those found in other approaches. Though this is not the place to give a comprehensive list of all of them let me mention some (see the excellent review by Wald in [19]):

- Early calculations within the Euclidean quantum gravity framework [20, 21].
- Blak hole entropy as entanglement entropy between degrees of freedom inside and outside the horizon [22, 23].
- Noether charge methods (generalizable to diffeomorphism invariant models) [24].
- Black hole entropy defined as the thermal entropy of the thermal atmosphere of the black hole. [25].
- Entropy in Sakharov's theory of induced gravity (the dynamical aspects of gravity arise from the collective excitations of massive fields with constraints introduced to cancel divergences and ensure that $\Lambda = 0$. [26]
- Entropy as a measure of the different shapes of the horizon [27].
- Entropy from conformal field theories [28].
- ...

Arguably string theories are the leading contenders in the quest for quantum gravity. As far as the understanding of black hole entropy is concerned string inspired models are well developed and have led to a number of unambiguous predictions (as the literature on the subject is vast let me refer the reader to the pioneering papers on the subject [29, 30, 31] and the pedagogical review [32]). At low energies string theories reduce to a 10-dim SUGRA theory with black hole solutions. One can alternatively consider a weak coupling limit of string theory where perturbative computations can be performed and map the weakly coupled states with black hole solutions (this is possible because there are conserved charges if D-branes are introduced). In the weak coupling limit one can use methods of standard statistical mechanics to compute the entropy and obtain in this way the entropy for the black holes to which these configurations can be mapped. For certain classes of extremal or almost extremal black holes one gets exactly the Bekenstein-Hawking law $S(A) = A/4$ and subleading corrections that agree with the Wald formula for extensions of General Relativity. In addition to the entropy computations, Hawking radiation has also been successfully studied in the setting provided by string theories. For non-extremal black holes the results available to date are not as satisfactory as those obtained for extremal black holes. Nonetheless, by using a *correspondence principle*, Horowitz and Polchinski [33] have given arguments that explain the proportionality of entropy and areas for generic black holes. In this situation, however, the $1/4$ factor appearing in the area law cannot be accounted for directly. At the present moment the problem of computing the entropy for "more physical" black holes is receiving the attention of some string theorists. In particular there is a lot of interest in showing that the area law (with the right proportionality factor) can be recovered for black hole configurations with less symmetry charges than the ones traditionally studied [34, 35] . It is fair to say that there has been steady progress in this direction and the results, up to date, suggest that the Bekenstein-Hawking law can be recovered for physical black holes.

I will conclude by saying that despite the fact that LQG is not a finished theory of quantum gravity the formalism as it stands today can be used to build a convincing model of semiclassical black holes. The expected results on the Bekenstein-Hawking law are recovered and also physically sensible logarithmic subleading corrections. In addition to this there are some tantalizing hints of other interesting phenomena in the behavior of the entropy. To the extent that these corresponds to generic features of the combinatorial problems involved in the actual computation of the entropy I think that they could be present in other approaches to quantum gravity. The search of such behaviours could be an interesting research project that could further illuminate black hole physics.

Acknowledgements The author wants to thank the organizers of the Spanish Relativity Meeting ERE 2010 for their kind invitation to participate and Eduardo J. S. Villaseñor for his careful reading of the manuscript. This work was supported by the Spanish MICINN Project FIS2009-11893 and the Consolider-Ingenio 2010 Program CPAN (CSD2007-00042).

References

[1] Smolin L 1995 *J. Math. Phys.* **36** 6417–6455 (*Preprint* gr-qc/9505028)
[2] Krasnov K V 1997 *Phys. Rev.* **D55** 3505–3513 (*Preprint* gr-qc/9603025)
[3] Rovelli C 1996 *Phys. Rev. Lett.* **77** 3288–3291 (*Preprint* gr-qc/9603063)
[4] Ashtekar A, Baez J, Corichi A and Krasnov K 1998 *Phys. Rev. Lett.* **80** 904–907 (*Preprint* gr-qc/9710007)
[5] Ashtekar A, Baez J C and Krasnov K 2000 *Adv. Theor. Math. Phys.* **4** 1–94 (*Preprint* gr-qc/0005126)
[6] Domagala M and Lewandowski J 2004 *Class. Quant. Grav.* **21** 5233–5244 (*Preprint* gr-qc/0407051)
[7] Corichi A, Borja E F and Diaz-Polo J 2007 *J. Phys. Conf. Ser.* **68** 012031 (*Preprint* gr-qc/0703116)
[8] Agullo I, Barbero G J F, Borja E F, Diaz-Polo J and Villaseñor E J S 2008 *Phys. Rev. Lett.* **100** 211301 (*Preprint* 0802.4077)
[9] Agullo I, Barbero G J F, Borja E F, Diaz-Polo J and Villasenor E J S 2010 *Phys. Rev.* **D82** 084029
[10] Engle J, Perez A and Noui K 2010 *Phys. Rev. Lett.* **105** 031302 (*Preprint* 0905.3168)
[11] Kaul R K and Majumdar P 1998 *Phys. Lett.* **B439** 267–270 (*Preprint* gr-qc/9801080)
[12] Ghosh A and Mitra P 2006 *Phys. Rev.* **D74** 064026 (*Preprint* hep-th/0605125)
[13] Sahlmann H 2007 *Phys. Rev.* **D76** 104050 (*Preprint* 0709.2433)
[14] Engle J, Noui K, Perez A and Pranzetti D 2010 *Phys. Rev.* **D82** 044050 (*Preprint* 1006.0634)
[15] Barbero G J F and Villaseñor E J S 2009 *Class. Quant. Grav.* **26** 035017 (*Preprint* 0810.1599)
[16] Barbero G J F and Villaseñor E J S 2008 *Phys. Rev.* **D77** 121502 (*Preprint* 0804.4784)
[17] Meissner K A 2004 *Class. Quant. Grav.* **21** 5245–5252 (*Preprint* gr-qc/0407052)
[18] Agullo I, Borja E F and Diaz-Polo J 2008 *Phys. Rev.* **D77** 104024 (*Preprint* 0802.3188)
[19] Wald R M 2001 *Living Rev. Rel.* **4** 6 (*Preprint* gr-qc/9912119)
[20] Gibbons G W and Hawking S W 1977 *Phys. Rev.* **D15** 2752–2756
[21] Iyer V and Wald R M 1995 *Phys. Rev.* **D52** 4430–4439 (*Preprint* gr-qc/9503052)
[22] Bombelli L, Koul R K, Lee J and Sorkin R D 1986 *Phys. Rev.* **D34** 373–383
[23] Callan Jr C G and Wilczek F 1994 *Phys. Lett.* **B333** 55–61 (*Preprint* hep-th/9401072)
[24] Wald R M 1993 *Phys. Rev.* **D48** 3427–3431 (*Preprint* gr-qc/9307038)
[25] 't Hooft G 1985 *Nucl. Phys.* **B256** 727
[26] Frolov V P and Fursaev D V 1997 *Phys. Rev.* **D56** 2212–2225 (*Preprint* hep-th/9703178)
[27] Sorkin R D 1996 (*Preprint* gr-qc/9701056)
[28] Carlip S 2009 *Lect. Notes Phys.* **769** 89–123 (*Preprint* 0807.4520)
[29] Sen A 1995 *Mod. Phys. Lett.* **A10** 2081–2094 (*Preprint* hep-th/9504147)
[30] Strominger A and Vafa C 1996 *Phys. Lett.* **B379** 99–104 (*Preprint* hep-th/9601029)
[31] Callan C G and Maldacena J M Prepared for ICTP Summer School in High-energy Physics and Cosmology, Trieste, Italy, 10 Jun - 26 Jul 1996
[32] Damour T 2004 (*Preprint* hep-th/0401160)
[33] Horowitz G T and Polchinski J 1997 *Phys. Rev.* **D55** 6189–6197 (*Preprint* hep-th/9612146)
[34] Emparan R and Horowitz G T 2006 *Phys. Rev. Lett.* **97** 141601 (*Preprint* hep-th/0607023)
[35] Horowitz G T and Roberts M M 2007 *Phys. Rev. Lett.* **99** 221601 (*Preprint* 0708.1346)

Journal of Physics: Conference Series **314** (2011) 012004 doi:10.1088/1742-6596/314/1/012004

Dynamics of the quantum vacuum: Cosmology as relaxation to the equilibrium state

F.R. Klinkhamer[1] **and G.E. Volovik**[2,3]

[1] Institute for Theoretical Physics, University of Karlsruhe, Karlsruhe Institute of Technology, 76128 Karlsruhe, Germany

[2] Low Temperature Laboratory, Aalto University, P.O. Box 15100, FI–00076 AALTO, Finland

[3] Landau Institute for Theoretical Physics RAS, Kosygina 2, 119334 Moscow, Russia

E-mail: `frans.klinkhamer@kit.edu; volovik@boojum.hut.fi`

Abstract. The behavior of the gravitating vacuum energy density in an expanding universe is discussed. A scenario is presented with a step-wise relaxation of the vacuum energy density. The vacuum energy density moves from plateau to plateau and follows, on average, the steadily decreasing matter energy density. The current plateau with a small positive value of the vacuum energy density (effective cosmological constant) may result from a still not equilibrated contribution of the light massive neutrinos to the quantum vacuum.

1. Introduction

There are many different contributions to the vacuum energy density. They can be separated into sub-Planckian and trans-Planckian contributions with respect to the energy scale $E_{\text{Planck}} \equiv [\hbar\, c^5/(8\pi\, G_N)]^{1/2} \approx 2.44 \times 10^{18}$ GeV. The sub-Planckian contributions are described by relativistic quantum fields propagating over a classical spacetime manifold. The trans-Planckian contributions come from the fundamental microscopic degrees of freedom of the "deep vacuum." In the perfect equilibrium Minkowski vacuum [1], the trans-Planckian degrees of freedom compensate the contribution of the sub-Planckian quantum fields to the gravitating vacuum energy density ρ_{vac} (effective cosmological constant Λ_{eff}).

The above discussion adopts the condensed-matter point of view, where the effective quantum fields emerge only at low energies and where the high-energy phenomena are determined by the fundamental (atomic) degrees of freedom. In a self-sustained equilibrium vacuum, the full compensation of the vacuum energy density occurs without fine-tuning [1, 2]: the microscopic degrees of freedom are automatically adjusted to the equilibrium state. The self-tuned nullification of the energy density in the ground state of an arbitrary equilibrium system, including the relativistic quantum vacuum, suggests a possible solution [3] of the main cosmological constant problem [4]. A brief review of this so-called q–theory approach to the cosmological constant problem is presented in Appendix A.

The q–theory approach to the cosmological constant problem differs from those approaches which consider only relativistic quantum fields and where the contributions of the different quantum fields, fermionic and bosonic, compensate each other. This fermion-boson compensation can be due to supersymmetry or to a special choice of the fermionic and bosonic content of the vacuum, relying on special relations between the masses of the scalar, spinor, and vector fields (see, e.g., Refs. [5, 6, 7, 8, 9]).

In the perfect equilibrium state, the individual contributions to the vacuum energy density cannot be distinguished. But the situation changes dramatically for the dynamical vacuum of an expanding universe [2]. Cosmology, then, corresponds to the process of relaxation towards the equilibrium vacuum state, with the vacuum energy density dropping from its initial large value [which can be of order $\rho_{\text{vac}} \sim (E_{\text{Planck}})^4$] to its present small value [$\rho_{\text{vac}} \approx (2\ \text{meV})^4$]. During the expansion of the Universe, the energy hierarchy of the different contributions to the vacuum energy density from the different quantum fields gives rise to a time sequence, since each epoch is characterized by one dominant contribution to the vacuum energy density and, thus, by one particular value of the effective cosmological constant. This suggests a step-wise relaxation of the effective cosmological constant, which, on average, follows the matter energy density.

The contribution of a quantum matter field to the vacuum energy density is given by the following expression (see, e.g., Refs. [5, 6, 7, 8, 9, 10, 11, 12]):

$$\rho_{\text{vac}}^{(\text{matter})}(M,\, E_{\text{uv}}) = c_4\,(E_{\text{uv}})^4 + c_2\,(E_{\text{uv}})^2\, M^2 + c_0\, M^4\, \ln\big[\,(E_{\text{uv}})^2/M^2\,\big] + \text{O}(M^4)\,, \qquad (1)$$

where M is the mass of the matter field considered and E_{uv} is the ultraviolet energy cutoff, which is typically assumed to be of the order of the energy scale E_{Planck} defined above. Different regularization schemes have been suggested, in order to obtain the coefficients c_n appearing in (1). Here, and in the following, natural units are used with $\hbar = c = k_B = 1$.

Cosmological phase transitions or crossovers lead to time-varying masses of the fermionic and bosonic fields and, thereby, affect the contributions of these fields to the effective cosmological constant. In this article, the focus will be on the fermionic degrees of freedom, whose number is larger than the one of the bosonic degrees of freedom in the standard model of elementary particle physics (cf. Sec. 19.3.2 of Ref. [13]).

It is natural to suppose that, for higher temperatures, the number of fermionic quantum fields with nonzero masses is smaller, thereby removing the corresponding negative contributions to the vacuum energy density (see Secs. 2–4 and Appendix B for details, in particular, for the reason of having a *negative* contribution). As a result, the vacuum energy density from fermionic mass contributions is typically large for high temperature and small for low temperature. In an expanding universe, this vacuum energy density (effective cosmological constant) then relaxes in a step-wise manner, following the appearance of masses of the fermionic matter fields (see Secs. 5–7 for further discussion). The step-wise relaxation continues until the vacuum energy density reaches the zero value of the perfect equilibrium state or until the vacuum energy density is frozen at a stage where the relaxation process becomes ineffective.

Several scenarios have been proposed for this final (frozen) stage and the corresponding remnant vacuum energy density. These scenarios rely, for example, on new TeV–scale electroweak physics [14, 15, 16] or nonperturbative QCD dynamics [17, 18, 19, 20, 21] with a remnant vacuum energy density given by, respectively, $\rho_{\text{vac},\infty} \sim (E_{\text{ew}})^8/(E_{\text{Planck}})^4$ or $\rho_{\text{vac},\infty} \sim (E_{\text{QCD}})^6/(E_{\text{Planck}})^2$. In Sec. 5, we add a scenario related to neutrino mass M_ν, which may give a remnant vacuum energy density of order $(M_\nu)^4$; see, e.g., Refs. [22, 23, 24] for related discussions. All these scenarios do not exclude each other. Different contributions to the vacuum energy density produce successive plateaux in the process of the relaxation of the effective cosmological constant and the one which survives (or is frozen) is responsible for the current plateau of the effective cosmological constant at the meV–scale level.

2. Fermion propagator in an interacting system and vacuum energy density ρ_{vac}

As a start, consider the contribution of a fermionic quantum field to the vacuum energy density. The discussion will be based on certain properties of the mass function in the fermion propagator.

This fermion propagator must apply to the virtual fermions of the quantum vacuum (the ground state in condensed-matter-physics language) which is a dense interacting system with Planck-scale energies.

The general form of the Green's function of a fermionic particle is

$$G(p) = \frac{Z(p^2)}{i\,\gamma^\mu\, p_\mu + M(p^2)}\,, \tag{2}$$

for a Euclidean metric, $p^2 \equiv p_\mu\, p_\nu\, \delta^{\mu\nu} = |\mathbf{p}|^2 + (p_0)^2$, and appropriate Dirac matrices, $\gamma^\mu\gamma^\nu + \gamma^\nu\gamma^\mu = 2\,\delta^{\mu\nu}$. The mass function $M(p^2)$ in (2) must disappear at large p^2, where interaction effects can be neglected. The simplest form of $M(p^2)$ satisfying that condition is

$$M(p^2) = \frac{a}{b + p^2}\,, \tag{3}$$

in term of parameters $a > 0$ and $b \geq 0$ with mass dimensions 3 and 2, respectively.

This two-parameter *Ansatz* reflects two important physical quantities of the mass function, which depend on the interaction strength characterizing the system. First, the parameter $b^{1/2}$ from model (3) gives the energy scale of the effective ultraviolet cutoff for mass-dependent effects. Second, the ratio of the parameters a and b represents the mass parameter at zero momentum, $M(0) = a/b = \left(a\, b^{-3/2}\right) b^{1/2}$. The fermion mass $\overline{M}$ at the pole of the Green's function (i.e., on mass shell, temporarily reverting to a Lorentzian metric) is determined by the equation $\overline{M}^{\,2}\left(b - \overline{M}^{\,2}\right)^2 = M^2(0)\, b^2$. For $a^2 \ll b^3$, the on-shell mass is simply given by $\overline{M} \approx M(0) = a/b$.

For charged leptons, the natural value of the cutoff is the electroweak energy scale, $b \sim (E_{\text{ew}})^2$. For neutrinos, the cutoff may be comparable with the neutrino mass itself (a heuristic argument based on the see-saw mechanism is given in Appendix B). For the analytic *Ansatz* (2), this gives $a^2 \sim b^3$ and the on-shell neutrino mass $\overline{M}_\nu$ typically differs from $M_\nu(0)$ but is still of order $M_\nu(0)$.

The fermionic contribution to the vacuum energy density is obtained from the following Euclidean effective action:

$$S_E = -V_4 \int \frac{d^4p}{(2\pi)^4}\, \text{tr}\, \ln\left[G(p)\, \widetilde{m}\,\right], \tag{4}$$

where V_4 is the Euclidean spacetime volume considered and $\widetilde{m}$ a fixed reference mass. The zeroth-order term in the gradient expansion gives the contribution of the fermionic field to the cosmological constant term,

$$S = -\int d^4x\, \sqrt{-\det[g(x)]}\;\, \rho_{\text{vac}}^{(\text{fermion})} + \ldots\,, \tag{5}$$

where, now, the action has been given for an arbitrary Lorentzian metric $g_{\mu\nu}(x)$ and the conventions of Ref. [3] have been used.

The fermionic contribution to the cosmological constant can be calculated by introducing the vierbein field in (4), $\gamma^a\, e_a^\mu\, p_\mu$, and taking the functional derivative of S_E with respect to e_a^μ. Consider the contribution from the denominator of the Green's function (2). In order to obtain a simple estimate of the mass effects, take the derivative of (4) with respect to a,

$$\frac{d\,S_E}{d\,a} \sim \int \frac{d^4p}{(2\pi)^4}\, \text{tr}\big[\,(b + p^2)^{-1}\, G\,\big] \sim \int \frac{d^4p}{(2\pi)^4}\, \frac{a}{a^2 + p^2\,(b + p^2)^2}\,, \tag{6}$$

leaving out V_4 for the sake of brevity. The case $b^3 \lesssim a^2$ gives $dS_E/da \sim a^{1/3}$ and $S_E \sim a^{4/3}$. The case $a^2 \lesssim b^3$ gives $S_E \sim a^2/b$. Combined, there is thus the following simple estimate of the fermion contribution to the vacuum energy density:

$$\rho_{\text{vac}}^{(\text{fermion})} \sim \begin{cases} -a^{4/3} & \text{for} \quad b^3 \lesssim a^2\,, \\ -a^2/b & \text{for} \quad b^3 \gtrsim a^2\,, \end{cases} \tag{7}$$

where the overall minus sign will be verified independently in the next section.

3. Fermion contributions to ρ_{vac} at the electroweak energy scale

In the electroweak standard model, the natural choice for the Green's function of quarks and leptons with masses $M \lesssim E_{\text{ew}}$ gives parameters $b \sim (E_{\text{ew}})^2$ and $a = M(0)\,b \lesssim (E_{\text{ew}})^3$ for the model mass function (3). According to (7), the contribution to the vacuum energy density from light fermions with $M \ll E_{\text{ew}}$ is of order $M^2\,(E_{\text{ew}})^2$ and the one of heavy fermions with $M \sim E_{\text{ew}}$ of order $(E_{\text{ew}})^4$. Hence, the general expectation is that the contribution of a charged fermionic field with mass $M \lesssim E_{\text{ew}}$ to the vacuum energy density is

$$\rho_{\text{vac}}^{(\text{fermion})}(M) \sim -M^2\,(E_{\text{ew}})^2\,, \tag{8}$$

taking over the minus sign from (7).

The previous estimates (7) and (8) are obtained from an analysis of the denominator of the Green's function (2), but the numerator gives similar results. The residue $Z(p^2)$ approaches unity as $p^2 \to \infty$. For finite values of p^2, however, the residue $Z(p^2)$ is again determined by two energy scales, $M(0)$ and E_{ew}.

Estimate (8), including the minus sign, can also be obtained by direct calculation of the mass contribution to the zero-point energy from a spin–$\frac{1}{2}$ Dirac field:

$$\int^{(E_{\text{cutoff}})} \frac{d^3\mathbf{p}}{(2\pi)^3}\,\Big(-\sqrt{|\mathbf{p}|^2+M^2}+|\mathbf{p}|\Big) \sim -M^2\,(E_{\text{cutoff}})^2\,, \tag{9}$$

where E_{cutoff} is the ultraviolet cutoff of this quadratically divergent integral (see, e.g., Ref. [5]). Since the Green's function of a massive standard-model fermion differs from the Green's function of a massless fermion only at energies below E_{ew}, it is the electroweak scale E_{ew} which provides the natural ultraviolet cutoff in (9) rather than the Planck scale: $E_{\text{cutoff}} = E_{\text{ew}}$.

A Planck-scale cutoff $E_{\text{cutoff}} = E_{\text{Planck}}$ in (9) would be relevant only if the mass of the fermion were fundamental or generated at the Planck energy scale, i.e., with $b \sim (E_{\text{Planck}})^2$ in the mass function (3). However, if the Green's function of a massive fermion differs from the Green's function of a massless fermions only at energies of order of mass scale (as discussed in the fourth paragraph of Sec. 2), then the cutoff in (9) is determined by the fermion mass itself, leading to $\rho_{\text{vac}}^{(\text{fermion})}(M) \sim -M^4$. We suggest that the last estimate holds for the neutrino contribution (see Sec. 5 for possible cosmological implications).

4. Fermion contributions to ρ_{vac} at the QCD energy scale

For the quarks of quantum chromodynamics (QCD), the natural choice of parameters in model (3) is $a \sim (E_{\text{QCD}})^3$ and $b = 0$. Such a choice effectively corresponds to the phenomenon of confinement, which leads to a singular behavior of the effective quark mass in the infrared limit $p \to 0$. Actually, this Green's function at $p_0 = 0$ has been used in Refs. [18, 19] to justify the presence of a running term $|H(t)|\,(E_{\text{QCD}})^3$ in the vacuum energy density [here, $H(t)$ is the Hubble parameter of the expanding universe] and to calculate the corresponding remnant vacuum energy density $\rho_{\text{vac},\infty} \sim (E_{\text{QCD}})^6/(E_{\text{Planck}})^2$.

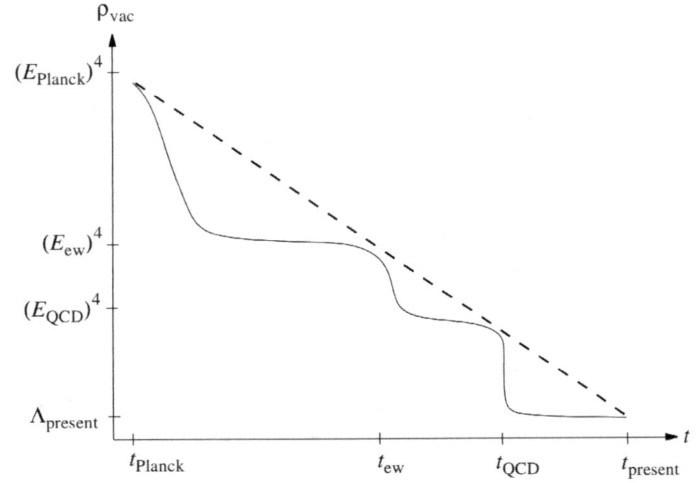

Figure 1. Sketch (as an approximate double-log plot) of the relaxation of the vacuum energy density during the evolution of the Universe. *Dashed curve*: relaxation according to the relation $< \rho_{\text{vac}}(t) > \sim (E_{\text{Planck}})^2/t^2$ from Ref. [2]. Only the envelope of the ultrafast (Planck-scale) oscillations of $\rho_{\text{vac}}(t)$ is shown, that is, $< \rho_{\text{vac}}(t) >$. The huge vacuum energy density $\rho_{\text{vac}} \sim (E_{\text{Planck}})^4$ at the initial moment $t_{\text{in}} \sim t_{\text{Planck}}$ relaxes to a zero value as $t \to \infty$. According to this estimate, the vacuum energy density at the present time passes through a value close to the one of the observed cosmological "constant": $\rho_{\text{vac}}(t_{\text{present}}) \sim (E_{\text{Planck}})^2/t^2_{\text{present}} \sim \Lambda^{(\text{obs})}_{\text{present}} \approx (2\ \text{meV})^4$. *Full curve*: dissipative processes and cosmological phase transitions lead to a modified behavior, with a step-wise decrease of the vacuum energy density towards zero. The vacuum energy density follows the matter energy density on average, but locally, on each plateau, the vacuum behaves as a medium with equation-of-state parameter $w \approx -1$.

However, the justification of the term $|H|\,(E_{\text{QCD}})^3$ is still problematic. Model (3) with $a \sim (E_{\text{QCD}})^3$ and $b = 0$ gives, in first approximation (7), a large vacuum energy density, $|\rho^{(\text{fermion})}_{\text{vac}}| \sim (E_{\text{QCD}})^4$. With $a > 0$, there is no infrared divergence in the integral (6) for $b = 0$ and the suggested $|H|\,(E_{\text{QCD}})^3$ correction does not appear. Still, this does not completely rule out a term of order $(E_{\text{QCD}})^6/(E_{\text{Planck}})^2$ in the vacuum energy density: a more elaborate model of the QCD vacuum energy density is needed, which properly takes into account the phenomenon of confinement. There is a hint from lattice simulations, that such a $(E_{\text{QCD}})^6/(E_{\text{Planck}})^2$ term is indeed possible [21].

5. Fermion contributions to ρ_{vac} at the mass scale of light neutrinos

At finite temperature T, the mass function $M(p^2)$ in (2) can be expected to depend also on T. Moreover, temperature may provide the relevant energy scale for the ultraviolet cutoff in (9), $E_{\text{cutoff}} \sim T$. As a result, the mass-dependent contribution becomes part of the thermal energy rather than the vacuum energy. This suggests a natural form of energy exchange between thermal fermions and vacuum fermions, i.e., between matter and vacuum. In any case, it can be expected that, with increasing temperature, the mass-dependent contributions to the vacuum energy density will be reduced or even disappear.

As the temperature of the adiabatically expanding universe decreases, more and more fermions become massive and add extra negative contributions (9) to the vacuum energy density. This implies that the total vacuum energy density ρ_{vac} decreases with time and that this change occurs in a step-like manner (Fig. 1), following, on the whole, the matter energy density.

According to this scenario, one of the plateaux in $\rho_{\text{vac}}(t)$ for $t \sim \bar{t}$ results from the lack of the neutrino-mass contribution to the vacuum energy density, $\rho_{\text{vac}}^{(\text{neutrino})}(M_\nu) \sim -(M_\nu)^4$. The lack of a negative neutrino contribution to the vacuum energy density corresponds, of course, to a positive value of $\rho_{\text{vac}}(\overline{t})$ at the time scale $\overline{t}$ considered. Contributions of fermionic matter fields with larger vacuum energy densities have already been released at higher temperatures and have been relaxed due to the fast oscillations of the microscopic q–type fields (see Appendix A.2).

Restrict the discussion temporarily to one neutrino flavor. Then, the neutrino contribution to the vacuum energy density can only be released after the temperature T_ν of the relic neutrino drops below the neutrino mass M_ν [which happens at a redshift of order $M_\nu/(3\times10^{-4}\,\text{eV}) \sim 10^2$ for $M_\nu \sim 0.05$ eV]. It is, however, very well possible that the contribution from the light neutrino is still not released at the present moment, because of the lack of an efficient equilibration mechanism between neutrino matter and neutrino vacuum in the current epoch. In that case, the excess of the vacuum energy density remains frozen at a positive value of order $(M_\nu)^4$. The lowest plateau in Fig. 1 can be expected to extend far beyond the present age of the Universe.

Turning to three neutrino flavors with substantial (near-maximal) mixing [13], a heuristic argument suggests the following expression for the frozen value of the vacuum energy density:

$$\rho_{\text{vac}}(t_{\text{present}}) \stackrel{?}{=} c_\nu \left(\text{tr}\left[(\widehat{M_\nu})^2\right]\right)\left(\det\left[(\widehat{M_\nu})^2\right]\right)^{1/3}, \tag{10}$$

where $\widehat{M_\nu}$ is the 3×3 neutrino mass matrix with eigenvalues $\{m_{\nu1},\, m_{\nu2},\, m_{\nu3}\}$ and the positive coefficient c_ν is assumed to be of order 1. Additional terms, possibly depending on the mixing angles, can be expected in (10). Note that neutrino contributions to the effective cosmological constant have been discussed in a number of recent articles (see, e.g., Refs. [22, 23, 24]).

Given the measured value $\rho_{\text{vac}}(t_{\text{present}}) \approx (2.3\text{ meV})^4$ for the left-hand side of (10) and the neutrino-oscillation data [13], there are three equations for the three neutrino masses. Take the neutrino-oscillation input to be $(m_{\nu3})^2 - (m_{\nu1})^2 = \pm 2.4\times10^{-3}\text{ eV}^2$ and $(m_{\nu2})^2 - (m_{\nu1})^2 = 7.7\times10^{-5}\text{ eV}^2$. Then, formula (10) with $c_\nu = 1$ gives the following neutrino mass spectra:

$$\left(m_{\nu1},\, m_{\nu2},\, m_{\nu3}\right)\Big|^{(c_\nu=1)} \stackrel{?}{=} \begin{cases} \left(2.793\times10^{-6},\, 8.775,\, 48.99\right)\times\text{meV}\,, \\ \left(48.99,\, 49.77,\, 1.783\times10^{-7}\right)\times\text{meV}\,. \end{cases} \tag{11}$$

These neutrino masses (close to the minimum values needed for the neutrino oscillations) lie outside the reach of the KATRIN tritium beta-decay detector with a 0.2–eV design sensitivity [25]. The same conclusion holds for the neutrino mass spectra resulting from formula (10) with $c_\nu \geq 10^{-8}$ [the $c_\nu = 10^{-8}$ spectra for both "normal" and "inverted" mass hierarchies have $m_{\nu1} \sim m_{\nu2} \sim m_{\nu3} \sim 0.2$ eV]. See Sec. 6 for additional comments.

6. Discussion

The main message of this article is that there is a hierarchy of different contributions to the vacuum energy density from different matter fields. In the context of an expanding universe, these contributions are step-by-step released with decreasing temperature of the Universe. Note that the few big steps of the vacuum energy density in Fig. 1 are only schematic, there may be many additional small steps. Some of these (small) steps of the vacuum energy density may even increase the vacuum energy density, if they result from the mass effects of bosonic fields [corresponding to (9) multiplied by -1]. But, as mentioned in Sec. 1, the present article focusses on the mass effects of fermionic fields, which typically give decreasing steps of the vacuum energy density as the Universe expands and cools.

The vacuum energy density roughly follows the matter energy density, but in a stepwise manner. This suggests a possible solution of the cosmic coincidence puzzle (cf. Ref. [14]), namely, why the observed cosmological "constant" is precisely of the order of the *present* matter energy density. Indeed, the step-wise relaxation of the vacuum energy density may solve the cosmic coincidence puzzle, because, during each epoch, the effective cosmological constant is related to the energy scale characterizing the given epoch and, thus, to the matter energy density.

It has been shown [2] that the huge initial vacuum energy density may continuously relax to a zero value as the age of the Universe goes to infinity. Moreover, it was found that this relaxing vacuum energy density passes through the observed numerical value ($\rho_{\text{vac},0} \sim 10^{-11}$ eV^4) for an age of the model universe of the order of the observed value ($t_0 \sim 10$ Gyr); see, in particular, Eq. (5.16) of Ref. [2]. This scenario gives a reasonable estimate of the present value of the vacuum energy density but contradicts the more detailed astronomical observations which are in favor of a genuine cosmological *constant*, at least, in the current epoch. The reason for the failure of the continuous-relaxation scenario may be that it is oversimplified, since it does not take into consideration the processes of radiation, dissipation, and cosmological phase transitions or crossovers. The new scenario with a step-wise decrease of the vacuum energy density may remove this discrepancy: the vacuum energy density follows the matter energy density on average, but locally, on each plateau, behaves as a medium with equation-of-state parameter $w \approx -1$.

The overall picture for the evolution of the vacuum energy density is, then, as follows. The initial relaxation dynamics of the vacuum energy density is dominated by microscopic (trans-Planckian) degrees of freedom [2], leading to $1/t^2$ decay from ultrafast oscillations or to exponential decay if radiation of matter fields or gravitational waves is taken into account. (This behavior is similar to that of Starobinsky inflation [26, 27], where the role of the microscopic degrees of freedom is played by heavy Planck-mass fields.) For later times, probably starting at the electroweak epoch, the vacuum dynamics and the energy exchange between vacuum and matter is dominated by the low-energy contributions of quantum fields and can be described by standard-model physics. The dynamics of the trans-Planckian degrees of freedom (e.g., the dynamics of the vacuum variable q from Ref. [2]) becomes irrelevant at late times, providing only a tiny response to perturbations occurring during the crossovers [15, 16]. The microscopic degrees of freedom have already played their main role: to fully compensate the vacuum energy density of quantum fields in the final equilibrium state, that is, the Minkowski vacuum.

In order to clarify this point about the main role of the microscopic degrees of freedom, consider, for example, the large vacuum energy density released by an intermediate cosmological phase transition. It is assumed that the chemical potential μ for the microscopic degrees of freedom takes its natural value μ_0, which corresponds to the true quantum vacuum of a global equilibrium (in the cosmological context, the final Minkowski state with massive fermions at zero temperature), and not a different value $\widetilde{\mu} \neq \mu_0$, which would correspond to a nonequilibrium state (in the cosmological context, the intermediate false-vacuum state with massless fermions at temperatures above the phase transition temperature considered). This value μ_0 either is fixed globally as an integration constant being conserved throughout the history of the Universe [1, 2] or emerges dynamically from an attractor-type solution [3]. This is the reason why the full curve in Fig. 1 (taking dissipative effects to have been operative at $t \gtrsim t_{\text{Planck}}$) does not approach a zero value of the vacuum energy density at times $t_{\text{Planck}} \ll t \ll t_{\text{ew}}$ but a finite positive value. Later, this vacuum energy density is reduced (released) at the transition to the next plateau.

If the vacuum energy density in the current epoch results from the lack of the neutrino contribution due to non-equilibration (as discussed in Sec. 5) and formula (10) holds with $c_\nu \sim 1$, then the 0.2–eV sensitivity of the KATRIN tritium beta-decay detector [25] does not suffice to detect the predicted nonzero (electron-type) neutrino mass $\lesssim 0.05$ eV from (11).

If, on the other hand, KATRIN does find a neutrino mass $\gtrsim 0.4$ eV, then this implies that the neutrino contribution to the vacuum energy density (10) [assuming $c_\nu \gtrsim 10^{-6}$] exceeds by several orders of magnitude the observed value of the current effective cosmological constant ($\rho_{\text{vac},0} \approx 3 \times 10^{-11}$ eV^4) and that this part of the vacuum energy density cannot be responsible for the presently observed value $\rho_{\text{vac},0}$. In this case, the neutrino contribution to $\rho_{\text{vac}}(t)$ must have dominated in one of the previous epochs and must have been released by now.

Assuming that the neutrino contribution has already equilibrated (as would be suggested by a positive result from KATRIN with $M_\nu \gtrsim 0.4$ eV), the remnant vacuum energy density $\rho_{\text{vac},\infty}$

Journal of Physics: Conference Series **314** (2011) 012004 doi:10.1088/1742-6596/314/1/012004

must be associated with other possible contributions such as $(E_{\text{ew}})^8/(E_{\text{Planck}})^4$ from Refs. [15, 16] or $(E_{\text{QCD}})^6/(E_{\text{Planck}})^2$ from Refs. [18, 19]. However, the last two explanations of $\rho_{\text{vac},\infty}$ may fare differently when compared with astronomical observations: the QCD explanation corresponds to a modified gravity theory and may already be ruled out by the astronomical observations, whereas the electroweak explanation corresponds to a standard ΛCDM model (now, with a calculated value of Λ) and appears to fit the current astronomical data well. Still, the electroweak explanation requires nonstandard physics at the TeV energy scale, which remains to be confirmed by particle-collider experiments.

All this also suggests that, in the past, there were additional plateaux in the effective cosmological constant associated with, for example, the mass M_e of the electron, the masses $M_{u,\,d}$ of the light (stable) quarks, the QCD energy scale, and the electroweak energy scale. These contributions to ρ_{vac} may be of order $(M_e)^2\,(E_{\text{ew}})^2$, $(M_{u,\,d})^2\,(E_{\text{ew}})^2$, $(E_{\text{QCD}})^4$, and $(E_{\text{ew}})^4$.

7. Conclusion

It has been argued, in this article, that cosmology can be viewed as the relaxation of the Universe towards the equilibrium vacuum state, with the vacuum energy density dropping from its initial Planck-scale value to its present meV–scale value. The initial relaxation dynamics of the vacuum is dominated by microscopic (trans-Planckian) degrees of freedom [1, 2], leading to $1/t^2$ decay of the average vacuum energy density (or to exponential decay from dissipative effects).

The dynamics of the vacuum energy density at later times is governed by contributions to the vacuum energy density from the relativistic quantum fields of the standard model. The energy hierarchy of the contributions of different quantum fields to the vacuum energy density leads to a step-wise relaxation with time of the effective cosmological constant (Fig. 1), which, on average, follows the matter energy density. Such a step-wise behavior of the vacuum energy density may solve the cosmic coincidence puzzle [14], because, in each epoch, the effective cosmological constant is related to the energy scale characterizing the given epoch and, thus, to the matter energy density. There are several scenarios for the origin of the latest plateau of the effective cosmological constant, including a possible contribution (10) from the masses of the neutrinos (more precisely, the positive effective cosmological constant would result from the lack of the negative mass-effect contribution of the neutrino fields, if these fields of the quantum vacuum still have not reached their equilibrium state).

Returning to a crucial point, the hierarchical structure of the quantum vacuum implies a complicated spectral function of the vacuum energy density. (The spectral function of the vacuum energy density has been introduced by Zeldovich [5] and its relation to the standard model of elementary particle physics has been discussed in, e.g., Refs. [7, 28, 29].) The individual contributions to the spectral function of the vacuum energy density cannot be resolved in the static Minkowski vacuum, since, in equilibrium, the contributions of the bosonic and fermionic quantum fields to the low-energy part of the spectral function are compensated by the trans-Planckian part of the spectrum [29]. The trans-Planckian degrees of freedom, which are responsible for the automatic nullification of the cosmological constant in the equilibrium Minkowski vacuum, play also an important role in the dynamics at the early (Planckian) stage of expansion of the Universe. At later stages, the dynamics of the quantum vacuum is primarily determined by the low-energy tail of the spectral function. The various hierarchies of low-energy scales give rise to different plateaux in the vacuum energy density, each of which resembles a genuine cosmological constant but has a dynamic origin nevertheless.

Acknowledgments

FRK thanks the Perimeter Institute for Theoretical Physics, Canada, and the Department of Physics, University of Toronto, Canada, for their hospitality in September, 2010. The work of GEV is supported in part by the Academy of Finland, Centers of Excellence Program 2006–2011.

Appendix A. Résumé of q–theory

One route to understanding the gravitational effects of the vacuum energy density goes under the name of "q–theory" [1, 2, 3]. The idea is to consider the macroscopic equations of a conserved microscopic variable q, whose precise nature need not be known. The goal of this appendix is to explain the basic logic of q–theory as an approach to the cosmological constant problem [4].

Appendix A.1. Statics

For a special type of vacuum variable q introduced in Ref. [1], the vacuum energy density $\rho_{\text{vac}}(q)$ entering the gravitational field equations has been found to differ from the vacuum energy density $\epsilon(q)$ appearing in the action:

$$\rho_{\text{vac}}(q) = \epsilon(q) - q\,\frac{d\epsilon(q)}{dq} \equiv \epsilon(q) - q\,\epsilon'(q)\,. \tag{A.1}$$

Here, q is a microscopic variable describing the physics of the deep (ultraviolet) vacuum, but its thermodynamics and dynamics are governed by macroscopic equations, because q is a conserved quantity. This quantity q is similar to the mass density in liquids, which describes a microscopic quantity – the number density of atoms – but obeys the macroscopic equations of hydrodynamics, because of particle-number conservation.

However, different from known liquids, the quantum vacuum is Lorentz invariant. The quantity q must, therefore, be Lorentz invariant, at least, in the equilibrium state,

$$q = q_0 = \text{constant}\,. \tag{A.2a}$$

The variable q naturally allows for a vanishing gravitating vacuum energy density (cosmological constant)

$$\rho_{\text{vac}}(q_0) = \epsilon(q_0) - q_0\,\epsilon'(q_0) = 0\,. \tag{A.2b}$$

Even though both terms in the middle expression of (A.2b) can be of order $(E_{\text{Planck}})^4$, they cancel exactly in the equilibrium state corresponding to Minkowski spacetime. As such, q–theory provides a possible solution [3] of the *main* cosmological constant problem [4], namely, why the energy scale of the observed cosmological constant is essentially zero compared to the known energy scales of elementary particle physics (e.g., $E_{\text{QCD}} \sim 10^2$ MeV and $E_{\text{ew}} \sim 1$ TeV).

Appendix A.2. Dynamics

For a particular realization of the q variable and a nontrivial q–dependence of the gravitational coupling parameter, it has been shown [2] that q becomes spacetime-dependent and so does the vacuum energy density $\rho_{\text{vac}}(q) = \frac{1}{2}\,(q-q_0)^2 + \mathrm{O}\big((q-q_0)^3\big)$.

In a spatially flat, isotropic, and homogeneous Robertson–Walker universe, the field equations then give a rapidly oscillating vacuum energy density $\rho_{\text{vac}}(t)$, which drops from a Planck-scale value to zero as $t \to \infty$. Specifically, the following behavior has been found [2]:

$$q(\tau)/q_0 - 1 \sim \tau^{-1}\,\sin\tau\,, \quad r_{\text{vac}}(\tau) \sim \tau^{-2}\,\sin^2\tau\,, \tag{A.3}$$

where the dimensionless cosmic time τ and the dimensionless vacuum energy density r_{vac} have been obtained by appropriate scalings with microscopic (Planckian) quantities such as q_0.

Appendix B. Neutrino mass and effective momentum cutoff

In this appendix, a heuristic argument is given for a low effective momentum cutoff of the mass effects from neutrino fields in an interacting system (in fact, the system relevant to the virtual particles populating the quantum vacuum up to Planck-scale energies).

Start from the see-saw mechanism for a single neutrino flavor (see, e.g., Ref. [30] and references therein), which gives a light neutrino mass, $M_\nu(0) = (M_D)^2/M_R$, for a right-handed Majorana scale M_R and a Dirac scale $M_D \sim E_{\text{ew}}$ with hierarchy $M_R \gg M_D$.

Adding linear momentum contributions to the standard see-saw mass matrix gives the following effective mass matrix:

$$\mathcal{M}_\nu^{(\mathrm{lin})}(p^2) = \begin{pmatrix} c_{11}\,p & M_D + c_{12}\,p \\ M_D + c_{12}\,p & M_R + c_{22}\,p \end{pmatrix}, \tag{B.1}$$

with $p \equiv \sqrt{p^2} \geq 0$ for a Euclidean metric [see below (2)] and nonnegative numerical coefficients c_{ij} of order 1. The c_{11} term in (B.1) corresponds to an effective Higgs-triplet contribution from the interactions, independent from mass terms. One eigenvalue of (B.1) is of order M_R. The absolute value of the other eigenvalue is

$$M_\nu(p^2) = (M_D)^2/M_R - c_{11}\,p + \mathrm{O}(p^2)\,, \tag{B.2}$$

for $p \ll M_D \ll M_R$. Observe that the two terms on the right-hand side of (B.2) have a relative minus sign, which is not the case for the individual entries of the matrix (B.1).

Equation (B.2) for $c_{11} \sim 1$ shows an effective momentum cutoff for mass effects of the order of the neutrino mass itself, as discussed in the fourth paragraph of Sec. 2. For the simple example considered in this appendix and small enough p^2, the *Ansatz* for the mass function would be nonanalytic in p^2, namely, $M_\nu(p^2) = \alpha/(\beta + p)$ with $\alpha \sim \beta^2$.

References

[1] Klinkhamer F R and Volovik G E 2008 *Phys. Rev. D* **77** 085015 (*Preprint* arXiv:0711.3170)
[2] Klinkhamer F R and Volovik G E 2008 *Phys. Rev. D* **78** 063528 (*Preprint* arXiv:0806.2805)
[3] Klinkhamer F R and Volovik G E 2010 *JETP Lett.* **91** 259 (*Preprint* arXiv:0907.4887)
[4] Weinberg S 1989 The cosmological constant problem *Rev. Mod. Phys.* **61**, 1
[5] Zeldovich Y B 1968 *Sov. Phys. Usp.* **11** 381
[6] Frolov V P and Fursaev D 1998 *Phys. Rev. D* **58** 124009 (*Preprint* arXiv:hep-th/9806078)
[7] Alberghi G L, Kamenshchik A Y, Tronconi A, Vacca G P and Venturi G 2008 *JETP Lett.* **88** 705 (*Preprint* arXiv:0804.4782)
[8] Barvinsky A O, Kamenshchik A Y, Kiefer C, Starobinsky A A and Steinwachs C F 2009 *J. Cosmol. Astropart. Phys.* JCAP12(2009)003 (*Preprint* arXiv:0910.1041)
[9] Froggatt C, Nevzorov R and Nielsen H B 2010 *PoS ICHEP2010* 442 (*Preprint* arXiv:1012.5121)
[10] Visser M 2002 *Mod. Phys. Lett. A* **17** 977 (*Preprint* arXiv:gr-qc/0204062)
[11] Akhmedov E K 2002 Vacuum energy and relativistic invariance *Preprint* arXiv:hep-th/0204048
[12] Ossola G and Sirlin A 2003 *Eur. Phys. J. C* **31** 165 (*Preprint* arXiv:hep-ph/0305050)
[13] Nakamura K *et al.* [Particle Data Group] 2010 *J. Phys. G* **37** 075021
[14] Arkani-Hamed N, Hall L J, Kolda C and Murayama H 2000 *Phys. Rev. Lett.* **85** 4434 (*Preprint* arXiv:astro-ph/0005111)
[15] Klinkhamer F R and Volovik G E 2009 *Phys. Rev. D* **80** 083001 2009 (*Preprint* arXiv:0905.1919)
[16] Klinkhamer F R 2011 *Phys. Rev. D* **84** 023011 2011 (*Preprint* arXiv:1101.1281)
[17] Schützhold R 2002 *Phys. Rev. Lett.* **89** 081302
[18] Klinkhamer F R and Volovik G E 2009 *Phys. Rev. D* **79** 063527 (*Preprint* arXiv:0811.4347)
[19] Klinkhamer F R 2010 *Preprint* arXiv:1005.2885
[20] Urban F R and Zhitnitsky A R 2010 *Nucl. Phys. B* **835** 135 (*Preprint* arXiv:0909.2684)
[21] Holdom B 2011 *Phys. Lett. B* **697**, 351 (*Preprint* arXiv:1012.0551)
[22] Capolupo A, Capozziello S and Vitiello G *Phys. Lett. A* **363** 53 (2007) (*Preprint* arXiv:astro-ph/0602467)
[23] Bauer F and Schrempp L 2008 *J. Cosmol. Astropart. Phys.* JCAP04(2008)006 (*Preprint* arXiv:0711.0744)
[24] Chitov G Y, August T, Natarajan A and Kahniashvili T 2011 *Phys. Rev. D* **83** 045033 (*Preprint* arXiv:0911.1728)
[25] Osipowicz A *et al.* 2001 *Preprint* arXiv:hep-ex/0109033; see also `http://www-ik.fzk.de/tritium/`
[26] Starobinsky A A 1980 *Phys. Lett.* **91 B** 99
[27] Vilenkin A 1985 *Phys. Rev. D* **32** 2511
[28] Kamenshchik A Y, Tronconi A, Vacca G P and Venturi G 2007 *Phys. Rev. D* **75** 083514 (*Preprint* arXiv:hep-th/0612206)
[29] Volovik G E 2009 *J. Phys.: Conf. Series* **174** 012007 (*Preprint* arXiv:0801.2714)
[30] King S F 2004 *Rept. Prog. Phys.* **67** 107 (*Preprint* arXiv:hep-ph/0310204)

Spanish Relativity Meeting (ERE 2010): Gravity as a Crossroad in Physics IOP Publishing
Journal of Physics: Conference Series **314** (2011) 012005 doi:10.1088/1742-6596/314/1/012005

Bottom up approach to Quantum Gravity

Enrique Álvarez
Instituto de Física Teórica UAM/CSIC and Departamento de Física Teórica
Universidad Autónoma de Madrid, E-28049–Madrid, Spain
E-mail

E-mail: enrique.alvarez@uam.es

Abstract. A general introduction is given to what can be predicated about quantum gravity once the lessons from the standard model of particle physics are taken into account. In particular, the effective lagrangian point of view is briefly commented upon.

1. Introduction

Can we assert some proposition about quantum gravity with some confidence?

The mass scale associated to this problem just by sheer dimensional analysis is Planck's mass, which is given in terms of Newton's constant, G by

$$m_p \equiv \sqrt{\frac{\hbar c}{8\pi G}} \sim G^{-1/2} \sim 10^{19} GeV \tag{1}$$

(we shall always work in units so that $\hbar = c = 1$). If we remember that $1\, GeV (= 10^3\, MeV)$ is the rough scale of hadronic physics (the mass and inverse Comptom wavelength of a proton, for example), this means that quantum gravity effects will only be apparent when we are able to explore concentrated energy roughly 10^{19} times bigger (or an scale distance correspondingly smaller; these two statements are supposed to be equivalent owing to Heisenberg's principle). To set the scale, the Large Hadron Collider works roughly at the $TeV (= 10^3\, GeV)$ scale, so there is a long way to go before reaching expected quantum gravity effects in accelerators.

If we want to get experimental information, we have to turn our attention towards Cosmology, or perhaps look for some clever precison experiment in the laboratory. Lacking any experimental clue, the only thing we can do is to think and try to look for logical (in)consistencies.

It has been repeatedly argued by many particle physicists that the practical utility of the answer to this question will not presumably be great. How would we know for sure beforehand?. There has always been a recurrent dream, exposed vehemently by Salam [18] that the inclusion of the gravitational interaction would cure many of the diseases and divergences of quantum field theory, through the inclusion in the propagator of terms of the type

$$e^{-\frac{1}{m_p^2 x^2}}$$

So that for example, the sum of tree graphs that leads to the Schwarzschild solution as worked out by Duff [8]

$$\frac{1}{r} + \frac{2M}{m_p^2 r^2} + \frac{4M^2}{m_p^4 r^3} + \ldots$$

would get modified to

$$\frac{1}{r}e^{-\frac{1}{m_p r}}+\frac{2M}{m_p^2 r^2}e^{-\frac{2}{m_p r}}+\ldots\sim\frac{1}{re^{-\frac{1}{m_p r}}-2\frac{M}{m_p^2}}$$

shifting the location of the horizon and eliminating the singularity at $r=0$. Nobody has been able to substantiate this dream so far.

At any rate quantum gravity is nevertheless a topic which has fascinated whole generations of physicists, just because it is so difficult. There seems to be a strong tension between the beautiful, geometrical world of General Relativity and the no less marvelous, less geometrical, somewhat mysterious, but very well tested experimentally, world of Quantum Mechanics.

As with all matters of principle we can hope to better understand both quantum mechanics and gravitation if we are able to clarify the issue.

The most conservative approach is of course to start from what is already known with great precision about the standard model of elementary particles associated to the names of Glashow, Weinberg and Salam. This can be called the *bottom-up approach* to the problem. In this way of thinking Wilson taught us that there is a working low energy effective theory, and some quantum effects in gravity can be reliably computed for energies much smaller than Planck mass. There are two caveats to this. First of all, we do not understand why the observed cosmological constant is so small: the natural value from the low energy effective lagrangian point of view ought to be much bigger. The second point is that one has to rethink again the lore of effective theories in the presence of horizons. We shall comment on both issues in due time.

There is not a universal consensus even on the most promising avenues of research from the opposite *top-down* viewpoint. Many people think that strings [17] are the best buy (I sort of agree with this); but it is true that after more than two decades of intense effort nothing substantial has come out of them. Others [16] try to quantize directly the Einstein-Hilbert lagrangian, something that is at variance with our experience in effective field theories. But it is also true that as we have already remarked, the smallish value of the observed cosmological constant also cries out of the standard effective theories lore.

2. Symmetries and observables

It is generally accepted that General Relativity, a generally covariant theory, is akin to a gauge theory, in the sense that the diffeomorphism group of the apace-time manifold, $Diff(M)$ plays a role similar to the compact gauge group in the standard model of particle physics. There are some differences though. To begin with, the group, $Diff(M)$ is too large; is not even a Lie group [13]. Besides, its detailed structure depends on the manifold, which is a dynamical object not given a priori. Other distinguished subgroups (such as the area-preserving diffeomorphisms,[2]) are perhaps also arguable for. Those leave invariant a given measure, such as the Lebesgue measure, $d^n x$.

It also seems clear that when there is a boundary of space-time, then the gauge group is restricted to the subgroup consisting in those diffeomorphisms that act trivially on the boundary. The subgroup that act not-trivially is related to the set of conserved charges. In the asympotically flat case this is precisely the Poincaré group, $SO(1,3)$ that gives rise to the ADM mass.

In the asympotically anti-de Sitter case, this is presumably related to the conformal group $SO(2,3)$.

It is nevertheless not clear what is the physical meaning of keeping constant the boundary of spacetime (or keeping constant some set of boundary conditions) in a functional integral of some sort.

Journal of Physics: Conference Series **314** (2011) 012005 doi:10.1088/1742-6596/314/1/012005

A related issue is that it is very difficult to define what could be *observables* in a diffeomorphism invariant theory, other than global ones defined as integrals of scalar composite operators $O(\phi_a(x))$ (where $\phi_a, a = 1 \ldots N$ parametrizes all physical fields) with the peudo-riemannian measure

$$\mathcal{O} \equiv \int \sqrt{|g|} d^4x O(\phi_a(x))$$

Some people claim that there are no local observables whatsoever, but only *pseudolocal* ones [10]; the fact is that we do not know. Again, the exception to this stems from keeping the boundary conditions fixed; in this case it is possible to define an S-matrix in the asymptotically flat case, and a conformal quantum field theory (CFT) in the asymptotically anti-de Sitter case. Unfortunatelly, the most interesting case from the cosmological point of view, which is when the space-time is asymptotically de Sitter is not well understood.

Incidentally, it is well known that the equivalence problem in four-dimensional geometries is undecidable [3]. In three dimensions Thurston's geometrization conjecture has recently been put on a firmer basis by Hamilton and Perelman, but it is still not clear whether it can be somehow implemented in a functional integral without some drastic restrictions. Those caveats should be kept in mind when reading the sequel.

A radically different viewpoint has recently been advocated by 't Hooft [19] by insisting in causality to be well-defined, so that the conformal class of the space-time metric should be determined by the physics, but not necessarily the precise point in a given conformal orbit. If we write the spacetime metric in terms of a unimodular metric and a conformal factor

$$g_{\mu\nu} = \omega^2(x)\hat{g}_{\mu\nu}$$

with

$$det\, \hat{g}_{\mu\nu} = 1$$

then the unimodular metric is in some sense intrisic and determines causality, whereas the conformal factor depends on the observer in a way dictated by *black hole complementarity.*

Finally, there is always the (in a sense, opposite) possibility that space-time (and thus diffeomorphism invariance) is not a fundamental physical entity in such a way that the appropiate variables for studying short distances are non geometrical. Something like that could happen in string theory, but our understanding of it is still in its infancy.

3. The Effective Lagrangian Approach to Quantum Gravity

The main purpose of this talk is to review the least ambitious approach possible to the topic. Actually, one can discuss what is the notion of causality when the spacetime metric is a fluctuating object, or whether there is an adequate notion of time to be used in writing down Scrödinger's equation for the Universe itself. But if our previous experience with the other interactions is to be of any relevance here, there ought to be a regime, experimentally accessible in the non too distant future, in which gravitons propagating in flat spacetime can be isolated. This is more or less unavoidable, provided gravitational waves are discovered experimentally, and the road towards gravitons should not be too different from the road that lead from the discovery of electromagnetic waves to the identifications of photons as the quanta of the corresponding interaction, a road that led from Hertz to Planck.

Any quantum gravity theory that avoids identifying gravitational radiation as consisting of large numbers of gravitons in a semiclassical state would be at variance with all we believe to know about quantum mechanics.

What we expect instead to be confirmed by observations somewhere in the future is that the number of gravitons per unit volume with frequencies between ω and $\omega + d\omega$ is given by Planck's formula

$$n(\omega)d\omega = \frac{\omega^2}{\pi^2}\frac{1}{e^{\frac{\hbar\omega}{kT}}-1}d\omega$$

It is natural to keep an open mind for surprises here, because it can be argued that gravitational interaction is not alike any other fundamental interaction in the sense that the whole structure of space-time ought to be presumably affected, but it cannot be denied that this is the most conservative approach and as such it should be explored first, up to its very limits, which could hopefully indicate further avenues of research.

From our experience then with the standard model of elementary particles, and assuming we have full knowledge of the fundamental symmetries of our problem, we know that we can parametrize our ignorance on the *fundamental* ultraviolet (UV) physics by writing down all local operators in the low energy fields $\phi_i(x)$ compatible with the basic symmetries we have assumed.

$$L = \sum_{n=0}^{\infty}\frac{\lambda_n(\Lambda)^n}{\Lambda^n}\mathcal{O}^{(n+4)}(\phi_i)$$

Here Λ is an ultraviolet (UV) cutoff, which restricts the contributions of large euclidean momenta (or small euclidean distances) and $\lambda_n(\Lambda)$ is an infinite set of dimensionless bare couplings.

Standard Wilsonian arguments imply that *irrelevant operators*, corresponding to $n > 4$, are less and less important as we are interested in deeper and deeper infrared (IR) (*low energy*) variables. The opposite occurs with *relevant operators*, corresponding to $n < 4$, like the masses, that become more and more important as we approach the IR. The intermediate role is played by the *marginal operators*, corresponding to precisely $n = 4$, and whose relevance in the I R is not determined solely by dimensional analysis, but rather by quantum corrections. The range of validity of any finite number of terms in the expansion is roughly

$$\frac{E}{\Lambda} << 1$$

where E is a characteristic energy of the process under consideration.

In the case of gravitation, we assume that general covariance (or diffeomorphism invariance) is the basic symmetry characterizing the interaction. We can then write

$$\begin{aligned} L_{eff} &= \lambda_0\Lambda^4\sqrt{|g|} + \lambda_1\Lambda^2 R\sqrt{|g|} + \lambda_2 R^2 + \frac{1}{2}g^{\alpha\beta}\nabla_\alpha\phi\nabla_\beta\phi\sqrt{|g|} + \\ &+\lambda_3\frac{1}{\Lambda^2}R^{\alpha\beta}\nabla_\alpha\phi\nabla_\beta\phi\sqrt{|g|} + \lambda_4\frac{1}{\Lambda^2}R^3\sqrt{|g|} + \lambda_5\phi^4\sqrt{|g|} + \\ &+\bar{\psi}\left(e^\mu_a\gamma^a\left(\partial_\mu - \omega_\mu\right)\psi - m\right)\psi + \frac{\lambda_5}{\Lambda^2}\bar{\psi}e^\mu_a\gamma^a R\left(\partial_\mu - \omega_\mu\right)\psi + \ldots \end{aligned} \quad (2)$$

where e^μ_a is the tetrad, such that

$$e^\mu_a e^\nu_\beta \eta^{\alpha\beta} = g^{\mu\nu}$$

$\eta^{\alpha\beta}$ being Minkowski's metric. The quantities ω_μ are the spin connection.

The need to recover General Relativity in the classical IR limit means

$$\lambda_1\Lambda^2 = -\frac{c^3}{16\pi G} \equiv -2M_p^2$$

This in turn, means that if

$$\lambda_0\Lambda^4$$

is to yield the observed value for the cosmological constant (which is of the order of magnitude of Hubble's constant, H_0^4, which is a very tiny figure when expressed in particle physics units, $H_0 \sim 10^{-33}\, eV$) then

$$\lambda_0 \sim 10^{-244}$$

This is one aspect of the cosmological constant problem; it seems most unnatural that the cosmological constant is observationally so small from the effective lagrangian point of view. I do not have anything new to say on this.

This expansion is fine as long as it is considered a low energy expansion. As Donoghue [7] has emphasized, even if it is true that each time that a renormalization is made there is a finite arbitrariness, there are physical predictions stemming from the non-local finite parts.

The problem is when energies are reached that are comparable to Planck's mass,

$$E \sim M_p.$$

Then all couplings in the effective Lagrangian become of order unity, and there is no *decoupling limit* in which gravitation can be considered by itself in isolation from all other interactions. This then seems the minimum prize one has to pay for being interested in quantum gravity; all couplings in the derivative expansion become important simultaneously. No significant differences appear when supergravity is considered.

In conclusion, it does not seem likely that much progress can be made by somehow quantizing Einstein-Hilbert's Lagrangian in isolation. To study quantum gravity means to study all other interactions as well.

On the other hand, are there any reasons to go beyond the standard model (SM)?

Yes there are some, both theoretical, and experimental. From the latter, and most important, side, both the existence of neutrino masses and dark matter do not fit into the SM. And from the former, abelian sectors suffer from Landau poles and are not believed to be UV complete; likewise the self-interactions in the Higgs sector appear to be a trivial theory. Also the experimental values of the particle masses in the SM are not natural from the effective lagrangian point of view.

The particle physics community has looked thoroughly for such extensions since the eighties: extra dimensions (Kaluza-Klein), supersymmetry and supergravity, technicolor, etc. From a given point of view, the natural culmination of this road is string theory

A related issue is the understanding of the so-called *semiclassical gravity*, in which the second memnber of Einstein's equations is taken as the expectation value of some quantum energy-momentum operator. It can be proved that this is the dominant $1/N$ approximation in case there are N identical matter fields (confer [11]). In spite of tremendous effort, there is not yet a full understanding of Hawking's emission of a black hole from the effective theory point of view (confer, for example, [12]). Another topic in which this approach has been extensively studied is Cosmology. Novel effects (or rather old ones on which no emphasis was put until recently) came from lack of momentum conservation and seem to point towards some sort of instability [14]; again the low energy theory is not fully understood; this could perhaps have something to do with the presence of horizons.

Coming back to our theme, and closing the loop, what are the prospects to make progress in quantum gravity?

Insofar as effective lagrangians are a good guide to the physics, there are only two doors open: either there is a ultraviolet (UV) attractive fixed point in coupling space, such as in Weinberg's *asymptotic safety* [20], or else, new degrees of freedom, like in string theory [17] exist in the UV. Even if Weinberg's approach is vindicated, the fact that the fixed point most likely lies at strong coupling combined with our present inability to perform analitically other than perturbative computations, means that lattice simulations should be able to cope with the integration over (a subclass of) geometries before physical predictions could be made with the techniques at hand at the present moment.

It is to be remarked that sometimes theories harbor the seeds of their own destruction. Strings for example, begin as theories in flat spacetime, but there are indications that space itself should be a derived, not fundamental concept. It is hoped that a simpler formulation of string theory exists bypassing the roundabouts of its historical development. This is far from being the case at present.

Finally, it is perhaps worth pointing out that to the extent that a purely gravitational canonical approach, as the ones based upon the use of Ashtekar [5] variables, makes contact with the classical limit (which is an open problem from this point of view) the preceding line of argument should still carry on.

It seems *unavoidable* with our present understanding, that any theory of quantum gravity should recover, for example, the prediction that there are quantum corrections to the gravitational potential given by [4]

$$V(r) = -\frac{Gm_1m_2}{r}\left(1 + 3\frac{G\,(m_1+m_2)}{r} + \frac{41}{10\pi}\frac{G\hbar}{r^2}\right)$$

(the second term is also a loop effect, in spite of the conspicuous absence of $\hbar$.) Similarly, and although this has been the subject of some controversy, it seems now established that there are gravitational corrections to the running of gauge couplings, first uncovered by Robinson and Wilczek [15] and given in standard notation by

$$\beta(g,E) = -\frac{b_0}{(4\pi)^2}g^3 - 3\frac{16\pi G}{(4\pi)^2\hbar c^3}gE^2$$

Sometimes these effects are dismissed as perturbative, and therefore trivial. This is not a healthy attitude.

Acknowledgments

I am grateful to Jaume Garriga, Tini Veltman and Enric Verdaguer for many discussions, as well as to John Donoghue, David Toms, and Sigurdur Helgason for useful correspondence. I also would like to thank the organizers of ERE 2010, C. Barceló and J.L. Jaramillo, for the kind invitation to attend this sucessful meeting.

This work has been partially supported by the European Commission (HPRN-CT-200-00148) as well as by FPA2009-09017 (DGI del MCyT, Spain) and S2009ESP-1473 (CA Madrid).

[1] E. Alvarez, "Quantum Gravity: A Pedagogical Introduction To Some Recent Results," Rev. Mod. Phys. **61** (1989) 561.

[2] E. Alvarez, "Can one tell Einstein's unimodular theory from Einstein's general JHEP **0503** (2005) 002 [arXiv:hep-th/0501146].

[3] E. Alvarez, "Some general problems in quantum gravity,"
"Some general problems in quantum gravity. 2. The Three-dimensional case," Int. J. Mod. Phys. D **2** (1993) 1 [arXiv:hep-th/9211050].

[4] N. E. JBjerrum-Bohr, J. F. Donoghue, B. R. Holstein, "Quantum gravitational corrections to the nonrelativistic scattering potential of two masses," Phys. Rev. **D67**, 084033 (2003). [hep-th/0211072].

[5] A. Ashtekar, "New Variables for Classical and Quantum Gravity," Phys. Rev. Lett. **57**, 2244 (1986).

[6] B. S. de Witt,*Dynamical Theory of Groups and Fields*, (Gordon and Breach, 1965)
Quantum theory of gravity, I,II,III Phys. Rev 162 (1967)1113,1195,1239.

[7] John F. Donoghue, *General Relativity as an effective field theory: The leading quantum corrections*, Phys. Rev. D50 (1994) 3874.
"Perturbative dynamics of quantum general relativity," arXiv:gr-qc/9712070.

[8] M. J. Duff, "Quantum Tree Graphs and the Schwarzschild Solution," Phys. Rev. D **7** (1973) 2317.
[9] R.P. Feynman, *Lectures on Gravitation*, (University of Bangalore Press,1997)
[10] S. B. Giddings, D. Marolf and J. B. Hartle, "Observables in effective gravity," Phys. Rev. D **74** (2006) 064018 [arXiv:hep-th/0512200].
[11] J. B. Hartle, G. T. Horowitz, "Ground State Expectation Value Of The Metric In The 1/n Or Semiclassical Approximation To Quantum Gravity," Phys. Rev. **D24**, 257-274 (1981).
[12] G. 't Hooft, "The Scattering matrix approach for the quantum black hole: An Overview," Int. J. Mod. Phys. **A11**, 4623-4688 (1996). [gr-qc/9607022].
[13] J. Milnor, "Remarks On Infinite Dimensional Lie Groups," *In *Les Houches 1983, Proceedings, Relativity, Groups and Topology, Ii*, 1007-1057*
[14] A. M. Polyakov, "De Sitter Space and Eternity," Nucl. Phys. B **797** (2008) 199 [arXiv:0709.2899 [hep-th]]. "Decay of Vacuum Energy," Nucl. Phys. B **834**, 316 (2010) [arXiv:0912.5503 [hep-th]]. E. Alvarez, R. Vidal, "Comments on the vacuum energy decay," [arXiv:1004.4867 [hep-th]].
[15] S. P. Robinson, F. Wilczek, "Gravitational correction to running of gauge couplings," Phys. Rev. Lett. **96**, 231601 (2006). [hep-th/0509050].
D. J. Toms, "Quantum gravitational contributions to quantum electrodynamics," [arXiv:1010.0793 [hep-th]].
[16] C. Rovelli, "Quantum Gravity," *Cambridge, UK: Univ. Pr. (2004) 455 p*
[17] Two standard references are
M. B. Green, J. H. Schwarz and E. Witten, "SUPERSTRING THEORY. VOL. 1: INTRODUCTION," *Cambridge, Uk: Univ. Pr. (1987) 469 P. (Cambridge Monographs On Mathematical Physics)*
"Superstring Theory. Vol. 2: Loop Amplitudes, Anomalies And Phenomenology," *Cambridge, Uk: Univ. Pr. (1987) 596 P. (Cambridge Monographs On Mathematical Physics)*
J. Polchinski, "String theory. Vol. 1: An introduction to the bosonic string," *Cambridge, UK: Univ. Pr. (1998) 402 p*
"String theory. Vol. 2: Superstring theory and beyond," *Cambridge, UK: Univ. Pr. (1998) 531 p*
A shorter reference is:E. Alvarez and P. Meessen, "String primer," JHEP **9902**, 015 (1999) [arXiv:hep-th/9810240].
[18] A. Salam, "Impact Of Quantum Gravity Theory On Particle Physics,"
[19] G. 't Hooft, "Quantum gravity without space-time singularities or horizons," arXiv:0909.3426 [gr-qc].
[20] S. Weinberg, *Ultraviolet divergences in quantum theories of gravitation*, in *General relativity: an Einstein centenary volume*, S.W. Hawking and W. Israel eds. (Cambridge University press,1979).
"Effective Field Theory, Past and Future," arXiv:0908.1964 [hep-th].

Spanish Relativity Meeting (ERE 2010): Gravity as a Crossroad in Physics IOP Publishing
Journal of Physics: Conference Series **314** (2011) 012006 doi:10.1088/1742-6596/314/1/012006

Simple model for quantum general relativity from loop quantum gravity

Carlo Rovelli

Centre de Physique Théorique de Luminy, [1] Case 907, F-13288 Marseille, EU

E-mail: rovelli@cpt.univ-mrs.fr

Abstract. New progress in loop gravity has lead to a simple model of a 'general-covariant quantum field theory'. I sum up the definition of the model in self-contained form, in terms accessible to those outside the subfield. I emphasize its formulation as a generalized topological quantum field theory with an infinite number of degrees of freedom, and its relation to lattice theory. I list the indications supporting the conjecture that the model is related to general relativity and UV finite.

1. The Model

A simple model has recently emerged in the context of loop quantum gravity. It has the structure of a generalized topological quantum field theory (TQFT), with an infinite number of degrees of freedom, local in sense of classical general relativity (GR). It can be viewed as an example of a "general-covariant quantum field theory". It is defined as a function of two-complexes and may have mathematical interest in itself. I present the model here in concise self-contained form.

The model has emerged from the unexpected convergence of many lines of investigation, including canonical quantization of GR in Ashtekar variables [1, 2, 3, 4, 5], Ooguri's [6] 4d generalization of matrix models [7, 8, 9, 10, 11], covariant quantization of GR on a Regge-like lattice [12, 13, 14], quantization of geometrical "shapes" [15, 16, 17, 18] and Penrose spin-geometry theorem [19]. The corresponding literature is intricate and long to penetrate. Here I skip all 'derivations' from GR, and, instead, list the elements of evidence supporting the conjectures that the transition amplitudes are finite and the classical limit is GR.

The model's dynamics is defined in Sec. 2. States and operators in Sec. 3 and 4. Sec. 5 reviews the evidence relating the model to GR, and some of its properties.

2. Feynman rules

The model is defined assigning transition amplitudes $Z_{\mathcal{C}}(h_l)$ with $h_l \in SU2$, to two-complexes $\mathcal{C}$ with boundary.

A two-complex (see Fig.1) is a finite set of F elements f denoted "faces", E elements e ("edges"), and V elements v ("vertices"), equipped with a boundary relation ∂ associating a cyclically ordered set of edges to each face and an ordered couple of vertices $\{s_e, t_e\}$ ("source" and "target") to each edge. Its boundary is a (possibly disconnected) graph Γ, whose L links

[1] Unité mixte de recherche du CNRS et des Universités de Provence, de la Méditerranée et du Sud.

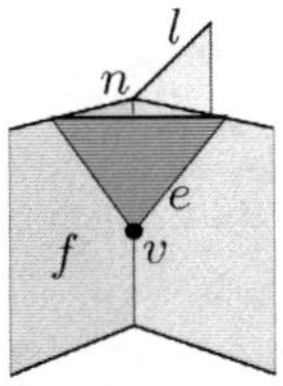

Figure 1. A two-complex with one bulk vertex.

l are edges of $\mathcal{C}$ bounding a single face and whose N nodes n are vertices of $\mathcal{C}$ bounding (links and) a single internal edge. $Z_\mathcal{C}(h_l)$ is defined as the integral obtained associating:

(i) Two group integrations to each internal edge (or one to each adjacent couple {internal edge, vertex})

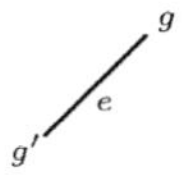

$$\longmapsto \int_{SL2C} dg_{es_e} \int_{SL2C} dg_{et_e} \tag{1}$$

(ii) A group integration to each couple of adjacent {face, internal edge}

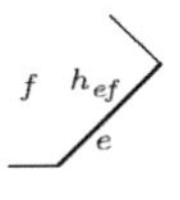

$$\longmapsto \int_{SU2} dh_{ef}\, \chi^{j_f}(h_{ef}) \tag{2}$$

$\chi^j(h)$ is the spin-j $SU2$ character of h.

(iii) A sum to each face f

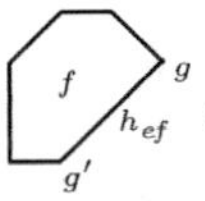

$$\longmapsto \sum_{j_f} d_{j_f}\, \chi^{\gamma(j_f+1),j_f}\Big(\prod_{e\in\partial f} g_{ef}^{\epsilon_{lf}}\Big). \tag{3}$$

where $g_{ef} := g_{es_e} h_{ef} g_{et_e}^{-1}$ for internal edges, and $g_{ef} = h_l \in SU2$ for boundary edges. d_j is $(2j+1)$. $\chi^{p,k}(g)$ is the $SL2C$ character in the unitary representation with (continuous and discrete) Casimir eigenvalues p and k. $\epsilon_{ef} = \pm 1$ according to whether the orientations (defined by ∂) of the edge e and the face f are consistent or not. γ is a fixed real parameter called Barbero-Immirzi parameter.

(iv) At each vertex, one of the integrals $\int_{SL2C} dg_{ev}$ in (1) (which is redundant) is dropped.

The resulting amplitude can be written compactly as

$$Z_\mathcal{C}(h_l) = \int_{(SL2C)^{2(E-L)-V}} dg_{ve} \int_{(SU2)^{\mathcal{V}-L}} dh_{ef}$$
$$\sum_{j_f} \prod_f d_{j_f}\, \chi^{\gamma(j_f+1),j_f}\Big(\prod_{e\in\partial f} g_{ef}^{\epsilon_{lf}}\Big) \prod_{e\in\partial f} \chi^{j_f}(h_{ef}) \tag{4}$$

where $\mathcal{V}$ is the sum of the valences of all the faces. This completes the definition of the model.

Journal of Physics: Conference Series **314** (2011) 012006 doi:10.1088/1742-6596/314/1/012006

For a two-complex without boundary, (4) reduces to the "partition function"

$$Z_{\mathcal{C}} = \int_{(SL2C)^{2E-V}} dg_{ve} \int_{(SU2)^{\mathcal{V}}} dh_{ef} \sum_{j_f} \prod_f d_{j_f} \qquad (5)$$
$$\chi^{\gamma(j_f+1),j_f}\Big(\prod_{e\in\partial f} (g_{es_e} h_{ef} g_{et_e}^{-1})^{\epsilon_{lf}}\Big) \prod_{e\in\partial f} \chi^{j_f}(h_{ef}).$$

(The sum defining the $SL2C$ character converges because the $SU(2)$ integral reduces it to a finite subspace.) A formulation more similar to the one common in the literature is in Sect. 4.2 (the one above is related to [20]).

Section 3 clarifies in which sense the $Z_{\mathcal{C}}(h_l)$ define a general covariant QFT, and Section 5 clarifies the relation with GR, and how these transition amplitudes can be used to compute physical quantities such as graviton's n-points functions or the evolution of a classical spacetime. Before going into this, however, I anticipate some comments on the intuitive physical interpretation of these quantities.

There are two related physical interpretations of the above equations, that can be considered. The first is as a concrete implementation of Misner-Hawking intuitive "sum over geometries"

$$Z = \int_{\text{Metrics/Diff}} Dg_{\mu\nu}\; e^{\frac{i}{\hbar}S[g_{\mu\nu}]}. \qquad (6)$$

As we shall see, indeed, the integration variables in (5) have a natural interpretation as 4d geometries (Sect. 4.2), and the integrand approximates the exponential of the Einstein-Hilbert action $S[g_{\mu\nu}]$ in the semiclassical limit (Sect.5). Therefore (5) gives a family of approximations of (6) as the two-complex is refined. But there is a second interpretation, compatible with the first but more interesting: the transition amplitudes (4), formally obtained sandwiching the sum over geometries (6) between appropriate boundary states, can be interpreted as terms in a generalized perturbative Feynman expansion for the dynamics of quanta of space (Sect. 4.1). In particular, (4) implicitly associates a vertex amplitude (given explicitly below in (21)) to each vertex v: this is the general-covariant analog for GR of the QED vertex amplitude

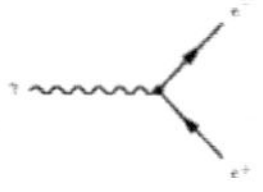

$$= e\,\gamma_\mu^{AB}\;\delta(p_1+p_2+k). \qquad (7)$$

Therefore the amplitudes (4) are a general covariant and background independent analog of Feynman graphs. These remarks on interpretation should become more clear in the last section.

The model has a euclidean version [13, 18], obtained replacing $SL2C$ with $SO4$, and can be written in a (euclidean or lorentzian) quantum deformed version, obtained by replacing $SO4$ and $SL2C$ with their q deformation (see [21]). The q-deformed version has not yet been sufficiently studied, but one might expect it to correspond to the the inclusion of a cosmological constant and its transition amplitudes (4) to be finite for appropriate values of q.

3. TQFT on manifolds with defects

Atiyah has provided a compelling definition of a general covariant QFT, by giving axioms for topological quantum field theory (TQFT) [22, 23]. In Atiyah scheme, a 4d TQFT is defined by the cobordisms between 3d manifolds. To each compact 3d manifold M_3 without boundaries is associated a finite dimensional Hilbert space $\mathcal{H}_{M_3}$, and to each 4d manifold M_4 with boundary ∂M_4 is associated a state $\psi_{M_4} \in \mathcal{H}_{\partial M_4}$. These satisfy natural composition axioms.

The model defined by (4) belongs to a simple generalization of Atiyah's TQFT, where: (i) boundary Hilbert spaces are not necessarily finite dimensional; (ii) 4d manifolds are replaced by two-complexes; (iii) 3d manifolds are replaced by graphs [24, 25, 26]. Graphs bound two-complexes in the same manner in which 3d manifolds bound 4d manifolds.

Consider a graph Γ, namely a set of L elements l called "links" and N elements n called "nodes", and a boundary relation ∂ associating to each link an ordered couple of nodes $\partial l = \{s_l, t_l\}$. Associate to each graph Γ the Hilbert space

$$\mathcal{H}_\Gamma = L_2[SU(2)^L/SU(2)^N] \tag{8}$$

where the L_2 is defined by the Haar measure and the "gauge" action of $SU(2)^N$ on the states $\psi(h_l) \in L_2[SU(2)^L]$ is

$$\psi(h_l) \to \psi(V_{s_l} h_l V_{t_l}^{-1}), \qquad V_n \in SU(2)^N. \tag{9}$$

If $\mathcal{C}$ is a two-complex bounded by the (possibly disconnected) graph Γ, then (4) defines a state in $\mathcal{H}_\Gamma$ which satisfies TQFT composition axioms [27]. Thus, the model defined above defines a generalized TQFT in the sense of Atiyah.[2]

In the next section I show (following [29]) that the states in this boundary space have a natural interpretation as 3-geometries, thanks to a beautiful theorem by Penrose.

4. Penrose metric operator

The boundary Hilbert space (8) has a natural interpretation as a space of quantum metrics, that was early recognized by Roger Penrose. The natural "momentum" operator on $L_2[SU(2)]$ is the derivative operator

$$L^i\psi(h) \equiv i\frac{d}{dt}\,\psi(he^{it\tau_i})\Big|_{t=0}, \tag{10}$$

where $i=1,2,3$ labels a hermitian basis $\vec{\tau}=\{\tau_i\}$ in the $su2$ algebra. The gauge invariant operator

$$G_{ll'} = \vec{L}_l \cdot \vec{L}_{l'} \tag{11}$$

where $\vec{L}_l = \{L^i_l\}$ is the derivative with respect to h_l and $s_l = s_{l'} := n$, is well defined on $\mathcal{H}_\Gamma$ and coincides with Penrose's metric operator. Penrose spin-geometry theorem then gives states in $\mathcal{H}_\Gamma$ a consistent interpretation as quantized 3-geometries. The metric operator $G_{ll'}$ determines the angle between the links l and l at the node n [19, 30, 31] (see Fig.2). The theorem states that these angles obey the dependency relations expected of angles in three dimensional space. A volume element associated to the node n can be defined in terms of Penrose metric operator, using standard relations between metric and volume element [32]. For instance, for a 4-valent node n, bounding the links $l_1, ..., l_4$ the volume operator V_n is given by

$$V_n^2 = |\vec{L}_{l_1} \cdot (\vec{L}_{l_2} \times \vec{L}_{l_3})|; \tag{12}$$

gauge invariance (9) at the node ensures that this definition does not depend on which triple of links is chosen. Analogously, the diagonal terms G_{ll} of the metric determines the area element A_l normal to the link l by

$$A_l^2 = \vec{L}_l \cdot \vec{L}_l. \tag{13}$$

[2] This generalization consists essentially in replacing manifolds M by "manifold with defects" $\tilde{M}$. A graph Γ is related to a 3d manifold with defects $\tilde{M}_3$ as follows. Take a cellular decomposition Δ of a (say, topologically trivial) 3d manifold M_3. Then $\tilde{M}_3$ is constructed removing the 1-skeleton Δ_1 of Δ from M_3, that is $\tilde{M}_3 = M_3 - \Delta_1$, and Γ is identified with $\Gamma = \Delta_1^*$, the 1-skeleton of the dual complex. Notice that Γ captures fully the fundamental group of $\tilde{M}_3$. Similarly, a two-complex $\mathcal{C}$ can be related to a 4d manifold with defects $\tilde{M}_4$ by $\tilde{M}_4 = M_4 - \Delta_2$ and $\mathcal{C} = \Delta_2^*$, namely removing the 2-skeleton of the cellular complex, and identifying $\mathcal{C}$ with the 2-skeleton of the dual complex. Now, recall that in Regge gravity curvature is concentrated on defects with codimension 2, and the holonomy of the Levi-Civita connections on the *flat* manifold with defects $M_n - \Delta_{n-2}$ captures entirely the geometry. Manifolds with codimension-2 defects (or graphs and two-complexes) are this natural carriers of *curved* Regge geometries. In [28], the space $\mathcal{H}_\Gamma$ is precisely constructed as the quantization of a space of flat $SU2$ connections on $\tilde{M}_3$, or equivalently a space of Regge metrics where curvature is on the defects.

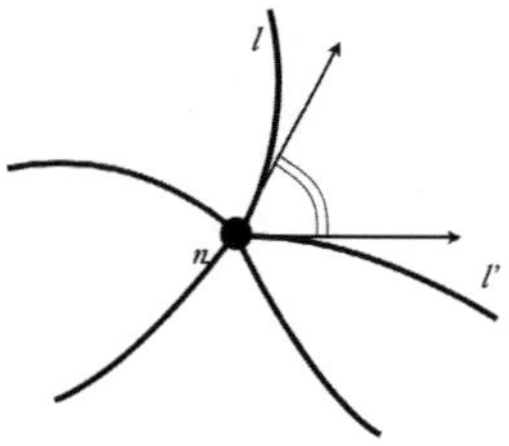

Figure 2. The angle defined by the Penrose metric operator on the graph.

The Area and Volume operators A_l and V_n form a complete set of commuting observables in $\mathcal{H}_\Gamma$, in the sense of Dirac. The spectrum of both operators can be computed [32]; it is discrete and it has a minimum step between zero and the lowest non-vanishing eigenvalue. In the case of the area, this gap is

$$a_0 = \frac{\sqrt{3}}{2}. \tag{14}$$

The orthonormal basis that diagonalizes the complete commuting commuting set of operators A_l, V_n is called the *spin-network* basis. This basis can be obtained via the Peter-Weyl theorem. It is labelled by a spin j_l for each link l and an $SU2$ intertwiner i_n for each node n [33, 34, 35], and defined by

$$\psi_{\Gamma,j_l,i_n}(h_l) = \big\langle \otimes_l \, d_{j_l} \, D^{j_l}(h_l) \; \big| \; \otimes_n \, i_n \, \big\rangle_\Gamma \tag{15}$$

where $D^{j_l}(h_l)$ is the Wigner matrix in the spin-j representation and $\langle\cdot|\cdot\rangle_\Gamma$ indicates the pattern of index contraction between the indices of the matrix elements and those of the intertwiners given by the structure of the graph.[3] A G-intertwiner, where G is a Lie group, is an element of a (fixed) basis of the G-invariant subspace of the tensor product $\otimes_l \mathcal{H}_{j_l}$ of irreducible G-representations —here those associated to the links l bounded by n. Since the Area is the $SU2$ Casimir, the spin j_l is easily recognized as the Area quantum number and i_n is the Volume quantum number.

Coherent states in $\mathcal{H}_\Gamma$ have been studied by a number of authors and are particularly useful in applications [36, 37, 38, 17, 39, 40, 41, 42, 43] (see also [44, 45, 46]).

4.1. Spin networks as quantum 3-geometries

The results above equip the boundary states of the model (4) with a geometrical interpretation: the spin network state ψ_{Γ,j_l,i_n} is interpreted as representing a granular space. Each node is a quantized "chunk", or "quantum" of space (see Fig.3); the graph gives the connectivity relations between these quanta; i_n is the quantum number of the volume of the n'th quantum of space; and j_l is the quantum number of the area of the elementary surface separating the adjacent nodes s_l and t_l.

Thus, the quantum states of the theory describe background-independent quantum excitations of the geometry of space. Physical space is built up, or "weaved up" [4] by such nets of atoms of space.

[3] Both tensor products live in

$$H_\Gamma \subset L_2[SU(2)^L] = \bigoplus_{j_l} \bigotimes_l \mathcal{V}_{j_l} \otimes \mathcal{V}_{j_l}^* = \bigoplus_{j_l} \bigotimes_n \bigotimes_{e\in\partial n} \mathcal{V}_{j_l}. \tag{16}$$

where $\mathcal{V}_j$ is the $SU2$ spin-j representation space, here identified with its dual.

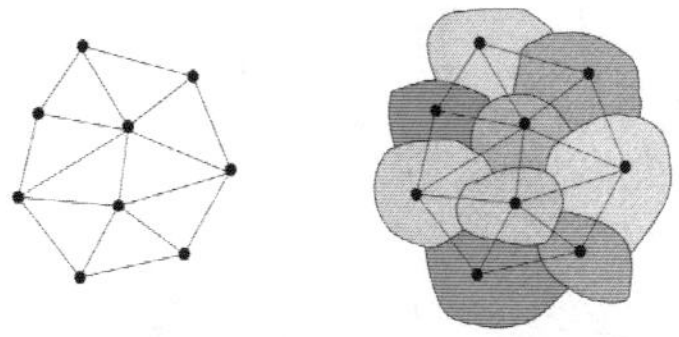

Figure 3. "Granular" space. Each node of the graph describes a "quantum" of space.

As in classical GR [47, 48], and unlikely in ordinary field theory, in this theory localization is only relative to the field itself. In this sense, the theory is profoundly different from ordinary *local* quantum field theory.

Two important comments about the length scale of the theory are in order. First, metric quantities are expressed here in natural units, without dimension-full parameters. To relate them to centimeters, we need the centimeters value of the minimal gap a_0, or equivalently the dimension-full expression of the operator L_l^i. Let's call L_{Pl} the unit of length in which all the equations above hold. L_{Pl} is a fundamental parameter of the theory, setting the scale at which the theory is defined, namely the scale of the quantum granularity of space.[4]

Second, the Hilbert space (8) is precisely the Hilbert space of lattice gauge theory, in the Kogut-Susskind [49] canonical formulation. The similarity with lattice gauge theory can be emphasized by rewriting (5) in the local form

$$Z_{\mathcal{C}} = \int dg_{ve} \int dh_{ef} \prod_f K_f(g_{ve}, h_{ef}) \tag{18}$$

where the "face amplitude" is

$$K_f(g_{ve}, h_{ef}) = \sum_j d_j \;\; \chi^{\gamma(j+1),j}\Big(\prod_{e\in\partial f} g_{ef}^{\epsilon_{lf}}\Big) \prod_{e\in\partial f} \chi^j(h_{ef}). \tag{19}$$

But there is a key difference between the physical interpretation in the two cases, which leads to a rather different dynamics. Lattice gauge theory assumes the lattice to be defined at a scale a, the "lattice spacing". This scale enters (indirectly) in the Hamiltonian and the physical theory is defined by appropriately taking the limit where $a \to 0$ and the number N of nodes of the lattice goes to infinity: $N \to \infty$. The lattice spacing is the imprint of the background metric. Here, instead, there is no background metric, and the lattice has no metrical significance whatsoever (as the coordinates of classical GR). It is the operator $G_{ll'}$ that has metric significance, and a metric emerges only in terms of expectation values and eigenvalues of such operator on the quantum states. Since geometrical operators have discrete eigenvalues and there are an Area and a Volume gaps, there is an intrinsic minimal scale (at the scale L_{Pl}), set by the quantum discreteness itself. It emerges in the same manner as the minimal scale in the energy of a quantum harmonic oscillator. The theory has no degrees of freedom at a smaller length scale. To capture the full theory, we only need to consider the $N \to \infty$ limit, namely arbitrary graphs, without any lattice spacing to be taken to zero.

[4] If we disregard radiative corrections, L_{Pl} can be related to $\hbar$ and the low-energy Newton constant G, using the classical limit of the theory. As we see later, indeed, the group elements U_l and the derivative operators L^i are recognized as the holonomy of the Ashtekar-Barbero connection and the inverse densitized triad. A quantum representation of the Poisson algebra of these is identical to the L_l^i, U_l operator algebra if $8\pi\gamma\hbar G = 1$. (The Newton constant and the Barbero-Immirzi parameter enter the action and hence the definition of the momentum; the Planck constant appears in promoting Poisson brackets to commutators.) Hence

$$L_{Pl} = 8\pi\gamma\hbar G, \quad \text{up to radiative corrections.} \tag{17}$$

The running of the Newton between the Planck scale and low-energy can modify this relation.

4.2. Transition amplitudes in terms of spinfoams

By explicitly performing all integrals in (4), and going to the spin network basis, it is not difficult to see that (4) can be rewritten in the form

$$Z_{\mathcal{C}}(j_l, i_n) = \sum_{j_f, i_e} \prod_f d_{j_f} \prod_v W_v(\sigma). \tag{20}$$

where i_e associates an $SU2$ intertwiner to each internal edge. A triple $\sigma = \{\mathcal{C}, j_f, i_e\}$ is called a *spinfoam*. The "vertex amplitude" $W_v(\sigma)$ turns out to be [12, 13, 14, 17, 18, 5, 50]

$$W_v(\sigma) = \mathrm{Tr} \prod_e I(i_e) \tag{21}$$

where the product is over the edges bounded by v and I is a map from $SU2$ intertwiners to $SL2C$ intertwiners defined as follows. Fix a subgroup $SU2$ of $SL2C$ and decompose the $SL2C$ irreducible representation $\mathcal{H}^{pk}$ into spin-j $SU2$ irreducibles $\mathcal{H}^{pk} = \oplus_j \mathcal{H}_j^{pk}$. Let Y_γ be the isomorphism $Y_\gamma : \mathcal{H}_j \to \mathcal{H}_j^{\gamma(j+1),j}$ sending a spin-j $SU2$ representation to the spin-j subspace of the unitary $SL2C$ representation with $p = \gamma(j+1), k = j$. Recall that $i_e \in \otimes_l \mathcal{H}_{j_l}$. Then I is defined by $I : \otimes_l \mathcal{H}_{j_l} \to P_{SL2C}[\otimes_l Y_\gamma \mathcal{H}_{j_l}]$ where P_{SL2C} is the projection on the $SL2C$ invariant subspace. The Trace Tr means that the $SL2C$ intertwiners are contracted among themselves in (21), following the pattern of index contraction formed by the graph surrounding the vertex. The expression (20) (or similar) is the one commonly found in the LQG literature. Notice that the QED vertex (7) too can be viewed as formed by intertwiners.

When Γ is disconnected, for instance if it is formed by two connected components, expression (20) defines transition amplitudes between the connected components. This transition amplitude can be interpreted as a quantum mechanical sum over histories. Slicing a two-complex, we obtain a history of spin networks, in steps where the graph changes at the vertices. The sum (20) can therefore be viewed as a Feynman sum over histories of 3-geometries, or a *sum over 4-geometries*. This is what connects the two intuitive physical pictures mentioned in Section 2: the particular geometries summed over can also be viewed as histories of interactions of quanta of space.

The amplitude of the individual histories is *local*, in the sense of being the product of face and vertex amplitudes. It is *locally Lorentz invariant* at each vertex, in the sense that the vertex amplitude (21) is $SL2C$ invariant: if we choose a different $SU2$ subgroup of $SL2C$ (in physical terms, if we perform a local Lorentz transformation), the amplitude does not change. The entire theory is *background independent*, in the sense that no fixed metric structure is introduced in any step of the definition of the model. The metric emerges only via the expectation value (or the eigenvalues) of the Penrose metric operator.

5. Relation with GR

A number of elements of evidence support the conjecture that the model is related to GR:

(i) The classical limit of the theory is given sending $\hbar \to 0$ at fixed value of boundary geometry. Since geometrical quantities are defined by spins j multiplied by powers of (17), the limit is the "large quantum numbers" $j \to \infty$ limit, as always in quantum theory. In other words, the classical limit of pure quantum gravity is also the large distance limit, as expected. The asymptotic expansion of the vertex (21) for high quantum numbers has been studied in detail and computed explicitly for five-valent vertices [51, 52, 53, 54]. The result is that it gives the Regge approximation of the Hamilton function of the spacetime region bounded by the 3-geometry determined by the spin network surrounding v. Since, in turn, the Regge action is known to be the Einstein-Hilbert action $S[g_{\mu\nu}]$ of a Regge geometry, we have that

$$W_v(\sigma) \sim e^{\frac{i}{\hbar} S[g_{\mu\nu}]}. \tag{22}$$

Accordingly, in the semiclassical regime the sum (5) truly reduces to a sum over geometries weighted by the exponential of the GR action, as in (6).

(ii) The Hilbert space and the operators of the theory match those obtained by a canonical quantization of GR using the Ashtekar variables and choosing Wilson loops as basic observables [2, 3, 55, 56, 57]. This convergence is the result that has sparked the interest in this model, a few years ago [12, 17, 18, 14]. A notable theorem states that under general assumptions —the key one being diff-invariance— this quantum kinematics is essentially unique [58, 59].[5]

(iii) GR's action can be written in the form [62]

$$S = \int (e \wedge e)^* \wedge F + \frac{1}{\gamma} \int e \wedge e \wedge F. \qquad (23)$$

The first term is the standard Einstein-Hilbert action $S[g_{\mu\nu}] = \int \sqrt{g} R$, written in first order form and in terms of a tetrad e and an $SL2C$ connection with curvature F. The second term is a parity violating term that does not affect the equations of motion and leads to the real Ashtekar variables. This action is the BF action

$$S_{BF} = \int B \wedge F \qquad (24)$$

where the two-form field B is restricted to the form $B = (e \wedge e)^* + \frac{1}{\gamma}(e \wedge e)$. A constraint on B forcing it to have this form is called "simplicity constraint". Now, (5) is as a modification of Ooguri's BF partition function [6]

$$\begin{aligned} Z_C &= \int_{G^{2E-V}} dg_{ve} \prod_f \delta\Big(\prod_{e\in\partial f} (g_{es_e} g_{et_e}^{-1})^{\epsilon_{lf}}\Big) \\ &= \int_{G^{2E-V}} dg_{ve} \sum_{j_f} \prod_f d_{j_f}^G \chi^{j_f}\Big(\prod_{e\in\partial f} (g_{es_e} g_{et_e}^{-1})^{\epsilon_{lf}}\Big), \end{aligned} \qquad (25)$$

obtained restricting the sum precisely to the states where such simplicity constraint hold [63, 64]. These constraints turn the (topologically invariant) BF partition function into the (non topologically invariant) partition function for GR. Because of the restriction in the representations summed over and the $SU2$ integrations, (5) relaxes the BF flatness condition implemented in (25) by the delta function on the holonomy around each face, turning local degrees of freedom on.

(iv) The model can be directly obtained via discretization and quantization of GR on a lattice [13, 18].

(v) It is possible to compute *particle*'s (graviton's) n-point functions from the model. n-point functions depend on the choice of a background. The background is introduced in the calculation via the choice of the boundary state. Coherent states in $\mathcal{H}_\Gamma$ give intrinsic and extrinsic [41] 3d-geometries, probed up to a given scale. Particle states over such geometries are obtained acting with the metric field operator on such states. (On the meaning of the notion of "particle" in this context see [69].) n-point functions for these particle states can then be computed perturbatively expanding the transition amplitudes in the number

[5] Alternatively, this Hilbert space can be obtained quantizing a space of the "shapes" of the geometry of solids figures (polyhedra) [15, 60, 51, 61, 40].

Journal of Physics: Conference Series **314** (2011) 012006 doi:10.1088/1742-6596/314/1/012006

of vertices [70, 71]. This technique allows in principle particle n-point functions to be computed at all orders, and therefore to compare the model with the standard perturbative quantum GR defined by conventional effective quantum field theoretical methods over flat space. The 2-point function has been computed in the euclidean theory to first order using this technique [65, 66] and the result is that it matches the one computed by expanding GR over a flat background, namely the free graviton propagator. Therefore the model can describes linearized gravitational waves.

(vi) A similar technique can be used to compute the cosmological evolution of homogeneous isotropic metrics (described by suitable coherent states). The result is that the (gravitational part) of the Friedmann equation has been derived from the model [67]. This indicates that the model may me consistent with the cosmological regime of classical GR.

All these facts converge in suggesting that the classical limit of the model is GR.

5.1. Physical amplitudes, expansion and divergences

Physical amplitudes. Consider the subspace of $\mathcal{H}_\Gamma$ where the spins j_l vanish on a subset of links. States in this subspace can be naturally identified with states in $\mathcal{H}_{\Gamma'}$, where Γ' is the subgraph of Γ where $j_f \neq 0$. Hence the family of Hilbert spaces $\mathcal{H}_\Gamma$ has a projective structure and the projective limit $\mathcal{H} = \lim_{\Gamma\to\infty} \mathcal{H}_\Gamma$ is well defined. $\mathcal{H}$ is the full Hilbert space of states of the theory. It describes an infinite number of degrees of freedom.[6]

In the same manner, two-complexes are partially ordered by inclusion: we write $\mathcal{C}' \leq \mathcal{C}$ if $\mathcal{C}$ has a sub-complex isomorphic to $\mathcal{C}'$. If the limit exist, we define

$$Z(h_l) = \lim_{\mathcal{C}\to\infty} Z_{\mathcal{C}}(h_l) \tag{26}$$

where the limit is in the sense of nets[7]. The transition amplitudes $Z(h_l)$ are defined on $\mathcal{H}$.

These same amplitudes can be defined *summing* over all two-complexes bounded by Γ

$$Z(j_l, i_n) = \sum_{\mathcal{C}} Z^*_{\mathcal{C}}(j_l, i_n). \tag{27}$$

where Z^* is defined by the same sum as Z, but excluding the $j_f = 0$ spins from the sum and including appropriate combinatorial factors. In spite of the apparent difference, these two definitions are equivalent [68], since the reorganization of the sum (26) in terms of the sub-complexes where $j_f \neq 0$ gives (27). The sum (27) can be viewed as the analog of the sum over all Feynman graphs in conventional QFT. Thus, the amplitudes (4) are families of approximations to the physical amplitudes (26).

A hint about the regime where this expansion is effective, namely where the complete sum is well approximated by its lowest terms (possibly renormalized, see below), is given by the fact that in the classical limit the vertex amplitude goes to the Regge action of large simplices. This indicates that the regime where the expansion is effective is around flat space; this is the hypothesis on which the calculations in items 5 and 6 above are based.

Divergences. There are *no* ultraviolet divergences, because there are no trans-Planckian degrees of freedom. However, there are potential large-volume divergences, coming from the sum over j. In ordinary Feynman graphs, momentum conservation at the vertices implies that the divergences are associated to closed loops. Here $SU(2)$ invariance at the edges implies that divergences are associated to "bubbles", namely subsets of faces forming a compact surface

[6] It has a structure similar to Fock space, with $\mathcal{H}_\Gamma$, which is a space of states with V quanta of space, being the analog to the Fock N-particle state.

[7] $\forall\epsilon\, \exists\mathcal{C}_\epsilon$ *s.t.* $|Z - Z_{\mathcal{C}}| \leq \epsilon\ \forall\mathcal{C} \geq \mathcal{C}_\epsilon$, where $\mathcal{C}$ and $\mathcal{C}_\epsilon$ have the same boundary.

without boundary [72, 73, 20, 75, 74]. Such large-volume divergences are well known in Regge calculus, and can be visualized as "spikes" of the 4-geometry.

Spikes are likely to be effectively regulated by going to the quantum group. It is commonly understood that the q-deformation amounts to the inclusion of a cosmological constant. This is consistent with the fact that q-deformed amplitudes are suppressed for large spins, correspondingly to the fact that the presence of a cosmological constant sets a maximal distance and effectively "puts the system in a box". Whether divergent or not, radiative corrections renormalize the vertex amplitude.

The second source of divergences is given by the limit (26). Less is known in this regard, but it is tempting to conjecture that this sum too could be regularized by the quantum deformation.

Scales. Equation (4) that defines the theory includes explicitly a single dimensionless parameter: γ. To this we add q in the q-deformed case, which determines the cosmological constant Λ in natural units; and the Planck scale, which enters the theory for the reason explained in Section 4. The model has therefore three parameters: L_P, which sets the minimal length scale, beyond which there are no degrees of freedom, Λ, which determines a maximal scale, and γ, which has analogies with the θ parameter in QCD, as evident from (23).

The transition amplitudes (4) can be coded into a generating functional. More precisely [76, 20], they can be seen as Feynman graphs of a generating auxiliary field theory, precisely as for the matrix models. From this perspective, a further dimensionless coupling constant λ can be naturally added to the theory as a coupling constant multiplying the vertex amplitude (21).

I close mentioning that strictly related to this theory is the ample literature on loop quantum cosmology [77, 78] and LQG black hole entropy [79, 81, 80], which has lead, respectively, to study the hypothesis of a quantum-gravity induced "Big-Bounce", and the hypothesis that the "quanta of space" described in Section 4 be the microstructure responsible for the Bekenstein-Hawking entropy.

Acknowledgments
I warmly thank Ilya Khrzhanovsky and the Kapitza Institute in Kharkov for the hospitality during October 1942, and in particularly Andrey Losev, for the engaging conversations during this visit, which have inspired this paper. Thanks to Matteo Smerlak, Eugenio Bianchi and Simone Speziale, for a careful reading of the first version of this paper and numerous suggestions.

References

[1] A. Ashtekar, "New Variables for Classical and Quantum Gravity," *Phys. Rev. Lett.* **57** (1986) 2244–2247.
[2] C. Rovelli and L. Smolin, "Knot Theory and Quantum Gravity," *Phys. Rev. Lett.* **61** (1988) 1155.
[3] C. Rovelli and L. Smolin, "Loop Space Representation of Quantum General Relativity," *Nucl. Phys.* **B331** (1990) 80.
[4] A. Ashtekar, C. Rovelli, and L. Smolin, "Weaving a classical geometry with quantum threads," *Phys. Rev. Lett.* **69** (1992) 237–240, `arXiv:hep-th/9203079`.
[5] W. Kaminski, M. Kisielowski, and J. Lewandowski, "Spin-Foams for All Loop Quantum Gravity," *Class. Quant. Grav.* **27** (2010) 095006, `arXiv:0909.0939`.
[6] H. Ooguri, "Topological lattice models in four-dimensions," *Mod. Phys. Lett.* **A7** (1992) 2799–2810, `arXiv:hep-th/9205090`.
[7] E. Brezin, C. Itzykson, G. Parisi, and J. B. Zuber, "Planar Diagrams," *Commun. Math. Phys.* **59** (1978) 35.
[8] F. David, "Planar diagrams, two-dimensional lattice gravity and surface models," *Nuclear Physics B* **257** (1985) 45–58.
[9] V. A. Kazakov, A. A. Migdal, and I. K. Kostov, "Critical Properties of Randomly Triangulated Planar Random Surfaces," *Phys. Lett.* **B157** (1985) 295–300.
[10] J. Ambjorn, B. Durhuus, and J. Frohlich, "Diseases of Triangulated Random Surface Models, and Possible Cures," *Nucl. Phys.* **B257** (1985) 433.
[11] D. J. Gross and A. A. Migdal, "Nonperturbative Two-Dimensional Quantum Gravity," *Phys. Rev. Lett.* **64** (1990) 127.

[12] J. Engle, R. Pereira, and C. Rovelli, "The loop-quantum-gravity vertex-amplitude," *Phys. Rev. Lett.* **99** (2007) 161301, arXiv:0705.2388.
[13] J. Engle, R. Pereira, and C. Rovelli, "Flipped spinfoam vertex and loop gravity," *Nucl. Phys.* **B798** (2008) 251–290, arXiv:0708.1236.
[14] J. Engle, E. Livine, R. Pereira, and C. Rovelli, "LQG vertex with finite Immirzi parameter," *Nucl. Phys.* **B799** (2008) 136–149, arXiv:0711.0146.
[15] A. Barbieri, "Quantum tetrahedra and simplicial spin networks," *Nucl. Phys.* **B518** (1998) 714–728, arXiv:gr-qc/9707010.
[16] J. W. Barrett and L. Crane, "Relativistic spin networks and quantum gravity," *J. Math. Phys.* **39** (1998) 3296–3302, arXiv:gr-qc/9709028.
[17] E. R. Livine and S. Speziale, "A new spinfoam vertex for quantum gravity," *Phys. Rev.* **D76** (2007) 084028, arXiv:0705.0674.
[18] L. Freidel and K. Krasnov, "A New Spin Foam Model for 4d Gravity," *Class. Quant. Grav.* **25** (2008) 125018, arXiv:0708.1595.
[19] R. Penrose, "Angular momentum: an approach to combinatorial spacetime," in *Quantum Theory and Beyond*, T. Bastin, ed., pp. 151–180. Cambridge University Press, Cambridge, U.K., 1971.
[20] J. B. Geloun, R. Gurau, and V. Rivasseau, "EPRL/FK Group Field Theory," arXiv:1008.0354.
[21] K. Noui and P. Roche, "Cosmological deformation of Lorentzian spin foam models," *Class. Quant. Grav.* **20** (2003) 3175–3214, arXiv:gr-qc/0211109.
[22] M. Atiyah, *Topological quantum field theory*, vol. 68. Publication mathematiques de l'I.H.E.S., 1988.
[23] M. Atiyah, *The geometry and physics of knots*. Cambridge University Press, 1990.
[24] J. C. Baez, "Spin foam models," *Class. Quant. Grav.* **15** (1998) 1827–1858, arXiv:gr-qc/9709052.
[25] J. C. Baez, "An introduction to spin foam models of BF theory and quantum gravity," *Lect. Notes Phys.* **543** (2000) 25–94, arXiv:gr-qc/9905087.
[26] L. Crane, "Holography in the EPRL Model,". http://arXiv.org/abs/1006.1248.
[27] E. Bianchi, D. Regoli, and C. Rovelli, "Face amplitude of spinfoam quantum gravity," arXiv:1005.0764.
[28] E. Bianchi, "Loop Quantum Gravity à la Aharonov-Bohm," arXiv:0907.4388.
[29] E. Bianchi, "Loop Quantum Gravity, Lectures at the XIX SIGRAV Conference on General Relativity and Gravitational Physics. Scuola Normale Superiore-Pisa." 9/2010.
[30] J. P. Moussouris, *Quantum models as spacetime based on recoupling theory*. PhD thesis, Oxford, 1983.
[31] S. A. Major, "Operators for quantized directions," *Class. Quant. Grav.* **16** (1999) 3859–3877, arXiv:gr-qc/9905019.
[32] C. Rovelli and L. Smolin, "Discreteness of area and volume in quantum gravity," *Nucl. Phys.* **B442** (1995) 593–622, arXiv:gr-qc/9411005.
[33] C. Rovelli and L. Smolin, "Spin networks and quantum gravity," *Phys. Rev.* **D52** (1995) 5743–5759, arXiv:gr-qc/9505006.
[34] J. Baez, "Spin Networks in Gauge Theory," *Adv. Math.* **117** (1996) no. 2, 253–272.
[35] J. Baez, "Spin Networks in Nonperturbative Quantum Gravity," in *The Interface of Knots and Physics*, L. Kauffman, ed., vol. 51 of *Proceedings of Symposia in Pure Mathematics*, pp. 197–203. American Mathematical Society, Providence, U.S.A., 1996.
[36] B. C. Hall, "Geometric quantization and the generalized Segal-Bargmann transform for Lie groups of compact type," *Communications in Mathematical Physics* **226** (2002) 233, arXiv:quant-ph/0012105.
[37] A. Ashtekar, J. Lewandowski, D. Marolf, J. Mourao, and T. Thiemann, "Coherent state transforms for spaces of connections," *J. Funct. Anal.* **135** (1996) 519–551, arXiv:gr-qc/9412014.
[38] T. Thiemann, "Complexifier coherent states for quantum general relativity," *Class. Quant. Grav.* **23** (2006) 2063–2118, arXiv:gr-qc/0206037.
[39] E. Bianchi, E. Magliaro, and C. Perini, "Coherent spin-networks," *Phys. Rev.* **D82** (2010) 024012, arXiv:0912.4054.
[40] E. Bianchi, P. Donà, and S. Speziale, "Polyhedra in loop quantum gravity," arXiv:1009.3402.
[41] L. Freidel and S. Speziale, "Twisted geometries: A geometric parametrisation of SU(2) phase space," arXiv:1001.2748.
[42] C. Rovelli and S. Speziale, "On the geometry of loop quantum gravity on a graph," arXiv:1005.2927.
[43] L. Freidel and S. Speziale, "From twistors to twisted geometries," arXiv:1006.0199.
[44] B. Dittrich and J. P. Ryan, "Phase space descriptions for simplicial 4d geometries," arXiv:0807.2806.
[45] D. Oriti and T. Tlas, "Encoding simplicial quantum geometry in group field theories," arXiv:0912.1546.
[46] V. Bonzom, "From lattice BF gauge theory to area-angle Regge calculus," *Class. Quant. Grav.* **26** (2009) 155020, arXiv:0903.0267.
[47] C. Rovelli, "What is observable in classical and quantum gravity?," *Class. Quant. Grav.* **8** (1991) 297–316.
[48] C. Rovelli, "GPS observables in general relativity," *Phys. Rev.* **D65** (2002) 044017, arXiv:gr-qc/0110003.

[49] J. Kogut and L. Susskind, "Hamiltonian formulation of Wilson's lattice gauge theories," *Phys. Rev. D* **11** (1975) 395–408.
[50] S. Alexandrov, "The new vertices and canonical quantization," arXiv:1004.2260.
[51] J. W. Barrett, R. J. Dowdall, W. J. Fairbairn, H. Gomes, and F. Hellmann, "A Summary of the asymptotic analysis for the EPRL amplitude," arXiv:0909.1882.
[52] J. W. Barrett, R. J. Dowdall, W. J. Fairbairn, F. Hellmann, and R. Pereira, "Lorentzian spin foam amplitudes: graphical calculus and asymptotics," arXiv:0907.2440.
[53] F. Conrady and L. Freidel, "Quantum geometry from phase space reduction," *J. Math. Phys.* **50** (2009) 123510, arXiv:0902.0351.
[54] E. Bianchi, E. Magliaro, and C. Perini, "Spinfoams in the holomorphic representation," arXiv:1004.4550.
[55] A. Ashtekar and J. Lewandowski, "Background independent quantum gravity: A status report," *Class. Quant. Grav.* **21** (2004) R53, arXiv:gr-qc/0404018.
[56] T. Thiemann, *Modern Canonical Quantum General Relativity.* Cambridge University Press, Cambridge, UK, 2007.
[57] C. Rovelli, *Quantum Gravity.* Cambridge University Press, Cambridge, UK, 2004.
[58] J. Lewandowski, A. Okolów, H. Sahlmann, and T. Thiemann, "Uniqueness of Diffeomorphism Invariant States on Holonomy-Flux Algebras," *Commun. Math. Phys.* **267** (2005) 703–733.
[59] C. Fleischhack, "Irreducibility of the Weyl algebra in loop quantum gravity," *Phys. Rev. Lett.* **97** (2006) 061302.
[60] J. W. Barrett and L. Crane, "A Lorentzian signature model for quantum general relativity," *Class. Quant. Grav.* **17** (2000) 3101–3118, arXiv:gr-qc/9904025.
[61] R. Pereira, *Spin foams from simplicial geometry* PhD thesis, Marseille, 2010.
[62] S. Holst, "Barbero's Hamiltonian derived from a generalized Hilbert-Palatini action," *Phys. Rev. D* **53** (1996) 5966–5969, arXiv:gr-qc/9511026.
[63] Y. Ding and C. Rovelli, "The volume operator in covariant quantum gravity," *Class. Quant. Grav.* **27** (2010) 165003, arXiv:0911.0543.
[64] Y. Ding and C. Rovelli, "Physical boundary Hilbert space and volume operator in the Lorentzian new spin-foam theory," *Class. Quant. Grav.* **27** (2010) 205003, arXiv:1006.1294.
[65] E. Alesci and C. Rovelli, "The complete LQG propagator: I. Difficulties with the Barrett-Crane vertex," *Phys. Rev.* **D76** (2007) 104012, arXiv:0708.0883.
[66] E. Bianchi, E. Magliaro, and C. Perini, "LQG propagator from the new spin foams," *Nucl. Phys.* **B822** (2009) 245–269, arXiv:0905.4082.
[67] E. Bianchi, C. Rovelli, and F. Vidotto, "Towards Spinfoam Cosmology," arXiv:1003.3483.
[68] C. Rovelli and M. Smerlak, "Summing over triangulations or refining the triangulation?" To appear.
[69] D. Colosi and C. Rovelli, "What is a particle?," *Class. Quant. Grav.* **26** (2009) 025002, arXiv:gr-qc/0409054.
[70] C. Rovelli, "Graviton propagator from background-independent quantum gravity," *Phys. Rev. Lett.* **97** (2006) 151301.
[71] E. Bianchi, L. Modesto, C. Rovelli, and S. Speziale, "Graviton propagator in loop quantum gravity," *Class. Quant. Grav.* **23** (2006) 6989–7028, arXiv:gr-qc/0604044.
[72] A. Perez and C. Rovelli, "A spin foam model without bubble divergences," *Nucl. Phys.* **B599** (2001) 255–282, arXiv:gr-qc/0006107.
[73] C. Perini, C. Rovelli, and S. Speziale, "Self-energy and vertex radiative corrections in LQG," *Phys. Lett.* **B682** (2009) 78–84, arXiv:0810.1714.
[74] V. Rivasseau and Z. Wang, "How are Feynman graphs resumed by the Loop Vertex Expansion?," arXiv:1006.4617.
[75] T. Krajewski, J. Magnen, V. Rivasseau, A. Tanasa, and P. Vitale, "Quantum Corrections in the Group Field Theory Formulation of the EPRL/FK Models," arXiv:1007.3150.
[76] D. Oriti, "The group field theory approach to quantum gravity: some recent results," arXiv:0912.2441.
[77] A. Ashtekar, "Loop Quantum Cosmology: An Overview," *Gen. Rel. Grav.* **41** (2009) 707–741, arXiv:0812.0177.
[78] A. Ashtekar, M. Campiglia, and A. Henderson, "Casting Loop Quantum Cosmology in the Spin Foam Paradigm," arXiv:1001.5147.
[79] A. Ashtekar, "Classical quantum physics of isolated horizons: A Brief overview," *Lect. Notes Phys.* **541** (2000) 50–70.
[80] J. Engle, A. Perez, and K. Noui, "Black hole entropy and SU(2) Chern-Simons theory," *Phys. Rev. Lett.* **105** (2010) 031302, arXiv:0905.3168.
[81] K. Krasnov and C. Rovelli, "Black holes in full quantum gravity," *Class. Quant. Grav.* **26** (2009) 245009, arXiv:0905.4916.

Journal of Physics: Conference Series **314** (2011) 012007 doi:10.1088/1742-6596/314/1/012007

Quantum Gravity phenomenology: achievements and challenges

S Liberati[1,2] and L Maccione[3]

[1] SISSA, via Bonomea, 265, 34136 Trieste, Italy
[2] INFN, Sezione di Trieste, Via Valerio 2, 34127 Trieste, Italy
[3] DESY, Theory Group, Notkestraße 85, 22607 Hamburg, Germany

E-mail: liberati@sissa.it

E-mail: luca.maccione@desy.de

Abstract. Motivated by scenarios of quantum gravity, Planck-suppressed deviations from Lorentz invariance are expected at observable energies. Ultra-High-Energy Cosmic Rays, the most energetic particles ever observed in nature, yielded in the last two years strong constraints on deviations suppressed by $O(E^2/M_{\rm Pl}^2)$ and also, for the first time, on space-time foam, stringy inspired models of quantum gravity. We review the most important achievements and discuss future outlooks.

1. Introduction

Quantum Gravity (QG) has posed a challenge to many theoretical physicists of the last generation and is at present far from understood. Although we do not yet have a single experiment or observation forcing us to introduce such a theory[1], we definitely need it, not only on philosophical grounds (reductionism as a driving force in physics), but also because we know that in physically relevant regimes (e.g. singularities in cosmology and in black holes...) our classical theory of gravitation fails to be predictive. However, when searching for QG, we have to tackle not only deep theoretical problems (e.g. the renormalizability of gravitational theories, the possible loss of unitarity in gravitational phenomena [1], the meaning of time in QG [2, 3]) but also the lack of observational and experimental guidance. The typical scale at which QG effects should become relevant is expected to be the one at which the gravitational action (the Einstein-Hilbert action for General Relativity) becomes of the order of the quantum of action $\hbar$. This happens at the so called Planck scale $M_{\rm Pl} \equiv \sqrt{\hbar c/G_N} \simeq 1.22 \times 10^{19}$ GeV$/c^2$ which corresponds to energies well above the capabilities of any Earth based experiment as well as any observationally accessible regime.

However, the situation may be better than it appears at first sight. In fact, models of gravitation beyond General Relativity and models of QG have shown that there can be several low energy "relic signatures" of these models, which would lead to deviation from the standard theory predictions in specific regimes.

[1] However, part of the gravitation theory community would remark that current cosmological observations (dark energy and dark matter issues) are definitely taking up this role.

Here we focus upon the phenomenology of violations of fundamental symmetries, given that a convenient way to perform high-precision tests is to look for experimental deviations from symmetries that are believed to hold *exactly* in nature and that could be broken by QG.

An example of such a fundamental symmetry is CPT invariance, which requires that physics be unchanged under the combination of charge conjugation (C), parity inversion (P) and time reversal (T). C connects particles and antiparticles, P represents a spatial reflection of physical quantities with respect to the coordinate origin and T reverses a physics reaction in time.

In Quantum Field Theory, Lorentz symmetry is intimately related to CPT symmetry. Indeed, one of the hypotheses of the well known "CPT theorem" is Lorentz invariance. If CPT is broken, then at least one of the hypotheses of the CPT theorem should also break down. It has been proven [4] that Lorentz symmetry is the failing assumption in the so called "anti-CPT theorem", which states that in any unitary, local, relativistic point-particle field theory CPT breaking implies Lorentz violation. Note however that the converse of this statement is not true: it is possible to violate Lorentz invariance while keeping CPT exact[2].

Thus, it is interesting to study both the theory and the phenomenology of Lorentz invariance violation (LV), which may yield a glimpse of QG. In recent years, attempts to place constraints on high-energy deviations from LI have mainly focused on modified dispersion relations for elementary particles. Indeed, specific hints of LV arose from various approaches to Quantum Gravity. Among the many examples are string theory tensor VEVs [6], space-time foam [7], semiclassical spin-network calculations in Loop QG [8], non-commutative geometry [9, 10, 11], some brane-world backgrounds [12] and condensed matter analogues of "emergent gravity" [13].

Lorentz symmetry breaking is not a necessary feature of QG, but it is clear that any possible LV effect connected with the Planck scale could provide an observational window into QG. However, to directly observe phenomena connected with $M_{\rm Pl}$ would require the center of mass energy of, e.g., a scattering process to be comparable to $M_{\rm Pl}$. This is 15 orders of magnitude larger than what the LHC can probe with its design center of mass energy of 14 TeV. On the other hand, if we are testing LI specifically, then also non-LI quantities can be important. The energy of the particle in some frame, or a cosmological propagation distance are widely discussed examples. These quantities can be so large as to effectively offset the $M_{\rm Pl}$ suppression to a physical observable, so that very small corrections are magnified. For this reason, they are called "windows on QG".

In order to correctly identify such "windows on QG" it is important to place them into a dynamical framework. A standard method is to study, within the context of Effective Field Theory (EFT), a Lagrangian containing the standard model fields and all LV operators of interest that can be constructed by coupling the standard model fields to new LV tensor fields that have non-zero vacuum expectation values [14, 15, 16, 17, 18, 19]. [3] A generic result of this procedure is the presence of modified dispersion relations for particles, of the form

$$E^2 - p^2 = m^2 + f(\vec{p}, M_{\rm Pl}; \mu) \;, \tag{1}$$

where m is the particle mass, E its energy, the function f represents the QG contribution and can depend generically on the momentum $\vec{p}$, on $M_{\rm Pl}$ and on some intermediate mass scale μ. For simplicity we assume that only boost invariance is broken, while rotations are preserved (see

[2] However, this theorem does not hold for theories that do not admit a local field theory formulation and that can therefore have unexpected properties [5].

[3] There are other approaches to either violate or modify Lorentz invariance, that do not necessarily yield a low energy EFT (see [20] and refs therein). However, these models do not easily lead to particle physics constraints as the dynamics of particles is less well understood. Therefore we do not consider them here. In particular, we remark here that ideas of deformation, rather than breaking, of the Lorentz symmetry (see, e.g., [21]) do not have an ordinary-EFT formulation, hence they cannot be tested with the procedures discussed here.

[22] for further comments on rotation breaking), so that f depends on $p = |\vec{p}|$, rather than on $\vec{p}$. Moreover, at $p \ll M_{\rm Pl}$ we can expand f so that Eq. (1) reads

$$E^2 - p^2 = m^2 + \sum_{n=1}^{N} \eta^{(n)} \frac{p^n}{M_{\rm Pl}^{n-2}}\,, \tag{2}$$

All renormalizable LV operators (corresponding to $n = 1, 2$) that can be added to the standard model are known as the (minimal) Standard Model Extension (mSME) [14]. These operators all have dimension three or four and can be further classified by their behavior under CPT. The CPT odd dimension five kinetic terms for QED were written down in [15] while the full set of dimension five operators were analyzed in [17]. The dimension five and six CPT even kinetic terms for QED for particles coupled to a non-zero background vector, which we are primarily interested in here, were analyzed in [18]. It is notable that SUSY forbids renormalizable operators for matter coupled to non-zero vectors [16] but permits nonrenormalizable operators.

Many of the operators in these various EFT parameterizations of LV have been very tightly constrained via direct observations (see [22, 23, 24] for extensive reviews). Moreover, higher dimension LV operators can be tightly constrained by EFT arguments [25] showing that they will generically induce via radiative corrections large dimension 3 and 4 operators in coupled particles if we assume no other relevant physics enters between the TeV and $M_{\rm Pl}$ energies.

This is a very powerful argument which applies basically to any Lorentz violating theory (see e.g. [26]) and should not be arbitrarily discounted. However, as the SUSY example above shows, this assumption can be a little dangerous as new physics above a TeV can change the hierarchy of terms. In particular SUSY would prohibit dimension 3 and 4 operators and once broken would add an extra $\mathcal{O}(\Lambda_{SUSY}/M_{\rm Pl})$-to-some-suitable-power suppression. We still do not know if SUSY can really do this job and it is not clear if this is the correct solution of this naturalness problem. However, given the present uncertainty on this, it would be nice, when possible, to constrain the dimension five and six LV kinetic terms directly via observation.

Mass dimension five CPT odd operators have been strongly constrained, both in QED and in the hadronic sector, using a wealth of observations spanning from the synchrotron spectrum of the Crab nebula (and its hard X-ray polarization) to the ultra-high-energy cosmic rays (UHECR), (see e.g. [24]). Here we shall consider explicitly terms coming from dimension five and six CPT even LV operators, so that the dispersion relations for protons, pions, and photons respectively, take the form [27]

$$E_p^2 = p^2 + m_p^2 + \eta_p \frac{p^4}{M_{\rm Pl}^2}, \quad E_\pi^2 = p^2 + m_\pi^2 + \eta_\pi \frac{p^4}{M_{\rm Pl}^2}, \quad \omega^2 = k^2 + \beta \frac{k^4}{M_{\rm Pl}^2} \tag{3}$$

The successful operation of the PAO has brought UHECRs to the interest of a wide community of scientists and has allowed to test fundamental physics (in particular Lorentz invariance in the QED sector) with unprecedented precision [28, 29, 30]. Given that the most important developments in the last years were achieved in the context of UHECR physics, we will review the role of UHECRs in the following.

2. Ultra-high-energy Cosmic Rays and LV

The UHECR constraints [31, 32, 33, 34, 35, 36, 27, 37] rely on the behavior of particle reaction thresholds with LV. What matters for threshold reactions in the presence of modified dispersion relations as in Eq. (2) is not the size of the LV correction compared to the absolute energy of the particle, but rather the size of the LV correction to the mass of the particles in the reaction. Hence the LV terms usually become important when their size becomes comparable to the mass of the heaviest particle. This criterion sets the presence of a critical energy E_{cr} above which LV effects are relevant in a given threshold reaction. If the LV term scales with energy as E^n, then $E_{cr} \sim \left(m^2 M_{\rm Pl}^{n-2}\right)^{1/n}$ [38]. According to this reasoning, the larger the particle mass the higher is the energy at which threshold LV effects come into play.

2.1. Constraints from the UHECR spectrum

The Cosmic Ray spectrum spans more than ten decades in energy (from $<$ 100 MeV to $>$ 10^{20} eV) with a power-law shape of impressive regularity $dN/dE \propto E^{-p}$. The spectral slope p has been measured as $p \simeq 2.7$ for 1 GeV $\lesssim E \lesssim 10^{15.5}$ eV, followed by a softening (the "knee") to $p \simeq 3.0$ for $10^{15.5}$ eV $\lesssim E \lesssim 10^{17.5}$ eV, a further steepening to $p \simeq 3.2$ (the "second knee") up to $E \simeq 10^{18.5}$ eV and a subsequent hardening (the "ankle") to again $p \simeq 2.7$ at $E \gtrsim 10^{18.5}$ eV [39, 40].

One of the most fascinating problems regarding CRs is at what energy the end-point to the CR spectrum occurs. A suppression to the spectrum is expected theoretically due to the interactions of UHECR protons with the Cosmic Microwave Background (CMB), leading to the production of charged and neutral pions, eventually dumping the energy of the UHECR protons into neutrinos and γ-rays. This reaction has a LI threshold energy $E_{\rm th} \simeq 5 \times 10^{19}$ $(\omega_b/1.3 \text{ meV})^{-1}$ eV (ω_b is the target photon energy). Therefore, at the present epoch, significant photo-pion production in a LI theory occurs only if the energy of the interacting proton is above a few 10^{19} eV, with the CR mean-free-path rapidly decreasing above this energy. Hence, it has long been thought to be responsible for a cut-off in the UHECR spectrum, the Greisen-Zatsepin-Kuzmin (GZK) cut-off [41, 42]. Moreover, trans-GZK particles arriving at Earth must be accelerated within the so called GZK sphere, whose radius is expected to be of the order of 100 Mpc at $\sim 10^{20}$ eV and to shrink down at larger energies. Experimentally, the presence of a suppression of the UHECR spectrum has been confirmed only recently with the observations by the HiRes detector [43] and the PAO [44]. Although the cut-off could be also due to the finite acceleration power of the UHECR sources, the fact that it occurs at roughly the expected energy favors a GZK explanation. The correlation results shown in [45] further strengthen this hypothesis.

It is this scenario where possible LV effects come into play. LV has two effects on UHECR propagation: it modifies standard reactions and allows new, normally forbidden reactions. In particular, in [27] it was considered

- $p + \gamma \to p + \pi^0$ $(n + \pi^+)$, which is modified by LV.
- $p \to p + \gamma$ and $p \to p + \pi$, which correspond respectively to photon and pion emission in vacuum and would be forbidden if LI were exact.

As a consequence of LV, the mean free path for the GZK reaction is modified. The propagated UHECR spectrum can therefore display features, like bumps at specific energies, suppression at low energy, recovery at energies above the cutoff, such that the observed spectrum cannot be reproduced. Moreover, the emission of Cherenkov γ-rays and pions in vacuum would lead to sharp suppression of the spectrum above the relevant threshold energy. After a detailed statistical analysis of the agreement between the observed UHECR spectrum and the theoretically predicted one in the presence of LV and assuming pure proton composition, the final constraints implied by UHECR physics are (at 99% CL) [27]

$$\begin{array}{lll} -10^{-3} \lesssim & \eta_p & \lesssim 10^{-6} \\ -10^{-3} \lesssim & \eta_\pi & \lesssim 10^{-1} \quad (\eta_p > 0) \quad \text{or} \quad \lesssim 10^{-6} \quad (\eta_p < 0)\,. \end{array} \tag{4}$$

2.1.1. Role of UHE nuclei UHECR constraints have relied so far on the hypothesis, not in contrast with any previous experimental evidence, that protons constituted the majority of UHECRs above 10^{19} eV. Recent PAO [46] and Yakutsk [47] observations, however, showed strong hints of an increase of the average mass composition with rising energies up to $E \approx$

$10^{19.6}$ eV, although still with large uncertainties. Hence, experimental data suggests that heavy nuclei can possibly account for a substantial fraction of UHECR on Earth.

One can assume that each individual nucleus has its own independent modified dispersion relation and make a further simplification by assuming that energy and momentum of the nucleus are the sum of energies and momenta of its constituents [38]. With this approximation (and also taking the masses of protons and neutrons to be equal), the dispersion relation for a nucleus of mass A and charge Z can be written as [48]

$$E_A^2 = (AE_1)^2 = (Ap_1)^2 + (Am_1)^2 + \frac{\eta}{A^2}\frac{(Ap_1)^4}{M_{\rm Pl}^2} = p_{A,Z}^2 + m_{A,Z}^2 + \frac{\eta_p}{A^2}\frac{p_{A,Z}^4}{M_{\rm Pl}^2}\,. \tag{5}$$

So now we have only one free parameter, η_p, for the nucleon, while for nuclei there are effective parameters of the form $\eta_A = \eta_p/A^2$. This phenomenological model guarantees that the correct dispersion relations are recovered when dealing with macroscopic objects [38], for which QG effects should be suppressed.

Assuming that current hints for a heavy composition at energies $E \sim 10^{19.6}$ eV [46] may be confirmed in the future, and that some UHECR is observed up to $E \sim 10^{20}$ eV [49], one could place a first constraint on the absence of spontaneous decay for nuclei which could not spontaneously decay without LV [48]. It will place a limit on $\eta_p < 0$, because in this case the energy of the emitted nucleon is lowered with respect to the LI case until it "compensates" the binding energy of the nucleons in the initial nucleus in the energy-momentum conservation.

An upper limit for $\eta_p > 0$ can instead be obtained from vacuum Cherenkov emission [48]. Assuming UHECR to be mainly iron at the highest energies the constraint is given by $\eta_p \lesssim 2\times10^2$ for nuclei observed at $10^{19.6}$ eV (and $\eta_p \lesssim 4$ for 10^{20} eV), while for He it is $\eta_p \lesssim 4\times10^{-3}$ (10^{-4}).

UHE nuclei suffer mainly from photodisintegration losses as they propagate in the intergalactic medium. Because photodisintegration is indeed a threshold process, it can be strongly affected by LV. According to [48], and in the same way as for the proton case, the mean free paths of UHE nuclei are modified by LV in such a way that the final UHECR spectra after propagation can show distinctive LV features. However, a quantitative evaluation of the propagated spectra has not been performed yet.

2.2. Constraints from UHE γ-rays

Photopion interactions of UHECR protons with the CMB lead to the production of neutral pions which subsequently decay into UHE γ-ray pairs. The PAO and the Yakutsk and AGASA experiments imposed limits on the presence of photons in the UHECR spectrum. In particular, the photon fraction is less than 2.0%, 5.1%, 31% and 36% (95% C.L) at $E = 10$, 20, 40, 100 EeV respectively [50, 51]. From the theoretical side, and bearing in mind the uncertainties related to source and propagation effects, it is well established that in a LI framework UHE photons are attenuated by pair production onto the CMB and Radio background during their travel to Earth, leading to their fraction in the total UHECR flux being reduced to less than 1% at 10^{19}eV and less than 10% at 10^{20} eV [52, 53]. It was shown in a framework with modified dispersion relations for both photons and e^+/e^- and standard energy/momentum conservation, that pair production could be effectively inhibited at high energy, due to the presence of an upper threshold [54],[4] and therefore the fraction of photons present in UHECRs on Earth would violate the present experimental upper limits. Hence, the *non* observation of a large fraction of UHE photons in UHECRs implies the constraint $|\xi| < \mathcal{O}(10^{-14})$ in the EFT framework [29, 30].

[4] An upper threshold is an energy above which it is not possible to simultaneously conserve energy and momentum in an interaction. If Lorentz symmetry is exact then upper thresholds do not exist, while they might well exist if it is violated [55].

2.2.1. Constraints on space-time foam models The recent detection of time delays on arrival of high energy γ-rays [56, 57] led to renewed interest of the astrophysics community in QG induced LV effects. The observed time delays can be explained, and are actually expected, in standard astrophysical scenarios hence they can be readily used to place constraints on LV models. However, time delays are naturally predicted also in generic LV QG models. It is now established that any LV model able to reproduce the observed delays and admitting an EFT formulation is in tension with other astrophysical observations (see e.g. [24]). Up to now, the only fully developed LV model able to explain the observed time delays has a string theory origin and does not admit an EFT formulation [58, 59, 60, 61, 62, 63, 64, 65]. Therefore, if observed time delays were due to such QG effects, the propagation of GeV photons over cosmological distances could not be described within EFT. Given that EFT is accurately verified with terrestrial accelerators up to $\sim$ 100 GeV, this would be a very striking and revolutionary conclusion.

In the model [58, 59, 60, 61, 62, 63, 64, 65] only purely neutral particles, such as photons or Majorana neutrinos, possess LV modified dispersion relations. For photons this has the form

$$E_\gamma^2 = p^2 - \xi \frac{p^3}{M} \, , \tag{6}$$

with the free parameter $\xi > 0$. Hence only subluminal photons are present in the theory, and photon propagation in vacuum is not birefringent. Due to stochastic losses in interactions with the D-brane foam, exact energy-momentum conservation during interactions does not hold. This last phenomenon is controlled by the free parameter ξ_I [59].

According to Eq. (6) photons with different energy travel at different speeds. Then, if a source at redshift $\bar{z}$ simultaneously emitted two photons at energy $E_1' \neq E_2'$, their time delay at Earth will be

$$\Delta t \simeq \xi \frac{\Delta E}{M} \frac{1}{H_0} \int_0^{\bar{z}} dz \frac{1+z}{\sqrt{\Omega_\Lambda + (1+z)^3 \Omega_{\rm M}}} \, , \tag{7}$$

where ΔE is the observed energy difference and the integral on redshift accounts also for redshift of the energy [66, 67, 68]. Time-of-flight constraints are then viable for this model, even though they lead at most to constraints on ξ, because ξ_I is not effective in this context.

Rather intriguingly, the FERMI Collaboration has recently reported the detection of delays on arrival of γ-ray photons emitted by distant GRBs, in particular GRB 080916C [56] and GRB 090510 [57] (see however [67] for an updated review). A thorough analysis of these delays in the energy range 35 MeV – 31 GeV allowed to place for the first time a conservative constraint of order $\xi \lesssim 0.8$ [57] on LV effects expressed as in Eq. (6). This is the best constraint so far available on the theory. On the other hand, FERMI results can be interpreted in terms of LV assuming $\xi \simeq 0.4$ and a possible evolution of the D-particle density with redshift [65].[5]

In order to constrain the D-brane model, the process of pair production, $\gamma\gamma \to e^+e^-$, can be exploited [69]. Indeed, according to [69] also in the D-brane model pair-production exhibits upper thresholds which, for values of the free parameters $\xi, \xi_I \gtrsim \mathcal{O}(10^{-12})$, are located at $E > 10^{19}$ eV. This would lead to UHECRs being constituted by a large fraction of photons, in contrast with experimental data [69]. Therefore, a limit $\xi, \xi_I \lesssim 10^{-12}$ is placed [69]. This also means that D-particle explanations of time delays in the GeV range are in conflict with data on the photon fraction in UHECRs (although some possible implementations of the model [62] were recently proposed [70] which would naturally evade the above constraints).

[5] Plausible astrophysical explanations of this phenomenon exist. No claim of a discovery of LV can be made on the basis of the data reported in [56, 57], where only LV constraints are discussed.

2.3. Foreseen constraints from UHE neutrinos

Neutrinos, with their tiny mass of order $m_\nu \simeq 0.01$ eV [71], are in principle the most suited particles to provide strong constraints on LV, at least for reactions involving *only* neutrinos. Despite the threshold being low for LV effects to kick in, neutrinos with ultra-high energy are necessary to achieve a signal, as they interact so weakly that the phase space for a LV reaction must be huge to generate an appreciable rate. This requirement implies that very large energies are needed.

If one neglects exotic sources of UHE neutrinos, the "cosmogenic" neutrino flux is created [72, 73, 74, 75] via the decay of charged pions produced by the interaction of primary nucleons with CMB photons above the GZK threshold. Violation of LI however introduces new phenomena in the propagation of UHE neutrinos. A detailed list can be found in [76], however we shall focuss here on the so called ν-splitting $\nu \to \nu\nu\bar{\nu}$ as it exclusively involves the neutrino sector and has a high enough rate to be seen at UH energies.

The effects of neutrino splitting on the UHE neutrino spectrum are twofold and can be understood qualitatively as follows.

Flux suppression at UH energies The splitting is effectively an energy loss process for UHE neutrinos. If the rate is sufficiently high, the energy loss length can be below 1 Mpc. Let us call $\bar{E}(\eta_\nu)$ the energy at which this happens. Then, being GZK neutrinos produced mainly at distances larger than 1 Mpc, we do not expect any neutrino to be detected at Earth with $E > \bar{E}$. The mere observation of neutrinos up to a certain energy $E_{\rm obs}$ would imply a constraint [76]

$$\eta_\nu^{(4)} \lesssim \left(\frac{E_{\rm obs}}{6 \times 10^{18}\ {\rm eV}}\right)^{-13/4} . \tag{8}$$

Flux enhancement at sub-UH energies Neutrinos lose energy by producing lower energy neutrinos. Eventually these neutrinos will become stable, either because their energy is below threshold, or because their lifetime is larger than their propagation time. Accordingly, an enhancement of the neutrino flux at energies below few $\times\ 10^{18}$ eV is expected [76].

Next generation neutrino detectors such as ANITA [77] and SuperEUSO [78, 79] are sensitive to neutrinos of energies $> 10^{19}$ eV. Further experiments, like the planned ARIANNA [80, 81] and IceRay [82], will cover the range $10^{17} \div 10^{20}$ eV. The scenarios described above can then be tested in the near future and constraints $\eta_\nu < 10^{-4}$ will be potentially cast according to Eq. (8).

3. Summary

QG phenomenology of Lorentz and CPT violations is a success story in physics. We have progressed in few years from almost no tests to tight, robust constraints on EFT models and some spacetime foam models. In summary for EFT with LV the situation is:

QED up to $\mathcal{O}(10^{-22})$ on $n = 2$, $\mathcal{O}(10^{-11})$ on $n = 3$, $\mathcal{O}(10^{-7})$ on $n = 4$

Hadrons up to $\mathcal{O}(10^{-31})$ on $n = 1$, $\mathcal{O}(10^{-27})$ on $n = 2$, $\mathcal{O}(10^{-14})$ on $n = 3$, $\mathcal{O}(10^{-6})$ on $n = 4$

Neutrinos up to $\mathcal{O}(10^{-27})$ on $n = 2$, $\mathcal{O}(10^{-14})$ on $n = 3$, expected $\mathcal{O}(10^{-4})$ on $n = 4$

Chances are high that improving observations in HE astrophysics will strengthen these constraints in a near future. Let us note however, that there is a noticeable missing voice in the above list, this is the gravitational sector. In particular, it would be important to cast in the future constraints which are purely gravitational, given that this framework seems to have both theoretical [83] as well as phenomenological reasons for being pursued [84][6]. We leave this

[6] Basically if LV operators are present only in the gravitational sector the induced LV operators in Standard Models particles are further suppressed by the smallness of the Newton constant provided the Lorentz breaking scale is much lower energy scale than the Planck one. This could be then a way out of the previously discussed naturalness problem.

for future studies.

4. Conclusions and Perspectives

Lorentz invariance of physical laws relies on only few assumptions: the principle of relativity, stating the equivalence of physical laws for non-accelerated observers, isotropy (no preferred direction) and homogeneity (no preferred location) of space-time, and a notion of precausality, requiring that the time ordering of co-local events in one reference frame be preserved [85, 86, 87]. In this sense a breakdown of Lorentz invariance does not necessarily imply a breakdown of the relativity principle. For this reason, it is worth exploring an alternative possibility that keeps the relativity principle but that relaxes one or more of the above postulates. Such a possibility can lead to the so-called very special relativity framework [88], which was discovered to correspond to the break down of isotropy and to be described by a Finslerian-type geometry [89, 90, 91]. Notice that in this example the generators of the new relativity group number fewer than the usual ten associated with Poincaré invariance. Specifically, there is an explicit breaking of the $O(3)$ group associated with rotational invariance. Finsler-type geometries have also been considered as a possible geometric framework for modified dispersion relations like Eq. (2) in [92] albeit the possibility to use them as the geometric counterpart of Minkowski spacetime for alternative special relativity groups seems hampered by structural problems (see e.g. [93] and references therein).

One may wonder whether there exist alternative relativity groups with the same number of generators as special relativity. Currently, we know of no such generalization in (commutative) coordinate space. However, it has been suggested that, in non-commutative spacetime, such a generalization is possible, and it was termed "doubly" or "deformed" (to stress the fact that it still has 10 generators) special relativity, DSR [94]. Unfortunately, the various DSR candidates face in general major problems regarding their physical interpretation and a working model is not yet available (see however [94] for recent attempts in new directions).

Finally, it is a logical, and rather simple, possibility that a Lorentz symmetry breakdown could be signaling an interpolation from a relativity group to another one, for example two special relativity groups characterized by different limit speeds (see [95] for an example in the so called analogue gravity context [13]) or between a Lorentzian and an Euclidean Poincaré group (see [96] for an explicit, analogue gravity inspired, example). Even more intriguingly it might be that a Lorentz invariant world could emerge from a non-relativistic system living in lower dimensions (e.g. the effective dimension of quantum gravity models seems to generically reduce to two at very short scales [97])

In conclusion, we should take the experience in constraining EFT with LV as a lesson that we can and we must challenge quantum/emergent gravity scenarios with the observational test. However, we cannot say yet "mission accomplished" for what regards testing possible deviations from local Lorentz invariance at small scales as we have started testing the most obvious, generic scenarios. New tests will probably require picking up more specific models for what lies beyond the Planck scale and will rely more heavily on these assumptions. This task will also probably require a better use and knowledge of the current (mainly astrophysical) data. Still we feel that any possibility to confront our ideas with reality should be pursued without hesitations, and that the path walked in these years should be followed. There cannot be any credible quantum gravity research without a rigorous quantum gravity phenomenology challenge.

References

[1] Hawking S 1976 *Phys.Rev.* **D14** 2460–2473
[2] Isham C 1995 (*Preprint* gr-qc/9510063)
[3] Butterfield J and Isham C 1998 To appear in The Arguements of Time, ed. J. Butterfield, Oxford University Press 1999 (*Preprint* gr-qc/9901024)
[4] Greenberg O W 2002 *Phys. Rev. Lett.* **89** 231602 (*Preprint* hep-ph/0201258)

[5] Chaichian M, Dolgov A D, Novikov V A and Tureanu A 2011 (*Preprint* 1103.0168)
[6] Kostelecky V A and Samuel S 1989 *Phys. Rev.* **D39** 683
[7] Amelino-Camelia G, Ellis J, Mavromatos N, Nanopoulos D and Sarkar S 1998 *Nature* **393** 763–765 (*Preprint* astro-ph/9712103)
[8] Gambini R and Pullin J 1999 *Phys. Rev.* **D59** 124021 (*Preprint* gr-qc/9809038)
[9] Carroll S M, Harvey J A, Kostelecky V A, Lane C D and Okamoto T 2001 *Phys. Rev. Lett.* **87** 141601 (*Preprint* hep-th/0105082)
[10] Lukierski J, Ruegg H and Zakrzewski W J 1995 *Ann. Phys.* **243** 90–116 (*Preprint* hep-th/9312153)
[11] Amelino-Camelia G and Majid S 2000 *Int. J. Mod. Phys.* **A15** 4301–4324 (*Preprint* hep-th/9907110)
[12] Burgess C, Cline J, Filotas E, Matias J and Moore G 2002 *J. High Energy Phys.* **2002** (*Preprint* hep-th/0201082)
[13] Barcelo C, Liberati S and Visser M 2005 *Living Rev. Rel.* **8** 12 (*Preprint* gr-qc/0505065)
[14] Colladay D and Kostelecký V A 1998 *Phys. Rev. D* **58** 116002 (*Preprint* hep-ph/9809521)
[15] Myers R C and Pospelov M 2003 *Phys. Rev. Lett.* **90** 211601 (*Preprint* hep-ph/0301124)
[16] Groot Nibbelink S and Pospelov M 2005 *Phys. Rev. Lett.* **94** 081601 (*Preprint* hep-ph/0404271)
[17] Bolokhov P A and Pospelov M 2008 *Phys. Rev. D* **77** 025022 (*Preprint* hep-ph/0703291)
[18] Mattingly D 2007 *From Quantum to Emergent Gravity: Theory and Phenomenology* (SISSA) (*Preprint* 0802.1561)
[19] Kostelecky V A and Mewes M 2009 *Phys. Rev. D* **80** 015020 (*Preprint* 0905.0031)
[20] Amelino-Camelia G 2008 (*Preprint* 0806.0339)
[21] Amelino-Camilia G 2002 *Int. J. Mod. Phys. D* **11** 35–60 (*Preprint* gr-qc/0012051)
[22] Mattingly D 2005 *Living. Rev. Rel.* **8** (*Preprint* gr-qc/0502097)
[23] Jacobson T, Liberati S and Mattingly D 2006 *Annals Phys.* **321** 150–196 (*Preprint* astro-ph/0505267)
[24] Liberati S and Maccione L 2009 *Ann. Rev. Nucl. Part. Sci.* **59** 245–267 * Brief entry * (*Preprint* 0906.0681)
[25] Collins J, Perez A, Sudarsky D, Urrutia L and Vucetich H 2004 *Phys. Rev. Lett.* **93** 191301 (*Preprint* gr-qc/0403053)
[26] Iengo R, Russo J G and Serone M 2009 *JHEP* **0911** 020 (*Preprint* 0906.3477)
[27] Maccione L, Taylor A M, Mattingly D M and Liberati S 2009 *J. Cosmol. Astropart. Phys.* **2009** 022 (*Preprint* 0902.1756)
[28] Galaverni M and Sigl G 2008 *Phys. Rev. Lett.* **100** 021102 (*Preprint* 0708.1737)
[29] Maccione L and Liberati S 2008 *J. Cosmol. Astropart. Phys.* **0808** 027 * Brief entry * (*Preprint* 0805.2548)
[30] Galaverni M and Sigl G 2008 *Phys. Rev. D* **78** 063003 (*Preprint* 0807.1210)
[31] Kifune T 1999 *Astrophys. J.* **518** L21–L24 (*Preprint* astro-ph/9904164)
[32] Aloisio R, Blasi P, Ghia P L and Grillo A F 2000 *Phys. Rev. D* **62** 053010 (*Preprint* astro-ph/0001258)
[33] Amelino-Camelia G and Piran T 2001 *Phys. Rev. D* **64** 036005 (*Preprint* astro-ph/0008107)
[34] Stecker F W and Scully S T 2005 *Astropart. Phys.* **23** 203–209 (*Preprint* astro-ph/0412495)
[35] Gonzalez-Mestres L 2009 *Nucl. Phys. Proc. Suppl.* **190** 191–197 * Brief entry * (*Preprint* 0902.0994)
[36] Scully S T and Stecker F W 2009 *Astropart. Phys.* **31** 220–225 (*Preprint* 0811.2230)
[37] Stecker F W and Scully S T 2009 *New J. Phys.* **11** 085003 * Brief entry * (*Preprint* 0906.1735)
[38] Jacobson T, Liberati S and Mattingly D 2003 *Phys. Rev. D* **67** 124011 (*Preprint* hep-ph/0209264)
[39] Gaisser T K and Stanev T 2005 *Nucl. Phys. A* **777** 98–110 (*Preprint* astro-ph/0510321)
[40] Bergman D R and Belz J W 2007 *J. Phys.* **G34** R359 (*Preprint* 0704.3721)
[41] Zatsepin G T and Kuz'min V A 1966 *JETP Letters* **4** 78
[42] Greisen K 1966 *Phys. Rev. Lett.* **16** 748–750
[43] Abbasi R *et al.* (HiRes) 2007 (*Preprint* astro-ph/0703099)
[44] Roth M (Pierre Auger) 2007 (*Preprint* arXiv:0706.2096[astro-ph])
[45] Abraham J *et al.* (Pierre Auger) 2007 *Science* **318** 938–943 (*Preprint* 0711.2256)
[46] Abraham J *et al.* (Pierre Auger Observatory Collaboration) 2010 *Phys.Rev.Lett.* **104** 091101 (*Preprint* 1002.0699)
[47] Glushkov A V, Makarov I T, Pravdin M I, Sleptsov I E, Gorbunov D S, Rubtsov G I and Troitsky S V 2008 *JETP Lett.* **87** 220–224 the Yakutsk EAS Array (*Preprint* astro-ph/0710.5508)
[48] Saveliev A, Maccione L and Sigl G 2011 * Temporary entry * (*Preprint* 1101.2903)
[49] Abraham J *et al.* (The Pierre Auger) 2010 *Phys. Lett.* **B685** 239–246 (*Preprint* 1002.1975)
[50] Abraham J *et al.* (Pierre Auger) 2008 *Astropart. Phys.* **29** 243–256 (*Preprint* 0712.1147)
[51] Rubtsov G I *et al.* 2006 *Phys. Rev.* **D73** 063009 (*Preprint* astro-ph/0601449)
[52] Sigl G 2007 *Phys. Rev.* **D75** 103001 (*Preprint* astro-ph/0703403)
[53] Gelmini G B, Kalashev O E and Semikoz D V 2007 *JCAP* **0711** 002 (*Preprint* 0706.2181)
[54] Galaverni M and Sigl G 2008 *Phys. Rev. Lett.* **100** 021102 (*Preprint* 0708.1737)
[55] Mattingly D, Jacobson T and Liberati S 2003 *Phys. Rev. D* **67** 124012 (*Preprint* hep-ph/0211466)

Journal of Physics: Conference Series **314** (2011) 012007 doi:10.1088/1742-6596/314/1/012007

[56] Abdo A, Ackermann M, Ajello M, Baldini L, Ballet J *et al.* 2009 *Science* **327** 1103–1106
[57] Ackermann M *et al.* (Fermi GBM/LAT Collaborations) 2009 *Nature* **462** 331–334 (*Preprint* 0908.1832)
[58] Ellis J R, Mavromatos N E and Nanopoulos D V 1992 *Phys. Lett.* **B293** 37–48 (*Preprint* hep-th/9207103)
[59] Ellis J R, Mavromatos N E and Nanopoulos D V 2001 *Phys. Rev.* **D63** 124025 (*Preprint* hep-th/0012216)
[60] Ellis J R, Mavromatos N E, Nanopoulos D V and Sakharov A S 2004 *Int. J. Mod. Phys.* **A19** 4413–4430 (*Preprint* gr-qc/0312044)
[61] Ellis J R, Mavromatos N E and Sakharov A S 2004 *Astropart. Phys.* **20** 669–682 (*Preprint* astro-ph/0308403)
[62] Ellis J R, Mavromatos N E and Nanopoulos D V 2008 *Phys. Lett.* **B665** 412–417 (*Preprint* 0804.3566)
[63] Ellis J, Mavromatos N E and Nanopoulos D V 2009 *Phys. Lett.* **B674** 83–86 (*Preprint* 0901.4052)
[64] Li T, Mavromatos N E, Nanopoulos D V and Xie D 2009 *Phys. Lett.* **B679** 407–413 (*Preprint* 0903.1303)
[65] Ellis J, Mavromatos N E and Nanopoulos D V 2009 (*Preprint* 0912.3428)
[66] Jacob U and Piran T 2008 *JCAP* **0801** 031 (*Preprint* 0712.2170)
[67] Amelino-Camelia G and Smolin L 2009 *Phys. Rev.* **D80** 084017 (*Preprint* 0906.3731)
[68] Ellis J R, Farakos K, Mavromatos N E, Mitsou V A and Nanopoulos D V 2000 *Astrophys. J.* **535** 139–151 (*Preprint* astro-ph/9907340)
[69] Maccione L, Liberati S and Sigl G 2010 *Phys.Rev.Lett.* **105** 021101 (*Preprint* 1003.5468)
[70] Ellis J, Mavromatos N E and Nanopoulos D V 2010 *Phys. Lett.* **B694** 61–64 (*Preprint* 1004.4167)
[71] Yao W M *et al.* (Particle Data Group) 2006 *J. Phys.* **G33** 1–1232
[72] Beresinsky V and Zatsepin G 1969 *Phys.Lett.* **B28** 423–424
[73] Stecker F W 1973 *Astrophys. Space Sci.* **20** 47–57
[74] Engel R, Seckel D and Stanev T 2001 *Phys. Rev.* **D64** 093010 (*Preprint* astro-ph/0101216)
[75] Semikoz D V and Sigl G 2004 *JCAP* **0404** 003 (*Preprint* hep-ph/0309328)
[76] Mattingly D M, Maccione L, Galaverni M, Liberati S and Sigl G 2010 *JCAP* **1002** 007 (*Preprint* 0911.0521)
[77] Gorham P W *et al.* (ANITA) 2009 *Phys. Rev. Lett.* **103** 051103 (*Preprint* 0812.2715)
[78] Petrolini A 2011 *Nucl. Instrum. Meth.* **A630** 131–135 (*Preprint* 0909.5220)
[79] Santangelo A and Petrolini A 2009 (*Preprint* 0909.5370)
[80] Barwick S W 2007 *J. Phys. Conf. Ser.* **60** 276–283 (*Preprint* astro-ph/0610631)
[81] Barwick S W 2009 *Nucl. Instrum. Meth.* **A602** 279–284
[82] Allison P *et al.* 2009 *Nucl. Instrum. Meth.* **A604** S64–S69 (*Preprint* 0904.1309)
[83] Horava P 2009 *Phys.Rev.* **D79** 084008 (*Preprint* 0901.3775)
[84] Pospelov M and Shang Y 2010 (*Preprint* 1010.5249)
[85] Brown H 2005 *Physical relativity: space-time structure from a dynamical perspective* Oxford Scholarship Online. Philosophy module (Clarendon Press) ISBN 9780199275830 URL http://books.google.it/books?id=T6IVyWiPQksC
[86] Liberati S, Sonego S and Visser M 2002 *Annals Phys.* **298** 167–185 (*Preprint* gr-qc/0107091)
[87] Sonego S and Pin M 2009 *J.Math.Phys.* **50** 042902 (*Preprint* 0812.1294)
[88] Cohen A G and Glashow S L 2006 *Phys.Rev.Lett.* **97** 021601 (*Preprint* hep-ph/0601236)
[89] Bogoslovsky G Y 2005 (*Preprint* math-ph/0511077)
[90] Bogoslovsky G 2006 *Phys.Lett.* **A350** 5–10 (*Preprint* hep-th/0511151)
[91] Gibbons G, Gomis J and Pope C 2007 *Phys.Rev.* **D76** 081701 (*Preprint* 0707.2174)
[92] Girelli F, Liberati S and Sindoni L 2007 *Phys.Rev.* **D75** 064015 (*Preprint* gr-qc/0611024)
[93] Skakala J and Visser M 2010 (*Preprint* 1008.0689)
[94] Amelino-Camelia G, Freidel L, Kowalski-Glikman J and Smolin L 2011 * Temporary entry * (*Preprint* 1101.0931)
[95] Fagnocchi S, Finazzi S, Liberati S, Kormos M and Trombettoni A 2010 *New J.Phys.* **12** 095012 (*Preprint* 1001.1044)
[96] Girelli F, Liberati S and Sindoni L 2009 *Phys.Rev.* **D79** 044019 (*Preprint* 0806.4239)
[97] Carlip S 2009 (*Preprint* 0909.3329)

Journal of Physics: Conference Series **314** (2011) 012008 doi:10.1088/1742-6596/314/1/012008

Gravitational fluctuations in de Sitter cosmology

Enric Verdaguer

Departament de Física Fonamental and Institut de Ciencies del Cosmos, Universitat de Barcelona, Av. Diagonal 647, 08028 Barcelona, Spain

E-mail: `enric.verdaguer@ub.edu`

Abstract. De Sitter spacetime plays an important role in cosmology: the geometry of most inflationary models is close to de Sitter spacetime and so will be the late-time behavior of the present universe with accelerated expansion. In linearized perturbation theory the metric fluctuations in a de Sitter cosmology describe very well the anisotropies of the microwave background and the observed large scale structure. Recently there has been some interest in the need to go beyond the linear approximation to include the effect of matter loops. This will allow testing perturbation theory in a de Sitter background, checking possible large back-reaction effects on the de Sitter geometry, and also eventually to discriminate between inflationary models that lead to similar results at tree level. Working in the framework of stochastic gravity, or equivalently in the large N expansion, one may derive the two-point correlations for the gravitational fluctuations incorporating the effects of matter loops. One may characterize the quantum gravitational fluctuations in a gauge invariant way by the two-point functions of the linearized Riemann tensor. This can be given in terms of the two-point linearized Einstein and Weyl tensors. Assuming minimally coupled scalar fields in a de Sitter invariant state, the two-point functions of the linearized Einstein tensor over the de Sitter background have been computed in terms of de Sitter invariant bi-tensors.

1. Introduction

De Sitter spacetime plays a key role in our understanding of the universe. On the one hand, the anisotropies of the cosmic microwave background and the observed large structure of the universe are successfully explained by a short early inflationary phase [1, 2]. On the other hand, the present universe is undergoing also a phase of accelerated expansion [3] which might be driven by a small non-vanishing cosmological constant. The geometry of most inflationary models as well as the late time behavior of our expanding universe are well approximated by the de Sitter geometry.

For this reason quantum gravitational perturbations around de Sitter and near-de Sitter spacetimes have been extensively studied in the context of inflationary models. The standard analysis is based on the linearized computation of the two-point function of the metric perturbations [4, 5]. But more recently the need for matter loop corrections has been emphasized by Weinberg [6], see also [7], who pointed out the importance of testing perturbation theory in cosmology given the great success of the tree level approximation. Another reason to include the effect of matter loops is to discriminate between different inflationary models that give very similar results at the tree level [8, 9]. A series of works including matter loop corrections in different models have been performed using different approaches and models [10, 11, 12] including stochastic gravity [13]. However, certain aspects such as the significance of possible large infrared effects, have not been completely settled yet. The standard theoretical framework assumes that the background dynamics of de Sitter and near-de Sitter spacetimes is not affected by large quantum corrections. However, there is a number of studies [14, 15, 16, 17, 18] that claim that one-loop corrections could give a significant back-reaction in inflationary models. It has also been argued that higher-order radiative corrections from graviton two-loops could give rise to a secular screening of the cosmological constant for pure gravity in de Sitter spacetime [19, 20]. Such nonlinear effects are dominated by infrared modes which are copiously generated by the exponential expansion and should be fairly insensitive to the short-distance behavior of quantum gravity. However, whether these large effects really take place is still under debate, mainly due to the gauge dependence of the results [21, 22, 23, 24, 25, 26, 27].

It is also worth noting that interacting field theories in de Sitter spacetime are usually plagued with infrared divergencies from loop diagrams in a perturbative expansion in the interaction parameter that leads to the breakdown of perturbation theory. An approach due to Starobinsky [28, 29] has been able to go beyond by non-perturbatively accounting for the infrared effects. It is based on the observation that the infrared part of the scalar field may be considered as a classical stochastic field with a probability distribution described by a Fokker-Planck equation, this is because the infrared modes outside the horizon undergo fast decoherence. In this way Starobinski and Yokoyama [30] were able to show that there is a de Sitter invariant state for a low-mass scalar field (that is, a mass smaller than the Hubble scale) with a quartic self-interaction. Also Riotto and Sloth [31] where able to evaluate the infrared finite correlation functions by resumming higher loop-diagrams non-perturbatively for $O(N)$-invariant scalar theory with quartic self-interactions; they use the Fokker-Planck equation and also a two-particle-irreducible formulation. These results are in agreement with a recent result by Marolf and Morrison [32] who prove by analytical continuation of the interacting Euclidean vacuum that for cubic and quartic interactions there is an interacting de Sitter invariant vacuum which is well behaved in the infrared and stable. Although for low masses the different terms of the perturbative series do not decrease as already noted in Ref. [33], where the similarities of the large fluctuations on superhorizon scales in de Sitter with some examples in condensed matter physics, near the vicinity of a critical point, are pointed out. Other authors, however, like Polyakov [34, 35] argue that interacting scalar fields in de Sitter are unstable and that this can lead to the decay, or screening, of the cosmological constant. This result is based on Feynman diagrams which are infrared divergent in the "in-out" formalism due to the exponential growth of the de Sitter volume element. The use of the "in-out" perturbation theory, however, has been known to lead to unstabilities which are not present in the "in-in" formalism [36, 37]; see also [38, 39] in relation to Polyakov's proposal.

Here, instead of a self-interacting scalar field we consider a free massive scalar field interacting with metric perturbations on a de Sitter background. To gain insight into the possibility of important quantum effects in de Sitter, it seems convenient to study quantum metric perturbations around de Sitter spacetime keeping de Sitter invariance always manifest. This may be achieved with the formalism of de Sitter-invariant bi-tensors developed by Allen and Jacobson [40]. The tree-level graviton two-point function has been computed using this formalism by several authors [41, 42, 43]. A conclusion that can be drawn from their results is that the two-point function of the linearized Riemann tensor ${R_{ab}}^{cd}$, which is gauge-invariant around de Sitter spacetime, decays like the fourth power of the physical distance between the two points [44]. Here, we summarize the main results obtained in Ref. [45] on the effects of matter loops on quantum gravitational fluctuations. A key ingredient for this calculation is the connected symmetrized two-point function of the stress tensor operator of the matter fields. Obtaining this object in de Sitter spacetime, in a de Sitter-invariant vacuum, and with the formalism of de Sitter-invariant bitensors, is one of the main results reported here.

We work in the framework of stochastic gravity. Stochastic gravity, which is based in

Journal of Physics: Conference Series **314** (2011) 012008 doi:10.1088/1742-6596/314/1/012008

the Einstein-Langevin equations, is a perturbative extension of semiclassical gravity which incorporates the stress tensor fluctuations [46, 47, 48, 49, 50], as a gravitational source. Moreover, in analogy with the results in some simple open quantum systems [51] and the results of stochastic gravity in Minkowski spacetime [52], it has been shown that the correlation functions of the metric fluctuations obtained in stochastic gravity reproduce the two-point metric fluctuations in the quantum theory of gravity interacting with N quantum fields to leading order in an $1/N$ expansion [53, 54]. Thus, stochastic gravity is a useful framework to study quantum metric fluctuations, and can be applied to the problem of cosmological perturbations to some advantage. It exactly reproduces the standard tree level results for cosmological perturbations when both the gravitational and matter field perturbations are linear and quantized [56, 57]. But since it incorporates the full effect of the matter fields it goes beyond the tree level by incorporating matter loop corrections [13]. Here we are concerned with the gravitational fluctuations in de Sitter spacetime including these corrections. More specifically, we consider fluctuations which are directly induced by the two-point functions of the stress tensor such as the linearized Einstein tensor fluctuations.

2. Stochastic gravity

General relativity is known to be a non-renormalizable quantum field theory, however, it makes perfect sense when it is regarded as an effective field theory [58, 59] which has a range of validity below a certain energy scale, often taken as the Planck mass. This low energy effective theory results from the separation of the low energy gravitational quantum effects from the high energy contributions. The action of the gravitational effective field theory is organized as a sum of all possible interactions which are consistent with general covariance and local Lorentz invariance in a series which has higher and higher powers of derivatives of the metric, as an energy expansion, with coefficients that must be determined by experiments. The lowest order terms are the cosmological constant and the Einstein-Hilbert terms.

Regarding quantum gravity as an effective field theory results in a perfectly well behaved quantum field theory. The semiclassical limit can be formally derived as the leading-order $1/N$ approximation of quantum gravity [60], interacting with N free quantum fields when Newton's gravitational constant G is rescaled to $\bar{G} = GN$. In this limit, in which $\bar{G}$ is kept fixed, one arrives after path integration at a theory where the gravitational field can be treated as a c-number and the quantum fields are fully quantized. If we call g_{ab} the metric tensor and $\hat{\phi}$ a scalar field (for notational simplicity we write just one scalar field) one arrives at the semiclassical Einstein equation for the metric g_{ab}:

$$G_{ab}[g] = \bar{\kappa}\langle T_{ab}[g]\rangle_{ren}, \tag{1}$$

where $\bar{\kappa} = \kappa N$ with $\kappa = 8\pi G = 8\pi/m_P^2 = 8\pi l_P^2$, where m_P and l_P are the Planck mass and Planck length, respectively, and $T_{ab} = T_{ab}[\hat{\phi}^2]$ is the stress tensor operator which is quadratic in the field operator $\hat{\phi}$. This operator, being the product of distribution valued operators, is ill defined and needs to be regularized and renormalized; the subscript ren means that the operator has been renormalized. The angle brackets on the right hand side mean that the expectation value of the stress tensor operator is computed in some quantum state, say $|\psi\rangle$, compatible with the geometry described by the metric g_{ab}. On the left hand side of equation (1), $G_{ab}[g]$ stands for the Einstein tensor for the metric g_{ab}. We also implicitly assume that the local geometric terms that are needed to renormalize the stress tensor operator are included on the right hand side; these are basically a term with a cosmological constant and two higher derivative terms quadratic in the curvature that can be derived from an action with the square of the Ricci scalar and another term with the square of the Weyl tensor, these terms come with two arbitrary renormalization parameters. These higher derivative terms are already present in

Journal of Physics: Conference Series **314** (2011) 012008 doi:10.1088/1742-6596/314/1/012008

the gravitational effective field theory. Thus, the effect of the high energy modes of the matter fields is to produce shifts in the coefficients of the effective field action up to terms quadratic in the curvature, which involve up to four derivatives of the metric tensor.

The purpose of stochastic gravity is to extend perturbatively the semiclassical theory in order to include the quantum fluctuations of the stress tensor. The physical observable that measures these fluctuations is the *noise kernel*: a four index bi-tensor defined as the expectation value of the anticommutator of the stress tensor operator T_{ab},

$$N_{abc'd'}(x,x') = \frac{1}{2}\langle\{T_{ab}(x), T_{c'd'}(x')\}\rangle - \langle T_{ab}(x)\rangle\langle T_{c'd'}(x')\rangle. \tag{2}$$

This expectation value is taken in the background metric g_{ab} which we assume to be a solution of the semiclassical equation (1). An important property of the symmetric bi-tensor, $N_{abc'd'}(x,x') = N_{c'd'ab}(x',x)$, is that it is finite because the ultraviolet divergences of T_{ab} are cancelled by the substraction of $\langle T_{ab}\rangle$. Since the operator T_{ab} is selfadjoint $N_{abc'd'}(x,x')$, which is the expectation value of an anticommutator, is real and positive semi-definite. This last property allows for the introduction of an auxiliary classical Gaussian stochastic tensor ξ_{ab} defined by

$$\langle\xi_{ab}(x)\rangle_s = 0, \quad \langle\xi_{ab}(x)\xi_{c'd'}(x')\rangle_s = N_{abc'd'}(x,x'). \tag{3}$$

This stochastic tensor is symmetric $\xi_{ab} = \xi_{ba}$ and divergenceless, $\nabla^a\xi_{ab} = 0$, as a consequence of the fact that the stress tensor operator is divergenceless. The subscript s means that the expectation value is just a classical stochastic average.

The idea now is to modify the semiclassical Einstein equation (1) by introducing a linear correction to the metric tensor g_{ab}, such as $g_{ab} + h_{ab}$, which accounts consistently for the fluctuations of the stress tensor. The simplest equation is obtained by adding to the right hand side of equation (1), but evaluated for the perturbed metric, the stochastic tensor just introduced. Substraction of the semiclassical equation (1) leads to the following equation,

$$G^{(1)}_{ab}[g+h] = \bar{\kappa}\langle T^{(1)}_{ab}[g+h]\rangle_{ren} + \bar{\kappa}\xi_{ab}[g]), \tag{4}$$

where we recall that the background metric g_{ab} is assumed to be a solution of equation (1). As indicated by the superscript (1) this stochastic equation must be thought of as a linear equation for the metric perturbation h_{ab} which will behave consequently as a stochastic field tensor. This equation is gauge invariant with respect to diffeomorphisms defined by any field on the background spacetime [47].

The stochastic equation (4) is known as the Einstein-Langevin equation, and predicts that the gravitational field has stochastic fluctuations over the background g_{ab}. It is linear in h_{ab}, thus its solutions can be written in terms of a solution of the homogeneous equation containing information on the initial conditions, $h^0_{ab}(x)$, and the auxiliary field $\xi^{cd}(x')$ with the retarded propagator of equation (4) with vanishing initial conditions, $G^{ret}_{abcd}(x,x')$. Form this and equations (3), the two-point correlations for the metric perturbations are:

$$\langle h_{ab}(x)h_{cd}(y)\rangle_s = \langle h^0_{ab}(x)h^0_{cd}(y)\rangle_s + \bar{\kappa}^2\int d^4x'd^4y'\sqrt{g(x')g(y')}G^{ret}_{abef}(x,x')N^{efgh}(x',y')G^{ret}_{cdgh}(y,y').$$

These two-point stochastic correlations, which are independent of the auxiliary stochastic tensor introduced above, are the most relevant physical observables.

It turns out that these stochastic correlations are connected to the correlations of the quantized metric fluctuations in the large N expansion. In fact, it can be shown that they correspond exactly to the symmetrized two-point quantum metric correlations obtained in the large N expansion: $\langle\{\hat{h}_{ab}(x), \hat{h}_{cd}(y)\}\rangle = 2\langle h_{ab}(x)h_{cd}(y)\rangle_s$, where $\hat{h}_{ab}(x)$ mean the quantum operator corresponding to the metric perturbations. This result was explicitly obtained in the Minkowski background in Ref. [52] where the two-point correlation in the stochastic context exactly agrees with the symmetrized part of the graviton propagator computed by Tomboulis [53] in the context of gravity interacting with N matter Fermion fields, where the graviton propagator is of order $1/N$. This result can be extended to an arbitrary background in the context of the large N expansion, a sketch of the proof with explicit details in the Minkowski background can be found in Ref. [54]. The connection between stochastic correlations and quantum correlations was studied in detail in the context of simpler open quantum systems [51].

3. Stress tensor correlation in de Sitter spacetime

3.1. De Sitter geometry and de Sitter invariant bi-tensors

The D-dimensional de Sitter spacetime, dS_D, may be defined as a submanifold embedded in $(D+1)$-dimensional Minkowski spacetime, whose points satisfy the equation

$$\eta_{AB} X^A(x) X^B(x) = H^{-2}, \tag{5}$$

where H is a parameter with dimensions of mass, η_{AB} the Minkowski metric, and $X^A(x)$ is the position vector in Minkowski spacetime of the point x. The hyperboloid is invariant under Lorentz transformations $X^A = \Lambda^A_B X^B$ and these define the isometries σ of dS_D: $X^A(\sigma(x)) = \Lambda^A_B X^B(x)$. Thus, the isometries of de Sitter form the Lorentz group $O(D,1)$. Another relevant isometry is the antipodal isometry $\alpha(x)$ defined by $X^A(\alpha(x)) = -X^A(x)$.

One may write the de Sitter metric in coordinates $(\tilde{\eta}, \chi, \theta, \varphi)$ which span a finite range $-\pi < \tilde{\eta} < 0$, $0 \le \chi < \pi$, with (θ, φ) the standard spherical coordinates. It is given by

$$ds^2 = \frac{1}{H^2 \sin^2 \tilde{\eta}} \left(-d\tilde{\eta}^2 + d\chi^2 + \sin^2 \chi \left(d\theta^2 + \sin^2 \theta d\varphi^2 \right) \right), \tag{6}$$

where the sections $\tilde{\eta} = const$ are three spheres. These coordinates cover the entire de Sitter spacetime. They represent a closed universe which contracts for $-\pi < \tilde{\eta} < -\pi/2$ and expand for $-\pi/2 < \tilde{\eta} < 0$. The de Sitter conformal diagram, for the conformal metric $(H^2 \sin^2 \tilde{\eta})\, ds^2$, may be represented in terms of the coordinates $(\chi, \tilde{\eta})$, each point of the conformal diagram represents a S^2 sphere of radius $\sin \chi$.

To introduce the formalism of the de Sitter invariant bi-tensors we first define the basic bi-scalar

$$Z(x, x') = \eta_{AB} X^A(x) X^B(x'), \tag{7}$$

where $x, x' \in dS_D$, and where we assume $H = 1$ for simplicity. The value of H will be restored in the final expressions. Note that Z is invariant under the de Sitter group, if σ is an isometry, then $Z(\sigma(x), \sigma(x')) = Z(x, x')$ and under the antipodal transformation $Z(\alpha(x), x') = -Z(x, x')$. This bi-scalar is related to the Minkowskian distance, defined by $d^2(x, x') = \eta_{AB}[X^A(x) - X^A(x')][X^B(x) - X^B(x')]$, which is the length of the straight line in the embedding $(D+1)$-Minkowski spacetime that joins x and x'. It follows that Z and d are related through

$$d^2 = 2(1 - Z). \tag{8}$$

Therefore, $Z = 1$ when x' is on the light cone of x, $Z > 1$ when the points are timelike separated, and $Z < 1$ when the points are spacelike separated. It should also be noted that $Z \in (-\infty, \infty)$, since the Minkowskian distance between two points on the hyperboloid can take any value.

Without loss of generality we can take the point x in the conformal diagram $(\chi = 0, \tilde{\eta} = -\pi/2)$ and consider the spacelike straight curve with $\tilde{\eta} = const$ connecting x and its antipodal $\alpha(x)$ which in that diagram is given by $(\chi = \pi, \tilde{\eta} = -\pi/2)$. This curve is a geodesic, because it is a straight line on a cylinder tangent to the hyperboloid. The Minkowskian distance between x and any other point x' in this geodesic satisfies $0 \leq d(x,x') \leq 2$. This implies

$$-1 \leq Z(x,x') \leq 1. \tag{9}$$

Any other spacelike geodesic passing through x can be obtained from this one by an isometry that leaves x invariant (it also leaves $\alpha(x)$ invariant). This is analogous to the case of the sphere, where any geodesic (meridian) passing through the north pole is obtained from a given one by a rotation that leaves the north pole (and south pole) invariant. This means that any point y' connected to x by a spacelike geodesic satisfies $y' = \sigma(x')$ for some x' in the above geodesic, and where σ is an isometry such that $\sigma(x) = x$. De Sitter invariance of Z then implies $Z(x,y') = Z(\sigma(x),\sigma(x')) = Z(x,x')$. Consequently, equation (9) holds for any point x' connected with x by a spacelike geodesic. In other words, spacelike separated points x, x' such that $Z(x,x') < -1$ cannot be joined by a geodesic. For timelike geodesics, however, there are theorems that guarantee the existence of geodesics between any two timelike separated points when the spacetime is globally hyperbolic [55].

A bi-tensor is said to be maximally symmetric if it is invariant under all the isometries of the manifold. The following bi-tensors are maximally symmetric [40] in dS_D: the distance, $\mu(x,x')$, along the shortest geodesic joining x and x', also called the geodesic distance; the unit vectors, $n_a(x,x')$ and $n_{a'}(x,x')$, tangent to the geodesic at the points x and x' respectively, pointing outward from it; the parallel propagator, $g_{ab'}(x,x')$, which parallel-transports a vector from x' to x along the geodesic; and the metric tensors, $g_{ab}(x,x')$ and $g_{a'b'}(x,x')$, at the points x and x' respectively. A theorem by Allen and Jacobson [40] states that any maximally symmetric bi-tensor is a linear combination of products of $n_a, n_{a'}, g_{ab}, g_{a'b'}$ and $g_{ab'}$, with coefficients that depend only on μ.

If $x(\lambda)$ is the shortest geodesic joining x and x', and its tangent vector is $v^a(\lambda)$, and we take $x(0) = x$ and $x(1) = x'$, the geodesic distance between x and x' is

$$\mu(x,x') = \int_0^1 d\lambda\, [g_{ab}(x(\lambda))v^a(\lambda)v^b(\lambda)]^{1/2}. \tag{10}$$

Note that $\mu(x,x')$ is a real number for spacelike separated points, and an imaginary number for timelike separated points. The sign prescription for the square root is chosen in such a way that $\mu(x,x') > 0$ for spacelike separated points, and $\mu(x,x') = i\xi(x,x')$, with $\xi(x,x') > 0$, for timelike separated points. The bi-tensors n_a and $n_{a'}$ are defined by the equations $n_a = \nabla_a \mu$, $n_{a'} = \nabla_{a'}\mu$. Allen and Jacobson's theorem can be applied to the computation of the covariant derivatives of $n_a, n_{a'}$ and $g_{ab'}$, which are also maximally symmetric bi-tensors. The corresponding μ-dependent coefficients were computed in Ref. [40].

For spacelike geodesics, $-1 \leq Z < 1$, the geodesic distance μ of dS_D is given by the equation

$$\mu = \cos^{-1} Z, \tag{11}$$

where $\cos^{-1}$ is the principal value of the inverse cosine, which is single-valued. For example, $\cos^{-1} 1 = 0$, and $\cos^{-1}(-1) = \pi$. This is easily seen for the geodesic with $\tilde{\eta} = const$ joining $(\chi = 0, \tilde{\eta} = -\pi/2)$ and $(\chi = \pi, \tilde{\eta} = -\pi/2)$. Other spacelike geodesics are obtained from this one by an isometry, and therefore, by de Sitter invariance of μ and Z, Eq. (11) is true for all spacelike geodesics. For spacelike separated points not joined by a geodesic one may still define μ by analytical continuation [40, 45]. The definition of n_a, $n_{a'}$ and of the parallel propagator can then be extended. A geometric interpretation of this analytic extension for spacelike separated points not joined by geodesics is given in Ref. [45]

Journal of Physics: Conference Series **314** (2011) 012008 doi:10.1088/1742-6596/314/1/012008

3.2. Noise kernel in de Sitter spacetime

Consider a quantum field in dS_D, whose stress tensor operator is T_{ab}, in a vacuum state $|0\rangle$ that we assume to be de Sitter invariant. We want to compute the noise kernel $N_{abc'd'}$ defined by Eq. (2). Since $|0\rangle$ is a de Sitter-invariant state, the expectation value of any observable must be de Sitter-invariant. Therefore, $N_{abc'd'}$ is a de Sitter-invariant bi-tensor and Allen and Jacobson's theorem may be used. On the other hand, $N_{abc'd'}(x,x') = N_{c'd'ab}(x',x)$, and, since the stress tensor is symmetric, we also have that $N_{abc'd'} = N_{bac'd'} = N_{abd'c'}$. The most general bi-tensor that is a linear combination of products of $n_a, n_{a'}, g_{ab}, g_{a'b'}$ and $g_{ab'}$, with coefficients that depend only on μ, and that satisfies the above constraints is

$$\begin{aligned} N_{abc'd'} &= P(\mu)n_a n_b n_{c'} n_{d'} + Q(\mu)(n_a n_b g_{c'd'} + n_{c'} n_{d'} g_{ab}) \\ &\quad + R(\mu)(n_a n_{c'} g_{bd'} + n_b n_{d'} g_{ac'} + n_a n_{d'} g_{bc'} + n_b n_{c'} g_{ad'}) \\ &\quad + S(\mu)(g_{ac'} g_{bd'} + g_{bc'} g_{ad'}) + T(\mu) g_{ab} g_{c'd'}. \end{aligned} \tag{12}$$

Moreover, the stress tensor conservation law implies $\nabla^a N_{abc'd'} = 0$ which imposes some differential equations for the coefficients P, Q, R, S, T. Also there are further restrictions on these coefficients imposed by the condition that the noise kernel should be unambiguous when x and x' are antipodal, see Ref. [45] for the details.

Now let us assume a free massive minimally coupled scalar field, $\xi = 0$, whose stress tensor is given by the coincidence limit, $T_{ab}(x) = \lim_{x''\to x} \mathcal{D}_{ab}(x,x'')\phi(x)\phi(x'')$, where $\mathcal{D}_{ab}$ is a second order differential operator [45]. The stress tensor two-point function of the free scalar field is obtained by substituting the previous expression in (2). Since the field ϕ is free, and $|0\rangle$ is a vacuum state, the four-point function appearing in this equation can be written in terms of the Wightman function, $G^+(x,x') = \langle 0|\phi(x)\phi(x')|0\rangle$,

$$\langle 0|\phi(x)\phi(x'')\phi(x')\phi(x''')|0\rangle = G^+(x,x'')G^+(x',x''') + G^+(x,x')G^+(x'',x''') + G^+(x,x''')G^+(x'',x').$$

This equation is obtained by expanding the field ϕ in terms of creation and annihilation operators. Note that, in the particular case where the product of fields on the left-hand side is time-ordered, this equation also follows from Wick's theorem. Substituting into (2) using the previous expression for the stress tensor, and taking the limit $x'' \to x$ and $x''' \to x'$, one finds $N_{abc'd'}$ in terms of G^+ [45].

We now use that the vacuum $|0\rangle$ is de Sitter-invariant, and impose that the points x and x' are spacelike separated. For spacelike separated points, the Wightman function is the expectation value of an observable, so that it must be a de Sitter-invariant biscalar. Thus it must have the form $G^+ = G(\mu)$. On the other hand, the Wightman function satisfies the Klein-Gordon equation. Imposing as boundary conditions that $G(\mu)$ diverges only at the coincidence limit, $\mu \to 0$, and that it does so in the same way as the standard Wightman function in Minkowski spacetime, the only solution [45] is $G(\mu) = c_{m,D} F\left(h_+, h_-; \frac{D}{2}; \frac{1+Z}{2}\right)$, where F denotes the hypergeometric function, $h_\pm$ and $c_{m,D}$ are constants, $2h_\pm = (D-1) \pm \sqrt{(D-1)^2 - 4m^2}$ and $c_{m,D}(4\pi)^{D/2}\Gamma(\frac{D}{2}) = \Gamma(h_+)\Gamma(h_-)$, and we are using that $Z = \cos\mu$, as follows from (11). One can see that there is a vacuum whose Wightman function for spacelike separated points is just that expression, which is the Bunch-Davies vacuum. Substituting $G(\mu)$ into the noise kernel, one finds that the noise kernel in the Bunch-Davies vacuum has the form (12), and the functions P, Q, R, S, T are explicitly given in terms of $G(\mu)$; see [45] for the details.

4. Quantum gravitational fluctuations

To characterize the gravitational fluctuations we may compute the two-point function of the linearized Riemann tensor. The Riemann tensor of the background de Sitter spacetime with suitably raised indices is given by ${R_{ab}}^{cd} = (R/6)\,\delta_{[a}^{[c}\delta_{b]}^{d]}$. Its linear perturbation is gauge invariant because its Lie derivative with respect to an arbitrary vector field vanishes. The Riemann tensor is entirely determined by the Ricci tensor (or equivalently the Einstein tensor) and the Weyl tensor. Both the linearized Einstein and the Weyl tensors over a de Sitter background are separately gauge invariant, and the Weyl tensor also vanishes in the background.

The previous results for the noise kernel can be directly used to get information on the quantum gravitational fluctuations in de Sitter spacetime. It turns out that the lowest-order contribution to the two-point function for the linearized Einstein tensor is of order $\bar{l}_P^4$ and comes directly from the one-loop contributions of the matter fields. Both from the stochastic gravity formalism [48, 49, 54, 50] or from the Einstein equation to linear order in the metric perturbations (including the matter fields as well as the metric perturbations). Using order reduction [61] one can see that the two higher derivative quadratic curvature tensors, which are implicit in the Einstein-Langevin equation (4), and which appear as counterterms in the renormalization of the stress tensor, both vanish in this case because they become zero for a de Sitter background. Thus only the linearized Einstein tensor survives, and one obtains the following result for the (connected) symmetrized two-point function of the Einstein tensor at that order:

$$\langle G^{a\,(1)}_{\ b}(x)\, G^{c'\,(1)}_{\ d'}(x')\rangle_c = 64\pi^2 \bar{l}_P^4 N^{-1}\, N^{a\ c'}_{\ b\ d'}(x,x'), \tag{13}$$

where $G^{a(1)}_{\ b}$ is the linearized Einstein tensor, and $N^{a\ c'}_{\ b\ d'}$ is the noise kernel. Therefore, the noise kernel in de Sitter gives the leading order contribution to the symmetrized two-point function of the linearized Einstein tensor. From the results for the noise kernel [45], for massless fields, $m = 0$, and long distances d, we have

$$\langle G^{a\,(1)}_{\ b}(x)\, G^{c'\,(1)}_{\ d'}(x')\rangle_c \sim \frac{\bar{l}_P^4 H^4}{N\, d^4}\,, \tag{14}$$

and for low mass fields, $m \ll H$, we have

$$\langle G^{a\,(1)}_{\ b}(x)\, G^{c'\,(1)}_{\ d'}(x')\rangle_c \sim \bar{l}_P^4 H^8 N^{-1}\,(Hd)^{-\frac{4m^2}{3H^2}}\,, \tag{15}$$

which implies very long range correlations.

The two-point function for the linearized Weyl tensor which is also gauge invariant gets contributions of order $\bar{l}_P^4$ and higher from matter loops, but also a non-vanishing tree-level contribution of order $\bar{l}_P^2$ even in the absence of matter loops, which was calculated by Kouris [44]. For spacelike separated points and long distances one has

$$\langle W^{(1)}_{abcd}(x) W^{(1)}_{a'b'c'd'}(x')\rangle_c \sim \frac{\bar{l}_P^2 H^2}{N\, d^4} + \mathrm{O}\left(\bar{l}_P^4 H^4\right). \tag{16}$$

The matter loop contributions of the order $\bar{l}_P^4$ have not been computed yet. When combined with the tree-level result for the correlations of the Weyl tensor in Ref. [44], our results for the correlations of the Einstein tensor completely characterizes the quantum fluctuations of the full Riemann tensor. It should be pointed out that the two-point function corresponding to the cross-correlation of the Einstein and Weyl tensors is of order $\bar{l}_P^4$ (rather than $\bar{l}_P^3$). This can be interpreted as the absence of correlations between the tree-level fluctuations of the Weyl tensor, with an amplitude of order $\bar{l}_P$, and the fluctuations of the Ricci tensor, with an amplitude of order $\bar{l}_P^2$. The two-point function for the linearized Weyl tensor given in the previous expression may be compared to the two-point function computed over a Minkowski background, given by

$$\langle W^{(1)}_{abcd}(x) W^{(1)}_{a'b'c'd'}(x')\rangle_c \sim \frac{\bar{l}_P^2}{N\, d^6}\,. \tag{17}$$

This result agrees also with the the two-point function for the linearized Weyl tensor in a de Sitter background at short distances.

Note that in contrast to the existing calculations of loop corrections to the spectrum of cosmological perturbations in inflationary models, where the self-interaction of the matter fields often plays an important role [10, 11], the only interaction vertices associated with the free fields that we have considered here are those corresponding to their gravitational interaction. Moreover, we do not consider a non-vanishing homogeneous background configuration for the scalar fields (which would give also a tree-level contribution to the stress-tensor two-point function [56]) and the usual treatment for the correlations of the scalar-type metric perturbations [6, 11, 12] cannot be directly applied. We believe that considering the correlations of the Riemann tensor and expressing them in terms of maximally invariant tensors is very useful in order to analyze whether de Sitter invariance is broken by loop corrections.

4.1. Acknowledgments

I am very grateful to the organizers of the Spanish Relativity Meeting 2010 on *Gravity as a Crossroad in Physics* for their kind invitation and for providing a very stimulating environment. I am also grateful to Guillem Pérez-Nadal and Albert Roura for their essential contribution to the work described here, and to Antonio Campos, Bei-Lok Hu, and Rosario Martin for many discussions during the last years on stochastic gravity. This work has been partially supported by the Research Projects MCI FPA2007-66665-C02-02, FPA2010-20807-C02-02, CPAN CSD2007-00042, within the program Consolider-Ingenio 2010, and AGAUR 2009-SGR-00168.

References

[1] Mukhanov V F 2005 *Physical Foundations of cosmology* (Cambridge: Cambridge University Press)
[2] Weinberg S 2008 *Cosmology* (New York: Oxford University Press)
[3] Perlmutter S. et al. 1999 *Astrophys. J.* **517** 565
[4] Starobinsky A A 1979 *JETP Lett.* **30**, 682
[5] Mukhanov V F, Feldman H A and Brandenberger R H 1992 *Phys. Rep.* **215**, 203.
[6] Weinberg S 2005 *Phys. Rev.* D **72** 043514.
[7] Senatore L and M. Zaldarriaga 2009 *arXiv:0912.2734 [hep-th].*
[8] Urakawa Y and Tanaka T 2009 *Prog. Theor. Phys.* **122** 779.
[9] Urakawa Y. and Tanaka T 2009 *Prog. Theor. Phys.* **122** 1207
[10] Sloth M S 2006 *Nucl. Phys.* B **748** 149
[11] Seery S 2007 *JCAP* **0711** 025
[12] Adshead P, Easther R and Lim E A 2009 *Phys. Rev.* D **80** 083521
[13] Urakawa Y and Maeda K 2008 *Phys. Rev* D **78** 064004
[14] Mukhanov V F, Abramo L R W and R. H. Brandenberger R H 1997 *Phys. Rev. Lett.* **78** 1624
[15] Abramo L R W, Brandenberger R H and Mukhanov V F 1997 *Phys. Rev.* D **56** 3248
[16] Abramo L R W and Woodard R P 1999 *Phys. Rev.* D **60** 044010
[17] Losic B and W. G. Unruh W G 2005 *Phys. Rev.* D **72** 123510
[18] Losic B and Unruh W G 2008 *Phys. Rev. Lett.* **101** 111101
[19] Tsamis N C and Woodard R P 1996 *Nucl. Phys.* B **474** 235
[20] Tsamis N C and Woodard R P 1997 *Ann. Phys. (NY)* **253** 1
[21] Abramo L R W and Woodard R P 2002 *Phys. Rev.* D **65** 043507
[22] Geshnizjani G and Brandenberger R H 2002 *Phys. Rev.* D **66** 123507
[23] Geshnizjani G and Brandenberger R H 2005 *JCAP* **0504**, 006
[24] Tsamis N C and Woodard R P 2005 *Class. Quantum Grav.* **22**, 4171
[25] Garriga J and Tanaka T 2008 *Phys. Rev.* D **77**, 024021
[26] Pérez-Nadal G, Roura A and Verdaguer E 2008 *Phys. Rev.* D **77** 124033
[27] Pérez-Nadal G, Roura A and Verdaguer E 2008 *Class. Quant. Grav.* **25** 154013
[28] Starobinsky A A 1986 *Field Theory, Quantum Gravity and Strings*, ed H J de Vega and N Sanchez, Lecture Notes in Physics, Vol. 246 (Springer-Verlag, Berlin) p 107
[29] Vilenkin A 1983 *Phys. Rev.* D **27** 2848
[30] Starobinsky A A and Yokoyama J 1994 *Phys. Rev.* D **50**, 6357
[31] Riotto A and Sloth M S 2008 *JCAP* **0804**, 030

[32] Marolf D and Morrison I 2010 *Preprint* arXiv:1006.0035[gr-qc]
[33] Burgess C P, Holman R, Leblond L and Shandera S 2010 *JCAP* **1010** 017
[34] Polyakov A M 2008 *Nucl. Phys.* B **797** 199
[35] Polyakov A M 2010 *Nucl. Phys.* B **834** 316
[36] Higuchi A and Cheong L Y 2009 *Class. Quant. Grav.* **26**, 135019
[37] Higuchi A 2009 *Class. Quant. Grav.* **29** 072001
[38] Alvarez E and Vidal R 2009 *JHEP* **10** 045
[39] Alvarez E and Vidal E 2010 *Prerint* arXiv:1004.4867 [hep-th]
[40] Allen B and Jacobson T 1986 *Commun. Math. Phys.* **103** 669
[41] Allen B and Turyn M 1987 *Nucl. Phys.* B **292** 813
[42] Antoniadis I and Mottola E 1991 *J. Math. Phys.* **32** 1037
[43] Higuchi A and Kouris S S 2001 *Class. Quantum Grav.* **18** 4317
[44] Kouris S S 2001 *Class. Quantum Grav.* **18** 4961
[45] Pérez-Nadal G, Roura A and Verdaguer E 2010 *JCAP* **1005** 036
[46] Campos A and Verdaguer E 1996 *Phys. Rev.* D **53** 1927
[47] Martín R and Verdaguer E 1999 *Phys. Lett.* B **465** 113
[48] Martín R and Verdaguer E 1999 *Phys. Rev.* D **60** 084008
[49] Hu B L and Verdaguer E 2003 *Class. Quantum Grav.* **20** R1
[50] Hu B L and Verdaguer E 2008 *Living Rev. Rel.* **11** 3
[51] Calzetta E, Roura A and Verdaguer E 2003 *Physica* A **319** 188
[52] Martín R and Verdaguer E 2000 *Phys. Rev.* D **61** 124024
[53] Tomboulis E 1977 *Phys. Lett.* B **70** 361
[54] Hu B L, Roura A and Verdaguer E 2004 *Phys. Rev.* D **70** 044002
[55] Wald R M (1984) *General Relativity* (Chicago: The University of Chicago Press)
[56] Roura A and Verdaguer E 2008 *Phys. Rev.* D **78** 064010
[57] Urakawa Y and Maeda K 2008 *Phys. Rev.* D **77** 024013
[58] Donoghue J F 1994 *Phys. Rev.* D **50** 3874
[59] Burgess C P 2004 *Living Rev. Rel.* **7** 5
[60] Hartle J B and Horowitz G T 1981 *Phys. Rev.* D **24** 257
[61] Flanagan E E and Wald R M 1996 *Phys. Rev.* D **54** 6233

Spanish Relativity Meeting (ERE 2010): Gravity as a Crossroad in Physics IOP Publishing
Journal of Physics: Conference Series **314** (2011) 012009 doi:10.1088/1742-6596/314/1/012009

On Critical Massive (Super)Gravity in adS_3

Eric A. Bergshoeff [1], **Olaf Hohm** [2], **Jan Rosseel** [1], **Ergin Sezgin** [3] **and Paul K. Townsend** [4]

[1] Centre for Theoretical Physics, University of Groningen, Nijenborgh 4, 9747 AG Groningen, The Netherlands
[2] Center for Theoretical Physics, Massachusetts Institute of Technology, Cambridge, MA 02139, USA
[3] George and Cynthia Woods Mitchell Institute for Fundamental Physics and Astronomy, Texas A&M University, College Station, TX 77843, USA
[4] Department of Applied Mathematics and Theoretical Physics, Centre for Mathematical Sciences, University of Cambridge, Wilberforce Road, Cambridge, CB3 0WA, U.K.

E-mail: E.A.Bergshoeff@rug.nl, ohohm@mit.edu, j.rosseel@rug.nl, sezgin@tamu.edu, P.K.Townsend@damtp.cam.ac.uk

Abstract.
We review the status of three-dimensional "general massive gravity" (GMG) in its linearization about an anti-de Sitter (adS) vacuum, focusing on critical points in parameter space that yield generalizations of "chiral gravity". We then show how these results extend to $\mathcal{N}=1$ super-GMG, expanded about a supersymmetric adS vacuum, and also to the most general 'curvature-squared' $\mathcal{N}=1$ supergravity model.

Massive gravity models in three spacetime dimensions (3D) have been intensively investigated over the past few years because of the possibility of finding a consistent 'toy' model of quantum gravity from which we might learn something useful. The oldest and simplest of these massive gravity models is "topologically massive gravity" (TMG), which is defined by the addition to the usual Einstein-Hilbert (EH) action of a Lorentz Chern-Simons (LCS) term [1]. TMG propagates a single massive spin-2 mode but this is a ghost unless the EH term has the non-standard sign. It is useful to define a sign σ such that $\sigma = 1$ yields the standard EH term, so that "non-standard" means $\sigma = -1$. The addition of a cosmological term allows the possibility of anti-de Sitter (adS) vacua, but then $\sigma = -1$ implies a negative mass for (BTZ) black holes. In the context of the $\mathrm{adS}_3/\mathrm{CFT}_2$ correspondence, this problem with the bulk theory translates to a negative central charge of the boundary conformal field theory (CFT). Taking $\sigma = 1$ allows positive central charges but at the cost of non-unitary propagation of the bulk spin-2 modes, which again implies [2, 3], although less directly, non-unitarity of the boundary CFT.

The classical equations of TMG depend on a mass parameter μ associated to the LCS term, and a length scale ℓ associated to the cosmological term, which we define such that ℓ is the radius of curvature in an adS vacuum. However, rescaling the metric is equivalent to rescaling μ and ℓ so the only parameter of the classical theory is the dimensionless product $\mu\ell$. Quantum corrections will introduce the additional dimensionless constant κ^2/ℓ, where $\kappa = \sqrt{16\pi G_3}$ is the gravitational coupling constant, but this appears classically only through an overall factor of ℓ/κ^2 in the on-shell action. In the equivalent approximation to the boundary CFT, the central charges take the form $c_\pm = (24\pi\ell/\kappa^2)f_\pm(\mu\ell)$ for dimensionless functions $f_\pm$. It was pointed

Published under licence by IOP Publishing Ltd

out in [4] that the parameter $\mu\ell$ can be tuned to a critical value, the "chiral point", at which either c_+ or c_- is zero. The bulk massive graviton mode then disappears from the spectrum, which suggests that the problem of a non-unitary bulk graviton for $\sigma = 1$ is circumvented by this "chiral gravity" theory. The definition of such a theory depends crucially on a choice of boundary conditions [5, 6]. Boundary conditions weaker than the standard Brown-Henneaux boundary conditions lead to well-defined logarithmic boundary CFTs. They also allow new logarithmic bulk modes, which is related to the observation of [7] that the bulk spin-2 mode is replaced by a bulk spin-1 mode at the chiral point.

An alternative, parity-preserving 3D massive gravity is "new massive gravity" (NMG), which is defined by the addition to the EH plus cosmological terms of a curvature-squared term constructed from the scalar [8]

$$G^{\mu\nu}S_{\mu\nu} \equiv R^{\mu\nu}R_{\mu\nu} - \frac{3}{8}R^2 \tag{1}$$

where the tensors G, S, R are, respectively, the Einstein, Schouten and Ricci tensors. This involves the introduction of a mass-squared parameter m^2 (which is positive for NMG but which we allow to be negative in the most general curvature-squared model). In an expansion about a Minkowski vacuum one finds that two spin-2 modes of mass m are propagated and, as for TMG, that perturbative unitarity requires $\sigma = -1$. As required by parity, the two spin-2 modes have opposite sign "helicities" s; i.e. opposite sign of the 3D Pauli-Lubanski pseudo-scalar divided by the mass; we call $|s|$ the "spin".

In the adS context, there is a family of classical NMG models parametrized by the dimensionless constant $m^2\ell^2$. In any parity-preserving 3D gravity theory with an adS vacuum, the central charges of the boundary CFT are equal, $c_+ = c_- = c$, and may be computed by the formula [9–11]

$$c = \frac{8\pi\ell}{\kappa^2} g_{\mu\nu} \frac{\partial L_{3D}}{\partial R_{\mu\nu}}\bigg|_{\text{adS vacuum}} \tag{2}$$

where L_{3D} is the 3D gravity Lagrangian of mass dimension two, i.e. without the factor of κ^{-2}. For NMG this gives [12–14]

$$c = \frac{24\pi\ell}{\kappa^2}\,\hat{\sigma}\,, \qquad \hat{\sigma} = \sigma + \frac{1}{2\ell^2 m^2}\,. \tag{3}$$

The expansion of the NMG action about the adS vacuum is greatly facilitated by starting with an alternative second-order version of the action with an auxiliary tensor field [8]. Diagonalization of the quadratic term of this action yields the sum of a linearized EH action with coefficient $\hat{\sigma}$, which is therefore the 'effective' EH coefficient in the adS background, and a spin-2 Fierz-Pauli (FP) action with coefficient $-1/\hat{\sigma}$ [13]. The linearized EH action propagates no modes so we get two massive gravitons from the FP action, but these are propagated unitarily only if $\hat{\sigma} < 0$. Hence there is a clash between unitary bulk gravitons and positive CFT central charges, as in TMG. There is also an analog of chiral gravity because there is a critical value of $\ell^2 m^2$ at which $c = 0$, and at which the quadratic action becomes equivalent to a spin-1 Proca action [13]. This critical NMG model is associated to a logarithmic CFT [15].

Both TMG and NMG are special cases of "general massive gravity" (GMG) [8], defined by the addition of *both* the LCS and the NMG curvature-squared term to the EH plus cosmological terms. Expanded about a Minkowski vacuum, this model propagates two spin-2 modes of opposite helicity but with different masses. In an adS vacuum there is a two-parameter family of classical GMG models parametrized by the constants $\mu\ell$ and $m^2\ell^2$. The central charges of the boundary CFT can be computed by a generalization of (2) to allow for parity-odd terms [16]. The result is

$$c_\pm = (24\pi\ell/\kappa^2) f_\pm\,, \qquad f_\pm = \hat{\sigma} \pm \frac{1}{\mu\ell}\,. \tag{4}$$

The no-ghost conditions follow from an expansion of the GMG action about an adS vacuum, but this 'off-shell' analysis is more subtle here, in part because the auxiliary tensor field 'trick' reduces the order in derivatives to three rather than two when the LCS term is present. Recent results in [17] indicate that the parameter ranges with positive central charge and positive energy of the graviton modes are mutually exclusive in GMG, as in TMG and NMG. Here we shall review, with some simplifications, the results of the on-shell analysis of [18] and explain how it extends to supergravity.

To facilitate our task, it will be convenient to first review the connection between a unitary irreducible representation (UIR) of the $SO(2,2)$ isometry group of adS$_3$ and fields on adS$_3$. Let $\varphi^{(s)}$ be a rank-$|s|$ ($|s| \geq 1$) totally-symmetric and tracefree field on adS$_3$ subject to the 'divergence-free' condition

$$\bar{D}^{\mu}\varphi^{(s)}_{\mu\nu_1\cdots\nu_{s-1}} = 0\,, \tag{5}$$

where $\bar{D}$ is the covariant derivative with respect to a background adS$_3$ metric $\bar{g}$. Let $\mathscr{D}(\eta)$ be the family of first-order linear differential operators, parametrized by a dimensionless constant η, that act on the space of such tensors according to the definition

$$\left[\mathscr{D}(\eta)\varphi^{(s)}\right]_{\mu_1\cdots\mu_s} = [\mathscr{D}(\eta)]_{\mu_1}{}^{\rho}\varphi^{(s)}_{\rho\mu_2\cdots\mu_s}\,, \qquad [\mathscr{D}(\eta)]_{\mu}{}^{\nu} = \ell^{-1}\delta^{\nu}_{\mu} + \frac{\eta}{\sqrt{|\bar{g}|}}\varepsilon_{\mu}{}^{\tau\nu}\bar{D}_{\tau}\,. \tag{6}$$

Despite appearances, the rank-$|s|$ tensor $\mathscr{D}(\eta)\phi^{(s)}$ is also traceless, totally symmetric and 'divergence-free'. Using the identity

$$\mathscr{D}(\eta)\mathscr{D}(-\eta)\varphi^{(s)} \equiv -\eta^2\left[\bar{D}^2 + \ell^{-2}\left(|s| + 1 - \eta^{-2}\right)\right]\varphi^{(s)} \tag{7}$$

we see that eigenfunctions of the covariant D'Alembertian $\bar{D}^2$ acting on spin-$|s|$ fields are linear combinations of solutions to the first-order equations

$$\mathscr{D}(\eta)\varphi^{(s)} = 0\,, \qquad \mathscr{D}(-\eta)\varphi^{(s)} = 0\,. \tag{8}$$

A physically acceptable[1] solution of either equation furnishes a UIR of $SO(2,2)$ labelled by its lowest weights (E_0, s), where $\ell^{-1}E_0$ is the lowest energy[2], which is real and satisfies the condition $E_0 \geq |s|$ [19–21]. The UIR furnished by a solution of (8) has lowest weights [22]

$$E_0 = 1 + \frac{1}{|\eta|}\,, \qquad s = \frac{|s|\eta}{|\eta|}\,. \tag{9}$$

We see that the sign of η gives the sign of the helicity s. The UIR's with $E_0 = |s|$ for $s \neq 0$ are called singleton irreps; their weight space is drastically reduced compared to a typical UIR, and they can be interpreted as describing modes that are confined to the 2D boundary of adS$_3$ [23]. More generally, the adS/CFT correspondence assigns to every UIR an operator in a dual 2D CFT with conformal weights $h_\pm = (E_0 \pm s)/2$, in which context the bound $E_0 \geq |s|$ translates into $h_\pm \geq 0$.

The above analysis applies for $|s| > 0$. For $s = 0$ we must consider the solutions of the second-order equation

$$\left(\bar{D}^2 - \mathscr{M}^2\right)\varphi = 0 \tag{10}$$

for constant $\mathscr{M}$. The solutions, if physically acceptable, furnish UIRs with $s = 0$ and

$$E_0 = 1 \pm \sqrt{1 + \ell^2\mathscr{M}^2}\,. \tag{11}$$

[1] i.e. nonsingular at the origin and normalizable with respect to the $SO(2,2)$ invariant measure [19,20].

[2] Or highest energy. It is not possible to distinguish between the two at the level of the equations of motion. This is why the unitarity of the irreducible representation is not sufficient for unitarity of the field theory.

Reality of E_0 implies the Breitenlohner-Freedman (BF) bound $\ell^2\mathscr{M}^2 \geq -1$. In addition, $E_0 \geq 0$ is required for a UIR. This condition allows both signs in (11) when $-1 \leq \ell^2\mathscr{M}^2 \leq 0$; otherwise only the plus sign is allowed.

Let us now see how all this applies to the 'Einstein' theory defined by the EH and cosmological terms alone. We write the metric as

$$g_{\mu\nu} = \bar{g}_{\mu\nu}\left(1 + \frac{1}{3}h\right) + H_{\mu\nu}\,, \qquad \bar{g}^{\mu\nu}H_{\mu\nu} = 0\,, \tag{12}$$

so that (H, h) are the tracefree and trace perturbations respectively. In the gauge

$$\bar{D}^\mu H_{\mu\nu} = 0\,, \tag{13}$$

the linearized field equations may now be written as

$$[\mathscr{D}(1)\mathscr{D}(-1)H]_{\mu\nu} = -\frac{1}{3}\left(\bar{D}_\mu\bar{D}_\nu - \frac{1}{3}\bar{g}_{\mu\nu}\bar{D}^2\right)h\,, \qquad \left(\ell^2\bar{D}^2 - 3\right)h = 0\,. \tag{14}$$

We have coupled equations for H and h but there is still a residual gauge invariance. If we write the vector gauge parameter of linearized diffeomorphisms as $\xi^T_\mu + \partial_\mu\xi$, where $\bar{D}^\mu\xi^T_\mu = 0$, then the gauge condition (13) is invariant provided that

$$\bar{D}_\mu\left(\ell^2\bar{D}^2 - 3\right)\xi = 0\,, \qquad \left(\ell^2\bar{D}^2 - 2\right)\xi^T_\mu = 0 \quad \Rightarrow \quad \left(\ell^2\bar{D}^2 + 2\right)\bar{D}_{(\mu}\xi^T_{\nu)} = 0\,. \tag{15}$$

The variation of h is $\delta h \sim \bar{D}^2\xi$ but this implies $\delta h \sim \xi$ for $\bar{D}^2\xi \sim \xi$. Since ξ is a residual gauge parameter, we may use it to set h to zero. This leaves us with the equation $\mathscr{D}(1)\mathscr{D}(-1)H = 0$, but the identity (7) implies that this is equivalent to

$$\left(\ell^2\bar{D}^2 + 2\right)H = 0\,. \tag{16}$$

This is clearly invariant under the residual gauge invariance, which suffices to remove all local degrees of freedom of H.

We are now in a position to discuss GMG. Linearized about an adS vacuum, the GMG equations are

$$[\mathscr{D}(1)\mathscr{D}(-1)\mathscr{D}(\eta_+)\mathscr{D}(\eta_-)H]_{\mu\nu} = -\frac{1}{3\ell^2}\left(\bar{D}_\mu\bar{D}_\nu - \frac{1}{3}\bar{g}_{\mu\nu}\bar{D}^2\right)h\,, \tag{17}$$

$$\frac{\Omega}{m^2}\left(\ell^2\bar{D}^2 - 3\right)h = 0\,, \tag{18}$$

where

$$\Omega \equiv \ell^2 m^2\hat{\sigma} - 1 \tag{19}$$

and [18]

$$\eta_\pm = \Omega^{-1}\left(-\frac{\ell m^2}{2\mu} \pm \sqrt{\frac{\ell^2 m^4}{4\mu^2} - \Omega}\right)\,. \tag{20}$$

We have retained the factor of Ω in the h-equation because it allows the $\Omega = 0$ equations to be obtained by taking the $\Omega \to 0$ limit. The region in parameter space with $\Omega > 0$ is divided from the region with $\Omega < 0$ by the curve $\Omega = 0$. To see the significance of this division, we observe that

$$\eta_+\eta_- = \Omega^{-1}\,. \tag{21}$$

This shows that the parameters $\eta_\pm$ have opposite signs in the $\Omega < 0$ region and the same sign in the $\Omega > 0$ region. The reality of $\eta_\pm$ and the UIR condition $|\eta_\pm| \leq 1$ translate into $m^2 c_\pm \leq 0$ for $\Omega < 0$ and $m^2 c_\pm \geq 0$, $\ell^2 m^4 \geq 4\mu^2 \Omega$ for $\Omega > 0$.

Provided that $\Omega \neq 0$, we may proceed as in the 'Einstein' case, eliminating h by a combination of its field equation and a residual gauge transformation, to arrive at the single 4th order equation

$$\mathscr{D}(1)\mathscr{D}(-1)\mathscr{D}(\eta_+)\mathscr{D}(\eta_-)H = 0 . \tag{22}$$

As long as $(\eta_+ - \eta_-)(|\eta_+| - 1)(|\eta_-| - 1) \neq 0$ the general solution is a linear combination of the two singleton modes and solutions of the first-order equations $\mathscr{D}(\eta_\pm)H = 0$. This state of affairs applies in the NMG limit $|\mu| = \infty$ since we then have

$$\eta_+ = -\eta_- = -1/\sqrt{-\Omega} . \tag{23}$$

Clearly, we must be on the $\Omega < 0$ side of the parameter space divide to take this limit, in which the bound $|\eta_\pm| \leq 1$ becomes $m^2\hat{\sigma} < 0$. As confirmed by the computation of the quadratic action in [13], this bound is the condition for the absence of tachyons. This computation also shows that $\hat{\sigma} < 0$ is the no-ghost condition, i.e. for unitarity of the quantum field theory. As $m^2 > 0$ for NMG, the absence of tachyons implies the absence of ghosts. It seems likely to us that this will remain true for GMG as long as $\Omega < 0$. However, it also seems likely that GMG has ghosts when $\Omega > 0$, even though it is certainly possible to satisfy the UIR bound $|\eta_\pm| < 1$, and hence avoid tachyons. This is because solutions of the equation $\mathscr{D}(\eta)\mathscr{D}(\eta')H = 0$ with $\eta\eta' > 0$ could be expected to arise naturally in the context of 5th order 3D gravity models, which necessarily propagate ghosts in a Minkowski vacuum [24]. To settle this issue we would need an 'off-shell' analysis like that presented for NMG in [13].

When $\Omega = 0$ the h-equation drops out. This leaves us with the H-equation, that can be shown to reduce to

$$[\mathscr{D}(\eta)\mathscr{D}(1)\mathscr{D}(-1)]_\mu{}^\rho \varepsilon_\rho{}^{\alpha\beta}\bar{D}_\alpha H_{\beta\nu} = 0 , \qquad \eta = -\frac{\mu}{\ell m^2} . \tag{24}$$

Unless $|\eta| = 1$, the solutions space is spanned by the singletons, the solution of $\mathscr{D}(\eta)H = 0$, and the solution of

$$\varepsilon_\rho{}^{\alpha\beta}\bar{D}_\alpha H_{\beta\nu} = 0 . \tag{25}$$

By acting on this equation with the operator $\varepsilon_\lambda{}^{\tau\mu}\bar{D}_\tau$, we see that the solution of (25) also obeys $(\ell^2\bar{D}^2 + 3)H = 0$. Comparing this with (7), we might naively conclude that it requires $\eta = \infty$ and hence $E_0 = 1$, which violates the spin-2 bound $E_0 \geq 2$. The resolution of this puzzle is subtle because of an additional gauge invariance; the relevant solutions of (25) are "partially massless" [25, 26].

Our main interest here is in the 'critical' GMG cases. For finite non-zero Ω these are those for which

$$(\eta_+ - \eta_-)(|\eta_+| - 1)(|\eta_-| - 1) = 0 . \tag{26}$$

When this condition holds, the solution space is no longer spanned by solutions of first-order equations of the form $\mathscr{D}(\eta)H = 0$. However, subject to appropriate boundary conditions, one finds that additional solutions appear. In the case that two of the η values coincide, these are logarithmic solutions of $\mathscr{D}^2(\eta)H = 0$ that do not solve $\mathscr{D}(\eta)H = 0$. In the case that three η values coincide we get additional doubly-logarithmic solutions of $\mathscr{D}^3(\eta)H = 0$ that solve neither $\mathscr{D}(\eta)H = 0$ nor $\mathscr{D}^2(\eta)H = 0$. A similar statement applies when $\Omega = 0$ but then the only critical case occurs when $\ell m^2 = |\mu|$, in which case we get a logarithmic solution of $\mathscr{D}^2(\pm 1)H = 0$. The catalog of critical points for $\Omega \neq 0$ is much richer[3]:

[3] The critical points of GMG have also been studied in [18]. Recently, properties of the logarithmic CFT dual to GMG at the critical points have been investigated in [17].

(i) $|\eta_+| = 1$ but $|\eta_-| \neq 1$ and $\Omega^{-1} \neq 0$. We find that $\eta_+ = \pm 1$ when $\ell\mu\hat{\sigma} = \mp 1$ and then $\eta_- = \mp 1/\left(1 \pm \ell m^2/\mu\right)$. We get no propagating graviton mode from $\mathscr{D}(\eta_+)H = 0$, just an extra logarithmic solution to $\mathscr{D}^2(\pm 1)H = 0$. In this case there is therefore a *single* propagating massive graviton of helicity $2\eta_-/|\eta_-|$. If $\Omega^{-1} \to 0$, then $m^2 \to \infty$, giving the TMG limit. In this case $\eta_- = 0$, and $|\ell\mu\sigma| = 1$.

(ii) $|\eta_-| = 1$ but $|\eta_+| \neq 1$ and $\Omega^{-1} \neq 0$. In this case $\eta_- = \pm 1$ when $\ell\mu\hat{\sigma} = \mp 1$ and then $\eta_- = \mp 1/\left(1 \pm \ell m^2/\mu\right)$. In this case there is a *single* propagating massive graviton of helicity $2\eta_+/|\eta_+|$. If $\Omega^{-1} \to 0$, then $m^2 \to \infty$, giving the TMG limit. In this case $\eta_+ = 0$, and $|\ell\mu\sigma| = 1$.

(iii) $\eta_+ = -\eta_-$ and $|\eta_\pm| = 1$. This case is realized by $\eta_- = -\eta_+ = 1$, which requires $|\mu| = \infty$, $m^2\hat{\sigma} = 0$. This is the critical limit of NMG. In this case (22) degenerates to

$$\mathscr{D}(1)^2\mathscr{D}(-1)^2 H = 0\,. \tag{27}$$

Although this propagates no gravitons, it does propagate two spin-1 modes [27]. One way to see this is to rewrite the fourth-order equation as the pair of second-order equations

$$\left(\ell^2\bar{D}^2 + 2\right) H = U\,, \qquad \left(\ell^2\bar{D}^2 + 2\right) U = 0\,, \tag{28}$$

where U is another symmetric traceless and 'divergence-free' tensor. Applying to the U-equation the same reasoning that we used above for the 'Einstein' case, we deduce that $U_{\mu\nu} = 2\bar{D}_{(\mu}A_{\nu)}$ for some vector field A satisfying $\left(\ell^2\bar{D}^2 - 2\right) A_\mu = 0$ and, since U is traceless, $\bar{D}^\mu A_\mu = 0$. In other words H satisfies

$$\left(\ell^2\bar{D}^2 + 2\right) H_{\mu\nu} = 2\bar{D}_{(\mu}A_{\nu)} \tag{29}$$

for a vector field such that

$$\left(\ell^2\bar{D}^2 - 2\right) A_\mu = 0\,, \qquad \bar{D}^\mu A_\mu = 0\,. \tag{30}$$

These equations for A, which also follow from taking the divergence and trace of (29), are Proca equations in the adS background. They are derivable from the Proca Lagrangian density

$$\mathscr{L}_{Proca} = -\tfrac{1}{4}F_{\mu\nu}F^{\mu\nu} - 2\ell^{-2}A^\mu A_\mu\,, \qquad F_{\mu\nu} = 2\partial_{[\mu}A_{\nu]}\,. \tag{31}$$

For singleton solutions the left-hand-side of (29) vanishes, implying that A is a Killing vector of the adS background. In contrast, the left-hand-side of (29) is non-vanishing for the logarithmic solutions of (27) and hence these must be related to non-trivial Proca modes. However, although a solution of the Proca equations furnishes a UIR of the adS_3 isometry group, with $(E_0, s) = (2, 1)$, a careful study of (29) shows that these solutions correspond to the first descendants of the logarithmic modes; see [28] for a detailed description of precisely such a descendant mode.

(iv) $\eta_+ = \eta_-$ but $|\eta_\pm| \neq 1$. This is possible only when $\Omega > 0$. In this case, $\eta_\pm = -2\mu/(m^2\ell)$. We get one propagating massive graviton from the solution to $\mathscr{D}(-2\mu/(m^2\ell)) = 0$ and an additional logarithmic solution of $\mathscr{D}^2(-2\mu/(m^2\ell)) = 0$.

(v) $\eta_+ = \eta_-$ *and* $|\eta_\pm| = 1$. This is possible only when $\Omega = 1$, which requires $2\mu = \mp m^2\ell$ according to whether $\eta_+ = \eta_- = 1$ or $\eta_+ = \eta_- = -1$. In this case we have, apart from singletons, only logarithmic and doubly logarithmic modes from $\mathscr{D}^3(1)H = 0$ or $\mathscr{D}^3(-1)H = 0$.

Observe that only cases (i)-(iv) are possible for $\Omega < 0$, which may be required if ghosts are to be avoided, for the reason given above. In these four cases we have

$$c_+ c_- = 0\,. \tag{32}$$

In other words, at least one of the two central charges of the boundary CFT is zero. Unless both are zero, "critical" can be interpreted as "chiral". As pointed out in [18], the "chiral GMG" models interpolate between chiral TMG, which is case (iii), and critical NMG, which is case (iv).

In the remainder of this paper, we shall explain how these results extend to the supergravity case. The $\mathscr{N}=1$ supersymmetric extensions of NMG and GMG have been found in recent work, along with all possible maximally supersymmetric vacua [27,29]. These constructions are complicated by the fact that the off-shell graviton multiplet contains not just the metric but also an additional scalar field S, which is auxiliary in 3D 'Einstein' supergravity [30] and in super-TMG [31] but which propagates in generic curvature-squared models. Even when the kinetic term for S is absent it will still propagate a scalar mode in adS vacua unless the coefficients of the cubic equation of motion for S are constants[4], and this condition defines a "super-GMG" model for which the bosonic action is precisely of GMG-form after elimination of S. Remarkably, *all* adS vacua of GMG correspond to supersymmetric vacua of super-GMG, so the spectrum in such vacua is determined by the GMG spectrum. Of course, the boundary CFT must now be a boundary SCFT, but it will still have the same central charge [32]. For these reasons, the results summarized above for GMG extend with no essential modifications to super-GMG. Unfortunately, this makes it unlikely that supersymmetry can help to resolve the boundary/bulk unitarity "clash" explained above.

Although the super-GMG model is the most interesting special case of the generic curvature squared 3D supergravity model constructed in [27], the results for the generic model provide us with the opportunity to explore some issues in the context of a model with more parameters. The bosonic Lagrangian is

$$e^{-1}\mathscr{L}_{bos} = -V(S) + f(S)R + \frac{1}{m^2}\left(R^{\mu\nu}R_{\mu\nu} - \frac{3}{8}R^2\right) + \frac{1}{8\tilde{m}^2}\left[R^2 - 16\,(\partial S)^2\right] + \frac{1}{\mu}e^{-1}\mathscr{L}_{LCS}\,, \tag{33}$$

where $\mathscr{L}_{LCS}$ is the Lorentz-Chern-Simons term, and

$$\begin{aligned} V(S) &= -MS + 2\sigma S^2 - \frac{1}{\check{\mu}}S^3 + \left(-\frac{3}{2m^2} + \frac{1}{\check{m}^2} + \frac{9}{2\tilde{m}^2}\right)S^4\,, \\ f(S) &= \sigma + \frac{1}{2\check{\mu}}S + \left(-\frac{1}{2m^2} + \frac{3}{10\check{m}^2} + \frac{3}{2\tilde{m}^2}\right)S^2\,. \end{aligned} \tag{34}$$

The model has six independent mass parameters $(M, m, \check{m}, \tilde{m}, \check{\mu}, \mu)$, which can be traded for dimensionless parameters using the gravitational coupling constant, and one discrete parameter $\sigma = \pm 1, 0$. Despite the notation, $(m^2, \tilde{m}^2, \check{m}^2)$ are allowed to take either sign.

The field equations following from (33) are given in [27]. The supersymmetric adS vacuum solutions of these equations have[5]

$$S = \bar{S} = -\ell^{-1}\,, \qquad M = -\ell^{-1}\left(4\sigma + \frac{2}{5\ell^2\check{m}^2}\right)\,. \tag{35}$$

[4] This condition may be relaxed slightly but we refer to [27] for a discussion of this point.
[5] We choose the sign of S so as to agree with the sign choice made in [27].

The central charges are given by (4) with $\hat{\sigma}$ given by

$$\hat{\sigma} = \sigma - \frac{1}{2\ell\tilde{\mu}} + \frac{3}{10\ell^2\breve{m}^2}\,. \tag{36}$$

The adS/CFT correspondence suggests that the spectrum of propagating gravitons will be independent of $\tilde{m}$ and will depend on the $\tilde{\mu}$ and $\breve{m}$ only through the parameter $\hat{\sigma}$. We shall verify this.

To expand the field equations about a supersymmetric adS vacuum, we write the metric as in (12) and we write

$$S = \bar{S} + \ell^{-1}s \tag{37}$$

for dimensionless scalar perturbation s. To present the results, we shall again make use of the parameters $\eta_\pm$ and Ω defined in (19) and (20) in terms of $\hat{\sigma}$, although it should be remembered that $\hat{\sigma}$ is now given by (36). In addition, it is useful to define the dimensionless parameter

$$a = m^2\left[-\frac{2}{m^2} + \frac{6}{5\breve{m}^2} + \frac{6}{\tilde{m}^2} - \frac{\ell}{\tilde{\mu}}\right]. \tag{38}$$

This agrees with the definition in [27] when restricted to the models (with $\tilde{m}^2 = \infty$) for which this parameter was defined there. We will also need the following (dimensionless) linear differential operators:

$$L = \ell^2\bar{D}^2 - 3\,, \qquad \tilde{L} = \frac{m^2}{\tilde{m}^2}L - \Omega\,. \tag{39}$$

Using these definitions, one finds that the h and s equations are, respectively,

$$\frac{1}{m^2}L\left(\tilde{L}h + 3as\right) = 0\,, \qquad \frac{1}{m^2}\left(\tilde{L} + 2a\right)s + \frac{a}{12m^2}Lh = 0\,, \tag{40}$$

and that the linearized H-equation is

$$\mathscr{D}(1)\mathscr{D}(-1)\mathscr{D}(\eta_+)\mathscr{D}(\eta_-)H = \Omega^{-1}J\,, \tag{41}$$

where

$$J_{\mu\nu} = \frac{1}{3\ell^2}\left(\bar{D}_\mu\bar{D}_\nu - \tfrac{1}{3}\bar{g}_{\mu\nu}\bar{D}^2\right)\left[\tilde{L}h + 3as\right]. \tag{42}$$

The integrability condition $\bar{D}^\mu J_{\mu\nu} = 0$ is satisfied as a consequence of the h-field equation.

Observe that we recover the GMG equations that we have already analyzed on setting $\tilde{m}^2 = |\tilde{\mu}| = \infty$ and $a = 0$. In that case, we argued that it was possible to set $h = 0$, and hence $J = 0$, by using a combination of the h-equation and a residual gauge transformation. It is no longer so clear that this argument still applies, but a simpler one is available, as long as $\Omega \neq 0$. We define the new symmetric traceless and 'divergence-free' perturbation

$$\tilde{H} = H - m^2\Omega^{-1}J\,. \tag{43}$$

Using the relation $\varepsilon_{(\mu}{}^{\alpha\beta}\bar{D}_{|\alpha}J_{\beta|\nu)} = 0$, we may deduce that

$$\mathscr{D}(1)\mathscr{D}(-1)\mathscr{D}(\eta_+)\mathscr{D}(\eta_-)\widetilde{H}_{\mu\nu} = 0\,. \tag{44}$$

The analysis of critical points for $\Omega \neq 0$ now proceeds *exactly* as for the GMG case. For $\Omega = 0$ the H-equation is given by

$$\left[\mathscr{D}(\eta)\mathscr{D}(1)\mathscr{D}(-1)\right]_\mu{}^\rho\varepsilon_\rho{}^{\alpha\beta}\bar{D}_\alpha H_{\beta\nu} = -\eta J_{\mu\nu}\,, \qquad \eta = -\frac{\mu}{\ell m^2}\,. \tag{45}$$

Note that by acting on this equation with the operator $\varepsilon_\lambda{}^{\tau\mu}\bar{D}_\tau$, we can arrive at the integrability condition $\mathscr{D}(\eta)\mathscr{D}(1)\mathscr{D}(-1)(\ell^2\bar{D}^2+3)H = 0$, an equation that also follows from (24). To summarize, the critical point structure of the generic curvature-squared supergravity model, expanded about a supersymmetric adS vacuum, is *identical* to that of the GMG model, with all dependence on the extra parameters absorbed into the effective EH coefficient $\hat{\sigma}$, as anticipated above.

The main difference of the generic case as compared to GMG is the possibility of additional scalar modes arising from the h and s equations (40). Here we shall consider some special cases.

- $\tilde{m}^2 = \infty$ *and* $a = 0$; this case was called "generalized GMG" in [27]; it reduces to the GMG case analysed above when $|\tilde{\mu}| = \infty$. For this case, and assuming non-zero finite m^2, the equations (40) become
$$\Omega L h = 0\,, \qquad (2a-\Omega)\,s = 0 \tag{46}$$
We see that $s=0$ (unless $\Omega = 2a$ but then it is undetermined, implying an 'accidental' gauge invariance that allows one to choose $s=0$). Provided $\Omega \neq 0$ we have $Lh = 0$ and residual gauge invariances allow us to set $h=0$. There are therefore no propagating scalars. One can show that the generalized super-NMG limit ($|\mu| = \infty$) is ghost-free for $\hat{\sigma} \leq 0$ [27].
- $\tilde{m}^2 = \infty$ but $a \neq 0$. In this case the S-equation of motion is algebraic but it still propagates modes in adS, for reasons explained in [27]. Here we verify this for the generic $\Omega \neq 0$ case. Under these conditions, the equations (40) become equivalent to (for $m^2 \neq 0$ and finite)
$$L\left(h - 3a\Omega^{-1}s\right) = 0\,, \qquad Ls = \frac{4\Omega\,(\Omega-2a)}{a^2}s \tag{47}$$
As the addition of a scalar to h does not change its residual gauge transformation, the first equation does not lead to propagating modes. The second one takes the form (10) with $\varphi = s$ and $\mathscr{M}^2 = 3 + 4\Omega\,(\Omega-2a)\,/a^2$. This equation therefore propagates a scalar mode that is non-tachyonic. However there is no guarantee that it is not a ghost.
- $a=0$ but $\tilde{m}^2 \neq \infty$. The equations (40) now reduce to
$$L\left(L - \frac{\Omega\tilde{m}^2}{m^2}\right)h = 0\,, \qquad \left(L - \frac{\Omega\tilde{m}^2}{m^2}\right)s = 0\,. \tag{48}$$
The first of these is a fourth-order equation for h that is not easy to analyse. The second equation takes the form (10) with $\varphi = s$ and $\ell^2\mathscr{M}^2 = 3 + \tilde{m}^2\Omega/m^2$. We therefore get a non-tachyonic scalar provided that $4 \geq -\tilde{m}^2\Omega/m^2$, although there is again no guarantee that it is not a ghost.

Acknowledgments

EB wishes to thank the organizers for a stimulating and inspiring atmosphere. PKT is supported by an EPSRC Senior Fellowship. The research of ES is supported in part by NSF grants PHY-0555575 and PHY-0906222. The work of O.H. is supported by the DFG–The German Science Foundation and in part by funds provided by the U.S. Department of Energy (DOE) under the cooperative research agreement DE-FG02-05ER41360.

References

[1] S. Deser, R. Jackiw and S. Templeton, "Topologically massive gauge theories", Annals Phys. **140** (1982) 372.
[2] K. Skenderis, M. Taylor and B. C. van Rees, "Topologically Massive Gravity and the AdS/CFT Correspondence," JHEP **0909** (2009) 045 [arXiv:0906.4926 [hep-th]].
[3] D. Grumiller and I. Sachs, "$AdS_3/LCFT_2$ – Correlators in Cosmological Topologically Massive Gravity," JHEP **1003** (2010) 012 [arXiv:0910.5241 [hep-th]].

[4] W. Li, W. Song and A. Strominger, "Chiral Gravity in Three Dimensions," JHEP **0804** (2008) 082 [arXiv:0801.4566 [hep-th]].
[5] D. Grumiller and N. Johansson, "Instability in cosmological topologically massive gravity at the chiral point," JHEP **0807** (2008) 134 [arXiv:0805.2610 [hep-th]].
[6] A. Maloney, W. Song and A. Strominger, "Chiral Gravity, Log Gravity and Extremal CFT," Phys. Rev. D **81** (2010) 064007 [arXiv:0903.4573 [hep-th]].
[7] S. Carlip, S. Deser, A. Waldron and D. K. Wise, "Cosmological Topologically Massive Gravitons and Photons," Class. Quant. Grav. **26** (2009) 075008 [arXiv:0803.3998 [hep-th]].
[8] E. A. Bergshoeff, O. Hohm and P. K. Townsend, "Massive Gravity in Three Dimensions," Phys. Rev. Lett. **102** (2009) 201301 [arXiv:0901.1766 [hep-th]]; "On massive gravitons in 2+1 dimensions," J. Phys. Conf. Ser. **229** (2010) 012005 [arXiv:0912.2944 [hep-th]]; "Gravitons in Flatland," arXiv:1007.4561 [hep-th].
[9] J. D. Brown and M. Henneaux, "Central Charges in the Canonical Realization of Asymptotic Symmetries: An Example from Three-Dimensional Gravity," Commun. Math. Phys. **104** (1986) 207.
[10] H. Saida and J. Soda, "Statistical entropy of BTZ black hole in higher curvature gravity," Phys. Lett. B **471** (2000) 358 [arXiv:gr-qc/9909061].
[11] P. Kraus and F. Larsen, "Microscopic Black Hole Entropy in Theories with Higher Derivatives," JHEP **0509** (2005) 034 [arXiv:hep-th/0506176].
[12] Y. Liu and Y. W. Sun, "Note on New Massive Gravity in AdS_3," JHEP **0904** (2009) 106 [arXiv:0903.0536 [hep-th]]; "Consistent Boundary Conditions for New Massive Gravity in AdS_3," JHEP **0905** (2009) 039 [arXiv:0903.2933 [hep-th]].
[13] E. A. Bergshoeff, O. Hohm and P. K. Townsend, "More on Massive 3D Gravity," Phys. Rev. D **79** (2009) 124042 [arXiv:0905.1259 [hep-th]].
[14] O. Hohm and E. Tonni, "A boundary stress tensor for higher-derivative gravity in AdS and Lifshitz backgrounds," JHEP **1004** (2010) 093 [arXiv:1001.3598 [hep-th]].
[15] D. Grumiller and O. Hohm, "$AdS_3/LCFT_2$ - Correlators in New Massive Gravity," Phys. Lett. B **686** (2010) 264 [arXiv:0911.4274 [hep-th]].
[16] P. Kraus and F. Larsen, "Holographic gravitational anomalies," JHEP **0601** (2006) 022 [arXiv:hep-th/0508218].
[17] D. Grumiller, N. Johansson, T. Zojer, "Short-cut to new anomalies in gravity duals to logarithmic conformal field theories," [arXiv:1010.4449 [hep-th]].
[18] Y. Liu and Y. W. Sun, "On the Generalized Massive Gravity in AdS_3," Phys. Rev. D **79** (2009) 126001 [arXiv:0904.0403 [hep-th]].
[19] R. Raczka, N. Limic and J. Niederle, "Discrete degenerate representations of the noncompact rotation groups,", J. Math. Phys. **7** (1966) 1861.
[20] P. Breitenlohner and D. Z. Freedman, "Stability In Gauged Extended Supergravity," Annals Phys. **144** (1982) 249.
[21] A. O. Barut and R. Raczka, "Theory Of Group Representations And Applications," (World Scientific, 1986).
[22] S. Deger, A. Kaya, E. Sezgin and P. Sundell, "Spectrum of D = 6, N = 4b supergravity on AdS(3) x S(3)," Nucl. Phys. B **536** (1998) 110 [arXiv:hep-th/9804166].
[23] A. M. Harun ar Rashid, C. Fronsdal and M. Flato, "Three D singletons and 2-D C.F.T," Int. J. Mod. Phys. A **7** (1992) 2193.
[24] E. A. Bergshoeff, O. Hohm and P. K. Townsend, "On Higher Derivatives in 3D Gravity and Higher Spin Gauge Theories," Annals Phys. **325** (2010) 1118 [arXiv:0911.3061 [hep-th]];
[25] S. Deser and R. I. Nepomechie, "Gauge Invariance Versus Masslessness In De Sitter Space," Annals Phys. **154** (1984) 396.
[26] S. Deser and A. Waldron, "Partial masslessness of higher spins in (A)dS," Nucl. Phys. B **607** (2001) 577 [arXiv:hep-th/0103198].
[27] E. A. Bergshoeff, O. Hohm, J. Rosseel, E. Sezgin and P. K. Townsend, "More on Massive 3D Supergravity," arXiv:1005.3952 [hep-th].
[28] G. Giribet, M. Kleban and M. Porrati, "Topologically Massive Gravity at the Chiral Point is Not Chiral," JHEP **0810** (2008) 045[arXiv:0807.4703 [hep-th]].
[29] R. Andringa, E. A. Bergshoeff, M. de Roo, O. Hohm, E. Sezgin and P. K. Townsend, "Massive 3D Supergravity," Class. Quant. Grav. **27** (2010) 025010 [arXiv:0907.4658 [hep-th]].
[30] P. S. Howe and R. W. Tucker, "Local Supersymmetry In (2+1)-Dimensions. 1. Supergravity And Differential Forms," J. Math. Phys. **19** (1978) 869;
[31] S. Deser and J. H. Kay, "Topologically Massive Supergravity," Phys. Lett. B **120** (1983) 97.
[32] M. Bañados, K. Bautier, O. Coussaert, M. Henneaux and M. Ortiz, "Anti-de Sitter/CFT correspondence in three-dimensional supergravity," Phys. Rev. D **58** (1998) 085020 [arXiv:hep-th/9805165].

New Horizons in Gravity: Dark Energy and Condensate Stars

Emil Mottola
Theoretical Div., Los Alamos National Laboratory, Los Alamos, NM 87545 USA
E-mail: emil@lanl.gov

Abstract.
Black holes are an apparently unavoidable prediction of classical General Relativity, at least if matter obeys the strong energy condition $\rho + 3p \geq 0$. However quantum vacuum fluctuations generally violate this condition, as does the eq. of state of cosmological dark energy $\rho = -p > 0$. When quantum effects are considered, black holes lead to a number of thermodynamic paradoxes associated with the Hawking temperature and assumption of black hole entropy, which are briefly reviewed. It is argued that the largest quantum effects arise from the conformal scalar degrees of freedom generated by the trace anomaly of the stress-energy tensor in curved space. At event horizons these can have macroscopically large backreaction effects on the geometry, potentially removing the classical event horizon of black hole and cosmological spacetimes, replacing them with a quantum phase boundary layer, where the effective value of the gravitational vacuum energy density can change. In the effective theory including the quantum effects of the anomaly, the cosmological term Λ becomes a dynamical condensate, whose value depends upon boundary conditions at the horizon. By taking a positive value in the interior of a fully collapsed star, the effective cosmological term removes any singularity, replacing it with a smooth dark energy de Sitter interior. The resulting gravitational vacuum condensate star (or *gravastar*) configuration resolves all black hole paradoxes, and provides a testable alternative to black holes as the final quantum mechanical end state of complete gravitational collapse. The observed Λ_{eff} dark energy of our universe likewise may be a macroscopic finite size effect whose value depends not on Planck scale or other microphysics but on the cosmological Hubble horizon scale itself.

1. Introduction: Classical Black Holes

Just a year after the publication of the field equations of General Relativity (GR), K. Schwarzschild found a simple, static, spherically symmetric solution of those equations, with the line element [1],

$$ds^2 = -f(r)\, d\tau^2 + \frac{dr^2}{h(r)} + r^2 \left(d\theta^2 + \sin^2\theta\, d\phi^2\right) , \tag{1}$$

where in this case

$$f(r) = h(r) = 1 - \frac{r_M}{r} = 1 - \frac{2GM}{rc^2} \,. \tag{2}$$

This Schwarzschild solution to the vacuum Einstein's equations, with vanishing Ricci tensor $R^a_{\ b} = 0$ and stress tensor $T^a_{\ b} = 0$ for all $r > 0$ describes an isolated, non-rotating object of total mass M.

Published under licence by IOP Publishing Ltd

Journal of Physics: Conference Series **314** (2011) 012010 doi:10.1088/1742-6596/314/1/012010

If the Schwarzschild solution (1)-(2) is taken seriously for $r < r_M = 2GM/rc^2$, the singularity at $r = 0$ is present in Einstein's theory for any mass $M > 0$, including the macroscopic mass of a collapsed star with the mass of the sun, $M_\odot \simeq 2 \times 10^{33}$ gm. or even that of supermassive objects with masses 10^6 to $10^9 M_\odot$. The collapse of such enormous quantities of matter to a single mathematical point at $r = 0$ certainly presents a challenge to the imagination, and one that it seems Einstein himself sought arguments to avoid [2]. The situation is scarcely more acceptable if the singularity is removed only by the intervention of quantum effects at the extremely tiny Planck length $L_{Pl} = (G\hbar/c^3)^{\frac{1}{2}} = 1.616 \times 10^{-33}$ cm.

On the other hand, unlike the central singularity of the Schwarzschild metric at $r = 0$, and despite the coordinate singularity at $r = r_M$, local scalar invariant quantities that can be constructed from the contractions of the Riemann curvature tensor remain finite as $r = r_M$. For example the fully contracted quadratic Riemann invariant

$$R^{abcd}R_{abcd} = \frac{12\, r_M^2}{r^6}\,, \tag{3}$$

which diverges at $r = 0$ remains finite at $r = r_M$. Also, although the time for an infalling particle to reach the horizon is infinite for any observer remaining fixed outside the horizon, the *proper time* measured by the particle itself during its infall remains finite as $r \to r_M$ [3, 4]. Since the line element (1) is non-singular for $0 < r < r_M$, the most straightforward possibility would seem to be to assume that this non-singular vacuum interior can be matched smoothly to the non-singular exterior Schwarzschild solution. This matching was achieved by the coordinate transformations and analytic continuation of the Schwarzschild solution found by Kruskal and Szekeres [5]. In the these new (T, X) coordinates, the Schwarzschild line element (1)-(2) becomes

$$ds^2 = \frac{4r_M^3}{r}\, e^{r/2r_M}\left(-dT^2 + dX^2\right) + r^2\left(d\theta^2 + \sin^2\theta\, d\phi^2\right)\,, \tag{4}$$

where r is to be regarded as the function of the Kruskal-Szekeres (T, X) defined implicitly by

$$\left(\frac{r}{r_M} - 1\right) e^{r/r_M} = X^2 - T^2\,. \tag{5}$$

Thus the (singular) transformation to Kruskal-Szekeres coordinates (T, X) has removed the coordinate singularity of the Schwarzschild line element (1)-(2).

An important point which is often left unstated is that this mathematical procedure of analytic continuation through the null hypersurface of an event horizon at $r = r_M$ actually involves a *physical assumption*, namely that the stress-energy tensor $T^a_{\ b}$ is vanishing there. Even in the purely classical theory of General Relativity, the hyperbolic character of Einstein's equations allows generically for stress-energy sources and hence metric discontinuities on the horizon which would violate this assumption. Additional physical information is necessary to determine what happens at the event horizon, and the correct matching of interior to exterior geometry depends on this physics, which may or may not be consistent with (complex) analytic continuation of coordinates around the singularity at $r = r_M$.

The static Schwarzschild solution of an isolated uncharged mass was generalized to include electric charge by Reissner & Nordström [6], and more interestingly for astrophysically realistic collapsed stars, to include rotation and angular momentum by Kerr [7]. The complete analytic extensions of the Reissner-Nordström and Kerr solutions were found as well [8]. The global properties of these analytic extensions are more complicated and arguably even more unphysical than in the Schwarzschild case. For slowly rotating black holes with angular momentum $J < GM^2/c$, there are an *infinite* number of black hole interior and asymptotically flat exterior regions, singularities, and *closed timelike curves* in the interior region(s), which violate causality on macroscopic distance scales [9]. Again these apparently unphysical features appear in GR only if the mathematical hypothesis of complex analytic extension and continuation through real coordinate singularities is assumed. Based on the Principle of Equivalence between gravitational and inertial mass, Einstein's theory possesses general coordinate invariance under all *regular* and *real* transformations of coordinates. It is the appending to classical General Relativity of the much stronger mathematical *hypothesis* of *complex* analytic continuation through *singular* coordinate transformations that leads to the global aspects of the Schwarzschild solution which may be unrealized in Nature. This analytic continuation is generally invalid if there are stress-tensor sources encountered at or before the breakdown of coordinates.

In the general black hole solution characterized by mass M, angular momentum J, and electric charge Q, one can define a quantity called the *irreducible mass* M_{irr} by the relation [10]

$$M^2 = \left(M_{irr} + \frac{Q^2}{4GM_{irr}}\right)^2 + \frac{c^2J^2}{4G^2M_{irr}^2}, \tag{6}$$

The differential form of (6) is [11, 12]

$$dE = dMc^2 = \frac{c^2}{8\pi G}\,\kappa\, dA + \Omega\, dJ + \Phi\, dQ \tag{7}$$

which is the Smarr formula for a Kerr-Newman rotating, electrically charged black hole, in which

$$\kappa = \frac{1}{M}\left[\frac{c^4}{4G} - \frac{4\pi^2 G}{c^4 A^2}\left(Q^4 + 4c^2J^2\right)\right], \tag{8a}$$

$$\Omega = \frac{4\pi J}{MA}, \tag{8b}$$

$$\Phi = \frac{Q}{M}\left[\frac{c^2}{2G} + \frac{2\pi Q^2}{Ac^2}\right], \tag{8c}$$

$$A = \frac{16\pi G^2}{c^4}M_{irr}^2 = \frac{4\pi G}{c^4}\left[2GM^2 - Q^2 + 2\sqrt{G^2M^4 - GM^2Q^2 - c^2J^2}\right], \tag{8d}$$

are the horizon surface gravity, angular velocity, electrostatic potential and area respectively. All dimensionful constants have been retained to emphasize that (7)-(8) are formulae derived from *classical* GR in which no $\hbar$ whatsoever appears. Notice also that the coefficient of dA in (7)

$$\left(\frac{\partial E}{\partial A}\right)_{J,Q} = \frac{c^2\kappa}{8\pi G} \tag{9}$$

has both the form and dimensions of a *surface tension.*

2. Quantum Black Holes and Thermodynamic Paradoxes

It has been shown that in any classical process the irreducible mass M_{irr} in (6), and therefore from (8d) the geometric area A of the horizon can never decrease [10, 13]. Since matter falling into a black hole would take its entropy with it, leading to an apparent violation of the Second Law of Thermodynamics, but classically A always increases, Bekenstein proposed that the black hole should itself be assigned an entropy proportional to A [14]. Since A does not have the units of entropy, it is necessary to divide the area by another quantity with units of length squared before multiplying by Boltzmann's constant k_B, to obtain an entropy. However, classical GR (without a cosmological term) contains no such quantity, G/c^2 being simply a conversion factor between mass and distance. Hence Bekenstein found it necessary for purely dimensional reasons to introduce Planck's constant $\hbar$ and the Planck length $L_{Pl} = (\hbar G/c^3)^{\frac{1}{2}} = 1.616 \times 10^{-33}$ cm into the discussion, proposing that the entropy of a black hole should be

$$S_{BH} = \gamma k_B \frac{A}{L_{Pl}^2} = \gamma \frac{16\pi G k_B}{\hbar c} M_{irr}^2 \,, \tag{10}$$

with γ a constant of order unity [14]. He showed that *if* such an entropy were assigned to a black hole, so that it is added to the entropy of matter, $S_{tot} = S_m + S_{BH}$, then this total generalized entropy would plausibly always increase. In fact, this is not difficult at all, and the generalized second law $\Delta S_{tot} \geq 0$ is usually satisfied by a very wide margin, simply because the Planck length L_{Pl} is so tiny, and the macroscopic area of a black hole measured in Planck units is so enormous. Hence even the small increase of mass and area caused by dropping into the black hole a modest amount of matter and concomitant loss of matter entropy $\Delta S_m < 0$ is easily overwhelmed by a great increase in S_{BH}, $\Delta S_{BH} \gg |\Delta S_m|$, guaranteeing that the generalized total entropy increases: $\Delta S_{tot} > 0$.

Soon after Bekenstein's proposal Hawking argued that black holes would also emit radiation at a characteristic temperature [15]

$$T_H = \frac{\hbar \kappa}{2\pi c k_B} \overset{J=Q=0}{=} \frac{\hbar c^3}{8\pi G k_B M} \,, \tag{11}$$

where the first equality is general, and the second equality applies only for a Schwarzschild black hole with $J = Q = 0$. With the temperature inversely proportional to its mass assigned to a black hole by this formula, if we interpret (7) as the first law of thermodynamics in the form

$$dE = dMc^2 = T_H \, dS_{BH} + \Omega \, dJ + \Phi \, dQ \,. \tag{12}$$

then the coefficient γ in (10) is fixed to be 1/4. Notice that this involves simply multiplying κ and dividing A in the *classical* Smarr formula (7) by $\hbar$.

Since the temperature $T_H < 0.1\mu K$ for a solar mass black hole is so exceedingly small, the prospects for testing the Hawking prediction experimentally are correspondingly remote. Nevertheless the formula (12) is simple and appealing, and it has been generally accepted since soon after Hawking's paper appeared. However, simultaneously and from the very beginning, a number of problems with this thermodynamic interpretation made their appearance as well.

Firsty, the entropy (10) is non-extensive both in not being proportional to the volume but to the area, and in being entirely independent of the number of particle species. This points to a basic problem in trying to account for the the entropy by a microcanonical counting of microstates according to Boltzmann's formula,

$$S = k_B \ln W(E) \,. \tag{13}$$

Normally one would expect the number of distinct microscopic states $W(E)$ at a given total energy E, and hence the entropy, to depend on the number of distinct particle species.

Secondly, tracing the thermal quanta at asymptotic temperature T_H backwards indicates that at late times they must have originated very close to the event horizon at $r = r_M$ with local energy

$$\hbar \, \omega_{loc} \sim \frac{k_B T_H}{f^{\frac{1}{2}}(r)} = \frac{k_B T_H}{\sqrt{1 - \frac{r_M}{r}}} \to \infty \,, \tag{14}$$

which is arbitrarily large, and in particular exceeds the Planck energy, at which the semi-classical approximation of a fixed classical background geometry would be expected to fail.

Thirdly, as pointed out by Hawking himself [16], a temperature inversely proportional to the $M = E/c^2$ implies that the heat capacity of a Schwarzschild black hole,

$$\frac{dE}{dT_H} = -\frac{8\pi G k_B M^2}{\hbar c} = -\frac{Mc^2}{T_H} < 0 \tag{15}$$

is *negative*, whereas in statistical mechanics the heat capacity of any system in stable equilibrium is related to the energy fluctuations about its mean value $\langle E\rangle$ by (at constant volume)

$$c_V = \left(\frac{d\langle E\rangle}{dT}\right)_V = \frac{1}{k_B T^2}\left\langle (E - \langle E\rangle)^2\right\rangle > 0\,, \tag{16}$$

and is necessarily positive [17]. A negative heat capacity can be formally obtained in the microcanonical treatment of certain non-relativistic gravitational systems such as globular clusters with very long relaxation times that have not yet reached a true equilibrium state [18, 19]. However whereas the relaxation time scale for a globular cluster due to two-body stellar encounters can be much larger than the dynamical time scale $\sqrt{R^3/GM}$ (with R a typical dimension of the cluster and M its mass), statistical fluctuations in the Hawking thermal flux occur on the dynamical time scale $2GM/c^3$ itself, which is very short ($\sim 10^{-5}$ sec for a solar mass black hole). This is indeed the time scale for an unstable mode around the equilibrium state to develop [20]. It is therefore by no means clear that equilibrium thermodynamics can be applied to black holes at all.

Finally, a cold quantum system generally has a low entropy since thermal excitations are suppressed at zero temperature. However in the limit $M \to \infty$ or $\hbar \to 0$, for which (11) gives $T_H \to 0$, (10) gives $S_{BH} \to \infty$, *i.e.* an *infinitely large* entropy at *absolute zero* temperature!

It is instructive to evaluate S_{BH} for typical astrophysical black holes. Taking as our unit of mass the mass of the sun, $M_\odot \simeq 2 \times 10^{33}$ gm., we have

$$S_{BH} \simeq 1.050 \times 10^{77}\, k_B \left(\frac{M}{M_\odot}\right)^2 . \tag{17}$$

This is truly an enormous entropy. For comparison, we may estimate the entropy of the sun as it is, a hydrogen burning main sequence star, whose entropy is given to good accuracy by the entropy of a non-relativistic perfect fluid. This is of the order Nk_B where N is the number of nucleons in the sun $N \sim M_\odot/m_N \sim 10^{57}$, times a logarithmic function of the density and temperature profile which may be estimated to be of the order of 20 for the sun. Hence the entropy of the sun is roughly

$$S_\odot \sim 2 \times 10^{58}\, k_B\,, \tag{18}$$

or nearly 19 orders of magnitude smaller than (17). Since (17) makes no reference to how a black hole is formed, we may in principle imagine forming one from the sun adiabatically by an arbitrarily slow contraction. At every time in this adiabatic process the entropy of the sun would remain (18). At the instant that the event horizon is reached this entropy would have to jump discontinuously somehow to (17). From Boltzmann's formula (13) this means that the number of microstates of a black hole must jump by $\exp(10^{19})$ at that instant at which the event horizon is reached, a truly staggering proposition. The loss of the information represented by this enormous jump in entropy is one form of the "information paradox," which is so serious that it led Hawking to suggest that quantum mechanics itself must be altered [21]. This is all the more strange when one considers that according to classical GR by a change to Kruskal-Szekeres coordinates (4) nothing at all is supposed to happen at the event horizon.

For all of these reasons the thermodynamic interpretation of (12) remains problematic. On the other hand, if the inserted $\hbar$ in (12) is simply cancelled and one returns to the differential

Smarr relation (7), derived from classical GR, these difficulties immediately vanish. One would only be left to explain in what sense the surface gravity κ is a surface tension.

If collapse to a black hole does not produce an equilibrium state, because of its negative heat capacity and rapidly growing unstable negative mode, the question then naturally arises of what is the final equilibrium state of complete gravitational collapse? In condensed matter physics at high density and low temperatures quantum effects always play an important role, and lead generally to a phase transition in which a macroscopic condensate is formed. In the proposal of a quantum phase transition producing a gravitational Bose-Einstein condensate (GBEC) in the interior of a fully collapsed object a physical surface and surface tension replaces the mathematical horizon of classical black hole, and removes all of the thermodynamic paradoxes of quantum black holes [22].

3. Quantum Effects at Horizons: The Role of the Trace Anomaly

The analytic continuation of the static Schwarzschild geometry to the black hole interior by Kruskal coordinates (4) assumes the complete absence of sources to Einstein's eqs. in the vicinity of the horizon. However, it has long been known that at least in certain states quantum stresses can be quite important there. In the natural "vacuum" state empty of all particles at infinity and corresponding to the static Schwarzschild time t, called the Boulware state $|B\rangle$ [23], the expectation value of the renormalized stress tensor [24]

$$\langle B|T^a_{\ b}|B\rangle_R \to -\frac{\pi^2}{90}\frac{\hbar c}{(4\pi r_M)^4}\left(1-\frac{r_M}{r}\right)^{-2} \mathrm{diag}\,(-3,1,1,1)\,, \tag{19}$$

becomes arbitrarily large as $r \to r_M$. Being a dimension four operator, the $\left(1-\frac{r_M}{r}\right)^{-2}$ behavior of $\langle B|T^a_{\ b}|B\rangle_R$ is a kinematic consequence of the blueshift (14). Clearly, if such a state were even close to being realized in practice, its stress-energy would act as a significant physical source for the semi-classical Einstein's equations,

$$R^a_{\ b} - \frac{R}{2}\,\delta^a_{\ b} + \Lambda\delta^a_{\ b} = 8\pi G\,\langle T^a_{\ b}\rangle_R\,. \tag{20}$$

and necessarily influence the background spacetime (1). The assumption that $T^a_{\ b} = 0$ on the horizon upon which the analytic continuation of coordinates in (4) critically depends would have to be re-evaluated, or discarded entirely, leading potentially to a very different interior solution.

Let us dispel at this point any worry that large quantum effects on the horizon violate coordinate invariance or the Equivalence Principle. It is only some classical notions about locality that are violated by such effects and in the same way that quantum mechanics itself usually violates them. The quantum effective action in a background field, be it electromagnetic or gravitational, is generally a non-local functional of the background. It is this non-locality which makes particle creation in electromagnetic fields or in curved spacetimes possible. For this same reason the stress tensor (19) derived from the one-loop quantum effective action depends on boundary conditions in the full spacetime, not simply local quantities such as the Riemann curvature (3), which are small on the horizon. Because of quantum mechanics matter behaves as waves and it is these virtual matter waves in the macroscopic quantum state, non-local and coherent on the horizon scale that lead to (19). In a *different* (Unruh) state it is exactly the same non-local wavelike nature of matter that leads to the Hawking particle creation process.

If there were but a single "vacuum" state of matter in a Schwarzschild background that showed evidence of large quantum effects near the horizon, it might be possible to argue that the single Boulware state $|B\rangle$ should be excluded as pathological. In fact, just the opposite is true. By considering the most important non-local terms in the quantum effective action, related to the conformal anomaly, it is possible to show that *most* states behave as in (19) [25, 26].

That the important quantum effects near horizons are associated with the conformal or trace anomaly is related to the fact that due to the kinematic blueshift (14) becoming greater than all other mass scales as the horizon is approached, all quantum fields, massless or not, become essentially conformal in the horizon region. Quantitatively, it has recently become possible to calculate these effects from the form of the quantum effective action due to the anomaly,

$$S_{anom} = \frac{b'}{2}\int d^4x\sqrt{-g}\left\{-(\Box\varphi)^2 + 2\left(R^{ab} - \frac{R}{3}g^{ab}\right)\nabla_a\varphi\nabla_b\varphi + \left(E - \frac{2}{3}\Box R\right)\varphi\right\} \tag{21}$$

$$+b\int d^4x\sqrt{-g}\left\{-(\Box\varphi)(\Box\psi) + 2\left(R^{ab} - \frac{R}{3}g^{ab}\right)\nabla_a\varphi\nabla_b\psi + \frac{1}{2}C_{abcd}C^{abcd}\varphi + \frac{1}{2}\left(E - \frac{2}{3}\Box R\right)\psi\right\}$$

in terms of curvature tensors and two new scalar field degrees of freedom φ and ψ [25].The form of this effective action is determined by general principles of covariance and the form of the trace anomaly, with only the coefficients b and b' dependent upon the number and spin of the underlying quantum fields [27]. The action (21) is clearly a spacetime scalar and therefore in that sense it is certainly consistent with the Equivalence Principle.

If (21) is varied with respect to the scalar fields φ and ψ, one obtains their (linear) eqs. of motion. Solving these linear differential eqs. formally in terms of the Green's functions of the relevant differential operator and inserting this formal solution for φ and ψ back into (21) allows one to rewrite the effective action in its original non-local but still covariant form. This non-local effective action of the trace anomaly in four dimensions is the analog of the two-dimensional anomaly effective action [28] which can also be put into a local form with the addition of a new local scalar degree of freedom [29].

The eqs. for the scalar fields $\varphi(r)$ and $\psi(r)$ derived from (21) can also be solved in the static spherically symmetric Schwarzschild background, and the results substituted into the the stress tensor also obtained from (21) by varying with respect to the metric. This stress tensor incorporating the one-loop quantum effects of the anomaly generally exhibits divergent behavior at the horizon, just as in (19), for a wide variety of states [25]. This verifies explictly the origin of such divergent behavior as the one-loop non-local quantum effective action of the anomaly, and the generic behavior of the divergences of the stress tensor near $r = r_M$ due to quantum effects. Let us remark also that it is by no means necessary for the state to have truly divergent behavior on the horizon as in (19) in order to have large quantum effects there. Any state in which $G\langle T^a_{\ b}\rangle_R$ becomes of order of the (small) classical curvature (23) near the horizon is already enough to produce significant backreaction effects on the geometry. To see this note that Einstein's equations (20) in the general static spherically symmetric metric (1) imply

$$\frac{1}{r}\frac{d}{dr}\left(\frac{h}{f}\right) = -8\pi G\,\frac{\rho + p}{f}\,, \tag{22}$$

while the Riemann curvature tensor has the component

$$R^{tr}_{\ \ tr} = \frac{h}{4}\left(\frac{f'^2}{f^2} - \frac{2f''}{f} - \frac{h'}{h}\frac{f'}{f}\right) \tag{23}$$

(where primes denote differentiation with respect to r). In a pure vacuum, $h(r) = f(r)$ and (23) becomes $-f''/2$ and hence remains finite. But from (22) if $\rho + p > 0$ in the horizon region where f or h vanishes, in general $h \neq f$ and the cancellation of the singularities in (23) will not occur. In that case even the Riemann curvature invariant can become large. The horizon region defined by one or more metric functions h or f approaching zero is extremely sensitive to matter source perturbations, and even "small" quantum vacuum polarization effects can have relatively large effects on the spacetime geometry in this region. The anomaly scalar fields $\varphi(r)$ or $\psi(r)$

behaving like $\ln f(r) \to \infty$ shows that non-local quantum effects can easily produce such large perturbations near the horizon.

In flat space one can understand the large quantum effects of the anomaly on the light cone (hence a null surface such as an event horizon of a black hole) as the result of the massless poles of the conformal φ and ψ scalar propagators, which describe correlated two-particle states of the underlying quantum fields at threshold [30]. Two-particle pairing at the Fermi surface is common in many-body physics, where it is responsible for superfluidity and superconductivity at low temperatures. In these non-relativistic quantum systems as the temperature is lowered or the pressure raised there is a phase transition to a condensate, in which a macroscopically large number of paired fermions or bosons collapse into a single coherent quantum state. In relativistic quantum field theory, anomalies (both axial and conformal) seem to be the only cases where two-particle pairing to form a spin zero bosonic state at the kinematic threshold $k^2 = 0$ of the Dirac sea (for a fermion pair) can occur consistent with Lorentz or general coordinate invariance. In this case, of course, it is the vacuum itself which becomes unstable. The strong effects of these anomalous quantum two-particle correlations near black hole horizons invites us then to consider the possibility that a phase transition in spacetime itself is possible, in which the black hole interior is replaced by a kind of low temperature gravitational vacuum condensate.

4. Gravitational Vacuum Condensate Stars

The region near the horizon $r = r_M$ in which quantum effects can be important is very narrow. The spatial extent of the region can be estimated by asking how close to r_M must r be for the quantum stresses (19) to give curvature corrections through Einstein's eqs. (20) of the same order as the classical Riemann curvature (23). This gives the estimate

$$|f(r_M \pm \delta r)| = \left|1 - \frac{r_M}{r_M \pm \delta r}\right| \simeq \frac{\delta r}{r_M} \equiv \epsilon \simeq \frac{M_{Pl}}{M} \simeq 10^{-38}\,\frac{M_\odot}{M}\,, \tag{24}$$

in solar mass units. Thus $\delta r \simeq L_{Pl}$. However because the physical distance is given by the line element (1) this corresponds to a physical distance

$$\ell \simeq \frac{\delta r}{f^{\frac{1}{2}}} \simeq \sqrt{L_{Pl}\, r_M} \simeq 3 \times 10^{-14}\sqrt{\frac{M}{M_\odot}}\ \text{cm}\,, \tag{25}$$

of the order of the diameter of an atomic nucleus for M of a few solar masses, which is tiny on astrophysical scales but still much larger than the microscopic Planck length. Well away from this boundary layer $f(r) = (1 - r_M/r) \to 1$, and quantum effects are of order ϵ^2, and hence completely negligible. Thus we expect the general conditions for an effective field theory treatment to hold everywhere except within this thin boundary layer of thickness ℓ, where the quantum effects are relatively large. Mathematically, this situation is similar to those encountered in hydrodynamic shock waves where flows can be accurately described by the continuum Navier-Stokes eqs. except for a thin boundary layer inside the shock front of order of the mean free path of the molecular constituents of the fluid where the continuum approximation breaks down. A second example from fluid mechanics of a boundary layer is the Prandtl layer of the flow of a fluid around an obstacle, such as a ship's bow or an airplane's wing, which is responsible for the full macroscopic drag on the body. In both of these situations one can obtain an accurate description in the continuum fluid description, provided one supplies higher derivative terms usually neglected and/or proper boundary conditions derived from conservation laws across the layer.

Away from the boundary layer on either side of it Einstein's eqs. (20) apply with negligible quantum effects. However, because the layer is a phase boundary, in which the quantum vacuum itself has changed its character, the parameters of the low energy description in terms of classical

General Relativity need not have the same values on opposite sides of the phase boundary. On general grounds the "latent heat" of the ground state of a system can change in a phase transition, and for the vacuum itself this is the cosmological term Λ in Einstein's eqs. (20). Moreover it has been shown that the fluctuations of the anomaly scalar degrees of freedom in (21) which are responsible for the large quantum effects near the horizon also give rise to the cosmological term becoming *dynamical*, varying in space and time [29, 31]. A non-zero Λ can also be interpreted thermodynamically as a zero entropy density condensate because of the Gibbs' relation

$$\rho + p = sT + \mu n\,. \tag{26}$$

Since $\Lambda > 0$ corresponds to a positive vacuum energy density with negative pressure $p = -\rho$, the entropy density s must vanish if there is no conserved number density or associated chemical potential $\mu = 0$. We are familiar with phenomenological low energy effective field theories such as the electroweak theory of the Standard Model where the vacuum energy and effective Λ changes due to the Higgs field developing a non-zero condensate $\langle\Phi\rangle \neq 0$ which is a pure quantum state with zero entropy, so that $W = 1$ and $S = 0$ in (13). In the bag model of hadrons there is a vacuum energy, the bag constant, associated with the rapid crossover (not a true phase transition) between the interior of a hadron where quarks and gluons are approximately free and the exterior where they cannot propagate. Note also that in all these cases the vacuum condensate eq. of state $p = -\rho$ with $\rho > 0$ *violates* the strong energy condition $\rho + 3p \geq 0$ which is used to prove the classical black hole singularity theorems [9].

Thus it is reasonable to assume that the effective value of Λ will change at the quantum phase transition and become non-zero in the interior of the quantum boundary layer at $r \simeq r_M$. Einstein's eqs. again apply within with this new value of Λ describing the interior vacuum condensate state. Just as the vacuum Einstein's eqs. possess a static, spherically symmetric solution for an isolated mass, namely the Schwarzschild solution (2), they possess a static, spherically symmetric vacuum solution for positive cosmological term, namely the de Sitter solution, (1) with [32]

$$f(r) = h(r) = 1 - \frac{r^2}{r_H^2}\,, \qquad \text{with} \qquad r_H \equiv \frac{1}{H} = \sqrt{\frac{3}{\Lambda}}\,. \tag{27}$$

Just as the exterior Schwarzschild geometry will attract all classical matter into its horizon at $r = r_M$, the de Sitter geometry will sweep out all classical matter outwards towards its horizon at $r = r_H$. Thus a strictly static solution of Einstein's eqs. is possible if both the exterior and condensate interior are free of other matter or stress energy sources. As the de Sitter horizon is approached from the interior there are again quantum effects of the anomaly action and stress tensor in generic quantum states that grow like $f^{-2}(r) = (1 - H^2r^2)^{-2}$ [25, 33]. Hence a globally static solution is possible only if the interior and exterior geometries are matched in their mutual near horizon regions: $r_H \simeq r_M$. This fixes the interior vacuum condensate energy density to

$$\rho_{cond} = \frac{3c^4H^2}{8\pi G} = \frac{3\,c^8}{32\pi G^3 M^2}\,, \tag{28}$$

so that the total mass of the interior condensate is

$$\frac{4\pi r_H^3}{3}\,\frac{\rho_{cond}}{c^2} = M\,, \tag{29}$$

the total mass of the Schwarzschild solution in the exterior. This static matching of the interior de Sitter vacuum condensate (27) to the exterior Schwarzschild geometry (2) is what we have called a gravitational vacuum condensate star or *gravastar* [22, 34].

A key point about matching an interior de Sitter solution to an exterior Schwarzschild solution is that this matching cannot be achieved across a null surface such as an horizon because such a null surface can contain radiation leading to arbitrary discontinuities of the metric functions $f(r)$ and $h(r)$. However, once the null horizon singularity of these functions is regulated by a *finite* boundary layer of order ℓ in (25), then the entire boundary layer is a *timelike* tube. In ref. [22] this was done by application of the Israel junction conditions, assuming a $p = \rho$ relativistic fluid eq. of state within the boundary layer [35]. A more accurate matching is now possible with the effective action (21) and stress tensor derived from it. The stability question can then be re-examined through the Lagrangian variational principle the effective action provides.

The gravastar proposal for the final equilibrium quantum ground state for complete gravitational collapse eliminates all of the paradoxes associated with black holes. Since the interior is described by the static patch of de Sitter spacetime (27), there is evidently no singularity and no place for information to disappear into. Quantum unitarity is preserved. Since the interior is a pure quantum vacuum state which can be regarded as a macroscopic condensate, it has zero entropy. Any entropy is associated with the fluctuations present in the quantum boundary layer at $r \simeq r_M = r_H$, and may be estimated to be of order [22]

$$S \sim \frac{\rho}{T}\ell A \sim k_B \left(\frac{M}{M_{Pl}}\right)^{\frac{3}{2}} \simeq 10^{57}\, k_B \left(\frac{M}{M_\odot}\right)^{\frac{3}{2}}\,, \tag{30}$$

which is within an order of magnitude or so of a typical stellar progenitor. The $M^{\frac{3}{2}}$ dependence on mass is consistent with estimates from ordinary thermodynamics for a relativistic star (see e.g. [36]). Hence there is no enormous mismatch of entropies to be explained as for (17), and no information paradox. There is also no Hawking radiation from a gravastar, the black hole event horizon having been replaced with a boundary layer or thin shell of finite thickness (25) and finite energy density. Although this energy density is enormous by terrestrial standards, it is of the same order as that of a neutron star, *i.e.* nuclear energy density, and far below Planck scale energy densities. Hence there is no trans-Planckian problem and the effective field theory of Einstein's equations (20) augmented by the quantum effects of the anomaly action (21) should be sufficient to describe the main macroscopic features of gravastars. The final state of a completely cooled gravastar as its temperature approaches absolute zero is a quantum stable ground state of zero entropy similar to a finite sized Bose-Einstein condensate of cold atoms at absolute zero, trapped by its own self-gravity.

As astrophysical objects gravastars would be cold, dark and compact, and therefore mimic classical black holes in almost every respect observationally. In complete isolation it would be virtually impossible to tell a gravastar from a black hole. The accretion of matter onto a gravastar would be similar to that of a black hole except at the very last stage when it interacts with the boundary layer. At the boundary layer, matter can be converted into pure energy and baryon and lepton number is likely also violated by anomalous processes in the Standard Model. Extensions to more realistic solutions with rotation, magnetic fields and taking into account Standard Model matter interactions through the anomaly effective action (21) are possible. Although the release of energy by matter striking the boundary layer can be very large, since the boundary layer thickness (25) is so small, any radiation from a *non-rotating* gravastar would be extremely redshifted out of γ-rays or X-rays to much lower energies. Thus the standard arguments for the absence of a surface [37] do not apply [38]. Gravitational waves should be generated at the natural oscillation frequencies of the physical thin shell, and perhaps would provide the cleanest evidence of the surface distinguishing a gravastar from a black hole. Once rotation, particularly rapid rotation is considered, the situation may be quite different, and the efficient conversion of matter into energy could conceivably provide the central powerhouse for Gamma Ray Bursters and some of the most energetic sources observed in the universe.

5. A New Model for Cosmological Vacuum Energy

The scalar φ and ψ fields couple to the metric particularly strongly on the cosmological Hubble horizon scale where they have their largest effect. Thus the fluctuations of the scalar degrees of freedom determined by the anomaly may lead to a phase transition to precisely the conformally invariant phase of gravity described by the fixed point $\Lambda = 0$ [29, 31] in the near vicinity of the horizon. Since the interior condensate phase has the eq. of state $p = -\rho$, exactly that of the cosmological dark energy observations suggest pervade our universe [39], and the gravastar solution automatically fixes the vacuum condensate energy density ρ_{cond} in terms of its size by (28), an interesting possibility raised by this solution is that the observable universe itself could be the interior of a gravastar. In that case the observed cosmological dark energy of our universe some 72% of the critical energy density ρ_{crit} defined by the present value of the Hubble parameter H_0 would be identified with the condensate energy density

$$\rho_{cond} \simeq (0.72)\,\rho_{crit} = (0.72)\,\frac{3c^4 H_0^2}{8\pi G} \simeq 6.5 \times 10^{-9}\ \mathrm{erg/cm^3}\,. \tag{31}$$

If our location is far from the bubble wall, it would be difficult to distinguish the cosmological model of a spherical bubble of vacuum energy condensate from the standard Friedmann-Robertson-Walker (FRW) model, although on the cosmological Hubble horizon scale $H_0^{-1} \simeq 4.2$ Gpc and of course globally it would be dramatically different, with a preferred origin and a physical surface. At the horizon the quantum fluctuations of the scalar degrees of freedom described by the anomaly action (21) would be large. In ref. [33] it was pointed out that these anomaly scalars couple to the metric near the horizon and generate fluctuations that are similar to those in inflationary models. Thus it may be that these fluctuations emanating from the horizon are what we observe in the Cosmic Microwave Background (CMB) anisotropies, without any need to introduce an inflaton. Moreover, since the fluctuations of the anomaly scalars lead to a conformally invariant phase of gravity [29, 31], the fluctuations should have a spectrum and statistics consistent with conformal invariance, including in their non-Gaussian features. Thus, one way of testing this cosmological model would be to detect the non-Gaussian bi-spectrum of the CMB with the shape predicted by conformal invariance [40, 41, 42].

The new cosmological model is then a kind of bubble or 'bag' of vacuum energy, containing these gravitationally coupled degrees of freedom, which is well described as approximately a de Sitter universe of finite size of the order of the Hubble radius. Like an ordinary bubble of gas, the pressure inside is determined by the size of the bubble and its surface tension, not any UV cutoff, and the interior pressure adjusts itself dynamically to the boundary conditions at the Hubble scale. In the effective field theory of Einstein gravity augmented by the anomaly action (21), the vacuum energy of infinite flat space is identically zero, and the residual small ρ_{cond} of (31) is a boundary effect of a macroscopic condensate. Then the essential cosmological problem becomes to explain why $\Omega_{cond} = \rho_{cond}/\rho_{crit} \simeq 0.72$ rather than unity, and how the vacuum energy and pressure adjusts itself dynamically at the boundary to maintain $\Omega_{cond} \approx 1$ over time.

Clearly additional details of the gravastar model incorporating matter and radiation need to be worked out and it then must pass the many successful tests of the usual FRW big bang model before it can be considered a realistic alternative cosmology. The essentially automatic incorporation of dark energy with the approximately correct value and solution of the cosmological dark energy problem would seem to make the model worth pursuing further.

Acknowledgments
I gratefully acknowledge all my collaborators on whose work this talk was based, and especially P. O. Mazur, with whom the hypothesis of gravitational vacuum condensate stars was developed. I also thank José Luis Jaramillo, Carlos Barceló, and all the organizers of the Spanish Relativity Meeting ERE2010 at which this talk was presented.

References

[1] Schwarzschild K 1916 *Sitzber. Preuss. Akad. Wiss. Berlin, Math.-Phys.* 189
[2] Einstein A 1939 *Ann. Math.* **40** 922
[3] Misner C W, Thorne K S and Wheeler J A 1973 *Gravitation* (San Francisco: Freeman)
[4] Weinberg S 1972 *Gravitation and Cosmology: Principles and Applications of the General Theory of Relativity* (New York: Wiley).
[5] Kruskal M D 1960 *Phys. Rev.* **119** 1743
Szekeres G 1960 *Publ. Mat. Debrecen.* **7** 285
[6] Reissner H 1916 *Ann. Phys. (Ger.)* **50** 106
Nordström G 1918 *Proc. Kon. Ned. Akad.* **20** 1238
[7] Kerr R P 1963 *Phys. Rev. Lett.* **11** 237
[8] Boyer R H and Lindquist R W 1967 *J. Math. Phys.* **8** 265
Carter B 1968 *Phys. Rev.* **174** 1559
[9] Hawking S W and Ellis G F R 1973 *The Large Scale Structure of Space-Time* (Cambridge: Cambridge Univ. Press)
[10] Christodoulou D 1970 *Phys. Rev. Lett.* **25** 1596
Christodoulou D and Ruffini R 1971 *Phys. Rev.* D **4** 3552
[11] Smarr L 1973 *Phys. Rev. Lett.* **30** 71
[12] Bardeen J M, Carter B and Hawking S W 1973 *Comm. Math. Phys.* **31** 161
[13] Hawking S W 1971 *Phys. Rev. Lett.* **26** 1344
[14] Bekenstein J D 1972 *N. Cimento Lett.* **4** 737; 1973 *Phys. Rev.* D **7** 2333; 1974 *Phys. Rev.* D **9** 3292
[15] Hawking S W 1974 *Nature* **248** 30; 1975 *Comm. Math. Phys.* **43** 199
[16] Hawking S W 1976 *Phys. Rev.* D **13** 191
[17] See Mazur P O 1996 *Acta Phys. Polon.* B **27** 1849 for further discussion of this point
[18] Lynden-Bell D 1967 *Mon. Not. R. Astron. Soc.* **136** 101;
Lynden-Bell D and Lynden-Bell R M 1977 *Mon. Not. R. Astron. Soc.* **181** 405
[19] Padmanabhan T 1990 *Phys. Rep.* **188** 285
[20] Gross D J, Perry M J, and Yaffe L G 1982 *Phys. Rev. D* **25** 330;
York J W 1986 *Phys. Rev.* D **33** 2092
[21] Hawking S W 1982 *Comm. Math. Phys.* **87** 395
[22] Mazur P O and Mottola E 2001 *Preprint* arXiv: gr-qc/0109035; 2004 *Proc. Nat. Acad. Sci.* **101** 9545
[23] Boulware D G 1975 *Phys. Rev.* D **11** 1404; 1976 *Phys. Rev.* D **13** 2169
[24] Christensen S M and Fulling S A 1977 *Phys. Rev.* D **15** 2088
[25] Mottola E and Vaulin R 2006 *Phys. Rev.* D **74** 064004
[26] For further discussion and extensive references see Mottola E 2010 *Acta Phys. Polon.* B **41** 2031
[27] Duff M 1977 *Nucl. Phys.* B **125** 334; 1984 *Nucl. Phys.* B **B234** 269
[28] Polyakov A M 1981 *Phys. Lett.* **B103** 207
[29] Antoniadis I, Mazur P O and Mottola E 2007 *New J. Phys.* **9** 11
[30] Giannotti M and Mottola E 2009 *Phys. Rev.* D **79** 045014
[31] Antoniadis I and Mottola E 1992 *Phys. Rev.* D **45** 2013
[32] de Sitter W 1917 *Mon. Not. R. Astron. Soc* **78** 3
Eddington A S 1924 *The Mathematical Theory of Relativity* (Cambridge: Cambridge Univ. Press)
[33] Anderson P R, Molina-Páris C and Mottola E 2009 *Phys. Rev. D* **80** 084005
[34] Similar ideas based on analogies with condensed matter systems have been expressed in Chapline G, Hohlfield E, Laughlin R B and Santiago D 2001 *Phil. Mag.* **B81** 235; 2003 *Int. Jour. Mod. Phys.* **A18**, 3587
[35] Israel W 1966 *N. Cimento* B **B44** 1; *N. Cimento* B **48** 463
[36] Zel'dovich Ya B and Novikov I D 1971 *Relativistic Astrophysics* Vol 1 (Chicago:Univ. of Chicago Press) [1996 *Stars and Relativity* (Mineola, NY: Dover)].
[37] Narayan R and Heyl J S 2002 *Astrophys.J.* **574** L139
Yuan Y F, Narayan R, and Rees M J 2004 *Astrophys.J.* **606** 1112
McClintock J E, Narayan R, and Rybicki G B 2004, *Astrophys.J.* **615** 402
[38] Abramowicz M A, Kluzniak W and Lasota J P 2002 *Astron. Astroph.*, **396** L31
[39] Riess A G *et. al.* 1998 *Astron. J.* **116** 1009 2004 *Astron. J.* **607** 665
Perlmutter S *et. al.* 1999 *Astrophys. J.* **517** 565
Tonry J L *et. al.* 2003 *Astrophys. J.* **594** 1
[40] Antoniadis I, Mazur P O and Mottola E 1997 *Phys. Rev. Lett.* **79** 14
[41] Mottola E in *2010 Cosmology: Proceedings of the XLVth Rencontres de Moriond* ed E Auge, J Dumarchez and J Tran Tranh Van (Vietnam: The Gioi Publishers) (*Preprint* arXiv:1103.1613)
[42] Antoniadis I, Mazur P O and Mottola E 2011 *Preprint* arXiv:1103.4164 submitted to *Phys. Rev.* D

Spanish Relativity Meeting (ERE 2010): Gravity as a Crossroad in Physics IOP Publishing
Journal of Physics: Conference Series **314** (2011) 012011 doi:10.1088/1742-6596/314/1/012011

The Frontiers of Observational Cosmology and the Confrontation with Theory

Malcolm Longair
Cavendish Laboratory, JJ Thomson Avenue, Cambridge CB3 0HE, UK
E-mail: msl1000@cam.ac.uk

Abstract. The current state of observational cosmology and the confrontation with theory is presented. The review is divided into the following sections:

- Basic observations on which the models are based.
- Testing the basic assumptions made in the construction of the standard cosmological models.
- Structure formation in the standard models
- Observational tests of the standard models - the confrontation with observation
- Basic problems and approaches to their solution
- Future challenges - the ESA EUCLID mission is given as an example.

1. Introduction

This brief summary of the current status of observational cosmology and the confrontation with theory is based upon the contents of my book *Galaxy Formation* to which reference should be made for many more details of the observations, their interpretation and the underlying astrophysical and cosmological theory (Longair 2008). For illustrative purposes, I use the following values of the cosmological parameters: $\Omega_0 = 0.3$, $\Omega_\Lambda = 0.7$, $h = H_0/100\,\mathrm{km\ s^{-1}\,Mpc^{-1}} = 0.7$. The dangerous bend sign indicates areas where care is needed.

2. Basic observations

Nowadays astrophysical cosmology begins with observations of the Cosmic Microwave Background Radiation as observed by the Cosmic Background Explorer (COBE) and the Wilkinson Microwave Anisotropy Probe (WMAP). The COBE observations showed that the spectrum of the background radiation is very precisely that of a black-body at temperature $T = 2.726$ K. A perfect dipole component is detected over the whole sky, corresponding to the motion of the Earth through a frame of reference in which the radiation would be perfectly isotropic. The COBE map of the whole sky had angular resolution of 7° while the more recent WMAP observations had a higher angular resolution of 0.3°. Both maps display the remarkable *isotopy* of the background cosmological signal over the whole sky once the Galactic foreground emissions have been subtracted. The large-scale distribution of radiation over the sky is isotropic to better than one part in 10^5 over the whole sky. Fluctuations in intensity are observed at about the level $\Delta T/T \sim 10^{-5}$ and these provide key information about cosmological parameters and the development of structure in the Universe. It is wholly convincing that this background radiation originated at a last scattering surface at a redshift $z \approx 1000$, or scale factor $a \approx 10^{-3}$, at the epoch of recombination of the primordial plasma.

The next issue is the homogeneity of the Universe. This has been established by large surveys of galaxies, starting with the local distribution determined by Geller and Huchra. Their famous picture shows that locally the galaxy distribution is highly inhomogeneous with huge walls and holes, or voids, on physical scales up to about 50–100 Mpc. The structure is 'sponge-like', the material of the sponge corresponding to the location of the galaxies. The important feature of this structure is that the material of the sponge is continuously connected as is the distribution of holes – this provides evidence for the gaussianity of the primordial fluctuations from which the structure formed. These structures have now been extended to five times greater distances at the present epoch by the 2dF and SDSS surveys, each containing over 200,000 galaxies. These surveys show the same type of cellular structure seen nearby and statistically, although the distribution of galaxies is highly irregular, the degree of irregularity is found to be the same at the present epoch out to redshifts $z \sim 0.25$. The power spectrum of the irregularities is of power-law form, $n(r) = n_0[1 + \xi(r)]$ with $\xi(r) \propto r^{-1.8}$ on scales up to about 10 Mpc; on larger scales, the amplitude of the power spectrum decreases more rapidly than this.

For future high precision cosmological studies, it should be noted that one consequence of the presence of this structure is that we observe the distant Universe through a distorting gravitational screen. These corrections are very small at the moment, but many of the future cosmological tests require exquisite control of all the uncertainties in the observations and then these distorting effects should not be ignored.

Finally, we need the redshift-distance relation for galaxies, the *Hubble diagram*. Suffice to say that all classes of galaxy follow the same Hubble's law $v = H_0 r$, where H_0 is Hubble's constant. Combined with the observed isotropy and homogeneity of the Universe, this means that the Universe as a whole is expanding uniformly.

3. Basic Assumptions

The standard models are based upon two assumptions and a theory of gravity:

- The *Cosmological Principle* states that we are not located at any special location in the Universe. Combined with the observations that the Universe is isotropic, homogeneous and uniformly expanding on a large scale, this leads to the *Robertson-Walker metric.*
- *Weyl's Postulate* is the statement that the world lines of particles meet at a singular point in the finite or infinite past. This solves the clock synchronisation problem and means that there is a unique world-line passing through every point in space-time.
- *General Relativity* enables us to relate the energy-momentum tensor T_{ab} of the cosmic fluid to the geometrical properties of space-time.

The Robertson-Walker metric can be written in the following form:

$$\mathrm{d}s^2 = \mathrm{d}t^2 - \frac{a^2(t)}{c^2}[\mathrm{d}r^2 + \Re^2 \sin^2(r/\Re)(\mathrm{d}\theta^2 + \sin^2\theta\,\mathrm{d}\phi^2)] \ . \tag{1}$$

t is *cosmic time* as measured by a clock carried by a fundamental observer and r is the *comoving radial distance coordinate* which is fixed to a galaxy for all time. The metric contains one unknown function $a(t)$, the *scale factor*, and the constant $\Re$, the radius of curvature of the geometry of the Universe at the present epoch. $a(t)$ is normalised to unity at the present epoch. It follows from this metric that the redshift $z = (\lambda_{\rm obs} - \lambda_{\rm em}/\lambda_{\rm em})$ is directly related to the scale-factor through the relation $a(t) = (1+z)^{-1}$, where $\lambda_{\rm em}$ is the emitted wavelength and $\lambda_{\rm obs}$ the observed wavelength.

A key test of the Robertson-Walker metric is that the same formula which describes the redshift of spectral lines should also apply to time intervals in the emitted and received reference frames. This test has now been extended to cosmological redshifts $z \approx 0.8$ using Type 1a supernova which have remarkably similar light curves, particularly when account is taken of the observed luminosity-width correlation. A clear time dilation effect has been observed which is exactly proportional to $(1 + z)$, as predicted by the Robertson-Walker metric.

A second test concenrs the radiation temperature of the Cosmic Microwave Background Radiation as a function of redshift. In the case of black-body radiation, the form of the spectrum is preserved as the Universe expands, but the radiation temperature $T_{\rm r}$ changes with redshift as $T_{\rm r} = T_0(1 + z)$. The temperature of the background radiation can be inferred from observations of fine structure lines in absorption lie systems in the spectra of distant quasars. Photons of the background radiation excite the fine-structure levels of the ground state of neutral carbon (CI) atoms and the relative strengths of the absorption lines originating from the ground and first excited states are determined by the temperature of the background radiation. This experiment requires very high quality absorption spectra of large redshift quasars. For the quasars Q1331+170 and QSO 0013-004, absorption systems at redshifts $z \sim 1.8$, the inferred radiation temperatures are about $7 - 8$ K, consistent with the expected relation. More recently, in the spectrum of the quasar PSS J1443+2724, there is an absorption system at redshift $z_{\rm abs} = 4.224$. The expected radiation temperature of the background radiation at this redshift is 14.2 K is consistent with the observed excitation of the ground and first fine-structure excited states of CI.

How good is general relativity? The present status is splendidly reviewed by Will (2006) in his on-line article. There is no evidence for departures from Einstein's Equivalence Principle at a level better than about one part in 10^{12}. Concerning the deviations from general relativity, the four classic tests show consistency with the standard theory in terms of the Parameterised Post-Newtonian (PPN) parameters β and γ. Specifically,

- The gravitational redshift of electromagnetic waves in a gravitational field has been measured using hydrogen masers in rocket payloads. These confirm the prediction of general relativity at the level of about 5 parts in 10^5.
- The Advance of the Perihelion of Mercury. Continued observations of Mercury by radar ranging have established the advance of the perihelion of its orbit to about 0.1% precision with the result $d\omega/dt = 42.98(1 \pm 0.001)$ arcsec per century. General relativity predicts a value of $d\omega/dt = 42.98$ arcsec per century.
- Gravitational deflection of electromagnetic waves by the Sun has been measured by Very Long Baseline Interferometry and the values found are $(1 + \gamma)/2 = 0.99992 \pm 0.00023$.
- The time delay of electromagnetic waves propagating through a varying gravitational potential, the Shapiro test, has been carried out by the Cassini spacecraft which found a time-delay corresponding to $(\gamma - 1) = (2.12.3) \times 10^{-5}$. Hence the coefficient $(1 + \gamma)/2$ must be within at most 0.0012 percent of unity.

The binary pulsar systems have proved to be very important in testing various aspects of general relativity. In binary pulsars such as PSR 1913+16, the pulsar is one of a pair of neutron stars in binary orbits about their common centre of mass. The radio pulses are assumed to be due to beams of radio emission from the poles of the magnetic field distribution. Many parameters of the binary orbit and the masses of the neutron stars can be measured with very high precision by accurate timing measurements. For this binary pulsar, the kinematics of the orbits are entirely consistent with general relativity and in addition the binary pulsar emits gravitational radiation leading to a speeding up of the stars in the binary orbit. The observed speeding up of the pulsars is entirely consistent with the predicted rate of gravitational radiation of the binary system. Even more impressive is the double pulsar system PSR J0737-3039 in which both neutron stars are pulsars. In addition, the period of the binary is only 2.4 hours and so the system enables general relativity to be tested with even higher precision. After only 2.5

years, the double pulsar enabled general relativity to be tested at a level similar to that of PSR 1913+16. In addition, the eclipse phenomenon observed in this double pulsar enabled the relativistic spin precession of PSR J0737-3039B to be measured with high statistical significance. The spin precession rate was found to be in agreement with general relativity.

Various solar system, astrophysical and cosmological tests have been made to find out if the gravitational constant has varied over cosmological time-scales. For binary pulsar data, the bounds are dependent upon the theory of gravity in the strong-field regime and on the neutron star equation of state. The big-bang nucleosynthesis bounds assume specific forms for time dependence of the gravitational constant. Lunar laser ranging provides the best limit of $\dot{G}/G \leq (4 \pm 9) \times 10^{-13}$ and so there can have been little variation in the value of the gravitational constant over the last 10^{10} years.

There has been some debate as to whether or not the fine-structure constant α has changed very slightly with cosmic epoch from observations of fine-structure lines in large redshift absorption line systems in quasars. The Australian observers reported a small decrease in the value of α, $\Delta\alpha/\alpha \approx 0.5 \times 10^{-5}$ at redshifts $1 - 2.5$ compared with the value at the present epoch. The ESO observers found little evidence for such a change.

4. The Growth of Small Density Perturbations

Within the background of the standard cosmological models of general relativity, a key role is played by the development of small perturbations with cosmic epoch. Many of the issues can be understood from the simplest solutions for a non-relativistic fluid on scales less than the cosmological horizon scale. The differential equations which come out of a simple perturbation analysis are:

$$\frac{\mathrm{d}\Delta}{\mathrm{d}t} = -\nabla \cdot \delta \boldsymbol{v} \ , \qquad (2)$$

$$\frac{\mathrm{d}^2\Delta}{\mathrm{d}\tau^2} + \left(\frac{\dot{a}}{a}\right)\frac{\mathrm{d}\Delta}{\mathrm{d}\tau} = \Delta(4\pi G \varrho_0 a^2 - k_\mathrm{c}^2 c_\mathrm{s}^2) \ , \qquad (3)$$

where $\Delta = \delta\varrho/\varrho_0$ is the *density contrast* and $c_\mathrm{s}^2 = \partial p/\partial\varrho$ is the adiabatic sound speed; k_c is the comoving wave number. Equation (2) relates the rate at which the density contrast develops to the peculiar velocity $\delta\boldsymbol{v}$ associated with the collapse of the perturbation. Equation (3) results in a dispersion relation for the growth rate of the small perturbations. The solutions correspond to the Jeans' instability in an expanding medium. Two limiting solutions are useful. In the critical Einstein-de Sitter model ($\Omega_0 = 1; \Omega_\Lambda = 0$), the perturbation grows linearly with scale factor, $\Delta \propto a = (1+z)^{-1}$. In the empty model ($\Omega_0 = 0; \Omega_\Lambda = 0$), there is no growth of the perturbations $\Delta =$ constant. In the more general case, the evolution of the density contrast depends upon both Ω_0 and Ω_Λ and, if the evolution of Δ with redshift could be measured observationally, this would provide an important way of determining cosmological parameters[1].

These solutions illuminate the origin of the basic problem of structure formation. Throughout the matter-dominated era, the growth rate of perturbations on large physical scales is $\Delta \propto a = (1+z)^{-1}$. Since galaxies and astronomers certainly exist at the present day $z = 0$, it follows that $\Delta \geq 1$ at $z = 0$ and so, at the last scattering surface, $z \approx 1,000$, fluctuations must have been present with amplitude at least $\Delta = 10^{-3}$. Therefore,

- The slow growth of density perturbations is the source of a fundamental problem in understanding the origin of galaxies – large-scale structures did not condense out of the primordial plasma by exponential growth of infinitesimal statistical perturbations.

[1] Examples of the evolution of the density contrast Δ with cosmic epoch are given in my book *Galaxy Formation* (2008).

Table 1.

Parameter	Definition	Standard value	Estimated value
h	Hubble's constant		0.72
$\omega_B = \Omega_B h^2$	baryon density parameter		0.0223
$\omega_D = \Omega_D h^2$	cold dark matter density parameter		0.127
Ω_Λ	dark energy density parameter		0.72
$\Omega_k = \Omega_\Lambda + \Omega_0$	space curvature		≈ 1
n_s	scalar spectral index	$n_s = 1$	≈ 1
τ	reionisation optical depth		0.09
σ_8	density variance in 8 Mpc spheres		0.74
w	dark energy equation of state	$w = -1$	
$f_\nu = \Omega_\nu/\Omega_D$	massive neutrino fraction	$f_\nu = 0$	
N_ν	number of relativistic neutrino species	$N_\nu = 3$	
$\Delta^2_\mathcal{R}$	amplitude of curvature perturbations		
r	tensor-scalar ratio		
A_s	amplitude of scalar power-spectrum		
$\alpha = \mathrm{d}\ln K/\mathrm{d}\ln k$	running of scalar spectral index	$\alpha = 0$	
A_{SZ}	SZ marginalization factor		
b	bias factor		
z_s	weak lensing source redshift		

- Because of the slow development of the density perturbations, we have the opportunity of studying the formation of structure on the last scattering surface at a redshift $z \approx 1,000$.

The methodology is therefore to follow the evolution of an initial power-spectrum of density perturbations from the early Universe when it was radiation dominated, through the epoch of equality of radiation and matter energy densities to the epoch of recombination at $z \approx 1000$ at which temperature perturbations are imprinted on the Cosmic Microwave Background Radiation. Typically the numerical simulations begin with a power-law spectrum in the early Universe on scales from 10^5 to $10^{20}\, M_\odot$. The power-law power spectrum of the form $P(k) \propto k$ is known as a *Harrison-Zeldovich power spectrum.*

To cut a long story short, these computations concerning the predicted power-spectrum of the distribution of galaxies and of fluctuations in the Cosmic Microwave Background Radiation have now become one of the most powerful approaches for the determination of cosmological parameters.

5. The Values of the Cosmological Parameters

The remarkable result of the last ten years is that the computations described in the last paragraph above and the observed power spectra are in highly satisfactory agreement with a number of quite independent ways of estimating the cosmological parameters. In Table 1, a list of cosmological parameters is given with estimates of their values which are consistent with observations of the power spectrum of the Cosmic Microwave Background Radiation and its polarisation. The estimates of these parameters are now known to better than 10% and in many cases with much greater precision. In the lower half of the table a list of additional cosmological parameters is given and these can be constrained or determined by future ground- and space-based observations.

Specifically,

- *Supernovae of Type 1a* Type 1a supernovae have dominated methods for extending the redshift-distance relation to large redshifts. They are brightest supernovae and are observed to have remarkably standard properties, particularly when corrections are made for the observed luminosity-width relation. It should be noted that the precise mechanism of the supernova explosion is not yet established. Despite this concern, the empirical evidence is impressive. The ESSENCE project has the objective of measuring the redshifts and distances of about 200 supernovae. The Supernova Legacy Project aims to obtain distances for about 500 supernovae. The observations are consistent with a finite and positive value of the cosmological constant, the best fitting values being $\Omega_\Lambda = 0.73$; the total density parameter in baryonic and cold dark matter form $\Omega_0 = \Omega_D + \Omega_B \approx 0.27$.
- *The power spectrum of galaxies from the Sloan Digital Sky Survey and the AAO 2dF galaxy survey* There is evidence for residual acoustic oscillations in the galaxy power spectrum corresponding to the first maxima observed in the power-spectrum of the Cosmic Microwave Background Radiation – these are referred to as *Baryonic Acoustic Oscillations.*
- *Mass density of the Universe from the infall velocities of galaxies into large scale structures* The two-dimensional correlation function for galaxies selected from the 2dF Galaxy Redshift Survey in the radial and tangential directions on the sky shows a flattening due to the infall of galaxies into large-scale density perturbations. The inferred overall density parameter is $\Omega_0 = 0.25$.
- *Formation of the Light Elements by Primordial Nucleosynthesis* The predicted primordial abundances of the light elements is a sensitive a function of the present baryon-to-photon ratio $\eta = 10^{10} n_B/n_\gamma = 274\Omega_B h^2$. Steigman finds a best fitting value $\Omega_B h^2 = 0.022\ ^{+0.003}_{-0.002}$. This can be compared with the independent estimate from the power-spectrum of fluctuations of the Cosmic Microwave Background Radiation $\Omega_B h^2 = 0.0224 \pm 0.0009$. The only concern is that recent estimates of the observed abundances of the light elements may differ slightly from the predictions of standard model.
- *Cosmic time scale from the theory of stellar evolution and nucleocosmochronology* Ages of the oldest globular clusters from studies of their Hertzsprung-Russell diagrams for have resulted in ages of typically $T_0 = 15 \pm 2.4$ (stat) $^{+4}_{-1}$ (syst) Gy by Bolte and $T_0 = (11.5 \pm 1.3)$ Gy by Chaboyer. An interesting case is the star CS 22892-052 which has iron abundance 1000 times less than the solar value. A number of species never previously observed in such metal-poor stars were detected, as well a single line of thorium. A lower limit to the age of the star is (15.2 ± 3.7) Gy. Schramm found a lower limit to the age of the Galaxy of 9.6 Gy from nucleocosmochronology. His best estimates of the age of the Galaxy are somewhat model-dependent, but typically ages of about (12 - 14) Gy.
- *The value of Hubble's constant* The controversies of the 1970s and 1980s have been resolved thanks to a very large effort by many observers to improve knowledge of the distances to nearby galaxies. The final result of the HST key project was $H_0 = 72 \pm 7(1\sigma)$ km s^{-1} Mpc^{-1}. From gravitational time delays a value $H_0 = 68 \pm 6$(stat) ± 8(syst) km s^{-1} Mpc^{-1} was found. The Sunyaev-Zeldovich effect in X-ray clusters of galaxies gave the result $H_0 = 76.9 \pm 4$(stat) ± 9(syst) km s^{-1} Mpc^{-1}.
- *The statistics of gravitational lenses* In the Cosmic Lens All Sky Survey (CLASS), a very large sample of flat spectrum radio sources was imaged by the Very Large Array, Very Long Baseline Array and the MERLIN long-baseline interferometer. 13 sources were multiply imaged out of a total sample of 8958 radio sources. The numbers are consistent with flat world models with density parameter $\Omega_0 = 0.31 \pm 0.25 \pm 0.12$(syst). For a flat universe with dark energy equation of state of the form $p = w\rho c^2$, they found an upper limit $w < -0.55 \pm 0.15$ consistent with $w = -1$.

The power-spectrum and polarisation of the Cosmic Microwave Background Radiation on their own provide excellent estimates of the values of the parameters in the top half of Table 1 but it is striking how well these now agree with the above independent estimates. The caution sign means that there are other parameters may play a role in determining the exact form of the spectrum of the temperature fluctuations of the Cosmic Microwave Background Radiation and these should not be ignored. Here are a few examples of the parameters which might impact the estimates of the values of the cosmological parameters included in Table 1.

- The value of w is found to be close to -1, which would correspond precisely to Einstein's cosmological constant into the field equations of general relativity. If the value differed from unity, this would indicate new physics beyond the standard model of Einsteinian gravity.
- We now know that the known types of standard neutrinos have non-zero rest mass and so at some level their presence must influence the predicted spectrum of the fluctuations in the Cosmic Microwave Background Radiation.
- The presence of primordial gravitational waves would influence the spectrum of perturbations on large angular scales. Already, constraints can be placed upon the tensor to scalar ratio r of these waves. Primordial gravitational waves with spectrum similar to that of scalar perturbations on large angular scales is a prediction of the standard inflationary picture of the early Universe. They could be detected through their characteristic B-mode quadrupole signature at low levels in the Cosmic Microwave Background Radiation, but it is a very demanding observation. In addition, it has to be detected in the presence of a similar B-mode quadrupole signature induced by gravitational lensing by large scale structures.
- A further concern is whether or not it is safe to assume that the initial power-spectrum from which large-scale structures formed was indeed of power-law form. It will be argued that the fact that the simplest inflationary models predict a Harrison-Zeldovich spectrum is compelling evidence that that picture should be taken seriously, but of course, there may well be good reasons why the actual Universe is more complex than that simple picture.

These are all issues which can be addressed by observation, but they are very challenging programmes at and beyond the current limits of technology. I have no doubt, however, that these will be carried out if the appropriate investment is made.

6. The Basic Problems

Granted these successes, there remain a number of fundamental questions.

- Why is the Cosmic Microwave Background Radiation so isotropic? At earlier cosmological epochs, the particle horizon scale $r \sim ct$ encompassed less and less mass and so the scale over which particles could be causally connected became smaller and smaller. On the last scattering layer at redshift $z = 1000$, the distance $r = 3ct$ corresponds to an angle $\theta = 2°$ on the sky. Regions of the sky separated by greater angular distances could not have been in causal communication. Why then is the sky isotropic to one part in 10^5 on all greater angular scales?
- Why is the Universe geometrically flat, $\Omega_k = 0$? According to the standard world models, if the Universe were set up with a value of the density parameter differing even slightly from the critical value $\Omega_0 = 1$, it would diverge very rapidly from this value at later epochs.
- Why is the Universe baryon-asymmetric with $N_\gamma/N_\mathrm{B} = 1.6 \times 10^9$? If photons were neither created nor destroyed, this ratio is conserved as the Universe expands. In the early Universe, baryon-antibaryon pair production took place with the result that there must have been a very small asymmetry in the baryon-antibaryon ratio in the very early Universe if we are to end up with the correct photon-to-baryon ratio at the present day.

- Why do the cosmological parameters take the values $\Omega_\Lambda = 0.72$, $\Omega_D = 0.23$, $\Omega_B = 0.05$. Furthermore, why is Ω_Λ about 10^{120} smaller than the value expected from quantum field theory? It is a surprise that Ω_Λ and Ω_0 are of the same order of magnitude at the present epoch. The matter density evolves with redshift as a^{-3}, while the dark energy density is unchanging with cosmic epoch. Why then do we live at an epoch when they have more or less the same values?
- What is the origin of the initial power spectrum of perturbations and why is it of power-law form with power spectrum $P(k) \propto k$? The amplitudes of the density perturbations which led to the formation of galaxies had to be of finite amplitude, $\Delta = \delta\rho/\rho \sim 10^{-5}$, on a very wide range of mass scales when they came through the horizon.

I have suggested five possible solutions to these problems.

- That is just how the Universe is – the initial conditions were set up that way.
- There are only certain classes of Universe in which intelligent life could have evolved. This approach involves the *Anthropic Cosmological Principle* according to which the Universe is as it is because we are here to observe it.
- The inflationary scenario for the early Universe.
- Seek clues from particle physics and extrapolate to the earliest phases of the Universe.
- Something else we have not yet thought of, what I sometimes refer to as the *Donald Rumsfeld model* – the things we don't know we don't know. This would certainly involve new physical concepts.

There is some merit in each of these approaches. My suspicion is that the answer almost certainly lies in the fifth bullet, but in the mean time the most promising approach seems to be the inflationary picture which surprisingly has the potential to provide answers to at least some of the basic problems.

In the standard inflationary scenario, the Universe went through a period of exponential expansion in its very early phases, in a typical realisation the expansion lasting for about 100 e-folding times from the time the Universe was about 10^{-34} s old. This has two effects. First, particles are driven beyond their local particle horizons. If there were sufficient e-folding times, particles which were originally in causal contact in the very early Universe and so could homogenise the material of the Universe, were driven far beyond their causal horizons by the present epoch, solving the horizon problem. Second, the inflationary expansion straightens out the geometry of the Universe, however complex it may have been in the early Universe. The Universe is driven to flat geometry by the end of the inflationary era. The inflationary scenario thus solves the flatness and homogeneity problems.

The considerations of the last paragraph may be thought of as inflation without physics. To put more physics into the model, we appeal to the properties of scalar fields, similar to those which may be responsible for the present exponential expansion of the Universe under the influence of the cosmological constant. Their equation of state can be written

$$\varrho_\phi = \frac{1}{2}\dot{\phi}^2 + V(\phi) \qquad ; \qquad p_\phi = \frac{1}{2}\dot{\phi}^2 - V(\phi) \ . \tag{4}$$

If the first terms in $\dot{\phi}^2$, the kinetic energy terms, are very small compared with the potential energy terms, we obtain a negative pressure equation of state $p_\phi = -\varrho_\phi c^2$. The form of the potential $V(\phi)$ has to be chosen very carefully so that the exponential expansion can last long enough. This is known as slow-roll inflation so that the homogenising effect of the expansion can encompass the whole Universe. Despite a huge amount of theoretical effort, there is not yet a genuinely physical theory of inflation. It looks at first as though we have not made much progress since we are putting in by hand the sort of potential which results in the Universe we observe today.

The inflationary picture has, however, a remarkable property which was not expected when it was first proposed. We need to consider the quantum fluctuations which are necessarily present in the scalar fields during the inflationary era in the very early Universe. Amazingly, in the simplest model of inflation, the fluctuations in the scalar field result in exactly the Harrison-Zeldovich power-spectrum, $P(k) \propto k$. In my book *Galaxy Formation*, I demonstrate how this result can be derived using simple concepts from elementary quantum mechanics. As remarked by Liddle and Lyth in their book *Cosmological Inflation and Large Scale Structure*,

> 'Although introduced to resolve problems associated with the initial conditions needed for the Big Bang cosmology, inflation's lasting prominence is owed to a property discovered soon after its introduction. It provides a possible explanation for the initial inhomogeneities in the Universe that are believed to have led to all the structures we see, from the earliest objects formed to the clustering of galaxies to the observed irregularities in the microwave background.'

7. Some Future Challenges

The considerations of this review result in a large number of observational and theoretical challenges. Here is a selection of these.

- The PLANCK mission of ESA, now in its second year in orbit, will result in an order of magnitude improvement in the determination of cosmological parameters, thus providing yet more precise estimates of these and providing yet more stringent tests of what has become the standard cosmological model.
- Is the physics of the standard model accurately understood at the 1% level? If the objective is to improve yet further the accuracy with which we know the cosmological parameters, say to the 1% level, we need to understand the physics of the standard model to better than 1%. This is a serious theoretical challenge.
- Does the evolution of galaxies and large scale structures really agree with the predictions of the standard Λ Cold Dark Matter model?
- Is the dark energy really the cosmological constant? Or, to put it another way, what is the value of w in the equation of state of the dark energy which drives the present exponential expansion of the Universe?
- Does the dark energy vary with cosmic epoch or is it a constant as expected if it were simply the cosmological constant?
- What clues can be derived from the Large Hadron Collider at CERN – what are the implications of detecting/not detecting Higgs particles? Are dark matter candidates found in the LHC experiments?
- The experiments to make direct detections of dark matter particles have made spectacular progress over the last few years, the present limits already excluding some of the potential supersymmetric candidates for the dark matter.

As an example of just how difficult some of these challenges are, let me give the example of the EUCLID mission of the European Space Agency which is one of the candidates for the next medium-sized mission. High precision dark energy measurements require a combination of two or more cosmological probes. Euclid aims to utilise the most promising dark energy probes: an all-sky survey of weak lensing and huge samples of galaxy redshifts to measure precisely the baryon acoustic oscillations in the distribution of galaxies. Individually, these are the two most powerful measures of dark energy. In combination they will control and measure a wide range

Journal of Physics: Conference Series **314** (2011) 012011 doi:10.1088/1742-6596/314/1/012011

of systematics effects which would effect either of them separately. The combination of these experiments will measure the two Newtonian potentials $\phi(k,z)$ and $\psi(k,z)$, which describe a perturbed Λ Cold Dark Matter Universe. In standard general relativity, $\phi(k,z) = \psi(k,z)$. The weak lensing and baryon acoustic oscillations measurements depend in different ways upon $\phi(k,z)$ and $\psi(k,z)$ - weak lensing depends upon both $\phi(k,z)$ and $\psi(k,z)$ whereas the baryon acoustic oscillations depends only on $\psi(k,z)$. Thus, Euclid enables us to test if General Relativity is the best description of gravity on the scale of the largest perturbations we can measure in the Universe.

The way in which the experiment will be carried out is that the weak lensing will reconstruct directly the distribution of the dark matter and the evolution of the growth rate of dark matter perturbations with redshift. The baryon acoustic oscillations act as standard rods, determine $P(k)$ and provide a measure of $H(z)$ and hence $w(z)$. They also map out the evolution of the baryonic component of the Universe. Together, these enable many systematic effects to be controlled, for example, intrinsic alignments in weak lensing and bias factors in baryon acoustic oscillations. The value of w will be measured to 1% an the variation of $w(z)$ with redshift to 10%. This is a very large undertaking, but it is only with efforts on this scale that the future challenges can be addressed.

References

[1] Liddle, A. R. and Lyth, D.H. (2009). *The Primordial Density Perturbation* Cambridge: Cambridge University Press. This book includes and supersedes the material from their earlier book *Cosmological Inflation and Large Scale Structure* (2000). Cambridge: Cambridge University Press.

[2] Longair, M. S. (2008). *Galaxy Formation.* Berlin: Springer-Verlag.

[3] Will, C. M. (2006). *Living Reviews in Relativity*, **9**, Online article: http://www.livingreviews.org/lrr-2006-3.

Spanish Relativity Meeting (ERE 2010): Gravity as a Crossroad in Physics IOP Publishing
Journal of Physics: Conference Series **314** (2011) 012012 doi:10.1088/1742-6596/314/1/012012

Loop Quantum Cosmology: A cosmological theory with a view

G A Mena Marugán

Instituto de Estructura de la Materia, CSIC, Serrano 121, 28006 Madrid, Spain.

E-mail: mena@iem.cfmac.csic.es

Abstract. Loop Quantum Gravity is a background independent, nonperturbative approach to the quantization of General Relativity. Its application to models of interest in cosmology and astrophysics, known as Loop Quantum Cosmology, has led to new and exciting views of the gravitational phenomena that took place in the early universe, or that occur in spacetime regions where Einstein's theory predicts singularities. We provide a brief introduction to the bases of Loop Quantum Cosmology and summarize the most important results obtained in homogeneous scenarios. These results include a mechanism to avoid the cosmological Big Bang singularity and replace it with a Big Bounce, as well as the existence of processes which favor inflation. We also discuss the extension of the frame of Loop Quantum Cosmology to inhomogeneous settings.

1. Introduction

Gravity remains as the only fundamental physical interaction which has not been satisfactorily described quantum mechanically. Leaving aside the belief that all interactions should be unified in a single theory, a strong motivation for a quantum theory of gravity comes from the singularity results of General Relativity [1]. In these classical singularities the predictability breaks down, indicating that the regime of applicability of Einstein's theory has been surpassed and that a new and more fundamental description is needed.

Apart from the ability to cure the classical singularities, one expects that a quantum theory of gravity would open a new window to physics, incorporating phenomena which originate in the quantum realm. Any candidate theory should make this compatible with the infrared behavior of General Relativity, so that no quantum effect alters much spacetime regions like those which we observe [2]. Explaining why our Universe is actually so (semi-)classical constitutes a real challenge to quantum gravity. Previous attempts to construct a quantum formalism for the metric fields (in geometrodynamics) employing a Wheeler-De Witt approach have proven unsuccessful. In the general theory, the approach finds functional analysis and interpretational obstacles which prevent further progress. On the other hand, in simple models where these obstacles can be circumvented, the classical singularities are not fully resolved (e.g., semiclassical states are peaked on trajectories where physical observables eventually diverge [3, 4]). Recently, a different approach to quantize General Relativity has been developed: the theory of Loop Quantum Gravity (LQG) [5, 6]. This theory is based on a nonperturbative canonical formalism of gravity and declares diffeomorphism invariance and background independence to be basic guidelines. Its application to simple cosmological models, generally homogeneous ones, gives rise to the new area of gravitational physics known as Loop Quantum Cosmology (LQC) [2, 7].

2. Loop Quantum Gravity

LQG is a canonical quantization of General Relativity, constructed starting from a Hamiltonian formulation of Einstein's theory. Let us begin by considering a globally hyperbolic spacetime. Introducing then a global foliation, the initial data for the construction of the Lorentzian spacetime metric are contained in the spatial 3-metric and the extrinsic curvature on the considered section of the foliation [8]. If one wants to introduce fermions in this framework, the spatial metric must be replaced with a triad, to which fermions couple directly. This coupling occurs in an internal su(2) index, and respects the invariance of the system under internal rotations, realized then as SU(2) gauge transformations. The spatial metric is recovered as the square of the triad, contracted in the internal indices by means of the Euclidean metric [the Killing-Cartan metric on su(2)]. A canonical set of variables to describe the gravitational phase space is formed, up to numerical factors, by the densitized triad (i.e., the triad multiplied by the square root of the determinant of the spatial metric), and the extrinsic curvature expressed in triadic form (i.e., the extrinsic curvature contracted with the triad in one spatial index).

At this stage, one can replace the extrinsic curvature by a connection valued 1-form, taking values on the algebra su(2). Classically, this Ashtekar-Barbero connection is a sum of the spin connection compatible with the co-triad and the triadic extrinsic curvature multiplied by an arbitrary positive number γ, known as the Immirzi parameter. By construction, this connection forms a canonical set with the densitized triad, modulo a factor $8\pi G\gamma$ in their Poisson brackets, where G is the Newton constant (from now on, we set the speed of light equal to the unity).

The introduction of the su(2)-connection allows one to incorporate nonperturbative techniques in the description of the system, similar to those developed in gauge field theories, like e.g. for Yang-Mills fields. In particular, the gauge invariant information about the connection is captured in the so-called Wilson loops, constructed from holonomies which describe the parallel transport of spinors around loops. We hence replace the connection by SU(2)-holonomies along (piecewise analytic) edges, where we understand that an edge is an embedding of the closed unit interval in the considered spatial manifold. This replacement involves a 1-dimensional smearing of the connections, and renders the information about them gauge invariant except for the effect of transformations at the end points of the edges. By joining a finite number of edges in those vertices to form a graph [5], and combining the holonomies there so that SU(2) invariance is respected everywhere, one obtains what is usually called a spin network. It is worth noticing that the construction of the holonomies, which contain a line integral of the connections, is made without recurring to any background structure.

Since the most relevant field divergences in our formalism come from the appearance of a 3-dimensional delta function on the Poisson brackets between the connection and the densitized triad, and we have already smeared the connection over one dimension, it seems natural to smear the triad similarly, but now over two dimensions. Given that the densitized triad is a (spatial) vector density, this smearing can be carried out again without employing any background structure. For any (piecewise analytic) surface, we can define the flux of the densitized triad through it, obtaining the desired smearing. Holonomies and fluxes form an algebra under Poisson brackets, which we regard from now on as our basic algebra of functions on the gravitational phase space. From this perspective, the quantization of General Relativity amounts to the representation of this algebra as an algebra of operators acting on a Hilbert space. In addition, we must take into account that the system is subject to a series of constraints, which must be imposed quantum mechanically. These are the Gauss [or SU(2)] constraint, the diffeomorphisms constraint, and the Hamiltonian or scalar constraint, which express the invariance of the system under SU(2) transformations, spatial diffeomorphisms, and time reparametrizations [5, 6].

An important result for the robustness of the predictions of LQG is a uniqueness theorem about the admissible representations of the holonomy-flux algebra, known as the LOST theorem (after the initials of its authors [9]). Specifically, this theorem states that there exists a unique cyclic representation of that algebra with a diffeomorphism invariant state (interpretable as a vacuum). In total, the choice of the algebra of elementary variables, motivated by background independence, and the status of spatial diffeomorphisms as a fundamental symmetry pick up a unique quantization, up to unitary equivalence and prior to the imposition of constraints.

To gain insight into the kind of quantization adopted in LQG, let us consider the so-called cylindrical functions: complex functions that depend on the connection only via the holonomies along a graph, formed by a finite numbers of edges. Completing this algebra of functions with respect to the norm of the supremum, we obtain a commutative C^*-algebra with identity, where the $*$-relation is provided by the complex conjugation [10, 11]. According to Gel'fand theory, this algebra is then (isomorphic to) the algebra of continuous functions on a certain compact space, called the spectrum. Smooth connections are dense in this space. Besides, the Hilbert space of the representation is just a space of square integrable functions on this Gel'fand spectrum with respect to a certain measure. The LOST theorem guaranties that there exists only one diffeomorphism invariant measure which supports not only a representation of the holonomies, but of the whole holonomy-flux algebra: the Ashtekar-Lewandowski measure, used to construct the representation in LQG. This representation turns out not to be equivalent to a standard one, and therefore leads to physical results which are different from those of other, conventional quantizations. In fact, the representation is not continuous; as a consequence, the connection cannot be defined as an operator valued distribution [5]. Finally, the resulting quantum geometry is discrete, with area and volume operators that have a point spectrum [5].

3. Loop Quantum Cosmology: the flat FRW model

As a paradigmatic example in LQC, let us now apply this type of loop quantization techniques to Friedmann-Robertson-Walker (FRW) cosmologies (namely homogeneous and isotropic spacetimes) with flat spatial sections of $\mathbb{R}^3$ topology and a matter content provided by a homogeneous massless scalar field ϕ, minimally coupled to the metric [3, 4]. We introduce a fiducial triad and an integration cell, adapted to it, to carry out all integrations and avoid in this way divergences due to the homogeneity and noncompactness of the spatial sections. We call V_0 the fiducial volume of this cell. It is possible to check that all physical results are indeed independent of these choices of fiducial elements [3, 4]. Besides, one can fix the gauge freedom so that both the densitized triad and the connection become diagonal. Given the isotropy, they are then totally specified by one single variable each, which we call p and c, respectively. These variables describe the geometry degrees of freedom, vary only in time, and form a canonical pair: $\{c, p\} = 8\pi\gamma G/3$. Classically, they are related with the scale factor a and its time derivative by the formulas $p = V_0^{2/3} a^2$ and $c = \gamma V_0^{1/3} \dot{a}$.

To retain all the gauge invariant information about the su(2)-connection, taking into account the homogeneity, it suffices to consider holonomies along (fiducial) straight edges. Similarly, triads are now smeared across (fiducial) squares. The fluxes are then totally determined by the variable p. Returning to the holonomies, it is easy to check that, for an edge of coordinate length $\lambda V_0^{1/3}$ in any fiducial direction, the matrix elements of the SU(2)-holonomy are linear combinations of exponentials of the form $e^{\pm i\lambda c/2}$. The corresponding configuration algebra is then the linear space of continuous and bounded complex functions in the real line ($c \in \mathbb{R}$), with elements of the form $f(c) = \sum_j f_j e^{i\lambda_j c/2}$. It is well known that the completion of this algebra with the supremum norm is just the Bohr C^*-algebra of almost periodic functions [11]. Its Gel'fand spectrum is the Bohr compactification of the real line, $\mathbb{R}_{\text{Bohr}}$. This compactification can be seen as the set of group homomorphisms from the group of real numbers (with the sum) to the multiplicative group $\mathbb{C}$ of complex numbers of unit norm. Indeed, for every real number c we have a homomorphism $x_c : \mathbb{R} \to \mathbb{C}$ of this kind, namely $x_c(\lambda) = e^{i\lambda c/2}$. Moreover, it is possible to see that the real line is actually dense in $\mathbb{R}_{\text{Bohr}}$, using the fact that our initial configuration algebra separates points $c \in \mathbb{R}$ [10, 11].

The operation $x\tilde{x}(\lambda) = x(\lambda)\tilde{x}(\lambda)$ provides a commutative group structure in $\mathbb{R}_{Bohr}$. Since the group $\mathbb{R}_{\rm Bohr}$ is compact, it has a (unique) invariant Haar measure. The functions on $\mathbb{R}_{\rm Bohr}$ consisting in the evaluation at a real point μ form an orthonormal basis in the corresponding Hilbert space of square integrable functions with the norm defined by that measure [11]. We designate each element in this basis with a ket $|\mu\rangle$. This basis allows us to pass from our configuration representation, in which holonomies act by multiplication, to a "momentum" representation in which the triad has a multiplicative action [10, 11]. Calling $N_\lambda = e^{i\lambda c/2}$, this "momentum" representation is given by $\hat{p}|\mu\rangle = (4\pi\gamma G/3)\mu|\mu\rangle$ and $\hat{N}_\lambda|\mu\rangle = |\mu+\lambda\rangle$ (we set $\hbar = 1$). Clearly, the basis $\{|\mu\rangle;\ \mu \in \mathbb{R}\}$ is uncountable, and therefore the Hilbert space is nonseparable. Nevertheless, normalizable states can get nonvanishing contributions only from a countable subset of states $|\mu\rangle$; otherwise their norm would not be finite. This "momentum" representation is the one usually employed in LQC. It is worth remarking that the representation fails to be continuous, owing to the discrete norm on the Hilbert space, $\langle\mu|\tilde{\mu}\rangle = \delta_\mu^{\tilde{\mu}}$. As a result, a connection operator does not truly exist, and the representation is inequivalent to the Wheeler-De Witt one, in total parallelism with the situation found in LQG for the general case.

Although homogeneity ensures that the diffeomorphisms constraint is satisfied, and the $SU(2)$ constraint has been removed by gauge fixing, the system is still subject to a Hamiltonian constraint, which must be imposed now quantum mechanically. In order to introduce a Hamiltonian constraint operator, there are essentially two building blocks which must be defined in terms of our elementary operators $\hat{p}$ and $\hat{N}_\lambda$. First, we need an operator to represent the phase space function $t(p) = \text{sign}(p)/\sqrt{|p|}$. This function contains all the p-dependence of the (nondensitized) triad of the model. Note that this triad diverges at the Big Bang, where the variable p vanishes. Correspondingly, the operator representing $t(p)$ cannot be defined exclusively from $\hat{p}$ using the spectral theorem: $\hat{p}$ has a point spectrum which contains the zero, and hence its inverse operator is not well defined. But it is possible to construct a regularized triad operator using commutators with holonomies, in addition to $\hat{p}$ [12]: $\widehat{t(p)} = 3(\hat{N}_{-\bar{\mu}}|\hat{p}|^{1/2}\hat{N}_{\bar{\mu}} - \hat{N}_{\bar{\mu}}|\hat{p}|^{1/2}\hat{N}_{-\bar{\mu}})/(4\pi\gamma G\bar{\mu})$. In principle, $\bar{\mu}$ may take any real value. It will be fixed later on in our discussion. The resulting operator is diagonal in the considered basis of μ-states. Furthermore, it turns out to be bounded from above, so that, in particular, the classical divergence disappears. Actually, our regularized operator annihilates the kernel of $\hat{p}$.

The other block that we need is the SU(2)-curvature operator. Recall that the connection operator is not well defined, therefore we cannot use it to construct the curvature. Nonetheless, it is possible to determine it using a square loop of holonomies. We use again edges of fiducial length $\bar{\mu}V_0^{1/3}$. Classically, the expression of the curvature would be recovered exactly in the limit of zero area, when $\bar{\mu}$ tends to zero. However, this limit is not well defined in LQC. The idea is to shrunk the square up to the minimum physical area Δ allowed in LQG, where the spectrum of the area operator is discrete [4]. This introduces a certain nonlocality in the formalism, and turns the parameter $\bar{\mu}$ into a state dependent quantity, since the physical area depends on the particular state under consideration. Explicitly, the relation that must be satisfied for each state is $\bar{\mu}^2|p| = \Delta$. At this stage, it is convenient to relabel the basis of μ-states as a basis of volume eigenstates, introducing an affine parameter v for the translation generated by $\bar{\mu}c/2$ [4]. By construction, we then get that $\hat{N}_{\bar{\mu}}|v\rangle = |v+1\rangle$. The parameter v is related with the physical volume of the fiducial cell, $V = p^{3/2}$, by the formula $v = \text{sign}(p)V/(2\pi\gamma G\sqrt{\Delta})$.

Finally, for the quantization of the matter field ϕ, we use the standard Schrödinger representation. So, the kinematical Hilbert space is the tensor product of the gravitational space of LQC and the standard one for matter. With a suitable factor ordering and choice of densitization [13], one then gets a Hamiltonian constraint of the form $\hat{H} = -6\hat{\Omega}^2 + \hat{P}_\phi^2$, where $\hat{P}_\phi$ is the momentum operator of the matter field, and acts by differentiation ($\hat{P} = -i\partial_\phi$). The gravitational part of the constraint is given by the operator $\hat{\Omega}^2$. Remarkably, this constraint leaves invariant the zero-volume state $|v=0\rangle$, as well as its orthogonal complement. Therefore, the analog of the classical singularity is removed in practice from the Hilbert space, and we can restrict all physical considerations to its complement. In this sense, the singularity gets resolved at a kinematical level. Moreover, the operator $\hat{\Omega}^2$ has an action of the following type: $\hat{\Omega}^2|v\rangle = f_+(v)|v+4\rangle + f(v)|v\rangle + f_-(v)|v-4\rangle$. Here, the real functions $f_+(v)$ and $f_-(v)$ have the outstanding property that they vanish in the respective intervals [-4,0] and [0,4] [13]. Thus, the action of $\hat{\Omega}^2$ preserves each of the subspaces of the gravitational Hilbert space obtained by restricting the label of the v-states to any of the semilattices $\mathcal{L}_\varepsilon^\pm = \{v = \pm(\varepsilon + 4n);\, n \in \mathbb{N}\}$, where $\varepsilon \in (0,4]$. Then, each of these semilattices provides a superselection sector. In each sector, the orientation of the triad is definite (v does not change sign) and $|v|$ has an strictly positive minimum, equal to ε. For concreteness, we choose sectors with $v>0$ from now on.

On the other hand, it is possible to show that, on each sector, the operator $\hat{\Omega}^2$ has a nondegenerate absolutely continuous spectrum equal to the positive real line [13, 14]. Recalling its action, this gravitational constraint operator might be understood as a second-order difference operator. But its eigenfunctions are entirely determined by their value at ε, point from which they can be constructed by solving the eigenvalue equation. In this sense, the gravitational constraint operator leads to a No-Boundary description, in which the eigenstates which encode the information about the quantum geometry are all determined without the need to introduce any boundary condition in the region around the origin. We also notice that, up to a global phase, these eigenfunctions $e_\delta^\varepsilon(v)$ are real, since so is the gravitational constraint operator.

With such eigenfunctions, one easily finds the solutions to the Hamiltonian constraint, which have the form $\psi(v,\phi) = \int_0^\infty d\delta\, e_\delta^\varepsilon(v)[\psi_+(\delta)e^{i\sqrt{6\delta}\phi} + \psi_-(\delta)e^{-i\sqrt{6\delta}\phi}]$. The scalar field ϕ plays the role of an emergent time. Then, physical states can be identified (e.g.) with the positive frequency solutions $\psi_+(\delta)$ that are square integrable over the spectral parameter $\delta \in \mathbb{R}^+$ [13]. A complete set of Dirac observables is formed by $\hat{P}_\phi$ and $|\hat{v}|_{\phi_0}$, the latter being defined by the action of the volume operator when the field equals ϕ_0. On the Hilbert space of physical states specified above, these observables are self-adjoint operators.

4. The Big Bounce

In the previous section we have completed the quantization of the flat FRW with a homogeneous massless scalar field. In order to analyze the physical predictions of this quantum theory, we will consider now the evolution of (positive frequency) quantum states with a semiclassical behavior. We study *Gaussian-like* states which, at an instant $\phi = \phi_0$ in the region of large emergent times $\phi_0 \gg 1$, are peaked on certain values of the elementary observables of the model, namely, the matter momentum, $P_\phi = P_\phi^0$, and the physical volume, $v = v^0$. We restrict our attention to states with large values of P_ϕ^0 and v^0 [3, 4]. The numerical analysis of the quantum evolution unveils an outstanding phenomenon in these states: the Big Bang singularity is resolved dynamically and is replaced by a bounce that connects the universe with another branch of the evolution, dictated again by the equations of General Relativity. This mechanism to elude the cosmological singularity is known as the Big Bounce [3, 4].

The numerical studies show that the considered semiclassical states remain peaked on a well defined trajectory during the whole evolution. On these states, the Big Bounce does not occur in a genuinely quantum region where one were to loose an effective notion of geometry and spacetime. The trajectory deviates from the one predicted by General Relativity only when the matter energy density ρ becomes of the order of one percent of a critical density, ρ_{crit}. This scale for the onset of corrections is of the Planck order and universal: it is the same for all the semiclassical states which suffer the bounce. Explicitly, the critical density is $\rho_{\text{crit}} = (\sqrt{3}/32\pi^2\gamma^3G^2) \approx 0.41\rho_{\text{Planck}}$, where ρ_{Planck} is the Planck density. For densities close to the critical one, i.e., in the regime close to the bounce, gravity behaves as a repulsive force owing to the effects of quantum geometry [2].

In addition, the trajectory followed by the peak of these states matches an effective dynamics, which has been deduced in detail (under certain assumptions on the family of states under consideration) using techniques of geometric quantum mechanics [15]. The agreement between the numerical simulations and the predictions of this effective dynamics is remarkable. In particular, the effective dynamics predicts a bounce precisely when the matter density reaches the critical value $\rho_{\rm crit}$. Further support to the role played by this critical density comes from the study of the operator which represents the matter density in the quantum theory. It is possible to prove that it has a bounded spectrum, the bound being given again by $\rho_{\rm crit}$. Then, the overview picture is that the emergence of important quantum geometry effects in this model is controlled by the value of the matter energy density. When this density approaches the Planck scale, quantum geometry phenomena enter the scene, preventing that it keeps on increasing and consequently avoiding the collapse into a cosmological singularity. It is worth pointing out that these quantum phenomena can be relevant even in regions which one would not consider to belong to the deep Planck regime. For instance, the volume v at the bounce is proportional to the value of the matter field momentum, which is conserved in the evolution. Hence, when the bounce occurs, the volume can be as large as desired.

The presence of a quantum bounce is actually generic in this quantum model, with implications that exceed the restriction to the discussed class of semiclassical states. We have already commented that the eigenfunctions of the gravitational constraint operator are real (up to a global phase). Studying the Wheeler-De Witt limit of this operator, one can prove that its eigenfunctions lead in fact to positive frequency solutions with ingoing and outgoing components of equal amplitude in this limit [13] (see [16] for the case of a specific superselection sector).

Furthermore, the Big Bounce mechanism is not restricted just to the flat FRW model with a massless scalar field, but is rather general. On the one hand, assuming the validity of the effective dynamical equations for other matter contents (assumption that is supported by the numerical analyses carried out so far), one can show that all strong singularities (á la Królak) are resolved in flat FRW for any kind of matter [17]. Only Type II and Type IV singularities may remain [17], but these singularities can be considered physically harmless, since geodesics can be extended beyond them (then, sufficiently strong in-falling detectors can survive these singularities). On the other hand, similar conclusions about the occurrence of the Big Bounce have been reached in other FRW models quantized in LQC. These include the flat model with negative cosmological constant [18], the closed model [19], the open model [20] (some problems of the treatment presented in that reference can be solved with the techniques of [21]), and the flat model with positive cosmological constant (recently studied by Ashtekar and Pawłowski), all of them with a homogenous scalar matter field present as well. For the flat FRW universe with negative cosmological constant and the closed FRW model, the classical evolution leads to a Big Crunch (i.e., the universe recollapses into a cosmological singularity). In the quantum theory, this Big Crunch is also resolved via a Big Bounce, like in the case of the Big Bang. In all cases, there exists an upper bound for the matter energy density, which is given again exactly by $\rho_{\rm crit}$, and the infrared regime shows an outstanding agreement with General Relativity.

In spite of some statements that have appeared in the literature of LQC [17], the effective equations for flat FRW in the presence of generic matter do not necessarily lead to an asymptotically de Sitter behavior if a vanishing or a divergent value of the scale factor a were to be approached in the evolution, without further assumptions. The confusion comes from the consideration of the identity $\ln(\rho/\rho_0) = \int_{a_0}^{a}[1+w(\tilde{a})]d\tilde{a}/\tilde{a}$, deduced from the conservation equation for the matter energy density, and where a_0 is a reference value for the scale factor, $\rho_0 = \rho(a_0)$, and $w(a)$ is the ratio between the pressure and the energy density of matter. In fact, the convergence of the above integral when $a \to \infty$ does not need that $w(a)$ tend to minus the unity [22] (the value that would correspond to a de Sitter regime). Besides, even if the energy density is required to be positive and bounded from above by the critical density, one may have a vanishing limit for it. Then, the considered integral would diverge in that limit, allowing for values of w different from minus one [22]. This situation might be found in the limit of null a. It is straightforward to see that these arguments invalidate the discussion in [17].

Similar conclusions about the resolution of cosmological singularities have been reached in other homogeneous scenarios which contain anisotropies. LQC has been implemented successfully to completion in Bianchi models of type I [14, 23, 24, 25, 26], type II [21], and type IX [27]. The analysis, complemented with numerical simulations in these models, confirms the Big Bounce scenario, though now there may exist bounces in different scale factors, since the spatial directions are not all equivalent in view of the anisotropy. Together with the BKL conjecture [28], these results suggest a generic resolution of spacelike singularities in LQC. Recall that the BKL conjecture says that spatial derivatives can be neglected against time derivatives when one approaches a spatial singularity, so that the dynamics at any point can be approximated locally by a homogeneous model (i.e., a Bianchi model).

The anisotropic model which has received more attention is Bianchi I. Recently, this model has been studied thoroughly with the prescription put forward by Ashtekar and Wilson-Ewing [24] to determine the lengths of the edges used to define the holonomies that enter the Hamiltonian constraint. Actually, it has been shown that this prescription is characterized by the requirement that the action of the corresponding holonomies produce a constant shift in the (absolute value of the) physical volume [25]. In addition, one can see that the initial value problem for the quantum evolution is well posed even in vacuo: an infinite though countable set of initial data on the section of minimum physical volume suffices to fix the solution to the constraint [26]. This allows one to identify the space of solutions and construct from it the Hilbert space of physical states. Finally, while the superselection sectors for the physical volume are the same as in flat FRW, the sectors for the anisotropies turn out to have a rather different structure. They are still discrete, and different triad orientations are not mixed, but these sectors are dense on the (positive) real line, instead of being formed by points separated by a constant distance [25].

Further support to the Big Bounce scenario in anisotropic models, using the prescription of [24] in the quantization, comes from the extrapolation of the effective dynamical equations, assuming their validity for Bianchi I. These effective equations guarantee that the directional Hubble rates, the expansion, and the shear scalar (of comoving observers) are all bounded from above in the evolution [29], preventing in this way the formation of dangerous singularities. Contrary to what one could naively expect [29], however, the bounded nature of these physical quantities cannot be extrapolated to the genuine quantum theory. The reason is that a bounded function on phase space (namely, one of our physical quantities in the effective theory) is not always represented by a bounded operator. One can ensure that the corresponding operator is bounded only when it can be defined in terms of a set of commuting elementary ones via the spectral theorem. But in generic situations this is not the case. Similarly, an unbounded function on phase space may be represented by a bounded operator. In addition, when superselection sectors enter the scene, the physically relevant spectra of our elementary operators do not coincide with the range of their classical analogs, therefore introducing limitations to the domain of applicability of the effective equations. For instance, the spectrum of the (absolute value of the) physical volume is bounded from below by a positive number on each of our superselection sectors. This invalidates the analysis of the limit of vanishing volume in the effective equations.

5. Inflation

The effects of quantum geometry in the *early universe* are important not only to elude cosmological singularities, but also to build up a satisfactory inflationary scenario. In standard cosmology, one generally needs a fine tuning of the initial conditions or the inflationary parameters in order to reach enough inflation to explain the observed universe, with at least 68 e-foldings. Recent results in LQC indicate that, on the contrary, the quantum phenomena that accompany the Big Bounce render natural an inflationary process with this number of e-foldings.

Let us consider a flat FRW model with an inflaton field with positive kinetic energy. The effective equations for this type of models in LQC imply a series of interesting properties [2, 30]. Firstly, the Hubble parameter is bounded from above. Besides, when the inflaton potential is nonnegative and bounded from below, the time derivatives of the inflaton and of the Hubble parameter turn out to be bounded from above in absolute value. In addition, for potentials which are unboundedly large when the inflaton field approaches plus/minus infinity, there exists an upper/lower bound on the value of this inflaton. On the other hand, and more remarkably, there always exists a phase of superinflation [31, 32] after the bounce, in which the Hubble parameter increases from zero up to its maximum value. This phenomenon of superinflation is robust in LQC, and appears even in the absence of potential. Nevertheless, one can show that the superinflation epoch alone does not yield sufficient e-foldings in generic situations.

For concreteness, we will focus our discussion on the inflationary potential $m^2\phi^2$, with a mass of the order of 10^{-6} in Planck units, which is the phenomenologically preferred value. We can calculate the probability of getting more than 68 e-foldings, starting with equiprobability for every unconstrained set of initial data. Taking the bounce as the most natural instant to define initial data, and imposing the Hamiltonian constraint, we obtain on the space of physical data a measure of the form $\sqrt{1-F_{\text{Bounce}}}\,d\phi_{\text{Bounce}}\,dv_{\text{Bounce}}$, where ϕ_{Bounce} and v_{Bounce} are the values of the field and the physical volume on the bounce section, respectively. Besides, F_{Bounce} is the fraction of the matter energy density which is due to the potential at that moment. On the other hand, we recall that $|\phi_{\text{Bounce}}|$ is bounded, because the studied potential grows unboundedly. Then, it is possible to see that, if $F_{\text{Bounce}} > 1.4\cdot 10^{-11}$, the superinflationary phase either provides the desired number of e-foldings by its own or supplements them by funneling the solutions to initial conditions such that there is a sufficiently long period of slow-roll inflation [30]. It is straightforward to compute that the relative probability for a value of F_{Bounce} in this range is greater than 0.99. These conclusions are not sensitive to reasonable changes, e.g., in the inflaton mass or the form of the potential. Therefore, the effective equations that arise in LQC solve the fine tuning problems for inflation.

6. Inhomogeneous models

Inhomogeneous models have been considered lately in the framework of LQC in order to extend the applicability of this quantization scheme beyond the simple FRW or Bianchi cosmological spacetimes and with the aim at obtaining predictions that, eventually, might be contrasted with observations. The quantum analysis of this type of models has been carried out adopting a hybrid approach [25, 26, 33, 34], which combines a loop quantization of the subspace of homogeneous solutions of the system (or rather of their geometry) and a Fock quantization of the matter fields and inhomogeneous gravitational waves contained in the model. This hybrid approach is based on the assumption that there exists a hierarchy of quantum phenomena, so that the most relevant effects of the loop quantum geometry are those that affect the sector of homogeneous degrees of freedom.

For a series of cosmological models, the ambiguity in the selection of a Fock quantization for the inhomogeneities can be removed by appealing to some recent theorems which guarantee the uniqueness of the choice of both a field description (among a reasonable set of possibilities) and a Fock representation for it [35, 36, 37, 38]. The uniqueness is ensured if the quantization respects certain conditions on the unitarity of the dynamics as well as the invariance of the vacuum under the spatial symmetries of the field equations. As a consequence, the predictions of the hybrid quantization are robust, in the sense that they are not affected by the typical ambiguities which plague the quantization of systems with an infinite number of degrees of freedom. Cosmological systems where these uniqueness results have been proven, and the hybrid quantization is at hand, include the Gowdy spacetimes [39] (with different spatial topologies and possibly containing scalar fields), and fields and perturbations around closed FRW models.

In the inhomogeneous cosmologies that have been studied in this way, the loop quantization of the homogeneous sector suffices to resolve the cosmological singularities at the kinematical level, in a similar manner as it happens for the flat FRW model. It is remarkable that, in spite of the field character of these systems, the quantization can been carried out to completion. Moreover, the Hilbert space of physical states that one attains with this hybrid approach is such that one recovers the Fock description of the inhomogeneities of the system, providing support to this conventional quantum treatment and proving its compatibility with the loop quantization.

In particular, the hybrid quantization process has been discussed in detail in the (vacuum) Gowdy model with 3-torus topology [25, 34], probably the simplest inhomogeneous cosmology. In this case, the effective dynamics that one obtains extrapolating the results deduced in FRW scenarios has been studied both numerically and (partially) analytically [40] (although the prescription used to determine the lengths of the edges for the holonomies is not that in [24]). The analysis confirms the presence of the Big Bounce, which happens in all the three spatial directions of this anisotropic model. The bounce occurs typically at values of the fluxes variables (i.e., the p-variables for each of the three spatial directions) which are at least a 13 per cent of those found in the absence of inhomogeneities [40]. This proves that the inhomogeneities do not have a drastic effect that could alter substantially the Big Bounce, e.g., driving the bounce into the deep Planck regime. In addition, the numerical studies have considered the change in amplitude of the inhomogeneous modes (which describe linearly polarized gravitational waves) between the two asymptotic regions of the effective trajectories, which correspond respectively to a contracting and an expanding universe. One can take a statistical average, disregarding phases in the mode decomposition of the inhomogeneities. Actually, it is possible to see that, in the sector of the space of solutions where the inhomogeneities dominate the bounce dynamics, the change in the amplitudes is antisymmetric with respect to the phase. Then, in average, the amplitudes are statistically preserved through the bounce. On the other hand, in the sector where the vacuum dynamics is approximately valid around the bounce, the change in the amplitudes is positive in average [40]. Although the scenario is not completely physical, since the considered inhomogeneities are not all those allowed in the most general cosmological setting, the found behavior may indicate a LQC mechanism that removes low amplitudes through the bounce.

7. Conclusion

The increasing attention paid recently to the quantization of cosmological systems using loop techniques has crystalized in the foundation of a new branch of gravitational physics. This new formalism of LQC allows one a rigorous control on the mathematical and interpretational aspects of quantum cosmology, providing significance and robustness to the predictions in an area where they are rarely falsifiable in a direct way. Moreover, in doing so, LQC has opened new views to the quantum phenomena of the *early universe*. It leads to a new paradigm for cosmology in which the Big Bang singularity is resolved and replaced with a Big Bounce. Remarkably, this Big Bounce respects the semiclassicality of the universe, connecting two branches whose asymptotic behavior is well described by General Relativity. On the other hand, LQC renders inflation a natural process, suppressing the need for a fine tuning in the initial conditions or parameters of the inflationary era. In addition, LQC might supply a mechanism to remove low amplitudes from the (primordial) inhomogeneities in cosmology. And it suggests new settings for the consideration of initial conditions on the inhomogeneities, which would not be imposed anymore in the region around the classical singularity. Further research is needed to explore the consequences in issues such as cosmological perturbations, primordial fluctuations, or the power spectrum of anisotropies in the cosmic background. These are exciting topics that can provide a suitable arena where the predictions of LQC might be compared with observations.

Acknowledgments
The author is grateful to the organizers of the ERE2010 for the warm atmosphere created during the conference. He thanks J. Cortez, G. Jannes, L. J. Garay, M. Martín-Benito, D. Martín-de Blas, J. Olmedo, and specially D. Jaramillo and J. M. Velhinho for enlightening conversations. Owing to space limitations, many authors that have contributed to the progress of LQC have not been cited explicitly. The list of names includes (but is not restricted to) K. Banerjee, G. Date, W. Kaminski, G. Khanna, G. M. Hossain, J. Lewandowski, C. Rovelli, S. Shankaranarayaran, L. Szulc, M. Varadarajan, F. Vidotto, and Yongge Ma. This work was supported by the MICINN Project No. FIS2008-06078-C03-03 and the Consolider Program CPAN No. CSD2007-00042.

References

[1] Hawking S and Ellis G F R 1973 *The Large Scale Structure of Space-Time* (Cambridge: Cambridge University Press)
[2] Ashtekar A 2010 *Preprint* arXiv:1005491
[3] Ashtekar A, Pawłowski T and Singh P 2006 *Phys. Rev. Lett.* **96** 141301
Ashtekar A, Pawłowski T and Singh P 2006 *Phys. Rev.* D **73** 124038
[4] Ashtekar A, Pawłowski T and Singh P 2006 *Phys. Rev.* D **74** 84003
[5] Ashtekar A and Lewandowski J 2004 *Classical Quantum Gravity* **21** R53
[6] Thiemann T 2007 *Modern Canonical Quantum General Relativity* (Cambridge: Cambridge University Press)
[7] Bojowald M 2008 *Living Rev. Rel.* **11** 4
[8] Wald R M 1984 *General Relativity* (Chicago: University of Chicago Press)
[9] Lewandowski J, Okolow A, Sahlmann H and Thiemann T 2006 *Comm. Math. Phys.* **267** 703
[10] Mena Marugán G A 2009 *AIP Conf. Proceedings* **1130** 89
[11] Velhinho J M 2007 *Classical Quantum Gravity* **24** 3745
[12] Ashtekar A, Bojowald M and Lewandowski J 2003 *Adv. Theor. Math. Phys.* **7** 233
[13] Martín-Benito M, Mena Marugán G A and Olmedo J 2009 *Phys. Rev.* D **80** 104015
[14] Martín-Benito M, Mena Marugán G A and Pawłowski T 2008 *Phys. Rev.* D **78** 064008
[15] Taveras V 2008 *Phys. Rev.* D **78** 064072
[16] Ashtekar A, Corichi A and Singh P 2008 *Phys. Rev.* D **77** 024046
[17] Singh P 2009 *Classical Quantum Gravity* **26** 125005
[18] Bentivegna E and Pawłowski T 2008 *Phys. Rev.* D **77** 124025
[19] Ashtekar A, Pawłowski T, Singh P and Vadersloot K 2007 *Phys. Rev.* D **75** 024035
[20] Vandersloot K 2007 *Phys. Rev.* D **75** 023523
[21] Ashtekar A and Wilson-Ewing E 2009 *Phys. Rev.* D **80** 123532
[22] Jaramillo D 2010 *Private Communication*
[23] Chiou D 2007 *Phys. Rev.* D **75** 024029
[24] Ashtekar A and Wilson-Ewing E 2009 *Phys. Rev.* D **79** 083535
[25] Garay L J, Martín-Benito M and Mena Marugán G A 2010 *Phys. Rev.* D **82** 044048
[26] Martín-Benito M, Mena Marugán G A and Wilson-Ewing E 2010 *Phys. Rev.* D **82** 084012
[27] Wilson-Ewing E 2010 *Phys. Rev.* D **82** 043508
[28] Belinskiĭ V A, Khlatnikov I M and Lifshitz E M 1970 *Adv. Phys.* **19** 525
[29] Corichi A and Singh P 2009 *Phys. Rev.* D **80** 044024
[30] Ashtekar A and Sloan D 2010 *Phys. Lett.* B **694** 108
[31] Bojowald M 2002 *Phys. Rev. Lett.* **89** 261301
[32] Singh P 2006 *Phys. Rev.* D **73** 063508
[33] Martín-Benito M, Garay L J and Mena Marugán G A 2008 *Phys. Rev.* D **78** 083516
[34] Mena Marugán G A and Martín-Benito M 2009 *Int. J. Mod. Phys.* A **24** 2820
[35] Corichi A, Cortez J, Mena Marugán G A and Velhinho J M 2006 *Classical Quantum Gravity* **23** 6301
[36] Cortez J, Mena Marugán G A and Velhinho J M 2007 *Phys. Rev.* D **75** 084027
[37] Cortez J, Mena Marugán G A and Velhinho J M 2010 *Phys. Rev.* D **81** 044037
[38] Cortez J, Mena Marugán G A, Olmedo J and Velhinho J M 2010 *JCAP* **10** 030
[39] Gowdy R H 1974 *Ann. Phys.* **83** 203
[40] Brizuela D, Mena Marugán G A and Pawłowski T 2010 *Classical Quantum Gravity* **27** 52001

Journal of Physics: Conference Series **314** (2011) 012013 doi:10.1088/1742-6596/314/1/012013

Towards modelling the central engine of short GRBs

José A. Font[1], Luciano Rezzolla[2], Bruno Giacomazzo[3,4], Luca Baiotti[5], David Link[6]

[1] Departamento de Astronomía y Astrofísica, Universitat de València, 46100 Burjassot (Valencia), Spain
[2] Max-Planck-Institut für Gravitationsphysik, Albert-Einstein-Institut, Potsdam-Golm, Germany
[3] Department of Astronomy, University of Maryland, College Park, MD, USA
[4] Gravitational Astrophysics Laboratory, NASA Goddard Space Flight Center, Greenbelt, MD USA
[5] Institute of Laser Engineering, Osaka University Yamada-oka 2-6, Suita, Osaka 565-0871, Japan
[6] Institut für Physik, Humboldt-Universität zu Berlin, Berlin, Germany

E-mail: `j.antonio.font@uv.es`

Abstract. Numerical relativity simulations of non-vacuum spacetimes have reached a status where a complete description of the inspiral, merger and post-merger stages of the late evolution of close binary neutron systems is possible. Determining the properties of the black-hole-torus system produced in such an event is a key aspect to understand the central engine of short-hard gamma-ray bursts (sGRBs). Of the many properties characterizing the torus, the total rest-mass is the most important one, since it is the torus' binding energy which can be tapped to extract the large amount of energy necessary to power the sGRB emission. In addition, the rest-mass density and angular momentum distribution in the torus also represent important elements which determine its secular evolution and need to be computed equally accurately for any satisfactory modelling of the sGRB engine. In this paper we summarize our recent results from fully general-relativistic simulations of the coalescence of unequal-mass binary neutron stars, whose evolution is followed through the inspiral phase, the merger and prompt collapse to a black hole, up until the appearance of a thick accretion disk, which is studied as it enters a regime of quasi-steady accretion. Our simulations show that large-scale, quasi-Keplerian tori with masses as large as $\sim 0.2\,M_{\odot}$ can be produced as the result of the inspiral and merger of binary neutron stars with unequal masses.

1. Introduction

Simulating the merger of binary neutron stars in general relativity has been a long-term major target of numerical relativity. Progress has been steady, owing to the difficulties imposed by the presence in such setting of strong gravitational and magnetic fields, shock waves, and matter motion at relativistic speeds. In addition, the multidimensional character of the scenario and the formation of curvature singularities as a plausible outcome of the merger, have also conspired to slow down progress. Nevertheless, considerable advances have been achieved in recent years due to major developments in several key aspects of the problem, such as the formulation of the equations and the incorporation of equations of state (EOS) from nuclear physics. Likewise, the use of high-resolution numerical methods and adaptive mesh refinement, along with increased computational resources, have also permitted to perform the simulations to significant levels of accuracy. A view of the state-of-the-art in the field is offered by the works of [1, 2, 3, 4, 5, 6, 7, 8, 9]

The characterization of the gravitational radiation produced in such mergers is one of the prime motivations behind the simulations. Such events are among the most promising sources of detectable gravitational radiation for laser interferometric detectors. Current estimates for the detection rate relative to first-generation detectors is ~ 1 event per $\sim 40-300$ years, increasing to an encouraging $\sim 10-100$ events per year for the advanced detectors [10]. The second motivation to simulate binary neutron star mergers is establishing whether the end-product of the merger can operate as the central engine of short-hard gamma-ray bursts (sGRBs) [11, 12]. Depending on the suitability of the model parameters, existing simulations show the formation of a stellar mass black hole surrounded by a hot disk. Driven by neutrino processes and magnetic fields, such a compact system may indeed be capable of launching a relativistic fireball with an energy of $\sim 10^{48}$ erg on a timescale of $0.1-1$ s [13].

In this paper we present a summary of our recent efforts in the simulation of binary neutron star mergers. Full details of these simulations were given in [9] to which the interested reader is addressed for a complete account of our study. In particular, we present here the results for only two models, and focus our discussion on the late-time dynamics of the torus formed after the merger of such two unmagnetized, equal- and unequal-mass neutron star binaries. As reported in [9] our simulations indicate that large-scale tori with masses as large as $\sim 0.2\,M_{\odot}$ can be produced as the result of the inspiral and merger of binary neutron stars with unequal masses. These tori are typically of large size, with quasi-Keplerian distribution of angular momenta, showing quasi-stationary evolutions and the absence of dynamical instabilities. Furthermore, the gravitational-wave emission reveals that the waveforms are sensitive to the mass ratio in the binary, both during the inspiral and after the merger, and could be used to extract important information on the structure and EOS of the progenitor stars.

2. Physical and computational framework

The basic set of equations we solve for the simulations reported in this work includes the equations describing the dynamics of the gravitational field (the Einstein equations) and the equations describing the dynamics of the matter. The Einstein equations are formulated following the conformal traceless BSSN formalism with a "1 + log" slicing condition and a "Gamma-driver" shift condition (the interested reader is addressed to [14] for a detailed discussion of the equations and gauges used). These equations are solved using the `CCATIE` code, a three-dimensional finite-differencing code. Correspondingly, the general-relativistic hydrodynamics equations are solved using the `Whisky` code [15], which adopts a flux-conservative formulation of the equations as presented in [16] and high-resolution shock-capturing schemes. For the simulations we employ an ideal fluid EOS, in which the pressure p is expressed as $p = \rho\epsilon(\Gamma-1)$, where ρ is the rest-mass density, ϵ is the specific internal energy, and Γ is the adiabatic exponent. The value chosen for the latter is $\Gamma = 2$. As in [3], the gravitational-wave signal is extracted using two independent methods, on the one hand a method based on the Newman-Penrose formalism and, on the other hand, a method based on the measurements of the non-spherical gauge-invariant perturbations of a Schwarzschild black hole. The hierarchy of the computational grid, one of the most important aspects of the simulations, is handled by the `Carpet` mesh refinement driver [17]. We use six refinement levels with a $2:1$ refinement factor for successive levels. The timestep on each grid is set by the Courant condition and, hence, by the spatial grid resolution for that level; the typical CFL coefficient is set to 0.35. The time evolution is carried out using a 4th-order–accurate Runge-Kutta integration algorithm. Boundary data for finer grids are calculated with spatial prolongation operators employing 3rd-order polynomials and with prolongation in time employing 2nd-order polynomials. Initially, the number of grid points across the star is ~ 100. The torus surrounding the black hole after

Journal of Physics: Conference Series **314** (2011) 012013 doi:10.1088/1742-6596/314/1/012013

the merger is usually not contained within the finest grid, but its high-density region is covered by the second finest grid with resolution $\Delta_2 = 0.3$. The whole grid is set up to be symmetric with respect to the (x, y) plane. The boundary conditions are chosen to be radiative for the metric, in order to prevent gravitational waves from scattering back into the grid, and static for the hydrodynamical variables.

We use quasi-equilibrium initial data generated with the multi-domain spectral-method code LORENE developed at the Observatoire de Paris-Meudon [18]. A full list of all considered models together with a selection of physical quantities defining them, *e.g.* baryon mass and ADM mass, orbital frequency and initial angular momentum, etc., is given in [9]. (The models of the lowest mass ratios in Ref. [9] were kindly provided by Dorota Gondek-Rosińska.) Here, as mentioned in the Introduction, we only focus in two representative models of the sample discussed in [9], as these highlight the differences in the evolution of equal-mass and unequal-mass binaries. Following the notation of [9] we label these two models M3.6q1.00 and M3.4q0.70, respectively. The first one is an equal-mass model (mass ratio $q = 1$) with total baryonic mass $M_{\rm tot} = 3.6 M_\odot$ and the second one is an extreme unequal-mass model ($q = 0.7$) with total baryonic mass $M_{\rm tot} = 3.4 M_\odot$.

3. Results

Figure 1 shows a selection of representative isodensity contours on the equatorial plane for both models. At the initial time (top panel), the stars are in their quasi-equilibrium configuration at a coordinate separation of 45 km. The asymmetry of model M3.4q0.70 before the evolution starts is already apparent. The more massive star is much more compact than its extended companion, which is deformed already at the initial distance by tidal forces. Along the evolution of model M3.6q1.00 (see [9]), matter is ejected in small amounts during the inspiral phase and in larger amounts during the merger phase, when shock waves resulting from the collision are much stronger. On the contrary, model M3.4q0.70 is characterized by the formation of large spiral arms extending well beyond the domain shown in the figure, which are responsible for the rapid ejection of matter. Therefore, for model M3.4q0.70, gravitationally bound matter travelling along the spiral arms away from the central object will form a more massive accretion torus around the central black hole than that formed in the case of an equal-mass, symmetric binary system.

For model M3.6q1.00, the mass and spin of the black hole measured at formation are $M = 2.56\ M_\odot$ and $a \equiv J/M^2 = 0.745$, values which barely change throughout the subsequent evolution of the system. These quantities could not be measured accurately for model M3.4q0.70 (see [9] for details). The recoil velocity imparted to the black hole at the end of the inspiral phase are respectively 0.28 km/s for model M3.6q1.00 and 15.82 km/s for model M3.4q0.70. While such recoil velocities are much smaller than those measured for binary black holes, they could still yield astrophysically interesting results being comparable or larger than the escape velocity from the core of a globular cluster, $v_{\rm esc} \sim 50$ km/s [19].

The middle panels of figure 1 correspond to the time when the system enters the regime of quasi-stationary accretion (QSA) shortly after the formation of the black hole, while the bottom two panels refer to the final time of the evolution. We define the onset of the QSA as the time when the condition $\dot{M}_{\rm tot}/M_{\rm tot} < 10^{-6} (Gc^{-3}\, M_\odot)^{-2}$ is satisfied for the first time, where $M_{\rm tot}$ is the total mass of the binary and $\dot{M}_{\rm tot}$ is the mass accretion rate, defined as $\dot{M}_{\rm tot} = \frac{d}{dt} \int \rho W \sqrt{\gamma}\, d^3x$, where W is the Lorentz factor and γ is the determinant of the spatial metric. In addition, figure 2 shows color-coded contours of the rest-mass density for both models in the (x, z) plane. Taken together, these two figures allow for a closer view of the morphological features of the forming disks, in particular, their spatial dimension and thickness. The morphological differences between both models are significant. The equal-mass model produces a highly symmetric, geometrically thin disk, similar to the ones already observed

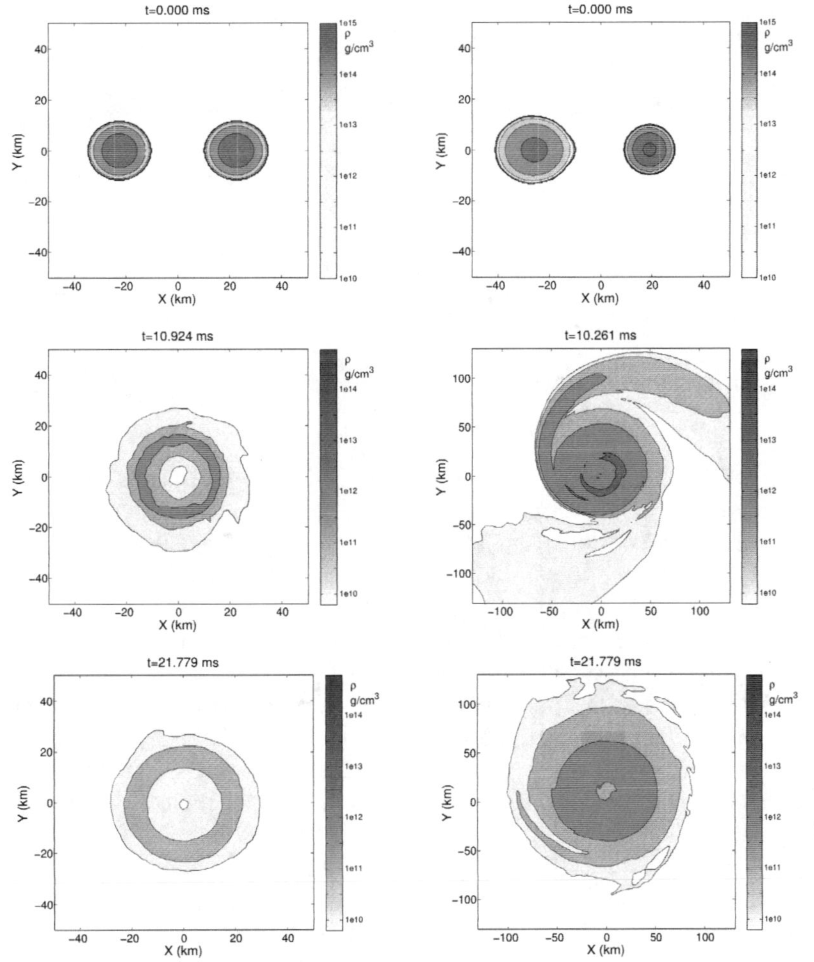

Figure 1. Isodensity contours for models M3.6q1.00 (left) and M3.4q0.70 on the (x, y) plane. The top panels show the binaries at $t = 0$ to highlight the asymmetry of the unequal-mass model. The times when the frames have been taken are shown on top of the plots while the color-code for the rest-mass density is indicated to the right of each plot. The middle panels show the morphology of the tori at the onset of the quasi-stationary accretion, while the bottom panels correspond to the end of the simulation. Note the change in the scale of the plots of the last two panels of model M3.4q0.70.

for other equal-mass initial data in [3]. The unequal-mass model, on the other hand, at the time of the onset of the regime of QSA is characterized by the presence of a large spiral arm, which has not yet been accumulated onto the central disk surrounding the formed black hole. The diameters of the disks and their vertical heights differ in a significant way between the two models: at the end of the evolution and using the $\rho = 10^{10}\,\mathrm{g/cm^3}$ isodensity contour as the reference value below which material is not considered part of the disk, our simulations yield disk diameters of $\sim$ 50 km for model M3.6q1.00 and $\sim$ 150 km for model M3.4q0.70. The corresponding vertical scale is $\sim$ 5 km and $\sim$ 35 km, respectively. While the tori differ in size by a factor $\sim$ 3, they differ by a factor $\sim$ 200 in mass while having comparable mean rest-mass densities.

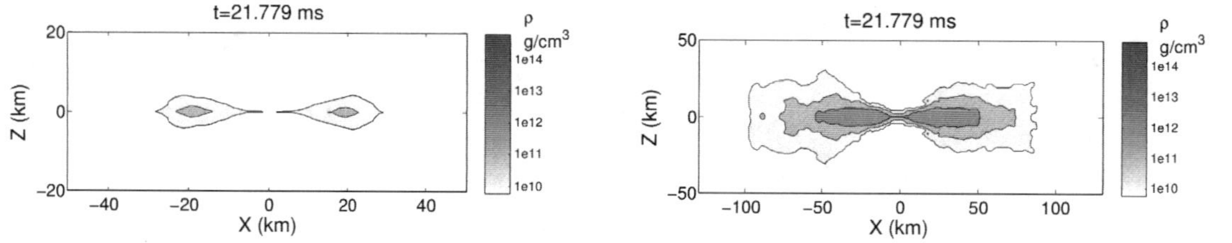

Figure 2. Isodensity contours for models M3.6q1.00 (left panel) and M3.4q0.70 showing the morphology of the tori at the end of the simulation on the (x, z) plane. Note that the disks in the two panels have very different lengthscales. The thickness of the disk for the unequal-mass binary is apparent.

The violent dynamics of the merger process is reflected on an associated large emission of gravitational radiation. This issue was discussed in [9], to which the interested reader is addressed for further specific details. The main result of our findings is that present gravitational-wave detectors are unlikely to detect any of the binaries of our sample if they are located at a distance of 100 Mpc and observed only during the final part of the inspiral. However, advanced detectors will be able to reveal these sources even at large distances and measure them with significant SNRs in the case of third-generation detectors such as the Einstein Telescope [20].

In agreement with previous simulations by [1], we have found that the smaller the mass ratio q of the binary the larger the mass of the resulting accretion disk. It should be noted that the definition of what constitutes a torus and, thus, what is its mass is rather arbitrary. This is due to the fact that, after the apparent horizon has formed, a substantial part of the rest-mass lies still outside, although it will accrete rapidly onto the black hole. Given this arbitrariness, we define the torus mass M_{tor} as the total rest mass outside the apparent horizon when the disk enters the QSA regime, a regime which is found in all models investigated in [9]. With this definition we find that while the equal-mass model produces a disk of barely $10^{-3} M_{\odot}$, model M3.4q0.70 produces significantly more massive tori with masses of about $0.2 M_{\odot}$. A phenomenological expression for the torus mass was derived in [9] (see also [1]), which showed that tori with masses up to $\sim 0.35 M_{\odot}$, are possible for mass ratios $q \sim 0.75 - 0.85$. These values are well above the typical lower bounds discussed in the context of short GRBs triggered by thermal energy deposition above the poles of the black hole by $\nu\bar{\nu}$ annihilation [21].

In [9] spacetime diagrams for observers comoving with the black hole were used to analyze the evolution of the tori and to gain some insight on their dynamics. An example is shown in figure 3, which displays the evolution of the color-coded rest-mass density embedded in a spacetime diagram with the $x - x_{\text{AH}}$ coordinate on the horizontal axis, where x_{AH} is the position of the apparent horizon, and the coordinate time t on the vertical axis. By comparing both diagrams it is again evident the stricking difference in the evolution of both models. On the one hand, the equal-mass model shows a global oscillatory movement with respect to the location of the black hole horizon. On the other hand, the spacetime diagram for the density evolution of model M3.4q0.70 shows very rapid expansions corresponding to the ejection of the large spiral waves mentioned before. As discussed by [9], most of this matter is still bound, as the criterion $u_t > -1$ is fulfilled, but it nevertheless reaches distances which are several hundreds of km away from the black hole, leading to tori that have spatial dimensions as large as $\sim$ 80 km. This matter will eventually fall back onto the tori, where it may lead to enhanced accretion and consequently to a new and *delayed* gamma-ray emission as the one recently observed in [22].

The stability of the tori resulting from binary neutron star mergers is a central issue for current models of GRBs. Any instability which might disrupt the black hole plus thick disk

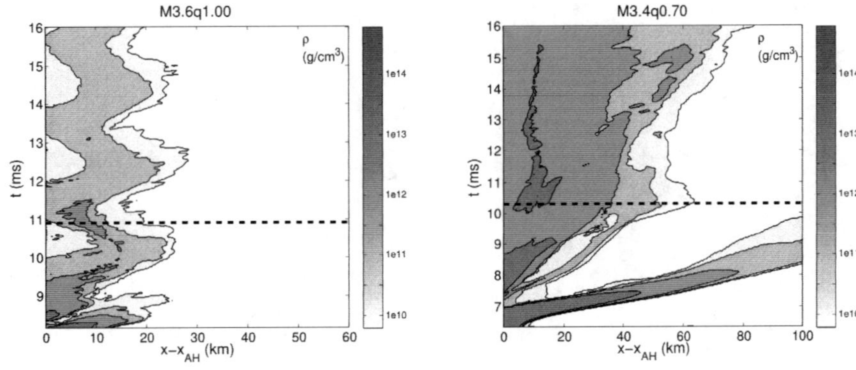

Figure 3. Evolution of the rest-mass density ρ for models M3.6q1.00 (left) and M3.4q0.70 along the positive x axis in a frame comoving with the BH. The panels show the color-coded rest-mass density embedded in a spacetime diagram with the $(x - x_{\rm AH})$ coordinate on the horizontal axis, being $x_{\rm AH}$ the position of the apparent horizon, and the coordinate time t on the vertical axis. For each model, the dotted horizontal line marks the onset of the regime of QSA.

system on timescales shorter than the duration of the energy release by the GRB source could pose a severe problem for the prevailing models. One such instability is the so-called runaway instability [23]. General relativistic hydrodynamical simulations of such instability were first performed by [24, 25, 26, 27], who treated the dynamics of the gravitational field in an approximate way and neglected the self-gravity of the torus. These investigations concluded that tori with constant distribution of specific angular momentum were unstable while non-constant (power-law) angular momentum disks were stable. More recently, in [28] the first simulations in full general relativity of marginally-stable self-gravitating tori in axisymmetry were presented. The purpose of this work was to evaluate the influence of the torus self-gravity on the runaway instability. The results of [28] indicate that tori are indeed stable irrespective of the angular momentum distribution. Our simulations, summarized in the spacetime diagrams of figure 3, which are not restricted to axisymmetry but are however constrained to much shorter timescales, reach the same conclusion: Self-gravitating tori around black holes, as those produced by the merger of binary neutron stars, are stable at least on the dynamical timescales investigated here. Further evidence has been recently collected by [29] where different types of non-axisymmetric instabilities of self-gravitating disks have been investigated. It is also worth pointing out that in the case of *magnetized* binary neutron star mergers, the resulting disks could be however unstable to the magneto-rotational instability [30].

The global oscillations of the equal-mass model shown in the left panel of the spacetime diagram of figure 3 are also found in the mass accretion rate, a quantity which is central to estimate the lifetime of the disk and which was shown in [9] to be subject to quasi-periodic oscillations as the torus moves in and out at about the radial epicyclic frequency. The mass flux of the unequal-mass model, on the other hand, is found to be rather constant in time, a fact which reflects a very different distribution of angular momentum in the tori. The profiles of the specific angular momentum $\ell \equiv -u_\phi/u_t$ along the x-axis are displayed in the left panel of figure 4, for the tori produced by the two models (model M3.6q1.00, blue lines extending to $\sim$ 20 km; model M3.4q0.70, red lines extending up to $\sim$ 70 km). The profiles are computed in a frame comoving with the black hole and for densities $\rho > 10^{10}\,{\rm g/cm^3}$. Quite clearly, the specific angular momentum decreases outward at all times for the equal-mass binary, while it increases outward for the unequal-mass one (although it was initially decreasing at the innermost

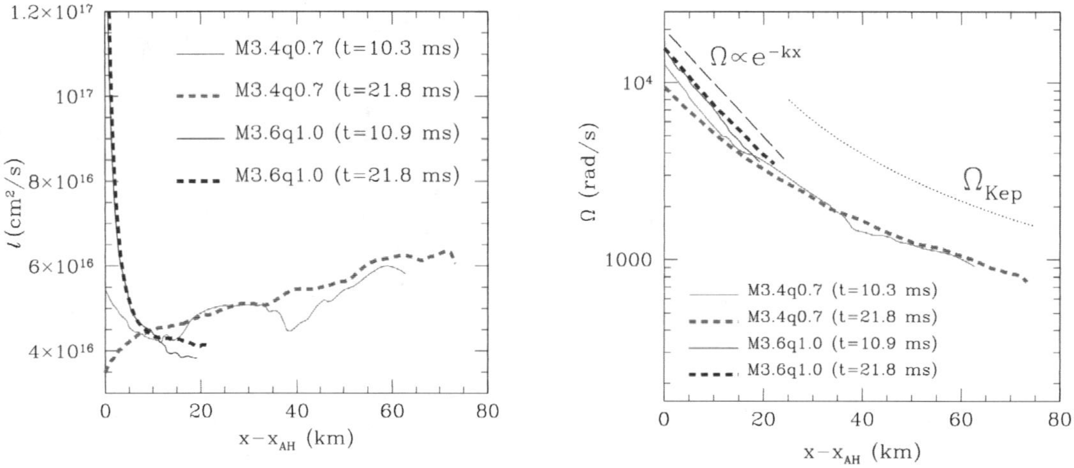

Figure 4. Profiles along the x-axis of the specific angular momentum (left) and angular velocity (right) of the tori produced by the two binaries considered. Different line types refer either to the onset of the QSA ($t \sim 10$ ms, thin solid lines) or to the end of the simulation ($t \sim 22$ ms, thick dashed lines). Left plot: Note the different profiles present in both models and that the specific angular momentum for the unequal-mass case increases outwards. Right plot: the dotted line indicates the reference Keplerian angular velocity, which matches very well the outer parts of the torus from the unequal-mass binary. The long-dashed line is an exponentially decaying profile, which instead reproduces well the profile for the equal-mass binary.

parts). Rayleigh's criterion for dynamical stability is clearly satisfied by model M3.4q0.70 but it is equally-clearly violated by model M3.6q1.00, which is nevertheless stable. The reason for this could be due to the fact that Rayleigh's criterion assumes stationary and purely azimuthal motion, a condition which is not satisfied for the unequal-mass binary, which shows instead large radial epicyclic oscillations.

Correspondingly, the right panel of figure 4 shows the profiles of the angular velocity Ω along the x-axis for the tori produced by our two models. While the equal-mass binary has an exponentially decaying profile (long-dashed line), $\Omega \propto \exp[-k(x - x_{\rm AH})] \sim \exp[-0.07(x - x_{\rm AH})]$, which does not change significantly with time, the unequal-mass binary reaches at the end of the simulation a profile which is, especially in the outer parts, essentially Keplerian, with $\Omega_{\rm Kep} \sim x^{-3/2}$ (dotted line). This feature explains the scaling of the specific angular momentum as $\ell \sim x^{1/2}$ and provides firm evidence that the tori produced in this case will be dynamically stable.

4. Summary

We have presented results from a subset of simulations of binary neutron star mergers reported in [9]. Our discussion has been focused on the late-time dynamics of the torus formed after the merger of two unmagnetized, equal- and unequal-mass neutron star binaries. The evolution of the neutron stars has been followed through the inspiral phase, the merger and prompt collapse to a black hole, up until the appearance of an accretion disk, which has been studied as it enters and remains in a regime of quasi-steady accretion.

We have found that the torus mass increases with the initial mass asymmetry of the neutron stars. Equal mass binaries with a total baryonic mass larger than 3.7 $M_\odot$ lead to prompt black hole formation and produce tori with radial extension of ~ 30 km and with masses $\sim 10^{-3}$ $M_\odot$. On the contrary, unequal-mass binaries produce significantly more massive tori with masses of about $0.2M_{\odot}$. The associated tori are also much more extended with typical sizes ~ 150 km. A phenomenological expression for the torus mass derived in [9] showed that tori with masses up to $\sim 0.35\ M_{\odot}$ are possible.

The torus dynamics is remarkably different for the equal- and unequal-mass binaries. While tori from equal-mass binaries exhibit a quasi-periodic form of accretion associated with the radial epicyclic oscillations of the disks, those from unequal-mass binaries exhibit a quasi-steady form of accretion. In addition, the analysis of the distribution of the angular momentum in the tori has shown no evidence for the development of non-axisymmetric instabilities. Angular momentum is transported outwards more efficiently for smaller values of the mass asymmetry, yielding Keplerian angular velocity distributions. As discussed in detail in [9], advanced gravitational wave detectors will be able to reveal these sources at large distances (~ 100 Mpc) and measure them with significant signal-to-noise ratios.

Acknowledgments

We are grateful to the Meudon group and to D. Gondek-Rosińska for producing some of the initial data used in the calculations reported in [9] which form the basis for the results of this paper. Additional gratitude goes to Gian Mario Manca, E. Schnetter and all the `Carpet/Cactus` developers. The computations were performed on the Damiana cluster at the AEI, on the MareNostrum cluster at the Barcelona Supercomputing Center and on the Ranger cluster at the Texas Advanced Computing Center through TERAGRID allocation TG-MCA02N014. This work was supported in part by the DFG grant SFB/Transregio 7, by "CompStar", a Research Networking Programme of the European Science Foundation, by the JSPS Postdoctoral Fellowship For Foreign Researchers, Grant-in-Aid for Scientific Research (19-07803), and by the Spanish Ministerio de Educación y Ciencia (AYA 2007-67626-C03-01).

References

[1] Shibata M and Taniguchi K 2006 *Phys. Rev. D* **73** 064027
[2] Anderson M *et al.* 2008 *Phys. Rev.* **D77** 024006
[3] Baiotti L, Giacomazzo B and Rezzolla L 2008 *Phys. Rev. D* **78** 084033
[4] Anderson M, Hirschmann E W, Lehner L, Liebling S L, Motl P M, Neilsen D, Palenzuela C and Tohline J E 2008 *Phys. Rev. Lett.* **100** 191101
[5] Baiotti L, Giacomazzo B and Rezzolla L 2009 *Class. Quantum Grav.* **26** 114005
[6] Liu Y T, Shapiro S L, Etienne Z B and Taniguchi K 2008 *Phys. Rev. D* **78** 024012
[7] Giacomazzo B, Rezzolla L and Baiotti L 2009 *MNRAS* **399** L164–L168
[8] Kiuchi K, Sekiguchi Y, Shibata M and Taniguchi K 2009 *Phys. Rev. D* **80** 064037
[9] Rezzolla, L, Baiotti, L, Giacomazzo, B, Link, D and Font, J.A. 2010, *Class. Quantum Grav.* **27** 114105
[10] Belczynski K, Taam R E, Kalogera V, Rasio F and Bulik T 2007 *Astrophysical Journal* **662** 504
[11] Narayan R, Paczynski B and Piran T 1992 *Astrophys. J.* **395** L83
[12] Zhang B and Meszaros P 2004 *Int. J. Mod. Phys. A* **19** 2385
[13] Piran T 2004 *Rev. Mod. Phys.* **76** 1143–1210
[14] Baiotti L, Hawke I and Rezzolla L 2007 *Class. Quantum Grav.* **24** S187–S206
[15] Baiotti L, Hawke I, Montero P J, Löffler F, Rezzolla L, Stergioulas N, Font J A and Seidel E 2005 *Phys. Rev. D* **71** 024035
[16] Banyuls F, Font J A, Ibáñez J M, Martí J M and Miralles J A 1997 *Astrophys. J.* **476** 221
[17] Schnetter E, Hawley S H and Hawke I 2004 *Class. Quantum Grav.* **21** 1465–1488
[18] Gourgoulhon E, Grandclément P, Taniguchi K, Marck J A and Bonazzola S 2001 *Phys. Rev. D* **63** 064029
[19] Webbink R F 1985 *Dynamics of Star Clusters* (*IAU Symposium* vol 113) ed J Goodman & P Hut pp 541–577
[20] Einstein Telescope `http://www.et-gw.eu`
[21] Oechslin R and Janka, H-Th 2006 *Astron. Astrophys.* **368**, 1489–1499
[22] Giuliani A and others 2010 *Astrophys. J.* **708** L84–L88
[23] Abramowicz M A, Calvani M and Nobili L 1983 *Nature* **302** 597–599
[24] Font J A and Daigne F 2002 *MNRAS* **334** 383–400
[25] Font J A and Daigne F 2002 *Astrophys.J* **581** L23–L26
[26] Zanotti O, Rezzolla L and Font J A 2003 *Mon. Not. Roy. Soc.* **341** 832

[27] Daigne F and Font J A 2004 *MNRAS* **349** 841–868
[28] Montero P J, Font J A and Shibata M 2010 *Phys. Rev. Lett.* **104** 191101
[29] Korobkin O, Abdikamalov E, Schnetter E, Stergioulas N and Zink B 2010 arXiv:1011.3010
[30] Giacomazzo B, Rezzolla L and Baiotti L 2010 *arXiv:1009.2468*

Journal of Physics: Conference Series **314** (2011) 012014 doi:10.1088/1742-6596/314/1/012014

Milli-Hertz Gravitational Waves: *LISA* and *LISA PathFinder*

H Araújo[5], P Cañizares[1,2], M Chmeissani[3], A Conchillo[1], M Díaz-Aguiló[2,4], E García-Berro[2,4], L Gesa[1], F Gibert[2], C Grimani[6], W Hajdas[7], D Hollington[5], I Lloro[1], A Lobo[1,2], I Mateos[1], M Nofrarias[8], C Puigdengoles[3], J Ramos-Castro[9], J Sanjuán[10], CF Sopuerta[1,2] and P Wass[5]

[1] Instituto de Ciencias del Espacio, *CSIC*, Campus *UAB*, Facultat de Ciències, Torre C-5 Parell, 2ª Planta, E-08193 Bellaterra (Cerdanyola del Vallès), Barcelona, Spain
[2] Institut d'Estudis Espacials de Catalunya (*IEEC*), Edifici *Nexus*, Gran Capità 2–4, 08034 Barcelona, Spain
[3] Institut de Física d'Altes Energies (*IFAE*), Edifici C, Universitat Autònoma de Barcelona, 08193 Bellaterra, (Barcelona), Spain
[4] Departament de Física Aplicada, Universitat Politècnica de Catalunya, Escola Politècnica Superior de Castelldefels, Avda. Canal Olímpic s/n, 08860 Castelldefels, Spain
[5] Blackett Laboratory, Imperial College London, Prince Consort Road, London SW7 2BW, UK
[6] Università degli Studi di Urbino, and *INFN* Florence, Dipartimento DiSBeF, Via Santa Chiara 27, 61029 Urbino, Italy
[7] Department of Particles and Matter, Paul Scherrer Institut, ODRA 120, 5232 Villigen, Switzerland
[8] Max-Planck-Institut für Gravitationsphysik (Albert-Einstein-Institut), Callinstrasse 38, D-30167 Hannover, Germany
[9] Departament d'Enginyeria Electrònica, *UPC*, Campus Nord, Edif. C4, Jordi Girona 1–3, 08034 Barcelona, Spain
[10] Department of Physics, University of Florida, NPB–22258 PO Box 118 440, Gainesville FL 32611-8440, USA

E-mail: `lobo@ieec.fcr.es`

Abstract. Ground based GW detectors are limited at their lower frequency band (1–10 Hz) by settlement gravity gradients and seismic noise, and their sensitivity peaks at around 100 Hz. Sources in this band are mostly short duration signals, and their rates uncertain. Going down to milli-Hertz frequencies significantly increases the number and types of available sources. *LISA* was planned with the idea to explore a likely richer region of the GW spectrum, beyond that accessible to ground detectors; the latter are however expected to produce the first GW observations. In this paper I will present the main *LISA* concepts; in particular, emphasis will be placed on *LISAPathFinder*, the *ESA* precursor of *LISA*, in which our research group in Barcelona is heavily involved.

1. Introduction

In the classical Newtonian theory of Gravity, gravitational fields do not propagate away from their sources, they rather instantly spread across the surrounding space out to infinite distances. This is of course incompatible with the causality principle set out in Einstein's Special Relativity

Published under licence by IOP Publishing Ltd

Theory [1], i.e., that there is no possibility that either energy or matter travel at speeds higher than the speed of light, c. Obviously, gravitational fields carry energy, hence they may not propagate to remote places instantly. The existence of Gravitational Waves (GW), or some sort of radiation gravitational field, must therefore exist in Nature if the causality principle, as stated above, is to hold true.

Einstein described GWs for the first time shortly after he completed his General Relativity theory [2]. He found that, according to his theory, GWs are small perturbations in the space-time geometry which, in an otherwise flat (Lorentz) space-time, travel at the speed of light. He also found they have quadrupole structure with only two polarisation degrees of freedom. Finally, he gave (approximate) formulae for the GW amplitudes and the energy loss rate to gravitational radiation by mechanical systems.

Strong confidence in Einstein's General Relativity theory predictions, endorsed by every piece of experimental and observational evidence so far, is a good motivation to search for GWs. Although this is of course a weak scientific argument, it did encourage the first research efforts to detect GWs. In 1974 the announcement by Russell Hulse and Joseph Taylor of the discovery of the binary pulsar PSR B1913+16 was meant to change the status of GWs from a theoretical conjecture into a real physical phenomenon. Later on, in 1993 Hulse and Taylor were awarded the Nobel Prize of Physics "for the discovery of a new type of pulsar, a discovery that has opened up new possibilities for the study of gravitation"; this indeed included the radiation of GWs from the binary system, hence we no longer need the above confidence argument to advocate the interest of GW detection research...

PSR B1913+16 was tracked until 2006, when the timing measurement precision reached the confusion limit with other noise sources —see reference [3] for technical details. The GW radiation effect is inferred from the secular shrinkage of the pulsar's orbit, normally expressed in the equivalent term of orbital period decay. This is seen to proceed, year after year, with an ever increasing coincidence with the predictions of General Relativity, the latest result [3] being that

$$\frac{\dot{P}_{\rm measured}}{\dot{P}_{\rm theory}} = 0.997 \pm 0.002 \tag{1}$$

i.e., theory and observations match to around 0.1% —a remarkable figure.

Quite a few other binary pulsars have been discovered since 1974, and all of them have proved to be the excellent gravity laboratories they were foreseen to be, in particular as regards the confirmation that binary systems emit gravitational radiation —see [4] for a review.

Table 1 provides a number of interesting parameters of the Hulse and Taylor binary pulsar. Let us consider a few consequences which will be useful further down in this review.

Character	Parameter	Value
	Orbital period	$P = 0.322997448930\,(4)$ days
Keplerian	Eccentricity	$e = 0.6171338\,(4)$
	Projected semi-major axis	$a\,\sin i = 2.3417725\,(8)$ light-sec
	Periastron precession	$\langle\dot{\varphi}\rangle = 4.226595\,(5)$ deg/year
Post-Keplerian	Time dilation + grav redshift	$\gamma = 0.0042919\,(8)$ sec
	Orbital period decay	$\dot{P} = \text{-}2.4184\,(9)\times 10^{-12}$
	Pulsar mass	$m_{\rm p} = 1.4414\,(2)\ M_\odot$
General Relativity	Companion mass	$m_{\rm c} = 1.3867\,(2)\ M_\odot$
	Orbit inclination	$\sin i = 0.73\,(4)$

Table 1. PSR B1913+16 parameters (drawn from reference [3])

Journal of Physics: Conference Series **314** (2011) 012014 doi:10.1088/1742-6596/314/1/012014

As already mentioned, General Relativity is strongly endorsed by the accuracy with which the results of experiments/observations match to this date its theoretical predictions. Concerning the binary pulsar, we can use the data in table 1 and the equations of General Relativity to *infer* the following for PSR B1913+16:

GW emission frequency	$f \sim 70\ \mu$Hz
Current GW emission amplitude	$h \sim 2\times10^{-23}$
Estimated lifetime	$\tau \sim 300$ million years

As we see, the GW emission happens at a very low frequency and with a very tiny amplitude. Nevertheless, as the binary looses energy to GWs, these two parameters (frequency and amplitude) grow, while the system size shrinks until the stars eventually merge in about 300 million years. Since this figure is much less than the age of the Universe, one reasonably expects that many other binaries exist which are older than PSR B1913+16, hence that they have entered the frequency band and sensitivity range of operating GW detectors and can be seen.

Since this review is about milli-Hz GW detection, the above gives at least an indicative idea of where GWs can surely be found. In the following sections we will present *LISA*, the first space-borne GW detector, and its expected scientific yield, i.e., the foreseen (and unforeseen) GW sources emitting at frequencies around 1 milli-Hz, a band where many are expected. We will also address briefly *LISA PathFinder* the technological precursor of *LISA*, in which we are involved.

2. The *LISA* concept and scientific scope

LISA (Laser Interferometer Space Antenna) is based on the same working principles as e.g. *VIRGO* and *LIGO*: a passing gravitational wave causes a distortion of the geometric properties of the space-time region it sweeps; electromagnetic waves circulating across that region are indeed also affected by the GWs, and this happens in the form of *phase shifts*. For example, the phase a beam of light at point B travelling from point A, say, is different if there is or there is not a GW in the intervening space. This fact can be taken advantage of to measure such phase shifts by interferometry, thence to detect GWs [5][1].

If laser links are established between pairs of masses which are in (nominally) pure geodesic motion, at least along the directions of the links, then phase shifts in the laser beams can be compared after the light is reflected in the remote masses and recombined in their origin by optical interferometry. This is how *LIGO* and *VIRGO* work, and also how *LISA* works —except that the latter has specific requirements due to its much larger size, and to its being in space. According to the theory [5], the phase shift difference between the interfering beams as they recombine after their round trips up and down the respective interferometer arms is

$$\delta\phi(t) \propto h(t)\,\sin\left(\frac{\Omega_{\rm GW}L}{c}\right) \tag{2}$$

where $h(t)$ is the amplitude of the incoming GW[2], $\Omega_{\rm GW} = 2\pi f_{\rm GW}$ its angular frequency, L the interferometer arm-length and c the speed of light in vacuum; the GW is assumed to impinge perpendicularly on the interferometer plane. The proportionality factor depends on the angle between the interferometer arms, and the signal is maximum when this angle is 90°. A clear

[1] This is more often worded in terms of the GW changing the distance between the points A and B, but the final result, i.e., how is the signal seen in the interferometer, is the same.

[2] Actually a combination of $h_+(t)$ and $h_\times(t)$ depending on the angles between the GW's axes and the laboratory axes.

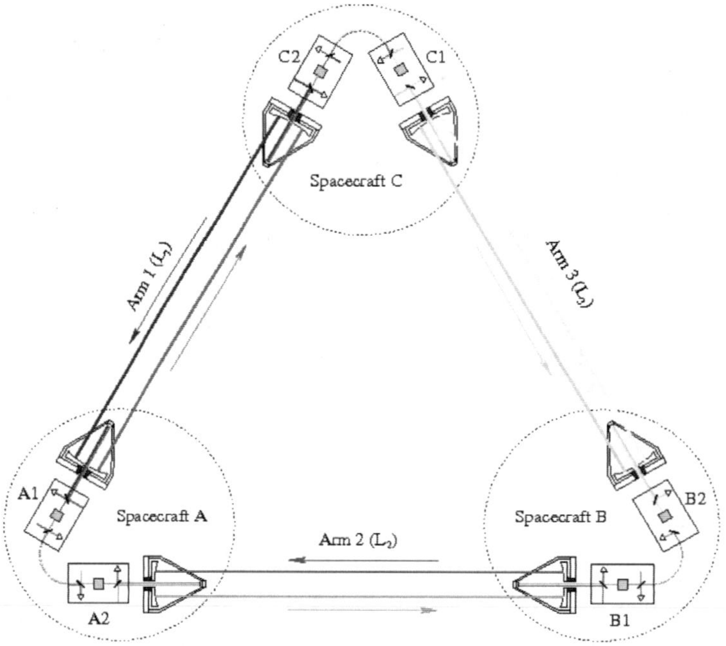

Figure 1. *LISA*'s equilateral triangle configuration. The distance between each pair of spacecraft is 5 million kilometres, which corresponds to a GW frequency of 15 milli-Hz — see equation (3). Each spacecraft houses two proof masses in nominally geodesic motion (free fall), and the relative distance variations in a frequency band around 1 mHz between corresponding proof masses in remote spacecraft is measured to picometre precision by means of laser interferometry.

and important consequence of equation (2) is that a maximum signal is obtained when the arm-length is tuned to the GW's wavelength as[3]

$$L = \frac{\lambda_{\rm GW}}{4} \tag{3}$$

This means if we want to optimise a detector for e.g. 1 kHz signals we need interferometer arms of 75 kilometres. This can be reduced by use of a multiple reflection system (or, equivalently, Fabry-Perot optical cavities), which can make sensitive detectors at 100 Hz with only 3–4 kilometre arm-lengths [6].

Astrophysical sources in lower frequency bands appear more likely, as powerful GW emission is bound to the motion of large masses, and these typically have long periods/characteristic time scales. If we wish to go down, say, to the milli-Hertz frequency range then we need arm-lengths in the order of tens of millions of kilometres. This is far beyond what can be done on earth, not only because of the arm-length constraint but also because the so called gravity gradient noise[4] at frequencies below ~1 Hz are extremely difficult to reduce [7].

In the early 90s, and following earlier studies during the previous 15 years or so to build a GW detector in space, an idea began to take shape which has basically survived to this date. This

[3] Note the relationship $\lambda_{\rm GW} = c/f_{\rm GW}$, which holds in the usual approximation that GWs are perturbations of flat space-time.

[4] This is due to the motion of masses around the detector settlement, whether soil vibrations, mass displacements, air winds, etc.

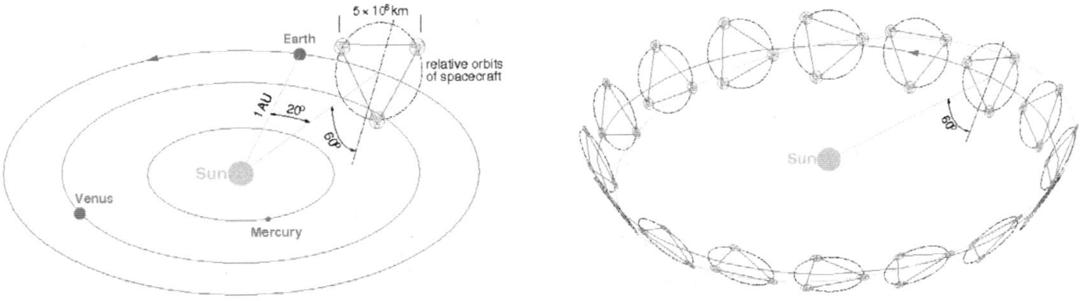

Figure 2. Left: *LISA*'s snapshot position in its heliocentric orbit. Note the tilt of the triangle plane relative to the ecliptic (60°) and the trailing angle behind the Earth (20°). Right: evolution of the constellation during one year. Note the normal to *LISA*'s plane describes a cone in the sky once a year, while the three satellites rotate clockwise around the barycentre of the triangle, also once per year.

is schematically shown in figure 1. The constellation of the three spacecraft will orbit the Sun at 1 AU, trailing the Earth by 20°, some 45 million km from us. The plane of the constellation is tilted by 60°, which ensures the best stability possible of the equilateral triangle formation across the year [8].

The intended sensitivity of *LISA* to GW signals is plotted in figure 3, along with that of *LIGO* for reference of the complementarity of these two instruments. The curves represent the (amplitude) spectral density of noise expressed in $\mathrm{Hz}^{1/2}$, i.e., they represent the instrumental noise as if it was a GW noise blurring the interesting signals. These are indicated (for *LISA*) as "Resolved galactic binaries" and "Coalescence of Massive Black Holes", which are very reliably

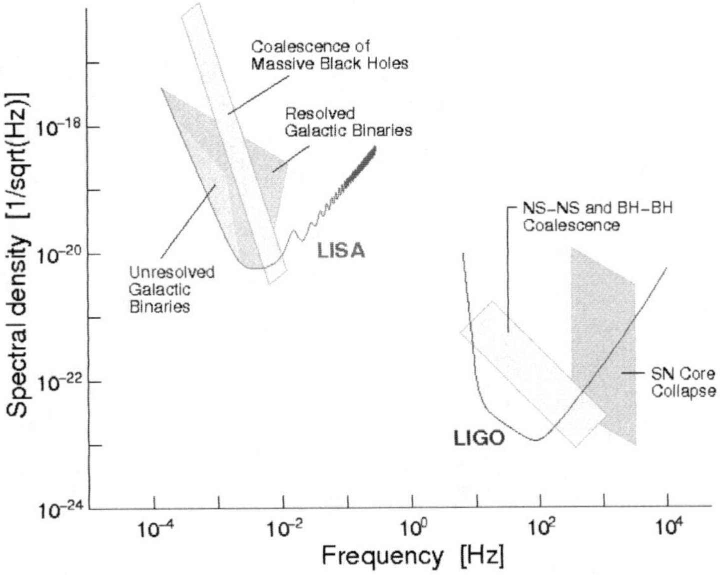

Figure 3. Amplitude spectral noise of *LISA* (left curve) and *LIGO* (right curve). *LISA* seems to be noisier, but what matters is the signals each detector can actually sense. Note *LISA* is contaminated by what are called *unresolved galactic binaries*: *LISA* is signal-dominated at frequencies roughly below 1 mHz.

predicted signals visible to *LISA*. In particular, a score or so of resolved galactic binaries will surely be detected by *LISA*, as their GW signals can be calculated and are seen to fall within the detectability region. These binaries are the so called *verification binaries* as they will provide a check of the performance of *LISA* and a calibration procedure, too. Another important type of sources for LISA is the capture and subsequent inspiral of a stellar compact object into a massive black hole at a galatic center, known as an extreme-mass-ratio inspiral (EMRI). Like every new instrument, though, we should be prepared for the unexpected, not just what we can at present predict on the basis of our current knowledge.

3. The *LISA* instrument

To reach the sensitivity levels shown in figure 3 is a real scientific and technological challenge: we are talking about picometre interferometry over a baseline of 5 million km, and measurements are to be done at milli-Hz frequencies, which is also a non trivial problem in electronic circuitry. In this section we review the most relevant issues which arise while developing *LISA*, as well as the solutions given to them.

First of all, for *LISA* to make meaningful GW measurements, all six proof masses must (ideally) be in *pure geodesic motion*, or *free fall*, i.e., no non-gravitational forces should act on them which would deviate them from geodesic trajectories. In the case of *LISA*, such trajectories are determined by the Sun, the planets, their moons, etc. *plus the gravitational waves themselves.* The latter however show up as *gravity gradients* which result in relative accelerations between pairs of masses which have a specific signature, e.g., they vary with time with some characteristic patterns in a certain frequency band. The geodesic motion of *LISA*'s proof masses, acting as interferometer mirrors, ensures that the GW induced phase shifts in the laser light is what is actually measured.

Such ideal situation is of course not feasible with real instruments: there are both *external* potential disturbances (e.g., solar radiation pressure, cosmic and solar ionising particles, etc.) and internal ones (temperature fluctuations, magnetic fields, local gravitation, etc.) which tend to deviate the proof masses from pure free fall. A requirement is thus defined which sets a limit on the amount of *noise* which can be tolerated for *LISA* to meet its scientific objectives. This is what we see in figure 3, left; translated into relative acceleration between proof masses in distant spacecraft, the requirement is in the order of

$$S_{\Delta a}^{1/2}(\omega) \lesssim 3\times 10^{-15}\ \mathrm{m\,s^{-2}\,Hz^{-1/2}} \qquad \text{for } LISA \tag{4}$$

leaving aside the frequency dependence which results in the V-shaped curve of figure 3. As we also see in the figure, *LISA* is sensitive to likely GW signals between 0.1 mHz and 1 Hz provided the requirement above is met in the frequency band.

3.1. LISA's drag-free system

Adverse conditions in the interplanetary medium must be avoided in the first place; the spacecraft hosting the proof masses provide this shielding, but this is not sufficient: the radiation pressure from the sunlight drags along the spacecraft with a tiny acceleration, so that the masses freely floating in their interiors would eventually crash against their walls, hence destroying the experiment.

To avoid this, a set of micro-thrusters is installed in the outer walls of each spacecraft which, upon detection of misplacement of the proof masses with respect to their centred position, are activated so that the spacecraft is gently driven back to the proof masses' geodesic trajectories. The activation scheme is derived from the readouts of the so called Gravitational Reference Sensor (GRS), which determines the position of the proof masses by a set of capacitive sensors: each mass is floating in the interior of a box called *electrode housing* whose interior walls have

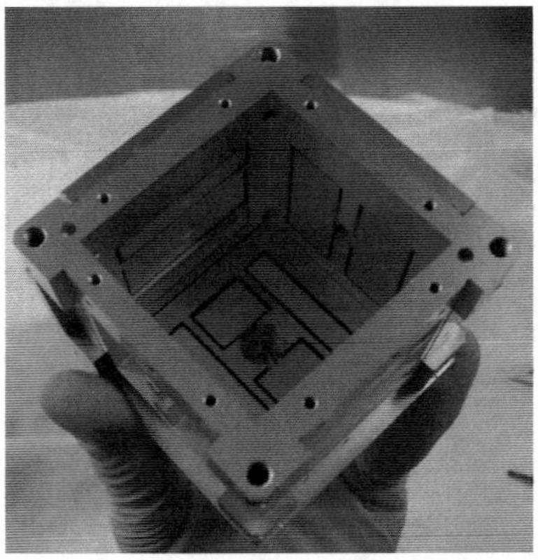

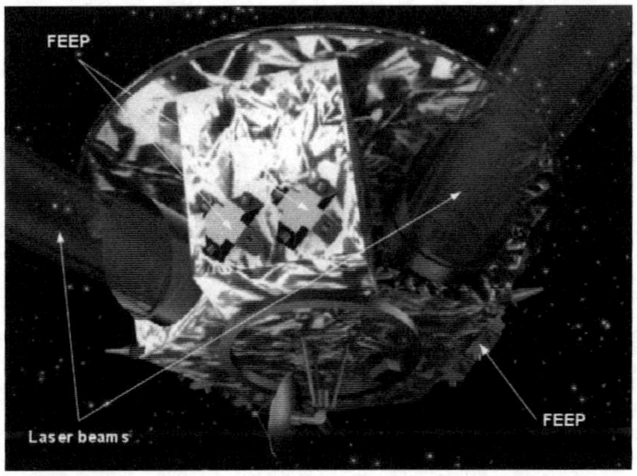

Figure 4. Left: interior of the electrode housing. The proof mass is a 2 kg Au+Pt alloy cube forming capacitors with each of the electrodes in the walls. Right: an artist's view of a *LISA* spacecraft from outside. Note two micro-thruster sets, dubbed FEEP in the drawing (for Field Effect Electric Propulsion, a thrust concept based on the emission and acceleration of ions from an on-board reservoir), which are suitably activated to counter the effect of proof mass misplacement relative to the housing, solidly linked to the satellite's structure.

electrodes making capacitors between them and the floating mass. A 100 kHz voltage is applied to these electrodes which enables a nanometre precision measurement of the gap size between the electrode and the proof mass. Figure 4 shows a real electrode housing and a drawing of micro-thrusters in a *LISA* spacecraft.

The combination of the GRS and the micro-thruster actuators is usually called the *drag-free* subsystem. This can also be partly driven by interferometer signals, one order of magnitude more sensitive than the capacitive sensors but only useful along the longitudinal degree of freedom in the direction of the remote spacecraft. This is an essential subsystem for *LISA*.

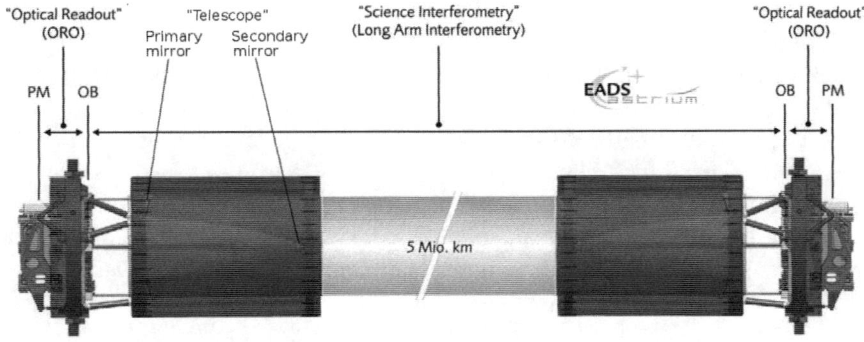

Figure 5. *LISA* interferometry: on the left and right we see (part of) the structure of two spacecraft, highlighting in particular the telescopes, used both to send and receive light to and from the other spacecraft. The diagram also shows the three step interferometry concept: the proof mass positions are first determined with respect to their host spacecraft (short range) then the distances between spacecraft are measured (long range), then all three measurements combined into a single one.

3.2. LISA interferometry

While the *drag-free* is necessary to make meaningful measurements, these measurements are effectively made by precision laser interferometry. In *LISA* the laser is infrared ($\lambda_{\rm laser} = 1.064\,\mu$m) and has a power of 2 watts. The light is sent to the remote spacecraft by means of a telescope (of the Cassegrain reflector type [9]), as it is collected on arrival, too —see figure 5. The distance between corresponding proof masses is determined in three steps: first the distance between a proof mass and its host satellite, which is a few centimetres, is measured with a local interferometer of the Mach-Zehnder type, identical to the one in *LISA PathFinder* [10] —see below. Next, the distance between remote spacecraft, in the order of 5 million kilometres is determined, and finally the distance between the remote spacecraft and, finally, the second proof mass position with respect to its host satellite is determined, like in the first step.

The *LISA PathFinder* interferometer has been extensively tested and is performing well above requirements, so it will be directly taken over into *LISA*. The most difficult part is therefore the second step, as the arm-length is not only extremely long, it is also varying with time. The reason is the *LISA* constellation is not totally stable, as we mentioned in section 2. The "instability" is not really an instability in the usual sense; in this context it refers to the fact that the relative positions between spacecraft varies during the year. This is reflected in three main effects: a) the distances between spacecraft vary (by about 60 000 km in one year, not very relevant as this happens at very low frequency —30 nano-Hz); b) the spacecraft move with respect to each other at velocities of 15 m/sec at most, which requires laser modulations in the range of 10 Mhz to compensate for the associated Döppler shifts; and c) the angles between arms vary by about $\pm 1.5°$.

The latter, together with the finite time the laser light takes to travel along one *LISA* arm (16.7 seconds) complicate the practical implementation of the interferometry: first of all, the telescope structure must be able to slightly and quietly pivot to compensate for the angle variations, and second, a *point ahead mechanism* is in charge of pointing the laser beam to the position where the other satellite will be 16.7 seconds later. While local interferometry can be accurately designed and tested on ground, the long arm interferometry is of course impossible to implement directly in the laboratory, so it is studied combining optic and electronic devices [11, 12].

3.3. Other LISA subsystems

The subsystems described in the last two sections are definitely the most important ones from the conceptual point of view. However many more are needed in the real implementation of the satellites: thermal insulators to preserve a very demanding temperature stability (temperature fluctuations affect the proof mass mechanical stability, the optical elements' properties, etc.), solar panels and batteries, on board computers to control the system workings, receive telecommands from ground and send telemetry, down-link antennas to make possible the communications, star-trackers to determine the satellite's positions (using also detailed sky maps), diagnostics sensors and calibrators, etc. Figure 6 provides a few graphical hints.

4. *LISA PathFinder*

To meet the requirements of *LISA* is definitely not trivial. But, even beyond that, the very mission concept is very unusual in space science and engineering, as the spacecraft and the payload are actually deeply integrated: the spacecraft is much more than a mere vehicle, as it is heavily controlled by the payload itself and needs to respond with high precision.

The drag-free subsystem described in the previous section is extremely demanding —see equation (4)—, orders of magnitude more than any drag-free previously flown, and cannot possibly be tested in Earth. In particular because neither a parabolic flight nor a drop tower can create a free fall environment for periods of time of several hours, as required by *LISA*, which has to work in the sub-mHz frequency range.

These circumstances, together with the major challenge *LISA* poses and its consequent high price, were the drivers to fly a technology precursor mission which would put to test in space the main subsystems of *LISA*. A successful such mission would help reduce the risks of flying from scratch a fully new concept mission and, hopefully, create instrumentation which can be (almost) directly be carried over into *LISA*.

This mission is *LISA PathFinder* (LPF) and is led by the European Space Agency (*ESA*), with the contributions of seven countries (Germany, Italy, United Kingdom, Spain, Switzerland, France and the Netherlands) to its payload.

The LPF concept is to squeeze one of the three *LISA* arms from 5 million kilometres to 30 cm (see figure 7, left), to fit it into a single spacecraft, and set up there a drag-free environment near the *LISA* frequency band and with slightly less demanding requirements. More specifically,

$$S^{1/2}_{\Delta a}(\omega) \lesssim 3\times 10^{-14}\ \mathrm{m\,s^{-2}\,Hz^{-1/2}} \qquad \text{for LPF} \tag{5}$$

in a frequency range between 1 mHz and 30 mHz. The margins in sensitivity and frequency band are intended to make LPF feasible while at the same time close enough to *LISA*, so that the final step to the large mission is affordable on the basis of what we learn from LPF.

4.1. Quick overview of LPF instruments

As already mentioned, one of the main subsystems in LPF is the *drag-free*. This is implemented in LPF by means of two cylindrical vacuum enclosures (VE), where the electrode housings (see figure 4, left panel), containing the proof masses, are hosted. This VE preserve pressure conditions at 10^{-5} Pascal, and this is done with so called getter pumps, which is a material that captures the particles which constantly escape from the various surfaces altogether (*outgassing*). The mass positioning signals are processed by the On-Board Computer (OBC) which activates the micro-thrusters as required.

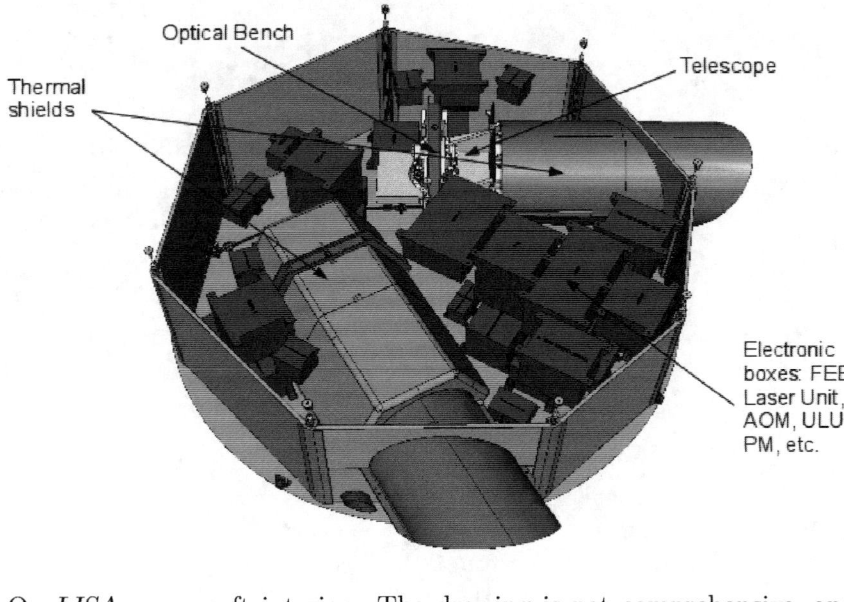

Figure 6. On *LISA* spacecraft interior. The drawing is not comprehensive, and just gives a qualitative flavour of the parts there.

Journal of Physics: Conference Series **314** (2011) 012014 doi:10.1088/1742-6596/314/1/012014

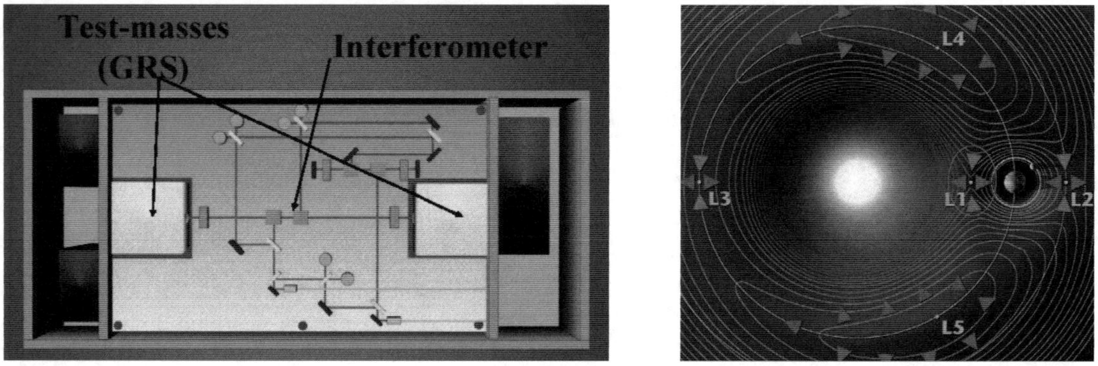

Figure 7. Left: schematics of the LPF concept: the two proof masses are enclosed in a single spacecraft, at 30 cm from each other and whose distance is measured by a Mach-Zehnder laser interferometer. Right: Lagrange points of the Sun-Earth system. LPF will orbit around L1 as shown in the diagram.

The space in between is occupied by the Optical Bench (OB), which contains various optical elements (mirrors, beam-splitters, photo-diodes and fibre injectors) to do the interferometry. The interferometer is a heterodyne Mach-Zehnder [13]; actually, up to four such interferometers are implemented in parallel in the LPF OB, each to measure a relevant signal: the main one is the differential channel, i.e., the distance between the two proof masses, the other three being the distance of one of the proof masses to the spacecraft structure and two more, one for frequency reference and the other for laser frequency stabilisation.

Another very important subsystem is the *Diagnostics Subsystem* which includes thermal,

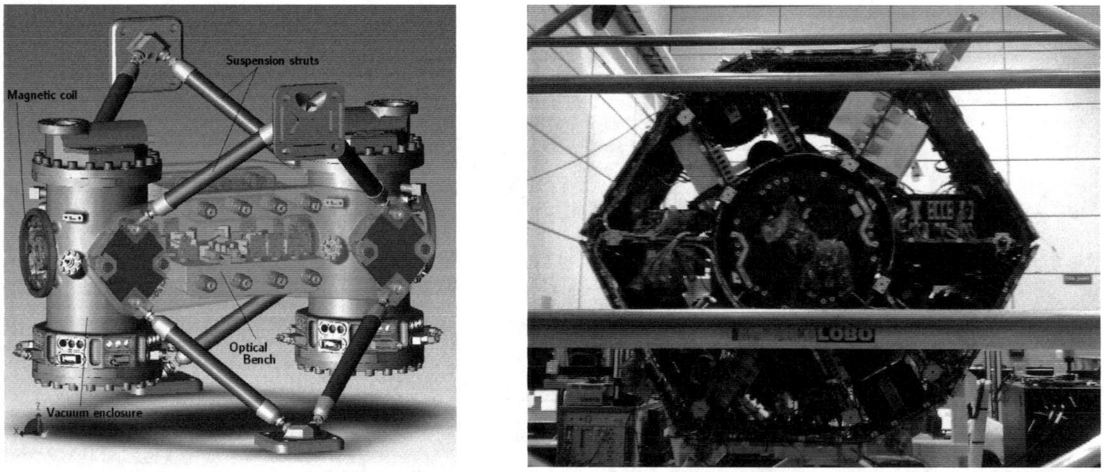

Figure 8. Left: The *LISA* Technology Package (LTP), an artist view. The proof masses are inside their Electrode Housings (see figure 4), and these inside the VEs, the vertical cylinders. In between, the OB with its optical elements and its vertical transparent walls; we also see eight suspension struts which hold the entire LTP into the spacecraft structure. The latter is shown in the right panel, without the top and bottom lids; the central circle contains the LTP (not seen because it is covered) while outside are all the electronics boxes, both LTP and platform boxes.

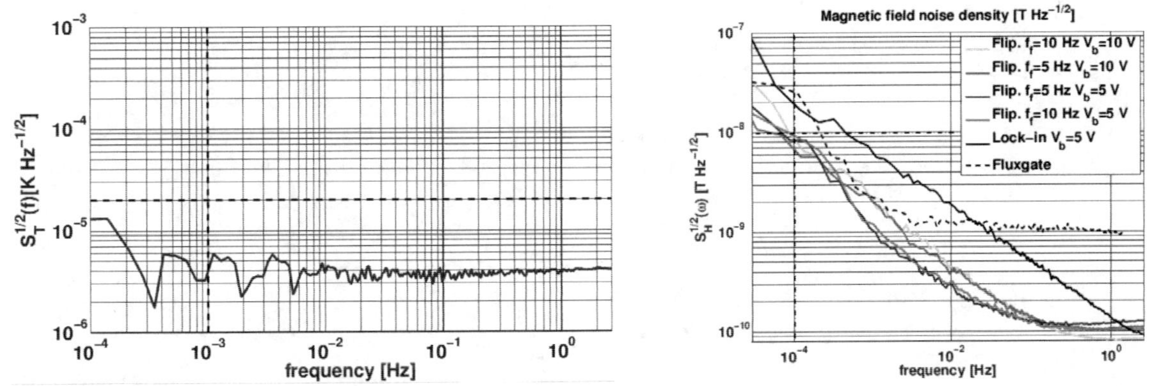

Figure 9. Left: Performance of the LPF thermal sensors and their electronics. The dashed lines indicate the LPF requirements, i.e., they are below and to the right of these lines. As can be seen, the spectral density is below 10^{-5} K $\mathrm{Hz}^{-1/2}$ down to 10^{-4} Hz, a remarkable achievement. Right: sensitivity of various magnetic sensors. The *fluxgate* (black dashed line) is the one which will fly in LPF, while the rest are AMR (Anisotropic Magneto-Resistance) in various configurations. Again, dash-dotted lines define the LPF requirements and we see that all magnetometers are compliant with it, but they are even good down to 10^{-4} Hz.

magnetic and charged particle detectors, along with calibration devices. This subsystem has been fully developed at *IEEC*, Barcelona. Apart from monitoring to high precision various disturbances in the LTP, the diagnostics subsystem provides a way to identify the real effect of those disturbances on the interferometer readout, i.e., to determine which part of the total system noise is of thermal or magnetic origin. Given the enormous stability required in the LTP, the diagnostics items have to be extremely sensitive, too. In Figure 9 we see the performance of the thermal sensors (NTC thermistors) and magnetic sensors. Deliberately, the plots stretch down to 10^{-4} Hz, one order of magnitude below the LPF requirement, but in the lower part of the *LISA* band. The results show that *LISA* is already within reach if the requirements on these diagnostics items do not differ significantly from those of LPF [14]. This is still an open question, but the plots definitely indicate we are on track. In addition, the development of tinier and yet sensitive magnetometers is a real must for *LISA*, as the current fluxgates have very serious setbacks: they have poor space resolution (because of their size), they must be kept far from the proof masses (due to magnetic back-action), and only a few can be installed (due to size and price). In this respect the research we are carrying forward at *IEEC* is very valuable for *LISA*, and the results so far look really encouraging [15].

The *LPF* Radiation Monitor (RM) is designed to obtain short term information on charged particles hitting the spacecraft and possibly reaching the proof masses, thereby distorting the measurements of the GRS. Due to the protection provided by the spacecraft and the VE, only particles with a primary energy above 70 MeV can go through[16], hence the RM has its particle detectors (PIN diodes) hidden in a copper shield of cubic shape with bevelled corners and a thickness of 6.4 mm —figure 10. By counting particles with those energies, it becomes possible and meaningful to study correlations between the RM readings and the charge management system of the LTP [17].

The DMU is the LTP computer. It controls all the diagnostics items and their interfaces, plus processes and sends information with various other subsystems, e.g., the phase-meter of the interferometer channels, caging mechanism, etc. In hardware it embodies three pairs of Printed Circuit Boards (PCB) —see figure 10—, duplication being implemented as redundancy for security requirements. The DMU requires software to execute its functionality, and this is

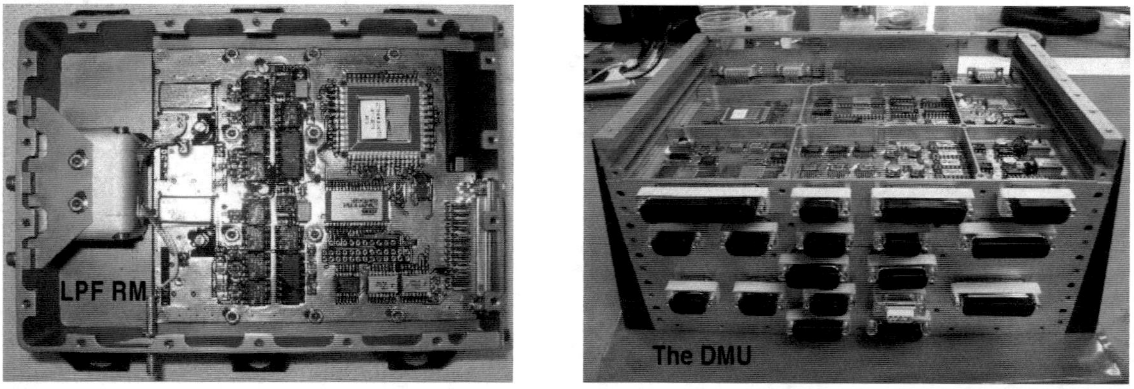

Figure 10. Left: the Radiation Monitor; the grey box on the left of the photo is a copper shield protecting the silicon sensors from particles with primary energies below 70 MeV. Right: Photo of the real DMU; the uncovered PCB is a Data Acquisition Unit (DAU), the other is missing in the picture as it was not installed at shoot time; below it are the other PCBs forming the DMU stack, all duplicated for security, like the DAU.

organised in two blocks: the Boot Software (BSW) and the Application Software (ASW). The former is a sort of mini-operating system (61 kbytes) which starts-up the system and takes care of several tasks, such as upgrading of the ASW in flight. The ASW receives telecommands from ground to drive the LTP experiment according to the general mission plan [18], and to send data and telemetry to the OBC, the main mission computer. The DMU software has been extensively tested, with consistently excellent results.

5. Conclusions

LISA is a complex and expensive mission, but it has a strong scientific case, as it will bring about a new way to observe the Universe. A way which may give us access to the unknown 96% of the Universe which does not shine in the electromagnetic channels of telescope Astronomy, but which does gravitate, hence perhaps generates GWs with an immense potential for discovery. In this brief review we have highlighted some of the features of *LISA* as currently conceived. At the time of writing (April–2011), however, new events are taking place which may change some of the statements made above regarding specific aspects of *LISA*'s performance; these have not been reflected in the text because they are still not consolidated. Nevertheless the main facts about *LISA* will definitely not affect either its scientific interest or its relevance.

Acknowledgments

Financial support from the Spanish Ministry of Innovation and Research (MICINN), contract ESP2007-61712, is gratefully acknowledged, as well as AYA08-04211-C02-01. CFS acknowledges support from the Ramón y Cajal Programme of the same Ministry, and a Marie Curie International Reintegration Grant (MIRGCT-2007-205005/PHY) within the 7th European Community Framework Programme, and PC acknowledges an FPU PhD Grant from MICINN. Part of this work was also supported by AGAUR, Generalitat de Catalunya.

References

[1] Albert Einstein *et al.* 1952 *The Principle of Relativity* (New York: Dover)
[2] Einstein A 1918 *Sitzungsberichte der Königlich Preussischen Akademie der Wissenschaften zu Berlin* 154
[3] Weisberg J and Taylor J 2010 Timing Measurements of the Relativistic Binary Pulsar PSR B1913+16 (*Preprint* `http://arxiv.org/abs/1011.0718`)

[4] Duncan R Lorimer 2008 Binary and Millisecond Pulsars, Living Reviews in Relativity, available on-line at http://relativity.livingreviews.org/Articles/lrr-2008-8
[5] Lobo A 1992 *Classical and Quantum Gravity* **9** 1385
[6] Stefan Hild 2009 How to listen to the Universe? Optimising future GW observatories for astrophysical sources, available on-line at http:www.sr.bham.ac.uk/~hild/presentations/nikhef_hild.pdf1
[7] Creighton T 2008 *Classical and Quantum Gravity* **25** 125011
[8] WM Folkner, F Hechler, TH Sweetser, MA Vincent, PL Bender 1997 *Classical and Quantum Gravity* **14** 1405
[9] Cassegrain reflector, see e.g. Wikipedia at http://en.wikipedia.org/wiki/Cassegrain_reflector
[10] G Heinzel et al 2004 *Classical and Quantum Gravity* **21** 581
[11] JI Thorpe RJ Cruz S S and Mueller G 2005 *Classical and Quantum Gravity* **22** 227
[12] H Halloin, O Jeannin, B Argence, V Bourrier, E de Vismes, P Prat 2010 LISA On Table: an Optical Simulator for LISA, Proceedings of the 2010 International Conference on Space Optics, Rhodes (Greece), available on-line at http://congrex.nl/icso/Papers/Session%2014b/FCXNL-10A02-2018762-1-Halloin_ICSO_Paper.pdf
[13] G Heinzel et al 2003 *Classical and Quantum Gravity* **20** S153
[14] J Sanjuán, J Ramos-Castro, A Lobo 2009 *Classical and Quantum Gravity* **26** 094009
[15] I Mateos, A Lobo, J Ramos-Castro, J Sanjuán, M Nofrarias 2009 *Journal of Physics: Conference Series* **154** 012005, available on-line at http://iopscience.iop.org/1742-6596/154/1/012005
[16] Araújo HM *et al* 2005 *Astroparticle Physics* **22** 451
[17] D Shaul, T Sumner and A Lobo 2008 Analysis of Data from the Radiation Monitor on LISA PathFinder Tech. rep. ICL and IEEC, S2-ICL-TN-3019
[18] L Gesa and A Conchillo 2010 ASW Requirements Tech. rep. IEEC and NTE, S2-NTE-SRD-3003

Journal of Physics: Conference Series **314** (2011) 012015 doi:10.1088/1742-6596/314/1/012015

Axially symmetric spacetimes: numerical and analytical perspectives

Sergio Dain

Facultad de Matemática, Astronomía y Física,
Universidad Nacional de Córdoba,
Ciudad Universitaria, (5000) Córdoba,
Argentina

E-mail: dain@famaf.unc.edu.ar

Abstract. Some new aspects of axially symmetric spacetimes are discussed. These results open the door for future interplay between analytical and numerical studies. The new developments are based on the role of the total mass in axial symmetry. Finally, a list of relevant open problems is presented. These problems can be hopefully solved with an interaction between numerical and analytical insights.

1. Introduction

In any physical theory, the presence of a symmetry reduces the degrees of freedom of the equations and hence it simplifies considerable the analysis. To study an isolated systems, the simplest models are the spherically symmetric ones. However, it is well known that for vacuum Einstein equations, due to Birkhoff's theorem, spherical symmetric spacetimes has no dynamics. The next possible model with symmetries in vacuum are axially symmetric spacetimes. It has been proved in [4] that no additional symmetry can be imposed to the spacetime if we want to keep the gravitational radiation and a complete null infinity. This result single out axially symmetric spacetimes as the only models for vacuum isolated, dynamical, system with symmetries. From this point of view, axially symmetric gravitational waves are the simplest possible waves emitted by isolated sources.

There exists many relevant physical models one can study in axial symmetry. In particular, for vacuum and in the strong field regime, we list the following

- Head-on collisions of two black holes.
- Rotating, non-stationary, black holes.
- Formation of black holes: weak cosmic censorship.
- Critical collapse of gravitational waves.

What are the difficulties of axial symmetry? To take advantage of the symmetry an adapted coordinate should be used. However, the reduced equations are formally singular at the axis. This singular behavior introduce a major difficulty in the analysis. In fact, it can be argued that this singular behavior near the axis is so complicated that the axially symmetric case is as hard as the full general case.

Journal of Physics: Conference Series **314** (2011) 012015 doi:10.1088/1742-6596/314/1/012015

What are the advantages of axial symmetry? The first obvious advantage for numerical computations is that axially symmetric spacetimes are less computationally expensive: only two effective spatial dimensions. Also, the number of equations and variables is reduced. The second advantage, which is also well known, is the conservation of angular momentum. The angular momentum is a quasilocal conserved quantity in axial symmetry (Komar integral of the Killing vector). That is, axially symmetric gravitational waves do not carry angular momentum. In particular no Penrose process and no superradiant scattering can occur for axially symmetric vacuum spacetimes. This represents an important simplification in the dynamics. The third advantage is the mass integral formula. The total mass can be written as a positive definite volume integral, as we will see. This is the main new ingredient that we would like to discuss in the following.

The purpose of this review is to summarize some new results for axially symmetric spacetimes with emphasis in the interplay between numerical and analytical studies. The plan of the article is the following. In section 2 we review the axially symmetric equations. The purpose of this section is to present the relevant maximal-isothermal gauge and the corresponding mass formula, which is valid only in this gauge. This formula represents the main new motivation to study axial symmetry. In section 3 we discuss recent results, both numerical and analytical which are based on the maximal-isothermal gauge and the mass formula. Finally, in section 4 we present relevant open problems.

2. Axial Symmetry

Consider a vacuum solution of Einstein's equations, i.e., a four dimensional manifold M with metric $g_{\mu\nu}$ (with signature $(-+++)$) such that the corresponding Ricci tensor vanishes

$$ {}^{(4)}\mathcal{R}_{\mu\nu} = 0. \tag{1} $$

Suppose, in addition, that the metric $g_{\mu\nu}$ admits a Killing field η^μ, that is η^μ satisfies the equation

$$ \hat{\nabla}_{(\mu}\eta_{\nu)} = 0, \tag{2} $$

where $\hat{\nabla}_\mu$ is the connection with respect to $g_{\mu\nu}$. Greek indices $\mu, \nu, \cdots$ denote four dimensional indices.

We define the square of the norm and the twist of η^μ, respectively, by

$$ \eta = \eta^\mu \eta^\nu g_{\mu\nu}, \quad \omega_\mu = \epsilon_{\mu\nu\lambda\gamma}\eta^\nu \hat{\nabla}^\lambda \eta^\gamma. \tag{3} $$

Using the field Eq. (1) it is possible to prove that

$$ \hat{\nabla}_{[\mu}\omega_{\nu]} = 0, \tag{4} $$

and hence ω_μ is locally the gradient of a scalar field ω

$$ \omega_\mu = \hat{\nabla}_\mu \omega. \tag{5} $$

Let $\mathcal{N}$ denote the collection of all trajectories of η^μ, and assume that it is a differential 3-manifold. We define the metric $h_{\mu\nu}$ on $\mathcal{N}$ by

$$ \eta g_{\mu\nu} = h_{\mu\nu} + \eta_\mu \eta_\nu. \tag{6} $$

The vacuum field equations (1) can be written in the following form on $\mathcal{N}$

$$\Box\eta = \frac{1}{\eta}(\nabla^a\eta\nabla_a\eta - \nabla^a\omega\nabla_a\omega), \tag{7}$$

$$\Box\omega = \frac{2}{\eta}\nabla^a\omega\nabla_a\eta, \tag{8}$$

$$^{(3)}R_{ab} = \frac{1}{2\eta^2}(\nabla_a\eta\nabla_b\eta + \nabla_a\omega\nabla_b\omega), \tag{9}$$

where ∇_a and $^{(3)}R_{ab}$ are the connexion and the Ricci tensor of h_{ab}, we have defined $\Box = \nabla_a\nabla^a$ and Latin indices $a, b\dots$ denote three dimensional indices on $\mathcal{N}$.

Up to this point, the only assumption we have made is that the spacetime admits a Killing vector field η^μ and that η^μ is not null, otherwise the metric h_{ab} is not defined. If the Killing field is timelike ($\eta < 0$) then the metric h_{ab} is Riemannian and the equations (7)–(9) are the stationary Einstein vacuum equations. On the other hand, when the Killing vector is spacelike ($\eta > 0$), the metric h_{ab} is a is a 3-dimensional Lorenzian metric (we chose the signature $(-++)$). In axial symmetry, the Killing vector η^μ is spacelike and its norm vanishes at the axis of symmetry. Hence, the equations are formally singular at the axis. This singular behavior at the axis represents the main difficulty to handle these equations.

In the Lorenzian case, Eq. (9) has the form of Einstein equations in three dimensions coupled with effective matter sources produced by η and ω. The effective matter Eqs. (7)–(8) imply that the energy-momentum tensor defined in terms of η and ω by

$$T_{ab} = \frac{1}{2\eta^2}(\nabla_a\eta\nabla_b\eta + \nabla_a\omega\nabla_b\omega) - \frac{1}{4\eta^2}h_{ab}(\nabla_c\eta\nabla^c\eta + \nabla_c\omega\nabla^c\omega), \tag{10}$$

is divergence free, i.e. $\nabla^a T_{ab} = 0$.

Eqs. (7)–(9) are purely geometric equations with respect to the metric h_{ab}. The essential point is that there exist no dynamical degrees of freedom in 3-dimensional gravity (the Weyl tensor vanishes) and hence all the dynamics is produced by the effective matter sources determined by (η, ω). In other words, the dynamics of vacuum axially symmetric gravitational waves has the behavior of matter moving in a lower dimensional space. Hence, it can be expected that the total mass of the waves can be computed as an "integral on the matter sources". As we will see in the following, this is precisely what happens and it leads to the mass formula in axial symmetry.

2.1. 2+1 decomposition

In order to formulate an initial value problem, we will perform an standard $2+1$ decomposition of Eqs. (7)–(9). Note that this is completely analogous to the $3+1$ decomposition of Einstein equations, in fact all the formulas are formally identical because the dimension does not appear explicitly in them (see, for example, [28], [27]).

Consider a foliation of spacelike, 2-dimensional slices S of the metric h_{ab}. Let t be an associated time function and let n^a be the unit normal vector orthogonal to S with respect to the metric h_{ab}. The intrinsic metric on S is denoted by q_{ab} and is given by

$$h_{ab} = -n_a n_b + q_{ab}. \tag{11}$$

Define the density μ by

$$\mu = 2\,{}^{(3)}R_{ab}n^a n^b + {}^{(3)}R, \tag{12}$$

and the current J_b by

$$J_b = -q_b^c n^a\, {}^{(3)}R_{ca}, \tag{13}$$

where ${}^{(3)}R = {}^{(3)}R_{ab}h^{ab}$ denotes the trace of ${}^{(3)}R_{ab}$. Then, using Eq. (9) we obtain

$$\mu = \frac{1}{2\eta^2}\left(\eta'^2 + \omega'^2 + |D\eta|^2 + |D\omega|^2\right), \tag{14}$$

$$J_A = -\frac{1}{2\eta^2}\left(\eta' D_A\eta + \omega' D_A\omega\right), \tag{15}$$

where D_A is the connexion with respect to q_{AB}. The prime denotes directional derivative with respect to n^a, that is

$$\eta' = n^a\nabla_a\eta = \frac{1}{\alpha}\left(\partial_t\eta - \beta^A D_A\eta\right) \tag{16}$$

where α is the lapse and β^A is the shift vector of the foliation. The indices $A, B, \cdots$ denote two dimensional indices on S. The constraints equations corresponding to (9) are given by

$${}^{(2)}R - \chi^{AB}\chi_{AB} + \chi^2 = \mu, \tag{17}$$

$$D^A\chi_{AB} - D_B\chi = J_B, \tag{18}$$

where ${}^{(2)}R$ is the Ricci scalar of q_{AB}, χ_{AB} is the second fundamental form of S and χ its trace

$$\chi = q^{AB}\chi_{AB}. \tag{19}$$

We use the following sign convention for the definition of χ_{AB}

$$\chi_{ab} = -q_a^c\nabla_c n_b = -\frac{1}{2}\pounds_n q_{ab}, \tag{20}$$

where $\pounds$ denotes Lie derivative. The evolution equations are given by

$$\partial_t q_{AB} = -2\alpha\chi_{AB} + \pounds_\beta q_{AB}, \tag{21}$$

$$\partial_t\chi_{AB} = \pounds_\beta\chi_{AB} - D_A D_B\alpha + \alpha\tau_{AB}, \tag{22}$$

where

$$\tau_{AB} = \chi\chi_{AB} + {}^{(2)}R_{AB} - {}^{(3)}R_{AB} - 2\chi_{AC}\chi_B^C. \tag{23}$$

and

$${}^{(3)}R_{AB} = \frac{1}{2\eta^2}(\partial_A\eta\partial_B\eta + \partial_A\omega\partial_B\omega). \tag{24}$$

The evolution equations (21)–(22) and the constraint equations (17)–(18) constitute a complete $2+1$ decomposition of the 3-dimensional Einstein Eq. (9). It remains to decompose the effective matter Eqs. (7)–(8). The result is the following

$$-\Sigma'' + \Delta_q\Sigma + D_A\Sigma\frac{D^A\alpha}{\alpha} + \Sigma'\chi = \frac{1}{\eta^2}\left(\omega'^2 - |D\omega|^2\right), \tag{25}$$

$$-\omega'' + \Delta_q\omega + D_A\omega\frac{D^A\alpha}{\alpha} + \omega'\chi = \frac{2}{\eta^2}\left(D_A\omega D^A\eta - \omega'\eta'\right), \tag{26}$$

where Δ_q is the Laplacian with respect to q_{AB}, i.e. $\Delta_q = D^A D_A$ and we have defined $\Sigma = \log\eta$.

The evolution equations (21)–(22) and (25)–(26) together with the constraint equations (17)–(18) represent the complete $2+1$ decomposition of the geometrical equations (7)–(9) in an arbitrary gauge. This set of equations presents an important feature which is peculiar of $2+1$ dimensions. Namely, the evolution equations (21)–(22) and the constraint equations (17)–(18)

are, roughly speaking, equivalent in the following sense. With an appropriate gauge choice (in the next section we will present an important example) the constraint equations (17)–(18) form an elliptic system that determines q_{AB} an χ_{AB} in terms of the effective matter sources η and ω. Hence, the evolution equations (25)–(26) together with the constraint equations (17)–(18) constitute a complete system of equations. In particular, a solution of this system will automatically satisfy the other evolution equations (21)–(22). Alternative, it is possible to solve the evolution equations (25)–(26) and (21)–(22). If the initial data satisfy the constraint equations at a given time, then the solution will satisfy the constraint equations for all times. Note that even when the gauge is fixed, we have different possibilities for constructing evolutions schemes. The first choice is called in the numerical literature "constrained evolution" and the second choice "free evolution". It is also possible to mix them to obtain "mixed evolution" schemes (see the discussion in [29]).

2.2. Gauge and mass formula

In this section we describe the maximal-isothermal gauge. In particular we review the mass formula valid in this gauge (see [15] for details). For the lapse, we impose the maximal condition on the 2-surfaces

$$\chi = 0. \tag{27}$$

Note that we are not imposing that the surfaces are maximal in the 3-dimensional picture as in [15]. The later condition is the one generally used (see, for example, [6] [30]), but the difference is only minor. In particular the mass formula is positive definite for both conditions. The one used here appears to be natural with respect to the rescaled metric h_{ab}. Eq. (27) implies the following well known equation for the lapse

$$\Delta_q \alpha = \alpha(\chi^{AB}\chi_{AB} + \mu_1), \tag{28}$$

where

$$\mu_1 = {}^{(3)}R_{ab}n^a n^b = \frac{1}{2\eta^2}\left(\eta'^2 + \omega'^2\right). \tag{29}$$

The maximal gauge (27) can be, of course, imposed in any dimensions and it is not related at all with axial symmetry. In contrast, the condition for the shift is peculiar for two space dimensions. The shift vector is fixed by the requirement that the intrinsic metric q_{AB} has the following form

$$q_{AB} = e^{2u}\delta_{AB}, \tag{30}$$

where δ_{AB} is a fixed (i.e. $\partial_t \delta_{AB} = 0$) flat metric in two dimensions. Then, using (27), we obtain that the trace free part of (21) is given by

$$2\alpha\chi_{AB} = (\mathcal{L}_q\beta)_{AB}, \tag{31}$$

where $\mathcal{L}_q$ is the conformal Killing operator in two dimensions with respect to the metric q_{AB}. Equation (31) is an elliptic first order system of equations for β^A.

The elliptic Eqs. (28) and (31) determine lapse and shift for the metric h_{ab} and hence they fix completely the gauge freedom in Eqs. (7)–(9). This gauge has associated a natural cylindrical coordinate system (t, ρ, z) for which the metric δ_{AB} is given

$$\delta = d\rho^2 + dz^2, \tag{32}$$

and the axis of symmetry is given by $\rho = 0$. The slices S are the half planes $\mathbb{R}^2_+$.

It is useful to introduce the following auxiliary functions. Instead of η we will use the function σ defined by

$$\eta = \rho^2 e^{\sigma}. \tag{33}$$

Note that for Minkowski $\eta = \rho^2$, that is, σ measure the non-flat part of the norm η. Also, instead of u it is convenient to work with the function q defined by

$$u = \log \rho + \sigma + q. \tag{34}$$

The important property of q is that it vanished at the axis. This is a consequence of the regularity conditions of the metric at the axis (see [20] for details).

The equations presented in section 2.1 can be explicitly written as partial differential equations in the coordinates (t, ρ, z) (see [20]). In particular the Hamiltonian constraint is given by

$$^{(3)}\Delta\sigma + \Delta q = -\frac{\epsilon}{4}, \tag{35}$$

where

$$\epsilon = \frac{e^{2u}}{\eta^2}\left(\eta'^2 + \omega'^2\right) + |\partial\sigma|^2 + \frac{|\partial\omega|^2}{\eta^2} + 2e^{-2u}\hat{\chi}^{AB}\chi_{AB}. \tag{36}$$

The hat on χ^{AB} means that the indices are moved with the flat metric δ_{AB}, ∂ denotes partial derivatives with respect to ρ and z, Δ is the 2-dimensional flat Laplacian, namely

$$\Delta = \partial_\rho^2 + \partial_z^2, \tag{37}$$

and $^{(3)}\Delta$ is the 3-dimensional flat Laplacian acting on axially symmetric functions, that is

$$^{(3)}\Delta = \Delta + \frac{\partial_\rho}{\rho}. \tag{38}$$

The positive scalar ϵ plays the role of the energy density of the gravitational waves in this gauge. The integral of ϵ over the slice S is given by

$$m = \frac{1}{16}\int_{\mathbb{R}^2_+} \epsilon\, \rho\, d\rho dz. \tag{39}$$

It is a non-trivial fact that this number m is precisely the total ADM mass of the spacetime. Moreover, the fall off properties of the solutions α and β of gauge conditions ensure that the mass is conserved along the evolution, namely

$$\frac{d}{dt}m = 0, \tag{40}$$

see [15] for details.

The mass formula (39) together with the conservation law (40) represent a relevant property of this gauge which is not present in other ones.

In any physical theory conserved quantities (in particular, conserved energies) are very important to control the evolution of the system. However, in General Relativity, the conserved mass appears as a boundary integral and not as a volume integral (as, for example, in the wave equation). Hence it is not possible to relate the mass with any norm of the fields to control the evolution of them (for the wave equation the energy is precisely the norm of the wave). The mass formula (39) for axially symmetric systems in the maximal-isothermal gauge represents a remarkable exception.

3. Results

3.1. Numerical results

Axially symmetric spacetimes has been studied numerically since the very beginnings of numerical relativity, see chapter 10.4 in [2] and references therein for an historical perspective. A very good reference for early results is the review article [3]. In particular, in this article the maximal and isothermal gauges are discussed. Another important review article is [26], where the symmetry reductions and $2+1$ decomposition mentioned in section 2 was used for first time in numerical relativity.

As we mentioned in the introduction, the difficulty introduced by the singular behavior at the axis is severe and that made axially symmetric evolution particularly difficult to handle. As a consequence of this "axisymmetric codes were practically abandoned as soon as computers became powerful enough to be able to handle full three-dimensional simulations in the early 1990s." (the quote is from [2]).

New insights in the study of the axial regularity problem in numerical relativity were presented in [23]. They introduced two main ingredients. The first one is that the behavior at the axis is treated as a singular boundary condition which are handled by introducing "ghost zones" outside the axis. The second one is that all the variables used are such that either they vanish at the axis or have vanishing derivative there, but not both. This cure the axis instabilities from the numerical point of view.

Later in [6] a new code was developed which also use similar techniques to handle the axis problem. This article is particularly relevant in our context because they use essentially the same equations and the same gauge conditions as presented in section 2. However, the mass formula (39) for this gauge was not known at that time and hence it was not used in that article.

Further studied were performed in [29] [31], where new evolutions schemes are proposed and the general case with twist is considered. In [32] the regularization procedure at the axis was performed for more general evolutions schemes. In [30] important issues concerning the maximal-isothermal gauge are studied, in particular a new choice of variable is proposed in order to guarantee the uniqueness of solutions of the relevant equations. I will come back to this in the next section.

In a very recent series of articles, the critical collapse of gravitational waves is studied [33] [34]. This code use a different gauge choice, namely harmonic coordinates. Finally, we mention [20] where the linear equations in the maximal-isothermal gauge were studied and boundary conditions compatible with the mass formula were presented.

3.2. Analytical results

The first analytical application of the mass formula is the proof of the following geometrical inequality

$$\sqrt{|J|} \leq m, \tag{41}$$

where J is the total angular momentum of the spacetime (see [14] [17] for details). Extension of this theorem, which remove technical assumptions and also simplify considerable its proof were presented in [8] [9]. The remarkable aspect of this proof if that extreme Kerr black hole appears as the minimizer of the mass integral. This suggests stability properties of these black holes. In particular, it suggests that it is perhaps easier to prove stability near an extreme black holes than a non-extreme one in axial symmetry (see the essay [16]). There exist important extensions to inequality (41) which include charges [11] [10].

The second geometrical inequality proved with a suitable quasi-local formulation of the mass formula is the following

$$8\pi|J| \leq A, \tag{42}$$

where A is the area and J is the angular momentum of the black hole horizon. See [18], [1] [21] for details. In the stationary case with matter and charge, it has been proved in [24]. Numerical evidences for the validity of this inequality has been presented in [25].

Inequality (42) ensures that the Christodoulou quasi local-mass of the black hole is a monotonically increasing quantity in the evolution (see the discussion in [18]).

Both geometrical inequalities are proven on the initial data, that is, the proofs do not make use of the evolution equations. In this gauge the equations reduce to a coupled hyperbolic–elliptic system which is formally singular at the axis. Due to the rather peculiar properties of the system, the local in time existence has proved to resist analysis by standard methods. To analyze the principal part of the equations, which may represent the main source of the difficulties, we study linear perturbation around the flat Minkowski solution in this gauge. In [22] we solved this linearized system explicitly in terms of integral transformations in a remarkable simple form. This representation is well suited to obtain useful estimates to apply in the non-linear case.

4. Open Problems

There exists two relevant open model problems for axially symmetric spacetimes:

(i) Black hole formation and critical collapse of axially symmetric gravitational waves (Numerical).

(ii) The stability of the Kerr black hole in axial symmetry (Mathematical).

These problems are expected to be difficult, they constitute the long term motivation of this study. In the following I will briefly discuss them in order to propose simpler open problems that can contribute to the resolution of 1 and 2 and at the same time are feasible to solve in the near future.

Most of the numerical work mentioned in section 3.1 is concerned with the vacuum equations and with the critical collapse of the gravitational waves. As we pointed out in the introduction, there exists other interesting physical systems in axial symmetry. For example, the collapse of an axially symmetric rotating star studied in [26]. However, if we restrict ourselves to the vacuum equations (as we do in this article) then probably the most relevant open problem to study in numerical axial symmetry is the critical collapse. And the reason is that this problem can not be handled, with present computers, without symmetry assumptions due to the high resolution needed. The subject has already a long history (see the review article [5]), and despite many efforts the problem still lacks definitive answer (see the recent article [34]).

Although there exist many possible gauge choices in axial symmetry, the mass formula suggests the maximal-isothermal gauge as privileged one. Also, constrained evolutions schemes appear to be better than free evolutions schemes (see the discussion in [30]). In the following we focus on the maximal-isothermal gauge in its constrained formulation. In the article [30] it was clearly pointed out one of the main problems of these equations. There are many elliptic equations and it is not clear a priori that they will always admit a unique solution along the evolution. A new set of variables was proposed in that article to ensure that the equations have the correct behavior at the linearized level to ensure uniqueness of solution. However, it is still not clear if the non-linear equations (notably, the Hamiltonian constraint) is always solvable. It is an open problem to see if these equations (or some variant of them) are well posed. I believe that this is an important problem that stress the interplay between analytical and numerical studies. Perhaps the resolution of the well-posedness question will lead us to select (or even discover) the correct evolution scheme.

Another question which is important in numerical evolution is the issue of outer boundary conditions on a finite domain. In the article [20] a set of boundary condition at the linear level were proposed. This boundary conditions have the property that the energy appears to leave the domain, at least for the class of initial data studied in this article. It is important to note that the mass formula provides a well-defined local measure of the gravitational energy. These boundary conditions have not yet been implemented for the complete equations.

The following two problems summarize the discussion above.

1.a. **Well-posedness for the maximal isothermal gauge.** Is the Hamiltonian constraint always solvable? Does the gauge breaks down in the strong field regime?

1.b. **Implementation of the radiation boundary conditions for the full equations.** How much they improve the evolution scheme? It is possible to prove (may be with some extra assumptions) that for such boundary conditions the energy always leaves the domain?

As we have seen, problem 1 is more numerically oriented, since it appears to be very difficult to understand critical phenomena with the current analytical tools. In contrast problem 2 is more analytical. In recent years there have been important new developments in the problem of the linear stability of the Kerr black hole from the analytical point of view. Most of this work is concerned with scalar waves on a fixed Kerr black hole background. We refer to the review articles [12] and [13] and reference therein. These works are not restricted to axial symmetry. The expectation is, of course, that axially symmetric spacetimes are simpler. In particular, axial symmetry provides extra geometrical structure that can give new insight into the equations. As we mentioned in section 3.2 much of the current evidences that this is the case come from the geometrical inequalities (41) and (42). These inequalities have been proved under some restricted conditions. In the case of the quasi-local inequality (42), the only restriction is that the initial data are assumed to be maximal. It is important to remove this restriction or to find counter examples. Numerics can play a relevant role in the search of possible counter examples. Also, it would be very interesting to generalize this inequality to include the charge. This is important not only because the charge is a relevant parameter of a black hole but also because the charge is well defined without any symmetry assumption (this is not the case of the quasi-local angular momentum). The natural question is whether a similar inequality holds without axial symmetry. This would involve an appropriate definition of quasi-local angular momentum. Another interesting problem is to study what happens in the case of equality in (42). A rigidity property has been proved in [21] that characterize the local geometry of the surface in that case. However, it is very likely that this kind of surface can only appear as asymptotic limit as the extreme Kerr initial data. A related question is to study small deformation of the extreme Kerr initial data, as the ones constructed in [19]. Do these initial conditions admit apparent horizons if the perturbation is small enough? This question is relevant because if the answer is negative then these initial conditions have the chance to provide complete Cauchy surfaces that are outside the black hole region, as in the extreme Kerr black hole.

Regarding the global inequality (41), the most relevant open problems is to prove it for multiple black holes and also to remove the maximal condition. The multiple black holes case is related with the uniqueness of the Kerr black hole among stationary black holes with disconnected horizon components.

All the previous problems are concerned with the initial conditions. The maximal-isothermal gauge will be useful to answer questions regarding the black hole stability if the mass formula can be used in some way to control the evolution. A natural first problem would be to recover the non-linear stability of Minkowski [7] in this gauge. The expectation is that the mass formula will provide a simpler (and different) kind of approach to this problem; although, of course, always restricted to axial symmetry. A second problem it is the study of axially symmetric perturbations of a black hole in this gauge. The expectation is that the mass formula can help to prove linear stability under axially symmetric perturbations of the Kerr black hole. The advantage of this problem in comparison with the non-linear stability of Minkowski in this gauge is that it is probably easier to deal with and also it will prove something new, since the linear stability of the Kerr black hole is still an open question.

We summarize this discussion in the following list of open problems in relation to problem 2.

2.a. **Geometrical inequalities.** Remove the maximal assumption. For the quasi-local inequality (42) include the charge and study the possible generalization of this inequality without any symmetry assumptions. Study the case of the equality: it is possible to have such a surface in an asymptotically flat initial data? Study the existence of minimal surfaces or apparent horizon for small perturbation of extreme Kerr initial data. For the global inequality (41), prove it for multiples black holes.

2.b. **Linear stability of the Kerr black hole in axial symmetry.** Use the mass conservation to control the linear evolution.

[1] Andrés Aceña, Sergio Dain, and María E Gabach Clément. Horizon areaangular momentum inequality for a class of axially symmetric black holes. *Classical and Quantum Gravity*, 28(10):105014, 2011, 1012.2413.

[2] Miguel Alcubierre. *Introduction to 3 + 1 numerical relativity*, volume 140 of *International Series of Monographs on Physics*. Oxford University Press, Oxford, 2008.

[3] J. M. Bardeen and T. Piran. General relativistic axisymmetric rotating systems: coordinates and equations. *Physics Reports*, 96:205–250, 1983.

[4] J. Bičák and A. Pravdová. Symmetries of asymptotically flat electrovacuum space–times and radiation. *Journal of Mathematical Physics*, 39(11):6011–6039, 1998.

[5] José M. Martín-García Carsten Gundlach. Critical phenomena in gravitational collapse. *Living Reviews in Relativity*, 10(5), 2007.

[6] Matthew W. Choptuik, Eric W. Hirschmann, Steven L. Liebling, and Frans Pretorius. An Axisymmetric Gravitational Collapse Code. *Class. Quant. Grav.*, 20:1857–1878, 2003, gr-qc/0301006.

[7] D. Christodoulou and S. Klainerman. *The Global Nonlinear Stability of the Minkowski Space*, volume 41 of *Princeton Mathematical series*. Princeton University Press, Princeton, NJ, 1993.

[8] Piotr T. Chrusciel. Mass and angular-momentum inequalities for axi-symmetric initial data sets I. Positivity of mass. *Annals Phys.*, 323:2566–2590, 2008, 0710.3680.

[9] Piotr T. Chruściel, Yanyan Li, and Gilbert Weinstein. Mass and angular-momentum inequalities for axi-symmetric initial data sets. II. Angular-momentum. *Ann. Phys.*, 323(10):2591–2613, 2008, arXiv:0712.4064.

[10] Piotr T. Chrusciel and Joao Lopes Costa. Mass, angular-momentum, and charge inequalities for axisymmetric initial data. *Class. Quant. Grav.*, 26:235013, 2009, 0909.5625.

[11] Joao Lopes Costa. Proof of a Dain inequality with charge. *Journal of Physics A: Mathematical and Theoretical*, 43(28):285202, 2010, 0912.0838.

[12] Mihalis Dafermos and Igor Rodnianski. Lectures on black holes and linear waves, 2008, 0811.0354.

[13] Mihalis Dafermos and Igor Rodnianski. The black hole stability problem for linear scalar perturbations. 2010, 1010.5137.

[14] Sergio Dain. Angular momemtum-mass inequality for axisymmetric black holes. *Phys. Rev. Lett.*, 96:101101, 2006, gr-qc/0511101.

[15] Sergio Dain. Axisymmetric evolution of Einstein equations and mass conservation. *Class. Quantum. Grav.*, 25:145021, 2008, 0804.2679.

[16] Sergio Dain. The inequality between mass and angular momentum for axially symmetric black holes. *International Journal of Modern Physics D*, 17(3-4):519–523, 2008, arXiv:0707.3118 [gr-qc].

[17] Sergio Dain. Proof of the angular momentum-mass inequality for axisymmetric black holes. *J. Differential Geometry*, 79(1):33–67, 2008, gr-qc/0606105.

[18] Sergio Dain. Extreme throat initial data set and horizon area-angular momentum inequality for axisymmetric black holes. *Phys. Rev. D*, 82(10):104010, Nov 2010, 1008.0019.

[19] Sergio Dain and Maria E. Gabach Clément. Small deformations of extreme Kerr black hole initial data. *Class. Quant. Grav.*, 28:075003, 2011, 1001.0178.

[20] Sergio Dain and Omar E. Ortiz. Well-posedness, linear perturbations, and mass conservation for the axisymmetric einstein equations. *Phys. Rev. D*, 81(4):044040, Feb 2010.

[21] Sergio Dain and Martin Reiris. Area - Angular momentum inequality for axisymmetric black holes, 2011, 1102.5215.

[22] Sergio Dain and Martin Reiris. Linear perturbations for the vacuum axisymmetric Einstein equations. *Annales Henri Poincare*, 12:49–65, 2011, 1005.5347.

[23] David Garfinkle and G. Comer Duncan. Numerical evolution of Brill waves. *Phys. Rev.*, D63:044011, 2001, gr-qc/0006073.

[24] Jörg Hennig, Carla Cederbaum, and Marcus Ansorg. A universal inequality for axisymmetric and stationary black holes with surrounding matter in the Einstein-Maxwell theory. *Commun. Math. Phys.*, 293:449–467, 2010, 0812.2811.

[25] Jose Luis Jaramillo, Nicolas Vasset, and Marcus Ansorg. A numerical study of Penrose-like inequalities in a family of axially symmetric initial data. 2007, 0712.1741.

[26] T. Nakamura, K. Oohara, and Y. Kojima. General Relativistic Collapse to Black Holes and Gravitational Waves from Black Holes. *Progress of Theoretical Physics Supplement*, 90:1–218, 1987.

[27] Alan Rendall. 3+1. `http://www.aei.mpg.de/~rendall/3+1.ps`, 2008.

[28] Alan D. Rendall. *Partial differential equations in general relativity*, volume 16 of *Oxford Graduate Texts in Mathematics*. Oxford University Press, Oxford, 2008.

[29] Oliver Rinne. *Axisymmetric Numerical Relativity.* PhD thesis, University of Cambridge, 2005, gr-qc/0601064.

[30] Oliver Rinne. Constrained evolution in axisymmetry and the gravitational collapse of prolate Brill waves. *Class. Quantum. Grav.*, 25:135009, 2008, 0802.3791.

[31] Oliver Rinne and John M. Stewart. A strongly hyperbolic and regular reduction of Einstein's equations for axisymmetric spacetimes. *Class. Quantum. Grav.*, 22:1143–1166, 2005, gr-qc/0502037.

[32] Milton Ruiz, Miguel Alcubierre, and Dario Nunez. Regularization of spherical and axisymmetric evolution codes in numerical relativity. *Gen. Rel. Grav.*, 40:159–182, 2008, 0706.0923.

[33] Evgeny Sorkin. An Axisymmetric generalized harmonic evolution code. *Phys.Rev.*, D81:084062, 2010, 0911.2011.

[34] Evgeny Sorkin. On critical collapse of gravitational waves. *Class.Quant.Grav.*, 28:025011, 2011, 1008.3319.

Journal of Physics: Conference Series **314** (2011) 012016 doi:10.1088/1742-6596/314/1/012016

Geometric invariants for initial data sets: analysis, exact solutions, computer algebra, numerics

Juan A. Valiente Kroon

School of Mathematical Sciences,
Queen Mary, University of London
Mile End Road, London E1 4NS, UK

E-mail: `j.a.valiente-kroon@qmul.ac.uk`

Abstract. A personal perspective on the interaction of analytical, numerical and computer algebra methods in classical Relativity is given. This discussion is inspired by the problem of the construction of invariants that characterise key solutions to the Einstein field equations. It is claimed that this kind of ideas will be or importance in the analysis of dynamical black hole spacetimes by either analytical or numerical methods.

1. Introduction

Gaining control over the dynamics of black hole spacetimes is one of the great challenges of modern classical General Relativity. This ambitious programme can only be successfully addressed by means of a “multidisciplinary” approach which combines ideas and notions from diverse areas of Relativity and beyond. In particular, very much of what we expect to be able to learn about the behaviour of dynamical black holes will depend on a successful interaction between the two strands of research generically known as *numerical* and *mathematical* Relativity. Broadly speaking, numerical Relativity can be understood as the endeavour of solving the Einstein field equations in computers by means of numerical schemes. Similarly, mathematical Relativity can be described as the analysis of solutions of the Einstein field equations by means of analytical methods.

In the last decade mainstream mathematical Relativity has focused on the analysis of the Einstein field equations by means of abstract methods to obtain a wide variety of existence and uniqueness theorems. There is, however, another line of research in mathematical Relativity which is of potential great value for the analysis of dynamical spacetimes: the construction of exact solutions and their characterisation.

The study of exact solutions of the Einstein field equations is a mature subject. The highlights of this programme can mostly be found in the monograph [1]. It is very unlikely that solutions of fundamental physical relevance will be found in the future. Given this state of affairs, it is natural to ask what can one do with this wealth of information that has been extracted from this programme. In this contribution I want to argue that key to exploit these ideas is bringing to the fore a closely related line of research —namely, the analysis of *invariant characterisations of spacetimes.*

The need for invariant characterisations for solutions to the Einstein field equations stems from the diffeomorphism invariance of the theory. Given two metrics, one can ask whether they are locally isometric —that is, if there is a coordinate that brings one metric to the other. The latter is usually known as the *Equivalence Problem*, and it was formally solved by Cartan who gave a prescription to answer the question —this prescription involves the computation and comparison of higher order derivatives of the Riemann tensor. Instead of using Cartan's procedure, one could also address the same problem by means of invariant characterisations — that is coordinate and gauge independent labelling of solutions to the Einstein field equations.

Here I want to argue that methods inspired from the invariant characterisation of spacetimes can prove extremely valuable in the analysis of black hole spacetimes. To see why, I turn my attention to some of the *paradigmatic* (open) problems in the analysis of black hole spacetimes. For this, let us first recall that the 2-parameter Kerr solution (which includes the Schwarzschild solution as special case) is the prototype of a rotating black hole spacetime in vacuum —its parameters correspond to the mass and angular momentum of the black hole. In this regard, it is strongly conjectured that all stationary vacuum black hole solutions are described by the Kerr family of solutions this problem is called the *uniqueness of the Kerr spacetime.* In principle, it should be possible to use the Kerr solution to model the exterior of a rotating massive body other than a black hole. This expectation, however, has not been realised and it is now conjectured that there is no perfect fluid configuration whose exterior is described by the Kerr solution —this problem is called the *non-existence of an interior for Kerr.* The Kerr spacetime is believed to describe, in some sense, the asymptotic state of the evolution of dynamical black hole spacetimes this is sometimes called the consensus view on black holes, and it provides the connection with Cosmic Censorship. Furthermore, black hole solutions which are, in some sense, close to a member of the Kerr family are expected to have the same global and asymptotic structure as Kerr black holes the so- called *non-linear stability of the Kerr spacetime.*

Although this list does not intend to be extensive, it helps to understand what is, in my opinion, the key problem in the study of dynamic black holes: *to understand the role of the Kerr spacetime in the space of solutions to the vacuum Einstein field equations.*

2. The Cauchy problem in General Relativity and Computer Algebra

In order to study generic properties of a wide classes of spacetimes, one needs a systematic procedure to construct and parametrise solutions to the Einstein field equations. The *Cauchy problem* in General Relativity —in which one prescribes some initial data on a some 3-dimensional hypersurface and then purports to reconstruct the spacetime arising from this data— provides a way of doing this. The Cauchy problem (and variations thereof) provide the starting point of not only analytical studies of the Einstein field equations but also of the numerical computation of solutions thereof.

In order to see what it is involved in the Cauchy problem, recall that the Einstein equations are tensorial equations valid in any coordinate system. As such, they cannot directly be identified as belonging to any of the three main families of partial differential equations (hyperbolic, elliptic, parabolic). In order to obtain partial differential equations one has to break the covariance of the equations by picking a direction along which to follow the evolution of the spacetime. Sometimes the peculiarities of the particular problem at hand suggest a natural way of doing this, but very often this choice is essentially arbitrary and only justified by its value in extracting information about the solution in question. The choice of of the direction of time (*time vector*) needs to be complemented with an election of coordinates. One then speaks of a gauge choice. More generally, and depending on the formulation of the Einstein field equations one is considering, a gauge choice must not be restricted only to a choice of time vector and coordinates, but can

Journal of Physics: Conference Series **314** (2011) 012016 doi:10.1088/1742-6596/314/1/012016

involve further considerations like the choice of a vector frame, a spin dyad or a conformal factor (conformal gauge).

Once the gauge has been fixed, one then proceeds to decompose the Einstein field equations along directions longitudinal and transverse to the time vector to obtain a 3+1 decomposition (or formulation) of the Einstein field equations. Generically, one finds that a certain subset of the resulting equations are intrinsic to the leaves of the foliation implied by the time vector —these are called *constraint equations*. The other part of the field equations gives describes the evolution of the various field quantities —the so-called *evolution or propagation equations*. Depending to the type of gauge being used, it may be necessary to supplement the propagation equations with evolution equations for the gauge. The justification behind the approach is a s follows:

(i) One first solves the constraint equations on some initial hypersurface.
(ii) On a second step one solves the evolution equations on some neighbourhood of the initial hypersurface.
(iii) Finally, one shows that by virtue of the propagation equations, the constraint equations are satisfied on the same neighbourhood where the propagation equations have a solution. This last step is usually called the *propagation* of the constraints.

The algebraic complexity of the Einstein field equations has made the analysis of 3+1 decompositions make the study of 3+1 decompositions of the Einstein field equations a topic where the modern capabilities of Computer Algebra systems have made an impact. One can go a step further: the use of Computer Algebra methods is a key ingredient if one is to fully exploit the potential of the Cauchy problem in General Relativity. The reason behind this assertion lies in the intrinsic nonuniqueness of 3+1 decompositions of the Einstein field equations. As it will be discussed in the sequel, the relevance of Computer Algebra resides not only in the ability to streamline or automatise tedious computations, but in the insight it provides on the structures underlying large and complex systems of equations.

3. Characterising initial data sets

One of the main challenges posed by the Cauchy problem in General Relativity is how to prescribe physically relevant data for the evolution equations. This task is made difficult by the fact that the data has to satisfy the constraint equations. Traditional approaches to the solution of the constraint equations have favoured analytical simplicity over physical considerations — like in the case of conformally flat initial data sets. It is only recently, that more physically realistic initial data sets to describe, say, dynamical black hole spacetimes have been brought to the fore.

As mentioned before, the Schwarzschild and Kerr spacetimes play have a fundamental role among the class of black hole solutions to the Einstein field equations. This point of view is supported on the theoretical side by what Penrose called the *establishment's point of view* — see e.g. [2]. From a more applied perspective, this point of view is strongly supported by the numerical simulations of dynamical black hole spacetimes. Hence, it seems of crucial importance to understand what it means that a certain initial data set for the Einstein field equations is data for the Schwarzschild or the Kerr spacetime. Notice that although the Schwarzschild and Kerr solutions have certain symmetries which in a certain sense almost fully characterise them, the existence of these symmetries can be hard to detect on a particular hypersurface of the spacetime.

I formulate the problem of characterising initial data sets for the Schwarzschild or Kerr spacetimes as follows: *to provide conditions on an initial data set for the Einstein field equations so that its development is isometric to the Schwarzschild or Kerr spacetimes.*

A priori, there seem to be many ways to address this problem. One can think of providing local or global conditions. Furthermore, the conditions can be given in terms of implicit conditions on the data or they can be provided in the form of an algorithm so that they can be explicitly verified. The choice of a particular approach depends very much on the particular applications one has in mind.

Before proceeding to discuss a strategy to attack the problem posed in previous lines, I want to point out that although this programme arose first as an attempt to bridge the gap between analytical and numerical considerations —where a resolution of the problem is of clear interest— it now seems to be also of relevance in providing alternative formulations (and hopefully alternative solutions!) to the paradigmatic problems of the mathematical theory of black holes. A further area of potential analytical and numerical interest is that of the construction of initial data sets for the Schwarzschild and Kerr spacetimes with particular geometric properties. For example, there have been attempts to construct initial data sets for the Kerr spacetime which are conformally flat —it now has been shown that these do not exist [3, 4]. Similarly, it is of interest to see whether it is possible to construct boosted initial data sets for the Schwarzschild spacetime which are maximal.

4. A strategy to characterise initial data sets for the Kerr spacetime

A consequence of the 4-dimensional nature of spacetime is that invariant characterisations of spacetimes are, in general, easier to obtain and formulate than characterisations of arbitrary slices of the spacetime. Accordingly, the approach to obtain a characterisation of data for the Kerr spacetime I want to advocate here is based on the following strategy:

(i) Look for a suitable characterisation of the Kerr spacetime.

(ii) Perform a 3+1 decomposition of the characterisation on an arbitrary slice of the spacetime. Some of the equations thus obtained will be intrinsic, while some others will describe the evolution of the characterisation. From the formed one obtains necessary conditions for an initial data set to be Kerrian.

(iii) Find further assumptions that ensure that the necessary conditions obtained in step (iii) propagate —that is, if they are satisfied on an initial hypersurface, then they are also satisfied at later times so that the spacetime characterisation of the Kerr spacetime is recovered. This step gives sufficient conditions.

Step (ii) in the aforementioned procedure is analogous to the 3+1 splitting of the Einstein field equations into constraints and evolution equations. In a natural way, Computer Algebra methods are of great value here. However, their relevance comes truly to the fore when analysing the interdependences of the various equations arising from the decomposition —ideally, one would like to obtain a minimal set of conditions. Step (iii) is, in turn, analogous to the problem of the propagation of the Einstein constraint equations. This is the most difficult part of the approach. It requires the construction of suitable evolution equations for the necessary conditions obtained in (iii). The use of Computer Algebra is essential in this particular step.

5. Characterising the Kerr spacetime

Several approaches to the problem of characterising the Kerr spacetime have been given in the literature. Each with their own advantages and disadvantages. Here we briefly mention some of these.

a. The Simon and Mars-Simon tensors. A convenient way of studying stationary solutions to the Einstein field equations is through the quotient manifold of the orbits of the stationary Killing vector. The Schwarzschild spacetime is characterised among all stationary solutions by the vanishing of the Cotton tensor of the metric of this quotient manifold —see e.g. [5]. In

Journal of Physics: Conference Series **314** (2011) 012016 doi:10.1088/1742-6596/314/1/012016

[6] a suitable generalisation of the Cotton tensor of the quotient manifold was introduced —the *Simon tensor*. The vanishing of the Simon tensor together with asymptotic flatness and non-vanishing of the mass characterises the Kerr solution in the class of stationary solutions. In [7, 8] a spacetime version of the Simon tensor was introduced —the so-called *Mars-Simon tensor*. The construction of this tensor requires the *a priori* existence of a Killing vector in the spacetime. Accordingly, it is tailored for the problem of the uniqueness of stationary black holes. The vanishing of the Mars-Simon tensor together with some global conditions (asymptotic flatness, non-zero mass, stationarity of the Killing vector) characterises the Kerr spacetime.

b. Characterisations using concomitants of the Weyl tensor. A concomitant of the Weyl tensor is an object constructed from tensorial operations on the Weyl tensor and its covariant derivatives. An invariant characterisation of the Kerr spacetime in terms of concomitants of the Weyl tensor has been obtained in [9]. This result generalises a similar result for the Schwarzschild spacetime given in [10]. These characterisations consist of a set of conditions on concomitants of the Weyl tensor, which if satisfied, characterise locally the Kerr/Schwarzschild spacetime. An interesting feature of the characterisation is that it provides expressions for the stationary and axial Killing vectors of the spacetime in terms of concomitants of the Weyl tensor. Unfortunately, the concomitants used in the characterisation are complicated, and thus, produce very involved expressions when performing a 3+1 split.

c. Characterisations by means of generalised symmetries. Generalised symmetries (sometimes also known as hidden symmetries) are generalisations of the Killing vector equation —like the Killing tensors and conformal Killing-Yano tensors. These tensors arise naturally in the discussion of the so-called Carter constant of motion and in the separability of various types of linear equations on the Kerr spacetime —see e.g. [11, 12, 13]. In particular, the existence of a conformal Killing-Yano tensor is equivalent to the existence of a valence-2 symmetric spinor satisfying the Killing spinor equation. An important property of the Schwarzschild and Kerr spacetimes is that they admit a Killing spinor. This Killing spinor generates, in a certain sense, the Killing vectors and Killing-Yano tensors of the exact solutions in question [14]. In [15, 16] it has been shown that for a spacetime which is neither conformally flat nor of Petrov type N, the existence of a Killing spinor associated to a Killing-Yano tensor together with the requirement of asymptotic flatness renders a characterisation of the Kerr spacetime. In a further analysis, it has been possible to remove the *a priori* assumptions on the Petrov type —see [17].

Although at first sight independent, the characterisations of the Schwarzschild and Kerr spacetimes described in the previous paragraphs are interconnected —sometimes in very subtle manners. This is not too surprising as all these characterisations make use in a direct or indirect manner of the fact that the Kerr spacetime is a vacuum spacetime of Petrov type D —see e.g. [1] for a discussion of the Petrov classification. The art in producing a useful characterisation of the Kerr spacetime lies in finding further conditions on type D spacetimes which are natural and simple to use.

6. Using Killing spinors to obtain a characterisation of Kerr initial data

In this section I want to briefly describe the key ideas behind a characterisation of Kerr data based on the notion of Killing spinors. This is a non-technical summary of the ideas exposed in [15, 16, 17, 18]. The reader is referred to these references for a detailed exposition of the topic.

The starting point of the construction is the observation that the existence of a Killing spinor in the Kerr spacetime allows to relate the Killing vectors of the spacetime with its curvature in a neat way. The reason for its importance can be explained in the following way: from a specific Killing spinor it is possible to obtain a Killing vector which in general will be complex. It turns out that for the Kerr spacetime this Killing vector is in fact real and coincides with the stationary Killing vector. It can be shown that the Kerr solution is the only asymptotically flat vacuum spacetime with these properties.

Given the aforementioned spacetime characterisation of the Kerr solution, the question now is how to make use of it to produce a characterisation in terms of initial data sets. For this, one has to encode the existence of a Killing spinor at the level of the data. The way of doing this was first discussed in [19] and follows the spirit of the well-known discussion of how to encode Killing vectors on initial data —see e.g. [20]. The use of Computer Algebra methods is instrumental to implement these ideas in an effective manner [1].

The conditions on the initial data that ensure the existence of a Killing spinor in its development are called the *Killing spinor initial data equations* and are, like the Killing initial data equations (KID equations), overdetermined. In [21], a procedure was given on how to construct equations which generalise the KID equations for time symmetric data. These generalised equations have the property that for a particular behaviour at infinity they always admit a solution. If the spacetime admits Killing vectors, then the solutions to the generalised KID equations with the same asymptotic behaviour as the Killing vectors are, in fact, Killing vectors. Therefore, one calls the solutions to the generalised KID equations *approximate symmetries*. The total number of approximate symmetries is equal to the maximal number of possible Killing vectors on the spacetime. A peculiarity of this procedure is that if the spacetime is not stationary, the approximate Killing vector associated to a time translation does not have the same asymptotic behaviour as a time translation[2].

The Killing spinor initial data equations consist of three conditions: one of them differential (the *spatial Killing spinor equation*)[3] and two *algebraic conditions*. Following the spirit of [21] we construct a generalisation of the spatial Killing spinor equation —*the approximate Killing spinor equation*. This equation is elliptic and of second order. This equation is the Euler-Lagrange equation of an integral functional —the L^2-norm of the exact spatial Killing spinor equation. For this equation it is possible to prove the following theorem:

Theorem 1. *For initial data sets to the Einstein field equations with suitable asymptotic behaviour, there exists a solution to the approximate Killing spinor equation with the same asymptotic behaviour as the Killing spinor of the Kerr spacetime.*

The proof of this theorem is based on an adaptation of standard arguments of the theory of systems of elliptic partial differential equations on non-compact manifolds. In particular it is necessary to make use of so-called weighted Sobolev spaces. Remarkably, some of the technical points of the proof require intensive Computer Algebra computations. Again, this type of methods show their effectiveness as a tool to investigate the properties of large systems tensorial and spinorial equations.

It should be pointed out that the conditions on the asymptotic behaviour of the initial data are rather mild and amount to requiring the data to be, in a sense, asymptotically Kerr data. The precise version of this theorem generalises the one discussed in [15] in that it allows for boosted data. This generalisation is only possible after a detailed analysis of the asymptotic solutions of the exact Killing spinor equation.

[1] The computations described in this contribution have been carried out in the suite `xAct` for `Mathematica`: www.xact.es.

[2] For a time translation it is understood a Killing vector which in some asymptotically Cartesian coordinate system has a leading term of the form ∂_t.

[3] The idea of using the spatial part of spinorial equations to characterise slices of particular spacetimes is not new. In [22] the spatial twistor equation has been used to characterise slices of conformally flat spacetimes. See also [23].

Journal of Physics: Conference Series **314** (2011) 012016 doi:10.1088/1742-6596/314/1/012016

The approximate Killing spinor discussed in the previous paragraphs can be used to construct a geometric invariant for the initial data set. This invariant is global and involves the L^2 norms of the Killing spinor initial data equations evaluated at the approximate Killing vector. It should be observed that part of the invariant satisfies a variational principle —this is a further difference with respect to the construction of [21]. As the initial data set is assumed to be asymptotically Euclidean, one expects its development to be asymptotically flat. This renders the desired characterisation of Kerr data and the following result.

Theorem 2. *Consider an initial data set for the vacuum Einstein field equations such that the L^2 norm of the Killing spinor initial data equations evaluated at the solution (with the same asymptotic behaviour as the Killing spinor of the Kerr spacetime) to the approximate Killing spinor equation vanishes. Then the initial data set is locally data for the Kerr spacetime. Furthermore, if the 3-manifold has the same topology as that of hypersurfaces of the Kerr spacetime, then the initial data set is data for the Kerr spacetime.*

The invariant described in this theorem is a non-negative number having the property that it vanishes if and only if the initial data set corresponds to data for the Kerr spacetime. Thus, the invariant measures the *non-Kerrness* of the initial data and condenses it in a single number. The question whether this invariant constitutes a good distance in the space of initial data sets is currently being investigated.

Keeping in mind possible applications to the numerical simulation of black hole spacetimes, it is worth pointing out that the construction of the invariant is fully amenable to a numerical implementation —the elliptic solvers that one would need to compute the solution to the approximate Killing spinor equation are, nowadays, standard technology.

In view of possible applications of the non-Kerrness to the problem of the uniqueness of stationary black holes and the non-linear stability of the Kerr spacetimes, it is important not only to consider hypersurfaces with two asymptotic ends, but also ones with an inner boundary having the topology of a 2-sphere. This boundary can represent the horizon of a black hole or the surface of a star. Generalisations of Theorems 1 and 2 for 3-dimensional hypersurfaces with the topology of the complement of an open ball in $\mathbb{R}^3$ have been given in [17]. The analysing leading to these generalisations is more technically involved than that of [15, 16] as it requires the verification of certain compatibility conditions on the boundary and the specification of the right boundary value for the elliptic problem. Remarkably, the theory of Killing spinors readily provides a formula for a Killing spinor candidate in terms of concomitants of the restriction of the Weyl tensor of the spacetime to the boundary of the hypersurface —a further connection with the theory of exact solutions to the Einstein field equations.

Further extensions (not yet fully analysed) of the ideas discussed in this section include 3-manifolds with a a cylindrical end (trumpet data) and hyperboloidal manifolds. There are reasons to suggest that the latter can be of particular importance in the analysis of evolution issues.

7. Concluding remarks

The purpose of the present contribution has been to highlight the potential that exist in carrying out research following an inclusive point of view that encompasses ideas from exact solutions, analytic methods, computer algebra techniques and numerical simulations.

A natural question concerning the invariants here discussed is that of their behaviour under time evolution. An input from numerical investigations can be of great utility to address this question. Conversely, it is conceivable that the feedback provided by this type of theoretical considerations can have a useful impact in numerical investigations of solutions to the Einstein field equations. If some type of constancy or monotonicity property could be established for the geometric invariant by either analytical or numerical methods, this would be a useful tool for studying non-linear stability of the Kerr spacetime and also in the numerical evolutions of black hole spacetimes.

Finally, it is worth pointing a further, rather intriguing connection of the geometric invariants I have been describing. Namely, in [18] it has been shown that the construction of [15, 16, 17] mimics several aspects of Witten's proof of of the *Mass Positivity Theorem* by spinorial methods —see [24]. The full implications of this connection still awaits a detailed consideration. As in previous analysis, it is very likely that a combination of various techniques coming from traditionally different areas of research in General Relativity will be instrumental to obtain key insights.

Acknowledgements
The research described in this contribution is part of an ongoing scientific collaboration with T Bäckdahl. We have benefited from the comments, insights and suggestions from a number of colleagues including: JL Jaramillo, A García-Parrado, C Lübbe, JMM Senovilla, M Mars, L Andersson, S Dain, N Kamran, PK Tod, G Ávila, A Aceña, S Hollands, PT Chruściel and R Beig. This project has been funded by an EPSRC Advanced Research fellowship. I thank the organisers of the ERE 2010 in Granada for the invitation to give a plenary talk.

[1] Stephani H, Kramer D, MacCallum M A H, Hoenselaers C and Herlt E 2003 *Exact Solutions of Einstein's Field Equations* (Cambridge University Press) second edition
[2] Penrose R 1973 *Ann. N. Y. Acad. Sci.* **224**
[3] Valiente Kroon J A 2004 *Phys. Rev. Lett.* **92** 041101
[4] Valiente Kroon J A 2004 *Class. Quantum Grav.* **21** 3237
[5] Friedrich H 2004 *50 years of the Cauchy problem in general relativity* ed Chruściel P T and Friedrich H (Birkhausser)
[6] Simon W 1984 *Gen. Rel. Grav.* **16** 465
[7] Mars M 1999 *Class. Quantum Grav.* **16** 2507
[8] Mars M 2000 *Class. Quantum Grav.* **17** 3353
[9] Ferrando J J and Sáez J A 2009 *Class. Quantum Grav.* **26** 075013
[10] Ferrando J J and Sáez J A 1998 *Class. Quantum Grav.* **15** 1323
[11] Carter B 1968 *Comm. Math. Phys.* **10** 280
[12] Kamran N and McLenaghan R G 1984 *Acad. Roy. Belg. Bull. Cl. Sci* **70** 596
[13] Penrose R and Rindler W 1986 *Spinors and space-time. Volume 2. Spinor and twistor methods in space-time geometry* (Cambridge University Press)
[14] Hughston L P and Sommers P 1973 *Comm. Math. Phys.* **33** 129
[15] Bäckdahl T and Valiente Kroon J A 2010 *Phys. Rev. Lett.* **104** 231102
[16] Bäckdahl T and Valiente Kroon J A 2010 On the construction of a geometric invariant measuring the deviation from kerr data In `arXiv:1005.0743[gr-qc]`
[17] Bäckdahl T and Valiente Kroon J A 2010 The "non-kerrness" of domains of outer communication of black holes and exteriors of stars In `arXiv:1010.2421[gr-qc]`
[18] Bäckdahl T and Valiente Kroon J A 2010 Approximate twistors and positive mass In `arXiv:1011.3712[gr-qc]`
[19] García-Parrado A and Valiente Kroon J A 2008 *J. Geom. Phys.* **58** 1186
[20] Beig R and Chruściel P T 1997 *Class. Quantum Grav.* **14** A83
[21] Dain S 2004 *Phys. Rev. Lett.* **93** 231101
[22] Tod K P 1984 *Gen. Rel. Grav.* **16** 435
[23] Beig R and Szabados L B 1997 *Class. Quantum Grav.* **14** 3091
[24] Witten E 1981 *Comm. Math. Phys.* **80** 381

The interior of axisymmetric and stationary black holes: Numerical and analytical studies

Marcus Ansorg

Theoretisch-Physikalisches Institut, Friedrich-Schiller-Universität Jena, Max-Wien-Platz 1, D-07743 Jena, Germany

E-mail: Marcus.Ansorg@uni-jena.de

Jörg Hennig

Dept. of Mathematics and Statistics, University of Otago, P.O. Box 56, Dunedin 9054, New Zealand

E-mail: jhennig@maths.otago.ac.nz

Abstract. We investigate the interior hyperbolic region of axisymmetric and stationary black holes surrounded by a matter distribution. First, we treat the corresponding initial value problem of the hyperbolic Einstein equations numerically in terms of a single-domain fully pseudo-spectral scheme. Thereafter, a rigorous mathematical approach is given, in which soliton methods are utilized to derive an explicit relation between the event horizon and an inner Cauchy horizon. This horizon arises as the boundary of the future domain of dependence of the event horizon. Our numerical studies provide strong evidence for the validity of the universal relation $A^+A^- = (8\pi J)^2$ where A^+ and A^- are the areas of event and inner Cauchy horizon respectively, and J denotes the angular momentum. With our analytical considerations we are able to prove this relation rigorously.

1. Introduction

Axisymmetric and stationary black hole space-times are characterized by the existence of two Killing vectors ξ and η. Outside the black hole, these vectors can always be linearly combined to form a time-like vector. In contrast, any such non-trivial linear combination, taken in some *interior* neighborhood of the event horizon $\mathcal{H}^+$, inevitably leads to a space-like vector. The corresponding Einstein equations, expressed in Boyer-Lindquist-type coordinates, are elliptic in the black hole's exterior and hyperbolic in its interior. It has been shown in [1] that for non-vanishing angular momentum J of the black hole, there always exists a regular boundary of the future domain of dependence of the event horizon, the 'inner Cauchy horizon' $\mathcal{H}^-$.

In this contribution, we investigate the interior region of axisymmetric and stationary black holes which are surrounded by a matter distribution. The Einstein field equations are considered in Boyer-Lindquist-type, as well as Weyl coordinates, where they possess a well-defined behavior at the horizons. Note that in Boyer-Lindquist-type coordinates, the mathematical form of the field equations is the same at $\mathcal{H}^+$ and $\mathcal{H}^-$.

In section 2 we apply a fully pseudo-spectral method and solve the degenerate hyperbolic Einstein equations as an initial value problem. We expand the field quantities with respect to both spatial and time coordinates in terms of Chebyshev polynomials. Starting at the past Cauchy horizon $\mathcal{H}^+$, we evolve the data into the black hole interior up to the future Cauchy horizon $\mathcal{H}^-$. The data, found in this manner at $\mathcal{H}^-$, obey the remarkable property

$$A^+A^- = (8\pi J)^2\,, \tag{1}$$

where A^+ and A^- are the areas of event and inner Cauchy horizon respectively. The numerical solutions possess the extreme accuracy known for spectral methods when applied to elliptic problems. In particular, the relation (1) was found to be valid up to 12 decimal digits.

In section 3 we review analytical work through which (1) was proved rigorously [1]. The corresponding techniques are based on 'Bäcklund transformations', a method known from soliton theory. We conclude this article with a discussion of generalized and related black hole properties in axisymmetry.

2. Numerical studies

In Boyer-Lindquist-type coordinates (R,θ,φ,t), the metric becomes singular at $R=\pm r_{\rm h}$,

$$\mathrm{d}s^2 = \hat\mu\left(\frac{\mathrm{d}R^2}{R^2-r_{\rm h}^2}+\mathrm{d}\theta^2\right)+\hat u\sin^2\theta(\mathrm{d}\varphi^2-\omega\mathrm{d}t)^2-\frac{4}{\hat u}(R^2-r_{\rm h}^2)\mathrm{d}t^2\,, \tag{2}$$

where $\pm r_{\rm h}$ denote the coordinate locations of the two horizons $\mathcal{H}^\pm$. The interior hyperbolic region of the black hole is characterized by

$$R\in(-r_{\rm h},r_{\rm h}),\qquad \theta\in[0,\pi]\,. \tag{3}$$

Here, the metric coefficients $\hat\mu,\omega,\hat u$ depend on R and θ only and are regular at $R=\pm r_{\rm h}$ (see [2, 3]). The corresponding hyperbolic Einstein equations read as follows:

$$(R^2-r_{\rm h}^2)\tilde u_{,RR}+2R\tilde u_{,R}+\tilde u_{,\theta\theta}+\tilde u_{,\theta}\cot\theta = 1-\frac{\hat u^2}{8}\sin^2\theta\left(\omega_{,R}^2+\frac{\omega_{,\theta}^2}{R^2-r_{\rm h}^2}\right), \tag{4}$$

$$(R^2-r_{\rm h}^2)(\omega_{,RR}+4\omega_{,R}\tilde u_{,R})+\omega_{,\theta\theta} = -\,\omega_{,\theta}(3\cot\theta+4\tilde u_{,\theta})\,, \tag{5}$$

$$\begin{aligned}(R^2-r_{\rm h}^2)\tilde\mu_{,RR}+R\tilde\mu_{,R}+\tilde\mu_{,\theta\theta} &= \frac{\hat u^2}{16}\sin^2\theta\left(\omega_{,R}^2+\frac{\omega_{,\theta}^2}{R^2-r_{\rm h}^2}\right)+R\tilde u_{,R}\\ &\quad -(R^2-r_{\rm h}^2)\tilde u_{,R}^2-\tilde u_{,\theta}(\tilde u_{,\theta}+\cot\theta),\end{aligned} \tag{6}$$

where $\tilde u := \frac12\ln\left(\hat u/\hat u_{\rm N}^+\right)$ and $\tilde\mu := \frac12\ln\left(\hat\mu/\hat u_{\rm N}^+\right)$ with $\hat u_{\rm N}^+ := \hat u(R=r_{\rm h},\theta=0)$[1]. Note that $\hat\mu$ does not enter in the equations (4, 5). An explicit solution of this system is given by the Kerr metric which describes a single rotating black hole in vacuum.

At $R=\pm r_{\rm h}$, the metric potentials have to obey the following boundary conditions (see [2, 3])

$$\mathcal{H}^\pm:\quad\left\{\begin{array}{l}\omega=\text{constant}=\omega^\pm\,,\\[2mm] \pm 2r_{\rm h}\,\tilde u_{,R}+\tilde u_{,\theta\theta}+\tilde u_{,\theta}\cot\theta = 1-\frac18\hat u^2\omega_{,R}^2\sin^2\theta\end{array}\right. \tag{7}$$

with the horizon angular velocities $\omega^\pm$.

[1] Here we choose the north pole value $\hat u_{\rm N}^+$ of $\hat u$ as normalization constant. It remains positive and finite even in the degenerate limit $r_{\rm h}\to 0$.

Expressions for the black hole angular momentum J and horizon areas $A^{\pm}$ have been discussed in [2, 3]. They are given by

$$J = \frac{1}{8\pi}\oint\limits_{\mathcal{H}^{\pm}} \eta^{a;b}\mathrm{d}S_{ab} = -\frac{1}{16}\int_0^{\pi} \hat{u}^2\omega_{,R}\big|_{\mathcal{H}^{\pm}} \sin^3\theta\,\mathrm{d}\theta, \tag{8}$$

$$A^{\pm} = 2\pi\int_0^{\pi}\sqrt{\hat{\mu}\hat{u}}\big|_{\mathcal{H}^{\pm}}\sin\theta\,\mathrm{d}\theta = 4\pi\hat{u}_{\mathrm{N}}^{\pm}, \tag{9}$$

where the 'north pole' values of $\hat{u}$ are denoted by

$$\hat{u}_{\mathrm{N}}^{\pm} = \hat{u}(R = \pm r_{\mathrm{h}}, \theta = 0).$$

Note that, for the area formula (9), we used the following relations:

$$\begin{array}{lcccl} \text{At the Horizons } \mathcal{H}^{\pm}\,(R=\pm r_{\mathrm{h}}) & : & \sqrt{\hat{\mu}\hat{u}} & = & \text{constant} \\ \text{At the Rotation axis } (\sin\theta = 0) & : & \hat{\mu} & = & \hat{u}\,. \end{array} \tag{10}$$

In our numerical scheme we solve the equations (4) and (5). The corresponding solutions provide us with the relevant information to compute J and $A^{\pm}$.

As a first step, we express the metric coefficients $\tilde{u}$ and ω in terms of auxiliary potentials U and Ω:

$$\begin{aligned} \tilde{u}(R,\theta) &= \tilde{u}^{+}(\theta) + (R - r_{\mathrm{h}})U(R,\theta) \\ \omega(R,\theta) &= \omega^{+} + (R - r_{\mathrm{h}})\omega_R^{+}(\theta) + (R - r_{\mathrm{h}})^2\Omega(R,\theta)\,. \end{aligned}$$

In this *ansatz*, we take into account the boundary condition $\omega(R = r_{\mathrm{h}},\theta) = \omega^{+}$. We prescribe freely a set of 'initial' data $\{\tilde{u}^{+}(\theta), \omega^{+}, \omega_R^{+}(\theta)\}$ which is then evolved from $R = r_{\mathrm{h}}$ into the black hole interior up to $R = -r_{\mathrm{h}}$ via the equations (4) and (5) (see figure 1). Note that the second boundary condition in (7) fixes the first 'time' derivative $\tilde{u}_{,R}(r_{\mathrm{h}},\theta)$ in terms of $\tilde{u}^{+}(\theta)$ and $\omega_R^{+}(\theta)$, and hence $\tilde{u}_{,R}(r_{\mathrm{h}},\theta)$ cannot be prescribed freely.

We have chosen a *fully* pseudo-spectral scheme [4] for the numerical integration of (4) and (5). In practice, we expand the functions U and Ω with respect to Chebyshev polynomials:

$$U \approx \sum_{j=0}^{n}\sum_{k=0}^{n} c_{jk}^{(U)} T_j\left(\frac{R}{r_{\mathrm{h}}}\right) T_k\left(\frac{2}{\pi}\theta - 1\right) \tag{11}$$

$$\Omega \approx \sum_{j=0}^{n}\sum_{k=0}^{n} c_{jk}^{(\Omega)} T_j\left(\frac{R}{r_{\mathrm{h}}}\right) T_k\left(\frac{2}{\pi}\theta - 1\right) \tag{12}$$

and consider the field equations on an $n \times n$-'spectral grid' with Gauss-Lobatto gridpoints (R_j, θ_k) (see figure 1),

$$R_j = r_{\mathrm{h}}\cos\frac{\pi j}{n-1}, \qquad \theta_k = \pi\sin^2\frac{\pi k}{2(n-1)} \qquad (j,k = 0,\ldots,n-1)\,. \tag{13}$$

The pseudo-spectral collocation point method provides an approximation of the values

$$U_{jk} = U\left(R_j,\theta_k\right), \quad \Omega_{jk} = \Omega\left(R_j,\theta_k\right) \tag{14}$$

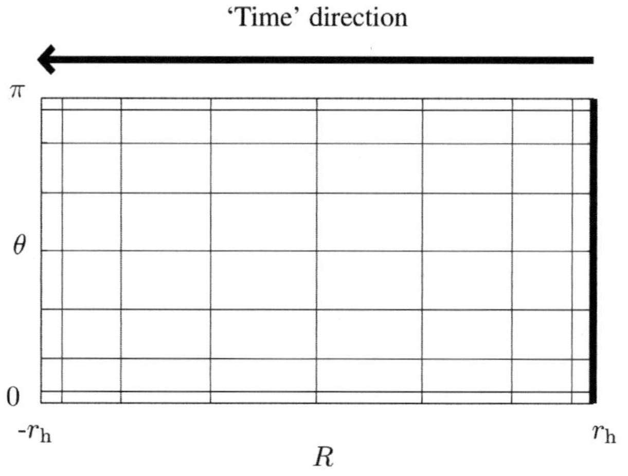

Figure 1. The interior hyperbolic black hole region with past $(R = r_\mathrm{h})$ and future $(R = -r_\mathrm{h})$ Cauchy horizons $\mathcal{H}^\pm$. The metric is determined as the solution to an initial value problem starting at $R = r_\mathrm{h}$ and evolving up to $R = -r_\mathrm{h}$. We indicate the location of the Gauss-Lobatto gridpoints on the spectral grid which is used in the pseudo-spectral collocation point method.

which we collect in the $2n^2$-dimensional vector

$$\mathbf{f} = \begin{pmatrix} U_{00} \\ \Omega_{00} \\ \vdots \\ U_{NN} \\ \Omega_{NN} \end{pmatrix}, \qquad N = n - 1\,. \tag{15}$$

From any such vector $\mathbf{f}$ the spectral coefficients $c_{jk}^{(U)}$ and $c_{jk}^{(\Omega)}$ of U and Ω can be computed. Moreover, the corresponding coefficients of first and second derivatives of U and Ω are easily determined. Returning from coefficient to physical space, we can build vectors $\mathbf{f}_{,R}, \ldots, \mathbf{f}_{,\theta\theta}$ containing the values of first and second derivatives $U_{,R}, \Omega_{,R}, \ldots, U_{,\theta\theta}, \Omega_{,\theta\theta}$ at all collocation points (R_j, θ_k).

Now consider the system (4, 5) at (R_j, θ_k) and insert the relevant components of $\mathbf{f}, \mathbf{f}_{,R}, \ldots, \mathbf{f}_{,\theta\theta}$ into these equations. We obtain a non-linear algebraic system of equations of the form

$$\mathbf{F}(\mathbf{f}) = 0\,, \tag{16}$$

where

$$\mathbf{F} = \begin{pmatrix} F_{00}^{(U)} \\ F_{00}^{(\Omega)} \\ \vdots \\ F_{NN}^{(U)} \\ F_{NN}^{(\Omega)} \end{pmatrix}. \tag{17}$$

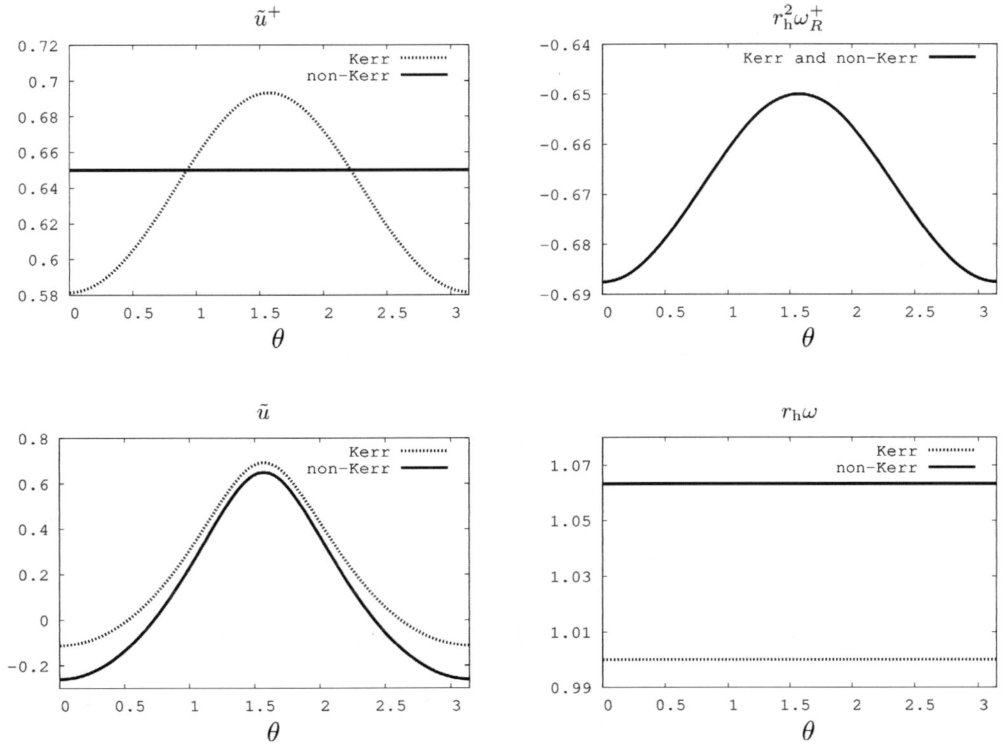

Figure 2. Initial data for Kerr and non-Kerr solution (upper panels) and corresponding values at the future Cauchy horizon $\mathcal{H}^-$ (lower panels).

Here, $F_{jk}^{(U)}$ and $F_{jk}^{(\Omega)}$ denote the difference of left and right hand sides of equations (4) and (5) respectively, taken at the collocation point (R_j, θ_k). The solution $\mathbf{f}$ of the discrete algebraic system (16) describes the spectral approximation of the solution to our hyperbolic initial value problem.

We find the vector $\mathbf{f}$ using a Newton-Raphson scheme,

$$\mathbf{f} = \lim_{m\to\infty} \mathbf{f}_m\,, \quad \mathbf{f}_{m+1} = \mathbf{f}_m - [\mathbf{J}(\mathbf{f}_m)]^{-1}\,\mathbf{F}\,(\mathbf{f}_m)\,, \tag{18}$$

where the Jacobian matrix is given by

$$\mathbf{J} = \frac{\partial \mathbf{F}}{\partial \mathbf{f}}\,. \tag{19}$$

Note that for the convergence of the scheme a 'good' initial guess $\mathbf{f}_0$ is necessary. For the first calculation, we take the corresponding coefficients of the Kerr solution for some specific parameters (say J and A^+). We may then depart from the Kerr solution and gradually approach some new solution with a non-Kerr initial data set $\{\tilde{u}^+(\theta), \omega^+, \omega_R^+(\theta)\}$.

In the following example we start with a Kerr black hole in which we prescribe the specific angular momentum $J/M^2 = 0.8$, where M denotes the black hole mass. As explained above, we gradually depart from this solution and approach an initial data set in which we replace the initial function $u^+(\theta)$ by a constant (we take $\tilde{u}^+(\theta) = 0.65$) but maintain ω^+ and $\omega_R^+(\theta)$ as previously determined for the Kerr case of $J/M^2 = 0.8$. The initial data $\tilde{u}^+(\theta)$ and $\omega_R^+(\theta)$, as well as the functions $\tilde{u}$ and ω, at $\mathcal{H}^-$, are displayed in figure 2. Here, the values at the inner Cauchy horizon $R = -r_h$ emerge from the pseudo-spectral calculation.

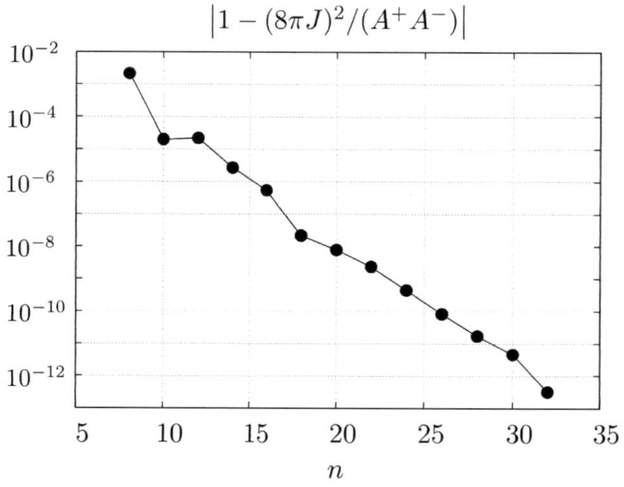

Figure 3. Numerical test of relation (1) for the non-Kerr example displayed in figure (2).

The numerical test of relation (1) is exhibited in figure 3. It shows a convergence plot of the expression $|1-(8\pi J)^2/(A^+A^-)|$ for our sample non-Kerr solution. We obtain a confirmation of the validity of (1) by 12 relevant decimal digits which is equivalent to the numerical round-off error.

3. Analytical studies

This section about the analytical study of the interior hyperbolic black hole region, including the rigorous derivation of the equality (1), is a brief summary of earlier work that has been presented in [1].

3.1. Weyl coordinates

As a first step towards a strict mathematical treatment in terms of *Bäcklund transformations*, we introduce canonical Weyl coordinates $(\rho, \zeta, \varphi, t)$ in a small exterior vacuum vicinity of the black hole[2]:

$$\rho^2 = 4(R^2 - r_{\rm h}^2)\sin^2\theta, \qquad \zeta = 2R\cos\theta. \tag{20}$$

The corresponding line element reads as follows:

$$\mathrm{d}s^2 = \mathrm{e}^{-2U}\left[\mathrm{e}^{2k}(\mathrm{d}\rho^2 + \mathrm{d}\zeta^2) + \rho^2\mathrm{d}\varphi^2\right] - \mathrm{e}^{2U}(\mathrm{d}t + a\mathrm{d}\varphi)^2\,. \tag{21}$$

Here, the metric potentials U, k and a are functions of ρ and ζ. The rotation axis ($\sin\theta = 0$) is given by:

$$\rho = 0, \qquad |\zeta| \geq 2r_{\rm h}\,, \tag{22}$$

and the exterior event horizon $\mathcal{H}^+$ is located at

$$\rho = 0, -2r_{\rm h} \leq \zeta \leq 2r_{\rm h}\,. \tag{23}$$

The event horizon is a degenerate surface in Weyl coordinates. As the interior region is characterized by $|R| < r_{\rm h}$, the corresponding Weyl coordinate ρ assumes imaginary values there. This means that the hyperbolic black hole region cannot be accessed in Weyl coordinates (see figure 4).

[2] We assume that for physically reasonable types of matter surrounding our stationary black hole, the immediate vicinity of the event horizon must be vacuum, see. e. g. discussion in [3].

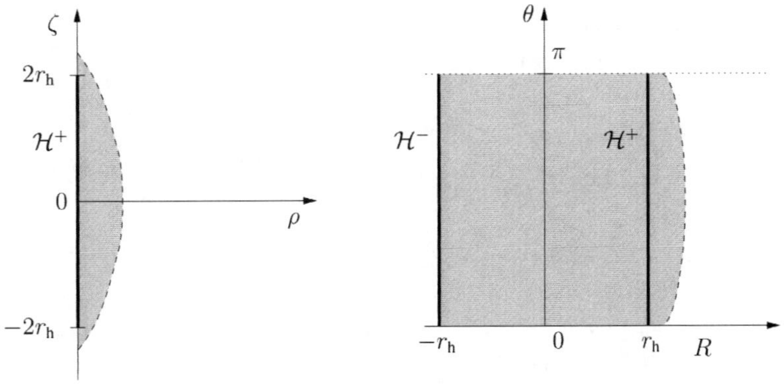

Figure 4. Portion of a black hole space-time in Weyl coordinates (left panel) and Boyer-Lindquist-type coordinates (right panel) (figure taken from [1]).

3.2. The Ernst equation

The complex Ernst potential f combines the metric functions U and b,

$$f = \mathrm{e}^{2U} + \mathrm{i}b\,, \tag{24}$$

where the twist potential b is related to the coefficient a via the relations:

$$a_{,\rho} = \rho\,\mathrm{e}^{-4U} b_{,\zeta}, \qquad a_{,\zeta} = -\rho\,\mathrm{e}^{-4U} b_{,\rho}. \tag{25}$$

In Boyer-Lindquist-type coordinates, this relation reads as:

$$a_{,R} = -2\sin\theta\,\mathrm{e}^{-4U} b_{,\theta}, \qquad a_{,\theta} = 2(R^2 - r_\mathrm{h}^2)\sin\theta\,\mathrm{e}^{-4U} b_{,R}. \tag{26}$$

The vacuum Einstein equations (which are valid in a vicinity of $\mathcal{H}^+$) are equivalent to the *Ernst equation* [5] which reads in Weyl coordinates as

$$(\Re f)\left(f_{,\rho\rho} + f_{,\zeta\zeta} + \frac{1}{\rho} f_{,\rho}\right) = f_{,\rho}^2 + f_{,\zeta}^2 \tag{27}$$

and in Boyer-Lindquist-type coordinates:

$$(\Re f)\left[(R^2 - r_\mathrm{h}^2) f_{,RR} + 2R f_{,R} + f_{,\theta\theta} + \cot\theta f_{,\theta}\right] = (R^2 - r_\mathrm{h}^2) f_{,R}^2 + f_{,\theta}^2. \tag{28}$$

Note that because of the degeneracy of $\mathcal{H}^+$ in Weyl coordinates, the potential f is, for $\rho = 0$, only a C^0-function in terms of ζ. In contrast, for a regular black hole, f is analytic with respect to the Boyer-Lindquist-type-coordinates R and θ.

3.3. Bäcklund transformation

The Ernst equation is, when written in Weyl coordinates, the integrability condition of an associated linear matrix problem [6, 7]. This is the great advantage of the Weyl coordinates with respect to the Boyer-Lindquist-type coordinates. The existence of this linear problem enables us to apply methods known from soliton theory. In this contribution, we are particularly interested in the so-called Bäcklund transformations through which new solutions from previously known ones are created [8]–[12]. As an example, the Kerr solution describing a rotating black hole in vacuum can be constructed from the flat Minkowski metric in this manner (see e. g. [10]).

In the following we use the Bäcklund transformation technique in order to write an arbitrary regular axisymmetric, stationary black hole solution f in terms of an auxiliary 'seed' potential f_0. Here, f_0 describes a space-time without a black hole but with a completely regular central vacuum region. More specifically, f_0 is characterized by the following properties:

(i) f_0 is defined in a vicinity of the axis section $\rho = 0, |\zeta| \le 2r_\mathrm{h}$.

(ii) In this vicinity, f_0 is an analytic function of ρ and ζ and an *even* function of ρ.

(iii) For $\rho = 0$ and $|\zeta| \le 2r_\mathrm{h}$, the values of f_0 in terms of the event horizon values of f are given by

$$f_0 = \frac{\mathrm{i}\left[2r_\mathrm{h}(b_\mathrm{N}^+ + b_\mathrm{S}^+) - (b_\mathrm{N}^+ - b_\mathrm{S}^+)\zeta\right] f + 4r_\mathrm{h} b_\mathrm{N}^+ b_\mathrm{S}^+}{4r_\mathrm{h} f - \mathrm{i}\left[2r_\mathrm{h}(b_\mathrm{N}^+ + b_\mathrm{S}^+) + (b_\mathrm{N}^+ - b_\mathrm{S}^+)\zeta\right]}, \tag{29}$$

where $b_\mathrm{N}^+ = b(\rho = 0, \zeta = 2r_\mathrm{h})$ and $b_\mathrm{S}^+ = b(\rho = 0, \zeta = -2r_\mathrm{h})$ denote the twist potential values at north and south poles of $\mathcal{H}^+$.

Now, from this Ernst potential f_0 the original potential f can be recovered by means of an appropriate Bäcklund transformation of the following form[3]:

$$f = \frac{\begin{vmatrix} f_0 & 1 & 1 \\ \bar{f}_0 & \alpha_1\lambda_1 & \alpha_2\lambda_2 \\ f_0 & \lambda_1^2 & \lambda_2^2 \end{vmatrix}}{\begin{vmatrix} 1 & 1 & 1 \\ -1 & \alpha_1\lambda_1 & \alpha_2\lambda_2 \\ 1 & \lambda_1^2 & \lambda_2^2 \end{vmatrix}}, \tag{30}$$

where

$$\lambda_j = \sqrt{\frac{K_j - \mathrm{i}\bar{z}}{K_j + \mathrm{i}z}}, \qquad j = 1, 2, \qquad K_1 = -2r_\mathrm{h}, \qquad K_2 = 2r_\mathrm{h} \tag{31}$$

with the complex coordinates $z = \rho + \mathrm{i}\zeta$, $\bar{z} = \rho - \mathrm{i}\zeta$, and α_1, α_2 are solutions to the Riccati equations

$$\alpha_{j,z} = -(\lambda_j\alpha_j^2 + \alpha_j)\frac{f_{0,z}}{f_0 + \bar{f}_0} + (\alpha_j + \lambda_j)\frac{\bar{f}_{0,z}}{f_0 + \bar{f}_0}, \tag{32}$$

$$\alpha_{j,\bar{z}} = -\left(\frac{1}{\lambda_j}\alpha_j^2 + \alpha_j\right)\frac{f_{0,\bar{z}}}{f_0 + \bar{f}_0} + \left(\alpha_j + \frac{1}{\lambda_j}\right)\frac{\bar{f}_{0,\bar{z}}}{f_0 + \bar{f}_0} \tag{33}$$

with

$$\alpha_j\bar{\alpha}_j = 1. \tag{34}$$

For a regular black hole, f is analytic with respect to R and $\cos\theta$ in an exterior vicinity of $\mathcal{H}^+$. Hence, we can expand it analytically into an *interior* vicinity of $\mathcal{H}^+$. A theorem by Chruściel (theorem 6.3 in [13]) assures that the potential f exists as a regular solution of the interior Ernst equation for all values [4]

$$(R, \cos\theta) \in (-r_\mathrm{h}, r_\mathrm{h}] \times [-1, 1]\,. \tag{35}$$

This region only excludes the Cauchy horizon $\mathcal{H}^-$ ($R = -r_\mathrm{h}$).

In the following derivation of the expression for f at the interior boundary $R = -r_\mathrm{h}$, a crucial role is played by the fact that f_0 is *even* in ρ. In terms of the Boyer-Lindquist-type coordinates, this means that f_0 is an analytic function of $(R^2 - r_\mathrm{h}^2)\sin^2\theta$ and $R\cos\theta$. The analytic expansion of f_0 into the region $R < r_\mathrm{h}$ retains this property. Hence, f_0 taken at the boundaries of the inner hyperbolic region, can be expressed in terms of f_0 taken at $R = r_\mathrm{h}$. Specifically we obtain

$$f_0(R = -r_\mathrm{h}\,, \cos\theta) = f_0(R = +r_\mathrm{h}\,, -\cos\theta) = f_0(\rho = 0, \zeta = -2r_\mathrm{h}\cos\theta)\,. \tag{36}$$

Also, it follows that f_0 is regularly defined in a sufficiently small vicinity of the boundary of the interior region, see figure 5.

From the values of f_0 at these boundaries we can construct f on $\mathcal{H}^-$ via the Bäcklund transformation, which is stated in the following.

[3] A bar denotes complex conjugation.

[4] We obtain Chruściel's form of the line element by substituting $R = r_\mathrm{h}\cos T$ and $\theta = \psi$.

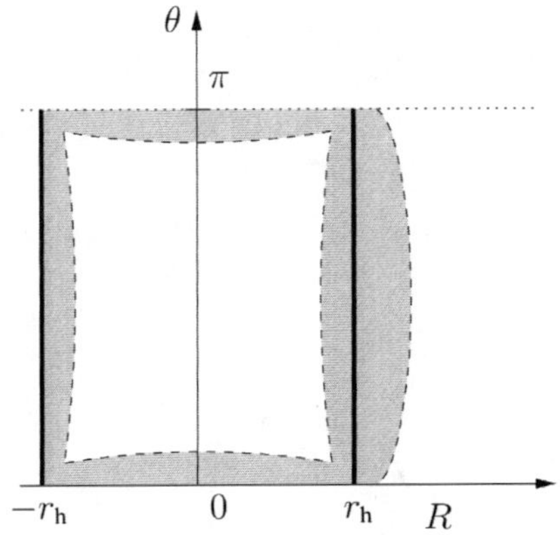

Figure 5. The seed function f_0 can be regularly defined in at least the grey areas (figure taken from [1]).

Theorem 3.1. *Any Ernst potential f of a regular, axisymmetric, and stationary black hole space-time with angular momentum $J \neq 0$, can be regularly extended into the interior of the black hole up to and including an interior Cauchy horizon, described by $R = -r_h$ in Boyer-Lindquist-type coordinates (R, θ). The values of f on the Cauchy horizon are given by*

$$f(R = -r_h, \cos\theta) = \frac{\mathrm{i}\,[\delta_1 + \delta_2 - (\delta_1 - \delta_2)\cos\theta]\, f_0(R = r_h, -\cos\theta) + 2\delta_1\delta_2}{2 f_0(R = r_h, -\cos\theta) - \mathrm{i}\,[\delta_1 + \delta_2 + (\delta_1 - \delta_2)\cos\theta]} \tag{37}$$

with

$$\delta_1 = \frac{b_S^+(b_N^+ - b_S^+) + 2b_N^+(b_{,\theta\theta})_N^+}{b_N^+ - b_S^+ + 2(b_{,\theta\theta})_N^+}, \qquad \delta_2 = \frac{b_N^+(b_N^+ - b_S^+) + 2b_S^+(b_{,\theta\theta})_N^+}{b_N^+ - b_S^+ + 2(b_{,\theta\theta})_N^+} \tag{38}$$

where the scripts '+' and 'N/S' indicate that the corresponding value of b or its second θ-derivative has to be taken at the event horizon's north or south pole respectively. The values of the seed solution f_0 for $R = r_h$ follow via (29) from f on the event horizon.

Note that, for $J \to 0$, the Cauchy horizon becomes singular. In this case we have $|\delta_{1/2}| \to \infty$ and consequently $f|_{\mathcal{H}^-}$ diverges.

By virtue of the above theorem we are able to write the inner Cauchy horizon area A^- completely in terms of Ernst expressions taken at the event horizon $\mathcal{H}^+$. As discussed in [1], the universal equality (1) arises in this manner. Moreover, the vanishing of A^- is obtained in the limit $J \to 0$, i. e. when the Cauchy horizon becomes singular.

4. Discussion

In this contribution we discussed the interior hyperbolic region of axisymmetric and stationary black holes which are surrounded by a matter distribution. We first looked at the corresponding degenerate hyperbolic Einstein equations in terms of fully pseudo-spectral methods, and confirmed the validity of relation (1) to high precision.

In the second part of the article, we used the Bäcklund technique in order to write the black hole metric in terms of an auxiliary seed potential f_0 that does not describe a black hole but a completely regular central region. We first derived simple relations between the values of f_0 at the boundaries of the interior region, and then carried these relations over to the original black hole metric by virtue of an appropriate Bäcklund transformation. A particular consequence of this relation is the universal equality (1).

Note that the key point for this method to work is the fact that the two horizons $\mathcal{H}^{\pm}$ are connected by specific axis sections. Indeed, the linear matrix problem associated to the Ernst equation (see discussion at the beginning of section 3.3) simplifies considerably for $\rho = 0$, that is, on the two horizons $\mathcal{H}^{\pm}$, and on the rotation axis ($\sin\theta = 0$). An alternative derivation, analyzing the linear matrix problem directly at the several sections where $\rho = 0$, was carried out in [14, 15] for the Einstein-Maxwell field. In that work, the corresponding steps yield the more general formula

$$(8\pi J)^2 + (4\pi Q^2)^2 = A^+A^-$$

where Q is the electric charge of the black hole.

It is interesting to remark that there are associated *inequalities* relating angular momentum and area of the black hole. In [16] (see also [17]) it was shown that for *subextremal* black holes (which possess trapped surfaces in every sufficiently small interior vicinity of the event horizon), the following inequalities hold:

$$A^- < \sqrt{(8\pi J)^2 + (4\pi Q^2)^2} < A^+.$$

In the case of pure Einsteinian gravity, this relation was proved in a different context, namely for non-stationary, axisymmetric, vacuum space-times [18]. In particular, the local inequality $A \geq 8\pi|J|$ was shown, where A and J are the area and angular momentum of any axially symmetric closed stable minimal surface in an axially symmetric maximal initial data.

Acknowledgments

We would like to thank S. Dain and P. T. Chruściel for illuminating discussions and R. Thompson for commenting on the manuscript.

References

[1] Ansorg M and Hennig J 2008 *The inner Cauchy horizon of axisymmetric and stationary black holes with surrounding matter* Class. Quant. Grav. **25** 222001
[2] Carter B 1973 *Black hole equilibrium states*, in *Black Holes* (Les Houches) ed C deWitt and B deWitt (London: Gordon and Breach) pp 57-214
[3] Bardeen J M 1973 *Rapidly rotating stars, disks, and black holes*, in *Black holes* (Les Houches) ed C deWitt and B deWitt (London: Gordon and Breach) pp 241-289
[4] Hennig J and Ansorg M 2009 *A Fully Pseudospectral Scheme for Solving Singular Hyperbolic Equations* Journal of Hyperbolic Differential Equations **6** 161
[5] Ernst F J 1968 *New Formulation of the Axially Symmetric Gravitational Field Problem* Phys. Rev. **167** 1175
[6] Neugebauer G 1979 *Bäcklund transformations of axially symmetric stationary gravitational fields* Journal of Physics A **12** L67
[7] Neugebauer G 1980 *A general integral of the axially symmetric stationary Einstein equations* Journal of Physics A **13** L19
[8] Harrison B K 1978 *Bäcklund Transformation for the Ernst Equation of General Relativity* Phys. Rev. Lett. **41** 1197
[9] Kramer D and Neugebauer G 1983 *Bäcklund transformations in General Relativity* in *Proc. of the Int. Seminar on Solutions of Einstein's Equations (Retzbach)* ed Hoenselaers C and Dietz W (Berlin: Springer-Verlag) pp 1-25
[10] Neugebauer G 1996 *Gravitostatics and Rotating Bodies* in *Proc. of the 46th Scottish Universities Summer School in Physics (Aberdeen)* ed Hall G S and Pulham J R (London: IOP) pp 61-81
[11] Ansorg M. 2001 *Differentially Rotating Disks of Dust: Arbitrary Rotation Law* Gen. Rel. Grav. **33** 309
[12] Ansorg M., Kleinwächter A., Meinel R. and Neugebauer G. 2002 *Dirichlet Boundary Value Problems of the Ernst Equation* Phys. Rev. D **65** 044006
[13] Chruściel P T 1990 *On space-times with $U(1) \times U(1)$ symmetric compact Cauchy surfaces* Ann. Physics **202** 100
[14] Ansorg M and Hennig J 2009 *Inner Cauchy horizon of axisymmetric and stationary black holes with surrounding matter in Einstein-Maxwell theory* Phys. Rev. Lett. **102** 221102
[15] Hennig J and Ansorg M 2009 *The inner Cauchy horizon of axisymmetric and stationary black holes with surrounding matter in Einstein-Maxwell theory: study in terms of soliton methods* Annales Henri Poincare **10** 1075
[16] Hennig J, Cederbaum C and Ansorg M 2010 *A universal inequality for axisymmetric and stationary black holes with surrounding matter in the Einstein-Maxwell theory* Comm. Math. Phys. **293** 449
[17] Hennig J, Ansorg M. and Cederbaum C 2008 *A universal inequality between angular momentum and horizon area for axisymmetric and stationary black holes with surrounding matter* Class. Quant. Grav. **25** 162002
[18] Dain S and Reiris M *Area - Angular momentum inequality for axisymmetric black holes* arXiv:1102.5215

Journal of Physics: Conference Series **314** (2011) 012018 doi:10.1088/1742-6596/314/1/012018

Session IV.A in Memory of Brian Edgar. Chairmans Introduction

This session is dedicated to the memory of Brian Edgar who died in June. Shortly before his death Brian submitted an abstract to the ERE meeting. At the meeting his talk was given by Jos Senovilla, a close friend and collaborator of Brian, with whom he discussed the topic extensively in May of this year. Brian was a frequent attendee at the ERE meetings in recent years and shared fruitful collaborations with several groups in Iberia and indeed worldwide. Brian was a dedicated teacher and supervisor and a very able and productive researcher. However to many of us, Brian was more than just a colleague, but rather a dear friend. It was an honour to have known him and he will be sorely missed. Brian would always find time to talk to students, whether his own or those of other people. Accordingly, as well as talks from a range of Brian's collaborators and friends, we have included a number of talks by students in this special session as we think Brian would have wanted.

Brian Edgar died aged 64 from a stroke on June 10, 2010. He is survived by his wife Adwoa, daughter Cara and son Neal. From 1987 right up until his death Brian worked at the Mathematics Department of Linkping University (LiU) in Sweden. He was born and brought up in Northern Ireland and had a marvellous broad Northern Ireland accent that many of us loved to hear. After graduating with a degree in Mathematics from Queens University in Belfast, Brian worked in the Cameroon on VSO (Voluntary Service Overseas) and then as a lecturer at the University of the Cape Coast, Ghana for a number of years before embarking his PhD studies at Kings College, London under the supervision of Clive Kilmister. Brian was an Effer, a name given by and to the PhD students who shared the office at room F of the old building of Kings College on Surrey Street from the mid 60s to mid 70s. This was a very exciting and productive period at Kings when many new ideas were developed on the mathematical aspects of general relativity by such luminaries as Felix Pirani, Hermann Bondi, Clive Kilmister, David Robinson and Roger Penrose (at nearby Birkbeck College) plus many frequent international visitors. After gaining his doctorate in 1977 Brian went back to a lectureship at Cape Coast, Ghana. It was there he met and married his wife Adwoa. When the economic and political climate in Ghana deteriorated, they moved to the University of Botswana in 1982. During his time in Gaborone, Brian supervised, in secret, a banned ANC activist, Mosibudi Mangena, who later became a minister in South African government. In 2001 the GR16 conference in Durban was officially opened by Mangena and Brian had the opportunity to meet up with his former student and exchange reminiscences of those times. Brian was a superb raconteur and he had many tales of this period and more generally of his experiences in Northern Ireland, Africa and Sweden. Brian moved to LiU in Sweden in 1987, becoming a full professor in 2000. He and Malcolm Ludvigsen inaugurated and developed the gravitational physics program at LiU. He was the supervisor a number of successful PhD and Masters students. Brian was a classical relativist, in the sense that his work was mainly on the geometrical aspects of Riemannian geometry in general relativity. Part of his work was devoted to the field of exact solutions and, in particular, the GHP formalism and its generalisations. An important outcome of these studies is the Edgar-Ludwig conformally flat pure radiation metrics. Latterly he was working with various international collaborators on

Lanczos potentials and dimensionally dependent identities in general relativity. At the time of his passing he was finishing one of such paper which appears in these proceedings. He was one of the main driving forces behind the formation of GRSweden (the gravitational section of the Swedish Physical Society) and served for several years as its chairman. Although he was a very active researcher, Brian was also an outstanding lecturer and was voted best teacher in 1990 by the mathematics students at LiU. His colourful personality and direct manner made him popular with his students, who held him in high, and somewhat amused, regard. Brian would always appear dressed as Father Christmas at the annual mathematics students Yuletide party; a role to which he was eminently suited with his booming voice and long white beard and hair. One of his interests was the development of applied mathematics programmes in developing countries, particularly Rwanda; a passion that was born during his time in Africa. At the time of his death, he was also working on a program for collaboration between the Swedish and Brazilian applied mathematical programmes. We will miss his scientific enthusiasm and his warm and generous personality.

Alan Barnes
barnesa@aston.ac.uk

Journal of Physics: Conference Series **314** (2011) 012019 doi:10.1088/1742-6596/314/1/012019

(Conformally) semisymmetric spaces and special semisymmetric Weyl tensors

S Brian Edgar[1] **and José M M Senovilla**[2]

[1]Deceased

[2] Física Teórica, Universidad del País Vasco, Apartado 644, 48080 Bilbao, Spain

E-mail: `josemm.senovilla@ehu.es`

Abstract. Semisymmetric spaces are a natural generalisation of symmetric spaces. For semisymmetric spaces in four dimensions with Lorentz signature, the Weyl tensor is easily seen (via spinors) to have a particularly simple quadratic property, which we call a special semisymmetric Weyl tensor. Using dimensionally dependent tensor identities, all (conformally) semisymmetric spaces are confirmed to have special semisymmetric Weyl tensors for all signatures in four dimensions. Furthermore, all Ricci-semisymmetric spaces with special semisymmetric Weyl tensors are shown to be semisymmetric for all signatures in four dimensions. Counterexamples demonstrate that these two properties have no direct generalisations in higher dimensions.

1. Introduction

It is known, for spaces in dimensions:

(i) $n \geq 5$ for all signatures [3, 9]

(ii) $n = 4$ for Lorentz signature (using spinors) [8]

that *semi-symmetry* [2, 14, 15] is equivalent to *conformal semi-symmetry* when the Weyl conformal tensor C_{abcd} is non-zero:

$$\nabla_{[a}\nabla_{b]}R_{cdef} = 0 \Longleftrightarrow \nabla_{[a}\nabla_{b]}C_{cdef} = 0 \quad (\text{if } C \neq 0).$$

Here, R_{abcd} is the Riemann tensor. The Lorentzian result was also implicit in [10, 11]. For dimensions $n = 4$ a single simple counterexample is known [5], which has proper Riemannian signature. Using *four-dimensional tensor identities* [7], it is possible to show precisely when the result fails for proper Riemannian signature, and for neutral signature [6].

2. The spinor inspiration

Using the spinor formalism with the standard notation [12, 13] for the curvature spinors, the semi-symmetric conditions $\nabla_{[a}\nabla_{b]}R_{cdef} = 0$ can be rewritten as:

$$0 = \Box_{AB}\Psi_{CDEF} = \mathrm{X}_{AB(C}{}^{G}\Psi_{DEF)G} \tag{1}$$

$$0 = \Box_{A'B'}\Psi_{CDEF} = \Phi_{A'B'(C}{}^{G}\Psi_{DEF)G} \tag{2}$$

$$0 = \Box_{AB}\Phi_{CDC'D'} = 2\mathrm{X}_{AB(C}{}^{E}\Phi_{D)EC'D'} + 2\Phi_{AB(C'}{}^{E'}\Phi_{|CD|D')E'} \tag{3}$$

where $X_{ABCD} = \Psi_{ABCD} + R\epsilon_{A(B}\epsilon_{C)D}/6$ and R is the scalar curvature. One must observe that (1) and (2) correspond to $\nabla_{[a}\nabla_{b]}C_{cdef} = 0$, while (3) corresponds to $\nabla_{[a}\nabla_{b]}R_{cd} = 0$ (this is the so-called "Ricci semi-symmetry" condition).

One can prove that [(1) and (2)] $\Longrightarrow$ (3) ($C \neq 0$) [8]. Similarly [(1) and (3)] $\Longrightarrow$ (2) [6]. There are some special cases having a zero Weyl tensor which have been fully discussed in [1], see also Åman's contribution to this volume.

It springs to mind the idea to concentrate solely on the expression (1), which is equivalent to

$$24\Psi_{AB(C}{}^{G}\Psi_{DEF)G} = -R\left(\varepsilon_{A(C}\Psi_{DEF)B} + \varepsilon_{B(C}\Psi_{DEF)A}\right).$$

Its tensor expression reads

$$C^{ef}{}_{k[a}C^{k}{}_{b]cd} + C^{ef}{}_{k[c}C^{k}{}_{d]ab} = \frac{R}{6}\left(\delta^{[e}_{[b}C_{a]}{}^{f]}{}_{cd} + \delta^{[e}_{[d}C_{c]}{}^{f]}{}_{ab}\right).$$

This can be straightforwardly generalised to all dimensions $n \geq 4$ as follows [6]:

Definition 1 (Edgar, 2010) *A Weyl tensor is said to be* special semi-symmetric *if it satisfies:*

$$C^{ef}{}_{k[a}C^{k}{}_{b]cd} + C^{ef}{}_{k[c}C^{k}{}_{d]ab} = \frac{2R}{n(n\quad 1)}\left(\delta^{[e}_{[b}C_{a]}{}^{f]}{}_{cd} + \delta^{[e}_{[d}C_{c]}{}^{f]}{}_{ab}\right).$$

3. The main result

Once the concept of special semi-symmetric Weyl tensor has been introduced, one can prove the following result.

Theorem 1 (Edgar, 2010) *All (conformally) semi-symmetric spaces have a special semi-symmetric Weyl tensor in four dimensions. This is not the case in dimensions higher than four.*

Remark: Observe that the result holds for arbitrary signature of the metric.
Proof. The conformal semisymmetric condition $\nabla_{[a}\nabla_{b]}C_{cdef} = 0$ reads

$$C^{ef}{}_{k[a}C^{k}{}_{b]cd} + C^{ef}{}_{k[c}C^{k}{}_{d]ab} - \frac{2}{n(n-1)}R\left(\delta^{[e}_{[b}C_{a]}{}^{f]}{}_{cd} + \delta^{[e}_{[d}C_{c]}{}^{f]}{}_{ab}\right)$$
$$= \frac{2}{n-2}\left(\tilde{R}^{[e}_{i}\delta^{f]}_{[a}C^{i}{}_{b]cd} + \tilde{R}^{[e}_{i}\delta^{f]}_{[c}C^{i}{}_{d]ab} - \tilde{R}^{[e}_{[a}C^{f]}{}_{b]cd} - \tilde{R}^{[e}_{[c}C^{f]}{}_{d]ab}\right)$$

where R_{ab} is the Ricci tensor and $\tilde{R}_{ab} \equiv R_{ab} - \frac{R}{n}g_{ab}$ its trace-free part. This relation has a non-trivial trace

$$0 = R^{ej}{}_{i[a}C^{i}{}_{b]cj} + R^{ej}{}_{i[c}C^{i}{}_{j]ab}$$

which can be split into two parts

$$0 = (n-1)C_{abi}{}^{(c}\tilde{R}^{d)i} - 2C_{i}{}^{(cd)}{}_{[a}\tilde{R}_{b]}{}^{i} - 2\delta^{(c}_{[a}C^{d)i}{}_{b]}{}^{j}\tilde{R}_{ij}, \tag{4}$$

$$C_{abij}C^{cdij} + 4C_{[a}{}^{ij[c}C_{b]ij}{}^{d]} - \frac{2R}{n}C_{ab}{}^{cd} =$$
$$\frac{1}{2(n-2)}\left((n-3)\tilde{R}^{i[c}C^{d]}{}_{iab} + \tilde{R}^{i}{}_{[a}C_{b]i}{}^{cd} + 2\tilde{R}_{ij}\delta^{[c}_{[a}C^{d]i}{}_{b]}{}^{j}\right). \tag{5}$$

A basic "fundamental dimensionally dependent identity" (fddi) [7] in 4 dimensions is

$$C_{[ab}{}^{[cd}\delta^{j]}_{i]} = 0 \tag{6}$$

from where one immediately deduces

$$0=9\tilde{R}^{i}{}_{j}C_{[ab}{}^{[cd}\delta^{j]}_{i]}=2\left(\tilde{R}^{i[c}C^{d]}{}_{iab}+\tilde{R}^{i}{}_{[a}C_{b]i}{}^{cd}+2\tilde{R}_{ij}\delta^{[c}_{[a}C^{d]i}{}_{b]}{}^{j}\right).$$

Thus, (5) becomes, for $n=4$,

$$C_{abij}C^{cdij}+4C_{[a}{}^{ij[c}C_{b]ij}{}^{d]}-\frac{R}{2}C_{ab}{}^{cd}=0\,. \tag{7}$$

Surprisingy, this implies by itself that the Weyl tensor is (for $n=4$) special semi-symmetric: use again the fddi (6) to build

$$C_{i}{}^{j}{}_{ef}C_{[ab}{}^{[cd}\delta^{j]}_{i]}+C_{i}{}^{j}{}_{ab}C_{[ef}{}^{[cd}\delta^{j]}_{i]}=0$$

which expands into

$$C_{efi[c}C_{d]}{}^{iab}+C_{cdi[e}C_{f]}{}^{iab}+C_{ij[c}{}^{[a}\delta^{b]}_{d]}C_{ef}{}^{ij}+C_{ij[e}{}^{[a}\delta^{b]}_{f]}C_{cd}{}^{ij}=0.$$

Managing this (and its trace) in a judicious manner [6], and using (7), one arrives at

$$C^{ef}{}_{k[a}C^{k}{}_{b]cd}+C^{ef}{}_{k[c}C^{k}{}_{d]ab}=\frac{R}{6}\left(\delta^{[e}_{[b}C_{a]}{}^{f]}{}_{cd}+\delta^{[e}_{[d}C_{c]}{}^{f]}{}_{ab}\right)$$

in the case $n=4$. This is precisely the condition given in Definition 1 for the Weyl tensor to be special semi-symmetric four dimensions. This proves the $n=4$ part of the theorem.

The result in higher dimensions follows because one can easily construct explicit counterexamples, showing that this property is exclusive to $n=4$. One particular simple counterexample is to consider a semi-riemmanian manifold decomposable into two submanifolds, each of which has constant curvature. It is then very simple to compute the Weyl tensor and check that the condition in Defintion 1 fails to hold for all $n>4$ [6]. ■

4. Final remarks: extension to pseudo-symmetric spaces

Actually, all the previous results hold for the more general classes of (conformally) *pseudo-symmetric* spaces. Pseudo-symmetric spaces are defined by (e.g. [4] and references therein)

$$R^{ef}{}_{k[a}R^{k}{}_{b]cd}+R^{ef}{}_{k[c}R^{k}{}_{d]ab}=L\left(\delta^{[e}_{[b}R_{a]}{}^{f]}{}_{cd}+\delta^{[e}_{[d}R_{c]}{}^{f]}{}_{ab}\right) \tag{8}$$

while conformally pseudo-symmetric spaces satisfy

$$R^{ef}{}_{k[a}C^{k}{}_{b]cd}+R^{ef}{}_{k[c}C^{k}{}_{d]ab}=\tilde{L}\left(\delta^{[e}_{[b}C_{a]}{}^{f]}{}_{cd}+\delta^{[e}_{[d}C_{c]}{}^{f]}{}_{ab}\right) \tag{9}$$

for some scalars L and $\tilde{L}$. Observe that they correspond respectively to

$$\nabla^{[e}\nabla^{f]}R_{abcd}=-L\left(\delta^{[e}_{[b}R_{a]}{}^{f]}{}_{cd}+\delta^{[e}_{[d}R_{c]}{}^{f]}{}_{ab}\right),$$
$$\nabla^{[e}\nabla^{f]}C_{abcd}=-\tilde{L}\left(\delta^{[e}_{[b}C_{a]}{}^{f]}{}_{cd}+\delta^{[e}_{[d}C_{c]}{}^{f]}{}_{ab}\right)$$

so that (8) implies (9) —for appropriate $\tilde{L}$. Then, analogously to the semi-symmetric case, one can split (9) by decomposing the Riemann tensor into Weyl plus Ricci terms and then check (for instance via spinors) that the Weyl tensor satisfies a particular quadratic relation for $n=4$. Generalizing it to arbitrary dimension, one can introduce the following definition

Definition 2 (Edgar, 2010) *A Weyl tensor is said to be* special pseudo-symmetric *if it satisfies:*

$$C^{ef}{}_{k[a}C^{k}{}_{b]cd} + C^{ef}{}_{k[c}C^{k}{}_{d]ab} = M\left(\delta^{[e}_{[b}C_{a]}{}^{f]}{}_{cd} + \delta^{[e}_{[d}C_{c]}{}^{f]}{}_{ab}\right)$$

for some scalar function M.

Clearly, the semi-symmetric Weyl tensor condition of Definition 1 is a special case with $n(n-1)M = 2R$.

Using exactly the same type of arguments —based again on the fddi (6) for $n = 4$—as in the proof of Theorem 1, one can then prove [6]

Theorem 2 (Edgar, 2010) *All (conformally) pseudo-symmetric spaces have a special pseudo-symmetric Weyl tensor in four dimensions. This is not the case in dimensions higher than four.*

Again, the result holds for arbitrary signature.

Acknowledgments

JMMS wants to express his deep gratitude to Brian Edgar —the true and only author of this contribution. Supported by grants FIS2010-15492 (MICINN), GIU06/37 (UPV/EHU), P09-FQM-4496 (J. Andalucía).

References

[1] Åman J 2010, Investigation of all Ricci semi-symmetric and all conformally semi-symmetric spacetimes, preprint arXiv:1006.5684
[2] Cartan É 1946, *Leçons sur la Géométrie des Espaces de Riemann*, 2nd ed. (Gauthier-Villars, Paris)
[3] Deszcz R and Grycak W 1989, On manifolds satisfying some curvature conditions *Colloquium Math.* **57** 89-92
[4] Defever F, Deszcz R, Verstraelen L, and Vrancken L 1994, On pseudo-symmetric spacetimes, *J. Math. Phys.* **35** 5908-5921
[5] Derdzinski A 1981, Examples de métriques de Kaehler et dEinstein autoduales sur le plan complexe, in *Géométrie riemannienne en dimension 4 (Séminaire Arthur Besse 1978/79)* (Cedic/Fernand Nathan Paris) pp 334-346
[6] Edgar S B 2010, preprint (unpublished)
[7] Edgar S B and Höglund A 2002, Dimensionally dependent tensor identities by double antisymmetrization *J. Math. Phys.* **43** 659-677
[8] Eriksson I and Senovilla J M M 2010, Note on (conformally) semi-symmetric spacetimes *Class. Quantum Grav.* **27** 027001
[9] Grycak W 1987, Riemannian manifolds with a symmetry condition imposed on the 2-nd derivative of the conformal curvature tensor, *Tensor N.S.* **46** 287-290
[10] Haesen S and Verstraelen L 2004, Classification of the pseudosymmetric space-times, *J.Math. Phys.* **45** 2343–2346.
[11] Haesen S and Verstraelen L 2005, Curvature and symmetries of parallel transport, in *Differential Geometry and Topology, Discrete and Computational Geometry*, Boucetta M and Morvan J.-M eds., (IOS Press, Nato Science Series) Vol. 197, chapter 8
[12] Penrose R and Rindler W 1984, *Spinors and spacetime vol 1.*, Cambridge University Press,
[13] Penrose R and Rindler W 1986, *Spinors and spacetime vol 2.*, Cambridge University Press, 1986.
[14] Szabó Z I 1982, Structure theorems on Riemannian spaces satisfying $R(X,Y)\cdot R = 0$. I. The local version, *J. Diff. Geom.* **17** 531-582
[15] Szabó Z I 1985, Structure theorems on Riemannian spaces satisfying $R(X,Y)\cdot R = 0$. II. Global versions, *Geom. Dedicata* **19** 65-108

Journal of Physics: Conference Series **314** (2011) 012020 doi:10.1088/1742-6596/314/1/012020

Investigation of all Ricci semi-symmetric and all conformally semi-symmetric spacetimes

Jan E Åman

Department of Physics, Stockholm University, SE-106 91 Stockholm, Sweden

E-mail: ja@fysik.su.se

Abstract.
We find all Ricci semi-symmetric as well as all conformally semi-symmetric spacetimes. Neither of these properties implies the other. However, we find that for spacetimes (3+1 dim) with nonzero Weyl tensor and nonzero tracefree part of the Ricci tensor conformal semi-symmetry and Ricci semi-symmetry is equivalent.

1. Introduction

Semi-symmetric spaces were introduced by Cartan [3] and are characterized by the curvature condition

$$\nabla_{[a}\nabla_{b]}R_{cdef} = 0, \tag{1}$$

where R_{cdef} denotes the Riemann tensor whilst round and square brackets enclosing indices indicate symmetrization and antisymmetrization, respectively.

A semi-Riemannian manifold is said to be *conformally semi-symmetric* if the Weyl tensor C_{abcd} satisfies

$$\nabla_{[a}\nabla_{b]}C_{cdef} = 0; \tag{2}$$

and *Ricci semi-symmetric* if the Ricci tensor satisfies

$$\nabla_{[a}\nabla_{b]}R_{cd} = 0. \tag{3}$$

The geometrical properties of semi-symmetric spaces were discussed in [11, 6, 8, 9, 4].

In this paper we find all semi-symmetric, Ricci semi-symmetric and all conformally semi-symmetric spacetimes. Eriksson and Senovilla [5] found all such non-conformally flat spacetimes.

It is an advantage to instead of tensors rather to use the Weyl spinor Ψ_{ABCD}, the curvature scalar spinor Λ, and the spinor $\Phi_{ABA'B'}$ for the tracefree part of the Ricci tensor. We also use the spinor $\mathrm{X}_{ABCD} = \Psi_{ABCD} + \Lambda\left(\varepsilon_{AC}\varepsilon_{BD} + \varepsilon_{AD}\varepsilon_{BC}\right)$. The spinor commutator $\Box_{AB}$ operating on spinors with one index is $\Box_{AB}\,\kappa_C = -\mathrm{X}_{ABC}{}^{E}\kappa_E$ and $\Box_{AB}\,\tau_{C'} = -\Phi_{ABC'}{}^{E'}\tau_{E'}$ [7].

In spinors (2) is equivalent to $\Box_{AB}\Psi_{CDEF} = 0$ and $\Box_{A'B'}\Psi_{CDEF} = 0$, or

$$\mathrm{X}_{AB(C}{}^{G}\Psi_{DEF)G} = 0, \tag{4}$$

$$\Phi_{A'B'(C}{}^{G}\Psi_{DEF)G} = 0. \tag{5}$$

2. Conditions for semi-symmetry

Calculation of the components of (4) in terms of Ψ_{ABCD} and Λ shows that it will not have 15 independent components but rather just 5 components and (4) can be replaced by its contraction over BC or

$$\Psi^{GH}{}_{(AD}\Psi_{EF)GH} - 2\Lambda\Psi_{ADEF} = 0. \tag{6}$$

We observe that the spinor commutator $\Box_{AB}$ yields 0 if operating on a scalar as the curvature scalar Λ, therefore only its effect on $\Phi_{ABA'B'}$, i.e. on the tracefree part of the Ricci tensor has to be considered. The condition for Ricci semi-symmetry (3) corresponds to spinor equations

$$\Box_{AB}\Phi_{CDC'D'} = -2\mathrm{X}_{AB(C}{}^{E}\Phi_{D)EC'D'} - 2\Phi_{AB(C'}{}^{E'}\Phi_{D')E'CD}, \tag{7}$$

$$\Box_{A'B'}\Phi_{CDC'D'} = -2\overline{\mathrm{X}}_{A'B'(C'}{}^{E'}\Phi_{D')E'CD} - 2\Phi_{A'B'(C}{}^{E}\Phi_{D)EC'D'}. \tag{8}$$

The condition (8) is however just the complex conjugate of (7) and will not give any additional conditions for Ricci semi-symmetry.

Equation (7) has three groups of two symmetric indices but will in fact not have 27 independent components as $\Box^{AB}\Phi_{ABC'D'} = 0$, it can be replaced by the fully symmetric spinors

$$\Box_{(AB}\Phi_{CD)C'D'} = -2\Psi_{(ABC}{}^{E}\Phi_{D)EC'D'} \tag{9}$$

$$\Box_{(A}{}^{F}\Phi_{C)FC'D'} = 4\Lambda\Phi_{ACC'D'} - \Psi^{EF}{}_{AC}\Phi_{EFC'D'} - 2\Phi^{E}{}_{A}{}^{F'}{}_{(C'}\Phi_{D')F'CE} \tag{10}$$

with 15 and 9 components respectively.

The various spacetimes are for simplicity studied in a frame where first Ψ_{ABCD} has been brought to a standard form depending on its Petrov type. Thereafter $\Phi_{ABA'B'}$ is brought to standard form depending on its Segre type [1, 10].

All of the above formulas have been implemented in CLASSI [2].

3. Main Results

The equations for Ricci semi-symmetric spacetimes (9) and (10) and Petrov Types **I**, **II** or **III** in standard frame require that $\Phi_{ABA'B'} = 0$, while spacetimes with a Λ-term only or vacuum ($\Lambda = 0$) are Ricci semi-symmetric.

For Petrov type **D** spacetimes (with only Ψ_2 nonzero) the conditions both for conformal and for Ricci semi-symmetry require all components of $\Phi_{ABA'B'}$ to be zero except for $\Phi_{11'}$. The conditions for conformal semi-symmetry reduce to $2\Lambda + \Psi_2 = 0$ and for Ricci semi-symmetry to $\Phi_{11'}(2\Lambda + \Psi_2) = 0$. Spacetimes of Segre type A1[(11)(1,1)] are therefore both Ricci and conformally semi-symmetric if they conform to $\Lambda = -\frac{1}{2}\Psi_2$ with arbitrary $\Phi_{11'}$. Spacetimes with Λ-term only are semi-symmetric if once again $\Lambda = -\frac{1}{2}\Psi_2$ but only Ricci semi-symmetric for other relations between Ψ_2 and Λ.

For Petrov type **N** spacetimes (with only Ψ_4 nonzero) the conditions for Ricci semi-symmetry as well as for conformal semi-symmetry require all components of $\Phi_{ABA'B'}$ to be zero except for $\Phi_{22'}$. The conditions here reduce to $\Lambda\Psi_4 = 0$ for conformal semi-symmetry and $\Lambda\Phi_{11'} = 0$ for Ricci semi-symmetry. They are therefore both Ricci and conformally semi-symmetric for arbitrary Ψ_4 and $\Phi_{22'}$ as long as $\Lambda = 0$ and also for vacuum spacetimes. Once again all spacetimes with Λ-term only are Ricci semi-symmetric.

3.1. Conformally flat spacetimes

This is the case not treated by Eriksson and Senovilla [5].

All conformally flat (Petrov type **0**) spacetimes are conformally semi-symmetric.

Calculations with CLASSI show that all conformally flat spacetimes of Segre type A1[(11)(1,1)] (only $\Phi_{11'}$ nonzero) with $\Lambda = 0$ are also Ricci semi-symmetric.

For conformally flat perfect fluids (Segre type A1[(111),1)], $\frac{1}{2}\Phi_{00'} = \Phi_{11'} = \frac{1}{2}\Phi_{22'}$) as well as for tachyons (Segre type A1[1(11,1)], $-\frac{1}{2}\Phi_{00'} = \Phi_{11'} = -\frac{1}{2}\Phi_{22'}$) the condition for Ricci semi-symmetry reduces to $\Lambda = \Phi_{11'}$. So perfect fluids with $\Lambda = \frac{1}{2}\Phi_{00'} = \Phi_{11'} = \frac{1}{2}\Phi_{22'}$ as well as tachyons with $\Lambda = -\frac{1}{2}\Phi_{00'} = \Phi_{11'} = -\frac{1}{2}\Phi_{22'}$ are also Ricci semi-symmetric.

All spacetimes of Segre type A3[(11,2)] with $\Lambda = 0$ (only $\Phi_{22'}$ nonzero) are also Ricci semi-symmetric.

All vacuum spacetimes and Λ-term only spacetimes are semi-symmetric.

For all other Segre types the conditions for Ricci semi-symmetry are not satisfied.

4. Summary

All conformally flat spacetimes are conformally semi-symmetric, all spacetimes with Λ-term only are Ricci semi-symmetric.

Semi-symmetric spacetimes are flat spacetimes, Petrov type **N** vacuum, Λ-term and Segre A1[(11)(1,1)] spacetimes of Petrov type **0** and of type **D** with $\Lambda = -\frac{1}{2}\Psi_2$, Segre type A3[(11,2)] with $\Lambda = 0$ of Petrov types **N** and **0**, as well as conformally flat perfect fluids and tachyons with $\Lambda = \Phi_{11'}$. See Table 1.

We find that *conformal semi-symmetry* and *Ricci semi-symmetry* is equivalent for spacetimes with nonzero Weyl tensor and nonzero tracefree part of the Ricci tensor.

Segre type / Petrov type	**I**	**II**	**III**	**D**	**N**	**0**
Vacuum	Ric s-s	Ric s-s	Ric s-s	Ric s-s	semi-sym	semi-sym
Λ-term A1[(111,1)], $\Lambda = -\frac{1}{2}\Psi_2$	Ric s-s	Ric s-s	$\nexists$	semi-sym	$\nexists$	$\nexists$
Λ-term A1[(111,1)] else	Ric s-s	Ric s-s	Ric s-s	Ric s-s	Ric s-s	semi-sym
A1[(11)(1,1)] $\Phi_{11'} \neq 0$, $\Lambda = -\frac{1}{2}\Psi_2$	-	-	$\nexists$	semi-sym	$\nexists$	$\nexists$
A1[(11)(1,1)] $\Phi_{11'} \neq 0$, Λ else	-	-	-	-	-	semi-sym
A3[(11,2)] $\Phi_{22'} \neq 0$, $\Lambda = 0$	-	-	-	-	semi-sym	semi-sym
A1[(111),1] perfect fluid, $\Lambda = \frac{1}{2}\Phi_{00'} = \Phi_{11'} = \frac{1}{2}\Phi_{22'}$	-	-	-	-	-	semi-sym
A1[1(11,1)] tachyon fluid, $\Lambda = -\frac{1}{2}\Phi_{00'} = \Phi_{11'} = -\frac{1}{2}\Phi_{22'}$	-	-	-	-	-	semi-sym
All other Ricci tensors	-	-	-	-	-	conf s-s

Table 1. Relations between Petrov type, Segre type and conformal semi-symmetry (conf s-s), Ricci semi-symmetry (Ric s-s) and semi-symmetry. A hyphen (-) indicates neither conformal nor Ricci semi-symmetry.

Acknowledgments

I thank Brian Edgar (1945-2010) for pointing out ref [5] and introducing me to semi-symmetric spacetimes. I thank Ingemar Bengtsson and José Senovilla for comments on the manuscript.

References

[1] Åman J E, d'Inverno R A, Joly G C and MacCallum M A H 1991 Quartic Equations and Classification of Riemann Tensors in General Relativity *General Relativity and Gravitation* **23** 1023-55

[2] Åman J E 2002 *Manual for CLASSI: Classification Programs for Geometries in General Relativity*, Department of Physics. Stockholm University, Technical Report, Provisional edition. Distributed with the sources for SHEEP and CLASSI

[3] Cartan É 1946 *Leçons sur la Géométrie des Espaces de Riemann* 2nd edn (Paris: Gauthier-Villars)

[4] Defever F, Deszcz R, Verstraelen L, Vrancken L 1994 On pseudo-symmetric spacetimes, *J. Math. Phys.* **35** 5908-21

[5] Eriksson I and Senovilla J M M 2010 Note on (conformally) semi-symmetric spacetimes *Class. Quantum Grav.* **27** 027001 [arXiv:0908.3246 [math.DG]]
[6] Haesen S and Verstraelen L 2004 Classification of the pseudosymmetric space-times *Journal of Mathematical Physics* **45** 2343-6
[7] Penrose R and Rindler W 1984 *Spinors and spacetime* vol 1 (Cambridge: Cambridge University Press)
[8] Senovilla J M M 2006 Second-order symmetric Lorentzian manifolds *AIP Conf. Proc.* **841** 370-7
[9] Senovilla J M M 2008 Second-order symmetric Lorentzian manifolds I: characterization and general results *Classical and Quantum Gravity* **25** 245011
[10] Stephani H, Kramer D, MacCallum M A H, Hoenselaers C, Herlt E 2003 *Exact Solutions to Einstein's Field Equations* 2nd edn (Cambridge: Cambridge University Press)
[11] Szabó Z I 1982 Structure theorems on Riemannian spaces satisfying $R(X,Y)\cdot R = 0$. I. The local version, *J. Diff. Geom.* **17** 531-82

Journal of Physics: Conference Series **314** (2011) 012021 doi:10.1088/1742-6596/314/1/012021

Second-order symmetric Lorentzian manifolds II: structure and global properties

O F Blanco[1], **M Sánchez**[1] **and J M M Senovilla**[2]

[1] Departamento de Geometría y Topología , Facultad de Ciencias, Universidad de Granada Campus Fuentenueva s/n, 18071 Granada, Spain

[2] Física Teórica, Universidad del País Vasco, Apartado 644, 48080 Bilbao, Spain

E-mail: oihane@ugr.es, sanchezm@ugr.es, josemm.senovilla@ehu.es

Abstract. We give a summary of recent results on the explicit local form of the second-order symmetric Lorentzian manifolds in arbitrary dimension, and its global version. These spacetimes turn out to be essentially a specific subclass of plane waves.

1. Introduction

Last year we presented the local explicit form of the *second-order symmetric* (or *2nd symmetric*[1] for short) spacetimes in four dimensions [3]. The techniques used for that purpose were specific of the dimension. Our aim now is to introduce new techniques that generalize the result to arbitrary dimension. Recall that 2nd symmetric spacetimes are characterized by the condition

$$\nabla_\rho \nabla_\sigma R^\alpha{}_{\beta\lambda\mu} = 0.$$

They were introduced by one of the authors in [17] as the simplest natural generalization of the *locally symmetric* spacetimes (those satisfying $\nabla_\sigma R^\alpha{}_{\beta\lambda\mu} = 0$). In the Riemannian [2],[7],[15] and Lorentzian [6] cases, the locally symmetric spaces are well known. However, 2nd symmetric spaces have been hardly considered in the literature, probably because the local de Rham decomposition implies that 2nd symmetry and local symmetry are equivalent in the proper Riemannian case [16],[22]. In the Lorentzian case such decomposition fails [23], and so does the equivalence. Thus, even though in arbitrary signature *semi-symmetric* spaces (those satisfying $\nabla_{[\rho}\nabla_{\sigma]} R^\alpha{}_{\beta\lambda\mu} = 0$, see for example [20],[21] for the Riemannian case and [12], [14] for the Lorentzian one) have always been regarded as the simple natural generalization of the locally symmetric ones, in indefinite signatures the 2nd symmetry is a simpler natural choice.

In this contribution we announce the following classification of *proper 2nd symmetric* spacetimes (here proper means that they are not locally symmetric, that is $\nabla_\sigma R^\alpha{}_{\beta\lambda\mu} \neq 0$) in arbitrary dimension [4]:

Theorem 1.1 *A n-dimensional proper 2nd symmetric spacetime (M, g) is locally isometric to a product spacetime $(M_1 \times M_2, g_1 \oplus g_2)$ where (M_2, g_2) is a non-flat Riemannian symmetric space*

[1] The notion of *3-symmetric space* (without ordinal), which is widely spread in the literature, was introduced by Gray [13] in a different context.

and (M_1, g_1) is a generalized Cahen-Wallach space of order 2, defined as $M_1 = \mathbb{R}^{d+2}$ $(d \geq 1)$ endowed with the metric

$$g_1 = -2du\left(dv + p_{ij}(u)x^i x^j du\right) + \delta_{ij}dx^i dx^j, \quad i,j = 2,\ldots,d+1$$

where $\{u, v, x^i\}$ are the Cartesian coordinates of $\mathbb{R}^{d+2}$, each function p_{ij} is affine, i.e., $p_{ij}(u) = \alpha_{ij}u + \beta_{ij}$ for some $\alpha_{ij}, \beta_{ij} \in \mathbb{R}$, and $\sum_{i,j=2}^{d+1}(\alpha_{ij})^2 \neq 0$.

If in addition (M, g) is geodesically complete and simply connected, then (M, g) is globally isometric to one of such products.

2. Sketch of the proof of theorem 1.1

As in the four-dimensional case, the starting point for the proof is [17, Theorem 4.2], which states when, in the Lorentzian case, the 2nd symmetry condition does not imply local symmetry. This can only happen when the spacetime possesses a null (non-vanishing) covariantly constant vector field. Therefore, it is locally isometric to a Brinkmann spacetime, whose metric can be written in an appropiate *Brinkmann chart* $\{u, v, x^i\}$ as [5],[24]:

$$ds^2 = -2du(dv + Hdu + W_i dx^i) + g_{ij}dx^i dx^j, \quad i,j \in \{2,\ldots,n-1\},$$

where H, W_i and $g_{ij} = g_{ji}$ are functions independent of v, otherwise arbitrary. With this data at hand, the steps of the proof are the following:

(i) A suitable choice of vector field basis simplifies the computation of the equations of 2nd symmetry.

(ii) As a first consequence, the leaves of the foliation $\overline{\mathcal{M}}$ generated by the submanifolds with constant values of u and v are locally symmetric Riemannian manifolds for any Brinkmann chart.

(iii) Using the integrability equations and (ii), one proves that the only non-vanishing components of the curvature in the basis (i) are $\nabla_0 R^1{}_{i0j}$.

(iv) Applying a generalization of the Eisenhart theorem [11], "the spatial part" g_{ij} is splits into two parts, one of them is Ricci-flat which, on using a result in [1], becomes actually flat.

(v) After lengthy computations, one can prove that there exists a Brinkmann chart such that the metric becomes:

$$\begin{aligned} ds^2 &= -2du\left(dv + \left(H_1(u,x^a) + H_2(u,x^{a'})\right)du + W_a(u,x^b)dx^a + W_{a'}(u,x^{b'})dx^{a'}\right) \\ &+ \delta_{ab}dx^a dx^b + g_{a'b'}dx^{a'}dx^{b'}, \end{aligned}$$

with $a, b \in \{2,\ldots,d+1\}$, $a', b' \in \{d+2,\ldots,n-1\}$, $g_{a'b'} \neq \delta_{a'b'}$.

(vi) At this stage, the equations can be reorganized in two blocks:
- one corresponding to the equations of local symmetry, whose solutions are already known [6] (see also [2]),
- the other one corresponding to the equations of 2nd symmetry for a plane wave, which are easily solvable.

To develop the proof, the introduction of some mathematical tools is required, namely:

(1) Some operators defined on the tensor bundle $T^r_s\overline{\mathcal{M}}$, denoted as D_0 (transverse operator) and $\overline{\nabla}$ (which arises from the covariant derivative on the leaves), and an adaptation of the exterior derivative to $\overline{\mathcal{M}}$.

(2) The curvature tensor $\overline{R}^s{}_{ijk}$ associated intrinsically to the foliation (i.e., to $\overline{\nabla}$), as well as its naturally associated Ricci tensor $\overline{R}_{ij}$ and scalar curvature $\overline{R}$.

(3) Some tensor fields on $\overline{\mathcal{M}}$ defined as projections of $R^{\alpha}{}_{\beta\lambda\mu}$ and $\nabla_\rho R^{\alpha}{}_{\beta\lambda\mu}$, as well as some algebraic lemmas on vector spaces with positive definite inner product.

(4) An appropriate adaptation of the Eisenhart theorem to Brinkmann spacetimes, as mentioned above. This involves the operators D_0 and $\overline{\nabla}$ and ensures the reducibility of the entire family of metrics g_{ij} —which depend on u— simultaneously.

Before dealing with the global part of the proof, we define

- *proper r-th symmetric* spacetimes as those satisfying $\nabla_{\rho_1}\dots\nabla_{\rho_r}R^{\alpha}{}_{\beta\lambda\mu} = 0$ but $\nabla_{\rho_1}\dots\nabla_{\rho_{r-1}}R^{\alpha}{}_{\beta\lambda\mu} \neq 0$,
- the generalized *Cahen-Wallach spacetimes of order r*, denoted by $CW_r^d(A)$, as the spacetimes $(\mathbb{R}^d, g_A)$, $d \geq 3$ with metric

$$g_A = -2du\left(dv + A_{ij}x^i x^j du\right) + \delta_{ij}dx^i dx^j,$$

where $A = (A_{ij})$ is a square symmetric matrix of dimension $d-2$ such that $A \equiv A(u) = A^{(r-1)}u^{r-1} + \dots + A^{(1)}u + A^{(0)}$, $A^{(r-1)} \neq 0$, and each $A^{(l)}$ is a square symmetric matrix of constants, $l = 1, \dots, r-1$. Observe that these spacetimes are proper rth symmetric [17], as well as analytic and geodesically complete [8].

Then, the global result follows directly from:

Proposition 2.1 *For any matrix A as above:*

- *the direct product of $CW_r^n(A)$ and a finite number of Riemannian symmetric (not necessarily simply connected) spaces is a proper r-th symmetric Lorentzian manifold, which is in fact geodesically complete and analytic, and*
- *if a complete Lorentzian manifold (M, g) is locally isometric to spaces of the form given above, then its universal covering is also of this type.*

Remark 2.2 The techniques used to solve the arbitrary dimensional case differ from the techniques used in four dimensions since in the latter case the Petrov classification of the Weyl tensor [19] was used. Such classification is not available in arbitrary dimension with the same simplicity, though some generalizations have already apeared in the literature (see [9],[10],[18]). Therefore, we had to refocuse the problem in arbitrary dimension using a different, and more powerful, viewpoint.

Acknowledgments

The three authors are partially supported by the grant P09-FQM-4496 (J. Andalucía), OFB and MS are partially supported by MTM2007-60731 (Spanish MICINN with FEDER funds), and JMMS is supported by grants FIS2010-15492 (MICINN) and GIU06/37 (UPV/EHU).

References

[1] Alekseevskii DV and Kimelfeld BN 1975 Structure of homogeneous Riemannian spaces with zero Ricci curvature (Russian) *Funkcional. Anal. i Prilozen.* **9** 2 5-11
[2] Besse A L *Einstein manifolds* (Springer, 1987)
[3] Blanco O F, Sánchez M and Senovilla J M M 2010 Complete classification of second-order symmetric spacetimes *J. Phys.: Conf. Ser.* **229** 012021
[4] Blanco O F, Sánchez M and Senovilla J M M Structure of second-order symmetric Lorentzian manifolds, in preparation
[5] Brinkmann H W 1925 Einstein spaces which are mapped conformally on each other *Math. Ann.* **94** 119-45
[6] Cahen M, Wallach F 1970, Lorentzian symmetric spaces *Bull. Amer. Math. Soc.* **76** 585-591
[7] Cartan É 1926 Sur une classe remarquable d'espaces de Riemann *Bull. Soc. Math. France* **54** 214-64
Cartan É 1927 Sur une classe remarquable d'espaces de Riemann *Bull. Soc. Math. France* **55** 114-34

[8] Candela A M, Flores J L and Sánchez M 2003 On general plane fronted waves. Geodesics *Gen. Relat. Gravit.* **35** 631-649.
[9] Coley A. 2008 Classification of the Weyl Tensor in Higher Dimensions and Applications *Class.Quant.Grav.* **25** 033001
[10] Coley A. Milson R. Pravda V. and Pravdov Ì A. 2004 Classification of the Weyl Tensor in Higher Dimensions *Class.Quant.Grav.* **21** L35-L42
[11] Eisenhart L P 1923, Symmetric tensors of the second order whose first covariant derivatives are zero *Trans. Amer. Math. Society* **25** 2 297-306
[12] Eriksson I and Senovilla J M M 2010 Note on (conformally) semi-symmetric spacetimes *Class. Quantum Grav.* **27** 027001
[13] Gray A 1972 Riemannian manifolds with geodesic symmetries of order 3 *J. Differential Geometry* **7**, 343-369
[14] Haesen S and Verstraelen L 2004 Classification of the pseudo-symmetric space-times, *J. Math. Phys.* **45** 2343-6
[15] Helgason S *Differential Geometry, Lie Groups, and Symmetric Spaces* (Academic Press, New York, 1978)
[16] Nomizu K and Ozeki H 1962 A theorem on curvature tensor fields *Proc. Nat. Acad. Sci. USA* **48** **206-7**
[17] Senovilla J M M 2008 2nd-order symmetric Lorentzian manifolds I: characterization and general results, *Class. Quantum Grav.* **25** 245011
[18] Senovilla J M M 2010 Algebraic classification of the Weyl tensor in higher dimensions based on its 'superenergy' tensor *Class. Quantum Grav.* **27** 222001
[19] Stephani H, Kramer D, MacCallum M, Hoenselaers C and Herlt E, *Exact Solutions of Einstein's Field Equations*, 2nd ed. (Cambrigde University Press., Cambrigde, 2003)
[20] Szabó Z I 1982 Structure theorems on Riemannian spaces satisfying $R(X,Y)\dot{R} = 0$: I. The local version *J. Differ. Geom.* **17** 531-82
[21] Szabó Z I 1985 Structure theorems on Riemannian spaces satisfying $R(X,Y)\dot{R} = 0$: I. The global versions *Geom. Dedicata* **19** 65-108
[22] Tanno S 1972 Curvature tensors and covariant derivatives *Ann. Mat. Pura Appl.* **96** 233-41
[23] Wu H 1967 Holonomy groups of indefinite metrics *Pacific J. Math.* **20** 351-92
[24] Zakharov V D 1973 *Gravitational Waves in Einstein's theory* (New York: Wiley)

Conformal Einstein Perfect Fluid Spacetimes

Norbert Van den Bergh

Department of Mathematical Analysis TW16, University of Ghent, Galglaan 2, Ghent, Belgium

E-mail: norbert.vandenbergh@ugent.be

Abstract. A review is given of results obtained in connection with perfect fluid spacetimes which are conformally related to Einstein spaces.

1. Introduction

It is well known that a conformal transformation of the metric,

$$\hat{g}_{ij} = e^{2\phi} g_{ij}, \tag{1}$$

induces a transformation of the curvature tensor,

$$e^{2\phi} \hat{R}^{ij}{}_{kl} = R^{ij}{}_{kl} - 4Y^{[i}_{[k}\delta^{j]}_{l]}, \tag{2}$$

with

$$Y_{ij} = \phi_{,i;j} - \phi_{,i}\phi_{,j} + \frac{1}{2} g_{ij} g^{mn} \phi_{,m}\phi_{,n}. \tag{3}$$

If the space-time $(\mathcal{M}, \hat{g})$ is Einstein,

$$\hat{R}_{ij} - \frac{1}{2}\hat{R}\hat{g}_{ij} + \Lambda \hat{g}_{ij} = 0, \tag{4}$$

and if the energy-momentum tensor of $(\mathcal{M}, g)$ is of the perfect fluid type,

$$R_{ij} - \frac{1}{2} R g_{ij} = (m + p) u_i u_j + p g_{ij}, \tag{5}$$

then $(\mathcal{M}, g)$ is called a conformally Einstein perfect fluid. This requires the existence of a solution ϕ of the over-determined system of equations

$$Y_{ij} = \frac{1}{2}(m + p) u_i u_j + \frac{1}{6}(m - \Lambda e^{2\phi}) g_{ij}. \tag{6}$$

When additionally $\Lambda = 0$ the space-time is called a conformally Ricci-flat perfect fluid. It is clear from both (5, 6) that the Λ-term introduces new physics, in the sense that it cannot be simply re-absorbed in pressure and energy-density. In combination with any of the following special situations,

- $\nabla\phi \parallel \boldsymbol{u}$
- the fluid is shear-free: $\sigma_{ij} = 0$
- the fluid is vorticity-free: $\omega_{ij} = 0$
- the fluid admits a barotropic equation of state and $\nabla\phi \perp \boldsymbol{u}$
- both $(\mathcal{M}, g)$ and $(\mathcal{M}, \hat{g})$ admit the same two-dimensional group of isometries,

the set (6) turned out to be very restrictive for the resulting space-times, the full classification of which was given in [5, 6, 7, 8]. In this contribution I redo the calculations, now including a possible non-zero Λ-term in the hope that more general classes of solutions might result. I also use the GHP-formalism [2] in order to present the calculations in a form which is as compact as possible, as well as to obtain a better insight in the relevant variables and equations. The null-tetrad is $\boldsymbol{e}_1, \boldsymbol{e}_2, \boldsymbol{e}_3, \boldsymbol{e}_4 \equiv \boldsymbol{m}, \bar{\boldsymbol{m}}, \boldsymbol{k}', \boldsymbol{k}$ with $\boldsymbol{k} \cdot \boldsymbol{k}' = -1$ and $\boldsymbol{m} \cdot \bar{\boldsymbol{m}} = 1$.

2. Basic equations

Aligning the $(\boldsymbol{k}', \boldsymbol{k})$-plane with the $(\boldsymbol{u}, \nabla\phi)$-plane, one has $\eth\phi\ (= \eth'\phi) = 0$. Introducing a $[1,1]$-weighted quantity Q (with $Q' = 1/Q$) allows one to write the fluid velocity as

$$\boldsymbol{u} = \frac{1}{\sqrt{2}}(Q\boldsymbol{k}' + Q'\boldsymbol{k}),$$

which implies the following expressions for the components of the Ricci-spinor:

$$\begin{cases} Q'^2\Phi_{00} = Q^2\Phi_{22} = 2\Phi_{11} = \frac{1}{4}(m+p) \\ R = m - 3p \\ \Phi_{01} = \Phi_{02} = \Phi_{12} = 0. \end{cases} \tag{7}$$

Herewith (and with $\mathcal{E} = \Lambda e^{2\phi}$) the basic equations (6) become:

$$\begin{cases} \text{Þ}^2\phi = (\text{Þ}\phi)^2 + \frac{1}{4}(m+p)Q^2 \\ \text{Þ}'\text{Þ}\phi = \frac{1}{12}(m+3p) + \frac{\mathcal{E}}{6}, \end{cases} \tag{8}$$

and

$$\begin{cases} \bar{\sigma}\text{Þ}'\phi + \sigma'\text{Þ}\phi = 0 \\ \kappa\text{Þ}'\phi - \bar{\pi}\text{Þ}\phi = 0 \\ \rho\text{Þ}'\phi + \bar{\rho}'\text{Þ}\phi = (\text{Þ}\phi)(\text{Þ}'\phi) + \frac{1}{6}(m - \mathcal{E}), \end{cases} \tag{9}$$

together with their primed versions. I also introduce a $[0,0]$-weighted quantity N (with $N' = 1/N$), determining the position of $\nabla\phi$ w.r.t. $\boldsymbol{k}$ and $\boldsymbol{k}'$:

$$N := Q'^2\text{Þ}\phi/\text{Þ}'\phi. \tag{10}$$

3. Naturally occurring subcases

When analysing the integrability conditions for the system (8,9) a number of subcases, each requiring a separate discussion, presents itself in a natural way:

3.1. $N - 1 = 0$

This is the case in which $\boldsymbol{u} \parallel \nabla\phi$: the vorticity then by definition vanishes and a fairly simple analysis along the lines of [6] implies the same is true for acceleration and shear: the only possible solutions are therefore the conformally flat FLRW space-times.

3.2. $N+3=0$ *(or, equivalently,* $N'+3=0$*)*

Again solutions turn out to be shear- and vorticity-free. However the presence of the Λ-term makes the analysis now considerably more complicated than in [7]. It turns out that the only allowed solutions belong to a shear-free subclass [1] of the conformally flat Stephani expanding universes[4], with $\Lambda < 0$ and with energy-density $m = -2\mu^2 Q^2 < 0$. These solutions admit at most one Killing vector.

3.3. $N^2+6N+1=0$

This particular case again gives rise to shear- and vorticity-free solutions. It was overlooked in the shear-free analysis of [5]-[8] (where it was claimed that any such non-conformally flat solutions could only arise for orthogonal $\nabla\phi$ and $\boldsymbol{u}$). However the resulting Petrov type D solutions again have a negative energy-density. The latter conclusion is based on an algebraic identity holding between m, N, $Þ'\phi$, Q and $\mathcal{E}$:

$$(\alpha_1 N + \alpha_2)m^2 + (\alpha_3 N + \alpha_4)m + (\alpha_5 N + \alpha_6) = 0,$$

with the α_i polynomials in $Þ'\phi$, Q and $\mathcal{E}$. An analysis of the roots of this quadratic in m involves the discriminant, which turns out to be proportional to the number $\zeta_1 + \zeta_2 N$, with

$$\zeta_1 = 1357297565629151284215396601095623063987021170856423988 1$$
$$\zeta_2 = 7910909947864917472994093612934674141882216009506134012 6.$$

Taking into account that the whole analysis up to this point has been carried out in exact arithmetic, this number is remarkably close to 0:

$$3 - 2\sqrt{2} - \frac{\zeta_1}{\zeta_2} \approx 0.28\ 10^{-112}\ !$$

Moreover one can show that $\frac{\zeta_1}{\zeta_2}$ is a so called "best" approximation of $3-2\sqrt{2}$ (meaning that there is no better rational approximation with a smaller denominator: this is related to the fact that, defining relatively prime p and q by $\frac{p}{q} = (3 - \frac{\zeta_1}{\zeta_2})/2$, one finds that p, q are solutions of the Pell-Fermat equation, $p^2 - 2q^2 = 1$, with $p - q\sqrt{2} = (1-\sqrt{2})^{148}$).

4. Shear-free case (and $N \notin \{1, -3, -1/3, -3 \pm 2\sqrt{2}\}$)

When N differs from any of the previous special values, a fairly easy analysis shows that shear-free solutions can only arise when $\nabla\phi \perp \boldsymbol{u}$ (i.e. $N=-1$) and when the expansion vanishes. Non-conformally flat solutions then necessarily require $\kappa = \Psi_0 = \Psi_1 = 0$, showing that the Petrov type is D.

When $Þ\phi \neq 0$ the $\Lambda \neq 0$ generalisations of the results in [5, 7] are obtained: solutions have a barotropic equation of state (although not a physically reasonable one) and are, for non-real ρ rotating LRS class I fluids, conformally related to the Cahen-Defrise metrics, whereas for real valued ρ they are (non-rotating) LRS class IIc or IIb (the acceleration can now vanish because of the presence of the Λ-term).

When $Þ\phi = 0$ a new non-rotating LRS IIc solution is obtained, belonging to the family of static fluid spheres (or their negatively curved or flat analogues): it is characterised by

$$\begin{cases} \delta \equiv \bar{\delta} \equiv 0 \\ D\phi = -\Delta\phi = k \text{ (constant)} \\ (D-\Delta)p = \frac{m+p}{2k}(p + 2\Lambda e^{2\phi} - 2k^2) \\ \Psi_2 = -\frac{2}{3}\Lambda e^{2\phi} \\ \Lambda < 0 \\ m = -2\Lambda e^{2\phi} - 6k^2 \end{cases} \tag{11}$$

and clearly becomes conformally flat (with negative energy density) in the $\Lambda = 0$ limit.

5. Vorticity-free case (with non-vanishing shear and $N \notin \{1, -3, -1/3, -3 \pm 2\sqrt{2}\}$)

When the vorticity vanishes and when none of the previous specifications applies, the $\Lambda \neq 0$ generalisations of the results in [6] are obtained: when $\nabla\phi \not\perp \boldsymbol{u}$ solutions are (pseudo-) spherically or plane symmetric expanding fluids, whereas for $\nabla\phi \perp \boldsymbol{u}$ one finds perfect fluid space-times admitting a 3-dimensional group of isometries acting transitively on 3-dimensional time-like hypersurfaces. The corresponding Einstein space is a $\Lambda \neq 0$ generalisation of the Bianchi VI_0 Ellis-MacCallum metric.

6. General situation

Currently the case with non-vanishing shear and vorticity is still being investigated. When $\nabla\phi \perp \boldsymbol{u}$ the consistency analysis of the system of equations looks tractable. The same cannot be said about the completely general case, which, even for $\Lambda = 0$ remains completely unsolved up to now. All *Ansätze* so far have led back to solutions in which the shear, or the vorticity, are 0.

References

[1] Barnes A 1973 *Gen. Rel. Grav.* **4** 105
[2] Geroch R, Held A and Penrose R 1973 *J. Math. Phys.* **14** 874
[3] Kozameh CN, Newman ET, Tod KP 1985 *Gen. Rel. Gravs.* **17** 343
[4] Stephani H 1967 *Comm. Math. Phys.* **4** 137
[5] Van den Bergh N 1986 *J. Math. Phys.* **27** 1076
[6] Van den Bergh N 1986 *Gen. Rel. Grav.* **18** 649
[7] Van den Bergh N 1986 *Lett. Math. Phys.* **11** 141
[8] Van den Bergh N 1988 *J. Math. Phys.* **29** 1451

Journal of Physics: Conference Series **314** (2011) 012023 doi:10.1088/1742-6596/314/1/012023

The role of special conformal Killing tensors in general relativity

Liselotte De Groote

Ghent University, Department of Mathematical Analysis, Galglaan 2, 9000 Ghent, Belgium

E-mail: `ldgroote@cage.ugent.be`

Abstract. Special conformal Killing tensors have a number of interesting properties, an overview of which is given in the Riemannian case. Some of these properties remain valid in pseudo-Riemannian manifolds. Presently, we are investigating the latter problem, where, in order to obtain a better insight, we first examine some specific cases, such as special conformal Killing tensors in conformally flat or Petrov type D space-times.

1. Introduction

A symmetric type $(0,2)$ tensor field L is a special conformal Killing tensor of a metric g if there exists a 1-form λ such that

$$L_{ij|k} = \frac{1}{2}\left(g_{jk}\lambda_i + g_{ik}\lambda_j\right), \tag{1}$$

where the rule denotes covariant differentiation with respect to the Levi-Civita connection of g. These solutions satisfy

$$L_{ij|k} + L_{jk|i} + L_{ki|j} = \lambda_i g_{jk} + \lambda_j g_{ki} + \lambda_k g_{ij}, \tag{2}$$

and are therefore conformal Killing tensors. Moreover, taking the trace of equation (1) on i and j shows that $\lambda = \mathrm{d}(\mathrm{tr}L)$, so L is of gradient type, and λ is in fact defined by the equation (1).

Special conformal Killing tensors have some very interesting properties. In the first place they can be used to construct Killing tensors. Furthermore the Nijenhuis torsion of any special conformal Killing tensor vanishes (actually we should refer here to the type $(1,1)$ tensor, obtained by raising an index with the metric):

$$L_l^k\left(L_{i|j}^l - L_{j|i}^l\right) - L_j^l L_{i|l}^k + L_i^l L_{j|l}^k = 0. \tag{3}$$

This has significant consequences for the eigenfunctions of L, when they are functionally independent. At each point x of the manifold M, $L(x)$ is a linear transformation of the tangent space T_xM, which is symmetric with respect to the scalar product defined by the metric. There is thus an orthogonal basis of T_xM with respect to which $L(x)$ is diagonal. The diagonal entries are then the eigenvalues of L. If L has functionally independent eigenvalues, n functions $u^1, u^2, ...u^n$ exist, such that at each point x of the manifold, $u^i(x)$ is an eigenvalue of $L(x)$ for each i, and the Jacobian matrix $(\partial u^i/\partial x^j)$ is everywhere non-singular. The u^i may then be taken as local coordinates. As a consequence of the fact that its torsion vanishes, L takes the form

$$L = \sum_{i=1}^{n} u^i \frac{\partial}{\partial u^i} \otimes \mathrm{d}u^i. \tag{4}$$

As they are orthogonal coordinates, $g_{ij} = 0$ for $i \neq j$.

2. Constructing Killing tensors

Vector fields K^μ generating isometries of a metric g are known as Killing vector fields. Killing tensors are a generalisation of Killing vectors. They are linked with integrals of geodesic motion and the theory of separation of variables. Killing tensors are symmetric type $(0, n)$ tensors which satisfy the Killing equation

$$\nabla_{(\sigma} K_{\mu_1 \ldots \mu_n)} = 0. \tag{5}$$

This equation is hard to solve, therefore methods are being used to construct Killing tensors from vectors and other tensors with special properties. We will only examine valence 2 Killing tensors.

Let us consider a (non-singular) special conformal Killing tensor L of metric g and take its cofactor tensor A, defined by

$$A_{ij} L^j_l = (\det L) g_{il}. \tag{6}$$

Then it is easy to see that A is a Killing tensor of g. To prove this take the covariant derivative of the equation (6) above, and use the SCK-equation (1) and the fact that L has vanishing torsion. Then

$$A_{ij|k} = (\det L)\left(\bar{L}_{ij}\bar{L}_{kl} - \frac{1}{2}\bar{L}_{ik}\bar{L}_{jl} - \frac{1}{2}\bar{L}_{il}\bar{L}_{jk}\right)\lambda^l, \tag{7}$$

where $\bar{L}$ is the inverse of L. This tensor clearly satisfies the Killing equation

$$A_{(ij|k)} = 0. \tag{8}$$

Also note that this Killing tensor has the same eigenvectors as L.

The special conformal Killing tensor equations (1) are linear, and obviously have the solution $L_{ij} = k g_{ij}$ for any constant k. So, the cofactor tensor of $aL_1 + bL_2$ is a Killing tensor for every constant a, b (or at least those for which $aL_1 + bL_2$ is non-singular). In particular, if L_{ij} is a special conformal Killing tensor, so is $L_{ij} + k g_{ij}$. By taking the coefficients of powers of k in the cofactor tensor of this special conformal Killing tensor, we obtain n valence 2 Killing tensors, simultaneously diagonal with L_{ij}, one of which is the metric itself. When the special conformal Killing tensor has simple eigenvalues, these Killing tensors are independent. Since the eigenvectors of $aL + kg$ are the same as the eigenvectors of L, these Killing tensors have the same eigenvectors.

It is important to note that the space of solutions of (1) is a finite-dimensional vector space. Therefore only a limited number of Killing tensors can be generated from special conformal Killing tensors. The maximal dimension of the space of solutions of (1), $\frac{1}{2}(n+1)(n+2)$, is achieved if and only if the space is of constant curvature [1].

3. Separation of variables

Related to the Killing tensor problem, is the question of separation of variables of classical differential equations. When a manifold admits a special conformal Killing tensor, the eigenfunctions of which are simple and functionally independent, these eigenfunctions are orthogonal separation coordinates for classical differential equations, such as the Hamilton-Jacobi equations for the geodesics of the manifold. To prove this, one has to use the fact that the Nijenhuis torsion of the special conformal Killing tensor vanishes. As we said in the introduction, the eigenvalues u^i of L can be used, when simple and functionally independent, as local orthogonal coordinates. These coordinates then are the orthogonal separation coordinates for the Hamilton-Jacobi equation for the geodesics of the manifold.

4. Projective equivalence

Two metrics are said to be projectively equivalent if they have the same geodesics upto reparametrization. This occurs if and only if a certain metric formed out of the two metric tensors is a special conformal Killing tensor.

Let h and g be two projectively equivalent metrics, then

$$L_{ij} = \left(\frac{\det h}{\det g}\right)^{1/(n+1)} g_{ik} g_{jl} h^{kl} \tag{9}$$

defines a special conformal Killing tensor L of metric g. Conversely, given L, a (non-singular) special conformal Killing tensor of g

$$h_{ij} = (\det L)^{-1} g_{ik} g_{jl} \bar{L}^{kl} \tag{10}$$

defines a metric projectively equivalent to g.

As mentioned before, the special conformal Killing tensors form a linear space, so if g admits an equivalent metric h, it admits a one-parameter family of equivalent metrics h^k. Indeed, starting with L, the special conformal Killing tensor generating h, we can consider all equivalent metrics determined by the pencil $L_k = L + kg$ (the constant $-k$ must not coincide with the eigenvalues of L).

Remark: if n is even, L in (9) is always well defined. If n is odd and $\det h / \det g$ is negative, we can consider $-h$ instead of h. This changes the sign of $\det h$ and makes L well defined.

For proofs and more information on special conformal Killing tensors in connection with projective equivalence, we refer the reader to [2]

5. Example: aligned Petrov type D and future work

We have examined the Petrov type D space-times, which admit an aligned special conformal Killing tensor, other than the trivial one (a constant times the metric). By aligned we mean that L has only a contribution along the principal null-directions of the tetrad. We choose the tetrad such that $\Psi_i = 0$, $i \neq 2$. It can be shown that the only possible special conformal Killing tensor in this case is a constant Killing tensor. Furthermore, we find that $\Psi_2 = -\frac{R}{12}$, R being the Ricci constant, $\Phi_{ij} = 0$ except Φ_{11}. If R is a constant, these solutions belong to the Bertotti-Robinson family.

Presently we are examining which Petrov type D metrics allow non-aligned special conformal Killing tensors. Also the cases of Petrov type N and conformally flat spaces are being looked at.

Acknowledgments

The author would like to thank Norbert Van den Bergh for his help and encouragement and Brian Edgar (1945-2010) for suggesting some papers.

[1] Crampin M 2007 *Differential geometric methods in mechanics and field theory* ed F Cantrijn et al (Ghent: Academia press) pp 57-70
[2] Bolsinov A V and Matveev V S 2003 *Jour. geometry and physics* **44** 489

Spanish Relativity Meeting (ERE 2010): Gravity as a Crossroad in Physics IOP Publishing
Journal of Physics: Conference Series **314** (2011) 012024 doi:10.1088/1742-6596/314/1/012024

A new special class of Petrov type D vacuum space-times in dimension five.

Alfonso García-Parrado Gómez-Lobo [1] **and Lode Wylleman** [2,3]

[1] Centro de Matemática, Universidade do Minho. 4710-057 Braga, Portugal
[2] Utrecht University, Department of Mathematics, Budapestlaan 6, 3584 CD Utrecht, the Netherlands
[3] Ghent University, Department of Mathematical Analysis IW16, Galglaan 2, 9000 Ghent, Belgium

E-mail: `alfonso@math.uminho.pt`

E-mail: `lwyllema@ugent.be`

Abstract. Using extensions of the Newman-Penrose and Geroch-Held-Penrose formalisms to five dimensions, we invariantly classify all Petrov type D vacuum solutions for which the Riemann tensor is isotropic in a plane orthogonal to a pair of Weyl aligned null directions [1].

1. The role of the Newman-Penrose (NP) and Geroch-Held-Penrose (GHP) formalisms in General Relativity

The NP formalism [1] consists of an explicit expansion of the Ricci and Bianchi identities in a *null tetrad*, yielding a set of equations (henceforth called NP equations) for the *spin coefficients* and *curvature scalars.* Under certain restrictions (vacuum, algebraic type of the Weyl and/or Ricci tensors, etc) one can choose a null tetrad in which some curvature scalars vanish bringing a simplification to the NP equations. In some cases it is possible to perform a complete study of the consistency conditions of the constraints and their higher order differentials, thus enabling the classification of the family of solutions satisfying the restrictions imposed. The GHP formalism [2] further exploits the invariance of the null tetrad directions under the group of boost and spin transformations in order to work with those variables which are *covariant* under this group. This reduces the number of variables and simplifies the equations. From these considerations, we deduce that the GHP formalism is best suited for situations in which the boost-spin group is the natural invariance group in the geometric set-up. An important example of this can be found in the case of Petrov type D solutions, where the two principal null directions of the Weyl tensor are naturally invariant under boosts, and their orthogonal 2-plane is invariant under spins (see [3, 4] for further details).

2. A generalisation of the NP and GHP formalisms to dimension five

The techniques described in the previous paragraph have been developed for the case of a four-dimensional (4D) space-time, but the ideas are general for any dimension as explained in [5, 6, 7]. In our work we analyse these ideas for the particular case of a 5D space-time. In this particular

[1] We dedicate this work to the memory of Professor S. B. Edgar (1945-2010).

case, one can still rely on spinor calculus to introduce the essential quantities of the formalisms, in the same way as in the 4D case. The basics of spinor calculus in a 5D space-time can be found in [8] and we stress that the rationale behind many of our definitions lies in the use of spinors, even though this will not be made explicit in the present work.

We start by introducing a *semi-null pentad* defined as follows

$$N \equiv \{l^a, n^a, m^a, \bar{m}^a, u^a\}, \qquad l^a n_a = -1\,, \quad m^a \bar{m}_a = 1\,, \quad u^a u_a = 1. \tag{1}$$

The *frame derivations* associated with each element of this frame are denoted as follows:

$$D \equiv l^a \nabla_a\,, \quad \Delta \equiv n^a \nabla_a\,, \quad \delta \equiv m^a \nabla_a\,, \quad \bar{\delta} \equiv \bar{m}^a \nabla_a\,, \quad \mathcal{D} \equiv u^a \nabla_a. \tag{2}$$

Next we need to define the *spin coefficients* and the *curvature scalars*. Some of them coincide with the quantities appearing in the standard 4D NP formalism (this is to be expected as we work in a frame obtained from the NP null tetrad by just adding the element u^a). Here we just enumerate the quantities.

- Twelve NP 4D spin coefficients: α, β, γ, ϵ, κ, λ, μ, ν, π, ρ, σ, τ.
- Ten complex 5D spin coefficients: ζ, η, θ, χ, ω, ϕ, ξ, υ, ψ, ς.
- Six real 5D spin coefficients: $\mathfrak{a}$, $\mathfrak{b}$, $\mathfrak{c}$, $\mathfrak{d}$, $\mathfrak{e}$, $\mathfrak{f}$.

$$2 \times 12 + 2 \times 10 + 6 = 50 \text{ real Ricci rotation coefficients.}$$

- Five complex 4D Weyl scalars: Ψ_0, Ψ_1, Ψ_2, Ψ_3, Ψ_4,
- Eleven complex 5D Weyl scalars: ${}^*\Psi_0$, ${}^*\Psi_1$, Ψ_1^*, ${}^*\Psi_2$, Ψ_2^*, ${}^*\Psi_3$, Ψ_3^*, Ψ_4^*, Ψ_{01}, Ψ_{02}, Ψ_{12}.
- Three real 5D Weyl scalars: Ψ_{00}, Ψ_{11}, Ψ_{22}.
- 4D NP Ricci scalars: Real: Φ_{00}, Φ_{11}, Φ_{22}, Complex: Φ_{01}, Φ_{02}, Φ_{12}.
- 5D trace-free Ricci scalars: Real: Ω, ${}^*\Phi_{01}$, ${}^*\Phi_{12}$, Complex: ${}^*\Phi_{02}$.
- Scalar curvature: Λ.

$$(2 \times 16 + 3) + (2 \times 4 + 6) + 1 = 50 \text{ independent real Riemann tensor components.}$$

The procedure is now to expand the Ricci and Bianchi identities in the semi-null pentad (1) and write them out in terms of the scalar quantities just introduced. The resulting equations are referred to as the 5D NP equations and they can be regarded as an *extension* of the well-known 4D NP equations. One can also compute the commutation relations of the operators defined in (2) in terms of the spin coefficients and again a set of equations is obtained which extends to dimension five the known commutation relations of the NP formalism.

Under the action of the boost-spin group the semi-null pentad transforms according to

$$l^a \mapsto z\bar{z}\, l^a\,, \quad n^a \mapsto \frac{n^a}{z\bar{z}}\,, \quad m^a \mapsto \frac{z}{\bar{z}} m^a\,, \quad \bar{m}^a \mapsto \frac{\bar{z}}{z} \bar{m}^a\,, \quad u^a \mapsto u^a \tag{3}$$

for some complex number $z \in \mathbb{C}$. A scalar quantity Q is said to be *weighted* if under (3) it transforms as

$$Q \mapsto z^p \bar{z}^q Q, \tag{4}$$

for some integers p and q (the pair (p, q) being called the weight of Q). It is not difficult to check that all the curvature scalars are weighted quantities whereas only a subset of the spin coefficients is weighted. Now, we can follow a procedure similar to the one presented in [2] and introduce new derivations *covariant* under boosts and spins. These new derivations turn out to be uniquely defined and are given by [8]

$$\begin{aligned}&þQ \equiv (D - p\epsilon - q\bar{\epsilon})Q\,, \quad þ'Q \equiv (\Delta - p\gamma - q\bar{\gamma})Q\,,\\ &ðQ \equiv (\delta - p\beta - q\bar{\alpha})Q\,, \quad ð'Q \equiv (\bar{\delta} - p\alpha - q\bar{\beta})Q\,, \quad \widehat{\mathcal{D}}Q \equiv (\mathcal{D} - p\theta - q\bar{\theta})Q\,,\end{aligned} \tag{5}$$

where Q is a quantity with weight (p, q) as in (4). They are referred to as *GHP operators.* Regarding the covariance under a boost-spin transformation (3) one specifically has

$$\begin{aligned}&þQ \mapsto z^{1+p}\bar{z}^{1+q}þQ\,, \quad þ'Q \mapsto z^{p-1}\bar{z}^{q-1}þQ\,,\\ &ðQ \mapsto z^{p-1}\bar{z}^{q+1}ðQ\,, \quad ð'Q \mapsto z^{p+1}\bar{z}^{q-1}ð'Q\,, \quad \widehat{\mathcal{D}}Q \mapsto \widehat{\mathcal{D}}Q\,.\end{aligned} \tag{6}$$

One can now extract the part of the NP equations which only involves derivatives of weighted quantities and obtain a set of equations which we call the 5D GHP equations. The remaining NP equations are absorbed in the commutator relations of the GHP operators.

3. The $\mathcal{A}$-class and its invariant classification

The *null alignment theory* [9] provides an algebraic classification of any tensor in a Lorentzian vector space of arbitrary dimension. For the Weyl tensor this implies a generalization of the 4D *Petrov types.* These are characterised by the *Weyl aligned null directions* (WANDS) and their *alignment order.* The WANDs generalise the notion of *principal null direction* of a 4D Weyl tensor. An important difference lies in the fact that the number of WANDs might be infinite or zero (the latter being in fact the generic case) in dimensions higher than four, whereas it is exactly four (counting multiplicity) in the 4D case. We refer the reader to [9] for further details.

In view of the complete classification of the 4D Petrov type D vacuum space-times (containing the well-known vacuum black hole solutions) [10, 11, 12, 13, 3] it is natural to endeavour the classification of 5D Petrov type D vacua (or *Einstein spaces*). By definition, these are space-times (possibly admitting a cosmological constant Λ) with vanishing trace-free Ricci tensor and exhibiting a pair of double WANDs spanned by l^a and n^a at each point. Taking these as the first two vectors of a semi-null pentad, the only surviving components of the Weyl tensor are the zero boost-weight Weyl scalars

$$\Psi_{11}\,, \quad \Psi_2\,, \quad \Psi_{02}\,, \quad \Psi_2^*\,, \quad {}^*\Psi_2. \tag{7}$$

It turns out that the resulting GHP equations and their consequences are rather unmanageable, although some general properties (even in general dimensions) were deduced in [14, 15]. In order to simplify the computations further, we restricted ourselves to the class $\mathcal{A}$, defined by the property that the Riemann (or Weyl) tensor is isotropic in a plane orthogonal to l^a and n^a. Taking the pentad vectors m^a and $\bar{m}^a$ along the complex null directions of this plane and u^a along its normal, only the $(0, 0)$-weighted Weyl scalars Ψ_{11} and Ψ_2 are possibly non-zero.

The procedure followed consists of imposing the conditions just described on the extended GHP equations and analyse the corresponding consistency conditions. The details of the integrability analysis will be presented in [16]. Here the possible cases are summarised in table 1. Each case is characterised by algebraic relations fulfilled by the quantities Ψ_{11} and Ψ_2. The case $\Psi_2 = 2\Psi_{11}$ has been completely integrated in [15] and only here is the pair of double WANDs not unique. The case $\Psi_{11} = 0$ is a direct generalization of the 4D solutions and can be fully integrated, just as the families characterized by the conditions h_1 and $\Psi_2 = -2\Psi_{11}$. All cases only depend on a number of invariantly defined continuous parameters ('global charges'). It is possible to obtain invariant information (even without performing the actual integrations) in the same spirit as e.g. [3]. The Karlhede bound refers to the number of covariant derivatives of the Riemann tensor needed in the invariant classification algorithm [13]. The symbol s (r) denotes the dimension of the isotropy (isometry) group.

Table 1. Characterizing relations for the subclasses of $\mathcal{A}$. The symbol h_1 denotes the condition $\Psi_2 = \bar{\Psi}_2$, $\Psi_2\bar{\Psi}_2 \neq 4\Psi_{11}^2 \neq 0$ and the symbol h_2 stands for the condition $\Psi_2\bar{\Psi}_2 = 4\Psi_{11}^2$, $\bar{\Psi}_2 \neq \Psi_2$.

Subclass	$\Psi_{11} = 0$	h_1	h_2	$\Psi_2 = 2\Psi_{11}$	$\Psi_2 = -2\Psi_{11}$
Karlhede bound	≤ 2	1	≥ 1	≤ 2	≤ 2
Global charges	≤ 4	2	≤ 3	≤ 1	≤ 1
s	≤ 2	2	≤ 2	3 or 4	3 or 4
r	≥ 2	6	≥ 3	7 or 9	7 or 9

4. Conclusion and further research

We have performed the invariant classification of a particular class $\mathcal{A}$ of five-dimensional vacuum metrics of Weyl-Petrov type D, possibly admitting a cosmological constant. Some issues have been left open in this work. Perhaps the most important one is a complete integration of the h_2 case in table 1, which would enable us to carry out a detailed study of the properties of the corresponding solutions. This would require the further development of integration techniques for the differential equations arising in the 5D GHP equations, in the line of those largely explored by Brian Edgar in the 4D case (see e.g. [17, 18, 3]). Another interesting avenue is to explore other Petrov type D subtypes different to the $\mathcal{A}$-class (the generalisation of our analysis to obtain the generic five-dimensional type D vacuum solution seems to be rather involved).

Acknowledgments

AGP is supported by the Research Centre of Mathematics of the University of Minho (Portugal) through the "Fundação para a Ciência e a Tecnologia" (FCT) Pluriannual Funding Program. LW is supported by a BOF Research Grant (UGent) and a FWO mobility grant.

References

[1] Newman E and Penrose R 1962 *J. Mathematical Phys.* **3** 566–578
[2] Geroch R, Held A and Penrose R 1973 *J. Mathematical Phys.* **14** 874–881
[3] Edgar S B, García-Parrado A and Martín-García J M 2009 *Classical Quantum Gravity* **26** 105022, 13
[4] Collins J M, d'Inverno R A and Vickers J A 1990 *Classical Quantum Gravity* **7** 2005–2015
[5] Pravda V, Pravdová A, Coley A and Milson R 2004 *Classical Quantum Gravity* **21** 2873–2897
[6] Ortaggio M, Pravda V and Pravdová A 2007 *Classical Quantum Gravity* **24** 1657–1664
[7] Durkee M, Pravda V, Pravdová A and Reall H S 2010 *Classical and Quantum Gravity* **27** 215010
[8] García-Parrado A and Martín-García J M 2009 *J. Math. Phys.* **50** 122504,26
[9] Milson R, Coley A, Pravda V and Pravdová A 2005 *Int. J. Geom. Methods Mod. Phys.* **2** 41–61
[10] Kinnersley W 1969 *J. Math. Phys.* **10** 1195
[11] Debever R, Kamran N and McLenaghan R G 1984 *J. Math. Phys.* **25** 1955
[12] García D A 1984 *J. Math. Phys.* **25** 1951
[13] Stephani H, Kramer D, MacCallum M A H, Hoenselaers C and Herlt E, *Exact Solutions to Einstein's Field Equations (Second Edition)* (Cambridge: Cambridge University Press, 2003)
[14] Pravda V, Pravdová A and Ortaggio M 2007 *Class. Quantum Grav.* **24** 4407
[15] Durkee M and Reall H S 2009 *Class. Quantum Grav.* **26** 245005
[16] García-Parrado Gómez-Lobo A and Wylleman L, *A special class of Petrov type D Einstein spaces in five dimensions*, in preparation
[17] Edgar S B and Ludwig G 1997 *Gen. Rel. Grav.* **29** 1309
[18] Edgar S B and Ramos M P M 2005 *Class. Quantum Grav.* **22** 791

Journal of Physics: Conference Series **314** (2011) 012025 doi:10.1088/1742-6596/314/1/012025

The complete solution of the conformastat electrovacuum problem

Guillermo A. González[1] **and Raül Vera**[2]

[1] Escuela de Física, Universidad Industrial de Santander, A. A. 678, Bucaramanga, Colombia
[2] Dept. Física Teórica, Universidad del País Vasco/Euskal Herriko Unibertsitatea, 644 PK, 48080 Bilbao, Basque Country, Spain

E-mail: guillego@uis.edu.co, raul.vera@ehu.es

Abstract. We find the complete solution of the Einstein-Maxwell field equations without sources for static spacetimes in which the space of Killing trajectories is conformally flat. The result is used to present an improved local characterisation of the Majumdar-Papapetrou class of solutions.

1. Introduction

We present the complete solution of the Einstein-Maxwell field equations without sources for conformastat spacetimes, i.e. the conformastat electrovacuum spacetimes. The only assumption made is that the electromagnetic field inherits the symmetry, thus being stationary.

Taking the definition in [1] as standard (see the original [2]) by conformastat spacetimes we mean the static subclass of the conformastationary spacetimes, which are, in turn, those stationary spacetimes with a conformally flat space of orbits. This class of spacetimes is important by its own right, since it includes the most relevant Schwarzschild solution and the Majumdar-Papapetrou class of metrics, solution of the Einstein-Maxwell equations without sources. But more interestingly, the study of conformastat electrovacuum spacetimes eventually provides us with some characterisation (or uniqueness) results for static charged black holes [3], because the resulting classes of solutions are tremendously restricted.

Conformastat *vacuum* spacetimes comprise three [4] out of the seven families that constitute the whole set of type D static vacuum spacetimes [5, 1] (Table 18.2 in [1]). These three families consist of the Schwarzschild solution together with its plane and hyperbolic counterparts.

In order to generalise this class of solutions Lukász *et al.* and Perjés tackled the problem of finding the whole set of conformastationary vacuum spacetimes in a series of three papers. Let us note that they refer to conformastationary spacetimes simply as "conformastat". In a first paper [6] the solutions possessing a functional relationship between the real and imaginary parts of the Ernst potential $\mathcal{E}$ were found to consist of three bi-parametric families of solutions. These three families are the stationary generalisations of the three conformastat vacuum solutions by the Ehlers transformation. In particular, the Schwarzschild solution generalises to the NUT solution. In two subsequent papers [7, 8], using purposely defined "Ernst coordinates", Perjés found that solutions with functionally independent real and imaginary parts of $\mathcal{E}$ do not exist. The final result is thus that the set of conformastationary vacuum spacetimes correspond to the three families presented in [6], which can be thought of being the NUT-type extensions of the conformastat vacuum solutions.

Another natural step to follow is the generalisation of the conformastat vacuum spacetimes by including a stationary electromagnetic field without sources.

To find the complete solution of the conformastat electrovacuum problem we first show how this problem and the conformastationary vacuum problem can both be treated within a common framework by using a suitable notation. This framework is then used to prove that non-trivial conformastat electrovacuum spacetimes must contain a functional relationship between the gravitational and electromagnetic potentials. The final part, which completes the study of conformastat electrovacuum spacetimes, consists in classifying and exploiting the necessary functional relationship.

The final result is that *all conformastat electrovacuum spacetimes either belong to the Majumdar-Papapetrou class or correspond to either the Bertotti-Robinson solution or the exterior Reissner-Nordström solution together with its plane and hyperbolic counterparts.* Furthermore, the procedure used for the classification provides an improved characterisation of the Majumdar-Papapetrou class. This is known to be the class of static electrovacuum spacetimes in which the usual rescaled induced metric in the space of orbits is flat. Here we find that it is sufficient to ask that the metric be conformally flat and with vanishing Ricci scalar.

2. Conformastationary and conformastat spacetimes

A stationary spacetime $(\mathcal{M}, g_{\mu\nu})$ is *locally* defined by the existence of a timelike Killing vector field ξ^μ whose space of orbits invariantly determines a differentiable 3-dimensional Riemannian manifold Σ_3. Local coordinates $\{t, x^a\}$ exist for which $\xi^\mu = \partial_t$ and such that the line-element reads [1]

$$ds^2 = -e^{2U}(dt + A_a dx^a)^2 + e^{-2U}\widehat{h}_{ab}dx^a dx^b, \tag{1}$$

where U, A_a and $\widehat{h}_{ab}$ do not depend on t. Applying the usual projection formalism (see e.g. [1]) we will think of U, A_a as objects living on $(\Sigma_3, \widehat{h}_{ab})$. Once U, A_a and $\widehat{h}_{ab}$ are given, the local geometry of the stationary spacetime $(\mathcal{M}, g_{\mu\nu})$ is fully specified by using (1). Let us use the first Latin indices $a, b, \ldots$ for objects defined on $(\Sigma_3, \widehat{h}_{ab})$. In this context a static spacetime is characterised by $A_a = 0$.

A *conformastationary* spacetime is a stationary spacetime whose space of orbits $(\Sigma_3, \widehat{h}_{ab})$ is conformally flat [1]. The intrinsic characterisation of a conformally flat 3-space is the vanishing of the York tensor density

$$Y_a{}^e \equiv \hat{\eta}^{bce}\left(2\hat{\nabla}_c\widehat{R}_{ba} - \frac{1}{2}\widehat{h}_{ab}\hat{\nabla}_c\widehat{R}\right) = 0, \tag{2}$$

where $\widehat{R}_{ab}$, $\hat{\eta}_{abc}$ and $\hat{\nabla}$ denote the Ricci tensor, the volume form and covariant derivative relative to $\widehat{h}_{ab}$. Note that $Y_{ae} = Y_{ea}$ and $Y_a{}^a = 0$.

Conformastat spacetimes are those conformastationary spacetimes which are, in fact, static, for which A_a can be set to zero.

3. Electrovacuum field equations

Let us first fix one basic assumption and some notation. First, we will restrict ourselves to Maxwell fields $F_{\alpha\beta}$ in $(\mathcal{M}, g_{\mu\nu})$ which inherit the stationary symmetry, i.e. for which $\mathcal{L}_\xi \boldsymbol{F} = 0$. The Einstein-Maxwell equations outside the sources imply (locally, at least) the existence of two complex scalars, $\Phi(x^a)$ the electromagnetic potential, and $\mathcal{E}(x^a)$ the Ernst potential. Then, the Einstein-Maxwell equations reduce to

$$\widehat{R}_{ab} = G_a\overline{G}_b + \overline{G}_a G_b - (H_a\overline{H}_b + \overline{H}_a H_b),$$

$$\hat{\nabla}^a H_a + \frac{1}{2}\overline{G}\cdot H - \frac{3}{2}G\cdot H = 0,$$
$$\hat{\nabla}^a G_a - \overline{H}\cdot H - (G-\overline{G})\cdot G = 0,$$

where $H_a \equiv (\mathrm{Re}\mathcal{E} + \Phi\overline{\Phi})^{-1/2}\Phi_{,a}$ and $G_a \equiv 1/2(\mathrm{Re}\mathcal{E} + \Phi\overline{\Phi})^{-1}(\mathcal{E}_{,a} + 2\overline{\Phi}\Phi_{,a})$, and the dot denotes the scalar product with respect to $\widehat{h}_{ab}$. The integrability conditions for the two potentials are given by $\mathrm{d}H = H\wedge \mathrm{Re}G$, $\mathrm{d}G = G\wedge\overline{G} + \overline{H}\wedge H$.

The metric is determined by the relations $e^{2U} = \mathrm{Re}\mathcal{E} + \Phi\overline{\Phi}$, $\mathrm{d}A_{ab} = 2e^{-4U}\hat{\eta}_{abc}\mathrm{Im}G^c$.

3.1. Vacuum stationary and electro-magnetostatic cases

The *vacuum* stationary case is defined by $\Phi = 0 \Rightarrow H_a = 0$ and hence the Einstein-Maxwell equations plus the integrability for the potentials read

$$\widehat{R}_{ab} = G_a\overline{G}_b + \overline{G}_a G_b, \qquad \hat{\nabla}^a G_a - (G-\overline{G})\cdot G = 0, \qquad \mathrm{d}G = G\wedge\overline{G}.$$

The *static* case corresponds to $G_a - \overline{G}_a = 0$. Then $G_a = U_{,a}$ and also $\overline{H}_a = e^{-2i\theta}H_a$ for some constant θ [4]. After defining $X_a \equiv e^{-i\theta}H_a = e^{-U}\Psi_{,a}$ where $\Psi \equiv e^{-i\theta}\Phi$ is real, and $\Sigma_a \equiv \frac{1}{2}(U_{,a} + jX_a)$ in the hypercomplex plane, where $j^2 = 1$ with conjugation $\breve{j} = -j$, the set of equations now read

$$\widehat{R}_{ab} = 4(\Sigma_a\breve{\Sigma}_b + \breve{\Sigma}_a\Sigma_b), \qquad \hat{\nabla}^a\Sigma_a - (\Sigma-\breve{\Sigma})\cdot\Sigma = 0, \qquad \mathrm{d}\Sigma = \Sigma\wedge\breve{\Sigma}.$$

4. Common framework

Denote by ι either i or j, so that $\iota^2 = \pm 1$ accordingly, and by $\tilde{}$ the general conjugation. Consider a "composed" vector field $\mathcal{Y}^a$ and a metric $\widehat{h}_{ab}$ that satisfy

$$\widehat{R}_{ab} = N(\mathcal{Y}_a\widetilde{\mathcal{Y}}_b + \widetilde{\mathcal{Y}}_a\mathcal{Y}_b), \qquad \hat{\nabla}^a\mathcal{Y}_a - (\mathcal{Y}-\widetilde{\mathcal{Y}})\cdot\mathcal{Y} = 0, \qquad \mathrm{d}\mathcal{Y} = \mathcal{Y}\wedge\widetilde{\mathcal{Y}}. \tag{3}$$

The *vacuum (stationary)* case is then recovered by using $N = 1$ and $\mathcal{Y}_a(=G_a)$ being complex, while the *static (electrovacuum)* case is recovered with $N = 4$ and $\mathcal{Y}_a(=\Sigma_a)$ being hypercomplex.

Regarding *conformastationarity*, the introduction of the 1-form $L \equiv \star(\mathcal{Y}\wedge\widetilde{\mathcal{Y}})$ allows us to write the condition (2) as

$$(\mathcal{Y}_a - \widetilde{\mathcal{Y}}_a)L^e + \hat{\eta}^{bce}(\widetilde{\mathcal{Y}}_b\hat{\nabla}_c\mathcal{Y}_a + \mathcal{Y}_b\hat{\nabla}_c\widetilde{\mathcal{Y}}_a) - \frac{1}{2}\widehat{h}_{ab}\hat{\eta}^{bce}\hat{\nabla}_c(\mathcal{Y}\cdot\widetilde{\mathcal{Y}}) = 0. \tag{4}$$

The proof involves solving the system (3)-(4) for $\mathcal{Y}_a$ and $\widehat{h}_{ab}$. In the following we present the main steps.

To solve the system one has to consider first the case $\mathcal{Y}\cdot\mathcal{Y} = 0$. In the static case $\mathcal{Y} = \Sigma$ is j-composed and $\Sigma\cdot\Sigma = 0$ implies in particular $G\cdot G + X\cdot X = 0$, which leads to the trivial case $G = X = 0$. However, in the vacuum case $\mathcal{Y} = G$ is complex and one can have, in principle, fields for which $G\cdot G = 0$. The study of these null fields was performed in [6], where it was proven that no null coformastationary vacuum spacetimes exist apart from the trivial case of flat spacetime.

Two different situations now arise depending on whether L_a vanishes or not. In the case $L_a \neq 0$ one can take the basis $\{L_a, \mathcal{Y}_a, \widetilde{\mathcal{Y}}_a\}$ and show that one of the components of (2) reads $L\wedge\mathrm{d}L = 0$, and therefore $L = \iota\chi\mathrm{d}\varphi$ for some real functions χ and (a nonconstant) φ. The integrability condition $\mathrm{d}\mathcal{Y} = \mathcal{Y}\wedge\widetilde{\mathcal{Y}}$ implies, in turn, that a composed function σ exists such that $\mathcal{Y} = (\sigma+\widetilde{\sigma})^{-1}\mathrm{d}\sigma$. The three potentials σ, $\widetilde{\sigma}$ and φ can be used as coordinates to write down the remaining equations and study the compatibility conditions, in particular for $N = 1$ and $N = 4$. A somewhat long calculation [3] shows that there are no possible solutions in this case.

The result so far is that the case $L_a \neq 0$ is empty, and therefore $L_a = 0$ necessarily, i.e. $\mathcal{Y}$ and $\widetilde{\mathcal{Y}}$ must be parallel. This translates in the vacuum and static cases to:

Table 1. The conformastat electrovacuum solutions.

$\widehat{R} = 0$	Majumdar-Papapetrou	
$\widehat{R} > 0$	Plane-symmetric field (plane Reissner-Nordström)	
	Bertotti-Robinson	
	Reissner-Nordström exterior	$M^2 - Q^2 > 0$
	hyperbolic Reissner-Nordström	
$\widehat{R} < 0$	Bertotti-Robinson	
	Reissner-Nordström exterior	$M^2 - Q^2 < 0$

Theorem 1 *Conformastationary vacuum spacetimes are always characterised by a functional relation between the potentials U and Ω.*

Theorem 2 *Conformastat electrovacuum spacetimes are always characterised by a functional relation between the potentials U and Φ.*

5. The complete solution

The study of the $L_a = 0$ case was presented in [9], but it turned out to be incomplete (see [3]). Indeed, the divergence equation for $\mathcal{Y}_a$ firstly fixes, for an arbitrary constant k, $e^{2U} = 1 - 2k\Psi + \Psi^2$, which can be rewritten in parametric form in terms of an auxiliary function V that satisfies $\widehat{\nabla}^2 V = 0$. Three cases arise depending on k, for which the corresponding Ricci equations read $k^2 = 1 : \widehat{R}_{ab} = 0$, $k^2 > 1 : \widehat{R}_{ab} = 2V_{,a}V_{,b}$, $k^2 < 1 : \widehat{R}_{ab} = -2V_{,a}V_{,b}$. The remaining equations for $\widehat{h}_{ab}$ and $V_{,a}$ are encoded in (2). All the resulting solutions [3] are presented in Table 1. The result mentioned in the introduction thus follows. Moreover one can state the following improved local characterisation of the Majumdar-Papapetrou class, known to be the static electrovacuum spacetimes with a flat $\widehat{h}_{ab}$ [1]. Here we have relaxed the requirement on $\widehat{h}_{ab}$ by showing that

Corollary 1 *The Majumdar-Papapetrou class of solutions are the static electrovacuum spacetimes with conformally flat $\widehat{h}_{ab}$ and $\widehat{R} = 0$.*

Acknowledgments

GAG wants to thank the warm hospitality of Dept. de Física Teórica, UPV/EHU, where this work was made. RV is funded by project IT-221-07 from the Basque Government, and thanks support from project FIS2007-61800 (MICINN).

References

[1] Stephani H, Kramer D, MacCallum M A H, Hoenselaers C and Herlt E (2003) *Exact solutions of Einstein's field equations. Second Edition*, Cambridge University Press, Cambridge
[2] Synge J L (1960) *Relativity: The General Theory*, North Holland, Amsterdam
[3] González G A and Vera R 2011 *Class. Quantum Grav.* **28** 025008
[4] Das A 1971 *J. Math. Phys.* **12** 1136-42
[5] Ehlers J and Kundt W (1962) "Exact solutions of the gravitational fields equations", *Gravitation*, ed. L. Witten, Wiley, New York, p. 49
[6] Lukács B, Perjés Z and Sebestyén Á 1983 *Gen. Rel. Grav.* **15** 511-22
[7] Perjés Z 1986 *Gen. Rel. Grav.* **18** 511-30
[8] Perjés Z 1986 *Gen. Rel. Grav.* **18** 531-47
[9] González G A and Vera R 2010 *J. Phys.: Conf. Ser.* **229** 012040

Journal of Physics: Conference Series **314** (2011) 012026 doi:10.1088/1742-6596/314/1/012026

Density perturbations in Kantowski-Sachs models with a cosmological constant

M Forsberg[1], M Bradley[1] and P K S Dunsby[2]
[1]Department of Physics, Umeå University, SE-901 87 Umeå, Sweden
[2]Department of Mathematics and Applied Mathematics, University of Cape Town, South Africa

E-mail: mats.forsberg@physics.umu.se, michael.bradley@physics.umu.se, peter.dunsby@uct.ac.za,

Abstract. The growth of density perturbations in Kantowski-Sachs cosmologies with a positive cosmological constant is studied, using the 1+3 and 1+1+2 covariant formalisms. For each wave number a closed system for eight scalars is obtained. These are formed from quantities that are zero on the background and hence are gauge invariant. As an example a numerical solution describing the evolution of density perturbations on a background that experiences a bounce is presented. Typically the density gradient in the bouncing directions experiences a local maxium at or slightly after the bounce.

1. Introduction

In this work we study the growth of density perturbations in perfect fluid Kantowski-Sachs universes with positive cosmological constant with special attention to the behaviour close to a bounce, using the 1+3 and 1+1+2 covariant splits of spacetime [1, 2, 3, 4]. As inhomogeneity variables we use the spatial gradients of the density, the expansion, the shear scalar and one more auxiliary scalar to close the system. These quantities are all covariantly defined and are also zero on the background [1, 5]. Hence they are gauge invariant [6]. By projecting along the preferred and orthogonal directions and taking divergences along the preferred and orthogonal directions respectively and making harmonic decompositions of the spatial derivatives, the system is reduced to a first order system of ordinary differential equations in time for eight scalar quantities for each wave number.

2. The 1+3 and 1+1+2 covariant formalisms

In [1, 2], to which the reader is referred for details, a covariant formalism for the 1+3 split of spacetimes with a preferred timelike vector, u^a, was developed. The projection operator onto the perpendicular 3-space is given by $h_a^b = g_a^b + u_a u^b$. With the help of this vectors and tensors can be covariantly decomposed in "spatial" and "timelike" parts. The covariant time derivative and projected spatial derivative are given by

$$\dot{\psi}_{a..b} \equiv u^c \nabla_c \psi_{a...b} \quad \text{and} \quad \tilde{\nabla}_c \psi_{a...b} \equiv h_c^f h_a^d ... h_b^e \nabla_f \psi_{d...e} \tag{1}$$

respectively. The covariant derivative of the 4-velocity, u^a, can be decomposed as

$$\nabla_a u_b = -u_a \dot{u}_b + \tilde{\nabla}_a u_b = -u_a \dot{u}_b + \frac{1}{3}\theta h_{ab} + \omega_{ab} + \sigma_{ab} \tag{2}$$

where the kinematic quantities of u^a, acceleration, expansion, shear and vorticity are defined by $\dot{u}_a \equiv u^b\nabla_b u_a$, $\theta \equiv \tilde{\nabla}_a u^a$, $\sigma_{ab} \equiv \tilde{\nabla}_{<a}u_{b>}$, and $\omega_{ab} \equiv \tilde{\nabla}_{[a}u_{b]}$ respectively. These quantities, together with the Ricci tensor (that in the present case is given by the energy density, μ, and the pressure, p) and the electric, $E_{ab} \equiv C_{acbd}u^c u^d$, and magnetic, $H_{ab} \equiv \frac{1}{2}\eta_{ade}C^{de}{}_{bc}u^c$, parts of the Weyl tensor, are then used as dependent variables. From the Ricci and Bianchi identities one obtains propagation equations in the u^a direction and constraints.

A formalism for a further split (1+2) with respect to a spatial vector n^a (with $u^a n_a = 0$) was developed in [3, 4]. Projections perpendicular to n^a are made with $N_a^b = g_a^b + u_a u^b - n_a n^b$. Derivatives along and perpendicular to n^a are given by

$$\hat{\psi}_{a...b} \equiv n^c\tilde{\nabla}_c\psi_{a...b} = n^c h_c^f h_a^d ... h_b^e \nabla_f \psi_{d...e} \quad \text{and} \quad \delta_c\psi_{a...b} \equiv N_c^f N_a^d ... N_b^e \tilde{\nabla}_f \psi_{d...e} \tag{3}$$

respectively. Similarly to the decomposition of $\nabla_a u_b$, $\tilde{\nabla}_a n_b$ and $\dot{n}_a$ can be decomposed into further 'kinematical' quantities of n^a. The set of equations is now decomposed into propagation equations in the u^a and n^a directions respectively and constraints.

3. Perturbations on Kantowski-Sachs

As backgrounds we use the Kantowski-Sachs cosmologies, [7], that are spatially homogeneous and locally rotationally symmetric (LRS). With zero vorticity the line-element can be written as

$$ds^2 = -dt^2 + a_1^2(t)dz^2 + a_2^2(t)\left(d\theta^2 + \sin^2\theta d\varphi^2\right) \tag{4}$$

with the 4-velocity of matter given by $u = \frac{\partial}{\partial t}$ and the direction of anisotropy by $n = \frac{1}{a_1}\frac{\partial}{\partial z}$. The expansion and the only independent component of the shear are given by

$$\theta = \frac{\dot{a}_1}{a_1} + 2\frac{\dot{a}_2}{a_2} \quad \text{and} \quad \Sigma \equiv \sigma_{nn} = \frac{2}{3}\left(\frac{\dot{a}_1}{a_1} - \frac{\dot{a}_2}{a_2}\right), \tag{5}$$

respectively.

Given an equation of state and the cosmological constant, Λ, the Kantowski-Sachs models are completely determined in terms of shear, Σ, expansion, θ, and energy density, μ. The electric part of the Weyl tensor is given algebraically as $\mathcal{E} = -\frac{2}{3}\mu - \frac{2}{3}\Lambda - \Sigma^2 + \frac{2}{9}\theta^2 + \frac{1}{3}\Sigma\theta$. The following evolution equations are obtained

$$\dot{\Sigma} = \frac{2}{3}(\mu + \Lambda) + \frac{1}{2}\Sigma^2 - \Sigma\theta - \frac{2}{9}\theta^2\,, \ \dot{\mu} = -\theta(\mu + p)\,, \ \dot{\theta} = -\frac{1}{3}\theta^2 - \frac{1}{2}(\mu + 3p - 2\Lambda) - \frac{3}{2}\Sigma^2\,. \tag{6}$$

The inhomogenities will be described by quantities that are zero on the background, and hence are gauge invariant [6]. The primary variable is the density gradient

$$\mathcal{D}_a \equiv \frac{a\tilde{\nabla}_a\mu}{\mu}, \tag{7}$$

where a is the average scale factor, defined from $\theta = 3\dot{a}/a$. The density fluctuations $\frac{\delta\mu}{\mu}$ on a dimensionless comoving length scale l_0 are related to the quantity $\mathcal{D}_a$ through $\frac{\delta\mu}{\mu} \sim (\mathcal{D}_a\mathcal{D}^a)^{1/2}l_0$. To close the system three auxillary quantities, that we choose as

$$\mathcal{Z}_a \equiv a\tilde{\nabla}_a\theta\,, \quad \mathcal{T}_a \equiv a\tilde{\nabla}_a\sigma^2 \quad \text{and} \quad \mathcal{S}_a \equiv a\tilde{\nabla}_a(\sigma^{ab}S_{ab})\,, \tag{8}$$

will be needed. Here S_{ab} is the traceless part of the 3-Ricci tensor. In the present case it can be rewritten as $S_{ab} = E_{ab} + \sigma_{<a}{}^c\sigma_{b>c} - \frac{1}{3}\theta\sigma_{ab}$ in terms of covariant quantities.

The gradients are then decomposed into components parallel and perpendicular to n^a like $\mathcal{D} \equiv \mathcal{D}_a n^a$ and $\mathcal{D}_{\bar{a}} \equiv \mathcal{D}_b N^b_{\bar{a}}$ respectively and equivalently for the other quantities in (8). To treat the spatial derivatives appearing in the equations we will do an harmonic decomposition. For this purpose it is suitable to get the spatial derivatives in the form of two Laplace-like operators,

$$\delta^2 \equiv \delta_a \delta^a \quad \text{and} \quad \hat{\Delta} \equiv n^a \tilde{\nabla}_a n^b \tilde{\nabla}_b \tag{9}$$

acting on our variables. To obtain this we define new variables $\hat{\mathcal{D}} \equiv n^a \tilde{\nabla}_a \mathcal{D}$ and $\not{\mathcal{D}} \equiv \delta^a \mathcal{D}_{\bar{a}}$, and similarly for the other variables. Finally, to remove some singular terms, $\mathcal{T}_a$ is redefined according to

$$\mathcal{T}_a = \Sigma^2 \tilde{\mathcal{T}}_a + \frac{\Sigma}{\tilde{S}} \mathcal{S}_a\,, \tag{10}$$

where $\tilde{S}$ to zeroth order is given by $\tilde{S} = -\frac{2}{3}\mu - \frac{2}{3}\Lambda - \frac{1}{2}\Sigma^2 + \frac{2}{9}\theta^2 = -\frac{2}{3}K < 0$ (K is the curvature of the 2-spheres $dt = dz = 0$). The system for the hat variables then becomes

$$\begin{aligned}
\dot{\hat{\mathcal{D}}} &= \left[\theta\left(\frac{p}{\mu} - \frac{1}{3}\right) - 2\Sigma\right]\hat{\mathcal{D}} - \left(1 + \frac{p}{\mu}\right)\hat{\mathcal{Z}} \\
\dot{\hat{\mathcal{Z}}} &= -\left(\theta + 2\Sigma\right)\hat{\mathcal{Z}} + \left[-\frac{1}{2}\mu + \frac{3}{2}\frac{\mu p'}{\mu + p}\left(\tilde{S} + \frac{3}{2}\Sigma^2\right)\right]\hat{\mathcal{D}} - 2\frac{\Sigma}{\tilde{S}}\hat{\mathcal{S}} - 2\Sigma^2\hat{\tilde{\mathcal{T}}} - \frac{\mu p'}{\mu + p}\hat{\Delta}\left[\hat{\mathcal{D}} + \not{\mathcal{D}}\right] \\
\dot{\hat{\tilde{\mathcal{T}}}} &= -\left(\frac{1}{3}\theta + 2\Sigma + \frac{\Sigma^3}{\tilde{S}}\right)\hat{\tilde{\mathcal{T}}} - \left(\frac{\Sigma^2}{\tilde{S}^2} + \frac{1}{\tilde{S}}\right)\hat{\mathcal{S}} - \left[\frac{\Sigma\mu}{\tilde{S}} + \frac{\mu p'}{\mu + p}\left(\theta - \frac{3}{2}\Sigma\right)\right]\hat{\mathcal{D}} + \left(1 + \frac{2}{3}\frac{\Sigma\theta}{\tilde{S}}\right)\hat{\mathcal{Z}} \\
&\quad + \frac{\mu p'}{\mu + p}\frac{1}{\tilde{S}}\left[\left(\frac{1}{2}\Sigma - \frac{1}{3}\theta\right)\hat{\Delta}\hat{\mathcal{D}} - \left(\Sigma - \frac{1}{6}\theta\right)\hat{\Delta}\left(\not{\mathcal{D}}\right)\right] - \frac{1}{\tilde{S}}\hat{\Delta}(\hat{\mathcal{Z}} - \frac{1}{2}\not{\mathcal{Z}}) + \frac{\Sigma}{\tilde{S}}\hat{\Delta}(\hat{\tilde{\mathcal{T}}} + \not{\tilde{\mathcal{T}}}) \\
&\quad + \frac{1}{\tilde{S}^2}\hat{\Delta}(\hat{\mathcal{S}} + \not{\mathcal{S}}) \\
\dot{\hat{\mathcal{S}}} &= \left[\mu\Sigma^2 + \frac{\mu p'}{\mu + p}\tilde{S}\left(\frac{5}{2}\theta\Sigma + \frac{3}{2}\tilde{S} - \frac{3}{2}\Sigma^2\right)\right]\hat{\mathcal{D}} - \left(\frac{2}{3}\theta\Sigma + \frac{5}{2}\tilde{S}\right)\Sigma\hat{\mathcal{Z}} + \left(\Sigma^4 + 2\tilde{S}^2\right)\hat{\tilde{\mathcal{T}}} + \\
&\quad \left(\frac{\Sigma^3}{\tilde{S}} - 2\theta\right)\hat{\mathcal{S}} + \Sigma\hat{\Delta}\hat{\mathcal{Z}} - \frac{1}{2}\Sigma\hat{\Delta}\left(\not{\mathcal{Z}}\right) - \Sigma^2\hat{\Delta}\hat{\tilde{\mathcal{T}}} + \frac{\mu p'}{\mu + p}\left[\left(-\tilde{S} + \frac{1}{3}\theta\Sigma - \frac{1}{2}\Sigma^2\right)\hat{\Delta}\hat{\mathcal{D}}\right. \\
&\quad \left. -\Sigma^2\hat{\Delta}\left(\not{\tilde{\mathcal{T}}}\right)\frac{1}{2}\left(\tilde{S} - \frac{1}{3}\theta\Sigma + 2\Sigma^2\right)\hat{\Delta}\left(\not{\mathcal{D}}\right)\right] - \frac{\Sigma}{\tilde{S}}\hat{\Delta}(\hat{\mathcal{S}} + \not{\mathcal{S}}).
\end{aligned} \tag{11}$$

A similar system is obtained for the variables $\not{\mathcal{D}}$, see [8].

By making a harmonic decomposition $\Psi = \sum_{k_\parallel, k_\perp} \Psi_{k_\parallel k_\perp} P_{k_\parallel} Q_{k_\perp}$ where

$$\hat{\Delta} P_{k_\parallel} = -\frac{k_\parallel^2}{a_1^2} P_{k_\parallel}\,, \quad \delta_a P_{k_\parallel} = \dot{P}_{k_\parallel} = 0 \quad \text{and} \quad \delta^2 Q_\perp = -\frac{k_\perp^2}{a_2^2} Q_\perp\,, \quad \hat{Q}_\perp = \dot{Q}_\perp = 0 \tag{12}$$

the system (11) and the corresponding system for the slashed variables are rewritten in terms of the constant comoving wave numbers $k_\parallel$ and $k_\perp$.

Due to the definitions of $\hat{\mathcal{D}}$ and $\not{\mathcal{D}}$ in terms of the spatial derivatives, the time dependences of the variables $\mathcal{D}_\parallel = a_1 \hat{\mathcal{D}}$ and $\mathcal{D}_\perp = a_2 \not{\mathcal{D}}$ give a better representation of the development of the relative density contrast $\frac{\delta\mu}{\mu}$.

4. Numerical solutions

An example of a numerical solution is given here. This background, where the equation of state is given by $p = \mu/3$ (radiation), starts in a state where the directions perpendicular to the direction of anisotropy have small and decreasing expansion rates that become almost neglible for a period of time. After this the expansion starts again to finally reach a constant value $\theta_\perp = \sqrt{\Lambda/3}$. The anisotropy direction starts from a contracting state, goes through a bounce and then starts expanding. Eventually its expansion also approaches $\theta_\| = \sqrt{\Lambda/3}$, giving a total expansion rate of $\theta = \sqrt{3\Lambda}$. The initial state is close to the critical point $_-X$ (contracting Kasner) and for an intermediate period the solution is close to another critical point $_+X$ (expanding Kasner), but for large times the solution approaches the sink point de Sitter $_+dS$. The evolutions of density, μ_0, expansion, θ_0, expansion in the anisotropy direction, $\theta_\|$, expansion in one of the perpendicular directions, $\theta_\perp$ and shear Σ_0 are depicted in figure 1.

In figure 2 the growth of the density gradient in the direction of anisotropy is shown for different values of the comoving wave numbers $k_\|$ and $k_\perp$. For $k = 0$ the density gradient in the direction of anisotropy reaches a small maximum after the bounce and after this a small minimum before it starts growing unboundedly. For higher values of the wave number k the density gradient in the anisotropy direction shows an oscillatory behaviour with an initally increasing amplitude that later on decreases, but does not fall of to zero. This behaviour with a local maximum in the density gradient at or slightly after the bounce in the bouncing direction seems to be typical. In the perpendicular directions (not depicted here) the gradient for the $k = 0$ case is roughly constant before it also starts growing. For higher k values the oscillations initially have an approximately constant amplitude that with time slowly starts growing.

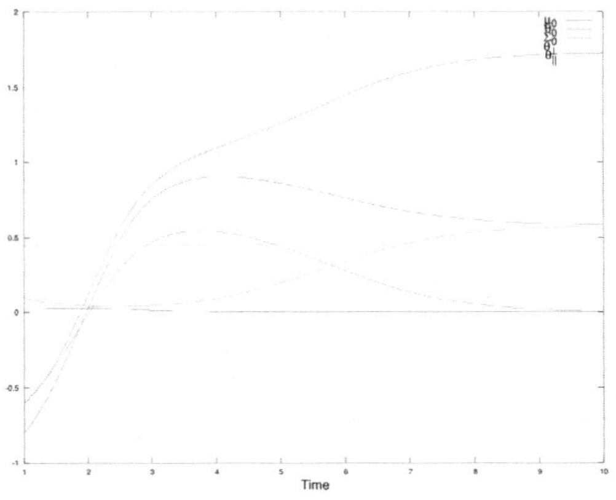

Figure 1. A radiation background that experiences a bounce. It starts close to $_-X$ and ends at $_+dS$. Initial values at $t_0 = 1$ are given by $\mu_0 = 0.02$, $\theta_0 = -0.6$, $\Sigma_0 = -0.6$, $\theta_\| = -0.8$ and $\theta_\perp = 0.1$.

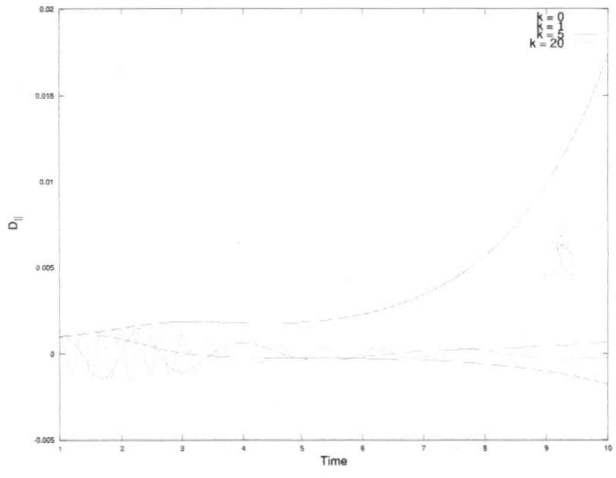

Figure 2. The growth of the density perturbations $\mathcal{D}_\|$ in the background given by figure 1 for the wave numbers $k_\|/a_{10} = k_\perp/a_{20} = 0, 1, 5$ and 20. Initially, at $t_0 = 1$, $\hat{\mathcal{D}} = \not{\mathcal{D}} = 0.001$. ($a_{10} = a_1(t_0)$, $a_{20} = a_2(t_0)$.)

References

[1] G.F.R Ellis and M. Bruni, Phys. Rev. D, **40**, 1804 (1989).
[2] G.F.R Ellis and H. van Elst, in M. Lachieze-Rey (ed.), *Theoretical and Observational Cosmology*, NATO Science Series, Kluwer Academic Publishers (1998), arXiv:gr-qc/9812046v5.
[3] C. A. Clarkson and R. Barrett, Class. Quantum Grav. **20**, 3855 (2003).
[4] C.A. Clarkson, Phys. Rev. D, **76**, 104034 (2007).
[5] M. Bruni, P.K.S. Dunsby and G.F.R. Ellis, Astrophys. J., **395**, 34 (1992).
[6] J.M. Stewart and M. Walker, Proc. R. Soc. London, **A341**, 49 (1974).
[7] R. Kantowski and R.K. Sachs, J. Math. Phys. **7**, 443 (1967).
[8] M. Bradley, P.K.S. Dunsby and M. Forsberg, preprint 2010.

Journal of Physics: Conference Series **314** (2011) 012027 doi:10.1088/1742-6596/314/1/012027

On the extremality and ultraspinning instability of Myers-Perry black holes

Jan E Åman, Narit Pidokrajt

Department of Physics, Stockholm University
SE-106 91 Stockholm, Sweden

E-mail: `ja@fysik.su.se`, `narit@fysik.su.se` (corresponding author)

Abstract. We use a contact geometric method to study Myers-Perry (MP) black holes in arbitrary dimensions with arbitrary angular momenta. We have shown that the MP black holes of dimension d with n equal nonzero spins and $2n \geqslant d-3$ all have extremal limits as expected and that we should classify MP black holes in three series depending on whether the value of $2n-d+3$ is 0, 1 or 2. For black holes with $2n < d-3$ the Ruppeiner curvature diverges although they have no extremal limits. In order to have an ultraspinning mode at least one spin of the MP black hole must be set to zero. Our result agrees with others in the literature where the authors are able to establish the minimum temperature surface on which the membrane phase of ultraspinning MP black holes occurs. We conjecture that the membrane phase of ultraspinning MP black holes is reached around the minimum temperature in the case $2n < d-3$ which is where the Ruppeiner curvature diverges.

Myers-Perry black holes, *i.e.* classical black hole solutions in more than four spacetime dimensions have been the subject of increasing attention in recent years due to the richness of the solutions themselves. Firstly, there is the possibility of rotation in N independent rotation planes with the rotation group $SO(d-1)$ with $N \equiv \lfloor \frac{d-1}{2} \rfloor$, where $\lfloor \ \rfloor$ denotes the integer part, and d denotes the number of spacetime dimensions. To each of these independent rotation planes there is an associated angular momentum parameter, J. Secondly, d plays a special role in MP solutions because rotation changes as we increase the number of dimensions in that there is a competition between the gravitational and centrifugal potentials. The radial fall-off of the Newtonian potential $-\frac{GM}{r^{d-3}}$ depends on the number of dimensions, whereas the centrifugal barrier $\frac{J^2}{M^2 r^2}$ does not as rotation is confined to a plane because it depends only on the magnitude of the position along the radial direction. It is readily seen that the competition between the two quantities is different for $d=4$, $d=5$, and $d \geqslant 6$. This suggests that the dimensionality will have dramatic consequences for the behaviour of black holes. For more detailed information on MP black holes and physics of black holes in higher dimensions (black rings and black objects and related issues), we refer the reader to a review by Emparan and Reall [2]. Topological classification of black hole species in higher dimensions is available in a recent review by Rodriguez [3]. Classical and thermodynamic stability of MP black holes has recently become a popular research topic following a discovery (conjecture) in 2003 by Emparan and Myers [4] that singly-spinning MP black holes will suffer ultraspinning instability. More precisely rapidly rotating MP black holes with a single spin are unstable. This type of instability would render the MP black hole behave more like a black membrane because the black hole's horizon topology is deformed along the rotational plane. Recent works in connection with the Emparan-Myers conjecture are *e.g.* [5]. It is believed that deeper information on the ultraspinning instablity of MP black hole would give rise to a more complete understanding of higher-dimensional physics itself. More recently ultraspinning instability of anti-de Sitter black holes has been studied by Dias *et al.* [6].

In this short article we report outcomes from our study on aspects of thermodynamics and extremal limits of MP black hole solutions—using a contact geometric method as a main tool. More precisely we employ the information geometric method of the Fisher-Rao type [7]. In information geometry one defines the Hessian matrix of a thermodynamic function (F) as a (information) metric tensor on the state space, namely $g^I_{ab} = \partial_a \partial_b F$. In this work F is either entropy (which gives Ruppeiner metric [8]) or internal energy/mass (which gives Weinhold metric [9]). Note also that the Ruppeiner and Weinhold geometries are conformally related with inverse temperature as a conformal factor, $g^R_{ab} = (1/T) g^W_{ab}$. This method has been widely used in the context of statistical mechanics for analyzing underlying interactions of physical systems. The salient feature of this is that for a system without underlying interactions such as the ideal gas the Ruppeiner geometry is flat, whilst the curvature singularity signals critical phenomena such as the phase transitions in the van der Waal gas [10]. In the past decade the information geometric method has also become an alternative and useful tool to analyze stability of black holes [11]. In [12] it is shown that the Ruppeiner geometric method is consistent with the Poincaré turning point method based on black hole/black ring transitions. The MP solutions we study are in d dimensions and with N number of angular momenta. More specifically we derive thermodynamic geometries (Ruppeiner and Weinhold) of the uncharged MP black hole solutions in an asymptotically flat spacetime with an arbitrary number of angular momenta. This work is an extension of the work done in [13] from a singly-spinning MP black holes to the multiply-spinning scenario.

In this work we have explicitly expressed the mass function of the MP black hole in terms of thermodynamic state parameters S and J_i.

$$M = \frac{d-2}{4} S^{\frac{d-3}{d-2}} \prod_i^n \left(1 + \frac{4J_i^2}{S^2}\right)^{\frac{1}{d-2}} . \tag{1}$$

From this equation we can derive a temperature of the MP black hole as

$$T = \frac{\frac{d-3}{4} \prod_i^n \left(1 + \frac{4J_i^2}{S^2}\right) - \frac{1}{2} \sum_k^n \frac{4J_k^2}{S^2} \prod_{i \neq k}^n \left(1 + \frac{4J_i^2}{S^2}\right)}{S^{\frac{1}{d-2}} \prod_i^n \left(1 + \frac{4J_i^2}{S^2}\right)^{\frac{d-3}{d-2}}} . \tag{2}$$

This equation can thus be used to obtain extremal limits (*i.e.* where $T = 0$) in terms of explicit thermodynamic parameters. It is worth pointing out that *in order to have an ultraspinning mode at least one spin of the MP black hole must be set to zero.* In such a mode the temperature will never reach zero, rather the temperature becomes minimal at what would otherwise be an extremal limit of the MP black hole had all the spins been turned on. In other words, there are no extremal limits for black holes that possess an ultraspinning mode. For MP black hole with spins all equal and non-zero we can express their extremal limits (bounds) in various coordinates as shown in Table 1.

We uncover new and intriguing geometric structures (see Fig. 1) of thermodynamic state spaces of MP black holes. The outcome of this investigation also allows us to conjecture that the membrane phase of ultraspinning MP black holes is reached after the MP black hole's temperature becomes minimal. This is in agreement with that found in [15] by Astefanesei, Rodriguez and Theisen. All the computations in this project were done using CLASSI [14].

d	n	Extr limits	Extr limits	Extr limits
4	1	$\frac{S}{J} \geqslant 2$	$\frac{J}{M^2} \leqslant 1$	$2 \leqslant \frac{S}{M^2} \leqslant 4$
5	1	$\frac{S}{J} \geqslant 0$	$\frac{J^2}{M^3} \leqslant \frac{2^4}{3^3}$	$0 \leqslant \frac{S^2}{M^3} \leqslant \frac{2^6}{3^3}$
5	2	$\frac{S}{J} \geqslant 2$	$\frac{J^2}{M^3} \leqslant \frac{2^2}{3^3}$	$\frac{2^4}{3^3} \leqslant \frac{S^2}{M^3} \leqslant \frac{2^6}{3^3}$
6	2	$\frac{S}{J} \geqslant \frac{2}{\sqrt{3}}$	$\frac{J^3}{M^4} \leqslant \frac{3^{\frac{3}{2}}}{2^7}$	$\frac{1}{2^4} \leqslant \frac{S^3}{M^4} \leqslant 1$
7	2	$\frac{S}{J} \geqslant 0$	$\frac{J^4}{M^5} \leqslant \frac{2^6}{5^5}$	$0 \leqslant \frac{S^4}{M^5} \leqslant \frac{2^{10}}{5^5}$
7	3	$\frac{S}{J} \geqslant \sqrt{2}$	$\frac{J^4}{M^5} \leqslant \frac{2^8}{3^3 5^5}$	$\frac{2^{10}}{3^3 5^5} \leqslant \frac{S^4}{M^5} \leqslant \frac{2^{10}}{5^5}$
8	3	$\frac{S}{J} \geqslant \frac{2}{\sqrt{5}}$	$\frac{J^5}{M^6} \leqslant \frac{5^{\frac{5}{2}}}{2^2 3^9}$	$\frac{2^3}{3^9} \leqslant \frac{S^5}{M^6} \leqslant \frac{2^6}{3^6}$
9	3	$\frac{S}{J} \geqslant 0$	$\frac{J^6}{M^7} \leqslant \frac{2^8}{7^7}$	$0 \leqslant \frac{S^6}{M^7} \leqslant \frac{2^{14}}{7^7}$
9	4	$\frac{S}{J} \geqslant \frac{2}{\sqrt{3}}$	$\frac{J^6}{M^7} \leqslant \frac{3^3}{7^7}$	$\frac{2^6}{7^7} \leqslant \frac{S^6}{M^7} \leqslant \frac{2^{14}}{7^7}$
10	4	$\frac{S}{J} \geqslant \frac{2}{\sqrt{7}}$	$\frac{J^7}{M^8} \leqslant \frac{7^{\frac{7}{2}}}{2^{27}}$	$\frac{1}{2^{20}} \leqslant \frac{S^7}{M^8} \leqslant \frac{1}{2^8}$
11	4	$\frac{S}{J} \geqslant 0$	$\frac{J^8}{M^9} \leqslant \frac{2^{10}}{3^{18}}$	$0 \leqslant \frac{S^8}{M^9} \leqslant \frac{2^{18}}{3^{18}}$
11	5	$\frac{S}{J} \geqslant 1$	$\frac{J^8}{M^9} \leqslant \frac{2^{18}}{3^{18} 5^5}$	$\frac{2^{18}}{3^{18} 5^5} \leqslant \frac{S^8}{M^9} \leqslant \frac{2^{18}}{3^{18}}$

Table 1. *Extremal limits of MP black holes in various coordinates for spins all equal and non-zero.*

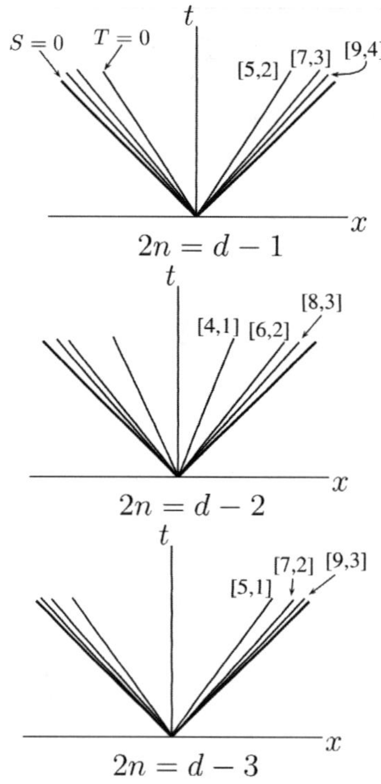

Figure 1. *Black hole solutions are presented in three series: Series 1 is when $2n = d - 1$, Series 2 when $2n = d - 2$, and Series 3 when $2n = d - 3$. For each series we have three cases as shown in the figures. Numbers in the brackets refer to $[d, n]$ with d the spacetime dimension, and n the number of nonzero spins. These graphs allow us to decide which MP black holes have genuine extremal limit in that only those that have wedge structures do possess such the limit.*

We have derived explicit thermodynamic functions of Myers-Perry black holes namely the mass formula in any dimension for the MP black holes in terms of entropy, S and angular momenta, J_i with an arbitrary number of angular momentum. We have also derived the Bekenstein-Hawking temperature and the specific heat of the MP black hole in general. The complexity of the thermodynamic metrics depends on the number of dimensions, d and the number of angular momenta, n. We establish extremal limits for the MP black holes with arbitrary n in any dimension, in other words we are able to establish the generalized Kerr bound (expressed as a function of explicit black hole state parameters) of multiply spinning Kerr black holes of equal spins in higher dimensions. We study thermodynamic geometries of the Myers-Perry black holes with arbitrary angular momenta in various dimensions and present the outcomes by drawing thermodynamic cone diagrams which capture the extremal limits of the black holes. Thermodynamic state space can be geometrically represented as a wedge embedded in Minkowski space. The opening angle of such the wedge is uniquely determined by the number of spacetime dimensions and the number of angular momenta (see Fig. 1). We believe that these results will be useful for the purpose of studying higher dimensional black holes as well as the information geometric methods themselves.

We conjecture that the membrane phase of ultraspinning MP black holes is reached at the minimum temperature in the case $2n < d - 3$ which is where the Ruppeiner curvature diverges.

Acknowledgements

Narit Pidokrajt kindly thanks Helge Ax:son Johnson Stiftelse for financial support for his participation in the ERE2010 meeting. He would also like to thank the KoF group, Fysikum, Stockholm University for its warm hospitality. We thank Roberto Emparan and Ingemar Bengtsson for useful discussions.

References

[1] Myers R C and Perry M J 1986 *Annals Phys. (N.Y.)* **172**, 304
[2] Emparan R and Reall H S 2008 Living Rev. Rel. **11**, 6
[3] Rodríguez M J 2010 On the black hole species (by means of natural selection) *Preprint arXiv:1003.2411*
[4] Emparan R and Myers R C 2003 *JHEP* **0309**, 025
[5] Emparan R, Harmark T, Niarchos V, Obers N A and Rodríguez M J 2007 *JHEP* **0710**, 110
Muruta K and Soda J 2008 *Prog. Theor. Phys.* **120**, 561
Dias O J C, Figueras P, Monteiro R, Santos J E and Emparan R 2009 *Phys. Rev. D* **80**, 111701
Dias O J C, Figueras P, Monteiro R, Reall H S and Santos J E 2010 *JHEP* **1005**, 076
Dias O J C, Figueras P, Monteiro R and Santos J E 2010 *Phys. Rev. D* **82**, 104025
[6] Dias O J C, Figueras P, Monteiro R and Santos J E 2010 Ultraspinning instability of anti-de Sitter black holes *Preprint arXiv: 1011.0996*
[7] Bengtsson I and Zyczkowski K 2006 *Geometry of Quantum States* (Cambridge University Press)
[8] Ruppeiner G 1995 *Rev. Mod. Phys.* **67**, 605
[9] Weinhold F 1975 *J. Chem. Phys.* **63**, 2479
[10] Brody D C and Hook D W 2008 *J. Phys. A* **42**, 023001
[11] Pidokrajt N 2009 *Information geometries in black hole physics*, PhD Thesis (Stockholm University, Sweden)
[12] Arcioni G and Lozano-Tellechea E 2005 *Phys. Rev. D* **72**, 104021
[13] Åman J E and Pidokrajt N 2006 *Phys. Rev. D* D **73**, 024017
[14] Åman J E 2002 *Manual for CLASSI: Classification Programs for Geometries in General Relativity*, (Department of Physics. Stockholm University, Technical Report), Provisional edition. Distributed with the sources for SHEEP and CLASSI.
[15] Astefanesei D, Rodriguez M J and Theisen S 2010 *JHEP* **1008**, 046

Spanish Relativity Meeting (ERE 2010): Gravity as a Crossroad in Physics IOP Publishing
Journal of Physics: Conference Series **314** (2011) 012028 doi:10.1088/1742-6596/314/1/012028

The Einstein field equations for cylindrically symmetric elastic configurations

I Brito[1], J Carot[2] and E G L R Vaz[3],

[1,3] Departamento de Matemática e Aplicações, Universidade do Minho, 4800-058 Guimarães, Portugal

[2] Departament de Física, Universitat de les Illes Balears, Cra Valdemossa pk 7.5, E-07122 Palma, Spain

E-mail: [1]`ireneb@math.uminho.pt`, [2]`jcarot@uib.cat`, [3]`evaz@math.uminho.pt`

Abstract. In the context of relativistic elasticity it is interesting to study axially symmetric space-times due to their significance in modeling neutron stars and other astrophysical systems of interest. To approach this problem, here, a particular class of these space-times is considered. A cylindrically symmetric elastic space-time configuration is studied, where the material metric is taken to be flat. The components of the energy-momentum tensor for elastic matter are written in terms of the invariants of the strain tensor, here chosen to be the eigenvalues of the pulled-back material metric. The Einstein field equations are presented and a condition confirming the existence of a constitutive function is obtained. This condition leads to special cases, in one of which a new system for the metric functions and an expression for the constitutive function are deduced. The new system depends on a particular function, which builds up the constitutive equation.

1. Introduction - Relativistic elasticity

General relativistic elasticity was formulated in the mid-twentieth century due to the necessity to study astrophysical problems such as deformations of neutron star crusts, which can be modelled by axially symmetric metrics. Relevant contributions to the theory of general relativistic elasticity were given, for example, by Carter and Quintana [1], Magli and Kijowski [2], Beig and Schmidt [3], Karlovini and Samuelsson [4]. The work here presented is based on Magli [5], [6] and Brito, Carot and Vaz [7], [8].

Consider a spacetime (M, g) filled with an elastic material, where M is a four-dimensional Hausdorff, simply connected manifold of class C^2 at least and g is a Lorentz metric with signature $(-,+,+,+)$. Let x^a, $a = 0,1,2,3$, denote the coordinates in M. The material space X is a three-dimensional manifold, whose points represent the particles of the material. The material space is equipped with a Riemannian metric γ, which is called the material metric. This metric measures the distance between particles, calculated in the locally relaxed state of the material. The material coordinates will be represented by y^A, $A = 1,2,3$. The configuration of the material is described by the configuration mapping $\Psi : M \longrightarrow X$, which gives rise to the rank three matrix $y^A_a = \dfrac{\partial y^A}{\partial x^a}$, the relativistic deformation gradient. The velocity field of the matter u^a, a future oriented, timelike unit vector field, which spans the one-dimensional Kernel of the relativistic deformation gradient, is defined by the conditions: $u^a y^A_a = 0$, $u^a u_a = -1$, and $u^0 > 0$. The pulled-back material metric $k_{ab} = y_a^A y_b^B \gamma_{AB}$ is orthogonal to the velocity field and can be used to measure the state of strain of the material. Thus, one can define the strain operator as $K^a{}_b = -u^a u_b + k^a{}_b$, and the relativistic strain tensor is then defined by $s_{ab} = \frac{1}{2}(k_{ab} - h_{ab}) = \frac{1}{2}(g_{ab} - K_{ab})$, where $h_{ab} = g_{ab} + u_a u_b$. The material is said to be in an unstrained state if $s_{ab} = 0$. Assuming that the internal energy of an elastic deformation, accumulated in an infinitesimal portion of the material, is invariant with respect to the space-time orientation of the material, the energy depends only on the invariants of the strain tensor. This energy is called the constitutive equation of the material and will be denoted by $v = v(I_1, I_2, I_3)$, where I_1, I_2, I_3 are scalar invariants constructed out of K_{ab}. Here we will use [6]

$$I_1 = \frac{1}{2}\left(\mathrm{Tr}K - 4\right), \quad I_2 = \frac{1}{4}\left[\mathrm{Tr}K^2 - (\mathrm{Tr}K)^2\right] + 3, \quad I_3 = \frac{1}{2}\left(\det K - 1\right). \tag{1}$$

These invariants can be written in terms of the eigenvalues of $K^a{}_b$.
The energy density ρ is defined by

$$\rho = \epsilon v(I_1, I_2, I_3), \tag{2}$$

where ϵ represents the particle number density. The energy density can also be rewritten as $\rho = \epsilon_0 \sqrt{\det K} v(I_1, I_2, I_3)$, where ϵ_0 is the particle number density of the relaxed material.

The energy-momentum tensor for elastic matter can be derived from the Lagrangian $\Lambda = \sqrt{-g}\rho$, which depends on y^A, y_a^A and x^a. The corresponding Euler-Lagrange equations are given by $\dfrac{\partial \Lambda}{\partial y^A} - \partial_a\left(\dfrac{\partial \Lambda}{\partial y_a^A}\right) = 0$. Using Noether's theorem one constructs the canonical energy-momentum tensor

$$T^a{}_b = \frac{1}{\sqrt{-g}}\frac{\partial \Lambda}{\partial y_a^A} y_b^A - \delta^a{}_b \Lambda, \tag{3}$$

which satisfies the conservation law $\nabla_a T^{ab} = 0$. The energy-momentum tensor can be rewritten in terms of the invariants of K_{ab}, see [6] for details, and the resulting expression can then be used to construct the Einstein field equations for elastic matter $G^a{}_b = 8\pi T^a{}_b$.

2. Cylindrically symmetric elastic configuration

Consider a static cylindrically symmetric space-time (M, g), whose metric g is given by the line-element

$$ds^2 = -e^{2\nu}dt^2 + e^{2\mu}dr^2 + e^{2\mu}dz^2 + e^{2\psi}d\phi^2. \tag{4}$$

The space-time coordinates are $x^a = (t, r, z, \phi)$ and ν, μ and ψ depend only on r. The associated material space X is assumed to be such that the configuration mapping ψ preserves the Killing vectors (KVs), so that if $\vec{\xi_A}$ are KVs in M, where $\vec{\xi_1} = \partial_t$, $\vec{\xi_2} = \partial_z$, $\vec{\xi_3} = \partial_\phi$, then $\psi_*(\vec{\xi_A}) = \vec{\eta_A}$ are also KVs in X. Therefore, the material metric γ is also cylindrically symmetric and it can be shown that the coordinates $y^A = (R, \zeta, \Phi)$ in X are defined by $R = R(r)$, $\zeta = z$ and $\Phi = \phi$. The material metric γ can be represented by the line-element $d\Sigma^2 = dR^2 + dz^2 + R^2 d\phi^2$, where $R = R(r)$. Here, we shall assume for simplicity that $R(r) = r$, so that γ takes the form

$$d\Sigma^2 = dr^2 + dz^2 + r^2 d\phi^2. \tag{5}$$

The velocity field of the matter turns out to be $u^a = \left(e^{-\nu(r)}, 0, 0, 0\right)$, and one can then construct the relevant tensors mentioned above, thus:

$$k^a{}_b = e^{-2\mu}\delta^a{}_1\delta^1{}_b + e^{-2\mu}\delta^a{}_2\delta^2{}_b + r^2 e^{-2\psi}\delta^a{}_3\delta^3{}_b. \tag{6}$$

and also

$$K^a{}_b = \begin{pmatrix} 1 & 0 & 0 & 0 \\ 0 & e^{-2\mu} & 0 & 0 \\ 0 & 0 & e^{-2\mu} & 0 \\ 0 & 0 & 0 & r^2 e^{-2\psi} \end{pmatrix}. \tag{7}$$

This operator has one eigenvalue equal to 1 and the other eigenvalues are

$$\eta = e^{-2\mu}, \quad \tau = r^2 e^{-2\psi}, \tag{8}$$

where η has algebraic multiplicity two. The invariants (1) can be expressed as

$$I_1 = \frac{1}{2}(2\eta + \tau - 3), \; I_2 = -\frac{1}{2}\left(\eta^2 + 2\eta\tau + 2\eta + \tau\right) + 3, \; I_3 = \frac{1}{2}\left(\eta^2\tau - 1\right). \tag{9}$$

The non-zero components of the energy-momentum tensor are then

$$\begin{aligned} T^0_{\ 0} &= -\rho, \; T^1_{\ 1} = -\rho + \frac{\partial\rho}{\partial I_3}\eta^2\tau - \frac{\partial\rho}{\partial I_2}(1+\eta+\tau)\eta + \frac{\partial\rho}{\partial I_1}\eta \\ T^2_{\ 2} &= T^1_{\ 1}, \; T^3_{\ 3} = -\rho + \frac{\partial\rho}{\partial I_3}\eta^2\tau - \frac{\partial\rho}{\partial I_2}(1+2\eta)\tau + \frac{\partial\rho}{\partial I_1}\tau. \end{aligned} \tag{10}$$

Using the fact that

$$\frac{\partial\rho}{\partial\eta} = \frac{\partial\rho}{\partial I_1} - (1+\eta+\tau)\frac{\partial\rho}{\partial I_2} + \eta\tau\frac{\partial\rho}{\partial I_3}, \quad \frac{\partial\rho}{\partial\tau} = \frac{1}{2}\frac{\partial\rho}{\partial I_1} - \left(\eta + \frac{1}{2}\right)\frac{\partial\rho}{\partial I_2} + \frac{1}{2}\eta^2\frac{\partial\rho}{\partial I_3},$$

the components of the energy-momentum tensor can be expressed simply as

$$T^0_{\ 0} = -\epsilon v, \; T^1_{\ 1} = T^2_{\ 2} = \epsilon\eta\frac{\partial v}{\partial\eta}, \; T^3_{\ 3} = 2\epsilon\tau\frac{\partial v}{\partial\tau}. \tag{11}$$

3. Einstein field equations

The Einstein field equations $G^a_{\ b} = 8\pi T^a_{\ b}$ for the cylindrically symmetric elastic configuration presented in the previous section can be written as

$$\frac{\mu'' + \psi'' + \psi'^2}{e^{2\mu}} = -\epsilon v\, 8\pi, \tag{12}$$

$$\frac{\mu'\nu' + \mu'\psi' + \nu'\psi'}{e^{2\mu}} = \epsilon\eta\frac{\partial v}{\partial\eta}\, 8\pi, \tag{13}$$

$$\frac{\nu'^2 + \nu'' + \psi'' + \psi'^2 + \nu'\psi' - \mu'\nu' - \mu'\psi'}{e^{2\mu}} = \epsilon\eta\frac{\partial v}{\partial\eta}\, 8\pi, \tag{14}$$

$$\frac{\nu'^2 + \nu'' + \mu''}{e^{2\mu}} = 2\epsilon\tau\frac{\partial v}{\partial\tau}\, 8\pi, \tag{15}$$

where primes denote derivatives with respect to r. Dividing (13) and (15) through by (12) leads to

$$\frac{\partial \ln v}{\partial E} = -\frac{\mu'\nu' + \mu'\psi' + \nu'\psi'}{\mu'' + \psi'' + \psi'^2} \quad \text{and} \quad \frac{\partial \ln v}{\partial T} = -\frac{1}{2}\frac{\nu'^2 + \nu'' + \mu''}{\mu'' + \psi'' + \psi'^2}, \tag{16}$$

respectively, where $E = \ln\eta = -2\mu$ and $T = \ln\tau = 2\ln r - 2\psi$. And since $T^2_{\ 2} = T^1_{\ 1}$, it follows that

$$2\mu'\nu' + 2\mu'\psi' - \nu'^2 - \nu'' - \psi'' - \psi'^2 = 0. \tag{17}$$

The integrability condition $\frac{\partial^2 \ln v}{\partial T \partial E} = \frac{\partial^2 \ln v}{\partial E \partial T}$ must be satisfied in order for a constitutive equation to exist. Therefrom one obtains

$$\frac{\partial}{\partial T}\left[-\frac{1}{\mu'}\left(\frac{1}{r} - \psi'\right)\right]\frac{\partial \ln v}{\partial T} = 0. \tag{18}$$

Here, we shall consider the particular case in which both factors vanish; i.e.:

$$\frac{\partial}{\partial T}\left[-\frac{1}{\mu'}\left(\frac{1}{r}-\psi'\right)\right]=0 \quad \text{and} \quad \frac{\partial \ln v}{\partial T}=0, \tag{19}$$

which implies

$$\psi=\ln(r)+k_0\mu+k_1 \quad \text{and} \quad T^3_3=0, \tag{20}$$

where k_0 and k_1 are constants. In order to avoid singularities at the axis of symmetry, one must have $e^{2\psi}=r^2L(r)$, where $L(r)\neq 0$ for $r=0$, (see [9]). Then, taking into account (17) and (20) and setting $k_0=1$, allows to write the metric functions as follows

$$\psi(r)=\ln(r)+\frac{1}{2}\ln(L), \quad \mu(r)=\frac{1}{2}\ln(L), \quad \nu(r)=-\frac{1}{4}\ln(L)\ +\ \text{constant}, \tag{21}$$

where the function $L=L(r)$ must satisfy the condition

$$6L'^2Lr-8L'''L^2r^2-8L''L^2r+16L''L'Lr^2-9L'^3r^2+8L'L^2=0, \tag{22}$$

as a consequence of the Einstein field equations (12)-(15). An example of a function $L(r)$ satisfying (22) is given by $L(r)=\exp\left(\frac{7-12r}{8r-7}\right)$. It can be easily shown that the Dominant Energy Condition is satisfied for $r>0.931855$. Since in the present case, where $T^3_3=0$, one has $\frac{\partial \ln v}{\partial r}=-2\mu'\frac{\partial \ln v}{\partial E}$, then, applying (16) one may also write the constitutive function v in terms of L as

$$v(r)=c\ \exp\left(\int\frac{L'^2}{-3L'^2r+4LL'+4L''Lr}dr\right), \tag{23}$$

where c is a constant.

Currently, the generic cases in which only one factor in (18) vanishes are under study, as well as the problem of the matching of these elastic spacetimes to vacuum or cosmological constant ones.

4. Acknowledgments

IB and EGLRV acknowledge financial support from FCT and CMAT. They also express their thanks and gratitude for the hospitality at the Universitat de les Illes Balears.

References

[1] Carter B and Quintana H 1972 *Proc. R. Soc.* **A 331** 57
[2] Magli G and Kijowski J 1992 *Gen. Rel. Grav.* **24** 139
[3] Beig R and Schmidt B 2003 *C*lass. Quantum Grav. **20** 889
[4] Karlovini M and Samuelsson L 2003 *Class. Quantum Grav.* **20** 3613
[5] Magli G 1993 *Gen. Rel. Grav.* **25** 441
[6] Magli G 1993 *Gen. Rel. Grav.* **25** 1277
[7] Brito I, Carot J and Vaz EGLR 2010 *Gen. Rel. Grav.* **42** 2357
[8] Carot J, Brito I and Vaz EGLR 2010 *J. Phys.: Conf. Series* **229** 012026
[9] Carot J, 2000 *Class. Quantum Grav.* **17** 2675

Journal of Physics: Conference Series **314** (2011) 012029 doi:10.1088/1742-6596/314/1/012029

The Hawking signal in density-density correlations in BECs

Alessandro Fabbri

Departamento de Física Teórica and IFIC, Universidad de Valencia-CSIC, C. Dr. Moliner 50, 46100 Burjassot, Spain

E-mail: afabbri@ific.uv.es

Abstract. We outline the derivation of the Hawking quanta-partner signal in correlations, and highlight the specific application to detect it in density-density correlations in BECs.

Black holes possess surprising quantum features. In the gravitational collapse of a star forming a black hole, Hawking [1] showed that the initial quantum vacuum state of matter fields gets distorted in such a way that our measurements, restricted to the exterior region, would detect a thermal state at the temperature $T_H = \frac{\kappa}{2\pi}$, where κ $(= 1/4M$ for Schwarzschild) is the horizon's surface gravity. A too tiny effect, unfortunately, since for solar mass black holes $T_H \sim 10^{-7}\ K$ is 7 orders of magnitude below the cosmic microwave background temperature.

The basic mechanism explaining the Hawking effect is pair-production in the near-horizon region, with one member escaping to infinity (the positive energy *Hawking quanta*) and the other falling into the black hole (the negative energy *partner*) [2]. We can see this effect by looking at correlations. The initial vacuum state contains local correlations that are transferred, in the course of the time evolution, to nonlocal correlations between the black hole and the external region (these are precisely the correlations Hawking quanta-partner). Such correlations, lost in the tracing operation implicit in any measurement performed in the external region, have a typical form that characterizes the Hawking effect. In the gravitational case, however, their study is of purely academic interest due to the impossibility of measuring them.

A completely different framework to test the existence and the detectability of the Hawking effect was proposed by Unruh [3], who used the mathematical equivalence between light propagation in a curved spacetime and sound propagation in inhomogeneous eulerian fluids to predict the production of a thermal flux of phonons whenever an acoustic horizon (i.e. a region separating subsonic and supersonic flow) is formed. The reasoning is the same as Hawking's, and rests on the fact that the nature of the effect is kinematical, i.e. it does not depend on the underlying dynamics but just on the details of wave propagation in a black-hole like geometry.

Unruh's motivation was to test the robustness of the Hawking effect against modifications of the relativistic (hydrodynamic in this case) theory at short distances, an analysis much more difficult to perform in the gravitational case (the so called transplanckian problem [4]).

Among the many systems proposed (see for instance [5]), Bose-Einstein condensates (BECs) offer particularly attractive experimental conditions [6]. Indeed, the huge difference between the Hawking temperature and that of the surrounding can be here reduced to just one order of magnitude. Still, it appears difficult to separate the thermal flux of the created Hawking-Unruh phonons from that of the (bigger) background.

The situation appears more promising if we look at the signature of this effect in the correlations, exploiting the crucial (and obvious) fact that in this context the acoustic origin of the horizon does not forbid correlation measurements between the exterior and the black hole.

We shall outline here a derivation of the Hawking effect for the case of a 2D conformal scalar field ϕ propagating in the two-dimensional (t, x) section of the BEC acoustic geometry

$$ds^2 = \frac{n}{mc(x,t)}\left[-(c^2(x,t) - v^2)dt^2 - 2vdtdx + dx^2 + dy^2 + dz^2\right] , \tag{1}$$

in which we consider a one-dimensional condensate with constant number density n and velocity v (< 0) along x (m is the mass of the atoms), and the only non-trivial quantity is the speed of sound $c(x, t)$, that can be tuned by varying the atom-atom scattering length.

The formation of the acoustic black hole is described by considering an initial spatially homogeneous configuration with $c_{in} = const. > |v|$ for $t < 0$ ('in' region) turning, as $t > 0$, into a nonhomogeneous one with $c(x) > |v|$ $(c(x) < |v|)$ for $x > 0$ $(x < 0)$ and an acoustic horizon (defined by $c(x) + v = 0$) at $x = 0$ ('out' region).

Our quantum field $\hat{\phi}$ is prepared in the state $|in\rangle$, which is the ground state of the decomposition

$$\hat{\phi} = \int_0^{+\infty} dw\left[\hat{a}_w^{u,in} u_w^{u,in} + \hat{a}_w^{v,in} u_w^{v,in} + h.c.\right] , \tag{2}$$

where $u_w^{u(v),in} = \frac{e^{iwu(v)^{in}}}{\sqrt{4\pi w}}$ $(u^{in} = t - \frac{x}{v+c_{in}},\ v^{in} = t + \frac{x}{c_{in}-v})$ are the positive frequency 'in' right-moving (left-moving) modes basis in the 'in' region. The modes are ortho-normalized according to the Klein-Gordon scalar product

$$(u_w, u_{w'}) = -i\int dx\left[u_w\left(\frac{\partial_t + v\partial_x}{c}\right)u_{w'}^* - u_{w'}^*\left(\frac{\partial_t + v\partial_x}{c}\right)u_w\right] \tag{3}$$

and $|in\rangle$ satisfies the condition $\hat{a}_w^{u(v),in}|in\rangle = 0$ for all w.

To find the physical consequences of the choice of $|in\rangle$ state in the 'out' region we need to evolve the 'in' modes and connect them with the 'out' modes basis. Moreover, being right-moving and left-moving modes decoupled we will only consider the u sector, which is the revelant one for our purposes.

A positive frequency 'in' mode

$$\phi_w(t < 0) = u_w^{u,in} = \frac{e^{-iwu^{in}}}{\sqrt{4\pi w}},\quad u^{in} = t - \frac{x}{v + c_{in}}, \tag{4}$$

is matched with a linear combination of both positive and negative frequency 'out' modes, $u_w^{u,out} = \frac{e^{-iwu^{out}}}{\sqrt{4\pi w}}$, $u^{out} = t - \int \frac{dx}{v+c(x)}$ (such modes are positive frequency outside the horizon $(x > 0)$ and negative-frequency inside $(x < 0)$)

$$\phi_w(t > 0) = \int_0^{+\infty} d\tilde{w}\left[\alpha_{w\tilde{w}} u_{\tilde{w}}^{u,out} + \beta_{w\tilde{w}} u_{\tilde{w}}^{u,out*}\right] \tag{5}$$

with Bogoliubov coefficients α and β. Such coefficients are given by the scalar products

$$\begin{aligned}\alpha_{w\tilde{w}} &= (\phi_w, u_{\tilde{w}}^{u,out}) = \frac{1}{2\pi}\sqrt{\frac{\tilde{w}}{w}}\int \frac{dx'}{v + c(x')} e^{-i\tilde{w}\int^{x'} \frac{dy}{v+c(y)} + \frac{iwx'}{v+c_{in}}},\\ \beta_{w\tilde{w}} &= -(\phi_w, u_{\tilde{w}}^{u,out*}) = \frac{1}{2\pi}\sqrt{\frac{\tilde{w}}{w}}\int \frac{dx'}{v + c(x')} e^{i\tilde{w}\int^{x'} \frac{dy}{v+c(y)} + \frac{iwx'}{v+c_{in}}} .\end{aligned} \tag{6}$$

It can be shown that the global solution (4)-(6) satisfies the two-dimensional Klein-Gordon equation $\nabla^2\phi = 0$ including the matching conditions at $t = 0$ ([ ] means variation across the jump at $t = 0$)

$$[\phi] = 0 \ , \quad \left[\frac{(\partial_t + v\partial_x)\phi}{c}\right] = 0 \ , \tag{7}$$

i.e.

$$\frac{e^{\frac{iwx}{v+c_{in}}}}{\sqrt{w}} = \int_0^{+\infty} \frac{d\tilde{w}}{\sqrt{\tilde{w}}} \left[\alpha_{w\tilde{w}} e^{i\tilde{w}\int^x \frac{dy}{v+c(y)}} + \beta_{w\tilde{w}} e^{-i\tilde{w}\int^x \frac{dy}{v+c(y)}}\right] , \tag{8}$$

$$\frac{\sqrt{w}e^{\frac{iwx}{v+c_{in}}}}{v+c_{in}} = \frac{1}{v+c(x)} \int_0^{+\infty} d\tilde{w}\sqrt{\tilde{w}} \left[\alpha_{w\tilde{w}} e^{i\tilde{w}\int^x \frac{dy}{v+c(y)}} - \beta_{w\tilde{w}} e^{-i\tilde{w}\int^x \frac{dy}{v+c(y)}}\right] \tag{9}$$

(the analysis of the four dimensional Klein-Gordon equation in the full acoustic geometry (1) is more involved because right-moving and left-moving modes are no more decoupled).

Eq. (5) allows us to relate the operators of the 'in' decomposition (2) with the 'out' one

$$\hat{\phi} = \int_0^{+\infty} d\tilde{w} \left[\hat{a}_{\tilde{w}}^{u,ext} u_{\tilde{w}}^{u,ext} + \hat{a}_{\tilde{w}}^{u,bh\dagger} u_{\tilde{w}}^{u,bh} + \hat{a}_{\tilde{w}}^{v,out} u_{\tilde{w}}^{v,out} + h.c.\right] \tag{10}$$

and in particular we get

$$\hat{a}_{\tilde{w}}^{u,ext} = \int dw \left[\alpha_{w\tilde{w}}^{(x>0)} \hat{a}_w^{u,in} + \beta_{w\tilde{w}}^{(x>0)*} \hat{a}_w^{u,in\dagger}\right], \hat{a}_{\tilde{w}}^{u,bh\dagger} = \int dw \left[\alpha_{w\tilde{w}}^{(x<0)} \hat{a}_w^{u,in} + \beta_{w\tilde{w}}^{(x<0)*} \hat{a}_w^{u,in\dagger}\right] . \tag{11}$$

Restricting to the exterior of the horizon it is not difficult to show, using the near-horizon expansion $c(x) \sim -v + \kappa x$ with κ the horizon's surface gravity to evaluate explicitly the main contribution to the Bogoliubov coefficients

$$\alpha_{w\tilde{w}}^{(x>0)} \sim \frac{1}{2\pi\kappa}\sqrt{\frac{\tilde{w}}{w}} \left[\frac{-iw}{\kappa(v+c_{in})}\right]^{\frac{i\tilde{w}}{\kappa}} \Gamma(-\frac{i\tilde{w}}{\kappa}) \ , \ \beta_{w\tilde{w}}^{(x>0)} \sim \frac{1}{2\pi\kappa}\sqrt{\frac{\tilde{w}}{w}} \left[\frac{-iw}{\kappa(v+c_{in})}\right]^{-\frac{i\tilde{w}}{\kappa}} \Gamma(\frac{i\tilde{w}}{\kappa}) \ , \tag{12}$$

that at late retarded times u the state $|in\rangle$ is thermally populated of Hawking-Unruh quanta

$$\langle in|\hat{a}_{\tilde{w}}^{u,ext\dagger} a_{\tilde{w}}^{u,ext}|in\rangle = \lim_{w'\to w} \int dw \beta_{w\tilde{w}}^{(x>0)} \beta_{w\tilde{w}'}^{(x>0)*} = \frac{\delta(0)}{e^{\frac{2\pi\tilde{w}}{\kappa}} - 1} \ . \tag{13}$$

The Dirac δ, associated to the use of continuum plane wave modes, would disappear if we considered wave-packets instead.

The possibility of exploring the interior region as well allows us to study the correlations between the emitted quanta and their partners inside the horizon. To determine the symmetric two-point function $\langle in|\{\hat{\phi}(t,x), \hat{\phi}(t',x')\}|in\rangle \equiv \frac{1}{2}\langle in|\hat{\phi}(t,x)\hat{\phi}(t',x') + \hat{\phi}(t',x')\hat{\phi}(t,x)|in\rangle$ for, say, $x < 0$ and $x' > 0$ we need also the Bogoliubov coefficients in the interior region

$$\alpha_{w\tilde{w}}^{(x<0)} \sim -\frac{1}{2\pi\kappa}\sqrt{\frac{\tilde{w}}{w}} \left[\frac{iw}{\kappa(v+c_{in})}\right]^{\frac{i\tilde{w}}{\kappa}} \Gamma(-\frac{i\tilde{w}}{\kappa}) = -e^{-\frac{\pi w}{\kappa}} \alpha_{w\tilde{w}}^{(x>0)} \ ,$$

$$\beta_{w\tilde{w}}^{(x<0)} \sim -\frac{1}{2\pi\kappa}\sqrt{\frac{\tilde{w}}{w}} \left[\frac{iw}{\kappa(v+c_{in})}\right]^{-\frac{i\tilde{w}}{\kappa}} \Gamma(\frac{i\tilde{w}}{\kappa}) = -e^{\frac{\pi w}{\kappa}} \beta_{w\tilde{w}}^{(x>0)} \ , \tag{14}$$

and we get (we select only the relevant u-sector contribution)

$$\langle in|\{\hat\phi(t,x),\hat\phi(t',x')\}|in\rangle =$$
$$\frac{1}{2}\int d\tilde{w}d\tilde{w}'dw\left[u^{u,bh}_{\tilde{w}}\alpha^{(x<0)}_{w\tilde{w}}+u^{u,bh*}_{\tilde{w}}\beta^{(x<0)}_{w\tilde{w}}\right](t,x)\left[u^{u,ext*}_{\tilde{w}'}\alpha^{(x>0)*}_{w\tilde{w}'}+u^{u,ext}_{\tilde{w}'}\beta^{(x>0)*}_{w\tilde{w}'}\right](t',x')+c.c.$$
$$=\frac{1}{2}\int d\tilde{w}d\tilde{w}'dw u^{u,bh}_{\tilde{w}}(t,x)u^{u,ext*}_{\tilde{w}'}(t',x')\left[\alpha^{(x<0)}_{w\tilde{w}}\alpha^{(x>0)*}_{w\tilde{w}'}+\beta^{(x<0)*}_{w\tilde{w}}\beta^{(x>0)}_{w\tilde{w}'}\right]+c.c.$$
$$=-\int\frac{d\tilde{w}}{4\pi\tilde{w}}\frac{\cos\tilde{w}(u^{bh}-u^{ext})}{\sinh\frac{\pi\tilde{w}}{\kappa}}=-\frac{1}{4\pi}\ln\cosh\frac{\kappa(u^{bh}-u^{ext})}{2}, \tag{15}$$

where $u^{bh}=t-\int^{x}\frac{dy}{v+c(y)}$ (with $x<0$) and $u^{ext}=t'-\int^{x'}\frac{dy'}{v+c(y')}$ $(x'>0)$.

Going specifically to the BEC case, the quantity of interest in the experiments is the one time, normalized correlation function of the density fluctuations $G^{(2)}(t;x,x')=\frac{1}{n^2}\langle\hat{n}_1(t,x)\hat{n}_1(t',x')\rangle|_{t=t'}$. In the hydrodynamical approximation $\hat{n}_1=-\frac{n}{mc^2}(\partial_t+v\partial_x)\hat\theta^1$ and the phase operator $\hat\theta_1$ is well approximated, up to a proportionality factor, by our field $\hat\phi$ ($\hat\theta^1\sim\sqrt{\frac{mc}{n}}\hat\phi$, see [7] for more details). If we consider points sufficiently far away so that the speed of sound is well approximated with its asymptotic values $c(x)\to c_l$, $c(x')\to c_r$ a straightforward calculation, using the crucial result (15), gives

$$G^{(2)}(t;x,x')=\frac{\kappa^2}{16\pi mn(c_rc_l)^{3/2}}\frac{c_rc_l}{(v+c_l)(v+c_r)}\frac{1}{\cosh^2[\frac{\kappa}{2}(\frac{x}{v+c_l}-\frac{x'}{v+c_r})]}\ . \tag{16}$$

These correlations characterize neatly the Hawking effect, namely the dependence on the surface gravity in the height and width and in particular the stationary peak at $\frac{x}{v+c_l}=\frac{x'}{v+c_r}$ identifying pairs of quanta created just inside and just outside the horizon at all times after its formation and traveling in opposite directions.

The robustness of these correlations has been numerically observed in an analysis carried out using the microscopic theory in [8], where it has been also shown that this effect is still present and clearly visible even in the presence of a (bigger) thermal background. Inserting numbers for existing experiments, we have correlations of the order of 10^{-3}, small but not negligible. Proposals to amplify the signal by a factor 10 by following the formation of the acoustic black hole with a period of free expansion may be important for the experimental search [9]. These facts, together with the vast technology available nowadays, indicate that density correlation measurements are very promising for an experimental verification of the Hawking-Unruh effect in the near future.

Acknowledgements: It is a pleasure to thank R. Balbinot, I. Carusotto and A. Recati for many useful and interesting discussions.

References

[1] S. W. Hawking, Nature **248**, 30 (1974); Commun. Math. Phys. **43**, 199 (1975).
[2] S. W. Hawking, 'The Quantum Mechanics Of Black Holes," Sci. Am. **236**, 34 (1977).
[3] W. G. Unruh, Phys. Rev. Lett. **46**, 1351 (1981).
[4] T. Jacobson, Phys. Rev. D **44**, 1731 (1991).
[5] C. Barcelo, S. Liberati and M. Visser, Living Rev. Rel. **8**, 12 (2005)
[6] L.J. Garay, J.R. Anglin, J.I. Cirac and P. Zoller, Phys. Rev. Lett. **85**, 4643 (2000)
[7] R. Balbinot, A. Fabbri, S. Fagnocchi, A. Recati and I. Carusotto, Phys. Rev. A **78**, 021603 (2008)
[8] I. Carusotto, S. Fagnocchi, A. Recati, R. Balbinot and A. Fabbri, New J. Phys. **10**, 103001 (2008)
[9] E. Cornell, talk presented at the worskshop "Towards the observation of Hawking radiation in condensed matter systems" (www.uv.es/workshopEHR), IFIC (Valencia), February 2009

On the robustness of acoustic black hole spectra

S Finazzi[1] **and R Parentani**[2]

[1] SISSA, via Bonomea 265, Trieste 34151, Italy; INFN sezione di Trieste
[2] Laboratoire de Physique Théorique, CNRS UMR 8627, Bât. 210, Université Paris-Sud 11, 91405 Orsay Cedex, France

E-mail: finazzi@sissa.it and renaud.parentani@th.u-psud.fr

Abstract.
We study the robustness of the spectrum emitted by an acoustic black hole by considering series of stationary flows that become either subsonic or supersonic, i.e. when the horizon disappears. We work with the superluminal Bogoliubov dispersion of Bose–Einstein condensates. We find that the spectrum remains remarkably Planckian until the horizon disappears. When the flow is everywhere supersonic, new pair creation channels open. This will be the subject of a forthcoming work.

1. Introduction

In the hydrodynamic approximation, i.e. for long wavelengths, the propagation of sound waves in a moving fuid is analogous to that of light in a curved spacetime [1, 2]. More precisely, a one dimensional flow defines the acoustic metric

$$ds^2 = -c^2dt^2 + (dx - v\,dt)^2 \,, \tag{1}$$

where v is the flow velocity and c the speed of sound. Assuming that the fluid flows from right to left ($v < 0$), a sonic horizon is present where $w = c + v$ crosses 0. This situation describes a black (white) hole horizon when $\kappa_{\rm K} \equiv \partial_x w$ evaluated at $w = 0$ is positive (negative). One thus expects that such system would emit a Hawking flux, i.e. a thermal flux of phonons at a temperature given by $T_{\rm H} = \kappa_{\rm K}/2\pi$ (in units where $k_B = \hbar = 1$).

However, since Hawking radiation relies on short wavelength modes [3, 4], the dispersion of sound waves, which is neglected in the hydrodynamic approximation, must be taken into account. Following the original work [5], this was done using analytical [6, 7, 8, 9] and numerical [10, 11, 12, 13] methods. Provided the dispersive scale ξ, the "healing length", is much smaller than the surface gravity scale c/κ, it was found that the thermal properties of the spectrum are extremely robust. In fact, so robust that it is very difficult to characterize and analyze the small deviations with respect to the standard flux.

Our aim is to complete the analysis of [12, 13] where it was shown that the spectrum remains Planckian whenever $T_{\rm H} \lesssim \omega_{\rm max}/15$, where $\omega_{\rm max}$ is a critical frequency which scales as $1/\xi$ but, more importantly, also depends on $w(x = -\infty)$, the asymptotic value of $c + v$. Since this result was obtained using symmetrical profiles like $w \sim D \tanh(\kappa x/D)$, one may wonder whether the robustness of the spectrum was not partially due to the symmetry of the flow w.r.t. the horizon, namely $w(-x) = -w(x)$. To investigate this question, we shall consider flows which are highly

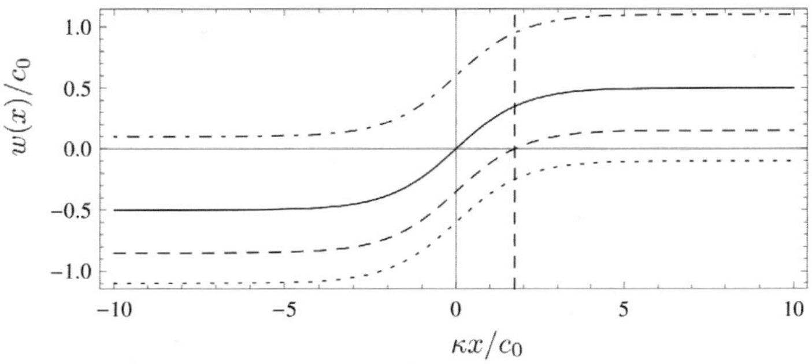

Figure 1. Four velocity profiles $w = c + v$ of Eq. (2) with the same height D but four values of D_1. For $D_1 = 0$ (solid line), one recovers the symmetric case of [12] with the Killing horizon (solid vertical line) at $x = 0$. For $-D < D_1 < 0$ (dashed line), the horizon (dashed vertical line) is shifted to $x > 0$. In the last two cases, there is no horizon: for $D_1 < -D$ (dotted line), the flow is everywhere supersonic, and for $D_1 > D$ (dotdashed line), it is subsonic.

asymmetric. In particular we shall study the spectrum in the interesting limiting cases where the horizon disappears because the flux becomes either sub or supersonic. In parallel with this work, we investigated in [14] other classes of profiles generalizing those of [12, 13].

2. Phonon spectra in asymmetric flows

We consider elongated condensates, stationary flowing along the longitudinal direction x. We assume that the transverse dimensions are small enough that the relevant phonon excitations are longitudinal. The system is thus effectively described by the $1+1$ dimension metric (1) where v and c depend only on x. We work with flows $w = v + c$ given by

$$\frac{w(x)}{c_0} = D_1 + D \tanh\left(\frac{\kappa x}{D c_0}\right). \tag{2}$$

D fixes the gap $[w(\infty) - w(-\infty)]/2c_0$, while D_1, which was taken to be 0 in [12, 13], fixes the asymmetry between sub and supersonic regimes. When $D_1 > D$, the flow is everywhere subsonic, and when $D_1 < -D$, it is supersonic, see Fig. 1. Instead, when $|D_1| < D$, there is a sonic (Killing) horizon where $w = 0$, localized at

$$x_{\rm H} = -\frac{D c_0}{\kappa} \operatorname{arctanh}\left(\frac{D_1}{D}\right). \tag{3}$$

It separates the sub ($x > x_{\rm H}$) from the supersonic region ($x < x_{\rm H}$). In such a flow, when ignoring dispersion, the spectrum of upstream phonons spontaneously emitted from the horizon would be very simple, and would strictly correspond to the Hawking radiation [1, 5]: It would follow a Planck law at Hawking temperature $T_{\rm H} = \kappa_{\rm K}/2\pi$. The surface gravity $\kappa_{\rm K}$ is here

$$\kappa_{\rm K} \equiv \partial_x (c+v)|_{x=x_H} = \kappa \left[1 - \left(\frac{D_1}{D}\right)^2\right]. \tag{4}$$

Taking dispersion into account, the spectrum becomes much more complicated. In this paper, we consider phonons in Bose–Einstein condensates (BEC) with the Bogoliubov dispersion

$$(\omega - vk)^2 = \Omega^2(k) = c^2 k^2 + \frac{\hbar^2 k^4}{4m^2} = c^2 k^2 \left(1 + \frac{\xi^2 k^2}{2}\right), \tag{5}$$

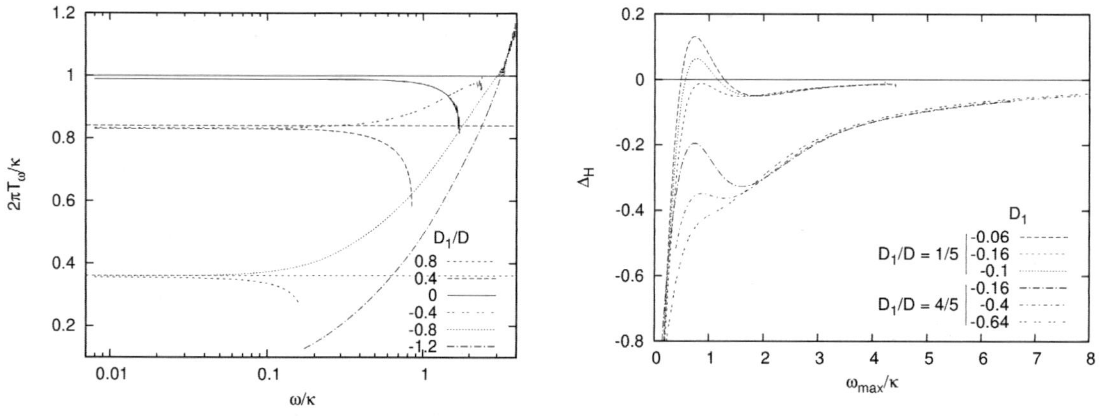

Figure 2. Left panel: T_ω for 6 values of D_1/D, with $D = 0.5$ and $\kappa\xi/\sqrt{2}c = 0.1$ fixed. The horizontal lines represent $T_{\rm H}(D_1)$ of Eq. (4), respectively for $|D_1| = 0, 0.2, 0.4$. For $D_1 < -D$ the curve is truncated at $\omega_{\rm min}$ as explained in the text. Right panel: $\Delta_{\rm H}$ as a function of $\omega_{\rm max}/\kappa$ for two values of D_1/D and various D_1. When $\omega_{\rm max}/\kappa$ is large enough, deviations from Planckianity are small and, more importantly, governed by the ratio D_1/D only.

where ω is the lab conserved frequency. In the sequel, following the numerical procedure of [13], we study the deviations of n_ω, the phonon spectrum, w.r.t. the standard Planck one, by varying the offset parameter D_1 while keeping fix all the other parameters.

To characterize these deviations, we use the temperature function T_ω defined by

$$n_\omega \equiv \frac{1}{\exp(\omega/T_\omega) - 1}, \tag{6}$$

which is constant when the spectrum is Planckian. This quantity thus allows a direct comparison with the Hawking temperature $T_{\rm H} = \kappa_{\rm K}/2\pi$ of Eq. (4). In Fig. 2, T_ω and $T_{\rm H}$ (horizontal lines) are plotted for various values of D_1. We distinguish four regimes:

- For $|D_1|$ sufficiently smaller than D, the spectrum is Planckian until ω approaches $\omega_{\rm max}$, where it vanishes. For $D_1 > 0$, $\omega_{\rm max}$ decreases since $|w(x = -\infty)|$ does so, and conversely for $D_1 < 0$. However for both signs, at low frequency T_ω closely follows $\kappa_{\rm K}/2\pi$ of (4).
- When $|D_1| \to D$, $\kappa_{\rm K}$ of (4) drops down to 0 and the sonic horizon disappears. In this critical regime, deviations from Planckianity appears even at low frequency.
- For $D_1 > D$, the flow is everywhere subsonic. There is no particle production because there are no negative norm modes with $\omega > 0$.
- For $D_1 < -D$, the flow is supersonic. A new critical frequency $\omega_{\rm min} < \omega_{\rm max}$ appears. For $0 < \omega < \omega_{\rm min}$, there are now 4 asymptotic in and out modes, and the scattering matrix is 4×4. Since the code of [13] is designed to the 3×3 case, this regime requires a new code [15]. However, for $\omega_{\rm min} < \omega < \omega_{\rm max}$ there are only 3 modes, and our code can handle this frequency band. In this range, as can be seen in the left panel, n_ω is close to the case where $D_1 > -D$, even though the spectrum is not at all Planckian.

Furthermore, we checked that the *shape* of the spectrum is hardly changed when varying the healing length ξ. Namely, spectra calculated with different values of ξ almost coincide when T_ω is plotted versus the rescaled quantity $\omega/\omega_{\rm max}(\xi)$. To identify the parameters governing the deviations from Planckianity we define

$$\Delta_{\rm H} \equiv \frac{f(\omega = T_{\rm H}) - f_{T_{\rm H}}(\omega = T_{\rm H})}{f_{T_{\rm H}}(\omega = T_{\rm H})}, \tag{7}$$

where $f(\omega) \equiv \omega n_\omega$ is the actual energy flux and $f_{T_{\rm H}}(\omega)$ is the thermal flux at temperature $T_{\rm H}$. In Fig. 2, right panel, $\Delta_{\rm H}$ is plotted as a function of $\omega_{\rm max}(\xi)/\kappa$. Various values of D_1 are used, while the ratio D_1/D is kept constant respectively for the three lower curves ($D_1/D = -1/5$), and for the three upper curves ($D_1/D = 4/5$). As expected, $\Delta_{\rm H}$ goes to 0 for large values of $\omega_{\rm max}/\kappa$. What we learn here is that the leading deviations from Planckianity do not depend separately on D_1 or D but only on their ratio.

3. Comments and conclusions

In this work, we have studied the robustness of the properties of black hole radiation in BEC by considering flows characterized by a high asymmetry between the sub and the supersonic region. We found that the spectrum remains Planckian until one approaches the critical cases ($|D_1| \to D$) where the surface gravity (4) vanishes.

In fact, for $0 < D_1 \to D$, since the fluid flows in the supersonic region with a velocity just above that of sound, the cutoff frequency $\omega_{\rm max}$ goes to zero. Hence the range of frequencies where the temperature T_ω is nearly constant shrinks and disappears in the limit. As expected, no emission of particles is found when the flow remains everywhere subsonic.

The situation is much more interesting in the opposite case, for $0 > D_1 \to -D$, when pushing the velocity of the subsonic region towards the speed of sound. In this regime, even though $\kappa_{\rm K}$ of (4) goes again to 0, $\omega_{\rm max}$ now increases. As a result, the low frequency plateau of T_ω at low temperature again shrinks, but the spectrum becomes more and more blue. This can be understood from the fact that more dispersive modes, i.e. modes with higher ω, have still their turning point near $x = 0$ where the gradient of w is $\kappa \gg \kappa_{\rm K}$.

When $D_1 < -D$, there is no sonic horizon since the flow is everywhere supersonic. Yet the spectrum is continuously deformed even though the Planck character is completely lost in the frequency range $\omega_{\rm min} < \omega < \omega_{\rm max}$, where $\omega_{\rm min}$ is a new critical frequency. For $\omega < \omega_{\rm min}$, new scattering channels open because there is a fourth asymptotic mode. Investigations of this case [15] are in progress.

To conclude, we stress that the thermal spectrum at the naïve surface gravity (4) provides a reliable approximation of the actual phonon spectrum in all cases but the critical ones where the sonic horizon disappears. In these extreme cases, the spectrum is no longer thermal and its properties are governed by the dispersive properties of field.

Acknowledgments

We are grateful to A. Coutant and S. Liberati for interesting discussions.

References

[1] Unruh W G 1981 *Phys. Rev. Lett.* **46** 1351–1353
[2] Barcelo C, Liberati S and Visser M 2005 *Living Rev. Rel.* **8** 12
[3] Jacobson T 1991 *Phys. Rev. D* **44** 1731–1739; 1993 *Phys. Rev. D* **48** 728–741
[4] Brout R, Massar S, Parentani R and Spindel P 1995 *Phys. Rept.* **260** 329–454
[5] Unruh W G 1995 *Phys. Rev. D* **51** 2827–2838
[6] Brout R, Massar S, Parentani R and Spindel P 1995 *Phys. Rev. D* **52** 4559–4568
[7] Corley S 1998 *Phys. Rev. D* **57** 6280–6291
[8] Himemoto Y and Tanaka T 2000 *Phys. Rev. D* **61** 064004
[9] Unruh W G and Schutzhold R 2005 *Phys. Rev. D* **71** 024028
[10] Corley S and Jacobson T 1996 *Phys. Rev. D* **54** 1568–1586
[11] Unruh W G 2007 *PoS* **QG-PH** 039
[12] Macher J and Parentani R 2009 *Phys. Rev. D* **79** 124008
[13] Macher J and Parentani R 2009 *Phys. Rev. A* **80** 043601
[14] Finazzi S and Parentani R 2010 (*Preprint* `arXiv:1012.1556[gr-qc]`)
[15] Coutant A, Finazzi S, Liberati S and Parentani R *in preparation*

Journal of Physics: Conference Series **314** (2011) 012031 doi:10.1088/1742-6596/314/1/012031

Horizon effects for surface waves in wave channels and circular jumps

G. Jannes[1], R. Piquet, J. Chaline, P. Maïssa, C. Mathis, G. Rousseaux

Université de Nice Sophia Antipolis, Laboratoire J.-A. Dieudonné, UMR CNRS-UNS 6621, Parc Valrose, 06108 Nice Cedex 02, France

E-mail: gil.jannes@unice.fr

Abstract. Surface waves in classical fluids experience a rich array of black/white hole horizon effects. The dispersion relation depends on the characteristics of the fluid (in our case, water and silicon oil) as well as on the fluid depth and the wavelength regime. In some cases, it can be tuned to obtain a relativistic regime plus high-frequency dispersive effects. We discuss two types of ongoing analogue white-hole experiments: deep water waves propagating against a counter-current in a wave channel and shallow waves on a circular hydraulic jump.

1. Introduction

Surface waves in classical fluids provide a natural and rich class of black/white hole analogues. Two familiar examples are the blocking of sea waves at a river mouth and the approximately circular jump created by opening the tap in a kitchen sink. We reproduce these two types of white hole analogues in controlled laboratory settings in order to study the associated horizon effects and their possible lessons for relativity (and vice versa: lessons from relativity for fluid mechanics). The river-mouth example corresponds to deep water waves propagating against a counter-current in a wave channel, while the kitchen-sink example corresponds to shallow waves on a circular hydraulic jump.

The general dispersion relation for capillary-gravity surface waves propagating against a counter-current of velocity U is

$$(\omega - Uk)^2 = \left(gk + \frac{\gamma}{\rho}k^3\right)\tanh(kH), \tag{1}$$

with g the gravitational constant, γ the surface tension, ρ the density, H the fluid depth, and as usual ω and k are the frequency and wavenumer, respectively. By developing the two extreme cases $kH \ll 1$ and $kH \gg 1$ one obtains the deep water and the shallow water limit which are applicable to the wave channel and the circular jump, respectively.

2. Deep water waves in a wave channel

The deep water case gives $(\omega - Uk)^2 \approx gk + \frac{\gamma}{\rho}k^3$. The gravity wave limit is obtained by neglecting the term in k^3, and has a white hole horizon when $U = \sqrt{g/k}$. The deep water case

[1] Speaker

Published under licence by IOP Publishing Ltd

with inclusion of capillarity possesses a tremendously rich phenomenology which includes not only white hole horizons, but also additional horizons such as a negative horizon (a blocking line for waves with a negative co-moving frequency) and a blue horizon (a blocking line for waves blue-shifted due to the effect of surface tension). A first series of experiments which led to the observation of negative-frequency waves was described in [1]. A theoretical development of the analogy with black/white hole physics as well as rainbow physics in the context of dynamical systems theory can be found in [2], while [3] contains an in-depth theoretical treatment of the dispersion relation and the associated horizon effects. Here, we limit ourselves to mentioning that there exist two possible scenarios in which the white hole horizon can be crossed:

- a double-bouncing scenario in which an incoming wave bounces back at the white hole horizon, then bounces forward again at the blue horizon, after which it is sufficiently blueshifted to cross the white hole horizon;
- a direct dispersive penetration in which an incoming wave of sufficiently high frequency penetrates directly through the white hole horizon, in spite of the presence of a counter-current which blocks all surface waves in the gravity wave limit.

It should be noted that in the gravity wave limit, the dispersion is "subluminal", since the group velocity $c = \sqrt{g/k}$ decreases with k. The white hole horizon is then a strict one-way membrane for pure gravity waves. The inclusion of surface tension is therefore crucial for both horizon penetration scenarios just mentioned. We refer to [3] for further details, and focus on the shallow-water case of the circular jump in the remainder of this text.

3. The circular hydraulic jump

When a vertical fluid jet impacts on a horizontal plate with a sufficient flow rate, it will form a thin layer near the impact zone, which expands radially and at a certain distance forms a sudden circular hydraulic jump.

3.1. *Mach cone experiment*

We have focused on the behaviour of surface waves propagating inward against the fluid flow of the jump itself, and in particular on the question of whether and where they are blocked. Since the propagation of surface waves on the circular jump can be described in terms of an effective Painlevé-Gullstrand metric, the location where the surface waves are blocked forms the hydrodynamical analogue of a white hole horizon. The above question could be answered by comparing the radial fluid velocity v_s^r at the surface and the propagation velocity c of the surface waves. However, given that there exist no satisfactory simultaneous measurements of both quantities, we have opted for a simpler alternative, which allows us to directly determine the ratio between them. Our method is based on the Mach cone well known in the case of sound waves, see Fig. 1. There, an object propagating at a speed V above the speed of sound c_s leaves behind an observable cone, the Mach cone. This is formed by the envelope of the subsequent wavefronts emitted by the object, which partially escape from each other. The half-angle θ of the cone obeys $\sin\theta = ct/|V|t = c/|V| = 1/M$, with M the Mach number, see Fig. 1. For an object propagating at a speed $V < c_s$, the subsequent wavefronts remain inside the previous ones and no Mach cone is formed.

Exactly the same argument can be applied to the case of an object standing still at the surface of a fluid flow. By comparing the fluid flow velocity v_s^r to the propagation speed of surface waves c, the following cases can occur:

- Supercritical region: $v_s^r > c \;\Rightarrow\; \theta \in \left[0, \frac{\pi}{2}\right]$
- Subcritical region: $v_s^r < c \;\Rightarrow\; \theta$ complex; Mach cone disappears
- Horizon: $v_s^r = c \;\Rightarrow\; \theta = \frac{\pi}{2}$

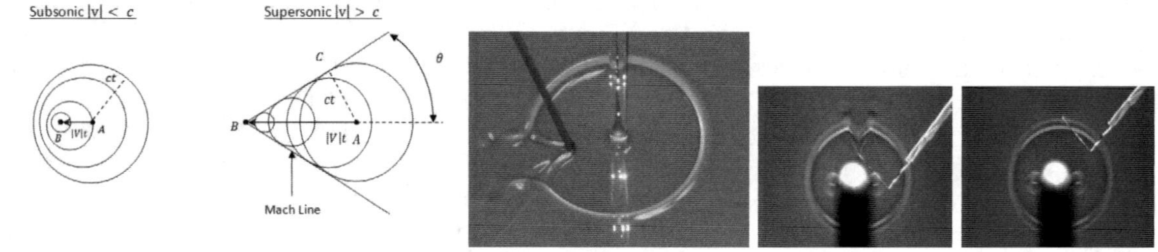

Figure 1. From left to right: Wavefronts emitted in subsonic/subcritical regime. Formation of Mach cone in supersonic/supercritical regime. Mach cone in circular jump (our experiments). Mach cone in circular jump, photo taken with high-speed camera. Disappearance of Mach cone just outside the jump.

Our experiment to demonstrate the presence of a hydrodynamic horizon is described in [4]. Essentially, we have pumped silicon oil through a steel nozzle onto a horizontal PVC plate. A needle was placed such as to penetrate the flow surface at varying distances from the centre of the circular jump, where the oil jet impacts on the PVC plate. For each position of the needle, we haved photographed the setup with a high-speed camera and measured the corresponding Mach angle θ. The resulting angles and the derived ratio v_r^s/c are shown in Fig. 2.

These results provide a clear proof that the circular hydraulic jump constitutes a two-dimensional hydrodynamic white hole: surface waves travelling at a velocity c from outside the jump are trapped outside the jump in precisely the same sense as light is trapped inside a gravitational black hole. The corresponding white-hole horizon is situated precisely at the radius of the jump itself. The following features of this hydrodynamic white hole are particularly striking, especially in comparison with other current or planned experiments, e.g. in optics or Bose-Einstein condensates:

- The white hole here is created "spontaneously". One only needs to arrange for a sufficient fluid flow rate, but no extraordinary engineering is required. In fact, it suffices in principle to open the tap in a kitchen sink to observe such a white hole. All other ingredients in the experimental setup (the choice of silicon oil, the precision of the pump etc) serve merely to make the experiment cleaner and free of perturbations and other undesired effects, but do not affect the essential point which is the creation of the hydrodynamic white hole.
- The white hole can be observed with the naked eye: the location of the white hole horizon is precisely where the fluid undergoes the characteristic jump.

Finally, let us go back to dispersion relation for surface waves and examine in which dispersive regime the circular jump typically lies.

3.2. Dispersion relation

Developing Eq. (1) for the case $kH \ll 1$ and truncating at $\mathcal{O}(k^4)$, one obtains

$$
\begin{aligned}
(\omega - Uk)^2 &\approx gHk^2 + \left(\frac{\gamma H}{\rho} - \frac{gH^3}{3}\right)k^4 + \mathcal{O}(k^6) \\
&= c^2k^2 + c^2\left(l_c^2 - \frac{H^2}{3}\right)k^4 + \mathcal{O}(k^6),
\end{aligned}
$$

where $l_c = \sqrt{\gamma/g\rho}$ is the capillary length ($l_c = 1.49$mm for the silicon oil in our experiments). Contrarily to the deep-water case of the wave channel, the shallow-water regime is relativistic

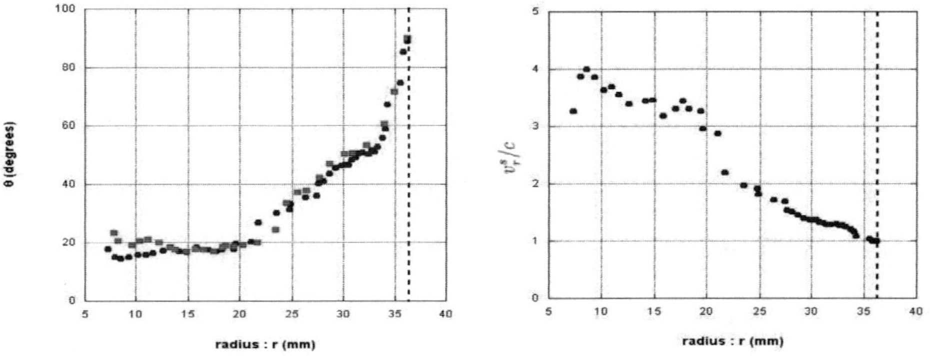

Figure 2. Left: Mach angle θ as a function of the distance r from the centre of the jump for two values of the external fluid height H (blue circles: $H = 0$mm, red squares: $H = 3$mm, respectively). Right: v_r^s/c as a function of r. The dashed vertical line represents the jump radius R_j. Experimental parameters: see [4].

at low values of k (the associated "relativistic speed" c is $c = \sqrt{gH}$, i.e. the surface wave velocity in the low-k or gravity limit where capillarity is negligible). Moreover, since the typical heights of the fluid inside the jump are very small (certainly smaller than $\sqrt{3}l_c$, both in our experiments and in other experiments reported in the literature), one is tempted to conclude that the circular jump should exhibit superluminal dispersion. Calculations of the the group velocity $c_g \equiv \partial\omega/\partial k$ from the complete dispersion relation confirm that the circular jump is superluminal (c_g increasing with k) for realistic experimental parameters. In such a superluminal regime, sufficiently high-frequency modes can penetrate across the horizon in the classically prohibited sense. The interior of the black/white hole is then no longer causally separated from the outside. Such superluminal horizon-crossing effects are considered in several scenarios for quantum gravity phenomenology. They are particularly interesting in the sense that they put the robustness of several aspects of black hole physics (and in particular, Hawking radiation) to the test [5]. This implies that it should be possible to test some of these issues associated to the robustness of semiclassical gravity with respect to transplanckian physics in the circular hydraulic jump.

We will report in the near future about further experiments to study the interaction between the circular jump and incoming surface waves.

References

[1] Rousseaux G, Mathis C, Maïssa P, Philbin T G and Leonhardt U 2008 *New J. Phys.* **10** 053015.
[2] Nardin J C, Rousseaux G and Coullet P 2009 *Phys. Rev. Lett.* **102** 124504.
[3] Rousseaux G, Maïssa P, Mathis C, Coullet P, Philbin T G and Leonhardt U 2010 *New J. Phys.* **12**.
[4] Jannes G, Piquet R, Maïssa P, Mathis C and Rousseaux G 2010 (*Preprint* arXiv:1010.1701).
[5] Barceló C, Garay L J and Jannes G 2009 *Phys. Rev.* **D79** 024016.

Journal of Physics: Conference Series **314** (2011) 012032 doi:10.1088/1742-6596/314/1/012032

Membrane Paradigm and Holographic Hydrodynamics

Christopher Eling[1], **Yasha Neiman**[2], **Yaron Oz**[2]

[1] SISSA, Via Bonomea 265, 34136 Trieste, Italy and INFN Sezione di Trieste, Via Valerio 2, 34127 Trieste, Italy

[2] Raymond and Beverly Sackler School of Physics and Astronomy, Tel-Aviv University, Tel-Aviv 69978, Israel

E-mail: `cteling@sissa.it`, `yashula@gmail.com`, `yaronoz@post.tau.ac.il`

Abstract.
We discuss recent work showing that in certain cases the membrane paradigm equations governing the dynamics of black hole horizons can be recast as relativistic conservation law equations. In the context of gauge/gravity dualities, these equations are interpreted as defining the viscous hydrodynamics of a holographically dual relativistic field theory. Using this approach, one can derive the viscous transport coefficients and the form of the entropy current for field theories dual to gravity plus matter fields.

1. Introduction

In the past decade there has been much interest in the hydrodynamics of relativistic conformal field theories (CFTs), largely due to the AdS/CFT correspondence between (quantum) gravitational theories on asymptotically Anti-de-Sitter (AdS) spacetimes in $(d+1)$ dimensions and CFTs in d dimensions [1]. In [2] it was shown explicitly that the d-dimensional CFT hydrodynamics equations are dual to the gravitational field equations describing the evolution of large scale perturbations of the $(d+1)$-dimensional black brane. This has been dubbed the fluid-gravity correspondence. It has been noticed that this correspondence naturally completes the picture of the old membrane paradigm [3, 4] of the 1980's, where black hole horizons are thought of as being analogous to fictitious viscous fluids. In this case, however, the dynamics of the entire spacetime is encoded holographically into the fluid living at the boundary of the spacetime and not at the event horizon [5].

On the other hand, at least in some circumstances, the horizon membrane paradigm can be modernized within the context of hydrodynamics as an effective field theory and the fluid-gravity correspondence. The starting point is an equilibrium solution containing a timelike Killing vector field and a stationary causal horizon. This solution is associated with a thermal state at uniform temperature. A general thermal state out of equilibrium is described by an inhomogeneous black hole, which is not an exact solution. However, when a hydrodynamic limit exists, we can work in a derivative expansion, assuming that there is no singularity at the horizon. In [6] we applied this approach to the event horizon of a boosted Einstein-Yang-Mills black brane in an asymptotically AdS spacetime. At lowest orders in derivatives, the set of Einstein and Maxwell equations projected into the horizon surface describe the hydrodynamics of a CFT with anomalous global non-abelian charges. Here we will briefly describe the formalism used in [6] and show how it can be applied to the simplest case of an uncharged black brane [7].

2. Horizon Geometry

In the bulk spacetime, we choose coordinates of the form $x^A = (r, x^\mu)$. The x^μ are coordinates on the horizon $\mathcal{H}$. r is a transverse coordinate, with $r = 0$ on the horizon. $\partial_A r$ is a null covector tangent to the $\mathcal{H}$. When raised with the metric, it gives a vector field $\ell^A = g^{AB}\partial_B r$ which is both normal and tangent to the horizon, and tangent to its null generators. In components, we have $\ell^A = (0, \ell^\mu)$.

The pullback of g_{AB} into $\mathcal{H}$ is the degenerate horizon metric $\gamma_{\mu\nu}$. Its null directions are the generating light-rays of $\mathcal{H}$, i.e. $\gamma_{\mu\nu}\ell^\nu = 0$. The Lie derivative of $\gamma_{\mu\nu}$ along ℓ^μ gives us the shear/expansion tensor, or "second fundamental form":

$$\theta_{\mu\nu} \equiv \frac{1}{2}\mathcal{L}_\ell \gamma_{\mu\nu}. \tag{1}$$

We can write a decomposition of $\theta_{\mu\nu}$ into a shear tensor $\sigma^{(H)}_{\mu\nu}$ and an expansion coefficient θ:

$$\theta_{\mu\nu} = \sigma^{(H)}_{\mu\nu} + \frac{1}{d-1}\theta\gamma_{\mu\nu}. \tag{2}$$

We found it is convenient raise indices with $(G^{-1})^{\mu\nu}$, which is the inverse of any matrix $G_{\mu\nu}$ of the form

$$G_{\mu\nu} = \lambda\gamma_{\mu\nu} - b_\mu b_\nu; \quad b_\mu \ell^\mu \neq 0. \tag{3}$$

Here we introduced the superfluous scalar field λ for later convenience: it will turn out that a matrix of the form (3) coincides at leading order with the metric $h_{\mu\nu} = \eta_{\mu\nu}$ associated with the hydrodynamic dual.

Since $\gamma_{\mu\nu}$ is degenerate, one cannot use it to define an intrinsic connection on the null horizon, as could be done for spacelike or timelike hypersurfaces. The bulk spacetime's connection does induce a notion of parallel transport in $\mathcal{H}$, but only along its null generators. This structure is not fully captured by $\gamma_{\mu\nu}$; instead, it is encoded by the extrinsic curvature, or 'Weingarten map' $\Theta_\mu{}^\nu$, which is the horizon restriction of $\nabla_A \ell^B$:

$$\Theta_\mu{}^\nu = \nabla_\mu \ell^\nu. \tag{4}$$

One can show that given an arbitrary $G_{\mu\nu}$ of the form (3), $\Theta_\mu{}^\nu$ can be written as:

$$\Theta_\mu{}^\nu = \lambda\theta_{\mu\rho}(G^{-1})^{\rho\nu} + c_\mu \ell^\nu; \quad c_\mu \ell^\mu = \kappa. \tag{5}$$

κ is the "surface gravity", which measures the non-affinity of ℓ^μ. The covector c_μ encodes the degrees of freedom in $\Theta_\mu{}^\nu$ which are independent of $\gamma_{\mu\nu}$. In the hydrodynamics, these degrees of freedom will roughly correspond to the velocity and temperature fields.

In their usual form, the Gauss-Codazzi equations involve the divergence of a hypersurface's extrinsic curvature. Our null horizon, however, does not possess an intrinsic connection, and a covariant divergence of $\Theta_\mu{}^\nu$ cannot be defined. The solution is to define a tensor *density* constructed out of $\Theta_\mu{}^\nu$ in the following manner [8]:

$$Q_\mu{}^\nu = v(\Theta_\mu{}^\nu - \kappa\delta^\nu_\mu), \tag{6}$$

where v is a scalar density equal to the horizon area density. It turns out that the horizon-intrinsic covariant divergence $\bar{\nabla}_\nu Q_\mu{}^\nu$ is uniquely defined. Using this divergence, the null Gauss-Codazzi equation can be written as

$$R_{\mu\nu}S^\nu = D^{(G)}_\nu \left(\lambda v\theta_{\mu\rho}(G^{-1})^{\rho\nu}\right) + v\theta\partial_\mu \ln\sqrt{\lambda} + c_\mu\partial_\nu S^\nu + 2S^\nu\partial_{[\nu}c_{\mu]} - v\partial_\mu\theta, \tag{7}$$

where we have defined the area entropy current $S^\mu = v\ell^\mu$. This is the form we we will use for calculations.

3. AdS Black Brane and Ideal Hydrodynamics

The vacuum Einstein equations with negative cosmological constant can be expressed as

$$R_{AB} + dg_{AB} = 0. \tag{8}$$

These equations have the homogeneous (boosted) black brane solution

$$ds^2_{(0)} = -2\ell_\mu dx^\mu dr - (r+R)^2 f \ell_\mu \ell_\nu dx^\mu dx^\nu + (r+R)^2 P_{\mu\nu} dx^\mu dx^\nu, \tag{9}$$

where $f = 1 - (\frac{R}{r})^d$. ℓ^μ is normalized $\eta_{\mu\nu}\ell^\mu\ell^\nu = -1$ and $P_{\mu\nu} = \eta_{\mu\nu} + \ell_\mu\ell_\nu$. The horizon quantities defined in the above section can be expressed in terms of the Bekenstein-Hawking entropy density $s = v^{(0)}/4 = R^{d-1}/4$ and the Hawking temperature $T = \kappa^{(0)}/2\pi = (4/\pi d)R$

$$S^{(0)\mu} = 4s\ell^\mu; \qquad \Theta_\mu^{(0)\nu} = -2\pi T \ell_\mu \ell^\nu; \qquad \gamma^{(0)}_{\mu\nu} = (4s)^{\frac{2}{d-1}} P_{\mu\nu}; \qquad c^{(0)}_\mu = -2\pi T \ell_\mu, \tag{10}$$

where the superscript zero anticipates inhomogeneous corrections. Consider again the configuration (9), but with (ℓ^μ, s) slowly varying functions of x^μ rather than constants. We will interpret this x^μ-dependence by treating these quantities as fields on the brane's horizon $\mathcal{H}$. Various quantities and equations can be expanded order by order in powers of the small ∂_μ derivatives. We use the symbolic small parameter ε to count these powers. We will refer to the power of ε involved as the "order" of a quantity or an equation. In general, (9) with x^μ-dependent parameters will not be a solution of the field equations.

Our analysis will take place on the event horizon $\mathcal{H}$ of the exact inhomogeneous solution. In the configuration (9) with inhomogeneous (ℓ^μ, s), the hypersurface $r = 0$ is null, and doesn't intersect the singularity. It is therefore an event horizon for the zeroth-order solution, and a zeroth-order approximation for the horizon of the corresponding exact solution. For the corrected solution, we will still use coordinates so that $r = 0$ on $\mathcal{H}$.

The Gauss-Codazzi equation (7) reads at first order,

$$c^{(0)}_\mu \partial_\nu S^{(0)\nu} + 2S^{(0)\nu}\partial_{[\nu} c^{(0)}_{\mu]} = O(\varepsilon^2). \tag{11}$$

Plugging in the zeroth-order values, and using the thermodynamic identities $\epsilon + p = Ts$ and $dp = sdT$, we rearrange the equation to get the conservation of a perfect fluid stress tensor

$$\partial_\nu T^\nu_\mu = \partial_\nu((\epsilon+p)\ell_\mu\ell^\nu + p\delta^\nu_\mu) = 0. \tag{12}$$

For the black brane one can show that $\epsilon = (d-1)p$, so the fluid is conformal, as expected.

4. Viscous Hydrodynamics

Now we want to consider the Gauss-Codazzi equations at second order. Obviously, before we write down these equations, we must consider the possible first-order corrections to the horizon ansatz (10). First of all, at viscous order in hydrodynamics there is naturally an ambiguity in definition of the fluid four-velocity. We make the (somewhat non-standard, but here natural) choice that the four-velocity is the direction of the entropy velocity; that is we fix the direction of the entropy velocity ℓ^μ and the magnitude of the entropy density s by requiring that they have the values in eq. (10) without corrections. Due to this condition, the correction to $\gamma_{\mu\nu}$ must be transverse to ℓ^μ and traceless with respect to $\eta_{\mu\nu}$. We will find that the precise form of $\gamma^{(1)}_{\mu\nu}$ is otherwise irrelevant to the constraint equations.

Let us now turn to the c_μ component of the extrinsic curvature. For our uncharged black brane, the most general correction to c_μ reads

$$c^{(1)}_\mu = \mathcal{A}\ell_\mu\partial_\nu\ell^\nu + \mathcal{B}\partial_\mu s, \tag{13}$$

where $\mathcal{A}$ and $\mathcal{B}$ are functions of s.

The requirement for the constraint equations to take the form of conservation laws will place restrictions on these functions. To start, we note that the horizon's first-order shear/expansion tensor can be derived directly from the zeroth-order metric via (1). Using the ideal equations, one finds that $\theta^{(1)} = 0$ and $\theta_{\mu\nu} = (4s)^{\frac{2}{d-1}}\pi_{\mu\nu}$. where $\pi_{\mu\nu}$ is the fluid shear tensor $P^{\rho}_{\mu}P^{\sigma}_{\nu}\partial_{(\rho}u_{\sigma)} - \frac{1}{d-1}P_{\mu\nu}\partial_{\rho}u^{\rho}$. Let us now evaluate the second order pieces of the Gauss-Codazzi equation (7). Plugging in (13), and using the ideal order hydrodynamics equations we find

$$\partial_{\nu}\left(4s\pi^{\nu}_{\mu}\right) + 4\partial_{\nu}\left(\mathcal{A}sP^{\nu}_{\mu}(\partial_{\rho}\ell^{\rho})\right) - 4\mathcal{A}(\partial_{\rho}\ell^{\rho})\partial_{\mu}s. \tag{14}$$

The first two terms can contribute to a relativistic conservation law equation. However, the last does not and therefore we must set $\mathcal{A} = 0$. As a result, at second-order, the Gauss-Codazzi equations have the form

$$\partial_{\nu}(T^{(0)\nu}_{\mu} + T^{(1)\nu}_{\mu}) = \partial_{\nu}\left((\epsilon + p)\ell_{\mu}\ell^{\nu} + p\delta^{\nu}_{\mu} - \frac{s}{2\pi}\pi^{\nu}_{\mu}\right) = O(\varepsilon^3), \tag{15}$$

which are the relativistic hydrodynamics equations for a conformal fluid with shear viscosity $\eta/s = 1/4\pi$.

5. Discussion

It is remarkable that in our analysis the dynamics of the null horizon does not encode an arbitrary hydrodynamic system: once the choices discussed above are made, the transport coefficients are uniquely determined. Specifically, the shear viscosity, the bulk viscosity (and in general, other quantities like conductivity, see [6]) are fixed by the null hypersurface equations to specific functions of state, while for a general fluid they may be arbitrary (under the restriction of positive semi-definiteness).

Note however that in relation to membrane paradigm picture of [5], our construction cannot capture all the dynamics of the theory at higher orders in the hydrodynamic ε expansion. Nevertheless, to viscous order we are able to obtain the hydrodynamics of the boundary fluid by working with the equations for the null horizon. The reason is that we parametrized the membrane equations in terms of the thermodynamic variables associated with the boundary fluid and required the metric $G_{\mu\nu}$ to match the boundary metric (in this case $\eta_{\mu\nu}$). In principle, one can also write the equations in terms of variables associated with a generic $r = const.$ hypersurface, thereby obtaining the hydrodynamics of those surfaces as described in [9].

References

[1] J. M. Maldacena, Adv. Theor. Math. Phys. **2**, 231 (1998) [Int. J. Theor. Phys. **38**, 1113 (1999)] [arXiv:hep-th/9711200]; O. Aharony, S. S. Gubser, J. M. Maldacena, H. Ooguri and Y. Oz, [arXiv:hep-th/9905111].
[2] S. Bhattacharyya, V. E. Hubeny, S. Minwalla and M. Rangamani, JHEP **0802**, 045 (2008) [arXiv:0712.2456 [hep-th]].
[3] T. Damour, these de doctorat d'etat, University of Paris VI, 1979 (unpublished); T. Damour and M. Lilley, arXiv:0802.4169 [hep-th];
[4] K. S. Thorne, R. H. Price and D. A. Macdonald, *Black holes: the membrane paradigm*, Yale University Press, 1986.
[5] V. E. Hubeny, M. Rangamani, S. Minwalla, and M. Van Raamsdonk, Int. J. Mod. Phys. **D17**, 2571-2576 (2009).
[6] C. Eling, Y. Neiman and Y. Oz, arXiv:1010.1290 [hep-th].
[7] C. Eling and Y. Oz, JHEP **1002**, 069 (2010) [arXiv:0906.4999 [hep-th]]; C. Eling, I. Fouxon, Y. Oz, Phys. Lett. **B680**, 496-499 (2009). [arXiv:0905.3638 [hep-th]].
[8] J. Jezierski, J. Kijowski and E. Czuchry, Phys. Rev. D **65**, 064036 (2002) [arXiv:gr-qc/0110018].
[9] I. Bredberg, C. Keeler, V. Lysov and A. Strominger, arXiv:1006.1902 [hep-th]; T. Faulkner, H. Liu, M. Rangamani, [arXiv:1010.4036 [hep-th]].

Journal of Physics: Conference Series **314** (2011) 012033 doi:10.1088/1742-6596/314/1/012033

Non-equilibrium Spacetime Thermodynamics, Entanglement viscosity and KSS bound

G.Chirco, C.Eling and S.Liberati

SISSA, Via Bonomea 265, 34136 Trieste, Italy and INFN Sezione di Trieste

E-mail: chirco@sissa.it, eling@sissa.it, liberati@sissa.it

Abstract. We propose a dual lower dimensional description of the vacuum state associated to a strongly coupled CFT living on Rindler wedge slice close to the horizon hypersurface. From this field theory, with a linear response approach, we show the possibility to derive an entanglement horizon viscosity via a holographic Kubo formula in terms of a two-point function of the stress tensor of matter fields in the bulk. The entanglement viscosity over entropy density ratio come out to satisfy the universal Kovtun-Son-Starinets (KSS) value $1/4\pi$ in four dimensions, suggesting the universal ratio may be a fundamental property of quantum entanglement.

1. Introduction

In the past decade, there has been significant interest in the idea of gravity as the thermodynamics of a quantum theory associated to some underlying microscopic structure of spacetime [1, 2, 3, 4, 5, 6]. A fundamental hint in this direction is given by Jacobson [1], who derive the Einstein equation as an equilibrium equation of state for a *spacetime thermal system* built on the thermodynamical properties of the Minkowski vacuum. The key point in the derivation is the possibility to characterize the Minkowski vacuum perceived by an uniformly accelerated observer as a canonical thermal state (Unruh effect), together with the idea that the local acceleration horizon, associated with the former observer, can be analogous to a tiny piece of a black hole event horizon, thereby allowing for an area-entropy proportionality assumption.

In this context, the coupling between geometry and matter is effectively provided by demanding a local equilibrium Clausius relation, $TdS = \delta Q$, as the main tool to relate the variation of entropy of the vacuum fields dS (horizon area deformation) to the perturbative effects of a local flux of matter energy δQ through the acceleration horizon.

More recently, in the attempt to understand whether and how the thermodynamical derivation of the Einstein equation could be generalized to allow for the higher curvature terms expected by the effective field theory, it has ben realized that the thermodynamical prescription requires a generalization of the local equilibrium condition to a more general entropy balance law. The new local equilibrium condition, $TdS = \delta Q + \delta N$ (generalized Clausius relation), can account for some extra irreversible entropy production terms due to the source contributions for the horizon area/entropy evolution which are quadratic in expansion and shear of the null geodesic bundle comprising the horizon.

Following the approach in [3, 4], one can separate the equilibrium and the non-equilibrium features associated with the gravitational dynamics, in relation with the activation and the

Journal of Physics: Conference Series **314** (2011) 012033 doi:10.1088/1742-6596/314/1/012033

propagation of the gravitational degrees of freedom entailed in the definition of the entropy functional, via the area-entropy proportionality relation. A deep investigation in this line has been carried on in a series of papers extending the thermodynamical derivation of the gravity field equations from General Relativity to generalized Brans-Dicke theories [5, 7]. These works show that the contributions leading to nonequilibrium spacetime thermodynamics can be actually related to the presence of heat fluxes associated to the purely gravitational/internal degrees of freedom of the theory. In particular, the irreversible interpretation for extra entropy contributions defined in terms of purely geometrical quantities, with relation to the notion of horizon viscosity, provided further hints toward the possible presence of a consistent underlying microscopic theory. The query for a viable description for the underlying microscopic theory is then a fundamental clue in this sense.

The fluctuation-dissipation theorem links viscous dissipation to fluctuations of a thermal equilibrium state. A natural attempt to interpret the viscous dissipation rate of a horizon would rely on the study the quantized gravitational fluctuations of the horizon shear, as already developed many years ago [9]. However, alternatively, one can take a different approach and associate the properties of vacuum fluctuations to the notion of quantum entanglement.

2. Horizon Shear Viscosity via Entanglement

In analogy to the black hole case, the equations governing the dynamics of a local Rindler causal horizon can be associated to those of a viscous fluid, according to the *membrane paradigm* approach [10, 11, 12]. This endows the local horizon system with some apparent hydrodynamic transport coefficients such as viscosities. In general, the relationship between the dynamics of a fluid and the dynamics of any black hole event horizon is just an analogy. In order for hydrodynamics to be a valid description, the characteristic wavelength and time scale of perturbations to the system must be much larger than the microscopic scale set by a correlation length (or mean free path). This basic criterion cannot be fulfilled even in the familiar example of a spherically symmetric Schwarzschild horizon. However, for the scale free Rindler horizon, the hydrodynamic limit exists.

The thermodynamic properties of the Rindler wedge are apparently encoded into a "pre-holographic" lower dimensional description associated with the horizon boundary. Namely, when a quantum state is restricted to a sub-region of the spacetime (in this case Minkowski vacuum state in the Rindler wedge), quantum fluctuations of this state have a dual, thermal description associated with the horizon boundary. Furthermore, an effective description of the large-scale dynamics of this vacuum thermal state is always provided by hydrodynamics. To this end, one can develop a simple Kubo-like formula for the viscosity induced on the horizon in terms of a two point stress-energy tensor correlation function for the quantum fields in the Rindler wedge [13]. In this line, one could expect to find a holographic "entanglement viscosity". This quantity, when cut off, should scale exactly with the entanglement entropy.

To test this hypothesis, we propose a microscopic Kubo-like formula for the shear viscosity associated with the fluid lower dimensional description of the vacuum thermal state, characterized by a conserved stress tensor operator [13],

$$\langle \hat{T}^{(d+1)\mu}{}_{\nu} \rangle = \int_{\ell_c}^{\infty} d\xi \, \langle \hat{T}_{(R)}{}^{\mu}_{\nu} \rangle = \int_{\ell_c}^{\infty} d\xi \, \kappa\xi \langle \hat{T}^{\mu}_{\nu} \rangle, \tag{1}$$

defined, by ansatz, as the radial integral of the bulk Minkowski vacuum stress energy tensor.

Now, in four spacetime dimensions the bulk energy density for a scalar field has the Planckian form

$$\epsilon(\xi) = \frac{\pi^2 T^4}{30} = \frac{1}{480\pi^2 \xi^4}. \tag{2}$$

From our ansatz (1), we find an energy density that appropriately scales like the area of the horizon boundary [14]

$$\epsilon^{2+1} = \frac{\kappa}{960\pi^2\ell_c^2}. \tag{3}$$

Using the Gibbs relation $\epsilon + P = sT$, and equation of state $\epsilon = 3P$ for the massless bulk scalar field, we find the entropy density s obeys

$$s = \frac{2\pi^3}{45}T^3 = \frac{1}{180\pi\xi^3}. \tag{4}$$

Integrating over ξ from ℓ_c to ∞ to find the effective area entropy yields

$$s = \frac{1}{360\pi\ell_c^2}, \tag{5}$$

which agrees with standard results in the literature for the brick wall/entanglement entropy [14, 15, 16].

If we apply the formalism of viscous hydrodynamics to this system, the shear viscosity should be given by the Kubo formula

$$\chi = \lim_{k^0\to 0}\frac{1}{k^0}ImG_R(k^0, \vec{k} = 0) \tag{6}$$

in terms of the effective stress tensor of the lower dimensional theory associated with the horizon,

$$\eta = \lim_{\omega\to 0}\frac{1}{\omega}\int d\tau d^dx e^{i\omega\tau}\theta(\tau)\langle[T_{xy}^{d+1}(\tau, x, y), T_{xy}^{d+1}(0)]\rangle, \tag{7}$$

where ω is a Rindler frequency. Using our ansatz that the lower dimensional densities are radial integrals of the bulk matter stress-tensor, we arrive at the following formula

$$\eta = \lim_{\omega\to 0}\frac{1}{\omega}\int_{\ell_c}^{\infty}d\xi'\int_{\ell_c}^{\infty}d\xi\int d\tau d^dx e^{i\omega\tau}\theta(\tau)\kappa^2\xi\xi'\langle[T_{xy}(\tau, x, y, \xi), T_{xy}(0, \xi')]\rangle. \tag{8}$$

which ends up to be a function of the correlator of the bulk stress energy tensor for the matter fields in the wedge.

As a test case of our viscosity formula, we consider the thermal state to consist of a free, minimally coupled scalar field in a four dimensional Rindler spacetime and we eventually get

$$\eta = \frac{1}{1440\pi^2\ell_c^2}, \tag{9}$$

which, as expected, is divergent in the limit $\ell_c \to 0$ and scales in ℓ_c as a $2+1$ quantity. Remarkably, the ratio of our shear viscosity to the entanglement entropy density is exactly the KSS ratio [17]. This suggests that the KSS ratio may be a fundamental holographic property of spacetime, rather than just of the AdS black hole solutions. The saturation of the universal KSS ratio for the Rindler acceleration horizon, in flat Minkowski spacetime, is indeed an interesting result, since gravity is absent. This is not in contrast with the ADS/CFT correspondence results [18, 19], where the KSS ratio, though apparently rooted in gravitational physics, curiously does not depend on the Newton constant.

3. Discussion

Our result suggests that the $1/4\pi$ ratio might be a fundamental property of quantum entanglement and its associated holography. It also provides support for the hypothesis that semi-classical gravity on macroscopic scales is induced or emergent as an effective theory of some lower dimensional, strongly coupled quantum system with a large number of degrees of freedom. In this picture, the $1/4\pi$ ratio is saturated in gauge theories with a Einstein gravity dual because 1) they have an area (BH) entropy and 2) in the large N limit the number of degrees of freedom diverges and gravity is turned off as the Newton constant goes zero.

In our formula for the viscosity, we consider the thermal state to consist of a free, minimally coupled scalar field in a four dimensional Rindler spacetime. One apparent problem with this choice is that the shear viscosity in an free field theory is typically ill-defined. In physical terms, shear viscosity measures the rate of transverse momentum diffusion between the elements of a fluid. Although the quasi-particle description in kinetic theory is not a good one in a strongly coupled system, we can gain some guidance by thinking of shear viscosity as a diffusion process. One can show that $\eta \sim \epsilon l_{\mathrm{mfp}}$, where l_{mfp} is the mean free path of the fluid. Since in a free field theory the mean free path diverges, η diverges as well. On the other hand, in our case the equivalence principle implies a field theory in Rindler space can be thought of as being in a constant gravitational field. Imposing a UV cutoff on this system seems to introduce gravitational dynamics. If the cutoff is placed near the Planck length (as we suspect) the gravitational dynamics is strongly coupled there. The idea is that the dominant effect in the relaxation of the vacuum thermal state is the strongly coupled gravitational interaction. This also seems to explain how there can be universality in the result for η.

References

[1] T. Jacobson, Phys. Rev. Lett. **75**, 1260 (1995) [arXiv:gr-qc/9504004].
[2] for a review, see T. Padmanabhan, Rept. Prog. Phys. **73**, 046901 (2010) [arXiv:0911.5004 [gr-qc]] and references therein
[3] C. Eling, R. Guedens and T. Jacobson, Phys. Rev. Lett. **96**, 121301 (2006) [arXiv:gr-qc/0602001];
[4] C. Eling, JHEP **0811**, 048 (2008) [arXiv:0806.3165 [hep-th]].
[5] G. Chirco and S. Liberati, Phys. Rev. D **81**, 024016 (2010) [arXiv:0909.4194 [gr-qc]].
[6] for alternative approaches, see E. Elizalde and P. J. Silva, Phys. Rev. D **78**, 061501 (2008) [arXiv:0804.3721 [hep-th]]; R. Brustein and M. Hadad, Phys. Rev. Lett. **103**, 101301 (2009) [arXiv:0903.0823 [hep-th]].
[7] G. Chirco, C. Eling and S. Liberati, Phys. Rev. D**83** 024032 (2011)
[8] R. Brustein, D. Gorbonos and M. Hadad, Phys. Rev. D **79**, 044025 (2009) [arXiv:0712.3206 [hep-th]]; also see E. Elizalde and P. J. Silva, Phys. Rev. D **78**, 061501 (2008) [arXiv:0804.3721 [hep-th]].
[9] P. Candelas and D. W. Sciama, Phys. Rev. Lett. **38**, 1372 (1977).
[10] T. Damour, these de doctorat d'etat, University of Paris VI, 1979 (unpublished); T. Damour, *Surface Effects in Black-Hole Physics* in *Proceedings of the Second Marcel Grossmann Meeting on General Relativity*, edited by R. Ruffini (North-Holland, Amsterdam, 1982), p. 587; also see T. Damour and M. Lilley, arXiv:0802.4169 [hep-th].
[11] R. H. Price and K. S. Thorne, Phys. Rev. D **33**, 915 (1986).
[12] K. S. Thorne, R. H. Price and D. A. Macdonald, *Black holes: the membrane paradigm*, Yale University Press, 1986.
[13] G. Chirco, C. Eling and S. Liberati, Phys. Rev. D**82** 024010 (2010)
[14] J. S. Dowker, Phys. Rev. **D** 18, 1856 (1978); J. S. Dowker, Class. Quant. Grav. **11**, L55 (1994) [arXiv:hep-th/9401159].
[15] L. Susskind and J. Uglum, Phys. Rev. D **50**, 2700 (1994) [arXiv:hep-th/9401070].
[16] G. 't Hooft, Nucl. Phys. B **256**, 727 (1985).
[17] P. Kovtun, D. T. Son and A. O. Starinets, Phys. Rev. Lett. **94**, 111601 (2005) [arXiv:hep-th/0405231].
[18] J. M. Maldacena, Adv. Theor. Math. Phys. **2**, 231 (1998) [Int. J. Theor. Phys. **38**, 1113 (1999)] [arXiv:hep-th/9711200].
[19] O. Aharony, S. S. Gubser, J. M. Maldacena, H. Ooguri and Y. Oz, Phys. Rept. **323**, 183 (2000) [arXiv:hep-th/9905111].

Journal of Physics: Conference Series **314** (2011) 012034 doi:10.1088/1742-6596/314/1/012034

A strained space-time

Angelo Tartaglia

Politecnico di Torino, corso Duca degli Abruzzi 24, 10129 Torino, Italy and INFN, Torino

E-mail: angelo.tartaglia@polito.it

Abstract. The paper outlines a theory where spacetime is treated as a physical four-dimensional continuum with properties similar to the ones of ordinary three-dimensional elastic continua. Two manifolds are compared: the first is flat; the second is curved and is thought as being obtained from the first by deformation. The intrinsic "rigidity" os spacetime implies that the deformation correspond to a strained state, where the strain tensor is given by half the difference between the metric tensor of the deformed and the undeformed manifolds. As it would happen in three dimensions the strain is associated with an elastic potential energy. The theory adds this deformation potential energy term to the standard spacetime Lagrangian density of General Relativity. Using the new Lagrangian density the theory is able to reproduce the accelerated expansion of the universe and is also consistent with the data from Big Bang Nucleosynthesis, the acoustic horizon of CMB and the structure formation after the recombination era.

1. Introduction

Spacetime has a fundamental role in General Relativity (GR) and interacts with matter/energy, which means that it is not a mathematical artifact but possesses physical properties on its own. This point is not always fully considered when trying and describing the behaviour of gravitating systems, especially at the cosmic scale. On the other hand we know that there are a few observed behaviours in the universe which is not easy to explain in terms of plain GR. This is the case, for instance, of the accelerated expansion discovered in 1998 [6] as well as of the missing mass or dark matter problem in galaxies and galaxy clusters. In order to solve the open problems a host of solutions have been envisaged, but often the different theories explore the mathematical side without bothering about the physical meaning of the additional terms or changes they introduce into the Lagrangian density of spacetime.

Here I am sketching an approach that finally leads to an additional term into the Lagrangian of empty spacetime, but that term is molded on known physics, drawing onto the analogy that is found between spacetime and three-dimensional material continua. If spacetime is a deformable Riemannian manifold we expect the induced curvature to be associated to a strain in the manifold, then to a corresponding deformation energy. This is the reason why we introduce into the Lagrangian density a potential deformation energy term, depending on the strain of the curved manifold. The analogy further suggests to make use of a generalized linear elasticity theory to find the explicit form of the deformation energy.

Once the new Lagrangian density has been written, we are able, as we shall see, to work out of it the accelerated expansion of the universe as well as consistent values for a number of cosmological tests, like the primordial nucleosynthesis, the acoustic scale of the CMB and the large scale structure formation.

2. The strained manifold

The easiest way to describe the Strained State Theory (SST) is to have recurse to an embedding of the physical four-dimensional manifold into a higher dimensional flat Riemannian space. Actually we may imagine to have a four-dimensional flat manifold, called the *reference manifold* and the real curved manifold (the *natural manifold*), both embedded into a (4+1)-dimensional Riemannian space. Ideally we may think as having an "initial" state which coincides with the reference manifold and successively evolves into the natural manifold by deformation due to the presence of matter/energy or of texture defects in the original continuum. This logical sequence has nothing to do with an evolution in time since time is part of the manifolds. The Lorentzian signature may or may not be present from the very beginning. If it is there the reference manifold is Minkowskian, otherwise it is Euclidean. The situation is graphically represented in fig. 1.

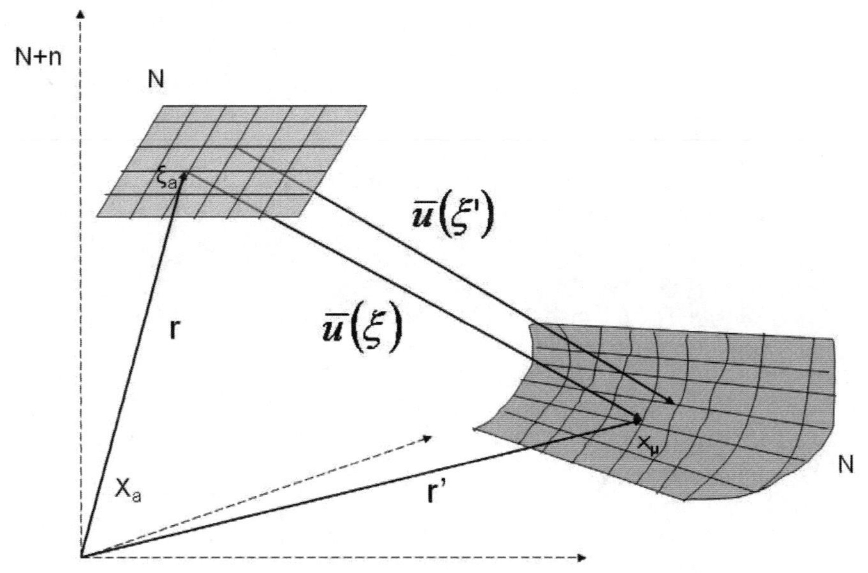

Figure 1. Schematic $(N+n)$-dimensional embedding (actually $4+1$) of a flat N-dimensional manifold and a curved N-dimensional one obtained by deformation from the first. The correspondence between line elements on the two manifolds (each one endowed with its own reference frame and coordinates) visualizes the strain induced in the curved one.

The (N+1)-dimensional vector field **u** visible in the figure establishes a one to one correspondence between events on the two N-dimensional submanifolds. In practice the **u**'s (the *displacement field*) define the distortion of the reference manifold into the natural one. On the basis of the established correspondence we may compare corresponding line elements on the two manifolds; each of them has its own 4-dimensional metric tensor: a general metric $g_{\mu\nu}$ for the natural manifold and a Minkowskian or Euclidean metric $\eta_{\mu\nu}$ for the reference manifold. The difference between the two tensors defines the strain tensor of the natural manifold according to:

$$\epsilon_{\mu\nu} = \frac{1}{2}(g_{\mu\nu} - \eta_{\mu\nu}) \tag{1}$$

The formal details of the theory may be found in [7].

2.1. The Lagrangian density

According to the model outlined in the previous section we can write the action integral of spacetime in the form:

$$S = \int [R + \frac{1}{2}(\lambda\epsilon^2 + 2\mu\epsilon_{\alpha\beta}\epsilon^{\alpha\beta}) + \kappa L_{matter}]\sqrt{-g}d^4x \tag{2}$$

The first term in the square brackets is of course the scalar curvature that plays the role of a "kinetic" term since it contains the derivatives of the dynamic coordinates, i.e. of the elements of the metric tensor; the second term is the potential deformation energy term; the third term is the Lagrangian density of ordinary matter/energy, coupled to geometry through $\kappa = 16\pi G/c^2$, as usual. The two paramters λ and μ are the Lamé coefficients of spacetime. The Lamé coefficients are the typical quantities that contain the elastic properties of a three-dimensional continuous material. Their number depends on the linearization of the theory and on the hypothesis of local isotropy of the manifold, but not on the number of dimensions, so we may use them also in our four-dimensional case.

Starting from eq. (2) and applying the variational principle we can solve various problems, finding the strained configuration of spacetime that minimizes the action integral consistently with the specific symmetry and constraints.

3. The Robertson-Walker symmetry

Studying an universe endowed with the typical Robertson-Walker symmetry we see that the Lagrangian density of empty spacetime becomes:

$$L = -6a(a\ddot{a}+\dot{a}^2)+\frac{1}{4}(\lambda+2\mu)a^3(1-b^2)^2+\frac{3}{8}(3\lambda+2\mu)\frac{(1-a^2)^2}{a}+\frac{3}{4}[\mu\frac{1-a^2}{a^2}-\lambda(1-b^2)]a(1-a^2) \tag{3}$$

The functions a and b depend on cosmic time τ only; a is the scale factor of the universe, b is a gauge function depending on the fact that we use the coordinates of the natural manifold to express the Minkowski metric tensor of the reference manifold and there are in principle an infinity of different ways to establish the correspondence between the two manifolds.

From the Lagrangian (3), after adding matter, we can obtain two equations, for a and b. Solving, we end up with the explicit form of the scale factor $a(\tau)$. Suppose the universe is filled up with dust and radiation, the result in terms of the Hubble function is:

$$H = \frac{\dot{a}}{a} = c\sqrt{\frac{B}{16}}\left\{3\left(1 - \frac{(1+z)^2}{a_0^2}\right)^2 + \frac{8\kappa}{3B}(1+z)^3\left[\rho_{m0} + \rho_{r0}(1+z)\right]\right\}^{1/2}. \tag{4}$$

Here $B = \frac{\mu}{4}\frac{2\lambda+\mu}{\lambda+2\mu}$ is the bulk modulus of spacetime; ρ_{m0} and ρ_{r0} are the present matter and radiation energy densities; a_0 is the present value of the scale factor; z is the redshift parameter.

4. Cosmological tests

Eq. (4) contains four parameters. Two of them, ρ_{m0} and ρ_{r0}, are connected with observation; the other two are typical of the SST. We may use the energy densities as constraints, then optimize the parameters of the theory in order to fit a number of cosmological quantities. First of all we have considered the dependence of the luminosity of SnIa supernovae on the redshift parameter [1], then we considered also the proportions between light elements in the primordial nucleosynthesis [4], the size of the sound horizon in the CMB [2], and the particle horizon at the recombination influencing the Large Scale Structure formation in the universe [3].

The results of this multiple optimization process are summarized in the numerical values that can be read in table 1 [5]; the maximum likelihood estimates are reported in brackets. It is $B_{a_0} = \frac{8}{9}\kappa\rho_{r0}a_0^4$.

Table 1. Estimated parameter values.

B $(10^{-52}\mathrm{m}^{-2})$	$\rho_{\mathrm{m}0}(10^{-29}\mathrm{g\ cm}^{-3})$	$B_{a_0}^{-1}(10^{52}\mathrm{m}^2)$
$2.22\ (2.22) \pm 0.06$	$0.260\ (0.258) \pm 0.009$	$0.011\ (0.009) \pm 0.006$

5. Conclusion

The theory (SST) we have described is based on the idea that spacetime, as a physical continuum, possesses properties similar to the ones of ordinary three-dimensional elastic materials, properly generalized to four dimensions and the Lorentzian signature. We have applied SST to the case of a Robertson-Walker universe with dust and radiation and we have optimized the parameters of the theory by means of four typical cosmological tests: primordial abundances of light elements, acoustic horizon in the CMB, large scale structure formation and the measured luminosity of type Ia supernovae. In order to evaluate whether our estimates are good or not we can compare our results with the ones obtained in the case of the standard ΛCDM theory, where dark energy is represented by a cosmological constant. The comparison has been made with standard statistical parameters and shows that the reliability of SST is more or less the same as the one of ΛCDM [5]. Our task now is to go on working out more implications and consequences of the SST for different symmetries and physical situations.

References

[1] D. Kowalski *et al.* *ApJ*, 686:749–778, 2008.
[2] K. M. Komatsu *et al.* Seven-year wilkinson microwave anisotropy probe (wmap) observations: Cosmological interpretation. arXiv:1001.4538, 2010.
[3] S. Cole *et al.* *MNRAS*, 362:505–534, 2005.
[4] Y. I. Izotov and T. X. Thuan. *Apjl*, 710:L67–L71, 2010.
[5] M. Sereno N. Radicella and A. Tartaglia. Cosmological constraints for the cosmic defect theory. Unpublished.
[6] A. G. Riess. *Astrophys. J.*, 116:1009.
[7] A. Tartaglia and N. Radicella. *Class. Quantum Grav.*, 27:035001, 2010.

Entropy bounds for uncollapsed matter

Gabriel Abreu and Matt Visser
School of Mathematics, Statistics and Operation Research
Victoria University of Wellington
Wellington, New Zealand.

E-mail: `Gabriel.Abreu@msor.vuw.ac.nz; Matt.Visser@msor.vuw.ac.nz`

Abstract. In any static spacetime the quasilocal Tolman mass contained within a volume can be reduced to a Gauss-like surface integral involving the flux of a suitably defined generalized surface gravity. By introducing some basic thermodynamics, and invoking the Unruh effect, one can then develop elementary bounds on the quasilocal entropy that are very similar in spirit to the holographic bound, and closely related to entanglement entropy.

1. Tolman mass and generalized surface gravity

Among the different notions of quasi-local mass, the Tolman mass has the advantage that when dealing with a horizonless object (like a star/planet/monster/gravastar/black star/quasi-black hole) in a static spacetime, it can ultimately be bounded in terms of quantities measurable in the surface of the object [1]; particularly the average surface gravity and the total area. In any static spacetime where the metric can be put into the form

$$\mathrm{d}s^2 = -e^{-2\Psi}\mathrm{d}t^2 + g_{ij}\mathrm{d}x^i\mathrm{d}x^j, \tag{1}$$

the Tolman mass [2] inside a region Ω is defined in terms of the diagonal components of the stress-energy tensor as follows:

$$m_T(\Omega) = \int_\Omega \sqrt{-g_4}\left(T_0^0 - T_i^i\right)\,\mathrm{d}^3x. \tag{2}$$

Here g_4 is the determinant of the $(3+1)$-dimensional metric. The Einstein equations then imply

$$m_T(\Omega) = \frac{1}{4\pi}\int_\Omega \sqrt{-g_4}\,R_0^0\,\mathrm{d}^3x, \tag{3}$$

which is a purely geometrical statement. Now, it is possible to rewrite the Tolman mass in terms of the three-acceleration of a specific set of fiducial observers (FIDOs). First consider the set of FIDOs with four-velocity parallel to the time-like Killing vector k^a,

$$V^a = \frac{k^a}{||k||} = \left(\sqrt{|g^{00}|},\,0,\,0,\,0\right). \tag{4}$$

Their four-acceleration is given by

$$A^a = V^b\nabla_b V^a = V^0\left(\partial_0 V^a + \Gamma^a{}_{c0}V^c\right) = \sqrt{|g^{00}|}\,\Gamma^a{}_{00}\sqrt{|g^{00}|} = |g^{00}|\,\Gamma^a{}_{00}. \tag{5}$$

That is, $A^a = \left(0;\, |g^{00}|\, \Gamma^i{}_{00}\right)$, since V is four-orthogonal to A.

On the other hand, consider the Raychaudhuri equation for this timelike congruence:

$$\frac{\mathrm{d}\theta}{\mathrm{d}s} = -R_{ab}\, V^a\, V^b + 2\,\omega^2 - 2\,\sigma^2 - \frac{1}{3}\theta^2 + \nabla_a A^a. \tag{6}$$

Here the expansion θ, shear expansion σ_{ab}, and vorticity ω_{ab} of the congruence are defined as usual. In terms of the spatial projection operator $h_{ab} = g_{ab} + V_a\, V_b$ we have:

$$\begin{aligned} \theta_{ab} &= h_{ac}\, \nabla^{(c} V^{d)}\, h_{db} \quad ; & \theta &= g^{ab}\, \theta_{ab} = \nabla_a\, V^a; \\ \sigma_{ab} &= \theta_{ab} - \frac{1}{3}\, h_{ab}\, \theta \quad ; & \sigma^2 &= \frac{1}{2}\, \sigma_{ab}\sigma^{ab}; \\ \omega_{ab} &= h_{ac} \nabla^{[c}\, V^{d]}\, h_{db} \quad ; & \omega^2 &= \frac{1}{2}\, \omega_{ab}\, \omega^{ab}. \end{aligned} \tag{7}$$

Then, since $k_{(a;b)} = 0$ in view of Killing's equation, we have

$$V_{(a;b)} = \nabla_{(a}\, \{k/||k||\}_{b)} = \frac{V_{(a}\, ||k||_{,b)}}{||k||}. \tag{8}$$

Therefore $\theta_{ab} = 0$, and consequently $\theta = 0$ and $\sigma_{ab} = 0$. Furthermore, we also have $V_a = -|g_{00}|^{-1/2}\ (1, 0, 0, 0)_a$, that is, in the notation of differential forms $V^\flat = e^\Psi\ dt$. This implies $\mathrm{d}V^\flat = \mathrm{d}\Psi \wedge V^\flat$ and consequently $\omega_{ab} = 0$ (the congruence of FIDOs is hypersurface orthogonal), implying $\omega = 0$. Thus in this situation the timelike Raychaudhuri equation implies

$$R_{ab}\, V^a\, V^b = \nabla_a A^a. \tag{9}$$

This finally reduces to the well known Landau–Lifshitz result [3],

$$R^0_0 = \frac{1}{\sqrt{-g_4}} \partial_i \left(\sqrt{-g_4}\, A^i\right). \tag{10}$$

Now using ordinary partial derivative integration by parts, for any 3–volume Ω the Tolman mass can be rewritten as

$$\begin{aligned} m_T(\Omega) &= \frac{1}{4\pi} \int_\Omega \partial_i \left(\sqrt{-g_4}\, A^i\right)\, \mathrm{d}^3x, \\ &= \frac{1}{4\pi} \int_\Omega \partial_i \left(\sqrt{g_3}\, e^{-\Psi}\, A^i\right)\, \mathrm{d}^3x, \\ &= \frac{1}{4\pi} \int_{\partial\Omega} \left(e^{-\Psi}\, A^i\right)\, \hat{n}_i\, \sqrt{g_2}\, \mathrm{d}^2x, \end{aligned} \tag{11}$$

where $\hat{n}$ is the unit normal, and $\sqrt{g_2}$ is the induced area measure on $\partial\Omega$.

We now consider a generalization of the surface gravity. One method to obtain the usual static surface gravity is by considering the magnitude of the four-acceleration of a suitable set of FIDOs, and renormalizing it using a suitable redshift factor. Here, instead we take the three spatial components of such four-acceleration to build a three-vector

$$\kappa^i = e^{-\Psi}\, A^i, \tag{12}$$

with magnitude $\kappa = e^{-\Psi} \sqrt{g_{ij} A^i\, A^j}$. This generalized surface gravity retains the *usual* physical interpretation of the surface gravity, but is not only valid for any event horizon present — this definition now also applies to FIDOs on the (arbitrary) boundary $\partial\Omega$. The Tolman mass is

$$m_T(\Omega) = \frac{1}{4\pi} \int_{\partial\Omega} \kappa^i\, \hat{n}_i\, \sqrt{g_2}\, \mathrm{d}^2x. \tag{13}$$

Furthermore, defining an average surface gravity over the boundary $\partial\Omega$,

$$\bar{\kappa}(\partial\Omega) \equiv \frac{\int_{\partial\Omega} \kappa \, \mathrm{d}(area)}{\int_{\partial\Omega} \mathrm{d}(area)}, \tag{14}$$

and total area $\mathscr{A}(\partial\Omega)$, and assuming there are no black holes regions, we finally have

$$m_T(\Omega) \leq \frac{\bar{\kappa}(\partial\Omega)\, \mathscr{A}(\partial\Omega)}{4\pi}. \tag{15}$$

This is a bound on the total Tolman mass of a horizonless object (such as a star, or a planet, or a monster [4], or a gravastar [5, 6, 7], or a black star [8, 9, 10], or a quasi–black hole [11, 12]) in a general static geometry, which only depends on quantities measurable on its surface.

2. Entropy bounds and the Unruh effect

To take thermodynamics into account, and calculate a bound on the entropy of these types of object, we now need to consider the Euler (Gibbs–Duhem) relation for the entropy density of the matter forming the horizonless object,

$$s = \frac{\rho + p - \mu\, n}{T}, \tag{16}$$

where (as usual) we have set $p = \frac{1}{3} tr\{T_{ij}\}$ and $\rho = T_{00}$. It is important to remember that the whole discussion in centered in some collection of ideal particles making up an uncollapsed object like a star/planet/monster/gravastar/black star/quasi-black hole. The total entropy due to this kind of matter inside any specific 3–volume is

$$S(\Omega) \;\; = \;\; \int_\Omega \sqrt{g_3}\, \frac{\rho + p + \mu\, n}{T}\, \mathrm{d}^3x. \tag{17}$$

The Tolman [2, 13], (Tolman–Ehrenfest [14]), and Tolman–Klein [15] equilibrium conditions

$$T\sqrt{-g_{00}} = T_\infty, \qquad \mu\sqrt{-g_{00}} = \mu_\infty, \tag{18}$$

imply

$$S(\Omega) = \frac{1}{T_\infty} \int_\Omega \sqrt{-g_4}\, (\rho + p)\; \mathrm{d}^3x - \frac{\mu_\infty}{T_\infty} N. \tag{19}$$

Here $N \equiv \int_\Omega \sqrt{g_3}\, n\, \mathrm{d}^3x$ is the total number of particles inside the 3–volume. Thermodynamic stability requires the chemical potential to be $\mu \geq 0$, then

$$S(\Omega) \leq \frac{1}{T_\infty} \int_\Omega \sqrt{-g_4}\; (\rho + p)\; \mathrm{d}^3x \leq \frac{1}{T_\infty} \int_\Omega \sqrt{-g_4}\; (\rho + 3p)\; \mathrm{d}^3x. \tag{20}$$

where we have assumed $p \geq 0$ throughout the interior. This last numerator is just equivalent to the definition of the Tolman mass. Hence

$$S(\Omega) \leq \frac{m_T(\Omega)}{T_\infty}. \tag{21}$$

We see that the total entropy is bounded by the Tolman mass divided by the temperature as measured at infinity. Furthermore from the previous discussion this entropy bound can be rephrased in terms of the average generalized surface gravity $\bar{\kappa}(\partial\Omega)$, and the total area $\mathscr{A}(\partial\Omega)$,

$$S(\Omega) \leq \frac{1}{4\pi\, T_\infty}\; \bar{\kappa}(\partial\Omega)\; \mathscr{A}(\partial\Omega). \tag{22}$$

Finally we invoke the only semiclassical notion we use throughout the calculations, to relate the surface gravity to a temperature, via the Unruh effect [16]. Due to the existence of the Unruh acceleration radiation phenomenon, it can be asserted that an observer on the boundary $\partial\Omega$ will measure a minimum local temperature of

$$T(x) \geq T_{\text{Unruh}}(x) = \frac{||A(x)||}{2\pi}, \tag{23}$$

which when redshifted to infinity implies

$$T_\infty \geq \max_{x \in \partial\Omega} \left\{ \frac{\kappa(x)}{2\pi} \right\}. \tag{24}$$

So the equilibrium temperature of the object confined inside $\partial\Omega$ satisfies

$$T_\infty \geq \frac{\bar{\kappa}(\partial\Omega)}{2\pi}. \tag{25}$$

Hence the entropy is bounded by *one half* of the total area $\mathscr{A}(\partial\Omega)$,

$$S(\Omega) \leq \frac{\mathscr{A}(\partial\Omega)}{2}. \tag{26}$$

3. Discussion

We have bounded the entropy of an object without horizons, in static spacetime, by half of its total area. Although the construction of this entropy bound is very minimalist, it is closely related to the usual entropy bounds [17, 18, 19, 20, 21, 22]. Specifically, it resembles the holographic bound [17], which corresponds to $S(\Omega) \leq \frac{1}{4}\mathscr{A}(\partial\Omega)$.

The fundamental reason for the coefficient mismatch between these two bounds is because we calculated the entropy of a uncollapsed distribution of matter, where the temperature has its normal interpretation as an intensive variable, and the Euler relation keeps its usual uncollapsed form. Further generalizations of this entropy bound, where a black hole region is taken into account, can be found in [1]. Overall, the importance of this entropy bound, for an uncollapsed distribution of matter, resides in its simplicity and its robustness.

References

[1] G. Abreu and M. Visser, Phys. Rev. Lett. **105**, 041302 (2010). [arXiv:1005.1132 [gr-qc]].
[2] R. C. Tolman, Phys. Rev. **35**, 8 (1930).
[3] L. D. Landau and E. M. Lifschitz, *The Classical Theory of Fields* (Pergamon Press, Oxford, 2000).
[4] R. D. Sorkin, R. M. Wald and Z. J. Zhang, Gen. Rel. Grav. **13**, 1127 (1981).
[5] P. O. Mazur and E. Mottola, Proc. Nat. Acad. Sci. **101**, 9545 (2004) [arXiv:gr-qc/0407075].
[6] M. Visser and D. L. Wiltshire, Class. Quant. Grav. **21**, 1135 (2004) [arXiv:gr-qc/0310107].
[7] C. Cattoen, T. Faber and M. Visser, Class. Quant. Grav. **22**, 4189 (2005) [arXiv:gr-qc/0505137].
[8] C. Barcelo, S. Liberati, S. Sonego and M. Visser, Phys. Rev. D **77**, 044032 (2008) [arXiv:0712.1130 [gr-qc]].
[9] C. Barcelo, S. Liberati, S. Sonego and M. Visser, AIP Conf. Proc. **1122**, 99 (2009) [arXiv:0909.4157 [gr-qc]].
[10] M. Visser, C. Barcelo, S. Liberati and S. Sonego, [arXiv:0902.0346 [gr-qc]].
[11] J. P. S. Lemos and O. B. Zaslavskii, Phys. Rev. D **81**, 064012 (2010) [arXiv:0904.1741 [gr-qc]].
[12] J. P. S. Lemos and O. B. Zaslavskii, Phys. Rev. D **78**, 024040 (2008) [arXiv:0806.0845 [gr-qc]].
[13] R. C. Tolman, *Relativity, thermodynamics, and cosmology,* (Dover, 1987).
[14] R. C. Tolman and P. Ehrenfest, Phys. Rev. **36**, 1791 (1930).
[15] O. Klein, Rev. Mod. Phys. **21** (1965), 531.
[16] W. G. Unruh, Phys. Rev. D **14**, 870 (1976).
[17] R. Bousso, Rev. Mod. Phys. **74**, 825 (2002), [arXiv:hep-th/0203101].
[18] J. D. Bekenstein, Phys. Rev. D **9**, 3292 (1974).
[19] J. D. Bekenstein, Phys. Rev. D **23**, 287 (1981).
[20] T. Padmanabhan, Class. Quant. Grav. **21**, 4485 (2004), [arXiv:gr-qc/0308070].
[21] T. Padmanabhan, Phys. Rev. D **81**, 124040 (2010), [arXiv:1003.5665 [gr-qc]].
[22] M. Srednicki, Phys. Rev. Lett. **71**, 666 (1993), [arXiv:hep-th/9303048].

Journal of Physics: Conference Series **314** (2011) 012036 doi:10.1088/1742-6596/314/1/012036

The generalized second law in universes with quantum corrected entropy relations

Ninfa Radicella and Diego Pavón

Departamento de Física, Universidad Autónoma de Barcelona, 08193 Bellaterra (Barcelona), Spain

E-mail: ninfa.radicella@uab.cat, diego.pavon@uab.es

Abstract. We apply the generalized second law of thermodynamics to discriminate among quantum corrections (whether logarithmic or power-law) to the entropy of the apparent horizon in spatially Friedmann-Robertson-Walker universes. We use the corresponding modified Friedmann equations along with either Clausius relation or the principle of equipartition of the energy to set limits on the value of a characteristic parameter entering the said corrections.

1. Introduction

As is well known, event horizons mimic black bodies and possess a nonvanishing temperature and entropy [1, 2]. Recently, it was demonstrated that cosmological apparent horizons are also endowed with thermodynamical properties, formally identical to those of event horizons [3].
The connection between gravity and thermodynamics was reinforced by Jacobson, who associated Einstein equations with Clausius' relation [4], and later on by Padmanabhan who linked the macroscopic description of spacetime, given by Einstein equations, to microscopic degrees of freedom, N, through the principle of equipartition of energy [5].
On the other hand, quantum corrections to the semi-classical entropy-law have been introduced in recent years, namely, logarithmic and power-law corrections. Logarithmic corrections, arises from loop quantum gravity due to thermal equilibrium fluctuations and quantum fluctuations [6],

$$S \propto \left[\frac{A}{4l_{pl}^2} + \alpha \ln \frac{A}{4l_{pl}^2}\right] . \tag{1}$$

On its part, power-law corrections appear in dealing with the entanglement of quantum fields in and out the horizon [7],

$$S \propto \frac{A}{4l_{pl}^2}\left[1 - K_\alpha A^{1-\alpha/2}\right] . \tag{2}$$

In the last two expressions, α denotes a dimensionless parameter whose value (in both cases) is currently under debate.

For the connection between horizons and thermodynamics to hold, these quantum entropy corrections must translate into modifications of the field equations of gravity, see e.g. [8, 3]. In any sensible physical context these modifications must fulfill the generalized second law (GSL)

Published under licence by IOP Publishing Ltd

Table 1. Expressions for $f(\alpha, H)$ and $g(\alpha, H)$.

	Logarithmic correction
Equipartition	$f(\alpha,H) = \frac{l_p^2\alpha}{2\pi c^2}H^2\left\{1-\frac{1}{2}\ln\left(\frac{\pi c^2}{l_p^2H^2}\right)\right\}$ $g(\alpha,H) = \frac{3l_p^2\alpha}{16\pi c^2}H^2\left\{1+\frac{2}{3}\ln\left(\frac{\pi c^2}{l_p^2H^2}\right)\right\}$
Clausius	$f(\alpha,H) = \frac{l_p^2\alpha}{2\pi c^2}H^2$ $g(\alpha,H) = \frac{3l_p^2\alpha}{16\pi c^2}H^2$
	Power-law correction
Equipartition	$f(\alpha,H) = -\alpha\frac{3-\alpha}{4-\alpha}(Hr_c)^{\alpha-2}$ $g(\alpha,H) = -\frac{3-\alpha}{4-\alpha}(Hr_c)^{\alpha-2}$
Clausius	$f(\alpha,H) = -\frac{\alpha}{2}(Hr_c)^{\alpha-2}$ $g(\alpha,H) = -(Hr_c)^{\alpha-2}$

of thermodynamics. The latter asserts that the entropy of the horizon plus the entropy of its surroundings must not decrease in time. As demonstrated by Bekenstein, this law is satisfied by black holes in contact with their radiation [9].

The aim of this work is to see whether the modified Friedmann equations coming from logarithmic corrections and from power-law corrections, in conjunction with Clausisus relation or the equipartition principle, are compatible with the generalized second law. This sets constraints on the parameter α introduced above, whose value is theory dependent and rather uncertain. We hope this may be of help in discriminating among quantum corrections, via a purely classical analysis.

2. Modified Friedmann equations

As said above, when quantum modifications to the entropy-area relation are incorporated it translates into modifications of Friedmann equations when computing either Clausius relation or the equipartition principle on the apparent horizon of a flat Robertson-Walker Universe. The radius of the horizon is given by $\tilde{r}_{AH} = \frac{c}{H}$, where $H = \dot{a}/a$ denotes the Hubble function. The detailed derivation can be found in [10], and the modified Friedmann equations read

$$H^2\left[1+g(\alpha,H)\right] = \frac{8\pi G}{3}\rho, \tag{3}$$

$$\dot{H}\left[1+f(\alpha,H)\right] = -4\pi G\left(\rho+\frac{P}{c^2}\right), \tag{4}$$

the explicit expressions of $f(\alpha, H)$ and $g(\alpha, H)$ depend on both the entropy corrections and the thermodynamical relation employed, as shown in Table 1.

As can be noted, the α parameter directly comes from quantum corrections to the entropy and consequently affects cosmological scenarios. Its value depends on the details of the quantum calculations, and for the time being there is not agreement on it. The following analysis determines in which intervals this parameter results compatible with the GSL.

3. The Generalized second law of Thermodynamics

Equipped with the entropy expressions (1) and (2), we set out to study whether the GSL is satisfied when the modified Friedmann equations (3) and (4) are employed.

Since the entropy depends on the area of the apparent horizon, $A_{AH} \propto H^{-2}$, it varies as $\dot{S}_H \propto F(H)\dot{H}$. Using eq.(4), it can be cast in terms of the Hubble parameter and the energy density and pressure of the fluid that fills the universe

$$\dot{S}_H = \mathcal{K}\frac{\mathcal{F}(H)}{H^3}\left(\rho + \frac{P}{c^2}\right), \tag{5}$$

where $\mathcal{K} = \frac{8\pi^2 c^5 k_B}{\hbar}$ and $\mathcal{F}(H)$ depends on the entropy corrections and the thermodynamic relation used to derive Friedmann equations, namely

$$\mathcal{F}(H) = \frac{1 + \frac{\alpha l_p^2 H^2}{\pi c^2}}{1 + f(\alpha, H^2)}, \tag{6}$$

for logarithmic entropy corrections, and

$$\mathcal{F}(H) = \frac{1 + \alpha (r_c H)^{\alpha-2}/2}{1 + f(\alpha, H)}, \tag{7}$$

for power-law corrections.
For the sake of clarity in what follows we split the analysis for the two classes of entropy corrections, but we will only consider perfect fluids assuming that the dominant energy condition (DEC) holds true (i.e., $\rho + P/c^2 > 0$) all along the expansion. Then, in view of eq.(5) the GSL is satisfied provided $\mathcal{F}(H)$ is non-negative which occurs only for some values of the parameter α.

3.1. Logarithmic entropy corrections

For convenience, we introduce the dimensionless variable $x = l_p^2/A_{AH}$ so that $x \simeq 1$ at the quantum regime and, provided $\dot{H} < 0$, it decreases as time goes on. We require $\mathcal{F}(\alpha, x)$ to be non-negative, if the GSL is to hold, as well as $\rho \geq 0$, not to deal with ghosts.
Left panel of fig.(1) depicts the regions in the plane (x, α) where $\mathcal{F} > 0$ is fulfilled and those in which $\mathcal{F} < 0$. We believe that the allowed values for α are just those such that the GSL holds throughout the expansion of the Universe. The upper bound on α is given by the local minimum of the dashed curve, that is $\alpha = 4e^3$, and the lower bound by the intersection of the dotted curve with the line $x = 1$ (i.e., when horizon area equals Planck's area), that is $\alpha = -1/4$ (not shown). Positive values of α seem to be largely favored, which is also consistent with some quantum calculations of the entropy corrections [11].
By using Clausius relation, instead, and adopting the above defined variable x, as seen in the right panel of Fig.1, the GSL is satisfied, $(\mathcal{F}(\alpha, x) \geq 0)$, for $\alpha > -1/(4x)$.

3.2. Power-law entropy correction

Before applying the GSL let us look at the Friedmann equations for power-law entropy correction. Inspection of eq.(2) shows that the values of $\alpha = 0$ and $\alpha = 2$ are special, in the sense that for $\alpha = 0$ there are no entropy corrections and the equations reduce to the corresponding general relativity expressions with a cosmological constant. Likewise $\alpha = 2$ represents just a renormalisation of Newton constant, G. Bearing this in mind, we start the analysis by introducing the dimensionless variable $x = (r_c H)^{-1}$, and identifying the crossover scale r_c with H_0^{-1}. Thus x tends to zero in the far past and its today value is $x_0 = 1$ (provided, again, that $\dot{H} < 0$).
Performing a similar analysis as before of the sign of the $\mathcal{F}$ function and on the energy density, as found in [10], constraints on the α parameter follow . These are shown in Fig. 2. In the case of Clausius relation, the GSL remains valid throughout the expansion so long as $-2 \leq \alpha < 2$ while equipartition of energy allows a wider range for the parameter α which can now lie in the range $3 \leq \alpha < 4$, as well.

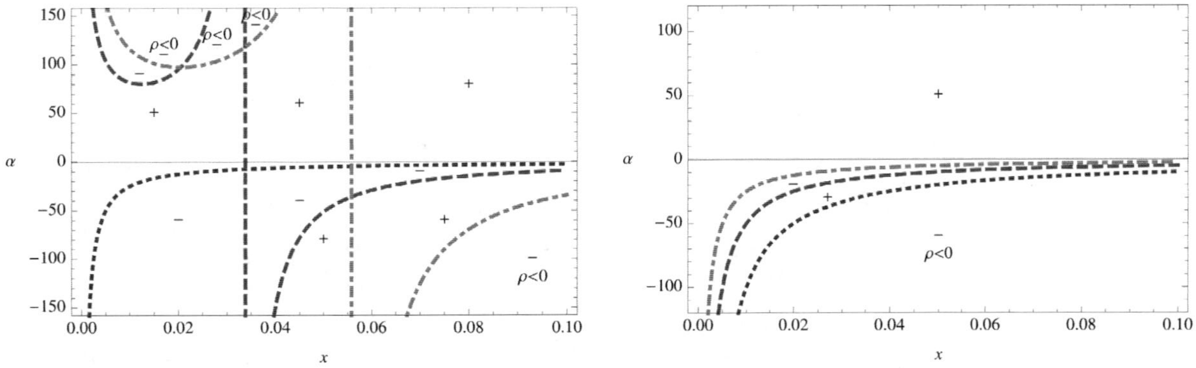

Figure 1. Plot of the sign of $\mathcal{F}$ depending on α along the Universe expansion. On the left panel the case of the logarithmic correction and equipartition principle is shown. The right panel corresponds to the case of using Clausius relation instead. The plots are restricted to the range $0 < x < 0.1$ but in the remaining region the curves behave monotonically.

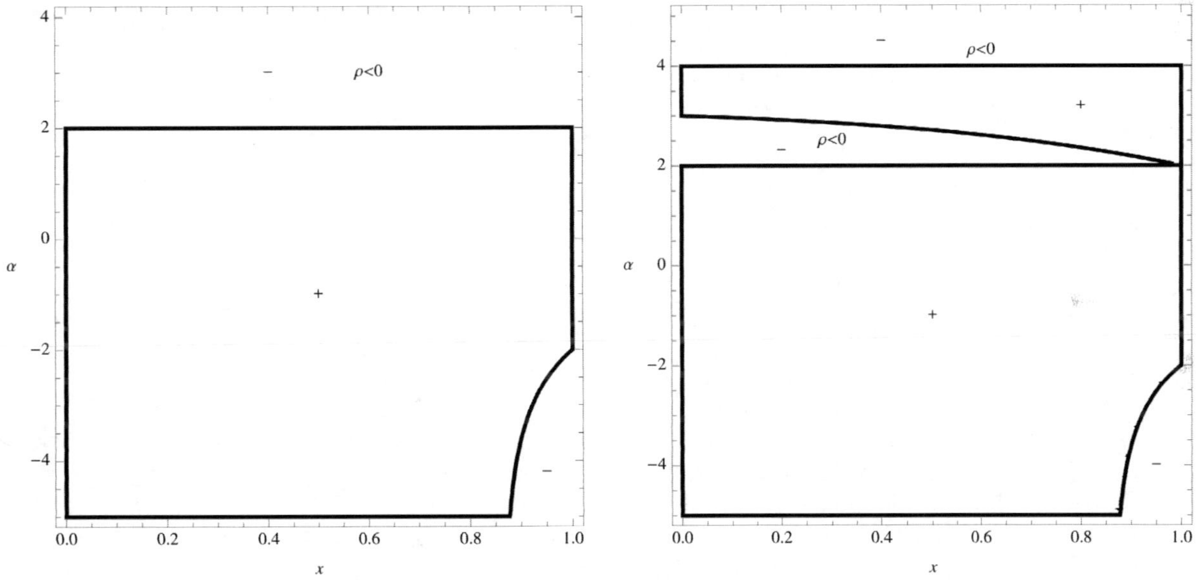

Figure 2. Plots of the sign of $\mathcal{F}$ depending on α along the Universe expansion. On the left panel the case of the power-law correction and the equipartition principle is shown. The right panel depicts the case when Clausius' relation is used instead.

References

[1] J.D. Bekenstein, *Phys. Rev. D*, 7:2333–2346, 1973.
[2] G.W. Gibbons and S.W. Hawking, *Phys. Rev. D*, 15:2738–2751, 1977.
[3] Rong-Gen Cai, Li-Ming Cao, and Ya-Peng . Hu, *Class. Quantum Grav.*, 26:155018, 2009.
[4] T. Jacobson, *Phys. Rev. Lett.*, 75:1260, 1995.
[5] T. Padmanabhan, arXiv:1003.5665v1 [gr-qc], 2010.
[6] K. A. Meissner, *Class. Quantum Grav.*, 21:5245–5252, 2004.
[7] Saurya Das, S. Shankaranarayanan, and Sourav Sur, *Phys. Rev. D*, 77:064013, 2008.
[8] J.E. Lidsey, *Class. Quantum Grav.*, 26:147001, 2009.
[9] J.D. Bekenstein, *Phys. Rev. D*, 9:3292, 1974.
[10] N. Radicella and D. Pavón Coloma, *Phys. Lett. B*, 691: 121, 2010.
[11] S. Hod, *Class. Quantum Grav.*, 21:L97, 2004.

Journal of Physics: Conference Series **314** (2011) 012037 doi:10.1088/1742-6596/314/1/012037

Thermal radiation from lorentzian traversable wormholes

Prado Martín-Moruno and Pedro F González-Díaz

Colina de los Chopos, Instituto de Física Fundamental,
Consejo Superior de Investigaciones Científicas, Serrano 121, 28006 Madrid, Spain

E-mail: pra@iff.csic.es

Abstract. In this contribution we show that lorentzian dynamic wormholes emit thermal phantom-like radiation. Analogously to as it occurs for black holes, the consideration of such radiation process allows the formulation of a wormhole thermodynamics which might help in the understanding of those objects.

1. Introduction

It is well known that the universe is currently undergoing a period of accelerating expansion and, in the framework of general relativity, this phenomenon is only understandable if dark energy dominates the universal dynamics. It is even possible that the dark fluid violates the null energy condition [1], being the equation of state parameter $w < -1$ (with $w = p/\rho$). As recent studies have shown, this strange fluid, dubbed phantom energy [2], has negative temperature [3].

On the other hand, lorentzian traversable wormholes are solutions of general relativity, describing short-cuts between two regions of the same universe or connections between two universes [4]. The problem with these solutions was that, in order to be traversable, they must be supported by some stuff violating the null energy condition, which had been called exotic matter. Nevertheless, this property is precisely the same which fulfils phantom energy. In fact, it has been shown that an inhomogeneous version of phantom energy can be the exotic stuff which is required to support wormholes [5].

Due to the possible existence of wormholes in our Universe, a deeper understanding of these objects is necessary. Therefore, in the present contribution, we consider the potential thermodynamical properties of lorentzian traversable wormholes [6]. The key point to derive such properties is the study of the radiation process.

2. Preliminaries.

In order to study the thermal radiation process, in this section we summarize some previous results necessary to characterize wormholes univocally.

2.1. Dark energy accretion onto black holes and wormholes.

Babichev, Dokuchaev and Eroshenko were the first to study the accretion of dark energy onto black holes phenomenon [7]. They used a test-fluid approach considering a Schwarzschild black hole and the dark energy described by an energy-momentum tensor of a perfect fluid, obtaining that the variation of the black hole mass with time takes the form [7]

$$\dot{M} = 4\pi D M^2 \left(p + \rho\right), \tag{1}$$

where D is a constant of order unity. This expression shows how the black hole mass, and with it its size, increases if the fluid fulfils the null energy condition, it decreases if $p + \rho < 0$, and remains constant in the cosmological constant case.

Soon after, Gonzalez-Diaz studied the dark energy accretion onto wormholes [8]. It can be seen that the application of the test-fluid approach to the Morris-Thorne wormhole [9] leads to

$$\dot{m} = -4\pi Q m^2 \left(p + \rho\right), \tag{2}$$

where Q is a positive constant. Therefore, the wormhole increases by the accretion of phantom energy, it decreases if the fluid fulfils the null energy condition, and it also remains constant in the cosmological constant case.

2.2. 2+2-formalism and its applications.

In order to study the possible thermodynamics of wormholes, we must consider a method based on local quantities since, by definition, there is no event horizon in these spacetimes. Therefore, we consider the 2+2 formalism developed by Hayward for spherically symmetric spacetimes [10]. The metric of these spacetimes can always be expressed, at least locally, in terms of the double null coordinates, $\xi^{\pm}$, as

$$\mathrm{d}s^2 = 2g_{+-}\mathrm{d}\xi^{+}\xi^{-} + r^2\mathrm{d}\Omega^2_{(2)}, \tag{3}$$

where $r(\xi^{+},\xi^{-})$ is the radius of the spheres of symmetry. The expansions along the null directions are $\Theta_{\pm} = 2\left(\partial_{\pm} r\right)/r$ and, since the sign of their product is an invariant, it can allow us to determine the kind of the spheres of symmetry. The sphere is trapped, marginal or untrapped, if this product is bigger, equal or less than zero. A marginal sphere (fixing $\Theta_{+} = 0$) can be future ($\Theta_{-} < 0$), bifurcating ($\Theta_{-} = 0$) or past ($\Theta_{-} > 0$) and outer ($\partial_{-}\Theta_{+} < 0$), degenerate ($\partial_{-}\Theta_{+} = 0$) or inner ($\partial_{-}\Theta_{+} > 0$). A trapping horizon is defined as the hypersurface foliated by marginal spheres and it has the same classification as the marginal spheres.

In spherically symmetric spacetimes one can consider the Kodama vector

$$k = \mathrm{curl}_2 r = -g^{+-}\left(\partial_{+} r \partial_{-} - \partial_{-} r \partial_{+}\right), \tag{4}$$

where the subscript 2 means the space normal to the spheres of symmetry. On the trapping horizon, this vector fulfils an equation similar to that of the temporal Killing vector in static spacestimes [10], therefore, a generalized surface gravity can be defined as [10]

$$\kappa = \frac{1}{2}\mathrm{div}_2\mathrm{grad}_2 r = g^{+-}\partial_{+}\partial_{-}r. \tag{5}$$

It can be seen that κ is positive, zero or negative if the trapping horizon is outer, degenerate or inner.

Hayward also found a generalized first law of thermodynamics for these spacetimes, which is [10]

$$L_z E = \frac{\kappa L_z A}{8\pi} + \omega L_z V, \tag{6}$$

where $z = z^{+}\partial_{+} + z^{-}\partial_{-}$ is the vector which generates the horizon, $E = r\left(1 - \partial^a r \partial_a r\right)/2$ is the Misner-Sharp energy and ω is an invariant. It can be noted that Eq. (6) suggests a relation between the entropy and the area, $S \propto A$.

Even more, as it has been shown [11], this method can be used to obtain the temperature of the thermal radiation for dynamical black holes, which is

$$T_{\mathrm{H}} = \frac{\kappa}{2\pi}. \tag{7}$$

Journal of Physics: Conference Series **314** (2011) 012037 doi:10.1088/1742-6596/314/1/012037

3. Wormholes characterization.

Using the Einstein equations for spherically symmetric spacetimes [10] and the definition of a trapping horizon, the following expression can be obtained

$$\text{sign}\left(\frac{z^+}{z^-}\right) = -\text{sign}\left[(p+\rho)_{\text{H}}\right]. \tag{8}$$

Therefore, a trapping horizon is space-like if the surrounding material has $p+\rho > 0$, null if $p+\rho = 0$, and time-like if $p+\rho < 0$. On the other hand, the evolution of the area can be written in terms of the expansions (one of which vanishes at the horizon) as

$$L_z A = \int \mu \left(z^+\Theta_+ + z^-\Theta_-\right). \tag{9}$$

The combination of Eqs. (8) and (9) and the definitions previously introduced imply that the black hole, characterized by a future outer trapping horizon, increases if $p+\rho > 0$ and decreases in the opposite case. Since this result is qualitatively the same as that obtained by Babichev et al.[7], we could consider that both methods describe just the same process, which is originated from a flow of the surrounding material into the hole.

Thus, dynamic wormholes must be characterized by past outer trapping horizons [6], in order to recover the results obtained following the accretion method [9] by using the 2+2 formalism. Of course, in the static case their trapping horizons will be bifurcating.

4. Wormholes thermal radiation and thermodynamics.

We consider in the present study a general spherically symmetric and dynamic wormhole which, therefore, is described through metric (3) with a past outer trapping horizon (fixing, without loss of generality, $\Theta_- = 0$). It can be seen that this metric can be consequently written in terms of the generalized retarded Eddington-Finkelstein coordinates, at least locally, as [6]

$$\mathrm{d}s^2 = -e^{2\Psi} C \mathrm{d}u^2 - 2e^{\Psi}\mathrm{d}u\mathrm{d}r + r^2\mathrm{d}\Omega^2, \tag{10}$$

where $\mathrm{d}u = \mathrm{d}\xi^-$, $\mathrm{d}\xi^+ = \partial_u\xi^+\mathrm{d}u + \partial_r\xi^+\mathrm{d}r$, and Ψ expressing the gauge freedom in the choice of the null coordinate u. Since $\partial_r\xi^+ > 0$, we have considered $e^{\Psi} = -g_{+-}\partial_r\xi^+ > 0$ and $e^{2\Psi}C = -2g_{+-}\partial_u\xi^+$. It can be seen that $C = 1 - 2E/r$, with E being the Misner-Sharp energy.

Although a traversable wormhole possesses, by definition, a classically allowed trajectory, we can think that the existence of a trapping horizon, characterized by a non-vanished generalized surface gravity, opens the possibility for an additional quantum traversing phenomenon through the wormhole. So, we consider the tunnelling probability along a classical forbidden trajectory, $\Gamma \propto \exp\left[-2\text{Im}\,(I)\right]$, where I is the action. Now, applying the method considered in Ref. [11], but taking in this case the metric (10) into account, it can be seen that Γ has a thermal form, $\Gamma \propto \exp\left(-\omega_\phi/T_H\right)$, being the temperature [6]

$$T_{\text{H}} = -\frac{\kappa}{2\pi}, \tag{11}$$

which is negative. Therefore, like in the black hole case, wormholes radiate matter of the same kind as their surrounding material, in this case, phantom energy.

Once calculated the temperature on the wormhole horizon, we can re-write the generalized first law, Eq. (6), as

$$L_z E = -T_{\text{H}} L_z S + \omega L_z V, \tag{12}$$

which implies the recovery of the usual relation between the entropy and the area, $S = A/4$.

At this point, we can formulate the three laws of wormhole thermodynamics [6].

- First law: the change in the gravitational energy of the wormhole is equal to the sum of the energy removed from the wormhole and the work done in the wormhole.
- Second law: the entropy of a wormhole can never decrease when the wormhole is placed in its most natural exotic environment.
- Third law: it is impossible to reach the absolute zero for surface gravity by means of any dynamical process.

This third law can be understood taking into account that if for some hypothetical process the surface gravity is reduced to vanish, then this process would also be changing the trapping horizon to be degenerate. Since a wormhole is characterized by an outer horizon, the resulting object would not be a wormhole and, therefore, these laws will not be applicable.

5. Conclusions and further comments.

It can be concluded that wormholes posses a well-defined thermodynamics and thermally radiate phantom energy.

Like in the black hole case, the radiation process would produce a decrease of the wormhole throat size, so decreasing the wormhole entropy, too. We expect that this violation of the second law would be only apparent, because it is the total entropy of the universe what should be meant to increase.

It is worth noticing that the initial conditions in the action integral could be chosen in such a way that the wormhole would radiate ordinary matter increasing its size in the process. But, if that case would be at all possible, the thermal radiation would be always thermodynamically forbidden in front of the accretion entropicaly favoured process.

Acknowledgments

P. M. M. thanks the financial support provided by the I3P framework of CSIC and the European Social Fund. This work was supported by MEC under Research Project No. FIS2008-06332/FIS.

References

[1] Mortlock D J and Webster R L 2000 The statistics of wide-separation lensed quasars *Mon. Not. Roy. Astron. Soc.* **319** 872 (*Preprint* astro-ph/0008081)

[2] Caldwell R R 2002 A Phantom Menace? *Phys. Lett.* B **545** 23 (*Preprint* astro-ph/9908168)

[3] Gonzalez-Diaz P F and Siguenza C L 2004 Phantom thermodynamics *Nucl. Phys.* B **697** 363 (*Preprint* astro-ph/0407421)

Saridakis E N, Gonzalez-Diaz P F and Siguenza C L 2009 Unified dark energy thermodynamics: varying w and the -1-crossing *Class. Quant. Grav.* **26** 165003 (*Preprint* astro-ph/0901.1213)

[4] Morris M S and Thorne K S 1988 *Am. J. Phys.* **56** 395

[5] Sushkov S V 2005 Wormholes supported by a phantom energy *Phys. Rev.* D **71** 043520 (*Preprint* gr-qc/0502084)

Lobo F S N 2005 Phantom energy traversable wormholes *Phys. Rev.* D **71** 084011 (*Preprint* gr-qc/0502099)

[6] Martin-Moruno P and Gonzalez-Diaz P F 2009 Lorentzian wormholes generalizes thermodynamics still further *Class. Quant. Grav.* **26** 215010 (*Preprint* gr-qc/0904.0099)

Martin-Moruno P and Gonzalez-Diaz P F 2009 Thermal radiation from Lorentzian traversable wormholes *Phys. Rev.* D **80** 024007 (*Preprint* gr-qc/0907.4055)

[7] Babichev E, Dokuchaev V and Eroshenko Yu 2004 Black hole mass decreasing due to phantom energy accretion *Phys. Rev. Lett.* **93** 021102 (*Preprint* gr-qc/0402089)

[8] González-Díaz P F 2004 Achronal cosmic future *Phys. Rev. Lett.* **93** 071301 (*Preprint* astro-ph/0404045)

[9] Gonzalez-Diaz P F 2006 Some notes on the big trip *Phys. Lett.* B **635** 1 (*Preprint* hep-th/0607137)

[10] Hayward S A 1998 Unified first law of black-hole dynamics and relativistic thermodynamics *Class. Quant. Grav.* **15** 3147 (*Preprint* gr-qc/9710089)

[11] Hayward S A, Di Criscienzo R, Vanzo L, Nadalini M and Zerbini S 2009 Local Hawking temperature for dynamical black holes *Class. Quant. Grav.* **26** 062001 (*Preprint* gr-qc/0806.0014)

Journal of Physics: Conference Series **314** (2011) 012038 doi:10.1088/1742-6596/314/1/012038

Towards the M(atrix) model action in an arbitrary 11D supergravity background. Progress report

Igor A. Bandos

Department of Theoretical Physics, University of the Basque Country UPV/EHU, P.O. Box 644, 48080 Bilbao, Spain and IKERBASQUE, Basque Foundation for Science, Bilbao, Spain

E-mail: igor_bandos@ehu.es, bandos@ific.uv.es

Abstract.
We are searching for the action principle for multiple M0–brane (multiple M-wave or mM0) system starting from the mM0 equations of motion obtained in the frame of superembedding approach. Surprisingly, the way from these equations to the action happens to be hampered by a problem which suggests a possible generalization of the action principle which we call "hierarchical action principle".

To our best knowledge, the M0–brane, which can be also called M-wave or 11D massless superparticle, was described for the first time in [1], where it was used to derive the action for D0-brane (Dirichlet particle, which is 10D type II massive superparticle) by generalized dimensional reduction. As far as the effective action for multiple M0-brane system is concerned, the construction of its purely bosonic limit starting from the 10D "dielectric brane action" by Myers [2] was the subject of [3]. However, as far as the 10D Myers action does not possesses neither supersymmetry nor Lorentz invariance, it should not be surprising that the Multiple M-wave or Multiple graviton action of [3] is also purely bosonic and is not Lorentz invariant.

Recently the *supersymmetric and* $SO(1,10)$ *Lorentz covariant equations of motion* for multiple M-wave system in an arbitrary supergravity background were derived in [4, 5] in the frame of superembedding approach [6, 7, 8, 9]. Probably these equations are approximate, but, if so, clearly going beyond the previously known approximations. The natural question is what is the action principle reproducing these set of equations? Surprisingly enough, construction of such an action meets a problem which we describe in this contribution. We also notice that possible resolution might consist in using a modification of the action principle which can be called "hierarchical action principle".

1. Action for single $mM0$ and equations for center of energy motion of $mM0$ system

The set of equations derived in [4, 5] is naturally split on the equations for the center of energy motion and the equations for the relative motion of the constituents of the mM0 system. The former subset of equations is obtained in [4, 5] as a consequence of the so-called superembedding equation, the basic equation of the superembedding approach to a single brane [6, 8], and formally coincide with the equations of motion for a single M0-brane (11D massless superparticle). These can be obtained from the action [10] (see [11, 12] for D=4 and D=3,6,10)

$$S_{M0} = \int_{W^1} \rho^{\#} \hat{E}^{=} := \int_{W^1} \rho^{\#} d\hat{Z}^M E_M^a(\hat{Z}) u_a^{=} \; . \tag{1}$$

In it W^1 is the superparticle worldline parametrized by proper time τ, $\hat{Z}^M(\tau) = (\hat{x}^\mu(\tau), \hat{\theta}^{\check{\alpha}}(\tau))$ are coordinate functions describing the embedding of this worldline into the target 11D superspace $\Sigma^{(11|32)}$ with local coordinates $Z^M = (x^\mu, \theta^{\check{\alpha}})$ (the set of which includes eleven bosonic coordinate x^μ, $\mu = 0, 1, \ldots, 9, 10$, and 32 fermionic coordinates $\theta^{\check{\alpha}}$, $\check{\alpha} = 1, \ldots, 32$; here 32 is the dimension of the Majorana spinor representation of $SO(1, 10)$), $u_a^{=} = u_a^{=}$ is a light–like vector variable, $u_a^{=} u^{a=} = 0$, $\hat{E}^{=} = \hat{E}^a u_a^{=}$, and $\hat{E}^a = d\hat{Z}^M E_M^a(\hat{Z})$ is the pull–back to W^1 of the bosonic supervielbein $E^a = dZ^M E_M^a(Z)$ of the target superspace. The detailed study of the action (1) can be found in [10] so that our presentation here will be brief and schematic.

The set of the 11D supervielbein forms $dZ^M E_M{}^A(Z) = (E^a, E^\alpha)$, including, in additional to the bosonic vectorial form E^a $(a = 0, 1, \ldots, 9, 10)$ also fermionic spinorial form $E^\alpha = dZ^M E_M^\alpha(Z)$ $(\alpha = 1, \ldots, 32)$, describes 11D supergravity background when obeys the set of superspace constraints [13, 14]. The most important of these constraints determine the bosonic torsion 2–form of $\Sigma^{(11|32)}$. This reads

$$T^a := DE^a = -iE^\alpha \wedge E^\beta \Gamma^a_{\alpha\beta} \, , \tag{2}$$

where $\Gamma^a_{\alpha\beta} = \Gamma^a_{\beta\alpha}$ are 11D Dirac matrices (see Appendix A), $\wedge$ denotes the exterior product of differential forms. The constraints (2) can be derived from the requirement that the M0-brane action (1) is invariant under the following local fermionic κ–symmetry transformations

$$i_\kappa \hat{E}^\alpha := \delta_\kappa \hat{Z}^M E_M^\alpha(\hat{Z}) = \kappa^{+q} v_q^{-\alpha} \, , \qquad i_\kappa \hat{E}^a = 0 \, , \quad \delta_\kappa u_a^{=} = 0 \, , \quad \delta_\kappa \rho^{\#} = 0 \, . \tag{3}$$

Here $\kappa^{+q} = \kappa^{+q}(\tau)$ is the κ–symmetry parameter, $q = 1, \ldots, 16$ can be considered as $SO(9)$ spinor index and $v_q^{-\alpha}$ is the set of 16 constrained 11D spinors $(\alpha = 1, \ldots, 32)$ which is related to the light–like vector $u_a^{=}$ by

$$v_q^- \Gamma_a v_p^- = u_a{}^{=} \delta_{qp} \, , \; (a) \qquad \delta_{qp} u_a^{=} = v_q^{-\alpha} \Gamma^a_{\alpha\beta} v_p^{-\beta} \, , \; (b) \qquad and \qquad v_q^{-\alpha} C_{\alpha\beta} v_p^{-\beta} = 0 \, . \; (c) \tag{4}$$

This set of constrained spinors can be also used to write the fermionic equations of motion of the M0 brane as follows

$$\hat{E}^\alpha := d\hat{Z}^M(\tau) E_M{}^\alpha(\hat{Z}(\tau)) = e^{+q} v_q^{-\alpha} \, . \tag{5}$$

The bosonic equations of motion can be also written in terms of the light–like vector $u_a^{=}$ and properly chosen covariant derivative. Their set includes the relations

$$\hat{E}^a := d\hat{Z}^M(\tau) E_M{}^a(\hat{Z}(\tau)) = \tfrac{1}{2} e^{\#} u_a^{=} \, , \; (a) \qquad Du_a^{=} = 0 \, , \; (b) \qquad Dv_q^{-\alpha} = 0 \, , \; (c) \, , \tag{6}$$

where D is an appropriate covariant derivative. In our worldline approach $D = d\tau D_\tau = e^{\#} D_{\#}$, where $e^{\#} = d\tau e_\tau^{\#}(\tau)$ is the einbein form, but in the superembedding formalism of [4, 5] it has also fermionic components $(D = e^{\#} D_{\#} + e^{+q} D_{+q})$ while (6) still holds. These covariant derivative enters as well in the equation for the Lagrange multiplies $\rho^{\#}$ which reads

$$D\rho^{\#} := d\rho^{\#} - 2\rho^{\#} \omega^{(0)} = 0 \, , \tag{7}$$

where $\omega^{(0)}$ is a certain $SO(1, 1)$ connection and, as above, $D = e^{\#} D_{\#}$.

To define the $SO(1, 10) \times SO(1, 1) \times SO(9)$ covariant derivative used in (6) and below it is convenient, to complete the light–like vector $u_a^{=}$ till the full *moving frame* containing, besides $u_a^{=}(\tau)$, also a complementary light-like vector $u_a^{\#}(\tau)$ and 9 normalized spacelike vectors $u_b^i(\tau)$ orthogonal to both of them and among themselves,

$$\begin{aligned} u_a^{=} u^{a\,=} &= 0 \, , \qquad & u_a^{\#} u^{a\,\#} &= 0 \, , \qquad & u_a^{\#} u^{a\,=} &= 2 \, , \\ u_a^{=} u^{a\,i} &= 0 \, , \qquad & u_a^{\#} u^{ai} &= 0 \, , \qquad & u_a^i u^{aj} &= -\delta^{ij} \, . \end{aligned} \tag{8}$$

With this vectors one can split Eq. (6a) on

$$\hat{E}^{=} := \hat{E}^a u_a^{=} = 0\,, \quad (a) \qquad \hat{E}^i := \hat{E}^a u_a^i = 0\,, \quad (b) \qquad \mathit{and} \qquad \hat{E}^{\#} := \hat{E}^a u_a^{\#} = e^{\#}\,.\,(c) \quad (9)$$

The latter equation defines the einbein induced by embedding of W^1 into $\Sigma^{(11|32)}$. The fermionic equations of motion for a single M0 brane, and also for the center of mass of the mM0 system in the superembedding description of [4, 5], can be written in the form of

$$\hat{E}_q^{-} := \hat{E}^{\alpha} v_{\alpha q}{}^{-} = 0\,. \qquad (10)$$

where $v_{\alpha q}^{-}$ is related to above $v_q^{-\beta}$ by $v_{\alpha q}^{-} = iC_{\alpha\beta} v_q^{-\beta}$. The equivalence of Eqs. (10) and (5) holds due to the constraint (4c). Further details on the worldline geometry induced by this embedding, on the $SO(1,10) \times SO(1,1) \times SO(9)$ covariant derivative D as well as on moving frame and spinor moving frame variables can be found in [10, 4].

The above equations describe as well the center of energy motion of the mM0 system as described by superembedding approach of [4, 5].

2. Lagrangian which might be used to produce the equation of relative motion of the $mM0$ constituents

The relative motion of the constituents of the system of N M0–branes was described in [4, 5] in terms of the nanoplet of anti-hermitian bosonic $N \times N$ matrix fields $\mathbf{X}^i$ with $SO(1,10)$ weight 2, the fact which can be expressed by writing $\mathbf{X}^i = \mathbf{X}^i_{\#} = \mathbf{X}^i_{++}$, by a 16-plet of fermionic $N \times N$ matrix fields $\mathbf{\Psi}_q$ with $SO(1,1)$ weight 3, $\mathbf{\Psi}_q = \mathbf{\Psi}_{\#+q} = \mathbf{\Psi}_{+++q}$,

$$\mathbf{X}^i(\tau) \in su(N)\,, \qquad i = 1, \ldots, 9\,, \qquad \mathbf{\Psi}_q(\tau) \in su(N)\,, \qquad q = 1, \ldots, 16\,, \qquad (11)$$

and by (auxiliary; actually, pure gauge) one dimensional $SU(N)$ connection $\mathbf{A}_{\#}$. These fields are inert under the 11D Lorentz $SO(1,10)$ transformations. The equations for $\mathbf{X}^i(\tau)$ and $\mathbf{\Psi}_q(\tau)$ obtained in [5] involve the projections $\hat{F}_{\#ijk} := F^{abcd} u_a{}^{=} u_b{}^i u_c{}^j u_d{}^k$, $\hat{T}_{\#i+q} := T_{ab}{}^{\beta}(\hat{Z}) v_{\beta q}^{-} u_a^{=} u_b^i$, $\hat{R}^{=i} := \hat{R}^{ab} u_a^{=} u_b{}^i = 0$ of the four form flux, gravitino field strength and curvature tensor; they can be reproduced by varying Lagrangian

$$\mathcal{L} := \mathcal{L}_{\#\#\#\#} = tr\left(\tfrac{1}{2} D_{\#}\mathbf{X}^i D_{\#}\mathbf{X}^i - \tfrac{1}{64}([\mathbf{X}^i, \mathbf{X}^j])^2 - 4i\mathbf{\Psi}_q D_{\#}\mathbf{\Psi}_q + 2i\mathbf{X}^i \mathbf{\Psi}_q \gamma^i_{qp} \mathbf{\Psi}_p\right) +$$
$$+\tfrac{1}{2}\hat{R}_{\#i\#j}\, tr(\mathbf{X}^i\mathbf{X}^j) + \tfrac{1}{12}\hat{F}_{\#ijk}\, tr(\mathbf{X}^i\mathbf{X}^j\mathbf{X}^k) + \tfrac{i}{3}\hat{F}_{\#ijk}\, tr(\mathbf{\Psi}\gamma^{ijk}\mathbf{\Psi}) - 2i\hat{T}_{\#i\,+q}\, tr(\mathbf{X}^i\mathbf{\Psi}_q) \quad (12)$$

with respect to Ψ_q, $A_{\#}$ and $\mathbf{X}^i$, and *omitting the complete derivatives.* This corresponds to the variation of action $S_{relative} = \int \mu^{\#\#\#\#} \mathcal{L}_{\#\#\#\#}$ constructed by integrating the Lagrangian (12) with a proper measure $\mu^{\#\#\#\#}$ allowing integration by parts, and omitting the boundary terms.

3. Problems with the measure for the mM0 action and possible relevance of the "hierarchic action principle"

The natural candidates for the measure with the appropriate $SO(1,1)$ weight (-8) are [1] $\hat{E}^{\#}(\rho^{\#})^3$ and $\hat{E}^{=}(\rho^{\#})^5$. However, the first one, when used in the $S_{mM0} = S_{M0} + S_{relative}$, would result in breaking of a number of gauge symmetries characteristics of the center of energy action S_{M0}. This forces us to consider the other candidate, $\mu^{\#\#\#\#} = \hat{E}^{=}(\rho^{\#})^5$, with which

$$S_{relative} = \int_{W^1} \hat{E}^{=}(\rho^{\#})^5 \mathcal{L}_{\#\#\#\#}\,. \qquad (13)$$

[1] Notice that in our notation the positive weight 1 with respect to $SO(1,1)$ group is denoted either by subindex $_+$ or by a superindex $^-$, so that one can write, for instance, $\mathbf{\Psi}_q = \mathbf{\Psi}_{\#+q} = \mathbf{\Psi}_{\#q}^{-} = \mathbf{\Psi}^{=-q}$, $\rho^{\#} = \rho^{++} = \rho_{--} = \rho_{=}$.

The variation of (13) does reproduce the equations of relative motion of the mM0 constituents as they were obtained in [5], but multiplied by $\hat{E}_\tau^=$. Then the problem is that $\hat{E}_\tau^=$ vanishes on the "center of energy mass shell" due to Eq. (9a), $\hat{E}^= = d\tau^= \hat{E}_\tau^= = 0$. Notice also that, when this equation is taken into account, the functional (13) vanishes itself so that the variational problem for $S_{relative}$ does not make sense on the center of energy mass shell.

These simple observations suggest the following (probably, looking a bit artificial) prescription of extracting the equations of [4, 5] from (13). First one does not set $\hat{E}_\tau^=$ equal to zero and vary the action (13) with respect to the relative motion variables $\mathbf{\Psi}_q$, $\mathbf{A}_\#$ and $\mathbf{X}^i$. Then one divide the equations thus obtained by $\hat{E}_\tau^=$ (thus compensating the $\hat{E}_\tau^=$ multipliers which are present in them) and, only after that, one uses the center of energy equations, including Eq. (9a), $\hat{E}_\tau^= = 0$. To resume, the equations for relative motion of mM0 constituents can be obtained as

$$\lim_{\hat{E}_\tau^= \mapsto 0} \frac{1}{\hat{E}_\tau^=} \frac{\delta S_{relative}}{\delta \mathbf{\Psi}_q} = 0 \, , \qquad \lim_{\hat{E}_\tau^= \mapsto 0} \frac{1}{\hat{E}_\tau^=} \frac{\delta S_{relative}}{\delta \mathbf{A}_\#} = 0 \, , \qquad \lim_{\hat{E}_\tau^= \mapsto 0} \frac{1}{\hat{E}_\tau^=} \frac{\delta S_{relative}}{\delta \mathbf{X}^i} = 0 \, . \tag{14}$$

Such a prescription of 'hierarchical action principle', formulated above for the relative motion, can be extended to the complete description of the multiple M0 system by the action

$$S_{mM0} = S_{M0} + S_{relative} = \int_{W^1} \hat{E}^= \rho^\# \left(1 + (\rho^\#)^4 \mathcal{L}_{\#\#\#\#} \right) \, , \tag{15}$$

where $\mathcal{L}_{\#\#\#\#}$ is defined in (12). Variation of (15) produces all the equations of center of energy motion of the mM0 system, including Eq. (9a) (when one assumes a generic situation with $\mathcal{L}_{\#\#\#\#} \neq -\frac{1}{5(\rho^\#)^4}$). Then the reason for the special rôle of Eq. (9a) is that the action (15) vanishes due to it, and the equations of the relative motion are obtained as in (14).

The selfconsistency of the above described *"hierarchical action principle"* requires additional study (hence the 'progress report' in the title). Notice that some generalizations of the action principle were proposed and used before. Let us mention the generalized action principle of the group manifold approach to supergravity [17], its p-brane counterpart [18] and the 'democratic action' [19] for 10D type II supergravity[2].

Aknowledgments The author is thankful to Dmitri Sorokin for useful discussion. The partial support from the research grants FIS2008-1980 from the Spanish MICINN and the Basque Government Research Group Grant ITT559-10 is greatly acknowledged.

References

[1] Bergshoeff E and Townsend P K 1997 Super-*D*-branes *Nucl.Phys.* **B490**, 145–162. (*Preprint* hep-th/9611173).
[2] Myers R C 1999 Dielectric-branes JHEP 12(1999)022 (*Preprint* hep-th/9910053).
[3] Janssen B and Lozano Y 2002 *Nucl. Phys.* **B643**, 399; **B658**, 281 (2003) (*Preprint* hep-th/0207199).
[4] Bandos I A 2010a *Phys. Lett.* **B687**, 258–263 (*Preprint* arXiv:0912.5125 [hep-th]).
[5] Bandos I A 2010b *Phys. Rev. Lett.* **105**, 071602 [1-4] (2010); Phys. Rev. **D82**, 105030 [1-19] (2010).
[6] Bandos I, Sorokin D, Tonin, Pasti P and Volkov D 1995 *Nucl. Phys.* **B446**, 79-118 (*Preprint* hep-th/9501113).
[7] Howe P S and Sezgin E 1996 *Phys. Lett.* **B390**, 133; **B394**, 62 (*Preprints* hep-th/9607227, hep-th/9611008).
[8] Sorokin D P 2000 Superbranes and superembeddings, *Phys. Rept.* **329**, 1-101 (*Preprint* hep-th/9906142).
[9] Bandos I A 2009 *Phys. Lett.* **B680**, 267–273 (2009) (*Preprint* 0907.4681 [hep-th]).
[10] Bandos I A 2008 *Phys. Lett.* B **659**, 388 *Nucl. Phys.* **B796**, 360 (*Preprint* 0707.2336, 0710.4342 [hep-th]).
[11] Bandos I A 1990 Sov. J. Nucl. Phys. **51**, 906–914 (1990).
[12] Bandos I A and Nurmagambetov A Yu 1995 Class. Quant. Grav. **12**, 1881-1892 (*Preprint* hep-th/9502143).
[13] Cremmer E and Ferrara S 1980 *Phys. Lett.* B **91**, 61.
[14] Brink L and Howe P S, *Phys. Lett.* B **91**, 384.
[15] Witten E 1996 *Nucl. Phys.* B **460**, 335 (*Preprint* hep-th/9510135).
[16] Banks T, Fischler W, Shenker S H and Susskind L 1997, *Phys. Rev.* D **55**, 5112-5128.
[17] Ne'eman Y and Regge T 1978 *Phys. Lett.* **B74**, 54; *Riv. Nuovo Cim.* **1N5**, 1.
[18] Bandos I A, Sorokin D P and Volkov D V 1995 *Phys. Lett.* **B352**, 269–275 (*Preprint* hep-th/9502141).
[19] Bergshoeff E, Kallosh R, Ortin T, Roest D and Van Proeyen A 2001 *Class. Quant. Grav.* **18**, 3359.
[20] Manvelyan R and Mkrtchyan R 2002 *Mod. Phys. Lett.* **A17**, 1393 (*Preprint* hep-th/0112233).

[2] An example of a more radical modification is the two-time 'multiaction' construction of [20].

Higher order first integrals and quantum gravitational anomalies

Anca Visinescu, Mihai Visinescu

Department Theoretical Physics, National Institute for Physics and Nuclear Engineering, Magurele MG-6, Bucharest, Romania

E-mail: avisin, mvisin@theory.nipne.ro

Abstract. Hidden symmetries on curved space-times are investigated in connection with higher rank Killing tensors. It is shown that at the quantum level the conformal Killing vectors and tensors do not in general produce operators that commute with the Klein-Gordon operator.

1. Introduction

The concept of symmetries is one of the key concepts in physics, Noether's theorem giving a correspondence between symmetries and conservation laws.

The evolution of a dynamical system is described in the entire phase-space and from this point of view it is natural to go in search of conserved quantities to genuine symmetries of the complete phase-space, not just the configuration one. Such symmetries are associated with higher rank symmetric Stackel-Killing (SK) tensors which generalize the Killing vectors. These higher order symmetries are known as *hidden symmetries* and the corresponding conserved quantities are quadratic, or, more general, polynomial in momenta. Also SK tensors play a pivotal role in Hamilton-Jacobi theory of separation of variables and the integrability of finite-dimensional Hamiltonian systems [1]. Another natural generalization of the Killing vectors is represented by the antisymmetric Killing-Yano (KY) tensors which in many respects are more fundamental than the SK tensors.

The conformal extension of the SK tensor equation determines the conformal Stackel-Killing (CSK) tensors which define first integrals of motion of the null geodesics. Investigations of the hidden symmetries of the higher dimensional space-times have pointed out the role of the conformal Killing-Yano (CKY) tensors to generate background metrics with black hole solutions.

Passing from the classical motions to the hidden symmetries of a quantized system it is necessary to investigate the corresponding quantum conserved quantities and separability of the equations of motion. Especially in the case of hidden symmetries it can appear *anomalies* representing discrepancies between the conservation laws at the classical level and the corresponding ones at the quantum level.

The plan of the paper is as follows. In Section 2 the generalized Killing equations are presented in connection with higher order conserved quantities. In Section 3 we discuss the special role of the KY tensors in generating higher order symmetries. In the next Section we describe the relationship between conformal symmetries and the corresponding quantum operators and analyze the existence of quantum gravitational anomalies. Finally, the last Section is devoted to conclusions.

2. Symmetries and conserved quantities

Let us consider the Hamilton function describing the geodesic motion in a curved n-dimensional (pseudo-)Riemmanian space $\mathcal{M}$ with the metric $\mathbf{g}$

$$H = \frac{1}{2} g^{\mu\nu} p_\mu p_\nu \,. \tag{1}$$

A conserved quantity of motion can be expanded as a power series in momenta

$$K = K_0 + \sum_{k=1}^{p} \frac{1}{k!} K^{\mu_1 \cdots \mu_k}(x) p_{\mu_1} \cdots p_{\mu_k} \,, \tag{2}$$

having a vanishing Poisson bracket with the Hamiltonian.

For the conservation of K its terms must satisfy

$$K^{(\mu_1 \cdots \mu_k ; \mu)} = 0 \,, \tag{3}$$

where a semicolon denotes the covariant differentiation corresponding to the Levi-Civita connection and round brackets denote full symmetrization over the indices enclosed. A symmetric tensor $K^{\mu_1 \cdots \mu_k}$ satisfying (3) is called a SK tensor of rank k. Let us note that for any geodesic γ with tangent vector $\dot{x}^\mu = p^\mu$

$$Q_K = K_{\mu_1 \cdots \mu_k} \dot{x}^{\mu_1} \cdots \dot{x}^{\mu_k} \,, \tag{4}$$

is constant along γ.

3. Killing-Yano tensors

KY tensors are a different generalization of Killing vectors which can be studied on a manifold. They were introduced by Yano [2] from a purely mathematical perspective and later on it turned out they have many interesting properties relevant to physics. In recent years the KY tensors are related to a multitude of different topics such as supersymmetries, index theorems, supergravity theories, and so on [3, 4].

A KY tensor $Y_{\mu_1 \cdots \mu_p}$ is antisymmetric satisfying the following equation:

$$Y_{\mu_1 \cdots \mu_{p-1} (\mu_p ; \nu)} = 0 \,. \tag{5}$$

The first connection with the symmetry properties of the geodesic motion is the observation that along every geodesic γ in $\mathcal{M}$, $Y_{\mu_1 \cdots \mu_{p-1} \nu} \dot{x}^\nu$ is parallel.

These two generalizations SK and KY of the Killing vectors could be related. Let $Y_{\mu_1 \cdots \mu_p}$ be a KY tensor, then the symmetric tensor field

$$K_{\mu\nu} = Y_{\mu\mu_2 \cdots \mu_p} Y_\nu{}^{\mu_2 \cdots \mu_p} \,, \tag{6}$$

is a SK tensor and it sometimes refers to this SK tensor as the associated tensor with $Y_{\mu_1 \cdots \mu_p}$.

Having in mind the special role of null geodesic for the motion of massless particles, it is convenient to look for conformal generalization of KY tensor. Let us mention also the remarkable role of the conformal Killing-Yano (CKY) tensors in the study of the properties of higher dimensional black holes (see e. g. [5, 6] and the cites contained therein). In what follows we limit ourselves to CKY tensors of rank 2 which satisfy [7]

$$Y_{\mu\nu;\lambda} + Y_{\lambda\nu;\mu} = \frac{2}{n-1} \left(g_{\lambda\mu} Y^\sigma{}_{\nu;\sigma} + g_{\nu(\mu} Y_{\lambda)}{}^\sigma{}_{;\sigma} \right) \,. \tag{7}$$

There is also a conformal generalization of the SK tensors, namely a symmetric tensor $K_{\mu_1\cdots\mu_p} = K_{(\mu_1\cdots\mu_p)}$ is called a CSK tensor if it obeys the equation

$$K_{(\mu_1\cdots\mu_p;\nu)} = g_{\nu(\mu_1}\tilde{K}_{\mu_2\cdots\mu_p)}\,, \tag{8}$$

where the tensor $\tilde{K}$ is determined by tracing the both sides of equation (8). Let us note that in the case of CSK tensors, the quantity (4) is constant only for null geodesics γ. There is also a similar relation between CKY and CSK tensors as in equation (6) [7].

4. Quantum gravitational anomalies

In order to find the necessary conditions for the existence of constants of motion in a first-quantized system we replace momenta by derivatives and look for operators commuting with the Hamiltonian:

$$\mathcal{H} = \Box = \nabla_\mu g^{\mu\nu}\nabla_\nu = \nabla_\mu\nabla^\mu\,, \tag{9}$$

corresponding to a free scalar particle and the covariant Laplacian is acting on scalars.

Many times the classical conserved quantities associated with SK tensors do not generally transfer to the quantized systems producing so-called quantum anomalies [8]. In what follows we shall analyze the quantum anomalies in the case of SK and CSK tensors and, to make things more specific, we confine ourselves to the case of tensors of rank 1 and 2.

For the beginning let us consider the conserved operator corresponding to a conformal Killing vector K^μ in the quantized system

$$\mathcal{Q}_V = K^\mu\nabla_\mu\,. \tag{10}$$

In order to identify a quantum gravitational anomaly we shall evaluate the commutator $[\Box, \mathcal{Q}_V]\Phi$ for $\Phi \in \mathcal{C}^\infty(\mathcal{M})$, solutions of the Klein-Gordon equation with the Klein-Gordon operator (9). A straightforward calculation gives

$$[\mathcal{H}, \mathcal{Q}_V] = \frac{2-n}{n}K_\lambda{}^{;\lambda\mu}\nabla_\mu + \frac{2}{n}K^\lambda{}_{;\lambda}\Box\,. \tag{11}$$

As it is expected, in the case of ordinary Killing vectors the r. h. s. of this commutator vanishes and there are no quantum gravitational anomalies. But for conformal Killing vectors, we confront with a quite different situation. Even if we evaluate the r. h. s. of (11) on solutions of the massless Klein-Gordon equation, $\Box\Phi(x) = 0$, the term $K_\lambda{}^{;\lambda\mu}\nabla_\mu$ survives. It is possible to find some particular spaces in which the r. h. s. of (11) vanishes, but in general the system is affected by quantum gravitational anomalies.

In the case of SK and CSK tensors of rank 2 the quantum analog of the classical conserved quantity (4) is

$$\mathcal{Q}_T = \nabla_\mu K^{\mu\nu}\nabla_\nu\,, \tag{12}$$

where $K^{\mu\nu}$ is a SK (3) or a CSK (8) tensor.

Now the evaluation of the commutator between $\mathcal{H}$ and the quantum operator (12) is more involved than in case of Killing vectors and the result of this tedious evaluation is [9, 10]:

$$\begin{aligned}[\Box, \mathcal{Q}_T] &= 2\left(\nabla^{(\lambda}K^{\mu\nu)}\right)\nabla_\lambda\nabla_\mu\nabla_\nu + 3\nabla_\tau\left(\nabla^{(\lambda}K^{\tau\nu)}\right)\nabla_\nu\nabla_\lambda + \left\{-\frac{4}{3}\nabla_\lambda\left(R_\tau{}^{[\lambda}K^{\nu]\tau}\right)\right.\\ &+ \left.\nabla_\lambda\left(\frac{1}{2}g_{\tau\sigma}(\nabla^\lambda\nabla^{(\tau}K^{\sigma\nu)} - \nabla^\nu\nabla^{(\tau}K^{\lambda\sigma)}) + \nabla_\mu\nabla^{(\lambda}K^{\mu\nu)}\right)\right\}\nabla_\nu\,. \end{aligned} \tag{13}$$

We mention that the last term is missing in the corresponding equation in [8]. Note also that the terms are arranged into groups with three, two and just one derivatives and consequently it is impossible to have compensations between them.

However in the case of SK tensors all the symmetrized derivatives vanish and we end up with a simpler result

$$[\Box, \mathcal{Q}_T] = -\frac{4}{3}\nabla_\mu (R_\sigma{}^{[\mu} K^{\nu]\sigma})\nabla_\nu \,. \tag{14}$$

Since the r. h. s. does not vanish identically, SK tensors exhibit quantum anomalies, i. e. the classical conservation law does not transfer to the quantum level.

Although in general SK tensors do not give quantum mechanical symmetries, there are a few notable conditions for which the r. h. s. of (14) vanishes. The simplest case in which this happens is obviously when the space is Ricci flat. A slightly more general case is when the space is Einstein, i. e. $R_{\mu\nu} \propto g_{\mu\nu}$ (that is, if the vacuum Einstein equations are fulfilled). In this case we get that the r. h. s. of the commutator involves $K^{[\mu\nu]}$ and consequently vanishes since SK tensors are symmetric. A more interesting and quite unexpected case is represented by SK tensors associated with KY tensors of rank 2 as in (6) [8], a situation which occurs for some spaces.

For CSK tensors the existence of some favorable conditions which could prevent the appearance of quantum anomalies is quite impossible. In the case of CSK tensors we could not simplify any more the commutator (13) since for them the symmetrized derivatives do not vanish. Even if we evaluate the commutator for CSK tensors associated with CKY tensors we do not obtain a cancellation of anomalies [10]. Therefore we are not able to find any favorable circumstances on the CSK tensors in order to achieve a conserved quantum operator.

5. Concluding comments

The (C)SK and (C)KY tensors are related to a multitude of different topics such as classical integrability of systems together with their quantization, supergravity, string theories, hidden symmetries in higher dimensional black-holes space-times, etc.

To conclude let us discuss shortly some problems that deserve a further attention. More elaborate examples working in a N-dimensional curved space-time and involving higher ranks ($k > 2$) SK tensors will be presented elsewhere. The concept of generalized (C)KY symmetry in the presence of a skew-symmetric torsion is more widely applicable and may become very powerful [11].

Acknowledgments

Support through CNCSIS program IDEI-571/2008 is acknowledged.

References

[1] Benenti S 1997 *J. Math. Phys.* **38** 6578
[2] Yano K 1952 *Ann. Math.* **55** 328
[3] Cariglia M 2004 *Class. Quantum Grav.* **21** 1051
[4] Visinescu M 2010 *Mod. Phys. Lett. A* **25** 341
[5] Krtouš P, Frolov V P and Kubizňák D 2008 *Phys. Rev. D* **78** 064022
[6] Houri T, Oota T and Yasui Y 2009 *Class. Quantum Grav.* **26** 045015
[7] Jezierski J 2002 *Class. Quantum Grav.* **19** 4405
[8] Carter B 1977 *Phys. Rev. D* **16** 3395
[9] Ianuş S, Visinescu M and Vîlcu G E 2009 *SIGMA* **5** 022
[10] Visinescu M 2010 *EPL Europhys. Lett.* **90** 41002
[11] Houri T, Kubizňák D, Warnick C M and Yasui Y 2010 *JHEP* **1007** 055

Journal of Physics: Conference Series **314** (2011) 012040 doi:10.1088/1742-6596/314/1/012040

Brane solutions and integrability: a status report

W. Chemissany [1], P. Fré [2], Jan Rosseel [3], A. S. Sorin [4], M. Trigiante [5], and T. Van Riet [6]

[1] Afdeling Theoretische Fysica, Katholieke Universiteit Leuven Celestijnenlaan 200D bus 2415, 3001 Heverlee, Belgium
[2] Dipartimento di Fisica Teorica, Universitá di Torino, & INFN - Sezione di Torino, via P. Giuria 1, I-10125 Torino, Italy
[3] Centre for Theoretical Physics, University of Groningen, Nijenborgh 4, 9747 AG Groningen, The Netherlands
[4] Bogoliubov Laboratory of Theoretical Physics, Joint Institute for Nuclear Research, 141980 Dubna, Moscow Region, Russia
[5] Dipartimento di Fisica Politecnico di Torino, C.so Duca degli Abruzzi, 24, I-10129 Torino, Italy
[6]Institutionen för Fysik och Astronomi, Box 803, SE-751 08 Uppsala, Sweden

E-mail: wissam@itf.fys.kuleuven.be, fre@to.infn.it, j.rosseel@rug.nl, sorin@theor.jinr.ru, mario.trigiante@polito.it,thomas.vanriet@fysast.uu.se

Abstract.
We review the status of the integrability and solvability of the geodesics equations of motion on symmetric coset spaces that appear as sigma models of supergravity theories when reduced over respectively the timelike and spacelike direction. Such geodesic curves describe respectively timelike and spacelike brane solutions. We emphasize the applications to black holes.

The construction of black hole solutions using sigma model techniques becomes very beneficial in supergravity theories with enough symmetries as first employed in [1]. This has later been extended to more general stationary p-brane solutions [2] and even time-dependent solutions (spacelike branes) [3–6], see [7] for the general framework. The sigma model technique reduces the problem to the one of geodesic motion on the appropriate moduli spaces. In the case of time-dependent solutions the moduli space is a Riemannian coset space G/H, where H is compact and in the stationary case a non-Riemannian coset space, G/H^*, where H^* is non-compact. In this status report we review the recent insights of [8,9,12–14] on the integrability and solvability of the geodesic equations of motion when the coset moduli spaces are symmetric. In these works we established the following results

- We made use of the group-theoretical structure of the target space to prove that the second order geodesic equations are integrable in the sense of Liouville [13], by explicitly constructing the correct number of conserved Poisson-commuting constants of motion from the underlying Poisson structure of the dual Borel algebras associated with G/H^* [9].
- We presented an integration method by the construction of a Lax algorithm that integrates the second order equations in one step [13], improving on the previously constructed two-step algorithms [8–10, 12] (see [4–6] for the case of Riemannian manifolds).
- In the case of black holes we showed that Liouville integrability implies that there always exists a description in terms of a (fake) superpotential $\mathscr{W}$, in the effective potential approach

(see e.g. [11,13,15–18] and refs therein). The correspondence between the effective potential and the sigma model approach was studied in depth in [14] by establishing an interesting generalization of Toda molecule equations.

In analogy with the Riemannian case [6], the geodesic equations on a symmetric space with indefinite signature can be recast into the Lax form [8,9]

$$\dot{L}(\tau) = [L(\tau), W(\tau)], \tag{1}$$

where the dot denotes the differentiation w.r.t τ. L and W are respectively the Lax operator and connection, being defined in terms of the pull-back Ω on the geodesic of the left-invariant 1-form on G/H^*, i.e., in terms of a coset representative $\mathbb{L}$,

$$\Omega = \mathbb{L}^{-1}\dot{\mathbb{L}} = L + W. \tag{2}$$

The corresponding geodesic action can thus be written as $S \sim \int d\tau \, \mathrm{Tr}(LL)$. Choosing the solvable gauge[1], one has

$$W \;\; = \;\; L_{>} - L_{<}, \tag{3}$$

where $L_{<(>)}$ denotes the upper-triangular (resp. lower-triangular) part of L. The Lax algorithm that we have devised to solve (1) is summarized by the following formula for the solution of the coset representative [13]

$$\left(\mathbb{L}(\tau)^{-1}\right)_{ij} \;\; \equiv \;\; \frac{1}{\sqrt{\mathfrak{D}_i(\mathscr{C})\mathfrak{D}_{i-1}(\mathscr{C})}}\mathrm{Det}\begin{pmatrix} \mathscr{C}_{1,1}(\tau) & \dots & \mathscr{C}_{1,i-1}(\tau) & (\mathscr{C}(\tau)\mathbb{L}(0)^{-1})_{1,j} \\ \vdots & \vdots & \vdots & \vdots \\ \mathscr{C}_{i,1}(\tau) & \dots & \mathscr{C}_{i,i-1}(\tau) & (\mathscr{C}(\tau)\mathbb{L}(0)^{-1})_{i,j} \end{pmatrix},$$

and we have defined

$$\mathscr{C}(\tau) := \mathrm{e}^{-2\tau L_0}, \tag{4}$$

$$\mathfrak{D}_i(\mathscr{C}) := \mathrm{Det}\begin{pmatrix} \mathscr{C}_{1,1}(\tau) & \dots & \mathscr{C}_{1,i}(\tau) \\ \vdots & \vdots & \vdots \\ \mathscr{C}_{i,1}(\tau) & \dots & \mathscr{C}_{i,i}(\tau) \end{pmatrix}, \quad \mathfrak{D}_0(\tau) := 1, \tag{5}$$

where L_0 and $\mathbb{L}(0)$ are, respectively, the initial values of the Lax matrix $L(\tau)$ and the coset representative $\mathbb{L}(\tau)$. The above equations generalize analogous equations written in [4–6] for the Lax operator in the case of Rimannian manifolds.

Having the Lax form does not yet imply the Liouville integrability of the system. By Liouville integrability we mean that there exist n functionally independent constants of motion (Hamiltonians $\mathscr{H}_i$) satisfying $\{\mathscr{H}_i, \mathscr{H}_j\} = 0$.

Let us summarize the proof of Liouville integrability of the full *second order* geodesic equations presented in detail in [13]. Consider the Noether charge matrix

$$Q = \mathbb{L}(\tau)L(\tau)\mathbb{L}(\tau)^{-1}, \quad Q = Q^A T_A, \quad L = Y^A T_A, \tag{6}$$

where T_A are the generators of the solvable algebra. One can verify the following Poisson brackets

$$\{Y_A, Y_B\} = -f_{AB}{}^{C} Y_c, \qquad \{Q_A, Q_B\} = f_{AB}{}^{C} Q_C. \tag{7}$$

[1] The solvable parametrization on G/H^* amounts to defining a coset representative $\mathbb{L} \in G/H^*$, namely, $\mathbb{L}(\phi) = \exp(\Phi^i T_i)$, where T_i's are the generators of the solvable algebra.

Note that Y_A and Q_A depend on the phase space variables of the Hamiltonian system. Denoting the dimension of the coset by n and the dimension of the symplectic leaves by $2h_O$, One can show that

$$2h_O = \mathrm{rank}(f_{AB}{}^C) \tag{8}$$
$$\#\text{Hamiltonians} = n - h_O \tag{9}$$
$$\#\text{Casimirs} = n - 2h_O \tag{10}$$

If we denote by

$$\mathscr{H}_a(Y_A), \quad a = 1, \cdots, h_O; \qquad \mathscr{H}_\ell(Y_A), \quad \ell = 1, \cdots, n - 2h_O \tag{11}$$

respectively the Hamiltonians on the leaves and the Casimirs, we can find the following $2(n-h_O)$ constants of motion in involution (using identity (7))

$$\mathscr{H}_a(Y_A), \qquad \mathscr{H}_\ell(Y_A), \qquad \mathscr{H}_a(Q_A), \qquad \mathscr{H}_\ell(Q_A), \tag{12}$$

where the $\mathscr{H}(Q)$'s are obtained by replacing Y_A by Q_A in the $\mathscr{H}(Y)$'s. One can demonstrate that each $\mathscr{H}_\ell(Q_A)$ is a Casimir on its own and can thus be expressed as a function of $\mathscr{H}_\ell(Y_A)$. This therefore implies that the total number of Hamiltonians in involution is $(n-h_O)+h_O = n$, the dimension of G/H^*, thereby proving Liouville integrability. The explicit construction and general form of the Hamiltonians are discussed in [13] for the case of $\mathrm{SL}(n,\mathbb{R})$. Amongst the $\mathscr{H}(Y)$'s are certain number of *polynomial* conserved constants of motion which have a simple interpretation in the context of black holes. One of them corresponds to the extremality parameter c. The remaining polynomial $\mathscr{H}(Y)$ keeps track of the regularity of the solution (they have to vanish for regular solutions). The Taub-NUT charge corresponds to one of the remaining non-polynomial $\mathscr{H}(Y)$, and electro-magnetic charges of the black hole solutions are functions of $\mathscr{H}(Q)$.

This has been shown to imply that the Hamilton-Jacobi formalism, HJ, can be applied. This allows us, formally, to rewrite the momenta P_i as functions of ϕ, the canonical coordinates (scalars parameterizing G/H^*), provided $J = \det(\partial\mathscr{H}_i/\partial P_j) \neq 0$ and conclude that (locally)

$$\dot{\phi}^i = g^{ij}(\phi)P_j(\phi, \mathscr{H}), \qquad P = d\mathscr{W}, \tag{13}$$

which is nothing else than the standard HJ equations. In this sense the function $\mathscr{W}$ is the *Hamilton's characteristic function* that solves the HJ equations

$$\mathscr{H}\left(\frac{\partial\mathscr{W}}{\partial\phi^i}, \phi^i\right) = \frac{1}{2}\frac{\partial\mathscr{W}}{\partial\phi^i}g^{ij}(\phi)\frac{\partial\mathscr{W}}{\partial\phi^j} = c^2, \tag{14}$$

where c is the affine velocity along the geodesic. In problems related to black holes, these results solve an open question about the existence of a fake superpotential for black hole solutions. We have proven the (local) existence of a fake superpotential for *all* stationary, spherically symmetric, black hole solutions to symmetric supergravity theories. In [13] we illustrated this in an explicit example, $\mathrm{SL}(3,\mathbb{R})$-model, in which we moreover have proven that the Jacobian J is always non-vanishing in the subspace of regular solutions, thus ensuring the global existence of $\mathscr{W}$.

As an application to black hole solutions, we have explicitly worked out in [13] the three findings mentioned in the introduction for the dilatonic black holes arising in Kaluza-Klein theories. Due to the simplicity of the dilatonic black hole model we are not able to understand the interpretation for the generic Hamiltonians as the model has only one non-polynomial $\mathscr{H}(Y)$, which is the Taub-NUT charge. Three of us are now trying to investigate the Hamiltonians in more involved models, such as the STU and G_2 models, which seems to make the physical interpretation more apparent.

Finally we want to end with a few interesting open problems that we would like to address in the future. First, it seems reasonable to extend the proof of Liouville integrability to the case of homogeneous non-symmetric spaces. Second, as already mentioned, it would be interesting to understand the physical interpretation of all constants of motion. As a further direction of investigation it might be useful to examine the role of the complete system of Hamiltonians in the context of radial quantization of black holes as a complete system of observables to uniquely define the quantum state.

Acknowledgments

W.C. is supported in part by the Natural Sciences and Engineering Research Council (NSERC) of Canada. The work of M.T. is supported by the government grant PRIN 36 2007. T.V.R. is supported by the Göran Gustafsson Foundation. The work of A.S. was partially supported by the RFBR Grants. W.C. wishes to thank the organizers for a stimulating and inspiring atmosphere. W.C. and T.V.R. also thank the university of Granada for hospitality.

References

[1] P. Breitenlohner, D. Maison and G. W. Gibbons, *Four-dimensional black holes from Kaluza-Klein theories*, Commun. Math. Phys. **120** (1988) 295.

[2] D. V. Gal'tsov and O. A. Rytchkov, "Generating branes via sigma-models," Phys. Rev. D **58**, 122001 (1998) [arXiv:hep-th/9801160].

[3] P. Fre, V. Gili, F. Gargiulo, A. S. Sorin, K. Rulik and M. Trigiante, "Cosmological backgrounds of superstring theory and solvable algebras: Oxidation and branes," Nucl. Phys. B **685**, 3 (2004) [arXiv:hep-th/0309237].

[4] P. Fre' and A. S. Sorin, "The arrow of time and the Weyl group: all supergravity billiards are integrable," arXiv:0710.1059 [hep-th].

[5] P. Fre and A. S. Sorin, "The Weyl group and asymptotics: All supergravity billiards have a closed form general integral," Nucl. Phys. B **815** (2009) 430.

[6] P. Fre and A. Sorin, *Integrability of supergravity billiards and the generalized Toda lattice equation*, Nucl. Phys. **B733** (2006) 334–355.

[7] E. Bergshoeff, W. Chemissany, A. Ploegh, M. Trigiante and T. Van Riet, *Generating Geodesic Flows and Supergravity Solutions*, Nucl. Phys. **B812** (2009) 343–401.

[8] W. Chemissany, J. Rosseel, M. Trigiante and T. Van Riet, *The full integration of black hole solutions to symmetric supergravity theories*, Nucl. Phys. **B830** (2010) 391–413.

[9] P. Fre and A. S. Sorin, "Supergravity Black Holes and Billiards and Liouville integrable structure of dual Borel algebras," JHEP **1003**, 066 (2010) [arXiv:0903.2559 [hep-th]].

[10] P. Fre and A. S. Sorin, "The Integration Algorithm for Nilpotent Orbits of G/H^* Lax systems: for Extremal Black Holes," arXiv:0903.3771 [hep-th].

[11] A. Ceresole, G. Dall'Agata, "Flow Equations for Non-BPS Extremal Black Holes," JHEP **0703** (2007) 110. [hep-th/0702088].

[12] W. Chemissany, P. Fre and A. S. Sorin, *The Integration Algorithm of Lax equation for both Generic Lax matrices and Generic Initial Conditions*, Nucl. Phys. **B833** (2010) 220–225.

[13] W. Chemissany, P. Fre, J. Rosseel, A. S. Sorin, M. Trigiante and T. Van Riet, "Black holes in supergravity and integrability," JHEP **1009**, 080 (2010) [arXiv:1007.3209 [hep-th]].

[14] W. Chemissany, J. Rosseel and T. Van Riet, "Black holes as generalised Toda molecules," Nucl. Phys. B **843**, 413 (2011) [arXiv:1009.1487 [hep-th]].

[15] J. Perz, P. Smyth, T. Van Riet and B. Vercnocke, *First-order flow equations for extremal and non-extremal black holes*, JHEP **03** (2009) 150.

[16] L. Andrianopoli, R. D'Auria, E. Orazi and M. Trigiante, "First Order Description of D=4 static Black Holes and the Hamilton-Jacobi equation," Nucl. Phys. B **833**, 1 (2010) [arXiv:0905.3938 [hep-th]].

[17] L. Andrianopoli, R. D'Auria, S. Ferrara and M. Trigiante, "Fake Superpotential for Large and Small Extremal Black Holes," JHEP **1008**, 126 (2010) [arXiv:1002.4340 [hep-th]].

[18] G. Bossard, Y. Michel and B. Pioline, *Extremal black holes, nilpotent orbits and the true fake superpotential*, JHEP **01** (2010) 038.

U(N) invariant dynamics for a simplified loop quantum gravity model

Enrique F. Borja
Institute for Theoretical Physics III, University of Erlangen-Nürnberg, Staudtstraße 7, D-91058 Erlangen (Germany).
Departamento de Física Teórica and IFIC, Centro Mixto Universidad de Valencia-CSIC. Facultad de Física, Universidad de Valencia, Burjassot-46100, Valencia (Spain).
E-mail: efborja@theorie3.physik.uni-erlangen.de

Jacobo Díaz-Polo
Institute for Gravitation and the Cosmos & Physics Department, Penn State, University Park, PA 16802-6300, U.S.A.
E-mail: jacobo@gravity.psu.edu

Iñaki Garay
Institute for Theoretical Physics III, University of Erlangen-Nürnberg, Staudtstraße 7, D-91058 Erlangen (Germany).
E-mail: igael@theorie3.physik.uni-erlangen.de

Etera R. Livine
Laboratoire de Physique, ENS Lyon, CNRS-UMR 5672, 46 Allée d'Italie, Lyon 69007, France.
E-mail: etera.livine@ens-lyon.fr

Abstract. The implementation of the dynamics in Loop Quantum Gravity (LQG) is still an open problem. Here, we discuss a tentative dynamics for the simplest class of graphs in LQG: Two vertices linked with an arbitrary number of edges. We use the recently introduced U(N) framework in order to construct SU(2) invariant operators and define a global U(N) symmetry that will select the homogeneous/isotropic states. Finally, we propose a Hamiltonian operator invariant under area-preserving deformations of the boundary surface and we identify possible connections of this model with Loop Quantum Cosmology.

1. Introduction

Loop quantum gravity (LQG) is now a well-established approach to quantum gravity [1]. It provides a non-perturbative mathematical formulation of the kinematical sector of the theory. The Hilbert space is generated by spin-networks: wave functions defined over oriented graphs whose edges are labeled by irreducible representations of the SU(2) group, and with intertwiners (SU(2) invariant tensors) on its vertices. Despite the several advances that have taken place in this field, one of the main challenges faced by the theory is the implementation of the dynamics. Our goal is to focus on a specific model in order to propose a suitable Hamiltonian for it.

Rovelli and Vidotto introduced a simple model based on a graph with 2 vertices and 4 edges [2]. They found that this model leads to a physical framework very similar to Loop Quantum Cosmology (LQC). We generalize this model and implement the LQG dynamics on a graph with 2 vertices joined by an arbitrary number N of edges [3].

We use the recently developed U(N) framework for SU(2)-intertwiners [4, 5, 6] in order to study the Hilbert space of spin networks on the 2-vertex graph. The operators of the U(N) formalism act on our Hilbert space and we identify a global U(N) symmetry generated by operators acting on the coupled system of the two vertices which reduces the full space of arbitrary spin network states to a space of homogeneous/isotropic states. Finally, we introduce global U(N)-invariant operators and use them to propose a U(N)-invariant Hamiltonian operator.

2. The U(N) framework

This framework was introduced in a series of papers [4, 5, 6]. Intertwiners with N legs are SU(2)-invariant states in the tensor product of N (irreducible) representations of SU(2). Then the basic tool used is the Schwinger representation of the $\mathfrak{su}(2)$ Lie algebra in terms of a pair of harmonic oscillators. Since we would like to describe the tensor product of N SU(2)-representations, we will need N copies of the $\mathfrak{su}(2)$-algebra and thus we consider N pairs of harmonic oscillators a_i, b_i with i running from 1 to N.

We can identify SU(2) invariant operators acting on pairs of (possibly equal) legs i, j [4, 6]:

$$E_{ij} = a_i^\dagger a_j + b_i^\dagger b_j, \quad (E_{ij}^\dagger = E_{ji}), \qquad F_{ij} = a_i b_j - a_j b_i, \quad (F_{ji} = -F_{ij}).$$

The operators E form a $\mathfrak{u}(N)$-algebra

$$[E_{ij}, E_{kl}] = \delta_{jk} E_{il} - \delta_{il} E_{kj},$$

and with the operators $F, F^\dagger$ form a closed algebra. This is why this formalism has been dubbed the U(N) framework for LQG. Notice that the diagonal operators give the energy on each leg, $E_{ii} = E_i$. Then the value of the total energy $E \equiv \sum_i E_i$ gives twice the sum of all spins $2 \times \sum_i j_i$, i.e twice the total area around the vertex in the context of LQG.

The E_{ij}-operators change the energy/area carried by each leg, while still conserving the total energy, while the operators F_{ij} (resp. $F_{ij}^\dagger$) will decrease (resp. increase) the total area E by 2:

$$[E, E_{ij}] = 0, \qquad [E, F_{ij}] = -2F_{ij}, \quad [E, F_{ij}^\dagger] = +2F_{ij}^\dagger.$$

This suggests to decompose the Hilbert space of N-valent intertwiners into subspaces of constant area:

$$\mathcal{H}_N = \bigoplus_{\{j_i\}} \text{Inv}\left[\otimes_{i=1}^N V^{j_i}\right] = \bigoplus_{J \in \mathbb{N}} \bigoplus_{\sum_i j_i = J} \text{Inv}\left[\otimes_{i=1}^N V^{j_i}\right] = \bigoplus_J \mathcal{H}_N^{(J)},$$

where V^{j_i} denotes the Hilbert space of the irreducible SU(2)-representation of spin j_i, spanned by the states of the oscillators a_i, b_i with fixed total energy $E_i = 2j_i$. Besides, it was proven [5] that each subspace $\mathcal{H}_N^{(J)}$ of N-valent intertwiners with fixed total area J carries an irreducible representation of U(N) generated by the E_{ij} operators.

Then the operators E_{ij} allow to navigate from state to state within each subspace $\mathcal{H}_N^{(J)}$. On the other hand, the operators F_{ij}, $F_{ij}^\dagger$ allow to go from one subspace $\mathcal{H}_N^{(J)}$ to the next $\mathcal{H}_N^{(J\pm1)}$, thus endowing the full space of N-valent intertwiners with a Fock space structure with creation operators $F^\dagger_{ij}$ and annihilation operators F_{ij}. It is worth to point out that the operators $E_{ij}, F_{ij}, F^\dagger_{ij}$ satisfy certain quadratic constraints [3] that look a lot like constraints on the multiplication of two matrices, one of them hermitian and the other antisymmetric [7].

3. The 2 vertex model

We consider the simplest class of non-trivial graphs for spin network states in Loop Quantum Gravity: a graph with two vertices linked by N edges, as shown in fig.1. This is a generalization of the simplest model (2 vertices and 4 links) introduced by Rovelli and Vidotto in [2] and which was shown to be related to models of quantum cosmology [8].

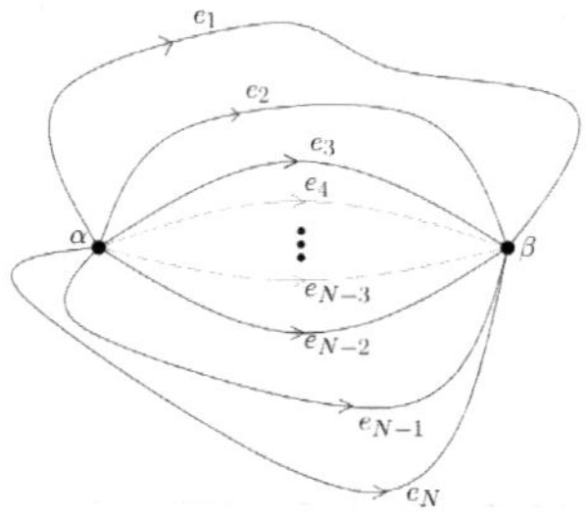

Figure 1. The 2-vertex graph with vertices α and β and the N edges linking them.

There are matching conditions [5] imposing that each edge carries a unique SU(2) representation, thus the spin on that edge must be the same as seen from α than from β i.e. $j_i^\alpha = j_i^\beta$. This translates into the fact that the oscillator energy for α on the leg i must be equal to the energy for β on its i-th leg:

$$\mathcal{E}_i \equiv E_i^{(\alpha)} - E_i^{(\beta)} = 0.$$

We can see that the matching conditions $\mathcal{E}_k$ generate a $\mathrm{U}(1)^N$ symmetry, so they are part of a larger $\mathrm{U}(N)$ symmetry algebra. Indeed, it is possible to introduce the more general operators:

$$\mathcal{E}_{ij} \equiv E_{ij}^{(\alpha)} - E_{ji}^{(\beta)} = E_{ij}^{(\alpha)} - (E_{ij}^{(\beta)})^\dagger,$$

that form a $\mathrm{U}(N)$ algebra and that reduce to the matching conditions in the case $i = j$.

Now, we can show [3] that by imposing the global $\mathrm{U}(N)$-invariance just defined on our 2-vertex system, we obtain a single state $|J\rangle$ for every total boundary area J. Thus, the $\mathrm{U}(N)$ invariance is restricting our system to states which are not sensitive to area-preserving deformations of the boundary between α and β. They are isotropic in the sense that all directions (i.e. all edges) are equivalent and the state only depends on the total boundary area, and they are homogeneous in the sense that the quantum state is the same at every point of space, i.e. at α and β.

In the following, we will restrict our study to this global $\mathrm{U}(N)$ invariant Hilbert space and we will propose a dynamics for this system. Studying the structure of the $\mathrm{U}(N)$ invariant operators acting on this space, we propose the simplest and most natural ansatz for a Hamiltonian operator:

$$H \equiv \lambda \sum_{ij} E_{ij}^{(\alpha)} E_{ij}^{(\beta)} + \left(\sigma \sum_{ij} F_{ij}^{(\alpha)} F_{ij}^{(\beta)} + \bar{\sigma} \sum_{ij} F_{ij}^{\alpha\dagger} F_{ij}^{\beta\dagger} \right),$$

where the coupling λ is real while σ can be complex a priori, so that the operator H is Hermitian. In fact, this is the most general $\mathrm{U}(N)$ invariant Hamiltonian (allowing only elementary changes in the total area), up to a renormalization by a E-dependent factor.

The action of this Hamiltonian over a state $|J\rangle$ is known, and we can also study the spectral properties of this operator. It turns out that it shares several mathematical analogies with the evolution operator used in Loop Quantum Cosmology [9, 10]. From this perspective, the 2-vertex model has a natural cosmological interpretation and exhibits a "big bounce" behavior avoiding the big bang singularity as in LQC.

4. Conclusions

The $\mathrm{U}(N)$ framework recently introduced in [4, 5, 6] represents a new and refreshing way to tackle several issues in the Loop Quantum Gravity framework.

In this work we have described a very simple model for LQG, based on the 2-vertex graph described above. We have the space of intertwiners for each vertex, plus some matching conditions coupling the two intertwiners by imposing that each edge i of the graph carries a unique spin label j_i as seen from both vertices. These matching conditions form the Cartan subalgebra of a larger "global" $\mathfrak{u}(N)$ algebra acting on both vertices, and we have seen that the invariance under this global $\mathrm{U}(N)$ symmetry implies the restriction to isotropic and homogeneous states $|J\rangle$ at the quantum level. This provides our $\mathrm{U}(N)$ symmetry with a very concrete physical interpretation. Indeed it could be viewed as a key step towards a full understanding between the general Loop Quantum Gravity (LQG) framework and the symmetry-reduced Loop Quantum Cosmology (LQC).

We have focused on the global $\mathrm{U}(N)$ invariant space of spin-network states. Using suitable $\mathrm{U}(N)$ invariant operators we have proposed a dynamics for this system. The defined $\mathrm{U}(N)$ invariant Hamiltonian allows a direct comparison with the Hamiltonian constraints used in LQC. This fact reinforces the possible connection between this kind of models and the Loop Quantum Cosmology framework.

Acknowledgments

This work was in part supported by the Spanish MICINN research grants FIS2008-01980, ESP2007-66542-C04-01 and FIS2009-11893. JD is supported by the NSF grant PHY0854743, The George A. and Margaret M. Downsbrough Endowment and the Eberly research funds of Penn State. IG is supported by the Department of Education of the Basque Government under the "Formación de Investigadores" program. EL is partially supported by the ANR "Programme Blanc" grants LQG-09 and by the ESF short visit travel grant 3595.

References

[1] T. Thiemann, *Modern Canonical Quantum General Relativity*, (Cambridge University Press, Cambridge, UK, 2007).
[2] C. Rovelli and F. Vidotto, *Stepping out of Homogeneity in Loop Quantum Cosmology*, Class.Quant.Grav.25 (2008) 225024 [arXiv:0805.4585]
[3] E.F. Borja, J. Diaz-Polo, I. Garay and E.R. Livine *Dynamics for a 2-vertex Quantum Gravity Model*, Class.Quant.Grav.27 (2010) 235010 [arXiv:1006.2451v2]
[4] F. Girelli and E.R. Livine, *Reconstructing Quantum Geometry from Quantum Information: Spin Networks as Harmonic Oscillators*, Class.Quant.Grav. 22 (2005) 3295-3314 [arXiv:gr-qc/0501075]
[5] L. Freidel and E.R. Livine, *The Fine Structure of* SU(2) *Intertwiners from* U(N) *Representations*, to appear in JMP 2010 [arXiv:0911.3553]
[6] L. Freidel and E.R. Livine, U(N) *coherent states for Loop Quantum Gravity*, arXiv:1005.2090
[7] E.F. Borja, L. Freidel, I. Garay and E.R. Livine *U(N) tools for Loop Quantum Gravity: The Return of the Spinor* [arXiv:1010.5451]
[8] M.V. Battisti, A. Marciano and C. Rovelli, *Triangulated Loop Quantum Cosmology: Bianchi IX and inhomogenous perturbations*, [arXiv:0911.2653]
[9] A. Ashtekar, T. Pawlowski, P. Singh and K. Vandersloot, *Loop quantum cosmology of k=1 FRW models*, Phys.Rev.D75 (2007) 024035 [arXiv:gr-qc/0612104]
[10] W. Kaminski and J. Lewandowski, *The flat FRW model in LQC: the self-adjointness*, Class.Quant.Grav.25 (2008) 035001 [arXiv:0709.3120]

Spanish Relativity Meeting (ERE 2010): Gravity as a Crossroad in Physics IOP Publishing
Journal of Physics: Conference Series **314** (2011) 012042 doi:10.1088/1742-6596/314/1/012042

Noncommutative Black Holes and the Singularity Problem [1]

C Bastos[1], O Bertolami[1,2], N C Dias[3,4] and J N Prata[3,4]

[1] Instituto de Plasmas e Fusão Nuclear, Instituto Superior Técnico, Avenida Rovisco Pais 1, 1049-001 Lisboa, Portugal

[2] Departamento de Física e Astronomia, Faculdade de Ciências da Universidade do Porto, Rua do Campo Alegre 687, 4169-007 Porto, Portugal

[3] Departamento de Matemática, Universidade Lusófona de Humanidades e Tecnologias, Avenida Campo Grande, 376, 1749-024 Lisboa, Portugal

[4] Grupo de Física Matemática, Universidade de Lisboa, Avenida Prof. Gama Pinto 2, 1649-003, Lisboa, Portugal

E-mail: cbastos@fisica.ist.utl.pt, orfeu.bertolami@fc.up.pt, ncdias@mail.telepac.pt, joao.prata@mail.telepac.pt

Abstract. A phase-space noncommutativity in the context of a Kantowski-Sachs cosmological model is considered to study the interior of a Schwarzschild black hole. Due to the divergence of the probability of finding the black hole at the singularity from a canonical noncommutativity, one considers a non-canonical noncommutativity. It is shown that this more involved type of noncommutativity removes the problem of the singularity in a Schwarzschild black hole.

1. Introduction

It is expected that a quantum theory of gravity may resolve the black hole (BH) singularity problem. However, as a quantum theory of gravity is beyond of reach, a quantum cosmology approach based on a minisuperspace approximation might provide some insight into problems of this nature. Assuming that the ultimate structure of space-time should be determined by quantum gravity and that should be noncommutative (NC), then noncommutativity in a quantum cosmological level should be considered. In fact, noncommutativity of space-time is shown to be relevant in the context of BHs physics [2]. One of the most interesting features of this noncommutativity is that it allows for a square integrable wave function, solution of the noncommutative Wheeler-DeWitt (WDW) equation.

In this contribution one presents a phase-space non-canonical NC extension of the Kantowski-Sachs (KS) cosmological model [1] and its application to study the singularity of a Schwarzschild BH [2]. In the interior of a Schwarzschild BH, for $r < 2M$, time and radial coordinates are interchanged ($r \leftrightarrow t$) and space-time is described by the metric:

$$ds^2 = -\left(\frac{2M}{t} - 1\right)^{-1} dt^2 + \left(\frac{2M}{t} - 1\right) dr^2 + t^2(d\theta^2 + \sin^2\theta d\varphi^2) \ . \tag{1}$$

[1] Based on a talk presented by CB at ERE 2010, Granada, Spain, 6th-10th September 2010.

That is, an isotropic metric turns into an anisotropic one, implying that the interior of a Schwarzschild BH can be described by an anisotropic cosmological space-time. Thus, the metric (1) can be mapped into the KS cosmological model [5], which, in the Misner parametrization, can be written as

$$ds^2 = -N^2dt^2 + e^{2\sqrt{3}\beta}dr^2 + e^{-2\sqrt{3}(\beta+\Omega)}(d\theta^2 + \sin^2\theta d\varphi^2) \ , \tag{2}$$

where Ω and β are scale factors, and N is the lapse function. The following identification for $t < 2M$,

$$N^2 = \left(\frac{2M}{t} - 1\right)^{-1} \quad , \quad e^{2\sqrt{3}\beta} = \left(\frac{2M}{t} - 1\right) \quad , \quad e^{-2\sqrt{3}\beta}e^{-2\sqrt{3}\Omega} = t^2 \ , \tag{3}$$

allows to map the metric Eq. (2) into the metric Eq. (1).

One considers that the two scale factors Ω and β, and the corresponding canonical conjugate momenta P_Ω and P_β do not commute, and

$$\begin{aligned}
\left[\hat{\Omega}, \hat{\beta}\right] &= i\theta\left(1 + \epsilon\theta\hat{\Omega} + \frac{\epsilon\theta^2}{1+\sqrt{1-\xi}}\hat{P}_\beta\right) \\
\left[\hat{P}_\Omega, \hat{P}_\beta\right] &= i\left(\eta + \epsilon(1+\sqrt{1-\xi})^2\hat{\Omega} + \epsilon\theta(1+\sqrt{1-\xi})\hat{P}_\beta\right) \\
\left[\hat{\Omega}, \hat{P}_\Omega\right] &= \left[\hat{\beta}, \hat{P}_\beta\right] = i\left(1 + \epsilon\theta(1+\sqrt{1-\xi})\hat{\Omega} + \epsilon\theta^2\hat{P}_\beta\right) ,
\end{aligned} \tag{4}$$

where θ, η and ϵ are positive constants and $\xi = \theta\eta < 1$, a quite physical condition [7]. The noncommutative WDW (NCWDW) equation of the KS cosmological model obtained in Ref. [3] is considered to describe the interior of the BH. The remaining commutation relations vanish. For $\epsilon \neq 0$ it implies that the noncommutative commutation and uncertainty relations are themselves position and momentum dependent. The case $\epsilon = 0$ corresponds to the canonical phase-space noncommutativity [4, 1, 2]. One can relate this NC algebra with the Heisenberg-Weyl algebra, via a suitable map. Suppose that $\left(\hat{\Omega}_c, \hat{P}_{\Omega_c}, \hat{\beta}_c, \hat{P}_{\beta_c}\right)$ obey the HW algebra, $\left[\hat{\Omega}_c, \hat{P}_{\Omega_c}\right] = \left[\hat{\beta}_c, \hat{P}_{\beta_c}\right] = i$. Then a suitable transformation would be:

$$\begin{aligned}
\hat{\Omega} &= \lambda\hat{\Omega}_c - \frac{\theta}{2\lambda}\hat{P}_{\beta_c} + E\hat{\Omega}_c^2 \ , \ \hat{\beta} = \lambda\hat{\beta}_c + \frac{\theta}{2\lambda}\hat{P}_{\Omega_c} \\
\hat{P}_\Omega &= \mu\hat{P}_{\Omega_c} + \frac{\eta}{2\mu}\hat{\beta}_c \ , \ \hat{P}_\beta = \mu\hat{P}_{\beta_c} - \frac{\eta}{2\mu}\hat{\Omega}_c + F\hat{\Omega}_c^2.
\end{aligned} \tag{5}$$

Here, μ, λ are real parameters such that $(\lambda\mu)^2 - \lambda\mu + \frac{\xi}{4} = 0 \Leftrightarrow 2\lambda\mu = 1 \pm \sqrt{1-\xi}$, and one chooses the positive solution given the invariance of the physics on the D map [4], and

$$E = -\frac{\theta}{1+\sqrt{1-\xi}}F \ , \ F = -\frac{\lambda}{\mu}\epsilon\sqrt{1-\xi}\left(1+\sqrt{1-\xi}\right) \ . \tag{6}$$

Of course, the inverse transformation can be obtained [3].

As in the canonical NC KS cosmological model [1, 2], one obtains the NCWDW equation considering the canonical quantization of the KS Hamiltonian, the operator $\hat{A}$ that commutes with the Hamiltonian and the D map in Eqs. (5), [3]

$$\begin{aligned}
&\mu^2\mathcal{R}'' - 48\exp\left(-\frac{2\sqrt{3}}{\mu}\Omega_c - 2\sqrt{3}E\Omega_c^2 + \frac{\sqrt{3}\theta a}{\mu\lambda}\right)\mathcal{R} - \frac{2\eta}{\mu}\left(\Omega_c + F\Omega_c^3\right)\mathcal{R} + \\
&+F^2\Omega_c^4\mathcal{R} + a^2\mathcal{R} + \left(\frac{\eta^2}{\mu^2} + 2aF\right)\Omega_c^2\mathcal{R} = 0 \ .
\end{aligned} \tag{7}$$

The dependence on β_c has completely disappeared and one is left with an ordinary differential equation for $\mathcal{R}(\Omega_c)$. Through the substitution, $\Omega_c = \mu z$, $\frac{d^2}{d\Omega_c^2} = \frac{1}{\mu^2}\frac{d^2}{dz^2}$ and $\mathcal{R}(\Omega_c(z)) := \phi_a(z)$ one obtains a second order linear differential equation, a Schrödinger-like equation:

$$-\phi_a''(z) + V(z)\phi_a(z) = 0 \ , \tag{8}$$

where the potential function, V(z), reads:

$$V(z) = -(\eta z - a)^2 - F^2\mu^4 z^4 - 2F\mu^2(\eta z - a)z^2 + 48\exp\left(-2\sqrt{3}z - 2\sqrt{3}\mu^2 E z^2 + \frac{\sqrt{3}\theta a}{\mu\lambda}\right). \tag{9}$$

Equation (8) depends explicitly on the noncommutative parameters θ, η, ϵ and the eigenvalue a. Notice that $E > 0$.

In this contribution one analyzes the singularity $t = 0$ in the Schwarzschild BH with this NCWDW equation, Eqs. (8) and (9). In fact, with the asymptotic solutions of this NCWDW equation one is able to study the singularity of the BH. In the canonical NC approach, these type of solutions vanish in the neighbourhood of $t = 0$, but the probability of finding the system at the singularity diverges [2]. However, using the non-canonical noncommutativity, one actually can remove the singularity of the Schwarzschild BH [3].

2. Singularity

To study the singularity of the BH, one needs to examine the asymptotic limit of the NCWDW. From Eq. (9) one finds that asymptotically ($z \to \infty$) the potential function is dominated by the term

$$V(z) \sim -F^2\mu^4 z^4, \tag{10}$$

which leads to $L^2(\mathbb{R})$ solutions of the Schrödinger equation (8). In fact, the Hamiltonian $H = -\frac{\partial^2}{\partial z^2} - F^2\mu^4 z^4$ has a continuous and real spectrum [6] in L^2. The asymptotic form of the eigenfunction corresponding to the eigenvalue zero and eigenvalue a of $\hat{A}$ is

$$\phi_a(z) \sim \frac{1}{z}\exp\left[\pm i\tfrac{F\mu^2}{3}z^3\right] \ . \tag{11}$$

Thus, one can write a general solution of the NCWDW equation as

$$\psi(\Omega_c, \beta_c) = \int C(a)\psi_a(\Omega_c, \beta_c)da \ , \tag{12}$$

where $C(a)$ are arbitrary complex constants and $\psi_a(\Omega_c, \beta_c)$ is of the form:

$$\psi_a(\Omega_c, \beta_c) = B\phi_a\left(\tfrac{\Omega_c}{\mu}\right)\exp\left[\tfrac{i\beta_c}{\mu}\left(a - \tfrac{\eta}{2\mu}\Omega_c\right)\right] \ . \tag{13}$$

Here $\phi_a(\Omega/\mu)$ is a solution of Eq. (8) and B is a normalization constant such that:

$$\int |\psi_a(\Omega_c, \beta_c)|^2 d\Omega_c = B^2 \int |\phi_a\left(\tfrac{\Omega_c}{\mu}\right)|^2 d\Omega_c = 1 \tag{14}$$

One fixes a constant β_c-hypersurface, given that the wave function is oscillatory in β_c. Thus, it is natural to consider a measure $d\xi = \delta(\beta - \beta_c)d\beta d\Omega_c$ in order to define the inner product in $L^2(\mathbb{R}^2)$ and the computation of probabilities. For the wave function Eq. (12) one then finds:

$$||\psi||_{L^2(\mathbb{R}^2, d\xi)} = \left(\int \psi(\Omega_c, \beta)\psi^*(\Omega_c, \beta)d\xi\right)^{1/2} = \left(\int C(a)C^*(a')\left(\int \psi_a(\Omega_c, \beta_c)\psi_{a'}^*(\Omega_c, \beta_c)d\Omega_c\right)da\,da'\right)^{1/2} . \tag{15}$$

The wave functions $\psi_a(\Omega_c, \beta_c)$ are solutions of a hyperbolic-type equation, and in general, they are not orthogonal to each other. So, one uses the Cauchy-Schwartz inequality to compute the $L^2(\mathbb{R}^2, d\xi)$-inner product between $\psi_a(\Omega_c, \beta_c)$ and $\psi^*_{a'}(\Omega_c, \beta_c)$. It follows that:

$$||\psi||_{L^2(\mathbb{R}^2,d\xi)} \leq \int |C(a)|\,||\psi_a(\Omega_c,\beta_c)||_{L^2(\mathbb{R}^2,d\xi)}da = \int |C(a)|da \qquad (16)$$

where, in the last step, the relation (14) was used. Thus, one finally concludes that for $C(a) \in L^1(\mathbb{R})$ the wave function Eq. (12) is squared integrable on constant β_c-hypersurfaces and the BH probability $P(r = 0, t = 0)$ at the singularity can now be calculated:

$$P(r=0,t=0) = \lim_{\tilde{\Omega}_c,\beta_c\to+\infty} \int_{\tilde{\Omega}_c}^{+\infty}\int_{-\infty}^{+\infty} |\psi(\Omega_c,\beta)|^2 d\xi = \lim_{\tilde{\Omega}_c,\beta_c\to+\infty} \int_{\tilde{\Omega}_c}^{+\infty} |\psi(\Omega_c,\beta_c)|^2\, d\Omega_c = 0\ . \qquad (17)$$

Notice that in the last step, it was used that $\psi(\Omega_c, \beta_c) \in L^2(\mathbb{R}^2, d\xi)$. Consequently, the integral in Ω_c vanishes, independently of the value of β_c.

3. Conclusions

In this contribution, a non-canonical phase-space NC extension of the KS cosmological model is used to study the singularity of a Schwarzschild BH. In contrast with the canonical NC studied in Ref. [2], this type of noncommutativity allows us to obtain square integrable solutions for the asymptotic NCWDW equation. In fact, this is the major property of the model. That is, the general solution of our NCWDW equation is square integrable on the constant β_c-hypersurfaces, and so one can use the same rules as in quantum mechanics to evaluate probabilities. In this case, one computes the probability of finding the BH at the singularity which is found to vanish.

Acknowledgments

The work of CB is supported by Fundação para a Ciência e a Tecnologia (FCT) under the grant SFRH/BPD/62861/2009. The work of OB is partially supported by the FCT grant PTDC/FIS/111362/2009. NCD and JNP were partially supported by the grants PTDC/MAT/69635/2006 and PTDC/MAT/099880/2008 of FCT.

References

[1] C. Bastos, O. Bertolami, N.C. Dias and J.N. Prata, Phys. Rev. D **78** (2008) 023516.
[2] C. Bastos, O. Bertolami, N.C. Dias and J.N. Prata, Phys. Rev. D **80** (2009) 124038.
[3] C. Bastos, O. Bertolami, N. Dias and J. Prata, Phys. Rev. **D 82** (2010) 041502 [R].
[4] C. Bastos, O. Bertolami, N.C. Dias and J.N. Prata, J. Math. Phys. **49** (2008) 072101
[5] P.R. Anderson, W.A. Hiscock and S. L. Larson, Phys. Rev. **D 56** (1997) 3571; R. Kantowski and R.K. Sachs, J. Math. Phys. **7** (1966) 443.
[6] B.L. Voronov, D.M. Gitman and I.V. Tyutin, *Self-adjoint differential operator associated with self-adjoint differential expressions*, arXiv: quant-ph/0603187.
[7] C. Bastos, O. Bertolami, N.C. Dias and J.N. Prata, *Entropic Gravity, Phase-Space Noncommutativity and the Equivalence Principle*, arXiv: 1010.4729 [hep-th].

Spanish Relativity Meeting (ERE 2010): Gravity as a Crossroad in Physics IOP Publishing
Journal of Physics: Conference Series **314** (2011) 012043 doi:10.1088/1742-6596/314/1/012043

Collision of gravitational shock waves in AdS and holography

Álvaro Dueñas-Vidal and Miguel A. Vázquez-Mozo.

Departamento de Física Fundamental, Universidad de Salamanca.
Plaza de la Merced s/n, E-37008 Salamanca, Spain.

E-mail: adv@usal.es, vazquez@usal.es

Abstract. We study numerically the formation of marginally closed trapped surfaces as the result of the collision of two shock gravitational waves in AdS in various dimensions ($D = 4, 5, 6, 7$ and 8). In all cases a critical value of the impact parameter is found above which no trapped surface of the type sought exists. We obtain a very simple scaling between the critical impact parameter and the energy of the incoming waves. The holographic implications of our results are discussed in the context of the AdS/CFT correspondence.

1. Introduction

The observed phenomenology at RHIC, and more recently at LHC, hints at strongly coupled quark-gluon plasma being produced in heavy ion collisions. This has motivated the use of AdS/CFT duality as a non-perturbative framework to describe the relevant physics involved. In this context it has been proposed that the collision of two gravitational waves propagating in AdS_5 could provide a good gravitational dual to model the formation of a quark-gluon plasma as the result of the collision of two energy lumps in the strongly-coupled $\mathcal{N} = 4$ supersymmetric Yang-Mills theory (SYM) on the boundary [1]. In spite of the difference between real QCD and $\mathcal{N} = 4$ SYM, the AdS/CFT calculations of some relevant observables show a good agreement with the experiments.

Here we study the collision of two gravitational shock waves in AdS_D. Before the collision takes place (i.e., $u < 0$ or $v < 0$), the bulk geometry in Poincare coordinates is given by

$$ds^2 = \frac{L^2}{z^2}\left[dz^2 - dudv + d\vec{x}^2 + \frac{z}{L}\Phi_+(z,\vec{x})\delta(u)du^2 + \frac{z}{L}\Phi_-(z,\vec{x})\delta(v)dv^2\right], \tag{1}$$

where L is the AdS_D scale and $\Phi_\pm(z,\vec{x})$ are the wave profiles. For Aichelburg-Sexl waves the profiles satisfy the Poisson-like equation

$$\left(\Box_{\mathbb{H}_{D-2}} - \frac{D-2}{L^2}\right)\Phi_\pm(z,\vec{x}) = -16\pi G_N \mu_\pm \left(\frac{z_\pm}{L}\right)\delta(z - z_\pm)\delta^{(D-3)}(\vec{x} - \vec{b}_\pm), \tag{2}$$

with $\mu_\pm$ the energy of the wave, and $\mathbb{H}_{D-2}$ the hyperbolic space transverse to the collision. Finally, $z_\pm$ and $\vec{b}_\pm$ are the impact parameters along the holographic and "field theory" coordinates respectively.

Journal of Physics: Conference Series **314** (2011) 012043 doi:10.1088/1742-6596/314/1/012043

The production of an apparent horizon as a result of the shock-wave collision can be interpreted as signaling the thermalization of the dual SYM plasma. A way to detect the formation of this apparent horizon while bypassing a full solution of the Einstein equations in the interaction region $u > 0$, $v > 0$ is to look for marginally outer closed surfaces lying on the region $\{u \leq 0, v = 0\} \cup \{u = 0, v \leq 0\}$ [2]. Notice that this region knows about the interaction between the waves because it includes the point $u = 0$, $v = 0$ common to both wavefronts.

Choosing appropriate coordinates $(U,V,Z,\vec{X})$ in which the null geodesics normal to the wavefronts are continuous, the closed trapped surface candidate $\mathcal{S} = \mathcal{S}_+ \cup \mathcal{S}_-$ is parametrized by two functions $\psi_\pm(Z,\vec{X})$ as

$$\begin{aligned} \mathcal{S}_+ &= \{(U,V,Z,\vec{X}) : U = 0, V = -\psi_+(Z,\vec{X}), \psi_+(Z,\vec{X}) \geq 0\}, \\ \mathcal{S}_- &= \{(U,V,Z,\vec{X}) : U = -\psi_-(Z,\vec{X}), V = 0, \psi_-(Z,\vec{X}) \geq 0\}. \end{aligned} \tag{3}$$

Imposing the marginally trapped surface condition to the shape functions $\Psi_\pm(z,\vec{x}) \equiv \frac{z}{L}\psi_\pm(z,\vec{x})$ we arrive at the boundary problem

$$\left(\Box_{\mathbb{H}_{D-2}} - \frac{D-2}{L^2}\right)(\Phi_\pm - \Psi_\pm) = 0, \quad \Psi_\pm(z,\vec{x})|_{\mathcal{C}} = 0, \quad g^{ab}\partial_a\Psi_+\partial_b\Psi_-\Big|_{\mathcal{C}} = 4. \tag{4}$$

where $\mathcal{C} = \mathcal{S}_+ \bigcap \mathcal{S}_-$. The first equation comes from the zero-convergence condition over the congruence of outer null geodesics normal to $\mathcal{S}_\pm$. The second one follows from the definition of $\mathcal{S}_\pm$ in (3). The last condition comes from the continuity of the congruence of null geodesics at the intersection $\mathcal{C}$ of the two branches.

2. Off-center collisions

The boundary problem (4) can be solved analytically in the case of head-on collisions [1]. Here, however, we will be interested in the situation where the to waves collide with a nonvanishing impact parameter, in which case the problem has to be tackled numerically. In particular, we consider off-center collisions where the impact parameter is along the "field theory coordinates", i.e. we take $z_+ = z_- = L$ in eq. (2).

The problem at hand has an $O(D-3)$ symmetry corresponding to rotations in $\mathbb{H}_{D-2}$ around the axis joining the two waves sources. For convenience we use radial coordinates where the metric of $\mathbb{H}_{D-2}$ reads

$$ds^2_{\mathbb{H}_{D-2}} = \frac{dr^2}{1 + r^2/L^2} + r^2 d\theta^2 + r^2\sin^2\theta d\Omega^2_{D-4}. \tag{5}$$

Without loss of generality we take the position of the wave sources in $\mathbb{H}_{D-2}$ at $r = \frac{b}{2}$, $\theta = 0$ and $r = \frac{b}{2}$, $\theta = \pi$. The intersection $\mathcal{C}$ is a $(D-3)$-dimensional hypersurface that in these coordinates is defined by a function $r = LG(\theta)$, while the symmetry of the problem implies that $\Psi_\pm(z,\vec{x}) = \Psi_\pm(r,\theta)$. This reduces the boundary problem (4) to a two-dimensional one. In addition, for the collision of two identical shock waves we have that $\mu_+ = \mu_-$, $\Phi_+(r,\theta) = \Phi_-(r,\theta) \equiv \Phi(r,\theta)$, and therefore by symmetry $\Psi_+(r,\theta) = \Psi_-(r,\theta) \equiv \Psi(r,\theta)$. Moreover $G(\theta) = G(\pi - \theta)$ and $\Psi(r,\theta) = \Psi(r,\pi-\theta)$.

Defining $H(r,\theta) \equiv \Phi(r,\theta) - \Psi(r,\theta)$, the boundary problem (4) reduces to

$$\begin{aligned} \left[\left(1 + \frac{r^2}{L^2}\right)\partial_r^2 + \frac{(D-3)L^2 + (D-2)r^2}{rL^2}\partial_r + \frac{1}{r^2}\partial_\theta^2 + \frac{D-4}{r^2\tan\theta}\partial_\theta - \frac{D-2}{L^2}\right] H &= 0, \\ \left[\left(1 + \frac{r^2}{L^2}\right)(\partial_r\Psi)^2 + \frac{1}{r^2}(\partial_\theta\Psi)^2\right]\Bigg|_{r=LG(\theta)} &= 4, \\ H\big|_{r=LG(\theta)} &= \Phi_+\left(LG(\theta),\theta\right). \end{aligned} \tag{6}$$

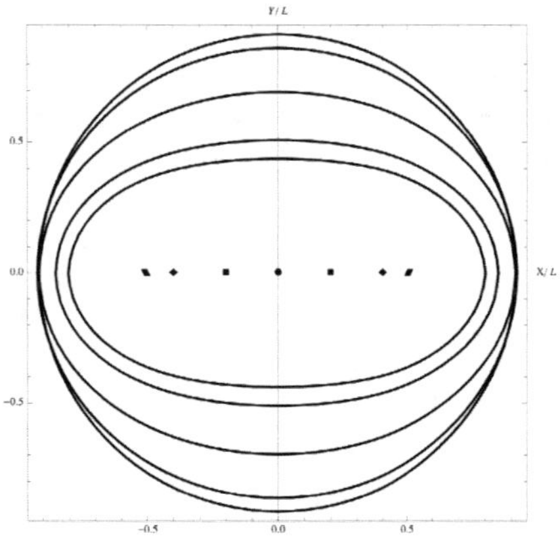

Figure 1. Some examples for the shape of $\mathcal{C}$ in $D = 7$ and $G_N \mu L^{3-D} = 1$ in the plane XY with $X = r\cos\theta$ and $Y = r\sin\theta$. The shock wave sources are marked on X axis. The inner curve corresponds to the critical impact parameter.

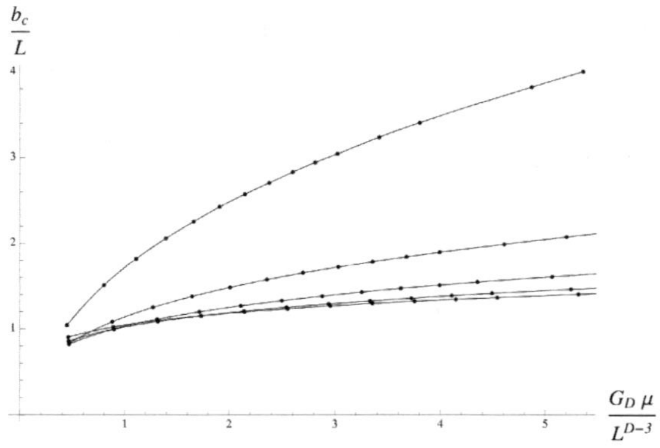

Figure 2. Critical impact parameter $\frac{b_c}{L}$ as a function of energy $\frac{G_N \mu}{L^{D-3}}$ for $D = 4$ (uppermost curve), 5, 6, 7, and 8 (lowest curve). The points corresponds to the numerical data while the continuous line is a fit using the scaling (7).

To solve it we use a finite difference method (with a 50×100 grid) combined with a trial-and-correction loop. First we give a trial shape for $\mathcal{C}$, $r = r_0$, and solve the partial differential equation with the second boundary condition in (6). Next, we use the solution to compute $T(\theta) = g^{ab}\partial_a\Psi\partial_b\Psi - 4$ on $\mathcal{C}$, and redefine the shape of $\mathcal{C}$ according to $r = r_0 + \epsilon T(\theta)$ with $\epsilon \ll 1$. The routine is repeated until $T(\theta) \ll 1$ and convergence is found. We have carried out the numerical analysis for $D = 4, 5, 6, 7$ and 8 with various values of the incoming energy. The details can be found in ref. [3]. For $D = 5$ we reproduce the results of [4].

In fig. 1 the sections of the closed trapped surfaces are shown for various values of the impact parameter in $D = 7$. Our results show that for all explored dimensions a critical impact parameter is always found. That is, for each energy there exists a value b_c such that no trapped surface is formed whenever $b > b_c$, whereas for $b = b_c$ a critical surface forms with finite size. In fig. 2 the values of b_c as a function of μ are plotted for various dimensions. The results are very well fit by the following scaling of the critical impact parameter with the energy

$$\frac{b_c}{L} \sim \left(\frac{G_N \mu}{L^{D-3}}\right)^{\frac{1}{D-2}}, \tag{7}$$

where the proportionality constant is of order one. For large values of the dimension this scaling implies the independence of the critical impact parameter with the shock wave energy.

From the boundary field theory point of view, the situation analyzed in this section corresponds to two relativistic energy lumps colliding with the same energy and impact parameter b. Assuming the existence of the trapped surface signals the eventual formation of an event horizon in the bulk, the critical behavior observed might be interpreted as indicating the absence of thermalization in the CFT plasma resulting from the collision for large enough values of the impact parameter.

3. Collisions between energy lumps of different size

In the field theory on the boundary the size of the incoming lumps is determined by the value of the holographic coordinate of the source as can be seen from

$$\frac{\int d^{D-2}x\,\vec{x}^{\,2}\langle T_{uu}\rangle_{\rm CFT}}{\int d^{D-2}x\langle T_{uu}\rangle_{\rm CFT}} = z_+^2, \tag{8}$$

and equivalently for the second wave replacing $\langle T_{uu}\rangle_{\rm CFT}$ by $\langle T_{vv}\rangle_{\rm CFT}$ and z_+ by z_-. Thus collisions with $z_+ \neq z_-$ are the dual descriptions for head-on collisions between energy lumps of different sizes.

The isometries of AdS can be exploited to link the computation of last section to collisions of gravitational waves with $z_+ \neq z_-$ [3]. In particular, using these symmetries the collision of two shock waves with $z_\pm = L(\sqrt{1+\beta^2}\pm\beta)^{-1}$ and $b_\pm = 0$ can be "rotated" into a symmetric one with $z''_\pm = L$ and $b''_\pm = \mp L\beta$, with β a real parameter. This is done in the following steps:

- An O(2) rotation of angle θ in the plane XY (cf. fig. 1). This moves the wave sources to

$$z'_\pm(\theta) = \frac{L}{\sqrt{1+\beta^2}\pm\beta\cos(\theta)}, \quad b'_\pm(\theta) = \frac{L\beta\sin\theta}{\sqrt{1+\beta^2}\pm\beta\cos(\theta)}. \tag{9}$$

  For $\theta = \frac{\pi}{2}$ the two sources have the same value of the holographic coordinate $z'_\pm = L(1+\beta^2)^{-\frac{1}{2}}$. Because of the UV/IR connection in AdS/CFT this rotation changes the energy of the shock waves to $\mu'_\pm = \mu_\pm\sqrt{1+\beta^2}(\sqrt{1+\beta^2}\pm\beta)^{-1}$.
- A longitudinal boost $\bar{u} = \lambda u$, $\bar{v} = \lambda^{-1}v$, with $\lambda = (\mu_+/\mu_-)(\sqrt{1+\beta^2}-\beta)(\sqrt{1+\beta^2}+\beta)^{-1}$. This results in the two shock waves having the same energy $\bar{\mu}_\pm = \sqrt{\mu_+\mu_-}$.
- Finally, a coordinates rescaling $z'' = z'\sqrt{1+\beta^2}$ and $\vec{x} = \vec{x}\,''\sqrt{1+\beta^2}$. The point of this last isometry is to eliminate the dependence of the holographic coordinate on the parameter β, so we are left with $z''_\pm = L$ and $b''_\pm = \mp L\beta$. The energy of the waves is now given by $\mu''_\pm = \sqrt{\mu_+\mu_-}(1+\beta^2)^{-\frac{1}{2}}$.

The analysis of the formation of closed trapped surfaces for the collision of unequal lumps can be now carried out by applying in reverse order this series of AdS isometries to the results of the previous section [3]. This leads to the conclusion that there is a critical value for the size difference $|z_+ - z_-|$ below which no trapped surface is formed. Physically this can be interpreted as indicating that a too small energy lump does not have enough degrees of freedom to induce thermalization after a collision with a big one. The fact that this depends only on the size difference is a consequence of conformal invariance.

Acknowledgments. A.D.-V. wishes to thank the organizers of the 2010 Spanish Relativity Meeting for the opportunity of presenting this work. A.D.-V. acknowledges support from a Castilla y León Regional Government predoctoral fellowship and Spanish Government Grant FIS2009-07238. M.A.V.-M. has been partially supported by Spanish Government Grants FPA2009-10612 and FIS2009-07238, Basque Government Grant IT-357-07 and Spanish Consolider-Ingenio 2010 Programe CPAN (CSD2007-00042).

References

[1] Gubser S S, Pufu S S and Yarom A 2008 *Phys. Rev.* **D78** 066014
[2] Penrose R 1974 Seminar at Cambridge University (unpublished).
[3] Dueñas-Vidal A and Vázquez-Mozo M A 2010 *J. High Energy Phys.* **21** 1007
[4] Lin S and Shuriak E 2009 *Phys. Rev.* **D79** 124015

Journal of Physics: Conference Series **314** (2011) 012044 doi:10.1088/1742-6596/314/1/012044

Classical GR as a topological theory with linear constraints[1]

Steffen Gielen

DAMTP, University of Cambridge, Centre for Mathematical Sciences, Wilberforce Road, Cambridge CB3 0WA, U.K.

Max Planck Institute for Gravitational Physics (Albert Einstein Institute), Am Mühlenberg 1, D-14476 Golm, Germany

E-mail: `sg452@cam.ac.uk, gielen@aei.mpg.de`

Abstract. We investigate a formulation of continuum 4d gravity in terms of a constrained topological (BF) theory, in the spirit of the Plebanski formulation, but involving only linear constraints, of the type used recently in the spin foam approach to quantum gravity. We identify both the continuum version of the linear simplicity constraints used in the quantum discrete context and a linear version of the quadratic volume constraints that are necessary to complete the reduction from the topological theory to gravity. We illustrate and discuss also the discrete counterpart of the same continuum linear constraints. Moreover, we show under which additional conditions the discrete volume constraints follow from the simplicity constraints, thus playing the role of secondary constraints. Our analysis clarifies how the discrete constructions of spin foam models are related to a continuum theory with an action principle that is equivalent to general relativity.

1. Motivation

As is well known, general relativity in three dimensions is topological, as there are no local degrees of freedom. Its first order formulation in terms of a connection ω^{AB} and a triad E^A is given by

$$S = \frac{1}{16\pi G}\int_\Sigma \epsilon_{ABC}E^A \wedge R^{BC}[\omega]\,. \qquad (1)$$

The quantisation of such a theory is rather well understood and different possible quantisation procedures are known. This is of course in stark contrast to general relativity in four dimensions, whose quantisation is to a large extent still an open problem. One may try to exploit what is known in 3D to write general relativity in 4D as a *constrained* topological theory with action

$$S = \int_\Sigma B^{AB} \wedge R_{AB}[\omega] + \lambda^\alpha C_\alpha[B]\,; \qquad (2)$$

one then needs to add constraints (the "simplicity constraints") which enforce $B^{AB} = \frac{1}{16\pi G}\epsilon^{AB}{}_{CD}E^C \wedge E^D$ (for some set of 1-forms E^A) to recover general relativity. The main obstacle in quantisation is then the identification and implementation of such constraints, which has indeed been a topic of much recent debate in the discrete setting of spin foam models [2].

[1] Based on a talk given at the Spanish Relativity Meeting (ERE2010), Granada, on 6 September 2010 [1].

If, for example as the starting point for loop quantum gravity, one wants to add a Holst term [3] to the first order connection form for general relativity, one instead needs to impose $\Sigma^{AB} = \frac{1}{8\pi\gamma G} E^A \wedge E^B$ on the two-form field Σ^{AB}, defined in terms of B^{AB} by

$$\Sigma^{AB} = \frac{1}{1 \pm \gamma^2} \left(B^{AB} - \frac{\gamma}{2} \epsilon^{AB}{}_{CD} B^{CD} \right) \tag{3}$$

(note that this is just a linear redefinition). We will focus on this case in the following and try to determine an appropriate set of constraints for Σ^{AB}.

In the traditional (Plebanski) formulation [4], one uses *quadratic constraints* for Σ^{AB},

$$\epsilon_{ABCD} \Sigma^{AB}_{ab} \Sigma^{CD}_{cd} = V \epsilon_{abcd}, \tag{4}$$

which have two *separate* sectors of solutions[2]:

$$\text{either } \Sigma^{AB} = \pm e^A \wedge e^B \quad \text{or } \Sigma^{AB} = \pm \frac{1}{2} \epsilon^{AB}{}_{CD} e^C \wedge e^D \tag{5}$$

for some set of 1-forms e^A. Classically, one can consistently remain within the "GR" sector [5]; quantum mechanically, the situation is less clear.

The discrete analogue of this construction has been used in spin foam models of quantum gravity: One introduces a triangulation of spacetime, integrates Σ^{AB} over triangles

$$\Sigma^{AB}_{ab}(x) \quad \Rightarrow \quad \Sigma^{AB}_{\triangle} \equiv \int_{\triangle} \Sigma^{AB} \in \mathfrak{so}(4) \simeq \Lambda^2 \mathbb{R}^4, \tag{6}$$

and imposes the constraint $\epsilon_{ABCD} \Sigma^{AB}_{\triangle} \Sigma^{CD}_{\triangle'} = 0$ if $\triangle = \triangle'$ or $\triangle$ and $\triangle'$ share an edge. It can be shown [6] that the remaining "volume" constraints can be replaced by the "closure constraint"

$$\sum_{\triangle \subset \blacktriangle} \Sigma^{AB}_{\triangle} = 0. \tag{7}$$

The quantum-mechanical implementation of these constraints led to the Barrett-Crane model [7]. The recently proposed new spin foam models [2], on the other hand, rely on the replacement of quadratic constraints on $\Sigma^{AB}_{\triangle}$ by *linear* constraints:

$$n_A(\blacktriangle) \Sigma^{AB}(\triangle) = 0 \quad \forall \triangle \subset \blacktriangle, \tag{8}$$

where $n_A(\blacktriangle)$ is normal to the tetrahedron $\blacktriangle$. These are stronger than the quadratic constraints; they restrict Σ^{AB} to the discrete analogue of $\Sigma^{AB} = \pm E^A \wedge E^B$. Here we present an extension of this construction, which was done in the discrete setting, to the *classical continuum theory*. We give a set of constraints whose solution implies that $\Sigma^{AB} = \pm E^A \wedge E^B$ for some E^A (ignoring the factor $8\pi\gamma G$ in the following). We then discretise these constraints and identify some which are sufficient at the discrete level. For details and proofs of our claims, see [8].

[2] This is under the additional assumption that $V \neq 0$! $V = 0$ configurations are not geometric at all, leading to additional potential difficulties in quantisation [5].

2. The construction

2.1. Continuum construction

Clearly, to find a continuum analogue of (8) that is derivable from an action, one has to introduce dynamical fields that play the analogue of the normal vectors n^A in (8). We introduce a basis of 1-forms e_a^A (i.e. assuming $\det(e_a^A) \neq 0$), which is dual to a basis of 3-forms by

$$n_{Adef} \equiv \epsilon_{ADEF} e_d^D e_e^E e_f^F, \quad e_A^c \sim \epsilon^{cdef} n_{Adef}. \tag{9}$$

Claim 1. For a basis of 3-forms n_A, the general solution to

$$n_{Adef} \Sigma_{ab}^{AB} = 0 \quad \forall \{a, b\} \subset \{d, e, f\} \tag{10}$$

is

$$\Sigma_{ab}^{AB} = G_{ab} e_a^{[A} e_b^{B]}, \tag{11}$$

where e^A is defined in terms of n_A as above. Note that $G_{ab} = G_{ab}(x)$ are functions on spacetime. One could try a linear redefinition $e_a^A = \lambda_a E_a^A$ to identify this general solution

$$\Sigma_{ab}^{AB} = G_{ab} e_a^{[A} e_b^{B]} \tag{12}$$

with $\Sigma^{AB} = \pm E^A \wedge E^B$, but this is not possible in general; one needs additional conditions.

In [8], we show that imposing the additional three constraints

$$\sum_b \sum_{\{a,f\} \not\subset \{b,e\}} n_{Abef} \Sigma_{ab}^{AB} = 0, \quad e \in \{0, 1, 2\} \text{ fixed}, \tag{13}$$

implies that $G_{ab}(x)$ is the same function $c(x)$ for all a, b, which can be absorbed by a pointwise rescaling $E^A = \sqrt{|c|} e^A$ to complete the identification $\Sigma^{AB} = \pm E^A \wedge E^B$. Note that not the field e^A we have introduced to define the constraints (10), but the rescaled E^A will play the role of an orthonormal tetrad determining physical lengths and angles. In particular, a completely degenerate triad $E^A = 0$ is possible since $c(x) = 0$ at some points is not excluded.

2.2. Discrete construction

The translation of the constraints into discretised variables is straightforward. Again one introduces a triangulation of spacetime and integrates Σ^{AB} over triangles and n_A over tetrahedra:

$$\Sigma_{ab}^{AB}, \; n_{Adef} \;\Rightarrow\; \Sigma^{AB}(\triangle), \; n_A(\blacktriangle). \tag{14}$$

Then the discrete analogue of the set of constraints (10) is (essentially by construction) the set of linear constraints used in the EPR(L)/FK models, $n_A(\blacktriangle)\Sigma^{AB}(\triangle) = 0 \quad \forall \triangle \subset \blacktriangle$. Perhaps more interestingly, the discrete version of the remaining constraints (13) would take the form

$$\sum_{\{i,j\} \not\ni \mathbf{A}} n_A(\blacktriangle_{\mathbf{i}}) \Sigma^{AB}(\triangle_{\mathbf{Aj}}) = 0, \tag{15}$$

where we label the tetrahedra in a 4-simplex by $\mathbf{A}, \mathbf{B}, \mathbf{C}, \mathbf{D}, \mathbf{E}$; there are five of these constraints per simplex (where $\mathbf{A}$ is replaced by $\mathbf{B}, \mathbf{C}, \mathbf{D}, \mathbf{E}$ respectively). These constraints play the role of the "volume" constraints which could be, in the case of quadratic constraints, replaced by a closure constraint relating different triangles in a tetrahedron. The situation here is completely analogous:

Result: The constraints (15) follow from the EPR(L)/FK linear constraints, the closure constraint on $\Sigma^{AB}(\triangle)$, plus an analogous "4D closure constraint"

$$n_A(\triangle_{\mathbf{A}}) + n_A(\triangle_{\mathbf{B}}) + n_A(\triangle_{\mathbf{C}}) + n_A(\triangle_{\mathbf{D}}) + n_A(\triangle_{\mathbf{E}}) = 0, \tag{16}$$

which can be given a clear geometric motivation, just as closure on Σ^{AB}: While the closure constraint (7) can be interpreted as the requirement for the triangles specified by $\Sigma^{AB}_{\triangle}$ to close up to form a tetrahedron, or alternatively as the discrete version of the field equation $\nabla_{[a}\Sigma^{AB}_{bc]} = 0$ integrated over an infinitesimal 3-ball, the four-dimensional constraint (16) can be interpreted as demanding that the five tetrahedra specified by $n^A(\triangle)$ close up to form a 4-simplex. The derivation of (16) from a continuum field equation is more subtle since one has to assume both $\nabla_{[a}\Sigma^{AB}_{bc]} = 0$ and the identification of Σ^{AB} with e^A to obtain the equation $\nabla_{[a}n^A_{bcd]} = 0$.

3. Summary

We have established a formulation of classical general relativity as a topological BF theory plus linear constraints which relied on the introduction of a basis of 3-forms at each point in addition to the two-form field Σ^{AB} and the connection ω^{AB}. This new continuum action is more closely related to current spin foam models, such as the EPR(L) and FK models, as discretising it leads to variables $\Sigma^{AB}(\triangle)$, $n_A(\triangle)$ which appear in these spin foam models, and are subject to the constraints used in these models, plus closure constraints on both Σ^{AB} and n_A. The latter is a new constraint which has been suggested before, but not received much attention in the literature so far. It suggests a new formulation for spin foam models in which the normals n_A carry information about the 3-volume and thus are given a fully geometric role, other than in the current models where one can set them to a fixed value. Other than searching for a modification of spin foam models in this direction, one should perform a detailed canonical analysis of the constraints used in our formulation. There are also indications for a close relation of our constructions both to current research in group field theories, as well as to the "edge simplicity" constraints introduced in [9]. We leave all this to future work.

Acknowledgments

This work was supported by EPSRC and Trinity College, Cambridge, partly through a Rouse Ball Travelling Studentship in Mathematics.

References

[1] Slides available as `http://www.iaa.es/ere2010/website/modules/tinyd0/content/talks/gielen.pdf`.
[2] Engle J, Pereira R and Rovelli C 2008 Flipped spinfoam vertex and loop gravity *Nuclear Physics* B **798** 251–290; Freidel L and Krasnov K 2008 A new spin foam model for 4D gravity *Classical and Quantum Gravity* **25** 125018
[3] Holst S 1996 Barbero's Hamiltonian derived from a generalized Hilbert-Palatini action *Physical Review* D **53** 5966–5969
[4] Plebański J F 1977 On the separation of Einsteinian substructures *Journal of Mathematical Physics* **18** 2511–2520
[5] Reisenberger M P 1999 Classical Euclidean general relativity from 'left-handed area = right-handed area' *Classical and Quantum Gravity* **16** 1357–1371
[6] Livine E R and Speziale S 2008 Solving the simplicity constraints for spinfoam quantum gravity *Europhysics Letters* **81** 50004
[7] Barrett J W and Crane L 1998 Relativistic spin networks and quantum gravity *Journal of Mathematical Physics* **39** 3296–3302
[8] Gielen S and Oriti D 2010 Classical general relativity as BF-Plebanski theory with linear constraints *Classical and Quantum Gravity* **27** 185017
[9] Dittrich B and Ryan J P 2008 Phase space descriptions for simplicial 4d geometries *Preprint* 0807.2806

Consistency of canonical formulation of Horava gravity

Chopin Soo

Department of Physics, National Cheng Kung University, Tainan, Taiwan

E-mail: cpsoo@mail.ncku.edu.tw

Abstract.
Both the non-projectable and projectable version of Horava gravity face serious challenges. In the non-projectable version, the constraint algebra is seemingly inconsistent. The projectable version lacks a local Hamiltonian constraint, thus allowing for an extra graviton mode which can be problematic. A new formulation (based on arXiv:1007.1563) of Horava gravity which is naturally *realized as a representation of the master constraint algebra (instead of the Dirac algebra)* studied by loop quantum gravity researchers is presented. This formulation yields a consistent canonical theory with first class constraints; and captures the essence of Horava gravity in retaining only spatial diffeomorphisms as the physically relevant non-trivial gauge symmetry. At the same time the *local* Hamiltonian constraint is equivalently enforced by the master constraint.

1. Introduction

Horava's theory of gravity [1] attempts to preserve unitarity by relinquishing space-time covariance, and improve renormalizability by including higher order *spatial derivatives*. Spatial diffeomorphism symmetry is retained at the fundamental level. A related development, in loop quantum gravity, is the application of the master constraint program [2, 3] to non-perturbative quantization of Einstein's theory. The master constraint algebra has the advantages of having structure constants rather than structure functions, and of spatial diffeomorphisms forming an ideal (thus allowing for the decoupling of the equivalent quantum Hamiltonian constraint from spatial diffeomorphism generators). In metric theories without full space-time covariance, departures of the constraint algebra from the Dirac algebra are to be expected. I present a new formulation (based on Ref. [4]) of Horava theory as a representation of the master constraint algebra,

$$\{\textstyle\int N^i H_i\, d^3x\, ,\int N'^j H_j\, d^3y\}_{\rm P.B.} = \int (\mathcal{L}_{\vec{N}} N'^i) H_i\, d^3x,$$
$$\{\textstyle\int N^i H_i\, d^3x\, , \mathbf{M}\}_{\rm P.B.} = 0\, ,\quad \{\mathbf{M}\, , \mathbf{M}\}_{\rm P.B.} = 0, \qquad (1)$$

which yields a consistent canonical theory with first class constraints. The *local* Hamiltonian constraint is equivalently implemented by the master constraint $\mathbf{M} := \int_\Sigma \frac{[H(x)]^2}{\sqrt{q(x)}}\, d^3x \approx 0.$

2. Horava gravity with detailed balance

In the canonical mould, the action for Horava gravity is $S = \int \pi^{ij}\dot{q}_{ij}\, d^3x\, dt - \int (NH + N^i H_i)\, d^3x\, dt$, wherein the super-Hamiltonian, with detailed balance [1], is $H = \frac{\kappa^2}{2}\frac{G_{ijkl}}{\sqrt{q}}[\pi^{ij}\pi^{kl} +$

$\frac{\delta W_{\rm T}}{\delta q_{ij}}\frac{\delta W_{\rm T}}{\delta q_{kl}}]$; and $H_i := 2q_{ij}\nabla_k \pi^{kj} \approx 0$ is the super-momentum constraint. It is assumed the space-time metric is of the Arnowitt-Deser-Misner form [5] $ds^2 = -N^2(cdt)^2 + q_{ij}(dx^i + N^i cdt)(dx^j + N^j cdt)$. The DeWitt supermetric, with deformation parameter λ which is not necessarily unity, is $G_{ijkl} = \frac{1}{2}(q_{ik}q_{jl} + q_{il}q_{jk}) - \frac{\lambda}{3\lambda-1}q_{ij}q_{kl}$; and $W_{\rm T}(q_{ij})$ is (up to 3rd order in spatial derivatives of the metric) the sum of a Chern-Simons action of the spatial affine connection and the spatial Einstein-Hilbert action with cosmological constant i.e. $W_{\rm T} = W_{\rm CS} + W_{\rm EH\Lambda} = \frac{1}{4w^2}\int \tilde{\epsilon}^{ikj}(\Gamma^l_{im}\partial_j\Gamma^m_{kl} + \frac{2}{3}\Gamma^l_{im}\Gamma^m_{jn}\Gamma^n_{kl})\,d^3x + \frac{\mu}{2}\int \sqrt{q}(R - 2\Lambda_W)\,d^3x$.

There is a striking feature in the *detailed balance condition* for Horava-Lifshitz theory: the Hamiltonian constraint can be succinctly expressed as [1]

$$H = \frac{\kappa^2 G_{ijkl}}{2\sqrt{q}}(\pi^{ij} + i\frac{\delta W_{\rm T}}{\delta q_{ij}})(\pi^{kl} - i\frac{\delta W_{\rm T}}{\delta q_{kl}}) =: \frac{\kappa^2}{2\sqrt{q}}G_{ijkl}Q^{ij}_+ Q^{kl}_- \approx 0; \tag{2}$$

wherein, as operators in the quantum theory, $\hat{Q}^{ij}_{\pm} := \hat{\pi}^{ij} \pm i\frac{\delta W_{\rm T}}{\delta q_{ij}} = e^{\pm\frac{W_{\rm T}}{\hbar}}\hat{\pi}^{ij}e^{\mp\frac{W_{\rm T}}{\hbar}}$. $Q_{\pm}$ are hermitian conjugates of each other if $W_{\rm T}$ is hermitian (classically real), and are separately hermitian if $W_{\rm T}$ is purely anti-hermitian (classically pure imaginary). In both cases, the classical H remains real. The superspace metric [7], $\delta S^2 \equiv G^{ijkl}\delta q_{ij}\delta q_{kl}$, has signature $({\rm sgn}[\frac{1}{3}-\lambda], +, +, +, +, +)$. So the corresponding Wheeler-DeWitt equation [7, 8] comes equipped with an 'intrinsic time' [7] for $\lambda > \frac{1}{3}$. Intriguingly, the emergent speed of light $c = \frac{\kappa^2\mu}{4}\sqrt{\frac{\Lambda_W}{1-3\lambda}}$, the cosmological constant $\Lambda = \frac{3}{2}\Lambda_W$, and Newton's gravitational constant $G = \frac{\kappa^2 c^3}{32\pi}$ can all be phenomenologically positive for $\lambda > \frac{1}{3}$ only if μ is pure imaginary and κ is real. Then H and the action is real only if w^2 is pure imaginary. This set of values renders $W_{\rm T}$ to be pure imaginary, and thus $Q_{\pm}$ become individually hermitian. States $\Psi_{Q_{\pm}} = Ie^{\pm\frac{W_{\rm T}}{\hbar}}$ with slowly varying I are semi-classical; and a pure imaginary $W_{\rm T}$ leads to real $\pi^{ij} = \mp i\frac{\delta W_{\rm T}}{\delta q_{ij}}$ solving the Hamilton-Jacobi equation with $\pm iW_{\rm T}$ as Hamilton functions.

3. Departures from the Dirac algebra

Both the projectable and non-projectable version of the theory face serious obstacles. In the projectable version, with the lapse function $N(t)$ dependent only on time, the theory has an integrated (rather than local) constraint $\int H\,d^3x = 0$. This produces a consistent first class constraint algebra. However, the absence of a local constraint $H(x) = 0$ implies the theory has an additional degree of freedom which can be pathological and phenomenologically problematic. In the Einstein-Hilbert theory the would-be pathological scalar mode is eliminated precisely by the local Hamiltonian constraint. Non-projectable Horava gravity with space-time dependent lapse function $N(x)$ is also problematic. $\pi_N(x) = 0$ leads to the secondary local constraint $H(x) = 0$. But for Horava gravity, $\{H(x), \int NH\,d^3y\}_{\rm P.B.} = (\Delta + \omega)N(x)$, wherein Δ contains spatial derivatives acting on N and ω does not [9, 10]. The only consistent solution is $N = 0$ [10]. Strictly speaking, $N = 0$ considered as a special case of a gauge-fixing condition $N = f$ results in $\{N(x) - f, \pi_N(y)\}_{\rm P.B.} = \delta(x - y)$ with second class constraints $\pi_N = N = 0$. All the constraints are formally stable under evolution. But this 'consistency', achieved at the cost of vanishing lapse function N, is dubious; moreover the space-time Arnowitt-Deser-Misner metric is degenerate whenever $N = 0$.

Dirac's algorithm for the analysis of constrained canonical systems [11] should reveal the true gauge symmetries of the theory. For Horava gravity, the requirement of vanishing lapse function from the algorithm seems to signal that *only* 3-dim. spatial diffeomorphism symmetry is physically relevant. Such theories cannot obey the Dirac algebra which is the

[1] After its initial discussion in Ref.[4], this particular form of the Hamiltonian constraint has also been investigated in Ref.[6].

hallmark of space-time covariance and the embeddability of hypersurface deformations, and from which Einstein's geometrodynamics can be uniquely recovered [12, 13]. There can be interesting modifications to the Dirac algebra in theories without full space-time covariance. For example, in ultra-local gravity with $H = \frac{2\kappa'}{\sqrt{q}} G_{ijkl}\pi^{ij}\pi^{kl}$, a strongly vanishing commutator, $\{\int NH\, d^3x, \int MH\, d^3y\}_{\text{P.B.}} = 0$ (even for $\lambda \neq 1$), replaces the usual commutation relation in the Dirac algebra [13]. When a scalar curvature term is added in deformations of Einstein-Hilbert theory with $\lambda \neq 1$ and $H = (\frac{2\kappa'}{\sqrt{q}} G_{ijkl}\pi^{ij}\pi^{kl} - \frac{\sqrt{q}}{2\kappa'} R)$, the corresponding Poisson bracket is

$$\{\int NH\, d^3x, \int N'H\, d^3y\}_{\text{P.B.}} = \int (N\nabla^i N' - N'\nabla^i N)[H_i - \frac{2(1-\lambda)}{3\lambda - 1}\nabla_i \pi]\, d^3x \tag{3}$$

With $\lambda \neq 1$ (the degenerate case [14] of $\lambda = \frac{1}{3}$ will not be discussed here), the consequent secondary constraint $Z_i := \nabla_i \pi = 0 \Leftrightarrow \pi = K(t)\sqrt{q}$ arises. The resultant condition for stability of $Z_i = 0$ is $W := [-\nabla^2 + R + \frac{2\kappa'^2\pi^2}{(3\lambda-1)q}]N = 0$. The constraint $H = 0$ allows us to write $R = \frac{4\kappa'^2}{q}\left(\overline{\pi}_{ij}\overline{\pi}^{ij} - \frac{1}{3(3\lambda-1)}\pi^2\right)$, wherein $\overline{\pi}^{ij} := \pi^{ij} - \frac{1}{3}q^{ij}\pi$ is the traceless part of the momentum. Together with $\pi = K\sqrt{q}$, the condition on N becomes

$$W = [-\nabla^2 + \frac{4\kappa'^2}{q}\left(\overline{\pi}_{ij}\overline{\pi}^{ij}\right) + \frac{16\kappa'^2 K^2}{3(3\lambda - 1)}]N = 0. \tag{4}$$

Since $-\nabla^2$ and $\frac{4\kappa'^2}{q}\overline{\pi}_{ij}\overline{\pi}^{ij}$ are both positive semi-definite operators, $W = 0$ can have non-vanishing solution for N only iff $\lambda < \frac{1}{3}$. For $\lambda < \frac{1}{3}$, the resultant theory (counting the 6 conjugate pairs (q_{ij}, π^{ij}), and $H_i = 0$ as first class and $H = 0, \tilde{\pi} = K\sqrt{q}$ as second class constraints) has $\frac{1}{2}[12 - 3(2) - 2(1)] = 2$ degrees of freedom. For $\lambda > \frac{1}{3}$, we are lead to the fact $N = 0$ is the only solution for $W = 0$. These conclusions differ from those of Ref. [15]. Although secondary constraints arise, a non-trivial solution exists for N for $\lambda < \frac{1}{3}$. However, for $\lambda \geq \frac{1}{3}$ (and for Horava gravity with local Hamiltonian constraint) only $N = 0$ is allowed.

4. Horava gravity as a master constraint theory

The Dirac algebra is not the Lie algebra of 4-dim. diffeomorphisms as structure *functions* are present in the commutator of two Hamiltonian constraints. But H_i and H constraints do generate 4-dim. diffeomorphisms on-shell. In a theory which possesses at the fundamental level only 3-dim. diffeomorphisms as gauge symmetry, we expect the constraints to generate, *on-shell, only spatial diffeomorphisms.* There is a formulation which precisely achieves this goal, and which at the same time gives rise to a condition equivalent to the local constraint $H(x) = 0$: Horava gravity realized as a master constraint theory. The master constraint $\mathbf{M} := \int_\Sigma \frac{[H(x)]^2}{\sqrt{q(x)}}\, d^3x$ is tailored to be invariant under spatial diffeomorphisms (with $H(x)$ being a scalar density of weight 1). For any real valued $H(x)$, the integrand is positive-semi-definite; and the master constraint equation, $\mathbf{M} = 0$, is mathematically equivalent to $H(x) = 0 \forall \vec{x}$ on the Cauchy surface Σ. Thus the master constraint equation replaces the infinite number of local restrictions $(H = 0)$ by a single global restriction. The canonical action for Horava gravity can be assumed to be

$$S = \int \pi^{ij}\dot{q}_{ij}\, d^3x\, dt - \frac{1}{\epsilon_o}\int N(t)\mathbf{M}\, dt + \int N^i H_i\, d^3x\, dt, \tag{5}$$

with H of the form in Eq.(2); ϵ_o has the physical dimension of energy density. The upshot is a consistent simple closed first class constraint algebra (as in Eq.(1)) with structure *constants.* Such a theory consistently generate equations of motion which are (on-shell) equivalent to spatial diffeomorphisms since $\{q_{ij}, N(t)\frac{\mathbf{M}}{\epsilon_o} + \int N^k H_k\, d^3x\}|_{\mathbf{M}=0\Leftrightarrow H=0} \approx \{q_{ij}, \int N^k H_k\, d^3x\}_{\text{P.B.}} = \mathcal{L}_{\vec{N}} q_{ij}$

(and similarly for π^{ij}). This formulation also realizes, in an explicit manner, the claim in Ref. [10] that time-reparamatrization symmetry of Horava gravity and freedom in the choice of $N(t)$ is on-shell trivial. Weak observables O should satisfy $\{O, \frac{N(t)}{\epsilon_0}\mathbf{M} + \int N^i H_i\, d^3x\}|_{\mathbf{M}=0 \Leftrightarrow H=0} \approx \{O, \int N^i H_i\, d^3x\}_{\text{P.B.}} \approx 0$, and are thus invariant with respect to the local gauge symmetry of spatial diffeomorphisms. In such theories, two configurations differing by four-dimensional, rather than spatial, coordinate transformations can be physically inequivalent. It is to be noted our formulation of Horava gravity *also requires the lapse function to depend only on t*.

'Troubles' in the constraint algebra of Horava gravity with *local* Hamiltonian constraint are to be expected of a canonical theory of 3-geometry which fundamentally possesses only spatial diffeomorphism invariance. Canonical Horava gravity can be consistently constructed as a master constraint theory, with the local Hamiltonian constraint equivalently enforced by the master constraint. Horava gravity is in fact an explicit, and highly non-trivial representation, of a master constraint theory with spatial diffeomorphism invariance. In not satisfying the Dirac algebra with local Hamiltonian constraint, Horava gravity is in fact more naturally associated with the master constraint theory than Einstein's General Relativity.

Acknowledgments

This work has been supported in part by the National Science Council of Taiwan under Grant Nos. NSC98-2112-M-006-006-MY3 and 98-2811-M-006-035. Beneficial interactions with J. Fernando Barbero G. are gratefully acknowledged.

References

[1] Horava P 2009 *Phys. Rev.* D **79** 084008
[2] Thiemann T 2006 *Class. Quantum Grav.* **23** 2211
[3] Dittrich B and Thiemann T 2006 *Class. Quantum Grav.* **23** 1025; 2006 *Class. Quantum Grav.* **23** 1067; 2006 *Class. Quantum Grav.* **23** 1089; 2006 *Class. Quantum Grav.* **23** 1043
[4] Soo C, Yang J and Yu H L 2010 New formulation of Horava-Lifshitz quantum gravity as a master constraint theory *Preprint* arXiv:1007.1563
[5] Arnowitt R L, Deser S and Misner C M 1959 *Phys. Rev.* **116** 1322
[6] Horava P, Melby-Thompson C M 2010 *Phys. Rev.* D **82** 064027
[7] DeWitt Bryce S 1967 *Phys. Rev.* **160** 1113
[8] Wheeler J A 1968 Superspace and the nature of quantum geometrodynamics, *Battelle Rencontres* ed C. M. DeWitt C M and Wheeler J A (New York: W. A. Benjamin)
[9] Li M and Pang Y 2009 JHEP08(2009)015
[10] Henneaux M, Kleinschmidt A and Gomez G L 2010 *Phys. Rev.* D **81** 0604002; 2010 Remarks on Gauge Invariance and First-Class Constraints *Preprint* arXiv:1004.3769
[11] Dirac P A M 1964 *Lectures in Quantum Mechanics* (New York: Yeshiva University Press)
[12] Hojman S, Kuchar K and Teitelboim C 1973 *Nature Phys. Sci.* **245** 97; 1976 *Ann. Phys.* **96** 88
[13] Teitelboim C 1980 The Hamiltonian structure of spacetime *General Relativity and Gravitation* vol 1, ed Held A (New York: Plenum)
[14] Pons J M and Talavara P 2010 *Phys. Rev.* D **82** 044011
[15] Bellorin J and Restuccia A 2010 On the consistency of the Horava Theory *Preprint* arXiv:1004.0055; 2010 Closure of the algebra of constraints for a non-projectable Horava model *Preprint* arXiv:1010.5531

A Theorem of Equivalence between TransverseDiff Theories and Scalar-Tensor Gravity

J.J. Lopez-Villarejo

European Organization for Nuclear Research, CERN CH-1211, Genève 23, Switzerland and Departamento de Física Teórica and Instituto de Física Teórica UAM/CSIC, E-28049-Madrid, Spain

E-mail: jj.lopezvillarejo@cern.ch; jjlopezvillarejo@gmail.com

Abstract. Transverse Diffeomorphism (TDiff) theories are well-motivated theories of gravity from the quantum perspective, which are based upon a gauge symmetry principle. The main contribution of this work is to firmly establish a correspondence between TransverseDiff and the better-known scalar-tensor gravity — in its more general form —, a relation which is completely analogous to that between unimodular gravity and General Relativity. We then comment on observational aspects of TDiff. In connection with this proof, we derive a very general rule that determines under what conditions the procedure of fixing a gauge symmetry can be equivalently applied before the variational principle leading to the equations of motion, as opposed to the standard procedure, which takes place afterwards; this rule applies to gauge-fixing terms without derivatives.

1. Introduction

Symmetry principles have a great relevance in modern physics theories. Specifically, there is a widespread point of view on General Relativity, stemming from its possible quantum interpretation, that considers it as the theory of spacetime diffeomorphism (Diff) invariance for spin-two massless particles: one starts from a linearized gravity action for a symmetric rank-two tensor in Minkowski space and, under the guidance of the Diff symmetry principle, one is "naturally" led to the full Einstein theory in the non-linear regime [9, 5] (see however [17]). Under this view, the *external* symmetry group of (active) diffeomorphisms acts as a *gauge* group which proves essential for eliminating the unphysical ill-defined polarization modes (ghosts) of the massless graviton.

On the other hand, it was already pointed out that the necessary and sufficient symmetry group for a consistent description of the massless graviton is not the full group of diffeomorphisms (Diff(M)) but rather a maximal subgroup of it [23]. Transformations belonging to this subgroup have been dubbed *transverse* (TDiff(M)) [1, 2], since the parameter describing the gauge transformation at this linear level is transverse, $\partial_\mu \xi^\mu = 0$, such as in the transformation law for the linearized metric:

$$h_{\mu\nu}(x) \to h_{\mu\nu}(x) + \partial_\mu \xi_\nu(x) + \partial_\nu \xi_\mu(x), \text{ with } \partial_\mu \xi^\mu(x) = 0. \tag{1}$$

At the non-linear level, one can maintain the same restriction $\partial_\mu \xi^\mu = 0$ on the symmetry group so that, for example, the transformation of the metric reads:

$$g_{\mu\nu}(x) \to g_{\mu\nu}(x) + \nabla_\mu \xi_\nu(x) + \nabla_\nu \xi_\mu(x), \quad \text{with } \partial_\mu \xi^\mu(x) = 0. \tag{2}$$

For a finite transformation, this corresponds to the subgroup of (finite) general coordinate transformations with Jacobian determinant equal to unity.

Following under this "quantum view", consider that the true symmetry of nature were TDiff, instead of Diff. Then, one is impelled to construct the most general action compatible with TDiff symmetry. The result is that, since $g \equiv -det(g_{\mu\nu})$ behaves as a scalar under this subgroup, one can add arbitrary functions of g to the usual terms of the GR Lagrangian [2]. Therefore, the most general Lagrangian invariant under transverse diffeomorphisms and of second order in derivatives reads:

$$S_{TDiff} = -\frac{1}{2\kappa^2}\int d^4x\sqrt{g}\left[f\left(\sqrt{g}-1\right)R + 2f_\lambda\left(\sqrt{g}-1\right)\Lambda + \frac{1}{2}f_k\left(\sqrt{g}-1\right)g^{\mu\nu}\partial_\mu\sqrt{g}\partial_\nu\sqrt{g}\right] + S_M \tag{3}$$

and the matter action may be taken to be of the form

$$S_M = \int d^4x\sqrt{g}L_{SM}\left[\psi_m, g_{\mu\nu};\, \sqrt{g}-1\right] \tag{4}$$

where we allow for an arbitrary extra-dependence on g in the conventional Standard Model Lagrangian; also, we have expressed this dependence on $\sqrt{g}-1$ instead of g for later convenience. As it stands, the action given by 3, 4 is clearly non-covariant, but is formulated in a specific set of coordinates — up to TDiff transformations of coordinates. The quantum ultraviolet behavior of such a theory was discussed in [3]; its observational signatures were addressed in [4] — we will have some comments to add under the light shed by our findings.

This way of proceeding from the TDiff symmetry principle is not, however, the standard route that followed from the work of Bij, Dam&Ng [23]. Indeed, these same authors started to develop what is known as *unimodular gravity* (recent proposals include [8, 21]). We compare and contrast these two approaches in section 2.

On another note, there is a significant lesson in the work by Kretschmann [12], when objecting to Einstein's views: every spacetime theory admits a generally covariant representation (see also [15, 16]). For TDiff, a representation of this kind can be found, e.g., in [18]. It introduces an absolute prior-geometric object [22], namely a background density $\tilde{g}$, so that all occurrences of g in the scalar Lagrangian appear in the covariant form $g/\tilde{g}$ (then again, when the coordinates employed are such that $\tilde{g}$ equals the unit scalar density we recover straightforwardly our action 3). We go one step further and introduce the notion of *physically-equivalent theories.* We find that, under certain simple boundary conditions, TDiff theory resembles a general scalar-tensor theory, to the same extent that General Relativity constrained to the harmonic gauge — or any other gauge — corresponds in its physical consequences to General Relativity in its covariant standard form (we are just "fixing the gauge"):

$$\delta\left(\int d^4x\sqrt{g}R\right) = 0 \quad \underset{\text{"physically equivalent"}}{\Longleftrightarrow} \quad \left[\delta\left(\int d^4x\sqrt{g}R\right) = 0\right]_{\partial_\mu(\sqrt{g}g^{\mu\nu})=0}$$

By *general* scalar-tensor theory we mean that the gravitational scalar is arbitrarily present all through the Lagrangian, i.e., not the standard presentation of scalar-tensor theory [11, 10, 6], which employs the principle of *universal coupling* in the matter Lagrangian or the equivalent concept of a *metric theory* of gravity (see for example [22, 24]), and which was developed to rescue the (Weak) Equivalence Principle (see however [7] regarding its convenience). The idea of this correspondence between TDiff and general scalar-tensor gravity has already been touched upon in previous studies [3, 4], and has even made use of, but now our aim is to definitively establish it; we address it in section 3. Finally, we conclude in section 4.

2. TDiff Gravity vs. Unimodular theories

The TransverseDiff symmetry at the linear level can be implemented in two different fashions in the Lagrangian, which give rise to two very different theories at the non-linear level.

The first route was explored in the aforementioned seminal paper [23]. One should note that the transversality condition in 1 is directly affecting the transformation rule of the trace of $h_{\mu\nu}$: $h^\mu_\mu \to h^\mu_\mu + \partial_\mu \xi^\mu$, such that this quantity is an invariant. Therefore, one can decide to restrict the value of the trace in the theory to have a fixed value, typically $h^\mu_\mu = 0$. In the full non-linear regime, this naturally translates into the condition $g \equiv -det(g_{\mu\nu}) = 1$, which explains the name *unimodular gravity*, but one could also fix the determinant of the metric to any arbitrary function.

Therefore, unimodular gravity is based upon a reduction of the functional space on which the Einstein-Hilbert action is defined:

$$g_{\mu\nu} \to \hat{g}_{\mu\nu}, \text{ with } det(\hat{g}_{\mu\nu}) = \epsilon_0$$

$$S_{EH}[g_{\mu\nu}] \equiv -\frac{1}{2\kappa^2}\int d^4x\sqrt{g}R[g_{\mu\nu}] \to S_{UG}[\hat{g}_{\mu\nu}] \equiv -\frac{1}{2\kappa^2}\int d^4x\epsilon_0 R[\hat{g}_{\mu\nu}] \tag{5}$$

where ϵ_0 is some fixed scalar density, usually taken unity. The symmetries of this action are the "volume preserving diffeomorphisms (VPD)", which respect $g(x) = g'(x)$ (and at the infinitesimal level $\delta_\xi g = \nabla_\mu \xi^\mu = \partial_\mu(\sqrt{g}\xi^\mu) = 0$). Note that these are not the same as 2 unless $\epsilon_0(x) = const$. We have 9 e.o.m.'s $\delta S/\delta \hat{g}_{\mu\nu} = 0$ (of which only 6 are independent, in virtue of the VPD symmetry) for 9 functions $\hat{g}_{\mu\nu}$. This theory resembles General Relativity with a gauge-fixing constraint on the metric. In fact, the only difference at the classical level is an arbitrary integration constant that appears, and which plays the role of a cosmological constant.

On the other hand, the approach that we are interested in, named "TDiff gravity" after the symmetry, is based upon application of a symmetry principle on the full space of metrics $g_{\mu\nu}$ (see 3 and 2). There are 10 e.o.m.'s: $\delta S/\delta g_{\mu\nu} = 0$ (only 7 independent in virtue of TDiff-based Bianchi identities), for 10 functions $g_{\mu\nu}$. This theory is connected to scalar-tensor theory with a gauge-fixing constraint, as we will see.

3. Correspondence between TDiff gravity and a general scalar-tensor theory.

Theorem. *Of correspondence between scalar-tensor and TDiff gravity.*

Let us have a TDiff theory of the general form 34, abbreviated as

$$S_{TDiff} = \int d^4x\sqrt{g(x)}\mathcal{L}(g_{\mu\nu}(x), \sqrt{g(x)} - 1;\ \psi(x)),$$

formulated in an open patch of coordinates B, with boundary ∂B, and where the $\psi(x)$ stands for a generic matter field. Consider now the corresponding scalar-tensor theory with action

$$S_{ST} = \int d^4x\sqrt{g(x)}\mathcal{L}(g_{\mu\nu}(x), \phi(x);\ \psi(x)),$$

where the explicit dependency on the determinant of the metric $\sqrt{g} - 1$ has been replaced by the new scalar field ϕ. Then, the set of (classical) solutions in both theories that fulfill that $\mathcal{L} \to 0$ at the boundary ∂B is the same, provided that we consider "physically equivalent" all solutions in a gauge orbit; specifically, we take the partial gauge fixing $\sqrt{g} - 1 = \phi$ on the scalar-tensor solutions:

$$\{(g_{\mu\nu}, \psi)|\ \delta S_{TDiff}[g_{\mu\nu}; \psi] = 0,\ \mathcal{L}|_{\partial B} = 0\} \equiv \{(g_{\mu\nu}, \phi, \psi)|\ \delta S_{ST}[g_{\mu\nu}, \phi; \psi] = 0,\ \mathcal{L}|_{\partial B} = 0\}|_{\phi=\sqrt{g}-1}$$

See [14] for the proof.

Let us make some comments following the theorem. First, a totally analogous theorem exists which connects GR and unimodular gravity 5. Indeed, the action for unimodular gravity can be written with the help of a Lagrange multiplier as ($\epsilon_0(x) = 1$)

$$S_{UG} = \int d^4x \left[\sqrt{g(x)}R(x) + \lambda(x)(\sqrt{g(x)} - 1)\right] \tag{6}$$

and, following a similar reasoning, one can establish the equivalence of the solutions of both theories when the Lagrangian goes to zero at the boundary. This result is already well-known, since the integration constant playing the role of a cosmological constant that appears in unimodular gravity — with respect to GR — is non-zero at the boundary.

Actually, a far more general conclusion can be drawn from our line of reasoning: *for any Lagrangian theory with a globally-defined gauge symmetry, it is equivalent to impose a gauge-fixing without derivatives before or after the calculation of variations, provided that we only consider (classical) solutions that make the Lagrangian vanish at the boundary of spacetime.* This adds up to the result in [19, 20] (Hamiltonian and Lagrangian treatment, respectively), that "the effect of plugging the gauge fixing constraint into the Lagrangian can be compensated by adding to the equations of motion for the reduced [restricted] theory some constraints that have disappeared as such along the process"

Second, this theorem somehow contradicts common beliefs of other authors, as in [18]

Third, we have expressed the explicit dependence on g of the TDiff theory by $\sqrt{g} - 1$ throughout the paper, so that it vanishes in Minkowski space with (canonical) cartesian coordinates. Whether the requirement $\mathcal{L} \underset{\partial B}{\to} 0$ is "natural" or too restrictive is left to the reader.

4. Conclusions

Let us summarize our results. We started by imposing a TDiff gauge symmetry group to the gravitational Lagrangian based on a symmetric rank-two tensor $g_{\mu\nu}$, as opposed to the standard Diff symmetry that leads us to General Relativity. Since the symmetry group is smaller, we thus get to a more general Lagrangian 34. Then, the theorem stated in section 3 tells us that, modulo some boundary conditions, one can view this theory just as a general scalar-tensor theory that has been presented in a particular gauge, namely $\phi = \sqrt{g} - 1$, but whose inherent symmetry group is again Diff (it is "physically-equivalent", in the sense of the introduction). At the end of the day, we have "gained" a scalar degree of freedom that was not postulated, but appears automatically due to the lack of symmetry. In a way, *the TDiff group can be thus considered to give rise to general-scalar tensor theory.* This present knowledge builds up from previous works [3, 4]. In addition, some collateral consequences of the theorem with a wider scope of application are presented at the end of section 3: these confront the question of "whether it makes a difference to impose the gauge-fixing before or after the calculus of variations"

Regarding the observational signatures of TDiff theories at the classical level, one can therefore rely on this equivalence to take advantage from the abundant literature that studies the bounds on the scalar-tensor theory. The same metric / universal coupling requirement that is usually imposed for scalar-tensor theory — which was mentioned in the introduction — can be applied directly in the TDiff Lagrangian 34 by considering only matter Lagrangians without explicit dependence on g.

I am grateful to Enrique Alvarez and Anton F. Faedo for our previous work on the subject, that opened this interesting perspective. I have largely benefited from discussion with Josep M. Pons, and from his comments. I also want to thank Daniel Zenhäusern for his interest and help with the project during my stay at EPFL, Lausanne. On the other hand, I missed some more support from the mathematics community that I consulted. This work has been partially supported by the European Commission (HPRN-CT-200-00148) as well as by FPA2009-09017 (DGI del MCyT, Spain) and S2009ESP-1473 (CA Madrid). The author is personally supported by a MEC grant, AP2007-00385.

[1] E. Alvarez, D. Blas, J. Garriga, and E. Verdaguer. Transverse Fierz-Pauli symmetry. *Nucl. Phys.*, B756:148–170, 2006.

[2] Enrique Alvarez and Anton F. Faedo. Unimodular cosmology and the weight of energy. *Phys. Rev.*, D76:064013, 2007.

[3] Enrique Alvarez, Anton F. Faedo, and J. J. Lopez-Villarejo. Ultraviolet behavior of transverse gravity. *JHEP*, 10:023, 2008.

[4] Enrique Alvarez, Anton F. Faedo, and J. J. Lopez-Villarejo. Observational constraints on transverse gravity. *JCAP*, 0907:002, 2009.

[5] David G. Boulware and Stanley Deser. Classical General Relativity Derived from Quantum Gravity. *Ann. Phys.*, 89:193, 1975.

[6] C. Brans and R. H. Dicke. Mach's principle and a relativistic theory of gravitation. *Phys. Rev.*, 124:925–935, 1961.

[7] Thibault Damour. Questioning the equivalence principle, arXiv:gr-qc/0109063v1.

[8] J. Earman. The cosmological constant, the fate of the universe, unimodular gravity, and all that. *Stud. Hist. Philos. Mod. Phys.*, 34:559–577, 2003.

[9] R.P. Feynman. *Lectures on Gravitation.* Caltech, 1962.

[10] M. Fierz. On the physical interpretation of P.Jordan's extended theory of gravitation. *Helv. Phys. Acta*, 29:128–134, 1956.

[11] P. Jordan. Schwerkaft und Weltall. *(Vieweg, Braunschweig, 1955).*

[12] E. Kretschmann. Uber den physikalischen Sinn der Relativitatspostulat, A Einsteins neue und seine ursprunglische Relativitatstheorie. *Annalen der Physik*, 53:575–614, 1993.

[13] D. L. Lee, A. P. Lightman, and W. T. Ni. Conservation laws and variational principles in metric theories of gravity. *Phys. Rev.*, D10:1685–1700, 1974.

[14] J. J. Lopez-Villarejo, TransverseDiff gravity is to scalar-tensor as unimodular gravity is to General Relativity. arXiv:1009.1023 [hep-th].

[15] J. D. Norton. General covariance, gauge theories, and the Kretschmann objection. In *Brading, K. (ed.) et al.: Symmetries in physics* 110- 123.

[16] J. D. Norton. General Covariance and the Foundations of General Relativity: Eight Decades of Dispute. *Reports on Progress in Physics*, 56:791–858, 1993.

[17] T. Padmanabhan. From gravitons to gravity: Myths and reality. *Int. J. Mod. Phys.*, D17:367–398, 2008.

[18] Yu. F. Pirogov. Unimodular metagravity vs. General Relativity with a scalar field. *Phys. Atom. Nucl.*, 73:134–137, 2010.

[19] J. M. Pons. Plugging the gauge fixing into the Lagrangian. *Int. J. Mod. Phys.*, A11:975–988, 1996.

[20] Josep M. Pons. Substituting fields within the action: consistency issues and some applications, arXiv:0909.4151 [hep-th].

[21] Lee Smolin. The quantization of unimodular gravity and the cosmological constant problem, Phys. Rev. **D**80, 084003, 2009.

[22] Kip S. Thorne, David L. Lee, and Alan P. Lightman. Foundations for a theory of gravitation theories. *Phys. Rev. D*, 7(12):3563–3578, Jun 1973.

[23] J. J. van der Bij, H. van Dam, and Yee Jack Ng. The exchange of massless spin two particles. *Physica*, 116A:307–320, 1982.

[24] Clifford M. Will. The confrontation between general relativity and experiment. *Living Reviews in Relativity*, 9(3), 2006.

Journal of Physics: Conference Series **314** (2011) 012047 doi:10.1088/1742-6596/314/1/012047

Quantum Gowdy model within the new loop quantum cosmology improved dynamics

M Martín-Benito[1,2], L J Garay[3,2], G A Mena Marugán[2]

[1]MPI für Gravitational Physics, Albert Einstein Institute, Am Mühlenberg 1, D-14476 Potsdam, Germany.

[2]Instituto de Estructura de la Materia, CSIC, Serrano 121, 28006 Madrid, Spain.

[3]Departamento de Física Teórica II, Universidad Complutense de Madrid, 28040 Madrid, Spain.

E-mail: mercedes@aei.mpg.de, luis.garay@fis.ucm.es, mena@iem.cfmac.csic.es

Abstract. The linearly polarized Gowdy T^3 model can be regarded as compact Bianchi I cosmologies with inhomogeneous modes allowed to travel in one direction. We study a hybrid quantization of this model that combines the loop quantization of the Bianchi I background, adopting the improved dynamics scheme put forward by Ashtekar and Wilson-Ewing, with a Fock quantization for the inhomogeneities. The Hamiltonian constraint operator provides a resolution of the cosmological singularity and superselects separable sectors. We analyze the complicated structure of these sectors. In any of them the Hamiltonian constraint provides an evolution equation with respect to the volume of the associated Bianchi I universe, with a well posed initial value problem. This fact allows us to construct the Hilbert space of physical states and to show that we recover the standard quantum field theory for the inhomogeneities.

1. Introduction

The Gowdy cosmologies represent a suitable arena to further develop loop quantum cosmology (LQC) [1] and include inhomogeneous settings, since they provide the simplest inhomogeneous cosmological spacetimes [2]. Here, we analyze the particular case of the Gowdy model with three-torus topology and linear polarization. The homogeneous sector of this model (given by the degrees of freedom that parametrize the subset of homogeneous solutions) coincides with the Bianchi I model with three-torus topology. In turn, the inhomogeneities describe a content of linearly polarized gravitational waves traveling in a single direction. In previous studies we carried out the quantization of this system by means of a hybrid procedure [3, 4]. More specifically, we applied a Fock quantization for the inhomogeneous sector, in order to deal with the field complexity, while for the homogeneous sector we employed the quantization of the Bianchi I model performed in LQC [5], with the main aim of achieving a quantum resolution of the cosmological singularity.

The quantization of the Bianchi I model in LQC is subject to several ambiguities. One of them concerns the representation of the curvature tensor of the homogeneous connection in the quantum theory. Different definitions of this object lead to different schemes of quantization. In this respect, Ashtekar and Wilson-Ewing have put forward the quantization of the Bianchi I model within a new *improved dynamics* scheme [6], which seems to be physically more appropriate than the scheme employed in [3, 4, 5]. In this note we summarize the hybrid quantization of the Gowdy model that results from adopting this new improved dynamics scheme in the representation of the homogeneous sector. We focus on the new features that the use of this scheme introduces, such as the complicated structure of the resulting sectors of superselection (aspect not analyzed in [6]) and the interpretation of the quantum Hamiltonian constraint as an evolution equation with respect to the volume of the associated Bianchi I universe. We will discus that this interpretation is valid inasmuch as the corresponding initial value problem is well posed: physical solutions are completely determined by a countable set of data given on an initial section of non-vanishing Bianchi I volume. Moreover, this fact allows us to identify the physical Hilbert space with the Hilbert space of initial data. For the details of this work we refer to [7].

2. Quantization of the model

The linearly polarized Gowdy T^3 cosmologies are globally hyperbolic spacetimes with three-torus spatial topology and with two axial and hypersurface orthogonal Killing vector fields ∂_σ and ∂_δ [2]. We use global coordinates adapted to the symmetries $\{t, \theta, \sigma, \delta\}$, with $\theta, \sigma, \delta \in S^1$. Then the spatial dependence occurs only in θ, and we can expand the field-like variables in Fourier series in that coordinate. After performing a partial gauge fixing [4, 7], the reduced phase space is described by two pairs of canonically conjugate point-particle variables (they do not depend on θ) and by one field ξ, together with its canonical momentum P_ξ. Two global constraints remain in the model, the spatial average of the θ-momentum constraint (which generates translations in S^1) and the spatial average of the densitized Hamiltonian constraint.

We call inhomogeneous sector the set of degrees of freedom given by the nonzero Fourier modes of ξ and those of P_ξ. To describe this sector we introduce creation and annihilation-like variables, denoted by $\{a_m^*, a_m\}$ $(m \in \mathbb{Z} - \{0\})$, defined as if ξ and P_ξ were a canonical pair for a free massless scalar field. In the quantum theory we promote them to annihilation and creation operators, with $[\hat{a}_m, \hat{a}_m^\dagger] = 1$. $\mathcal{F}$ will denote the corresponding Fock space. The generator of translations on the circle only involves this inhomogeneous sector. It is not difficult to obtain its quantum counterpart in terms of the basic operators $\hat{a}_m$ and $\hat{a}_m^\dagger$. The states that verify this constraint form a Fock subspace $\mathcal{F}_p \subset \mathcal{F}$ [4, 7].

In turn, the homogeneous sector is formed by the rest of degrees of freedom: the two point-particle variables and the zero mode of ξ, together with its momentum. This sector coincides with the phase space of the Bianchi I model. We describe it in the Ashtekar-Barbero formalism. Using a diagonal gauge, the nontrivial components of the densitized triad are $p_i/4\pi^2$, with $i = \theta, \sigma, \delta$, and $c_i/2\pi$ are those of the $su(2)$ connection. They satisfy $\{c_i, p_j\} = 8\pi G\gamma\delta_{ij}$, where γ is the Immirzi parameter and G is the Newton constant. In order to represent this sector in the quantum theory we choose the kinematical Hilbert space of the Bianchi I model constructed in LQC [5, 6], that we call $\mathcal{H}_{\text{kin}}^{\text{BI}}$. We recall that, on $\mathcal{H}_{\text{kin}}^{\text{BI}}$, the operators $\hat{p}_i$ have a discrete spectrum equal to the real line. The corresponding eigenstates, $|p_\theta, p_\sigma, p_\delta\rangle$, form an orthonormal basis (in the discrete norm) of $\mathcal{H}_{\text{kin}}^{\text{BI}}$. Owing to this discreteness, there is no well defined operator representing the connection, but rather its holonomies. They are computed along straight edges in the fiducial directions. The so-called improved dynamics prescription states that, when writing the curvature tensor in terms of holonomies, we have to evaluate them along edges with a certain minimum dynamical (state dependent) length $\bar{\mu}_i$. We use the specific improved dynamics prescription put forward in [6]: all the elementary operators which represent the matrix elements of the holonomies, called $\hat{\mathcal{N}}_{\bar{\mu}_i}$, produce a constant shift in the Bianchi I volume of the compact spatial section. In order to simplify the analysis, it is convenient to relabel the basis states in the form $|v, \lambda_\sigma, \lambda_\delta\rangle$, where v is proportional to the volume and such that the operators $\hat{\mathcal{N}}_{\pm\bar{\mu}_i}$ cause a shift on it equal to ± 1, and the parameters λ_i are all equally defined in terms of the corresponding parameters p_i, and verify that $v = 2\lambda_\theta\lambda_\sigma\lambda_\delta$.

The Hamiltonian constraint of the Gowdy model is formed by the Hamiltonian constraint of the Bianchi I model plus the coupling term which involves both homogeneous and inhomogeneous sectors. Out of the basic operators $\hat{p}_i$, $\widehat{\mathcal{N}_{\bar{\mu}_i}}$, $\hat{a}_m$, and $\hat{a}_m^\dagger$, we represent it as an operator $\widehat{\mathcal{C}}$ well defined on a dense domain of the kinematical Hilbert space $\mathcal{H}_{\text{kin}}^{\text{BI}} \otimes \mathcal{F}$ [7]. We choose a very suitable symmetric factor ordering such that $\widehat{\mathcal{C}}$ decouples the states of $\mathcal{H}_{\text{kin}}^{\text{BI}} \otimes \mathcal{F}$ with support on $v = 0$, namely, the states with vanishing homogeneous volume. Moreover, our operator $\widehat{\mathcal{C}}$ does not relate states with different orientation of any of the eigenvalues of the operators $\hat{p}_i$. Owing to this property we can then restrict the homogeneous sector (also in the domain of definition of $\widehat{\mathcal{C}}$) to e.g. the space spanned by the states $|v > 0, \lambda_\sigma > 0, \lambda_\delta > 0\rangle$. We call the resulting Hilbert space $\mathcal{H}_{\text{kin}}^{\text{BI},+}$. Remarkably, for this restriction of the homogeneous sector we do not need to impose any particular boundary condition or appeal to any parity symmetry. Note that both $\mathcal{H}_{\text{kin}}^{\text{BI}}$ and $\mathcal{H}_{\text{kin}}^{\text{BI},+}$ are non-separable Hilbert spaces, feature which is not desirable for a physical theory. This problem is overcome by the very action of the Hamiltonian constraint operator. Indeed, $\widehat{\mathcal{C}}$ defined on (a dense domain in) $\mathcal{H}_{\text{kin}}^{\text{BI},+} \otimes \mathcal{F}$ turns out to leave invariant some separable Hilbert subspaces that provide superselection sectors. Specifically, the homogeneous sector of these superselection sectors is the space spanned by the states $|v = \varepsilon + 4n, \lambda_\sigma = \lambda_\sigma^\star \omega_\varepsilon, \lambda_\delta = \lambda_\delta^\star \omega_\varepsilon\rangle$. Here $\varepsilon \in (0, 4]$ and $n \in \mathbb{N}$; therefore v takes support on semilattices of constant step equal to 4 starting in a minimum non-vanishing value ε. In addition, $\lambda_a^\star$ $(a = \sigma, \delta)$ is some positive real number and ω_ε runs over the subset of $\mathbb{R}^+$ given by

$$\left\{ \left(\frac{\varepsilon - 2}{\varepsilon} \right)^z \prod_k \left(\frac{\varepsilon + 2m_k}{\varepsilon + 2n_k} \right)^{p_k} \right\},$$

where $m_k, n_k, p_k \in \mathbb{N}$, and $z \in \mathbb{Z}$ when $\varepsilon > 2$, while $z = 0$ otherwise. One can check that indeed this set is dense in $\mathbb{R}^+$ and countable [7]. Therefore, any of these sectors provide separable Hilbert spaces contained in $\mathcal{H}_{\text{kin}}^{\text{BI}}$. We denote them as $\mathcal{H}_{\varepsilon, \lambda_\sigma^\star, \lambda_\delta^\star} = \mathcal{H}_\varepsilon \otimes \mathcal{H}_{\lambda_\sigma^\star} \otimes \mathcal{H}_{\lambda_\delta^\star}$. Note that the fact that the states with support on $v = 0$ are removed means that there is no analog in our quantum theory of the cosmological singularities, classically requiring that some p_i vanish. We thus solve these singularities already at the kinematical level in a very simple way.

We then restrict the study to any of these superselection sectors and define the homogeneous terms of the Hamiltonian constraint operator $\widehat{\mathcal{C}}$ on the span of the states $|\varepsilon + 4n, \lambda_\sigma^\star \omega_\varepsilon, \lambda_\delta^\star \omega_\varepsilon\rangle$. The Hamiltonian constraint turns out to provide a difference equation in the parameter v [7]. Therefore, if we regard v as an internal time, the constraint can be interpreted as an evolution equation in it. The condition that this constraint imposes is very complicated and the solutions can be obtained only formally. This is in part owing to the fact that one of the inhomogeneous terms creates in every step of the evolution an infinite number of particles (for a detailed discussion see [3, 4, 7]). Nonetheless, thanks to the precise structure of the superselection sectors for the variables λ_a we can argue that a set of countable data given on the initial section $v = \varepsilon$ completely determines the physical solutions. This issue is non-trivial. Actually, both the separability of the support of λ_a and the fact that it is dense in $\mathbb{R}^+$ have been essential to show it (the details can be found in [8]). At the end of the day, the initial value problem is well posed and it is indeed feasible to make the above interpretation of the constraint as an evolution equation in v.

We can thus identify solutions with initial data. Furthermore, we can construct a(n over) complete set of observables acting on the space of initial data and impose reality conditions on them [7, 8]. This determines a unique inner product that endows the space of initial data with a Hilbert structure. Then, we finally impose the S^1 symmetry and identify the resulting Hilbert space with the physical Hilbert space. The result is that the physical Hilbert space is $H_{\lambda_\sigma^\star} \otimes \mathcal{H}_{\lambda_\delta^\star} \otimes \mathcal{F}_p$, with $H_{\lambda_\sigma^\star} \otimes \mathcal{H}_{\lambda_\delta^\star}$ being the physical Hilbert space of the Bianchi I model in LQC [8].

3. Conclusions

This work summarizes the quantization in the framework of LQC of the Gowdy T^3 model with linear polarization, which is an inhomogeneous cosmological spacetime with spatial dependence in a single direction θ. In order to perform it, first, at the classical level, we fix the gauge partially. The reduced system is subject to two global constraints: the spatial average of the θ-momentum constraint and the spatial average of the densitized Hamiltonian constraint. These constraints are imposed à la Dirac in the quantum theory. Moreover, they are represented by means of a hybrid approach, which combines the polymeric techniques of LQC with those of the Fock quantization. The former ones, applied in the representation of the homogeneous sector of the system, allow us to resolve the cosmological singularity in the quantum theory, while the Fock quantization for the inhomogeneities successfully deal with the field complexity.

Employing this hybrid approach we obtain well defined operators for the constraints. For the Hamiltonian constraint operator, we choose a very convenient symmetric factor ordering, such that it allows us to restrict the study of the homogeneous sector to suitable, separable superselection sectors. In any of them, the Hamiltonian constraint provides a difference equation in the volume of the homogeneous sector of the Gowdy model, which coincides with the Bianchi I model. We have interpreted this volume as an internal time and the Hamiltonian constraint as an evolution equation in that time. We have argued that this interpretation is valid, inasmuch as the corresponding initial value problem is well posed. In fact, a countable dense set of initial data completely specifies the physical solutions, and then we can characterize the physical Hilbert space as a Hilbert space of initial data. The physical inner product is fixed by demanding reality conditions on a complete set of observables and imposing the S^1 symmetry. We conclude that the physical Hilbert space is the physical Hilbert space of the Bianchi I model in LQC tensor product a Fock space $\mathcal{F}_p$.

It is important to note that, if the model is totally deparametrized and quantized as in standard quantum field theory, it admits a unique Fock quantization with unitary dynamics and such that the vacuum is invariant under S^1 translations [9, 10, 11]. The physical Hilbert space for the inhomogeneities, namely $\mathcal{F}_p$, is actually equivalent to that obtained for them in the standard quantization of the deparametrized model. Therefore we recover the standard description of the inhomogeneities, which can be seen as degrees of freedom propagating over the polymerically quantized Bianchi I background. This result, although expected, is not trivial since homogeneous and inhomogeneous sectors have been quantized with very different methods and they are coupled in a very involved way by the Hamiltonian constraint.

Acknowledgments

This work was in part supported by the Spanish MICINN Project No. FIS2008-06078-C03-03 and the Consolider-Ingenio Program CPAN No. CSD2007-00042. M. M-B has been supported by CSIC and the European Social Fund under the grant I3P-BPD2006.

References

[1] Bojowald M 2008 *Living Rev. Rel.* **11** 4
[2] Gowdy R H 1974 *Ann. Phys.* **83** 203
[3] Martín-Benito M, Garay L J and Mena Marugán G A 2008 *Phys. Rev.* D **78** 083516
[4] Mena Marugán G A and Martín-Benito M 2009 *Int. J. Mod. Phys.* A **24** 2820
[5] Martín-Benito M, Mena Marugán G A and Pawłowski T 2008 *Phys. Rev.* D **78** 064008
[6] Ashtekar A and Wilson-Ewing E 2009 *Phys. Rev.* D **79** 083535
[7] Garay L J, Martín-Benito M and Mena Marugán G A 2010 *Phys. Rev.* D **82** 044048
[8] Martín-Benito M, Mena Marugán G A and Wilson-Ewing E 2010 *Phys. Rev.* D **82** 084012
[9] Corichi A, Cortez J and Mena Marugán G A 2006 *Phys. Rev.* D **73** 084020
[10] Corichi A, Cortez J, Mena Marugán G A and Velhinho J M 2006 *Classical Quantum Gravity* **23** 6301
[11] Cortez J, Mena Marugán G A and Velhinho J M 2007 *Phys. Rev.* D **75** 084027

Journal of Physics: Conference Series **314** (2011) 012048 doi:10.1088/1742-6596/314/1/012048

Further improvements in the understanding of LQC

J Olmedo[1], **M Martín-Benito**[2,1], **G A Mena Marugán**[1]

[1] Instituto de Estructura de la Materia, CSIC, Serrano 121, 28006 Madrid, Spain.

[2] MPI für Gravitational Physics, Albert Einstein Institute, Am Mühlenberg 1, D-14476 Potsdam, Germany.

E-mail: `olmedo@iem.cfmac.csic.es`, `mercedes@aei.mpg.de`, `mena@iem.cfmac.csic.es`

Abstract. Loop Quantum Cosmology provides a successful quantization of isotropic and homogeneous flat universes with a massless, homogeneous scalar field as the matter content. Here, we propose a new ordering for the Hamiltonian constraint operator that facilitates the quantization of this model and makes the physical consequences much more transparent. In particular, our constraint is such that, in the gravitational sector, the zero volume state decouples, allowing us to get rid of the cosmological singularity already at a kinematical level, as well as to introduce a consistent densitization procedure for the constraint. Furthermore, the typical discretization of the spatial volume is achieved in superselection sectors which prove to be most suitable, with support on semilattices and where the basic functions that codify all the relevant information about the geometry have the expected Wheeler-DeWitt limit of standing waves. Thanks to these properties, we can demonstrate that the quantum bounce is generic for any physical state and superselection sector.

1. Introduction

Loop Quantum Cosmology (LQC) is an approach to construct a quantum description of homogeneous spacetimes using loop quantization techniques [1]. In the particular situation of a homogeneous, isotropic spacetime with flat spatial curvature coupled to a homogeneous, massless scalar field, the complete quantization was first studied in [2, 3, 4, 5]. Other systems with less symmetry have also been quantized using these techniques, including e.g. homogeneous Bianchi I models [6, 7].

In spite of the fact that the above works about the isotropic flat model succeeded in incorporating the effects of quantum geometry, some aspects of the quantization needed a deeper analysis. For instance, a rigorous densitization procedure of the Hamiltonian constraint was missing, and it was not clear whether and how one could attain superselection sectors that were simpler and with better properties. Actually, in previous studies of Bianchi I spacetimes in LQC [6], these questions and problems had been naturally solved. Inspired by those studies, we have adopted a new ordering for the representation of the gravitational Hamiltonian constraint of the isotropic case as an operator in the corresponding kinematic Hilbert space [8]. On the one hand, our constraint operator decouples the state which is the counterpart of the classical singularity in that Hilbert space (namely, the zero volume state), so that we can just remove it. Thanks to this, we are able to define a map between the solutions of the original Hamiltonian constraint and those of the (simpler) densitized one. On the other hand, the constraint (together with the physical observables) superselects sectors in the gravitational part of the kinematical Hilbert space with support on semilattices (where the zero volume state is not present) that are contained e.g. in the positive real line. The simple structure of these sectors considerably simplifies the analysis of the model. Indeed, when we consider the gravitational part of the quantum Hamiltonian constraint, we can fully determine the form and behavior of its generalized eigenfunctions. These admit a type of no-boundary description, and have the expected asymptotic limit, which agrees with the Wheeler-DeWitt (WDW) limit of the model. All these properties allow us to prove that the quantum bounce is generic and occurs for all superselection sectors.

2. The quantum Hamiltonian constraint

The phase space of the model is described by two pairs of conjugate variables. One pair describes the matter content: the homogeneous massless scalar field ϕ and its momentum p_ϕ, with $\{\phi, p_\phi\} = 1$. On the other hand, the geometry is represented in terms of a densitized triad and an Ashtekar-Barbero $su(2)$-connection. In the homogeneous and isotropic situation under analysis, and in a diagonal gauge, each of them is specified by a single coefficient, which we call p for the triad and c for the connection, with $\{c, p\} = 8\pi G\gamma/3$. Here, G is the Newton constant and γ the Immirzi parameter. Note that, owing to the homogeneity and in order to avoid divergences, one must restrict all the spatial integrals to a given cell, whose volume with respect to a Euclidean fiducial metric will be called V_0.

We introduce a standard quantization of the homogeneous scalar field. Regarding the geometry, we employ a loop quantization procedure. In LQC, and using the fiducial structure, one considers as basic variables fluxes of densitized triads through square surfaces, which can be identified essentially with p, and holonomies of the connection along straight lines of fiducial length $\bar{\mu}V_0^{1/3}$. The matrix elements of these holonomies, denoted by $\mathcal{N}_{\bar{\mu}}$, are given by exponential functions of ic. Besides, for these holonomies, we adopt an "improved dynamics" prescription. Then, the considered fiducial length must satisfy that $\bar{\mu}^{-1} = \sqrt{p/\Delta}$ (for more details see [4]). Here, Δ is the minimum non-zero area allowed quantum mechanically. The gravitational part of the configuration algebra Cyl_S is the algebra of almost periodic functions of c, which are identified with kets $|v\rangle$. The Cauchy completion with the discrete norm $\langle v'|v\rangle = \delta_{v'v}$ provides the gravitational part of the kinematical Hilbert space, $\mathcal{H}_{\text{kin}}^{\text{grav}}$. The action of the basic operators on the basis states $|v\rangle$ is given by $\hat{p}|v\rangle = \text{sign}(v)(2\pi\gamma G\hbar\sqrt{\Delta}|v|)^{2/3}|v\rangle$ and $\hat{\mathcal{N}}_{\bar{\mu}}|v\rangle = |v+1\rangle$. Here, $\hbar$ is the Planck constant. Using these basic operators, we can promote the classical Hamiltonian constraint into a quantum one. In particular, the $SU(2)$-curvature can be described in terms of holonomies forming a closed loop whose physical area coincides with the minimum area Δ (see [3, 4]). At this stage, we exploit the freedom in the factor ordering to introduce a new operator realization of the constraint, firstly studied in [6] and discussed in detail in [8]. Our Hamiltonian constraint is given by

$$\hat{C} := \left[\widehat{\frac{1}{V}}\right]^{1/2}\left(-\frac{6}{\gamma^2}\widehat{\Omega}^2 + 8\pi G\hat{p}_\phi^2\right)\left[\widehat{\frac{1}{V}}\right]^{1/2}, \tag{1}$$

where the gravitational part is the square of

$$\widehat{\Omega} := \frac{1}{4i\sqrt{\Delta}}\left[\widehat{\frac{1}{\sqrt{|p|}}}\right]^{-1/2}\widehat{\sqrt{|p|}}\left[\left(\hat{\mathcal{N}}_{2\bar{\mu}} - \hat{\mathcal{N}}_{-2\bar{\mu}}\right)\widehat{\text{sign}(p)} + \widehat{\text{sign}(p)}\left(\hat{\mathcal{N}}_{2\bar{\mu}} - \hat{\mathcal{N}}_{-2\bar{\mu}}\right)\right]\widehat{\sqrt{|p|}}\left[\widehat{\frac{1}{\sqrt{|p|}}}\right]^{-1/2}. \tag{2}$$

The inverse volume operator $[\widehat{1/V}] := [\widehat{1/\sqrt{|p|}}]^3$ is regularized employing the so-called Thiemann's trick (see, e.g. [2]).

2.1. Densitization procedure

Our constraint operator $\hat{C}$ annihilates the zero volume state $|v = 0\rangle$ in the gravitational sector, and its action leaves invariant the orthogonal complement of $|v = 0\rangle$, denoted by $\widetilde{\mathcal{H}}_{\text{kin}}^{\text{grav}}$.

Then, the zero volume state has no contribution to the non-trivial solutions of the Hamiltonian constraint, and we can thus restrict the study to $\widetilde{\mathcal{H}}_{\text{kin}}^{\text{grav}}$. To this extent, the classical singularity is kinematically resolved. Since we have removed the kernel of the inverse volume operator, we can use $\widehat{[1/V]}^{-1}$ to define a bijection between the solutions of $\hat{C}$ and the solutions of its densitized version [8], given by

$$\hat{\mathcal{C}} := \left[\widehat{\frac{1}{V}}\right]^{-1/2} \hat{C} \left[\widehat{\frac{1}{V}}\right]^{-1/2} = -\frac{6}{\gamma^2}\widehat{\Omega}^2 + 8\pi G\hat{p}_\phi^2. \tag{3}$$

Let us call $\widetilde{\text{Cyl}}_S$ the linear span of the states $|v\rangle$ with $v \neq 0$, whose Cauchy completion with respect to the discrete norm gives $\widetilde{\mathcal{H}}_{\text{kin}}^{\text{grav}}$. Since in general the solutions to the Hamiltonian constraint are not normalizable states in the kinematic Hilbert space, one should expect them to belong to a bigger space, like e.g. the algebraic dual of $\widetilde{\text{Cyl}}_S$, namely $\widetilde{\text{Cyl}}_S^*$. In that space, the map between the solutions $(\psi|$ of the constraint $\hat{C}$ and those of $\hat{\mathcal{C}}$, $(\psi'| = (\psi|\widehat{[1/V]}^{1/2}$, is indeed a bijection. Using $\hat{\mathcal{C}}$, we can directly recognize two commuting Dirac observables, $\widehat{\Omega}^2$ and $\hat{p}_\phi^2$, a fact that simplifies the resolution of the constraint.

2.2. Properties of $\widehat{\Omega}^2$

The gravitational constraint operator has an action of the form [8]

$$\widehat{\Omega}^2|v\rangle = F_-(v)|v+4\rangle + G(v)|v\rangle + F_+(v)|v-4\rangle, \tag{4}$$

so that it only connects states $|v\rangle$ separated by 4 units in the label v. Here, $F_\pm(v)$ and $G(v)$ are certain real functions such that $F_+(v) = 0$ if $v \in [-4, 0)$ and $F_-(v) = 0$ if $v \in (0, 4]$. In total, $\widehat{\Omega}^2$ only relates states with support on semilattices of the form $\mathcal{L}_{\tilde{\varepsilon}}^{\pm} := \{v = \pm(\tilde{\varepsilon} + 4n),\, n \in \mathbb{N}\}$, where $\tilde{\varepsilon} \in (0, 4]$. We note that the gravitational Hilbert subspaces corresponding to states with support on any of these semilattices are all separable. In the following, we will restrict the analysis of the quantum geometry to just one of these separable Hilbert spaces, e.g. one with $v > 0$. In [8] we demonstrated that the operator $\widehat{\Omega}^2$ defined in any of these superselection sectors turns out to be essentially self-adjoint, with a non-degenerate absolutely continuous spectrum equal to the positive real line. We will use the parameter $\lambda \in [0, \infty)$ to denote the generalized eigenvalues of $\widehat{\Omega}^2$.

2.3. Basis of generalized eigenfunctions

The generalized eigenstates $|e_\lambda^{\tilde{\varepsilon}}\rangle = \sum_{v \in \mathcal{L}_{\tilde{\varepsilon}}^+} e_\lambda^{\tilde{\varepsilon}}(v)|v\rangle$ are completely determined by a piece of initial data $e_\lambda^{\tilde{\varepsilon}}(\tilde{\varepsilon})$, which can be chosen positive. The resulting eigenfunctions are then real, since $\widehat{\Omega}^2$ is a real operator. The elements of this basis must be normalized to the Dirac delta, condition that fixes the norm of the initial datum. For large v, our operator tends to a differential operator which also possesses an absolutely continuous spectrum, but 2-fold degenerated. Its generalized eigenfunctions, denoted by $\underline{e}_{\pm\omega}(v)$, are given (up to a factor $1/\sqrt{v}$) by plane waves in the logarithm of v, each of them interpretable as expanding or contracting waves in the volume (once the dependence in the matter field is taken into account and the discussion restricted to positive-frequency solutions). In the regime $v \to \infty$, then, the eigenstates $e_\lambda^{\tilde{\varepsilon}}(v)$ must be linear combinations of these plane waves. Since the combination has to be real:

$$e_\lambda^{\tilde{\varepsilon}}(v) \to r\left\{\exp\left[i\phi_{\tilde{\varepsilon}}(\omega)\right]\underline{e}_\omega(v) + \exp\left[-i\phi_{\tilde{\varepsilon}}(\omega)\right]\underline{e}_{-\omega}(v)\right\}, \tag{5}$$

where $\underline{e}_\omega(v) = \exp\left(i\omega \ln v/\alpha\right)/\sqrt{2\pi\alpha v}$, $\omega = \sqrt{\lambda}$ and $\alpha = 4\pi\gamma G\hbar$. In addition, r is a real number and $\phi_{\tilde{\varepsilon}}(\omega)$ is certain real function (that in general varies slowly with ω).

The asymptotic limit of the eigenfunctions $e_\lambda^{\tilde{\varepsilon}}(v)$ has a clear physical interpretation: they behave as standing waves, with exactly the same contribution of incoming and outgoing flux.

2.4. *Comparison and comments about different constraint operators*

The operators chosen to represent the gravitational constraint in [3, 4, 5], denoted there as $\widehat{\Theta}$, have some qualitative differences with respect to ours. On the one hand, the superselection sectors in those works, in general, are lattices instead of semilattices. The absolutely continuous spectrum of $\widehat{\Theta}$ is 2-fold degenerate. The criterion suggested there to construct the generalized eigenfunctions needs the union of two lattices. Numerical analyses showed that, on each lattice and in the regime of semiclassical states peaked around classical trajectories, they behave as *almost* standing waves for $|v| \to \infty$, with different limits in each lattice (the phase of the almost standing waves is in principle different in each lattice). Even though this difference does not affect the norm of the considered physical states, the scenario changes drastically for more general regimes where the behavior as standing waves cannot be assured or the overlapping between contracting and expanding branches becomes non-negligible. Therefore, our proposal ensures a good behavior in regimes where the others commented cannot. On the other hand, there are some specific situations in which the operator $\widehat{\Theta}$ has essentially the same properties as $\widehat{\Omega}^2$. In more detail, let us consider the operator $\widehat{\Theta}$ of [4] (which is also one of the proposals in [3]) defined in the symmetric superselection sector that contains the $|v = \pm 2\rangle$ states. One can straightforwardly check that, in the difference eigenvalue equation corresponding to this constraint operator, some of the coefficients vanish if $v = \pm 2$. Something similar happens in the simplified model of [5] for $v = 0$. This fact leads to a decoupling between semiaxes. But this is so only in these exceptional situations, whereas the results of [8], confirming the standing wave behavior responsible of the quantum bounce, are valid in all circumstances and superselection sectors.

3. Conclusions

We have analyzed the flat Friedmann-Lemaître-Robertson-Walker model coupled to a massless scalar field in the context of LQC. We have proposed a new Hamiltonian constraint operator, based on a different factor ordering, which leads to a quantum resolution of the cosmological singularity already at the kinematical level. Employing this fact, we have been able to carry out a rigorous densitization of the constraint, facilitating the subsequent analysis of the system. Remarkably, the basic elements that codify the information about the geometry have support on semilattices, emerging from an initial section with non-zero volume. In this sense, they provide a no-boundary description. Furthermore, they behave as standing waves on macroscopic regions of the universe. These two properties indicate that any incoming flux evolves, through a quantum region, to outgoing flux and vice versa, for any physical state and in all superselection sectors. In other words, the intrinsic properties of the gravitational part of the constraint, inherited by its eigenfunctions, allow us to prove that the quantum bounce is generic. We have compared also the properties of our Hamiltonian constraint operator with those of previous proposals. Our operator is the only one which guaranties the phenomenon of a generic quantum bounce, independently of the particular state and superselection sector considered.

4. References

[1] Bojowald M 2008 *Living Rev. Rel.* **11** 4
[2] Ashtekar A, Bojowald M and Lewandowski J 2003 *Adv. Theor. Math. Phys.* **7** 233
[3] Ashtekar A, Pawłowski T and Singh P 2006 *Phys. Rev. Lett.* **96** 141301
Ashtekar A, Pawłowski T and Singh P 2006 *Phys. Rev.* D **73** 124038
[4] Ashtekar A, Pawłowski T and Singh P 2006 *Phys. Rev.* D **74** 84003
[5] Ashtekar A, Corichi A and Singh P 2008 *Phys. Rev.* D **77** 024046
[6] Martín-Benito M, Mena Marugán G A and Pawłowski T 2008 *Phys. Rev.* D **78** 064008
Martín-Benito M, Mena Marugán G A and Pawłowski T 2009 *Phys. Rev.* D **80** 084038
[7] Ashtekar A and Wilson-Ewing E 2009 *Phys. Rev.* D **79** 083535
[8] Martín-Benito M, Mena Marugán G A and Olmedo J 2009 *Phys. Rev.* D **80** 104015

Journal of Physics: Conference Series **314** (2011) 012049 doi:10.1088/1742-6596/314/1/012049

Spinfoam Cosmology
quantum cosmology from the full theory

Francesca Vidotto

Centre de Physique Théorique de Luminy[1] , Case 907, F-13288 Marseille, EU;
Dipartimento di Fisica Nucleare e Teorica, Università degli Studi di Pavia, and
Istituto Nazionale di Fisica Nucleare, Sezione di Pavia, via A. Bassi 6, I-27100 Pavia, EU.

E-mail: `vidotto@cpt.univ-mrs.fr`

Abstract. Quantum cosmology is usually studied quantizing symmetry-reduced variables. Is it possible, instead, to define quantum cosmology starting from the full quantum gravity theory? In Loop Quantum Gravity (LQG), it is possible to cut the degrees of freedom in a suitable way, in order to define a cosmological model. Such a model provides a tool for describing general fluctuations of the quantum geometry at the bounce that replaces the initial singularity. I focus on its simplest version, a "dipole" formed by two tetrahedra. This has been shown to describe a universe with anisotropic and inhomogenous degrees of freedom. Its dynamics can be given using the spinfoam formalism. I briefly review the present state of this approach.

1. Introduction

I review here the main ideas of a new approach to quantum cosmology called *spinfoam cosmology*. In order to do this, I need to start by recalling what is usually meant by *quantum cosmology*. Where does the idea of quantum cosmology come from?

The original problem was the quantization of gravity. When the Wheeler-DeWitt equation was written down, it immediately appeared to be an intractable equation. Brice DeWitt and Charles Misner realized that one could study the quantization problem in a much simpler context: the finite dimensional dynamical system defined by cosmology. In cosmology the symmeties assumed reduce the gravitational field to a system with one (or few) degrees of freedom (DOF). These are far easier to quantize than the full theory, of course. In this way we arrive to the Wheeler-DeWitt equation for cosmology, and to all the following minisuperspace models. The application of Loop ideas to this program, in the last decade, has lead to Loop Quantum Cosmology (LQC)[1], with its beautiful results (first of all singularity resolution [2, 3, 4]) and therefore to a remarkable realization of the research program of DeWitt and Misner.

In the meanwhile, however, there have been main developments in the quantization of the full theory. Now the original problem of "how to quantize gravity" is rather well defined and LQG gives us several tool [5]. Therefore today we have the possibility to follow a different road to quantum cosmology: we can try to study quantum cosmology starting from the full quantum gravity theory.

[1] Unité mixte de recherche (UMR 6207) du CNRS et des Universités de Provence (Aix-Marseille I), de la Méditerranée (Aix-Marseille II) et du Sud (Toulon-Var); laboratoire affilié à la FRUMAM (FR 2291).

What can we obtain by exploring this road? From a technical point of view, this could help us understanding how LQC is embedded in the full theory. But what I find more tantalizing is the possibility to have a tool to describe the quantum geometry at the bounce, namely in a regime where a single DOF could not be enough to capture the geometry. The other DOF of this model provide a natural way to include inhomogeneities [6, 7]. One long term hope is to obtain a meaningful setting to describe the seeds of structure formation.

Here I describe a strategy to address these questions. LQC is imbedded in the full theory as an approximation. Therefore I discuss the approximations in the full theory. I review some basic features of the present model of LQG, describing at each step the approximation taken, and relating these approximations with cosmology. Let me start from the approximation that is at the base of the birth of modern cosmology, as we can see it from this prospective.

2. Embedding the cosmological principle in the graph Hilbert space

Modern cosmology is based on the cosmological principle, that says that "the dynamics of a homogeneous and isotropic space approximates well our universe". The presence of inhomogeneities can be disregarded in a first order approximation, where we consider the dynamics as described on the scale of the scale factor, namely the size of the universe. Thus our approximation depends on the scale factor, it is not just a large scale approximation: it depends on the ratio between the scale factor and the interaction that we want to consider. If we consider the dynamics of the whole universe, this ratio gives 1, and an unique DOF is concerned. We can then recover the full theory adding these DOF one by one. We obtain an approximate dynamics of the universe, with a finite number of DOF.

Graph truncation In LQG there is a natural way to implement the cosmological expansion. This will become clear considering the kinematic. In LQG the theory is defined on a direct sum of *graph spaces* $\mathcal{H}_\Gamma = L^2[SU(2)^L/SU(2)^N]$, where the states are restricted to be the ones invariant under $SU(2)^N$ gauge transformations. Each $\mathcal{H}_\Gamma$ is an Hilbert space defined on a given graph Γ, and all the subgraphs and the graphs obtained by a permutation will be contained in such a space. The graph captures the large scale DOF obtained averaging the metric over the faces of a cellular decomposition formed by N cells. Working with graphs can be thought as taking an averaging of the metric on the faces of the cellular decomposition. These graphs can be characterized simply in terms of the number of nodes N, the number of links L and their adjacency.

A simple example of graph that can be chosen is the "dipole" $\Delta_2^* =$  with 2 nodes and 4 links. It is particularly nice because its dual Δ_2 has a direct classical interpretation being a triangulation of a 3-sphere, formed by two tetrahedra glued together. This graph can be very useful in order to do calculation, so we will make use of it later, but our arguments can be generalized for all the graphs.

The truncation of the Hilbert space of LQG to the Hilbert space on a given single graph can be put in correspondence to the expansion in modes in cosmology, truncated at a finite order. This defines an approximated kinematics of the universe, inhomogeneous but truncated at a finite number of cells. One can recover the full theory adding the DOF one by one. We start from the cosmological DOF, namely a single variable given by the volume, and then adding more DOF corresponds to add more nodes and links on the graph considered. To obtain an homogeneous and isotropic space, one have to consider a regular graph.

Coherent states On a graph space we can define an over-complete basis of wave packets [8]

$$\psi_{H_\ell}(U_\ell) = \int_{SU(2)^N} dg_n \bigotimes_{\ell\in\Gamma} K(g_{n_{source(\ell)}} U_\ell \, g^{-1}_{n_{target(\ell)}} H_\ell^{-1}) \quad \text{where } g_n \in SU(2) \ \forall \text{ nodes } n$$

and $\forall$ link ℓ, the heat kernel K peaks the holonomies U_ℓ on the $SL(2,\mathbb{C})$ element H_ℓ. This

Journal of Physics: Conference Series **314** (2011) 012049 doi:10.1088/1742-6596/314/1/012049

$H_\ell = g_{\vec{n}}\, e^{-i(\xi+i\eta)\frac{\sigma_3}{2}}\, g_{\vec{n}'}^{-1}$ determines two normals to the faces of the cellular decomposition $\vec{n}$ and $\vec{n}'$, and the variables ξ and η, associated respectively to the extrinsic curvature and the area of a face. These states are superposition of spinnetwork states, but they are peaked on a given geometry. In particular, it is possible to choose to peak them on a homogeneous and isotropic geometry. This corresponds to take a regular cellular decomposition. In particular, for the dipole, this corresponds to fix the normals and take all the spins being the same, so that we reduce to an unique holomorphic variable: $z_\ell = \xi_\ell + i\eta_\ell \;\; \rightarrow \;\; z = \alpha c + i\beta p$. Geometry is determined by a couple (c, p) in the past and (c', p') in the future.

3. Transition amplitude between an initial and a final universe

Given these states, to which we have associated a cosmological interpretation, we can ask what is the transition amplitude between an initial and a final state. In LQG this transition amplitude is given as a sum over spinfoams $\sum_\sigma$, a product of "face amplitude" [9] and a product of vertex amplitudes $\prod_f d_f(\sigma)$, that reads

$$\langle W \mid \psi \rangle = \sum_\sigma \prod_f d_f(\sigma) \prod_v W_v(\sigma) \; .$$

The form of this amplitude has been defined in [10, 11, 12, 13]. Here I focus on how to use it to calculate the transition amplitude associated to the transition from a 3-sphere to a 3-sphere. We use the dipole as initial and final boundary graph, and we choose the coherent states corresponding to a homogeneous isotropic geometry. We want to calculate this transition at the first order. This corresponds to take a spinfoam with only one vertex. For each vertex

$$W(H_\ell) = \int_{SO(4)^N} dG_n \prod_\ell P_t(H_\ell\,,\, G_{n_{source(\ell)}} G^{-1}_{n_{target(\ell)}}) \qquad \text{where } \; G_n \in SO(4) \;\; \forall \text{ nodes } n$$

and

$$P_t(H, G) = \sum_j (2j+1) e^{-2t\hbar j(j+1)} \mathrm{Tr}\left[D^{(j)}(H) Y^\dagger D^{(j^+,j^-)}(G) Y \right] \; .$$

$D^{(j)}$ and $D^{(j^+,j^-)}$ are respectively the Wigner matrix of the spin-j representation of $SU(2)$ and of the spin-(j^+, j^-) representation of $SO(4)$. These matrices have different dimensions, and the map Y glues them together.

Finally a further approximation is needed to obtain the semiclassical limit: we consider the case of large spins j, equivalent to a large-distant expansion. The calculation is now well-defined and can be performed [14], and the result is this transition amplitude:

$$W(z, z') = C\; zz' e^{-\frac{z^2 + (z')^2}{\hbar}} \; .$$

Now, what does this transition amplitude mean physically? Our transition amplitude can be used directly to study the semiclassical behavior of the theory: in fact, semiclassical trajectories will be peaked on this amplitude (this can be seen numerically [15]).

Effective Hamiltonian We note that the transition amplitude is annihilated by the operator

$$\hat{H}(z, \bar{z})\; W(z', z) = \frac{3}{8\pi G} \left(\frac{\hat{z}^2 - \hat{\bar{z}}^2 - 3t\hbar}{4\alpha\beta\gamma} \right)^2 W(z', z) = \frac{3}{8\pi G} \left(\frac{4i\,\alpha\beta\, cp - 3t\hbar}{4\alpha\beta\gamma} \right)^2 = 0 \; .$$

In the large-distance approximation, where p is larger than the Planck scale, one can disregard the term with $\hbar$, and we can use the fact that $p = a^2$ and c corresponds to $\gamma\dot{a}$, thus $\hat{H} \sim \frac{3}{8\pi G}\, a^4\, \dot{a}^2$. We can divide by the volume in order to obtain the proper physical dimensions (since this is a constrain, it does not change the dynamics). We obtain $\hat{H} \sim \frac{3}{8\pi G}\, a\, \dot{a}^2$ that is the usual Hamiltonian constraint for the FLRW dynamics.

This results shows that it is possible to derive the Friedmann equation from the full covariant loop gravity, and give a further proof of the robustness of the vertex used, since it gives a good semiclassical limit.

Journal of Physics: Conference Series **314** (2011) 012049 doi:10.1088/1742-6596/314/1/012049

4. Summary, comments and further developments

To compute the transition amplitude we have used the expansions in the graph, the vertex and for large-distance. We have clarified their role in selecting a geometry corresponding to some cosmology, and given two cosmological boundary states, we have seen that the corresponding transition amplitude leads to Friedmann dynamics in the semiclassical limit.

I want to stress the fact that the boundary states chosen are homogeneous and isotropic, but the Hilbert space in which they live, which is the Hilbert space of the dipole graph, contains more DOF than the isotropic and homogeneous ones. One of the advantage of this procedure, indeed, is to provide a model that take into account naturally the inhomogeneous and anisotropic DOF, and that can be use to describe the quantum geometry at the bounce, namely in a regime where we are interested to go beyond a single DOF. The main motivation in studying spinfoam cosmology is that it could provide the right setting to describe the quantum geometry at the bounce. In particular I want to understand if quantum fluctuations of the geometry affect structure formation. In order to do this, one has to relax the assumptions of homogeneity and isotropy. In spinfoam cosmology is possible to choose how many degrees of freedom one wants to describe, by computing the dynamics on the appropriate graph. For this reason it would be interesting to investigate further the relation between the graph and the physical interpretation of the degrees of freedom involved. It could turn out to be very useful the to study this also in the $U(N)$ formalism recently introduced [16, 17].

In Spinfoam cosmology the next step to check the validity of the theory beyond the trivial flat solution is to include matter and/or a cosmological constant. One can include matter at an effective level, of course, but the coupling of matter with the quantum gravity is still an open task. On the other hand, there are already some developed ideas that can be followed to add a cosmological constant in spinfoam.

Cosmology provides a privileged framework to study the relation between the canonical and the covariant LQG. There are several issues that can be explored. I would like to mention in particular the presence of ambiguities in the regularization of the Hamiltonian constraint. In LQC has been introduced an "improved dynamics" [20, 21], where the regularization is fixed by requiring a good semiclassical behavior. This choice can be made more robust if the resulting Hamiltonian will match with the one obtained in the covariant framework [14].

References

[1] Ashtekar A 2009 *Gen. Rel. Grav.* **41** 707–741 (*Preprint* 0812.0177)
[2] Bojowald M 2001 *Phys. Rev. Lett.* **86** 5227–5230 (*Preprint* gr-qc/0102069)
[3] Singh P 2009 *Class. Quant. Grav.* **26** 125005 (*Preprint* 0901.2750)
[4] Ashtekar A 2009 *J.Phys.Conf.Ser.* **189** 012003 (*Preprint* 0812.4703)
[5] Rovelli C 2010 (*Preprint* 1010.1939)
[6] Rovelli C and Vidotto F 2008 *Class.Quant.Grav.* **25** 225024 (*Preprint* 0805.4585)
[7] Battisti M V, Marciano A and Rovelli C 2010 *Phys.Rev.D* **81** 064019 (*Preprint* 0911.2653)
[8] Bianchi E, Magliaro E and Perini C 2009 (*Preprint* 0912.4054)
[9] Bianchi E, Regoli D and Rovelli C 2010 (*Preprint* 1005.0764)
[10] Livine E R and Speziale S 2007 *Phys. Rev.* **D76** 084028 (*Preprint* 0705.0674)
[11] Engle J, Livine E, Pereira R and Rovelli C 2008 *Nucl. Phys.* **B799** 136–149 (*Preprint* 0711.0146)
[12] Freidel L and Krasnov K 2008 *Class. Quant. Grav.* **25** 125018 (*Preprint* 0708.1595)
[13] Kaminski W, Kisielowski M and Lewandowski J 2009 (*Preprint* 0909.0939)
[14] Bianchi E, Rovelli C and Vidotto F 2010 *Phys.Rev.D* **82** 084035 (*Preprint* 1003.3483)
[15] *http://www.iaa.es/ere2010/website/modules/tinyd0/content/talks/vidotto.pdf*
[16] Freidel L and Livine E R 2010 (*Preprint* 1005.2090)
[17] Borja E F, Diaz-Polo J, Garay I and Livine E R 2010 (*Preprint* 1006.2451)
[18] Bahr B and Dittrich B 2009 *Phys.Rev.D* **80** 124030 (*Preprint* 0907.4323)
[19] Bahr B and Dittrich B 2010 *New J.Phys.* **12** 033010 (*Preprint* 0907.4325)
[20] Ashtekar A, Pawlowski T and Singh P 2006 *Phys. Rev.* **D74** 084003 (*Preprint* gr-qc/0607039)
[21] Corichi A and Singh P 2008 *Phys.Rev.D* **78** 024034 (*Preprint* 0805.0136)

Journal of Physics: Conference Series **314** (2011) 012050 doi:10.1088/1742-6596/314/1/012050

Some physical consequences of an exact vacua distribution in the Bousso-Polchinski Landscape

César Asensio and Antonio Seguí
Departamento de Física Teórica, Universidad de Zaragoza, Spain
E-mail: casencha@unizar.es

Abstract. The Bousso-Polchinski (BP) Landscape is a proposal for solving the Cosmological Constant Problem. The solution requires counting the states in a very thin shell in flux space. We find an exact formula for this counting problem which has two simple asymptotic regimes, one of them being the method of counting low Λ states given originally by Bousso and Polchinski. We finally give some applications of the extended formula: a robust property of the Landscape which can be identified with an effective occupation number, an estimator for the minimum cosmological constant and a possible influence on the KKLT stabilization mechanism.

1. Introduction

The *cosmological constant problem* [1, 2] is the smallness of the observed value[1] $\Lambda_{\text{obs}} = 1.5 \times 10^{-123}$ [3, 4] when compared with naive expectations from particle physics. An attempt for a solution is proposed in the Bousso-Polchinski Landscape [5], in which a large amount J of quantized fluxes of charges $\{q_j\}_{i=1,\cdots,J}$ leads to an effective cosmological constant

$$\Lambda = \Lambda_0 + \frac{1}{2}\sum_{j=1}^{J} n_j^2 q_j^2 \,. \tag{1}$$

In (1), Λ_0 is a negative number of order -1, and the integer J-tuple $(n_1, \cdots, n_J)$ characterizes each of the vacua of the Landscape, which is a finite subset (yet an enormous one) of an infinite lattice comprising the nodes with cosmological constant smaller than some value $\Lambda_1 = \mathcal{O}(1)$. For large J and incommensurate charges $\{q_j\}$ this model contains states of small Λ. The problem arises now as how to count them.

Each state in the Bousso-Polchinski Landscape can be viewed as a node of a lattice in flux space surrounded by a cell of volume $\operatorname{vol} Q = \prod_{i=1}^{J} q_i$. On the other hand, each value of the cosmological constant $\Lambda_0 \leq \Lambda \leq \Lambda_1$ defines a ball $\mathcal{B}^J(r)$ in flux space of radius $r = \sqrt{2(\Lambda - \Lambda_0)}$. The BP counting argument [5, 6] consists of computing the number of states inside a ball of radius r by

$$\Omega_J(r) = \frac{\operatorname{vol} \mathcal{B}^J(r)}{\operatorname{vol} Q} \,. \tag{2}$$

Nevertheless, as the authors of [5] point out, this argument is not valid when any of the charges q_i exceed $R_0/\sqrt{J}$. Therefore, we will propose an exact counting formula which is reduced to (2) in the appropiate regime.

[1] We use reduced Planck units in which $8\pi G = \hbar = c = 1$.

2. The BP Landscape degeneracy

2.1. The exact representation

We start with the number of nodes in the lattice inside a sphere in flux space of radius r. This magnitude is called $\Omega_J(r)$ above. It can be given in terms of the characteristic function of an interval I, $\chi_I(t) = 1$ if $t \in I$ and 0 otherwise:

$$\Omega_J(r) = \sum_{\lambda\in\mathcal{L}} \chi_{[0,r]}(\|\lambda\|)\,. \tag{3}$$

Expression (3) is exact and finite, and it is equivalent to directly counting the nodes (the "brute-force" counting method), hence it cannot be used in order to obtain numbers as in (2).

The density of states associated to (3) is $\omega_J(r) = \frac{\partial\Omega_J(r)}{\partial r} = 2r\sum_{\lambda\in\mathcal{L}}\delta(r^2 - \|\lambda\|^2)$ which will be called the "BP Landscape degeneracy". The counting function $\Omega_J(r)$ is a stepwise monotonically non-decreasing function, and thus its derivative $\omega_J(r)$ is a sum of Dirac deltas. We can express these Dirac deltas as integrals in complex plane along a vertical line γ crossing the positive real axis in complex plane. We obtain

$$\omega_J(r) = \frac{2r}{2\pi i}\int_\gamma e^{sr^2}\left[\prod_{j=1}^{J}\vartheta(sq_j^2)\right]\mathrm{d}s\,. \tag{4}$$

The sum is hidden in the function $\vartheta(s) = \sum_{n\in\mathbb{Z}} e^{-sn^2}$, valid for $\operatorname{Re} s > 0$.

The integration of (4) with the initial condition $\Omega_J(0) = 1$ gives $\Omega_J(r)$.

2.2. The large distance (or BP) regime

Now we will turn to the approximate evaluation of $\omega_J(r)$. For this purpose we need the asymptotic behavior of ϑ function. There is a middle regime where the asymptotic regimes are not accurate enough, and we have computed numerically all the quantities.

The first case is $s \to 0$. In this regime, we make the integration contour pass near the origin, where ϑ has a singularity. Assuming that the main contribution to the integral will come from this region, we can replace ϑ by its asymptotic value when $s \to 0$ and we obtain

$$\omega_J(r) \approx \frac{2r}{2\pi i}\int_\gamma e^{sr^2}\left[\prod_{j=1}^{J}\sqrt{\frac{\pi}{q_j^2 s}}\right]\mathrm{d}s = \frac{\pi^{\frac{J}{2}}}{\operatorname{vol}Q}\frac{2r}{2\pi i}\int_\gamma e^{sr^2}\frac{\mathrm{d}s}{s^{\frac{J}{2}}} = \frac{2\pi^{\frac{J}{2}}}{\Gamma(\frac{J}{2})}\frac{r^{J-1}}{\operatorname{vol}Q}\,. \tag{5}$$

Equation (5) is the derivative of (2), that is, BP count. It is valid for large r distances, $h = \frac{Jq^2}{r^2} < \frac{2}{e} \approx 0.736$.

2.3. The small distance regime

In this case we are in the regime in which the asymptotic expansion of ϑ for large values of its argument is valid. We can estimate the integral using the saddle point approximation. Unfortunately, we cannot solve the saddle point equation in closed form for arbitrary charges. Nevertheless, in the simplest case in which all charges are equal $q_1 = \cdots = q_J = q$, we obtain

$$\omega_J(r) = \frac{(2h-2)^{\frac{J}{h}}}{q\sqrt{2\pi h}}\left(\frac{h}{h-1}\right)^{J+\frac{1}{2}}\,. \tag{6}$$

The saddle point and the asymptote are in the same region if $h = \frac{Jq^2}{r^2} > 1 + \frac{e^2}{2} \approx 4.694$.

3. Applications

3.1. Number of states in the Weinberg Window

The number of states of positive cosmological constant bounded by a small value Λ_ε is the number of nodes of the lattice in flux space whose distance to the origin lies in the interval $[R, R_\varepsilon]$, where $R = \sqrt{2|\Lambda_0|}$ and $R_\varepsilon = \sqrt{2(\Lambda_\varepsilon - \Lambda_0)} \approx R + \frac{\Lambda_\varepsilon}{R}$ so that the width of the shell is $\varepsilon = \frac{\Lambda_\varepsilon}{R}$. If Λ_ε is the width of the anthropic range Λ_{WW} (the so-called Weinberg Window), then the number of states in it is

$$\mathcal{N}_{\mathrm{WW}} = \frac{\omega_J(R)}{R}\Lambda_{\mathrm{WW}}\,. \tag{7}$$

Computation of $\omega_J(R)$ should be done along the lines of the previous section.

3.2. Typical number of non-vanishing fluxes

The set of nodes inside a thin shell of width $\varepsilon = R_\varepsilon - R$ above radius R will be called Σ_ε. We will assume that ε is smaller than the charges q_i so that (7) is valid but $\mathcal{N}_\varepsilon \gg 1$.

We find that the typical number of non-vanishing components of a state drawn randomly from Σ_ε is J for the cases $J = 2, 3$. We wonder whether it happens for all J. We will answer this question by computing the fraction of states in the shell having a fixed fraction α of non-vanishing components.

When a state λ is selected at random from Σ_ε with uniform probability, α becomes a discrete random variable taking values in the $[0, 1]$ interval. Assuming equal charges, its probability distribution is given by (see [7] for details)

$$P(\alpha) = \frac{2R}{\omega_J(R)}\binom{J}{\alpha J}\frac{1}{2\pi i}\int_\gamma e^{\phi(s,\alpha)}\,\mathrm{d}s \quad \text{with} \quad \phi(s,\alpha) = sR^2 + \alpha J \log\left[\vartheta(q^2 s) - 1\right]. \tag{8}$$

It can be seen that $P(\alpha)$ is locally Gaussian around its peak $\alpha^*(h)$, with standard deviation $\sim 1/\sqrt{J}$. $\alpha^*(h)$ is the typical number of non-vanishing fluxes in the shell Σ_ϵ (and essentially also in the whole Landscape). Its computation must be done numerically, either using the saddle point method on (8), or by statistical sampling, see [7]. The results are plotted in figure 1.

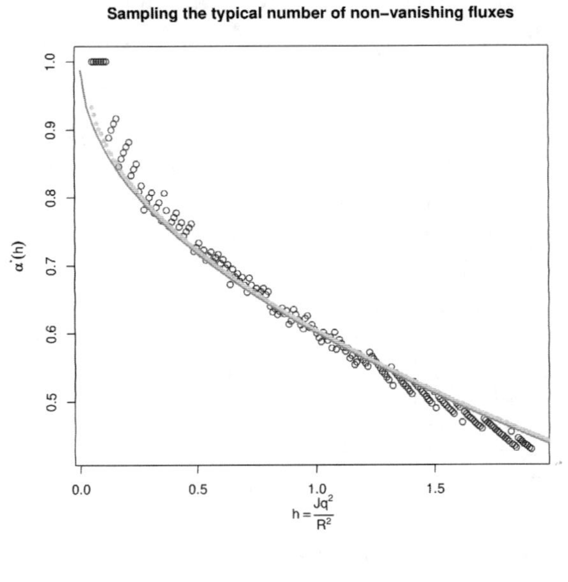

Figure 1. Samples of the typical number of non-vanishing fluxes. Two sampling methods have been used. The saddle point solution is also shown (red line).

3.3. Estimating the minimum positive cosmological constant

We can roughly estimate the explicit dependence of the minimum positive cosmological constant with respect to the parameters of the Landscape. We will call Λ^* the actual minimum value, and Λ_ε the corresponding estimator.

In the case of incommensurate charges, the symmetry degeneracy of a state is $2^{J\alpha^*}$, so that we have $\Lambda_\varepsilon \approx \frac{2^{J\alpha^*}R}{\omega_J(R)}$. We can check this estimate with brute-force data for low J and we find a good agreement in the statistical sense.

3.4. A possible influence on the KKLT mechanism

The Giddings-Kachru-Polchinski model [8] is a more realistic approach to the true string theory Landscape, and it can be endowed with a mechanism for fixing the compactification moduli, the so-called KKLT mechanism [9]. In this model, moduli are stabilized by the presence of fluxes and corrections to the superpotential coming from localized branes.

As far as we know, there is no combination between the BP Landscape and the KKLT mechanism, in the sense that there is no known realistic model in which all moduli are fixed and a large amount of three-cycles lead to an anthropic value of the cosmological constant. Let us assume that such a model will be built in the near future. If the $\alpha^*(h)$ curve discussed in the section 3.2 can be generalized, that is, the typical occupation number of the fluxes is different from 1, then there will be a finite fraction $1-\alpha^*$ of three-cycles with vanishing flux. This fraction of vanishing fluxes can spoil the stabilization mechanism.

4. Conclusions

We have developed an exact formula for counting states in the Bousso-Polchinski Landscape which reduces to the volume-counting one in certain (BP) regime. Numeric computations and brute-force searches have been carried out to check the results of our analytic approximations, and we have found remarkable agreement in all explored regimes.

In particular, we have discovered a robust property of the BP Landscape, the typical fraction of non-vanishing fluxes $\alpha^*(h)$, which reveals the structure of the lattice inside a sphere for large J as the union of hyperplane portions of effective dimension near $J\alpha^*$. This result is important in computing degeneracies, which are used in estimating the minimum cosmological constant, and it could be an obstacle for a realistic implementation of the KKTL moduli stabilization mechanism.

Acknowledgments

We would like to thank Pablo Diaz, Concha Orna and Laura Segui for carefully reading this manuscript. We also thank Jaume Garriga for useful discussions and encouragement. This work has been supported by CICYT (grant FPA-2006-02315 and grant FPA-2009-09638) and DGIID-DGA (grant 2007-E24/2). We thank also the support by grant A9335/07 and A9335/10 (Física de alta energía: Partículas, cuerdas y cosmología).

References

[1] Weinberg S 1989 *Rev. Mod. Phys.* **61**, 1-23
[2] Bousso R 2008 *Gen. Rel. Grav.* **40** 607 (*Preprint* 0708.4231 [hep-th])
[3] Perlmutter S *et al.* 1999 *Astrophys. J.* **517** 565-586 (*Preprint* astro-ph/9812133)
[4] Riess A G *et al.* 1998 *Astron. J.* **116** 1009-1038 (*Preprint* astro-ph/9805201)
[5] Bousso R and Polchinski J 2000 *JHEP* **06** 006 (*Preprint* hep-th/0004134)
[6] Clifton T, Shenker S and Sivanandam N 2007 *JHEP* **0709** 034 (*Preprint* 0706.3201 [hep-th])
[7] Asensio C and Segui A 2010 *Preprint* 1003.6011 [hep-th]
[8] Giddings S B, Kachru S and Polchinski J 2002 *Phys. Rev.* D **66** 106006 (*Preprint* hep-th/0105097)
[9] Kachru S, Kallosh R, Linde A and Trivedi S 2003 *Phys. Rev.* D **68** 046005 (*Preprint* hep-th/0301240)

Journal of Physics: Conference Series **314** (2011) 012051 doi:10.1088/1742-6596/314/1/012051

The effects of quantum field renormalization on the predictions of inflation for the CMB anisotropies

I. Agullo

Institute for Gravitation and the Cosmos, Physics Department, Penn State, University Park, PA 16802-6300, USA.

E-mail: agullo@gravity.psu.edu

Abstract. In single-field, slow-roll inflationary models scalar and tensorial (Gaussian) perturbations are usually characterized by the so called power spectrum in momentum space. Even though these power spectra are finite and well define in momentum space, typical ultraviolet divergences in quantum field theory appear when these quantities are expressed in position space. The requirement of a finite variance in position space forces the introduction of regularization technics in quantum field theory in an expanding universe. The regularization process has an important impact on the predicted scalar and tensorial power spectra for wavelengths that today are at observable scales.

1. Introduction

Inflation nowadays constitutes an important piece of the standard cosmological model [1]. It was originally introduced as a mechanism able to alleviate the classical problems of the big bang model [2]. However, it was soon realized [3] that inflation is able to provide an elegant quantum mechanical mechanism to account for the origin of small inhomogeneities in the early universe, essential for explaining the structures that we see today. In the simplest models of inflation, a scalar field slowly rolling down its potential causes an accelerated expansion of the background geometry. The rapidly changing spacetime geometry is able to amplify the vacuum fluctuations [4] of the inflaton field itself and of purely tensorial modes (gravitational waves), generating a primordial spectrum of scalar and tensor perturbations. These perturbations acquire classical properties during their evolution and provide the initial conditions for classical cosmological perturbations. The potential detection of the effects of primordial tensorial metric fluctuations in future high-precision measurements of the CMB anisotropies would constitute a highly non-trivial test of the inflationary paradigm. Therefore, it is particularly important to scrutinize from all point of views the predictions of inflation for the tensorial and scalar power spectra. In this respect, it was pointed out in [5] (see also [6]) that quantum field renormalization significantly modifies the amplitude of quantum fluctuations, and hence the corresponding power spectra, in de Sitter inflation. The analysis was further improved in [7, 8] to understand how the basic testable predictions of (single-field) slow-roll inflation could be affected by renormalization. In this talk we summarize our approach.

Let us assume that $\varphi(\vec{x}, t)$ represents a perturbation obeying a free field wave-equation on the inflationary background $ds^2 = -dt^2 + a^2(t)d\vec{x}^2$, where $a(t)$ is a quasi-exponential expansion factor. At the quantum level, this field is expanded as $\varphi(\vec{x},t) = (2\pi)^{-3/2}\int d^3k[\varphi_k(t)a_{\vec{k}} + \varphi_k^*(t)a_{\vec{k}}^\dagger]e^{i\vec{k}\vec{x}}$, where the creation and annihilation operators satisfy the canonical commutation relation. The mode functions $\varphi_k(t)$ are required to satisfy the adiabatic condition [9]. The power spectrum for this perturbation, $\Delta_\varphi^2(k,t)$, is usually defined in terms of the Fourier transform of the variance of the field [1]

$$\langle \hat{\varphi}_{\vec{k}}(t)\hat{\varphi}_{\vec{k}'}^\dagger(t)\rangle = \delta^3(\vec{k}+\vec{k}')\frac{2\pi^2}{k^3}\Delta_\varphi^2(k,t) \ , \tag{1}$$

where $\hat{\varphi}_{\vec{k}}(t) \equiv \varphi_k(t)a_{\vec{k}}$. The field modes $\varphi_k(t)$ describe a perturbation field characterized, in momentum space, by a zero mean $\langle\hat{\varphi}_{\vec{k}}(t)\rangle = 0$ and the variance $\Delta_\varphi^2(k,t)$. The advantage of working in momentum space resides in the fact that different modes fluctuate independently of each other, as explicitly displayed by the presence of the delta function in (1). This way, the quantum field is regarded as an infinite collection of independent oscillators, each with a different value of $\vec{k}$. In position space the perturbation is also characterized by a zero mean $\langle\varphi(\vec{x},t)\rangle = 0$ and a variance $\langle\varphi^2(\vec{x},t)\rangle = (2\pi)^{-3}\int d^3kd^3k'\langle\hat{\varphi}_{\vec{k}}(t)\hat{\varphi}_{\vec{k}'}^\dagger(t)\rangle e^{i(\vec{k}+\vec{k}')\vec{x}}$. This variance is formally related to the power spectrum by

$$\langle\varphi^2(\vec{x},t)\rangle = \int_0^\infty \frac{dk}{k}\Delta_\varphi^2(k,t) \sim \frac{1}{4\pi^2}\int_0^\infty dk(\frac{k}{a^2} + \frac{\dot{a}^2}{a^2k} + ...) \ , \tag{2}$$

where the large k behavior of the integrand is shown. As it is well-known in quantum field theory, the above expectation value quadratic in the field φ is ultraviolet divergent (quadratic and logarithmically). The quadratic divergence corresponds to the usual contribution from vacuum fluctuations in Minkowski space and can be eliminated by standard renormalization in flat spacetime. The logarithmic divergence, however, appears as a consequence of the non trivial spacetime expansion. Because the different k-modes fluctuate independently of each other, one could be tempted to get rid of this logarithmic ultraviolet divergence by simply eliminating the modes with $k > aH$ and leaving the rest unaffected. If one eliminates this divergence using a window function in this way, as is usual for random fields, then one obtains $\Delta_\varphi^2(k) \approx H^2/4\pi^2$, where $\Delta_\varphi^2(k)$ is defined by the quantity $\Delta_\varphi^2(k,t)$ evaluated a few Hubble times after the "Hubble exit time" t_k (defined by $a(t_k)/k = H(t_k)$), since this is the time scale at which the modes become time independent and behave as classical perturbations. However, one should take into account that the field fluctuations are quantum in nature and, therefore, one should consider the subtle points of quantum field theory (QFT) regarding the ultraviolet divergences.

Even though free quantum field theory is usually regarded as an infinite set of independent harmonic oscillators (one for each k-mode), there are fundamental holistic aspects of QFT that can not be properly understood in terms of independent modes. Renormalization is the hallmark of the holistic aspect of QFT [10]. This is clear in the fact that, although the renormalization schemes in QFT in curved spacetimes are based on the ultraviolet behavior of the theory, the infrared sector is also affected by renormalization, leading potentially to observable consequences. Therefore, the logarithmic divergence in (2) should be dealt with by renormalization in curved spacetimes. We propose that in the standard definitions of the spectrum $\Delta_\varphi^2(k,t)$, as given in (2), one should replace the unrenormalized $\langle\varphi^2(\vec{x},t)\rangle$ by the renormalized variance, $\langle\varphi^2(\vec{x},t)\rangle_{ren}$. Writing $\tilde{\Delta}_\varphi^2(k,t)$ for the spectrum defined in this way, the definition in (2) is replaced by the corresponding renormalized expression

$$\langle\varphi^2(\vec{x},t)\rangle_{ren} = \int_0^\infty \frac{dk}{k}\tilde{\Delta}_\varphi^2(k,t) \ . \tag{3}$$

Since the power spectrum is defined in momentum space, the natural scheme is renormalization in momentum space, so we define

$$\langle \varphi^2(\vec{x},t)\rangle_{ren} = \frac{4\pi}{(2\pi)^3}\int_0^\infty k^2 dk(|\varphi_k(t)|^2 - C_k(t)) \ , \tag{4}$$

where $C_k(t)$ represents the renormalization subtraction terms. Adiabatic renormalization [11, 9] provides a natural expression for $C_k(t)$. Moreover, the Bunch-Parker renormalization in momentum space [12, 9] (which turns out to be equivalent, when translated to position space, to the DeWitt-Schwinger proper time renormalization) provides another answer for $C_k(t)$. When these schemes are applied to the field perturbations arising from slow-roll inflation, which should be considered as massless free fields with a second-order adiabatic term encoding the dependence on the inflationary potential, the resulting expressions for $C_k(t)$ coincide, thus defining a unique expression for the spectrum $\tilde{\Delta}^2_\varphi(k,t)$. The holistic nature of QFT is then explicitly realized through (3-4); although the counterterms are fully determined by the ultraviolet behavior of the modes, the long wavelength sector, and hence the new $\tilde{\Delta}^2_\varphi(k,t)$, is significantly affected by the subtractions. In the slow-roll scenario, when H slowly decreases with time, the effects of renormalization have a non-trivial impact on $\tilde{\Delta}^2_\varphi(k,t)$ when this quantity is evaluated a few n Hubble times after the Hubble exit time t_k ($n > 1$ but $n\epsilon \ll 1$). We obtain for the tensorial and scalar spectra

$$\begin{aligned}
\tilde{\Delta}^2_t(k,n) &\approx \frac{8}{M_P^2}\left(\frac{H(t_k)}{2\pi}\right)^2 \epsilon(t_k)(2n-3/2) \\
\tilde{\Delta}^2_s(k,n) &\approx \frac{1}{2M_P^2\epsilon(t_k)}\left(\frac{H(t_k)}{2\pi}\right)^2 (3\epsilon(t_k)-\eta(t_k))(2n-3/2) \ ,
\end{aligned} \tag{5}$$

where $\epsilon = M_P^2/2(V'/V)^2$ and $\eta = M_P(V''/V)$ are the standard slow-roll parameters associated with the inflaton potential energy V. It is important to note that the subtraction terms introduce a time dependence, characterized by n, in the power spectra parameterizing the (unknown) time at which the modes exhibit classical behavior. However, the spectral indices and the tensor-to-scalar ratio r are not sensitive to the unknown parameter n, in the same way as they do not depend on the scale of inflation $H(t_k)$

$$\begin{aligned}
\tilde{n}_s - 1 \equiv \frac{d\ln\tilde{\Delta}^2_s}{d\ln k} = -6\epsilon + 2\eta + \frac{(12\epsilon^2 - 8\epsilon\eta + \xi)}{3\epsilon-\eta} \quad &; \quad \tilde{n}_t \equiv \frac{d\ln\tilde{\Delta}^2_t}{d\ln k} = 2(\epsilon-\eta) \\
\tilde{r} \equiv \frac{\tilde{\Delta}^2_t}{\tilde{\Delta}^2_s} = 16\epsilon\frac{\epsilon}{3\epsilon-\eta} \ ,
\end{aligned} \tag{6}$$

where ξ is another slow roll parameter: $\xi \equiv M_P^4(V'V'''/V^2)$. This parameter can be reexpressed in terms of ϵ, η and the running of the tensorial index $\tilde{n}'_t \equiv d\tilde{n}_t/d\ln k$ as $\tilde{n}'_t = 8\epsilon(\epsilon-\eta) + 2\xi$.

The expression (5) and (6) for the power spectra and spectral indices show that renormalization has a significant impact on the predictions of slow-roll inflation. It is important to notice that, in contrast to the standard approach, the renormalized tensorial power spectrum is not uniquely related to the Hubble parameter and now depends on the combination $H(t_k)/\epsilon(t_k)$. Another important difference concerns the consistency relation that relates the tensor-to-scalar ratio r to the spectra indices. This relation, that reads $r \equiv \Delta^2_t/\Delta^2_s = -8n_t$ if renormalization is not considered, becomes now

$$\tilde{r} = 4(1-\tilde{n}_s-\tilde{n}_t) + \frac{4\tilde{n}'_t}{\tilde{n}_t^2 - 2\tilde{n}'_t}\left(1-\tilde{n}_s - \sqrt{2\tilde{n}'_t + (1-\tilde{n}_s)^2 - \tilde{n}_t^2}\right) \ . \tag{7}$$

This expression is much more involved that the standard one allowing a wider range of possibilities. For instance, an exact scale-invariant tensorial power spectrum, $\tilde{n}_t = 0$, is now compatible with a non-zero ratio $r \approx 0.12 \pm 0.06$, which is forbidden by the standard prediction. Also, the values of $\tilde{n}_t$ are not constrained to be negative in contrast with the unrenormalized result. Additionally, the running of $\tilde{n}_t$ can not be determine form the values of n_s and n_t, and it is now an independent quantity that needs to be measured independently. This aspect could make more challenging the experimental verification of the consistency condition of (single-field) slow roll inflation.

An important question is related to the time at which the subtraction terms employed in the renormalization procedures should be evaluated. A definitive answer requires a deep analysis of the process in which the quantum perturbation acquire classical properties and its relation with quantum field renormalization. However, the fact that we find observable differences offers a deep way to experimentally probe this question.

Acknowledgements. This work has been supported by the PHY-0854743 NSF grant and Eberly research funds at Penn State University, and has been also partially supported by the spanish grant FIS2008-06078-C03-02. The author thanks J. Navarro-Salas, G. J. Olmo and L. Parker for collaboration of the content of this talk.

References

[1] Lyth D.H. and Liddle A.R. *The primordial density perturbation*, Cambridge University Press, (2009); Weinberg S., *Cosmology*, Oxford University Press, (2008); Dodelson S., *Modern Cosmology*, Academic Press, (2003).
[2] Guth A., *Phys. Rev.* D**23**, 347 (1981). Starobinsky A.A., *Phys. Lett.* B**91**, 99 (1980). Linde, A.D., *Phys. Lett.* B**108**, 389 (1982); *Phys. Lett.* B**129**, 177 (1983). Albrecht A. and Steinhardt P. J., *Phys. Rev. Lett.* **48**,1220 (1982). Sato, K., *Mon. Not. Roy. Astron. Soc.* **195**, 467 (1981).
[3] Mukhanov V.F. and Chibisov G.V., *JETP Letters* **33**, 532(1981). Hawking, S. W., *Phys. Lett.* B**115**, 295 (1982). Guth A. and Pi, S.-Y., *Phys. Rev. Lett.* **49**, 1110 (1982). Starobinsky, A. A., *Phys. Lett.* B**117**, 175 (1982). Bardeen, J.M., Steinhardt, P.J. and Turner, M.S., *Phys. Rev.*D**28**, 679 (1983).
[4] Parker L., *Phys. Rev. Lett.* **21** 562 (1968); *Phys. Rev.* **183**, 1057(1969).
[5] Parker L., *Amplitude of perturbations from inflation*, hep-th/0702216.
[6] Agullo I., Navarro-Salas J., Olmo G.J. and Parker L., *Phys. Rev. Lett.* **101**, 171301 (2008).
[7] Agullo I., Navarro-Salas J., Olmo G.J. and Parker L., *Phys. Rev. Lett.* **103**, 061301 (2009); *Gen. Rel. Grav.* **41**, 2301 (2009).
[8] Agullo I., Navarro-Salas J., Olmo G.J. and Parker L., *Phys. Rev.* D**81**, 043514 (2010).
[9] Parker L. and Toms D.J., *Quantum field theory in curved spacetime: quantized fields and gravity*, Cambridge University Press, (2009).
[10] Hollands S. and Wald R., arXiv:gr-qc/0405082v1.
[11] Parker L. and Fulling S.A., *Phys. Rev. D* **9** 341 (1974).
[12] Bunch T.S. and Parker L. *Phys. Rev. D* **20** 2499 (1979).

Journal of Physics: Conference Series **314** (2011) 012052 doi:10.1088/1742-6596/314/1/012052

Stochastic Inflation in Compact Extra Dimensions

Larissa Lorenz

Institute of Mathematics and Physics, Centre for Cosmology, Particle Physics and Phenomenology, Louvain University, 2 Chemin du Cyclotron, 1348 Louvain-la-Neuve, Belgium

E-mail: larissa.lorenz@uclouvain.be

Abstract. While moving down the potential on its classical slow roll trajectory, the inflaton field is subject to quantum jumps, which take it up or down the potential at random. In "stochastic inflation", the impact of these quantum jumps is modeled by smoothing out the field over (at least) Hubble-patch sized domains and treating fluctuations on smaller scales as noise. The inflaton thus becomes a stochastic process whose values at a given time are calculated using its probability distribution. We generalize this approach for non-canonic kinetic terms of Dirac Born Infeld (DBI) type and investigate the resulting modifications of the field's trajectory. Since models of DBI inflation arise from string-inspired scenarios in which the scalar field has a geometric interpretation, we insist that field value restrictions imposed by the model's string origin must be respected at the quantum level.

1. Introduction

Recent advances in string theory have lead to the construction of explicit models of inflation in which the scalar field ϕ is, for example, associated with the position of a $(p+1)$-dimensional hypersurface (a so-called Dp-brane) embedded into the ten-dimensional string background spacetime [1]. For $p = 3$, the brane's world volume corresponds to our four-dimensional Universe, while the six extra dimensions are compactified on a suitable internal manifold [2]. It is hoped that string theory can bring us closer to establishing inflation's theoretical foundations, while the cosmological parameters derived from "brane inflation" may be the only potentially observable consequences of string theory for decades to come. Due to the D3's Dirac Born Infeld (DBI) kinetic term, these models provide a "stringy" realization of k-inflation [3], giving rise to a rich and non-standard phenomenology.

While inflationary predictions for the properties of the cosmic microwave background evidently apply inside our observable patch of the Universe, the stochastic approach to inflation [4] seeks to understand the behavior of the quantum inflaton field on *much* larger scales. In this work [5], we extend the stochastic treatment applied to standard inflation in Refs. [6, 7] to DBI scenarios. There, the impact of the inflaton's stochastic character on scales beyond the Hubble patch was (in the Gaussian approximation) entirely expressed in terms of integrals of the Hubble parameter $H(\phi)$ and its derivatives. Here, we find that the influence of the modified kinetic term is readily absorbed into the same formalism if the so-called Lorentz factor $\gamma(\phi)$ [8], which measures the dynamics' deviation from the usual $\dot{\phi}^2/2$ (with $\gamma \approx 1$ in the standard and $\gamma \to \infty$ in the "ultrarelativistic" case), is treated on an equal footing with $H(\phi)$.

2. Stochastic inflation

2.1. The standard case

The notion of the inflaton as a stochastic process $\varphi[\xi]$ arises as follows: the field $\phi(\boldsymbol{x},t)$ is coarse-grained on a given (large) scale L, meaning that a spatially smooth field $\varphi(t)$ is constructed from ϕ's long-wavelength Fourier modes only, while fluctuations $\delta\phi(\boldsymbol{x},t)$ on smaller scales are interpreted as stochastic "noise". Explicitly, this decomposition reads

$$\phi(\boldsymbol{x},t) = \varphi(t) + \int \frac{\mathrm{d}^3\boldsymbol{k}}{(2\pi)^{3/2}}\,\Theta\left(k - L^{-1}\right)\left[a_{\boldsymbol{k}}\,\delta\phi_{\boldsymbol{k}}(t)\,\mathrm{e}^{-i\boldsymbol{k}\cdot\boldsymbol{x}} + a^\dagger_{\boldsymbol{k}}\,\delta\phi^*_{\boldsymbol{k}}(t)\,\mathrm{e}^{i\boldsymbol{k}\cdot\boldsymbol{x}}\right], \tag{1}$$

where $L^{-1} = \varepsilon a H$ with ε a small parameter. The smoothed $\varphi(t)$ then obeys a Langevin-type equation, which is obtained from the slow-roll[1] Friedmann and Klein Gordon equations plus a noise term $\xi(t)$ due to the small-scale fluctuations (a prime denotes φ-derivatives),

$$\dot\varphi = -\frac{2}{\kappa}\,H' + \frac{H^{3/2}}{2\pi}\,\xi(t)\,, \qquad \text{where} \quad \langle\xi(t)\rangle = 0\,, \quad \langle\xi(t)\,\xi(t')\rangle = \delta(t-t')\,. \tag{2}$$

In Ref. [6], this equation was solved using a perturbative expansion in the noise, *i.e.* by writing $\varphi(t) = \varphi_{\rm cl}(t) + \delta\varphi_1(t) + \delta\varphi_2(t) + \ldots$, with $\delta\varphi_1(t) \propto \mathcal{O}(\xi)$, $\delta\varphi_2(t) \propto \mathcal{O}(\xi^2)$ *etc.* By sorting the resulting terms on both sides of Eq. (2) by orders of ξ, evolution equations for $\delta\varphi_1$ and $\delta\varphi_2$ are identified, which can be solved by a time integration of H, its φ-derivatives and the noise. While the explicit time-dependence of $\xi(t)$ is unknown, its stochastic properties (2) can be used to obtain the mean values $\langle\varphi_1^2\rangle$ and $\langle\varphi_2\rangle$ in terms of φ-space integrals of H, H' *etc.* only. These mean values enter into the probability density function (PDF) $P_{\rm c}(\varphi) = \langle\delta(\varphi - \varphi[\xi])\rangle$, which describes the probability for the stochastic process $\varphi[\xi]$ to assume a given value φ at time t inside a single coarse-grained domain. In the Gaussian approximation, one has

$$P_{\rm c}(\varphi) = \frac{1}{\sqrt{2\pi\left\langle\delta\varphi_1^2\right\rangle}}\,\exp\left[-\frac{(\varphi - \varphi_{\rm cl} - \langle\delta\varphi_2\rangle)^2}{2\left\langle\delta\varphi_1^2\right\rangle}\right], \tag{3}$$

where $\varphi_{\rm cl}$ is the inflaton's value on the *classical slow-roll* trajectory. Hence, since $\langle\varphi_1^2\rangle$ and $\langle\varphi_2\rangle$ are functions of $\varphi_{\rm cl}$ (and the initial field value $\varphi_{\rm in}$ at the onset of inflation), one can use them to calculate the *quantum-corrected* mean $\langle\varphi\rangle = \varphi_{\rm cl} + \langle\delta\varphi_2\rangle$ along with the standard deviation. Ref. [6] carried out this calculation for several model classes, illustrating how quantum jumps significantly modify the classical trajectory in some cases.

Let us add two further remarks: first, the PDF of Eq. (3) considers $\varphi[\xi]$ inside one L^3 volume. However, since $L \propto aH$ itself depends on the inflaton's classical value inside it, different domains should be "weighted" with respect to their size. This is achieved by calculating a volume-corrected distribution $P_{\rm v}(\varphi)$, which gives a $\langle\varphi\rangle_{\rm v} \neq \langle\varphi\rangle$ [6]. Second, the perturbative expansion evidently only holds if the noise-induced corrections become smaller and smaller at higher orders in ξ. In Ref. [7], this condition was quantified and used as an intrinsic criterion for the regime of validity of the above ansatz.

2.2. The DBI case

We showed [5] that the results of Refs. [6, 7] are straightforwardly generalized to the case of string-inspired DBI inflation, provided the coarse-graining scale in Eq. (1) is adapted and the DBI Langevin equation is used. It can be shown that the DBI analogue of Eq. (2) is obtained by replacing $H' \to H'/\gamma$ in the classical drift term[2], where the Lorentz factor is given by

[1] These read $\dot\varphi \approx -V'/(3H)$ and $H^2 \approx (\kappa/3)\,V$, where $\kappa \equiv 8\pi/m_{\rm Pl}^2$.
[2] This can be shown using the "DBI slow-roll" equations $\dot\varphi \approx -V'/(3H\gamma)$ and $H^2 \approx (\kappa/3)V$.

$\gamma(\varphi) = \left[1 + 4H'^2/(\kappa^2 T)\right]^{1/2}$. (For the DBI domain size, one has $L_{\rm DBI}^{-1} = \varepsilon a \gamma H$.) Here, the so-called warping function $T(\varphi)$ –like the potential $V(\varphi)$– is *a priori* a free function but in a concrete model of brane inflation, both V and T are derived from the 10d string background [1, 2]. Using the same perturbative expansion in ξ, one can again solve for $\langle \delta\varphi_1^2 \rangle$ and $\langle \delta\varphi_2 \rangle$ in terms of integrals along φ's (DBI-)classical trajectory, which read

$$\langle \delta\varphi_1^2 \rangle = \frac{\kappa}{2}\left(\frac{H'}{2\pi\gamma}\right)^2 \int_{\varphi_{\rm cl}}^{\varphi_{\rm in}} {\rm d}\psi \left[\frac{H(\psi)\gamma(\psi)}{H'(\psi)}\right]^3 , \tag{4}$$

$$\langle \delta\varphi_2 \rangle = \frac{(H'/\gamma)'}{2\,(H'/\gamma)}\langle \delta\varphi_1^2 \rangle + \frac{H'/\gamma}{4\pi m_{\rm Pl}^2}\left[\left(\frac{\gamma^2 H^3}{H'^2}\right)_{\varphi_{\rm in}} - \left(\frac{\gamma^2 H^3}{H'^2}\right)_{\varphi_{\rm cl}}\right] , \tag{5}$$

respectively. [The results of Ref. [6] are reproduced for $\gamma \equiv 1$ in Eqs. (4,5).] Evidently, the only input needed to carry out the integral (4) are the potential and the warping, which (in the slow-roll limit) completely determine H and γ. In Ref. [5], we considered several examples, but here we focus on the case of "UV brane inflation" (see below) with $V(\varphi) = V_0 + (m^2/2)\varphi^2$ and $T(\varphi) = \varphi^4/\lambda$. In the limit where the constant V_0 dominates, Eq. (4) can be integrated (taking on board that $\varphi < m_{\rm Pl}$ always, a consequence of the string-geometric background [9]), giving

$$\langle \delta\varphi_1^2 \rangle_{\rm UV} \simeq \frac{16}{15\sqrt{2}}\left(\frac{m}{m_{\rm Pl}}\right)^2 \frac{\beta^{3/2}}{\alpha^{1/2}}\frac{m_{\rm Pl}^3}{\varphi_{\rm cl}} , \quad \langle \delta\varphi_2 \rangle_{\rm UV} \simeq -\frac{4}{15\sqrt{2}}\left(\frac{m}{m_{\rm Pl}}\right)^2 \frac{\beta^{3/2}}{\alpha^{1/2}}\frac{m_{\rm Pl}^3}{\varphi_{\rm cl}^2} , \tag{6}$$

where $\alpha \equiv 12\pi m_{\rm Pl}^2/(\lambda m^2)$ and $\beta \equiv V_0/(m^2 m_{\rm Pl}^2)$ are parameter combinations. The definition of the inflaton's Gaussian PDF (3) remains unchanged in the DBI case. Hence, for a given α, β, m and $\varphi_{\rm in}$, Eqs. (6) give the mean $\langle\varphi\rangle = \varphi_{\rm cl} + \langle\delta\varphi_2\rangle$ for any $\varphi_{\rm cl}$ on the DBI slow-roll trajectory.

3. Geometric consistency in compact extra dimensions

In the 10d string theory picture, the scenario of UV brane inflation corresponds to a D3-brane inside a 6d "warped throat" (*i.e.* cone-shaped) geometry, where a radial direction r extends above a 5d angular base. Up to renormalization, the D3's position along r is the inflaton, which therefore has a limited range $\phi \in [\phi_0, \phi_{\rm UV}]$, with ϕ_0 the "tip" of the cone and $\phi_{\rm UV}$ at its base (where it is joined onto a more complicated compact 6d geometry). Classically, the inflaton value $\varphi_{\rm cl}$ in this scenario decreases during inflation (*i.e.* the D3 starts out close to the UV end of the throat, moving towards the tip). What, however, follows for the quantum-corrected $\langle\varphi\rangle$? According to Eq. (6), $\langle\delta\varphi_2\rangle$ is negative and since $\langle\varphi\rangle = \varphi_{\rm cl} + \langle\delta\varphi_2\rangle$, this bears the imminent risk of $\langle\varphi\rangle$ decreasing below ϕ_0. Translated back into the D3's r-coordinate, the quantum jumps hence seem to take the brane "outside of spacetime". While a similar effect was observed in Ref. [6], where the inflaton developed a non-zero probability for negative field values under the influence of the noise, the geometric origin of a stringy inflaton renders the issue far more severe.

As advocated in Ref. [5], this problem is avoided if reflecting or absorbing walls are installed at certain positions in φ-space. While Ref. [10] considered the case of one reflecting wall, we derived the PDF for a stochastic (DBI or standard) inflaton field between two absorbing walls at ϕ_0 and $\phi_{\rm UV}$. (For UV brane inflation, this is interpreted as the D3 being "trapped" inside the warped throat.) To this end, we repeatedly applied the familiar method of images from electrodynamics, introducing an infinite number of "image branes" of alternating charge. The final PDF for this configuration is $P_{\rm w}(\varphi) = \left(2\pi\langle\delta\varphi_1^2\rangle\right)^{-1/2}\tilde{P}(\varphi)$, where

$$\tilde{P}(\varphi) = \sum_{n=-\infty}^{+\infty}\left(\exp\left\{-\frac{\left[\varphi - \phi_{\rm m} + 2n\,(\phi_{\rm UV} - \phi_0)\right]^2}{2\langle\delta\varphi_1^2\rangle}\right\} - \exp\left\{-\frac{\left[\varphi + \phi_{\rm m} - 2\phi_0 + 2n\,(\phi_{\rm UV} - \phi_0)\right]^2}{2\langle\delta\varphi_1^2\rangle}\right\}\right)$$

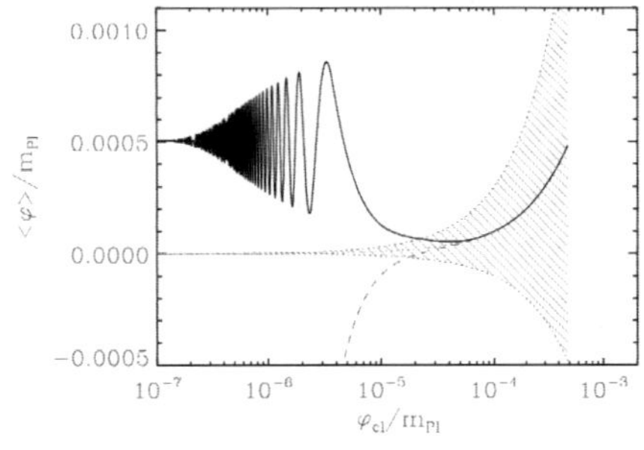

Figure 1. Evolution of a DBI inflaton for the shape of V and T cited in the text and $\alpha = 38$, $\beta = 3.7$, $m \simeq 2.19 \times 10^{-7} m_{\rm Pl}$ and $\varphi_{\rm in} = 5 \times 10^{-4} m_{\rm Pl}$. The green dotted line is the *classical* trajectory, while the red dashed line is the *quantum* mean $\langle\varphi\rangle = \varphi_{\rm cl} + \langle\delta\varphi_2\rangle$. The black solid line corresponds to $\langle\varphi\rangle_{\rm w}$ for absorbing walls at $\phi_0 = 10^{-5} m_{\rm Pl}$ and $\phi_{\rm UV} = 10^{-3} m_{\rm Pl}$. The hatched area is the validity regime of the perturbative treatment.

with $\phi_{\rm m} = \langle\varphi\rangle$ the former mean found with the "no walls" PDF (3). The true mean field value in the presence of the walls then is $\langle\varphi\rangle_{\rm w} \propto \int {\rm d}\varphi\, P_{\rm w}(\varphi)\, \varphi$ and was calculated in Ref. [5]. Interestingly, it shows oscillatory behavior, as illustrated by the example of Fig. 1. As the inflaton approaches the throat's tip, it seems that $\langle\varphi\rangle_{\rm w}$ stabilizes at a position above ϕ_0. There is, however, potential concern about the validity of the perturbative approach.

4. Conclusions

In brane inflation, the scalar field ϕ corresponds to the position of a D-brane within the 6d compact string geometry. On the one hand, this enriches the inflationary phenomenology due to the brane's non-standard kinetic term. On the other hand, the inflaton's ten-dimensional origin leaves a geometric legacy to the four-dimensional effective theory: the values of ϕ should respect the field space limitations imposed by the finite size of the extra dimensions. Here, we generalized several results of stochastic inflation to DBI dynamics: where, in the standard case, integrals of $H(\phi)$ (and derivatives) were sufficient, the DBI case is described accurately only if the Lorentz factor $\gamma(\phi)$ is included alongside $H(\phi)$. We obtained the Gaussian probability distribution function $P_{\rm c}(\varphi)$ for the smoothed-out field φ, which allowed us to calculate the (second order noise-corrected) mean field value $\langle\varphi\rangle$ to be compared with the classical one $\varphi_{\rm cl}$.

We then argued that, to avoid conflict with the underlying 10d string theory picture, even the quantum jumps superimposed on the inflaton's DBI trajectory should not take φ out of the geometrically admitted range. We proposed to implement this restriction by using suitable PDF boundary conditions in field space, *i.e.* by installing walls where the compact dimension associated with φ "ends". As an example, we considered a quadratic inflaton potential together with a quartic warping and imposed absorbing boundary conditions at both ends of the (radial direction of the) extra dimensions. As seen in Fig. 1, the trajectory then changes significantly for a chosen parameter set. Our hope is that, as the shape of the warping and the potential are better understood from string theory, these results may shed light on the possibly complicated path of the brane (that is, the inflaton) inside the six-dimensional compact space.

References

[1] Dvali G R and Tye S H H 1999 *Phys. Lett.* **B450** 72–82 (*Preprint* hep-ph/9812483)
[2] Kachru S *et al.* 2003 *JCAP* **0310** 013 (*Preprint* hep-th/0308055)
[3] Garriga J and Mukhanov V F 1999 *Phys. Lett.* **B458** 219–225 (*Preprint* hep-th/9904176)
[4] Starobinsky A A and Yokoyama J 1994 *Phys. Rev.* **D50** 6357–6368 (*Preprint* astro-ph/9407016)
[5] Lorenz L, Martin J and Yokoyama J 2010 *Phys. Rev.* **D82** 023515 (*Preprint* 1004.3734)
[6] Martin J and Musso M 2006 *Phys. Rev.* **D73** 043516 (*Preprint* hep-th/0511214)
[7] Martin J and Musso M 2006 *Phys. Rev.* **D73** 043517 (*Preprint* hep-th/0511292)
[8] Alishahiha M, Silverstein E and Tong D 2004 *Phys. Rev.* **D70** 123505 (*Preprint* hep-th/0404084)
[9] Baumann D and McAllister L 2007 *Phys. Rev.* **D75** 123508 (*Preprint* hep-th/0610285)
[10] Chandrasekhar S 1943 *Rev. Mod. Phys.* **15** 1–89

Journal of Physics: Conference Series **314** (2011) 012053 doi:10.1088/1742-6596/314/1/012053

Reduced phase space quantization of FRW universe

Fumitoshi Amemiya and Tatsuhiko Koike

Department of Physics, Keio University, 3-14-1 Hiyoshi, Kohoku-ku, 223-8522 Yokohama, Japan

E-mail: famemiya@rk.phys.keio.ac.jp, koike@phys.keio.ac.jp

Abstract. A gauge-invariant quantum theory of the flat Friedmann-Robertson-Walker (FRW) universe with dust is studied in terms of the Ashtekar variables. We use the reduced phase space quantization which has following advantages: (i) fundamental variables are all gauge invariant, (ii) there exists a physical time evolution of gauge-invariant quantities, so that the problem of time is absent and (iii) the reduced phase space can be quantized in the same manner as in ordinary quantum mechanics. Analyzing the dynamics of a wave packet, we show that the classical initial singularity is replaced by a big bounce in quantum theory.

1. Introduction

One of the motivations of quantum cosmology is to shed light on quantum nature of the initial singularity. However, there exists potential problems that have not been completely resolved yet. A problem is about what should be interpreted as observables in classical and quantum gravity [1, 2]. A canonical formulation of general relativity (GR) is a constrained system with first-class constraints in which the spacetime diffeomorphisms are interpreted as gauge transformations. In gauge theories, only gauge-invariant quantities are observables. However, there are technical and conceptual difficulties in the realization of the idea especially in GR. In many works, gauge-variant quantities are used as observables. This issue must be seriously considered especially in quantum gravity because it is substantially related to the problem of time [3].

In this paper, we shall construct and analyze a gauge-invariant quantum theory of the flat FRW universe with the Brown-Kuchař dust [4] in terms of the Ashtekar variables [5, 6]. We use the reduced phase space quantization method where the so-called relational formalism [7, 8] is used to construct the classical reduced phase space spanned by gauge-invariant quantities, and then the system is quantized in the same manner as in ordinary quantum mechanics. The quantization gives a possible resolution to the problem of time. As for the dynamics of the universe, we consider the motion of a wave packet and evaluate the expectation value of the scale factor. It is shown that the expectation value has a non-zero minimum, that is, the initial singularity is replaced by a big bounce in quantum theory. The remarkable point is that the big bounce mixes the states representing right-handed and left-handed systems. See [9, 10] for details of the work. In this paper we adopt the unit in which $c = 1$.

2. Reduced phase space of Friedmann-Robertson-Walker universe with dust

In the Ashtekar formulation [5, 6], the variables (A^i_a, E^a_i) form a canonically conjugate pair where A^i_a is a SU(2) connection and E^a_i is an orthonormal triad with density weight 1. In the flat FRW

Journal of Physics: Conference Series **314** (2011) 012053 doi:10.1088/1742-6596/314/1/012053

model, the Ashtekar variables can be written in terms of only one independent components $\tilde{c}$ and $\tilde{p}$ [11],

$$A_a^i = \tilde{c}(t)\omega_a^i, \quad E_i^a = \tilde{p}(t)X_i^a, \tag{1}$$

where ω^i are bases of left invariant one-forms and X_i are invariant vector fields dual to the one-forms. These variables have relations to the scale factor a such that

$$|\tilde{p}| = a^2, \quad \tilde{c} = \mathrm{sgn}(p)\frac{\gamma}{N}\dot{a}, \tag{2}$$

where γ is the so-called Barbero-Immirzi parameter, N is the lapse function and the dot denotes the derivative with respect to t. Note that, while the scale factor is restricted to be nonnegative, $\tilde{p}$ ranges over the entire real line, carrying an orientation of triads determined by the sign of $\tilde{p}$. We here consider a compact universe to avoide the divergence of the three-space integral and in particular we only consider the case of three-dimensional torus, where we take a cube of coordinate range $0 \leq x, y, z \leq V^{\frac{1}{3}}$, and identify the opposite faces.

If we define new variables as $p := V^{\frac{2}{3}}\tilde{p}$ and $c := V^{\frac{1}{3}}\tilde{c}$, the total action for gravity plus the Brown-Kuchař dust [4] is written as

$$S_{\mathrm{tot}} = \int dt \left[\frac{3}{\kappa\gamma}p\dot{c} + P_T\dot{T} - NH_{\mathrm{tot}}\right], \tag{3}$$

where $\kappa = 8\pi G$, T is the proper time measured along the particle flow lines when the equations of motion hold, P_T is its conjugate momentum and the Hamiltonian constraint takes the form

$$H_{\mathrm{tot}} = H_{\mathrm{grav}} + H_{\mathrm{dust}} = -\frac{3}{\kappa\gamma^2}c^2\sqrt{|p|} + P_T = 0. \tag{4}$$

The key observation of the relational formalism [7, 8] to define gauge-invariant quantities is as follows. Take two gauge-variant functions F and T on the phase space, and choose one of the functions T as a clock. Then, the value of F at $T = \tau$ is gauge-invariant even if F and T themselves are gauge variant. Suppose a phase space has a $2n$-dimension ($n \geq 2$), and there are canonical coordinates $(q^a, p_a, a = 1, \cdots, n)$ such that $\{q^a, p_b\} = \delta_b^a$. We will denote a first-class constraint by H and a phase space point by $y = (q^a, p_a)$. Under the gauge transformation generated by H, a point y is mapped to $y \mapsto \alpha_H^t(y)$, where t is a gauge parameter. That is, $\alpha_H^t(y)$ is a gauge flow generated by H starting from y. Then we can define the gauge-invariant quantity $O_F^\tau(y)$ as

$$O_F^\tau(y) := F(\alpha_H^t(y))|_{T(\alpha_H^t(y))=\tau}. \tag{5}$$

A constraint equation $H = 0$ is said to be of deparametrized form if it is written as $H(q^a, T, p_a, P_T) = P_T + h(q^a, p_a) = 0$ with some phase space coordinates $\{q^a, T; p_a, P_T\}$. In the deparametrized theories, the reduced phase space is spanned by the gauge-invariant quantities $\left(O_{q^a}^\tau(y), O_{p_a}^\tau(y)\right)$ associated with q^a and p_a with the simple symplectic structure $\left\{O_{q^a}^\tau(y), O_{p_b}^\tau(y)\right\} = \delta_b^a$. The physical Hamiltonian H_{phys} is obtained by replacing q^a and p_a in $h(q^a, p_a)$ with $O_{q^a}^\tau(y)$ and $O_{p_a}^\tau(y)$: $H_{\mathrm{phys}}\left(O_{q^a}^\tau(y), O_{p_a}^\tau(y)\right) := h\left(O_{q^a}^\tau(y), O_{p_a}^\tau(y)\right)$. The Hamiltonian generates the time evolution of the gauge-invariant quantity associated with a function F which depends only on q^a and p_a: $\frac{\partial O_F^\tau(y)}{\partial\tau} = \{H_{\mathrm{phys}}, O_F^\tau(y)\}$.

In the present case, it is natural to choose the function T as the clock variable. Then, the reduced phase space is coordinatized by the gauge-invariant quantities $C(\tau) := O_c^\tau(y)$ and $P(\tau) := O_p^\tau(y)$ associated with c and p with very simple symplectic structure

$$\{C(\tau), P(\tau)\} = \frac{\kappa\gamma}{3}. \tag{6}$$

Moreover, we can obtain the physical Hamiltonian H_{phys} by replacing c and p in $H_{\text{grav}}(c,p)$ with C and P,

$$H_{\text{phys}} = -\frac{3}{\kappa\gamma^2}C(\tau)^2\sqrt{|P(\tau)|}. \tag{7}$$

3. Quantization

In this section, we shall quantize the system on the reduced phase space obtained in the previous section. Now the physical variables are operators and the Poisson bracket $\{\bullet,\bullet\}$ is replaced with the commutation relation $(1/i\hbar)[\bullet,\bullet]$. Thus (6) becomes the canonical commutation relation

$$[\hat{C},\hat{P}] = \frac{i\kappa\gamma\hbar}{3}. \tag{8}$$

Let us choose the ordinary Schrödinger representation in which the operators $\hat{P}$ and $\hat{C}$, respectively, act on a wave function $\Psi(P)$ in the following way: $\hat{P}\Psi(P) = P\Psi(P)$, $\hat{C}\Psi(P) = \frac{i\hbar\kappa\gamma}{3}\frac{\partial\Psi(P)}{\partial P}$. As a concrete example, we choose the following operator ordering for the Hamiltonian,

$$\hat{H}_{\text{phys}} = -\frac{3}{\kappa\gamma^2}\sqrt{|\hat{P}|}\hat{C}^2. \tag{9}$$

Then the Schrödinger equation takes the simplest form,

$$i\hbar\frac{\partial\Psi}{\partial\tau} = \frac{\kappa\hbar^2}{3}\sqrt{|P|}\frac{\partial^2\Psi}{\partial P^2}. \tag{10}$$

Since the present Hamiltonian is different from the ordinary kinematical term, we choose the Hilbert space as $\mathcal{H} = L^2(\mathbb{R},|P|^{-\frac{1}{2}}dP)$ in order to make the Hamiltonian (9) Hermitian up to surface term. $\hat{H}_{\text{phys}}$ is somewhat singular at the origin $P=0$, it is indeed self-adjoint in $\mathcal{H}$.

4. Dynamics of the universe

Let us now analyze the dynamics of a wave packet. The procedure is as follows. First, we prepare an initial wave packet $\Psi(P,0)$ at some nonzero P. Then, we numerically evolve it backward in time by the Schrödinger equation (10) and evaluate the expectation value of $|P|$ as a function of the internal time $\tilde{\tau}$. Here we consider $|P|$ because both the positive and negative P correspond to the universe of the same size with different orientation of triads. For simplicity, we here choose the initial wave function as a Gaussian wave packet

$$\Psi(P,0) = C_0\exp\left(-\frac{(P-P_0)^2}{2\sigma^2} - ik_0P\right), \tag{11}$$

where C_0 is the normalization constant. Figs. 1 show the absolute value of the wave function as a function of P and τ, and the expectation value of $|P|$ is plotted as a function of the time τ.

We can see from Fig. 1 (a) that a part of the wave packet is reflected and the rest is transmitted at the origin. We here remind that the sign of P determines an orientation of triads, which correspond to a right-handed and left-handed systems respectively. Thus, the result indicates that if the present state of the universe is in a right-handed system, the past state is in superposition of the states of a right-handed and left-handed systems. As for the expectation value of $|P|$, Fig. 1 (b) indicates that the expectation value never goes to zero and bounces at a nonzero minimum. That is, the initial singularity is replaced by a big bounce in the present model.

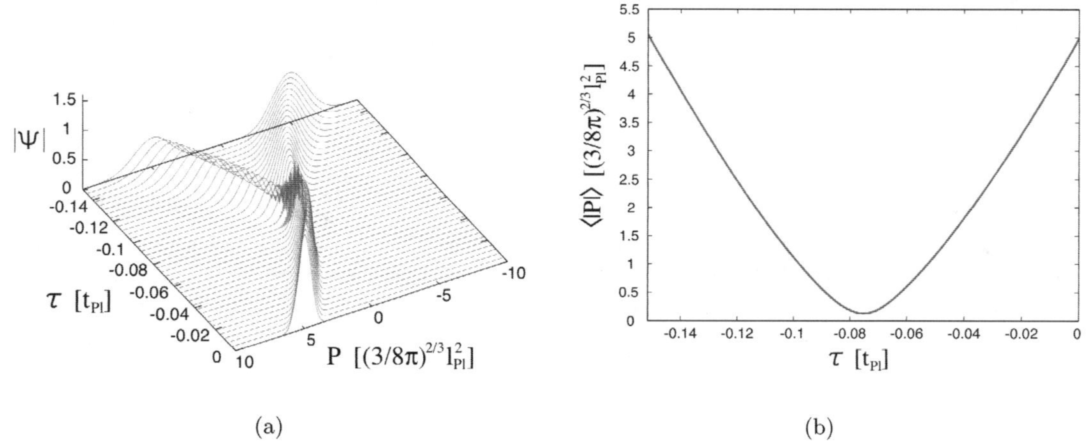

Figure 1. Fig.(a) shows the absolute value of the wave function as a function of τ and P. Fig.(b) shows the expectation values of $|P|$ as a functions of τ.

5. Conclusions

A gauge-invariant quantum theory of the flat FRW universe with dust has been studied in terms of the Ashtekar variables. We have first constructed the classical reduced phase space of the system by using the relational formalism and then have quantized the reduced system. The advantages of the quantization method are as follows: (i) fundamental variables are gauge-invariant quantities, (ii) a natural time evolution of the gauge-invariant quantities exists, so that the problem of time is absent and (iii) the reduced phase space can be quantized in the same manner as in ordinary quantum mechanics because there are no constraints in the reduced phase space. In the obtained quantum theory, we have analyzed the dynamics of a wave packet and have shown that the expectation value of P has a non-zero minimum, that is, the initial singularity is replaced by a big bounce in quantum theory. The interpretation of the wave packet is that if the present state of the universe is in a right-handed system, the past state has been in a superposition of the states of a right-handed and left-handed systems.

Achnowledgement

This work was supported in part by Global COE Program "High-Level Grobal Cooperation for Leading-Edge Platform on Access Spaces (C12)".

References

[1] P. G. Bergmann, Rev. Mod. Phys. **33** 510 (1961).
[2] C. Rovelli, Class. Quant. Grav. **8**, 1895 (1991); Phys. Rev. D **65**, 124013 (2002).
[3] For a comprehensive review on the problem of time, see e.g. C. J. Isham, arXiv:gr-qc/9210011.
[4] J. D. Brown and K. V. Kuchař, Phys. Rev. D **51**, 5600 (1995).
[5] A. Ashtekar, Phys. Rev. Lett. **57**, 2244 (1986); Phys. Rev. D **36**, 1587 (1987).
[6] J. F. Barbero G., Phys. Rev. D **51**, 5507 (1995).
[7] B. Dittrich, Gen. Rel. Grav. **39**, 1891 (2007).
[8] T. Thiemann, Class. Quant. Grav. **23**, 1163 (2006).
[9] F. Amemiya and T. Koike, Phys. Rev. D **80**, 103507 (2009).
[10] F. Amemiya and T. Koike, Phys. Rev. D **82**, 104007 (2010).
[11] A. Ashtekar, M. Bojowald and J. Lewandowski, Adv. Theor. Math. Phys. **7**, 233 (2003).

Journal of Physics: Conference Series **314** (2011) 012054 doi:10.1088/1742-6596/314/1/012054

Non-singular Universes à la Palatini

Gonzalo J. Olmo[1]

Departamento de Física Teórica and IFIC, Centro Mixto Universidad de Valencia - CSIC. Facultad de Física, Universidad de Valencia, Burjassot-46100, Valencia, Spain and Instituto de Estructura de la Materia, CSIC, Serrano 121, 28006 Madrid, Spain

E-mail: gonzalo.olmo@uv.es

Abstract. It has recently been shown that f(R) theories formulated in the Palatini variational formalism are able to avoid the big bang singularity yielding instead a bouncing solution. The mechanism responsible for this behavior is similar to that observed in the effective dynamics of loop quantum cosmology and an f(R) theory exactly reproducing that dynamics has been found. I will show here that considering more general actions, with quadratic contributions of the Ricci tensor, results in a much richer phenomenology that yields bouncing solutions even in anisotropic (Bianchi I) scenarios. Some implications of these results will be discussed.

1. Introduction.

In the last years it has been shown that though many $f(R)$ models have the ability to produce late-time cosmic acceleration and fit well the background expansion history, they are not in quantitative agreement with the structure and evolution of cosmic inhomogeneities (see [1] and references therein). Additionally, the fact that matter is concentrated in discrete structures like atoms causes the modified dynamics to manifest also in laboratory experiments[2, 3, 4], which confirms earlier suspicions on the viability of such models according to their corresponding Newtonian and post-Newtonian properties [5]. This is a very disturbing aspect of the models with infrared corrections, which demands the consideration of a microscopic description of the sources and prevents the use of macroscopic, averaged representations of the matter. A careful analysis of this point put forward the existence of non-perturbative effects induced by the Palatini dynamics in a number of contexts [6, 2, 7, 3]. Though this certainly is an undesired property of infrared-corrected models, it could become a very useful tool for models with corrections at high curvatures. Can we construct singularity-free cosmological models that recover GR at low curvatures using the non-perturbative properties of Palatini theories? As we report here, ultraviolet-corrected Palatini models turn out to be very efficient at removing the big bang cosmic singularity in various situations of interest.

2. Non-singular $f(R)$ cosmologies.

The recent interest in the dynamics of the early-universe in Palatini theories arose, in part, from the observation that the effective equations of loop quantum cosmology [8] (LQC) could be exactly reproduced by a Palatini $f(R)$ Lagrangian [9]. In LQC, non-perturbative quantum gravity effects lead to the resolution of the big bang singularity by a quantum bounce without

[1] Published under license in Journal of Physics: Conference Series by IOP Publishing Ltd.

introducing any new degrees of freedom. Though fundamentally discrete, the theory admits a continuum description in terms of an effective Hamiltonian that in the case of a homogeneous and isotropic universe filled with a massless scalar field leads to the following Friedmann equation

$$3H^2 = 8\pi G\rho\left(1-\frac{\rho}{\rho_{crit}}\right) , \tag{1}$$

where $\rho_{crit} \approx 0.41\rho_{Planck}$. At low densities, $\rho/\rho_{crit} \ll 1$, the background dynamics is the same as in GR, whereas at densities of order ρ_{crit} the non-linear new matter contribution forces the vanishing of H^2 and hence a cosmic bounce. This singularity avoidance seems to be a generic feature of loop-quantized universes [10].
Palatini $f(R)$ theories share with LQC the fact that the modified dynamics that they produce is not due to the existence of new dynamical degrees of freedom but rather to non-linear effects produced by the matter sources, which contrasts with other approaches to quantum gravity and to modified gravity. This allows to relate Eq.(1) with the corresponding $f(R)$ equation

$$3H^2 = \frac{f_R\left(\kappa^2\rho + (Rf_R - f)/2\right)}{\left(f_R - \frac{12\kappa^2\rho f_{RR}}{2(Rf_{RR}-f_R)}\right)^2} . \tag{2}$$

Taking into account the trace of the field equations (see [1] for details), which for a massless scalar becomes $Rf_R - 2f = 2\kappa^2\rho$ and implies that $\rho = \rho(R)$, one finds that a Palatini $f(R)$ theory able to reproduce the LQC dynamics (1) must satisfy the differential equation

$$f_{RR} = -f_R\left(\frac{Af_R - B}{2(Rf_R - 3f)A + RB}\right) \tag{3}$$

where $A = \sqrt{2(Rf_R - 2f)(2R_c - [Rf_R - 2f])}$, $B = 2\sqrt{R_c f_R(2Rf_R - 3f)}$, and $R_c \equiv \kappa^2\rho_c$. Imposing suitable boundary conditions, one can show [9] that the function

$$\frac{df}{dR} = -\tanh\left(\frac{5}{103}\ln\left[\left(\frac{R}{12\mathcal{R}_c}\right)^2\right]\right) \tag{4}$$

faithfully captures the LQC dynamics from low curvatures up to the non-perturbative bouncing region. Though the function (4) implies that the LQC Lagrangian is an infinite series, which is a manifestation of the non-local properties of the quantum geometry, one can find non-singular cosmologies with a finite number of terms. In fact, a simple quadratic Lagrangian of the form $f(R) = R + R^2/R_P$ does exhibit non-singular solutions for certain equations of state [11] depending on the sign of R_P. To be precise, if $R_P > 0$ the bounce occurs for sources with $w = P/\rho > 1/3$. If $R_P < 0$, then the bouncing condition is satisfied by $w < 1/3$.

Besides avoiding the development of curvature singularities, bouncing cosmologies can solve the horizon problem [12], which makes them interesting as a substitute for inflation. However, it has been found [13] that $f(R)$ models that develop a bounce when the condition $f_R = 0$ is met turn out to exhibit singular behavior of inhomogeneous perturbations in a flat, dust-filled universe. Further insight on the robustness of the bounce under perturbations was obtained in [14] studying the properties of $f(R)$ theories in anisotropic spacetimes of Bianchi-I type, for which $ds^2 = -dt^2 + \sum_{i=1}^3 a_i^2(t)(dx^i)^2$. In space-times of this type with a perfect fluid, one can derive a number of useful analytical expressions. In particular, one finds that the expansion $\theta = \sum_i H_i$ and the shear $\sigma^2 = \sum_i \left(H_i - \frac{\theta}{3}\right)^2$ are given by

$$\sigma^2 = \frac{\rho^{\frac{2}{1+w}}}{f_R^2}\frac{(C_{12}^2 + C_{23}^2 + C_{31}^2)}{3} , \quad \frac{\theta^2}{3}\left(1+\frac{3}{2}\Delta_1\right)^2 = \frac{f + \kappa^2(\rho + 3P)}{2f_R} + \frac{\sigma^2}{2} \tag{5}$$

where the constants $C_{ij} = -C_{ji}$ set the amount and distribution of anisotropy and satisfy the constraint $C_{12} + C_{23} + C_{31} = 0$. In the isotropic case, $C_{ij} = 0$, one has $\sigma^2 = 0$ and $\theta^2 = 9H^2$, with H^2 given by Eq.(2). Now, since homogeneous and isotropic bouncing universes always require the condition $f_R = 0$ at the bounce [14], a glance at (5) indicates that the shear diverges as $\sim 1/f_R^2$. This shows that any isotropic $f(R)$ bouncing model will develop divergences when anisotropies are present. Moreover, one can check by direct calculation that the Kretschman scalar $R_{\mu\nu\sigma\rho}R^{\mu\nu\sigma\rho} = 4(\sum_i(\dot{H}_i + H_i^2)^2 + H_1^2H_2^2 + H_1^2H_3^2 + H_2^2H_3^2)$ diverges at least as $\sim 1/f_R^4$, which is a clear geometrical pathology and signals the presence of a physical singularity.

3. Nonsingular cosmologies beyond $f(R)$.

The previous section provides reasons to believe that Palatini $f(R)$ models are not able to produce a fully satisfactory and singularity-free alternative to GR. Though the homogeneous and isotropic case greatly improves the situation with respect to GR, the existence of divergences when anisotropies and inhomogeneities are present spoil the hopes deposited on this kind of Lagrangians. To the light of these results, we explored new Palatini theories [14] to determine if the introduction of new geometrical invariants in the gravitational action could avoid the problems that appear in the $f(R)$ models. We considered Lagrangians of the form $f(R, Q)$, with $Q = R_{\mu\nu}R^{\mu\nu}$. Using the particular Lagrangian

$$f(R, R_{\mu\nu}R^{\mu\nu}) = R + a\frac{R^2}{R_P} + \frac{R_{\mu\nu}R^{\mu\nu}}{R_P} \ , \tag{6}$$

it was found that completely regular bouncing solutions exist for both isotropic and anisotropic homogeneous cosmologies filled with a perfect fluid. In particular, we found that for $a < 0$ the interval $0 \leq w \leq 1/3$ is always included in the family of bouncing solutions, which contains the dust and radiation cases. For $a \geq 0$, the fluids yielding a non-singular evolution are restricted to $w > \frac{a}{2+3a}$, which implies that the radiation case $w = 1/3$ is always nonsingular. For a detailed discussion and classification of the non-singular solutions depending on the value of the parameter a and the equation of state w see [14].

The field equations that follow from the Lagrangian (6) when $R_{\mu\nu}$ is assumed symmetric were derived in [15] and take the form

$$f_R R_{\mu\nu} - \frac{f}{2}g_{\mu\nu} + 2f_Q R_{\mu\alpha}R^\alpha{}_\nu = \kappa^2 T_{\mu\nu} \ , \ \nabla_\beta \left[\sqrt{-g}\left(f_R g^{\mu\nu} + 2f_Q R^{\mu\nu}\right)\right] = 0 \tag{7}$$

where $f_R \equiv \partial_R f$ and $f_Q \equiv \partial_Q f$. The connection equation in (7) can be solved introducing an auxiliary metric $h_{\alpha\beta}$ such that it takes the form $\nabla_\beta\left[\sqrt{-h}h^{\mu\nu}\right] = 0$, which implies that $\Gamma^\rho_{\mu\lambda}$ can be written as the Levi-Civita connection of $h_{\mu\nu}$. When the matter sources are represented by a perfect fluid, $T_{\mu\nu} = (\rho + P)u_\mu u_\nu + Pg_{\mu\nu}$, one can show that $h_{\mu\nu}$ and $h^{\mu\nu}$ are given by [15]

$$h_{\mu\nu} = \Omega\left(g_{\mu\nu} - \frac{\Lambda_2}{\Lambda_1 - \Lambda_2}u_\mu u_\nu\right) \ , \ h^{\mu\nu} = \frac{1}{\Omega}\left(g^{\mu\nu} + \frac{\Lambda_2}{\Lambda_1}u^\mu u^\nu\right) \tag{8}$$

where $\lambda = \sqrt{\kappa^2 P + \frac{f}{2} + \frac{f_R^2}{8f_Q}}$, $\Lambda_1 = \sqrt{2f_Q}\lambda + \frac{f_R}{2}$, $\Lambda_2 = \sqrt{2f_Q}\left[\lambda \pm \sqrt{\lambda^2 - \kappa^2(\rho + P)}\right]$, and $\Omega = \left[\Lambda_1(\Lambda_1 - \Lambda_2)\right]^{1/2}$. The metric field equation in (7) takes the following form

$$R_{\mu\nu}(h) = \frac{1}{\Lambda_1}\left[\frac{(f + 2\kappa^2 P)}{2\Omega}h_{\mu\nu} + \frac{\Lambda_1\kappa^2(\rho + P)}{\Lambda_1 - \Lambda_2}u_\mu u_\nu\right] \ . \tag{9}$$

In this expression, the functions f, Λ_1, and Λ_2 are functions of the density ρ and pressure P. In particular, in a universe filled with radiation, for which $R = 0$, the function Q in our

quadratic model (6) boils down to $Q = \frac{3R_P^2}{8}\left[1 - \frac{8\kappa^2\rho}{3R_P} - \sqrt{1 - \frac{16\kappa^2\rho}{3R_P}}\right]$. This expression recovers the GR value at low curvatures, $Q \approx 4(\kappa^2\rho)^2/3 + 32(\kappa^2\rho)^3/9R_P + \ldots$ but reaches a maximum $Q_{max} = 3R_P^2/16$ at $\kappa^2\rho_{max} = 3R_P/16$. At ρ_{max} the shear also takes its maximum allowed value, namely, $\sigma^2_{max} = \sqrt{3/16}R_P^{3/2}(C_{12}^2 + C_{23}^2 + C_{31}^2)$, which is always finite, and the expansion vanishes producing a cosmic bounce regardless of the amount of anisotropy.
It should be noted that the choice of a symmetric Ricci tensor in the analysis of $f(R,Q)$ bouncing cosmologies presented above is not arbitrary. As shown in [16], the antisymmetric part of the Ricci tensor introduces new dynamical degrees of freedom in the form of a massive vector field (see also [17] for a related result). If one looks for a framework suitable for the description of the effective dynamics of LQC (including anisotropies) and, more generally, of other theories of quantum geometry not involving new degrees of freedom, it seems natural to impose constraints on the spectrum of possible Lagrangians to avoid new propagating fields. In this sense, it should be noted that the $f(R,Q)$ theories discussed here are also able to reproduce [18] other aspects of the expected phenomenology of quantum gravity at the Planck scale. In particular, without imposing any a priori phenomenological structure, the quadratic Palatini model (6) predicts an energy-density dependence of the metric components that closely matches the structure conjectured in models of Doubly (or Deformed) Special Relativity and Rainbow Gravity [19]. This confirms that Palatini theories represent a new and powerful framework to address different aspects of quantum gravity phenomenology.

Work supported by the Spanish grants FIS2008-06078-C03-02, FIS2008-06078-C03-03, and the Consolider Programme CPAN (CSD2007-00042).

References

[1] G. J. Olmo, Int. J. Mod. Phys. **D**, to appear (2011); A. De Felice, S. Tsujikawa, Living Rev. Rel. **13**, 3 (2010); T.P. Sotiriou and V. Faraoni, Rev. Mod. Phys. 82, 451-497 (2010).
[2] G. J. Olmo, Phys. Rev. Lett. **98**, 061101 (2007).
[3] G. J. Olmo, Phys. Rev. **D 77**, 084021(2008).
[4] B. Li, D. F. Mota and D. J. Shaw, Phys. Rev. **D 78**, 064018 (2008); Class. Quant. Grav. **26**, 055018 (2009).
[5] G. J. Olmo, Phys. Rev. Lett. **95**, 261102 (2005); Phys. Rev. **D72**, 083505 (2005).
[6] E. E. Flanagan, Phys. Rev. Lett.**92**, 071101 (2004).
[7] E. Barausse, T.P. Sotiriou, and J.C. Miller, *Class.Quant.Grav.* 25,062001(2008); E. Barausse, T.P. Sotiriou, and J.C. Miller, *Class.Quant.Grav.* 25,105008(2008).
[8] M. Bojowald, *Living Rev. Rel.* **8**, 11 (2005); A. Ashtekar, *Nuovo Cim.* **122 B**, 135 (2007); A. Ashtekar, T. Pawlowski and P. Singh, *Phys. Rev. Lett.* **96**, 141301 (2006); *Phys. Rev.* **D** 73, 124038 (2006); *Phys. Rev.* **D** 74, 084003 (2006); L. Szulc, W. Kaminski, J. Lewandowski, *Class.Quant.Grav.* **24**, 2621 (2007); A. Ashtekar, T. Pawlowski, P.Singh, K. Vandersloot, *Phys.Rev.* **D** 75, 024035 (2007); K. Vandersloot, *Phys.Rev.* **D** 75, 023523 (2007).
[9] G.J.Olmo and P.Singh, JCAP 0901, 030 (2009).
[10] P. Singh, Class. Quant. Grav. **26**, 125005 (2009).
[11] C. Barragan, G. J. Olmo, H. Sanchis-Alepuz, Phys. Rev. **D80**, 024016 (2009); C. Barragan, G. J. Olmo, H. Sanchis-Alepuz, [arXiv:1002.3919 [gr-qc]]; G. J. Olmo, AIP Conf. Proc. **1241**, 1100-1107 (2010), [arXiv:0910.3734 [gr-qc]].
[12] M.Novello and S.E. Perez Bergliaffa, *Phys.Rep.* 463, 127-213 (2008);
[13] T.Koivisto, Phys. Rev. **D 82**, 044022 (2010).
[14] C. Barragan and G. J. Olmo, Phys. Rev. **D 82**, 084015 (2010).
[15] G. J. Olmo, H. Sanchis-Alepuz, S. Tripathi, Phys. Rev. **D80**, 024013 (2009).
[16] H. A. Buchdahl, J. Phys. A: Math. Gen. 12, 1235 (1979); V. Vitagliano, T. P. Sotiriou and S. Liberati, Phys. Rev. D **82**, 084007 (2010).
[17] D. N. Vollick, arXiv:gr-qc/0601016.
[18] G. J. Olmo, *Palatini actions and quantum gravity phenomenology*, to appear (2011).
[19] J. Magueijo and L. Smolin, Phys.Rev.Lett.**88**, 190403(2002); Phys.Rev. D**67**, 044017(2003); J.Magueijo and L.Smolin, Class. Quant. Grav.**21**, 1725 (2004).

Journal of Physics: Conference Series **314** (2011) 012055 doi:10.1088/1742-6596/314/1/012055

Cosmological solutions in F(R) Hořava-Lifshitz gravity

Diego Sáez-Gómez

Consejo Superior de Investigaciones Científicas, ICE(CSIC-IEEC), Campus UAB Facultat de Ciències, Torre C5-Parell-2a pl, E-08193 Bellaterra (Barcelona) Spain

Abstract. At the present work, it is studied the extension of $F(R)$ gravities to the new recently proposed theory of gravity, the so-called Hořava-Lifshitz gravity, which provides a way to make the theory power counting renormalizable by breaking Lorentz invariance. It is showed that dark energy can be well explained in the frame of this extension, just in terms of gravity. It is also explored the possibility to unify inflation and late-time acceleration under the same mechanism, providing a natural explanation the accelerated expansion.

1. Introduction to F(R) Hořava-Lifshitz gravity

Since observational data suggests the well known fact that probably the Universe is immersed in an accelerating phase, a lot of models have been proposed in order to explain such unusual behavior of gravity at large scales. The first and still most popular model was the cosmological constant, which gives a negative pressure and then, an acceleration on the expansion. Nevertheless, the fine tuning problem on the model has made that another kind of candidates has been seriously taken into account. One of them suggests the possibility of the modification of gravitational field equations such that at large scales, gravity behaves different than in local systems, producing accelerated expansion and even open the possibility to explain, at the same time, the inflationary epoch in a natural way. One of this theories is the so-called $F(R)$ gravity, which just generalized the Hilbert-Einstein action to more complex ones (for a recent review see Ref. [1]). On the other hand, recently a new gravitational theory has emerged, which suggests to be power-counting renormalizable but breaking explicitly Lorentz invariance [2] by means of different scaling properties on the coordinates,

$$x^i = bx^i\,, \quad t = b^z t\,. \tag{1}$$

where z is a dynamical critical exponent that makes the theory to be renormalizable for $z = 3$ in a $3+1$ spacetime dimensions. This new theory, already known as Hořava-Lifshitz gravity, does not explain the dark energy epoch in its original form as it also occurs in General Relativity (GR). Then, generalizations of the original action has been suggested in the same way as $F(R)$ gravity from GR in order to explain the dark energy and even the inflationary epochs under just a gravitational mechanism. Let us briefly review this extension called $F(R)$ Hořava-Lifshitz gravity. Giving a general metric written in the ADM decomposition in a 3+1 spacetime dimensions,

$$ds^2 = -N^2 dt^2 + g^{(3)}_{ij}(dx^i + N^i dt)(dx^j + N^j dt)\,, \tag{2}$$

The action considered here can be written as [3],

$$S = \frac{1}{2\kappa^2}\int dtd^3x\sqrt{g^{(3)}}NF(\tilde{R}),$$

$$\tilde{R} = K_{ij}K^{ij} - \lambda K^2 + R^{(3)} + 2\mu\nabla_\mu(n^\mu\nabla_\nu n^\nu - n^\nu\nabla_\nu n^\mu) - L^{(3)}(g_{ij}^{(3)}), \tag{3}$$

where K_{ij} is the extrinsic curvature, and $R^{(3)}$ is the spatial scalar curvature. Note that, apart of the gravitational coupling κ, there are two new constants (λ, mu), which account for the violation of the full diffeomorphism. Original Hořava-Lifshitz gravity is given by $F(\tilde{R}) = \tilde{R}$, where the term in front of μ becomes a total derivative and is omitted. The term $L^{(3)}(g_{ij}^{(3)})$ is chosen to be $L^{(3)}(g_{ij}^{(3)}) = E^{ij}G_{ijkl}E^{kl}$ with G_{ijkl} is the generalized De Witt metric, and the expression for E_{ij} is constructed to satisfy the "detailed balance principle", being defined through variation of an action (for more details see Ref. [2]).
Here, we are interested to study solutions of a spatially flat FRW Universe, described by the metric,

$$ds^2 = -N^2dt^2 + a^2(t)\sum_{i=1}^{3}\left(dx^i\right)^2. \tag{4}$$

If we also assume the projectability condition, N can be taken to be just time-dependent and, by using the foliation-preserving diffeomorphisms, it can be fixed to be unity, $N = 1$. For the metric (4), the scalar $\tilde{R}$ is given by $\tilde{R} = \frac{3(1-3\lambda+6\mu)H^2}{N^2} + \frac{6\mu}{N}\frac{d}{dt}\left(\frac{H}{N}\right)$, and assuming the FRW metric (4), the second FRW equation can be obtained by varying the action with respect to the spatial metric $g_{ij}^{(3)}$, which yields

$$0 = F(\tilde{R}) - 2(1-3\lambda+3\mu)\left(\dot{H}+3H^2\right)F'(\tilde{R}) - 2(1-3\lambda)\dot{\tilde{R}}F''(\tilde{R})$$

$$+2\mu\left(\dot{\tilde{R}}^2F^{(3)}(\tilde{R}) + \ddot{\tilde{R}}F''(\tilde{R})\right) + \kappa^2p_m, \tag{5}$$

here $\kappa^2 = 16\pi G$, p_m is the pressure of a perfect fluid that fills the Universe, and $N = 1$. Note that this equation becomes the usual second FRW equation for convenient $F(R)$ gravity (4), by setting the constants $\lambda = \mu = 1$. Assuming the projectability condition, variation over N of the action (3) yields the first FRW equation

$$0 = F(\tilde{R}) - 6\left[(1-3\lambda+3\mu)H^2 + \mu\dot{H}\right]F'(\tilde{R}) + 6\mu H\dot{\tilde{R}}F''(\tilde{R}) - \kappa^2\rho_m, \tag{6}$$

Hence, starting from a given $F(\tilde{R})$ function, and solving Eqs. (5) and (6), a cosmological solution can be obtained.

2. ΛCDM model in F(R) Hořava-Lifshitz gravity

Let us now discuss some cosmological solutions of $F(\tilde{R})$ Hořava-Lifshitz gravity, specially the solution that reproduces the ΛCDM model. The first FRW equation, given by (6) with $C = 0$, can be rewritten as a function of the number of e-foldings $\eta = \ln\frac{a}{a_0}$, instead of the usual time t. This technique has been developed in Ref. [5] for convenient $F(R)$ gravity, where it was shown that any $F(R)$ theory can be reconstructed for a given cosmological solution. Here, we extend such formalism to the Hořava-Lifshitz $F(R)$ gravity. By using the energy conservation equation, and assuming a perfect fluid with equation of state (EoS) $p_m = w_m\rho_m$, the energy density yields

Journal of Physics: Conference Series **314** (2011) 012055 doi:10.1088/1742-6596/314/1/012055

$\rho_m = \rho_0 a^{-3(1+w_m)} = \rho_0 a_0^{-3(1+w_m)} e^{-3(1+w_m)\eta}$. And since $\frac{d}{dt} = H\frac{d}{d\eta}$ and $\frac{d^2}{d\eta^2} = H^2\frac{d^2}{d\eta^2} + H\frac{dH}{d\eta}\frac{d}{d\eta}$, the FRW eq. (6) takes the form

$$0 = F(\tilde{R}) - 6\left[(1-3\lambda+3\mu)G + \frac{\mu}{2}G'\right]\frac{dF(\tilde{R})}{d\tilde{R}} + 18\mu\left[(1-3\lambda+6\mu)GG' + \mu GG''\right]\frac{d^2F(\tilde{R})}{d^2\tilde{R}}$$
$$-\kappa^2\rho_0 a_0^{-3(1+w)} e^{-3(1+w)\eta}\,, \tag{7}$$

where the primes denote derivatives with respect to η, and we have rewritten the Hubble parameter as $G(\eta) = H^2(\eta)$. The scalar curvature is now written as $\tilde{R} = 3(1-3\lambda+6\mu)G + 3\mu G'$. Hence, for a given cosmological solution $H^2 = G(\eta)$, one can resolve Eq. (7), and the $F(\tilde{R})$ that reproduces such solution is obtained.

Let us consider the solution that reproduces the ΛCDM model. In this case the Hubble parameter takes the form,

$$H^2 = G(\eta) = H_0^2 + \frac{\kappa^2}{3}\rho_0 a^{-3} = H_0^2 + \frac{\kappa^2}{3}\rho_0 a_0^{-3} e^{-3\eta}\,. \tag{8}$$

Then, by introducing the solution (8) in the equation (7), and by performing a change of variable $x = \frac{\tilde{R} - 9\mu H_0^2}{3H_0^2(1+3(\mu-\lambda))}$, and after some algebraic manipulation, it yields the differential equation,

$$0 = x(1-x)\frac{d^2F}{dx^2} + (\gamma - (\alpha+\beta+1)\,x)\frac{dF}{dx} - \alpha\beta F - 3H_0^2(1+3(\mu-\lambda))x - 3(1-3\lambda+9\mu)H_0^2\,, \tag{9}$$

with the set of parameters (α, β, γ) being given by

$$\gamma = -\frac{1}{2}\,, \quad \alpha + \beta = \frac{1-3\lambda-\frac{3}{2}\mu}{3\mu}\,, \quad \alpha\beta = -\frac{1+3(\mu-\lambda)}{6\mu}\,. \tag{10}$$

The complete solution of Eq. (9) is a Gauss' hypergeometric function plus a linear term and a cosmological constant coming, namely

$$F(\tilde{R}) = C_1 F(\alpha, \beta, \gamma; x) + C_2 x^{1-\gamma} F(\alpha-\gamma+1, \beta-\gamma+1, 2-\gamma; x) + \frac{1}{\kappa_1}\tilde{R} - 2\Lambda\,, \tag{11}$$

where C_1 and C_2 are constants, $\kappa_1 = 3\lambda - 1$ and $\Lambda = -\frac{3H_0^2(1-3\lambda+9\mu)}{2(1-3\lambda+3\mu)}$. In this case, the solution (11) behaves similarly to the classical $F(R)$ theory, except that now the parameters of the theory depend on (λ, μ), which are allowed to vary. We can conclude that the cosmic evolution described by the ΛCDM model given by the Hubble parameter (8) is well reproduced by this class of theories. Note that for an special choice of the parameters $\mu = \lambda - \frac{1}{3}$, which decouples the scalar mode included in all $F(\tilde{R})$ models from the metric tensor (for more details see Ref. [3]), the solution reduces to $F(\tilde{R}) = \frac{1}{\kappa_1}\tilde{R} - 2\Lambda$, which is equivalent to a cosmological constant.

Nevertheless, other class of solutions for dark energy (as phantom models) can well be reproduced in this class of theories as it was showed in Ref. [4].

3. Unifying inflation with dark energy

Here we extend the class of models called viable $F(R)$ gravity, which pass the local gravity tests, and can reproduce dark energy and even inflation, to the Hořava-Lifshitz gravity. We consider the action, $F(\tilde{R}) = \tilde{R} + f(\tilde{R})$, assuming that the term $f(\tilde{R})$ becomes important at cosmological scales, while for scales compared with the Solar system one the theory becomes linear on $\tilde{R}$. As an example, we consider the following function,

$$f(\tilde{R}) = \frac{\tilde{R}^n(\alpha\tilde{R}^n - \beta)}{1 + \gamma\tilde{R}^n}\,, \tag{12}$$

where (α, β, γ) are constants and $n > 1$. This theory reproduces the inflationary and cosmic acceleration epochs in standard $F(R)$ gravity, which is also the case in the present theory, as will be shown. During inflation, it is assumed that the curvature scalar tends to infinity. In this case the model (12), behaves as $\lim_{\tilde{R}\to\infty} F(\tilde{R}) = \alpha\tilde{R}^n$. Then, by solving the FRW equation (6), this kind of function yields a power-law solution of the type

$$H(t) = \frac{h_1}{t}, \quad \text{where} \quad h_1 = \frac{2\mu(n-1)(2n-1)}{1 - 3\lambda + 6\mu - 2n(1 - 3\lambda + 3\mu)}. \tag{13}$$

This solution produces acceleration during the inflationary epoch if the parameters of the theory are properly defined. The acceleration parameter is given by $\frac{\ddot{a}}{a} = h_1(h_1 - 1)/t^2$, thus, for $h_1 > 1$ the inflationary epoch is well reproduced by the model (12). On the other hand, the function (12) has a minimum at $\tilde{R}_0$, given by

$$\tilde{R}_0 \sim \left(\frac{\beta}{\alpha\gamma}\right)^{1/4}, \qquad f'(\tilde{R}) = 0, \qquad f(\tilde{R}) = -2\Lambda \sim -\frac{\beta}{\gamma}, \tag{14}$$

where we have imposed the condition $\beta\gamma/\alpha \gg 1$. Then, at the current epoch the scalar curvature acquires a small value which can be fixed to coincide with the minimum (14), such that the FRW equations (5) and (6) yield

$$H^2 = \frac{\kappa^2}{3(3\lambda - 1)}\rho_m + \frac{2\Lambda}{3(3\lambda - 1)} \qquad \dot{H} = -\kappa^2\frac{\rho_m + p_m}{3\lambda - 1}, \tag{15}$$

which look very similar to the standard FRW equations in GR, except for the parameter λ. As at the current epoch the scalar $\tilde{R}$ is small, the theory is in the IR limit where the parameter $\lambda \sim 1$, and the equations approach the usual ones for $F(R)$ gravity. Hence, the FRW equations (15) reproduce the behavior of the well known ΛCDM model with no need to introduce a dark energy fluid to explain the current universe acceleration, as well as the inflationary epoch.

Hence, we can conclude that the extension $F(R)$ Hořava-Lifshitz gravity can well reproduce the dark energy and inflationary epochs with no need of any exotic field. It has been showed that viable models as (12) can be extended to this new class of gravity, and adjusted to reproduce the whole expansion history. However, a deeper analysis has to be performed in order to obtain more information on this class of theories (see Ref. [6]). Even the problems on the scalar graviton can be resolved by introducing a U(1) symmetry in $F(R)$ Hořava-Lifshitz gravity (see Ref. [7]).

Acknowledgements
I would like to thank Emilio Elizalde, Sergei Odintsov and Josef Klusoň for useful discussions. I acknowledge an FPI grant, project FIS2006-02842, provided by Ministry of Science (Spain).

References
[1] S. Nojiri, S. D. Odintsov, arxiv:1011.0544 [gr-qc]
[2] P. Hořava, Phys. Rev. D **79**, 084008 (2009) [arXiv:0901.3775 [hep-th]].
[3] M. Chaichian, S. Nojiri, S. D. Odintsov, M. Oksanen and A. Tureanu, Class. Quantum Grav. **27** 185021 (2010) [arXiv:1001.4102 [hep-th]].
[4] E. Elizalde, S. Nojiri, S. D. Odintsov, D. Sáez-Gómez, Eur. Phys. J. C **70** 351 (2010) [arxiv:1006.3387 [hep-th]]
[5] S. Nojiri, S. D. Odintsov and D. Saez-Gomez, Phys. Lett. B **681**, 74 (2009) [arXiv:0908.1269 [hep-th]].
[6] D. Sáez-Gómez, arxiv:1011.2090 [hep-th]
[7] J. Klusoň S. Nojiri, S. D. Odintsov, D. Sáez-Gómez, arxiv:1012.0473 [hep-th]

Journal of Physics: Conference Series **314** (2011) 012056 doi:10.1088/1742-6596/314/1/012056

$f(G)$ modified gravity and the energy conditions

Nadiezhda M. García, Francisco S. N. Lobo, José P. Mimoso

Centro de Astronomia e Astrofísica da Universidade de Lisboa,
Campo Grande, Ed. C8 1749-016 Lisboa, Portugal

E-mail: nadiezhda@cosmo.fis.fc.ul.pt, flobo@cii.fc.ul.pt, jpmimoso@cii.fc.ul.pt

Tiberiu Harko

Department of Physics and Center for Theoretical and Computational Physics,
The University of Hong Kong, Pok Fu Lam Road, Hong Kong

E-mail: harko@hkucc.hku.hk

Abstract. Motivated by string/M-theory predictions that scalar field couplings with the Gauss-Bonnet invariant, G, are essential in the appearance of non-singular early time cosmologies, we discuss the viability of an interesting alternative gravitational theory, namely, modified Gauss-Bonnet gravity, and present the viability bounds arising from the energy conditions. In particular, we consider a specific realistic form of $f(G)$ analyzed in the literature that accounts for the late-time cosmic acceleration and that has been found to cure the finite-time future singularities present in the dark energy models, and further examine the respective viability of the specific $f(G)$ model imposed by the weak energy condition.

1. Introduction

A central theme in cosmology is the perplexing fact that the Universe is undergoing an accelerating expansion [1]. Several candidates, responsible for this expansion, have been proposed in the literature, in particular, dark energy models and modified gravity. In this context, a renaissance of modified theories of gravity, such as $f(R)$ gravity [2], has recently been verified in an attempt to explain the late-time accelerated expansion of the Universe. An interesting alternative theory is modified Gauss-Bonnet gravity, or $f(G)$ gravity, where $f(G)$ is a general function of the Gauss-Bonnet term [3, 4]. Specific realistic models of $f(G)$ gravity were constructed to account for the late-time cosmic acceleration [4, 5], and it is these forms of $f(G)$ that we consider in this work. The respective constraints of the parameters of the models were also analyzed in [5]. It was shown that by taking into account higher-order curvature corrections the finite-time future singularities in $f(G)$-gravity are cured. In this context, we further consider the constraints imposed by the energy conditions and verify whether the parameter range of the specific models considered in [5] are consistent with the energy conditions [6]. More specifically, we define generalized energy conditions for $f(G)$ modified theories of gravity, and consider their realization for flat Friedmann cosmological models. In particular, we analyze whether the weak energy condition is satisfied by particular choices of $f(G)$ which were advocated in Refs. [4, 5] as leading to viable models.

2. Gravitational field equations and the energy conditions

2.1. Field equations

The action of modified Gauss-Bonnet gravity is given by

$$S = \int d^4x\sqrt{-g}\left[\frac{R}{2\kappa} + f(G)\right] + S_M(g^{\mu\nu},\psi)\,, \tag{1}$$

where $\kappa = 8\pi$. $S_M(g^{\mu\nu},\psi)$ is the matter action, in which matter is minimally coupled to the metric $g_{\mu\nu}$ and ψ collectively denotes the matter fields. The Gauss-Bonnet invariant is defined as $G \equiv R^2 - 4R_{\mu\nu}R^{\mu\nu} + R_{\mu\nu\alpha\beta}R^{\mu\nu\alpha\beta}$.

In this work, we consider the flat FRW background, so that the gravitational field equations for $f(G)$ modified gravity are provided by the following form

$$\rho_{\rm eff} = \frac{3}{\kappa^2}H^2, \qquad p_{\rm eff} = -\frac{1}{\kappa^2}\left(2\dot{H} + 3H^2\right), \tag{2}$$

where the overdot denotes a derivative with respect t; $\rho_{\rm eff}$ and $p_{\rm eff}$ are the effective energy density and pressure, respectively, defined as

$$\rho_{\rm eff} = \rho + \frac{1}{2\kappa^2}\Big[-f(G) + 24H^2\left(H^2+\dot{H}\right)f'(G) - 24^2H^4\left(2\dot{H}^2 + H\ddot{H} + 4H^2\dot{H}\right)f''(G)\Big] \tag{3}$$

$$\begin{aligned} p_{\rm eff} &= p + \frac{1}{2\kappa^2}\Big\{f(G) - 24H^2\left(H^2+\dot{H}\right)f'(G) + (24)8H^2\Big[6\dot{H}^3 + 8H\dot{H}\ddot{H} + 24\dot{H}^2H^2 \\ &\quad + 6H^3\ddot{H} + 8H^4\dot{H} + H^2\dddot{H}\Big]f''(G) + 8(24)^2H^4\left(2\dot{H}^2 + H\ddot{H} + 4H^2\dot{H}\right)^2 f'''(G)\Big\}\,. \end{aligned} \tag{4}$$

For the flat FRW metric the curvature scalar and the Gauss-Bonnet invariant are given by $R = 6(2H^2 + \dot{H})$ and $G = 24H^2(H^2 + \dot{H})$, respectively.

2.2. Energy conditions

The energy conditions arise when one refers to the Raychaudhuri equation for the expansion. Note that the Raychaudhuri equation is a purely geometric statement, and as such it makes no reference to any gravitational field equations. The condition for attractive gravity reduces to $R_{\mu\nu}k^\mu k^\nu \geq 0$ and in general relativity, through the Einstein field equation, one can write the above condition in terms of the stress-energy tensor given by $T_{\mu\nu}k^\mu k^\nu \geq 0$. Now, in $f(G)$ modified theories of gravity the field equation can be written as the following effective gravitational field equation $G_{\mu\nu} \equiv R_{\mu\nu} - \frac{1}{2}R\,g_{\mu\nu} = T^{\rm eff}_{\mu\nu}$. Thus, the positivity condition, $R_{\mu\nu}k^\mu k^\nu \geq 0$, provides the following form for the null energy condition, $T^{\rm eff}_{\mu\nu}k^\mu k^\nu \geq 0$, through the modified gravitational field equation. As the Raychaudhuri equation holds for any geometrical theory of gravitation, we will maintain its physical motivation, namely, the focussing of geodesic congruences, along with the attractive character of gravity to deduce the energy conditions in the context of $f(G)$ modified gravity.

Thus, using the modified (effective) gravitational field equations the null energy condition (NEC), weak energy condition (WEC), strong energy condition (SEC) and the dominant energy condition (DEC) are given by

$$\text{NEC} \Longleftrightarrow \rho_{\rm eff} + p_{\rm eff} \geq 0\,, \tag{5}$$

$$\text{WEC} \Longleftrightarrow \rho_{\rm eff} \geq 0 \text{ and } \rho_{\rm eff} + p_{\rm eff} \geq 0\,, \tag{6}$$

$$\text{SEC} \Longleftrightarrow \rho_{\rm eff} + 3p_{\rm eff} \geq 0 \text{ and } \rho_{\rm eff} + p_{\rm eff} \geq 0\,, \tag{7}$$

$$\text{DEC} \Longleftrightarrow \rho_{\rm eff} \geq 0 \text{ and } \rho_{\rm eff} \pm p_{\rm eff} \geq 0\,, \tag{8}$$

respectively.

In standard mechanics terminology the first four time derivatives of position are referred to as velocity, acceleration, jerk and snap. In a cosmological setting, in addition to the Hubble parameter $H = \dot{a}/a$, it is appropriate to define the deceleration, jerk, and snap parameters as

$$q = -\frac{1}{H^2}\frac{\ddot{a}}{a}\,, \qquad j = \frac{1}{H^3}\frac{\dddot{a}}{a}\,, \qquad \text{and} \quad s = \frac{1}{H^4}\frac{\ddddot{a}}{a}\,. \tag{9}$$

In terms of the above parameters, we consider the following definitions

$$\dot{H} = -H^2(1+q), \qquad \ddot{H} = H^3(j+3q+2), \qquad \dddot{H} = H^4(s-2j-5q-3), \tag{10}$$

respectively.

Using these definitions, the weak energy condition takes the following form:

$$\begin{aligned} \rho_{\text{eff}} + p_{\text{eff}} &= \rho + p + \frac{96}{k^2}\Big\{ -(6q^3+27q^2+21q+8qj+9j-s)f''(G) \\ &\quad +24[4(q^2+2q+1)H^2+2q^2+7q+j+4]f'''(G)\Big\}H^8 \geq 0. \qquad (11) \\ \rho_{\text{eff}} &= \rho + \frac{1}{2k^2}\Big[-f(G) - 24H^4 q f'(G) - (24)^2 H^8(2q^2+3q+j)f''(G)\Big] \geq 0. \end{aligned}$$

3. Specific $f(G)$ models

Here, we consider a realistic model of $f(G)$ gravity, which have been found to reproduce the current acceleration [4]:

$$f(G) = \frac{a_1 G^n + b_1}{a_2 G^n + b_2}\,, \tag{12}$$

where a_1, a_2, b_1, b_2 and n are constants. In the following, we always assume $n > 0$.

In addition to accounting for the late-time cosmic acceleration, it was also found that the specific case of Eq. (12) analyzed in [5] cured the four types of finite-time future singularities emerging in the late-time accelerating era. Rather than exhaustively analyze all of the cases, we consider a specific case that does indeed prove that in addition to curing the finite-time future singularities it satisfies the weak energy condition.

For simplicity, in the example analyzed we consider vacuum, i.e., $\rho = p = 0$. The WEC constraints, i.e., $\rho_{\text{eff}} \geq 0$ and $\rho_{\text{eff}} + p_{\text{eff}} \geq 0$, are respectively given by

$$\begin{aligned} &-[a_1(-24qH^4)^n + b_1][a_2(-24qH^4)^n + b_2] + n(-24qH^4)^n(a_1b_2 - a_2b_1) \\ &+(24)^2 nH^8(-24qH^4)^{n-2}[a_2(n+1)(-24qH^4)^n + b_2(1-n)]\frac{(a_1b_2-a_2b_1)(2q^2+3q+j)}{a_2(-24qH^4)^n+b_2} \geq 0 \end{aligned} \tag{13}$$

$$\begin{aligned} &n(-24qH^4)^n(-b_1a_2+a_1b_2)\{(6q^3+27q^2+21q+8qj+9j-s)[n(a_2^2(-24qH^4)^{2n}-b_2^2) \\ &+(a_2(-24qH^4)^n+b_2)^2] + (4H^2+8H^2q+4H^2q^2+2q^2+7q+j+4)[4a_2b_2(1-n^2)(-24qH^4)^n \\ &+a_2^2(n^2+3n+2)(-24qH^4)^{2n} + b_2^2(n^2-3n+2)]q^{-1}H^{-4}\} \geq 0\,. \end{aligned} \tag{14}$$

The constraints provided by the inequalities (13)-(14) are too complicated to find exact analytical expressions for the parameter ranges of the constants a_1, a_2, b_1, b_2, and n, so we consider specific values for some of the parameters. In particular, we impose the following values $a_1 = -1$, $b_1 = -1$, and $a_2 = 2$, and plot the WEC as a function of b_2 and n, which is depicted in Fig. 1. The latter does indeed prove that the specific form of $f(G)$ given by Eq. (12) considered in [5] is consistent with the WEC inequalities.

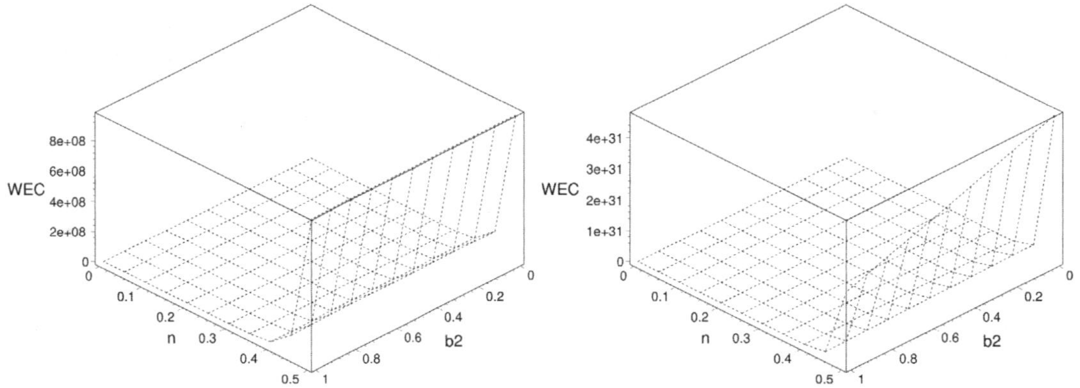

Figure 1. The plots depict the weak energy condition for the specific form of $f(G) = \frac{a_1 G^n + b_1}{a_2 G^n + b_2}$. The left plot corresponds to $\rho_{\rm eff} \geq 0$; the right plot corresponds to $\rho_{\rm eff} + p_{\rm eff} \geq 0$. We have considered the values $a_1 = -1$, $b_1 = -1$, and $a_2 = 2$. The plots show that the weak energy condition are satisfied for the parameter range considered. See the text for details.

4. Conclusion

In this work, we discussed the viability of an interesting alternative gravitational theory, namely, modified Gauss-Bonnet gravity or $f(G)$ gravity. We considered a specific realistic form of $f(G)$ analyzed in the literature that accounts for the late-time cosmic acceleration and that cured the finite-time future singularities [4, 5]. The general inequalities imposed by the energy conditions were outlined and using the recent estimated values of the Hubble, deceleration, jerk and snap parameters we have shown the viability of the above-mentioned forms of $f(G)$ imposed by the weak energy condition.

References

[1] Perlmutter, S. *et al.* 1999, Astrophys. J., 517, 565; Riess, A. G. *et al.* 1998, Astron. J., 116, 1009.

[2] Carroll, S.M., Duvvuri, V., Trodden, M., and Turner, M.S. 2004, Phys. Rev. D, 70, 043528; Copeland, E.J., Sami, M and Tsujikawa, S. Int. J. Mod. Phys. D **15**, 1753 (2006); Nojiri, S and Odintsov, S.D, Int. J. Geom. Meth. Mod. Phys. **4**, 115 (2007); Lobo, F.S.N. arXiv:0807.1640 [gr-qc]; De Felice, A. and Tsujikawa, S., Living Rev. Rel. **13**, 3 (2010); Nojiri, S. and Odintsov, S.D., arXiv:1011.0544 [gr-qc]; Bertolami, O. *et al*, Phys. Rev. D **75**, 104016 (2007); Boehmer, C.G., *et al*, Phys. Rev. D **76**, 084005 (2007); Boehmer, C.G., *et al*, Astropart. Phys. **29**, 386 (2008); Boehmer, C.G., *et al*, JCAP **0803**, 024 (2008); Bertolami, O., *et al*, Phys. Rev. D **78**, 064036 (2008); Harko, T. and Lobo, F.S.N., arXiv:1007.4415 [gr-qc]; Harko, T. and Lobo, F.S.N., Eur. Phys. J. C **70**, 373 (2010).

[3] Cognola, G., Gastaldi, M. and Zerbini, S., Int. J. Theor. Phys. **47**, 898 (2008); Nojiri, S. and Odintsov, S.D., Phys. Lett. B **631**, 1 (2005); Nojiri, S. and Odintsov, S.D., Int. J. Geom. Meth. Mod. Phys. **4**, 115 (2007); Cognola, G., Elizalde, E., Nojiri, S. and Odintsov, S.D., and Zerbini, S., Phys. Rev. D **73**, 084007 (2006); Cognola, G., Elizalde, E., Nojiri, S. and Odintsov, S.D., and Zerbini, S., Phys. Rev. D **75**, 086002 (2007); Li, B., Barrow, J.D., and Mota, D.F., Phys. Rev. D **76**, 044027 (2007); De Felice, A., and Tsujikawa, S., Phys. Lett. B **675**, 1 (2009); Alimohammadi, M., and Ghalee, A., Phys. Rev. D **79**, 063006 (2009); Boehmer, C.G., and Lobo, F.S.N., Phys. Rev. D **79**, 067504 (2009); Zhou, S.Y., Copeland, E.J., and Saffin, P.M., JCAP **0907**, 009 (2009); Goheer, N., Goswami, R., Dunsby, P.K.S., and Ananda, K., Phys. Rev. D **79**, 121301 (2009); Sadeghi, J., Setare, M.R., and Banijamali, A., Phys. Lett. B **679**, 302 (2009); De Felice, A., and Tsujikawa, S., arXiv:0907.1830 [hep-th]; Boehmer, C.G., Hollenstein, L., Lobo, F.S.N., and Seahra, S.S., arXiv:1001.1266 [gr-qc].

[4] Nojiri, S., Odintsov, S.D., and Tretyakov, P.V., Prog. Theor. Phys. Suppl. **172**, 81 (2008).

[5] Bamba, K., Odintsov, S.D., Sebastiani, L., and Zerbini, S., [arXiv:0911.4390 [hep-th]].

[6] Garcia, N.M., Harko, T., Lobo, F.S.N., and Mimoso, J.P., arXiv:1011.4159 [gr-qc].

Spanish Relativity Meeting (ERE 2010): Gravity as a Crossroad in Physics IOP Publishing
Journal of Physics: Conference Series **314** (2011) 012057 doi:10.1088/1742-6596/314/1/012057

Unified models of the cosmological dark sector

W Zimdahl[1], **W S Hipólito-Ricaldi**[2] **and H E S Velten**[1]

[1] Universidade Federal do Espírito Santo, Departamento de Física, Av. Fernando Ferrari, 514, Campus de Goiabeiras, CEP 29075-910, Vitória, Espírito Santo, Brazil

[2] Universidade Federal do Espírito Santo, Departamento de Ciências Matemáticas e Naturais, CEUNES Rodovia BR 101 Norte, km. 60, CEP 29932-540, São Mateus, Espírito Santo, Brazil

E-mail: winfried.zimdahl@pq.cnpq.br;hipolito@ceunes.ufes.br;velten@cce.ufes.br

Abstract. We model the cosmological substratum by a viscous fluid that is supposed to provide a unified description of the dark sector and pressureless baryonic matter. In the homogeneous and isotropic background the *total* energy density of this mixture behaves as a generalized Chaplygin gas. The perturbations of this energy density are intrinsically non-adiabatic and source relative entropy perturbations. The resulting baryonic matter power spectrum is shown to be compatible with the 2dFGRS and SDSS (DR7) data. A joint statistical analysis, using also Hubble-function and supernovae Ia data, shows that, different from other studies, there exists a maximum in the probability distribution for a negative present value $q_0 \approx -0.53$ of the deceleration parameter. Moreover, different from other approaches, the unified model presented here favors a matter content that is of the order of the baryonic matter abundance suggested by big-bang nucleosynthesis.

1. Introduction

According to the prevailing interpretation, our Universe is dynamically dominated by a cosmological constant Λ (or a dynamical equivalent, called dark energy (DE)) which contributes more than 70% to the total cosmic energy budget. More than 20% are contributed by cold dark matter (CDM) and only about 5% are in the form of conventional, baryonic matter. Because of the cosmological constant problem in its different facets, including the coincidence problem, a great deal of work was devoted to alternative approaches in which a similar dynamics as that of the ΛCDM model is reproduced with a time varying cosmological term, i.e., the cosmological constant is replaced by a dynamical quantity. Both dark matter (DM) and DE manifest themselves so far only through their gravitational interaction. This provides a motivation for approaches in which DM and DE appear as different manifestation of one single dark-sector component. The Chaplygin-gas model and its different generalizations realize this idea. Unified models of the dark sector of this type are attractive since one and the same component behaves as pressureless matter at high redshifts and as a cosmological constant in the long time limit. While the homogeneous and isotropic background dynamics for the (generalized) Chaplygin gas (GCG) is well compatible with the data, the study of the perturbation dynamics resulted in problems which apparently ruled out all Chaplygin-gas type models except those that are observationally almost indistinguishable from the ΛCDM model. To circumvent this problem, nonadiabatic perturbations were postulated and designed in a way to make the effective sound speed vanish. But this amounts to an ad hoc procedure which leaves open the physical origin of nonadiabatic perturbations. There exists, however, a different type of unified models of

Journal of Physics: Conference Series **314** (2011) 012057 doi:10.1088/1742-6596/314/1/012057

the dark sector, namely viscous cosmological models, which are intrinsically nonadiabatic [1]. In the homogeneous and isotropic background a one-component viscous fluid shares the same dynamics as a GCG. Now, what is observed in the redshift surveys is not the spectrum of the dark-matter distribution but the baryonic matter spectrum. Including a baryon component into the perturbation dynamics for a universe with a Chaplygin-gas dark sector, there appears the new problem that the unified Chaplygin-gas scenario itself is disfavored by the data. It is only if the unified scenario with a fixed pressureless (supposedly) baryonic matter fraction of about 0.043 (according to the results from WMAP and primordial nucleosynthesis) is *imposed* on the dynamics, that consistency with the data is obtained [2]. If the pressureless matter fraction is left free, its best-fit value is much larger than the baryonic fraction. In fact it becomes even close to unity, leaving only a small percentage for the Chaplygin gas, thus invalidating the entire scenario. In other words, a Chaplygin-gas-based unified model of the dark sector is difficult to reconcile with observations. One may ask now, whether the status of unified models can again be remedied by replacing the Chaplygin gas by a viscous fluid. It is exactly this question that we are going to investigate in the present paper. It is our purpose to study cosmological perturbations for a two-component model of baryons and a viscous fluid, where the latter represents a one-component description of the dark sector [3]. We shall show that such type of unified model is not only consistent for a fixed fraction of the baryons but also for the case that the matter fraction is left free. Our analysis demonstrates that the statistically preferred value for the abundance of pressureless matter is compatible with the mentioned baryon fraction 0.043 that follows from the synthesis of light elements.

2. The two-component model

The cosmic medium is assumed to be describable by an energy-momentum tensor T^{ik} which splits into a matter part T^{ik}_M and viscous fluid part T^{ik}_V,

$$T^{ik} = \rho u^i u^k + p\left(g^{ik} + u^i u^k\right) , \qquad T^{ik} = T^{ik}_M + T^{ik}_V , \tag{1}$$

with

$$T^{ik}_M = \rho_M u^i_M u^k_M + p_M\left(g^{ik} + u^i_M u^k_M\right) , \qquad T^{ik}_V = \rho_V u^i_V u^k_V + p_V\left(g^{ik} + u^i_V u^k_V\right) , \tag{2}$$

where the subscript "M" stands for matter and the subscript "V" stands for viscous. The total cosmic fluid is characterized by a four velocity u^m while u^i_M represents the four velocity of the matter part and u^i_V represents the four velocity of the viscous fluid. Energy-momentum conservation is supposed to hold separately for each of the components, i.e., $T^{ik}_{M\,;i} = T^{ik}_{V\,;i} = 0$. Let the matter be pressureless, i.e. $p_M = 0$ and the viscous fluid be characterized by a bulk viscous pressure $p_V = p = -\zeta\Theta$, where $\zeta =$ const and $\Theta = u^i_{;i}$ is the fluid expansion. Under this condition the total pressure coincides with the pressure of the viscous component. In terms of the present value q_0 of the deceleration parameter the Hubble rate H can be written as [1]

$$H = \frac{H_0}{3}\left[1 - 2q_0 + 2\left(1 + q_0\right) a^{-\frac{3}{2}}\right] , \tag{3}$$

where H_0 is the present value of H. The expression (3) coincides with the Hubble rate of a specific ($\alpha = -1/2$) GCG. Since $\rho_M = \rho_{M0} a^{-3}$, we have $\rho_V = \rho - \rho_{M0} a^{-3}$. It is the total energy density that behaves as a GCG, not the component V. This type of unified model differs from unified models in which the total energy density is the sum of a GCG and a baryon component. Only if the baryon component is ignored, both descriptions coincide. Consequently, in the homogeneous and isotropic background, a generalized Chaplygin gas with $\alpha = -1/2$ can be seen as a unified description of the cosmic medium, consisting of a separately conserved matter component and a bulk viscous fluid with $\zeta =$ const, where the latter itself represents a unified model of the dark sector.

3. Perturbations

The nonadiabaticity of the system as a whole is characterized by

$$\frac{\hat{p}}{\rho+p} - \frac{\dot{p}}{\dot{\rho}}\frac{\hat{\rho}}{\rho+p} = 3H\frac{\dot{p}}{\dot{\rho}}\left(\frac{\hat{\rho}}{\dot{\rho}} - \frac{\hat{\Theta}}{\dot{\Theta}}\right) , \tag{4}$$

where a dot represents a derivative with respect to the cosmic time and a caret denotes a first-order perturbation variable. Quantities without a caret are background quantities. The combination (4) is governed by the dynamics of the total energy-density perturbation $\hat{\rho}$ and by the perturbations $\hat{\Theta}$ of the expansion scalar, which is also a quantity that characterizes the system as a whole. The behavior of these quantities is described by the energy-momentum conservation for the entire system and by the Raychaudhuri equation, respectively. Both of these equations are coupled to each other. The remarkable point is that these quantities and, consequently, the total energy density perturbation, are independent of the two-component structure of the medium. The reason is the direct relation $\hat{p} = -\zeta\hat{\Theta}$ between the pressure perturbations and the perturbations of the expansion scalar. It is convenient to describe the perturbation dynamics in terms of gauge invariant quantities which represent perturbations on comoving hypersurfaces, indicated by a superscript c. These are defined as (v is the velocity potential, defined by $\hat{u}_\mu = v_{,\mu}$)

$$\frac{\hat{\rho}^c}{\dot{\rho}} \equiv \frac{\hat{\rho}}{\dot{\rho}} + v \ , \qquad \delta \equiv \frac{\hat{\rho}^c}{\rho} . \tag{5}$$

Then, in linear order,

$$\delta'' + f(a)\,\delta' + g(a)\,\delta = 0 \ , \tag{6}$$

where $\delta' \equiv \frac{d\delta}{da}$ and the coefficients $f(a)$ and $g(a)$ are given by explicitly known background quantities [1, 3]. The relative entropy perturbations are defined by

$$S_{MV} \equiv \frac{\hat{\rho}_M}{\rho_M} - \frac{\hat{\rho}_V}{\rho_V + p_V} \tag{7}$$

and obey the equation

$$S''_{VM} + r(a)S'_{VM} + s(a)S_{VM} = c(a)\delta' + d(a)\delta \ . \tag{8}$$

Again, the coefficients $r(a)$, $s(a)$, $c(a)$ and $d(a)$ are combinations of analytically given background variables [3]. The quantity relevant for the observations is the fractional perturbation $\delta_M \equiv \frac{\hat{\rho}^c_M}{\rho_M}$ of the energy density of the baryons, given by

$$\delta_M = \frac{1}{\gamma}\left[\delta - \frac{\rho_V + p}{\rho}S_{VM}\right] . \tag{9}$$

At early times, i.e. for small scale factors $a \ll 1$, the equation (6) has the asymptotic form $\delta'' + \frac{3}{2a}\delta' - \frac{3}{2a^2}\delta = 0$ independent of q_0 and for all scales. Eq. (8) then reduces to $S''_{VM} + \frac{3}{2a}S'_{VM} = 0$ with the solution $S_{VM} = \text{const} = 0$. Consequently, there are neither nonadiabatic contributions to the total energy-density fluctuations nor relative entropy perturbations and we have purely adiabatic perturbations $\delta_M = \delta$ at $a \ll 1$. This allows us to match the initial conditions for equations (6) and (8) with those of the ΛCDM model at early times.

The results of our statistical analysis are shown in Fig. 1.

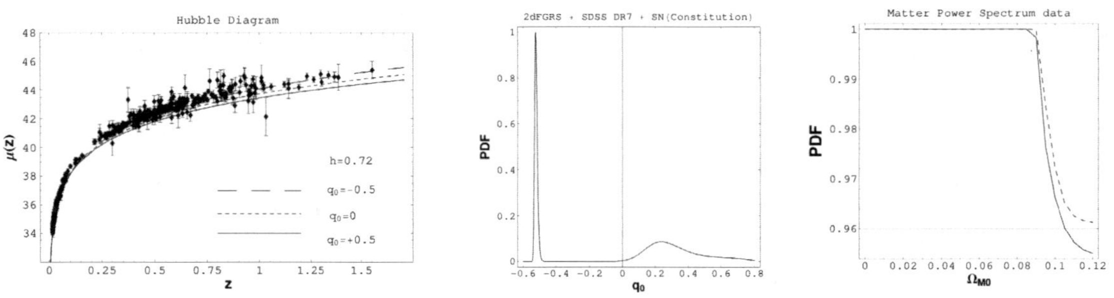

Figure 1. Left panel: Hubble diagram, center panel: probability distribution function (PDF) for q_0, based on a joint analysis of matter power spectrum and SN Ia data. The right panel shows the PDF for the pressureless component. For more details see [3].

4. Discussion and Conclusions

The approach presented here is based on the fact that the two-component system of a bulk viscous fluid and a separately conserved baryon component behaves in the background as a generalized Chaplygin gas with $\alpha = -\frac{1}{2}$. While the baryon component may be considered dynamically negligible in the background, the situation is different on the perturbative level, since the observed matter agglomerations are related to baryonic density fluctuations. These fluctuations are obtained from a combination of the nonadiabatic total energy density perturbations and relative entropy perturbations in the two-component system where the former source the latter. The probability distribution for the deceleration parameter has a maximum at $q_0 \approx -0.53$ which partially removes the degeneracy of previous studies which, taken at face value, were incompatible with an accelerated expansion and thus in obvious tension with results for the background. Perhaps still more important is the test of the unified model itself. Many investigations on approaches with a unified dark sector fix the pressureless matter component to be that of the favored (by the WMAP data) baryon fraction and then check whether or not the resulting dynamics can reproduce the observations. But this does not say anything on how probable the division of the total cosmic substratum into roughly 96% of a dark substance and roughly 4% of pressureless matter is. To decide this question, one has to consider the pressureless matter fraction as a free parameter and to find out which abundance is actually favored by the data. Our analysis revealed that the matter fraction probability is indeed highest for values smaller than roughly 8%. This is a result in favor of the unified viscous model. We recall that a corresponding analysis for a Chaplygin gas results in values close to unity [2] which seems to rule out such type of approaches. The present viscous model, on the other hand, remains an option for a unified description of the dark sector, at least as far as the matter power spectrum is concerned.

Acknowledgments

Support by CNPq and FAPES is gratefully acknowledged.

References

[1] Hipólito-Ricaldi W S, Velten H E S and Zimdahl W 2009 *JCAP* **06** 016
[2] Fabris J C, Velten H E S and Zimdahl W 2010 *Phys. Rev. D* **81**, 087303
[3] Hipólito-Ricaldi W S, Velten H E S and Zimdahl W 2010 *Phys. Rev. D* **82** 063507

Journal of Physics: Conference Series **314** (2011) 012058 doi:10.1088/1742-6596/314/1/012058

Holographic dark energy at the Ricci scale

Iván Durán and Diego Pavón

Departament de Física, Universitat Autònoma de Barcelona, 08193 Bellaterra, Spain

E-mail: ivan.duran@uab.cat, diego.pavon@uab.es

Abstract. We consider a holographic cosmological model in which the infrared cutoff is fixed by the Ricci's length and dark matter and dark energy do not evolve separately but interact non-gravitationally with one another. This substantially alleviates the cosmic coincidence problem as the ratio between both components remains finite throughout the expansion. We constrain the model with observational data from supernovae, cosmic background radiation, baryon acoustic oscillations, gas mass fraction in galaxy clusters, the history of the Hubble function, and the growth function. The model shows consistency with observation.

1. Introduction

Holographic models of late acceleration have become of fashion, among other things, because they link the dark energy density to the cosmic horizon, a global property of the universe, and have a close relationship to the spacetime foam [1], [2]. For a summary motivation of holographic dark energy, see section 3 in [3]. The cosmological model we present here assumes a spatially flat homogeneous and isotropic universe dominated by dark matter (DM) and dark energy (DE) (subscripts M and X, respectively), the latter obeying the holographic relationship

$$\rho_X = \frac{3M_P^2\, c^2}{L^2}, \tag{1}$$

where c^2 is a dimensionless parameter that we will take as constant -though strictly speaking it may slowly vary with time [4]-, and $L = (\dot{H} + 2H^2)^{-1/2}$ denotes the Ricci's lenght. We adopt the latter because it corresponds to the size of the maximal perturbation leading to the formation of a black hole [5]. For models that identify L with Hubble's length, H^{-1}, see [6] and references therein.

The other main assumption is that DE and DM interact non-gravitationally with each other according to

$$\dot{\Omega}_M - \left(1 - \frac{2\Omega_X}{c^2}\right)(1-\Omega_X)\, H \;=\; QH\,, \tag{2}$$

$$\dot{\Omega}_X + \left(1 - \frac{2\Omega_X}{c^2}\right)(1-\Omega_X)\, H \;=\; -QH\,. \tag{3}$$

Here Ω_M and Ω_X stand for the fractional densities of the components, and the interaction term

$$Q = -\frac{r_f}{(1+r_f)^2}\left(1 + r_f - \frac{2}{c^2}\right), \tag{4}$$

Published under licence by IOP Publishing Ltd

Journal of Physics: Conference Series **314** (2011) 012058 doi:10.1088/1742-6596/314/1/012058

is such that ratio $r \equiv \Omega_M/\Omega_X$ evolves from a constant value at early times to a final, finite, value -denoted as r_f- at late times (see [7] for details and Fig. 1). This clearly alleviates the coincidence problem (namely, "why are the densities of matter and dark energy of the same order precisely today?" [8]), something beyond the reach of the ΛCDM model. A consequence of the model is that the equation of state of dark energy $w \equiv p_X/\rho_X$ varies with expansion as shown in Fig. 2.

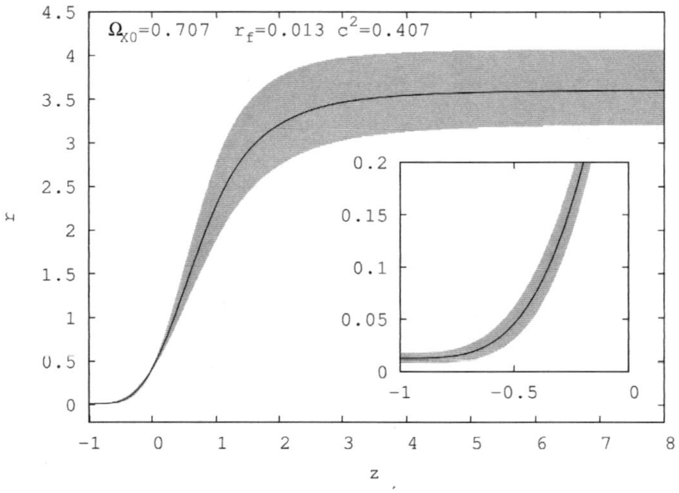

Figure 1. Plot of the ratio r between the energy densities vs. redshift for the best-fit model. As the inset shows, $r_f \equiv r(z)$ does not vanish when $z \to -1$. In this, and the next figure, the red swath indicates the region obtained by including the 1σ uncertainties of the constrained parameters used in the calculation.

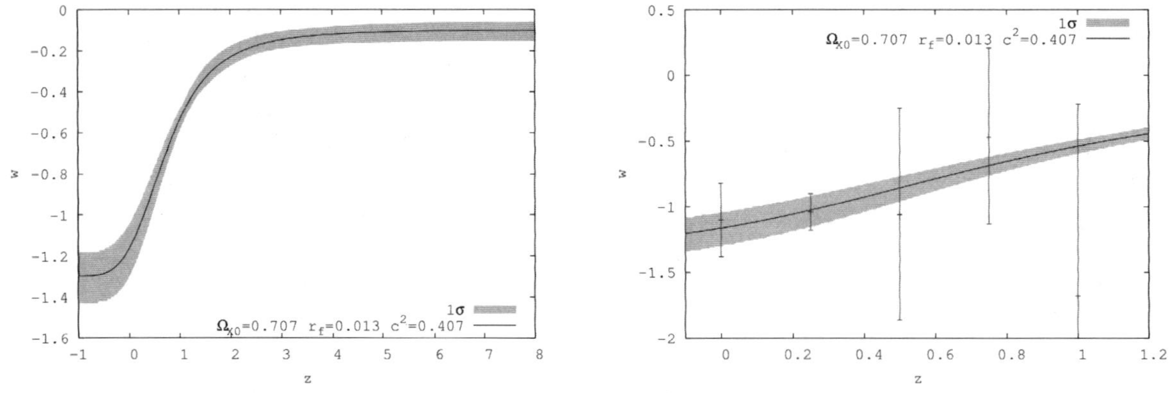

Figure 2. The equation of state parameter vs. redshift up to $z = 8$ (left panel), and up to $z = 1.2$ only (right panel) for the best fit holographic model. At high redshifts w approaches the equation of state of non-relativistic matter and at low redshifts it does not depart significantly from -1. The observational data are taken from [9]. The error bars indicate a 2σ uncertainty.

Table 1. χ^2 values of the best-fit holographic model and the best fit ΛCDM model.

Model	χ^2_{sn}	χ^2_{cmb}	χ^2_{bao}	χ^2_{x-rays}	χ^2_H	χ^2_{gf}	$\chi^2_{\rm total}$	$\chi^2_{\rm total}/dof$
Holographic	543.70	0.01	1.20	41.79	9.57	1.06	597.34	0.96
ΛCDM	542.87	0.05	1.13	41.59	8.73	0.43	594.80	0.96

2. Contrasting the model with observation

The model has four free parameters (Ω_{X0}, c^2, r_f, and H_0) which we constrained with observational data from supernovae (557 data points [10]), cosmic background radiation [11], baryon acoustic oscillations (at $z = 0.35$ [12] and $z = 0.2$ [13]), gas mass fraction in galaxy clusters (42 data points from x-ray measurements [14]), the Hubble function history (15 data points [15], [16], [17], and [18]), and the growth function (5 data points from Table 2 in [19]). (See [7] for details).

Data from the growth function -which correspond to matter perturbations within the horizon- are particularly relevant because they can more easily discriminate between cosmological models with a similar history of the Hubble function. For the model under consideration the growth function of matter, $f \equiv d\ln\delta_M/d\ln a$, is governed by

$$f' + f^2 + \left(\frac{1}{c^2(1+r)} + Q\frac{1+r}{r}\right) f - \frac{3r^3 + 2Q^2(1+r)^3 - 2Qr(2+r-r^2)}{2r^2(1+r)} = 0\,. \qquad (5)$$

In the non-interacting limit, $Q \to 0$, this expression reduces to the corresponding one of the Einstein-de Sitter cosmology, namely: $f' + f^2 + [2 + (\dot{H}/H^2)]f = 3\Omega_M/2$.

The best-fit values of the free parameters are found to be: $\Omega_{X0} = 0.707 \pm 0.009$, $c^2 = 0.407^{+0.033}_{-0.028}$, $r_f = 0.013^{+0.006}_{-0.005}$, and $H_0 = 71.8 \pm 2.9$ km/s/Mpc. This together Table 1 and figures 3 and 4 summarizes our findings. It is worthy of note that the non-interacting case, $Q = 0$ (which implies $r_f = 0$ via Eq. (4)), lies over 2σ away from the best fit value. This seems to be a generic feature of holographic models.

Table 1 shows the partial, total, and total χ^2 over the number of degrees of freedom of the holographic model along with the corresponding values for the ΛCDM model. The latter has just two free parameters, Ω_{M0} and H_0. Their best-fit values after constraining the model to the same sets of data are $\Omega_{M0} = 0.266 \pm 0.006$, and $H_0 = 71.8 \pm 1.9$ km/s/Mpc.

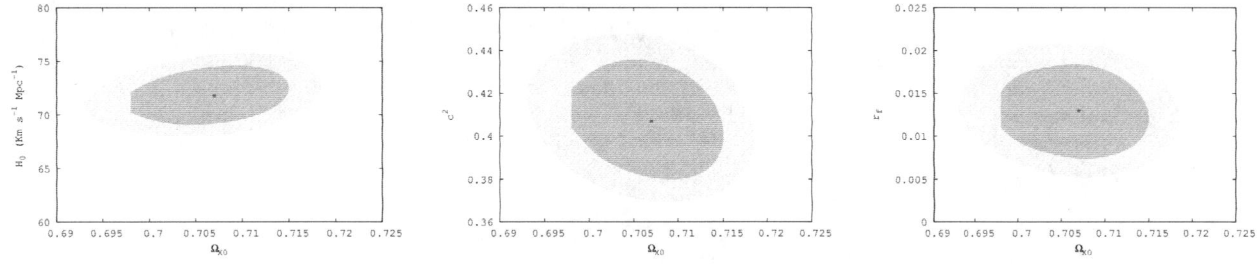

Figure 3. Panels from left to right show the 68.3% and 95.4% confidence contours for the pairs of free parameters (Ω_{X0}, H_0), (Ω_{X0}, c^2), (Ω_{X0}, r_f), respectively, obtained by constraining the holographic model with SN Ia+CMB-shift+ BAO+x-ray+H(z)+growth function data. The solid point in each panel locates the best fit values.

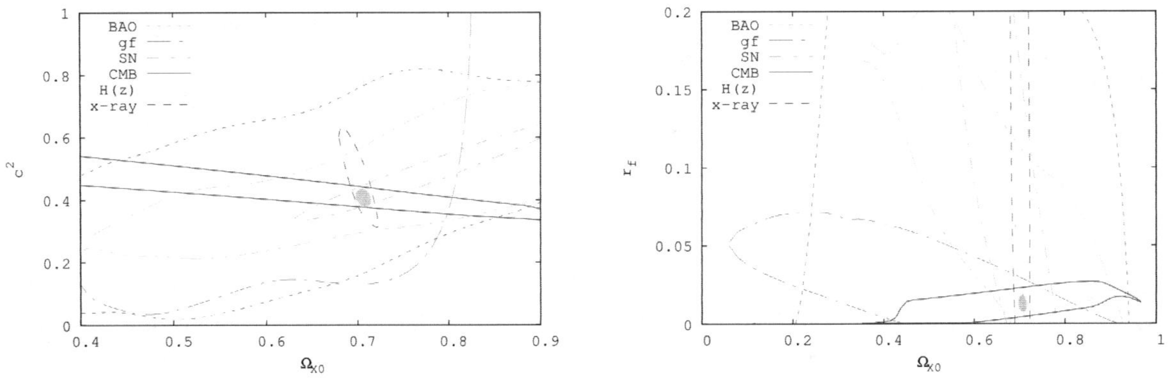

Figure 4. Left panel: Probability contours for SN Ia, CMB, BAO, x-ray, $H(z)$, and growth function, in the (Ω_{X0}, c^2) plane. The joined constraint uses $\chi^2_{\rm total} = \chi^2_{sn} + \chi^2_{cmb} + \chi^2_{bao} + \chi^2_{x-rays} + \chi^2_{Hubble} + \chi^2_{gf}$. Right panel: Idem for the (Ω_{X0}, r_f) plane.

3. Conclusions

The statistical analysis sketched above shows that the ratio χ^2_{total}/dof is lower than unity for the holographic interacting model of section 1, whereby it is compatible with observation. However, as Table 1 shows, the ΛCDM model fits better the same sets of data. Yet, the latter cannot explain the cosmic coincidence problem while the former can.

Acknowledgments

ID was funded by the "Universidad Autónoma de Barcelona" through a PIF fellowship. This research was partly supported by the Spanish Ministry of Science and Innovation under Grant FIS2009-13370-C02-01, and the "Direcció de Recerca de la Generalitat" under Grant 2009SGR-00164.

References

[1] Ng YJ 2001 *Phys. Rev. Lett.* **86** 2946
[2] Arzano M, Kephart TW and Ng YJ 2007 *Phys. Lett.* B **649** 243
[3] Zimdahl W and Pavón D 2007 *Class. Quantum Grav.* **24** 5461
[4] Radicella N and Pavón D 2010 *J. Cosmol. Astropart. Phys.* JCAP10(2010)005
[5] Brustein R 2008 "Cosmological Entropy Bounds" in *String Theories and Fundamental Interactions* (Lecture Notes in Physics vol 737) eds M Gasperini and J Maharana (Heidelberg: Springer) pp 619-659
[6] Durán I, Pavón D and Zimdahl W 2010 *J. Cosmol. Astropart. Phys.* JCAP07(2010)018
[7] Durán I and Pavón D 2011 *Phys. Rev.* D, in the press (*Preprint* arXiv:1012.2986 [astro-ph.CO])
[8] Steinhardt PJ 1997 "Cosmological Challenges for the 21st Century" in *Critical Problems in Physics* eds VL Fitch *at al* (Princeton: Princeton University Press) pp 123-146
[9] Serra P *et al* 2009 *Phys. Rev.* D **80** 121302
[10] Amanullah R *et al* 2010 *Astrophys. J.* (in the press) *Preprint* arXiv:1004.1711
[11] Komatsu E *et al* 2009 *Astrophys. J. Suppl.* **180** 330
[12] Eisenstein DJ *et al* 2005 *Astrophys. J.* **633** 560
[13] Percival WJ *et al* 2010 *Mon. Not. R. Astron. Soc.* **401** 2148
[14] Allen SW *et al* 2008 *Mon. Not. R. Astron. Soc.* **383** 879
[15] Riess AG *et al* 2009 *Astrophys. J.* **699** 539
[16] Gaztañaga E, Cabré A, and Hui L 2009 *Mon. Not. R. Astron. Soc.* **399** 1663
[17] Simon J, Verde L and Jiménez R 2005 *Phys. Rev.* D **71** 123001
[18] Stern D, Jiménez R, Verde L, Kamionkowski M and Stanford SA 2010 *J. Cosmol. Astropart. Phys.* JCAP02(2010)008
[19] Gong Y 2008 *Phys. Rev.* D **78** 123010

The parameter of the dark energy equation of state for high redshifts.

Ariadna Montiel and Nora Bretón.
Physics Dept., Centro de Investigación y de Estudios Avanzados del I.P.N. (México City)
E-mail: amontiel@fis.cinvestav.mx, nora@fis.cinvestav.mx

Abstract. We study the behavior of the parameter $w(z)$ of the dark-energy equation of state, $p_x = w(z)\rho_x$, as function of the redshift data from GRBs, to check its deviations from its most accepted value of -1. To this end we first find a reasonable calibration for the GRB in order to extract the luminosity distance d_L as a function of the redshift. Then we proceed to calculate the Hubble function $H(z)$ and $w(z)$.

1. Introduction

The present accumulation of high quality cosmological data and their detailed analyzes have suggested that recently our universe has entered into an accelerated expansion epoch. The responsible of such acceleration presumably is the 70 % of the total content of the matter and energy of the universe, the so called dark energy. Therefore, one of the most important questions in cosmology is to elucidate the nature of dark energy.

The main observations that point to this state of the universe are the related to the luminosity of the supernovae Ia (SN Iae): distant SN Iae are dimmer on average by 0.2 mag than what expected in an Einstein-de Sitter model, fact that translates in they being farther away than expected. These results imply that we live in an accelerated phase of the expansion of the universe.

The cause and rate of the acceleration is of crucial importance and the dark energy equation of state should be analyzed with all possible data. Supernovae cannot be detected out past redshift 1.7, but Gamma-ray bursts (GRBs) offer a means to extend the Hubble measurements to redshifts as high as 8 [1]. The reason is that GRBs are visible across much larger distances than supernovae. GRBs are now known to have several light-curve and spectral properties from which the luminosity of the burst can be calculated. Correlations between various properties of the emissions have been used to determine their distances, however a straightforward use of GRBs as cosmological probes has been discouraged by the intrinsic faintness of the nearby events, therefore, the extrapolation of their correlations to low redshift events needs the introduction of a model. However, if one assumes that once leaving their source, light rays should propagate in the same way throughout the cosmic spacetime, then we can calibrate GRBs at low redshift using the known data for SNIa.

In this contribution we use a 24 sample of the GRB data given in [2] to analyze the behavior of $w(z)$ assuming a FRW spacetime with a dark fluid matter with an equation of state (EoS) $p_x = w(z)\rho_x$; in a flat universe, as suggested by WMAP.

2. $w(z)$ parameter of the dark energy EoS

Let us consider the Friedmann equation for a flat spacetime,

$$\dot{H} = -4\pi G(\rho + p), \quad (1)$$

where $H(z) = \dot{a}/a$ is the Hubble parameter, ρ and p denote the total energy density and pressure of all the species present in the universe at a given epoch. We shall assume that $\rho = \rho_m + \rho_x$, with ρ_m being the (dark and barionic) matter and ρ_x the dark energy density. Defining the dimensionless fractional densities Ω_i by

$$\Omega_i = \frac{8\pi G}{3H^2}\rho_i, \quad (2)$$

assuming that matter is modeled by dust, then $p_m = 0$ and $\rho_m = \rho_{0m}(1+z)^3$, taking into account that $\dot{H} = -(1+z)HH'$, considering the redshift dependence of the Hubble parameter, $dz = -(1+z)Hdt$ and the prime denoting derivative respect to redshift, we arrive to the following expression for $w(z)$

$$w(z) = -1 + \left[1 - \frac{H_0^2}{H^2}\Omega_{0m}(1+z)^3\right]^{-1}\left[\frac{2H'}{3H}(1+z) - \frac{H_0^2}{H^2}\Omega_{0m}(1+z)^3\right] \quad (3)$$

where the subindex 0 means the present value of the function. Moreover $H(z)$ can be deduced from observational data through its relationship to the luminous distance d_L,

$$H(z) = \left\{\frac{d}{dz}\left(\frac{d_L(z)}{c(1+z)}\right)\right\}^{-1}. \quad (4)$$

Note that so far our assumptions are a flat FRW spacetime and dust modeling matter. The concordance principle is implicitly assumed as well, $\Omega_x + \Omega_m = 1$. With these considerations we constrain the dark energy equation of state, as far as the uncertainties in the GRBs measurements allow.

3. Calibrating GRBs

Based in the data of Type Ia Supernovae [3], Kodama *et al* [4], used the following calibration to the luminosity distance

$$\frac{d_L}{10^{27}\text{cm}} = 14.57z^{1.02} + 7.16z^{1.76}. \quad (5)$$

With Eq. (5) and the 69 GRBs observational data, we can use Eq. (3) and infer if the dark energy effective EoS parameter w is close to -1 (cosmological constant) or varies in time (quintessence).

However with the Kodama calibration and assuming $H_0 = 70\text{kms}^{-1}\text{Mpc}^{-1}$ and $\Omega_{0m} = 0.31$, using Eqs.(3) and (4), we obtained a physically unreasonable $w(z)$, diverging at redshift $z \sim 1$ (see Fig.2). We tried to repair this inconsistency by finding out another calibration for the GRBs extracted at low redshifts directly from Supernovae data.

Considering the empirical formula of the luminosity distance of Type Ia supernovae from Riess *et al* (2007) [3] with $0.359 < z < 1.755$,

$$\frac{d_L}{10^{27}cm} = 16.85z + 4.97z^2. \quad (6)$$

We apply Eq.(6) to 16 GRBs with the redshift $z < 1.755$ in our sample to obtain $E_\gamma = 4\pi d_L^2 S_{\text{bolo}}(1+z)^{-1}F_{\text{beam}}$, where S_{bolo} is the bolometric fluence estimated in $1-10,000$keV energy range in GRB rest frame and F_{beam} is a beaming factor related with the jet opening

Journal of Physics: Conference Series **314** (2011) 012059 doi:10.1088/1742-6596/314/1/012059

angle. The used sample was obtained from Table 4 in [2], they are the 27 reported objects with their respective $S_{\rm bolo}$ and $F_{\rm beam}$; from this set we removed three objects: GRBs 050505, 050904 and 060210 because of their large uncertainty in $E_{\rm peak}$.

Moreover, the typical spectrum of the prompt emission of GRBs can be expressed as exponentially connected broken power-law, so called Band function. Then we can determine spectral peak energy $E_{\rm peak}$, $E_{\rm peak} = (1+z)E_{\rm p,obs}$, corresponding to the photon energy at maximum in νF_ν spectra. The prompt emission property is related to $E_{\rm peak}$, as a $E_{\rm peak} - E_\gamma$ relation found by L. Amati *et al* [5] (see also [6]) that connects $E_{\rm peak}$ with the isotropic equivalent energy E_γ. We verify that the Amati relation is fulfilled by the 16 GRBs with $z < 1.755$,

$$\frac{E_\gamma}{10^{52}erg} = 3.41 \times 10^{-6}\left(\frac{E_{\rm peak}}{1{\rm keV}}\right)^{1.63}. \tag{7}$$

Then, we apply this Amati relation to 8 GRBs with high redshifts, $z < 4.3$, to determine the luminosity distance as a function of z.

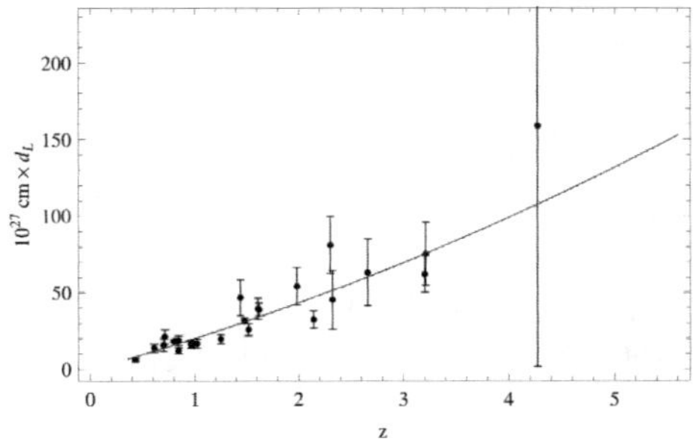

Figure 1. Luminous distance adjustment for the GRB's data from [2], derived using the Amati relation [5]; the best fit is $10^{27}{\rm cm} \times d_L = 18.55z + 1.55z^2$ with Rsquare $R^2 = 0.65$. The corresponding $w(z)$ is plotted in Fig.2

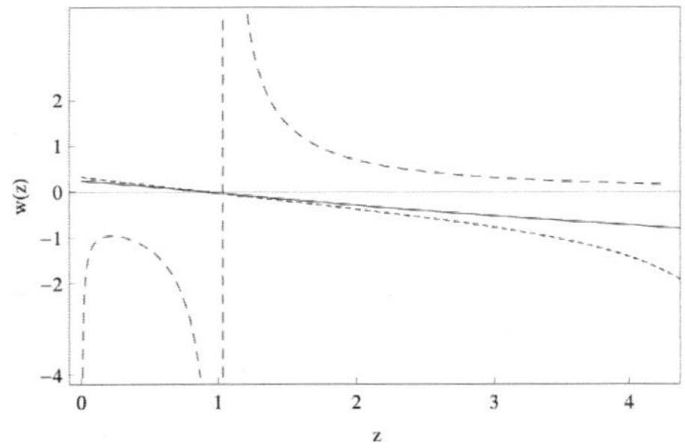

Figure 2. The plot for $w(z)$, calculated as a function of the redshift using Eqs.(3) and (10), with $\Omega_{0m} = 0.32$ (dotted) and $\Omega_{0m} = 0.1$ (continuous). The dashed curve that diverges represents $w(z)$ derived from the empirical luminous distance for SNIae [3], Eq. (5).

4. Applying the calibrated relation to high redshifts

For each GRB with $z = z^i$ we have the observed $S_{\rm bolo}$ in units of erg/cm^{-2}, the dimensionless $F_{\rm beam}$ and the observed $E_{\rm p,obs}$ in keV. Then, using Eq.(7) and $E_\gamma = 4\pi d_L^2 S_{\rm bolo}(1+z)^{-1}F_{\rm beam}$, the luminosity distance can be derived as

$$d_L(z^i) = 10^{23}{\rm cm}\sqrt{\frac{3.41}{4\pi F^i_{\rm beam}S^i_{\rm bolo}}(E^i_{\rm p,obs})^{0.815}(1+z^i)^{1.315}}. \tag{8}$$

With the error propagation calculated according to

$$(\Delta d_L(z))^2 = \left(\frac{\partial d_L(z)}{\partial S_{\rm bolo}}\sigma_{S_{\rm bolo}}\right)^2 + \left(\frac{\partial d_L(z)}{\partial F_{\rm beam}}\sigma_{F_{\rm beam}}\right)^2 + \left(\frac{\partial d_L(z)}{\partial E_{\rm p,obs}}\sigma_{E_{\rm p,obs}}\right)^2 \tag{9}$$

where σ_i is the uncertainty related to each data, as reported in [2]. From Eq. (8) and the data of GRBs, we obtained Fig. 1 for the luminosity distance with $0.359 < z < 4.3$ and the empirical formula given by

$$\frac{d_L}{10^{27}cm} = 18.55z + 1.55z^2 \tag{10}$$

Using this empirical rule for the luminosity distance as function of z, and using Eqs. (3)-(4) we derived the dark-energy equation-of-state parameter $w(z)$ that is plotted in Fig. 2.

Note also that we have used $H_0 = 49.9$km s^{-1}Mpc^{-1} that corresponds to the present value ($z = 0$) of the Hubble parameter given by our fit, Eqs. (4) and (10),

$$\begin{aligned} H(z) &= \left\{\frac{d}{dz}\left(\frac{d_L(z)}{c(1+z)}\right)\right\}^{-1} = \frac{(1+z)^2}{a+bz+cz^2}\text{km}s^{-1}\text{Mpc}^{-1}. \\ w(z) &= -1 + \frac{(2/3)(1+z)[2a+b(z-1)-2cz] - H_0^2\Omega_{0m}(a+bz+cz^2)^3}{[1+z-H_0^2\Omega_{0m}(a+bz+cz^2)^2](a+bz+cz^2)}, \end{aligned} \tag{11}$$

where the values of the constants obtained from our fitting Eq. (10), are $a = 0.02, b = 0.003, c = 0.002$.

5. Conclusions

The aim of this work is to determine the functional dependence of the dark-energy equation-of-state parameter in terms of the redshift, $w(z)$, from observational data coming from the GRBs.

First we find the best calibration between the observational data of GRBs, obtaining the luminosity distance as function of redshift, $d_L(z)$, Eq. (10). Then we obtain the corresponding Hubble function and then $w(z)$. Our estimation for $H_0 = 49.91$kms^{-1}Mpc^{-1} has the correct order of magnitude.

We noticed that Eq.(3) for $w(z)$ is not valid for any redshift: There is a critical redshift z_c where the factor $[1-(H_0^2/H(z)^2)\Omega_{0m}(1+z)^3]$ becomes zero, and then $w(z)$ diverges; the value of z_c depends on the chosen Ω_{0m} and H_0. With our calibration of GRBs and the derived fit for the luminous distance, the obtained $w(z)$ does not diverge in the considered redshift range, $z < 4.3$. However, the behavior of $w(z)$ depends strongly on the Ω_{0m} assumed: this is the influence of dark matter on dark energy [7]; $w(z)$ is closer to -1 as the considered Ω_{0m} is smaller (see Fig.2).

Unfortunately the GRBs data have still such uncertainties that do not allow us to obtain definite conclusions. Qualitatively our results are in agreement with related works in [4], [8] and [9]. In spite that GRBs data by themselves are unable to strongly constraint cosmological parameters, if in the future they can be considered as standard candles, they will be a complement to the SN Ia data sets and potentially very significant.

Acknowledgments

We gratefully acknowledge discussions with Ruth Lazkoz. A. M. acknowledges the Organizing Committee of ERE2010 for partial support to attend the Conference. Partial financial support of Conacyt Project 49182-F is also acknowledged.

References

[1] Salvaterra R *et al*, 2009 Nature, **461** 1258
[2] Schaefer B. E., 2007, Astrophys. J. **660** 16.
[3] Riess *et al*, 2007, Astrophys. J., **659** 98-121.
[4] Kodama Y *et al.* 2008 Mon. Not. R. Astron. Soc. **391**, L1-L4.
[5] Amati L., *et al.*, 2008 Mon. Not. R. Astron. Soc. **391**, 577-584.
[6] Ghirlanda, G. *et al.* 2004, Astrophys. J. , **616** 331.
[7] Maor I., Brustein R., McMahon J., Steinhardt P.J., 2002, Phys. Rev. D, **65**, 123003
[8] Capozziello, S., Izzo, L., 2009 Nuclear Phys. B, **194**, 206-211.
[9] Samushia L., Ratra B., 2010 Astrophys. J., **714**, 1347-1354,

Spanish Relativity Meeting (ERE 2010): Gravity as a Crossroad in Physics IOP Publishing
Journal of Physics: Conference Series **314** (2011) 012060 doi:10.1088/1742-6596/314/1/012060

Late-time cosmic acceleration: Dark gravity

Francisco S. N. Lobo

Centro de Astronomia e Astrofísica da Universidade de Lisboa, Campo Grande, Ed. C8
1749-016 Lisboa, Portugal

E-mail: flobo@cii.fc.ul.pt

Abstract. A central theme in cosmology is the perplexing fact that the Universe is undergoing an accelerating expansion. The latter, one of the most important and challenging current problems in cosmology, represents a new imbalance in the governing gravitational field equations. Several candidates, responsible for this expansion, have been proposed in the literature, in particular, dark energy models and modified gravity, amongst others. In this paper, we explore the possibility that the late-time cosmic acceleration is due to infra-red modifications of Einstein's theory of General Relativity, and review some of the modified theories of gravity that address this intriguing and exciting problem facing modern physics.

1. Introduction

Modern astrophysical and cosmological models are faced with two severe theoretical difficulties, that can be summarized as the dark energy and the dark matter problems. Relative to the former, high-precision observational data have confirmed with startling evidence that the Universe is undergoing a phase of accelerated expansion [1]. Several candidates, responsible for this expansion, have been proposed in the literature, in particular, dark energy models and modified gravity, amongst others. The cause of this acceleration still remains an open and tantalizing question. Nevertheless, a very promising way to explain these major problems is to assume that at large scales Einstein's theory of General Relativity breaks down, and a more general action describes the gravitational field. The Einstein field equation of General Relativity was first derived from an action principle by Hilbert, by adopting a linear function of the scalar curvature, R, in the gravitational Lagrangian density.

However, there are no a priori reasons to restrict the gravitational Lagrangian to this form, and indeed several generalizations of the Einstein-Hilbert Lagrangian have been proposed, such as involving an arbitrary function of the scalar invariant, $f(R)$ [2]. Recently, a renaissance of $f(R)$ modified theories of gravity has been verified in an attempt to explain the late-time accelerated expansion of the Universe. In particular, it was shown that the cosmic acceleration can indeed be explained in the context of $f(R)$ gravity [3]. In addition to this, in considering alternative higher-order gravity theories, one is liable to be motivated in pursuing models consistent and inspired by several candidates of a fundamental theory of quantum gravity. Indeed, motivations from string/M-theory predict that scalar field couplings with the Gauss-Bonnet invariant G are important in the appearance of non-singular early time cosmologies. It is also possible to apply these motivations to the late-time Universe in an effective Gauss-Bonnet dark energy model [4] (many aspects of Gauss-Bonnet gravity have been analyzed in the literature [5]).

Still in the context of modified gravity, an interesting possibility is the existence of extra dimensions. It is widely believed that string theory is moving towards a viable quantum gravity theory, and one of the key predictions of string theory is precisely the existence of extra spatial dimensions. In the brane-world scenario, motivated by recent developments in string theory, the observed 3-dimensional universe is embedded in a higher-dimensional spacetime [6]. The Dvali-Gabadadze-Porrati (DGP) models [7] achieve a covariant infra-red modification of General Relativity, where it is possible for extra-dimensional gravity to dominate at low energies, via a brane induced gravity effect. The generalization of the DGP models to cosmology lead to late-accelerating cosmologies [8], even in the absence of a dark energy field [9]. This exciting feature of "self acceleration" may help towards a new resolution to the dark energy problem, although this model deserves further investigation as a viable cosmological model [10]. While the DGP braneworld offers an alternative explanation to the standard cosmological model, for the expansion history of the universe, it offers a paradigm for nature fundamentally distinct from dark energy models of cosmic acceleration, even those that perfectly mimic the same expansion history. It is also fundamental to understand how one may differentiate this modified theory of gravity from dark energy models.

In this work, we review several modified theories of gravity, namely, $f(R)$ gravity and the DGP braneworld model, briefly exploring some of their interesting properties and characteristics, and in particular, focussing mainly on the late-time cosmic acceleration.

2. Modified theories of gravity: Late-time cosmic acceleration

2.1. $f(R)$ modified theories of gravity

A promising avenue that has been extensively investigated recently are the $f(R)$ modified theories of gravity, where the standard Einstein-Hilbert action is replaced by an arbitrary function of the Ricci scalar R.

The action for the $f(R)$ modified theories of gravity is given by

$$S = \frac{1}{2\kappa}\int d^4x\sqrt{-g}\; f(R) + S_M(g^{\mu\nu},\psi)\,, \tag{1}$$

where $\kappa = 8\pi G$. $S_M(g^{\mu\nu},\psi)$ is the matter action, in which matter is minimally coupled to the metric $g_{\mu\nu}$ and ψ collectively denotes the matter fields. Varying the action with respect to $g^{\mu\nu}$, provides the following field equation

$$FR_{\mu\nu} - \frac{1}{2}f\, g_{\mu\nu} - \nabla_\mu\nabla_\nu F + g_{\mu\nu}\nabla_\alpha\nabla^\alpha F = \kappa\, T^{(m)}_{\mu\nu}\,, \tag{2}$$

where $F \equiv df/dR$.

It is interesting to note that $f(R)$ gravity may lead to an effective dark energy, without the need to introduce a negative pressure ideal fluid. Taking into account the flat FRW metric and a perfect fluid description for matter, we verify that the gravitational field equation, Eq. (2), provides the generalised Friedmann equations in the following form [11]:

$$\left(\frac{\dot a}{a}\right)^2 = \frac{\kappa}{3}\rho_{\rm tot}\,, \qquad \left(\frac{\ddot a}{a}\right) = -\frac{\kappa}{6}\left(\rho_{\rm tot} + 3p_{\rm tot}\right), \tag{3}$$

where $\rho_{\rm tot} = \rho + \rho_{(c)}$ and $p_{\rm tot} = p + p_{(c)}$, and the curvature stress-energy components, $\rho_{(c)}$ and $p_{(c)}$, are defined as

$$\rho_{(c)} = \frac{1}{\kappa F(R)}\left\{\frac{1}{2}\left[f(R) - RF(R)\right] - 3\left(\frac{\dot a}{a}\right)\dot R F'(R)\right\}, \tag{4}$$

$$p_{(c)} = \frac{1}{\kappa F(R)}\left\{2\left(\frac{\dot a}{a}\right)\dot R F'(R) + \ddot R F'(R) + \dot R^2 F''(R) - \frac{1}{2}\left[f(R) - RF(R)\right]\right\}, \tag{5}$$

respectively. The late-time cosmic acceleration is achieved if the condition $\rho_{\rm tot} + 3p_{\rm tot} < 0$ is obeyed, which follows from Eq. (3).

For simplicity, consider the absence of matter, $\rho = p = 0$. Now, taking into account the equation of state $\omega_{\rm eff} = p_{(c)}/\rho_{(c)}$, with $f(R) \propto R^n$ and a generic power law $a(t) = a_0(t/t_0)^\alpha$ [11], the parameters $\omega_{\rm eff}$ and α are given by

$$\omega_{\rm eff} = -\frac{6n^2 - 7n - 1}{6n^2 - 9n + 3}, \qquad \alpha = \frac{-2n^2 + 3n - 1}{n - 2}, \tag{6}$$

respectively, for $n \neq 1$. Note that a suitable choice of n can lead to the desired value of $\omega_{\rm eff} < -1/3$, achieving the late-time cosmic acceleration.

Other forms of $f(R)$ have also been considered in the literature, for instance those involving logarithmic terms [5]. These models also yield acceptable values for the effective equation of state parameter, resulting in the late-time cosmic acceleration.

2.2. DGP brane gravity and self-acceleration

One of the key predictions of string theory is the existence of extra spatial dimensions. In the brane-world scenario, motivated by recent developments in string theory, the observed 3-dimensional universe is embedded in a higher-dimensional spacetime [6]. One of the simplest covariant models is the Dvali-Gabadadze-Porrati (DGP) braneworld model, in which gravity leaks off the $4D$ Minkowski brane into the $5D$ bulk at large scales. The generalization of the DGP models to cosmology lead to late-accelerating cosmologies [8], even in the absence of a dark energy field [9].

The $5D$ action describing the DGP model is given by

$$S = \frac{1}{2\kappa_5}\int d^5x\sqrt{-g}\,{}^{(5)}R + \frac{1}{2\kappa_4}\int d^4x\sqrt{-\gamma}\,{}^{(4)}R - \int d^4x\sqrt{-\gamma}\,\mathcal{L}_m, \tag{7}$$

where $\kappa_5 = 8\pi G_5$ and $\kappa_4 = 8\pi G_4$. The first term in the action is the Einstein-Hilbert action in five dimensions for a five-dimensional bulk metric, g_{AB}, with a five-dimensional Ricci scalar ${}^{(5)}R$; the second term is the induced Einstein-Hilbert term on the brane, with a four-dimensional induced metric γ on the brane; and $\mathcal{L}_m$ represents the matter Lagrangian density confined to the brane.

The transition from $4D$ to $5D$ behavior is governed by a cross-over scale, r_c, given by $r_c = \frac{\kappa_5}{2\kappa_4}$. Gravity manifests itself as a 4-dimensional theory for characteristic scales much smaller than r_c; for large distances compared to r_c, one verifies a leakage of gravity into the bulk, consequently making the higher dimensional effects important. Thus, the leakage of gravity at late times initiates acceleration, due to the weakening of gravity on the brane.

Taking into account the FRW metric, and considering a flat geometry, the modified Friedmann equation is given by

$$H^2 - \frac{\epsilon}{r_c}H = \frac{8\pi G}{3}\rho, \tag{8}$$

where $\epsilon = \pm 1$, and the energy density satisfies the standard conservation equation, i.e., $\dot\rho + 3H(\rho + p) = 0$. For scales $H^{-1} \ll r_c$, the second term is negligible, and Eq. (8) reduces to the general relativistic Friedmann equation, i.e., $H^2 = 8\pi G\rho/3$. The second term becomes significant for scales comparable to the cross-over scale, $H^{-1} \geq r_c$. Self-acceleration occurs for the branch $\epsilon = +1$, and the modified Friedmann equation shows that at late times in a CDM universe characterized by a scale factor $\rho \propto a^{-3}$, the universe approaches a de Sitter solution $H \to H_\infty = \frac{1}{r_c}$. Thus, one may achieve late-time acceleration if the H_0 is of the order of H_∞. Note that the late-time acceleration in the DGP model is not due to the presence of a negative pressure, but simply due to the weakening of gravity on the brane as a consequence of gravity leakage at late times.

Although the weak-field gravitational DGP behaves as $4D$ on scales smaller than r_c, linearized DGP gravity is not described by General Relativity [12]. It is also interesting to note that despite the fact that the expansion history of the DGP model and General Relativity are the same, the structure formation in both are essentially different [13]. Combining these features provides the possibility of distinguishing the DGP model from dark energy models in General Relativity. Another interesting aspect of this model is that the self-accelerating branch in the DGP model contains a ghost at the linearized level [12, 14]. The presence of the ghost implies a negative sign for the kinetic term, resulting in negative energy densities, consequently leading to the instability of the spacetime. We refer the reader to Ref. [12], and references therein, for more details on the DGP model.

3. Conclusion

The standard model of cosmology is remarkably successful in accounting for the observed features of the Universe. However, there remain a number of fundamental open questions at the foundation of the standard model. In particular, we lack a fundamental understanding of the acceleration of the late universe. Recent observations of supernovae, together with the WMAP and SDSS data, lead to the remarkable conclusion that our universe is not just expanding, but has begun to accelerate [1]. What is the so-called 'dark energy' that is driving the acceleration of the universe? Is it a vacuum energy or a dynamical field ("quintessence")? Or is the acceleration due to infra-red modifications of Einstein's theory of General Relativity? How is structure formation affected in these alternative scenarios? What will the outcome be of this acceleration for the future fate of the universe? Deciding between these possible sources of the cosmic acceleration will be one of the major objectives in cosmology in the next decade with several surveys and experiments to address the nature of dark energy. All of these aspects present an extremely fascinating aspect for future theoretical research.

References

[1] Perlmutter, S. *et al.* 1999, Astrophys. J., 517, 565; Riess, A. G. *et al.* 1998, Astron. J., 116, 1009.
[2] Copeland, E.J., Sami, M and Tsujikawa, S. Int. J. Mod. Phys. D **15**, 1753 (2006); Nojiri, S and Odintsov, S.D, Int. J. Geom. Meth. Mod. Phys. **4**, 115 (2007); Lobo, F.S.N. arXiv:0807.1640 [gr-qc]; De Felice, A. and Tsujikawa, S., Living Rev. Rel. **13**, 3 (2010); Nojiri, S. and Odintsov, S.D., arXiv:1011.0544 [gr-qc]; Bertolami, O. *et al*, Phys. Rev. D **75**, 104016 (2007); Boehmer, C.G., *et al*, Phys. Rev. D **76**, 084005 (2007); Boehmer, C.G., *et al*, Astropart. Phys. **29**, 386 (2008); Boehmer, C.G., *et al*, JCAP **0803**, 024 (2008); Bertolami, O., *et al*, Phys. Rev. D **78**, 064036 (2008); Harko, T. and Lobo, F.S.N., arXiv:1007.4415 [gr-qc]; Harko, T. and Lobo, F.S.N., Eur. Phys. J. C **70**, 373 (2010).
[3] Carroll, S.M., Duvvuri, V., Trodden, M., and Turner, M.S. 2004, Phys. Rev. D, 70, 043528.
[4] Nojiri, S., Odintsov, S.D., and Sasaki, M. 2005, Phys. Rev. D, 71, 123509; Nojiri, S., Odintsov, S.D., and Tretyakov, P.V., Prog. Theor. Phys. Suppl. **172**, 81 (2008); Bamba, K., Odintsov, S.D., Sebastiani, L. and Zerbini, S., [arXiv:0911.4390 [hep-th]].
[5] Nojiri, S., and Odintsov, S.D 2007. Int. J. Geom. Meth. Mod. Phys., 4, 115; Cognola, G., Elizalde, E., Nojiri, S., Odintsov, S.D., and Zerbini, S. 2006, Phys. Rev. D, 73, 084007; Boehmer, C.G., and Lobo, F.S.N., Phys. Rev. D **79**, 067504 (2009); Garcia, N.M., et al., arXiv:1011.4159 [gr-qc].
[6] Maartens, R. 2004, Living Rev. Rel. 7, 7; Randall, L., and Sundrum, R. 1999, Phys. Rev. Lett., 83, 3370; Randall, L., and Sundrum, R. 1999, Phys. Rev. Lett., 83, 4690.
[7] Dvali, G., Gabadadze, G., and Porrati, M. 2000, Phys. Lett. B, 485, 208.
[8] Deffayet, C. 2001, Phys. Lett. B, 502, 199.
[9] Brown, R.A. et al. 2005, J. Cosmol. Astropart. Phys., 11, 008.
[10] Lue, A. 2006, Phys. Rep., 423, 1.
[11] Capozziello, S., Carloni, S., and Troisi, A. 2003, Recent Res. Dev. Astron. Astrophys., 1, 625.
[12] Koyama, K. 2007, Class. Quant. Grav., 24, R231.
[13] Koyama, K., and Maartens, R. 2006, J. Cosmol. Astropart. Phys., 0601, 016.
[14] Koyama, K. 2005, Phys. Rev. D, 72, 123511.

Spanish Relativity Meeting (ERE 2010): Gravity as a Crossroad in Physics IOP Publishing
Journal of Physics: Conference Series **314** (2011) 012061 doi:10.1088/1742-6596/314/1/012061

w-singularities in cosmological models

Leonardo Fernández-Jambrina
ETSI Navales, Universidad Politécnica de Madrid, Arco de la Victoria s/n, E-28040-Madrid, Spain
E-mail: leonardo.fernandez@upm.es

Abstract. Recently a new type of cosmological singularity has been postulated for infinite barotropic index w in the equation of state $p = w\rho$ of the cosmological fluid, but vanishing pressure and density at the singular event. Apparently the barotropic index w would be the only physical quantity to blow up at the singularity. In this talk we would like to discuss the strength of such singularities and compare them with other types. We show that they are weak singularities.

1. Introduction

Until the observation of accelerated expansion of our universe, the only types of singularities that were considered to appear in Friedman-Lemaître-Robinson-Walker (FLRW) cosmological models were singularities in the form of a big bang or a big crunch, for which the density of the matter content of the spacetime blows up [1], as well as the scale factor of the universe. However, since the observational value of the barotropic index of the universe nowadays (the quotient between the pressure and the density of the content of the universe, $w = p/\rho$) is around minus one, there is a possibility of violation of the energy conditions and new types of singularities could appear. In fact, such singularities appear naturally in gravitational theories which try to cope with the observed phenomena.

Many of these new singularities are included in a descriptive classification due to Nojiri, Odintsov and Tsujikawa (N.O.T. in the following), which takes into account which physical quantities (scale factor of the universe a, Hubble ratio H, pressure p or density ρ) are singular [2]:

- Big bang / crunch: Vanishing a, divergent H, ρ and p.
- Type I: "Big rip": Infinite a, ρ and p [3].
- Type II: "Sudden": Finite a, H and ρ, divergent $\dot{H}$ and p [4].
- Type III: "Big freeze": finite a, infinite H, ρ and p [5].
- Type IV: Finite a, H, $\dot{H}$, ρ and p, but infinite higher derivatives of a.

The strength of these singularities [6, 7] has been checked in [8, 9]. Summarizing, big rip singularities are strong whereas sudden [10], big freeze and type IV singularities are weak. In this sense, the latter cannot be taken as the final stage of the universe, since the spacetime could be extended continuously beyond the singularity.

Out of this scheme, directional singularities for which some observers experience an infinite curvature, though the curvature scalars vanish at the singularity, have been found in some theories [11].

Also out of this classification, in [12] a FLRW cosmology is shown in which only the barotropic index w becomes infinite at $t = t_s$,

$$\begin{aligned} a(t) &= \frac{a_s}{1-\frac{3\gamma}{2}\left(\frac{n-1}{n-\frac{2}{3\gamma}}\right)^{n-1}} + \frac{1-\frac{2}{3\gamma}}{n-\frac{2}{3\gamma}} \frac{n a_s}{1-\frac{2}{3\gamma}\left(\frac{n-\frac{2}{3\gamma}}{n-1}\right)^{n-1}} \left(\frac{t}{t_s}\right)^{\frac{2}{3\gamma}} \\ &+ \frac{a_s}{\frac{3\gamma}{2}\left(\frac{n-1}{n-\frac{2}{3\gamma}}\right)^{n-1}-1} \left(1-\frac{1-\frac{2}{3\gamma}}{n-\frac{2}{3\gamma}}\frac{t}{t_s}\right)^n, \end{aligned} \tag{1}$$

for $\gamma > 0$ and $n \neq 1$ a natural number, whereas the scale factor, $a(t_s) = a_s$, and all its derivatives remain finite and the density and the pressure vanish at t_s. In [13] similar singularities are obtained but with non-vanishing pressure.

This new type of singularities has been dubbed barotropic index w-singularities [12], or simply w-singularities, or even type V [14] singularities, remarking the difference between them and type IV singularities, since no higher derivatives of the scale factor blow up at the time of the singularity t_s.

In this talk we would like to show a characterization of barotropic index w-singularities in FLRW cosmologies in terms of the form of the scale factor. More details may be found in [15].

In the next section we derive a characterization of barotropic index w-singularities in terms of the exponents of the power expansion in time of the scale factor of the universe. A discussion of the results is included in the last section.

2. Characterization of barotropic index w-singularities

In gravitational theories such as general relativity, the content of the universe is depicted as a perfect fluid of density ρ and pressure p, depending on just the time coordinate t, due to the homogeneity and isotropy of FLRW spacetimes. This means that there is an equation of state for the perfect fluid, $p = p(\rho)$, and the ratio of pressure and density is defined as the barotropic index of the fluid, $w = p/\rho$. This index evolves with time except for power-law cosmologies, which have a constant barotropic index.

In flat FLRW cosmologies the barotropic index is written in terms of the scale factor of the universe and its derivatives,

$$\rho = 3\left(\frac{\dot a}{a}\right)^2, \qquad p = -\left(\frac{\dot a}{a}\right)^2 - \frac{2\ddot a}{a}. \tag{2}$$

$$w = -\frac{1}{3} - \frac{2}{3}\frac{a\ddot a}{\dot a^2}, \tag{3}$$

and if we assume that it may be expanded as a generalized power expansion of time around the time of the singularity [16, 8]

$$a(t) = c_0(t_s - t)^{\eta_0} + c_1(t_s - t)^{\eta_1} + \cdots, \qquad \eta_0 < \eta_1 < \cdots, \qquad c_0 > 0, \tag{4}$$

where η_0, η_1,... are real numbers, the barotropic index can be expanded accordingly,

$$w(t) = w_0(t_s - t)^{\zeta_0} + w_1(t_s - t)^{\zeta_1} + \cdots,$$

Table 1. Singularities in barotropic index.

η_0	η_1	η_2	w_s
Nonzero	(η_0, ∞)	(η_1, ∞)	Finite
Zero	$(0, 1)$	(η_1, ∞)	Infinite
	1	$(1, 2)$	Infinite
	1	$[2, \infty)$	Finite
	$(1, \infty)$	$[\eta_1, \infty)$	Infinite

and the exponents and coefficients of this expansion can be written in terms of the former ones.

This gives rise to a hierarchy of cases and subcases depending on the values of η_0 and η_1 [15] that are consigned in table 1. We notice that infinite barotropic indices appear just for vanishing η_0 exponent, except for $\eta_1 = 1$ with $\eta_2 \geq 2$.

The requirement of finite derivatives of the scale factor implies that all exponents η_i must be natural numbers. Hence, we are left just with the subcase of infinite barotropic index with $\eta_0 = 0$ and $\eta_1 \geq 2$.

From (2) and (4) we learn that vanishing density requires $\eta_1 > 1$ and vanishing pressure requires $\eta_1 > 2$ [15].

Hence we see that an infinite barotropic index with finite derivatives of the scale factor of the universe requires a vanishing density at the time of the singularity but not a vanishing pressure.

This may be summarized in the following statement:

Theorem: A FLRW cosmological model has an infinite barotropic index at a finite time t_s with finite derivatives of the scale factor of the universe if and only if $a(t)$ can be expanded as a Taylor series at t_s with a null linear term,

$$a(t) = c_0 + \sum_{n=2}^{\infty} c_n (t_s - t)^n. \tag{5}$$

Furthermore if we require a vanishing pressure either (barotropic index w-singularity), the quadratic term c_2 is also to be zero.

3. Conclusions

We have shown a simple characterization of barotropic index w-singularities in FLRW cosmological models in terms of Taylor expansions of the scale factor of the universe. The necessary and sufficient condition is just the vanishing of the linear term of the expansion. And if we consider singularities with vanishing pressure, then both the linear and quadratic terms must be zero.

Most exotic singularities are weak in the sense that when we consider finite objects they are not disrupted by tidal forces on experiencing the singularity, since the spacetime can be extended continuously beyond the singular event [8].

In that reference, the strength of the singularities was checked in terms of the exponents of the expansion (4) of the scale factor of the universe. Though barotropic index w-singularities were not considered, the results obtained there are general enough to be applied to cosmological models with vanishing linear term. In fact, it was shown that singularities in cosmological models with $\eta_0 = 0$ and η_1 larger or equal than one were weak [17] with both Tipler [6] and Królak's [7] criteria.

Table 2. Strength of singularities in FLRW cosmological models.

η_0	η_1	η_2	$\{\eta_i\}$	**Tipler**	**Królak**	**N.O.T.**
$(-\infty, 0)$	(η_0, ∞)	(η_1, ∞)	I	Strong	Strong	I
0	$(0, 1)$	(η_1, ∞)	S	Weak	Strong	III
	1	$(1, 2)$	S	Weak	Weak	II
		$[2, \infty)$	S	Weak	Weak	IV
	$(1, 2)$	(η_1, ∞)	S	Weak	Weak	II
	$[2, \infty)$	(η_1, ∞)	S	Weak	Weak	IV
	2	$[3, \infty)$	N	Weak	Weak	w with $p_s \neq 0$
	$[3, \infty)$	$[\eta_1 + 1, \infty)$	N	Weak	Weak	w with $p_s = 0$
$(0, \infty)$	(η_0, ∞)	(η_1, ∞)	I	Strong	Strong	Crunch

Hence barotropic index w-singularities are weak and cannot be considered as a final stage for the universe, as it happens with other exotic singularities but big rip.

In table 2 we summarize our results on exotic singularities and their strength in terms of the exponents of the power expansion of the scale factor of the universe around the time of the singularity.

Singularities in the derivatives of the scale factor of the universe appear only for non-natural exponents in the expansion. This is the reason for including a column $\{\eta_i\}$ which deals with that fact.

The letter I (independent) means no additional condition on the exponents of the expansion around the time of the singularity. S (some) states that there must be at least one non-natural exponent in order to have a singularity in one of the derivatives. Finally, N (natural) means that all exponents are natural.

Acknowledgments
The author wishes to thank the University of the Basque Country for their hospitality and facilities to carry out this work.

References
[1] Hawking S W and Ellis G F R 1973 *The Large Scale Structure of Space-time* (Cambridge: Cambridge University Press)
[2] Nojiri S, Odintsov S D and Tsujikawa S 2005 *Phys. Rev. D* **71** 063004
[3] Caldwell R R, Kamionkowski M and Weinberg N N 2003 *Phys. Rev. Lett.* **91** 071301
[4] Barrow J D 2004 *Class. Quant. Grav.* **21** L79
[5] Bouhmadi-López M, Gonzalez-Díaz P F and Martín-Moruno P 2008 *Phys. Lett. B* **659** 1
[6] Tipler F J 1977 *Phys. Lett. A* **64** 8
[7] Królak A 1986 *Class. Quant. Grav.* **3** 267
[8] Fernández-Jambrina L and Lazkoz R 2006 *Phys. Rev. D* **74** 064030
[9] Fernández-Jambrina L and Lazkoz R 2009 *Phys. Lett. B* **670** 254
[10] Fernández-Jambrina L and Lazkoz R 2004 *Phys. Rev. D* **70** 121503
[11] Fernández-Jambrina L 2007 *Phys. Lett. B* **656** 9
[12] Dąbrowski M P and Denkiewicz T 2009 *Phys. Rev. D* **79** 063521
[13] Yurov A V 2010 *Phys. Lett. B* **689** 1
[14] Kiefer C 2010, *Journ. Phys. Conf. Ser.* **222** 012049
[15] Fernández-Jambrina L 2010 *Phys. Rev. D* **82** 124004
[16] Cattoën C and Visser M 2005 *Class. Quant. Grav.* **22** 4913
[17] Clarke C J S and Królak A 1985 *Journ. Geom. Phys.* **2** 127

Journal of Physics: Conference Series **314** (2011) 012062 doi:10.1088/1742-6596/314/1/012062

Gravitational waves in hybrid quintessential inflationary models

Paulo M Sá[1] and Alfredo B Henriques[2]

[1] Departamento de Física, Faculdade de Ciências e Tecnologia, Universidade do Algarve, Campus de Gambelas, 8005-139 Faro, Portugal

[2] Centro Multidisciplinar de Astrofísica – CENTRA and Departamento de Física, Instituto Superior Técnico, UTL, Av. Rovisco Pais, 1049-001 Lisboa, Portugal

E-mail: pmsa@ualg.pt, alfredo.henriques@ist.utl.pt

Abstract. The generation of primordial gravitational waves is investigated within the hybrid quintessential inflationary model. Using the method of continuous Bogoliubov coefficients, we calculate the full gravitational-wave energy spectrum. The post-inflationary kination period, characteristic of quintessential inflationary models, leaves a clear signature on the spectrum, namely, a sharp rise of the gravitational-wave spectral energy density Ω_{GW} at high frequencies. For appropriate values of the parameters of the model, Ω_{GW} can be as high as 10^{-12} in the MHz–GHz range of frequencies.

In this article, we investigate the generation of gravitational waves within the hybrid quintessential inflationary model [1]. Gravitational waves of cosmological origin [2] are of the utmost importance, since they have the potential of providing us with unique information about the very early Universe, not directly available by any other means. In particular, the detection of these waves would allow for a better understanding of several issues relevant both for cosmology and high-energy physics, like inflation, preheating and reheating mechanisms, post-inflationary phase transitions, topological defects of grand unified theories and string theory.

The hybrid quintessential inflationary model, proposed recently by Bastero-Gil *et al.* [3], is a modification of the original quintessential inflationary model of Peebles and Vilenkin [4]. In the original model, the unification of inflation and dark energy within a single framework is achieved with a scalar field ϕ playing both roles of inflaton and quintessence. Since, in such model, the potential of the scalar field ϕ has no minimum, reheating cannot proceed in the usual way, namely, by the decay of the scalar field into quanta of other fields. Instead, reheating proceeds by gravitational particle production, a quite inefficient mechanism which may lead to cosmological problems associated with large isocurvature fluctuations and overproduction of gravitinos and moduli fields [5]. Within the hybrid model, the usual reheating mechanism is recovered by introducing another scalar field χ, coupled to the original inflaton/quintessence field ϕ, with a hybrid-like potential [3]

$$U(\phi,\chi) = \frac{1}{2}g^2\chi^2(\phi^2 - m^2) + \frac{1}{4}\lambda_\chi\chi^4 + \begin{cases} \lambda_\phi(\phi^4 + M^4), & \text{for} \quad \phi < 0, \\ \lambda_\phi M^8(\phi^4 + M^4)^{-1}, & \text{for} \quad \phi \geqslant 0, \end{cases} \tag{1}$$

where g, m, λ_χ, λ_ϕ and M are constants. Recent measurements of the cosmic microwave background radiation and large-scale structure constrain the value of λ_ϕ to be of the order of

10^{-13} [6], while agreement with the measured value of today's dark-energy density [7] requires M to be of the order of $10^{-14}\, m_{\rm p}$. The requirement that the scalar field χ responds quickly enough to changes in the potential $U(\phi,\chi)$ and that it does not influence significantly the evolution of the inflaton/quintessence field ϕ imposes restrictions on the values of g, m and λ_χ, namely,

$$g\left(\frac{m}{m_{\rm p}}\right)^2 \gtrsim 0.2\,\lambda_\phi^{1/2} \quad \text{and} \quad \frac{g^4}{\lambda_\chi}\left(\frac{m}{m_{\rm p}}\right)^4 \lesssim 0.001\,\lambda_\phi, \tag{2}$$

where $m \ll m_{\rm p}$ [1, 3].

For $|\phi| \geqslant m$ the potential $U(\phi,\chi)$ has just one minimum at $\chi = 0$, while for $|\phi| < m$ it has two minima located at $\chi = \pm\sqrt{g^2(m^2-\phi^2)/\lambda_\chi}$.

Within the hybrid quintessential inflationary model the evolution of the Universe can be divided into four stages. During the first stage ($\phi \leqslant -m$), the potential energy of the scalar field ϕ dominates the evolution, yielding a period of chaotic inflation, while the field χ remains at rest at the minimum of the potential U, $\chi_{\rm min} = 0$. Enough inflation is guaranteed by choosing, as initial condition, $\phi_i \lesssim -5m_{\rm p}$. Inflation is followed by a period of kination, during which the scalar field ϕ behaves as stiff matter with equation of state $p = \rho$. During the second stage of evolution, which takes place for $-m < \phi < m$, the origin of the potential becomes unstable for the scalar field χ. As a result, this field rolls down toward one of the new minima of the potential, $\chi_{\rm min} = \pm\sqrt{g^2(m^2-\phi^2)/\lambda_\chi}$, and starts to oscillate around it, acquiring kinetic energy. The Universe continues to undergo a period of kination. During the third stage of evolution ($\phi \geqslant m$), the scalar field χ returns to the minimum of the potential, located at $\chi = 0$, and oscillates around it. The transference of energy from this field to a radiation fluid is achieved by the introduction of a phenomenological dissipative coupling, proportional to the mass of χ, namely [1, 8],

$$\Gamma_\chi = \mu g\sqrt{\phi^2 - m^2}. \tag{3}$$

In the above expression, μ is the proportionality constant, which is constrained to be [1, 3]

$$0.2\,\lambda_\phi^{1/6}\left(\frac{m_{\rm p}}{m}\right)^{2/3}\frac{g^{5/3}}{\lambda_\chi} \ll \mu < \mu_{\rm crit}, \tag{4}$$

where $\mu_{\rm crit}$ is a critical value above which the oscillatory motion of the scalar field χ becomes over damped. The scalar field χ is assumed to decay away completely during this stage of evolution, reheating the Universe to a temperature of the order of 10^{14} GeV. Gradually, kination gives place to a radiation-dominated Universe. Finally, during the fourth stage of evolution, the Universe undergoes, successively, a period of radiation-domination, matter-domination and accelerated expansion, the latter being induced by the field ϕ, which also played the role of inflaton in the first stage of evolution.

Integrating numerically the equations of motion, corresponding to the above mentioned stages of evolution, with appropriate initial conditions and for the allowed values of the parameters of the model, one obtains the scale factor a and its first and second time derivatives as functions of time (see, for more details, Ref. [1]). The knowledge of these quantities, from the beginning of the inflationary period to the present time, allows us to determine the full gravitational-wave energy spectrum, i.e., the gravitational-wave spectral energy density parameter, $\Omega_{\rm GW}$, as a function of today's value of the angular frequency of the gravitational waves, ω_0. The method used to obtain this spectrum is the method of continuous Bogoliubov coefficients, first applied to particle production in an expanding Universe [9] and then extended to the case of gravitons [10]. For applications of this method to several cosmological models see Ref. [11].

Journal of Physics: Conference Series **314** (2011) 012062 doi:10.1088/1742-6596/314/1/012062

The gravitational-wave spectral energy density parameter is given by

$$\Omega_{\rm GW} = \frac{8\hbar G}{3\pi c^5 H_0^2}\omega_0^4\beta_0^2, \tag{5}$$

where $\hbar$ is the reduced Planck constant, G is the Newton constant, c is the speed of light, H is the Hubble parameter and β is a Bogoliubov coefficient. The subscript "0" denotes quantities evaluated at the present time. The squared Bogoliubov coefficient β^2, which gives the number of gravitons, can be expressed as $\beta^2 = (X - Y)^2/4$, where $X(t)$ and $Y(t)$ are functions of time, determined from the system of differential equations

$$\dot{X} = -i\omega_0\frac{a_0}{a}Y, \tag{6}$$

$$\dot{Y} = -\frac{i}{\omega_0}\frac{a}{a_0}\left[\omega_0^2\left(\frac{a_0}{a}\right)^2 - \frac{\ddot{a}}{a} - \left(\frac{\dot{a}}{a}\right)^2\right]X. \tag{7}$$

In the above equations, a_0 is today's value of the scale factor and a dot denotes a derivative with respect to the cosmic time t.

The gravitational-wave spectrum for the hybrid quintessential inflationary model, obtained applying the above mentioned procedure, is shown in Fig. 1. The sharp rise of the gravitational-wave spectral energy density parameter $\Omega_{\rm GW}$ in the high-frequency region of the spectrum is due to the existence of a kination period, occurring between the end of the inflationary era and the beginning of the radiation-dominated era. This peak, which is firmly located in the MHz–GHz range of frequencies, is an unavoidable feature of quintessential inflationary models [12].

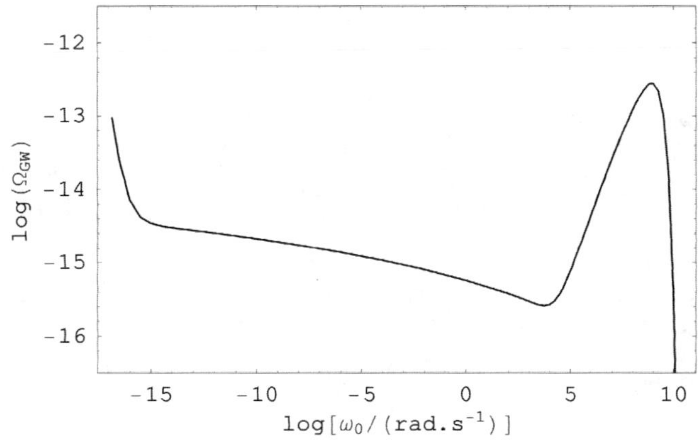

Figure 1. Gravitational-wave spectrum for $\lambda_\phi = 10^{-13}$, $M = 3.16\times10^{-14}\,m_{\rm p}$; $g = 6.3\times10^{-4}$, $m = 0.01\,m_{\rm p}$, $\lambda_\chi = 1$ and $\mu = 0.1$.

The duration of the kination period, and hence the height of the high-frequency peak, depends on the values of the parameters of the hybrid-like potential g, m and λ_χ, as well as on the value of the dissipation parameter μ [1]. Namely, if g and m increase (λ_χ decreases), the kination period becomes shorter and the peak smaller. This can be understood as follows. Let us consider, for example, the parameter g. An increase of g (for fixed values of the other parameters) leads to an increase of the value of χ at the minimum of the potential during the second stage of evolution. This implies that more energy is acquired by this field as it oscillates around the minimum of the potential. Consequently, more energy is available to be transferred, during the third stage of evolution, from the scalar field χ to the radiation fluid, leading to a higher reheating temperature. This, in turn, implies that, after the complete decay of χ, the period of time required for the energy density of radiation to become greater than the kinetic energy of the scalar field ϕ is shorter. As a result, the radiation-dominated era begins earlier, the kination period is shorter and the high-frequency peak of the spectrum is smaller. For the dissipation parameter μ, the duration of the kination period, as well as the height of the peak, increases as μ increases. Indeed, for higher values of μ, the energy transfer from the scalar field χ to the radiation fluid proceeds faster. As a consequence, the energy density of radiation begins to decrease as a^{-4} sooner, implying that the energy density of radiation is smaller after the complete decay of the field χ. Therefore, the time it takes for this energy density to dominate the dynamics of the Universe is longer, implying that the kination period is also longer and the peak bigger.

For the allowed values of the parameters of the model, the high-frequency peak of the gravitational-wave spectrum can be as high as $\Omega_{\mathrm{GW}} \simeq 10^{-12}$. Microwave-cavity and interferometric detectors, presently under development [13], are far from being sensitive enough to detect the primordial gravitational waves predicted by the hybrid quintessential inflationary model. However, one can hope that, in a not-so-distant future, a detection would be possible, allowing us to dramatically improve our understanding about processes taking place in the very early Universe.

Acknowledgments

This work was supported in part by *Fundação para a Ciência e a Tecnologia*, Portugal.

References

[1] Sá P M and Henriques A B 2010 *Phys. Rev.* D **81** 124043
[2] Grishchuk L P 1974 *Sov. Phys. JETP* **40** 409
Starobinskii A A 1979 *JETP Lett.* **30** 682
Abbott L F and Harari D D 1986 *Nucl. Phys.* B **264** 487
Allen B 1988 *Phys. Rev.* D **37** 2078
Sahni V 1990 *Phys. Rev.* D **42** 453
Grishchuk L P and Solokhin M 1991 *Phys. Rev.* D **43** 2566
Allen B 1997 *Proc. of the Les Houches School on Astrophysical Sources of Gravitational Waves (Les Houches, France, 1995)*, ed Marck J-A and Lasota J-P (Cambridge University Press) p 373
[3] Bastero-Gil M, Berera A, Jackson B M and Taylor A 2009 *Phys. Lett.* B **678** 157
[4] Peebles P J E and Vilenkin A 1999 *Phys. Rev.* D **59** 063505
[5] Felder G, Kofman L and Linde A 1999 *Phys. Rev.* D **60** 103505
[6] Smith T L, Kamionkowski M and Cooray A 2006 *Phys. Rev.* D **73** 023504
[7] Komatsu E *et al.* 2010 Seven-Year Wilkinson Microwave Anisotropy Probe (WMAP) Observations: Cosmological Interpretation *Preprint* arXiv:1001.4538v2 [astro-ph.CO]
[8] Yokoyama J and Maeda K 1988 *Phys. Lett.* B **207** 31
[9] Parker L 1969 *Phys. Rev.* **183** 1057
[10] Henriques A B 1994 *Phys. Rev.* D **49** 1771
Moorhouse R G, Henriques A B and Mendes L E 1994 *Phys. Rev.* D **50** 2600
Mendes L E, Henriques A B and Moorhouse R G 1995 *Phys. Rev.* D **52** 2083
[11] Henriques A B 2004 *Class. Quantum Gravity* **21** 3057
Henriques A B 2007 *Class. Quantum Gravity* **24** 6431 (E)
Sá P M and Henriques A B 2008 *Phys. Rev. D* **77** 064002
Sá P M and Henriques A B 2009 *Gen. Relativ. Gravit.* **41** 2345
Henriques A B, Potting R and Sá P M 2009 *Phys. Rev.* D **79** 103522
[12] Giovannini M 1999 *Phys. Rev.* D **60** 123511
[13] Bernard Ph, Gemme G, Parodi R and Picasso E 2001 *Rev. Sci. Instrum.* **72** 2428
Cruise A M and Ingley R M J 2006 *Class. Quantum Gravity* **23** 6185
Akutsu T *et al.* 2008 *Phys. Rev. Lett.* **101** 101101

Journal of Physics: Conference Series **314** (2011) 012063 doi:10.1088/1742-6596/314/1/012063

Fitting formulae for photon spectra from WIMP annihilation

J. A. R. Cembranos[1], A. de la Cruz-Dombriz[1,2], A. Dobado[1], R. Lineros[3] and A. L. Maroto[1]

[1] Departamento de Física Teórica I, Universidad Complutense de Madrid, E-28040, Spain.
[2] Department of Mathematics and Applied Mathematics, University of Cape Town, South Africa.
[3] IFIC, CSIC-Universidad de Valencia, Ed. Institutos, Apdo. correos 22085, E-46071, Spain.

E-mail: `dombriz at fis.ucm.es`

Abstract. Annihilation of different dark matter (DM) candidates into Standard Model (SM) particles could be detected through their contribution to the gamma ray fluxes that are measured on the Earth. The magnitude of such contributions depends on the particular DM candidate, but certain imprints of produced photon spectra may be analyzed in a model-independent fashion. In this work we provide the fitting formulae for the photon spectra generated by WIMP annihilation into quarks, leptons and gauge bosons channels in a wide range of WIMP masses.

1. Introduction

According to present observations of large scale structures, CMB anisotropies and light nuclei abundances, DM cannot be accommodated within the SM of elementary particles. Indeed, DM is not only a required component on cosmological scales, but it is also needed to provide a satisfactory description of rotational speeds of galaxies, orbital velocities of galaxies in clusters, gravitational lensing of background objects by galaxy clusters and the temperature distribution of hot gas in galaxies and clusters of galaxies. The experimental determination of the DM nature will require the interplay of collider experiments and astrophysical observations. Concerning the latter, DM searches use to be classified in direct or indirect ones (see [1] and references in the introduction section in [2]). Concerning direct detection, the elastic scattering of DM particles from nuclei should lead directly to observable nuclear recoil signatures although the weak interactions between DM and the standard matter makes DM direct detection extremely difficult.

On the other hand, DM might be detected indirectly, by observing their annihilation products into standard model particles. Thus, even if WIMPs (Weakly Interacting Massive Particles) are stable, two of them may annihilate into ordinary matter such as quarks, leptons and gauge bosons. Their annihilation in different places (galactic halo, Sun, etc.) produce cosmic rays which could be discriminated from the background through distinctive signatures. A cascade process follows the WIMPs annihilation. In the end, the potentially observable stable particles would be neutrinos, gamma rays, positrons and antimatter that may be observed through different devices. Among them, neutrinos and gamma rays have the advantage of maintaining their original direction due to their null electric charges.

Journal of Physics: Conference Series **314** (2011) 012063 doi:10.1088/1742-6596/314/1/012063

Concerning gamma rays, photon fluxes in specific (mainly SUSY) DM models are usually obtained by software packages such as DarkSUSY and micrOMEGAs based on PYTHIA Monte Carlo event generator [3] after having fixed a WIMP mass for the particular model under consideration.

This communication precisely focuses on providing the fitting functions for the photon production coming from quarks (including t quark), leptons and gauge bosons annihilations channels. This would allow to apply the results to alternative DM candidates for which software packages have not been developed. In addition, our investigation determine the dependence of such spectra on the WIMP mass in a model independent way.

On the other hand, the information about channel contribution and mass dependence can be very useful in order to identify gamma-ray signals for specific WIMP candidates and may also provide relevant information about the photon energy distribution when SM particle pairs annihilate.

Let us remind that the γ-ray flux from the annihilation of two WIMPs of mass M into two SM particles coming from all possible annihilation channels (labelled by the subindex i) is given by:

$$\frac{\mathrm{d}\,\Phi_\gamma^{\mathrm{DM}}}{\mathrm{d}E_\gamma} = \frac{1}{4\pi M^2}\sum_i \langle\sigma_i v\rangle \frac{\mathrm{d}N_\gamma^i}{\mathrm{d}E_\gamma} \times \frac{1}{\Delta\Omega}\int_{\Delta\Omega}\mathrm{d}\Omega \int_{\mathrm{l.o.s.}} \rho^2[r(s)]\;\mathrm{d}s\;, \tag{1}$$

where $\langle\sigma_i v\rangle$ holds for the thermal averaged annihilation cross-section of two WIMPs into two (i^{th} channel) SM particles, $\mathrm{d}N_\gamma^i/\mathrm{d}E_\gamma$ is the photon spectra in this i^{th} channel and ρ is the DM density. The integral is performed along the line of sight (l.o.s.) to the target and averaged over the detector solid angle $\Delta\Omega$.

2. Procedure

We have used the particle physics PYTHIA software [3] to obtain our results. The WIMPs annihilation is usually split into two separated processes: The first describes the annihilation of WIMPs and its SM output. The second one considers the evolution of the obtained SM unstable products. Due to the typical velocity dispersion of DM, we expect most of the annihilations to happen quasi-statically. This fact allows to state that by considering different center of mass (CM) energies for the obtained SM particles pairs from WIMP annihilation process, we are indeed studying different WIMP masses, i.e. $E_{\mathrm{CM}} \simeq 2\,M$. The procedure to obtain the photon spectra is thus straightforward: For a given pair of SM particles which are produced in the WIMP annihilation, we count the number of photons. The number of simulated collisions in each bin was fixed in order to provide suitable statistics in the number of produced photons. For instance, for the high energy bins many collisions are required to get a significant number of photons, whereas for low-intermediate energy, many photons are usually produced even for a small number of collisions.

3. Results

According to the extensive PYTHIA simulations performed for electroweak gauge bosons, leptons and quarks channels, three different fitting functions were required to fit the photon spectra: one for light quarks and leptons, one for gauge bosons (W and Z) and a final one for t quark. Those expressions depended on both WIMP mass dependent and independent parameters. Concerning the WIMP mass independent parameters, their values nevertheless do depend on the considered annihilation channel whereas for WIMP mass dependent ones, their evolutions with WIMP mass are parametrized by continuous and smooth curves as seen in [2]. Figure 1 shows two of these fits. The resulting expressions for the fitting functions are the following:

3.1. Light quarks and leptons

For $q\bar{q}$ (except the $t\bar{t}$ studied separately in section 3.3), $\tau^+\tau^-$ and $\mu^+\mu^-$ channels, the most general formula needed to reproduce the $\mathrm{d}N_\gamma^i/\mathrm{d}E_\gamma$ simulations may be written as:

$$\frac{\mathrm{d}N_\gamma}{\mathrm{d}x} = \frac{a_1}{x^{1.5}} \exp\left(-b_1 x^{n_1} - b_2 x^{n_2} - \frac{c_1}{x^{d_1}} + \frac{c_2}{x^{d_2}}\right) + q \ln\left[p(1-x^l)\right] \frac{x^2 - 2x + 2}{x} \tag{2}$$

where the logarithmic term takes into account the final state radiation through a Weizsäcker-Williams expression. Strictly speaking, the p parameter in the Weizsäcker-Williams term in the previous formula is $(M/m_{particle})^2$ where $m_{particle}$ is the mass of the charged particle that emits radiation. However in our approach, it will be a free parameter to be fitted since the radiation comes from many possible charged particles, which are produced along the decay and hadronization processes. Therefore all the bremsstrahlung effects were encapsulated in a single Weizsäcker-Williams-like term by using p and q parameters. l parameter is equal to one except for the $\mu^+\mu^-$ channel.

In fact, for the $\mu^+\mu^-$ channel, the above expression (2) becomes simpler since the exponential contribution is absent. The total gamma rays flux for this channel is thus well fitted by:

$$\frac{\mathrm{d}N_\gamma}{\mathrm{d}x} = q \ln\left[p(1-x^l)\right] \frac{x^2 - 2x + 2}{x} \tag{3}$$

For both leptons channels the covered WIMP mass range was from 25 to $5 \cdot 10^4$ GeV. Concerning the mass dependent parameters, they are n_1 and p in expression (2) for $\tau^+\tau^-$ channel (the rest of parameters are mass independent) and p, q and l in expression (3) for μ lepton. For $q\bar{q}$ channels, no general rule about mass (in)dependence of parameters can be settled, but specific results channel by channel are presented in [2].

Let us finally mention that the gamma rays from e^-e^+ pairs, the only contribution is that coming from bremsstrahlung. Therefore, the previous expression (3) is also valid with $q = \alpha_{\mathrm{QED}}/\pi$, $p = (M/m_{e^-})^2$ and $l \equiv 1$, parameters choice that obviously corresponds to the well-known Weizsäcker-Williams formula.

3.2. W and Z gauge bosons

For the W^+W^- and ZZ channels, the parametrization used to fit the Monte Carlo simulation is:

$$x^{1.5}\frac{\mathrm{d}N_\gamma}{\mathrm{d}x} = a_1 \exp\left(-b_1\, x^{n_1} - \frac{c_1}{x^{d_1}}\right) \left\{\frac{\ln[p(j-x)]}{\ln p}\right\}^q \tag{4}$$

This expression differs from expression (2) in the absence of the additive logarithmic contribution that acquires nonetheless a multiplicative behaviour. The exponential contribution is also quite simple with only one positive and one negative power laws. Moreover, a_1, n_1 and q parameters are WIMP mass independent (but they have different values depending on the studied gauge boson channel). The rest of parameters, i.e., b_1, c_1, d_1, p and j turned out to be both channel and mass dependent. The covered WIMP mass range was from 100 to 10^4 GeV for both channels. Nonetheless, at masses higher than 1000 GeV, no significant change in the photon spectra for both channels was observed.

3.3. t quark

Unlike the rest of the $q\bar{q}$ channels, the parametrization given by (2) is not valid when photon spectra are studied in the $t\bar{t}$ channel. For this one, the parametrization used to fit the simulations

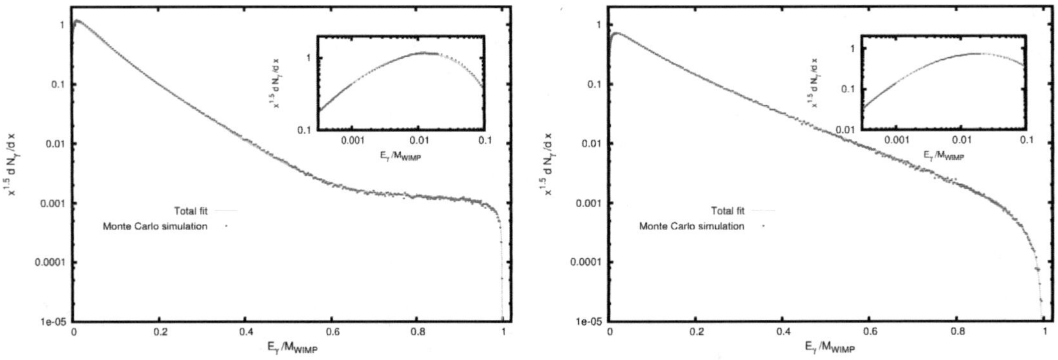

Figure 1. Photon spectra for $M_{WIMP} = 1$ TeV in the $b\bar{b}$ (left) and W^+W^- (right) channels. Dotted points are PYTHIA simulations and lines correspond to the proposed fitting functions.

is:

$$x^{1.5}\frac{\mathrm{d}N_\gamma}{\mathrm{d}x} = a_1 \exp\left(-b_1\, x^{n_1} - \frac{c_1}{x^{d_1}} - \frac{c_2}{x^{d_2}}\right)\left\{\frac{\ln[p(1-x^l)]}{\ln p}\right\}^q \tag{5}$$

In this formula, the exponential contribution is more complicated than the one in expression (4), with one positive and two negative power laws. Again, the additive logarithmic contribution is absent and acquires a multiplicative behaviour. Notice the exponent l in the logarithmic argument, required to provide correct fits in this channel.

The covered WIMP mass range was from 200 to 10^5 GeV. Nevertheless, at masses higher than 1000 GeV no significant change in the gamma-ray spectra was observed. The mass dependent parameters for this channel are b_1, n_1, c_2, p, q and l whereas a_1, c_1, d_1 and d_2 are mass independent.

4. Conclusions

We have presented the model-independent fitting functions for the photon spectra coming from WIMPs pair annihilation into Standard Model particle-antiparticle pairs for all the phenomenologically relevant channels. Explicit calculations for all studied channels [2] are available at the website http://teorica.fis.ucm.es/%7EPaginaWeb/downloads.html.

5. Acknowledgments

This work was supported by MICINN (Spain) project numbers FIS 2008-01323 and FPA 2008-00592, FPA2008-00319, CAM/UCM 910309, and Consolider-Ingenio MULTIDARK CSD2009-00064. AdlCD was also supported by the National Research Foundation (South Africa). RL was also supported by the EC contract UNILHC PITN-GA-2009-237920 and PROMETEO/2009/091 (Generalitat Valenciana).

References

[1] K. Sigurdson and M. Kamionkowski, PRL **92**, 171302 (2004); J. A. R. Cembranos *et al.* , PRL **90**, 241301 (2003); J. A. R. Cembranos and L. E. Strigari, PRD **77**, 123519 (2008); J. A. R. Cembranos, PRL **102**, 141301 (2009).

[2] J. A. R. Cembranos, A. de la Cruz-Dombriz, A. Dobado, R. A. Lineros, A. L. Maroto, arXiv(hep-ph): http://arxiv.org/abs/1009.4936.

[3] T. Sjostrand, S. Mrenna and P. Skands, JHEP05 (2006) 026 (LUTP 06-13, FERMILAB-PUB-06-052-CD-T).

Journal of Physics: Conference Series **314** (2011) 012064 doi:10.1088/1742-6596/314/1/012064

Over spinning a black hole?

Mariam Bouhmadi-López[1], Vitor Cardoso[1,2], Andrea Nerozzi[1] and Jorge V. Rocha[1]

[1] CENTRA, Dept. de Física, Instituto Superior Técnico, Av. Rovisco Pais 1, 1049 Lisboa, Portugal

[2] Department of Physics and Astronomy, The University of Mississippi, University, MS 38677, USA

E-mail: mariam.bouhmadi@ist.utl.pt, vitor.cardoso@ist.utl.pt, andrea.nerozzi@ist.utl.pt, jorge.v.rocha@ist.utl.pt

Abstract. A possible process to destroy a black hole consists on throwing point particles with sufficiently large angular momentum into the black hole. In the case of Kerr black holes, it was shown by Wald that particles with dangerously large angular momentum are simply not captured by the hole, and thus the event horizon is not destroyed. Here we reconsider this gedanken experiment for black holes in higher dimensions. We show that this particular way of destroying a black hole does not succeed and that Cosmic Censorship is preserved.

1. Introduction

Black holes can be formed through the collapse of matter, through sufficiently high-energy collisions of particles or quantum fluctuations in the early universe. Basically any process capable of confining a large portion of matter in a small enough space. Once formed, black holes are hard to kill. Quantum processes aside, no known classical mechanism can destroy a black hole. One of such processes was considered by Wald [1] many years ago and revisited recently [2, 3, 4, 5, 6]. It consists in throwing a point particle at a (four-dimensional) Kerr black hole with just the right angular momentum to spin the black hole up in such a way that eventually the horizon is disrupted. Indeed, the angular momentum of Kerr black holes is bounded by $J \leq M^2$, thus *if it were possible* for the black hole to capture particles of high enough angular momenta, then one might exceed this bound, possibly creating a naked singularity. Wald showed this cannot happen, as the potentially dangerous particles (i.e., those with large enough angular momentum) are never captured by the black hole [1].

The purpose of this short letter is to extend Wald's analysis to other spacetimes, in particular the Myers-Perry family of rotating black holes in higher dimensions [7]. Such an analysis is interesting because it allows one to test Cosmic Censorship in a very simple, yet realistic scenario.

2. Effective potential for radial motion of a test particle

We are interested in analysing the motion of a test particle in a D-dimensional rotating black hole. More precisely, we will be considering an asymptotically flat stationary black hole described by the Myers-Perry solution [7]. Unlike Kerr black hole, where there is only one possible rotation axis, and therefore there is only one angular momentum parameter, in higher dimensions there are several choices of rotation axis and there is a multitude of angular momentum parameters,

Journal of Physics: Conference Series **314** (2011) 012064 doi:10.1088/1742-6596/314/1/012064

each referring to a particular rotation plane [7]. Here, we will focus on two possible situations: (i) a black hole with all the angular momenta equal or (ii) a singly spinning black hole. In addition, we focus exclusively on the intuitively most dangerous process: particles falling in along the equator. Therefore, we use the effective "2+1" dimensional metric along the equatorial plane. It is as well easy to define the conserved energy, E, and the angular momentum L (per unit test-particle mass m_0 in the case of time-like geodesics [1]) associated to the time-like and rotational Killing vectors of the background geometry.

The equatorial motion in the geometry just described can be reduced to the radial equation $\dot{r}^2 = V_r$ [8, 9]. For a single rotational plane

$$r^2 V_r = \left[r^2 E^2 + \frac{M}{r^{D-3}} (aE - L)^2 + (a^2 E^2 - L^2) - \delta_1 \Delta_D \right] , \tag{1}$$

where $\Delta_D = r^2 + a^2 - M r^{5-D}$. Similar equations can be written when all angular momenta are equal. For instance, for even $D = 2(d+1)$ we find,

$$r^3 V_r = M \left(r^2 + a^2 \right)^{(4-D)/2} \left[r^2 \delta_1 + (L - aE)^2 \right] + r \left[(r^2 + a^2)(E^2 - \delta_1) - L^2 \right] , \tag{2}$$

while for odd $D = 2d + 1$ we obtain

$$r^2 V_r = M \left(r^2 + a^2 \right)^{(3-D)/2} \left[r^2 \delta_1 + (L - aE)^2 \right] + \left[(r^2 + a^2)(E^2 - \delta_1) - L^2 \right] . \tag{3}$$

On the previous equations $\delta_1 = 1, 0$ for timelike and null geodesics, respectively, and a dot stands for derivative respect to the proper time of the particle. The parameters a and M are related to the angular momentum and mass of the black hole (cf. Ref. [9])

The radial motion is completely governed by the potential V_r. If there are turning points outside the event horizon, then a particle coming from infinity can not reach the event horizon. Thus, the analysis we want to make is to study the maximum value of L for which there are either no turning points, or all of them lie inside the event horizon.

3. Spinning-up a black hole by throwing point particles

Now, we address the main issue of this paper: can we spin-up a black hole with mass $\mathcal{M}_0$ and angular momentum $\mathcal{J}_0$ in general D spacetime dimensions? For that, we throw in a particle of mass m_0 with angular momentum $\delta\mathcal{J} = m_0 L$ and energy $\delta\mathcal{M} = m_0 E$, such that $\delta\mathcal{M} \ll \mathcal{M}_0$ and $\delta\mathcal{J} \ll \mathcal{J}_0$. Upon absorption of this particle, the dimensionless spin of the black hole

$$j \equiv \frac{\mathcal{J}}{\mathcal{M}^{\frac{D-2}{D-3}}} , \tag{4}$$

changes to

$$j = j_0 + \delta j , \tag{5}$$

where the subscript stands for initial parameters of the black hole and

$$\delta j = \frac{m_0}{\mathcal{M}_0} \left(\frac{L}{\mathcal{M}_0^{\frac{1}{D-3}}} - E j_0 \frac{D-2}{D-3} \right) . \tag{6}$$

[1] For massless particles, the quantities E and L may be regarded as the energy and angular momentum.

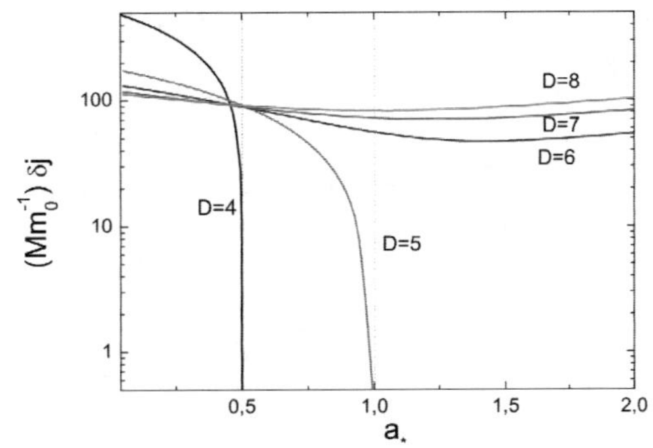

Figure 1. This figure shows the maximum increase in spin, $\frac{M}{m_0}(\delta j)_{\max}$ caused by a particle with $E/m_0 = 100$ falling into a Myers-Perry black hole with a single rotation parameter. The dimensionless rotation parameter a_* is defined as $a_* \equiv \frac{a}{M^{1/(D-3)}}$. Notice that it is not possible to spin-up an extremal black hole (for $D = 4, 5$, the extremal value is marked with a dotted line).

3.1. Single rotation parameter

As in four dimensions, in $D = 5$ we can also spin the black hole to the extremal limit and not further than that [10, 9]. What about general D? We have numerically searched for the critical angular momentum, and computed δj in Eq. (6). The results, which are summarised in Fig. 1, are clear: neutral black holes in four and five spacetime dimensions with a single rotation cannot be spun-up past extremality. For larger D, there is no extremal limit, and the black holes can be spun-up to an arbitrarily high rotation.

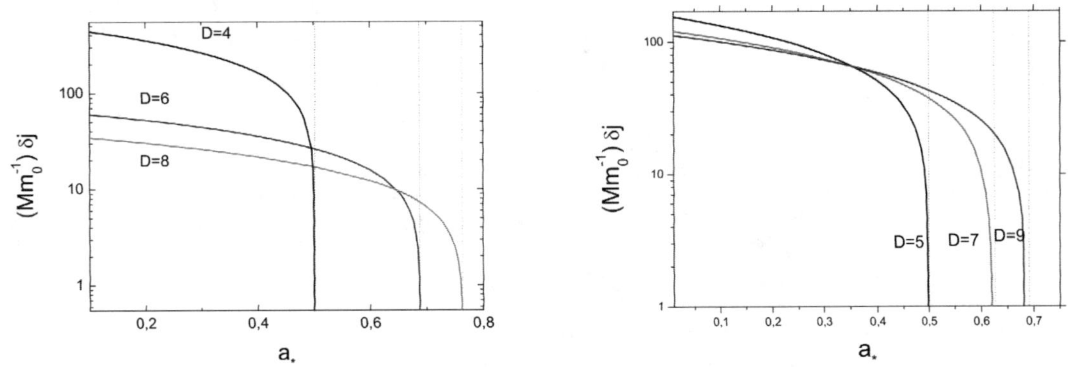

Figure 2. This figure shows the maximum increase in spin, $\frac{M}{m_0}(\delta j)_{\max}$ caused by d particles with $E/m_0 = 100$ falling into a Myers-Perry black hole with all rotation parameters equal. The left panel considers even D, the right panel refers to odd spacetime dimensions D. It is not possible to spin-up an extremal black hole, marked with a dotted line. The dimensionless rotation parameter $a_* \equiv \frac{a}{M^{1/(D-3)}}$.

3.2. All angular momenta equal

As we are dealing with a Myers-Perry black hole with equal angular momenta, by throwing only one point particle, we would end up with a Myers-Perry black hole with different angular momenta where the condition of all angular momenta being equal does not apply and we could conclude erroneously a violation of the cosmic censorship. How can we avoid this situation? Instead of throwing in one test particle, we consider d particles following similar geodesics along the d orthogonal rotation planes. The final black hole will also have all angular momenta equal. In this situation the expression (6) should read

$$\delta j = \frac{m_0}{\mathcal{M}_0}\left(\frac{L}{\mathcal{M}_0^{\frac{1}{D-3}}} - dEj_0\frac{D-2}{D-3}\right), \tag{7}$$

and the results are presented in Fig. 2. In full analogy with the singly spinning case in $D = 5$ in which the spin is bounded, we cannot exceed the extremal limit by throwing in test particles.

4. Conclusions

We have shown that a D-dimensional Myers-Perry black hole is immune to the throwing of point particles: in the geodesic approximation employed here, particles which are captured by the black hole have an angular momentum which is sufficiently low so as to be harmless; in fact sufficiently low that they are never able to spin-up the geometry past the extremal value.

Acknowledgements

V.C. is supported by a "Ciência 2007" research contract and by Fundação Calouste Gulbenkian through a short-term scholarship. M.B.L., A.N. and J.V.R. acknowledge financial support from *Fundação para a Ciência e Tecnologia* (FCT)-Portugal through the fellowships SFRH/BPD/26542/2006, SFRH/BPD/47955/2008 and SFRH/BPD/47332/2008, respectively. This work was partially funded by *Fundação para a Ciência e Tecnologia* (FCT)-Portugal through projects PTDC/FIS/64175/2006, PTDC/ FIS/098025/2008, PTDC/FIS/098032/2008 and CERN/FP/109290/2009. The authors thankfully acknowledge the computer resources, technical expertise and assistance provided by the Barcelona Supercomputing Center - Centro Nacional de Supercomputación.

References

[1] R. Wald, "Gedanken experiments to destroy a black hole," Ann. Phys. **82**, 548 (1974).
[2] G. E. A. Matsas and A. R. R. da Silva, Phys. Rev. Lett. **99**, 181301 (2007) [arXiv:0706.3198 [gr-qc]].
[3] G. E. A. Matsas, M. Richartz, A. Saa, A. R. R. da Silva and D. A. T. Vanzella, Phys. Rev. D **79**, 101502 (2009) [arXiv:0905.1077 [gr-qc]].
[4] S. Hod, Phys. Rev. Lett. **100**, 121101 (2008) [arXiv:0805.3873 [gr-qc]].
[5] V. E. Hubeny, Phys. Rev. D **59**, 064013 (1999) [arXiv:gr-qc/9808043].
[6] T. Jacobson and T. P. Sotiriou, arXiv:0907.4146 [gr-qc].
[7] R. C. Myers and M. J. Perry, Annals Phys. **172**, 304 (1986).
[8] V. Cardoso, A. S. Miranda, E. Berti, H. Witek and V. T. Zanchin, Phys. Rev. D **79**, 064016 (2009) [arXiv:0812.1806 [hep-th]].
[9] M. Bouhmadi-López, V. Cardoso, A. Nerozzi and J. V. Rocha, Phys. Rev. D **81**, 084051 (2010) [arXiv:1003.4295 [gr-qc]].
[10] H. Elvang, R. Emparan and A. Virmani, JHEP **0612**, 074 (2006) [arXiv:hep-th/0608076].

Spanish Relativity Meeting (ERE 2010): Gravity as a Crossroad in Physics IOP Publishing
Journal of Physics: Conference Series **314** (2011) 012065 doi:10.1088/1742-6596/314/1/012065

Electrically charged black hole solutions in generalized gauge field theories

J. Diaz-Alonso, D. Rubiera-Garcia

LUTH, Observatoire de Paris, CNRS, Université Paris Diderot. 5 Place Jules Janssen, 92190 Meudon, France, and
Departamento de Fisica, Universidad de Oviedo. Avda. Calvo Sotelo 18, 33007 Oviedo, Asturias, Spain

E-mail: joaquin.diaz@obspm.fr ; diego.rubiera-garcia@obspm.fr

Abstract. We summarize the main features of a class of *anomalous* (asymptotically flat, but non Schwarzschild-like) gravitational configurations in models of gravitating non-linear electrodynamics (G-NED) whose Lagrangian densities are defined as arbitrary functions of the two field invariants and constrained by several physical admissibility conditions. This class of models and their associated electrostatic spherically symmetric black hole (ESSBH) solutions are characterized by the behaviours of the Lagrangian densities around the vacuum and at the boundary of their domain of definition.

1. Preliminaries

For several decades a great deal of attention has been focussed on G-NED as generalizations of the Reissner-Nordström (RN) solution of the Einstein-Maxwell field equations, whose more popular example is the gravitating Born-Infeld theory [1]. Some other G-NED models supporting asymptotically Schwarzschild-like solutions have been also considered in the literature [2]. In addition to such classes of models (and gravitational configurations) there are others leading to ESSBH solutions approaching asymptotic flatness slower than the Schwarzschild field [3]. In this work we shall analyze the full class of physically admissible G-NEDs supporting this kind of *asymptotically anomalous* solutions.

The action determining the dynamics of a G-NED is defined as

$$S = S_G + S_{NED} = \int d^4x\sqrt{-g}\left[\frac{R}{16\pi G} - \varphi(X,Y)\right], \tag{1}$$

where $\varphi(X,Y)$ is a (unspecified) function of the two field invariants $X = -\frac{1}{2}F_{\mu\nu}F^{\mu\nu} = \vec{E}^2 - \vec{H}^2, Y = -\frac{1}{2}F_{\mu\nu}F^{*\mu\nu} = 2\vec{E}\cdot\vec{H}$, which are built from the field strength tensor $F_{\mu\nu}$ and its dual $F^*_{\mu\nu} = \frac{1}{2}\epsilon_{\mu\nu\alpha\beta}F^{\alpha\beta}$ and where the electric $\vec{E}$ and magnetic $\vec{H}$ fields are defined in the usual way. The associated energy-momentum tensor in flat space is obtained as

$$T_{\mu\nu} = 2F_{\mu\alpha}\left(\frac{\partial\varphi}{\partial X}F^{\alpha}_{\nu} + \frac{\partial\varphi}{\partial Y}F^{*\alpha}_{\nu}\right) - \varphi(X,Y)\eta_{\mu\nu}. \tag{2}$$

Consistently with previous conventions we shall call *physically admissible* the models satisfying the minimal requirements of (i) definiteness, continuity and regularity of $\varphi(X,Y)$ in their domain of definition of the $X-Y$ plane, assumed to be open, connected and including the vacuum, (ii) parity invariance $\varphi(X,Y)=\varphi(X,-Y)$, (iii) positive definite character of the energy, endorsed by the necessary and sufficient condition

$$\rho \geq \left(\sqrt{X^2+Y^2}+X\right)\frac{\partial\varphi}{\partial X}+Y\frac{\partial\varphi}{\partial Y}-\varphi(X,Y)\geq 0. \quad (3)$$

The associated *flat-space* field equations lead, for the electrostatic spherically symmetric (ESS) solutions ($E(\vec{r})=E(r)\vec{r}/r, \vec{H}=0$), to the first-integral

$$r^2\frac{\partial\varphi}{\partial X}\Big|_{Y=0}E(r)=Q, \quad (4)$$

where Q is an integration constant, identified as the electric charge. This equation allows the complete determination of the ESS solutions for admissible models. Assuming a field behaviour of the form $E(r\to 0)\sim r^p$ we obtain three possible cases: ultraviolet divergent (**UVD**) when $p\leq -1$, for which the integral of the energy diverges around $r=0$; **A1** when $-1<p<0$, for which the field diverges at $r=0$ but the integral of energy converges there; **A2** when $p=0$ (and $E(r\to 0)\sim a-br^\sigma$ with $\sigma>0$) for which the central field is finite and the integral of energy converges there. Concerning the asymptotic behaviour of the solutions, assumed to be of the form $E(r\to\infty)\sim r^q$, we are lead to four possible cases: infrared divergent (**IRD**) when $-1\leq q<0$, for which the fields vanish as $r\to\infty$ but the integral of energy diverges there; **B1** ($-2<q<-1$), **B2** ($q=-2$) and **B3** ($q<-2$) cases, for which the integral of energy converges asymptotically. The twelve combinations of these central and asymptotic behaviours exhaust the full class of admissible NEDs. Excluding the models leading to UVD and IRD behaviours we obtain six classes exhausting the set of admissible models supporting finite-energy ESS solutions. The minimal coupling to gravity of the nine classes of models obtained by excluding the IRD behaviour was extensively analyzed in Ref.[4]. This analysis determines and classifies the possible G-ESS configurations, which are reduced to *extreme BHs* (EBH), *one or two-horizons BHs*, *extreme and non-extreme black points* and *naked singularities* (NS), depending on the values of Q and the ADM mass coming from the integration of the field equations. All these configurations behave asymptotically as the Schwarzschild field. As we shall see, the three classes of models supporting ESS solutions with IRD behaviour lead to asymptotically anomalous gravitational configurations with similar structures at finite r.

2. Asymptotically anomalous gravitational configurations

Let us write the general expression of the G-ESS solutions associated to the action (1). The symmetry of the source $T_0^0=T_1^1$ leads to the static spherically symmetric line element [4]

$$ds^2=\lambda(r)dt^2-\lambda^{-1}(r)dr^2-r^2d\Omega^2, \quad (5)$$

where $d\Omega^2=d\theta^2+\sin^2\theta d\varphi^2$. With the line element (5) both the first-integral (4) and $X=\vec{E}^2(r,Q)$ have the same expression in flat space and in the gravitational context, allowing an immediate translation of all results obtained in the former case to the latter one.

Let us analyze three classes of admissible models supporting asymptotically IRD solutions with central field behaviours A1, A2 or UVD. Now the integral of energy diverges at $r\to\infty$ as $r^{q+1}, (-1\leq q<0)$ but converges at the center in the A1 and A2 cases, allowing the definition of the *interior integral of energy* $\varepsilon_{in}(r,Q)=\int_0^r R^2T_0^0(R,Q)dR$ which, owing to the admissibility conditions, is a **monotonically increasing and convex** function of r. The general solution of the Einstein equations for the metric reads in these cases

$$\lambda(r,Q,C)=1+\frac{C}{r}-\frac{2\varepsilon_{in}(r,Q)}{r}, \quad (6)$$

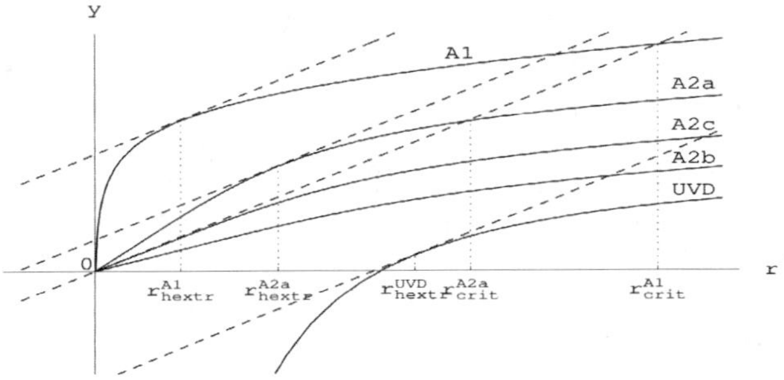

Figure 1. : Interior integral of energy for the different families of NED models with asymptotically IRD field behaviours, cut by the beam of straight lines $(r+C)/2$.

where C is an integration constant. The radii of the horizons (zeroes of (6)) are obtained as solutions of $\frac{r+C}{2} = \varepsilon_{in}(r,Q)$ and thus are given by the intersections between the curves $\varepsilon_{in}(r,Q)$ and the beam of straight lines $(r+C)/2$ corresponding to different values of C and Q (see Fig.1). Through this procedure we are led to the following structures for the configurations:

2.1. A1-IRD

In this case $\varepsilon_{in}(r,Q)$ has a divergent slope as $r \sim 0$ and the above method leads to five classes of configurations. In particular there is a value $C = C_{extr}(Q)$, corresponding to the straight line tangent to the curve $\varepsilon_{in}(r,Q)$, defining the radius $r_{hextr}(Q)$ of a EBH configuration (for which the horizon is degenerate, i.e. $\lambda(r) = \lambda'(r)|_{r=r_h} = 0$). These values are obtained as

$$8\pi r^2_{hextr}(Q) T^0_0(r_{hextr}, Q) = 1 \ ; \ C_{extr}(Q) = 2\varepsilon_{in}(r_{hextr}(Q), Q) - r_{hextr}(Q). \tag{7}$$

As easily seen in Fig.1, for $C > C_{extr}(Q)$ there are not cut points and the associated configurations are NS, while for $0 < C < C_{extr}(Q)$ they are two-horizons BHs. On the other hand, for $C < 0$ there is always a single cut point corresponding to BHs with a single (non-extreme) horizon. In the critical case ($C = 0$) the metric diverges to $-\infty$ around $r = 0$ as

$$\lambda(r) = 1 - \frac{32\pi Q}{(2-p)(p+1)} r^p + \cdots, \tag{8}$$

and increases monotonically to $\lambda(r \to \infty) = 1$ exhibiting a single-horizon as well. Consequently the central singularity is timelike for $C > 0$ and spacelike for $C \leq 0$. As $C \to 0^{\pm}$ in Eq.(6) the metric function converges to the critical one for $r > 0$ but differs at the center by δ-like terms. At large r, for all these IRD cases the asymptotic behaviour of the metric (6) approaches flatness as $\lambda(r \to \infty) - 1 \sim -r^q$ (for $-1 < q < 0$, and as $\sim \ln(r)/r$ for $q = -1$) and is not Schwarzschild-like. Thus the ADM mass is not defined in the asymptotically "anomalous" cases.

2.2. A2-IRD

From Fig.1 we see that similar configurations as in case A1 are obtained for $C \neq 0$ while for the critical configuration, $C = 0$, the metric at the center, which behaves as

$$\lambda(r \to 0, C = 0, Q, a, b, \sigma) \sim 1 - 16\pi Q a + \frac{32\pi b Q}{(\sigma+1)(2-\sigma)} r^{\sigma} + \chi r^2, \tag{9}$$

(χ: integration constant) is finite. Its sign there depends on the quantity $16\pi Q a$. In the **A2a** case the critical metric takes the value at the center $\lambda(0) = 1 - 16\pi Q a > 0$ and its slope there is

$$\left.\frac{\partial \lambda}{\partial r}\right|_{r\to 0} \sim \frac{32\pi b Q \sigma}{(\sigma+1)(2-\sigma)} r^{\sigma-1} + 2\chi r. \tag{10}$$

This implies that the slope of the metric at the center for the critical configuration is non-negative and depends on σ, becoming divergent for $\sigma < 1$, vanishing for $\sigma > 1$ and taking a finite value for $\sigma = 1$. As a consequence the critical metric, which is always monotonically increasing, leads to timelike NS. On the other hand, in case **A2b** corresponding to $\lambda(0) = 1 - 16\pi Qa < 0$ the critical configuration exhibits a single-horizon BH with a spacelike singularity at the center. The case **A2c** with $\lambda(0) = 1 - 16\pi Qa = 0$ leads to extreme black points [5].

2.3. UVD-IRD

In this case $\varepsilon_{in}(r, Q)$ cannot be defined, but now the Einstein equations can be integrated as

$$\lambda(r, Q, D) = 1 + \frac{D}{r} - \frac{2\varepsilon(r, Q, 0)}{r}, \qquad (11)$$

where $\varepsilon(r, Q, \Gamma) = 4\pi \int r^2 T_0^0(r, Q) + \Gamma$ and the arbitrary constant Γ has been absorbed in the constant D. This function is monotonically increasing and convex, exhibiting a vertical asymptote at $r = 0$ and diverging with vanishing slope as $r \to \infty$. Following the same procedure as in the previous cases, the radii of the horizons are obtained from the cutting points of the function $\varepsilon(r, Q, 0)$ with the beam of straight lines $(r + D)/2$, corresponding to different values of D. There is again a tangent line to the curve $\varepsilon(r, Q, 0)$, determining the radius of an EBH configuration through the same equation (7). For $D > D_{hextr}(Q)$ the configurations are NS, while two-horizons BHs appears for $D < D_{hextr}(Q)$, as can be seen from Fig.1.

3. Discussion

The analysis of the asymptotically anomalous class of NED models considered here, together with the one of the asymptotically Schwarzschild-like class studied in [4], exhausts the family of G-ESS solutions of physically admissible NEDs. For these families, aside from NS, EBHs and two-horizon BHs already present in the RN case, there are in addition single-horizon BHs and (in case A2) extreme and non-extreme black points [5]. Thermodynamics of the different gravitating ESS structures outlined here, making use of the fact that NED models satisfy both the zeroth and first laws of BH thermodynamics [6], is currently in progress.

Let us finally point out that the present analysis can be extended to non-abelian gauge fields. In this case the lagrangian density is still given by Eq.(1), where the field invariants are obtained by using the usual prescription in the calculation of the traces over the gauge group generators, leading to $X = -\frac{1}{2}F_{\mu\nu a}F_a^{\mu\nu} = \vec{E}_a^2 - \vec{H}_a^2, Y = -\frac{1}{2}F_{\mu\nu a}F_a^{*\mu\nu} = 2\vec{E}_a \cdot \vec{H}_a$, $a = 1 \cdots N$ (however, other prescriptions are possible). This problem can be reduced, for ESS fields ($\vec{E}^a(r) \neq 0, \vec{H}^a = 0$), to the abelian one in a correspondence between ESS fields of both theories associated to the same form of the Lagrangian density, solving the gravitating problem for the former in terms of the solution of the latter. This extension will appear in a forthcoming publication.

References

[1] Garcia A, Salazar H and Plebanski J F 1984 *Nuovo. Cim.* **84** 65
Demianski 1986 *Found. Phys.* **16** 187
Breton N 2003 *Phys. Rev.* D **67** 124004
Dey T K 2004 *Phys. Lett.* B **595** 484
[2] de Oliveira H P 1994 *Class. Quantum. Grav.* **11** 1469
Soleng H H 1995 *Phys. Rev.* D **52** 6178
Yajima H and Tamaki T 2001 *Phys. Rev.* D **63** 064007
[3] Hassaine M and Martinez C 2007 *Phys. Rev.* D **75** 027502
Hassaine M and Martinez C 2008 *Class. Quant. Grav.* **25** 195023
[4] Diaz-Alonso J and Rubiera-Garcia D 2010 *Phys. Rev.* D **81** 064021
[5] Diaz-Alonso J and Rubiera-Garcia D 2010 *Phys. Rev.* D **82** 085024
[6] Rasheed D A 1997 *Preprint* arXiv:hep-th/9702087

Spanish Relativity Meeting (ERE 2010): Gravity as a Crossroad in Physics IOP Publishing
Journal of Physics: Conference Series **314** (2011) 012066 doi:10.1088/1742-6596/314/1/012066

The behaviour of geodesics in constant-curvature spacetimes with expanding impulsive gravitational waves

Robert Švarc and Jiří Podolský

Institute of Theoretical Physics, Faculty of Mathematics and Physics,
Charles University in Prague, V Holešovičkách 2, 180 00 Prague 8, Czech Republic

E-mail: robert.svarc@mff.cuni.cz and podolsky@mbox.troja.mff.cuni.cz

Abstract. Motion of free test particles in spacetimes of constant curvature with an expanding impulsive gravitational wave is completely described. Explicit formulae which identify the particle positions and velocities "in front of" and "behind" the impulse are derived for a general impulse of this type. As a particular example, the effect of spherical impulsive wave generated by a snapped cosmic string is analyzed and visualized.

1. Introduction

In the classic work [1], Penrose described the construction of expanding spherical impulsive gravitational waves using his "cut and paste" method. Continuous metric form which represents this class of spacetimes was found later by Hogan [2]-[4] and, in the general case, by Podolský and Griffiths [5]. Motion of free test particles in flat background space with these waves was studied by Podolský and Steinbauer [6]. Recently we generalised these results to an arbitrary cosmological constant Λ in [7], where we described the behaviour of general C^1 geodesics in detail.

2. Spacetimes with expanding impulsive waves

Expanding impulsive gravitational waves of a spherical shape can be constructed using the Penrose "cut and paste" method, i.e., by "cutting" a constant-curvature spacetime along a null cone and "glueing it together" with a suitable warp. Continuous metric form which describes such a family of spacetimes can be written as

$$\mathrm{d}s^2 = \frac{2\left|\frac{V}{p}\mathrm{d}Z + U\Theta(U)p\bar{H}\mathrm{d}\bar{Z}\right|^2 + 2\mathrm{d}U\mathrm{d}V - 2\epsilon\mathrm{d}U^2}{\left[1+\frac{1}{6}\Lambda U(V-\epsilon U)\right]^2}\,, \tag{1}$$

where $p = 1+\epsilon Z\bar{Z}$, $\epsilon = -1, 0, +1$, $H = \frac{1}{2}\left[h'''/h' - \frac{3}{2}(h''/h')^2\right]$, and $h(Z)$ is an arbitrary complex function of the complex spatial coordinate Z (see [2]-[5] for more details). The impulse is located on the null hypersurface $U = 0$ which is a sphere expanding with the speed of light. Relations of this continuous line element to the conformally flat metric form of the background

$$\mathrm{d}s^{\pm 2} = \frac{-\mathrm{d}t^{\pm 2} + \mathrm{d}x^{\pm 2} + \mathrm{d}y^{\pm 2} + \mathrm{d}z^{\pm 2}}{\left[1 + \frac{\Lambda}{12}(-t^{\pm 2} + x^{\pm 2} + y^{\pm 2} + z^{\pm 2})\right]^2}\,, \quad \text{i.e.,} \quad \mathrm{d}s^{\pm 2} = \frac{2\mathrm{d}\eta^{\pm}\mathrm{d}\bar{\eta}^{\pm} - 2\mathrm{d}\mathcal{U}^{\pm}\mathrm{d}\mathcal{V}^{\pm}}{[1 + \frac{1}{6}\Lambda(\eta^{\pm}\bar{\eta}^{\pm} - \mathcal{U}^{\pm}\mathcal{V}^{\pm})]^2} \tag{2}$$

with $\eta^{\pm} \equiv \frac{1}{\sqrt{2}}(x^{\pm} + \mathrm{i}y)$, $\mathcal{U}^{\pm} \equiv \frac{1}{\sqrt{2}}(t^{\pm} - z^{\pm})$, $\mathcal{V}^{\pm} \equiv \frac{1}{\sqrt{2}}(t^{\pm} + z^{\pm})$ [which is Minkowski ($\Lambda = 0$), de Sitter ($\Lambda > 0$) or anti-de Sitter ($\Lambda < 0$) spacetime] are given by two different transformations, namely

$$\mathcal{V}^{+} = AV - DU\,, \qquad \mathcal{U}^{+} = BV - EU\,, \qquad \eta^{+} = CV - FU\,, \tag{3}$$

"in front of" the impulse ($U > 0$), and

$$\mathcal{V}^{-} = \frac{V}{p} - \epsilon U\,, \qquad \mathcal{U}^{-} = \frac{Z\bar{Z}}{p}V - U\,, \qquad \eta^{-} = \frac{Z}{p}V\,, \tag{4}$$

"behind" the impulse ($U < 0$), where the functions in (3) are given by

$$\begin{aligned}
A &= \frac{1}{p|h'|}\,, \qquad B = \frac{|h|^2}{p|h'|}\,, \qquad C = \frac{h}{p|h'|}\,,\\
D &= \frac{1}{|h'|}\left\{\frac{p}{4}\left|\frac{h''}{h'}\right|^2 + \epsilon\left[1 + \frac{Z}{2}\frac{h''}{h'} + \frac{\bar{Z}}{2}\frac{\bar{h}''}{\bar{h}'}\right]\right\}\,,\\
E &= \frac{|h|^2}{|h'|}\left\{\frac{p}{4}\left|\frac{h''}{h'} - 2\frac{h'}{h}\right|^2 + \epsilon\left[1 + \frac{Z}{2}\left(\frac{h''}{h'} - 2\frac{h'}{h}\right) + \frac{\bar{Z}}{2}\left(\frac{\bar{h}''}{\bar{h}'} - 2\frac{\bar{h}'}{\bar{h}}\right)\right]\right\}\,,\\
F &= \frac{h}{|h'|}\left\{\frac{p}{4}\left(\frac{h''}{h'} - 2\frac{h'}{h}\right)\frac{\bar{h}''}{\bar{h}'} + \epsilon\left[1 + \frac{Z}{2}\left(\frac{h''}{h'} - 2\frac{h'}{h}\right) + \frac{\bar{Z}}{2}\frac{\bar{h}''}{\bar{h}'}\right]\right\}\,.
\end{aligned} \tag{5}$$

3. Refraction formulae

Suppose C^1 geodesics $Z = Z(\tau)$, $U = U(\tau)$ and $V = V(\tau)$ in the continuous coordinates of (1) and denote the positions and velocities at the interaction time τ_i when $U = 0$ as $Z_i \equiv Z(\tau_i)$, $V_i \equiv V(\tau_i)$, $\dot{U}_i \equiv \dot{U}(\tau_i)$, $\dot{V}_i \equiv \dot{V}(\tau_i)$, $\dot{Z}_i \equiv \dot{Z}(\tau_i)$. We may apply the above two different transformations (4) and (3) in the half-spaces "behind" and "in front of" the impulse to obtain

$$\begin{aligned}
\mathcal{V}_i^{-} &= \frac{V_i}{p}\,, & \dot{\mathcal{V}}_i^{-} &= -\frac{\epsilon V_i}{p^2}(Z_i\dot{\bar{Z}}_i + \bar{Z}_i\dot{Z}_i) + \frac{\dot{V}_i}{p} - \epsilon\dot{U}_i\,,\\
\mathcal{U}_i^{-} &= \frac{|Z_i|^2}{p}V_i\,, & \dot{\mathcal{U}}_i^{-} &= \frac{V_i}{p^2}(Z_i\dot{\bar{Z}}_i + \bar{Z}_i\dot{Z}_i) + \frac{|Z_i|^2}{p}\dot{V}_i - \dot{U}_i\,,\\
\eta_i^{-} &= \frac{Z_i}{p}V_i\,, & \dot{\eta}_i^{-} &= \frac{V_i}{p^2}(\dot{Z}_i - \epsilon Z_i Z_i\dot{\bar{Z}}_i) + \frac{Z_i}{p}\dot{V}_i\,,
\end{aligned} \tag{6}$$

and

$$\begin{aligned}
h(Z_i) &= \frac{\eta_i^{+}}{\mathcal{V}_i^{+}}\,, & \dot{Z}_i &= \frac{p^2}{V_i}\left(\bar{C}_{,\bar{Z}}\,\dot{\eta}_i^{+} + C_{,\bar{Z}}\,\dot{\bar{\eta}}_i^{+} - A_{,\bar{Z}}\dot{\mathcal{U}}_i^{+} - B_{,\bar{Z}}\,\dot{\mathcal{V}}_i^{+}\right),\\
V_i = \frac{\mathcal{U}_i^{+}}{B} &= \frac{\mathcal{V}_i^{+}}{A} = \frac{\eta_i^{+}}{C}\,, & \dot{V}_i &= D\dot{\mathcal{U}}_i^{+} + E\,\dot{\mathcal{V}}_i^{+} - \bar{F}\,\dot{\eta}_i^{+} - F\,\dot{\bar{\eta}}_i^{+} + 2\epsilon\,\dot{U}_i\,,\\
& & \dot{U}_i &= \frac{1}{V_i}\left(\bar{\eta}_i^{+}\dot{\eta}_i^{+} + \eta_i^{+}\dot{\bar{\eta}}_i^{+} - \mathcal{V}_i^{+}\dot{\mathcal{U}}_i^{+} - \mathcal{U}_i^{+}\dot{\mathcal{V}}_i^{+}\right)\,,
\end{aligned} \tag{7}$$

respectively. Combining (6) with (7), it is now straightforward to express the interaction parameters behind the impulse as specific functions of the parameters in front of the impulse. In the conformally flat coordinates of (2), for the positions of test particles we obtain

$$\begin{aligned} x_i^- &= |h_i'|\frac{Z_i+\bar{Z}_i}{h_i+\bar{h}_i}x_i^+ , & y_i^- &= |h_i'|\frac{Z_i-\bar{Z}_i}{h_i-\bar{h}_i}y_i^+ , \\ z_i^- &= |h_i'|\frac{|Z_i|^2-1}{|h_i|^2-1}z_i^+ , & t_i^- &= |h_i'|\frac{|Z_i|^2+1}{|h_i|^2+1}t_i^+ , \end{aligned} \tag{8}$$

and for their velocities we get

$$\begin{aligned} \dot{x}_i^- &= a_x\dot{x}_i^+ + b_x\dot{y}_i^+ + c_x\dot{z}_i^+ + d_x\dot{t}_i^+ , & \dot{y}_i^- &= a_y\dot{x}_i^+ + b_y\dot{y}_i^+ + c_y\dot{z}_i^+ + d_y\dot{t}_i^+ , \\ \dot{z}_i^- &= a_z\dot{x}_i^+ + b_z\dot{y}_i^+ + c_z\dot{z}_i^+ + d_z\dot{t}_i^+ , & \dot{t}_i^- &= a_t\dot{x}_i^+ + b_t\dot{y}_i^+ + c_t\dot{z}_i^+ + d_t\dot{t}_i^+ , \end{aligned} \tag{9}$$

where the coefficients a_p, b_p, c_p, d_p are (somewhat complicated) functions of Z_i and $h_i \equiv h(Z_i)$, see [7]. If we define natural angles which characterize the positions and velocity directions in the (x,z) and (y,z) planes as

$$\tan\alpha^\pm \equiv \frac{x_i^\pm}{z_i^\pm} , \qquad \tan\beta^\pm \equiv \frac{\dot{x}_i^\pm}{\dot{z}_i^\pm} , \qquad \tan\gamma^\pm \equiv \frac{y_i^\pm}{z_i^\pm} , \qquad \tan\delta^\pm \equiv \frac{\dot{y}_i^\pm}{\dot{z}_i^\pm} , \tag{10}$$

the expressions (8) identifying the positions can be written as

$$\tan\alpha^- = \frac{(|h_i|^2-1)\,\mathrm{Re}\,Z_i}{(|Z_i|^2-1)\mathrm{Re}\,h_i}\tan\alpha^+ , \qquad \tan\gamma^- = \frac{(|h_i|^2-1)\,\mathrm{Im}\,Z_i}{(|Z_i|^2-1)\mathrm{Im}\,h_i}\tan\gamma^+ , \tag{11}$$

while, using (9), the inclinations of the velocity vector on both sides of the impulse are related by

$$\begin{aligned} \tan\beta^- &= \frac{v_z^+(a_x\tan\beta^+ + b_x\tan\delta^+ + c_x) + d_x}{v_z^+(a_z\tan\beta^+ + b_z\tan\delta^+ + c_z) + d_z} , \\ \tan\delta^- &= \frac{v_z^+(a_y\tan\beta^+ + b_y\tan\delta^+ + c_y) + d_y}{v_z^+(a_z\tan\beta^+ + b_z\tan\delta^+ + c_z) + d_z} , \end{aligned} \tag{12}$$

where $v_z^+ \equiv \dot{z}_i^+/\dot{t}_i^+$. These are explicit and general refraction formulae for motion of free test particles influenced by any expanding impulsive gravitational wave.

4. Impulses generated by snapped cosmic strings

We may apply these general results to an important particular family of spacetimes in which the expanding spherical impulse is generated by a snapped cosmic string (identified by a deficit angle in the region $U > 0$ in front of the impulse). This coresponds to the choice of the function $h(Z) = Z^{1-\delta}$, where $\delta \in [0, 1)$ is a real constant which characterizes the deficit angle $2\pi\delta$.

For simplicity, we consider here the ring of static test particles in the (x^+, z^+) plane, i.e., $y_i^+ = 0$, $\dot{x}^+ = \dot{y}^+ = \dot{z}^+ = 0$. The resulting motion remains in the (x^-, z^-) plane and general refraction formulae (11) and (12) reduce to

$$\tan\alpha^- = \frac{\sinh((1-\delta)r)}{\sinh r}\tan\alpha^+ , \qquad \tan\beta^- = \frac{\delta(1-\frac{\delta}{2})\cosh((1-\delta)r)}{(1-\frac{\delta}{2})^2\sinh(\delta r) + \frac{\delta^2}{4}\sinh((2-\delta)r)} , \tag{13}$$

where $r = \frac{1}{1-\delta}\log\left(\tan\frac{\alpha^+}{2}\right)$. These functions are drawn in Fig. 1. Using the above relations we can also study the deformation of such (initially static) ring of test particles, as shown in Fig. 2. We observe that the test particles are "dragged" along the z-direction by the impulse generated by moving "endpoints" of the snapped string (for more details see [7]).

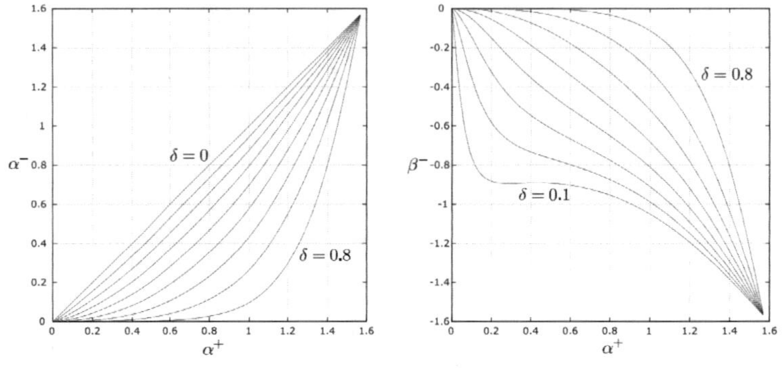

Figure 1. The function $\alpha^-(\alpha^+)$ which determines the displacement of the position of an initially static particle when it crosses the impulse generated by a snapped cosmic string (left). The function $\beta^-(\alpha^+)$ determines the dependence of the velocity vector inclination behind the impulse on the particle's position in front of the impulse (right). Here $\delta = 0.1, 0.2, \ldots, 0.8$.

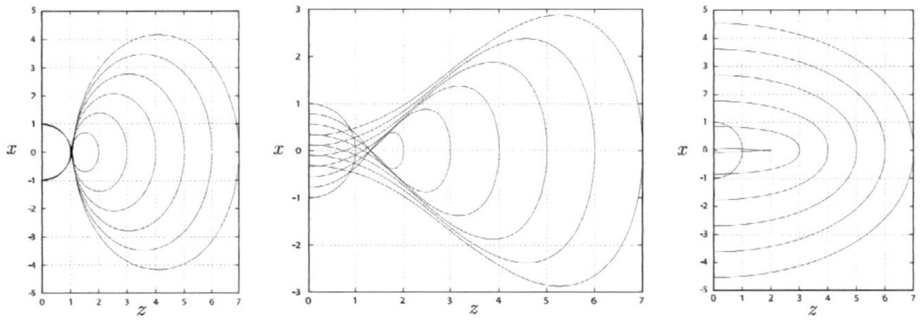

Figure 2. Time sequences which show the deformation of the ring of test particles (indicated here by initial semi-circles of unit radius for $\alpha^+ \in [-\frac{\pi}{2}, \frac{\pi}{2}]$) caused by different impulses generated by snapped cosmic strings with $\delta = 0.005$ (left), $\delta = 0.2$ (middle) and $\delta = 0.8$ (right).

5. Conclusions

We presented the framework for a description of the influence of expanding spherical impulsive gravitational waves on free test particles in spacetimes of a constant curvature. This generalizes the previous results [6] obtained for Minkowski background to any cosmological constant Λ. We derived the general refraction formulae (11) and (12) characterizing the shift of positions and the change of velocity vectors of the particles. More details can be found in the work [7].

Acknowledgments

R. Š. was supported by the grants GAČR 205/09/H033, SVV 261301 and GAUK 259018.
J. P. was supported by the grant GAČR 202/08/0187.

References

[1] Penrose R 1972 *General Relativity* ed L O'Raifeartaigh (Oxford: Clarendon Press) 101
[2] Hogan P A 1992 *Phys. Lett.* A **171** 21
[3] Hogan P A 1993 *Phys. Rev. Lett.* **70** 117
[4] Hogan P A 1994 *Phys. Rev.* D **49** 6521
[5] Podolský J and Griffiths J B 1999 *Class. Quantum Grav.* **16** 2937
[6] Podolský J and Steinbauer R 2003 *Phys. Rev.* D **67** 064013
[7] Podolský J and Švarc R 2010 *Phys. Rev.* D **81** 124035

Journal of Physics: Conference Series **314** (2011) 012067 doi:10.1088/1742-6596/314/1/012067

Uniqueness of static spacetimes with marginally outer trapped surfaces

Alberto Carrasco[1] **and Marc Mars**[2]

Dept. Física Fundamental, Univ. de Salamanca, Pl. de la Merced s/n, 37008 Salamanca, Spain

E-mail: [1]acf@usal.es, [2]marc@usal.es

Abstract. Marginally outer trapped surfaces (MOTS) are routinely used as quasi-local replacements for black holes and it is expected that MOTS and event horizons should coincide in stationary and static spacetimes. In this short notice we present a uniqueness theorem of static spacetimes containing MOTS.

1. Introduction

The concept of black hole involves, by its very definition, global properties of a spacetime. In an evolutive approach (as in a numerical calculation, for instance) the complete knowledge of the evolved spacetime is not available a priori (and in many cases this remains true even after the evolution has been carried out). Thus, the concept of black hole cannot be immediately used and a suitable replacement thereof is required.

Trapped surfaces (and their various relatives) are widely believed to provide such natural replacements. In particular, marginally outer trapped surfaces (MOTS) are considered to be suitable quasi-local versions of the event horizon. It is clear however, that both concepts are a priori very different and that a careful study of the relationship between them is necessary. In this respect, Ben-Dov has proven [1] that the boundary of the spacetime region containing weakly outer trapped surfaces (defined below) coincides with the event horizon in one of the simplest dynamical black hole configurations, namely the Vaidya spacetime. Nevertheless, from a 3+1 perspective the outermost MOTS in a given slice and the intersection of the event horizon with this slice do not coincide in general (consider for instance a spherical Cauchy foliation of the Vaidya spacetime, see e.g. [2]).

In any case, it is expected that in a stationary situation, MOTS should always lie on the event horizon. There is no proof of this fact, however. A possible approach to this problem goes through studying black uniqueness theorems (in four dimensions) when the assumption of black hole is replaced by the existence of a MOTS. In this work, we present a uniqueness theorem for static spacetimes with MOTS provided some conditions are satisfied (Theorem 3 below)

2. Definitions

A **static Killing initial data set** (static KID) $(\Sigma, g, K; \vec{\xi}\,)$ is a spacelike hypersurface (possibly with boundary $\partial\Sigma$) embedded in a static spacetime $(M, g^{(4)})$ with a static Killing field $\vec{\xi}$. Let us define $\lambda = -g^{(4)}(\vec{\xi}, \vec{\xi})$. Throughout this text a **surface** S is a 2-dimensional, orientable, embedded submanifold S in Σ which may have boundary ∂S provided $\partial S \subset \partial\Sigma$.

We will assume that $(\Sigma, g, K; \vec{\xi})$ has a chosen asymptotically flat end Σ_0^∞. Then, a surface $S \subset \Sigma$ is **bounding** if $\Sigma \setminus S$ splits into two disjoint sets Ω_1 and Ω_2 such that

(i) Ω_1 is connected, $\partial\Omega_1 = S$ and $\overline{\Omega_1} = \mathcal{K} \cup \hat{\Sigma}_0^\infty$ with $\mathcal{K}$ compact and $\hat{\Sigma}_0^\infty$ is an asymptotically flat end of Σ_0^∞, and

(ii) $\Omega_2 = \emptyset$ if and only if $S = \partial\Sigma$.

The sets Ω_1 and Ω_2 will be called the **exterior** and the **interior** of S in Σ, respectively.

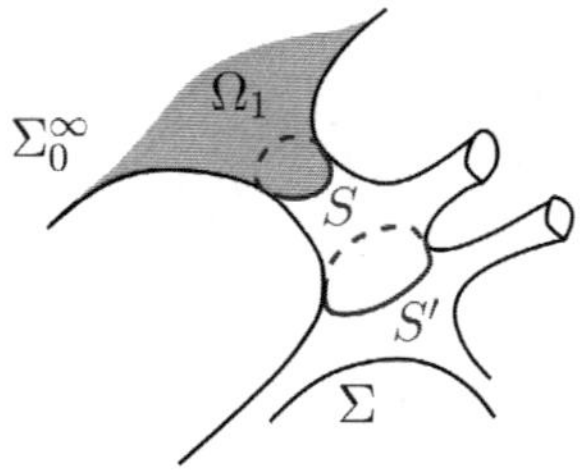

Figure 1. In short, a bounding surface S separates the manifold Σ in two sets such that one of them, Ω_1 (the exterior region, in gray color), is connected and contains Σ_0^∞ and no other end or boundary. The surface S' is not bounding because the exterior region including Σ_0^∞ contains an additional boundary.

Using the fact that S is orientable, it can be easily proven that a surface S admits two vectors fields $\{\vec{l}_+, \vec{l}_-\}$ which are null, future-directed and form a basis for the normal bundle of S in M at every point $p \in S$. We will take $\vec{l}_+$ and $\vec{l}_-$ to satisfy $g^{(4)}(\vec{l}_+, \vec{l}_-) = -2$. Whenever S is bounding, we will choose $\vec{l}_+$ to be the direction whose orthogonal projection onto Σ points into the exterior of S in Σ. Let us denote by θ^+ and θ^- the null expansions of S along $\vec{l}_+$ and $\vec{l}_-$, respectively.

A closed (i.e. compact and without boundary) bounding surface S is a:

- **Weakly outer trapped surface** (WOTS) if $\theta^+ \leq 0$.
- **Marginally outer trapped surface** (MOTS) if $\theta^+ = 0$.
- **Past weakly outer trapped surface** (past WOTS) if $\theta^- \geq 0$.

Finally, the **trapped region** T^+ is the union of the interiors of all WOTS in Σ and, similarly, the **past trapped region** T^- is the union of all the interiors of all past WOTS in Σ.

3. Motivation and statement of the result

Uniqueness theorems of black holes, which are cornerstones in General Relativity, have been proven in the stationary and the static cases for several matter models, such as e.g. vacuum and electrovacuum, under certain technical assumptions. To date, the most powerful method to prove uniqueness of static black holes is the so-called *doubling method* of Bunting and Masood-ul-Alam, which was used for the first time in [3] to show that the Schwarzschild family exhausts the set of static and vacuum non-degenerate black hole spacetimes, independently of whether the event horizon is a priori connected of not, thus improving the original uniqueness proof by Israel [4]. Although, as mentioned above, black holes require global hypotheses for their definition, in the static case these conditions can be translated into restrictions on the hypersurface Σ orthogonal to the static Killing vector. Thus, the theorem by Bunting and Masood-ul-Alam can be expressed as follows.

Journal of Physics: Conference Series **314** (2011) 012067 doi:10.1088/1742-6596/314/1/012067

Theorem 1 (Bunting and Masood-ul-Alam, 1987, [3]) *Let $(\Sigma, g, K; \vec{\xi})$ be a connected, asymptotically flat, static and vacuum KID. Assume time-symmetry (i.e. $K = 0$, $\vec{\xi} \perp \Sigma$), $\partial\Sigma \neq \emptyset$ and $\lambda|_\Sigma \geq 0$ with $\lambda = 0$ only on $\partial\Sigma$.*

Then (int Σ, g) (where int Σ is the interior of Σ) is isometric to the $\{t = 0, r > 2m\}$ slice of the Schwarzschild spacetime of mass m, for some $m > 0$.

The hypothesis of non-degeneracy of the event horizon, which in the theorem above appears implicitly in the fact that Σ has non-empty boundary and has no other ends besides the asymptotically flat one, was dropped by Chruściel in [5] after a suitable application of the doubling method. It is important to remark that the doubling method also works for other matter models. The key common feature that allows one to use the doubling method successfully in all these cases is the existence of a surface on Σ where $\lambda = 0$ and in whose exterior $\lambda > 0$. The existence of such a surface is guaranteed by the presence of an event horizon in the spacetime.

Since MOTS are expected to coincide with event horizons in the static case, it is natural to ask oneself whether the uniqueness theorems for static black holes can be extended to static spacetimes containing MOTS. A first answer to this question was given by Miao who proved the following theorem for the vacuum case.

Theorem 2 (Miao, 2005, [6]) *Let $(\Sigma, g, K; \vec{\xi})$ be a time-symmetric, asymptotically flat, static, vacuum KID with compact minimal (i.e. with vanishing mean curvature) boundary.*

Then (int Σ, g) is isometric to the $\{t = 0, r > 2m\}$ slice of the Schwarzschild spacetime of mass m, for some $m > 0$.

It is important to remark that in the time-symmetric setting all MOTS are bounding minimal surfaces and viceversa. In this work we try to generalize Miao's theorem in two directions. First, we will allow any matter model for which the doubling method gives uniqueness and, second, we will relax the hypothesis of time-symmetry. The proof by Miao relies strongly on the vacuum field equations so the approach will require completely new methods. This will require in particular a proper understanding of MOTS in static spacetimes.

As a consequence of the vacuum field equations, the set $\{\lambda = 0\}$ in a time-symmetric slice is a totally geodesic (in particular, minimal) boundary. Thus, recalling Theorem 1, Miao's theorem can be rephrased as follows: *No minimal surface can penetrate into the exterior region where the Killing vector is timelike in any time-symmetric, asymptotically flat, static and vacuum static initial data set.* Thus, Miao's result can be regarded as a confinement result for MOTS in time-symmetric and vacuum slices. Motivated by this fact, in [7] we proved that in a static KID, not necessarily time-symmetric, satisfying the null energy condition (which requires that the Einstein tensor contracted twice with any null vector gives a non-negative quantity), *no MOTS can penetrate in the exterior region where the static Killing is timelike* provided some technical hypotheses hold. The key tool which allowed us to achieve this result was a powerful theorem by Andersson and Metzger [8] which, roughly speaking, states that the boundary of the trapped region ∂T^+ is a MOTS and, in fact, the outermost MOTS. Our result in [7] extends Theorem 2 as a confinement result for MOTS. However, Miao's theorem also has a uniqueness part. This aspect can also be generalized as follows

Theorem 3 *Let $(\Sigma, g, K; \vec{\xi})$ be a static KID with a chosen asymptotically flat end Σ_0^∞ and satisfying the null energy condition. Suppose that Σ possesses a WOTS (in particular, MOTS) S. Assume:*

(i) $T^- \subset T^+$;

(ii) $H_1(\Sigma, \mathbb{Z}_2) = 0$ (i.e. the first homology group of Σ with coefficients in $\mathbb{Z}_2$ is trivial);

(iii) The spacetime $(M, g^{(4)})$ containing Σ is such that all the Killing prehorizons are embedded (a Killing prehorizon is, by definition, a null and injectively immersed submanifold of M where the Killing vector $\vec{\xi}$ is tangent and different from zero [9]).

(iv) The matter model is such that the Bunting and Masood-ul-Alam doubling method for time-symmetric slices gives uniqueness of black holes.

Then $(\Sigma \setminus T^+, g, K)$ *is a slice of such a unique spacetime.*

Let us end this short communication by briefly commenting on the hypotheses of Theorem 3. To that aim, it will be necessary to give some hints on the proof. Let U be the connected component of $\{\lambda > 0\}$ containing Σ_0^∞. Basically, the proof consists in showing that $\partial U = \partial T^+$, which is equivalent to asserting that $\lambda|_{\Sigma\setminus T^+} \geq 0$ and $\lambda = 0$ only in $\partial(\Sigma \setminus T^+)$. Once this is shown, hypothesis *(iv)* implies uniqueness. The proof of $\partial U = \partial T^+$ is by contradiction: Assume $\partial U \neq \partial T^+$ and then construct a WOTS outside ∂T^+ (which is impossible by the theorem by Andersson and Metzger [8] mentioned above).

The construction of the WOTS outside ∂T^+ is very specific and cannot cope with situations where hypothesis *(i)* does not hold. Nevertheless, we do not know whether condition *(i)* is really necessary for the validity of the theorem. In any case, in dynamical situations of strong gravitational collapse, it is to be expected that MOTS do form during the evolution, but no past WOTS are present at any time (since this would imply typically the presence of a singularity in the past). In this sense, hypothesis *(i)* seems reasonable. It is also important to mention that condition *(i)* is satisfied identically when the initial data set is time symmetric.

The WOTS outside ∂T^+ is constructed by joining suitable pieces of surfaces with $\lambda = 0$. The resulting set is proven to be injectively immersed, smooth, compact, without boundary and have vanishing null expansion. However, it may still be non-bounding. The topological condition *(ii)* is imposed to guarantee that this immersed submanifold is, in addition, bounding so that it splits the manifold into disjoint pieces. Hypothesis *(iii)* excludes the possibility that this immersed submanifold is non-embedded. It may well be the case that non-embedded prehorizons cannot exists under any circumstances but this is not known (see however [9] for a result in this direction under global spacetime assumptions). It should be remarked that the non-existence of non-embedded prehorizons is automatically satisfied if the spacetime is analytic.

Acknowledgments

We are very grateful to Miguel Sánchez Caja for useful comments. We acknowledge financial support under the projects FIS2006-05319 (Ministerio de Ciencia y Tecnología) and P09-FQM-4496 (Junta de Andalucía).

References

[1] Ben-Dov I. 2007 Outer trapped surfaces in Vaidya spacetimes *Phys. Rev. D* **75** 064007 (15pp)
[2] I. Bengtsson, J.M.M. Senovilla, 2009 A Note on trapped Surfaces in the Vaidya Solution, *Phys. Rev. D* **79** 024027 (6 pp)
[3] Bunting G and Masood-ul-Alam A K M 1987 Nonexistence of multiple black holes in asymptotically euclidean static vacuum space-time *Gen. Rel. Grav.* **19** 147-54
[4] Israel W 1967 Event horizons in static vacuum space-times *Phys. Rev* **164** 397-9
[5] Chruściel P T 1999 The classification of static vacuum space-times containing an asymptotically flat spacelike hypersurface with compact interior *Class. Quantum Grav.* **16** 661-87
[6] Miao P 2005 A remark on boundary effects in static vacuum initial data sets *Class. Quantum Grav.* **22** L53-9
[7] Carrasco A and Mars M 2008 On marginally outer trapped surfaces in stationary and static spacetimes *Class. Quantum Grav.* **25** 055011(19pp)
[8] Andersson L and Metzger J 2009 The area of horizons and the trapped region *Commun. Math. Phys.* **290** 941-72
[9] Chruściel P T and Galloway G J 2010 Uniqueness of static black-holes without analyticity *Preprint* gr-qc/1004.0513v1

Journal of Physics: Conference Series **314** (2011) 012068 doi:10.1088/1742-6596/314/1/012068

Closed trapped surfaces in higher dimensional self-similar Vaidya solution

Masahiro Shimano[1] **and T. Harada**[2]

Department of Physics, Rikkyo University, Toshima, Tokyo 171-8501, Japan

E-mail: [1]`shimano@rikkyo.ac.jp` and [2]`harada@rikkyo.ac.jp`

Abstract. Although an event horizon defines a black hole, it is difficult to explain a boundary of dynamical black holes. A trapped surface is a candidate of the boundary of dynamical black holes, and is studied in various spacetimes and settings. Usually, there is no trapped surface in a Minkowski region, however, Bengtsson and Senovilla showed an interesting result as follows: in a four-dimensional self-similar Vaidya spacetime, they considered non-spherical trapped surfaces and showed that trapped surfaces can extend into the Minkowski region, if and only if a mass function rises fast enough [1]. We extend this result in a higher-dimensional spacetime, because recently studying higher-dimensional black holes is significant in the context of large extra dimensions or TeV-scale gravity. In this paper we investigate a higher-dimensional Vaidya spacetime with a self-similar mass function. We match two kinds of $(n + 1)$-surfaces and construct trapped surfaces extended into the Minkowski region by using Bengtsson and Senovilla's way. Moreover, we demonstrate that there is no naked singularity, if the spacetime has trapped surfaces as above. These results might become a foothold to define the boundary of dynamical black holes in higher-dimensional spacetimes.

1. Introduction

A boundary of a region in a spacetime that cannot be observed from infinity is called event horizon. The event horizon defines a boundary between an inside and an outside of a black hole. Moreover, this horizon has a teleological property: an entire future history of the spacetime must be known before a position of the event horizon can be determined. Since the event horizon is defined at future timlike infinity, a shape of the event horizon does not change. However, some quantum effect might deform the boundary of dynamical black holes. It might be impossible to explain the boundary of dynamical black holes with the event horizon.

Eardley conjectured that the boundary of the region which contains marginally outer trapped surfaces coincide with the event horizon [2]. A surface called outer trapped surface is the closed spacelike two-surface (in four-dimensional case) and whose outer null expansion is negative. This conjecture is very interesting, because we can translate the event horizon with outer trapped surface, i.e., we can define the event horizon constructively. In a four-dimensional Vaidya spacetime, Ben-Dov showed that Eardley's conjecture is true [3]. However, the outer trapped surface cannot be considered in the general spacetime, because it is defined only in asymptotically flat spacetimes [4]. Moreover, the outer trapped surface only consider an outer null ray. There remains a possibility of which we can observe an inner null ray. Since the black hole is an invisible region, to define this region we should consider a notion of which both null rays cannot be observed.

Journal of Physics: Conference Series **314** (2011) 012068 doi:10.1088/1742-6596/314/1/012068

A surfaces called trapped surface defines the boundary of an invisible region. The trapped surface is a closed spacelike two-surface (in four-dimensional case) and whose both null expansions are negative. Since the black hole cannot be observed even from infinity, its boundary might be defined with the trapped surface. Usually, there is no trapped surface in a Minkowski spacetime. However, in the four-dimensional Vaidya spacetime it was showed that trapped surfaces can extend into the Minkowski region. Numerical results of Schnetter and Krishnan showed that an outer boundary of trapped surfaces can extend into the Minkowski region [5]. Moreover, Bengtsson and Senovilla considered the self-similar Vaidya spacetime, and they analytically showed that trapped surfaces can extend into the Minkowski region, if and only if a mass function rises fast enough [1].

Recently, higher-dimensional scenarios with large [6] and warped [7] extra dimensions were proposed to resolve the hierarchy between the gravitational and electroweak interactions. One of the most striking predictions of such scenarios is productions of a large number of miniblack holes in high-energy particle collisions [8]. Therefore, studying higher-dimensional black holes is significant in the context of above scenarios.

In this paper, we extend Bengtsson and Senovilla's way into a higher-dimensional Vaidya spacetime with a self-similar mass function. We match two kinds of $(n+1)$-surfaces (where $n = D-3$) and show that trapped surfaces can extend into the Minkowski region. Moreover, we demonstrate that there is no naked singularity, if the spacetime has trapped surfaces as above.

2. D dimensional Vaidya spacetime

We consider a D-dimensional Vaidya solution [9]

$$ds^2 = -\left(1 - \frac{2m}{nr^n}\right) dv^2 + 2dvdr + r^2 d\Omega_{n+1}^2 \tag{1}$$

where $n = D-3$, $D \geq 4$, and $d\Omega_{n+2}^2$ is a line element of of a unit (n+2)-sphere. We choose a mass function $m(v)$ such as

$$m = \begin{cases} 0 \quad , & v \leq 0 \\ \mu v^n \quad , & 0 \leq v \leq M^{1/n}/\mu \ , \\ M \quad , & v \geq M^{1/n}/\mu \end{cases} \tag{2}$$

where μ and M are constants, respectively. There is a radial influx of null fluid in an initially empty region of a D-dimensional Minkowski region. The region in $0 \leq v \leq M^{1/n}/\mu$ is a D-dimensional Vaidya region with the self-similar mass function, and the region in $v \geq M^{1/n}/\mu$ is a D-dimensional Schwarzschild region. We call a boundary between the Minkowski region and the Vaidya region MV-boundary, and also call the boundary between the Vaidya region and the Schwarzschild region VS-boundary. We know that a naked singularity will occur if and only if the mass function satisfies the following condition [10]:

$$0 < \mu < \left[\frac{n}{2(n+1)}\right]^{n+1} . \tag{3}$$

2.1. Two types of trapped surfaces

In order to construct trapped surfaces extended into the Minkowski region, we consider two kinds of $(n+1)$-surfaces as in Table. 1. We call the surface in which r and v are the function of ρ, and in which $\theta_i = \pi/2$ *constant inclination surface* (CIS). Similarly, we call the surface in which θ_i and v are the function of ρ, and in which $r = r_0$ *constant radius surface* (CRS).

Table 1. The type of $(n+1)$-surfaces (where $i, j = 1, \cdots n$)

	v	r	θ_i	θ_{n+1}	$\theta_{j\neq i}$
$\text{CIS}_{(i)}$	$V(\rho)$	$R(\rho)$	$\pi/2$	$0 \leq \theta_{n+1} \leq 2\pi$	$0 \leq \theta_j \leq \pi$
$\text{CRS}_{(i)}$	$V(\rho)$	r_0	$\Theta_i(\rho)$	$0 \leq \theta_{n+1} \leq 2\pi$	$0 \leq \theta_j \leq \pi$

2.2. Closed trapped surfaces

We consider the following types of $(n+1)$-surfaces in each region, respectively:

- The Minkowski region: We consider CISs which are a topological disk given by the hyperboloid $v = t_0 + r - \sqrt{r^2 + k^2}$, where t_0 and k are constants. We have chosen $t = v - r$ and $R = \rho$. These surfaces have negative both null expansions automatically.
- The Vaidya region: We consider CISs in which V and R satisfy the relation $dV/dR = a/(b - X)$,where $X = V/R$, a and b are constants, and we have chosen $R = \rho$. If a satisfies $a > b^2/4$ and $a > (n/\mu)^{1/n}$, both null expansions of these surfaces are negative.
- The Schwarzschild region: We consider CRSs which is a capping disk defined by $\Theta_i^2 + \left(V\,(n/(\gamma M))^{1/n} - \sigma_0\right)^2 = \pi^2/4$, where $\gamma = nR^n/M$ and σ_0 is a constant. Both null expansions of this surface are negative, if γ satisfies the following inequalities which are different in each CRSs: In the case of $i = 1$ the inequality is given by

$$n\sqrt{\frac{2}{\gamma} - 1}\left(\frac{1}{\gamma} - 1\right) > \frac{2}{\pi}. \tag{4}$$

We solve numerically this inequality with respect to γ, and plot an upper bound of γ in Figure. 1. The upper bound of γ approaches one for infinitely large dimension. Therefore,

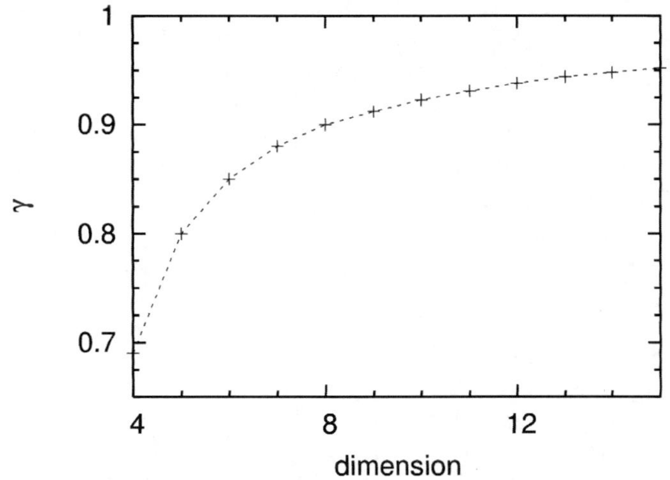

Figure 1. The upper bound of γ, where "+" is the upper bound of γ for each dimensions. We have connected each "+" by a dashed line. The dashed line approaches one for infinitely large dimension.

γ is less than one in any dimensions. On the other hand, in the case of $i \neq 1$ the inequalities are given by

$$n\sqrt{\frac{2}{\gamma}-1}\left(\frac{1}{\gamma}-1\right) > \frac{2}{\pi}\prod_{k=1}^{l-1}\sin\theta_k, \tag{5}$$

where $l = a, \cdots, n$. Since θ_k exists in the range of $0 \le \theta_k \le \pi$, the right hand side in Eq. (5) exists in the range of zero to $2/\pi$. In this case γ is also less than one in any dimensions. In both cases both null expansions are negative, if γ is less than one.

Matching these surfaces we construct trapped surfaces extended into the Minkowski region. To do this we impose the following matching conditions into parameters: On the MV-boundary if the mass function satisfies

$$\mu > \frac{1}{4}\left(\frac{n}{\gamma}\right)^{1/n}, \tag{6}$$

surfaces are matched. On the other hand, On the VS-boundary if γ is less than one, surfaces are matched. Because conditions for getting negative both null expansions are not inconsistent to matching conditions, we are able to construct trapped surface extended into the Minkowski region.

2.3. Naked singularity

We shall consider the naked singularity in this spacetime. Combining Eqs. (6) and (3) we get the condition with respect to γ such as $\gamma > n\left[(2(n+1)/n)^{n+1}/4\right]^n$.If γ satisfies this inequality, the spacetime has the naked singularity. However, this inequality is inconsistent to $\gamma < 1$. Therefore, if trapped surfaces given by above discussions exist in the spacetime, there is no naked singularity.

3. Conclusion

We have considered the higher-dimensional self-similar Vaidya spacetime. By using Bengtsson and Senovilla's way we have constructed trapped surfaces extended into the Minkowski region. Moreover, we have shown that there is no naked singularity, if the spacetime has trapped surfaces constructed above. Although in the higher-dimensional case there have been many CISs and CRSs compared to four-dimensional case, a feature of trapped surfaces constructed in this study and the possibility of the naked singularity have been same to four-dimensional case.

The event horizon is the surface which satisfies $\gamma = 2$, on the other hand, trapped surfaces constructed in this study satisfies $\gamma < 1$. Therefore, trapped surfaces are far from the event horizon. In higher-dimensional spacetime it is possible to consider many kinds of $(n+1)$-surfaces without CISs and CRSs. By using these $(n+1)$-surfaces, we could construct trapped surfaces which exist near the event horizon. Moreover, similar to this study, if we extend the Ben-Dov's way into the higher-dimensional Vaidya spacetime, we may show that Eardley's conjecture is true in this spacetime.

References

[1] I. Bengtsson, and J. M. M. Senovilla, Phys. Rev. D **79**, 024027 (2009).
[2] D. M. Eardley, Phys. Rev, D **57**, 2299 (1998).
[3] I. Ben-Dov, Phys. Rev, D **75**, 064007 (2007).
[4] S. W. Hawking and G. F. R. Ellis, *Large Scale Structure of Spacetime* (Cambridge University Press, Cambridge, 1972).
[5] E. Schnetter and B. Krishnan, Phys. Rev, D **73**, 021502(R) (2006).
[6] N. Arkani-Hamed, S. Dimopoulos and G. R. Dvali, Phys. Lett. B **429**, 263 (1998).
[7] L. Randall and R. Sundrum, Phys. Rev. Lett. **83**, 3370 (1999).
[8] P. C. Argyres, S. Dimopoulos and J. March-Russell, Phys. Lett. B **441**, 96 (1998).
[9] B. R. Iyer and C. V. Vishveshwara, Paramana **32**, 749 (1989).
[10] S. G. Ghosh and N. Dadhich, Phys. Rev, D **64**, 047501 (2001).

Journal of Physics: Conference Series **314** (2011) 012069 doi:10.1088/1742-6596/314/1/012069

Principles of Einstein–Finsler Gravity and Cosmology

Sergiu I. Vacaru

Department of Science, University Al.I. Cuza at Iaşi (UAIC),
54 Lascar Catargi str., Iaşi, Romania, 700107

E-mail: sergiu.vacaru@uaic.ro, Sergiu.Vacaru@gmail.com

Abstract. We analyze the foundations of Finsler gravity theories with metric compatible connections constructed on nonholonomic tangent bundles, or (pseudo) Riemannian manifolds. There are considered "minimal" modifications of Einstein gravity (including theories with violation of local Lorentz invariance) and shown how the general relativity theory and generalizations can be equivalently re–formulated in Finsler like variables. We focus on Finsler branes solutions and perspectives in modern cosmology.

1. Introduction

The bulk of classical gravitational effects are described experimentally in the framework of General Relativity (GR). Recently, it was proposed that certain exceptions in theoretical cosmology may be related to the dark matter and dark energy problems and some approaches were formulated for Finlser gravity models. It is also expected that small corrections with violations of equivalence principle and local Lorentz invariance (in brief, LV) have to be considered in low energy limits within general approaches to quantum gravity (QG).[1]

There are two general classes of Finsler type gravity theories: The first one originates from E. Cartan works on Finsler geometry [8], see further geometrical developments and applications in [1, 4, 5, 7, 9]. Such most related to "standard physics" constructions were elaborated for the metric compatible Cartan and canonical distinguished connections (in brief, d–connections). This class of Einstein–Finsler gravity (EFG) models can be elaborated following geometric and physical principles which are very similar to those in GR but with Cartan–Finsler type connections used instead of the Levi–Civita connection. In the second class of theories, the Finsler connections (for instance, the Berwald/ Chern d–connections) are not metric compatible [10, 9, 3]. This results in non–standard physical theories with ambiguities in definition of Dirac operators, quantum/noncommutative generalizations of gravity models etc [7].

The goals, and structure, of this paper, summarizing certain our recent results [4, 5, 7], are to study Finsler modifications of GR derived from MDR of classical gravity and QG theories with LV and formulate certain fundamental principles and modifications of axioms of GR including the EFG theories (section 2) and speculate on Finsler/–brane models with acceleration of Universe (section 3). Conclusions are presented in section 4.

[1] We cite a series of works and reviews of results on Finsler geometry methods and modifications of gravity and related string/brane gravity models and QG phenomenology [1, 2, 3, 4], noncommutative geometry and deformation/two connection quantization [5], nonholonomic Ricci flows and cosmology [6, 7].

2. Principles of Einstein–Finsler Gravity

2.1. Motivations and fundamental geometric/physical objects

We consider four important motivations to study Finsler modifications of gravity theories and formulations of GR in almost Kähler – Finsler variables [2, 7]:

(i) For deformations of Minkovski spacetimes resulting in nonlinear energy–momentum relations, $E^2 = p^2c^2 + m_0^2c^4 + \varphi(E,p)$, where $E \sim \frac{\partial}{\partial t}, p_{\widehat{i}} \sim \frac{\partial}{\partial x^{\widehat{i}}}$, $\omega \sim \frac{\partial}{\partial t}$ $k_{\widehat{i}} \sim \frac{\partial}{\partial x^{\widehat{i}}}$, $(x^1 = ct, x^2, x^3, x^4); \widehat{i}, \widehat{j}... = 2,3,4$; (for instance, $i,j,... = (1,\widehat{i}),(1,\widehat{j})$), we obtain modified dispersion relations (MDR) $\omega^2 = c^2[g_{\widehat{i}\widehat{j}}k^{\widehat{i}}k^{\widehat{j}}]^2(1 - q_{\widehat{i}_1\widehat{i}_2...\widehat{i}_{2r}}y^{\widehat{i}_1}...y^{\widehat{i}_{2r}}/r[g_{\widehat{i}\widehat{j}}k^{\widehat{i}}k^{\widehat{j}}]^{2r})$. To such MDR, we can associate a fundamental Finsler function $F(x^i, \beta y^j) = \beta F(x^i, y^j), \beta > 0$, in tangent space TV, and a nondegenerated Hessian, ${}^Fg_{ij}(x^i, y^j) = \frac{1}{2}\frac{\partial F^2}{\partial y^i \partial y^j}$, when

$$ds^2 = F^2 \approx -(cdt)^2 + g_{\widehat{i}\widehat{j}}(x^k)y^{\widehat{i}}y^{\widehat{j}}[1 + \frac{1}{r}q_{\widehat{i}_1\widehat{i}_2...\widehat{i}_{2r}}(x^k)y^{\widehat{i}_1}...y^{\widehat{i}_{2r}}/(g_{\widehat{i}\widehat{j}}(x^k)y^{\widehat{i}}y^{\widehat{j}})^r] + O(q^2). \quad (1)$$

Various scenarios in modern cosmology with dark energy/matter, QG phenomenology with LV, corrections from string/brane gravity, noncommutative gravity models etc are aimed to solve some important theoretical problems and explain existing experimental data [2, 1, 7].

(ii) The Einstein equations in GR, and modifications, can be integrated "almost in general" forms, for arbitrary finite dimension, in terms of Finsler like connections; imposing nonholonomic constraints, we generate solutions for the Levi–Civita connection [4, 3].

(iii) Rewriting GR, equivalently, in almost Käehler – Finsler variables, the theory can be quantized following methods of deformation/ geometric and A–brane quantization when a "two–connection" geometric renormalization procedure can be performed [5].

(iv) Finsler metrics and connections are induced via nonholonomic and/or noncommutative Ricci flows of Einstein spacetimes and generalizations [6].

Claim: A (pseudo) Finsler fundamental function F^2 (1) and its Hessian ("vertical" metric in TV) ${}^Fg_{ij}(x^k, y^a)$ do not state completely a geometric model. A Finsler geometry/ gravity $(F : \mathbf{N}, \mathbf{g}, \mathbf{D})$ is defined by three fundamental geometric objects generated by any $F(x,y)$:

- a nonlinear connection (N–connection) structure (i.e. a nonholonomic distributions with conventional h- / v–splitting ${}^F\mathbf{N} = \{N_i^a(x,y)\} : TTV = hTV \oplus vTV)$ determined canonically by the Euler–Lagrange equations for $L = F^2$ treated equivalently as semi–spray equations (in such a case, $\mathbf{N} = {}^c\mathbf{N}$);[2]
- a distinguished metric (d–metric), ${}^F\mathbf{g} = h\mathbf{g} \oplus v\mathbf{g}$; up to frame transforms, it can be parametrized in N–adapted form, ${}^F\mathbf{g} = \{{}^F\mathbf{g}_{\alpha\beta}\} = {}^Fg_{ij}(u)dx^i \otimes dx^j + {}^Fg_{ab}(u)\ {}^c\mathbf{e}^a \otimes {}^c\mathbf{e}^b$;
- a distinguished connection (d–connection; a N–adapted linear connection preserving $\mathbf{N}$ under parallelism), ${}^F\mathbf{D} = (hD, vD)$, which also determined by ${}^Fg_{ij}$ and N_i^a.

A Finsler geometry can be modelled on a tangent bundle $\mathbf{V} = TV$ to a manifold V, or on a nonholonomic manifold $\mathbf{V} = (\mathcal{V}, \mathcal{D})$, where $\mathcal{D}$ is a nonholonomic distribution, for instance, defining a N–connection structure [3, 7, 4]. If $\mathcal{V}$ is a (pseudo) Riemannian spacetime with a conventional nonholonomic $2+2$ splitting, we can introduce Finsler variables in GR.[3]

[2] there are N–elongated (co) frames (respectively, partial derivatives and differential operators), $\mathbf{e}_\nu = (\mathbf{e}_i = \partial_i - N_i^a\partial_a, e_a = \partial_a)$ and $\mathbf{e}^\mu = (e^j, \mathbf{e}^b = dy^b + N_j^b dy^b)$, for h- and v–coordinates $u^\alpha = (x^i, y^a); i,j,... = 1,2,...n$ and $a,b,... = n+1, n+2, ...n+n$ (on local formulas see details in [4, 7]);

[3] For any given (Finsler) d–connection $\mathbf{D} = \{\mathbf{\Gamma}^\alpha{}_{\beta\gamma}\}$, there are computed in standard forms (see details in [9, 10, 3, 4, 7]) the N– adapted coefficients of physically important tensors: torsion, $\mathbf{T}^\alpha{}_{\beta\gamma} = \{T^i{}_{jk}, T^i{}_{ja}, T^a{}_{ji}, T^a{}_{bi}, T^a{}_{bc}\}$, curvature, $\mathbf{R}^\alpha{}_{\beta\gamma\delta} = \{R^i{}_{hjk}, R^a{}_{bjk}, R^i{}_{jka}, R^c{}_{bka}, R^i{}_{jbc}, R^a{}_{bcd}\}$, Ricci tensor, $\mathbf{R}_{\alpha\beta} := \mathbf{R}^\tau{}_{\alpha\beta\tau} = \{R_{ij}, R_{ia}, R_{ai}, R_{ab}\}$, and scalar curvature, ${}^s\mathbf{R} := \mathbf{g}^{\alpha\beta}\mathbf{R}_{\alpha\beta} = R + S$, for $R = g^{ij}R_{ij}$ and $S = h^{ab}R_{ab}$. In Finsler like geometries, the well known Levi–Civita connection ∇, ($\mathbf{T}[\nabla] = 0$ and $\nabla\mathbf{g} = 0$), is not used (it does not preserve the h-/v– splitting).

Journal of Physics: Conference Series **314** (2011) 012069 doi:10.1088/1742-6596/314/1/012069

Theorem 1: Any (pseudo) Finsler / Riemannian geometry, for a chosen N–connection $\mathbf{N}$, can be modelled equivalently by geometric data $(\mathbf{g}, \nabla)$ and/or $(\mathbf{N}, {}^F\mathbf{g}, {}^g\mathbf{D})$ if ${}^g\mathbf{D} = {}^g\nabla + {}^g\mathbf{Z}$, when the connections and distortion tensor ${}^g\mathbf{Z}$ are determined by $\mathbf{g} = {}^F\mathbf{g}$.

Example: There is the Cartan (Finsler type) d–connection $\tilde{\mathbf{D}} = (h\tilde{D}, v\tilde{D})$ uniquely defined by the conditions $\tilde{\mathbf{D}}\ {}^F\mathbf{g} = 0$ and $T^i_{\ jk} = T^a_{\ bc} = 0$.[4] For instance, an EFG theory on TV, or $\mathbf{V}$, can be constructed similarly to GR when $\tilde{\mathbf{D}}$ is taken instead of ∇.

2.2. Principles and axioms of EFG; fundamental field equations

A model of EFG with $\mathbf{N} \sim {}^F\mathbf{N}$ and metric compatible $\mathbf{D}$ uniquely defined by $\mathbf{g} \sim {}^F\mathbf{g}$ (on TV, or $\mathbf{V}$) is derived similarly to GR and the standard particle model following such principles:

- **Generalized equivalence principle** for free fall and universality of gravitational redshift and related **principle of general covariance** with "mixing" of Finsler variables.
- **Generalized Mach principle** for the quantum energy/motion encoded via $(\mathbf{N}, \mathbf{g}, \mathbf{D})$ in aether spacetime models with anisotropic dependence on "velocity" type (nonholonomically constrained) variables y^a. This principle is less clear formulated both in GR and EFG.
- We can formulate a **constructive–axiomatic approximation** (as generalized Ehlers–Pirani –Schild, EPS, axioms) with a generalized paradigm for "Lorentzian 4–manifold" in GR extended on nonholonomic tangent bundles/ manifolds, for a corresponding model of EFG constructed in a form to be compatible with experimental data.
- Any **motion equation and conservation laws** are derived from nonholonomc Bianchi identities for $\mathbf{D}$, with unique deformations $\nabla_i T^{ij} = 0 \rightarrow \mathbf{D}_\alpha \Upsilon^{\alpha\beta} \neq 0$, for generalized stress–energy tensors $\Upsilon^{\alpha\beta}$ derived following a N–adapted variational/ covariant calculus.
- Generalized **Einstein–Finsler/ Yang–Mills / Dirac** etc equations are formulated for certain data $(\mathbf{N}, \mathbf{g}, \mathbf{D})$. For zero distortions ${}^g\mathbf{Z}$, we get standard constructions similar to those for $(\mathbf{g}, \nabla)$ in GR. The Einstein–Finsler (gravitational) field equations split as

$$R_{ij} - \frac{1}{2}(R+S)g_{ij} = \Upsilon_{ij},\ R_{ab} - \frac{1}{2}(R+S)h_{ab} = \Upsilon_{ab},\ R_{ai} = \Upsilon_{ai},\ R_{ia} = -\Upsilon_{ia}. \quad (2)$$

Remark: The equations (2) can be integrated in very general forms and quantized, for instance, for the canonical/Cartan d–connection [4, 5, 7]. Imposing nonholonomic constraints, we can generate exact solutions in GR and/or quantum models with the Levi–Civita connection.

3. Finsler Branes and Cosmological Solutions

Diagonal brane configurations on TV can be generated by ansatz

$$\mathbf{g} = \phi^2(y^5)\eta_{\alpha\beta}du^\alpha \otimes du^\beta - (l_P)^2\, \overline{h}(y^5)[\, dy^5 \otimes dy^5 + dy^6 \otimes dy^6 \pm dy^7 \otimes dy^7 \pm dy^8 \otimes dy^8], \quad (3)$$

where $\eta_{\alpha\beta} = diag[\pm 1]$ and $\alpha, \beta, ... = 1,2,3,4$; with a "length constant" l_P.[5] Such a metric is a solution of (2) if $\phi^2(y^5) = \frac{3\epsilon^2 + a(y^5)^2}{3\epsilon^2 + (y^5)^2}$ and $l_P\sqrt{|\overline{h}(y^5)|} = \frac{9\epsilon^4}{[3\epsilon^2+(y^5)^2]^2}$, where a is an integration constant and the width of brane is $\epsilon^2 = 40M^4/3\Lambda$, with some fixed integration parameters

[4] $\tilde{\mathbf{D}}$ is also a canonical almost symplectic connection used in deformation/ A–brane quantization (and two connection renormalization) of Einstein and generalized Finsler geometries [5]; its N–adapted coefficients are computed $\tilde{\mathbf{\Gamma}}^\alpha_{\gamma\tau} = \{(\tilde{L}^i_{jk} = \frac{1}{2}\ {}^F g^{ir}(\mathbf{e}_k\ {}^F g_{jr} + \mathbf{e}_j\ {}^F g_{kr} - \mathbf{e}_r\ {}^F g_{jk}),\ \tilde{C}^a_{bc} = \frac{1}{2}\ {}^F g^{ad}(e_c\ {}^F g_{bd} + e_c\ {}^F g_{cd} - e_d\ {}^F g_{bc})\}$.

[5] There are used indices ${}^1\alpha = (\alpha, 5, 6)$ and ${}^2\alpha = ({}^1\alpha, 7, 8)$, respectively, for 6–d and 8–d models; indices ${}^2\alpha,\ {}^2\beta, ...$ run values $1,2,3,4,5,...,m$ where $m \geq 2$; ${}^1a = 5,6$ and ${}^2a = 7,8$; coordinates y^5, y^6, y^7, y^8 are velocity type (for instance, y^5 has a finite maximal value y^5_0 on TV because the light velocity is finite). The anholonomic deformation method allows us to construct solutions for any convenient signature; for instance, (non) homogeneous cosmological metrics can be constructed for $u^2 = t$, or $u^3 = t$.

when $\frac{\partial^2\phi}{\partial(y^5)^2}\ |_{y^5=\epsilon}=0$ and $l_P\sqrt{|\overline{h}(y^5)|}\ |_{y^5=0}=1$; this states the conditions that on diagonal branes the Minkowski metric is 6–d or 8–d (for TV). For a "bulk" cosmological constant Λ, the nonzero components of stress–energy tensor are chosen $\Upsilon^{\beta}{}_{\delta}=\Lambda-M^{-(m+2)}\overline{K}_1(y^5)$, $\Upsilon^5{}_5=\Upsilon^6{}_6=\Lambda-M^{-(m+2)}\overline{K}_2(y^5)$, for a fundamental mass scale M on TV, $\dim TV=8$; $\overline{K}_1$ and $\overline{K}_2$ must satisfy certain compatibility conditions with the field equations and conservation laws [7].

We can construct *off–diagonal Finsler brane* trapping solutions (with cosmological scenarios for $h_3(x^i,y^3=v=t)$) deforming the metric (3) to ansatz with nontrivial N–connection coefficients: $N_i^3=w_i(x^k,v), N_i^4=n_i(x^k,v)$; $N_\alpha^5={}^1w_\alpha(x^k,v,y^5)$; $N_\beta^6={}^1n_\beta(x^k,v,y^5)$; $N^7{}_{1_\alpha}={}^2w\ {}_{1\alpha}(x^k,v,y^7)$; $N^8{}_{1_\beta}={}^2n\ {}_{1\beta}(x^k,v,y^7)$. Such EFG spaces are characterized by metrics

$$\mathbf{g}=g_idx^i\otimes dx^i+h_a\mathbf{e}^a\otimes\mathbf{e}^a+(l_P)^2(\overline{h}/\phi^2)[\ {}^qh\ {}_{1_a}\ \mathbf{e}\ ^{1_a}\otimes\ \mathbf{e}\ ^{1_a}+\ {}^qh\ {}_{2_a}\ \mathbf{e}\ ^{2_a}\otimes\ \mathbf{e}\ ^{2_a}] \qquad (4)$$

where $\mathbf{e}^3=dy^3+w_idx^i, \mathbf{e}^4=dy^4+n_idx^i$, $\mathbf{e}^5=dy^5+\ {}^1w_idx^i$, $\mathbf{e}^6=dy^6+\ {}^1n_idx^i$, $\mathbf{e}^7=dy^7+\ {}^2w_idx^i$, $\mathbf{e}^8=dy^8+\ {}^2n_idx^i$. Explicit formulas for coefficients (see [7]) are determined by "effective/polarized" cosmological constants induced by sources in (2), $\widetilde{\mathbf{\Upsilon}}^{\beta}{}_{\delta}=diag[\widetilde{\mathbf{\Upsilon}}^1{}_1=\widetilde{\mathbf{\Upsilon}}^2{}_2=\widetilde{\mathbf{\Upsilon}}_2, \widetilde{\mathbf{\Upsilon}}^3{}_3=\widetilde{\mathbf{\Upsilon}}^4{}_4=\widetilde{\mathbf{\Upsilon}}_4$, $\widetilde{\mathbf{\Upsilon}}^5{}_5=\widetilde{\mathbf{\Upsilon}}^6{}_6=\widetilde{\mathbf{\Upsilon}}_6, \widetilde{\mathbf{\Upsilon}}^7{}_7=\widetilde{\mathbf{\Upsilon}}^8{}_8=\widetilde{\mathbf{\Upsilon}}_8]$ and for corresponding classes of generating and integration functions.

4. Conclusions

The main goal of this work was to show how basic principles and axioms for the GR theory can be extended on nonholonomic tangent bundles/ manifolds. We provided a self–consistent scheme for formulating metric compatible Finsler gravity models in a form most closed to the standard theory of particle physics. Following the anholonomic deformation method, there were constructed new classes of cosmological off–diagonal solutions for Finsler branes (4).

We found that metric compatible Finsler connections allows us to realize trapping gravitational configurations with well–defined physical properties for a range of parameters. Such brane/trapping effects of QG and/or LV depend on the mechanism of Finsler type gravitational and matter fields interactions on tangent bundle TV over a spacetime V in GR and it is expected that they may be detected in TeV physics, or via modifications in modern cosmology scenarios and astrophysics.

Acknowledgement: I'm grateful to Organizers ERE2010 and M. Sanchez Caja for supporting my participation at Spanish Relativity Meeting, Granada (September 6-10, 2010).

References

[1] Vacaru S., J. Math. Phys. **37,** 508 (1996); Ann. Phys. (NY) **256,** 39 (1997); Nucl. Phys. B, **434,** 590 (1997).
[2] Perlick V., GRG **38,** 365 (2006); Girelli F., Liberati S. and Sindoni L., Phys. Rev. D **75,** 064015 (2007); Gibbons G. W., Gomis J., Pope C. N., Phys. Rev. D **76** (2007) 081701; Skakala J., Visser M., J. Phys. Conf. Ser. **189** (2009) 012037; Lämmerzahl C., Lorek D., Dittus H., GRG **41,** 1345 (2009); Kouretsis A. P., Stathakopoulos M., Stavrinos P. C., Phys. Rev. D **82** (2010) 064035; Mavromatos N. E, arXiv: 1011.3526.
[3] Vacaru S., Stavrinos P., Gaburov E. and Gonţa D., *Clifford and Riemann- Finsler Structures in Geometric Mechanics and Gravity,* Selected Works, Differential Geometry – Dynamical Systems (Geometry Balkan Press, 2006); www.mathem.pub.ro/dgds/mono/va-t.pdf & arXiv: gr-qc/0508023.
[4] Vacaru S., JHEP **4,** 009 (2001); IJGMMP **4,** 1285 (2007); **5,** 473 (2008); IJTP **49,** 884 (2010).
[5] Vacaru S., Phys. Lett. B **498,** 74 (2001); J. Math. Phys. **46,** 042503 (2005); CQG **27,** 105003 (2010); Phys. Lett. A **372,** 2949 (2008); IJGMMP **6,** 873 (2009); **7,** 713 (2010).
[6] Vacaru S., IJMPA **21,** 4899 (2006); J. Math. Phys. **49,** 043504 (2008); **50,** 073503 (2009).
[7] Vacaru S., Phys. Lett. B **690,** 224 (2010); IJTP **49,** 2753 (2010); arXiv: 1004.3007; 1008.4912; 1010.5457.
[8] Cartan E., Les Espaces de Finsler (Paris, Herman, 1935).
[9] Miron R. and Anastasiei M., The Geometry of Lagrange Spaces: Theory and Applications, FTPH no. **59** (Kluwer Academic Publishers, Dordrecht, Boston, London, 1994).
[10] Bao D., Chern S. -S., and Shen Z., An Introduction to Riemann–Finsler Geometry. Graduate Texts in Math., 200 (Springer–Verlag, 2000).

Spanish Relativity Meeting (ERE 2010): Gravity as a Crossroad in Physics IOP Publishing
Journal of Physics: Conference Series **314** (2011) 012070 doi:10.1088/1742-6596/314/1/012070

Double shell stars as source of the Kerr metric in the CMMR approximation

Cuchí, J. E.[1], **Molina, A.**[2], **Ruiz, E.**[1]
[1]Dpto. Física Fundamental. Universidad de Salamanca
[2]Dpto. de Física Fonamental. Universitat de Barcelona
E-mail: jecuchi@usal.es, alfred.molina@ub.edu, eruiz@usal.es

Abstract. The Cabezas, Martín, Molina and Ruiz (CMMR) method allows us to build global analytic solutions of Einstein's equations for stationary isolated and rigidly rotating perfect fluid solutions. We start from a double approximation, postminkowskian and slow rotation, and end up getting a matched global solution from the inner and outer metrics. The metrics this way obtained have some uses. In particular, we will show the application of the scheme to the equation of state $\mu + (1 - n)p = \mu_0$ and how it can be applied to a build a source with two concentric comoving shells of fluid with different μ_0. We will also analyse the conditions under which this configuration can be a source of the Kerr metric.

1. Introduction. The CMMR approximation and previous applications

In Cabezas *et al.*[1], some of us described a new analytic approximation method to deal with the problem of obtaining interior solutions of stationary and axisymmetric perfect fluid configurations and their matched asymptotically flat vacuum exterior. To solve the Einstein's equations iteratively we use the postminkowskian approximation. Their general solution to each order gives rise to an infinite expansion in tensorial spherical harmonics that can be cut considering slow rotation. Using this, we have built and studied the global metrics of a compact ball of constant density, a polytropic fluid[2] and also one with equation of state(EOS) $\mu + (1 - n)p = \mu_0$. This last case has also been used to establish a relation between the r_0 constant of Wahlquist's metric and the physical parameters of the source, as well as to confirm that there is no asymptotically flat exterior for it[3, 4]. We have also faced the open problem of the existence of a suitable source for the Kerr metric. There have been studies of sources with singularities, and some clues[5] from the local behaviour of the interior near the surface indicating possible perfect fluid candidates, as well some others indicating the opposite[6]. Nevertheless, the common stream tends to believe that no reasonable physical interior exists. We have checked all our previously built metrics and none of them can generate a Kerr exterior.

With this idea in mind, we start building the first order solution for a rotating ball of perfect fluid with constant energy density μ_1 (core) surrounded by a corotating thick shell with energy density μ_2 (crust) and again perfect fluid. Then, we will match it to an asymptotically flat vacuum spacetime and compare the resulting values of the multipole moments with the Kerr ones. After that, we will check how much would the results change if we introduced a singular term in the core metric.

2. The spacetime considered

We want to find the metric for three different zones. All of them are stationary and axisymmetric. Being the material content vacuum or perfect fluid, we can adapt coordinates to the symmetries so that the metrics have Papapetrou's structure[7].

The fluid is rigidly rotating (rotational speed as seen from the infinite ω is constant) everywhere. The first spacetime, the core of the perfect fluid, has no convective motion so its 4-velocity is $u^\alpha = \psi_1\,(\xi^\alpha + \omega\,\zeta^\alpha)$, with $\boldsymbol{\xi}$ and $\boldsymbol{\zeta}$ the timelike and the closed orbits-spacelike Killings, respectively. Its EOS is $\mu_1 + (1-n_1)p = \epsilon_1$ and its surface can be written as $r_I = r_i\left[1 + S_i\,\Omega^2 P_2(\cos\theta)\right]$ using the slow rotation to cut its expansion in Legendre polynomials. Here r_i is the value of r when $\omega = 0$ and S_i a constant to be determined. The crust of the fluid is also non-convective, its EOS is $\mu_2 + (1-n_2)p = \epsilon_2$ and its surface $r_\Sigma = r_s\left[1 + S_s\,\Omega^2 P_2(\cos\theta)\right]$. Solving Euler's equations, the pressure and energy densities are

$$
\begin{aligned}
p_i &= \frac{\epsilon_i}{n_i}\left[-1 + k_i\left(\frac{\psi_i}{\psi_{Ii}}\right)^{n_i}\right] \\
\mu_i &= \frac{\epsilon_i}{n_i}\left[1 + k_i(n_i-1)\left(\frac{\psi_i}{\psi_{Ii}}\right)^{n_i}\right]
\end{aligned}
\qquad i = 1, 2; \tag{1}
$$

and k_i constants to adjust imposing $p_1|_{r_I} = p_2|_{r_I}$ and $p_2|_{r_\Sigma} = 0$.

The outer vacuum region shares the symmetries and we impose asymptotical flatness upon it.

3. Building the global solution

We use harmonic Cartesian-like coordinates $\{x^a\}$, although some results are expressed in the orthonormal spherical associated cobasis $\{\omega^a\}$. We choose the parameter controlling the postminkowskian approximation as $\lambda = \frac{4}{3}\pi\epsilon_i r_0^2$ and the slow rotation one as $\Omega = \lambda^{-1/2}\omega r_0$, with r_0 an arbitrary length. We can now start to solve the field equations. Decomposing the exact metric $\mathbf{g}$ as $\mathbf{g}(\lambda,\Omega) = \boldsymbol{\eta} + \mathbf{h}(\lambda,\Omega)$, with $\boldsymbol{\eta}$ the flat metric, the system to solve is, to first order in λ,

$$
\begin{cases}
\triangle h_{\alpha\beta} = -16\pi t_{\alpha\beta}\,, \\
\partial^k\left[h_{k\alpha} - \dfrac{1}{2}h\,\eta_{k\alpha}\right] = 0\,,
\end{cases} \tag{2}
$$

where $t_{\alpha\beta}$ comes from the energy-momentum tensor.

The general vacuum solution is

$$
\mathbf{h} = \mathbf{h}_{\rm part} + 2\sum_{n=0}^{\infty}\frac{M_n}{r^{n+1}}\left(\mathbf{T}_n + \mathbf{D}_n\right) + 2\sum_{n=1}^{\infty}\frac{J_n}{r^{n+1}}\,\mathbf{Z}_n + \sum_{n=0,2}\frac{A_n}{r^{n+3}}\mathbf{E}_{n+2} + \frac{B_2}{r^3}\,\mathbf{F}_2\,. \tag{3}
$$

Here M_n and J_n are multipole moments; A_n and B_n are pure gauge constants; $\mathbf{D}_0 \equiv \delta_{ij}dx^i\!\otimes dx^j$, $\mathbf{T}_2 \equiv P_2(\cos\theta)\,\omega^t\otimes\omega^t$, $\mathbf{D}_2 \equiv P_2(\cos\theta)\,\delta_{ij}dx^i\!\otimes dx^j$, $\mathbf{T}_0 \equiv \omega^t\otimes\omega^t$ and $\mathbf{Z}_1 \equiv P_1^1(\cos\theta)\,(\omega^t\otimes\omega^\varphi + \omega^\varphi\otimes\omega^t)$ are tensorial spherical harmonics, as well as $\mathbf{E_n}$, $\mathbf{F_n}$ though the last ones have a more cumbersome definition that can be essentially found in [1].

The crust solution is

$$
\begin{aligned}
\mathbf{h} = \mathbf{h}_{\rm part} &+ \sum_{n=0}^{\infty}\tilde{m}_n\,r^n\left(\mathbf{T}_n + \mathbf{D}_n\right) + \sum_{n=1}^{\infty}\tilde{j}_n\,r^n\,\mathbf{Z}_n + \sum_{n=0}^{\infty}\tilde{a}_n r^n\mathbf{E}^*{}_n + \sum_{n=0}^{\infty}\tilde{b}_n\,r^n\,\mathbf{F}^*{}_n \\
&+ 2\sum_{n=0}^{\infty}\frac{\tilde{M}_n}{r^{n+1}}\left(\mathbf{T}_n + \mathbf{D}_n\right) + 2\sum_{n=1}^{\infty}\frac{\tilde{J}_n}{r^{n+1}}\,\mathbf{Z}_n + \sum_{n=0,2}\frac{\tilde{A}_n}{r^{n+3}}\mathbf{E}_{n+2} + \frac{\tilde{B}_2}{r^3}\,\mathbf{F}_2\,,
\end{aligned} \tag{4}
$$

and the one in the core

$$\mathbf{h} = \mathbf{h}_{\text{part}} + \sum_{n=0}^{\infty} m_n\, r^n\, (\mathbf{T}_n + \mathbf{D}_n) + \sum_{n=1}^{\infty} j_n\, r^n\, \mathbf{Z}_n + \sum_{n=0}^{\infty} a_n r^n \mathbf{E}^*{}_n + \sum_{n=0}^{\infty} b_n\, r^n\, \mathbf{F}^*{}_n\,. \tag{5}$$

We need to parametrize the multipole moments and gauge constants. As explained in [1], based on symmetry and newtonian results arguments, their assumed expressions are

$$\left.\begin{array}{ll} M_n \to \lambda\Omega^n \dfrac{\bar{M}_n}{r_0^n}\,, & J_n \to \lambda^{\frac{3}{2}}\Omega^n \dfrac{\bar{J}_n}{r_0^n}\,, \\ A_n \to \lambda\Omega^n \dfrac{\bar{A}_n}{r_0^{n+3}}\,, & B_2 \to \lambda\Omega^2 \dfrac{\bar{B}_n}{r_0^3}\,. \end{array}\right\} \quad \text{where } \bar{X} = \bar{X}^{(1)} + \lambda\bar{X}^{(2)} + \ldots \tag{6}$$

and $X^{(n)}$ can still contain Ω terms. The infinite series can now be cut using the slow rotation approximation. We will retain only terms up to $\mathcal{O}(\Omega^3)$. The particular solution $\mathbf{h}_{\text{part}}$ in each case is easy to find, and the last step towards a global metric is the matching of spacetimes.

Considering that we are interested in the multipole moments and they are coordinate independent, imposing Lichnerowicz matching conditions $[g_{\alpha\beta}]_\Sigma = [\partial_\alpha g_{\beta\gamma}]_\Sigma = 0$ is not restrictive. We get

$$\bar{M}_0^{(1)} = \frac{{r_i}^3(\epsilon_1-\epsilon_2)+{r_s}^3\epsilon_2}{{r_0}^3}, \tag{7}$$

$$\begin{aligned} \bar{J}_1^{(1)} &= \frac{2\left({r_i}^5(\epsilon_1-\epsilon_2)+{r_s}^5\epsilon_2\right)}{5{r_0}^5} + \Omega^2 \times \\ &\times \frac{-2{r_s}^2\left(5{r_i}^8(\epsilon_1-\epsilon_2)^2 - 8{r_i}^5{r_s}^3(\epsilon_1-\epsilon_2)\epsilon_2 - {r_s}^8\epsilon_2(2\epsilon_1+3\epsilon_2)\right)}{3{r_0}^5\left(9{r_i}^5(\epsilon_1-\epsilon_2)\epsilon_2 - 2{r_s}^5\epsilon_2(2\epsilon_1+3\epsilon_2) - 5{r_i}^3{r_s}^2\left(2{\epsilon_1}^2+\epsilon_1\epsilon_2-3{\epsilon_2}^2\right)\right)}, \end{aligned} \tag{8}$$

$$\bar{M}_2^{(1)} = \frac{{r_s}^2\left(5{r_i}^8(\epsilon_1-\epsilon_2)^2 + 8{r_i}^5{r_s}^3(\epsilon_1-\epsilon_2)\epsilon_2 + {r_s}^8\epsilon_2(2\epsilon_1+3\epsilon_2)\right)}{{r_0}^5\left(9{r_i}^5(\epsilon_1-\epsilon_2)\epsilon_2 - 2{r_s}^5\epsilon_2(2\epsilon_1+3\epsilon_2) - 5{r_i}^3{r_s}^2\left(2{\epsilon_1}^2+\epsilon_1\epsilon_2-3{\epsilon_2}^2\right)\right)}, \tag{9}$$

$$\begin{aligned} \bar{J}_3^{(1)} &= \frac{1}{7{r_0}^7\left(9{r_i}^5(\epsilon_1-\epsilon_2)\epsilon_2 - 2{r_s}^5\epsilon_2(2\epsilon_1+3\epsilon_2) - 5{r_i}^3{r_s}^2\left(2{\epsilon_1}^2+\epsilon_1\epsilon_2-3{\epsilon_2}^2\right)\right)} \times \\ &\times \Big[2{r_s}^2\left(5{r_i}^{10}(\epsilon_1-\epsilon_2)^2 + 5{r_i}^7{r_s}^3(\epsilon_1-\epsilon_2)\epsilon_2\right. \\ &\qquad \left. + 3{r_i}^5{r_s}^5(\epsilon_1-\epsilon_2)\epsilon_2 + {r_s}^{10}\epsilon_2(2\epsilon_1+3\epsilon_2)\right)\Big]. \end{aligned} \tag{10}$$

4. Matching, Kerr and singular solutions

Our vacuum metric can be used to describe Kerr metric near infinity choosing suitable values of the multipole moments[8], but we have to check if it is still possible with the values imposed by the matching. Kerr multipole moments are given by $M_n^{Kerr} = m(ia)^n$ where m and a are the standard parameters of Kerr in Boyer-Lindquist coordinates. If we relate them with ours we have

$$M_0^{Kerr} = m \quad \xrightarrow{ours} \quad \lambda r_0 \bar{M}_0 \tag{11}$$

$$M_1^{Kerr} \equiv J_1^{Kerr} = ma \quad \xrightarrow{ours} \quad \lambda^{3/2}\Omega r_0^2 \bar{J}_1 \tag{12}$$

$$M_2^{Kerr} = -ma^2 \quad \xrightarrow{ours} \quad \lambda\Omega^2 r_0^3 \bar{M}_2 \tag{13}$$

$$M_3^{Kerr} \equiv J_3^{Kerr} = ma^3 \quad \xrightarrow{ours} \quad \lambda^{3/2}\Omega^3 r_0^4 \bar{J}_3 \tag{14}$$

Using the relations (11) and (12), we can build $-ma^2 = -\lambda^2\Omega^2 r_0^3 \bar{J}_1^2/\bar{M}_0$, a quantity of order $\mathcal{O}(\lambda^2)$. Also, relation (13) gives $M_2^{\text{Kerr}} = -ma^2 = \lambda\Omega^2 r_0^3(\bar{M}_2^{(1)} + \lambda\bar{M}_2^{(2)} + \ldots)$. Comparing this two results we can see that $\bar{M}_2^{(1)}$ must be zero. A similar analysis shows that the same holds for $\bar{J}_3^{(1)}$. This two conditions are only compatible with $\epsilon_1 = \epsilon_2 = 0$, no fluid at all.

We wonder now what would happen if we manually introduced a singular term $\frac{A}{r}$ in the core solution. If we do it in the way

$$\mathbf{h} = \mathbf{h}_{\text{part}} + \sum_{n=0}^{\infty}\left(m_n\, r^n + \frac{A}{r}\right)(\mathbf{T}_n + \mathbf{D}_n) + \sum_{n=1}^{\infty} j_n\, r^n\, \mathbf{Z}_n + \sum_{n=0}^{\infty} a_n r^n \mathbf{E}^*{}_n + \sum_{n=0}^{\infty} b_n\, r^n\, \mathbf{F}^*{}_n\,, \quad (15)$$

then, repeating all the process, the multipole moments after the matching are

$$\bar{M}_0^{(1)} = \frac{{r_s}^3 \epsilon_2}{{r_0}^3}\,, \qquad \bar{M}_2^{(1)} = \frac{{r_s}^5\left({r_0}^5 + {r_i}^5(\epsilon_1 - \epsilon_2)\right)}{3{r_0}^{10}}\,,$$
$$\bar{J}_1^{(1)} = \frac{2{r_s}^5 \epsilon_2}{5{r_0}^5} + \Omega^2 \frac{-20{r_0}^8{r_s}^2\epsilon_2 + 2{r_i}^5{r_s}^5(-\epsilon_1 + \epsilon_2)}{15{r_0}^{10}}\,, \quad \bar{J}_3^{(1)} = \frac{2{r_i}^5{r_s}^7(\epsilon_1 - \epsilon_2) + 20{r_0}^8{r_i}^2{r_s}^2\epsilon_2}{35{r_0}^{12}}\,. \quad (16)$$

Once again, $\bar{M}_2^{(1)}$ and $\bar{J}_3^{(1)}$ must be zero, but in this case this leads us to

$$\epsilon_1 = -\frac{10{r_0}^8 - {r_i}^3{r_s}^5}{10{r_0}^3{r_i}^5}\,, \qquad \epsilon_2 = \frac{{r_s}^5}{10{r_0}^3{r_i}^2}\,, \quad (17)$$

thus, the double shell + singularity configuration is compatible with Kerr to this order.

5. Remarks

We have built within the CMMR approximation the global solution for an asymptotically flat vacuum containing a source composed of two thick shells of perfect fluid up to order $\mathcal{O}(\lambda, \Omega^3)$. We have analysed its multipole moments discarding it as a possible source of the Kerr metric and compared it with the case in wich a singular term is allowed in the core zone. This last case has been shown compatible with Kerr up to this order.

6. Acknowledgments

This work was supported by projects FIS2009-07238 and FIS2007-63034 from the Spanish Ministerio de Educación y Ciencia. JEC thanks Junta de Castilla y León for predoctoral fellowship EDU/1165/2007.

References

[1] Cabezas J A, Martín J, Molina A and Ruiz E 2007 *General Relativity and Gravitation* **39** 707–736

[2] Martín J, Molina A and Ruiz E 2008 *Classical and Quantum Gravity* **25** 105019

[3] Cuchí J, Martín J, Molina A and Ruiz E 2010 *Journal of Physics: Conference Series* vol 229 (IOP Publishing) p 012032

[4] Cuchí J E, Gil-Rivero A, Molina A and Ruiz E 2008 *Spanish Relativity Meeting 2007. Relativistic Astrophysics and Cosmology* vol 30 ed Oscoz A, Mediavilla E and Serra-Ricart M (EAS Publications Series) pp 311–314 URL `http://dx.doi.org/10.1051/eas:0830051`

[5] Roos W 1976 *General Relativity and Gravitation* **7** 431–444 ISSN 0001-7701

[6] Wolf T and Neugebauer G 1992 *Classical and Quantum Gravity* **9** L37

[7] Papapetrou A 1966 *Ann. Inst. H. Poincare A* **4** 83–105

[8] Beig R and Simon W 1981 *Proceedings of the Royal Society of London. Series A, Mathematical and Physical Sciences* **376** 333–341 ISSN 0080-4630

Journal of Physics: Conference Series **314** (2011) 012071 doi:10.1088/1742-6596/314/1/012071

Linear perturbations of a Schwarzschild black hole by thin disc

P. Čížek

Institute of theoretical physics, Faculty of Mathematics and Physics, Charles University

E-mail: `ciz@matfyz.cz`

Abstract. The article written by C.M.Will (1974) describes a method how to obtain a perturbation of an (originally) Schwarzschild metric by a slowly rotating light ring. This scheme is revisited in order to adapt it to the case of a thin disc.It turns out for some class of mass distribution on the disc bounded between some radii the solution (or at least its decomposition into spherical harmonics) can be found explicitly.

1. Introduction

Many of the very compact active astrophysical objects (for example active galactic nuclei and X-ray binaries) are believed to contain disc-like structures. However these very compact objects generates strong and inhomogenous gravitational field so Newton's theory is not sufficient to describe them. On the other hand general relativity is non-linear and the simple superposition cannot be used. Recently these and even more complicated (for example time dependant) fields are successfully treated numerically. However so far the only known "analytical" solutions of this problem contains a very high degree of symmetry, usually they are axisymmetric and stationary. Even in this "symmetric" model the equations are quite complicated so it is a bit surprising in the *static* case the superposition can be used and the equations solved explicitly (Weyl class of solutions).

The fundamental treatment of *stationary* axisymmetric thin discs can be found in the paper by Bardeen [1]. It was succeeded by the article by Will [2], which remains a major reference in this subject even now.

2. Basic concepts

Considering axisymmetric stationary case we can write metric in form

$$ds^2 = -e^{2\nu}dt^2 + r^2B^2e^{-2\nu}\sin^2\theta(d\varphi - \omega dt)^2 + e^{2\zeta-2\nu}\left(dr^2 + r^2d\theta^2\right), \tag{1}$$

where ν, ω, B and ζ stands for an unknown metric functions of coordinates r and θ.

The energy-momentum tensor of thin dust disc can expressed

$$T^\alpha_\beta = \sigma e^{2\nu-2\zeta}u^\alpha u_\beta r^{-1}\delta(\cos\theta), \tag{2}$$

where σ describes surface mass density and u^α is four-velocity of the fluid. It is convenient to write four-velocity in the form $u^\alpha = \frac{e^{-\nu}}{1-v^2}(1,0,0,\Omega)$, where $\Omega = \frac{d\phi}{dt}$ is coordinate angular velocity and $v = r\sin\theta B e^{-2\nu}(\Omega-\omega)$ linear velocity[1] of the fluid.

It can be shown function B does not contain physically relevant information. As long as it satisfies the equation $\nabla\cdot(r\sin\theta\nabla B)=0$(where the divergence $(\nabla\cdot)$ and gradient (∇) operators are the same as in the virtual Euclidean space-time with spherical coordinates r, θ and ϕ) it can be chosen arbitrary[2] and the solutions for different B will only differ in the choice of coordinates.

The important pair Einstein equations is

$$\nabla\cdot(B\nabla\nu) - \frac{1}{2}r^2\sin^2\theta e^{-4\nu}\nabla\omega\cdot\nabla\omega = 4\pi B\sigma\frac{1+v^2}{1-v^2}\frac{1}{r}\delta(\cos\theta), \tag{3}$$

$$\nabla\cdot\left(r^2\sin^2\theta B^3 e^{-4\nu}\nabla\omega\right) = -16\pi B^2\sigma e^{-2\nu}\frac{v}{1-v^2}\delta(\cos\theta) \tag{4}$$

and the last two independent equations can be used to express ζ using known ν and ω.

Now let us assume the background metric is perturbed by a thin dust disc. The metric function can be written in form $\nu = \nu_0 + \nu_1 + \cdots$ where the lower index corresponds with the power of the disc mass it contains (and analogously for the functions ω and ζ). In the subsequent text we will work only with the linear perturbations.

The Schwarzschild black hole in the isotropic coordinates takes form

$$ds^2 = -\left(\frac{2r-M}{2r+M}\right)^2 dt^2 + \left(1+\frac{M}{2r}\right)^4\left(dr^2 + r^2 d\theta^2 + r^2\sin^2\theta d\phi^2\right), \tag{5}$$

so the background metric functions are

$$B = 1-\frac{M^2}{4r^2}, \qquad \nu_0 = \ln\left(\frac{2r-M}{2r+M}\right), \qquad \omega_0 = 0, \qquad \zeta_0 = \ln\left(1-\frac{M^2}{4r^2}\right). \tag{6}$$

The linear perturbation of this black hole by disc will lead us to the equations

$$\frac{\partial}{\partial x}\left[(x^2-1)\frac{\partial\nu_1}{\partial x}\right] + \frac{1}{\sin\theta}\frac{\partial}{\partial\theta}\left[\sin\theta\frac{\partial\nu_1}{\partial\theta}\right] = 4\pi r\sigma\frac{1+v^2}{1-v^2}\delta(\cos\theta), \tag{7}$$

$$\frac{\partial}{\partial x}\left[(x+1)^4\frac{\partial\omega_1}{\partial x}\right] + \frac{(x+1)^3}{x-1}\frac{1}{\sin^3\theta}\frac{\partial}{\partial\theta}\left[\sin^3\theta\frac{\partial\omega_1}{\partial\theta}\right] = -\frac{\pi}{4M^2}\frac{(2r+M)^3}{2r-M}\sigma\frac{v}{1-v^2}\delta(\cos\theta), \tag{8}$$

where $x = \frac{r}{M}\left[1+M^2/(4r^2)\right]$ is new radial coordinate. Considering the particles are moving on geodesics their velocity can be written[3] as $v = (x-1)^{-1/2}$.

To solve these equations we will use expansion into the Legendre (for the function ν_1) and Gegenbauer (ω_1) polynomials[4]. So let us define

$$\nu_1 = \sum_{n=0}^{\infty}\nu_{1\,n}(x)P_n(\cos\theta), \qquad \omega_1 = \sum_{n=0}^{\infty}\omega_{1\,n}(x)T_n^{3/2}(\cos\theta). \tag{9}$$

[1] With respect to the zero angular momentum observer.
[2] With the exception $B = \frac{1}{r\sin\theta}$, which represents plane waves solutions instead of stationary axisymmetric ones.
[3] In the zeroth order of expansion. We need not to consider first order since the velocity is used in the source term of the equations, which is clearly proportional to the disc mass solely because of σ.
[4] It can be shown the series converges absolutely anywhere besides the axis. On the axis ν_1 converges absolutely, however in the case of ω_1 it is still not clear. The conjecture is ω_1 converges, but not absolutely.

Moreover we will write the r.h.s. of Einstein equations using these polynomials

$$4\pi r\sigma\frac{1+v^2}{1-v^2}\delta(\cos\theta)=\sum_{n=0}^{\infty}R_nP_n(\cos\theta),\text{ where }R_n=(4n+2)\pi P_n(0)x\chi(x)\quad\text{and}\quad(10)$$

$$\frac{-\pi}{4M^2}\frac{(2r+M)^3}{2r-M}\frac{\sigma v}{1-v^2}\delta(\cos\theta)=\sum_{n=0}^{\infty}S_nT_n^{3/2}(\cos\theta),\text{ where }S_n=\frac{\pi T^{3/2}(0)(2n+3)(x+1)^{3/2}\chi(x)}{-2M(n+1)(n+2)}\quad(11)$$

In both expressions $\chi(x)=r\sigma/(x-2)$ stands for the new function used to describe mass density. Also we can see $R_n=S_n=0$ when n is odd.

After substitution into linearized Einstein equations and their separation we will obtain

$$\frac{\mathrm{d}}{\mathrm{d}x}\left[(x^2-1)\frac{\mathrm{d}\nu_{1\,n}}{\mathrm{d}x}\right]-n(n+1)\nu_{1\,n}=R_n,\quad(12)$$

$$\frac{\mathrm{d}}{\mathrm{d}x}\left[(x+1)^4\frac{\mathrm{d}\omega_{1\,n}}{\mathrm{d}x}\right]-n(n+3)\frac{(x+1)^3}{x-1}\omega_{1\,n}=S_n.\quad(13)$$

The base of fundamental system of the first equation is the n-th Legendre polynomial $P_n(x)$ and the Legendre function of the second kind $Q_n(x)$. Analogously we can obtain fundamental system of the second equation. It consists of the polynomial $F_n(x)$ and the "function of the second kind" $G_n(x)$ defined as

$$F_n(x)={}_2\mathrm{F}_1\left(-n,n+3,4;(x+1)/2\right),\qquad G_n(x)=F_n(x)\int_x^{\infty}\frac{\mathrm{d}\xi}{[F_n(\xi)]^2(\xi+1)^4},\quad(14)$$

where ${}_2\mathrm{F}_1\left(a,b,c;z\right)$ denotes hypergeometric function. Assuming asymptotic flatness and regularity of the perturbation one can construct corresponding Green functions

$$\mathcal{G}_n^{\nu}(x,x')=-P_n[\min(x,x')]Q_n[\max(x,x')]\text{ and }\mathcal{G}_n^{\omega}(x,x')=-F_n[\min(x,x')]G_n[\max(x,x')].\quad(15)$$

These can be used to calculate first perturbation, however the integrals are quite complicated.

In special case, where $\chi(x)$ is an even polynomial, particular solution can be found explicitly. $\nu_{1\,n}^{\mathrm{Part}}$ is exactly

$$x\chi(x)=\sum_{l=0}^{m}c_{2l+1}P_{2l+1}(x)\quad\Rightarrow\quad\nu_{1\,n}^{\mathrm{Part}}=\sum_{l=0}^{m}\frac{c_{2l+1}}{2(l+1)(2l+1)-n(n+1)}P_{2l+1}(x),\quad(16)$$

where m and c_j are some constants. The case of $\omega_{1\,n}^{\mathrm{Part}}$ is slightly more difficult, but it takes form

$$\omega_{1\,n}^{\mathrm{Part}}=\frac{A_n(x)}{\sqrt{x+1}}+C_nF_n(x)\ln\left(\frac{\sqrt{x+1}+\sqrt{2}}{\sqrt{x+1}-\sqrt{2}}\right),\quad(17)$$

where $A_n(x)$ stands for a suitable polynomial and C_n is some constant.

However we would like to obtain solution with disc spanned between some radii $r_{\min}$ and $r_{\max}$ and we want the solution to be regular on the horizon and asymptotically flat. Its uniqueness is clear from the uniqueness of the Green functions. It can be also described by

$$\nu_{1\,n}=\begin{cases}A_<P_n(x) & \text{for}\quad x\le x_{\min}\\ A_=P_n(x)+B_=Q_n(x)+\nu_{1\,n}^{\mathrm{Part}}(x) & \text{for}\quad x_{\min}\le x\le x_{\max}\\ B_>Q_n(x) & \text{for}\quad x\ge x_{max}\end{cases},\quad(18)$$

where $A_<$, $A_=$, $B_=$ and $B_>$ are constants obtained by the the requirement of continuity of metric functions and their first derivatives at the radii of the rims of the disc (and analogously we can express $\omega_{1\,n}$).

3. Numerical examples

Let us present some numerical examples of the procedure described above. In both cases we take $M=1$, $r_{\min}=5$ and $r_{\max}=6$ and we consider only first 30 orders in the "harmonic" expansion (i.e. $P_n(\cos\theta)$ and $T_n^{3/2}(\cos\theta)$).

Journal of Physics: Conference Series **314** (2011) 012071 doi:10.1088/1742-6596/314/1/012071

The first one (Fig. 1 and 2) deals with $\chi = (40 - x^2) \cdot 10^5$ and the second one (Fig. 3 and 4) with $\chi = 10^{-7}$.

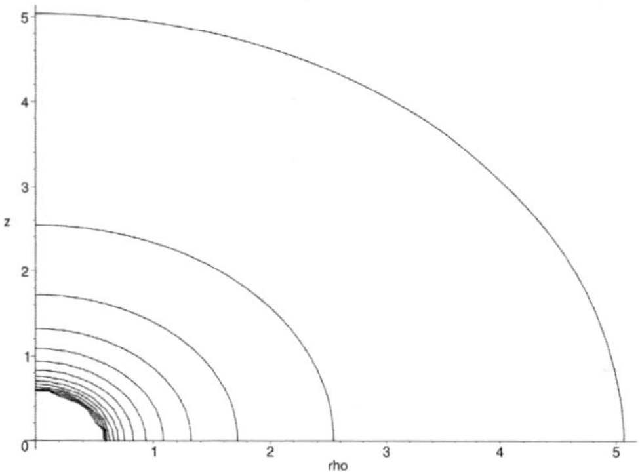

Figure 1. Gravitational potential $\nu = \nu_0 + \nu_1$. The lines corresponds with the potential values $-1.5, -1.4, \cdots, -0.1$.

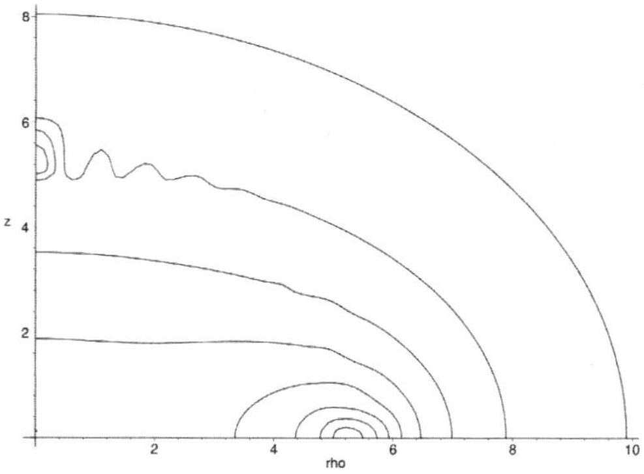

Figure 2. Dragging $\omega = \omega_0 + \omega_1$. The lines corresponds with the values $8 \cdot 10^{-6}, 7 \cdot 10^{-6}, \cdots, 1 \cdot 10^{-6}$.

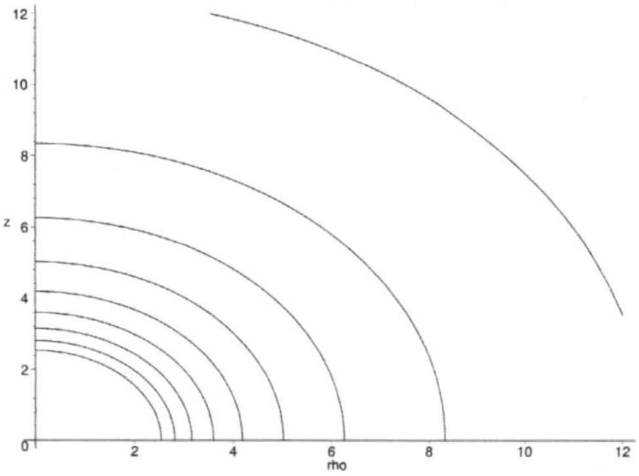

Figure 3. Gravitational potential $\nu = \nu_0 + \nu_1$. The lines corresponds with the potential values $-0.20, -0.18, \cdots, -0.04$.

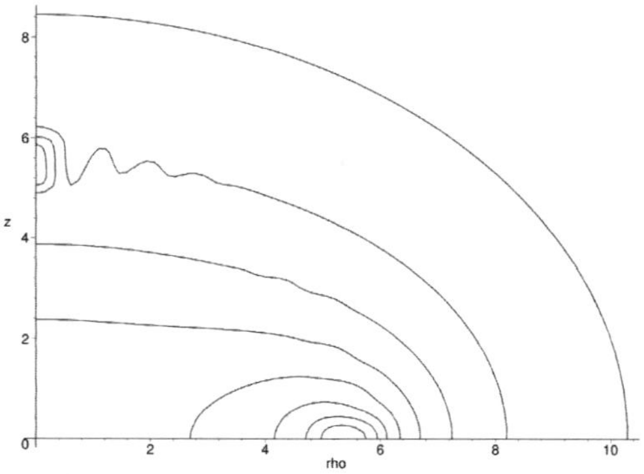

Figure 4. Dragging $\omega = \omega_0 + \omega_1$. The lines corresponds with the values $10 \cdot 10^{-9}, 9 \cdot 10^{-9}, \cdots, 1 \cdot 10^{-9}$.

4. Conclusion

Analogously to the paper by Will [2] we were able to obtain procedure calculating the first order perturbations of Schwarzchild black hole by a thin dust disc for some class of matter distributions. However mentioned procedure is not suited well for second or higher order expansion in the mass because of the problems associated with multiplication of spherical harmonics. Moreover it is not well suited for the numerical calculations (as we can see in the numerical examples).

5. Acknowledgements

I thank Oldřich Semerák for ideas and many enlightening conversations concerning this topic and Marcus Ansorg for the remarks concerning convergence of the series. Beside that I would like to acknowledge financial support from the grants GAUK-86508 and GACR-205/09/H033.

References

[1] Bardeen J.M., Wagoner R.V. 1971 Relativistic disks. I. Uniform rotation *Ap.J.* **167** 359-423

[2] Will C.M. 1974 Perturbation of a slowly rotating black hole by a stationary axisymmetric ring of matter. I. equilibrium configurations *Ap.J.* **191** 521-531

Journal of Physics: Conference Series **314** (2011) 012072 doi:10.1088/1742-6596/314/1/012072

Electromagnetic and Gravitational Invariants

L. Filipe O. Costa*, Carlos A. R. Herdeiro†, Lode Wylleman‡

*Centro de Física do Porto, Universidade do Porto, Porto, Portugal
†Dep. de Física da Universidade de Aveiro and I3N, 3810-183 Aveiro, Portugal
‡Faculty of App. Sciences TW16, Ghent University, Galglaan 2, 9000 Ghent, Belgium

E-mail: filipezola@fc.up.pt, herdeiro@ua.pt, lwyllema@gmail.com

Abstract. The curvature invariants have been subject of recent interest in the context of the experimental detection of the gravitomagnetic field, namely due to the debate concerning the notions of "extrinsic" and "intrinsic" gravitomagnetism. In this work we explore the physical meaning of the curvature invariants, dissecting their relationship with the gravitomagnetic effects.

1. Introduction

There is an ongoing debate concerning the question if there is a fundamental difference between the "translational" gravitomagnetism claimed to have been detected with Lunar Laser Raging [1], and (to an high accuracy) in the observations [2] of the binary pulsar PSR 1913 +16, and the gravitomagnetic field produced by the rotation of the Earth, detected in the LAGEOS Satellites data [3] (whose detection was also the primary goal of the Gravity Probe-B mission). These phenomena have been dubbed, respectively, "extrinsic" and "intrinsic" gravitomagnetism, such distinction being based [4, 5] on the quadratic curvature invariants $\mathbf{R}\cdot\mathbf{R}$, $\star\mathbf{R}\cdot\mathbf{R}$, and driven by their formal analogy with the invariants $\mathbf{F}\cdot\mathbf{F}$, $\star\mathbf{F}\cdot\mathbf{F}$ of the Maxwell tensor. In this work [6], starting from previous knowledge on the classification of purely electric/magnetic spacetimes [7, 8], and using insight from the *formal* analogy $\{\mathbf{F}\cdot\mathbf{F},\ \star\mathbf{F}\cdot\mathbf{F}\} \leftrightarrow \{\mathbf{R}\cdot\mathbf{R},\ \star\mathbf{R}\cdot\mathbf{R}\}$, we explain the invariant structure of the relevant gravitational fields. Then, using the gravito-electromagnetic analogy based on tidal tensors [9] as a *physical* guiding principle, we clarify the physical meaning of the curvature invariants and the implications on the motion of test particles.

2. Electromagnetic and gravitational scalar invariants — algebraic meaning

Wrt any unit timelike 4-vector u^α, the Maxwell tensor splits irreducibly into the two spatial vectors $(E^u)^\alpha \equiv F^{\alpha\beta}u_\beta$ and $(B^u)^\alpha \equiv \star F^{\alpha\beta}u_\beta$ (c.f. Eq. (16) of [10]), which are, respectively, the electric and magnetic fields as measured by an observer of 4-velocity u^α. Both $(E^u)^\alpha$ and $(B^u)^\alpha$ depend on the observer 4-velocity u^α, but combining them one can construct the two quadratic (i.e., second order) scalar invariants (e.g. [11]):

$$E^\alpha E_\alpha - B^\alpha B_\alpha = -\frac{1}{2}F_{\alpha\beta}F^{\alpha\beta} \equiv -\frac{1}{2}\star\mathbf{F}\cdot\mathbf{F}; \qquad E^\alpha B_\alpha = -\frac{1}{4}F_{\alpha\beta}\star F^{\alpha\beta} \equiv -\frac{1}{4}\star\mathbf{F}\cdot\mathbf{F}\,. \quad (1)$$

They have the following physical interpretation: (i) if $\star\mathbf{F}\cdot\mathbf{F} \neq 0$ then for all observers u^α one has $(E^u)^\alpha \neq 0 \neq (B^u)^\alpha$; (ii) if $\star\mathbf{F}\cdot\mathbf{F} = 0$ and $\mathbf{F}\cdot\mathbf{F} > 0$ (< 0) then observers u^α do exist wrt $(B^u)^\alpha = 0$ $((E^u)^\alpha = 0)$; (iii) the case $\mathbf{F}\cdot\mathbf{F} = \star\mathbf{F}\cdot\mathbf{F} = 0$ (but $F^{\alpha\beta} \neq 0$) corresponds to pure electromagnetic radiation.

On the other hand, the Riemann tensor *in vacuum* becomes the Weyl tensor, which analogously is completely decomposed (c.f. Eq. (30) of [10]) in its electric part $\mathbb{E}_{\alpha\beta} = R_{\alpha\mu\beta\nu}u^\mu u^\nu$ and magnetic part $\mathbb{H}_{\alpha\beta} = \star R_{\alpha\mu\beta\nu}u^\mu u^\nu$ wrt u^α. Moreover, $\mathbb{E}_{\alpha\beta}$ and $\mathbb{H}_{\alpha\beta}$ form quadratic invariants *formally* analogous to (1):

$$\mathbb{E}^{\alpha\gamma}\mathbb{E}_{\alpha\gamma} - \mathbb{H}^{\alpha\gamma}\mathbb{H}_{\alpha\gamma} = \frac{1}{8}R_{\alpha\beta\gamma\delta}R^{\alpha\beta\gamma\delta} \equiv \frac{1}{8}\mathbf{R}\cdot\mathbf{R} \qquad \mathbb{E}^{\alpha\gamma}\mathbb{H}_{\alpha\gamma} = \frac{1}{16}R_{\alpha\beta\gamma\delta}\star R^{\alpha\beta\gamma\delta} \equiv \frac{1}{16}\star\mathbf{R}\cdot\mathbf{R}, \quad (2)$$

which are usually combined into $I \equiv (\mathbf{R}\cdot\mathbf{R} + i\star\mathbf{R}\cdot\mathbf{R})/8$. However, whereas the invariants (1) are the only two independent scalar invariants of $F^{\alpha\beta}$, in the case of $R_{\alpha\beta\gamma\delta}$ there also two independent *cubic* invariants, given by $A \equiv R^{\alpha\beta}{}_{\lambda\mu}R^{\lambda\mu}{}_{\rho\sigma}R^{\rho\sigma}{}_{\alpha\beta}/16$ and $B \equiv R^{\alpha\beta}{}_{\lambda\mu}R^{\lambda\mu}{}_{\rho\sigma}\star R^{\rho\sigma}{}_{\alpha\beta}/16$, and combined into $J \equiv A - iB$. It turns out [7, 8] that one basically obtains formally equivalents statements (i)-(iii) as above by replacing $\mathbf{F}$ by $\mathbf{R}$, *provided that the condition* $M \equiv I^3/J^2 - 6 \geq 0$ (real or infinite) *is added to* (ii). Further comments and a more detailed treatment hereof shall be given in [6].

3. Interpretation of the invariant's structure for relevant setups

For astrophysical applications we will be interested essentially in the gravitational far field of non-spinning and spinning bodies obeying criterion $M \geq 0$, which thus have an electromagnetic counterpart. Now the important point is that $\mathbb{E}_{\alpha\gamma}$ and $\mathbb{H}_{\alpha\gamma}$ also obey transformation laws in a change of observer/frame which exhibit a degree of similarity with the transformation of the electromagnetic fields that allows for invariant structure of a gravitational setup to be understood comparing with the analogous electromagnetic setup. The magnetic part of the Riemann tensor $(\mathbb{H}^U)_{\alpha\gamma}$ as measured by a given observer $U^\alpha = U^0(1, \vec{v})$, can be obtained, in terms of the tensors $(\mathbb{E}^u)_{\alpha\gamma}$ and $(\mathbb{H}^u)_{\alpha\gamma}$ measured by $u^\alpha = u^0\delta^\alpha_0$, from decomposition (30) of [10], yielding [13], to first order in v, Eq. (3i) below, which exhibits suggestive similarities with its electromagnetic counterpart (3ii):

$$\overleftrightarrow{\mathbb{H}}(U) \simeq \overleftrightarrow{\mathbb{E}}(u) \times \vec{v} - \vec{v} \times \overleftrightarrow{\mathbb{E}}(u) + \overleftrightarrow{\mathbb{H}}(u) \quad (i) \quad \vec{B}(U) \simeq -\vec{v} \times \vec{E}(u) + \vec{B}(u) \quad (ii) \quad (3)$$

where $(\overleftrightarrow{\mathbb{E}} \times \vec{v})_{kl} \equiv \epsilon^{ij}{}_l \mathbb{E}_{ik} v_j$, $(\vec{v} \times \overleftrightarrow{\mathbb{E}})_{kl} \equiv \epsilon^{ij}{}_k v_i \mathbb{E}_{lj}$.

The examples of interest will be discussed in detail in [6]. Consider first a single point charge; for a generic observer U^α, we have $\vec{B}(U) \neq 0$; but as one can read from (3ii), always $\vec{B} \perp \vec{E}$, since there are observers (those comoving with the charge) for which $\vec{B}$ vanishes everywhere; and the invariant $\star\mathbf{F}\cdot\mathbf{F} = 0$ reflects that. Analogous arguments explain the vanishing of $\star\mathbf{R}\cdot\mathbf{R} = 0$ in Schwarzschild spacetime.

If we consider a system of two charges in arbitrary motion, then in general $\star\mathbf{F}\cdot\mathbf{F}\neq 0$, there is no frame where $\vec{B}=0$ *globally*, which we relate with the fact that there is no observer for which *both* particles are at rest (i.e., the currents involved cannot be made to vanish by a Lorentz boost). However, as explicitly shown in [6], in the special case of planar motion, *in the plane of the motion* we have $\star\mathbf{F}\cdot\mathbf{F}=0$, meaning that *therein* (and only therein) there are observers, *whose 4-velocity* U^α *varies from point to point,* relative to which $\vec{B}$ vanishes, which arises from an exact cancellation between the magnetic field produced by the motion of the two charges relative to the observer U^α. It should be emphasized that even in the plane of the motion $\vec{B}$ does not globally vanish in any rigid frame (as the velocity field U^α for which $B^\alpha=0$ is a shearing congruence), which is a clearly distinct situation from the case of a single charge, despite both being characterized by $\star\mathbf{F}\cdot\mathbf{F}\neq 0$. As shall be shown in [6], the analogous gravitational two body systems (which are studied to Post Newtonian order) exhibit an analogous invariant structure: $\star\mathbf{R}\cdot\mathbf{R}\neq 0$ in the most general case, telling us that in general there are no observers for which $\mathbb{H}_{\alpha\beta}=0$; and for planar motion, which is the case of the Earth-Sun system discussed in [1] we have $\star\mathbf{R}\cdot\mathbf{R}=0$ in the orbital plane, and $\star\mathbf{R}\cdot\mathbf{R}\neq 0$ elsewhere. This structure can be explained by precisely the same reasoning as in the electromagnetic case, since the tensor $\mathbb{H}_{\alpha\beta}$ is linear to Post Newtonian order and therefore to this accuracy a superposition principle applies.

One of the key points of this work is that the very same arguments are shown to apply to the case of a spinning body. The structure of the electromagnetic invariants of the field of a spinning charge is similar to the one of a system of charges in planar motion: $\star\mathbf{F}\cdot\mathbf{F}=0$ in the equatorial plane, $\star\mathbf{F}\cdot\mathbf{F}\neq 0$ elsewhere (explicit expressions will be given in [6]). A spinning body can be viewed as an assembly of translating elements; and there is no observer for which every part of it is at rest[1]; hence (except in the equatorial plane) it is not possible to make B^α vanish. In the equatorial plane, again by exactly the same principle from the case of $N\geq 2$ sources in planar motion, there are observers for which $B^\alpha=0$ (*not* the observers co-rotating with the source, despite rotating in the same sense; these observers have a velocity that decays as r^{-1}, which is the same asymptotic dependence as case of the two body system in planar motion mentioned above, c.f. [6]). The analogy with the gravitational case again holds: the curvature invariants for Kerr spacetime have the structure $\star\mathbf{R}\cdot\mathbf{R}\neq 0$ elsewhere, and $\star\mathbf{R}\cdot\mathbf{R}=0$ (exactly!) in the equatorial plane, where, likewise, there are observers for which $\mathbb{H}_{\alpha\beta}=0$; their angular velocity (which again depends on r) asymptotically matches its electromagnetic counterpart up to a factor of 2 [6, 12]. Hence, from the point of view of the structure of the invariants, whilst clearly distinguishing between the fields produced by a *single* translating body from the ones of a rotating body (which is in agreement with [5]), we found no fundamental distinction between the latter and the

[1] In a co-rotating frame indeed the whole spinning body would be seen to be at "rest"; but a (rigid) rotating frame consists of a congruence of observers *all with different* 4-velocities U^α, whilst having the same angular velocity. $\vec{B}$ does *not* vanish in the co-rotating frame, even though there are no currents therein; the reason being (taking the perspective of the rotating frame) that the vorticity of the congruence also contributes as a source for $\vec{B}$ (see Eqs. (17)-(20) of [10]). It is crucial to realize that unlike the translation of a single body, which can always be made to vanish by a simple boost, rotation (like the motion of $N\geq 2$ translating bodies in general) imprints itself in intrinsic properties (like the intrinsic angular momentum) that cannot be made to vanish by any change of frame.

ones from a system of translating bodies in planar motion. This is one point that (we hope) may shed some light on the ongoing debate.

It should also be remarked that, as the examples above make clear, whereas $\star\mathbf{F}\cdot\mathbf{F}\neq 0$ ($\star\mathbf{R}\cdot\mathbf{R}\neq 0$) signals intrinsic magnetic field (curvature), the case $\star\mathbf{F}\cdot\mathbf{F}=0$ ($\star\mathbf{R}\cdot\mathbf{R}=0$ with $M\geq 0$) does not distinguish static solutions, e.g. Coulomb field of a point charge (Schwarzschild solution), where B^{α} ($\mathbb{H}_{\alpha\beta}$) vanishes globally in a rigid frame, from setups where B^{α} ($\mathbb{H}_{\alpha\beta}$) can only be made to vanish for a a class of observers specific to each point (i.e., forming shearing congruences), as is the case of the equatorial plane of a spinning charge (equatorial plane of Kerr spacetime).

4. Dynamical implications of the invariants. Gravitomagnetism.

In spite of the insight it provides to interpret the structure of the gravitational invariants, the analogy with the electromagnetic invariants is purely formal (see Sec. 4 of [14]), and should not be used as a physical guiding principle: in one case we are dealing with quantities built on electromagnetic *fields* E^{α}, B^{α}; in the other case with gravitational *tidal tensors* $\mathbb{E}_{\alpha\beta}$, $\mathbb{H}_{\alpha\beta}$, which, as shown in [9], play in gravity analogous dynamical roles *not* to E^{α}, B^{α}, but to the *electromagnetic tidal tensors* (which are quantities one order higher in differentiation). This distinction is crucial, because the effects involved may actually be opposite: $\star\mathbf{F}\cdot\mathbf{F}=0$, telling us that there there are observers for which $B^{\alpha}=0$, means that magnetic dipoles carried by them do not undergo Larmor precession, but in general they will feel a force (since the magnetic *tidal* field does not vanish [12, 6]); by contrast what $\star\mathbf{R}\cdot\mathbf{R}=0$ (with $M\geq 0$) tells us is that there is a class of observers carrying gyroscopes that to not feel gravitational force (see Eq. (28) of [9]), but in general they "precess" (since it is the gravitomagnetic *tidal tensor* $\mathbb{H}_{\alpha\beta}$ that vanishes, not the so called [1, 4, 5] "gravitomagnetic *field*"). In [6] this will be shown explicitly for the physical systems considered in the previous section. To conclude, curvature invariants *do not* tell us about the "gravitomagnetic *field*" itself, nor the Lense Thirring effect (gravitomagnetic field and gyroscope "precession" are actually artifacts of the reference frame, thus cannot be manifest in invariant quantities). Curvature invariants tell us about the gravitomagnetic *tidal* field (magnetic curvature), and the appropriate probe to measure it is the *force* (*not the precession*!) on a gyroscope.

References

[1] M. Soffel *et al*, *PRD* **78** (2008) 024033; T. W. Murphy *et at*, *PRL* **98** (2007) 071102
[2] K. Nordtvedt, *Int. J. Theoretical Physics* **27** (1988) 2347
[3] I. Ciufolini, E. Pavlis, *Nature* **431** (2004) 958;
[4] I. Ciufolini, New Astronomy **15** (2010) 332
[5] I. Ciufolini, J. A. Wheeler, "Gravitation and Inertia," Princeton Ser. in Phys. (1995); Sec. 6.11.
[6] Costa L. F. O., Herdeiro C. A. R., Wylleman L. (work in progress)
[7] C. McIntosh, R. Arianhod, S. T. Wade, C. Hoenselaers, Class. Quant. Grav. **11** (1994) 1555
[8] L. Wylleman, N. Van den Bergh, *Phys. Rev. D.* **74** (2006) 084001
[9] Costa L F O and Herdeiro C A R 2008 *Phys. Rev. D* **78** 024021
[10] R. Maartens, B. Bassett, Class. Q. Grav. **15** (1998) 705
[11] L. Bel, Cahiers de Physique, **16** (1962) 59; eng. trans. Gen. Rel. Grav., **32** (2000) 2047.
[12] Costa L F O, Herdeiro C A R, Natário J, Zilhão M, in preparation.
[13] W. Campbell, T. Morgan, Am. J. Phys. **44** (1976) 356.
[14] L. F. Costa, C. A. R. Herdeiro, [arXiv:gr-qc/0612140v2]

Journal of Physics: Conference Series **314** (2011) 012073 doi:10.1088/1742-6596/314/1/012073

Polarization modes for strong-field gravitational waves

Bethan Cropp and Matt Visser

School of Mathematics, Statistics, and Operations Research
Victoria University of Wellington
Wellington, New Zealand

E-mail: bethan.cropp@msor.vuw.ac.nz, matt.visser@msor.ac.nz

Abstract. Strong-field gravitational plane waves are often represented in either the Rosen or Brinkmann forms. These forms are related by a coordinate transformation, so they should describe essentially the same physics, but the two forms treat polarization states quite differently. Both deal well with linear polarizations, but there is a qualitative difference in the way they deal with circular, elliptic, and more general polarization states. In this article we will describe a general algorithm for constructing arbitrary polarization states in the Rosen form.

1. Brinkmann form

Consider the general *pp* spacetime geometry [1, 2, 3, 4, 5]

$$\mathrm{d}s^2 = -2\,\mathrm{d}u\,\mathrm{d}v + H(u,x,y)\,\mathrm{d}u^2 + \mathrm{d}x^2 + \mathrm{d}y^2. \tag{1}$$

It is then a standard result that the only nonzero component of the Ricci tensor is

$$R_{uu} = -\frac{1}{2}\left\{\partial_x^2 H(u,x,y) + \partial_y^2 H(u,x,y)\right\}. \tag{2}$$

Restricting attention to vacuum plane waves [6] gives us the form:

$$\mathrm{d}s^2 = -2\,\mathrm{d}u\,\mathrm{d}v + \left\{[x^2-y^2]\,H_+(u) + 2xy\,H_\times(u)\right\}\,\mathrm{d}u^2 + \mathrm{d}x^2 + \mathrm{d}y^2. \tag{3}$$

In this form of the metric the two polarization modes are explicitly seen to decouple. By choosing $H_+(u)$ and $H_\times(u)$ appropriately we can construct any general polarization state.

2. Rosen form

The "most general" form of the Rosen metric is [1, 7]

$$\mathrm{d}s^2 = -2\,\mathrm{d}u\,\mathrm{d}v + g_{AB}(u)\,\mathrm{d}x^A\,\mathrm{d}x^B, \tag{4}$$

where $x^A = \{x, y\}$. The only non-zero component of the Ricci tensor is [1]:

$$R_{uu} = -\left\{\frac{1}{2}\,g^{AB}\,g''_{AB} - \frac{1}{4}\,g^{AB}\,g'_{BC}\,g^{CD}\,g'_{DA}\right\}. \tag{5}$$

Though relatively compact, because of the implicit matrix inversions this is a grossly nonlinear function of the metric components. In particular, in this form of the metric the + and × linear polarizations do not decouple in any obvious way [8].

3. Linear Polarization

Consider the strong-field gravity wave metric in the + linear polarization. That is, set $g_{xy} = 0$. It is found most useful to put the resulting metric in the form [8]

$$ds^2 = -2\,du\,dv + S^2(u)\,\left\{e^{+X(u)}\,dx^2 + e^{-X(u)}dy^2\right\}. \tag{6}$$

Then

$$R_{uu} = -\frac{1}{2}\left\{4\,\frac{S''}{S} + (X')^2\right\}. \tag{7}$$

In vacuum we have the general vacuum wave for + polarization in the form

$$ds^2 = -2\,du\,dv{+}S^2(u)\,\left\{\exp\left(2\int^u\sqrt{-S''/S}\,du\right)\,dx^2 + \exp\left(-2\int^u\sqrt{-S''/S}\,du\right)dy^2\right\}. \tag{8}$$

From the + polarization, by rotating the x–y plane through a fixed but arbitrary angleΘ_0, we can easily deal with linear polarization modes along any desired axis.

4. Arbitrary Polarization

Now take an arbitrary, possibly u dependent, polarization and consider the following metric ansatz [8]:

$$\begin{aligned} ds^2 \;&=\; -2\,du\,dv + S^2(u)\,\Big\{[\cosh(X(u)) + \cos(\theta(u))\sinh(X(u))]dx^2 \\ &\quad + 2\sin(\theta(u))\sinh(X(u))dx\,dy + [\cosh(X(u)) - \cos(\theta(u))\sinh(X(u))]dy^2\Big\}. \end{aligned} \tag{9}$$

Note setting $\theta(u) = \Theta_0$ corresponds to linear polarization. The vacuum field equations imply

$$4\,\frac{S''}{S} + (X')^2 + \sinh^2(X(u))\,(\theta')^2 = 0. \tag{10}$$

Let us introduce a dummy function $L(u)$ and split this into the two equations

$$4\,\frac{S''}{S} + (L')^2 = 0, \qquad (L')^2 = (X')^2 + \sinh^2(X(u))\,(\theta')^2. \tag{11}$$

The first of these equations is just the equation you would have to solve for a pure + (or in fact any linear) polarization. The second of these equations can be rewritten as

$$dL^2 = dX^2 + \sinh^2(X)\,d\theta^2, \tag{12}$$

and is the statement that L can be interpreted as distance in the hyperbolic plane H_2.

Compare this to Maxwell electromagnetism, where polarizations can be specified by

$$\vec{E}(u) = E_x(u)\,\hat{x} + E_y(u)\,\hat{y}, \tag{13}$$

with no additional constraints. Thus an electromagnetic wavepacket of arbitrary polarization can be viewed as an arbitrary "walk" in the (E_x, E_y) plane. We could also go to a magnitude-phase representation (E,θ) where

$$\vec{E}(u) = E(u)\cos\theta(u)\,\hat{x} + E(u)\sin\theta(u)\,\hat{y}. \tag{14}$$

So an electromagnetic wavepacket of arbitrary polarization can also be viewed as an arbitrary "walk" in the (E,θ) plane, where the (E,θ) plane is provided with the natural Euclidean metric

$$dL^2 = dE^2 + E^2\,d\theta^2. \tag{15}$$

In contrast for gravitational waves in the Rosen form we are now dealing with an arbitrary "walk" in the hyperbolic plane, H_2. Furthermore, because of the nonlinearity of general relativity, there is still one remaining differential equation to solve.

5. Circular Polarization

As an important example, we consider strong-field circular polarization. Circular polarization corresponds to a fixed distortion X_0 with a linear advancement of $\theta(u)$:

$$\theta(u) = \Omega_0\, u; \qquad X(u) = X_0. \tag{16}$$

Then [8]

$$\begin{aligned} ds^2 &= -2\,du\,dv + S^2(u)\,\Big\{[\cosh(X_0) + \cos(\Omega_0\, u)\sinh(X_0)]dx^2 \\ &\quad + 2\sin(\Omega_0 u)\sinh(X_0)dx\,dy + [\cosh(X_0) - \cos(\Omega_0 u)\sinh(X_0)]dy^2\Big\}. \end{aligned} \tag{17}$$

So the only nontrivial component of the Ricci tensor is

$$R_{uu} = -\frac{1}{2}\left\{4\,\frac{S''}{S} + \sinh^2(X_0)\,\Omega_0^2\right\}. \tag{18}$$

Solving the vacuum equations gives

$$S(u) = S_0\,\cos\left\{\frac{\sinh(X_0)\,\Omega_0\,(u-u_0)}{2}\right\}. \tag{19}$$

This now describes a spacetime that has good reason to be called a strong-field circularly polarized gravity wave. Note the weak-field limit corresponds to $X_0 \ll 1$ so for an arbitrarily long interval in u we have $S \approx S_0$, and without loss of generality we can set $S \approx 1$. Then, as expected, we obtain [8]

$$ds^2 \approx -2\,du\,dv + dx^2 + dy^2 + X_0\,\Big\{\cos(\Omega_0\,u)[dx^2 - dy^2] + 2\sin(\Omega_0\,u)\,dx\,dy\Big\}. \tag{20}$$

6. Decoupling the general form

Let us return to considering the metric in general Rosen form

$$ds^2 = -2\,du\,dv + g_{AB}(u)\,dx^A\,dx^B, \tag{21}$$

where x^A, x^B represent any arbitrary number of dimensions ($d_\perp \geq 2$) transverse to the (u, v) plane. It is easy to check that the only non-zero component of the Ricci tensor is still

$$R_{uu} = -\left\{\frac{1}{2}\,g^{AB}\,g''_{AB} - \frac{1}{4}\,g^{AB}\,g'_{BC}\,g^{CD}\,g'_{DA}\right\}. \tag{22}$$

Let us now decompose the $d_\perp \times d_\perp$ matrix g_{AB} into an "envelope" $S(u)$ and a unit determinant related to the "direction of oscillation" [8]. That is, let us take

$$g_{AB}(u) = S^2(u)\,\hat{g}_{AB}(u), \tag{23}$$

where $\det(\hat{g}) \equiv 1$. (A related discussion can be found in [9].) Calculating the various terms of the Ricci tensor using the relations

$$[\hat{g}^{AB}\,\hat{g}'_{AB}] = 0, \qquad [\hat{g}^{AB}\,\hat{g}''_{AB}] - [\hat{g}^{AB}\,\hat{g}'_{BC}\,\hat{g}^{CD}\,\hat{g}'_{DA}] = 0, \tag{24}$$

it is found that [8]

$$R_{uu} = -d_\perp\,\frac{S''}{S} - \frac{1}{2}\,[\hat{g}^{AB}\,\hat{g}'_{BC}\,\hat{g}^{CD}\,\hat{g}'_{DA}]. \tag{25}$$

Note that we have now succeeded in decoupling the determinant ($\det(g) = S^{2d_\perp}$; effectively the "envelope" $S(u)$ of the gravitational wave) from the unit-determinant matrix $\hat{g}(u)$.

Now consider the set $SS(I\!R, d_\perp)$ of all unit determinant real symmetric matrices, and on that set consider the Riemannian metric

$$\mathrm{d}L^2 = \mathrm{tr}\left\{[\hat{g}]^{-1}\,\mathrm{d}[\hat{g}]\,[\hat{g}]^{-1}\,\mathrm{d}[\hat{g}]\right\}. \tag{26}$$

Then

$$R_{uu} = -\frac{1}{2}\left\{2d_\perp\,\frac{S''(u)}{S(u)} + \left(\frac{\mathrm{d}L}{\mathrm{d}u}\right)^2\right\}. \tag{27}$$

This means an arbitrary polarization vacuum Rosen wave is an arbitrary walk in $SS(I\!R, d_\perp)$, with distance along the walk $L(u)$ related to the envelope function $S(u)$ as in the discussion above.

7. Discussion

Arbitrary polarizations, while trivial in the Brinkmann form, are difficult to implement in the Rosen form. To address this puzzle we have re-analyzed the Rosen strong-field gravity wave in terms of an "envelope" function and two freely specifiable functions. The vacuum field equations can be interpreted in terms of a single differential equation governing the "envelope", coupled with an arbitrary walk in polarization space. In particular we have indicated how to construct a circularly polarized Rosen form gravity wave, and how to generalize this central idea beyond (3+1) dimensions. Further detailed calculations and discussions of these Rosen form polarizations can be found in [8].

Acknowledgments

This research was supported by the Marsden Fund administered by the Royal Society of New Zealand.

References

[1] Hans Stephani, Dietrich Kramer, Malcolm MacCallum, Cornelius Hoenselaers, and Eduard Herlt, *Exact Solutions of Einstein's Field Equations.* (Cambridge University Press, Cambridge, 2003). See especially section 24.5.

[2] Roger Penrose, "A remarkable property of plane waves in general relativity", Rev. Mod. Phys. **37** (1965) 215–220. doi:10.1103/RevModPhys.37.215.

[3] Roger Penrose, "Any spacetime has a plane wave as a limit". In *Differential Geometry and Relativity*, edited by M. Cahen and M. Flato. (Kluwer/Riedel, Dordrecht, 1976). Pages 271–275.

[4] J. D. Steele, "On generalised *pp* waves",
http://web.maths.unsw.edu.au/~jds/Papers/gppwaves.pdf
[Retrieved 14 April 2010].

[5] J. B. Griffiths and J. Podolský, *Exact space-times in Einstein's general relativity*, (Cambridge University Press, England, 2009).

[6] H. W. Brinkmann, "Einstein spaces which are mapped conformally on each other", Mathematische Annalen **18** (1925)119. doi:10.1007/BF01208647.

[7] A. Einstein and N. Rosen, "On gravitational waves", J. Franklin Inst. **223** (1937) 43.

[8] Cropp B and Visser M 2010 "General polarization modes for the Rosen gravitational wave" *Class. Quant. Grav.* **27** 165022.

[9] L. D. Landau and E. M. Lifschitz, *The classical theory of fields*, [fourth revised English edition], (Pergamon Press, 1980).

Quasi-normal frequencies: Semi-analytic results for highly damped modes

Jozef Skakala and Matt Visser

School of Mathematics, Statistics and Operations Research,
Victoria University of Wellington,
New Zealand

E-mail: jozef.skakala@msor.vuw.ac.nz, matt.visser@msor.vuw.ac.nz

Abstract. Black hole highly-damped quasi-normal frequencies (QNFs) are very often of the form $\omega_n =$ (offset) $+ in$ (gap). We have investigated the genericity of this phenomenon for the Schwarzschild–deSitter (SdS) black hole by considering a model potential that is piecewise Eckart (piecewise Pöschl–Teller), and developing an analytic "quantization condition" for the highly-damped quasi-normal frequencies. We find that the $\omega_n =$ (offset) $+ in$ (gap) behaviour is *common but not universal*, with the controlling feature being whether or not the ratio of the surface gravities is a rational number. We furthermore observed that the relation between rational ratios of surface gravities and periodicity of QNFs is very generic, and also occurs within different analytic approaches applied to various types of black hole spacetimes. These observations are of direct relevance to any physical situation where highly-damped quasi-normal modes are important.

1. Introduction

Black hole quasi-normal modes (QNMs) become (after some suitable damping timescale) the dominant contribution arising from arbitrary perturbations of any black hole spacetime. Thus they form the black hole's characteristic "sound". They are solutions of Regge–Wheeler (axial), or Zerelli (polar) equation, with outgoing radiation boundary conditions. The Regge–Wheeler (Zerelli) equation becomes a time-independent Schrödinger-like equation for the "eigenvalues" ω_n^2, where the ω_n are non-real quasi-normal frequencies. The real parts of the QNM frequencies are spaced symmetrically with respect to the imaginary axis, and with suitable conventions the imaginary part is always positive. The background black hole spacetime is encoded in the Regge–Wheeler (Zerelli) potential, which typically shows exponential fall-off as one moves towards the event horizon(s). The rate of this fall-off is given by the horizon's surface gravity.

Originally, interest was focused on the fundamental lowest damped modes. One of the methods to calculate them was to approximate the real potential by the analytically solvable Pöschl–Teller potential [1]. To obtain the fundamental modes one needs to fit the Pöschl–Teller potential by the real potential's peak height and peak curvature.

Interest in the highly damped modes ($Im(\omega)$ being large) significantly increased after Hod's conjecture about their relation to black hole thermodynamics was formulated [2]. There are good physics arguments to expect that the highly damped modes depend qualitatively on the behavior of the tails of the potential. (The wave-pocket formed from such modes spreads quickly to the region of the tails of the potential.) Thus approximating the tails of the potential by analytically

Journal of Physics: Conference Series **314** (2011) 012074 doi:10.1088/1742-6596/314/1/012074

solvable potentials is expected to bring considerable *qualitative* understanding regarding the behavior of highly damped QNMs. Perhaps surprisingly, such an approximation is both tractable and informative, at least for the Schwarzschild spacetime and Schwarzschild–de Sitter spacetime. Detailed calculations are contained in [3, 4] and the basic results are summarized in this paper.

2. Schwarzschild spacetime

There is a wide agreement regarding the behavior of asymptotic (highly damped) quasi-normal modes in the case of Schwarzschild asymptotically flat spacetime [5, 6]. They are described by the formula (see for instance [7, 8, 9]):

$$\omega = \frac{\ln 3}{2\pi}\kappa + i\kappa\left(n + \frac{1}{2}\right) + O(n^{-1/2}). \tag{1}$$

In this case the tails of the Regge–Wheeler (Zerelli) potential are approximated by the exactly solvable piecewise smooth potential:

$$V(x) = \begin{cases} V_{0-}\ \mathrm{sech}^2(\kappa x) & \text{for} \quad x < 0. \\ V_{0+}/(x+a)^2 & \text{for} \quad x > 0, \quad a > 0; \end{cases} \tag{2}$$

where κ is the surface gravity of the black hole horizon. For $x > 0$ the solutions giving the outgoing radiation (for $V_{0+} \geq -1/4$) are [3, 4]

$$\psi_+(x) = C_+ \sqrt{\frac{x+a}{\omega}}\left[J_{\alpha_+}(\omega(x+a)) - e^{-i\alpha\pi} J_{-\alpha_+}(\omega(x+a)\right], \qquad (x > 0), \tag{3}$$

where the $J_{\alpha+}(x)$ are Bessel functions with

$$\alpha_+ \doteq \sqrt{1 + 4V_{0+}}. \tag{4}$$

On the other hand, for $x < 0$ we have [3, 4]

$$\psi_-(x) = C_-\, e^{i\omega x}\, {}_2F_1\left(\frac{1}{2} + \alpha_-, \frac{1}{2} - \alpha_-, 1 + i\frac{\omega}{\kappa}, \frac{1}{1 + e^{-2\kappa x}}\right), \qquad (x < 0), \tag{5}$$

where ${}_2F_1(...)$ is the hypergeometric function with

$$\alpha_- = \sqrt{\frac{1}{4} - \frac{V_{0-}}{\kappa^2}} \tag{6}$$

The equation for QNM modes is obtained from the junction condition:

$$\frac{\psi_+'(0)}{\psi_+(0)} = \frac{\psi_-'(0)}{\psi_-(0)}. \tag{7}$$

The derivatives of both hypergeometric and Bessel functions can be rewritten in terms of hypergeometric and Bessel functions themselves. Furthermore, at the specific point of interest, the hypergeometric functions can themselves be rewritten in terms of Gamma functions, but it is still difficult to look for solutions of (7) without taking some approximations [3, 4]. Fortunately from the beginning of the problem we were interested only in highly damped modes, where $Im(\omega) \gg 0$ and $Im(\omega) \gg |Re(\omega)|$. In this case we can approximate both Bessel functions and also Gamma functions (after playing with some Gamma identites we can apply the Stirling formula) by trigonometric functions. Then (7) simplifies to the extremely simple condition:

$$\sin\left(\frac{\pi\omega}{\kappa}\right) = 0. \tag{8}$$

The solution of this is: $\omega = i\kappa n$, qualitatively recovering the result (1) at leading order.

Journal of Physics: Conference Series **314** (2011) 012074 doi:10.1088/1742-6596/314/1/012074

3. Schwarzschild–deSitter spacetime

In this situation the exactly solvable potential approximating the tails of Regge–Wheeler (Zerelli) potential is [3, 4] (see also [10]):

$$V(x) = \begin{cases} V_{0-} \operatorname{sech}^2(\kappa_- x) & \text{for} \quad x < 0; \\ V_{0+} \operatorname{sech}^2(\kappa_+ x) & \text{for} \quad x > 0. \end{cases} \tag{9}$$

Here $\kappa_\pm$ again denotes the surface gravities of the two horizons (black hole and cosmological). The solutions of the Regge–Wheeler (Zerelli) equation are

$$\psi_\pm(x) = C_\pm \, e^{\mp i\omega x} \, {}_2F_1\left(\frac{1}{2} + \alpha_\pm, \frac{1}{2} - \alpha_\pm, 1 + i\frac{\omega}{\kappa_\pm}, \frac{1}{1 + e^{-2\kappa_\pm x}}\right), \tag{10}$$

where

$$\alpha_\pm = \sqrt{\frac{1}{4} - \frac{V_{0\pm}}{\kappa_\pm^2}} \tag{11}$$

Now after applying the junction condition (7) we can proceed in the same way as previously. (Rewriting derivatives of hypergeometric functions in terms of hypergeometric functions, evaluating the hypergeometric functions at the particular point of interest in terms of Gamma functions, using Gamma function identities and the Stirling approximation for $Im(\omega) \gg 0$ and $Im(\omega) \gg |Re(\omega)|$.) After some calculations [3, 4] we obtain:

$$\cos(-i\pi\omega[\kappa_+^{-1} - \kappa_-^{-1}]) - \cos(-i\pi\omega[\kappa_+^{-1} + \kappa_-^{-1}]) = 2\,\cos(\pi\alpha_+)\cos(\pi\alpha_-). \tag{12}$$

The solutions of equation (12) have numerous interesting properties. In particular:

Theorem 3.1 *Take the case when $\kappa_+/\kappa_- = p_-/p_+ \in \mathbf{Q}$. Take $\kappa_* = p_\pm \kappa_\pm$ ($p_\pm \in \mathbf{N}$), hence $\kappa_* = lcm(\kappa_+, \kappa_-)$. Then the solutions of (12) are given by equi-spaced families of quasi-normal modes:*

$$\omega_{an} = \omega_{0a} + ing\kappa_* \qquad (n \in \mathbf{N}). \tag{13}$$

Here $g = 1$ if $p_+ \cdot p_-$ is odd and $g = 2$ if $p_+ \cdot p_-$ is even. The $\omega_a = \frac{\kappa_* \ln(z_a)}{\pi}$ are related to roots z_a of the specific polynomial:

$$z^{2[p_+ + p_-]} - z^{2p_+} - z^{2p_-} - 4\cos(\pi\alpha_+)\cos(\pi\alpha_-)\, z^{[p_+ + p_-]} + 1. \tag{14}$$

Theorem 3.2 *Assume that (12) has some set of solutions of the form $\omega_n = \omega_0 + in \cdot (gap)$ ($n \in \mathbf{N}$). Then the ratio κ_+/κ_- must be rational.*

Thus existence of infinite equi-spaced sets of solutions of the equation (12) is *equivalent* to the rational ratio of surface gravities. Whenever the ratio is rational, the generic number of different equi-spaced families is $g(p_+ + p_-)$, as it is related to the number of roots of the polynomial (14).

4. Discussion

We have presented strong physics arguments why our model is likely to give good *qualitative* estimates for the QNFs of the real Regge-Wheeler potential of the SdS spacetime. But "*qualitative*" means that only the periodicity with the correct gap structure is recovered, the modes do not have to necessarily *quantitatively* match. (Specifically, the ω_0 mode might differ.) Now, since different families of modes are distinguished only by different ω_{0a} the following question can be asked: Does the multi-family splitting transfer from our approximate model to the real SdS case? The answer to this question is suggested by the following statement (which can be exactly proven):

Assume that the gap structure is of the form given in (3.1). Assume also that the number of families is a bounded function of the variables κ_+, κ_- $(\kappa_+/\kappa_- \in \mathbf{Q})$. *Then for every point* $\tilde{\kappa}_+, \tilde{\kappa}_-$ *it holds that maximally one QNF can be a continuous function of* κ_+, κ_- *variables at the point* $\tilde{\kappa}_+, \tilde{\kappa}_-$.

This statement strongly suggests (since there are good reasons to expect that QNMs are continuous functions of surface gravities) that if the gap structure transfers to the real physical situation, also the basic features of multi-family splitting must transfer. Furthermore [3] we are able to show that the limit $\kappa_+ \to 0$ gives in our model the same results as the case $V_{0+} = 0$. In both cases we obtain the QNFs of Schwarzschild spacetime, which is the expected result.

A complementary set of semi-analytic results for QNFs is obtained by using monodromy techniques [5, 7, 11, 12, 13]. All the equations derived by this method (for Schwarzschild, Schwarzschild–de Sitter, Reissner–Nordström, Reissner–Nordström–de Sitter) are of the form [14]

$$\sum_{n=1}^{N} A_n \exp\left(i\pi\omega\sum_{j=1}^{H}\frac{B_{nj}}{\kappa_j}\right) = 0. \qquad (15)$$

As a result of this, the theorem 3.1 can be completely generalized for all the known analytic results, and the theorem 3.2 is very generic as well (it holds with very rare and mostly unphysical exceptions) [14].

5. References

[1] V. Ferrari and B. Mashhoon, "New approach to the quasinormal modes of a black hole," Phys. Rev. D **30** (1984) 295

[2] S. Hod, "Bohr's correspondence principle and the area spectrum of the quantum black holes", Phys. Rev. Lett **81** (1998) 4293 [arXiv:gr-qc/9812002]

[3] J. Skakala and M. Visser, "Semi-analytic results for quasi-normal frequencies," JHEP **1008**, (2010) 061 [arXiv:1004.2539 [gr-qc]].

[4] J. Skakala and M. Visser, "Highly-damped quasi-normal frequencies for piecewise Eckart potentials," Phys. Rev. D **81** (2010) 125023 [arXiv:1007.4039 [gr-qc]].

[5] E. Berti, V. Cardoso and A. O. Starinets, "Quasinormal modes of black holes and black branes," Class. Quant. Grav. **26** (2009) 163001 [arXiv:0905.2975 [gr-qc]].

[6] H. P. Nollert, "Quasinormal modes: the characteristic 'sound' of black holes and neutron stars," Class. Quant. Grav. **16** (1999) R159.

[7] L. Motl and A. Neitzke, "Asymptotic black hole quasinormal frequencies," Adv. Theor. Math. Phys. **7** (2003) 307 [arXiv:hep-th/0301173].

[8] A. J. M. Medved, D. Martin and M. Visser, "Dirty black holes: Quasinormal modes," Class. Quant. Grav. **21** (2004) 1393 [arXiv:gr-qc/0310009].

[9] A. J. M. Medved, D. Martin and M. Visser, "Dirty black holes: Quasinormal modes for squeezed horizons," Class. Quant. Grav. **21** (2004) 2393 [arXiv:gr-qc/0310097].

[10] V. Suneeta, "Quasinormal modes for the SdS black hole: An analytical approximation scheme," Phys. Rev. D **68** (2003) 024020 [arXiv:gr-qc/0303114].

[11] J. Natario and R. Schiappa, "On the classification of asymptotic quasinormal frequencies for *d*-dimensional black holes and quantum gravity," Adv. Theor. Math. Phys. **8** (2004) 1001 [arXiv:hep-th/0411267].

[12] V. Cardoso, J. Natario and R. Schiappa, "Asymptotic quasinormal frequencies for black holes in non-asymptotically flat spacetimes", J. Math. Phys. **45** (2004) 4698 [arXiv: hep-th/0403132v3].

[13] A. Ghosh, S. Shankaranarayanan and S. Das, "High frequency quasi-normal modes for black holes with generic singularities. II: Asymptotically non-flat spacetimes," Class. Quant. Grav. **23** (2006) 1851 [arXiv:hep-th/0510186].

[14] J. Skakala and M. Visser, "Generic master equations for quasi-normal frequencies", JHEP **1007**, (2010) 070 [arXiv:gr-qc/1009.0080]

Journal of Physics: Conference Series **314** (2011) 012075 doi:10.1088/1742-6596/314/1/012075

Time-domain modelling of Extreme-Mass-Ratio Inspirals for the Laser Interferometer Space Antenna

Priscilla Canizares[1], **and Carlos F. Sopuerta**[2]

[1,2] Institut de Ciències de l'Espai (CSIC-IEEC). Facultat de Ciències, Campus UAB, Torre C5 parells. Bellaterra, 08193 Barcelona, Spain.

E-mail: [1] pcm@ieec.uab.es, [2] sopuerta@ieec.uab.es

Abstract. When a stellar-mass compact object is captured by a supermassive black hole located in a galactic centre, the system losses energy and angular momentum by the emission of gravitational waves. Subsequently, the stellar compact object evolves inspiraling until plunging onto the massive black hole. These EMRI systems are expected to be one of the main sources of gravitational waves for the future space-based Laser Interferometer Space Antenna (LISA). However, the detection of EMRI signals will require of very accurate theoretical templates taking into account the gravitational self-force, which is the responsible of the stellar-compact object inspiral. Due to its potential applicability on EMRIs, the obtention of an efficient method to compute the scalar self-force acting on a point-like particle orbiting around a massive black hole is being object of increasing interest. We present here a review of our time-domain numerical technique to compute the self-force acting on a point-like particle and we show its suitability to deal with both circular and eccentric orbits.

1. Introduction

It is known that the majority of the galactic nuclei harbour a (super)massive black hole (MBH). Such a MBH can gravitationally capture a stellar-mass compact object (SCO), with mass in the range $m = 1 - 50M_{\odot}$, which eventually will perform a slow inspiral due to the emission of gravitational radiation until it plunges onto the MBH. These events are usually known as Extreme-Mass-Ratio Inspirals. EMRIs with MBHs with masses in the range $M = 10^4 - 10^7 M_{\odot}$ are on the main targets of the future space-based gravitational-wave observatory LISA. One main issue is that EMRI signals will be buried in the LISA instrumental noise and the gravitational-wave foreground produced by compact galactic binaries inside the LISA band. In order to pull out these signals from the noise, we will have to cross-correlate the LISA data stream with a bank of EMRI waveform templates. These templates should be accurate enough to capture the physical relevant parameters of the system. For this reason, we need to include very precisely the gravitational *back-reaction* of the SCO on its own trajectory when modelling EMRIs. Due to the extreme mass ratios involved, an EMRI can be modeled within the framework of black hole perturbation theory, where the SCO is pictured as a point particle moving in the spacetime geometry of the MBH and the gravitational back reaction is described as the action of a local force, the *self-force*.

As a testbed for the development of numerical techniques to compute the self-force, we use a simplified EMRI system, in which the SCO is a scalar charged particle, with charge q, which orbits a non-rotating Schwarzschild MBH [1] and whose inspiral is driven by a self-force generated by

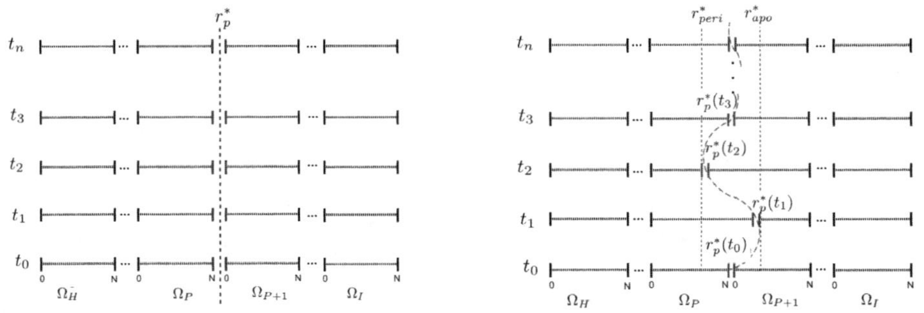

Figure 1. One-dimensional spatial grid for circular orbits (left) and eccentric orbits (right).

a scalar field, Φ. This simplified model contains all the essential ingredients of the gravitational problem. The scalar field Φ generated by the particle satisfies a wave equation in the MBH geometry, and the *scalar* self-force acting on the particle is given by: $F^{\mu} = q(g^{\mu\nu}+u^{\mu}u^{\nu})\,(\nabla_{\nu}\Phi)|_{\gamma}$, where $u^{\mu} = dz^{\mu}/d\tau$ is the particle's velocity , and τ is proper time.

The spherical symmetry of the MBH spacetime implies that the retarded field can be decomposed into decoupled scalar spherical harmonics modes:

$$\Phi(x) = \sum_{\ell=0}^{\infty}\sum_{m=-\ell}^{\ell} \Phi^{\ell m}(t,r)Y^{\ell m}(\theta,\phi)\,. \tag{1}$$

However, due to the singular character of the point particle, the sum over all the retarded field modes diverges at the particle location and, in order to obtain a self-force with a truly physical meaning, the field modes must be regularised. In order to do so, we use the *mode sum* regularisation scheme [2, 3, 4, 5, 6, 7], which gives an analytical expression for the singular contribution of the retarded field, $\Phi^{\rm S}$, at the particle location. Then, by subtracting the singular field modes from the retarded ones we obtain a smooth and differentiable field, $\Phi^{\rm R} = \Phi - \Phi^{\rm S}$, which generates the scalar self-force:$F^{\mu} = q(g^{\mu\,\alpha} + u^{\mu}u^{\alpha})\nabla_{\alpha}\Phi^{\rm R}$. Here, we review our work on the development of new techniques for the numerical computation of the retarded field, Φ. We also show the results we have obtained for the self-force after applying the mode-sum scheme.

2. Avoiding singularities: The Particle-without-Particle Formulation

In order to avoid the presence of the singularity associated with the point particle and all the related issues related with its numerical resolution we have proposed a new time-domain computational technique: *the Particle-without-Particle* (PwP) scheme. This technique was introduced in [8, 9] for the case of a charged scalar point particle orbiting in a circular orbit around a non-rotating MBH. In a subsequent work [10], the procedure was adapted to the case of eccentric orbits (see also [11]).

The main feature of our numerical scheme consists in splitting the one-dimensional spatial computational domain, parameterized in terms of the radial tortoise coordinate r^*, into two regions or subdomains, one at the left and one at the right of the particle: $r^* \in [-\infty,\infty] = [-\infty, r^*_p] \cup [r^*_p, \infty]$, where r^*_p denotes the particle radial location. Based on this division of the computational domain, all the variables associated with our problem can be written as: $\boldsymbol{U}(t,r^*) = \boldsymbol{U}_-(t,r^*)\Theta(r^*_p(t)-r^*)+\boldsymbol{U}_+(t,r^*)\Theta(r^*-r^*_p(t))$, where Θ is the Heaviside step function. In this way, by defining the jump of the variables as $[\boldsymbol{U}]_p \equiv \lim_{r^*\to r^*_p} \boldsymbol{U}_+(t,r^*) - \lim_{r^*\to r^*_p} \boldsymbol{U}_-(t,r^*)$

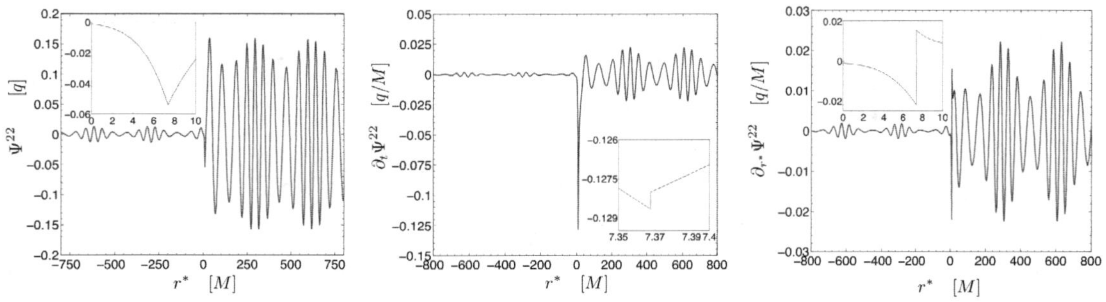

Figure 2. Snapshots of the evolution of the scalar field variable $\Psi = r\,\Phi$ (left), its time derivative $\partial_t\Psi$ (center), and its radial derivative $\partial_{r*}\Psi$ (right), for a particle in an eccentric orbit with orbital parameters $(e = 0.2, p = 7\,M)$. The internal frames show the ability of our method to resolve the different variables at the particle location, despite of the jumps in the time and radial derivatives. The snapshots have been taken after an evolution time $t_{\text{final}} - t_{\text{initial}} = 860M$.

(see [12, 9, 10]) we can obtain the jumps (matching conditions) at the particle location. Using this information we evolve the $1 + 1$ wave-type equations for each mode inside each subdomain (which are homogeneous equations thanks to the PwP scheme). The solutions obtained are communicated through the boundaries by enforcing the analytic jumps. We have developed two practical methods for enforcing the jumps: (i) The *penalty* method, and (ii) the direct communication of the characteristic fields (see [10]). This is the essence of the PwP scheme, i.e. replacing the particle by a set of boundary conditions.

In our implementation of the PwP scheme, the spatial domain is discretized using the PseudoSpectral Collocation (PSC) method. In this way, each subdomain is discretized independently using a *Lobatto-Chebyshev* grid and the variables are expanded in a basis of Chebyshev polynomials (see [13, 9]). One of the most interesting properties of the PSC method is that it provides exponential convergence for smooth functions. The implementation for the circular case has no special difficulties as the particle is always at the same radial position (see Figure 1) and the domain structure is static and given by $\Omega = \bigcup_{a=1}^{D} \Omega_a$, where $\Omega_a = \left[r^*_{a,\mathrm{L}}, r^*_{a,\mathrm{R}}\right]$, and the particle is placed at the interface of two subdomains, say Ω_a and Ω_{a+1}, with $r^*_p = r^*_{a,\mathrm{R}} = r^*_{a+1,\mathrm{L}} = \text{const}$. However in the eccentric case, the radial position of the particle changes with time. In this case, we keep the particle at the interface by using a time-dependent mapping between the spectral domain, $[-1, 1]$, and the corresponding physical subdomains, so that the particle is always mapped to the boundaries of the spectral domain (see Figure 1).

3. Discussion

Here we have summarized our work on the development of new time-domain scheme for the computation of the self-force [9, 10], the PwP scheme. We have shown it is a suitable scheme by presenting results on the simplified case of a scalar point particle in circular and eccentric orbits around a non-rotating MBH. In Figure 2 we show snapshots of the evolution of the different variables used in our implementation both for circular and eccentric orbits. This figure shows that the computational scheme can resolve in a very smooth way the retarded field and its derivatives at the particle location. In particular it can resolve jumps in the derivatives of the scalar field, due to the particle (this is how the particle appears in our PwP scheme). Then, by applying the mode-sum regularization scheme we obtain the self-force components. In Table 1 we show the self-force results obtained for the particle in a circular orbit, $(e = 0.0, p = 6)$, and also

Table 1. Self-Force components obtained for $\ell_{max} = 20$ and 24 subdomains

	Circular Orbit Case	(e = 0.0, p = 6 M)
$\partial_t \Phi^R$	$\partial_r \Phi^R$	$\partial_\phi \Phi^R$
$3.608899 \cdot 10^{-4}$	$1.677117 \cdot 10^{-4}$	$-5.304114 \cdot 10^{-3}$
	Eccentric Orbit Case	**(e = 0.2, p = 7 M)**
$\partial_t \Phi^R$	$\partial_r \Phi^R$	$\partial_\phi \Phi^R$
$4.278105 \cdot 10^{-4}$	$1.946249 \cdot 10^{-4}$	$-5.961118 \cdot 10^{-3}$

in a eccentric orbit with orbital parameters: $(e = 0.2, p = 7M)$. In the circular case, the radial component of the self-force, the only one that needs to be regularized, agrees within a $1 \cdot 10^{-4}\%$ with the results obtained by [14] in the frequency domain. In Table 1 we also present results obtained for the eccentric orbit just mentioned. These results have been obtained by employing between 20 to 24 subdomains and 50 collocation points per domain. The average time for a full self-force calculation with $\ell_{max} = 20$ (which involves the calculation of 231 harmonic modes) in a computer with two Quad-Core Intel Xeon processors at 2.27GHz is around 10 to 15 minutes. See [11] for a detailed description of how to tune the method for performance and efficiency. In another recent work [15] we discussed the question of initial conditions and how to avoid the presence of spurious solutions that can affect self-force computations.

Acknowledgments
PCM is supported by a predoctoral FPU fellowship of the Spanish Ministry of Science and Innovation. CFS acknowledges support from the Ramón y Cajal Programme of the Ministry of Education and Science of Spain and by a Marie Curie International Reintegration Grant (MIRG-CT-2007-205005/PHY) of the European Community (FP7). We acknowledge the computational resources provided by the Centre de Supercomputació de Catalunya and the Centro de Supercomputación de Galicia (ICTS 121 and 175).

References
[1] Poisson E 2004 *Living Rev. Relativity* **7** 6 (*Preprint* gr-qc/0306052)
[2] Barack L and Ori A 2000 *Phys. Rev.* **D61** 061502 (*Preprint* gr-qc/9912010)
[3] Barack L 2000 *Phys. Rev.* **D62** 084027 (*Preprint* gr-qc/0005042)
[4] Barack L 2001 *Phys. Rev.* **D64** 084021 (*Preprint* gr-qc/0105040)
[5] Barack L and Ori A 2002 *Phys. Rev.* **D66** 084022 (*Preprint* gr-qc/0204093)
[6] Detweiler S, Messaritaki E and Whiting B F 2003 *Phys. Rev.* **D67** 104016 (*Preprint* gr-qc/0205079)
[7] Haas R and Poisson E 2006 *Phys. Rev.* **D74** 044009 (*Preprint* gr-qc/0605077)
[8] Canizares P and Sopuerta C F 2009 *J. Phys. Conf. Ser.* **154** 012053 (*Preprint* 0811.0294)
[9] Canizares P and Sopuerta C F 2009 *Phys. Rev.* **D79** 084020 (*Preprint* 0903.0505)
[10] Canizares P, Sopuerta C F and Jaramillo J L 2010 *Phys. Rev.* **D82** 044023 (*Preprint* 1006.3201)
[11] Canizares P and Sopuerta C F 2011 (*Preprint* 1101.2526)
[12] Sopuerta C F and Laguna P 2006 *Phys. Rev.* **D73** 044028 (*Preprint* gr-qc/0512028)
[13] Boyd J P 2001 *Chebyshev and Fourier Spectral Methods* 2nd ed (New York: Dover)
[14] Diaz-Rivera L M, Messaritaki E, Whiting B F and Detweiler S 2004 *Phys. Rev.* **D70** 124018 (*Preprint* gr-qc/0410011)
[15] Jaramillo J L, Sopuerta C F and Canizares P 2011 (*Preprint accepted* 1101.2324)

Journal of Physics: Conference Series **314** (2011) 012076 doi:10.1088/1742-6596/314/1/012076

Collapse of rotating very massive stellar core to a black hole and a disk

Yuichiro Sekiguchi

Yukawa Institute for Theoretical Physics, Kyoto University, Kyoto 606-8502

E-mail: sekig@yukawa.kyoto-u.ac.jp

Abstract. Recent studies of modeling a progenitor of long gamma-ray bursts (LGRBs) suggest that progenitors of LGRBs might have a core with higher entropy than that of ordinary presupernove. Based on the above suggestion, we performed fully general relativistic, two-dimensional simulations of collapse of *higher entropy* core to a black hole and an accretion disk taking account of important microphysics. The initial core is simply modeled by a equilibrium configuration with constant entropy and electron fraction. We clarified that the formation dynamics depends even qualitatively on the amount of angular momentum. As a novel feature, we discovered that accretion disks formed after collapse can be convectively unstable.

1. Introduction

Gamma-ray bursts (GRBs) have been one of the outstanding problems in astrophysics since their discovery. Although progenitors of LGRBs have not been fully clarified yet, a number of progenitor models of LGRBs have been proposed (for a review, see[1]). Some authors [2, 3, 4, 5] proposed *interacting binary* models, in which the tidal force in a close binary keeps a helium star in synchronous, rapid rotation. Fryer and Heger [6] proposed a *binary-merger* model and showed that a merger of two helium cores during the common-envelope inspiral phase can produce a rapidly rotating core which may produces a LGRB. On the other hand, it has been showed that a single star can fulfill the requirements of the collapsar models if it is initially rapidly rotating ($\gtrsim 50\%$ of the Keplerian velocity at the equatorial surface) and of low metallicity ($Z/Z_{\odot} \lesssim 0.1$) (chemically homogeneous evolution model [7, 8, 9]).

All of these progenitor cores are anomalous in the sense that they are quite different from those of ordinary supernovae. Qualitatively speaking, the progenitor models have larger angular momentum and a *higher entropy* core (and hence, the core is more massive). The chemically homogeneous models predict a well-mixed, larger core with higher central entropy. It is also expected that the object formed after the binary merger will have a higher entropy, if the mass ratio of merging stars is not far from unity [10, 11]. Such a massive stellar core may collapse directly to a black hole without producing any explosion. In this article, we perform fully general relativistic simulations of collapse of rapidly rotating, higher entropy core, taking into account detailed microphysics.

2. Setting

Recently, we developed a fully general relativistic hydrodynamic code implementing a realistic equation of state, self-consistent electron and positron captures, and neutrino cooling by a

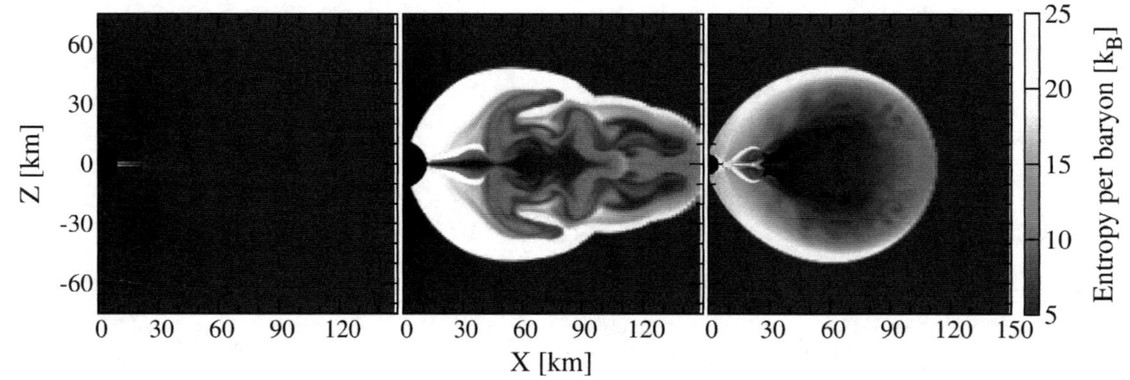

Figure 1. Contours of entropy per baryon for the slowly (left panel), moderately (middle panel), and rapidly (right panel) rotating models at selected timeslices.

general relativistic leakage scheme [12, 13]. We follow this paper for hydrodynamic equations for which the readers may refer to the details.

We construct approximate initial models in the following manner. We first calculate a spherical equilibrium configuration with a constant electron fraction and with a constant entropy per baryon ($Y_e = 0.5$ and $s = 8k_B$ in this article). We set the central density to be $\rho_c \approx 10^8$ g/cm^3. Following Nakazato et al. [14], we define the boundary of the iron core to be where the temperature is 5×10^9 K. Then the mass and the radius of the iron core are $M_{\rm iron} \approx 13 M_\odot$ and $R_{\rm iron} \approx 7000$ km. We adopt a wider region of $R_{\rm init} \approx 14000$ km for simulation.

At the current status, little is known about the angular momentum distribution in the progenitor core. Therefore we add the rotation profile according to $\Omega(\varpi) = \Omega_0 \exp[-R_c^2/2(\varpi^2 + R_c^2)] \exp[-\varpi^2/R_0^2]$, where $\varpi = \sqrt{x^2 + y^2}$. Ω, R_0 and R_c are parameters which control the degree of differential rotation. We fix the values of R_0 and R_c as $R_0 = R_{\rm init}/5$ and $R_c = R_{\rm init}/8$ and vary Ω_0 as 0, 0.4, 0.5 and 0.6 (hereafter referred to as spherical, slowly rotating, moderately rotating, and rapidly rotating models).

3. Results

The general feature of collapse until black hole formation can be summarized as follows. Gravitational collapse is triggered by the photo-dissociation of heavy nuclei and the electron capture. As the collapse proceeds the pressure is dominated by thermal gas pressure, and a weak bounce occurs at a subnuclear density forming a weak shock wave which is to be stalled quickly. Because the weak bounce is weak and ram pressure is strong, a black hole is formed soon after the bounce. Soon after the black hole formation neutrino luminosities decrease drastically because the main neutrino-emission region is swallowed by the black hole. After the black hole formation, on the other hand, the dynamics and feature of neutrino emission depend on the amount of angular momentum.

3.1. Slowly Rotating Model

In the slowly rotating model, a geometrically thin (but optically thick) accretion disk is formed soon after the black hole formation. A part of the material that falls onto the disk produces shock waves in the inner part of the disk (see the left panel of Figure 1). Thermal energy generated at the shock is not efficiently stored and the disk remains geometrically thin for a long time.

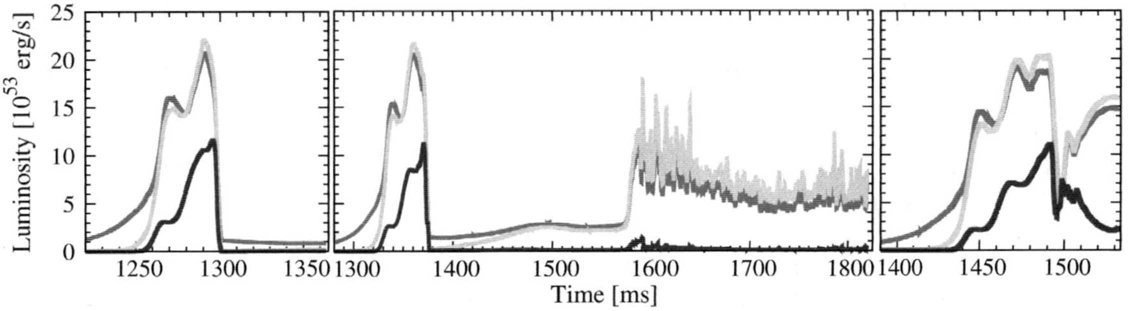

Figure 2. Time evolution of neutrino luminosities for the slowly (left panel), moderately (middle panel), and rapidly (right panel) rotating models. The red, green and blue curves correspond to luminosities of electron neutrinos, electron anti-neutrinos, and μ- and τ-neutrinos, respectively.

The left panel of Figure 2 plots the time evolution of neutrino luminosities for the slowly rotating model. It is found that the geometrically thin accretion disk emits $\approx 10^{53}$ erg/s by neutrinos. The efficiency of neutrino emission is very small as $L_{\nu,\rm tot}/(\dot{M}_{\rm BH}c^2) \approx 0.002$–$0.003$, which indicates that only small amount of material experiences the shock heating, and that most of energy is advected into the black hole before released by neutrinos.

3.2. Moderately Rotating Model

The feature of collapse is similar to that of slowly rotating model until formation of a geometrically thin disk. However, as the material with higher specific angular momentum in the outer region falls onto the disk, the density and mass of the disk increases. As the thermal energy is stored, the disk height H increases according to the balance relation $(P_{\rm disk} - P_{\rm ram})/H \sim GM_{\rm BH}\rho_{\rm disk}H/R^3_{\rm disk}$. The density and the temperature inside the disk increase and the ram pressure decreases. Consequently, H increases to be $\sim R_{\rm disk}$. The approximate balance relation is now written as $P_{\rm disk} - P_{\rm ram} \sim GM_{\rm BH}\rho_{\rm disk}/H$. Because the binding due to the gravitational force by the black hole decreases as H increases, the disk expands forming shock waves [15]. The neutrino opacities decrease as the disk expands, and accordingly, the cooling timescale becomes shorter. Then, the shock wave is stalled and the disk relaxes to a new geometrically thick state. Note that neutrino luminosities are significantly enhanced after the thick torus formation.

After the formation of the geometrically thick torus, convective motions are excited in the torus and near the shocked region (see the middle panel of Figure 1). The origin of the convection is explained as follows. The shock heating is more efficient in an inner part of the disk because kinetic energy of infalling material is larger. On the other hand, the neutrino cooling is less efficient in an inner part of the disk because of its lager optical depth. Thus, regions of negative entropy gradient along the radial direction near the equatorial plane are developed. Such configurations are known to be unstable to convection.

As a natural consequence of the convective activities of the accretion flow, neutrino luminosities vary violently in time (see the middle panel of Figure 2). If GRBs are driven by the pair annihilation of neutrinos and anti-neutrinos, such time-variability may be associated with the observed time-variability of GRB light curves.

3.3. Rapidly Rotating Model

In the rapidly rotating model, the disk formation process is qualitatively different from that in the moderately rotating model: A geometrically thick torus is formed *immediately* after the black hole formation because the pressure gradient and the angular momentum of the fluid near the equator are large enough that it retains an orbit outside the ISCO. Shock waves formed at the weak bounce is not swallowed into the black hole and a torus-shaped standing accretion shock remains around the black hole.

After the black hole formation, the luminosities decrease slightly. However, a dense torus surrounding the black hole is formed in a short time scale resulting in large total luminosity of $\sim 3 \times 10^{54}$ erg/s (see the right panel of Figure 2).

Convective motions are also observed in the rapidly model as in the moderately rotating model. However, strong large-scale circulations, appeared in the moderately rotating model, do not occur in the rapidly rotating model, although small-scale convective activities are driven (see the right panel in Figure 1). This is due to the stabilizing effect of the epicyclic frequency. Due to the absence of large-scale convective modes, effects of the convection on neutrino luminosities are likely to be minor. Indeed, no violent time-variability is observed after the thick torus formation.

4. Summary

We performed axisymmetric simulations of massive stellar core collapsing to a system composed of a rotating black hole and surrounding disk in full general relativity.

In the moderately rotating model, a geometrically thin accretion disk is first formed around the black hole and shocks are formed on its surface. As the thermal energy is stored in the disk, the disk expands to be a geometrically thick accretion torus. After the thick torus formation, convective activities set in because a region with negative entropy gradient emerges in the inner part of the torus. As a consequence the neutrino luminosities show violent time-variability with $L_\nu \sim 10^{54}$ erg/s.

The evolution of the accretion disk and neutrino emissivity depend strongly on the degree of initial rotation. In the slowly rotating model, the disk remains geometrically thin for a long time, and hence, the neutrino emissivity also remains relatively small ($\sim 10^{53}$ erg/s). In the rapidly rotating model, by contrast, a geometrically thick torus is immediately formed after the black hole formation, and luminosities of neutrinos emitted from the torus are as high as 10^{54} erg/s even at its formation. However, the convection is suppressed by the stabilizing epicyclic mode due to the rapid rotation and no violent time-variability is observed in the neutrino luminosities.

The author thanks to T. Shiromizu and T. Fukushige for their grateful aids. Numerical computations were performed on the NEC SX-9 at the data analysis center of NAOJ and on the NEC SX-8 at YITP in Kyoto University. This work is supported by the Grant-in-Aid for Scientific Research (21018008, 21105511, 21340051).

References

[1] Fryer C L et al. 2007 *Publ. Astron. Soc. Pacific* **119** 1211
[2] Izzard R G et al. 2004 *Mon. Not. Roy. Astron. Soc.* **348**, 1215
[3] Podsiadlowski et al. 2004 *Astrophys. J.* **607**, L17
[4] van den Heuvel E P J and Yoon S C 2007 *Astrophys. Space Sci.* **311** 177
[5] Cantiello M et al., 2007 *Astron. Astrophys.* **465**, L29
[6] Fryer C L and Heger A 2005 *Astrophys. J.* **623**, 302
[7] Yoon S C and Langer N 2005 *Astron. Astrophys.* **443** 643
[8] Yoon S C, Langer N and Norman C 2006 *Astron. Astrophys.* **460** 199
[9] Woosley S E and Heger A 2006 *Astrophys. J.* **637** 914
[10] Suzuki T K et al. 2007 *Astrophys. J.* **668** 435
[11] Gaburov E, Lombardi J C and Portegies Zwart S 2008 *Mon. Not. Roy. Astron. Soc.* **383** L5
[12] Sekiguchi Y 2010 *Class. Quant. Grav.* **27** 114107
[13] Sekiguchi Y 2010 *Prog. Theor. Phys.* **124** 331
[14] Nakazato K, Sumiyoshi K and Yamada S 2007 *Astrophys. J.* **666** 1140
[15] Sekiguchi Y and Shibata M 2007 *Prog. Theor. Phys.* **117** 1029

Influence of thermonuclear effects on the collapse of supermassive stars

Pedro J. Montero[1], **H.-T. Janka**[1], **E. Müller**[1], **B. Müller**[1]

1 Max-Planck-Institute für Astrophysik, Karl-Schwarzschild-Str. 1, 81748, Garching bei München, Germany

E-mail: `montero@mpa-garching.mpg.de`

Abstract.
We present results of general relativistic simulations of collapsing supermassive stars using the two-dimensional general relativistic numerical code Nada, which solves the Einstein equations written in the BSSN formalism and the general relativistic hydrodynamic equations with high resolution shock capturing schemes. These numerical simulations use a tabulated equation of state which includes effects of radiation and gas pressure, and those associated with the electron-positron pairs. We also take into account the effect of thermonuclear energy released by hydrogen and helium burning. We find that objects with mass $\approx 5 \times 10^5 M_\odot$ and initial metallicity greater than $Z_{CNO} \approx 0.004$ do explode if non-rotating, while the threshold metallicity for an explosion is reduced to $Z_{CNO} \approx 0.002$ for objects uniformly rotating.

1. Introduction

The observation of luminous quasars detected at redshifts higher than 6 in the Sloan Digital Sky Survey (SDSS) implies that SMBHs with masses $\sim 10^9 M_\odot$, which are believed to be the engines of such powerful quasars, were formed within the first billion years after the Big Bang (see [1] for a recent review). However, it is still an open question how SMBH seeds are formed and grow to reach such high masses in such a short amount of time [2].

A number of different routes based on stellar dynamical processes, hydrodynamical processes or a combination of both have been suggested. One of the theoretical scenarios proposes that if sufficient primordial gas in massive halos is unable to cool below $T_{vir} \approx 10^4 K$, it may lead to the formation of a supermassive star (SMS), with mass $\geq 5 \times 10^4 M_\odot$, which would eventually collapse to form a SMBH. This route assumes that fragmentation and cooling is suppressed by the presence of sufficiently strong UV radiation that prevents the formation of molecular hydrogen [3]. If isentropic SMSs [4, 5] form, it is expected that their quasi-stationary evolution of cooling and contraction will drive the stars to the onset of a general relativistic gravitational instability that would cause their gravitational collapse, and possibly also leads to the formation of a SMBH [6].

Fuller et al. [7] performed simulations of non-rotating SMSs in the mass range 10^5-$10^6 M_\odot$ with post-Newtonian corrections, an equation of state including $e^{\pm}$-pairs (which reduce the adiabatic index during collpase), and nuclear reactions describing hydrogen burning by the CNO cycles and the break-out from the β-limited hot CNO (HCNO) cycle. They found that SMSs with masses $M \gtrsim 10^5 M_\odot$ and initial metallicities $Z_{CNO} \lesssim 0.005$ do not explode, whereas SMSs with initial metallicities $Z_{CNO} \gtrsim 0.005$ do explode.

General relativistic simulations of collapsing supermassive stars [8, 9, 10, 11] neglecting the effects of nuclear burning (justified in the case of SMSs with masses exceeding $10^6 M_\odot$) indicate that about 90% of the initial mass would end up in the SMBH with a spin parameter $J/M^2 \approx 0.75$.

It is still an open question whether hydrogen burning by the β-limited HCNO cycle and its break-out via the $^{15}O(\alpha,\gamma)^{19}Ne$ reaction (rp-process) can halt the gravitational collapse of rotating SMSs with non-zero initial metallicities and generate enough thermal energy to lead to an explosion. To address this issue, we perform a series of general relativistic hydrodynamic simulations with a microphysical tabulated equation of state accounting for contributions from radiation, $e^{\pm}$-pairs, and baryonic matter. We also take into account the net thermonuclear energy released by the nuclear reactions involved in hydrogen and helium burning. The numerical simulations have been carried out with the Nada code [12].

2. Basic equations

We follow the so-called BSSN formulation of the Einstein equations [13, 14]. BSSN makes use of a conformal 3-metric, $\tilde{\gamma}_{ij} = e^{-4\phi}\gamma_{ij}$ and the conformal-traceless extrinsic curvature $\tilde{A}_{ij} = e^{-4\phi}(K_{ij} - \gamma_{ij}K/3)$, with the conformal factor ϕ satisfying $e^{4\phi} = \gamma^{1/3} \equiv \det(\gamma_{ij})^{1/3}$. In addition to evolution equations for $\tilde{\gamma}_{ij}$ and $\tilde{A}_{ij}$, there are evolution equations for the conformal factor ϕ (or $\chi \equiv e^{-4\phi}$), the trace of the extrinsic curvature K, and the "conformal connection functions" $\tilde{\Gamma}^i$ [14]. The lapse α and the shift vector β^i are computed using the so-called "moving puncture gauge".

The GRHD equations are the local conservation laws of momentum and energy, encoded in the stress-energy tensor $T^{\mu\nu}$, and of the matter density, J^μ (the continuity equation), i.e., $\nabla_\mu T^{\mu\nu} = 0$ and $\nabla_\mu J^\mu = 0$, where ∇_μ stands for the 4-dimensional covariant derivative. Following [15] the GRHD equations are written in flux-conservative form in cylindrical coordinates.

In order to avoid the complexity and stiffness from the solution of a nuclear reaction resulting network coupled to the hydrodynamic evolution, we follow an approximate method to take into account the net effect of the nuclear reactions of interest on the dynamics of the collapsing SMSs. We compute the nuclear energy release by hydrogen burning (through pp-chain, cold and HCNO cycles, and its break-out via the $^{15}O(\alpha,\gamma)^{19}Ne$ reaction) and helium burning (through 3-α reaction) as a function of rest-mass density, temperature and mass fractions of hydrogen (X_H), helium (X_{He}) and CNO metallicity (Z_{CNO}). This nuclear energy generation rate is added as a source term in the right-hand-side of the energy evolution equation.

3. Numerical approach and initial data

The evolution equations are integrated by the method of lines, for which we use a strong stability-preserving (SSP) Runge-Kutta algorithm of fourth-order. Since it is not possible to follow the gravitational collapse of a SMS from the early stages to the phase of black hole formation with a uniform Cartesian grid (computationally too demanding), we adopt a "regridding" procedure [10].

All initial models are chosen such that they are gravitationally unstable, i.e. their central rest-mass density is slightly larger than the critical central density required for the onset of the collapse. A list of the six different SMSs we have considered is given in Table 1. Models S1.a, S1.b and S1.c represent spherically symmetric SMSs with a gravitational mass of $M = 5 \times 10^5 M_\odot$, whereas R1.a, R1.b and R1.c are uniformly rotating initial models with the same gravitational mass. The rotating initial models are computed with the Lorene code [16]. We introduce a perturbation to trigger the gravitational collapse by reducing the pressure by about 2%. In order to determine the threshold metallicity required to halt the collapse and produce an explosion we carried out several numerical simulations for each initial model with different values of the initial metallicity as listed in Table 1.

Table 1. Main properties of the initial models studied and their fate. From left to right the columns show: model name, gravitational mass, initial central rest-mass density, ratio of rotational to gravitational energy $T/|W|$, initial central temperature, metallicity, and the outcome of the evolution.

Model	$M_{\rm grav}$ $[M_\odot]$	ρ_c $[g/cm^3]$	$T/\|W\|$	Temp [K]	Initial metallicity Z_{CNO}	Fate
S1.a	5.0×10^5	2.42×10^{-2}	0	5.8×10^7	2.0×10^{-3}	BH
S1.b	5.0×10^5	2.42×10^{-2}	0	5.8×10^7	4.0×10^{-3}	Explosion
S1.c	5.0×10^5	2.42×10^{-2}	0	5.8×10^7	5.0×10^{-3}	Explosion
R1.a	5.0×10^5	4.0×10^{-1}	0.0088	1.3×10^8	5.0×10^{-4}	BH
R1.b	5.0×10^5	4.0×10^{-1}	0.0088	1.3×10^8	8.0×10^{-4}	Explosion
R1.c	5.0×10^5	4.0×10^{-1}	0.0088	1.3×10^8	2.0×10^{-3}	Explosion

4. Results

First we consider a gravitationally unstable spherically symmetric SMS with a gravitational mass of $M = 5 \times 10^5 M_\odot$ (S1.a, S1.b and S1.c), which corresponds to a model extensively discussed in [7], and therefore allows for a comparison with the results presented here. In [7] it was found that an unstable non-rotating SMSs with $M = 5 \times 10^5 M_\odot$ and initial metallicity $Z_{CNO} = 2 \times 10^{-3}$ collapses to a BH, while a model with an initial metallicity $Z_{CNO} = 5 \times 10^{-3}$ explodes due to the nuclear energy released by the HCNO cycle. They also found that the central density and temperature at the thermal bounce (moment where collapse is reversed to explosion) are $\rho_{c,b} = 3.16$ g/cm^3 and $T_{c,b} = 2.6 \times 10^8$ K, respectively.

The left panels in Figure 1 show the increase in the central rest-mass density (upper panel) and temperature (lower panel) for model S1.a ($Z_{CNO} = 2 \times 10^{-3}$) with dashed lines until the point where an apparent horizon is found, indicating the formation of a BH. The solid lines represent the central rest-mass density and temperature evolution for model S1.c ($Z_{CNO} = 5 \times 10^{-3}$). They show that a thermal bounce occurs (at approximately $t \sim 9 \times 10^5$ s) entirely due to the HCNO cycle. The rest-mass density at bounce is $\rho_{c,b} = 2.9$ g/cm^3 and the temperature $T_{c,b} = 2.56 \times 10^8$ K, thus up to a few percent in agreement with [7]. Furthermore, we also find that model S1.b with metallicity $Z_{CNO} = 4 \times 10^{-3}$ explodes.

The evolutionary tracks for the central density and temperature of the rotating models R1.a, R1.b and R1.c are also shown in Figure 1. We refer with a dashed line to model R1.a with an initial metallicity $Z_{CNO} = 5 \times 10^{-4}$, which collapses to a BH, and with a solid line to model R1.c with $Z_{CNO} = 2 \times 10^{-3}$, which explodes due to the energy released by the HCNO cycle. Model R1.b with a lower metallicity of $Z_{CNO} = 8 \times 10^{-4}$ is a marginal case that explodes when the central temperature is high enough to trigger burning by the rp-process. As a result of the kinetic energy stored in the rotation of models R1.b and R1.c, the critical metallicity needed to lead to an explosion decreases relative to the non-rotating case. We observe that rotating models with initial metallicities up to $Z_{CNO} = 5 \times 10^{-4}$ do not explode even via the rp-process, which dominates at temperatures higher than $T \approx 5 \times 10^8$ K and increases the hydrogen burning rate by $200 - 300$ times relative to that of the HCNO cycle. We note that the time scale for the collapse and bounce phase is reduced because rotating models are more compact and have higher initial central density and temperature than spherical ones at the onset of the gravitational instability.

The right panel in Figure 1 shows the total nuclear energy generation rate in erg/s for the exploding models as a function of time during the collapse and subsequent bounce. The major contribution to the nuclear energy generation is due to hydrogen burning by the HCNO cycle. As expected the maximum nuclear energy needed to produce an explosion is lower in the rotating models.

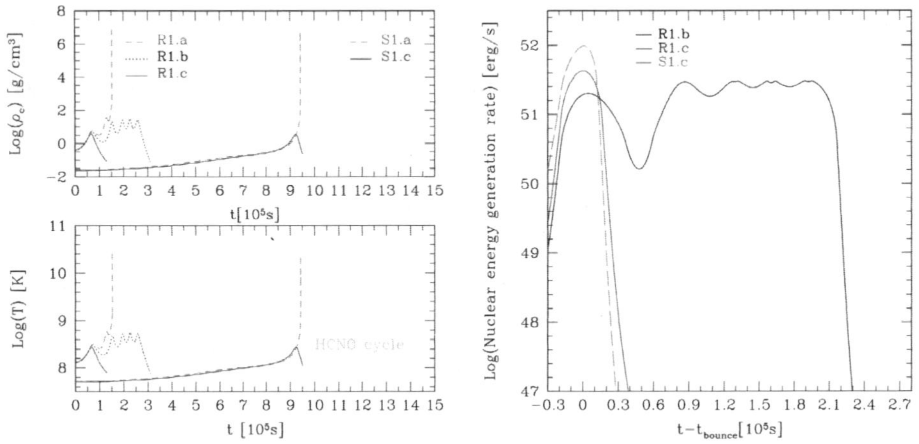

Figure 1. Time evolution of the central rest-mass density for all initial models (upper panel), and of the central temperature (lower panel). Horizontal dotted lines mark the temperature range where nuclear energy in primarily released by the HCNO cycle. The right panel shows the total nuclear energy generation rate $[erg/s]$ for the exploding models around the time of bounce.

5. Summary

We present results of general relativistic simulations of collapsing supermassive stars using the two-dimensional general relativistic numerical code Nada. We employed a tabulated equation of state which includes effects of radiation, $e^{\pm}$-pairs, and gas pressure. We took into account the effects of thermonuclear energy release by hydrogen and helium burning, and we determined the critical initial metallicity at which the collapse to a black hole is prevented by a thermonuclear-powered explosion of the rotating and non-rotating SMSs with mass $M = 5 \times 10^5 M_\odot$.

Acknowledgments

Work supported by the Deutsche Forschungsgesellschaft (DFG) through its Transregional Centers SFB/TR 7, and SFB/TR 27, and the Excellence Cluster EXC153).

References

[1] X. Fan, New Astronomy Reviews, 50, 665 (2006)
[2] M.J. Rees, in Black Holes in Binaries and Galactic Nuclei, ed. L. Kaper, E. P. J. van de Heurel, and P.A. Woudt, New York Springer, 351 (2001)
[3] K. Omukai, R. Schneider and Z. Haiman, ApJ **686**, 801 (2008).
[4] I. Iben, ApJ **138**, 1090 (1963).
[5] F. Hoyle and W. A. Fowler, MNRAS **125**, 169 (1963).
[6] W.S. Fowler, Rev.Mod.Phys., **36**, 545 (1964).
[7] G. M. Fuller, S. E. Woosley and T.A. Weaver, ApJ **307**, 675 (1986).
[8] S. L. Shapiro and S. A. Teukolsky, ApJ **234**, L177 (1979).
[9] F. Linke, J. A. Font, H.-T. Janka, E. Müller and P. Papadopoulos, A&A, **376**, 568 (2001)
[10] M. Shibata and S. L. Shapiro, ApJ **572** ,L39(2002).
[11] M. Saijo and I. Hawke, Phys. Rev. D **80** , 064001 (2009).
[12] P. J. Montero, J. A. Font and M. Shibata, Phys. Rev. D **78**, 064037 (2008).
[13] M. Shibata and T. Nakamura, Phys. Rev. D **52**, 5428 (1995).
[14] T. W. Baumgarte and S. L. Shapiro, Phys. Rev. D **59**, 024007 (1999).
[15] M. Shibata, Phys. Rev. D **67**, 024033 (2003).
[16] URL http://www.lorene.obspm.fr

Journal of Physics: Conference Series **314** (2011) 012078 doi:10.1088/1742-6596/314/1/012078

Gravitational waves in Fully Constrained Formulation in a dynamical spacetime with matter content

Isabel Cordero-Carrión[1,2], Pablo Cerdá-Durán[1] and José María Ibáñez[2]

[1] Max-Planck-Institut für Astrophysik, Karl-Schwarzschild-Str. 1, D-85741, Garching, Germany

[2] Departamento de Astronomía y Astrofísica, Universidad de Valencia, C/ Dr. Moliner 50, E-46100 Burjassot, Valencia, Spain

E-mail: chabela@mpa-garching.mpg.de, cerda@mpa-garching.mpg.de, jose.m.ibanez@uv.es

Abstract. We analyze numerically the behaviour of the hyperbolic sector of the Fully Constrained Formulation (FCF) (Bonazzola et al. 2004). The numerical experiments allow us to be confident in the performances of the upgraded version of the CoCoNuT code (Dimmelmeier et al. 2005) by replacing the Conformally Flat Condition (CFC), an approximation of Einstein equations, by FCF. First gravitational waves in FCF in a dynamical spacetime with matter content will be shown.

1. Introduction

The 3+1 formalism (see, e.g., [1]) is commonly used in most numerical codes which solve Einstein equations in spacetimes containing astrophysical compact objects. The Fully Constrained Formalism (FCF) [2, 3, 4] is a constrained evolution formulation of full Einstein equations. FCF is a natural generalization of the Conformally Flat Condition (CFC) [5, 6], an approximation of Einstein equations. FCF includes a hyperbolic system governing the gravitational radiation. We present numerical simulations of this system and compare some gravitational waves extracted with other methods.

2. Formalism

Given an asymptotically flat spacetime $(\mathcal{M}, g_{\mu\nu})$ we consider a $3+1$ splitting by spacelike hypersurfaces Σ_t. Latin (greek) indices go from 1 (0) to 3. $\gamma_{\mu\nu}$ denotes the 3-metric on Σ_t, $K_{\mu\nu}$ the extrinsic curvature, N the lapse function and β^i the shift vector. We denote $K = \gamma^{ij} K_{ij}$. A flat metric f_{ij} is introduced. We define $\gamma := \det \gamma_{ij}$ and $f := \det f_{ij}$. We introduce the conformal decomposition $\gamma_{ij} = \psi^4 \tilde{\gamma}_{ij}$. We define $h^{ij} := \tilde{\gamma}^{ij} - f^{ij}$. Maximal slicing, $K = 0$, and the so-called generalized Dirac gauge, $\mathcal{D}_k \tilde{\gamma}^{ki} = 0$, where $\mathcal{D}_k$ is the Levi–Civita connection associated with f_{ij}, are chosen. Einstein equations become a coupled elliptic-hyperbolic system: the elliptic sector acts on ψ, N, and β^i, and the hyperbolic one on h^{ij} [2]. If $h^{ij} = 0$ is imposed, CFC is recovered.

We introduce the decomposition $\hat{A}^{ij} = \psi^{10} \left(K^{ij} - \frac{1}{3} K \gamma^{ij} \right)$ motivated by the local uniqueness properties of the elliptic equations shown in [4]. We rewrite the equation for h^{ij} as a first order

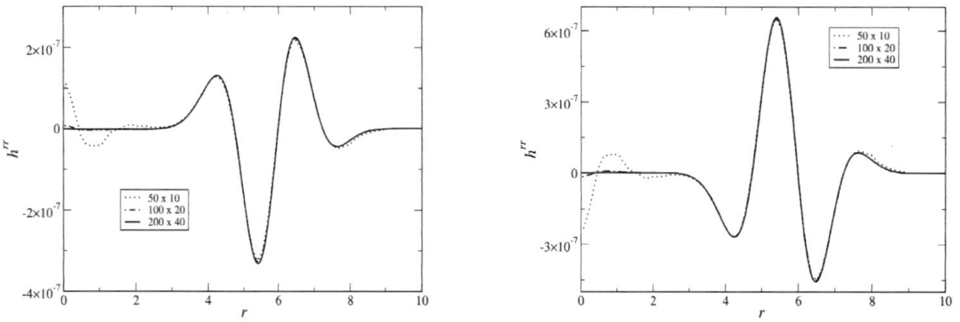

Figure 1. Radial profile of h^{rr} at $t = 6$. Right (left) panel corresponds to the equator (pole). Dotted, dashed-dotted and solid lines corresponds to $n_r = 50$ and $n_\theta = 100$ (radial and angular number of grid points), $n_r = 100$ and $n_\theta = 20$, and $n_r = 200$ and $n_\theta = 40$, respectively.

evolution system for the tensors $(h^{ij}, \hat{A}^{ij}, w^{ij}_k)$, where $w^{ij}_k := \mathcal{D}_k \tilde{\gamma}^{ij}$.

3. Numerical simulations

We perform the numerical evolution of the tensors $(h^{ij}, \hat{A}^{ij}, w^{ij}_k)$ and the evolution of matter (only in the last test), with the CoCoNuT code [7, 8]. Some basic elements of the code are: 4th order scheme in space and time; axisymmetry and symmetry with respect to the equatorial plane; spherical orthonormal coordinates; a Sommerfeld condition at the outer boundary; and Kreiss-Oliger dissipative term.

3.1. Teukolsky waves

We evolve firstly axisymmetric Teukolsky waves [9], with an amplitude of 10^{-5}. They are linearized wave equation solutions in vacuum, satisfy the Dirac gauge and are traceless (linear approximation of unit determinant). The background is flat, i.e., $N = \psi = 1$, and $\beta^i = 0$.

We display in Fig. 1 the radial profile of the component h^{rr} at $t = 6$, at the equator and at the pole, with different values of the number of radial and angular points. The analytical solution is recovered: velocity and amplitude of the wave, and its decay with the radius. We obtain an order of convergence close to the implemented 4th order for the tensor h^{ij}.

3.2. Equilibrium configuration of rotating neutron stars

We consider an axisymmetric and uniformly rotating neutron star in equilibrium. The initial data have been obtained from LORENE [10]. The models have non-vanishing β^i, N, ψ and matter fields, which are all kept fixed during the evolution of the tensor h^{ij}. It is possible to compute CFC and FCF initial models. During the evolution of a non equilibrium configuration we will consider the CFC approximation for the elliptic equations and the matter fields. The initial model is a neutron star with a 550 Hz rotation frequency, a 1.6 $M_\odot$ baryon mass and a 12.86 km coordinate equatorial radius. Details about the radial logarithmic grid used, and the chosen outer boundary and extracting radius for the gravitational waves can be found in [11].

Using the CFC approximation for the initial data for N, ψ and β^i, we find deviations in the components of the tensor h^{ij} during the evolution, which decay as $1/r$ (all the components should decay as $1/r^3$). It can be observed in Fig. 2, where we display the radial profile of the component h^{rr} in logarithmic scale at the equator for different times. The left panel in Fig. 2 corresponds to our control simulation, in which we have kept the same accuracy as the one used (for the elliptic sector) in [12]. We have improved the accuracy of the elliptic sector, in order

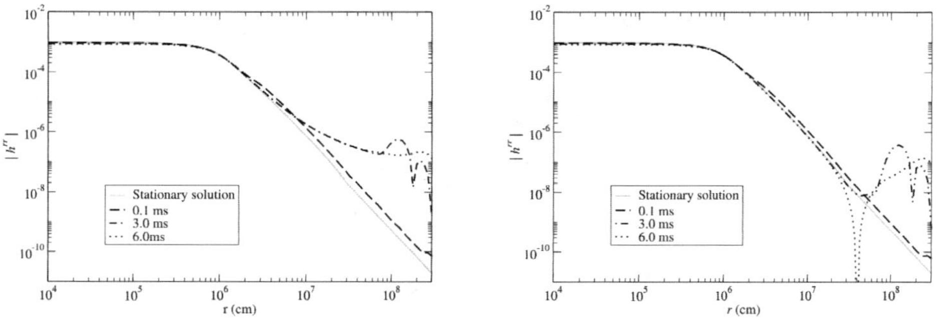

Figure 2. Radial profile of $|h^{rr}|$ with lower (higher) resolution in left (right) panel. Grey lines correspond to the stationary solution (as reference). Dashed, dot-dashed and dotted lines correspond to the evolution of the initial stationary data at 0.1, 3.0 and 6.0 ms, respectively.

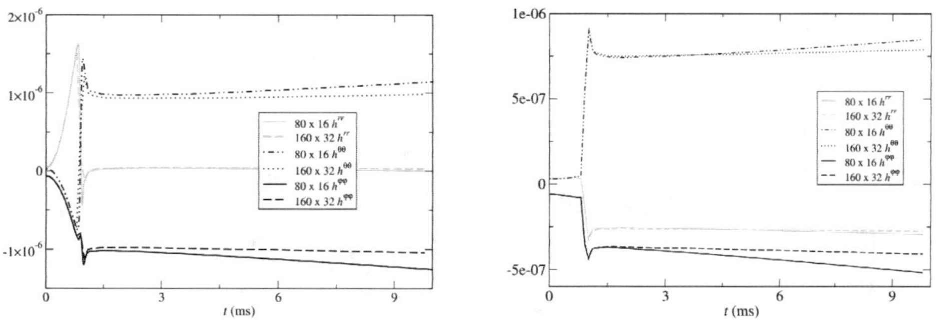

Figure 3. Evolution in time of the diagonal components of the h^{ij} tensor, at coordinate radius $r = 2.5 \times 10^7$ cm, using the CFC approximation (FCF formulation) in the left (right) panel.

to reduce the off-set, in the simulation showed in the right panel, and get the $1/r^3$ profile. The waves are generated initially when the CFC approximation is recovered in the code from the FCF initial data. Both simulations have 80 (16) radial (angular) grid points in the matter domain. The first reason for these deviations is, therefore, the accuracy in the numerical solutions of the elliptic equations.

We study now the influence of the accuracy of the evolution of the tensors $(h^{ij}, \hat{A}^{ij}, w_k^{ij})$, keeping the CFC approximation for the elliptic equations and the same accuracy in their numerical solutions. In Fig. 3 we plot the evolution in time of the diagonal components of the tensor h^{ij} at $r = 2.5 \times 10^7$ cm, for two different resolutions, in the CFC (FCF) case in the left (right) panel. Lower resolution corresponds to previous numerical grid, while the higher one has 160 (32) radial (angular) grid points in the already mentioned matter domain. In the CFC case, a deviation appears in $h^{\theta\theta}$ and $h^{\varphi\varphi}$, not corrected with resolution.

We compare simulations using CFC and FCF solutions of the elliptic equations. In the last case, a small deviation which decays as $1/r$ will appear, due to the numerical accuracy in the solutions of the elliptic equations, which is observed in h^{rr}. However, the largest errors occur in $h^{\theta\theta}$ and $h^{\varphi\varphi}$, and in the FCF case the absolute error is reduced around $1/4$. We conclude that the numerical accuracy in the solution of the elliptic equations, firstly, and the terms including the h^{ij} tensor in their sources, secondly, are the main reasons for the off-set in the gravitational wave signal, that we will compute in the non stationary simulations.

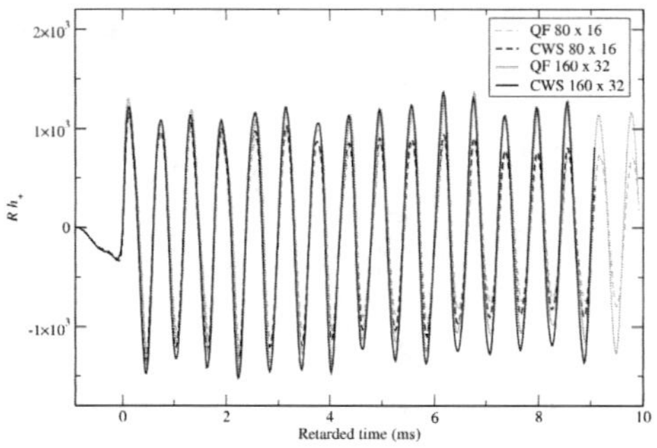

Figure 4. Gravitational wave from a perturbed rotating neutron star, at $r = 2.5 \times 10^6$cm. Real part of Weyl scalar corrected with the numerical stationary off-set (CWS, black lines), and the quadrupole formula (QF, orange lines), are plotted in terms of retarded time. Dashed (solid) lines correspond to lower (higher) resolution.

3.3. Perturbed equilibrium configuration of rotating neutron star

We evolve a perturbed neutron star. Hydrodynamic equations governing the matter evolution and elliptic equations are solved in the CFC approximation. In Fig. 4, we plot the real part of Ψ_4 Weyl scalar corrected with the numerical off-set of the stationary simulation, at $r = 2.5 \times 10^6$cm, and the gravitational wave obtained within the quadrupole formula, for the already used resolutions. We observe an extremely well agreement. The amplitude of the wave is damped during the evolution due to numerical errors, but it is corrected with resolution.

4. Conclusions

CoCoNuT is a general-relativistic magneto-hydrodynamic numerical code which uses the CFC approximation. Next step is the replacement of the CFC approximation by the FCF. We show that the hyperbolic sector works properly through several tests and compare the extracted gravitational waveforms with the quadrupole formula.

Acknowledgments

I. C.-C. acknowledges support from the Alexander von Humboldt Foundation. Work partially supported by the Spanish Ministry of Education and Science, Grant AYA2007-67626-C03-01. This work was also supported by the Collaborative Research Center on Gravitational Wave Astronomy of the Deutsche Forschungsgesellschaft (DFG SFB/Transregio 7).

References

[1] Alcubierre M 2008 *Introduction to 3+1 numerical relativity* (Oxford: Oxford University Press)
[2] Bonazzola S, Gourgoulhon E, Grandclément P and Novak J 2004 *Phys. Rev. D* **70** 104007
[3] Cordero-Carrión I, Ibáñez J M, Gourgoulhon E, Jaramillo J L and Novak J 2008 *Phys. Rev. D* **77** 084007
[4] Cordero-Carrión I, Cerdá-Durán P, Dimmelmeier H, Jaramillo J L, Novak J and Gourgoulhon E 2009 *Phys. Rev. D* **79** 024017
[5] Isenberg J A 2008 *Int. J. Mod. Phys. D* **17** 265
[6] Wilson J R and Mathews G J 1989 *Relativistic hydrodynamics* (Frontiers in numerical relativity) (Cambridge: Cambridge University Press)
[7] http://www.mpa-garching.mpg.de/hydro/COCONUT
[8] Dimmelmeier H, Novak J, Font J A, Ibáñez J M and Müller E 2005 *Phys. Rev. D* **71** 064023
[9] Teukolsky S A 1982 *Phys. Rev. D* **26** 745
[10] http://www.lorene.obspm.fr/
[11] Cordero-Carrión I, Cerdá-Durán P and Ibáñez J M 2010 *J. Phys. Conf. Ser.* **228** 012055.
[12] Dimmelmeier H, Font J A and Mueller E 2002 *Astron. Astrophys.* **393** 523

Journal of Physics: Conference Series **314** (2011) 012079 doi:10.1088/1742-6596/314/1/012079

Hydromagnetic instabilities and magnetic field amplification in core collapse supernovae

P Cerdá-Durán[1], M Obergaulinger[1,2], M A Aloy[3], J A Font[3] and E Müller[1]

[1] Max-Planck-Institut für Astrophysik, Karl-Schwarzschild-st. 1, 85748 Garching bei München, Germany
[2] Racah Institute of Physics, The Hebrew University, Jerusalem 91904, Israel
[3] Departamento de Astronomía y Astrofísica, Universidad de Valencia, 46100 Burjassot (Valencia), Spain

E-mail: cerda@mpa-garching.mpg.de

Abstract. Some of the most violent events in the universe, the gamma ray burst, could be related to the gravitational collapse of massive stellar cores. The recent association of long GRBs to some class of type Ic supernova seems to support this view. In such scenario fast rotation, strong magnetic fields and general relativistic effects are key ingredients. It is thus important to understand the mechanism that amplifies the magnetic field under that conditions. I present global simulations of the magneto-rotational collapse of stellar cores in general relativity and semi-global simulations of hydromagnetic instabilities under core collapse conditions. I discuss effect of the magneto-rotational instability and the magnetic field amplification during the collapse, the uncertainties in this process and the dynamical effects in the supernova explosion.

1. Introduction

There is strong evidence of the presence of strong magnetic fields in at least some of the compact objects resulting from the collapse of massive stars: the existence of magnetars, strongly magnetized neutron stars, as inferred from SGR and AXP observations [1], and the association of long-GRBs with some type Ic supernovae [2], which supports the collapsar model [3]. In this picture, after the collapse of rapidly rotating cores of massive stars, the magnetic field is amplified extracting energy from the rotation necessary to power the GRB.

Among the different mechanisms to amplify the magnetic field in this scenario the most promising is the magneto-rotational instability (MRI) [4]. In a simplified scenario, accretion discs, the condition for this instability to operate is that the angular velocity decreases outwards, $\partial_r \Omega^2 < 0$, where Ω is the angular velocity of the fluid. In this case, any magnetic field becomes unstable and grows exponentially in the time-scale of the rotation period. The typical length-scale of the fastest growing mode is $\lambda \sim 2\pi c_{\rm A}/\Omega$, being $c_{\rm A}$ the Alfvén speed. The more general criterion can be found in [5, 6].

For the typical conditions in core collapse supernovae, the outer layers of the proto-neutron star (PNS) fulfill the instability condition with a growth time of a few milliseconds [7] for the fastest rotating progenitors. This time-scale is comparable with the dynamical evolution time of the PNS, and therefore we expect the MRI to have a strong influence in the evolution of the PNS. However the length-scale of the fastest growing mode in this case is $\lambda \sim 100(B/10^{11}G)$ m.

Resolving this small length-scale in a multidimensional numerical simulation including the collapse of the iron core ($\sim$ 2000 km) to the PNS ($\sim$ 30 km) is computationally challenging and so far only models with unrealistically high initial magnetic field have been able to study the development of the MRI in core collapse [8, 9]. In this paper we present results of some of these global core collapse simulations in general relativity. As an alternative approach to the problem of the smallness of the length scales we present semi-global simulations of the MRI.

2. General relativistic global simulations

We have performed simulations of the collapse of rotating stellar cores using the general relativistic magneto-hydrodynamics (GRMHD) code CoCoNuT [9]. The code solves the evolution of the GRMHD equations using state-of-the-art Godunov-type schemes combined with a constrained transport scheme for the magnetic field. This technique allows for energy-momentum conservation, the preservation of the magnetic flux, and the correct behaviour of shocks and discontinuities. The metric is treated in the CFC approximation [10, 11] and is numerically evolved using spectral methods [12]. We use realistic stellar evolution progenitor models [13], a microphysical equation of state [14] and the deleptonization prescription of [15]. The initial magnetic field configuration is a purely-poloidal dipole-like field with three different initial strengths: 0, 10^{10} and 10^{12} G. We use a differentially rotating angular velocity profile for the progenitor with sufficient angular momentum to form a neutron star with a few milliseconds period. The progenitor core is unstable and it collapses helped by the lose of pressure support due to the deleptonization.

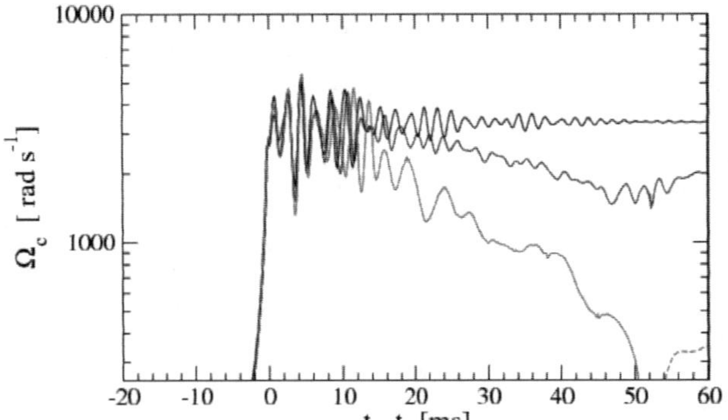

Figure 1. Time evolution of the central angular velocity during a global GRMHD simulation of the collapse of a rotating stellar core, without magnetic field (black line), with 10^{10} G (blue line) and with 10^{12} G (red line) initial field. t_b is the bounce time.

As the central density reaches the nuclear matter density, nuclear forces dominate the pressure and the equation of state stiffens. This leads to the formation of a shock wave which travels outwards and stalls around a few 100 km radius. The central region forms a compact PNS of about 30 km. In the model with highest magnetic field, thanks to the high resolution used in this region ($\sim$ 100 m), we were able to resolve the fastest growing mode of the MRI and the formation of channel flows. That leads to an exponential spin down of the PNS in a few ten milliseconds (see figure 1). In the case of weaker magnetic field, 10^{10} G, the spin down is slower. We suspect that in this case the spin-down could be comparable to the strongly magnetized case if we were able to resolve numerically the fastest growing mode of the MRI. Unfortunately the numerical resolution requirements increase linearly with the magnetic field and for this model it would mean increasing the numerical resolution by a factor 100, which is computationally not affordable nowadays. More details about these simulations can be found in [9].

3. Semi-global simulations

Instead of trying to increase the resolution in global simulations we have decided to simulate only a small box with the conditions present in the PNS. The box is about 1 km size which is

Journal of Physics: Conference Series **314** (2011) 012079 doi:10.1088/1742-6596/314/1/012079

sufficiently small so that we can afford the resolution needed to study the MRI, and at the same time is sufficiently large to study the effect of stratification. We use the Newtonian ideal MHD code AENUS [6] which uses high order Godunov-type schemes (up to 9th order) and constrained transport for the magnetic field evolution.

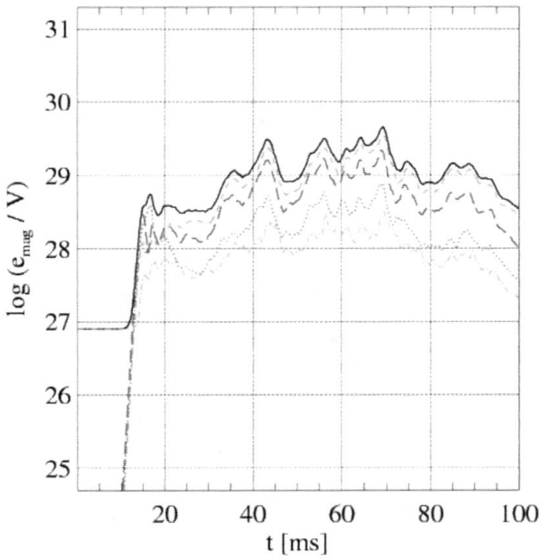

Figure 2. Time evolution of the average Maxwell stress components (different colors) in a typical semi-global 3D simulation. The exponential growth terminates after about 15 ms and a highly turbulent saturated state is formed.

The evolution of the magnetic field in these simulations is characterized by three phases as seen in figure 2: exponential growth, termination and saturation. The exponential growth phase can be described using the linear theory [5]. In this phase channel flows grow and amplify the magnetic stresses in the box. We have checked that the growth rates of the instability agree with theory as long as the fastest growing mode is spatially resolved by at least a few points. This holds in all the regimes of the MRI from the pure magneto-shear instability case to the magneto-convective limit. The termination phase occurs as the channel flows are destroyed by parasitic instabilities, mainly tearing modes. Despite of using an ideal MHD code, tearing modes appear due to the presence of numerical resistivity. We have obtained scaling relations between the magnetic stress at termination and the number of cells per channel flow, which is related to the numerical resistivity of the code. The computation shows that lower numerical resistivity produces higher termination values, due to the slower growth rate of the parasitic instabilities. The last regime is the saturation. In this regime 2 and 3 dimensional results differ considerably. In 2D simulations the magnetic stresses increase after the termination forming broader channel flows until they fill completely the box. In the 3D case (see figure 2) a quasi-stationary level, slightly higher than the termination point, is reached. The final state is highly turbulent with reappearance of coherent flow patterns in some cases. More details about these simulations can be found in [6].

4. Conclusions

We conclude that the MRI probably plays a extremely important role in the collapse of rapidly rotating stellar cores, even for weakly magnetized progenitors. However the numerical requirements to solve the GRMHD equations resolving all necessary length scales at the same time is unaffordable with present supercomputers. The only possible approach to this problem is to perform semi-global simulations. Those simulations show that it is possible to amplify the magnetic field under the conditions present in the PNS. However the saturation level, which ultimately sets the real influence in the global dynamics, depends strongly on numerical resistivity. Therefore we need numerical studies with physical resistivity in order to solve the question of the influence of the MRI in the core collapse scenario. Furthermore, large qualitative differences were found between 2D and 3D simulations. This implies that we cannot rely in the less computationally expensive 2D simulations to study the MRI and future simulations will have to be full three dimensional.

Acknowledgments

This work was supported by the Collaborative Research Center on Gravitational Wave Astronomy of the Deutsche Forschungsgesellschaft (DFG SFB/Transregio 7) and by the Spanish Ministry of Education and Science, Grant AYA2007-67626-C03-01.

References

[1] Kouveliotou C, Dieters S, Strohmayer T, et al. 1998 *Nature* **393** 235
[2] Kelly P L, Kirshner R P, and Pahre M 2008 *Astrophys J* **687** 1201
[3] Woosley S E, Langer N and Weaver T A 1993 *Aastrophys J* **411** 823
[4] Balbus S A and Hawley J F 1991 *Astrophys J* **376** 214
[5] Balbus S A 1995 *Astrophys J* **453** 380
[6] Obergaulinger M, Cerdá-Durán P, Müller E and Aloy M A 2009 *Astronom Astrophys* **498** 241
[7] Cerdá-Durán P, Font J A and Dimmelmeier H 2007 *Astron Astrophys* **474** 169
[8] Obergaulinger M, Aloy M A and Müller E 2006 *Astron Astrophys* **450** 1107
[9] Cerdá-Durán P, Font J A, Antón L and Müller E 2008 *Astron Astrophys* **492** 937
[10] Isenberg J A 2008 *Int J Mod Phys* D **17** 265
[11] Wilson J R and Mathews G J 1989 Relativistic hydrodynamics (Frontiers in numerical relativity) (Cambridge: Cambridge University Press)
[12] Dimmelmeier H, Font J A and Müller E. 2002 *Astronom Astrophys* **388** 917
[13] Woosley S E, Heger A and Weaver T A 2002 *Rev Mod Phys* **74** 1015
[14] Shen H, Toki H, Oyamatsu K and Sumiyoshi K 1998 *Prog Theor Phys* **100** 1013
[15] Liebendörfer M 2005 *Astrophys J* **633** 1042

Spanish Relativity Meeting (ERE 2010): Gravity as a Crossroad in Physics IOP Publishing
Journal of Physics: Conference Series **314** (2011) 012080 doi:10.1088/1742-6596/314/1/012080

Gravitational-wave signatures in successful vs. failed core-collapse supernovae

Kei Kotake

Division of Theoretical Astronomy, National Astronomical Observatory of Japan, 2-21-1, Osawa, Mitaka, Tokyo, 181-8588, Japan

E-mail: kkotake@th.nao.ac.jp

Abstract. In this contribution, we discuss possible gravitational-wave (GWs) signatures emitted from core-collapse supernovae that do or do not produce explosions. For the former case, we study properties of GWs based on the three-dimensional (3D) supernova simulations, which demonstrate the neutrino-driven explosions aided by the standing accretion shock instability (SASI). By taking into account the effects of stellar rotation, we find that the gravitational waveforms from neutrinos in models that include rotation exhibit a common feature otherwise they vary much more stochastically in the absence of rotation. We point out that a recently proposed future space interferometers like Fabry-Perot type DECIGO would permit the detection of these signals for a Galactic supernova. For the black-hole forming supernovae, we study the GW emission based on a long-term special relativistic magnetohydrodynamic simulation in the light of collapsar model of long-duration gamma-ray bursts (GRBs). We find that the GWs from anisotropic neutrino emission illuminated by accretion disk become as high as the GWs contributed from matter motions of accreting material. These signals, possibly visible to the BBO-class detectors for a hundred Megaparsec distance scales, may give us an important probe into the central engines of GRBs.

1. Introduction

The successful detection of neutrinos from SN1987A paved the way for *Neutrino Astronomy*, an alternative to conventional astronomy by electromagnetic waves. Core-collapse supernovae are now expected to be opening yet another astronomy, *Gravitational-Wave Astronomy*. Currently long-baseline laser interferometers such as LIGO, VIRGO, GEO600, and TAMA300 are operational (e.g., [1] for a recent review). For these detectors, core-collapse supernovae have been proposed as one of the most plausible sources of gravitational waves (GWs) (see, for example, [2, 3] for recent reviews).

Traditionally, most of the theoretical predictions of GWs have focused on the bounce signals (e.g., see [4, 5, 2, 3] and references therein). However recent stellar evolution calculations suggest that rapid rotation assumed in most of the previous studies is not canonical for progenitors with neutron star formations [6]. Besides the rapid rotation of the core, convective matter motions and anisotropic neutrino emission in the much later postbounce phase are expected to be the primary GW sources with comparable amplitudes to the bounce signals. Thus far, various physical ingredients for producing asphericities and the resulting GWs in the postbounce phase have been studied, such as the roles of pre-collapse density inhomogeneities [7], moderate rotation of the iron core [8], g-modes pulsations [9] of protoneutron stars (PNSs), and the standing accretion shock instability (SASI) (e.g., [10, 11]).

However, most of them have been based on two-dimensional (2D) simulations that assume axisymmetry. Then, the growth of SASI (e.g., [12]) and the large-scale convection, both of which are now considered to generically develop in the postbounce phase and to help the neutrino-driven explosions [15], develop along the symmetry axis preferentially, thus suppressing the anisotropies in explosions. There are several three-dimensional (3D) studies so far (e.g., [7]), however little has been studied about the GW emission from these models.

By performing 3D simulations that demonstrate the neutrino-driven core-collapse supernovae aided by SASI, we pointed out in our previous study [13] that the gravitational waveforms vary much more stochastically in 3D than 2D because the explosion anisotropies depend sensitively on the growth of the SASI which develops chaotically in all directions. In the former half of this contribution, we study the effects of rotation on the stochastic nature of the GW signals (section 2). In the latter half, we explore the GW signature emitted from the black-hole (BH) forming core-collapse supernovae in the light of collapsar simulations of long-duration GRBs (section 3).

2. GWs from 3D exploding supernovae with rotation

Figure 1 shows the gravitational waveforms for a typical 3D model without (left panel) and with rotation (right panel). For the rotating model, we give a uniform rotation on the flow advecting from the outer boundary of the iron core as in [14], whose specific angular momentum is assumed to agree with recent stellar evolution models [6]. Comparing these panels, one can clearly see the sudden rise in the GW amplitudes seen from the equator (right panel) after around 500 ms. By changing the initial angular momentum as well as the input neutrino luminosities from the protoneutron star, we computed fifteen 3D models. The GW features mentioned above are found to be common to the other rotating models.

Figure 2 illustrates a typical snapshot of the flow fields for the rotating model when the spiral SASI modes have already entered the non-linear regime, seen from the pole (top) or from the equator (bottom), respectively. From the left panel, one may guess the presence of the sloshing modes that happen to develop along the rotational axis (z-axis) at this epoch. It should be emphasized that the dominance of $h_{\nu,+}^{\rm equ}$ observed in the current 3D simulations have nothing to do with the one found in our previous 2D studies [11]. Free from the 2D axis effects, the major axis of the SASI changes stochastically with time, and the flow patters behind the standing shock simultaneously change in every direction. As a result, the sloshing modes can make only a small contribution to the GW emission. The remaining possibility is that the spiral flows clearly seen in the right panel should be a key importance to understand the GW feature mentioned above. In fact, by analyzing the matter distribution on the equatorial plane, we find that the compression of matter is more enhanced in the vicinity of the equatorial plane due to the growth of the spiral SASI modes, leading to the formation of the spiral flows circulating around the spin axis with higher temperatures. As a result, the neutrino emission seen parallel to the spin axis becomes higher than the ones seen from the other direction. Remembering again that the lateral-angle (θ) dependent function of the GW formulae (e.g., equation (9) in [11]) is positive near the north and south polar caps, the dominance of the polar neutrino luminosities leads to make the positively growing feature in $h_{\nu,+}^{\rm equ}$. By performing the spectral analysis of the gravitational waveforms, we find it very difficult to detect these neutrino-originated signals with slower temporal evolution ($\sim> O(10)$ms) by ground-based detectors whose sensitivity is limited mainly by the seismic noises at such lower frequencies. However these signals may be detectable by the recently proposed future space interferometers like Fabry-Perot type DECIGO ([16]). Contributed by the neutrino GWs in the lower frequency domains, the total GW spectrum tends to become rather flat over a broad frequency range below $\sim$ 100 Hz. These GW features obtained in the context of the SASI-aided neutrino-driven mechanism are different from the ones expected in the other candidate supernova mechanism, such as the MHD mechanism (e.g., [17])

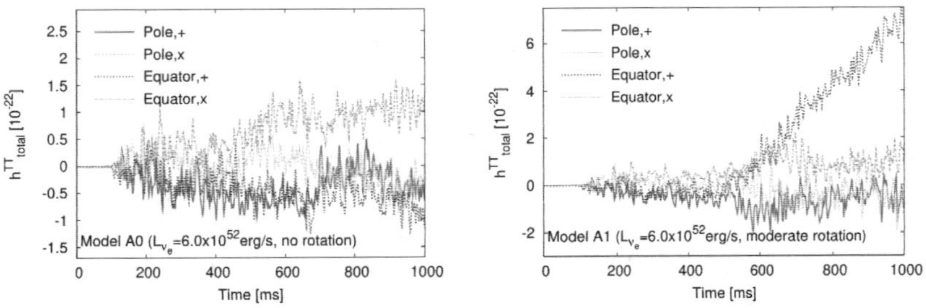

Figure 1. Gravitational waveforms from the sum of neutrinos and matter motions for models without (left panel), and with rotation (right panel). The increasing trend (blue line) in the right panel comes from the neutrino contribution to the total GW amplitudes. The time is measured from the epoch when the neutrino luminosity is injected from the surface of the neutrino sphere. For models presented here, the rotational flow is adjusted to advect to the PNS surface at around $t = 400$ ms. The supernova is assumed to be located at the distance of 10 kpc.

and the acoustic mechanism [9]. Therefore the detection of such signals is expected to provide an important probe into the long-veiled explosion mechanism.

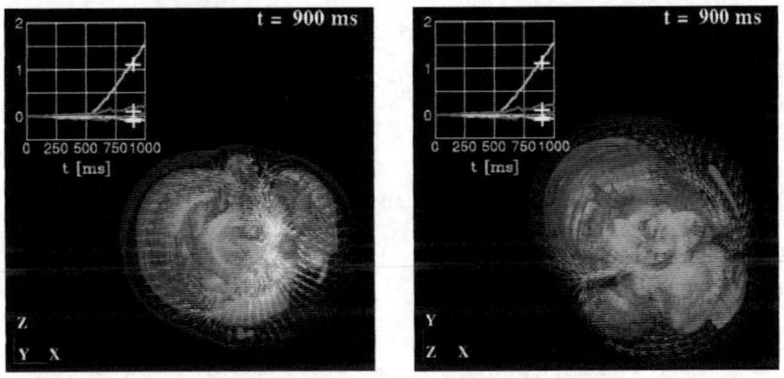

Figure 2. Partial cutaway of the entropy isosurfaces and the velocity vectors on the cutting plane for the rotating model. Left and right panels are for the equatorial and polar observer, respectively. The insets show the gravitational waveforms with '+' on each curves representing the time of the snapshot. Note that the colors of the curves are taken to be the same as the top panel of Figure 1.

3. GWs from collapsar evolution

To explore the GW emission from BH forming supernovae, we estimate the GW signals based on our long-term collapsar simulation in special relativistic magnetohydrodynamics (e.g., [18]). The left panel of Figure 3 is a snapshot taken from the collapsar simulation, showing a clear accretion-disk and BH system with the polar funnel regions along the spin axis of the disk. The dynamically rotating accretion disk is the primary source of the GWs from matter motions, while the anisotropy of neutrinos coming out from the surface of the accretion disk (white line in the left panel) gives rise to the neutrino-originated GWs. The middle panel shows the contribution to the total GW amplitudes (red line) from neutrinos (light blue line) and from matter motions (green line + purple line). As well known, the sign of the GW amplitudes from oblately deforming object is negative. This is the reason of the negative growth of the matter GWs. On the other hand, the positively growing feature of the neutrino GWs is due to the excess of neutrino emission parallel to the rotational axis. From the accretion disk that has a ring-like structure, the neutrino emission parallel to the spin axis becomes highest. This argument is essentially the same as the one in the previous section. From the right panel, it can be seen that the BBO-class detectors in the next generation would permit the detection of these signals, depending on the initial angular momentum of precollapse progenitor (indicated by J0.6 and J0,8 in the panel (see, [18] for more details), for a 100 Mpc distance scale. If the long-duration GRBs are originated from the so-called collapsar, as it is receiving quite some interest for more than decade, the detection of these signals may provide us an important clue to understand their long-veiled central engines.

It is gratefully acknowledged that the results presented in this article are the outcome of the collaborations with W. Iwakami Nakano, N. Ohnishi, S. Harikae, and T. Takiwaki. This study was supported in part by the Grants-in-Aid for the Scientific Research from the Ministry of Education, Science and Culture of Japan (Nos. 19540309 and 20740150).

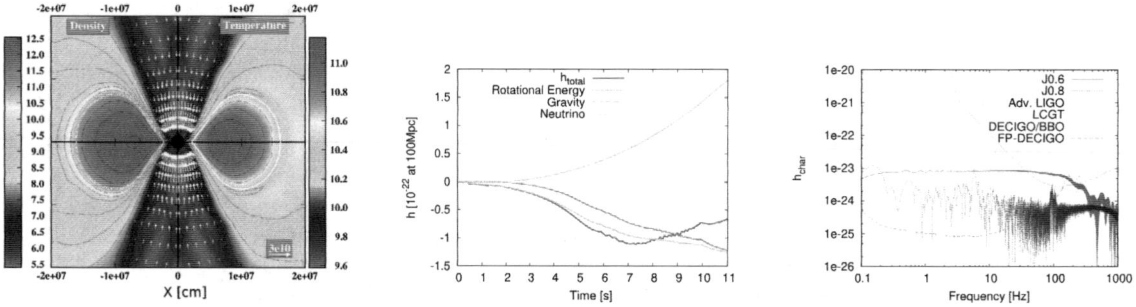

Figure 3. Left panel is the snapshot at 9.1 s after the onset of gravitational collapse for model J0.8 when the accretion disk is in a stationary state (see [18] for more detail). The logarithmic density (in, left-half) and temperature (in K, right-half) are shown. The white solid line denotes the area where the density is equal to 10^{11}, representing the surface of the accretion disk. The central black circle ($\approx 4M_\odot$) represents the inner boundary of our computations. The middle and right panel is the gravitational waveform and spectrum for the model, respectively (see text for more detail). The distance to the source is taken to be 100 Mpc.

References

[1] Hough, J., & Rowan, S. 2005, Journal of Optics A: Pure and Applied Optics, 7, 257
[2] Kotake, K., Sato, K., & Takahashi, K. 2006, Reports of Progress in Physics, 69, 971
[3] Ott, C. D 2009, Classical and Quantum Gravity, 26, 063001
[4] Kotake, K., Yamada, S., & Sato, K. 2003, Phys. Rev. D, 68, 044023
[5] Kotake, K., Yamada, S., Sato, K., Sumiyoshi, K., Ono, H., & Suzuki, H. 2004, Phys. Rev. D, 69, 124004
[6] Heger, A., Woosley, S. E., & Spruit, H. C. 2005, Astrophys. J., 626, 350
[7] Fryer, C. L., Holz, D. E., & Hughes, S. A. 2004, ApJ, 609, 288
[8] Müller, E., Rampp, M., Buras, R., Janka, H.-T., & Shoemaker, D. H. 2004, Astrophys. J., 603, 221
[9] Ott, C. D., Burrows, A., Dessart, L., & Livne, E. 2006, Physical Review Letters, 96, 201102
[10] Kotake, K., Ohnishi, N., & Yamada, S. 2007, Astrophys. J., 655, 406
[11] Kotake, K., Iwakami, W., Ohnishi, N., & Yamada, S. 2009, Astrophys. J, 704, 951
[12] Ohnishi, N., Kotake, K., & Yamada, S. 2006, Astrophys. J., 641, 1018
[13] Kotake, K., Iwakami, W., Ohnishi, N., & Yamada, S. 2009, Astrophys. J., 697, L133
[14] Iwakami, W., Kotake, K., Ohnishi, N., Yamada, S., & Sawada, K. 2009, Astrophys. J., 700, 232
[15] Marek, A., & Janka, H.-T. 2009, Astrophys. J., 694, 664
[16] Kawamura, S., et al. 2006, Classical and Quantum Gravity, 23, 125
[17] Takiwaki, T., & Kotake, K. 2010, arXiv:1004.2896
[18] Harikae, S., Takiwaki, T., & Kotake, K. 2009, Astrophys. J., 704, 354

Spanish Relativity Meeting (ERE 2010): Gravity as a Crossroad in Physics IOP Publishing
Journal of Physics: Conference Series **314** (2011) 012081 doi:10.1088/1742-6596/314/1/012081

Non-axisymmetric Torsional Oscillations of Relativistic Stars

Hajime Sotani[1] **and Kostas D. Kokkotas**[2]

[1]Division of Theoretical Astronomy, National Astronomical Observatory of Japan, 2-21-1 Osawa, Mitaka, Tokyo 181-8588, Japan

[2]Theoretical Astrophysics, University of Tübingen, Tübingen 72076, Germany

E-mail: sotani@astro.auth.gr

Abstract. We examine the non-axisymmetric Alfvén oscillations of relativistic stars with a strong dipole magnetic fields, where we have done the 2D simulation. We find that the spectrum of non-axisymmetric axial-type Alfvén oscillations is discrete as well as the axisymmetric polar Alfvén oscillations, in contrast to the spectrum of axisymmetric axial Alfvén oscillations which is continuum. We also show the dependence of Alvén frequency on the strength of magnetic field and find that such frequencies are smaller than those of the axisymmetric axial Alfvén oscillations. That is, to explain the observed evidence of quasi-periodic oscillations in the soft gamma repeaters, the non-axisymmetric axial oscillations could also play an important role.

1. Introduction

One of the most promising candidates of magnetars is the soft gamma repeaters (SGRs), which rarely emit quite strong gamma-rays called as "giant flares". Up to now, at least three giant flares have been detected in three different SGRs. Through the timing analysis of the decaying tail followed after the initial strong sharp burst in the giant flares, the existence of quasi-periodic oscillations (QPOs) is revealed. The specific frequencies of these QPOs are in the range from tens Hz up to a few kHz [1]. The many researchers are considering that these QPOs come from the oscillations of magnetars and the observations of QPOs in SGRs might be first evidences to detect the oscillations of neutron stars directly. To explain these QPO frequencies in SGRs theoretically, many attempts have been done. However, most of them are based on the axisymmetric oscillations, e.g., the oscillations in the crust region [2, 3] and the Alfvén oscillations [4, 5, 6, 7]. On the other hand, the non-axisymmetric oscillations on the relativistic magnetar are not really acknowledged, in spite of the fact that consideration of non-axisymmetric oscillations is more realistic and more natural. There are a few studies about such oscillations, but those are mainly based on the Newtonian framework [8, 9]. In practice, considering the non-axisymmetric oscillations is more complicated because the axial type oscillations are coupled with the polar type oscillations due to the presence of magnetic field even for a non-rotating magnetar. This coupling could be important for the observation of the gravitational waves, if one stands on the statement that the polar oscillations would be induced from the axial oscillation. Thus in this article we consider only the axial type non-axisymmetric oscillations as a first step, where we neglect the coupling with polar type ones. In order to know the feature of axial type non-axisymmetric oscillations, we will perform the 2D numerical simulations about the perturbation equations of relativistic magnetar models, which have a global dipole magnetic field.

2. Equilibrium Models

We neglect the stellar deformation from spherical symmetry due to the magnetic pressure, because the magnetic energy is much smaller than the gravitational binding energy. Thus the equilibrium of the non-rotating neutron star becomes static spherical symmetry. The magnetic field inside the star is superimposed on this stellar model, where we focus only on the dipole magnetic field with ideal MHD approximation. This magnetic field is given as [10].

3. Perturbation Equations

We restrict attention to non-axisymmetric axial-type perturbations with relativistic Cowling approximation, for which the perturbed matter quantities should be zero except for the θ and ϕ components of perturbed four-velocity, δu^θ and δu^ϕ. The perturbation equations describing δu^θ and δu^ϕ can be derived from the θ and ϕ components of the linearized equation of motion, which corresponds to the equation (35) in [3]. On the other hand, the perturbations of magnetic fields are determined from the linearized induction equation, which is corresponding to the equation (37) in [3]. Now, we can replace the variables δu^θ and δu^ϕ with

$$\delta u^\theta = im\mathcal{U}_1(t,r,\theta)e^{im\phi}/\sin\theta, \quad \text{and} \quad \delta u^\phi = \mathcal{U}_2(t,r,\theta)e^{im\phi}/\sin\theta. \tag{1}$$

Then the perturbation equations to solve can be written as

$$\begin{bmatrix} \mathcal{A}_{00} & \mathcal{A}_{01} \\ \mathcal{A}_{10} & \mathcal{A}_{11} \end{bmatrix} \begin{bmatrix} \partial_t^2\mathcal{U}_1 \\ \partial_t^2\mathcal{U}_2 \end{bmatrix} = \begin{bmatrix} \mathcal{F}_1 \\ \mathcal{F}_2 \end{bmatrix}, \tag{2}$$

where the coefficients $\mathcal{A}_{00}$, $\mathcal{A}_{01}$, $\mathcal{A}_{10}$, and $\mathcal{A}_{11}$, and the source terms $\mathcal{F}_1$ and $\mathcal{F}_2$ are some functions of the background properties.

Since the angular dependences for the axial perturbations can be generally described as $\delta u^\theta \sim \partial_\phi Y_{\ell m}/\sin\theta$ and $\delta u^\phi \sim (\partial_\theta Y_{\ell m})/\sin\theta$, the θ dependences of the variables $\mathcal{U}_1$ and $\mathcal{U}_2$ become $\mathcal{U}_1 \sim P_\ell^{|m|}(\cos\theta)$ and $\mathcal{U}_2 \sim \partial_\theta P_\ell^{|m|}(\cos\theta)$, where $P_\ell^m(\cos\theta)$ is the associated Legendre polynomial of order ℓ and m. Thus the boundary conditions for $\mathcal{U}_1$ and $\mathcal{U}_2$ are

$$\mathcal{U}_1 = 0 \;\text{ and }\; \partial_\theta\mathcal{U}_2 = 0 \qquad \text{for } m = 1, \tag{3}$$

$$\mathcal{U}_1 = 0 \;\text{ and }\; \mathcal{U}_2 = 0 \qquad \text{for } m \neq 1 \tag{4}$$

on the axis ($\theta = 0$), while

$$\mathcal{U}_1 = 0 \;\text{ and }\; \partial_\theta\mathcal{U}_2 = 0 \qquad \text{as } (\ell + m) \text{ is odd}, \tag{5}$$

$$\partial_\theta\mathcal{U}_1 = 0 \;\text{ and }\; \mathcal{U}_2 = 0 \qquad \text{as } (\ell + m) \text{ is even} \tag{6}$$

on the equatorial plane ($\theta = \pi/2$). On the other hand, with respect to the radial direction, since in the vicinity of stellar center the variables $\mathcal{U}_1$ and $\mathcal{U}_2$ are expanded as $\mathcal{U}_1 \sim C_1 r^{\ell-1} + \cdots$ and $\mathcal{U}_2 \sim C_2 r^{\ell-1} + \cdots$, where C_1 and C_2 are some constant, we can impose the boundary conditions at $r = 0$ as $\mathcal{U}_1 = \mathcal{U}_2 = 0$. The boundary conditions at the stellar surface are the zero traction conditions, which lead that $\partial_r\mathcal{U}_1 = \partial_r\mathcal{U}_2 = 0$.

4. Numerical Results

In order to construct magnetar models, we adopt a polytropic equation of state (EOS) called EOS A (see [11] for more details). Especially, we will focus on the stellar model with $M = 1.4M_\odot$. For simplicity, we consider only the pure poloidal magnetic component in this article. It should be noticed that the perturbed variables $\mathcal{U}_1$ and $\mathcal{U}_2$ should be the real variables if the stellar magnetic field has only poloidal component. To evolve the perturbation equations (2), we adopt

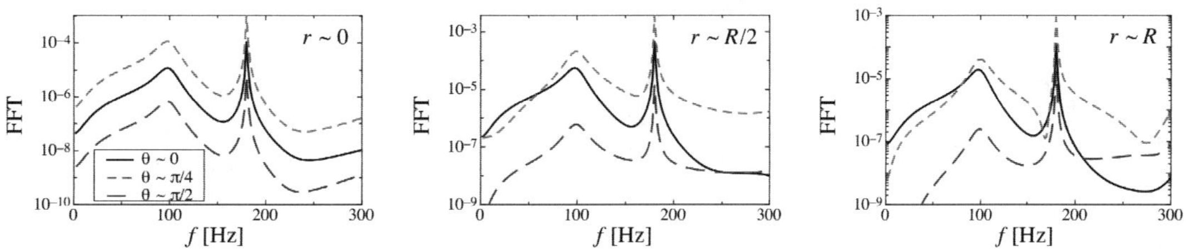

Figure 1. The FFT of the perturbation function $\mathcal{U}_2$.

the iterated Crank-Nicholson scheme together with a fourth-order Kreiss-Oliger dissipation. The region to calculate in this article is restricted in the two-dimensional region of (r, θ) with 100×40 grid points, where r and θ are in the range of $0 \le r \le R$ and $0 \le \theta \le \pi/2$, respectively. After obtaining the evolutions of perturbed variables, one can extract the specific frequencies for each spacial position by performing a fast Fourier transformation (FFT) of the time-depending perturbed variables.

First, to see the spectrum feature for the non-axisymmetric axial oscillations, we check the position of peak frequencies in FFT with different spacial position in the star. Now, figure 1 shows the FFT of perturbation variables $\mathcal{U}_2$ at various points inside the star for a stellar model with $B = 5 \times 10^{16}$ Gauss, where the value of m is fixed to be $m = 2$ in the evolution. In this figure three lines are corresponding to the different angular positions as $\theta \simeq 0$, $\pi/4$, and $\pi/2$, while three different panels correspond to the different radial positions as $r \simeq 0$, $R/2$, and R. From this figure, one can observe two peaks, whose frequencies are 98 and 180 Hz, and one can say that the positions of peak frequencies in FFT are independent of the observer inside the star. The FFT for $\mathcal{U}_1$ is also similar picture to figure 1. That is, the spectrum of non-axisymmetric axial Alfvén oscillation could not be continuum but discrete.

Figure 2 shows the dependence of the frequencies of non-axisymmetric axial Alfvén oscillations on the magnetic field strength, where the labels of (2,2) and (4,2) denote the values of (ℓ, m) and B_{16} is defined as $B_{16} = B/10^{16}$. These Alfvén modes are corresponding to the peak frequencies in the above FFTs. In the figure, the marks are shown our numerical results while the lines are corresponding to the fitting formula as mentioned the below. Obviously one can see that the frequencies are proportional to the magnetic field strength. With our numerical results, we can obtain the fitting formula as

$$f_{22}/B_{16} = 19.31 \quad \text{and} \quad f_{42}/B_{16} = 35.99, \tag{7}$$

where f_{22} and f_{42} denote the frequencies of Alfvén modes with $(\ell, m) = (2, 2)$ and $(4, 2)$, respectively. With these fitting formulae, the oscillation frequencies for the stellar model with $B = 4 \times 10^{15}$ Gauss become $f_{22} = 7.7$ Hz and $f_{42} = 14.4$ Hz. On the other hand, the minimum frequency of the axisymmetric axial Alfvén oscillation for the same stellar model is around 15 Hz [4]. That is, the non-axisymmetric axial oscillations can be realized the lower oscillation rather than the axisymmetric axial oscillations. This result could be quite important to explain theoretically the observed evidence of quasi-periodic oscillations in the SGR, because it has been difficult to explain the lower part of frequencies (see for the discussion in [3, 4]). We are considering that the non-axisymmetric axial Alfvén oscillations could play an important role in the oscillations of magnetar and probably it might be possible to explain clearly the observed evidence with these type of oscillations. To fit the possible stellar model with the observations in SGRs, however, it is necessary to produce more oscillation frequencies with different value of (ℓ, m) for the stellar models constructed with different EOSs. Since now we find that the non-axisymmetric axial oscillations become discrete spectrum, one can safely study the calculation

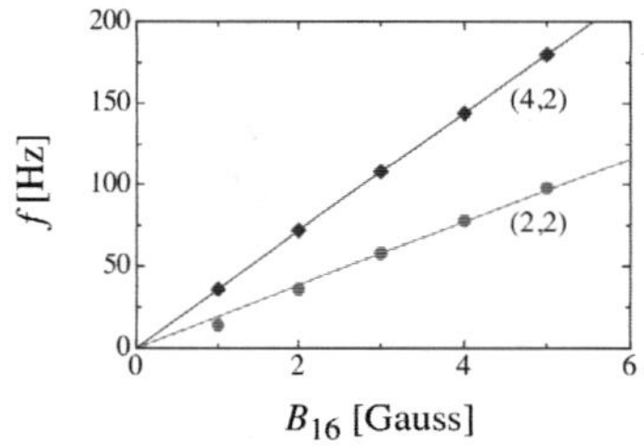

Figure 2. Frequencies of the Alfvén modes as functions of magnetic field strength.

to determine the specific frequencies by using modal decomposition as the study by [2, 8] and this method could be better to determine the frequencies of stellar oscillations. As a future work, we will do this and find the most possible magnetar model to explain the observations.

5. Summary and Discussion

To examine the non-axisymmetric axial-type oscillations on magnetar, we derive the perturbation equations for an arbitrary magnetic distribution inside the star, where we neglect the effect of the solid crust near the stellar surface. For simplicity, we consider the magnetar models with the pure poloidal magnetic field and with this stellar model the perturbation variables can reduce the real ones. Performing the time-evolution of two-dimensional perturbation equations, we estimate the Alfvén oscillations. In order to know the spectrum feature for these type oscillations, we check the dependence of positions of peak frequencies in FFTs upon the observer position inside the star. Then we observe that the positions of peak frequencies in FFTs are independent of the observers, i.e., we can conclude that these type oscillations become discrete spectrum as well as the axisymmetric polar Alfvén oscillations. Additionally we see the dependence of frequencies on the magnetic field strength and find that the frequencies are obviously proportional to the magnetic field strength. The frequencies of the non-axisymmetric axial oscillation become smaller than those of the axisymmetric axial oscillations. In order to explain the observation of specific frequencies of the quasi-periodic oscillations in SGRs, these smaller frequencies could be play an important role and as a future work we will find the most possible magnetar model by determining systematically more frequencies with different values of ℓ, m and with different EOSs. At the end, we make a comment about the introduction of the solid crust. As shown in the previous articles such as [3], the oscillation of solid curst region becomes discrete spectrum. Thus even if considering this region in the magnetar model, we can expect that there exist both frequencies of non-axisymmetric axial Alfvén oscillations and the crustal oscillations. Anyway taking into account the existence of solid crust region is also important future work.

References

[1] Watts A L and Strohmayer T E 2006 *Adv. Space Res.* **40** 1446
[2] Lee U 2007 *Mon. Not. R. Astron. Soc.* **374** 1015
[3] Sotani H, Kokkotas K D and Stergioulas N 2007 *Mon. Not. R. Astron. Soc.* **375** 261
[4] Sotani H, Kokkotas K D and Stergioulas N 2008a *Mon. Not. R. Astron. Soc.* **385** L5
[5] Colaiuda A, Beyer H R and Kokkotas K D 2009 *Mon. Not. R. Astron. Soc. 396* 1441
[6] Cerda-Duran P, Stergioulas N and Font J A 2009 *Mon. Not. R. Astron. Soc.* **397** 1607
[7] Sotani H and Kokkotas K D 2009 *Mon. Not. R. Astron. Soc.* **395** 1163
[8] Lee U 2009, *Preprint* arXiv:0911.3966 [astro-ph]
[9] Lander S K, Jones D I and Passamonti A *Preprint* arXiv:0912.3480 [astro-ph]
[10] Sotani H, Colaiuda A and Kokkotas K D 2008b *Mon. Not. R. Astron. Soc.* **385** 2161
[11] Sotani H and Kokkotas K D 2004 *Phys. Rev.* D **70** 084026

Journal of Physics: Conference Series **314** (2011) 012082 doi:10.1088/1742-6596/314/1/012082

Non-linear preheating after inflation and gravitational wave production

Mar Bastero-Gil

Departamento de Física Teórica y del Cosmos, Universidad de Granada, Granada-18071, Spain

E-mail: mbg@ugr.es

Abstract. We present the evolution of the *full* set of Einstein equations during preheating after inflation, for a generic supersymmetric model of hybrid inflation. Preheating of the scalar metric fluctuations does affect the evolution of vector and tensor modes, and in particular they do enhance the induced stochastic background of gravitational waves during preheating.This gives an energy density in general *an order of magnitude* larger than that obtained by evolving the tensors fluctuations in an homogeneous background metric. This enhancement can improve the expectations for detection by planned gravitational waves observatories.

1. Introduction

Background (CMB) measurements [1], are consistent with an early period of inflation, which gives rise to the primordial curvature perturbation which seeds the large scale structure observed today. Inflation should be followed by a reheating period, during which the inflationary vacuum energy is converted into radiation. During the first stages of reheating, the evolution of the system may be dominated by non-perturbative effects as those of preheating, i.e., parametric amplification of quantum field fluctuations in a background of oscillating fields [2]. It does also enhance the tensor perturbations, sourced by the field anisotropic stress-energy tensor, giving rise to a stochastic background of gravitational waves (GW) [3].

In most of the studies of preheating of GW, fields and tensor fluctuations are evolved in a background Friedmann-Robertson-Walker (FRW) metric. However, beyond linear perturbation, tensors are also seeded by scalar and vector components of the metric. [4]. Given that metric and fields are non-linearly coupled through the Einstein equations, the parametric amplification of field fluctuations will be rapidly transferred to all metric perturbations. In [5] we showed that for the case of the scalar metric perturbations. And preheating of the scalar metric perturbations can then affect the amplification of tensors. Here we show that this is indeed the case for hybrid models, independently of model parameters [6].

2. Gravity waves from preheating

We have integrated the full set of Einstein equations, with the stress-energy tensor provided by the fields in hybrid inflation. Einstein equations are written in the so-called BSSN (Baumgarte-Shapiro-Shibata-Nakamura) formalism [7]. The spatial metric is given by $\gamma_{ij} = exp(4\beta)\tilde{\gamma}_{ij}$, with $\det\tilde{\gamma}_{ij} = 1$. The metric dynamical variables are then β and $\tilde{\gamma}_{ij}$, the trace of the extrinsic curvature $K = \tilde{\gamma}^{ij}K_{ij}$, and its traceless part $\tilde{A}_{ij} = exp(-4\beta)(K_{ij} - \gamma_{ij}K/3)$. We choose to work in the synchronous gauge, where we have $K = 6\dot{\beta}$, with $\langle K\rangle/3 = H(t)$ being the average expansion rate, and $\langle e^{2\beta}\rangle = a(t)$ the average scale factor ("$\langle\cdot\rangle$" denotes spatial average.). The spatial metric $\tilde{\gamma}_{ij}$ encodes the two transverse and traceless degrees of freedom of the gravity waves, plus one additional scalar mode and 2 vector degrees of freedom, with $\dot{\tilde{\gamma}}_{ij} = 2\tilde{A}_{ij}$.

The set of equations of motion (EOM) are then given by the Klein-Gordon equations for the fields, and those for the metric variables [7] given by:

$$\ddot{\beta} + 2\dot{\beta}^2 = \frac{1}{6m_P^2}(V - 2T) - \frac{1}{24}\dot{\tilde{\gamma}}_{ij}\dot{\tilde{\gamma}}^{ij}\,, \quad (1)$$

$$\ddot{\tilde{\gamma}}_{ij} + K\dot{\tilde{\gamma}}_{ij} = 2e^{-4\beta}(M_{ij}^{TF} - R_{ij}^{TF}) + \dot{\tilde{\gamma}}_j^k\dot{\tilde{\gamma}}_{ki}\,, \quad (2)$$

where V is the potential for supersymmetric hybrid inflation, $V(\phi,\chi) = V_0 + g^2\chi^4/4 + g^2(\Phi^2 - \phi_c^2)\chi^2 + m_\phi^2\Phi^2/2$, and $T = (\dot{\Phi}^2 + \dot{\chi}^2)/2$ is the kinetic energy of the fields; Φ is the inflaton field and χ the waterfall field which triggers the phase transition at the end of inflation once the inflaton goes below the critical value ϕ_c (both taken to be real scalar fields). The field dependent source is $M_{ij} = m_P^{-2}(\partial_i\Phi\partial_j\Phi + \partial_i\chi\partial_j\chi)$, R_{ij} is the Ricci tensor of the metric γ_{ij}, and the superscript "TF" denotes the trace-free part of the tensor. m_P is the reduced Planck mass.

The system is placed in a finite and discrete 3D comoving box of length L and N sites per spatial dimension. The procedure introduces a comoving ultraviolet cut-off in both space and momentum. For the problem of preheating after inflation, one tries to optimize the choice of the ratio $O(N/L)$ to have an ultraviolet comoving momentum cut-off still larger than the preheating cut-off by the end of the simulation. This means that already at the start of the simulation our comoving box is smaller than the observable universe, but the relevant physical modes for preheating are all included.

We start the simulations some fraction of e-fold $\Delta N_e = 0.05$ after the end of inflation, with the background inflaton field still close to the critical point, $\langle\Phi\rangle = \phi_c\exp(-\eta_\phi\Delta N_e)$, and its background velocity given by the slow-roll conditions The background values for the waterfall field are set to zero at this point. Classical inflaton field fluctuation in a spatial lattice with periodic boundary conditions are expanded as usual in Fourier modes Φ_k, with an initial vacuum amplitude $|\Phi_k(0)| \simeq 1/\sqrt{2\omega_k}$, where $\omega_k = \sqrt{k^2 + m_\phi^2}$. The remaining variables must be chosen to satisfy the Einstein constraint equations, the momentum and the hamiltonian constraint, at $t = 0$. This is the well known initial-value problem in general relativity [8]. We choose initially vanishing tensors and vectors, i.e., $\tilde{\gamma}_{ij}(0) = \delta_{ij}$, such that the momentum constraint reduces to $2\partial_i K = -(3/m_P^2)(\dot{\Phi}\partial_i\Phi + \dot{\chi}\partial_i\chi)$. In order to fulfill this equation, we set the waterfall field as $\chi(0) = \Phi(0) - \langle\Phi(0)\rangle$, and $\dot{\chi}(0) = \langle\dot{\Phi}(0)\rangle - \dot{\Phi}(0)$. This allows to solve the momentum constraint for the initial value of the expansion rate fluctuations, and use the hamiltonian constraint to fix the fluctuations of the scale factor.

Preheating in hybrid inflation models has been extensively studied in the literature [9]. The parametric amplification of the fluctuations takes place first through an spinodial instability for the fields, during which the lower modes are quickly amplified. After a few oscillations, the amplitude of the fields has decayed enough to be out of the spinodial region, and tachyonic preheating ends. We just follow the evolution of the fields and metric variables up to the end of the resonance, before we loose the ultraviolet cut-off for the field modes. Metric variables will follow the same pattern of parametric amplification than the fields, with $|\beta|^2 \sim m_P^{-2}|\Phi|^2$. Keeping only the leading β terms in the Ricci tensor, we have:

$$R_{ij}^{TF} \simeq [-4\partial_i\beta\partial_j\beta + 2\partial_i\partial_j\beta]^{TF}\,, \quad (3)$$

and thus R_{ij}^{TF} becomes comparable to the field contribution M_{ij}^{TF} in Eq. (2).

The metric variable $\tilde{A}_{ij}$ is traceless but non-transverse, i.e., it contains more degrees of freedom than those two corresponding to GW. The traceless and transverse (TT) components

Journal of Physics: Conference Series **314** (2011) 012082 doi:10.1088/1742-6596/314/1/012082

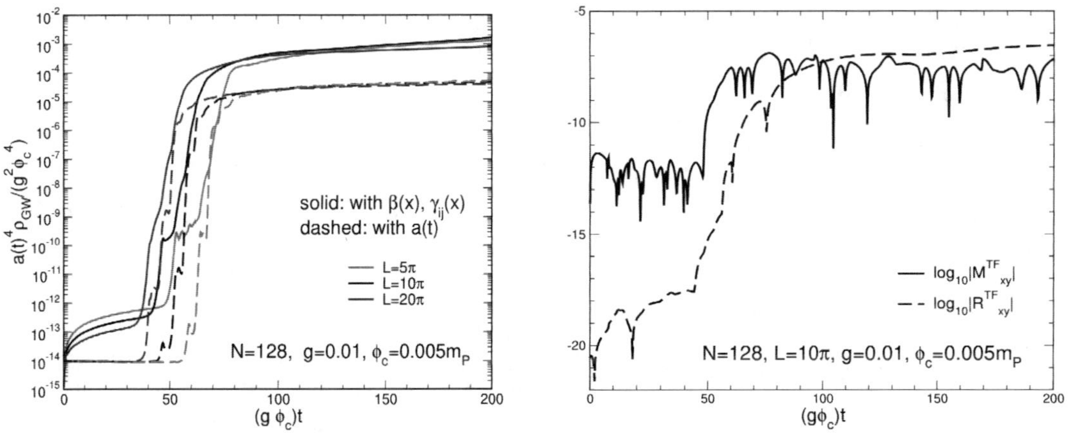

Figure 1. LHS plot: evolution of the energy density of GW, ρ_{GW}, for different values of the box size. Solid lines are the result of integrating the Einstein equations, dashed lines show the evolution of ρ_{GW} without including other metric fluctuations. RHS plot: comparison of the evolution of the absolute values of the source terms M_{xy}^{TF} and R_{xy}^{TF} in Eq. (3).

are projected by using the operator [3] $\Lambda_{ij,lm}(\hat{k}) = P_{il}(\hat{k})P_{jm}(\hat{k}) - P_{ij}(\hat{k})P_{lm}(\hat{k})/2$, where $P_{ij} = \delta_{ij} - \hat{k}_i\hat{k}_j$ and $\hat{k}_i = k_i/k$. The energy density of the GW is then $\rho_{GW} = m_P^2\langle \tilde{A}_{ij}(t,x)\tilde{A}^{ij}(t,x)\rangle^{TT} = m_P^2 \int d^3k |\tilde{A}_{ij}^{TT}(t,k)|^2$ [3]. On the LHS in Fig. 1, we have plotted ρ_{GW}, normalized to the initial vacuum energy $\rho_i = g^2\phi_c^4$. We have taken as parameter models $g = 0.01$, $\phi_c = 0.005 m_P$ and $\eta_\phi = 0.05$. The value of ρ_{GW} does not depend on the value of the coupling, which can be rescaled out from all the equations, but it does depend on the value of ϕ_c [3], which sets the scale for the field source term such that $\rho_{GW}/\rho_i \propto \phi_c^2$. We have included the results for different choices of the box size, to show that the final value of ρ_{GW} does not depend on the choice of L as far as we have all the relevant modes to start and end tachyonic preheating. With a larger comoving box we have more modes in the low momentum regime, and then tachyonic resonance starts slightly sooner, as can be seen in the plot. In this figure we have compared the results obtained when integrating the full Einstein equations (solid lines) with those obtained when integrating the tensor modes in a FRW background metric (dashed lines). The former are always roughly *an order of magnitude larger* due mainly to the contribution of the scalar modes of the metric fluctuations in Eq. (3). On the RHS in Fig. 1, we show that both source terms, M_{ij}^{TF} and the leading term R_{ij}^{TF} in Eq. (3), are of the same order by the end of the resonance. Scalar metric fluctuations start growing immediately after inflation due to the increase in the kinetic energy of the field (see Eq. (1)), and this effect leads the initial growth of the tensor perturbations.

In Fig. 2 we show the spectrum of GW per logarithmic frequency interval, $\rho_{GW}(k) = d\rho_{GW}/d\ln k$, at different times until the end of the tachyonic resonance. Comparing with the spectrum obtained in a FRW metric, the spectrum is enhanced and the peak is shifted towards lower values at around $k \simeq g\phi_c$. This effect at low momenta is again due to the scalar metric perturbations, which spectrum peaks below $g\phi_c$.

3. Summary

In summary, we have shown that non-linear effects due to metric perturbations enhance the amplitude of the GW stochastic background by an order of magnitude with respect to the calculations in a FRW background metric. Taking into account that $\rho_{GW}(k)$ is redshifted like radiation after the resonance, and assuming that entropy is conserved from reheating onwards, its

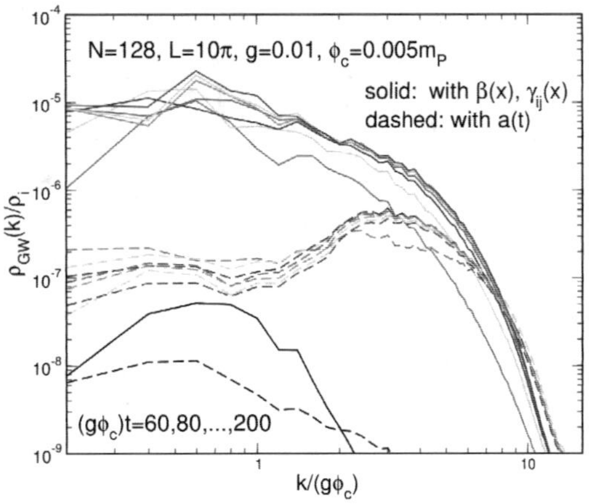

Figure 2. Spectrum $\rho_{GW}(k)/\rho_i$ per logarithmic k interval, for $(g\phi_c)t = 60$ to 200 in intervals of 20 from bottom to top. Solid lines are the result of integrating the Einstein equations, dashed lines show the result of the integration in a FRW metric.

maximum present day value normalized by the critical density today is given by $h^2\Omega_{GW}^{peak} \simeq 5.5 \times 10^{-9}\left(\frac{\phi_c/m_P}{0.005}\right)^2\left(\frac{T_{RH}^4}{\rho_i}\right)^{1/3}$, where T_{RH} is the reheating temperature. This amplitude is within the reach of the future GW observatory Advanced-LIGO for $\phi_c > 0.005m_P$, or BBO for $\phi_c < 0.005m_P$ [10]. However, today's values for the frequency are $f = 6.4\times 10^{10}\sqrt{g}(k/(g\phi_c))(T_{RH}/\rho_i^{1/4})^{1/3}$ Hz, while the operating frequency range for Advanced-LIGO is $1 - 10^3$ Hz, and BBO will operate in the range $10^{-3} - 10^2$ Hz. Thus, to have the GW spectrum within the observable range, the coupling should be at most of the order of $g \simeq 10^{-14}$. The numerical simulations so far can resolve the spectrum for the typical frequency of the resonance, $k \simeq g\phi_c$ and above, but not the infrared (IR) tail for subhorizon and superhorizon modes. The behavior of the GW spectrum in this IR range and how they are affected by non-linear metric effects is still an open question. A tail of subhorizon modes at the time of preheating rising slower than k^3 could be detected by BBO, although f_{peak} were in the range of 10^5 Hz.

Acknowledgments

We acknowledge for computing resources the IDRIS-CNRS (Babel) facility, and the cluster UGRGRID in Granada [11].

References

[1] E. Komatsu et al., arXiv:1001.4538.
[2] Y. Shtanov, J. Traschen and R. Brandenberger, Phys. Rev. **D 51** (1995)5438; L. Kofman, A. Linde and A. A. Starobinsky, Phys. Rev. **D 56** (1997) 3258; D. I. Kaiser, Phys. Rev. **D56** (1997) 706.
[3] R. Easther, J. T. . Giblin and E. A. Lim, Phys. Rev. Lett. **99** (2007) 221301; J. García-Bellido and D. G. Figueroa, Phys. Rev. Lett. **98** (2007) 061302; J. F. Dufaux, A. Bergman, G. N. Felder, L. Kofman and J. P. Uzan, Phys. Rev. D **76** (2007) 123517.
[4] S. Mollerach, D. Harari and S. Matarrese, Phys. Rev. **D 69** (2004) 063002; K. N. Ananda, C. Clarkson and D. Wands, Phys. Rev. D **75**, 123518 (2007).
[5] M. Bastero-Gil, M. Tristram, J. F. Macias-Perez and D. Santos, Phys. Rev. D **77** (2008) 023520.
[6] M. Bastero-Gil, J. F. Macias-Perez and D. Santos, Phys. Rev. Lett. **105** (2010) 081301.
[7] M. Shibata and T. Nakamura, Phys. Rev. **D52** (1995) 5428; T. W. Baumgarte and S. L. Shapiro, Phys. Rev. **D59** (1999) 024007.
[8] J. W. York, Phys. Rev. Lett. **26** (1971) 1656; P. Laguna, H. Kurki-Suonio and R. A. Matzner, Phys. Rev. **D 44** (1991) 3077.
[9] M. Bastero-Gil, S. F. King and J. Sanderson, Phys. Rev. **D60** (1999) 103517; G. N. Felder et al., Phys. Rev. Lett **87** (2001) 011601; G. N. Felder, L. Kofman and A. D. Linde, Phys. Rev. **D 64** (2001) 123517.
[10] LIGO URL: http://www.ligo.caltech.edu; BBO URL: http://universe.nasa.gov/new/program.bbo.html.
[11] IDRIS URL: http://www.idris.fr; UGRGRID URL: http://ugrgrid.ugr.es.

Spanish Relativity Meeting (ERE 2010): Gravity as a Crossroad in Physics IOP Publishing
Journal of Physics: Conference Series **314** (2011) 012083 doi:10.1088/1742-6596/314/1/012083

Bulk flows from clusters of galaxies.

F Atrio-Barandela[1], A Kashlinsky[2], H Ebeling[3], D Kocevski[4] and A Edge[5]

[1] Física Teórica, Universidad de Salamanca, 37008 Salamanca, Spain
[2] SSAI and Observational Cosmology Laboratory, Code 665, Goddard Space Flight Center, Greenbelt MD 20771, USA
[3] Institute for Astronomy, University of Hawaii, 2680 Woodlawn Drive, Honolulu, HI 96822, USA
[4] Department of Physics, University of California at Davis, 1 Shields Avenue, Davis, CA 95616, USA
[5] Department of Physics, University of Durham, South Road, Durham DH1 3LE, UK

E-mail: atrio@usal.es

Abstract. The kinematic Sunyaev-Zeldovich effect produces a measurable temperature anisotropy in the cosmic microwave background that can be used to measure the bulk motion of clusters of galaxies. We discuss our recent measurement of motions on scales of $\sim 300-800$ Mpc with amplitude and direction similar to the cosmological dipole, its systematics, uncertainties and the correlation with X-ray luminosity. The latter is a crucial test that supports the existence of the flow. We discuss the cosmological implications if the flow extends out to the cosmological horizon.

1. Introduction

In the local matter rest frame, the recession velocity of a galaxy of redshift z is: $cz = H_o r + v_r$. The peculiar velocity $\vec{v}$ is a local departure from the Hubble expansion $H_o r$. From redshift and distance measurements we can determine only its projection along the line of sight, v_r. Peculiar velocities play important role in understanding the large scale matter distribution in the Universe and have been subject of intense research. Uncertainties on distance indicators makes it difficult to measure peculiar velocities of individual galaxies. An alternative statistic is to measure the center of mass motion of galaxies within a given region, known as bulk flow. Even this statistic is difficult to determine from galaxy surveys beyond a scale of 50Mpc/h (see Strauss & Willick 1995 for a review). Nevertheless, the motion of the Local Group is well known. It has an amplitude of 627 ± 22 km/s and direction $(273^0 \pm 3^0, 30^0 \pm 3^0)$ in galactic coordinates. It has been determined from the Cosmic Microwave Background (CMB) dipole. This motion is supposed to be generated by local inhomogeneities but due to the lack of reliable distance estimators, is not clear what scales contribute. Kocevski & Ebeling (2006) found that most of the peculiar velocity of the Local Group was due to over-densities at $> 150h^{-1}$Mpc, a result difficult to accommodate in the standard ΛCDM model.

2. Peculiar velocities from the kinematic Sunyaev-Zeldovich effect.

Alternative methods to measure peculiar velocities was proposed in Kashlinsky and Atrio-Barandela (2000) and Atrio-Barandela, Kashlinsky and Mücket (2004). Clusters of galaxies are the largest astronomical objects bound gravitationally. The hot Intra Cluster gas is highly ionized and the free electrons scatter the CMB radiation. When the cluster moves in the isotropic CMB frame, the Doppler shift of the radiation produces a dipole pattern in the cluster rest frame. When the CMB photons cross the cluster, they get thermalize through Compton scattering with the electrons within the potential well. This effect produces a secondary anisotropy of amplitude $\Delta T/T_o(\hat{n}) = -\tau(\hat{n}\cdot\vec{v}_{cl})/c$. This component, termed kinematic Sunyaev-Zeldovich (kSZ), depends on the cluster velocity v_{cl}, the electron density along the line of sight τ and the direction of observation $\vec{n}$. This signal is subdominant in most clusters with respect to the anisotropy generated by the thermal motion of the electrons within the potential well of the cluster. This effect is known as thermal Sunyaev-Zeldovich (tSZ) effect. In the Rayleigh-Jeans regime, frequencies at which the WMAP satellite operates, it produces a decrement in the CMB temperature.

Kashlinsky and Atrio-Barandela (2000) showed that a bulk flow motion of N_{cl} clusters will produce a dipole pattern on the CMB at cluster locations with amplitude:

$$a_{1m} = 1\mu K \frac{v_{bulk}}{300km/s} \pm 3\mu K \left[\frac{N_{cl}}{1000}\right]^{1/2} \pm 0.6\mu K \left[\frac{N_{pixels}}{10000}\right]^{1/2} \pm 0.2\mu K \left[\frac{N_{cl}}{1000}\right]^{1/2} \tag{1}$$

The signal will be overshadowed by three competing effects: intrinsic CMB, noise and tSZ dipole due to the inhomogeneous cluster distribution on the sky. For typical cluster parameters, the expected signal of $\sim 1\mu$K would be dominated by the sampling variance of the intrinsic CMB signal at the cluster positions. Since the CMB signal is spatially correlated, Kashlinsky and Atrio-Barandela (2000) suggested to filter the data to remove the intrinsic CMB signal.

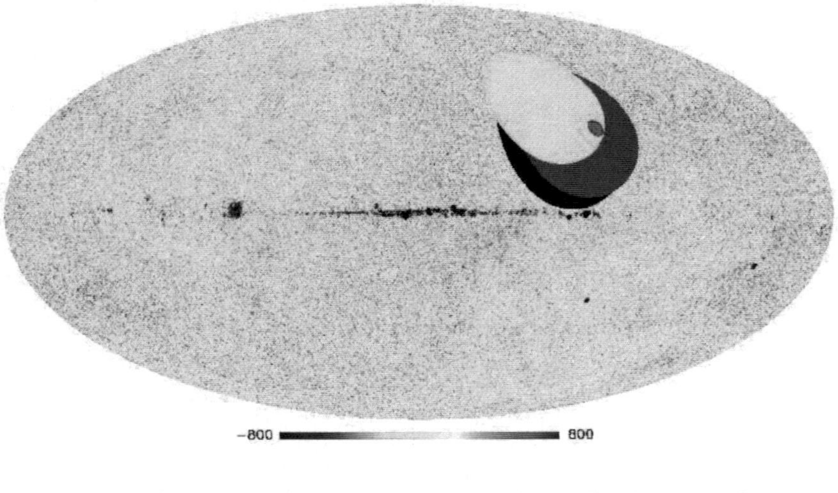

Figure 1. WMAP W1 Differencing Assembly. The three overlapping circles show the direction and 1σ uncertainty in the direction of the dark flow for three subsets of the scout cluster catalog: all clusters have X-ray luminosities in the Rossat broadband larger than 2×10^{44} erg/s, and redshifts smaller than 0.16 (dark blue) and 0.2 (blue) and 0.25 (cyan). The red spot indicates the direction of the CMB dipole.

3. The Dark Flow.

To measure bulk flows using the kSZ effect as described in the previous section it is necessary to have a large cluster sample. We are currently using the SCOUT catalog, described in Kashlinsky et al (2010). The same analysis could be also carried out using public data, as indicated in Kashlinsky, Atrio-Barandela and Ebeling (2010). The dipole evaluated on a CMB map at cluster positions would be dominated by the intrinsic CMB signal, as indicated in equation (1). In Kashlinsky et al (2008) and (2009) we showed this measurement could be carried out by adequately filtering the data. A description of the filter, the cluster catalog, the cosmological implications and a detailed description of the technical issues involved was also given. The dipole at cluster locations was measured at zero monopole, implying that the tSZ did not generate a significant dipole due to the inhomogeneous cluster distribution. As shown in Atrio-Barandela et al (2008), this is the result of clusters being well described by a Navarro, Frenk and White (1997) profile. If the baryons are in hydrostatic equilibrium, the temperature of the gas falls with distance from the core. When we integrate the signal including the cluster outskirts, the tSZ contribution drops and the kSZ signal becomes dominant.

Later, an improved analysis using WMAP 5yr data was carried out in Kashlinsky et al (2010). To summarize the results, in Figure 1 we represent the direction of the flow as derived for the most luminous clusters in the SCOUT catalog. The width of the circle corresponds to the 1σ error bar. The red spot indicates the direction of the CMB dipole. The direction is very consistent within the three subsamples and with the direction of the measured dipole.

In Atrio-Barandela et al (2010) we showed the filter to be optimal in the sense that its efficiency was limited by cosmic variance.

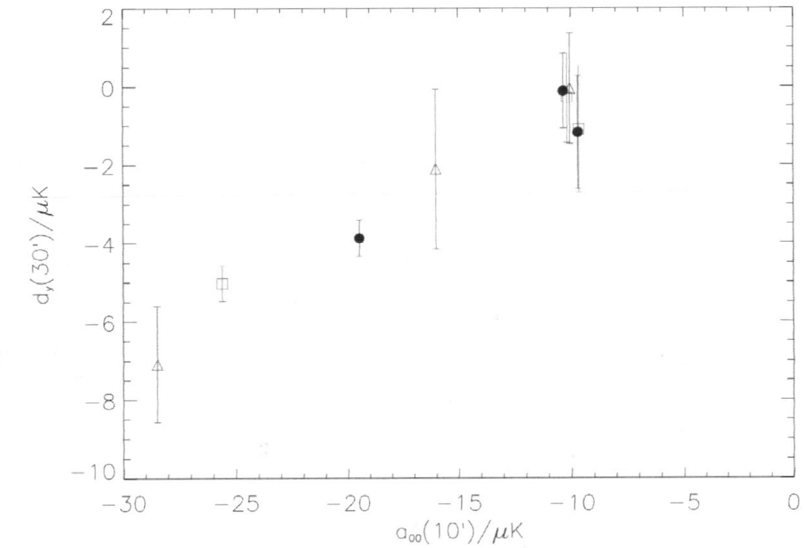

Figure 2. Y-component of the dipole evaluated at cluster positions on discs of size 30' versus the monopole evaluated on the original WMAP W-band on discs of size 10'. Clusters are bin by X-ray luminosity in three intervals: $L_X \in [1.0, 1.5], [1.5, 2.0]$ and $L_X > 2$ (in units of 10^{44}erg/s). The blue triangles correspond to clusters with redshifts $z \leq 0.16$, the red squares to $z \leq 0.2$ and the full green circles to $z \leq 0.25$. The error bars are obtained from the dispersion within the 4 W Differencing Assemblies.

The evidence we have provided in favor of the measured dipole being associated with the kSZ effect can be summarized as follows: (a) the dipole originates at cluster locations, (b) it is present when the number of clusters grow from ~ 120 clusters to more than 700 and, as indicated above, (c) the dipole is measured when the tSZ monopole was effectively zero. In Figure 2 we represent the measured y-component for clusters subsamples bin by X-ray luminosity, as a function of the measured tSZ monopole on the *unfiltered* four W Differencing Assemblies (DA). When computing the tSZ signal, all clusters are assigned an angular size of $10'$. The dipole is computed on *filtered* W DA with apertures of $30'$, where we have verified the tSZ monopole is close to zero. The error bars are computed from the dispersion within the W DA. Figure 2 shows a clear correlation between the measured dipole and the tSZ effect. The latter is a proxy for X-ray luminosity, since a more luminous cluster produces a larger tSZ effect. Since our filtering does not use information about cluster positions or their luminosities, this correlation can not be an artifact of our data processing or some unknown systematic effect present in WMAP data. It argues strongly in favor of the signal being real. The differences found between the subsamples are arguably due to clusters at higher redshifts having their tSZ and kSZ effects more strongly diluted by the beam.

4. Cosmological Implications.

As the data shows no signs of converging, one could assume that the scale of the flow extends all the way up to the observable horizon. Equivalently, seen from the matter rest frame the cosmological dipole is intrinsic, i.e., the matter and isotropic CMB frames do not coincide. An intrinsic dipole of amplitude $\Delta T/T_o \sim 10^{-3}$ can not be generated during inflation without generating a quadrupole of similar amplitude (Kashlinsky, Tkachev and Frieman 1994). However, an isocurvature pre-inflationary remnant of amplitude $\delta \sim 1$ and wavelength $L \sim 10^3 cH_o^{-1}$ could produce such anisotropy:

$$D \sim \frac{v}{c} \sim \left(\frac{cH_o^{-1}}{L}\right) \delta_L \sim 10^{-3} \qquad Q \sim \left(\frac{cH_o^{-1}}{L}\right)^2 \delta_L \sim 10^{-6} \tag{2}$$

without generating a significant quadrupole. In this context, our local motion could be generated by a pre-inflation perturbation of wavelength much larger than the current horizon. In this respect, the concordance ΛCDM model would be valid on sub-horizon scales, but one could not longer assume that the matter rest frame and the isotropic CMB frame were equivalent.

Acknowledgments
This work was partially funded by NASA NNG04G089G, 09-ADP09-0050 and FIS2009-07238, CSD 2007-00050, GR-234 grants from Spanish Ministerio de Educación y Ciencia and Junta de Castilla y León.

References

Atrio-Barandela F, Kashlinsky A, Ebeling H, Kocevski D and Edge A 2010 *Astrophys. J.* **719** 77
Atrio-Barandela F, Kashlinsky A, Kocevski D and Ebeling H 2008 *Astrophys. J.* **675** L57
Atrio-Barandela F, Kashlinsky A and Mücket J 2004 *Astrophys. J.* **601** L111
Kashlinsky A and Atrio-Barandela F 2000 *Astrophys. J.* **536** L67
Kashlinsky A, Atrio-Barandela F and Ebeling H 2010 Measuring the dark flow with public X-ray cluster data. *Preprint* ArXiv:1012.3214
Kashlinsky A, Atrio-Barandela F, Kocevski D and Ebeling H 2008 *Astrophys. J.* **686** L49
Kashlinsky A, Atrio-Barandela F, Kocevski D and Ebeling H 2009 *Astrophys. J.* **691** 1479
Kashlinsky A, Atrio-Barandela F, Ebeling H, Edge A and Kocevski D 2010 *Astrophys. J.* **712** L81
Kashlinsky A, Tkachev I and Frieman J 1994 *Phys. Rev. Lett.* **73** 1582
Kocevski D D and Ebeling H 2006 *Astrophys. J.* **645** 1043
Navarro J F, Frenk C S and White S D M 1997 *Astrophys. J.* **490** 493
Strauss M and Willick J A 1995 *Phys. Rep.* **261** 271
Sunyaev R and Zel'dovich I 1972 *Comments on Astrophys. Space Phys.* **4** 173

Journal of Physics: Conference Series **314** (2011) 012084 doi:10.1088/1742-6596/314/1/012084

Black hole candidates and the Kerr bound

Tomohiro Harada[1] and Rohta Takahashi[2]

[1]Department of Physics, Rikkyo University, Toshima, Tokyo 175-8501, Japan

[2]Cosmic Radiation Laboratory, the Institute of Physical and Chemical Research, 2-1 Hirosawa, Wako, Saitama 351-0198, Japan

E-mail: harada@rikkyo.ac.jp

Abstract. The specific angular momentum of a Kerr black hole must not be larger than its mass in the geometrical units. The observational confirmation of this bound which we call a Kerr bound directly suggests the existence of a black hole. On the other hand, the violation of this bound may suggest the existence of a superspinning object which might be suggested from a string theory argument. In order to investigate observational testability of this bound by using the X-ray energy spectrum of black hole candidates, we calculate the energy spectra from an optically thick and geometrically thin accretion disk of a superspinning object which is described by a Kerr metric but whose specific angular momentum is larger than its mass, and then compare the spectra of this object with those of a black hole. Based on this calculation, we present that the observational confirmation of the Kerr bound is very hard but the violation of it may be detectable if only the continuum X-ray spectrum of the disk is available.

1. Introduction

Black hole candidates will be directly observed by radio interferometers by X-ray satellites and gravitational wave detectors in the near future. Although these candidates are most likely to be black holes, it is not obvious to confirm that. In principle, to prove that black hole candidates are really black holes, we have to show that the observed data cannot be explained by anything else. In this relation, there is an argument that if we can show that a sufficiently large amount of mass, say, above the model-independent maximum mass of neutron stars $\sim 3M_{\odot}$, is compactified within the size comparable with its gravitational radius, general relativity assures that it is a black hole. It turns out however that this argument crucially depends on the cosmic censorship hypothesis [1, 2]. The singularity theorems predict the existence of spacetime singularities in generic spacetimes, where the spacetime curvature is singular in typical cases. The cosmic censorship hypothesis is an unproven hypothesis about the properties of spacetimes which are solutions to the Einstein equation with physically reasonable matter fields and with generic initial data. It claims that spacetime singularities formed in physical gravitational collapse are covered by horizons and hidden from any or distant observers. Thus, the cosmic censorship hypothesis ensures the future predictability of the classical theory including general relativity. On the other hand, since quantum gravity should replace classical general relativity in the Planckian regime, singular spacetimes in classical theory might be modified to regular ones in some sense. If this is true, the cosmic censorship hypothesis for the classical theory should be reconsidered in the new context. One of the interesting possibilities is that naked-singular solutions of the classical Einstein equation are physically significant in the low-curvature region but largely modified to be smooth spacetimes around singularities in the classical solutions due to quantum gravity

Journal of Physics: Conference Series **314** (2011) 012084 doi:10.1088/1742-6596/314/1/012084

effects. This article is based on Takahashi and Harada [3]. We use the geometrical units, where $c = G = 1$.

The Kerr metric is a solution to the vacuum Einstein equation. This solution is stationary, axisymmetric and asymptotically flat. The metric is given by

$$\begin{aligned} ds^2 &= -\left(1-\frac{2Mr}{\Sigma}\right)dt^2 - \frac{4aMr\sin^2\theta}{\Sigma}dtd\phi + \frac{\Sigma}{\Delta}dr^2 + \Sigma d\theta^2 \\ &\quad + \left(r^2+a^2+\frac{2Mra^2\sin^2\theta}{\Sigma}\right)\sin^2\theta d\phi^2, \end{aligned}$$

where $\Sigma = r^2+a^2\cos^2\theta$, $\Delta = r^2-2Mr+a^2$, and M and a are the mass and the Kerr parameter, respectively. We put $a_* \equiv a/M$ as the nondimensional Kerr parameter. The Kerr solution has a curvature singularity at $r=0$ and $\theta=\pi/2$, where $\Sigma=0$ and this is called a ring singularity. For $|a_*| \le 1$, the solution has an event horizon at $r=r_H$, where r_H is a larger root of $\Delta=0$. We can see that the ring singularity is covered by an event horizon. For $|a_*|>1$, there is no event horizon and the ring singularity is therefore not covered by a horizon. A singularity which is not covered by a horizon is called a naked singularity. Therefore, the condition $|a_*| \le 1$ must be satisfied for the singularity to be covered and we call this condition the Kerr bound on the Kerr parameter.

Here we should mention an important example where the naked singularity is remedied due to quantum gravity effects in string theory [4]. There is a solution called the BMPV solution of supergravity in (4+1) dimensions. This solution describes a black hole in some region of the parameter space as a sub-extremal case but has a naked singularity in other region as a super-extremal case. This naked singularity is actually excised by a domain wall of strings and D-branes, if stringy effects are taken into account. Although the Kerr solution has not yet been dealt with in the same context, we can assume that the naked singularity in the super-extremal Kerr solution is excised as well due to stringy effects and this regular object is called a superspinar. In this connection, we should also mention that it is suggested that we might overspin a sub-extremal black hole to a super-extremal object by plunging a test body [5].

2. The standard accretion disk in the general Kerr spacetime

We newly invent the consistent accretion disk model for the superspinar. We generalise the general relativistic standard accretion disk model around a black hole [6] to a superspinar. We adopt the following assumptions. The disk is axisymmetric, steady-state, geometrically thin and optically thick. The angular momentum of the fluid is transferred by the viscous torque and the dissipative energy is used to thermalise the disk. The local thermal equilibrium holds in the disk fluid. The inner edge is torque free and given by the innermost stable circular orbit (ISCO). Then, we find the solution of the fluid equation in the background Kerr metric. Given a Kerr spacetime, the accretion flow is fixed by two parameters, the accretion rate $\dot{M}$ and the radius of the outer edge $r_{\rm max}$. The accretion rate must be sub-Eddington for the consistency of the model. The result is not sensitive to the choice of $r_{\rm max}$ if it is chosen to be sufficiently large. Figure 1(a) shows the present accretion disk model schematically.

It is well known that the Kerr black hole has an ISCO. As a_* is increased from 0 to 1, $r_{\rm ISCO}$, the radius of the ISCO, decreases from $6M$ to M, while $r_{\rm H}$, the horizon radius, decreases from $2M$ to M. In fact, the Kerr spacetime has an ISCO even for $a_*>1$. As a_* is increased from 1, $r_{\rm ISCO}$ continues to decrease to 2/3 at $a_* = 4\sqrt{2}/(3\sqrt{3}) \simeq 1.09$, takes a minimum there and turns to increase. $r_{\rm ISCO}$ is greater than its Schwarzschild value $6M$ for $a_* > 8\sqrt{6}/3 \simeq 6.53$. This is shown in Fig. 1(b).

The efficiency ϵ of emission can be calculated as $1-E_{\rm ISCO}$, where $E_{\rm ISCO}$ is the specific energy of the particle orbiting the ISCO. Figure 2(a) shows the efficiency as a function of the

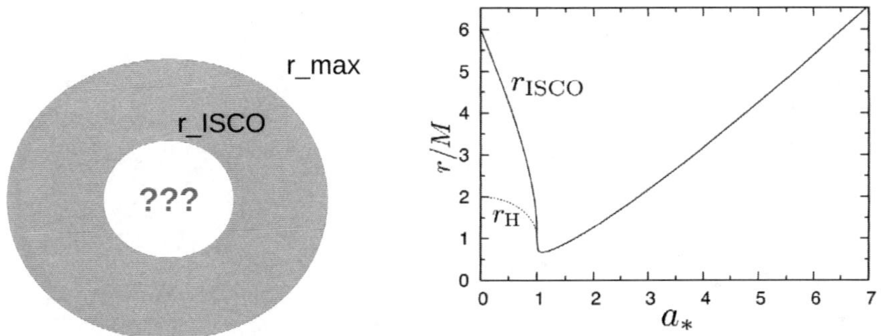

(a) The accretion disk model (b) The radius of the ISCO

Figure 1. (a) The schematic figure of the accretion disk model in the general Kerr spacetime and (b) the radius of the ISCO as a function of the Kerr parameter [3].

spin parameter a_*, where only the short-dashed line and the solid line are relevant to the present accretion disk model. The efficiency is ~ 5.7 % for the Schwarzschild black hole $a_* = 0$ and increases to ~ 42 % for $a_* = 1^-$. Then, it jumps to the maximum value $\sim 157.7\%$. The efficiency is larger than 100% for $1 < a_* \lesssim 1.09$. It then decreases to ~ 42 % for $a_* = 5/3 \simeq 1.67$ and continues to decrease as a_* is increased further. Note that the efficiency becomes ~ 5.7 % again for $a_* \simeq 6.53$. Figure 2(b) shows the temperature profiles of the accretion disks. We can see that there appears a peaky spike near the inner edge for $1 < a_* \lesssim 1.09$ due to the high efficiency of the disk. We should note that for a superspinar with $1.09 \lesssim a_* \lesssim 6.53$, there is a black hole counterpart with very similar temperature structure.

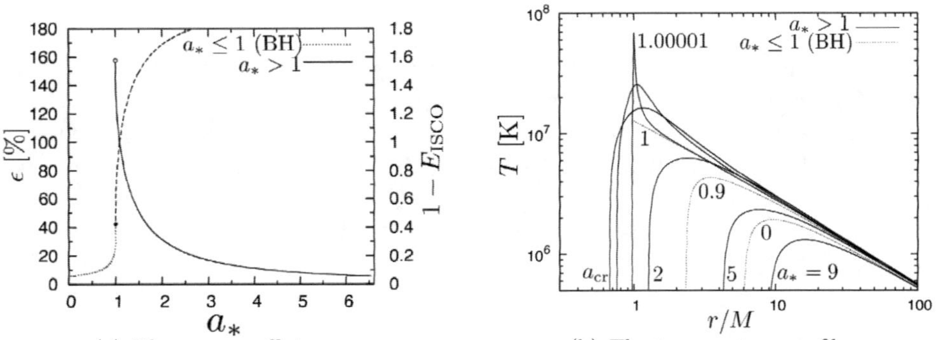

(a) The energy efficiency (b) The temperature profile

Figure 2. (a) The energy efficiency $\epsilon = 1 - E_{\rm ISCO}$ as a function of the Kerr parameter and (b) the temperature profiles of the accretion disks for the different values of the Kerr parameter [3].

3. The X-ray spectrum from the accretion disk

In addition to the dynamics and thermodynamics of the accretion disk, we assume that the dissipative energy due to the viscous torque is fully converted to the locally black-body radiation. We solve the general relativistic radiative transfer, including all kinematical general relativistic effects through the spacetime metric and the motion of the fluid. We neglect the absorption,

Journal of Physics: Conference Series **314** (2011) 012084 doi:10.1088/1742-6596/314/1/012084

emission and reflection by the surrounding gases for simplicity. We assume there is no emission and no absorption by the superspinar at the centre.

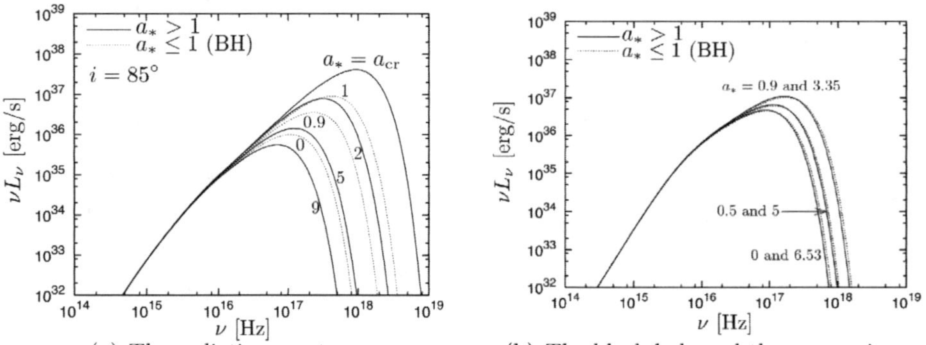

(a) The radiation spectra (b) The black hole and the superspinar

Figure 3. (a) The spectra of the emission from the disk with the different values of the Kerr parameter and (b) the demonstration of the similarity of the spectra between the black hole and the superspinar counterpart [3].

Figure 3(a) shows the predicted energy spectra for the different spin parameter values. For $1 \lesssim a_* \lesssim 1.67$, the spectrum extends to higher energy compared to black holes. For $a_* \gtrsim 6.53$, the photon energy is lower compared to black holes. However, surprisingly, any black hole has its superspinar counterpart which gives a very similar X-ray spectrum. This is demonstrated in Fig. 3(b). In contrast, for a superspinar with $1 \lesssim a_* \lesssim 1.67$ or $a_* \gtrsim 6.53$, there is no black hole counterpart.

4. Summary

In the present settings, it is very challenging to distinguish black holes from superspinars only by the X-ray spectrum observation. In contrast, some of the superspinars can be clearly distinguished from black holes. In other words, we can potentially find the violation of the Kerr bound by the X-ray spectrum observation. The present disk and radiation model might be too simple. It is interesting to study the observational testability of the Kerr bound with other accretion models, such as Radiatively Inefficient Accretion Flow (RIAF). See Takahashi and Harada [3] for further details.

References

[1] Penrose R., 1969, Riv. Nuovo Cim. **1**, 252

[2] Penrose R., 1979, in *General Relativity and Einstein Centenary Survey*, edited by S. W. Hawking and W. Israel (Cambridge University Press, Cambridge, England, 1979)

[3] Takahashi R., Harada T., 2010 Class. Quant. Grav. **27**, 075003

[4] Gimon E. G., Horava P., 2009, Phys. Lett. B**672**, 299

[5] Jacobson T., Sotiriou T. P., 2009, Phys. Rev. Lett. **103**, 141101; **103**, 209903(E)

[6] Page D. N., Thorne K. S., 1974, Astrophys. J. **191**, 499

Journal of Physics: Conference Series **314** (2011) 012085 doi:10.1088/1742-6596/314/1/012085

Toward restrictions on boson stars as black hole mimickers

F. S. Guzmán

Instituto de Física y Matemáticas, Universidad Michoacana de San Nicolás de Hidalgo. Edificio C3, Cd. Universitaria, 58040 Morelia, Michoacán, México.

E-mail: guzman@ifm.umich.mx

Abstract. The status of boson stars as black hole mimickers is presented among other mimickers. We focus on the analysis of the emission spectrum of a simple accretion disk model. We describe the free parameters that allow a boson star to become a black hole mimicker and present an example of a particular astrophysical case.

1. Introduction

One of the most important problems in relativistic astrophysics is that of the nature of black hole candidates (BHCs). It is commonly assumed that black hole solutions are correspond to BHCs. Nevertheless there are various other types of solutions of Einstein's equations sourced by different matter models that provide similar behavior to high energy observations in the vicinity of BHCs. These solutions are known as black hole mimickers. Among those solutions there are wormholes [1, 2], gavastars [2, 3, 4, 5], brane world solutions [6], naked singularities [7] and boson stars [8, 9, 10, 11]. In fact, one important issue is to determine whether or not BHCs have an event horizon [12].

In order to estimate observable differences between black holes and mimickers is focused on the one hand to the search of gravitational wave fingerprints of the BHCs and their perturbative properties. In the first place perturbations to Schwarzschild black holes are very well studied (for a recent review see e.g. [13]). On the other hand, black hole mimickers' perturbations have not been studied as much, however some results are in turn, for instance, the stability of the whole set of wormhole solutions usually thought of as mimickers is not clear, instead the instability of wormhole solutions has been shown for basic wormhole solutions [14, 15, 16, 17], destroying previous hope on the possibility that these are stable as shown in [18] for particular types of perturbations. Concerning gravastars, stability has been explored and has been found that there are stability regions [3, 19, 20]. About boson star solutions the stability of the solutions has actually been exhaustively studied, for instance using perturbative methods [21, 22], catastrophe theory [23] and full non-linear numerical relativity, both in spherical symmetry [22, 24, 25] and full 3D [26], and the stable branches of solutions are well known, and it has also been studied the fundamental quasinormal modes and tails of boson stars [27].

The study of boson stars has been pushed forward up to the binary boson star collision [28] and the non-linear evolution of perturbed boson stars [29], in both cases considering the system as a source of gravitational waves. The study of gravitational wave signatures has also been studied in order to tell between a black hole and a gravastar [4]. Instead, in the case of wormholes e.g. simple solutions supported by a phantom scalar field would not stand the perturbative analysis and there is no hope for a study of a binary system, because as shown in [16] the life time of these solutions is rather short and they should either collapse and form black holes or explode before they could merge.

On the other hand, a yet more common strategy to find observable differences among BHCs is associated to the study of the electromagnetic -observable- emission spectrum of geometrically thin optically thick disk models, since they are based on the study of time-like geodesics on a fixed background space-time and assume very simple conditions on the accreting material [6, 7, 8, 9, 10, 11].

2. Boson stars

Boson stars (BSs) are spherically symmetric stationary solutions to Einstein's field equations minimally coupled to a complex scalar field sourced by the stress energy tensor $T_{\mu\nu} = \frac{1}{2}[\partial_\mu\phi^*\partial_\nu\phi + \partial_\mu\phi\partial_\nu\phi^*] - \frac{1}{2}g_{\mu\nu}[\phi^{*,\alpha}\phi_{,\alpha} + V(|\phi|^2)]$, with $V(|\phi|) = m^2|\phi|^2 + \lambda|\phi|^4$ where m is associated to the mass of the boson and λ to the self-interaction of the boson system and the scalar field obeys the Klein-Gordon equation. The condition to find BSs solutions is that the scalar field has a harmonic dependence on time $\phi(r,t) = \phi_0(r)e^{-i\omega t}$, where r is the radial coordinate, t the coordinate time and ω is a characteristic frequency that determines the central density of the solution [10]. This condition implies that the stress-energy tensor becomes time-independent, which in turn implies that the geometry of the space-time is also time-independent. Using the rescaling $\tilde{\phi}_0 = \sqrt{4\pi}\phi_0$, $\tilde{r} = mr$, $\tilde{t} = \omega t$, $\tilde{\alpha} = \frac{m}{\omega}\alpha$ and $\Lambda = \frac{2\lambda}{8\pi m^2}$, one constructs the boson star solutions like those shown in Fig 1a. for various values of Λ. Each point of each curve represents a BS solution. Those to the left of the maxima are stable whereas those to the right are unstable [30].

3. The accretion disk model

The disk model around a given BHC is defined as follows. It corresponds to a geometrically thin, optically thick, steady accretion disk. The power per unit area generated by such a disk rotating around a central object is given by [31] and used also in the study of disks around gravastars [5] and naked singularities [7] reads $D(r) = \frac{\dot{M}}{4\pi r}\frac{\alpha}{a}\left(-\frac{d\Omega}{dr}\right)\frac{1}{(E-\Omega L)^2}\int_{r_i}^{r}(E-\Omega L)\frac{dL}{dr}dr$, where $\dot{M}$ is the accretion mass rate, E, L, Ω are the energy per unit mass, angular momentum per unit mass and angular velocity of test particles the disk is made of and r_i is the inner edge of the disk. For black holes this radius is assumed to be at the ISCO ($r = 6M$) of the hole. For BSs we choose $r_i = 0$ based on two considerations: i) BSs allow circular orbits in the whole spatial domain and thus there is no kinematical restriction to choose $r_i = 0$ and ii) it has been shown that the luminosity of the disk for BSs considering this inner radius never reaches the value of Eddington luminosity in the whole spatial domain even for high accretion rates and thus there are no radiation pressure effects imposing a restriction on the inner edge location [9]. Now, assuming it is possible to define a local temperature the Steffan-Boltzmann law is used so that $D(r) = \sigma T^4$, where $\sigma = 5.67 \times 10^{-5}\ erg\ s^{-1}\ cm^{-2}\ K^{-4}$ is the Steffan-Boltzmann constant. Now, considering the disk emits as a black body, we use the dependence of T on the radial coordinate and therefore the luminosity $L(\nu)$ of the disk for each frequency ν and the flux $F(\nu)$ can be calculated using the expression for the black body spectral distribution: $L(\nu) = 4\pi d^2 F(\nu) = \frac{16\pi h}{c^2}\cos(\vartheta)\nu^3\int_{r_i}^{r_f}\frac{rdr}{e^{h\nu/kT}-1}$, where d is the distance to the source, r_i and r_f indicate the location of the inner and outer edges of the disk, $h = 6.6256 \times 10^{-27}\ erg\ s$ is the Planck constant, $k = 1.3805 \times 10^{-16}\ erg\ K^{-1}$ is the Boltzmann constant and ϑ is the disk inclination. The algorithm to construct the emission spectrum for such a model of accretion disk around a BS and a BH is as follows: 1) Define the space-time functions a and α by choosing one of the equilibrium configurations in Fig. 1a and calculate M; 2) Define the metric of the equivalent BH through $\alpha_{BH} = \sqrt{1-2M/r}$ and $a_{BH} = 1/\sqrt{1-2M/r}$; 3) Calculate the angular velocity, angular momentum and energy of a test particle for both space-times $\Omega_{BS,BH}$, $L_{BS,BH}$, $E_{BS,BH}$; 4) Use such quantities to calculate the power emitted in both cases $D_{BS}(r)$ and $D_{BH}(r)$; 5) Calculate the temperature of the disk in both cases $T_{BS}(r) = (D_{BS}(r)/\sigma)^{1/4}$ and $T_{BH}(r) = (D_{BH}(r)/\sigma)^{1/4}$; 6) Use such temperature to integrate the luminosity $L_{BS}(\nu)$ and $L_{BH}(\nu)$ for several values of ν.

4. Mimicker configurations

The free parameters of BSs that one can play with in order to produce a spectrum that mimics that due to the presence of a black hole are m and Λ. As an example in Fig. 1b the power spectrum of a black hole is compared with two spectra due to two different BS configurations of the same mass but with different self-interaction Λ, one of which is a mimicker. It can be noticed how the spectrum due to one of the BSs lies on top of that due to the black hole. This is the effect that a black hole mimicker produces. For each value of Λ it is possible to find mimickers by changing m. For instance, in Fig. 1a stars indicate the configurations corresponding to black hole mimickers for various values of Λ.

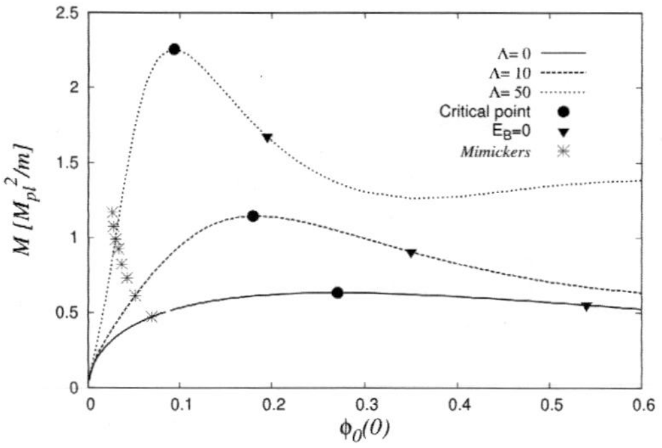

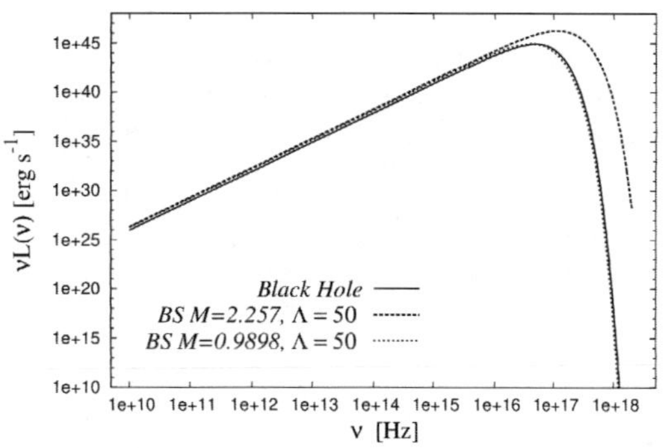

Figure 1. In the left panel it is shown a set of BSs configurations; each point of each curve corresponds to a BS configuration. Those configurations to the left of the maxima are stable, those between the maxima and the The mimicker configurations are indicated with stars for values of $\Lambda = 0, 10, 20, 30, 40, 50, 60, 70$. It is expected that for every value of Λ there is a mimicker configuration that lies near the ones indicated for the values shown here. We show the mass of the inverted triangles are unstable with negative binding energy and collapse into black holes, whereas those to the right of the inverted triangles are unstable configurations that instead of collapsing explode [30]. In the right panel we show an example of spectrum generated by a disk around BS and around a Schwarzschild black hole. The mimicker configuration used corresponds to a BS configuration with $M = 0.9898M_{pl}^2/m$, $\Lambda = 50$ and boson mass $m = 5.25 \times 10^{-22}$GeV, and works as a mimicker of a black hole with mass $M = 10M_{\odot}$ with an accretion rate parameter $2 \times 10^{-12}M_{\odot}/yr$. The spectra due to the black hole and to the mimicker are on top of each other, and for reference the spectrum due to a non-mimicker configuration is also shown.

5. Discussion

A summary of boson stars as black hole mimickers has been presented. It has been found that it is possible to find a boson star configuration that may supplant a black hole based on the emission spectrum of an accretion disk model. Nevertheless there are items that have to be addressed if boson stars are to still be considered black hole mimickers. The most important one is the question of whether or not boson stars are capable of trapping baryons, which eventually would form a mechanical surface that would behave like that of a baryonic compact object as assumed in studies of mimickers [12]. The fact that boson star space-times are regular everywhere indicates that particles (e.g. baryons) following geodesics would truly be accreted and retained on top of the boson star space-time; this would imply that there is nothing like accretion of matter, and instead, that the region of the boson star is a place where particles fall into and are expelled out, because it has been shown [10] that particles following radial trajectories would cross the origin and get out, and when test particles have a non-zero impact parameter there is an infinite potential barrier that forbids test particles to remain near the center of the boson star. However this is not a definitive answer, instead it is necessary to consider the self-gravity of baryons and their contribution to the geometry. With that information at hand it will be possible to determine whether or not boson stars are still potential black hole mimickers and then proceed to other more stringent astrophysics related tests like gravitational lensing and gravitational radiation.

Acknowledgments. This research is partly supported by grants: CIC-UMSNH-4.9 and CONACyT 106466.

References

[1] Tiberiu Harko, Zoltan Kovacs, Francisco S.N. Lobo, Phys. Rev. D 79: 064001, 2009.
[2] Jose P.S. Lemos, Oleg B. Zaslavskii Phys. Rev. D 78: 024040, 2008.
[3] Matt Visser, David L. Wiltshire Class. Quant. Grav. 21 (2004) 1135-1152.
[4] Cecilia B. M. H. Chirenti, Luciano Rezzolla, Class. Quant. Grav. 24: 4191-4206, 2007.
[5] Tiberiu Harko, Zoltan Kovacs, Francisco S.N. Lobo, arXiv: 0905.1355 [gr-qc].
[6] C.S.J. Pun, Z. Kovacs, T. Harko Phys.Rev. D78:084015, 2008.
[7] Z. Kovacs, T. Harko, Phys.Rev.D82:124047,2010.
[8] D. F. Torres, S. Capozziello and G. Lambiase, Phys. Rev. **D 62**, 104012 (2000).
[9] D. F. Torres, Nucl. Phys. **B 26**, 377 (2002).
[10] F. S. Guzmán, Phys. Rev. D 73:021501(R),2006.
[11] F. S. Guzmán and J. M. Rueda-Becerril, Phys. Rev. D 80:084023,2009.
[12] Ye-Fei Yuan, Ramesh Narayan, Martin J. Rees, Astrophys.J.606:1112-1124,2004. Ronald A. Remillard, Dacheng Lin, Randall L.Cooper, Ramesh Narayan, Astrophys.J. 646 (2006) 407-419. Jeffrey E. McClintock, Ramesh Narayan, George B. Rybicki, Astrophys.J. 615 (2004) 402-415.
[13] A. Nagar and L. Rezzolla, Class. Quantum Grav. **22** (2005) R167-R192.
[14] Hisa-aki Shinkai and Sean A. Hayward, Phys. Rev. D66, 044005:2002.
[15] J. A. González, F. S. Guzmán, O. Sarbach, Class. Quant. Grav. 26:015010, 2009.
[16] J. A. González, F. S. Guzmán, O. Sarbach, Class. Quant. Grav. 26:015011, 2009.
[17] J. A. González, F. S. Guzmán, O. Sarbach, Phys.Rev.D80:024023,2009.
[18] C. Armendariz-Picon, Phys.Rev. D65 (2002) 104010.
[19] Benedict M.N. Carter Class.Quant.Grav. 22 (2005) 4551-4562.
[20] Cecilia B. M. H. Chirenti, Luciano Rezzolla, Phys.Rev.D78:084011,2008.
[21] M. Gleiser, Phys. Rev. **D 38**, 2376 (1988).
[22] S. H. Hawley and M. W. Choptuik, Phys. Rev. **D 62**, 104024 (2000).
[23] F. V. Kusmartsev, et al., Phys. Rev. D 43, 3895 - 3901 (1991).
[24] E. Seidel and W-M. Suen, Phys. Rev. **D 42**, 384 (1990).
[25] J. Balakrishna, E. Seidel and W-M. Suen, Phys. Rev. **D 58**, 104004 (1998).
[26] F. S. Guzmán, Phys. Rev. **D 70**, 044033 (2004).
[27] F. D. Lora-Clavijo, A. Cruz-Osorio, F S Guzmán, Phys.Rev.D82:023005,2010.
[28] C. Palenzuela, I. Olabarrieta, L. Lehner, S. Liebling, Phys. Rev. **D 75**, 064005 (2007). C. Palenzuela, L. Lehner, S. Liebling, Phys. Rev. **D 77**, 044036 (2008).
[29] J. Balakrishna, R. Bondarescu, G. Daues, F. S. Guzmán, E. Seidel. Class. Quantum Grav. **23** (2006) 2631-2652.
[30] F. S. Guzmán, Rev. Mex. Fis. **55**, 321-326 (2009).
[31] Don N. Page and Kip S. Thorne, Astrophys.J. 191 (1974) 499-506.

Journal of Physics: Conference Series **314** (2011) 012086 doi:10.1088/1742-6596/314/1/012086

Quasinormal modes for the charged Vaidya metric

Cecilia Chirenti [1] **and Alberto Saa** [2]

[1] Centro de Matemática, Computação e Cognição, UFABC, 09210-170 Santo André, SP, Brazil
[2] Departamento de Matemática Aplicada, UNICAMP, 13083-859 Campinas, SP, Brazil

Abstract. The scalar wave equation is considered in the background of a charged Vaidya metric in double null coordinates (u, v) describing a non-stationary charged black hole with varying mass $m(v)$ and charge $q(v)$. The resulting time-dependent quasinormal modes are presented and analyzed. We show, in particular, that it is possible to identify some signatures in the quasinormal frequencies from the creation of a naked singularity.

1. Introduction

Even though macroscopically charged black holes are not expected to be produced in realistic astrophysical scenarios, we can obtain fascinating information on the structure and behavior of Einstein equations and general relativity with the study of these objects. Here, we used the charged Vaidya metric to model time-dependent black holes with an electric charge, and study the behavior of the quasinormal modes in a non-stationary background.

The charged Vaidya metric[1, 2, 3] is an asymptotic flat and spherically symmetric solution of Einstein–Maxwell equations in the presence of a radial flow of a null-type fluid. For our purposes, it is more convenient to consider the charged Vaidya metric in double-null coordinates, as proposed by Waugh and Lake in [4]. In double-null coordinates (u, v, θ, ϕ), the charged Vaidya metric reads

$$ds^2 = -2f(u,v)dudv + r^2(u,v)d\Omega^2\,, \tag{1}$$

where $d\Omega^2$ stands for the metric on the unity sphere (θ, ϕ). The Einstein-Maxwell equations imply that the functions $f(u, v)$ and $r(u, v)$ and $h(u, v)$ obey

$$f(u,v) = 2B(v)r_{,u}(u,v) \tag{2}$$

and

$$r_{,v}(u,v) = -B(v)\left(1 - \frac{2m(v)}{r} + \frac{q^2(v)}{r^2}\right), \tag{3}$$

where $B(v)$, $m(v)$ and $q(v)$ are arbitrary integration functions. With the choice of $B(v)$ = constant, one can interpret $m(v)$ and $q(v)$, respectively, as the total mass and the total charge of the solution. Both functions are only required to be monotonic, but a further condition must be imposed due to the weak energy condition in the case where both $\dot{m}$ and $\dot{q}$ are non null [5].

The solutions of Eq. (3) corresponds to the congruence of ingoing and outgoing null geodesics of the metric (1). We adopt here the semi-analytical method proposed in [6, 7] to construct such geodesics. In Fig. 1 we present as an example the null geodesics obtained for a case with $\dot{m} = \dot{q} = 0$, that is, a Reissner-Nordström (RN) black hole, in this parametrization. Fig. 2 presents the same solutions for a time dependent background. For an interesting analysis of the time dependent horizons in the Vaidya metric, see [8].

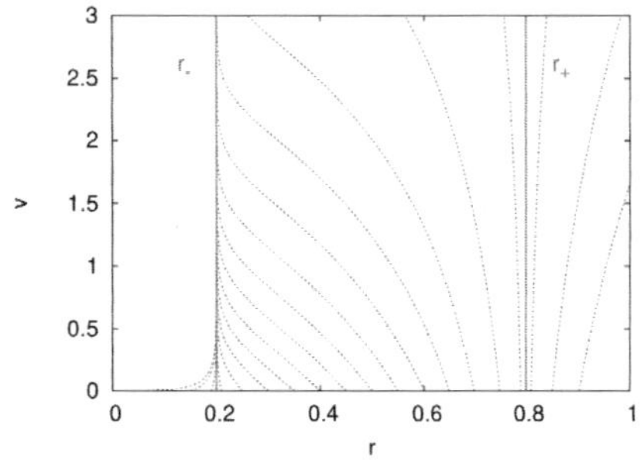

Figure 1. Example of null-geodesics obtained from eq. (3) for a RN black hole, with $B = -1/2$, $m = 0.5$, $q = 0.4$, $n = 4$ and $\Lambda = 0$ and taking $r(u, v = 0) = -u/2$. Here r_+ and r_- are defined as usual as $r_\pm = m \pm \sqrt{m^2 - q^2}$.

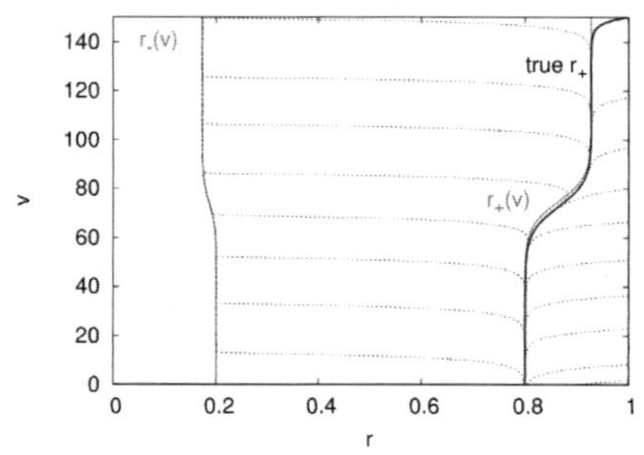

Figure 2. Same as in Figure 1, but this time for a time dependent charged black hole, with the choices $B = -1/2$, $m(v) = 1.0$ and $q(v) = \frac{1}{2}(q_2 - q_1)[1 + \tanh \rho(v - v_m)] + q_1$, with $q_1 = 0.5$, $q_2 = 0.7$, $\rho = 0.1$ and $v_m = 90$.

2. Scalar perturbations

The evolution of a massless scalar field in this background is given by the Klein-Gordon equation:

$$\frac{1}{\sqrt{-g}} \left(\sqrt{-g} g^{\mu\nu} \Psi_{,\nu}\right)_{,\mu} = 0 . \tag{4}$$

Decomposing the field $\Psi(u, v, \theta, \phi)$ in spherical harmonics

$$\Psi(u, v, \theta, \phi) = \sum_{\ell,m} r^{-1} \varphi(u, v) Y_{\ell m}(\theta, \phi) , \tag{5}$$

and using eqs. (2) and (3), we find

$$\varphi_{,uv} + V(u, v)\varphi = 0 , \tag{6}$$

with the potential $V(u, v)$ given by

$$V(u, v) = \frac{1}{2} f(u, v) \left(\frac{\ell(\ell + 1)}{r^2} + \frac{2m(v)}{r^3} - \frac{2q^2(v)}{r^4} \right) . \tag{7}$$

For more details on the numerical algorithm, initial conditions and wave extraction process for Eq. (6) we refer the reader to refs. [9, 10].

3. Numerical results

We first tested our code for static Schwarzschild and Reissner-Nordström configurations, with very good agreement with values found in the literature [11, 12].

In Fig. (3) we present the time variation in the $\omega_r \times \omega_i$ plane of the real and imaginary parts of the QNM of the scalar perturbation with $\ell = 2$ for a RN black hole with constant mass and a time dependent charge function $q(v)$. The frequencies change in a similar way with the time variation of the mass or the charge, but are typically more sensitive to the charge. The specific case depicted in Fig. (3) corresponds to a charge variation where q/m goes from 0.7 to 0.8, where we see how the variation goes from one stationary state to another through a non-adiabatic trajectory. The relevant timescale here for characterizing mass or charge changes which are adiabatic or not is essentially the same one of the $q = 0$ case studied in [10]: QNM modes will relax and follow an adiabatic trajectory in the $\omega_r \times \omega_i$ plane whenever $r''_+(v) < \omega_i(v)$.

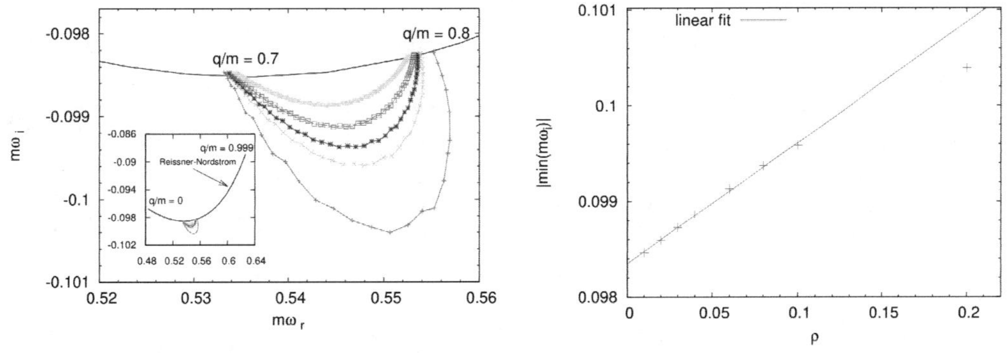

Figure 3. Left plot: quasinormal modes in the $\omega_r \times \omega_i$ plane, showing the transition between stationary states for the $\ell = 2$ scalar perturbation of a black hole with constant mass $m = 0.5$ and $q(v) = \frac{1}{2}(q_2 - q_1)[1 + \tanh \rho(v - v_m)] + q1$, with $q_1 = 0.7$, $q_2 = 0.8$, $v_m = 75$ and different values of ρ. Right plot: the minima of ω_i obtained in the transition shown in the left plot for different values of ρ, to quantify the non-adiabatic behavior.

We would like also to use our formalism and numerical setup to probe the formation of a naked singularity in the spacetime. In order to do so and avoid numerical problems as $q \to m$, we choose $m = m_0$ and $q(v) = (q_2 - q_1)[1 + \tanh \rho(v - v_m)] + q_1$, taking $q_2 = m_0$. This implies that the RN black hole will be (rigorously) extremal only for $v \to \infty$, but allows us to probe configurations very near this end state at finite times. By varying the parameter ρ, we can control the speed with which the black approaches the extremal state. The results are shown in Fig. (4). In the left plot we show how the results depend on the manner in which $q(v) \to m$. We remind here that an extremal RN black hole is not a naked singularity, and its QNMs are well defined as the appropriated limit of the non-extremal case [13, 14, 15]. The results in the right plot of Fig. 4 show what happens when we try to artificially create a naked singularity. We chose configurations with constant charge q and time dependent mass functions with final mass m_2 approaching and smaller than q. The results for ω_r are shown in the right plot of Fig. 4 and are truncated when $m(v) = q$ since the simulation crashes at this point, as expected.

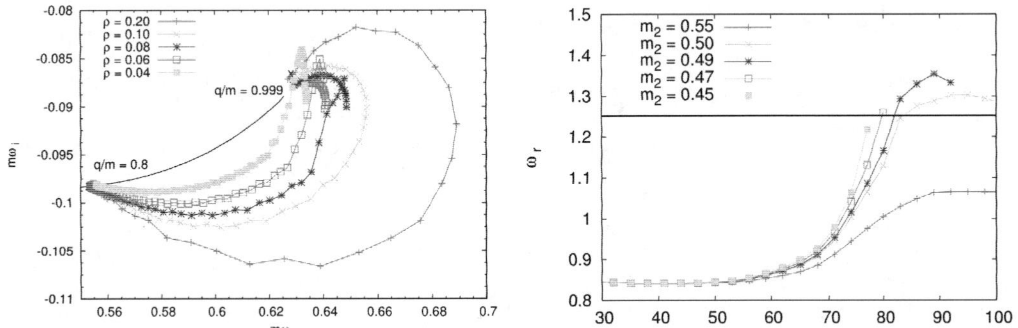

Figure 4. Left plot: same as in the left plot shown in Fig. (3), but this time with $q_2 = 0.5$ so that the RN black hole becomes extreme as $v \to \infty$. Right plot: $\omega_r(v)$ of a scalar perturbation with $\ell = 2$ obtained for "artificial" background configurations with m_2 approaching and smaller than q; $\omega_i(v)$ presents a similar behavior.

4. Final remarks

Our results confirm, for the charged case, the same non-stationary behavior corresponding to some sort of inertia for QNM frequencies identified in [9]. In particular, for situations with $r''_+(v) > \omega_i(v)$, the QNM frequencies will not follow in the $\omega_r \times \omega_i$ plane a trajectory corresponding to the instantaneous frequency associated to a RN black hole of a given q/m ratio, and a certain inertial behavior for the QNM can be identified. Moreover, we see that the faster the change in $m(v)$ or $q(v)$ is, the bigger the deviation from the stationary regime will be (see Fig. (3)). These points are now under investigation, and a more complete analysis will appear soon [16].

As for the $q = 0$ case considered in [9], an interesting extension of this work would be analysis of the highly damped QNM, the so-called overtones. Since for such overtones the ratio $|\omega_i/\omega_r|$ is always larger than for the fundamental ($n = 0$) QNM considered here, including, for sufficient large n, cases for which $|\omega_i/\omega_r| > 1$, it would be interesting to check if the non-stationary inertial behavior could somehow be attenuated for $n > 0$. We notice that the numerical analysis presented here cannot be extended directly to the $n > 0$ case since one cannot identify the $n > 0$ frequencies with sufficient accuracy. We believe this could be attained, in principle, by means of the WKB approximation.

Acknowledgments

This work was supported by CNPq, FAPESP and the Max Planck Society. It is a pleasure to thank Rodrigo Panosso Macedo and Luciano Rezzolla for useful discussions.

References

[1] Krori K D and Barua J 1974 Exterior solution for a charged radiating sphere in general relativity (stars) *J. Phys. A: Math. Gen.* **7** 2125
[2] Chatterjee S, Bhui B and Banerjee A 1990 Higher-dimensional Vaidya metric with an electromagnetic field *J. Math. Phys.* **31** 2208
[3] Fayos F, Martin-Prats M M and Senovilla J M M 1995 On the extension of Vaidya and Vaidya-Reissner-Nordström spacetimes *Class. Quantum Grav.* **12** 2565
[4] Waugh B and Lake K 1986 Double-null coordinates for the Vaidya metric *Phys. Rev. D* **34** 2978
[5] Ori A 1991 Charged null fluid and the weak energy condition *Class. Quant. Grav.* **8** 1559
[6] Girotto F and Saa A 2004 Semi-analytical approach for the Vaidya metric in double-null coordinates *Phys. Rev. D* **70** 084014
[7] Saa A 2007 N-dimensional Vaidya metric with cosmological constant in double-null coordinates *Phys. Rev. D* **75** 124019
[8] Booth I and Martin J 2010 On the proximity of black hole horizons: lessons from Vaidya *Preprint* arXiv:1007.1642v3 [gr-qc]
[9] Abdalla E, Chirenti C B M H and Saa A 2006 Quasinormal modes for the Vaidya metric *Phys. Rev. D* **74** 084029
[10] Abdalla E, Chirenti C B M H and Saa A 2007 Quasinormal mode characterization of evaporating mini black holes *JHEP* **10** 086
[11] Konoplya R A 2003 Gravitational quasinormal radiation of higher-dimensional black holes *Phys. Rev. D* **68** 124017
[12] Konoplya R A 2002 Decay of charged scalar field around a black hole: quasinormal modes of R-N, R-N-AdS and dilaton black holes *Phys. Rev. D* **66** 084007
[13] Onozawa H, Mishima T, Okamura T and Ishihara H 1996 Quasinormal modes of maximally charged black holes *Phys. Rev. D* **53** 7033
[14] Berti E 2004 "Black hole quasinormal modes: hints of quantum gravity?" *Proc. of Int. Workshop on Dynamics and Thermodynamics of Black Holes and Naked Singularities* (Milan) p 145 (*Preprint* gr-qc/0411025v1)
[15] Daghigh R G and Green M 2008 The Highly Damped Quasinormal Modes of Extremal Reissner-Nordström and Reissner-Nordström-de Sitter Black Holes *Class.Quant.Grav.* **25** 055001
[16] Chirenti C B M H and Saa A Nonstationary regime for quasinormal modes of the charged Vaidya metric, *to appear*

Journal of Physics: Conference Series **314** (2011) 012087 doi:10.1088/1742-6596/314/1/012087

Chaotic geodesic motion around a black hole and disc

Petra Suková

Institute of Theoretical Physics, Faculty of Mathematics and Physics, Charles University, V Holesovickach 2, 180 00 Praha 8, Czech Republic

E-mail: lvicekps@seznam.cz

Abstract. We study the chaotic features of time-like free motion around a Schwarzschild black hole, induced by the presence of a static, axially and reflectionally symmetric thin disc or ring. We describe the field by an exact (Weyl-type) solution of Einstein's equations and use Poincaré sections and time-series analysis in order to recognise overall tendencies in the evolution of phase portrait as well as changes that can be traced down to single orbits and their segments.

1. Introduction

Black holes assumed in galactic nuclei certainly dominate their surroundings gravitationally. However, even in non-active nuclei like that in our Galaxy, there is enough mass (in stars and gas) to influence *its own* (and each others') motion in a non-negligible way. One of tiny but robust effects is that the geodesic motion, which is integrable in the field of an *isolated* Schwarzschild or Kerr black hole, becomes chaotic when some additional matter is present. Actually, even if the additional matter does not break the space-time symmetries of the centre, the geodesic motion in the resulting field is no longer completely integrable, so it may become chaotic. This is important for the long-term evolution of the system and may have direct observational consequences, namely it can be recognised in electromagnetic and gravitational radiation emitted by the orbiting particles.

Here we consider exact static and axially symmetric space-times generated by a Schwarzschild black hole surrounded by a thin, reflectionally symmetric disc or ring, in order to show how the dynamics of free timelike motion changes from regular to chaotic (and possibly back) when parameters of the system are changed. The information revealed on Poincaré sections is supplemented by and compared with that provided by time series of phase variables. In particular, we illustrate whether and how the overall tendencies in the evolution of geodesic dynamics correspond to changes taking place on the level of individual orbits.

2. Weyl space-times

Static and axially symmetric vacuum regions of space-time can be described, in Weyl coordinates (t, ρ, z, ϕ), by the Weyl metric

$$ds^2 = -e^{2\nu(\rho,z)}dt^2 + \rho^2 e^{-2\nu(\rho,z)}d\phi^2 + e^{2\lambda(\rho,z)-2\nu(\rho,z)}(d\rho^2 + dz^2)\,. \qquad (1)$$

The Einstein equations yield the Laplace equation for ν, hence this function (representing gravitational potential) superposes linearly. The other unknown function λ can be obtained by line integration of

$$\lambda = \int_{\text{axis}}^{\rho,z} \rho \left[(\nu_{,\rho}^2 - \nu_{,z}^2) \mathrm{d}\rho + 2\nu_{,\rho}\nu_{,z} \mathrm{d}z \right] . \tag{2}$$

We have specifically considered the superposition of an (initially) Schwarzschild black hole with three different types of additional source — the inverted Morgan-Morgan counter-rotating discs [1], the discs with power-law density profile [2] and the Bach-Weyl rings [3]. All these sources have finite mass and the disc families were shown to contain physically realistic cases within non-negligible parameter ranges. The only serious limitation of the Weyl solutions is the lack of rotation. Indeed, the systems we would like to simulate (astrophysical black holes with accretion discs or rings) are likely to rotate rapidly rather than being static. The *exact* inclusion of rotation would also be interesting because it has been indicated on approximate solutions in the literature that dragging effects can alter (probably attenuate rather than amplify) the chaotic features significantly. Unfortunately, the superposition of sources is much more difficult in a non-static (stationary) case and no solution of such a type exists that could be submitted to a similar extensive numerical study.

The Weyl space-times are stationary (even static) and axially symmetric, so the energy at infinity per unit mass $E = -g_{tt}u^t$ and the angular momentum at infinity per unit mass $L = g_{\phi\phi}u^\phi$ conserve along the geodesics. In addition, one has normalization of four-velocity, but there is no other isolating integral in general, so the (four-dimensional) motion is not completely integrable and may become chaotic.

The phenomenon of chaos has been detected and studied in many systems, notably in physics, mathematics, biology and economics. It is of special interest in systems of general relativity where the configuration space is itself described by a non-linear theory. Immediately a question thus arises whether and how general relativity contributes to chaos and whether this contribution stems mainly from special relativistic dynamics or rather from dragging effects and spatial curvature. These questions have also practical aspects, because the general relativistic description of a particular system of sources is much more difficult than in Newton's theory and it is often necessary to resort to approximations like pseudo-Newtonian potentials or weak-field limit. It has already been shown in the literature that geodesic dynamics clearly distinguishes between these approximations and full general relativistic system, which indicates that the approximations typically fail to simulate the long-term behaviour correctly.

3. Results

There are many methods how to diagnose and classify chaos. However, in general relativity the usage of some of them is not easy or does not seem to have good sense, because they are usually applied in a particular coordinate system while in relativity coordinates (time, in particular) do not have an invariant meaning. Such queries mainly occur when trying to quantify "sensitive dependence on initial conditions", namely to compare how fast nearby orbits diverge in various directions. In particular, this requires to compare certain tangent vectors (namely velocities of the neighbouring particles) at different points of the manifold which is not a unique concept as one can use different ways of how to transport the quantities to the same point. It is only to be clarified how to compute, in general relativity, correctly and efficiently the various exponents proposed as a measure of the orbital divergence. It has been proved [4], fortunately, that the sign of the mostly used *Lyapunov* exponents is independent of coordinates, although their value is otherwise not invariant. Recently several new suggestions of how to practically compute the Lyapunov exponents have appeared [5]. We plan to try them in the future.

On the other hand, some of the methods can be adopted to general relativity quite straightforwardly. In particular, the Poincaré surfaces of section are very efficient in revealing the properties of phase space for systems where one is able to find the whole trajectories. Quite naturally, we began the study of geodesics in the fields of black holes surrounded by

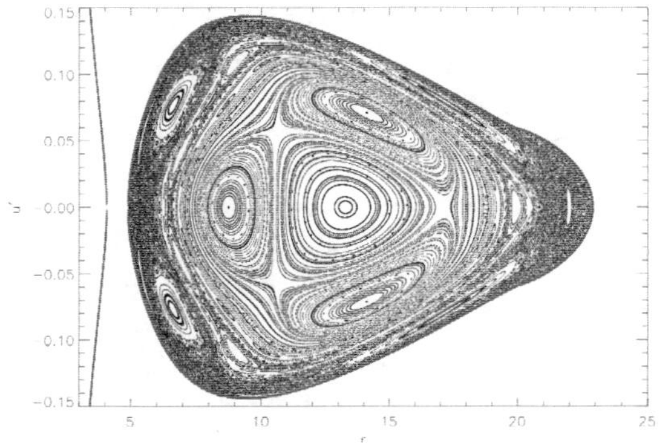

Figure 1. Poincare's surface of section (equatorial plane) for geodetic motion ($E = 0.954$, $L = 3.75M$) around BH with first iMM disc ($r = 20M$, $\mathcal{M} = 0.5M$).

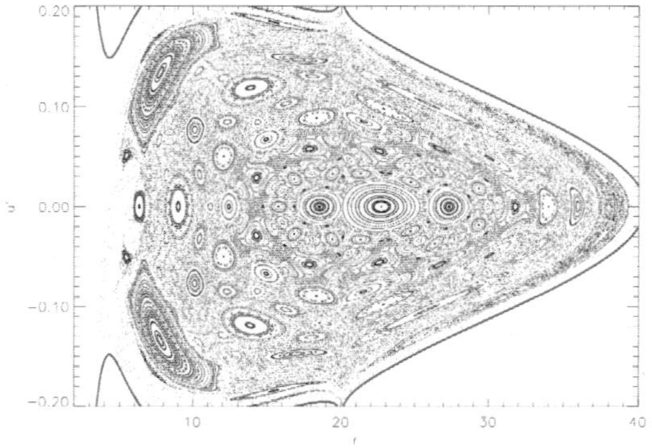

Figure 2. Poincare's surface of section (equatorial plane) for geodetic motion ($E = 0.977$, $L = 3.75M$) around BH with Bach-Weyl ring ($r = 20M$, $\mathcal{M} = 0.08M$).

discs or rings by plotting the Poincaré sections [6]; two examples for different external sources are given in figure 1 and figure 2. However, it is problematic to compare the Poincaré-section method directly with astrophysical observational data, because the latter mostly do not contain information about all three directions. The data are usually represented by time series of dynamical quantities (e.g. projection of position onto the celestial sphere). In order to decipher them, one tries to recognise their components showing different degrees of regularity, in particular, to decompose the "signal" that is probable to originate from a deterministic system from what appears to be just noise. There exist various routines that can provide the degree of regularity/stochasticity of the time series. We specifically tested the technique proposed in [7] on the data computed for our hole-disc system.

The method lies in analysis of the trajectory determined by a given time series $a(t)$ in a 3D space whose axes are $a(t)$, $a(t-\tau)$ and $a(t-2\tau)$, where $a(t-\tau)$ and $a(t-2\tau)$ are "clones" of the original series once and twice delayed by a given time shift τ. The space thus obtained is divided into m x m x m boxes and the passages of the trajectory through the boxes is recorded in terms of a normalised directional vector pointing from the point where the trajectory enters a particular box to the point where it leaves the box. The vectors provided by passages through each particular (j-th) box are then summed into a "total j-th box vector" V_j and this vector's norm is computed and divided by the number of passages through the j-th box n_j (it is assumed that the time series is long enough to yield many transits through the box). In autonomous deterministic systems the dynamics is given by equations of motion which depend on position in phase space, so when the boxes are small enough, the tangent vectors recorded for a given box point in similar directions, so their normalised sum $|V_j/n_j|$ is close to one. On the contrary, for random time series, the norm of this average tangent vector decreases with n_j as the average displacement per step for a random walk in a 3D space, i.e., for (large) n_j, as

$$\bar{R}^3_{n_j} = \frac{4}{(6\pi)^{1/2}} n_j^{-\frac{1}{2}}. \tag{3}$$

For a suitably chosen τ, a deterministic signal yields $|V_j/n_j|$ close to one for arbitrary n_j, whereas for a random signal the value falls off as $n_j^{-1/2}$.

Finally, one can compute a weighted average of $|V_j/n_j|$ over all traversed boxes for a given τ,

$$\bar{\Lambda}(\tau) = \Big\langle \frac{(V_j/n_j)^2 - (R^3_{n_j})^2}{1 - (R^3_{n_j})^2} \Big\rangle. \tag{4}$$

Journal of Physics: Conference Series **314** (2011) 012087 doi:10.1088/1742-6596/314/1/012087

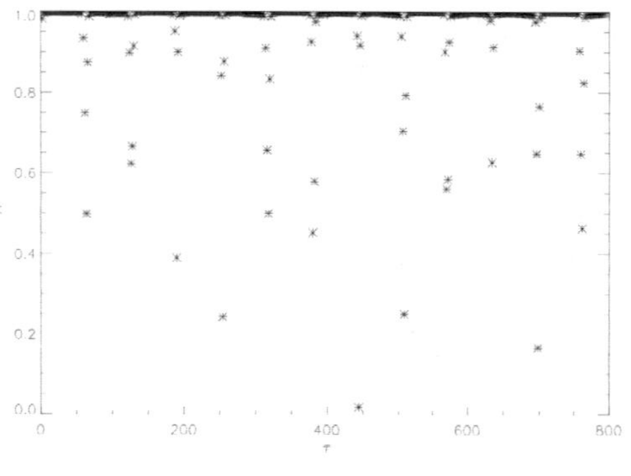

Figure 3. Weighted average of directional vectors.

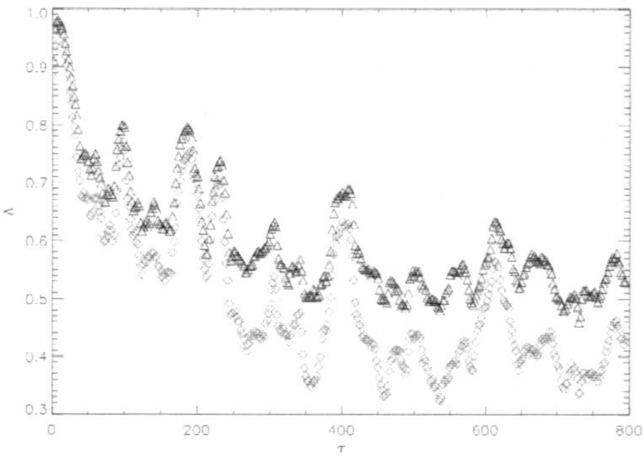

Figure 4. Weighted average of directional vectors ($m = 20$ – red, $m = 32$ – black).

The behaviour of this quantity in dependence on τ is again very different for regular and random trajectories, and we were curious to check what it does for chaotic ones. For regular trajectories the value proved to remain close to one for increasing τ except for certain specific values related to the period of motion. For chaotic trajectories the value turns out to decrease with τ, with some peaks occurring on the curve. The course of $\bar{\Lambda}$ for regular trajectory is plotted in figure 3 while for the chaotic one it is given in figure 4. Both these trajectories have the same global parameters (first inverted Morgan-Morgan disc with $r = 20M$, $\mathcal{M} = 0.5M$ and constants of the motion $E = 0.953$, $L = 3.75M$) but different initial conditions.

The value of $\bar{\Lambda}(t)$ depends on the sampling frequency of the time series (on how dense the data points are) and also on the box size, but the latter dependence (on m) is weak within the range where the elementary boxes are small enough to capture the folds of the trajectory and at the same time large enough to catch at least several points of the trajectory. This is shown in figure 4, where the red data points are computed for $m = 20$ and the black ones are computed for $m = 32$ and their behaviour is very similar.

4. Conclusions

We have shown on Poincaré surfaces of section that in the system of a static black hole surrounded by a thin static and axially symmetric dics or ring the geodesic flow could be chaotic. As usual in the weakly perturbed systems, the chaotic layers spread over the phase space from the unstable periodic points and separatrices between regular islands. We checked the degree of individual-orbit chaoticity by analysing the corresponding time series — we computed their power spectra and also applied the method of computing the weighted average of directional transit vectors suggested by [7].

Acknowledgments

The work was supported by the grants GAUK-86508, GACR-205/09/H033 and SVV 261301.

References

[1] Semerák O and Záček M 2000 *Class. Quantum Gravit.* B **17** 1613
[2] Semerák O 2004 *Class. Quantum Grav.* **21** 2203
[3] Bach R and Weyl H 1922 *Math. Zeit.* **13** 134
[4] Motter A E 2003 *Phys. Rev. Lett.* **91** 231101-1
[5] Stachowiak T and Szydlowski M 2010 *Preprint* arXiv:1008.3368v1
[6] Semerák O and Suková P 2010 *Mon. Not. R. Astron. Soc.* **404** 545
[7] Kaplan D T and Glass L 1992 *Phys. Rev. Lett.* **68** 427

New model of relativistic slowly rotating neutron stars with surface layer *crust*: application to giant *glitches* of Vela Pulsar

L. M. González-Romero and J. L. Blázquez-Salcedo

Depto. Física Teórica II, Facultad de Ciencias Físicas, Universidad Complutense de Madrid, 28040-Madrid, Spain.

E-mail: mgromero@fis.ucm.es, joseluis.blazquez@fis.ucm.es

Abstract.
Introducing a surface layer of matter on the edge of a neutron star in slow rigid rotation, we analyze, from an intrinsic point of view, the junction conditions that must be satisfied between the interior and exterior solutions of the Einstein equations. In our model the core-*crust* transition pressure arise as an essential parameter in the description of a configuration. As an application of this formalism, we describe giant *glitches* of the Vela pulsar as a result of variations in the transition pressure, finding that these small changes are compatible with the expected temperature variations of the inner crust during *glitch* time

1. Introduction

Many pulsars show sudden spin jumps, *glitches*, superimposed to the gradual spin down due to the continued loss of angular momentum suffered by the star. The study of the properties of the star during the *glitch* time is essential to understand the structure of the neutron star that models the pulsar [1, 2]. In Vela pulsar, giant *glitches* with relative period variation of the order of 10^{-6} have been observed. Many mechanisms have been proposed to explain *glitches*.

The models that deal with *glitches* are associated with the layer structure of neutron stars. Two regions of the star can be differentiated: the core and the *crust*. It is thought that the *crust* region has a solid crystalline structure similar to a metal [3]. Some theories propose that the *glitch* is triggered by the rupture of the *crust* as a consequence of the tensions on the *crust* that try to adequate the ellipticity of the *crust* to the changing angular velocity of the star.

Because of the large density gradient close to the transition between the core and the *crust*, most of the crustal matter resides in the shells in the inner part of the *crust* [4]. Furthermore, it is noteworthy that the dynamical properties of the neutron star depend strongly on the transition pressure between the star's core and the *crust* as have been pointed up by [5] and [6]. Because most of the matter of the *crust* is found in the shells near the transition region between the core of the star and the inner *crust*, this is the region where the properties of the *crust* are relevant [4]. Hence, we will treat the *crust* as a surface layer that envelops the star's core. In the next section we will describe how our model of slowly rotating neutron star with a surface layer *crust* is constructed.

2. Construction of models of slowly rotating neutron stars with surface *crust*.

We assume that the neutron star is in permanent rigid rotation. The star rotates with constant angular velocity Ω around an axis, so the resulting space-time is axisymmetric and stationary. We assume that the rotation of the star is sufficiently slow and, hence, the Hartle-Thorne perturbative solution can be used [7, 8]. To second order in the angular velocity of the star Ω, appropriate coordinates can be chosen so that the metric can be written as follows $ds^2 = -e^{2\psi(r)}[1+2h(r,\theta)]dt^2 + e^{2\lambda(r)}[1+2m(r,\theta)]dr^2 + r^2[1+2k(r,\theta)]\left[d\theta^2 - \sin^2\theta(d\phi - \omega(r)dt)^2\right]$.

We have to solve the Einstein equations in the interior of the star, where the matter is described by a perfect fluid tensor with equation of state $\rho = \rho(p)$ and in the outer region, where space is empty. Also, we have to impose appropriate boundary conditions in the surface of the star. It is very desirable to introduce a new radial coordinate R adapted to the surface of the star [7, 8].

We assume that the core of the star is surrounded by a thin surface layer (simplified *crust* model). Then we find that in the surface of the star $R = a$, the stress-energy tensor is non null and it can be written as $T_c^{\mu\nu}(R) = \rho_c(\theta)u_c^\mu u_c^\nu(a)$, i.e. the surface layer can be described as a perfect fluid with its own angular velocity Ω_c. The surface energy density can be written up to second order as $\rho_c(\theta) = \varepsilon + \delta\varepsilon(\theta)$. We must impose the usual junction conditions on the interior and exterior solutions, but taking into account the introduction of a surface layer of matter on the border of the star. We use the intrinsic formulation of these conditions [9] [10]: the first fundamental form must be continuous across the surface of the star, and the second fundamental form presents a discontinuity given by the stress-energy tensor in the surface of the star. In order to use these junction conditions in the slow rotation approximation, we expand them in terms of Ω, finding the following results for each one of the metric functions:

Order zero: we obtain important conditions for the mass function and for the pressure of transition between core and *crust*

$$M_{ext} = M_{int} + 4\pi a^2\sqrt{1 - \frac{2M_{int}}{a}}\varepsilon - 8\pi^2 a^2\varepsilon^2 \quad (1)$$

$$\frac{M_{ext}}{a^2\sqrt{1-2M_{ext}/a}} - \frac{M_{int} + 4\pi a^3 p_{int}}{a^2\sqrt{1-2M_{int}/a}} = 4\pi\varepsilon \quad (2)$$

The equation (1) gives the total mass of the star to zero order (M_{ext}) as the addition of three terms: the core mass ($M_{int} \equiv 4\pi\int_0^a \rho R^2 dR$), the *crust* mass, and a negative bounding energy term. We interpret (2) as an equation giving the surface density of energy ε in terms of the total mass of the star, the interior mass, and the core-*crust* transition pressure p_{int}. This condition is used to obtain the radius of the star.

First order: we obtain continuity of the rate of rotation of the inertial frames $\omega(a)_{ext} = \omega(a)_{int} \equiv \omega(a)$, and an expression for the total angular momentum:

$$J = a^6 e^{\lambda(a)_{ext}}(e^{-\lambda(a)_{int}}\left[\partial_R\overline{\omega}(a)\right]_{int} + 16\pi\varepsilon\overline{\omega_c})/6 \quad (3)$$

where $\overline{\omega_c} = \Omega_c - \omega(a)$. It can be seen [7, 8] that the first term of expression (3) has the same sign as Ω (in our case always positive). However, the second term depends on the sign of Ω_c. Hence, contra-rotating configuration are possible (the core and the *crust* contra-rotate), and even, we could find certain critical configuration in which the total angular momentum is null.

Second order: we obtain $\Delta[r^*(a,\theta)] = 0$ where r^* is defined as in [7, 8], so both the mean radius and the eccentricity are continuous. Also, conditions for the second order perturbation of the mass function and surface density ($\delta\varepsilon$) are obtained.

Finally the last of the second order conditions fixes the value of the angular velocity of the *crust* $\Omega_c = \omega(a) \pm (\Omega - \omega(a))$. We obtain two possible configurations for the same star core and transition pressure: one with the *crust* co-rotating with the interior fluid and with identical velocity, and a special configuration with contra-rotating *crust*.

Journal of Physics: Conference Series **314** (2011) 012088 doi:10.1088/1742-6596/314/1/012088

3. Some properties of the model using realistic equations of state for the core

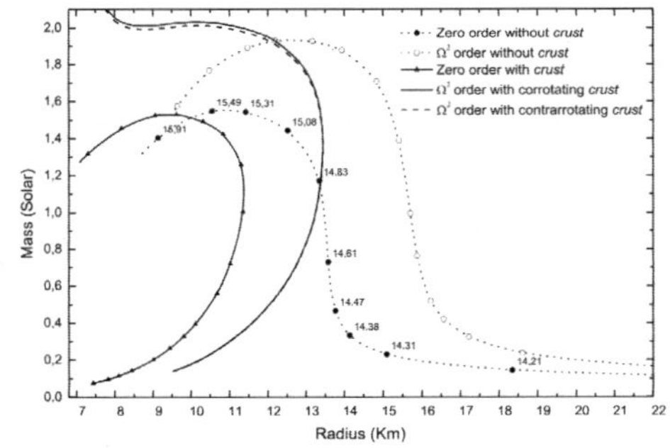

Figure 1. Mass vs radius for static/rotating configurations with a 10% *crust* (BPS)

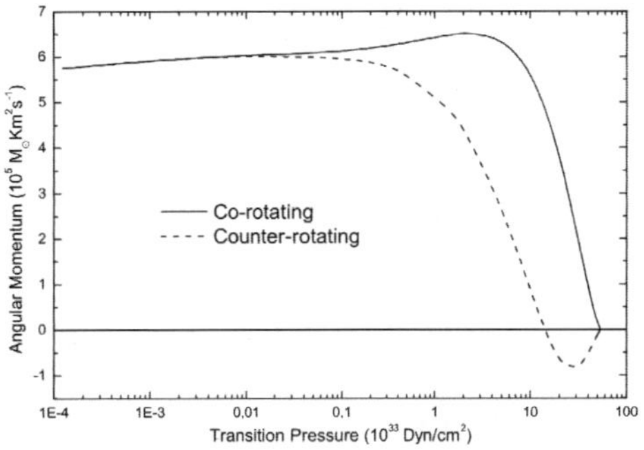

Figure 2. Angular momentum vs transition pressure for fixed ρ_c (BPS).

To obtain a model of a neutron star with a surface layer *crust* in permanent slow rotation, for a given equation of state of the core, we have to fix three parameters. We choose the total mass of the star, the central density, and the core-*crust* transition pressure. Then, we integrate the Einstein equations inside the star (using a realistic equation of state for the core) and we match this solution with the exterior solution found in [7, 8] using the junction conditions explained above. We have performed our simulations using an equation of state for high pressures from [11, K=240 MeV, $\frac{m^*}{m} = 0.78$ and $x_\sigma = 0.6$] and the BPS equation of state for the lower pressures. Similar results have been obtained for the analytical fit of the Sly equation presented in [3].

We present some of our results in figures 1 and 2. In fig. 1, we show multiple configurations for both static stars and rotating stars, with a *crust* of approximately 10% of the total mass and mass shedding angular velocity. On the static configurations without *crust* we write the value $\log 10(\rho_c)$. Note that the mass-radius behavior of the configurations with surface *crust* resembles those of the quark stars [11]. In fig. 2 we show the angular momentum for mass shedding angular velocity as a function of the core-*crust* transition pressure. In the counter rotating configuration we find that if the core-*crust* transition pressure is large enough, the angular momentum of the inner *crust* region could be equal or even higher than the angular momentum of the star core, resulting in a null or negative total angular momentum.

4. Core-*crust* transition pressure evolution in giant *glitches* of the Vela pulsar

On a typical pulsar, the rupture of the solid *crust* produces a release of energy 10^{41}-10^{43} ergs to the inner part of the *crust*. Several works study the thermal response of a neutron star after a *glitch* [12, 13], obtaining that increases in the temperature of the transition region of the order of $10^6 K$. This small rise of the temperature will produce an small but sensitive increase in the transition pressure [14]. We propose that the *glitch* can be explained by changes in the equation of the state of the star in the transition region between the *crust* and the core. In our model, these changes are represented by variations of the transition pressure due to the increase of the temperature produced by the deposition of the energy, coming from the rupture of the *crust*.

Using the observational data from the ATNF Pulsar Catalogue [15] we can study the giant glitches of Vela pulsar using our model. We assume that during the *glitch* neither the total mass nor the central density of the star are changed. Hence, in our description, the changes in the pulsar during this epoch are due to readjustment of the core-*crust* transition pressure, i.e., a variation of the equation of state in transition region between the core and the *crust*. Assuming this, we can give a description of the effects of these *glitches* over the neutron star using our model of slowly rotating neutron star with superficial *crust*. We assume that the star has a total

Date (MJD)	$\delta\Omega(10^{-6})$	$\delta p_{int}(10^{-10})$	$\delta ecc(10^{-6})$	$\delta Q(10^{-6})$
40289	2.34	0.70	2.36	4.73
41192	2.05	0.58	1.97	3.94
43693	3.06	0.93	3.15	6.31
45192	2.05	0.58	1.96	3.93
48457	2.72	0.81	2.76	5.51
51559	3.09	0.93	3.14	6.28
53959	2.62	0.82	2.74	5.48

Table 1. Date of each giant *glitch* of the Vela pulsar in modified Julian days and relative increases of angular velocity, transition pressure, eccentricity and quadrupolar moment

mass of $1.44M_{\odot}$, a central density of $1.279 \cdot 10^{15} gcm^{-3}$, and a transition pressure before the *glitch* of $3.751 \cdot 10^{33} dyn/cm^2$. We calculate the resulting configuration after the *glitch* varying the core-*crust* transition pressure until we reach an equivalent configuration (same mass and central density) with the resulting angular velocity just after the *glitch*. As it can be seen in table 1, our model predicts a relative transition pressure increase of the order of 10^{-10} for the considered *glitches*. This small variation in the equation of state on the surface of the star causes a redistribution of other properties of the star. The eccentricity of the star increases order 10^{-6}. Also, a movement of matter on the surface of the star, from the poles to the equator, produce an increase of the quadrupolar momentum (10^{-6}) which may cause gravitational radiation.

Using a estimation for the thermal pressure variations it can be seen that the corresponding increase of the inner *crust* temperature is order $10^6 K$ [14], which agree with the results obtained in the works previously commented. So the relative core-*crust* transition pressure changes we obtain in our model can be understood as variations in the equation of state of the transition region between the core and the *crust* of the neutron star, resulting from the deposition of energy of 10^{41}-$10^{43} erg$ in the outer layers of the neutron star during the *glitch* time.

In conclusion, we propose a new model for the giant *glitches* of the Vela pulsar: The tensions on the *crust* trying to adequate the ellipticity of the *crust* to the changed angular velocity produce the rupture of the *crust*. Then, there is a sudden energy deposition in the inner *crust* layers, which causes a rise of its temperature, resulting in an increase of the core-*crust* transition pressure of the neutron star (a change in the equation of the state in the core-*crust* transition region is produced). Hence, the properties of the neutron star (angular velocity, eccentricity, surface matter distribution and quadrupolar moment) change and a *glitch* is generated/observed.

The present work has been supported by Spanish Ministry of Science Project FIS2009-10614. The authors wish to thank F. Navarro-Lérida for valuable discussions.

References

[1] Lorenz C. P., Ravenhall D. G., Pethick C. J., 1993, *Phys. Rev. Lett.* **70**, 379
[2] Link B., Epstein R. I., Lattimer J. M., 1999, *Phys. Rev. Lett.* **83** 3362
[3] Haensel P., Potekhin A. Y., 2004, *Astron. Astrophys.* **428** 191 [arXiv:astro-ph/0408324]
[4] Pethick C. J., Lorenz C. P., Ravenhall D. G., 1995, *Nucl. Phys. A* **584** 675
[5] Lattimer J. M., Prakash M., 2001, *Astrophys. Journal* **550** 426
[6] Cheng K. S., Yuan Y. F., Zhang J. L., 2002, *Astrophys. Journal* **564** 909
[7] Hartle J. B., 1967, *Astrophys. Journal* **150** 1005
[8] Hartle J. B., Thorne K. S., 1968, *Astrophys. Journal* **153**, 807
[9] Misner C. W., Thorne K. S., Wheeler J. A., 1973, *Gravitation*, Freeman, San Francisco
[10] González-Romero L. M., 2003, *Phys. Rev. D* **67**, 064011
[11] Glendening N. K., 2000, *Compact stars* Astronomy & Astrophysics library, Springer, New York
[12] Hirano S. *et al* , 1997, *Astrophys. Journal* **491**, 286
[13] Van Riper K. A., Epstein I., Miller, G. S., 1991, *Astrophys. Journal* **381** L47
[14] Prakash M. *et al* , 1997, *Phys. Rep.* **280** 1
[15] Manchester, R. N. *et al* , 2005, *Astrophys. Journal* **129**, 1993

Cosmic magnetic fields and dark energy in extended electromagnetism

Jose Beltrán Jiménez[†,‡] **and Antonio L. Maroto**[†]

[†]Departamento de Física Teórica, Universidad Complutense de Madrid, 28040, Madrid, Spain
[‡] Institute de Physique Théorique, Université de Genève, 24 quai E. Ansermet, 1211 Genève 4, Switzerland

E-mail: Jose.Beltran@unige.ch, maroto@fis.ucm.es

Abstract. We discuss an extended version of electromagnetism in which the usual gauge fixing term is promoted into a physical contribution that introduces a new scalar state in the theory. This new state can be generated from vacuum quantum fluctuations during an inflationary era and, on super-Hubble scales, gives rise to an effective cosmological constant. The value of such a cosmological constant coincides with the one inferred from observations as long as inflation took place at the electroweak scale. On the other hand, the new state also generates an effective electric charge density on sub-Hubble scales that produces both vorticity and magnetic fields with coherent lengths as large as the present Hubble horizon.

1. Introduction

Out of the four known fundamental interactions in nature, two of them are particularly interesting in cosmological contexts due to their long range action, namely: gravity and electromagnetism. In fact, the behaviour of these two interactions at large scales is far from clear. On the gravitational sector we find the intriguing problem of the cosmic acceleration. Although a cosmological constant provides a simple and accurate description of it, from a theoretical point of view it would be even more desirable to have a fundamental explanation for the tiny value of such a constant. In this sense, several modifications of the gravitational interaction on cosmological scales have been proposed in the literature [1]. Concerning the electromagnetic sector, the unknown origin of the μG magnetic fields observed in galaxies and clusters [2] and, more remarkably, the very recent claim of detection of extra galactic magnetic fields [3] still lacks of a satisfactory explanation.

In this work we will consider the potential role of a modified electromagnetic theory in the dark energy problem [5, 6, 8] and how this modified electromagnetism can generate magnetic fields at large scales. Thus, we shall explore the interesting possibility of finding a link between dark energy and the origin of cosmic magnetic field.

2. Extended electromagnetism without the Lorenz condition

In the covariant quantization of the electromagnetic field, the starting point is the modified Maxwell action including a gauge fixing term, that, in a curved spacetime, reads:

$$S = \int d^4x\sqrt{-g}\left[-\frac{1}{4}F_{\mu\nu}F^{\mu\nu} + \frac{\xi}{2}(\nabla_\mu A^\mu)^2 + J_\mu A^\mu)^2\right] \tag{1}$$

which leads to the modified Maxwell equations:

$$\nabla_\nu F^{\mu\nu} + \xi \nabla^\mu (\nabla_\nu A^\nu) = J^\mu. \tag{2}$$

Taking the divergence of this equation, we obtain:

$$\Box(\nabla_\nu A^\nu) = 0 \tag{3}$$

The quantization in the Gupta-Bleuler formalism and in Minkowski spacetime proceeds as follows [4]: One works with the four polarizations of A_μ and then imposes the weak Lorenz condition on the physical states $|\phi\rangle$ so that $(\partial_\mu A^\mu)^{(+)}|\phi\rangle = 0$ to get rid of the unphysical polarizations. With this condition, the expected value of any physical observable only depends on the the transverse degrees of freedom, because the temporal and longitudinal polarizations contribute with opposite signs and, since the Lorenz condition imposes that there must be the same number of temporal and longitudinal photons in any physical state, they cancel each other. However, in an expanding universe, we see from (3) that $\nabla_\nu A^\nu$ can be excited from vacuum quantum fluctuations because it behaves as a massless scalar field and, therefore, the Lorenz condition is violated. The reason for such a violation is that the expansion excites a different number of temporal and longitudinal photons so the aforementioned cancellation does not occur anymore [5, 6].

In order to avoid the difficulties found when quantizing in the covariant formalism in an expanding universe, let us explore the possibility that the fundamental theory of electromagnetism is given by the modified action (1), where we allow the $\nabla_\mu A^\mu$ field to propagate. This theory still has the residual gauge symmetry $A_\mu \to A_\mu + \partial_\mu\theta$ provided $\Box\theta = 0$. Thus, having removed one constraint, the theory contains one additional degree of freedom and the general solution for the modified equations can be written as:

$$\mathcal{A}_\mu = \mathcal{A}_\mu^{(1)} + \mathcal{A}_\mu^{(2)} + \mathcal{A}_\mu^{(s)} + \partial_\mu\theta \tag{4}$$

where $\mathcal{A}_\mu^{(i)}$ with $i = 1, 2$ are the two transverse modes of the massless photon, $\mathcal{A}_\mu^{(s)}$ is the new scalar state, which is the mode that would have been eliminated if we had imposed the Lorenz condition and, finally, $\partial_\mu\theta$ is a purely residual gauge mode, which can be eliminated by means of a residual gauge transformation in the asymptotically free regions, in a completely analogous way to the elimination of the A_0 component in the Coulomb quantization. The fact that Maxwell's electromagnetism could contain an additional scalar mode decoupled from electromagnetic currents, but with non-vanishing gravitational interactions, was already noticed in a different context in [7].

The evolution of the new mode is given by (3), so that on super-Hubble scales, $|\nabla_\mu \mathcal{A}_k^{(s)\mu}| =$ *const.* which, as shown in [5], implies that the field contributes as a cosmological constant in (1). Notice that, as seen in (3), the new scalar mode is a massless free field and it is possible to calculate the corresponding power spectrum generated during inflation, $P_{\nabla A}(k) = 4\pi k^3 |\nabla_\mu \mathcal{A}_k^{(s)\mu}|^2$. In the super-Hubble limit, we get in a quasi-de Sitter inflationary phase characterized by a slow-roll parameter ϵ:

$$P_{\nabla A}(k) = \frac{9H_{k_0}^4}{16\pi^2}\left(\frac{k}{k_0}\right)^{-4\epsilon} \tag{5}$$

where H_{k_0} is the Hubble parameter when the k_0 mode left the horizon [5]. Notice that this result implies that $\rho_A \sim (H_{k_0})^4$. The measured value of the cosmological constant then requires $H_{k_0} \sim 10^{-3}$ eV, which corresponds to an inflationary scale $M_I \sim 1$ TeV. Thus we see that the cosmological constant scale can be naturally explained in terms of physics at the electroweak scale. This is one of the most relevant aspects of the present model in which, unlike existing dark energy theories based on scalar fields, dark energy can be generated without including any potential term or dimensional constant.

Since the field amplitude of the scalar state remains frozen on super-Hubble scales, there is no modification of Maxwell's equation on those scales. However, as the amplitude starts decaying once the mode enters the horizon in the radiation or matter eras, the ξ-term in (2) generates an effective current which can produce magnetic fields on cosmological scales, as we will show in the following.

By passing, we notice that, in Minkowski spacetime, the theory (1) is completely equivalent to standard QED because, although non-gauge invariant, the corresponding effective action is equivalent to the standard BRST invariant effective action of QED [6]. This prevents from potential unobserved effects in accelerators because both theories lead to the same phenomenology in flat spacetime, being distinguishable only in curved backgrounds.

On the other hand, despite the fact that the homogeneous evolution in the present case is the same as in ΛCDM, the effective cosmological constant generated by the new scalar state fluctuates so that the evolution of metric perturbations could be different. We have calculated the evolution of metric, matter density and electromagnetic perturbations [8]. The propagation speeds of scalar, vector and tensor perturbations are found to be real and equal to the speed of light, so that the theory is classically stable. Moreover, the three physical states carry positive energy so it is also quantum-mechanically stable. On the other hand, it is possible to see that all the post-Newtonian parameters [9] agree with those of General Relativity, i.e. the theory is compatible with all the local gravity constraints for any value of the homogeneous background electromagnetic field at the same level of accuracy as General Relativity [5, 10]. Concerning the evolution of scalar perturbations, we find that the only relevant deviations with respect to ΛCDM appear on large scales $k \sim H_0$ and that they depend on the primordial spectrum of electromagnetic fluctuations. However, the effects on the CMB temperature and matter power spectra are compatible with observations except for very large primordial fluctuations [8].

3. Generation of cosmic magnetic fields

It is interesting to note that the ξ-term can be seen, at the equations of motion level, as a conserved current acting as a source of the usual Maxwell field. To see this, we can write $-\xi\nabla^\mu(\nabla_\nu A^\nu) \equiv J^\mu_{\nabla\cdot A}$ which, according to (3), satisfies the conservation equation $\nabla_\mu J^\mu_{\nabla\cdot A} = 0$ and we can express (2) as:

$$\nabla_\nu F^{\mu\nu} = J^\mu_T \tag{6}$$

with $J^\mu_T = J^\mu + J^\mu_{\nabla\cdot A}$ and $\nabla_\mu J^\mu_T = 0$. Physically, this means that, while the new scalar mode can only be excited gravitationally, once it is produced it will generally behave as a source of electromagnetic fields. Therefore, the modified theory is described by ordinary Maxwell equations with an additional "external" current. For an observer with four-velocity u^μ moving with the cosmic plasma, it is possible to decompose the Faraday tensor in its electric and magnetic parts as: $F_{\mu\nu} = 2E_{[\mu}u_{\nu]} + \frac{\epsilon_{\mu\nu\rho\sigma}}{\sqrt{g}}B^\rho u^\sigma$, where $E^\mu = F^{\mu\nu}u_\nu$ and $B^\mu = \epsilon^{\mu\nu\rho\sigma}/(2\sqrt{g})F_{\rho\sigma}u_\nu$. Due to the infinite conductivity of the plasma, Ohm's law $J^\mu - u^\mu u_\nu J^\nu = \sigma F^{\mu\nu}u_\nu$ implies $E^\mu = 0$. Therefore, in that case the only contribution would come from the magnetic part. Thus, from Maxwell's equations, we get, for comoving observers in a FLRW metric (see also [11]):

$$\vec{\omega}\cdot\vec{B} = \rho^0_g \tag{7}$$

where $\vec{v} = d\vec{x}/d\eta$ is the conformal time fluid velocity, $\vec{\omega} = \vec{\nabla}\times\vec{v}$ is the fluid vorticity, $\rho^0_g = -\xi\partial_0(\nabla_\mu A^\mu)$ is the effective charge density today whose power spectrum can be obtained

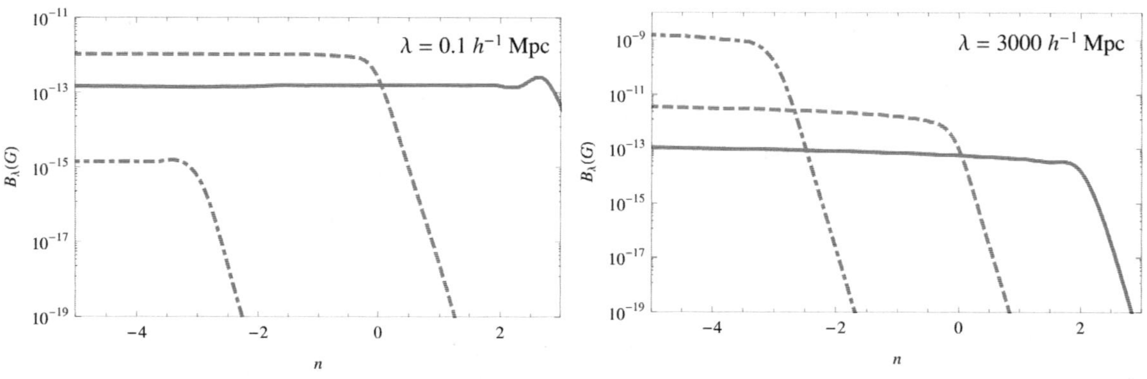

Figure 1. Lower limits on the magnetic fields generated on galactic scales (left panel) and Hubble horizon scales (right panel) in terms of the magnetic spectral index n for different values of the vorticity spectral index m. Dot-dashed blue for $m = 0$, dashed green for $m \simeq -3$ and full red for $m \simeq -5$.

from (5) (see [12] for more details), and the $\vec{B}$ components scale as $B_i \propto 1/a$ as can be easily obtained from $\epsilon^{\mu\nu\rho\sigma}F_{\rho\sigma;\nu} = 0$ to the lowest order in $|\vec{v}|$. Thus, the presence of the non-vanishing cosmic effective charge density necessarily creates both magnetic field and vorticity. Due to the presence of the effective current, we find that vorticity grows as $|\vec{\omega}| \propto a$, from radiation era until present.

Using (7), it is possible to translate the existing upper limits on vorticity coming from CMB anisotropies [11] into *lower* limits on the amplitude of the magnetic fields generated by this mechanism [12]. In Fig. 1 we show the corresponding limits on the magnetic field at galactic and Hubble horizon scales for different spectral indices of magnetic field (n) and vorticity (m). We see that this mechanism allows to generate relatively strong magnetic fields on scales as large as the Hubble horizon and act as seeds for a galactic dynamo or even play the role of primordial fields and account for observations just by amplification due to the collapse and differential rotation of the protogalactic cloud.

4. Discussion

We have discussed and extended electromagnetic theory in which we do not need to impose the Lorenz condition. In order to quantise the theory we introduce an additional scalar state which can be excited by gravitational fields. Indeed, fluctuations of such a state during an inflationary era at the electroweak scale generates an effective cosmological constant on super-Hubble scales with the correct value. This theory is free form both classical and quantum instabilities and is consistent with all local gravity tests at the same level as General Relativity, CMB and large scale structure observations. On the other hand, the sub-Hubble modes of the new state generated during inflation acts as an effective electromagnetic current so it can produce cosmological magnetic fields all the way to the Hubble horizon (but not beyond). This allows to establish an important link between the problems of dark energy and cosmic magnetic fields. In fact, also non-minimal couplings have been considered in [13] where a potential relationship between angular momentum and magnetic fields has been explored.

Acknowledgements: This work has been supported by MICINN (Spain) project numbers FIS 2008- 01323 and FPA 2008-00592, CAM/UCM 910309, MEC grant BES-2006-12059 and MICINN Consolider-Ingenio MULTIDARK CSD2009-00064. J.B. also wishes to thank support from the Norwegian Council under the YGGDRASIL project no 195761/V11.

References

[1] S. M. Carroll, V. Duvvuri, M. Trodden and M. S. Turner, Phys. Rev. D **70**, 043528 (2004); G. R. Dvali, G. Gabadadze and M. Porrati, Phys. Lett. B **485**, 208 (2000).
[2] L. M. Widrow, Rev. Mod. Phys. **74** (2002) 775; R. M. Kulsrud and E. G. Zweibel, Rept. Prog. Phys. **71** (2008) 0046091; P. P. Kronberg, Rept. Prog. Phys. **57** (1994) 325.
[3] A. Neronov and I. Vovk, Science **328** (2010) 73; F. Tavecchio, et al., arXiv:1004.1329; S. Ando, A. Kusenko, Astrophys. J. **722** (2010) L39
[4] C. Itzykson and J.B. Zuber, *Quantum Field Theory*, McGraw-Hill (1980); N.N. Bogoliubov and D.V. Shirkov, *Introduction to the theory of quantized fields*, Interscience Publishers, Inc. (1959).
[5] J. Beltrán Jiménez and A.L. Maroto, JCAP **0903** (2009) 016; Int. J. Mod. Phys. D **18** (2009) 2243; AIP Conf. Proc. **1241**, 1033 (2010).
[6] J. Beltrán Jiménez and A.L. Maroto, Phys. Lett. B **686** (2010) 175
[7] S. Deser, *Ann. Inst. Henri Poincaré*, **16**: 79 (1972)
[8] J. Beltrán Jimenez, T. S. Koivisto, A. L. Maroto and D. F. Mota, JCAP **0910** (2009) 029
[9] C. Will, *Theory and experiment in gravitational physics*, Cambridge University Press, (1993)
[10] J. B. Jimenez and A. L. Maroto, JCAP **0902** (2009) 025
[11] C. Caprini, S. Biller and P. G. Ferreira, JCAP **0502** (2005) 006
[12] J. Beltrán Jiménez and A.L. Maroto, arXiv:1010.3960 [astro-ph.CO].
[13] J. Beltrán Jiménez and A.L. Maroto, arXiv:1010.4513 [astro-ph.CO]. *Accepted for publication in* JCAP.

Spanish Relativity Meeting (ERE 2010): Gravity as a Crossroad in Physics IOP Publishing
Journal of Physics: Conference Series **314** (2011) 012090 doi:10.1088/1742-6596/314/1/012090

An infinite family of relativistic magnetized finite thin disks.

Antonio C. Gutiérrez-Piñeres [1] **and Guillermo A. González** [2]

[1]Facultad de Ciencias Básicas, Universidad Tecnológica de Bolívar, Cartagena, Colombia
[2]Universidad Industrial de Santander, A. A. 678, Bucaramanga, Colombia

E-mail: acgutierrez@unitecnologica.edu.co

E-mail: guillego@uis.edu.co

Abstract. An infinite family of relativistic finite thin disk model with magnetic field is presented. The model is obtained for solving the Einstein-Maxwell equations for static spacetimes by means of the Horský-Mitskievitch generating conjecture. The vacuum limit of these obtained solutions is the well known Morgan and Morgan solution. The obtained expressions are simply written in terms of oblate spheroidal coordinates. The mass of the disks are finite and the energy-momentum tensor agrees with all the energy conditions. The magnetic field and the circular velocity are evaluated explicitly. All the physical quantities obtained shown an acceptable behavior

1. Einstein-Maxwell Equations and Thin Disks

The vacuum Einstein-Maxwell equations, in geometrized units such that $c = 8\pi G = \mu_0 = \epsilon_0 = 1$, can be written as

$$G_{ab} = T_{ab}, \quad F^{ab}{}_{;b} = 0, \tag{1}$$

with the electromagnetic energy-momentum tensor given by

$$T_{ab} = F_{ac}F_b{}^c - \frac{1}{4}g_{ab}F_{cd}F^{cd}, \quad F_{ab} = A_{b,a} - A_{a,b}, \tag{2}$$

where A_a is the electromagnetic four potential.

By using the distributional approach [1, 2] and by choosing the electromagnetical potential as $A_a = A(r,z)\delta_a^0$ we can write the surface energy-momentum tensor and the current density of a source like a disk in the canonical form as

$$S^{ab} = \epsilon V^a V^b + p W^a W^b, \quad j^a = jW^a. \tag{3}$$

Here ϵ and p are the energy density and the azimuthal pressure of the disk, respectively, and we have used the orthonormal tetrad $e_{(b)}^{\ a} = \{V^a, W^a, X^a, Y^a\}$, with

$$V^a = e^{-\Phi}\delta_0^a, \quad W^a = \frac{e^{\Phi}}{r}\delta_1^a, \quad X^a = e^{\Phi-\Lambda}\delta_2^a, \quad Y^a = e^{\Phi-\Lambda}\delta_3^a, \tag{4}$$

and we have defined $\mu = \epsilon + p$ as the mass density on the surface of the disk. The energy-momentum tensor S^{ab} and the current density j^a can be interpreted as the superposition of two counterrotating fluids (see [3] and [4]). In order to do this we cast

$$S^{ab} = \epsilon_+ U_+^a U_+^b + \epsilon_- U_-^a U_-^b , \qquad j^a = \sigma_+ U_+^a + \sigma_- U_-^a , \tag{5}$$

where

$$\epsilon_+ = \epsilon_- = (\epsilon - p)/2, \qquad \sigma_+ = -\sigma_- = \frac{je^{\Phi}}{2r}\sqrt{\frac{\epsilon}{p} - 1}, \tag{6a}$$

are the energy densities and charge densities of the two counterrotating fluids. The counterrotating velocity vectors are given by

$$U_\pm^a = \frac{V^a \pm UW^a}{\sqrt{1-U^2}}, \qquad U^2 = \frac{p}{\epsilon} \leq 1, \tag{7}$$

with U the counterrotating tangential velocity. Therefore we have two counterrotating charged fluids with equal energy densities and equal but opposite charge densities.

2. The Horský-Mitskievitch conjecture and the WLP metric

The line element of a static vacuum spacetime can be written in the Weyl-Lewis-Papapetrou form

$$ds^2 = -f(r,z)^2\, e^{2\Phi_s} dt^2 + \frac{e^{-2\Phi_s} r^2}{f(r,z)^2} d\varphi^2 + f(r,z)^2 e^{2(\Lambda_s - \Phi_s)}(dr^2 + dz^2), \tag{8}$$

which admits two Killing vectors

$$\boldsymbol{\xi_t} = -f(r,z)^2 e^{2\Phi_s} \boldsymbol{dt}, \qquad \boldsymbol{\xi_\varphi} = \frac{r^2 e^{-2\Phi_s}}{f(r,z)^2} \boldsymbol{d\varphi}, \tag{9}$$

with $f(r,z)$ an arbitrary function and $\Phi_s = \Phi_s(r,z)$ and $\Lambda_s = \Lambda_s(r,z)$, being (r,z,φ) the usual cylindrical coordinates. The Killing vectors $\boldsymbol{\xi}$ and the electromagnetic four-potential $\boldsymbol{A}$ satisfy the equations

$$^\star\boldsymbol{d}^\star\boldsymbol{d\xi} = 0, \qquad ^\star\boldsymbol{d}^\star\boldsymbol{dA} = 0. \tag{10}$$

Then, Killing vector $\boldsymbol{\xi_\varphi}$ induces the four-potential

$$\boldsymbol{A} = \frac{qr^2}{fe^{2\Phi_s}} \boldsymbol{d\varphi}. \tag{11}$$

It can be verified through standard calculations that the sourceless Einstein-Maxwell equations are fulfilled if

$$f(r,z) = 1 + c_1 f_1(r,z), \tag{12}$$

where the function f_1 must be a solution of the differential equation

$$G^{(a)}{}_{(a)} = -R = 0, \tag{13}$$

arising from the fact that for electrovacuum spacetimes the Einstein tensor, $G_{(a)(b)}$, is traceless. Naturally, if $c_1 = 0$ we arrive to a solution of the vacuum Einstein equation.

As in the reference [5] we can choose $c_1 = q^2$ and $f_1 = r^2 e^{-2\Phi_s}$, then we have the four-potential

$$\mathbf{A} = \frac{qr^2}{q^2r^2 + e^{2\Phi_s}} \boldsymbol{d\varphi}. \tag{14}$$

Using equations (3) and (4) we found the surface energy density and the azimuthal pressure

$$\epsilon = \frac{4E^2(r)e^{3\Phi_s - \Lambda_s}(1 - r\Phi_{s,r})\Phi_{s,z}}{(e^{2\Phi} + q^2r^2)^3}, \qquad p = \frac{4rE(r)e^{3\Phi_s - \Lambda_s}[2q^2r + E(r)\Phi_{s,r}]\Phi_{s,z}}{(e^{2\Phi_s} + q^2r^2)^3}, \tag{15}$$

the surface mass density and the surface current density

$$\mu = \frac{4E(r)e^{3\Phi_s - \Lambda_s}\Phi_{s,z}}{(e^{2\Phi_s} + q^2r^2)^2}, \qquad j = \frac{4qre^{4\Phi_s - \Lambda_s}\Phi_{s,z}}{(e^{2\Phi_s} + q^2r^2)^2}, \tag{16}$$

and the circular velocity

$$U^2 = U_s^2 \left[1 + \frac{2q^2r}{E(r)\Phi_{s,r}}\right], \tag{17}$$

where $E(r) = e^{2\Phi_s} - q^2r^2$, and

$$U_s^2 = \frac{r\Phi_{s,r}}{1 - r\Phi_{s,r}} \tag{18}$$

is the circular velocity of the unmagnetized source, i.e. $q = 0$.

Let's restrict the previous general model to obtain a solution of the Einstein-Maxwell equations describing an infinite family of finite static magnetized thin disks. In order to do this, we use as a seed solution (Φ_s, Λ_s), the well known Morgan and Morgan metric disk (ϕ, λ) [6],

$$\phi(x, y) = -\sum_{n=0}^{\infty} C_{2n} q_{2n}(x) P_{2n}(y),$$

where $P_n(y)$ are the usual Legendre polynomials and $q_n(x) = i^{n+1}Q_n(ix)$. $Q_n(z)$ being the Legendre functions of second kind (see [7, 8]). The x and y are the oblate spheroidal coordinates related with the cylindrical coordinates by the relations [8]

$$r^2 = a^2(1 + x^2)(1 - y^2), \qquad z = axy,, \tag{19a}$$

where $0 \leq x < \infty$ and $-1 \leq y < 1$. The disk has coordinates $x = 0$, $0 \leq y^2 < 1$ and, on crossing the disk, y changes sign but does not change in the absolute value. We use the constants C_{2n} as determined by [9]. Then, these solutions correspond to the magnetized version of the well known Morgan-Morgan disks [6]. On the other hand, from the Eq.(15) the energy density of the magnetized disk can be written as

$$\epsilon_n = \frac{4E^2e^{3\phi_n - \lambda_n}[y + (1 - y^2)\phi_{n,y}]\phi_{n,x}}{ay^2[e^{2\phi_n} + q^2a^2(1 - y^2)]^3}, \qquad E(y) = e^{2\phi_n} - q^2a^2(1 - y^2), \tag{20}$$

whereas from the equations (15) and (16) the pressure, p, and mass density, μ, in the surface of the disk are

$$p_n = \frac{4E(1 - y^2)e^{3\phi_n - \lambda_n}\phi_{n,x}[2q^2a^2y - \phi_{n,y}E]}{ay^2[e^{2\phi_n} + q^2a^2(1 - y^2)]^3}, \qquad \mu_n = \frac{4Ee^{3\phi_n - \lambda_n}\phi_{n,x}}{ay[e^{2\phi_n} + q^2a^2(1 - y^2)]^2}, \tag{21}$$

respectively. Whereas, from the equations (16) and (17) we have for the surface current density and circular velocity

$$j = \frac{4q(1-y^2)^{1/2}e^{4\phi_n - \lambda_n}\phi_{n,x}}{y[e^{2\phi_n} + q^2a^2(1-y^2)]^2}, \tag{22}$$

and

$$U_n^2 = U_s^2\left[1 - \frac{2q^2a^2y}{E\phi_{n,y}}\right], \qquad U_s^2 = -\frac{(1-y^2)\phi_{n,y}}{y+(1-y^2)\phi_{n,y}}, \tag{23}$$

being U_n^2 the circular velocity of the unmagnetized source, i.e. $q = 0$.

The non zero components of the magnetic field in the surface of the disk are

$$B_z = -\frac{2qa(1-y^2)^{1/2}e^{2\phi_n}[y+(1-y^2)\phi_{n,y}]}{y[e^{2\phi_n} + q^2a^2(1-y^2)]^2}, \qquad B_r = -\frac{2qa(1-y^2)e^{2\phi_n}\phi_{n,x}}{y[e^{2\phi_n} + q^2a^2(1-y^2)]^2}, \tag{24}$$

here we have used the magnetic potential of the equation (11) and all the quantities are evaluated on the disk.

We have presented an infinite family of new exact solutions of the vacuum Einstein-Maxwell equations for static and axially symmetric spacetimes. The solutions describe an infinite family of magnetized finite thin disks, the magnetized version of the family of Morgan and Morgan relativistic thin disks [6]. Unlike the Morgan and Morgan seed solution, generated spacetimes are not asymptotically flat. Nevertheless, their energy densities and mass densities are positive everywhere and well behaved. In such way their energy-momentum tensor are in fully agreement with all the energy conditions. The first member, $n = 1$, of the family of the magnetized Morgan-Morgan disk has a singularity at the rim of the disk where the Kretschmann invariant becomes infinite, although its mass density is finite everywhere. Whereas all the $n > 1$ disks are regular. This property in the curvature is, of course, inherited from the seed Morgan and Morgan metric disk. On the other hand, as all the metric functions of the solutions were explicitly computed, these are the first fully integrated exact solutions for such kind of magnetized thin disk sources. Moreover, the method outlined may serve as a guideline to find more physical acceptable solutions in future works. Furthermore their importance as a new family of exact solutions of the Einstein-Maxwell vacuum equations.

Acknowledgments
I (A.C.G-P.) want to thank COLCIENCIAS. I am very grateful to Nathalie R. A., who was a strong motivating force through this project.

References

[1] A. Lichnerowicz, C.R. 1971 *Acad. Sci.* **273**, 528
[2] A. H. Taub 1980 *J. Math. Phys.* **21**, 1423
[3] G. A. González and O. A. Espitia 2003 *Phys. Rev.* D **68**, 104028
[4] G. García-Reyes and G. A. González 2004 *Phys. Rev.* D **70**, 104005
[5] L. Richtereck, J. Novotný and J. Horský 2000 *Czech. J. Phys.* **50**, 925
[6] T. Morgan and L. Morgan 1969 *Phys. Rev.* **183**, 1097
[7] Arfken G. and Weber H 2005 *Mathematical Methods for Physicists* (Academic Press, New York
[8] P. M. Morse and H. Fesbach 1953 *Methods of Theoretical Physics* (McGraw-Hill)
[9] G. A. González and J. I. Reina 2006 *Mon. Not. R. Astron. Soc.* **371**, 1873

Vacuum Spacetimes of Embedding Class Two

Alan Barnes

School of Engineering & Applied Science, Aston Umiversity, Birmingham B4 7ET, UK

E-mail: `barnesa@aston.ac.uk`

Abstract. Doubt is cast on the much quoted results of Yakupov that the torsion vector in embedding class two vacuum space-times is necessarily a gradient vector and that class 2 vacua of Petrov type III do not exist. The first result is equivalent to the fact that the two second fundamental forms associated with the embedding necessarily commute and has been assumed in most later investigations of class 2 vacuum space-times. Yakupov stated the result without proof, but hinted that it followed purely algebraically from his identity: $R_{ijkl}C^{kl} = 0$ where C_{ij} is the commutator of the two second fundamental forms of the embedding.

From Yakupov's identity, it is shown that the only class two vacua with non-zero commutator C_{ij} must necessarily be of Petrov type III or N. Several examples are presented of non-commuting second fundamental forms that satisfy Yakupovs identity and the vacuum condition following from the Gauss equation; both Petrov type N and type III examples occur. Thus it appears unlikely that his results could follow purely algebraically. The results obtained so far do not constitute definite counter-examples to Yakupov's results as the non-commuting examples could turn out to be incompatible with the Codazzi and Ricci embedding equations. This question is currently being investigated.

1. Introduction

A spacetime is said to be of *embedding class N* if it can be (locally) isometrically embedded in a flat pseudo-Riemannian manifold of dimension $4 + N$, but in no flat manifold of *smaller* dimension. Thus a spacetime is of embedding class 2 if it can be minimally embedded in a flat space of signature $(+1, -1, -1, -1, e_1, e_2)$ where $e_i = \pm 1$. It is well-known that there are no vacuum spacetimes of embedding class 1 and any spacetime is of embedding class 6 or less [1].

The embedding of a spacetime of class two is determined by two second fundamental forms Ω_{ab} and Λ_{ab} plus a torsion vector t_a satisfying

$$\text{Gauss equations :} \quad R^{ab}{}_{cd} = 2e_1\Omega^a_{[c}\Omega^b_{d]} + 2e_2\Lambda^a_{[c}\Lambda^b_{d]} \tag{1}$$

$$\text{Codazzi equations :} \quad \Omega^a_{[b;c]} = -e_2\Lambda^a{}_{[b}t_{c]} \tag{2}$$

$$\Lambda^a_{[b;c]} = e_1\Omega^a{}_{[b}t_{c]} \tag{3}$$

$$\text{Ricci equations :} \quad t_{[a;b]} = -\Omega^c{}_{[a}\Lambda_{b]c} \tag{4}$$

The two normals to the embedded spacetime in the enveloping six-dimensional manifold (and so the two second fundamental forms) are only determined up to a rotation and reflections if $e_1e_2 = 1$ or up to a boost and reflections if $e_1e_2 = -1$. Thus, for $e_1e_2 = 1$:

$$\tilde{\mathbf{\Omega}} = \pm(\cos\theta\mathbf{\Omega} + \sin\theta\mathbf{\Lambda}) \qquad \tilde{\mathbf{\Lambda}} = \pm(-\sin\theta\mathbf{\Omega} + \cos\theta\mathbf{\Lambda}) \tag{5}$$

whereas, for $e_1e_2 = -1$:

$$\tilde{\mathbf{\Omega}} = \pm(\cosh\theta\mathbf{\Omega} + \sinh\theta\mathbf{\Lambda}) \qquad \tilde{\mathbf{\Lambda}} = \pm(\sinh\theta\mathbf{\Omega} + \cosh\theta\mathbf{\Lambda}) \tag{6}$$

Under this change of normals the torsion vector transforms as $\tilde{t}_a = t_a + \theta_{,a}$. Thus, if the torsion vector is a gradient, (equivalently, from (4), if $\mathbf{\Omega}$ & $\mathbf{\Lambda}$ commute), the torsion vector may be set to zero and the two Codazzi equations (2, 3) decouple.

Yakupov [2] proved a useful identity for class 2 vacua:

$$R^{ab}{}_{cd}C^{cd} = 0 \quad \text{where} \quad C_{ab} = \Omega^e{}_{[a}\Lambda_{b]e}. \tag{7}$$

Note $\mathbf{C}$ is the commutator of $\mathbf{\Omega}$ and $\mathbf{\Lambda}$. Subsequently, he stated *without proof* two results [3]:

- For class 2 vacua, the commutator C_{ab} must vanish and the torsion vector t_a is a gradient.
- There are no class 2 vacua of Petrov type III.

For a summary of Yakupov's work in English, see [4]. This summary, and Yakupov himself, give a few hints, but no details, regarding the proof of these two results. As far as one can ascertain the proof was purely algebraic following from the Gauss equations (1), vacuum conditions (8) and the identity (7). Most subsequent work (e.g. [5] & [6]) has assumed $C_{ab} = 0$ and, in particular, van den Bergh [6] confirmed the non-existence class two vacua of Petrov type III subject to this assumption.

2. Embedding Class Two Vacua

Contracting the Gauss equation (1) the vacuum condition becomes:

$$e_1(\Omega^a_c\Omega^c_b - \omega\Omega^a_b) + e_2(\Lambda^a_c\Lambda^c_b - \lambda\Lambda^a_b) = 0 \tag{8}$$

where ω and λ are the traces of $\mathbf{\Omega}$ and $\mathbf{\Lambda}$ respectively. Let us call $\mathbf{\Omega^2} - \omega\mathbf{\Omega}$ and $\mathbf{\Lambda^2} - \lambda\mathbf{\Lambda}$ the *Ricci squares* of $\mathbf{\Omega}$ and $\mathbf{\Lambda}$ respectively. Thus, the Ricci squares of $\mathbf{\Omega}$ and $\mathbf{\Lambda}$ differ by at most a sign and so *a fortiori* have the same Segré type. Moreover, from (8), it follows that the Ricci square $\mathbf{\Omega^2} - \omega\mathbf{\Omega}$ and $\mathbf{\Lambda}$ commute and so do $\mathbf{\Lambda^2} - \lambda\mathbf{\Lambda}$ and $\mathbf{\Omega}$. However, this does not imply that $\mathbf{\Omega}$ and $\mathbf{\Lambda}$ themselves commute. As will be seen below, it is easy to find many examples where the two second fundamental forms $\mathbf{\Omega}$ and $\mathbf{\Lambda}$ satisfy the vacuum conditions, but do not commute.

All of these examples were found to satisfy Yakupov's identity automatically. At first this was surprising as the published proof [2] used the Gauss, Codazzi and Ricci equations plus the Ricci identities. However, it was quickly realised that the identity can be proved using *only* the Gauss equations (1) and vacuum conditions (8). As the first step in his proof, Yakupov derived the following equations from the Codazzi equations (2, 3) using the Ricci identities:

$$\Omega_{a[b;cd]} = \Omega^e_{[b}R_{cd]ea} = 2e_2t_{[c;d}\Lambda_{b]a} \qquad \Lambda_{a[b;cd]} = \Lambda^e_{[b}R_{cd]ea} = -2e_1t_{[c;d}\Omega_{b]a} \tag{9}$$

However, the second equality in each of these equations (with $t_{[a;b]}$ replaced by the commutator C_{ab}) can be derived by substituting for R_{abcd} using the Gauss equation (1) in the expressions $\Omega^e_{[b}R_{cd]ea}$ and $\Lambda^e_{[b}R_{cd]ea}$. The rest of the proof uses only the Gauss and vacuum equations and follows identical lines to Yakupov's (with $t_{[a;b]}$ replaced throughout by the commutator C_{ab}). Thus (7) is a *purely algebraic constraint* and Yakupov's results [3] cannot follow purely algebraically; if these potential counter-examples are to be excluded, it is necessary to consider *integrability conditions* derived from the Codazzi and Ricci equations.

From the identity (7) it follows that C_{ab}, if non-zero, is an eigenbivector of the Riemann tensor with *zero* eigenvalue. This immediately excludes Petrov types II and D. Furthermore Brans [7] proved that Petrov type I vacuum spaces with a zero eigenvalue do not exist. His proof used the Newman-Penrose formalism and made extensive use of the Bianchi identities. Thus, *the only class two vacua with non-zero commutator C_{ab} must have Petrov type N or III.*

3. Construction of Non-Commuting $\boldsymbol{\Omega}$ and $\boldsymbol{\Lambda}$

If the Ricci square of $\boldsymbol{\Omega}$ (and so that of $\boldsymbol{\Lambda}$) is non-degenerate, then so are $\boldsymbol{\Omega}$ and $\boldsymbol{\Lambda}$ and, moreover, they have the same invariant subspace structure and so commute. Thus it is only necessary to consider cases where the Ricci squares are degenerate, but where $\boldsymbol{\Omega}$ and $\boldsymbol{\Lambda}$ are not (and cases where they are less degenerate than their Ricci squares). Furthermore $\boldsymbol{\Omega}$ and $\boldsymbol{\Lambda}$ must have different invariant subspace structures if they are not to commute. The cases where the Ricci square is degenerate are summarised below.

If a second fundamental form has Segré type $[2,1,1]$ with eigenvalues λ, μ and ν respectively, its Ricci square has degenerate Segré type:

$[(1,1),1,1]$	if $\mu+\nu=0$
$[2,(1,1)]$	if $\lambda=0$
$[(1,1),(1,1)]$	if $\lambda=\mu+\nu=0$

If it has Segré type $[3,1]$ with eigenvalues λ and μ respectively, its Ricci square has degenerate Segré type:

$[(2,1),1]$	if $\lambda+\mu=0$
$[(3,1)]$	if $\lambda=0$

If it has Segré type $[1,1,1,1]$ with eigenvalues λ, μ, ν and σ respectively, its Ricci square has degenerate Segré type:

$[(1,1),1,1]$	if $\nu+\sigma=0$
$[1,1,(1,1)]$	if $\lambda+\mu=0$
$[(1,1),(1,1)]$	if $\lambda+\mu=\nu+\sigma=0$

If it has Segré type $[Z,\bar{Z},1,1]$ with eigenvalues λ, $\bar{\lambda}$, μ and ν respectively, its Ricci square has degenerate Segré type:

$[(1,1),1,1]$	if $\mu+\nu=0$
$[Z,\bar{Z},(1,1)]$	if $\lambda+\bar{\lambda}=0$
$[(1,1),(1,1)]$	if $\lambda+\bar{\lambda}=\mu+\nu=0$

There are at least 18[1] possible non-commuting combinations of Segré types compatible with the vacuum conditions (8). Most of these lead to Riemann tensors of Petrov type O (flat space) or type I (which are excluded by Brans' theorem). However, examples exist where the Riemann tensor has Petrov type III and N respectively; these are presented in the next two sections.

4. A Type III Example

The second fundamental forms $\boldsymbol{\Omega}$ and $\boldsymbol{\Lambda}$ both have Segré type $[211]$ with Ricci squares of type $[(21)1]$ (or both $[(11)11]$ with Ricci squares $[(111)1]$ if $\beta=0$) and take the form:

$$\Omega_{ab}=\beta\ell_a\ell_b+\lambda(\ell_a n_b+n_a\ell_b+y_a y_b)-\mu x_a x_b \qquad \Lambda_{ab}=\beta\ell_a\ell_b+\lambda(\ell_a\tilde{n}_b+\tilde{n}_a\ell_b+y_a y_b)-\mu\tilde{x}_a\tilde{x}_b$$

where(ℓ^a,n^a,x^a,y^a) is a half-null tetrad and $\tilde{n}^a$ and $\tilde{x}^a$ are related to (ℓ^a,n^a,x^a) by a null rotation about ℓ^a leaving y^a fixed:

$$\tilde{x}^a=x^a+\alpha\ell^a \qquad \tilde{n}^a=n^a+\alpha x^a+\tfrac{1}{2}\alpha^2\ell^a$$

To satisfy (8) the normals to the spacetime must have opposite causal characters (i.e. $e_1=-e_2$). The commutator is given by $C_{ab}=2\alpha(\lambda-\mu)^2\ell_{[a}x_{b]}$ and the Riemann tensor has the form:

$$R^{ab}{}_{cd}=4e_1\alpha\lambda(\lambda-\mu)\left(\alpha(\ell^{[a}x^{b]}\ell_{[c}x_{d]}-\ell^{[a}y^{b]}\ell_{[c}y_{d]})+\ell^{[a}x^{b]}\ell_{[c}n_{d]}+\ell^{[a}n^{b]}\ell_{[c}x_{d]}-\ell^{[a}y^{b]}x_{[c}y_{d]}-x^{[a}y^{b]}\ell_{[c}y_{d]}\right)$$

The Riemann tensor clearly has Petrov type III with ℓ^a being the repeated principal null vector, since $R^{ab}{}_{c[d}\ell^c\ell_{e]}=0$, but $R^{ab}{}_{cd}\ell^c\neq 0$.

[1] The precise number depends on whether degenerate Segré types are counted separately and whether use is made of the freedom (5, 6) in $\boldsymbol{\Omega}$ and $\boldsymbol{\Lambda}$ to show certain combinations of different Segré types are, in fact, equivalent

5. A Type N Example

The second fundamental forms $\mathbf{\Omega}$ and $\mathbf{\Lambda}$ have Segré types [211] and [(11)11] respectively with Ricci squares of type [(11)(11)] and take the form:

$$\Omega_{ab} = \beta\ell_a\ell_b - \lambda(x_ax_b - y_ay_b) \qquad \Lambda_{ab} = -\mu\tilde{x}_a\tilde{x}_b - \nu\tilde{y}_a\tilde{y}_b$$

where(ℓ^a, n^a, x^a, y^a) is a half-null tetrad and $(\tilde{x}^a, \tilde{y}^a)$ is an orthonormal dyad given by:

$$\tilde{x}^a = \cos\theta x^a + \sin\theta y^a \qquad \tilde{y}^a = -\sin\theta x^a + \cos\theta y^a$$

The vacuum conditions (8) are satisfied provided that $e_1\lambda^2 = e_2\mu\nu$. The commutator is given by $C_{ab} = 2\lambda(\mu - \nu)\sin 2\theta x_{[a}y_{b]}$. The Riemann tensor has the form:

$$R^{ab}{}_{cd} = -4e_1\beta\lambda\left(\ell^{[a}x^{b]}\ell_{[c}x_{d]} - \ell^{[a}y^{b]}\ell_{[c}y_{d]}\right)$$

This clearly has Petrov type N with principal null vector ℓ^a since $R^{ab}{}_{cd}\ell^c = 0$.

6. Conclusions

The examples presented in the previous two sections are potential counter-examples to Yakupov's result [3] that the commutator of the two second fundamental forms in a vacuum spacetime of embedding class two must vanish. In fact, after allowing for the freedom in the choice of the normals to the embedded spacetime (5, 6), they are essentially the only possible counter-examples.

The results obtained so far do not constitute definite counter-examples to Yakupov's results as they could turn out to be incompatible with the integrability conditions arising from the Codazzi and Ricci embedding equations (2-4). A preliminary investigation of these equations for the type III spacetimes reveals that they must be of Kundt's class [8] with the null vector ℓ^a being twist-free, shear-free and geodesic and having zero expansion (i.e. the Newman-Penrose quantities ρ, σ and κ all vanish), in agreement with a theorem of Collinson [9].

Similar techniques to those in §2–3 may be employed to investigate Einstein spaces (i.e. $R_{ab} = \kappa g_{ab}$) of embedding class two. In this case Yakupov's identity takes the form

$$C^{ab}{}_{cd}C^{cd} = \tfrac{4}{3}\kappa C_{ab} \tag{10}$$

Here, again, it appears that there exist many examples in which the commutator of the second fundamental forms is non-zero contrary to Yakupov's assertion [3]. All Petrov types seem to be possible except types III, N and O which are immediately excluded by (10).

A full investigation of the integrability conditions for both the vacuum and the Einstein space cases is in progress and will be presented elsewhere.

Acknowledgments

The author wishes to thank Norbert van den Bergh for useful discussions and encouragement regarding this work.

References

[1] Eisenhart L P 1949 *Riemannian Geometry* (Princeton, Princeton Univ. Press) pp 187–200
[2] Yakupov M Sh 1968 On Einstein spaces of second embedding class *Sov. Phys., Doklady* **13**, 585–7
[3] Yakupov M Sh 1973 Einstein spaces of class 2 (in Russian) *Grav. Teor. Otnos., Univ. Kazan* **9** 109–14
[4] Stephani H, Kramer D, MacCallum M A H, Hoenselaers C and Herlt E 2003 *Exact Solutions to Einstein's Field Equations (2nd ed.)* (Cambridge UK, Cambridge Univ. Press) chapter 37, pp 598–9
[5] Hodgkinson D E 1984 Empty space-times of embedding class 2 *Gen. Rel. Grav.* **16** 569–87
[6] Van den Bergh N 1996 Lorentz- and hyperrotation-invariant classification of symmetric tensors and the embedding class 2 problem *Class. Quantum Grav.* **13** 2817–37
[7] Brans C H 1975 Some restrictions on algebraically general vacuum metrics *J. Math. Phys.* **16** 1008–10
[8] Kundt W 1961 The plane-fronted gravitational waves *Z. Phys.* **163** 77–86 (see also chapter 31 of [4]).
[9] Collinson C D 1962 Empty space-times of embedding class two *J. Math. Phys.* **7** 608–11

Spanish Relativity Meeting (ERE 2010): Gravity as a Crossroad in Physics IOP Publishing
Journal of Physics: Conference Series **314** (2011) 012092 doi:10.1088/1742-6596/314/1/012092

Singularity theorems assuming trapped submanifolds of arbitrary dimension

Gregory J Galloway[1] **and José M M Senovilla**[2]

[1]Department of Mathematics, University of Miami, Coral Gables, FL, U.S.A.
[2] Física Teórica, Universidad del País Vasco, Apartado 644, 48080 Bilbao, Spain

E-mail: josemm.senovilla@ehu.es

Abstract. New singularity theorems are proven in Lorentzian manifolds of arbitrary dimension n if they contain closed trapped submanifolds of arbitrary co-dimension. The timelike or null convergence conditions must be generalized to a condition on sectional curvatures, or tidal forces, which reduces to the former in the cases of co-dimension 1, 2 or n. Applications to higher dimensional theories and to the case of trapped circles are briefly mentioned.

1. Introduction

The classical Hawking-Penrose singularity theorem [6] is

Theorem 1 (Hawking and Penrose) *Spacetime (in 4 dimensions) is causal geodesically incomplete if the strong-energy, causality and generic conditions hold and if there is one of the following:*

- *a closed achronal set without edge,* (co-dimension 1)
- *a closed trapped surface,* (co-dimension 2)
- *a point with re-converging light cone.* (co-dimension 4)

What about co-dimension 3 —a closed spacelike curve? In order to include this missing case we need a unification of the concept of trapping for submanifolds of arbitrary co-dimension. This is given by the mean curvature vector $\vec{H}$ of the submanifold ζ [7, 8, 9]. Recall that, for any vector field $\vec{n}$ orthogonal to ζ, $\theta(\vec{n}) \equiv n_\mu H^\mu$ is called the expansion of the submanifold along $\vec{n}$. A spacelike ζ is said to be future-trapped if $\vec{H}$ is timelike and future-pointing on all ζ. This is equivalent to having $\theta(\vec{n}) < 0$ for all possible future normal vectors $\vec{n}$.

2. Existence of focal points

The only intermediate result that is not immediately available in order to prove singularity theorems —such as theorem 1— based on trapped submanifolds ζ of arbitrary dimension is the existence of points focal to ζ. To that end, we use the following notation:

- n_μ: *future-pointing* normal to the spacelike submanifold ζ
- $\vec{e}_A$: vector fields tangent to ζ. They obviously satisfy $n_\mu e^\mu_A = 0$
- γ: geodesic curve tangent to n^μ at ζ

- u: affine parameter along γ, setting $u = 0$ at ζ
- N^μ: geodesic vector field tangent to γ such that $N_\mu|_{u=0} = n_\mu$
- $\vec{E}_A$: vector fields defined by parallelly propagating $\vec{e}_A$ along γ, and such that $\vec{E}_A|_{u=0} = \vec{e}_A$. Observe that $N_\mu E^\mu_A = 0$ for all u
- $P^{\nu\sigma} \equiv \gamma^{AB} E^\nu_A E^\sigma_B$ (at $u = 0$ this is the projector to ζ), where γ^{AB} is the inverse of the first fundamental form $\gamma_{AB} \equiv g_{\mu\nu} e^\mu_A e^\nu_B$ of ζ in the spacetime. Here $g_{\mu\nu}$ denotes the Lorentzian metric in the manifold.

Proposition 1 *Let ζ be a spacelike submanifold of co-dimension m in a Lorentzian manifold of dimension n, and let n_μ be a future-pointing normal to ζ. If $\theta(\vec{n}) \equiv (m-n)c < 0$ and the curvature tensor satisfies the inequality*

$$\boxed{R_{\mu\nu\rho\sigma} N^\mu N^\rho P^{\nu\sigma} \geq 0} \tag{1}$$

along γ, then there is a point focal to ζ along γ at or before $\gamma|_{u=1/c}$, provided γ arrives that far.

The proof can be found in [4].

Remarks:

(i) Spacelike hypersurfaces: $m = 1$. There is a unique timelike orthogonal direction n_μ. Then $P_{\mu\nu} = g_{\mu\nu} - (N_\rho N^\rho)^{-1} N_\mu N_\nu$ and (1) reduces to

$$R_{\mu\nu} N^\mu N^\nu \geq 0.$$

This is the *timelike convergence condition* along γ [5].

(ii) Spacelike 'surfaces': $m = 2$. Now, there are two linearly independent null normals on ζ, say n_μ and ℓ_μ. Define L_μ parallelly propagating ℓ_μ on γ. Then, $P_{\mu\nu} = g_{\mu\nu} - (N_\rho L^\rho)^{-1}(N_\mu L_\nu + N_\nu L_\mu)$ and again (1) reduces to

$$R_{\mu\nu} N^\mu N^\nu \geq 0$$

This is the *null convergence condition* along γ [5].

2.1. The curvature condition (1)

For co-dimension $m > 2$, the interpretation of condition (1) can be given physically in terms of *tidal forces*, or geometrically in terms of *sectional curvatures* [4].

The sectional curvature $k(n, e)$ relative to the plane spanned by $\vec{n}$ and $\vec{e}$ ($n_\mu e^\mu = 0$) is defined by [1, 2, 7, 9]

$$R_{\mu\nu\rho\sigma} n^\mu e^\nu n^\rho e^\sigma = k(n, e)(n_\rho n^\rho)(e_\rho e^\rho)$$

if n_μ is not null, and by [1]

$$R_{\mu\nu\rho\sigma} n^\mu e^\nu n^\rho e^\sigma = -k(n, e)(e_\rho e^\rho)$$

if it is null. Observe that the second of these coincide with the first for the case of a *unit* timelike normal. (There is a different sign with respect to the literature [1, 4]).

Hence (1) states that: *the sum of the $n - m$ sectional curvatures relative to a set of mutually orthogonal planes aligned with n_μ is non-positive, and remains so along γ.* These planes are timelike (respectively null) for a timelike (resp. null) normal n_μ.

In physical terms, this is a statement about the attractiveness of the gravitational field on average. The tidal force in directions initially tangent to ζ is attractive on average.

3. New singularity theorems

We concentrate on the generalization of the more elaborated Hawking-Penrose theorem, but many other theorems can be similarly generalized, see [4].

Recall that for any set ζ, $E^+(\zeta) \equiv J^+(\zeta)\backslash I^+(\zeta)$, using the standard notation for the causal $J^+(\zeta)$ and chronological $I^+(\zeta)$ futures of ζ, see e.g. [1, 6, 5, 7, 10, 12]. For the proofs of the following statements we refer the reader to [4].

Proposition 2 *If $(\mathcal{V}, g)$ is strongly causal and there is a closed f-trapped submanifold ζ of arbitrary co-dimension $m > 1$ such that condition (1) holds along every null geodesic emanating orthogonally from ζ, then either $E^+(E^+(\zeta) \cap \zeta)$ is compact, or the spacetime is null geodesically incomplete, or both.*

Thus, $\eta \equiv E^+(\zeta) \cap \zeta$ provides the compact achronal set in the standard Hawking-Penrose Lemma [6, 5, 12], the achronality of η being a direct consequence of the achronality of $E^+(\zeta)$. (The case with $m = 1$ is not included as it is trivial. If ζ is an achronal hypersurface, then $E^+(\zeta) = \zeta$ and so this set is compact).

The generalization of the Hawking-Penrose theorem follows from the above.

Theorem 2 *If the chronology, generic and strong energy conditions hold and there is a closed f-trapped submanifold ζ of arbitrary co-dimension such that condition (1) holds along every null geodesic emanating orhogonally from ζ then the spacetime is causal geodesically incomplete.*

Remark: For co-dimension $m = 1$ there are no null geodesics orthogonal to ζ so that there is no need to assume (1). For co-dimension $m = 2$ the condition (1) is actually included in the strong energy condition as explained after Proposition 1. Something analogous occurs for co-dimension $m = n$. These three cases cover the original Hawking-Penrose theorem, which was valid in arbitrary dimension (but only for co-dimension $m = 1, 2, n$).

4. Selected applications

The main obvious application of these theorems is to higher dimensional spacetimes, and thereby they can be relevant in Kaluza-Klein, string, supergravity, or M-type theories. For instance, in dimension $n = 11$, say, there are now 10 inequivalent possibilities for the boundary condition in the singularity theorems. This can be of special relevance in connection with the compactified extra-dimensions.

As an example, consider the arguments by Penrose [11] on the *classical* instability of compactified spatial extra-dimensions, which seem to imply that some sort of singularities may develop within a tiny fraction of a second. These arguments can be extended and probably clarified or reinforced by using our generalized Theorems.

Even in the traditional case of $n = 4$, the new theorems have applications to the cases with closed trapped curves. These are curves whose proper acceleration vector is timelike. An obvious relevant example is the case of spacetimes with whole cylindrical symmetry [13]

$$ds^2 = -A^2 dt^2 + B^2 d\rho^2 + F^2 d\varphi^2 + E^2 dz^2, \tag{2}$$

where $\partial_\varphi, \partial_z$ are spacelike commuting Killing vectors, whence all the functions depend only on t and ρ, and the axis is located at $\rho \to 0$. The coordinate φ is closed with standard periodicity 2π. The cylinders given by constant values of t and ρ are geometrically preferred; however, they are *not* compact in general.

On the other hand, the spacelike curves defined by constant values of t, ρ and z are certainly *closed*. Their mean curvature vector is proportional to dF. Thus, the causal character of the gradient of the norm of the circular Killing vector ∂_φ characterizes the trapping of these closed circles. Many results on incompleteness of geodesics can be derived thereof. There also appears a new type of horizon, defined as the set of points where dF is null, acting as the boundary which separates the trapped from the untrapped circles.

Finally, an important remark is that the curvature condition (1) in Proposition 1 can be weakened. As happens in the case of proper Riemannian spaces [3], it is sufficient that it holds on the average in the sense that only a milder version, using the integral along the geodesic, is needed. This is discussed in [4].

Acknowledgments

Supported by grants FIS2004-01626 (MICINN), GIU06/37 (UPV/EHU), P09-FQM-4496 (J. Andalucía), and DMS-0708048 (NSF).

References

[1] Beem J K, Ehrlich P E and Easley K L 1996 *Global Lorentzian Geometry* (Pure and Applied Math. 202, Marcel Dekker, New York)
[2] Eisenhart L P 1966 *Riemannian Geometry*, (Princeton University Press, Princeton).
[3] Galloway G J 1981, "Some results on the occurrence of compact minimal submanifolds" *Manuscripta Math.* **35** 209–219
[4] Galloway G J and Senovilla J M M 2010 "Singularity theorems based on trapped submanifolds of arbitrary co-dimension" *Class. Quantum Grav.* **27** 152002
[5] Hawking S W and Ellis G F R 1973, *The large scale structure of spacetime* (Cambridge Univ. Press, Cambridge)
[6] Hawking S W and Penrose R 1970 "The singularities of gravitational collapse and cosmology" *Proc. Roy. Soc. London A* **314** 529-548
[7] Kriele M 1999 *Spacetime* (Springer-Verlag, Berlin)
[8] Mars M and Senovilla J M M 2003, "Trapped surfaces and symmetries" *Class. Quantum Grav.* **20** L293–L300.
[9] O'Neill B 1983 *Semi-Riemannian geometry* (Academic Press, New York)
[10] Penrose R 1972, *Techniques of Differential Topology in Relativity*, Regional Conference Series in Applied Math. **7** (SIAM, Philadelphia)
[11] Penrose R 2003, "On the instability of extra space dimensions" in *The future of theoretical physics and cosmology*, Gibbons G W, Shellard E P S, and Rankin S J, eds. (Cambridge Univ. Press, Cambridge)
[12] Senovilla J M M 1998, "Singularity Theorems and Their Consequences", *Gen.Rel.Grav.* **30** 701-848
[13] Stephani H, Kramer D, MacCallum M A H, Hoenselaers C, and Herlt E 2003 *Exact Solutions to Einstein's Field Equations* 2nd ed. (Cambridge University Press, Cambridge)

New examples of marginally trapped surfaces and tubes in warped spacetimes

Miguel Ortega

Department of Geometry and Topology, University of Granada, 18071 Granada (Spain)

E-mail: miortega@ugr.es

Abstract. By using the classical Hopf map, we construct another fibration that allows us to obtain examples of marginally trapped tori and marginally outer trapped tubes (MOTT) which are foliated by tori, all of them embedded in a closed Friedman-Lemaître-Robertson-Walker 4-spacetime. In addition, we show examples of MOTTs with any causal character.

1. Preliminaries

We consider a 4-dimensional spacetime (M^4, g), a compact, without boundary, embedded spacelike surface S in (M^4, g). We choose two normal, future-pointing lightlike vector fields $\overrightarrow{k}$, $\overrightarrow{l}$ such that $g(\overrightarrow{l}, \overrightarrow{k}) = -1$. We call $A_{\overrightarrow{k}}$ and $A_{\overrightarrow{l}}$ the associated shape operators with $\overrightarrow{k}$ and $\overrightarrow{l}$ respectively. We recall that S is called *Marginally Outer Trapped Surface* (MOTS for short) when one of the expansions vanish, i. e., $\mathrm{trace}(A_{\overrightarrow{l}}) = 0$ and $\mathrm{trace}(A_{\overrightarrow{k}}) \neq 0$ everywhere, or viceversa. We should point out that the mean curvature vector $\vec{H}$ of a MOTS always satisfies $\|\vec{H}\| = 0$, but the converse does not hold. Actually, we are going to solve this equation in general, and then we will add topological hypothesis to obtain MOTS.

Next, we recall that a *Marginally Outer Trapped Tube* (MOTT for short) is a 3-dimensional smooth manifold $\mathcal{G}$ which admits a foliation by surfaces $\{S_\lambda : \lambda \in \Lambda\}$ such that there is a smooth immersion $\Phi : \mathcal{G} \to M^4$ satisfying:

(i) each $\Phi(S_\lambda)$ $(\lambda \in \Lambda)$ is a MOTS in M^4,

(ii) $\Phi(S_\lambda) \cap \Phi(S_\mu) = \emptyset$ for any $\lambda \neq \mu$.

Also, it is worth pointing out that the causal character of the MOTT may vary from point to point.

Our target is to obtain new examples MOTS and MOTT in a closed Friedman-Lemaître-Robertson-Walker 4-spacetime. We will obtain some partial results firstly, by using CMC surfaces in $\mathbb{S}^3$, and secondly, by using the classical Hopf map.

2. Using CMC surfaces in $\mathbb{S}^3$

Let us consider a smooth function $f : I \subset \mathbb{R} \to (0, \infty)$, $t \in I$, and a 3-dim. Riemannian manifold (M^3, g_3). With them, we construct the Generalized-Robertson-Walker 4-spacetime $\overline{M}_1^4 = I \times M^3$ with line element $\overline{g}_4 = -dt^2 + f^2 g_3$, where f is playing the rôle of the scale factor. Next, we

Journal of Physics: Conference Series **314** (2011) 012093 doi:10.1088/1742-6596/314/1/012093

consider a surface S and an immersion $\varphi : S \to M^3$. For a fixed $t_o \in I$, we define

$$\begin{array}{ccccc} S & \xrightarrow{\varphi} & M^3 & \xrightarrow{\psi} & \overline{M}_1^4 \\ & & p & \mapsto & (t_0, p) \end{array} \Rightarrow \phi := \psi \circ \varphi$$

being ϕ an immersion of S in $\overline{M}_1^4$ in the $t = t_o$ slice.

From the composition $\phi = \psi \circ \varphi$ it is clear that the mean curvature vector associated with ϕ is a linear combination of the mean curvature vectors associated with ψ and φ. Thus, if $\vec{H}_\phi$ and $\vec{H}_\varphi$ stand for the mean curvature vectors associated with ϕ and φ, respectively, one obtains

$$\vec{H}_\phi(p) = \frac{\vec{H}_\varphi(p)}{f^2(t_o)} + \frac{f'(t_o)}{f(t_o)} \partial_t|_{(t_o,p)}, \quad p \in S. \tag{1}$$

Theorem 1 *A surface $\phi : S \to \bar{M}^4$ contained in a t_0-slice of $(\overline{M}_1^4, -dt^2 + f^2 g_3)$ satisfies $\|\vec{H}_\phi\| = 0$ if, and only if, $\varphi : S \to M^3$ has constant mean curvature with $\|\vec{H}_\varphi\| = |f'(t_0)|$.*

Corollary 1 *There exist MOTS with arbitrary genus in closed ($M^3 = \mathbb{S}^3$) FLRW spacetimes.*

Proof: In [1] and [3], the authors show the existence of compact, with arbitrary genus, embedded surfaces in the (standard) round 3-sphere with (small) constant mean curvature. Then, by considering one of such surfaces, we construct one of our embeddings in a closed FLRW 4-spacetime at a suitable time t_0, and we only have to resort to Theorem 1.

3. Examples of MOTT in closed FLRW foliated by tori with different causality

We are interested in constructing some MOTT in closed FLRW spacetimes foliated by tori with different causality. Let $\mathbb{C}$ be the complex numbers, with $i = \sqrt{-1}$, $|z|$ the modulus of $z \in \mathbb{C}$, $\overline{z}$ its complex conjugate. We consider the round 3-sphere $\mathbb{S}^3 = \{(z, w) \in \mathbb{C}^2 : |z|^2 + |w|^2 = 1\}$, with standard metric g_3. Now, we recall the embedded torus with constant mean curvature C_u in $\mathbb{S}^3$ given by

$$C_u := \left\{(z_1, z_2) \in \mathbb{S}^3 \subset \mathbb{C}^2 : |z_1| = \cos(u),\ |z_2| = \sin(u)\right\},$$

$u \in (0, \pi/2)$, with mean curvature $\|\vec{H}_u\| := |2\cot(2u)|$. We define an auxiliary function

$$h : I \to (0, \pi/2),\ h(t) = \frac{1}{2}\operatorname{arccot}\left(\frac{f'(t)}{2}\right) = \frac{\pi}{4} - \frac{1}{2}\arctan\left(\frac{f'(t)}{2}\right),$$

and now, the embedding

$$\begin{gathered} \phi : I \times \mathbb{S}^1 \times \mathbb{S}^1 \to -I \times_f \mathbb{S}^3, \\ \phi(t, e^{i\theta}, e^{i\nu}) = \left(t, e^{i\theta}\cos(h(t)), e^{i\nu}\sin(h(t))\right). \end{gathered}$$

For each $t \in I$, the map $\phi(t, -, -) : \mathbb{S}^1 \times \mathbb{S}^1 \to \overline{M}_1^4$, is an embedding of a torus in the t-slice, with constant mean curvature $\|\vec{H}_u\| = \left|2\cot(2u)\ |_{u=h(t)}\right| = |f'(t)|$. By Theorem 1, each torus is a MOTS, and therefore, ϕ is a MOTT. The induced line element on $I \times \mathbb{S}^1 \times \mathbb{S}^1$ is

$$\phi^* \bar{g}_4 = \bar{g}_4(\phi_t, \phi_t)dt^2 + (f(t)\cos(h(t)))^2\, d\theta^2 + (f(t)\sin(h(t)))^2\, d\nu^2.$$

Clearly, the causal character depends only on ϕ_t, so we are going to study it by defining the function

$$z(t) := \bar{g}_4(\phi_t, \phi_t) = -1 + \left(\frac{f(t) f''(t)}{4 + f'(t)^2}\right)^2.$$

Next, we exhibit four different examples, showing that any causal character can actually happen.

(i) Given $a, b > 0$ such that $a^2 = 4 + b^2$, define the function $f : I = \mathbb{R} \to (0, \infty)$, $f(t) = a\cosh(t) + b\sinh(t)$. Then, $z(t) \equiv 0$. Therefore, ϕ_t is everywhere lightlike.

(ii) Define the function $f : (-1, 1) \to (0, \infty)$, $f(t) = \frac{2}{1-t^2}$. By simple computations, we obtain $z(t) \geq 3$, for any $t \in (-1, 1)$, and therefore ϕ_t is always spacelike.

(iii) Take real constants $c_1, c_2 > 0$. Then, the function $f : \mathbb{R} \to (0, \infty)$, $f(t) = \frac{4+c_1^2}{4c_2}t^2 + c_1 t + c_2$ is well-defined. A simple computation shows $z(t) \equiv -3/4$. This implies that ϕ_t is everywhere timelike.

(iv) Given the function $f : \mathbb{R} \to (0, \infty)$, $f(t) = 3 + \cos(2t)$. A straightforward computation gives

$$z(t) := -1 + \left(\frac{f(t)f''(t)}{4 + f'(t)^2}\right)^2 = -1 + \frac{4\cos^2(2t)(3 + \cos(2t))^2}{(3 - \cos(4t))^2}.$$

Finally, it is easy to check $z(0) = 15$ and $z(\pi/4) = -1$. In this case, the causal character changes *with time*.

4. Using the Hopf map

We regard now the standard 2-round sphere of radius 1/2 as $\mathbb{S}^2(1/2) = \{(z, x) \in \mathbb{C} \times \mathbb{R} : |z|^2 + x^2 = 1/4\}$, with standard line element g_2. The classical `Hopf map` is

$$\pi : \mathbb{S}^3 \to \mathbb{S}^2(1/2), \quad \pi(z, w) = \left(z\overline{w}, \frac{1}{2}|z|^2 - \frac{1}{2}|w|^2\right).$$

This map has many marvellous properties. Among them, we recall the following:

(i) π is a Riemannian submersion.

(ii) For each $(z, a) \in \mathbb{S}^2(1/2)$, then $\pi^{-1}\{(z, a)\}$ is a closed geodesic in $\mathbb{S}^3$.

Now, we extend it to a new submersion as follows:

$$\overline{\pi} : (-I \times_f \mathbb{S}^3, -dt^2 + f^2 g_3) \to (-I \times_f \mathbb{S}^2(1/2), -dt^2 + f^2 g_2), \ \overline{\pi}(t, p) = (t, \pi(p)).$$

Next, we consider a curve $\alpha : J \to -I \times_f \mathbb{S}^2(1/2)$, and its `pullback` $\overline{\pi}^*(\alpha) = J \times \mathbb{S}^1 \to -I \times_f \mathbb{S}^3$. We point out that the geometric elements of α determine the properties of the mean curvature vector of $\overline{\pi}^*\alpha$. For instance, if α is embedded and open/closed, then $\overline{\pi}^*(\alpha)$ is an embedded cylinder/torus in $-I \times_f \mathbb{S}^3$. In addition, these surfaces may not be contained in a single t-slice.

To make computations, we ask α to be a unit spacelike Frenet curve. Next, let β be a `horizontal lift` of α, i. e., $\overline{\pi} \circ \beta = \alpha$ and β' is orthogonal to $\ker(\overline{\pi})_*$ (i. e., β' is horizontal.) Also, for each $e^{i\theta} \in \mathbb{S}^1$, the map

$$\overline{\Gamma}_\theta : -I \times_f \mathbb{S}^3 \to -I \times_f \mathbb{S}^3, \quad \overline{\Gamma}_\theta(t, (z, w)) = \left(t, (e^{i\theta} z, e^{i\theta} w)\right)$$

is an isometry, so we can construct a parametrization of the surface $\overline{\pi}^*(\alpha)$ by

$$\phi : \overline{\pi}^*(\alpha) = J \times \mathbb{S}^1 \to -I \times_f \mathbb{S}^3, \quad \phi(s, \theta) = \overline{\Gamma}_\theta(\beta(s)).$$

If $\overline{g}_3 = -dt^2 + f^2 g_2$ is the line element, since α is a unit Frenet curve, we can compute its Frenet apparatus $\{T = \dot{\alpha}, N, B\}$ and κ, τ, with Frenet equations

$$\nabla_T T = \epsilon_2 \kappa N, \quad \nabla_T N = \kappa T + \epsilon_3 \tau B, \quad \nabla_T B = -\epsilon_2 \tau N,$$

where $\epsilon_2 = \overline{g}_3(N, N)$, $\epsilon_3 = \overline{g}_3(B, B)$, $\epsilon_2 = -\epsilon_3 = \pm 1$, and $\{T, N, B\}$ is a positive basis along α. As a summary, we have the following commutative diagram:

$$\begin{array}{ccc} \overline{\pi}^*(\alpha) = J \times \mathbb{S}^1 & \xrightarrow{\phi} & (-I \times_f \mathbb{S}^3, \overline{g}_4 = -dt^2 + f^2 g_3) \\ \downarrow & \overset{\beta}{\nearrow} & \downarrow \overline{\pi} \\ J & \xrightarrow{\alpha} & (-I \times_f \mathbb{S}^2(1/2), \overline{g}_3 = -dt^2 + f^2 g_2). \end{array}$$

Let $\tilde{N}$ and $\tilde{B}$ be horizontal lifts of N and B, resp., along β.

Lemma 1 *The mean curvature vector of ϕ is given by*

$$\vec{H}_\phi = \frac{\epsilon_2}{2}\left(\kappa + \frac{f'}{f}\overline{g}_3(\partial_t, N)\right)(\overline{\Gamma}_\theta)_*\tilde{N} + \frac{\epsilon_3}{2}\left(\frac{f'}{f}\overline{g}_3(\partial_t, B)\right)(\overline{\Gamma}_\theta)_*\tilde{B}$$

Proposition 1 *The mean curvature vector $\vec{H}_\phi$ satisfies $\|\vec{H}_\phi\| = 0$ if, and only if,*

$$\left(\kappa + \frac{f'}{f}\overline{g}_3(\partial_t, N)\right)^2 - \left(\frac{f'}{f}\overline{g}_3(\partial_t, B)\right)^2 = 0.$$

5. Final remarks

In [2], we obtained an open embedded surface with null mean curvature vector, and crossing two regions, one expanding and one collapsing. As a result, we pose the following open problem: to find an explicit MOTS in the 4-dim closed FLRW spacetime, which is not contained in any t-slice, from a closed curve in the toy model $-I \times_f \mathbb{S}^2(1/2)$.

6. Conclusions

- There exist MOTS in closed FLRW 4-spacetimes embedded in t_0-slices with arbitrary topology.
- This leads to MOTT in closed FLRW 4-spacetimes with any causal character. Our examples are foliated by tori, but other topologies are also possible.
- From a curve in a (toy model) closed FLRW 3-spacetime $\gamma : J \to (-I \times_f \mathbb{S}^2(1/2), -dt^2 + f^2 g_2)$, it is possible to construct embedded cylinders and tori in the closed FRLW 4-spacetime $(-I \times_f \mathbb{S}^3, -dt^2 + f^2 g_3)$ with some control of the mean curvature vector.
- Problem: to construct such a tori which is also a MOTS.

Acknowledgments

Partially supported by the Spanish MEC-FEDER Grant MTM2007-60731 and the Junta de Andalucia Regional Grant P09-FQM-4496 with FEDER funds.

References

[1] A. Butscher, F. Pacard, *Doubling Constant Mean Curvature Tori in* $\mathbb{S}^3$, Ann. Scuola Norm. Sup. Pisa Cl. Sci., (5) Vol. V (2006), 611-638.

[2] J. L. Flores, S. Haesen, M. Ortega, *Class. Quantum Grav.* **27** (2010) 145021 (18pp)

[3] H. B. Lawson, *Complete Minimal Surfaces in* $\mathbb{S}^3$, Annals of Math. **92** (1970) 335-374.

[4] B. O'Neill, Semi-Riemannian geometry. With applications to relativity. *Pure and Applied Mathematics, 103. Academic Press, Inc.*, New York, 1983

Spanish Relativity Meeting (ERE 2010): Gravity as a Crossroad in Physics IOP Publishing
Journal of Physics: Conference Series **314** (2011) 012094 doi:10.1088/1742-6596/314/1/012094

Horizon-entropy increase laws for spherically symmetric horizons in Brans-Dicke theory.

Alex B. Nielsen
Max-Planck-Institut für Gravitationsphysik,
Am Mühlenberg 1, D-14476 Golm,
Germany
E-mail: alex.nielsen@aei.mpg.de

Abstract. We derive the horizon-entropy increase law for spherically symmetric quasi-local horizons in Brans-Dicke theory. The quasi-local horizons used do not marginally trap null rays, and hence are not apparent horizons or foliated by marginally trapped surfaces, but instead have instantaneously constant gravitational entropy in the outgoing null direction. The relation derived has a very direct comparison with the horizon-entropy increase law for event horizons.

Perhaps one of the first to suggest black hole thermodynamics be applied to quasi-local horizons such as apparent horizons rather than event horizons was Hajicek [1] who conjectured that Hawking radiation originates from the region close to the apparent horizon independently of whether an event horizon exists or not. This idea was further examined by Hiscock [2] who proposed identifying the entropy with one quarter the area of the apparent horizon and Collins [3] who obtained a $T\mathrm{d}S = \mathrm{d}Q$ like relation for apparent horizons. An important contribution was made by Hayward [4] who defined an outer condition for apparent horizons and was able to show that the area of a future outer trapping horizon is non-decreasing if the null energy condition is satisfied on the horizon. These ideas received further attention when it was shown that the microstates of black hole entropy can be counted in loop quantum gravity for a constant area apparent horizon [5] and in analogue models where an event horizon is not necessary for the production of Hawking radiation [6]. The first tentative observations of this effect in analogue models may already have been observed [7].

Here we consider the horizon-entropy increase law for quasi-local horizons. We will consider the case of spherically symmetric horizons in a spherically symmetric spacetime using a Brans-Dicke action. In this case horizon generators can be written as $r^a = Bl^a + Cn^a$ with l^a and n^a the ingoing and outgoing radial null vectors respectively and $B > 0$ and C spherically symmetric functions on the horizon. Generalisations to non-spherically symmetric situations in more general theories can be found in [8].

Brans-Dicke theory is the prototype alternative theory of gravity with scalar and tensor modes. The action is given by

$$S = \int \mathrm{d}^4x\sqrt{-g}\left[\frac{1}{16\pi}\left(\phi R - \frac{\omega}{\phi}\nabla^a\phi\nabla_a\phi\right) + \mathcal{L}_{matter}\right]. \tag{1}$$

ω is the Brans-Dicke parameter and we assume the scalar field satisfies $\phi > 0$. Variation of this action with respect to the metric gives the gravitational field equations, which, when contracted twice with a null vector l^a give

$$R_{ab}l^a l^b = \frac{\omega}{\phi^2}\left(l^a \nabla_a \phi\right)^2 + \frac{8\pi}{\phi} T_{ab} l^a l^b + \frac{l^a l^b \nabla_a \nabla_b \phi}{\phi}\,, \tag{2}$$

where T_{ab} is the energy-momentum tensor of the matter fields. The last term arises from an integration by parts in the variation of the ϕR term. Because of this last term the sign of $R_{ab}l^a l^b$ is indeterminate. Even in situations where the null energy condition holds, $T_{ab}l^a l^b \geq 0$, the null curvature condition $R_{ab}l^a l^b \geq 0$ may not be satisfied. This is the condition that appears in the Raychaudhuri equation and on which the results of Hawking [9] and Hayward [4] depend. It was noticed for example in [10] that the apparent horizon can appear outside the event horizon in dynamical black hole collapse models in Brans-Dicke gravity.

Under a conformal transformation Brans-Dicke theory can be put into the form of Einstein gravity plus a scalar field non-minimally coupled to matter. Transforming

$$g_{ab} \to \tilde{g}_{ab} = \phi(x) g_{ab}\,, \tag{3}$$

transforms the action to

$$S = \int \mathrm{d}^4x \sqrt{-\tilde{g}} \left[\frac{1}{16\pi}\left(\tilde{R} - \frac{(3+\omega)}{2\phi^2}\tilde{\nabla}^a\phi\tilde{\nabla}_a\phi\right) + \frac{\mathcal{L}_{matter}}{\phi^2}\right]. \tag{4}$$

In this Einstein frame the contracted Ricci tensor is

$$\tilde{R}_{ab}l^a l^b = \frac{(3+\omega)}{2\phi^2}\left(l^a\tilde{\nabla}_a\phi\right)^2 + \frac{8\pi}{\phi^2}T_{ab}l^a l^b\,. \tag{5}$$

The theory automatically satisfies the null curvature condition in the Einstein frame if it satisfies the null energy condition. In the Einstein frame the apparent horizon appears inside the event horizon [10]. The reason why the apparent horizon can be outside the event horizon in one frame, but inside the event horizon in another is because the location of the trapping horizon is not invariant under a conformal transformation. The conformal transformation leaves the spacetime coordinates, causal structure and null rays unchanged. The location of the event horizon remains the same. But the area of a two-dimensional surface will change as

$$A \to \tilde{A} = \phi A\,. \tag{6}$$

In spherical symmetry the expansion of a congruence of rays normal to a spherically symmetric surface can be found by calculating the change of the area. Under a conformal transformation the expansion changes as

$$\theta_l \to \tilde{\theta}_l = \frac{l^a\nabla_a\tilde{A}}{\tilde{A}} = \theta_l + \frac{l^a\nabla_a\phi}{\phi}\,. \tag{7}$$

The vanishing of θ_l for a given surface is therefore not necessarily invariant under a conformal transformation. And thus the location of an apparent horizon satisfying $\theta_l = 0$ is not necessarily invariant.

A general result for the stationary entropy of a diffeomorphism invariant gravity theory was given in [11].

$$S_g = -2\pi \int_H \frac{\partial \mathscr{L}}{\partial R_{abcd}}\hat{\varepsilon}_{ab}\hat{\varepsilon}_{cd}\sqrt{q}\,\mathrm{d}^2x + \text{higher derivative terms}, \tag{8}$$

where $\hat{\varepsilon}_{ab}$ is the antisymmetric binormal form for the surface H, $\hat{\varepsilon}_{ab} = l_a n_b - n_a l_b$ and $\mathscr{L}$ is the full Lagrangian density. For ordinary Einstein gravity this formula gives the familiar result $S_g = A/4$ and for Brans-Dicke theory it gives the gravitational entropy as $S_g = \phi A/4$.

Consider now the following conditions

$$\begin{aligned} l^a\nabla_a S_g &= 0\,, \\ n^a\nabla_a S_g &< 0\,, \\ n^a\nabla_a\left(l^c\nabla_c S_g\right) &< 0\,. \end{aligned} \tag{9}$$

The conditions relate to how S_g is changing in the null directions normal to the horizon surface. In Brans-Dicke theory the first condition would be satisfied where $\theta_l + l^a\nabla_a\phi/\phi = 0$. Therefore these conditions are not satisfied by apparent horizons in Brans-Dicke theory. In ordinary Einstein gravity however, these would reduce to the requirements on the null expansions for a trapping horizon given in [4] since, in this case, $S_g = A/4$. The variation of the generalised entropy is now

$$r^a\nabla_a S_g = Cn^a\nabla_a S_g. \tag{10}$$

Since we require the tangent r^a to generate evolution along the generalised trapping horizon, on which $l^a\nabla_a S_g = 0$, we have $r^a\nabla_a(l^b\nabla_b S_g) = 0$ and thus

$$C = -\frac{Bl^a\nabla_a\left(l^b\nabla_b S_g\right)}{n^c\nabla_c\left(l^d\nabla_d S_g\right)}\,. \tag{11}$$

Putting this together with the Raychaudhuri equation and the equations of motion (2), the final result for the change of S_g along the horizon is then

$$r^a\nabla_a S_g = \frac{A\phi Bn^a\nabla_a S_g}{n^c\nabla_c\left(l^d\nabla_d S_g\right)}\left(\frac{\theta_l^2}{2} + \frac{\omega+1}{\phi^2}\left(l^a\nabla_a\phi\right)^2 + \frac{8\pi}{\phi}T_{ab}l^al^b\right)\,. \tag{12}$$

Since these horizons satisfy $\theta_l + l^a\nabla_a\phi/\phi = 0$ the term $\frac{1}{2}\theta_l^2 + \frac{\omega+1}{\phi^2}\left(l^a\nabla_a\phi\right)^2$ is guaranteed positive if $\omega > -3/2$. For a Brans-Dicke theory of this form and matter obeying the null energy condition $T_{ab}l^al^b \geq 0$, the generalised entropy is guaranteed to be non-decreasing on horizons satisfying the conditions (9).

A similar result can be derived for causal horizons such as event horizons. Because causal horizons are generated by null rays we have $r^a = l^a$ and immediately

$$\begin{aligned} l^b\nabla_b\left(l^a\nabla_a S_g\right) &= A\phi\Bigg[\left(\theta_l + \frac{l^a\nabla_a\phi}{\phi}\right)^2 + \kappa\left(\theta_l + \frac{l^a\nabla_a\phi}{\phi}\right) \\ &\quad -\frac{\theta_l^2}{2} - \frac{\omega+1}{\phi^2}\left(l^a\nabla_a\phi\right)^2 - \frac{8\pi}{\phi}T_{ab}l^al^b\Bigg]\,. \end{aligned} \tag{13}$$

This is now the second derivative of the horizon-entropy. But if the causal horizon settles down to a Killing horizon at late times, then it can be shown that the first derivative of the horizon-entropy cannot ever be negative [8]. Causal horizons will not coincide in general dynamical situations with the quasi-local horizons considered here [12].

Similar results can be extended to other theories such as scalar-tensor and $f(R)$ theories and non-spherically symmetric horizons [8]. These models cover a wide variety of cases from low-energy effective string theory actions, Kaluza-Klein compactifications of higher dimensions and quantum loop corrections. The quasi-local horizons used coincide with trapping horizons in Einstein gravity, but are generally different in all these other theories and models. Even if one restricts attention purely to Einstein gravity, the use of modified theories shows that the instantaneous constancy of gravitational entropy, rather than the focussing of light rays, is what is required to prove the horizon-entropy increase law. This has obvious implications for our understanding of entropy of the gravitational field and its relation to horizons.

The gravitational entropy used here was first derived for purely stationary systems and its extension to dynamical situations is known to contain some ambiguities [11]. The extension of other gravitational-thermodynamic properties to dynamical situations also contains some ambiguities [13, 14, 15]. But we have explicitly applied this horizon-entropy to dynamical situations here and found that a horizon-entropy increase law can be derived for different gravity models and for both quasi-local and causal horizons.

The definition also impacts our understanding of black holes. Ideally we would like to define a black hole in a way that indicates the presence of certain properties that are absent from objects such as neutron stars. Such properties might be the production of Hawking radiation transporting energy from the gravitational field to null infinity or regions with large deviations from classical metric geometry or some measure of black hole thermodynamics with horizon-entropy [16]. Black hole thermodynamics needs a surface in order to compute an area to equate with entropy. The non-uniqueness of trapping horizons [17] will also apply to the surfaces proposed here. But the horizon-entropy will only be conformally invariant for certain types of surfaces, typically requiring the scalar field to be constant on the horizon slice, such as the spherically symmetric surfaces considered here. This may pick out a preferred horizon for computing the horizon-entropy, or it may be that there is no unique definition of such a surface. The physical effects we commonly associate with black hole spacetimes may not be easily associated to a single surface [18], in which case the implications of the area-entropy relation and related holographic principles will need to be revisited.

Acknowledgments
The author is very grateful for support from the Alexander von Humboldt Foundation and hospitality at the Max Planck Institute for Gravitational Physics in Potsdam-Golm.

References
[1] P. Hajicek, Phys. Rev. D **36** (1987) 1065.
[2] W. A. Hiscock, Phys. Rev. D **40** (1989) 1336.
[3] W.Collins, Phys. Rev. D **45** (1992) 495.
[4] S. A. Hayward, Phys. Rev. D **49** (1994) 6467.
[5] A. Ashtekar, J. Baez, A. Corichi and K. Krasnov, Phys. Rev. Lett. **80** (1998) 904
[6] M. Visser, Int. J. Mod. Phys. D **12**, 649 (2003) [arXiv:hep-th/0106111].
[7] S. Weinfurtner et al., Phys. Rev. Lett. **106** (2011) 021302 [arXiv:1008.1911 [gr-qc]].
[8] V. Faraoni and A. B. Nielsen, arXiv:1103.2089 [gr-qc].
[9] Hawking and Ellis *"The Large scale structure of space-time,"* CUP, Cambridge, 1973
[10] M. A. Scheel, S. L. Shapiro and S. A. Teukolsky, Phys. Rev. D **51** (1995) 4236
[11] R. M. Wald, Phys. Rev. D **48** (1993) 3427 [arXiv:gr-qc/9307038].
[12] A. B. Nielsen, Class. Quant. Grav. **27** (2010) 245016 [arXiv:1006.2448 [gr-qc]].
[13] A. B. Nielsen and J. H. Yoon, Class. Quant. Grav. **25** (2008) 085010 [arXiv:0711.1445 [gr-qc]].
[14] A. B. Nielsen and D. h. Yeom, Int. J. Mod. Phys. A **24** (2009) 5261 [arXiv:0804.4435 [gr-qc]].
[15] M. Pielahn, G. Kunstatter and A. B. Nielsen, arXiv:1103.0750 [gr-qc].
[16] A. B. Nielsen, Gen. Rel. Grav. **41** (2009) 1539 [arXiv:0809.3850 [hep-th]].
[17] A. B. Nielsen, M. Jasiulek, B. Krishnan and E. Schnetter, arXiv:1007.2990 [gr-qc].
[18] C. Barcelo, S. Liberati, S. Sonego and M. Visser, Phys. Rev. Lett. **97** (2006) 171301

Conformal Yano-Killing Tensors in General Relativity

Jacek Jezierski

Department of Mathematical Methods in Physics, Faculty of Physics, University of Warsaw, ul. Hoża 74, 00-682 Warsaw, Poland

E-mail: `Jacek.Jezierski@fuw.edu.pl`

Abstract. How CYK tensors appear in General Relativity?

- Geometric definition of the asymptotic flat spacetime: *strong asymptotic flatness*, which guarantees well defined total angular momentum [2, 3, 4]
- Conserved quantities – asymptotic charges ($\mathscr{I}$, i^0) [2, 3, 4, 5, 6, 9]
- Quasi-local mass and "rotational energy" for Kerr black hole [5]
- Constants of motion along geodesics and symmetric Killing tensors [5, 6]

Spacetimes possessing CYK tensor [10]:

- Minkowski (quadratic polynomials) [5]
- (Anti-)deSitter (natural construction) [7, 8, 9]
- Kerr (type D spacetime) [5]
- Taub-NUT (new symmetric conformal Killing tensors) [6]

Other applications:

- Symmetries of Dirac operator
- Symmetries of Maxwell equations

The Conformal Yano-Killing tensor Q (as a 2-form), in n-dimensional spacetime with metric $g_{\mu\nu}$, is a solution to the following equation:

$$Q_{\lambda\kappa;\sigma} + Q_{\sigma\kappa;\lambda} = \frac{2}{n-1}\left(g_{\sigma\lambda}Q^{\nu}{}_{\kappa;\nu} + g_{\kappa(\lambda}Q_{\sigma)}{}^{\mu}{}_{;\mu}\right) . \tag{1}$$

This is a natural generalization of conformal Killing covector field for higher rank differential $p-$forms. For any p, CYK tensor is the element of the kernel of the twistor operator $Q \to \mathcal{T}_{wist}Q$ defined as follows:

$$\forall X \;\; \mathcal{T}_{wist}Q(X) := \nabla_X Q - \frac{1}{p+1}X \lrcorner \mathrm{d}Q + \frac{1}{n-p+1}g(X)\wedge \mathrm{d}^*Q \, . \tag{2}$$

The CYK tensor has several nice properties like: covariance with respect to the conformal rescaling of the metric tensor (Th. 2 in [5]), its divergence is a Killing vector (Th. 4 in [5]), and its Hodge dual is a CYK tensor (Th. 5 in [5]).

On the other hand, for the 4-dimensional closed Riemannian manifold the integrability conditions of CYK equation imply that overdetermined equation (1) is equivalent to harmonic-like equation (Th. 9 in [5])

$$\nabla^{\mu}\nabla_{\mu}Q_{\lambda\kappa} = R^{\sigma}{}_{\kappa\lambda\mu}Q_{\sigma}{}^{\mu} + R_{\sigma[\lambda}Q_{\kappa]}{}^{\sigma} \, . \tag{3}$$

In gravity we shall concentrate on the case $p = 2$. For $n = 4$ the contraction of the 2-form Q with spin-2 field (Weyl tensor) gives another 2-form which fulfills Maxwell equations. This observation enables one to propose a geometric definition of the *asymptotically flat spacetime* (Def. 3 and 4 in [4]): *strong asymptotic flatness*, which guarantees well defined total angular momentum [2, 3, 4]. The global conserved quantities – asymptotic charges – are well defined at spatial infinity if the corresponding *asymptotic* CYK tensor exists. For Schwarzschildean spacetimes there exists a full set of ACYK tensors at spatial infinity [3]. Moreover, ACYK tensor gives some insight in the behaviour of gravitational charges at null infinity [4]. In particular, the obstruction for the existence of angular momentum at $\mathscr{I}$ one can describe in terms of the corresponding ACYK tensor. Recall (see Sec. 5.6 in [1]) that in Bondi-Sachs coordinates (u, x, x^A) the space-time metric takes the form:

$$g = -xVe^{2\beta}du^2 + 2e^{2\beta}x^{-2}dudx + x^{-2}h_{AB}\left(dx^A - U^A du\right)\left(dx^B - U^B du\right) . \tag{4}$$

The extra condition for finiteness of angular momentum at $\mathscr{I}^+$ is related with the term (Sec. 3.2 in [4]):

$$x^{-1}\tilde{W}^{\mu\lambda\alpha\beta}\tilde{Q}_{\alpha\beta} = \begin{cases} x^{-2}\tilde{W}^{uxAx}v_{,A} + O(1) & \text{for boost} \\ x^{-2}\tilde{W}^{uxAx}\varepsilon_A{}^B\hat{v}_{,B} + O(1) & \text{for rotation} \end{cases} , \tag{5}$$

where $v = \frac{X_k}{r}$, $\hat{v} = \varepsilon_{klm}\frac{X^m}{r}$ are dipole functions on a sphere S^2, which is parameterized by coordinates x^A, $r^2 = \sum_k (X_k)^2$. Moreover, $\tilde{W}$ and $\tilde{Q}$ are conformally rescaled tensors, Weyl and CYK respectively, by appropriate power of the conformal factor x. The component (5) becomes finite if we assume that dipole part[1] $\text{dip}\left(\tilde{W}^{uxAx}\right) = O(x^2)$. It also means that the charges $\int_S W^{\mu\lambda}{}_{\alpha\beta}Q^{\alpha\beta}dS_{\mu\lambda}$, built from asymptotically constant CYK tensors Q, are *obstructions* for the existence of angular momentum at $\mathscr{I}^+$ (ofam-charges). The asymptotic conditions which guarantee finiteness of the four-momentum at future null infinity are as follows: energy — $\text{mon}\left(x^{-1}\tilde{W}^{uxux}\right) = O(1)$[2], linear momentum — $\text{dip}\left(x^{-1}\tilde{W}^{uxux}\right) = O(1)$ and they are fulfilled for any asymptotically flat spacetime satisfying the usual asymptotic conditions: $\tilde{W}^{\mu}{}_{\lambda\alpha\beta} = O(x)$, $\tilde{Q}^{\alpha\beta\lambda} = O(1) = \tilde{Q}^{\alpha\beta}$. However, for angular momentum the asymptotic condition $\text{dip}\left(x^{-2}\tilde{W}^{uxAx}\right) = O(1)$ is an extra assumption which is fulfilled for standard asymptotics of Bondi metrics (4) but it is not obvious.

The construction of global charges in General Relativity has a long history. In the case of asymptotically flat spacetime (asymptotically Minkowskian) the concept of asymptotic CYK tensor led to the strong asymptotic flatness condition at spatial infinity and to the construction of charges in terms of the Weyl tensor which are free from supertranslation ambiguity contrary to the "superpotentials" based on asymptotic Killing vector fields.

There are several interesting examples for the existence of the exact (and not only asymptotic) solutions of equation (1) in the physically important spacetimes like Minkowski, Kerr, Taub-NUT, deSitter or anti-deSitter. It is rather surprising that from geometrical point of view it is easier to find in (anti-)deSitter than in Minkowski spacetime the maximal set of 20 CYK 2-forms [8, 9]. More precisely, the assumption of Theorem 1 in [7] that the traceless part of extrinsic curvature of $M \subset N$ vanishes and the trace is constant on M is easily fulfilled for AdS spacetime M embedded in ambient flat 5-dimensional spacetime N. Hence, a pullback of constant CYK

[1] The dipole part of a vector field on S^2 is its orthogonal projection onto the six-dimensional space of conformal vector fields which is simultaneously a "first" eigenspace (with unit eigenvalue) for Laplace–Beltrami operator Δ

[2] By mon we denote monopole part of a scalar field on S^2 i.e. its orthogonal projection onto the one-dimensional kernel of Laplace–Beltrami operator Δ

tensors in N to a submanifold M gives all solutions of a CYK equation in AdS spacetime [8]. We can write the general form of a solution of equation (1) in Minkowski spacetime:

$$Q^{\mu\nu} = q^{\mu\nu} + 2x^{[\mu}u^{\nu]} + \varepsilon^{\mu\nu}{}_{\kappa\lambda}v^{\kappa}x^{\lambda} + 4x^{[\mu}k^{\nu]\lambda}x_{\lambda} + k^{\mu\nu}x_{\lambda}x^{\lambda}\,, \tag{6}$$

where $q^{\mu\nu}$ and $k^{\mu\nu}$ are any constant skew-symmetric tensors, and u^{μ} and v^{μ} are any constant vectors. Moreover, in Kerr spacetime the existing pair of CYK tensors:

$$Q_{\mathrm{Kerr}} = r\sin\theta \mathrm{d}\theta \wedge \left[\left(r^2+a^2\right)\mathrm{d}\phi - a\mathrm{d}t\right] + a\cos\theta \mathrm{d}r \wedge \left(\mathrm{d}t - a\sin^2\theta \mathrm{d}\phi\right) \tag{7}$$

$$*Q_{\mathrm{Kerr}} = a\cos\theta\sin\theta \mathrm{d}\theta \wedge \left[\left(r^2+a^2\right)\mathrm{d}\phi - a\mathrm{d}t\right] + r\mathrm{d}r \wedge \left(a\sin^2\theta \mathrm{d}\phi - \mathrm{d}t\right) \tag{8}$$

defines quasi-local mass and the dual mass in a geometric way. Obviously, the dual mass (as a topological charge) vanishes.

The property, that the symmetrized product $K_{\mu\nu} := Q_{\lambda(\mu}P_{\nu)}{}^{\lambda}$ of CYK tensors $Q_{\mu\nu}$ and $P_{\mu\nu}$ is a conformal (symmetric) Killing tensor (Th. 3 in [5]), enables one to construct symmetric conformal Killing tensors. Some of them seems to be not known [5, 6]. Those objects are useful in the construction of constants of motion along geodesics. More precisely, the geodesic equation $g^{\mu\nu}p_{\mu}\nabla_{\nu}\, p = 0$ implies that for a null covector p and for any conformal Killing tensor K we have $g^{\mu\nu}p_{\mu}\nabla_{\nu}(K^{\mu\nu}p_{\mu}p_{\nu}) = 0$, i.e. $K^{\mu\nu}p_{\mu}p_{\nu}$ is a constant of motion along null geodesics.

Let $W_{\mu\sigma\alpha\beta}$ be a spin-2 field (Weyl tensor) and by "$*$" we denote its dual $W^*{}_{\mu\nu\alpha\beta} := \frac{1}{2}W_{\mu\nu\rho\sigma}\varepsilon^{\rho\sigma}{}_{\alpha\beta}$. A super-tensor

$$2\mathcal{T}_{\mu\nu\alpha\beta\gamma\delta} := W_{\mu\sigma\alpha\beta}W_{\nu}{}^{\sigma}{}_{\gamma\delta} + W_{\mu\sigma\gamma\delta}W_{\nu}{}^{\sigma}{}_{\alpha\beta} + W^*{}_{\mu\sigma\alpha\beta}W^*{}_{\nu}{}^{\sigma}{}_{\gamma\delta} + W^*{}_{\mu\sigma\gamma\delta}W^*{}_{\nu}{}^{\sigma}{}_{\alpha\beta}\,,$$

described in [4] and firstly proposed by Öktem [11], is an example of the object which is "above" the Bel–Robinson tensor, i.e. its trace $g^{\beta\delta}\mathcal{T}_{\mu\nu\alpha\beta\gamma\delta}$ gives the original Bel–Robinson tensor. Theorem 8 in [4] enables one to define (with the help of CYK 2-form) "quadratic charge" as a 3-volume integral. Quite unusual application of this idea is given in [5], where the spinning energy of the rotating black hole is given in terms of the corresponding CYK tensor.

We summarize some relationships by the following diagram:

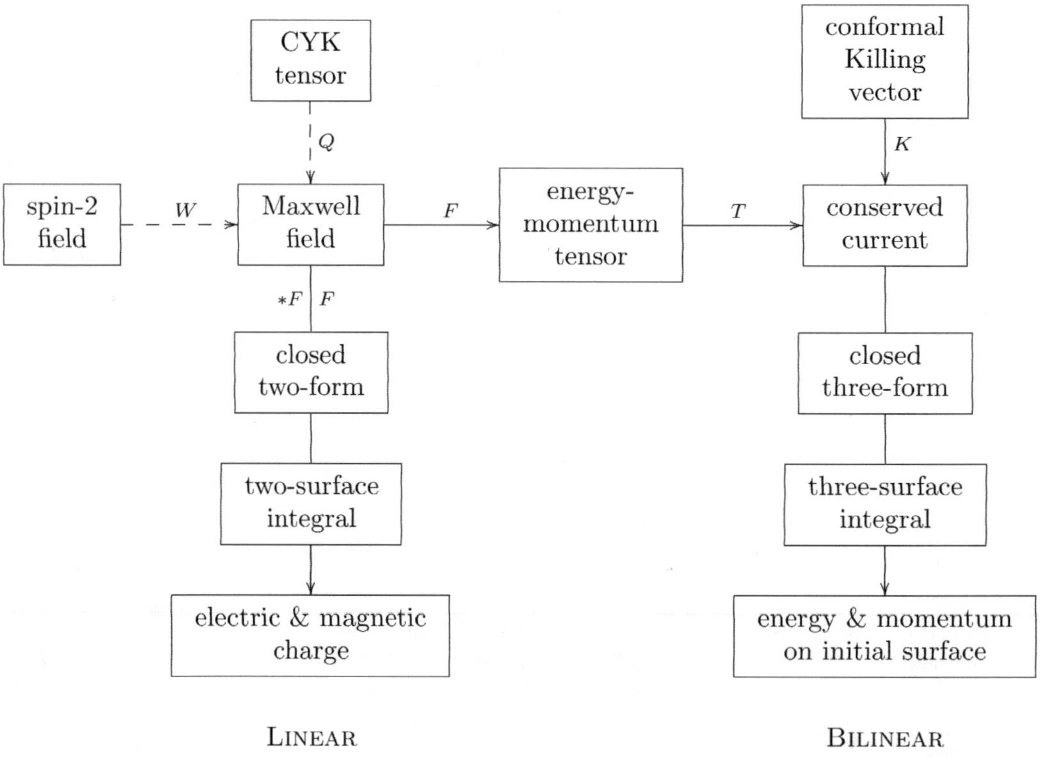

The left column corresponds to the expressions which are linear with respect CYK 2-form Q and the right one is bilinear in terms of CYK tensor.

Finally, in [7] the *asymptotic* AdS spacetime and its ACYK tensors are analyzed. The construction of (linear) asymptotic charges leads to well defined global charges. In particular, the exact CYK tensor $*\tilde{Q}_{\rm NUT\text{-}AdS} := \mathbf{l}dr \wedge [dt + 2\mathbf{l}(\cos\theta - 1)d\phi] + r(r^2 + \mathbf{l}^2)\sin\theta d\theta \wedge d\phi$ produces dual mass charge in NUT-AdS spacetime [9]. The 2-form $\tilde{F}(*\tilde{Q})$ is closed, defines dual mass unambiguously, and its value gives the interpretation of the NUT parameter $\mathbf{l}$.

We would like to stress that for the usual asymptotically flat spacetimes the existence of ACYK tensor gives a restriction on the asymptotics and leads to the notion of *strong asymptotic flatness*. On the other hand, in the standard *asymptotic* AdS spacetime [7, 9]

$$\tilde{g} = \tilde{g}_{\mu\nu} dz^\mu \otimes dz^\nu = \frac{l^2}{z^2}\left(dz \otimes dz + h_{ab} dz^a \otimes dz^b\right) \tag{9}$$

the 3-metric h obeys an asymptotic condition: $h = h_{ab} dz^a \otimes dz^b = \overset{(0)}{h} + z^2 \overset{(2)}{h} + z^3 \chi + O(z^4)$. The metric $\overset{(0)}{h}_{ab}$ and canonically conjugated momenta $\pi^{ab} := 3l^2 \sqrt{-\det \overset{(0)}{h}} \left(\chi^c{}_c \overset{(0)}{h}{}^{ab} - \chi^{ab}\right)$ (in A.D.M. densitized form) is a canonical data on a conformal boundary with respect to the symplectic structure $3l^2 \int_{\mathscr{I}} \delta \left[\sqrt{-\det \overset{(0)}{h}} \left(\chi^c{}_c \overset{(0)}{h}{}^{ab} - \chi^{ab}\right)\right] \wedge \delta \overset{(0)}{h}_{ab}$. Moreover, the boundary values at scri for any ACYK tensor Q in terms of the corresponding asymptotic Killing vector field L are given by ${}^{[AB]}Q = \overset{(0)}{h}(L^{AB}) \wedge dz$, $*{}^{[AB]}Q = L^{AB} \rfloor \mathrm{vol}(\overset{(0)}{h})$, where $\mathrm{vol}(\overset{(0)}{h})$ is a canonical volume 3-form on $\mathscr{I}$. They enable one to construct global charges which are unambiguously defined. Standard asymptotic AdS spacetime always admits maximal set of ACYK tensors in contradistinction to the asymptotically Minkowskian spacetime. In other words, there is no difference between standard asymptotic AdS spacetime and strongly asymptotic AdS spacetime.

[1] Chruściel P, Jezierski J and Kijowski J 2002 *A Hamiltonian framework for field theories in the radiating regime*, Springer Lecture Notes in Physics, Monographs, vol. **70**

[2] Jezierski J 1995 The Relation between Metric and Spin–2 Formulations of Linearized Einstein Theory *General Relativity and Gravitation* **27** 821–843

[3] Jezierski J 1997 Conformal Yano–Killing tensors and asymptotic CYK tensors for the Schwarzschild metric *Classical and Quantum Gravity* **14** 1679–1688

[4] Jezierski J 2002 CYK Tensors, Maxwell Field and Conserved Quantities for Spin-2 Field *Classical and Quantum Gravity* **19** 4405–4429 (*Preprint* gr-qc/0211039)

[5] Jezierski J, Łukasik M 2006 Conformal Yano-Killing tensor for the Kerr metric and conserved quantities *Classical and Quantum Gravity* **23** 2895–2918 (*Preprint* gr-qc/0510058v2)

[6] Jezierski J, Łukasik M 2007 Conformal Yano-Killing tensors for the Taub-NUT metric *Classical and Quantum Gravity* **24** 1331–1340 (*Preprint* gr-qc/0610090v1)

[7] Jezierski J 2008 Asymptotic conformal Yano-Killing tensors for asymptotic anti-de Sitter spacetimes and conserved quantities *Acta Physica Polonica B* **39** 75–114 (arXiv: 0705.4370v1 [gr-qc])

[8] Jezierski J 2008 A simple construction of conformal Yano-Killing tensors in anti-de Sitter spacetime, in ERE 2007 Relativistic Astrophysics and Cosmology, EAS Publications Series, Vol. 30, Edited by A. Oscoz, E. Mediavilla and M. Serra-Ricart (EDP Sciences, Les Ulis), pp. 249–252

[9] Jezierski J 2008 Conformal Yano-Killing tensors in anti-de Sitter spacetime *Class. Quantum Grav.* **25** 065010

[10] Jezierski J, Łukasik M 2009 Conformal Yano-Killing tensors in Einstein spacetimes *Reports on Mathematical Physics* **64** 205–221

[11] Öktem F 1968 On an Extension of Electromagnetic and Gravitational Tensors and of the Bel-Robinson Tensor, *Il Nuovo Cimento* **58 B** 167–176

[12] Tachibana S 1969 On conformal Killing tensor in a Riemannian space, *Tohoku Math. J.* (2) **21** 56–64

[13] Kashiwada T 1968 On conformal Killing tensor *Natur. Sci. Rep. Ochanomizu Univ.* **19** 67–74

[14] Tachibana S and Kashiwada T 1969 On the integrability of Killing–Yano's equation *J. Math. Soc. Japan* **21** 259–265

[15] Yano K 1952 Some remarks on tensor fields and curvature *Ann. Math.* **55** 328–347

Journal of Physics: Conference Series **314** (2011) 012096 doi:10.1088/1742-6596/314/1/012096

Algebraic Rainich conditions for the fourth rank tensor V

Lau Loi So

Department of Physics, National Central University, Chung Li 320, Taiwan

Abstract. Algebraic conditions on the Ricci tensor in the Rainich-Misner-Wheeler unified field theory are known as the Rainich conditions. Penrose and more recently Bergqvist and Lankinen made an analogy from the Ricci tensor to the Bel-Robinson tensor $B_{\alpha\beta\mu\nu}$, a certain fourth rank tensor quadratic in the Weyl curvature, which also satisfies algebraic Rainich-like conditions. However, we found that not only does the tensor $B_{\alpha\beta\mu\nu}$ fulfill these conditions, but so also does our recently proposed tensor $V_{\alpha\beta\mu\nu}$, which has many of the desirable properties of $B_{\alpha\beta\mu\nu}$. For the quasilocal small sphere limit restriction, we found that there are only two fourth rank tensors, $B_{\alpha\beta\mu\nu}$ and $V_{\alpha\beta\mu\nu}$, which form a basis for good energy expressions. Both of them have the completely trace free and causal properties, these two form necessary and sufficient conditions. Surprisingly either completely traceless or causal is enough to fulfill the algebraic Rainich conditions.

1. Introduction

In 1925 Rainich proposed a unified field theory for source-free electromagnetism and gravitation [1]. Misner and Wheeler [2] 32 years later proposed a geometrically unified theory, based on the Rainich idea, now called the Rainich-Misner-Wheeler theory. The necessary conditions are called the Rainich conditions. The algebraic Rainich conditions refer to the Ricci tensor, but this tensor can be replaced by other tensors. Penrose [3] and more recently Bergqvist and Lankinen [4, 5] made an analogy from the Ricci tensor to the Bel-Robinson tensor $B_{\alpha\beta\mu\nu}$, a certain fourth rank tensor quadratic in the Weyl curvature, which also satisfies algebraic Rainich-like conditions. We found that not only does the tensor $B_{\alpha\beta\mu\nu}$ fulfill these conditions, but so also does our recently proposed tensor $V_{\alpha\beta\mu\nu}$, which has many of the desirable properties of $B_{\alpha\beta\mu\nu}$. We define the Bel-Robinson tensor as follows

$$B_{\alpha\beta\mu\nu} := R_{\alpha\lambda\mu\sigma}R_{\beta}{}^{\lambda}{}_{\nu}{}^{\sigma} + R_{\alpha\lambda\nu\sigma}R_{\beta}{}^{\lambda}{}_{\mu}{}^{\sigma} - \frac{1}{8}g_{\alpha\beta}g_{\mu\nu}\mathbf{R}^2, \tag{1}$$

where $\mathbf{R}^2 = R_{\rho\lambda\sigma\tau}R^{\rho\lambda\sigma\tau}$. Here we propose [6] an unique alternative 4th rank tensor for the non-negative gravitational energy in the quasilocal limit

$$V_{\alpha\beta\mu\nu} =: B_{\alpha\beta\mu\nu} + W_{\alpha\beta\mu\nu} \neq B_{\alpha\beta\mu\nu}, \tag{2}$$

in general, where

$$W_{\alpha\beta\mu\nu} := \frac{3}{2}(R_{\alpha\mu\lambda\sigma}R_{\beta\nu}{}^{\lambda\sigma} + R_{\alpha\nu\lambda\sigma}R_{\beta\mu}{}^{\lambda\sigma}) + \frac{1}{8}(g_{\alpha\mu}g_{\beta\nu} + g_{\alpha\nu}g_{\beta\mu} - 2g_{\alpha\beta}g_{\mu\nu})\mathbf{R}^2, \tag{3}$$

Analog of the electric E_{ab} and magnetic part H_{ab} are defined in terms of the Weyl tensor

$$E_{ab} := C_{a0b0}, \quad H_{ab} := *C_{a0b0}, \quad a, b = 1, 2, 3. \tag{4}$$

The positive gravitation energy density for both $B_{\alpha\beta\mu\nu}$ and $V_{\alpha\beta\mu\nu}$ are

$$B_{0000} \equiv V_{0000} = E_{ab}E^{ab} + H_{ab}H^{ab} \geq 0. \tag{5}$$

In the small region $P_\mu \propto B_{\mu000} = V_{\mu000} = (E^2_{ab} + H^2_{ab},\ 2\epsilon_{cab}E^{ad}H^b{}_d)$. Since we want $P^0 \geq |\vec{P}| \geq 0$, then $E^2_{ab} + H^2_{ab} \geq 2\epsilon_{cab}E^{ad}H^b{}_d$, we call this the causal condition.

2. Algebraic Rainich conditions

In 1925 Rainich proposed a unified field theory for the geometrization of the electromagnetic field. The basic idea is from the Einstein-Maxwell field equations:

$$R_{\alpha\beta} - \frac{1}{2}g_{\alpha\beta}R = -8\pi\left(F_{\alpha\lambda}F_\beta{}^\lambda - \frac{1}{4}g_{\alpha\beta}F_{\rho\lambda}F^{\rho\lambda}\right). \tag{6}$$

where $R_{\alpha\beta}$ is the 2nd rank Ricci tensor and $F_{\alpha\beta}$ is the electromagnetic field tensor. Taking the trace of (6), one finds $R^\alpha{}_\alpha = R = 0$. Then rewrite (6)

$$R_{\alpha\beta} = -8\pi\left(F_{\alpha\sigma}F_\beta{}^\sigma - \frac{1}{4}g_{\alpha\beta}F_{\rho\sigma}F^{\rho\sigma}\right). \tag{7}$$

Using this, one can obtain $R_{\alpha\lambda}R_\beta{}^\lambda = \frac{1}{4}g_{\alpha\beta}R_{\rho\lambda}R^{\rho\lambda}$. Therefore the algebraic Rainich conditions can be written as

$$R_{\alpha\lambda}R_\beta{}^\lambda = \frac{1}{4}g_{\alpha\beta}R_{\rho\lambda}R^{\rho\lambda}, \qquad R^\alpha{}_\alpha = 0, \qquad R_{00} \geq 0, \tag{8}$$

where the last condition is to ensure positive energy.

3. Fourth rank Algebraic Rainich conditions

Penrose replaced the Ricci tensor by stress tensor $T_{\alpha\beta}$ using the Einstein field equation $R_{\mu\nu} - \frac{1}{2}g_{\mu\nu}R = \kappa T_{\mu\nu}$. Rewrite the algebraic Rainich conditions as follows

$$T_{\alpha\lambda}T_\beta{}^\lambda = \frac{1}{4}g_{\alpha\beta}T_{\rho\lambda}T^{\rho\lambda}, \qquad T^\lambda{}_\lambda = 0, \qquad T_{00} \geq 0, \tag{9}$$

Furthermore, analog from 2nd to 4th rank tensor, the Bel-Robinson tensor is the first tensor satisfying the algebraic Rainich conditions [3, 4],

$$B_{\alpha\lambda\sigma\tau}B_\beta{}^{\lambda\sigma\tau} = \frac{1}{4}g_{\alpha\beta}B_{\rho\lambda\sigma\tau}B^{\rho\lambda\sigma\tau}, \quad 0 = B^\rho{}_{\rho\sigma\tau} = B^\rho{}_{\sigma\rho\tau} = ..., \quad B_{0000} \geq 0, \tag{10}$$

where $B_{\alpha\lambda\sigma\tau} \equiv B_{(\alpha\lambda\sigma\tau)}$. Moreover, our recently proposed $V_{\alpha\beta\mu\nu}$ also satisfies these algebraic Rainich conditions

$$V_{\alpha\lambda\sigma\tau}V_\beta{}^{\lambda\sigma\tau} = \frac{1}{4}g_{\alpha\beta}V_{\rho\lambda\sigma\tau}V^{\rho\lambda\sigma\tau}, \quad 0 = V^\rho{}_{\rho\sigma\tau} = V^\rho{}_{\sigma\rho\tau} = ..., \quad V_{0000} \geq 0, \tag{11}$$

where $V_{\alpha\lambda\sigma\tau} \equiv V_{(\alpha\lambda)(\sigma\tau)} \equiv V_{\sigma\tau\alpha\lambda}$ and $V^\rho{}_{\rho\sigma\tau} \equiv 0 \equiv V^\rho{}_{\sigma\rho\tau}$, but $V_{\alpha\lambda\sigma\tau} \neq V_{(\alpha\lambda\sigma\tau)}$. We will show in section (IV) that the first condition indicated in (10) and (11) is an identity.

4. Two basic components of the fourth rank tensor

Edgar and Wingbrant [7] found the one-quarter identity

$$R_{\alpha\xi\lambda\kappa}R_{\sigma}{}^{\xi}{}_{\tau}{}^{\kappa}R_{\beta\pi}{}^{\lambda}{}_{\gamma}R^{\sigma\pi\tau\gamma}=\frac{1}{4}g_{\alpha\beta}R_{\rho\xi\lambda\kappa}R_{\sigma}{}^{\xi}{}_{\tau}{}^{\kappa}R^{\rho}{}_{\pi}{}^{\lambda}{}_{\gamma}R^{\sigma\pi\tau\gamma}, \tag{12}$$

$$R_{\alpha\xi\lambda\kappa}R_{\sigma}{}^{\xi}{}_{\tau}{}^{\kappa}R_{\beta\pi}{}^{\tau}{}_{\gamma}R^{\sigma\pi\lambda\gamma}=\frac{1}{4}g_{\alpha\beta}R_{\rho\xi\lambda\kappa}R_{\sigma}{}^{\xi}{}_{\tau}{}^{\kappa}R^{\rho}{}_{\pi}{}^{\tau}{}_{\gamma}R^{\sigma\pi\lambda\gamma}. \tag{13}$$

We found that the above two are not the only cases that satisfy the one-quarter property, but for any combination

$$M_{\alpha\lambda\sigma\tau}M_{\beta}{}^{\lambda\sigma\tau}=\frac{1}{4}g_{\alpha\beta\mu\nu}M_{\rho\lambda\sigma\tau}M^{\rho\lambda\sigma\tau}, \tag{14}$$

where $M_{\alpha\lambda\sigma\tau}$ is any tensor quadratic in the Riemann curvature, non-vanishing in vacuum. For any 4th rank tensor, Deser [8] stated that there are only two basic components. For example, $(R_{\rho\lambda\sigma\tau}R^{\rho\lambda\sigma\tau})^2$ and $(R_{\rho\lambda\sigma\tau} * R^{\rho\lambda\sigma\tau})^2$. Iihoshi [9] found

$$4B_{\rho\lambda\sigma\tau}B^{\rho\lambda\sigma\tau}=(R_{\rho\lambda\sigma\tau}R^{\rho\lambda\sigma\tau})^2+(R_{\rho\lambda\sigma\tau} * R^{\rho\lambda\sigma\tau})^2. \tag{15}$$

Moreover, we found

$$2B_{\rho\lambda\sigma\tau}B^{\rho\lambda\sigma\tau}=(R_{\rho\lambda\sigma\tau}R^{\rho\lambda\sigma\tau})^2-2R_{\rho\lambda\xi\kappa}R_{\sigma\tau}{}^{\xi\kappa}R^{\rho\lambda}{}_{\mu\nu}R^{\sigma\tau\mu\nu}, \tag{16}$$

which does not contain '2' at the last term in [8].

5. Completely traceless imply positivity or causal

There are 4 basic quadratic Riemann tensors in the small region limit [10], we use our own notation

$$\tilde{B}_{\alpha\beta\mu\nu}=R_{\alpha\lambda\mu\sigma}R_{\beta}{}^{\lambda}{}_{\nu}{}^{\sigma}+R_{\alpha\lambda\nu\sigma}R_{\beta}{}^{\lambda}{}_{\mu}{}^{\sigma},\quad \tilde{K}_{\alpha\beta\mu\nu}=R_{\alpha\lambda\beta\sigma}R_{\mu}{}^{\lambda}{}_{\nu}{}^{\sigma}+R_{\alpha\lambda\beta\sigma}R_{\nu}{}^{\lambda}{}_{\mu}{}^{\sigma},$$
$$\tilde{S}_{\alpha\beta\mu\nu}=R_{\alpha\mu\lambda\sigma}R_{\beta\nu}{}^{\lambda\sigma}+R_{\alpha\nu\lambda\sigma}R_{\beta\mu}{}^{\lambda\sigma},\quad \tilde{T}_{\alpha\beta\mu\nu}=-\frac{1}{8}g_{\alpha\beta}g_{\mu\nu}R_{\rho\lambda\sigma\tau}R^{\rho\lambda\sigma\tau}. \tag{17}$$

We let

$$A_{\alpha\beta\mu\nu}:=a_1\tilde{B}_{\alpha\beta\mu\nu}+a_2\tilde{S}_{\alpha\beta\mu\nu}+a_3\tilde{K}_{\alpha\beta\mu\nu}+a_4\tilde{T}_{\alpha\beta\mu\nu}, \tag{18}$$

and consider all the traces. It turns out because of the symmetry there are only two distinct traces

$$0=2A^{\alpha}{}_{\alpha\mu\nu}=(a_1+a_2-a_4)g_{\mu\nu}\mathbf{R}^2, \tag{19}$$
$$0=8A^{\alpha}{}_{\mu\alpha\nu}=(a_1-2a_2+3a_3-a_4)g_{\mu\nu}\mathbf{R}^2, \tag{20}$$

where a_1 to a_4 are constants. The solutions are $a_4=a_1+a_2$ and $a_2=a_3$. Consider the linear combination of these 4 fundamental tensors, we found $B_{\alpha\beta\mu\nu}$ and $V_{\alpha\beta\mu\nu}$ are the unique tensors that have the completely traceless property and imply positivity

$$\begin{aligned}A_{\alpha\beta\mu\nu} &= a_1(\tilde{B}_{\alpha\beta\mu\nu}+\tilde{T}_{\alpha\beta\mu\nu})+a_2(\tilde{S}_{\alpha\beta\mu\nu}+\tilde{K}_{\alpha\beta\mu\nu}+\tilde{T}_{\alpha\beta\mu\nu})\\ &= a_1B_{\alpha\beta\mu\nu}+a_2V_{\alpha\beta\mu\nu},\end{aligned} \tag{21}$$

using $B_{\alpha\beta\mu\nu}=\tilde{B}_{\alpha\beta\mu\nu}+\tilde{T}_{\alpha\beta\mu\nu}$ and $V_{\alpha\beta\mu\nu}=\tilde{S}_{\alpha\beta\mu\nu}+\tilde{K}_{\alpha\beta\mu\nu}+\tilde{T}_{\alpha\beta\mu\nu}$.

6. Positive or causal implies completely traceless

Counter example for satisfying positive energy but not fulfilling completely traceless,

$$X_{\alpha\beta\mu\nu} = R_{\alpha\lambda\beta\sigma} R_\mu{}^\lambda{}_\nu{}^\sigma, \quad X_{0000} = E_{ab}E^{ab} \geq 0, \quad X^\alpha{}_{\mu\alpha\nu} = \frac{1}{4} g_{\mu\nu} \mathbf{R}^2 \neq 0. \tag{22}$$

Returning back to causal, consider the energy-momentum integral in a quasilocal small sphere with constant time evolution of the hypersurface. The fourth rank tensor $X_{\alpha\beta\mu\nu}$ needs to be symmetric at the last two indices because of the small sphere limit

$$\begin{aligned} N^\mu P_\mu &= \int_{t=0} N^\mu X^0{}_{\mu\xi\kappa} x^\xi x^\kappa \eta_0 = \int_{t=0} N^\mu X^0{}_{\mu ij} x^i x^j dV = \int_{t=0} N^\mu X^0{}_{\mu l}{}^l \frac{r^2}{3} dV \\ &= N^\mu X^0{}_{\mu l}{}^l \frac{4\pi r^5}{15} = N^\mu \left(X^0{}_{\mu\alpha}{}^\alpha - X^0{}_{\mu 0}{}^0 \right) \frac{4\pi r^5}{15} = N^\mu X^0{}_{\mu 00} \frac{4\pi r^5}{15}, \end{aligned} \tag{23}$$

where we made the assumption that $X_{0\mu\alpha}{}^\alpha$ vanishes and fulfills causal. Consider the requirement for the energy-momentum being future pointing and non-spacelike (i.e., causal) in the small sphere limit :

$$\begin{aligned} A_{\mu 0 l}{}^l &= (-2a_1 - 4a_2 + 2a_3 + 3a_4) E_{ab}E^{ab} + (4a_1 + 4a_2 - 3a_4) H_{ab}H^{ab} \\ &\quad + (2a_1 + 2a_3) \epsilon_{cab} E^{ad} H^b{}_d. \end{aligned} \tag{24}$$

Causal requires the magnitude of $E_{ab}E^{ab}$ and $H_{ab}H^{ab}$ to be the same and the energy is greater than or equal to the momentum. After some simple algebra, we found

$$A_{\mu 0 l}{}^l = a_1 B_{\mu 000} + a_3 V_{\mu 000} = (a_1 + a_3) B_{\mu 000}, \tag{25}$$

where we require $a_1 + a_3 \geq 0$. Hence, the completely traceless and causal properties form necessary and sufficient conditions. This means we can further simplify the algebraic Rainich conditions for a fourth rank tensor; as far as the quasilocal small sphere limit is concerned, we only need the completely trace free condition or positivity (i.e., causal). This is an interesting result which is valid in the quasilocal small sphere region.

7. Conclusion

Either completely traceless or causal is enough to fulfill the algebraic Rainich conditions.

[1] Rainich G Y 1925 *Trans. Amer. Math. Soc.* **27** 106
[2] Misner C W and Wheeler J A 1957 *Ann. of Phys.* **2** 525
[3] Penrose R and Rindler W 1984 *Spinors and spacetime* (Cambridge U.P., Cambridge) Vol. 1
[4] Bergqvist G and Lankinen P 2004 *Class. and Quantum Grav* **21** 3499
[5] Bergqvist G and Lankinen P 2005 *Proc. R. Soc. A* **461** 2181
[6] L. L. So, arXiv:1005.0674 [gr-qc].
[7] Edgar S B and Wingbrant O 2003 *J. Math. Phys.* **44** 6140
[8] Deser S Preprint arXiv:gr-qc/9901007
[9] Iihoshi M and Ketov S V 2008 *Advances in High Energy Physics* Article ID 521389
[10] Deser S, Franklin J S and Seminaea D 1999 *Class. Quantum Grav.* **16** 2815

Spanish Relativity Meeting (ERE 2010): Gravity as a Crossroad in Physics IOP Publishing
Journal of Physics: Conference Series **314** (2011) 012097 doi:10.1088/1742-6596/314/1/012097

Late-time behaviour of the Einstein-Vlasov system with Bianchi I symmetry

Ernesto Nungesser
Max-Planck-Institut für Gravitationsphysik
Albert-Einstein-Institut
Am Mühlenberg 1
14476 Potsdam, Germany

E-mail: ernesto.nungesser@aei.mpg.de

Abstract.
The late-time behaviour of the Einstein-dust system is well understood for homogeneous spacetimes. For the case of Bianchi I we have been able to show that the late-time behaviour of the Einstein-Vlasov system is well approximated by the Einstein-dust system assuming that one is close to the unique stationary solution which is the attractor of the Einstein-dust system.

1. Introduction

The Einstein-Vlasov system with Bianchi symmetry is a system of integro-differential equations on a Lorentzian manifold M which has the structure $M = I \times G$ where I is an interval and G a Lie group. It can be seen as a kind of transport equation, say $Xf = 0$, coupled to other ordinary differential equations. There exists certain limiting process such that the transport equation is transformed into an ordinary differential equation. This happens if the function f becomes a distribution. The resulting system is called Einstein-dust system and the late-time behavior is well understood. It has been studied by the theory of dynamical systems and for almost all Lie groups the ω-limit set is known [1]. The strategy to study the late-time behaviour of the Einstein-Vlasov system consists in trying to show that the late-time behavior is well approximated by the Einstein-dust system. Physically this makes sense since one might expect a decay of the velocity dispersion for expanding cosmological models. The result presented in the following shows that this is true for case of the Lie group $\mathbb{R}^3$ and assuming that one is close to the unique stationary solution which is the attractor of the Einstein-dust system.

2. Einstein-Vlasov system with Bianchi I symmetry

2.1. Einstein equations with Bianchi I symmetry

A Bianchi spacetime is defined to be a spatially homogeneous spacetime whose isometry group possesses a three-dimensional subgroup G that acts simply transitively on the spacelike orbits. They can be classified by the structure constants of the Lie algebra associated to the Lie group. We will only consider the simplest case where the structure constants vanish, i.e. the case of

Bianchi I where the metric has the following form using Gauss coordinates:

$$^4g = -dt^2 + g_{ab}(t)dx^a dx^b. \tag{1}$$

In terms of coordinate expressions

$$\begin{aligned} \rho &= T^{00} \\ j_a &= T^0_a \\ S_{ab} &= T_{ab} \end{aligned}$$

where ρ, j_a and $T_{\mu\nu}$ are the energy density, matter current and energy momentum tensor respectively. Using the 3+1 decomposition of the Einstein equations as made in [2] the basic equations are

$$\dot{g}_{ab} = -2k_{ab} \tag{2}$$
$$\dot{k}_{ab} = Hk_{ab} - 2k_{ac}k^c_b - 8\pi(S_{ab} - \frac{1}{2}g_{ab}\,\mathrm{tr}\,S) - 4\pi\rho g_{ab} \tag{3}$$
$$-k_{ab}k^{ab} + H^2 = 16\pi\rho \tag{4}$$
$$T_{0a} = 0. \tag{5}$$

where k_{ab} is the second fundamental form, H its trace and a dot above a letter denotes a derivative with respect to the cosmological (Gaussian) time t. It has been assumed that the cosmological constant vanishes.

2.2. Vlasov equation with Bianchi I symmetry

For the matter model we will take the point of view of kinetic theory. This means that we have a collection of particles (in a cosmological context the particles are galaxies or clusters of galaxies) which are described statistically by a non-negative distribution function $f(x^\alpha, p^\alpha)$ which is the density of particles at a given spacetime point with given four-momentum. We will assume that all the particles have equal mass (one can relax this condition if necessary, see [3]) . We want that our matter model is compatible with our symmetry assumption, so we will also assume that f does not depend on x^a. In addition to that we will assume that there are no collisions between the particles. In this case the distribution function satisfies the Vlasov equation (See (3.38) of [2]):

$$\frac{\partial f}{\partial t} + 2k^a_b p^b \frac{\partial f}{\partial p^a} = 0. \tag{6}$$

where f is defined on the set determined by the equation

$$-(p^0)^2 + g_{ab}p^a p^b = -m^2$$

called the mass shell. For a given Bianchi I geometry the Vlasov equation can be solved explicitly with the result that if f is expressed in terms of the covariant components p_i then it is independent of time. The non-trivial components of the energy momentum tensor are:

$$\rho = \int f_0(p_i)(m^2 + g^{cd}p_c p_d)^{\frac{1}{2}}(\det g)^{-\frac{1}{2}}dp_1 dp_2 dp_3 \tag{7}$$
$$S_{ab} = \int f_0(p_i)p_a p_b(m^2 + g^{cd}p_c p_d)^{-\frac{1}{2}}(\det g)^{-\frac{1}{2}}dp_1 dp_2 dp_3 \tag{8}$$

For this kind of matter all the energy conditions hold. In particular $\rho \geq \mathrm{tr}\,S \geq 0$. Our system of equations consists of the equations (2)-(8).

3. Central results

The assumption that the spacetime is close to isotropic is expressed by assuming that the quantity

$$F = \frac{\sigma_{ab}\sigma^{ab}}{H^2}. \tag{9}$$

is small. Here σ_{ab} denotes the trace-free part of the second fundamental form. The quantity F is related to the so called *shear parameter*, which is bounded by the cosmic microwave background radiation and is a dimensionless measure of the anisotropy of the Universe (See chapter 5.2.2 of [1]). The other assumption which is needed is that the spacetime is close to "dust-like". This is expressed by assuming that the absolute value of the momenta of the particles is bounded. We define P as the supremum of the absolute value of the momenta at a given time t:

$$P(t) = \sup\{|p| = (g^{ab}p_a p_b)^{\frac{1}{2}} | f(t,p) \neq 0\} \tag{10}$$

The main result is the following:

Theorem 1. *Consider any C^∞ solution of the Einstein-Vlasov system with Bianchi I-symmetry and with C^∞ initial data. Assume that $F(t_0)$ and $P(t_0)$ are sufficiently small. Then at late times one can make the following estimates:*

$$\begin{aligned} H(t) &= -2t^{-1}(1+O(t^{-1})) && (11)\\ P(t) &= O(t^{-\frac{2}{3}+\epsilon}) && (12)\\ F(t) &= O(t^{-2}) && (13) \end{aligned}$$

where ϵ is a small and strictly positive constant. These estimates imply that the spacetime isotropizes and that asymptotically there is a dust-like behaviour. The proof is based on a bootstrap argument. See [4] for the details. Using these estimates one can obtain more detailed information about the behaviour of the metric. In analogy to the Kasner solution one can define the *generalized Kasner exponents* for the non-vacuum case. Let λ_i be the eigenvalues of k_{ij} with respect to g_{ij} we define

$$p_i = \frac{\lambda_i}{H} \tag{14}$$

as the generalized Kasner exponents. Having analyzed carefully the dust case with small data we could conclude that:

Theorem 2. *Consider the same assumptions as in the previous theorem. Then*

$$p_i = \frac{1}{3} + O(t^{-1}) \tag{15}$$

and

$$\begin{aligned} g_{ab} &= t^{+\frac{4}{3}}[\mathcal{G}_{ab} + O(t^{-2})] && (16)\\ g^{ab} &= t^{-\frac{4}{3}}[\mathcal{G}^{ab} + O(t^{-2})] && (17) \end{aligned}$$

where $\mathcal{G}_{ab}$ and $\mathcal{G}^{ab}$ are independent of t.

Journal of Physics: Conference Series **314** (2011) 012097 doi:10.1088/1742-6596/314/1/012097

4. Conclusions and Outlook

The result can be seen as a generalization of theorem 5.4 of [5]. We obtain the same the result, but a) we also obtain how fast the expressions converge b) we obtain an asymptotic expression for the spatial metric c) we do not assume additional symmetries. However we used a different kind of restriction namely the small data assumptions. For the study of more complicated Lie groups a starting point can be [6] - [9] where the late-time asymptotics are obtained assuming an additional symmetry. Another direction of generalization is the study of the asymptotics towards the initial singularity. In [10] and [11] the case of two fluids has been studied which leads to the non-diagonal case. For the Vlasov case already the non-LRS case has been analyzed in [12]. Already in this case surprising new features like the existence of heteroclinic networks arose. Finally it would be interesting concerning the direction of the initial singularity to show that also in the Vlasov case the off-diagonal components of the metric tend to constants and thus are not important for the dynamics. See [13] for the importance of this behaviour and [14] for consequences in a quantum version.

Acknowledgments

The author would like to thank Alan D. Rendall and Michael Koehn for a lot of helpful discussions.

References

[1] Wainwright J Ellis G F R 1997 *Dynamical Systems in Cosmology* (Cambridge University Press)

[2] Rendall A D 2008 *Partial Differential Equations in General Relativity* (Oxford University Press)

[3] Calogero S and Heinzle J M 2010 Oscillations toward the singularity of LRS Bianchi type IX cosmological models with Vlasov matter *SIAM J. Appl. Dyn. Syst.* 9, 1244-1262

[4] Nungesser E 2010 Isotropization of non-diagonal Bianchi I spacetimes with collisionless matter at late times assuming small data *Class. Quantum Grav.* 27 235025

[5] Rendall A D 1996 The initial singularity in solutions of the Einstein-Vlasov system of Bianchi type I *J. Math Phys.* 37, 438-451.

[6] Rendall A D and Tod K P 1998 Dynamics of spatially homogeneous solutions of the Einstein-Vlasov equations which are locally rotationally symmetric *Class. Quantum Grav.* 16, 1705-1726

[7] Rendall A D and Uggla C 2000 Dynamics of spatially homogeneous locally rotationally symmetric solutions of the Einstein-Vlasov equations *Class. Quantum Grav.* 17, 4697-4713

[8] Rendall A D 2002 Cosmological models and centre manifold theory *Gen. Rel. Grav.* 34, 1277-1294

[9] Calogero S and Heinzle J M 2009 Bianchi Cosmologies with Anisotropic Matter: Locally Rotationally Symmetric Models *Preprint arXiv* 0911.0667v1

[10] Sandin P and Uggla C 2008 Bianchi type I models with two tilted fluids *Class. Quantum Grav.* 25, 225013

[11] Sandin P 2009 Tilted two-fluid Bianchi type I models. *Gen. Rel. Grav.* 41, 2707-2724

[12] Heinzle J M and Uggla C 2006 Dynamics of the spatially homogeneous Bianchi type I Einstein-Vlasov equations *Class. Quantum. Grav.* 23, 3463-3490

[13] Damour T, Henneaux M and Nicolai H 2003 Cosmological Billiards *Class. Quantum Grav.* 20 R145-R200

[14] Kleinschmidt A, Koehn M and Nicolai H 2009 Supersymmetric quantum cosmological billiards *Phys. Rev.* D 80:061701

Spanish Relativity Meeting (ERE 2010): Gravity as a Crossroad in Physics IOP Publishing
Journal of Physics: Conference Series **314** (2011) 012098 doi:10.1088/1742-6596/314/1/012098

Causality and entropic arguments pointing to a null Big Bag hypersurface

E Minguzzi

Dipartimento di Matematica Applicata, Università degli Studi di Firenze, Via S. Marta 3, I-50139 Firenze, Italy

E-mail: `ettore.minguzzi@unifi.it`

Abstract. I propose a causality argument in order to solve the homogeneity (horizon) problem and the entropy problem of cosmology. The solution is based on the replacement of the spacelike Big Bang boundary with a null boundary behind which stays a chronology violating region. This solution requires a tilting of the light cones near the null boundary and thus it is based more on the behavior of the light cones and hence on causality than on the behavior of the scale factor (expansion). The connection of this picture with Augustine of Hippo famous philosophical discussion on time and creation is mentioned.

1. Introduction

The usual mathematical approach to cosmology starts with the Einstein equations,

$$G_{ab} + \Lambda g_{ab} = 8\pi T_{ab}, \qquad T_{ab} = (\rho + p)u_a u_b + p g_{ab},$$

passes through the assumption of the cosmological principle, namely through the assumption that all the observers moving with the cosmological flow (i.e. covariant velocity u) should see an isotropic universe as we do, which implies $g = -dt^2 + a(t)^2[\mathrm{d}r^2/(1-kr^2) + r^2\mathrm{d}\Omega^2]$, where $k = -1, 0, +1$. Finally, it arrives at the Friedmann equation $H^2 + k/a^2 = (\Lambda + 8\pi\rho)/3$, which together with the equation of state $p = p(\rho)$ should determine the dynamics of the large scale geometry of the Universe (the FLRW model). Most work of modern cosmology revolves around these two equations but as I shall argue in this work, observations point to a different model according to which the cosmological principle holds only at late times and cannot be extrapolated to the beginning of the Universe. Indeed, I will argue that the mathematical framework above removes completely the non-conformal dynamics, that is, the tilting of the light cones, which is fundamental in order to solve in a very simple way the homogeneity and entropy problems of cosmology.

2. The homogeneity (horizon, isotropy) problem

The first suggestion comes from the homogeneity problem. The cosmic microwave background radiation (CMB) is the remnant of the decoupling between light and matter which happened at about $z = 1000$ when the baryon density decreased so much that the mean free path of light became infinite and light could travel freely to reach us today. Observations show that it has a quite good black body spectrum at 2.7K with very high isotropy: variations are of the

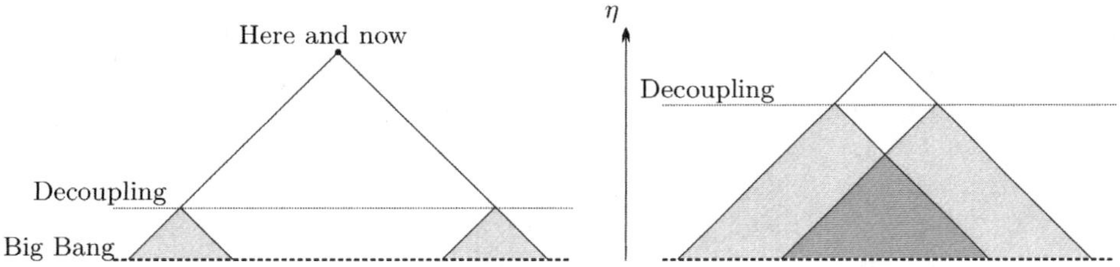

Figure 1. The conformal diagram of an FLRW universe which displays the homogeneity problem and the inflation solution to it. The Big Bang is a spacelike hypersurface in the conformal completion.

order of 10^{-5} for angles as large as $80°$. This spherical thermal photograph of the sky tells us that the intersection of our past light cone with the decoupling hypersurface was a region very homogeneous in temperature. If the cosmological principle holds from the time of decoupling to our days, this means that the whole Universe at decoupling time was thermally homogeneous. This conclusion is puzzling because according to the FRLW universe the opposite portions of the night sky that we observe today had no past in common at decoupling time, so it is not clear why their temperature should have been related. It is useful to rewrite the metric in terms of the conformal time $\eta(t) = \int_0^t \frac{dt}{a(t)}$ since it makes the coordinate speed of light constant (Fig. 1).

The theory of inflation solves the homogeneity problem by claiming that the scale factor a grows enormously in the first instants of the Universe. The idea is that if a patch of space expands so much to include the whole surface we see today, then it should be natural to observe homogeneity. This argument works only if homogeneity is assumed at a different scale, actually at a much smaller scale, prior to inflation, namely if the initial patch is considered homogeneous.

This criticism has been moved to inflation by several authors, as rather that solving the problem of homogeneity, inflation seems to replace a type homogeneity assumption with another [1]. R. Penrose argues that inflation may well prove to be correct but not for the initial arguments moved in its favor [1].

Mathematically, inflation pushes up the decoupling time in the conformal diagram. In this way the past cones of two decoupling events seen from us today intersect and could influence each other. The idea is that they have roughly the same temperature because there has been time for thermalization before decoupling. This solution based on thermalization increases the entropy difference from the Big Bang to the decoupling time and in particular makes worse the problem of justifying why the entropy of the Big Bang state was so small - the cosmological entropy problem.

3. The cosmological entropy problem

As Penrose clarifies one has to distinguish between matter and gravity when talking about homogeneity and order [2]. The second law of thermodynamics which we know to hold would not hold if the Universe were already at a state of maximum entropy. We are lead to the idea that either the Universe we see was generated by a large entropy fluctuation around the equilibrium or, more reasonably, that the entropy was small already at the beginning of the Universe. According to Penrose's argument in the coarse graining of the phase space the Universe was located in a very small cell at the Big Bang, a fact which made possible the increase of entropy (by the way, Penrose argues that the entropy has nothing to due with the size of the Universe). This idea can be quantified by comparing the observed entropy of the Universe with its maximum that would be reached after every baryon enters a black hole and using Bekenstein-Hawking formula $S_{BH} = \frac{kc^3}{4g\hbar}A$ which relates the area of the black hole with its entropy. This calculation shows that the cell at decoupling is about one part of $10^{10^{123}}$ than that at end of the Universe. This figure gives an idea of how peculiar was the the Big Bang. According to Penrose the Big Bang was special because the gravity degrees of freedom were in a special state (the CMB shows that on the contrary matter and radiation were in a hot disordered state). Looking for some mathematical expression of the gravitational entropy he observes that the Weyl tensor $C_{ab}{}^{dc} = R_{ab}{}^{dc} - 2R_{[a}{}^{[c}g_{b]}{}^{c]} + \frac{1}{3}Rg_{[a}{}^{c}g_{b]}{}^{d}$, increases when the gravitational entropy increases, indeed an homogeneous matter configuration through the action of gravity will tend to clump and the Weyl tensor would increase. He proposes that the Universe at the Big Bang had to have a vanishing or very small Weyl tensor, in such a way that the gravitational degrees of freedoms were frozen. This is Penrose's Weyl tensor hypothesis [2, 3]. (In order to solve this problem Penrose proposes a cyclic picture of the Universe. His proposal differs from mine.) Even if one could prove observationally that the Big Bang satisfied the Weyl tensor hypothesis, one would still be faced with the problem: if the Big Bang is made by causally unrelated events on the boundary how could it be special?

4. A simple causality solution: The Big Bang hypersurface is null

The solution I propose to these two problems, simple as it is, is that the Big Bang in the conformal completion diagram should not be spacelike but rather lightlike and generated by lightlike geodesics. The mechanism I propose works particularly well if the Big Bang hypersurface is also compact thus I will consider only this case. The idea is that every two points at the Big Bang share the same future, that is, *future distinction* [4] fails at the Big Bang. This is accomplished not through a particular behavior of the conformal factor $a(t)$ but rather of the light cones which tilt near the Big Bang hypersurface as they have to be tangent to it. Exact solutions of the Einstein equations that exhibit this kind of behavior exist, e.g. the Taub-NUT and the Λ-Taub-NUT metrics.

Thus, any two points at the last scattering hypersurface see the whole Big Bang and thus it is clear that they have roughly the same temperature. However, in this picture the null Big Bang hypersurface is generated by lightlike geodesics and is compact, thus it is made by closed or almost closed causal curves. In practice the events in the Big Bang hypersurface are in (almost) causal contact and hence the matter and radiation content is highly homogeneous already at the Big Bang. There is no need to assume thermalization from Big Bang to decoupling time in order to account for the homogeneity of the CMB radiation. In particular the problem of the specialness of the Big Bang is made less severe.

5. The role of rigidity

This picture for the beginning of the Universe solves naturally also the cosmological entropy problem. Indeed, if $\mathcal{B}$ is the Big Bang hypersurface then $\mathcal{B}$ is the boundary of the Universe that generates from it $\mathcal{B} = \partial I^+(\mathcal{B})$. Since $\mathcal{B}$ is an achronal boundary, the lightlike geodesics that run on it must be achronal and hence cannot have conjugate points. However, $\mathcal{B} = H^-(S)$ where S is a Cauchy hypersurface on the future of $\mathcal{B}$ (for simplicity we are making some causality assumptions) and because of a theorem due to Hawking [4, Lemma 8.5.5], since $\mathcal{B}$ is compact it is future null complete. Further, by using the area theorem [4, p.297] [5] one obtains that $\theta = R_{ab}K^aK^b = 0$ on the geodesics of $\mathcal{B}$ where K^a is the tangent vector.

Consider the equations for the expansion θ (measuring the divergence of the transverse section to the flow) and for the shear σ_{mn} (measuring the deformation) of the geodesics running on such hypersurface [4, Sect. 4.2]

$$\frac{\mathrm{d}\theta}{\mathrm{d}v} = -R_{ab}K^aK^b - 2\sigma^2 - \frac{1}{2}\theta^2, \qquad \frac{\mathrm{d}\sigma_{mn}}{\mathrm{d}v} = -C_{m4n4} - \theta\sigma_{mn} - \sigma_{mp}\sigma_{pn} + \delta_{mn}\sigma^2, \quad \text{(sum over p)}$$

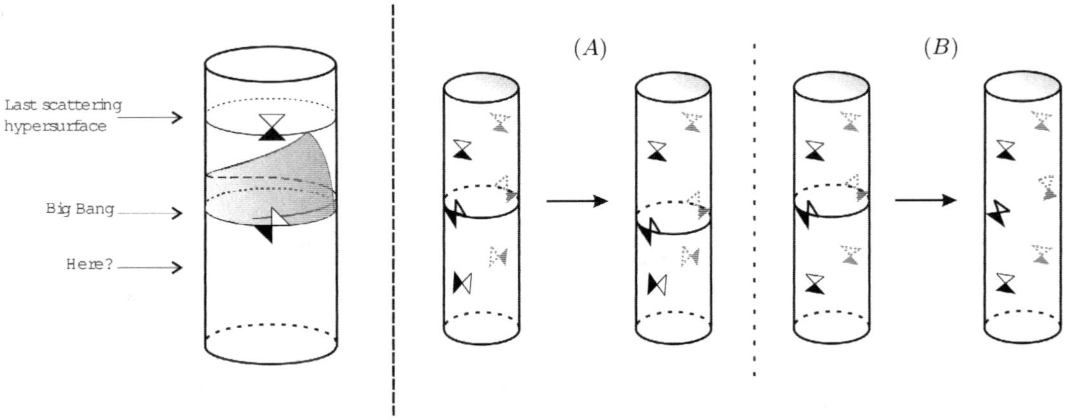

Figure 2. The chronology violating region behind the Big Bang (case A) guarantees the Big Bang stability and hence that of our solution to the homogeneity and entropy problems.

where $m, n = 1, 2$, and v is the affine parameter and a base adapted to the congruence has been chosen. From the first equation we obtain that the shear vanishes and from the second that $C_{m4n4} = 0$ on $\mathcal{B}$. The proposal implies that several components of the Weyl tensor vanish at the Big Bang and hence that the geometry there is very special. Penrose's Weyl tensor hypothesis is then at least supported if not proved.

6. The chronology violating set and the connection with Augustine's cosmology

One can also ask if the causal structure (but not the metric, as one expects a singularity due to the singular behavior of the conformal factor at the Big Bang) of the spacetime can be continued before the Big Bang $\mathcal{B}$. We argue that if so then there must be a chronology violating region $\mathcal{C}$ before the Big Bang. Indeed, in this case the achronal hypersurface of the Big Bang would be stable under small perturbations of the metric as $\mathcal{B}$ would be the slice between the non-chronological region $\mathcal{C}$ and the chronological region $M\backslash\mathcal{C}$ of spacetime (Fig. 2).

I have proved a theorem [6] which shows that under weak conditions the chronological region $M\backslash(\mathcal{B} \cup \mathcal{C})$ is actually stably causal [4] and hence admits a time function $t : M \to \mathbb{R}$, namely a function which increases over every causal curve. The physical content of this theorem recalls Augustine of Hippo's conception of time and creation [7, 6] according to which God ($\mathcal{C}$) created the whole Universe ($M = I^+(\mathcal{C})$) and time (t) began to exist at the act of creation ($\mathcal{B}$) and throughout the created world ($M\backslash(\mathcal{B} \cup \mathcal{C})$), but still outside God which indeed, in the middle age philosophical thought, was usually considered atemporal [8, 6].

References

[1] Penrose R 2005 *The road to reality: A complete guide to the laws of the Universe* (New York: A. A. Knopf)
[2] Penrose R 1979 *Singularities and time-asymmetry* vol General relativity: An Einstein centenary survey (Cambridge: Cambridge University Press) pp 581–638
[3] Goode S W 1991 *Class. Quantum Grav.* **8** L1–L6
[4] Hawking S W and Ellis G F R 1973 *The Large Scale Structure of Space-Time* (Cambridge: Cambridge University Press)
[5] Chruściel P T, Delay E, Galloway G J and Howard R 2001 *Ann. Henri Poincaré* **2** 109–178
[6] Minguzzi E Can God find a place in physics? St. Augustine's philosophy meets general relativity preprint: 0909.3876
[7] St Augustine 1992 *Confessions* World's Classics (Oxford: Oxford University Press) translated by H. Chadwick
[8] Craig W L and Smith Q 1993 *Theism, Atheism and Big Bang Cosmology* (Oxford: Oxford University Press)

Spanish Relativity Meeting (ERE 2010): Gravity as a Crossroad in Physics IOP Publishing
Journal of Physics: Conference Series **314** (2011) 012099 doi:10.1088/1742-6596/314/1/012099

On the non-linear stability of scalar field cosmologies

Artur Alho[♭], **Filipe C. Mena**[♭] **and Juan A. Valiente Kroon**†

[♭]Centro de Matemática, Universidade do Minho, 4710-057 Braga, Portugal

†School of Mathematical Sciences, Queen Mary, University of London, London E1 4NS, U.K.

E-mail: aalho@math.uminho.pt, fmena@math.uminho.pt, jav@maths.qmul.ac.uk

Abstract. We review recent work on the stability of flat spatially homogeneous and isotropic backgrounds with a self-interacting scalar field. We derive a first order quasi-linear symmetric hyperbolic system for the Einstein-nonlinear-scalar field system. Then, using the linearized system, we show how to obtain necessary and sufficient conditions which ensure the exponential decay to zero of small non-linear perturbations.

1. Introduction

The asymptotic stability of cosmological spacetimes is an important problem of mathematical relativity. This problem can be addressed by considering small perturbations of a given background solution and investigating whether they asymptotically decay in time. Most of the approaches to this question have been limited to the use of linear or higher-order truncated perturbation theory, and thus, they never take fully into account the non-linearity of the Einstein-Field-Equations (EFEs). In [2], Friedrich extended his frame representation of the vacuum EFEs to the case of matter sources consisting of perfect fluids. This system implies a *first-order quasi-linear symmetric hyperbolic (FOSH) system.* FOSH systems are of the form

$$\mathbf{A}^0(\mathbf{u})\partial_t\mathbf{u} = \mathbf{A}^j(\mathbf{u})\partial_j\mathbf{u} + \mathbf{B}(\mathbf{u})\mathbf{u}, \qquad j = 1,2,3, \tag{1}$$

where $\mathbf{u} = \mathbf{u}(\boldsymbol{x}, t)$ is a smooth vector-valued function of dimension s with domain in $\Sigma \times [0,T]$ where Σ is a spacelike 3-dimensional manifold, and $\mathbf{A}^j$, $\mathbf{B}$ denote smooth $s \times s$ matrix valued-functions such that $\mathbf{A}^0$, $\mathbf{A}^j$ are symmetric and $\mathbf{A}^0$ is positive definite. The operators ∂_t and ∂_j stand for the partial derivatives with respect to the coordinate $t \in [0,T]$ and the spatial coordinates x^j in Σ, respectively. A natural way of performing a stability analysis (see also [3]) is to consider a sequence of smooth initial data $\mathbf{u}_0^\varepsilon$ for the EFEs satisfying the constraints equations on a Cauchy hypersurface Σ. The sequence is assumed to depend continuously on the parameter ε in such a way that the limit $\varepsilon \to 0$ renders the data of the reference solution $\mathring{\mathbf{u}}_0$. In particular, one can write the full solution to the EFEs as the Ansatz $\mathbf{u}^\varepsilon = \mathring{\mathbf{u}} + \varepsilon\breve{\mathbf{u}}$, where $\breve{\mathbf{u}}$ is a (non-linear) perturbation whose size is controlled by the parameter ε. Using this splitting in equation (1), we are led to consider the following initial value problem for the non-linear perturbations:

$$\begin{aligned} &\left[\mathring{\mathbf{A}}^0(\mathring{\mathbf{u}}) + \varepsilon\breve{\mathbf{A}}^0(\breve{\mathbf{u}},\varepsilon)\right]\partial_t\breve{\mathbf{u}} = \left[\mathring{\mathbf{A}}^j(\mathring{\mathbf{u}}) + \varepsilon\breve{\mathbf{A}}^j(\breve{\mathbf{u}},\varepsilon)\right]\partial_j\breve{\mathbf{u}} + \left[\mathring{\mathbf{B}}(\mathring{\mathbf{u}}) + \varepsilon\breve{\mathbf{B}}(\breve{\mathbf{u}},\varepsilon)\right]\breve{\mathbf{u}}, \\ &\breve{\mathbf{u}}(\boldsymbol{x},0) = \breve{\mathbf{u}}_0(\boldsymbol{x}). \end{aligned} \tag{2}$$

where $\mathring{\mathbf{A}}^\mu$ ($\mu = 0, 1, 2, 3$) and $\mathring{\mathbf{B}}$ are defined as the non-vanishing terms with $\epsilon = 0$. When the linearised system ($\epsilon = 0$) of (2) has constant coefficients, then for small enough ϵ, global existence in time t, and exponential decay to zero of a solution to (2) can be obtained —see e.g. [4]. Roughly, this follows from requiring that the eigenvalues of the non-principal part of the linearised system ($\mathring{\mathbf{B}}$) have negative real part. For the type of systems being considered here, where the matrices $\mathring{\mathbf{B}}$, $\mathring{\mathbf{A}}^\mu$ are not constant but depend smoothly on time, these methods can be easily generalised —see also [3].

2. The Einstein-Friedrich-nonlinear scalar field system

Using Friedrich's formulation of EFEs for the Einstein-Euler system [2], we obtained the following symmetric hyperbolic reduction for the Einstein-non-linear scalar field system —see [1] for details:

$$
\begin{aligned}
\partial_t\phi &= \psi,\\
\partial_t\psi &= -\psi\chi - \frac{d\mathcal{V}}{d\phi},\\
2\partial_t\chi_{(bd)} - 2\,e'^{\,0}_{(b}\partial_t a_{d)} - 2\,e'^{\,j}_{(b}\partial_j a_{d)} &= \frac{2}{3}\left(\mathcal{V}(\phi) - \psi^2\right)h_{bd} - 2\chi^{p}{}_{(d}\chi_{b)p} + 2a_b a_d + 2E_{bd}\\
&\quad - \left(\gamma^p{}_{bd} + \gamma^p{}_{db}\right)a_p\\
\partial_t a_c - e'^{\,0}_{p}\partial_t\chi_c{}^p - e'^{\,j}_{p}\partial_j\chi_c{}^p &= \left(\frac{2}{\psi}\frac{d\mathcal{V}}{d\phi} + \chi\right)a_c - \chi_c{}^p a_p - \gamma^q{}_{cp}\chi_q{}^p + \gamma^p{}_{qp}\chi_c{}^q,\\
2\partial_t E_{bd} - 2\epsilon^{pa}{}_{(b|}e'^{\,0}_{a}\partial_t B_{p|d)} - 2\epsilon^{pa}{}_{(b|}e'^{\,j}_{a}\partial_j B_{p|d)} &= -\psi^2\left(\chi_{(bd)} - \frac{1}{3}\chi h_{bd}\right) - 4\chi E_{bd} + 10\chi^q{}_{(b}E_{d)q}\\
&\quad - 2h_{bd}\chi^{qp}E_{qp} + 4a_a B_{p(b}\epsilon_{d)}{}^{pa} - 2\gamma^q{}_{pa}B_{q(d}\epsilon_{b)}{}^{pa}\\
&\quad - 2\epsilon^{pa}{}_{(d}\gamma^q{}_{b)a}B_{pq},\\
2\partial_t B_{bd} - 2\epsilon^{ap}{}_{(d|}e'^{\,0}_{a}\partial_t E_{|b)p} - 2\epsilon^{ap}{}_{(d}e'^{\,j}_{|a}\partial_{j|}E_{b)p} &= -2\chi B_{bd} + 6\chi^q{}_{(b}B_{d)q} + 2\chi_{ac}B_{pq}\epsilon^{pa}{}_{(b}\epsilon_{d)}{}^{qc}\\
&\quad - 4a_a E_{p(b}\epsilon_{d)}{}^{pa} - 2\gamma^q{}_{pa}E_{q(b}\epsilon_{d)}{}^{ap}\\
&\quad - 2\epsilon^{ap}{}_{(b}\gamma^q{}_{d)a}E_{pq},\\
\partial_t e'^{\,0}_{b} &= -\chi_b{}^c e'^{\,0}_{c} + a_b,\\
\partial_t e'^{\,i}_{b} &= -\chi_b{}^c e'^{\,i}_{c},\\
\partial_t\gamma'^a{}_{bd} &= B_{dp}\epsilon^{pa}{}_b - \chi_d{}^p\gamma'^a{}_{bp} + 2h^{ap}\chi_{d[p}a_{b]},
\end{aligned}
\tag{3}
$$

and the following result:

Theorem 1. *The Einstein-Friedrich-nonlinear scalar field (EFsf) system consisting of the equations in* (3) *forms a quasi-linear first-order symmetric hyperbolic (FOSH) system for the scalar field* (ϕ)*, its momentum-density* (ψ)*, the spatial frame coefficients* ($e'^{\,i}_{b}, e'^{\,0}_{b}$)*, the connection coefficients* ($\chi_{(ab)}, a_c, \gamma'^a{}_{bd}$)*, and the electric and magnetic parts of the Weyl tensor* ($E_{(bd)}, B_{(bd)}$)*, relatively to the slices of constant time t, as long as the quadratic form*

$$\sum_{a=1,2,3}\theta^a{}_i\theta^a{}_j - \frac{\partial_i\phi}{\psi}\frac{\partial_j\phi}{\psi}, \qquad i, j = 1, 2, 3,$$

with $\theta^a{}_i$ the spatial co-frame coefficients, is positive definite.

3. Stability Analysis

In this section we use the symmetric hyperbolic system of last section to show that, for some classes of potentials, the evolution of sufficiently small nonlinear perturbations of a Friedmann-

Robertson-Walker background with a scalar field with positive potential, asymptotic exponential decay to zero in time.

3.1. *The background solution*

As it is well known, the metric of a Friedman-Robertson-Walker (FRW) spacetime —i.e. a spatially homogeneous and isotropic spacetime— can be written as

$$\mathrm{d}s^2 = -\mathrm{d}t^2 + \left(\frac{a(t)}{\omega}\right)^2 \delta_{ij}\mathrm{d}x^i\mathrm{d}x^j,$$

where $a(t)$ is the Robertson-Walker scale factor, $\omega = 1 + \frac{k}{4}\delta_{ij}x^ix^j$, $\partial_i\omega = kx_i$, and $k = -1, 0, 1$ is the curvature of the spatial hypersurfaces. Since the metric is conformal flat, it follows that $\mathring{E}_{bd} = \mathring{B}_{bd} = 0$. The gauge conditions [1] for the frame are satisfied if $\mathring{e}_0{}^\mu = \delta_0{}^\mu$, $\mathring{e}_b{}^\mu = (\omega/a)\delta_b{}^\mu$ where $b = 1, 2, 3$. Thus, the spatial connection coefficients are given by $\mathring{\gamma}^c{}_{bd} = (k/2a^2)\left(h_{db}x^c - h_d{}^c x_b\right)$, with $x^\mu = (\omega/a)\delta^\mu{}_c x^c$. The remaining non-vanishing connection coefficients are $\mathring{\gamma}^0{}_{bd} = \mathring{\chi}_{db} = Hh_{bd}$, $\mathring{\gamma}^b{}_{0d} = \mathring{\chi}_d{}^b = Hh_d{}^b$, where $H(t) \equiv \dot{a}/a$ is the so-called *Hubble function* and $\dot{}$ denotes differentiation with respect to t. In particular $\chi_{[bd]} = a_b = 0$ and $\chi_{(bd)} = 0$ for $b \neq d$, and for such background metrics $\mathring{\chi} = 3\dot{a}/a \equiv 3H$. Thus, in the case of a FRW cosmology the Einstein-scalar field system (3) reduces to the evolution equations

$$\frac{\mathrm{d}\mathring{\phi}}{\mathrm{d}t} = \mathring{\psi}, \quad \frac{\mathrm{d}\mathring{\psi}}{\mathrm{d}t} = -3H\mathring{\psi} - \frac{\mathrm{d}\mathcal{V}}{\mathrm{d}\mathring{\phi}} \quad \text{and} \quad \frac{\mathrm{d}H}{\mathrm{d}t} = -H^2 - \frac{1}{3}\mathring{\psi}^2 + \frac{1}{3}\mathcal{V}(\mathring{\phi}), \tag{4}$$

subject to the Friedmann-scalar field constraint equation $H^2(t) - \frac{1}{6}\mathring{\psi}^2(t) - \frac{1}{3}\mathcal{V}(\mathring{\phi}) = -k/a^2$.

3.2. *Linearised evolution equations*

In order to perform the linearisation procedure we compute $\left.\frac{\mathrm{d}\mathbf{u}^\epsilon}{\mathrm{d}\epsilon}\right|_{\epsilon=0}$ and drop all (nonlinear) terms of coupled perturbations. In this way, we obtain the following system

$$\begin{aligned}
\partial_t\breve{\phi} &= \breve{\psi},\\
\partial_t\breve{\psi} &= -\left(\frac{\mathrm{d}^2\mathcal{V}}{\mathrm{d}\mathring{\phi}^2}\right)\breve{\phi} - 3H\breve{\psi} - \mathring{\psi}\breve{\chi},\\
2\partial_t\breve{\chi}_{(bd)} - 2\left(\frac{\omega}{a}\right)\delta_{(d}{}^j\partial_j\breve{a}_{b)} &= \frac{2}{3}\left(\left(\frac{\mathrm{d}\mathcal{V}}{\mathrm{d}\mathring{\phi}}\right)\breve{\phi} - 2\mathring{\psi}\breve{\psi}\right)h_{bd} - 4H\breve{\chi}_{(bd)} + 2\breve{E}_{bd}\\
&\quad - \frac{k}{a^2}\left(h_{bd}x^p\breve{a}_p - x_{(b}\breve{a}_{d)}\right),\\
\partial_t\breve{a}_c - \left(\frac{\omega}{a}\right)\delta_p{}^j\partial_j\breve{\chi}_c{}^p &= \left(2H + \frac{2}{\mathring{\psi}}\frac{\mathrm{d}\mathcal{V}}{\mathrm{d}\mathring{\phi}}\right)\breve{a}_c + \left(\frac{\mathrm{d}H}{\mathrm{d}t}\right)\breve{e}_c{}^0 + \frac{k}{2a^2}x_c\breve{\chi} - \frac{3k}{2a^2}x^q\breve{\chi}_{(qc)},\\
2\partial_t\breve{E}_{bd} - 2\left(\frac{\omega}{a}\right)\epsilon^{pa}{}_{(b|}\delta_a{}^j\partial_j\breve{B}_{p|d)} &= -\mathring{\psi}^2\left(\breve{\chi}_{(bd)} - \frac{h_{bd}}{3}\breve{\chi}\right) - 2H\breve{E}_{bd} + \frac{k}{a^2}x_p\epsilon^{pq}{}_{(b}\breve{B}_{d)q},\\
2\partial_t\breve{B}_{bd} - 2\left(\frac{\omega}{a}\right)\epsilon_{(d|}{}^{ap}\delta_a{}^j\partial_j\breve{E}_{p|b)} &= -2H\breve{B}_{bd} + \frac{k}{a^2}x_p\epsilon^{qp}{}_{(b}\breve{E}_{d)q},\\
\partial_t\breve{e}_b{}^0 &= -H\breve{e}_b{}^0 + \breve{a}_b,\\
\partial_t\breve{e}_b{}^i &= -H\breve{e}_b{}^i - \left(\frac{\omega}{a}\right)\delta_c{}^i\breve{\chi}_b{}^c,\\
\partial_t\breve{\gamma}^a{}_{bd} &= -H\breve{\gamma}^a{}_{bd} - \frac{k}{2a^2}\left(x^a\breve{\chi}_{db} - x_b\breve{\chi}_d{}^a\right)\\
&\quad + H\left(\delta^a_d\breve{a}_b - h_{bd}\breve{a}^a\right) + \breve{B}_{dp}\epsilon^{pa}{}_b.
\end{aligned} \tag{5}$$

As a consequence, the linearised system is of the following form

$$\mathring{\mathbf{A}}^0\partial_t\breve{\mathbf{u}} - \mathring{\mathbf{A}}^j(t,\boldsymbol{x})\partial_j\breve{\mathbf{u}} = \mathring{\mathbf{B}}(t,\boldsymbol{x})\breve{\mathbf{u}}.$$

Since we are considering perturbations over a Friedmann-Robertson-Walker background with flat spatial sections, the linearized matrices $\mathring{\mathbf{A}}^\mu$ and $\mathring{\mathbf{B}}$ are functions of cosmic time t, only. In this case the characteristic polynomial of $\mathring{\mathbf{B}}$ has the following form

$$\begin{aligned}&(\lambda+H)^{21}\times(\lambda+2H)^3\times\left(\lambda^2+6H\lambda+2\mathring{\psi}^2+8H^2\right)^3\times\\&\times\left(\lambda^2-\left(H+2\frac{\mathring{\mathcal{V}}'}{\mathring{\psi}}\right)\lambda-\left(\frac{\mathrm{d}H}{\mathrm{d}t}+2H^2+2H\frac{\mathring{\mathcal{V}}'}{\mathring{\psi}}\right)\right)^3\times f(\lambda)\end{aligned} \tag{6}$$

where f is a polynomial of degree 8 in λ with coefficients depending on the background quantities $(\mathring{\mathcal{V}}'', \mathring{\mathcal{V}}', \mathring{\psi}, H)$. Here $\mathring{\mathcal{V}}'$ and $\mathring{\mathcal{V}}''$ denote, respectively, the first and second derivatives of the potential with respect to $\mathring{\phi}$. In order to obtain conditions which ensure that the characteristic polynomial has eigenvalues with negative real part, we can make use of the *Liénard-Chipart theorem* —see e.g. [5]. This theorem states that a polynomial $f(z) = a_0z^n + a_1z^{n-1} + a_2z^{n-2} + \cdots + a_n,\quad a_0 > 0$, with real coefficients, has roots with negative real part if and only if *all coefficients of f are positive* and the *Hurwitz determinants* defined by

$$\delta_0 \equiv 1, \qquad \delta_l \equiv det\begin{pmatrix} a_1 & a_3 & a_5 & \cdots & a_{2l-1}\\ a_0 & a_2 & a_4 & \cdots & a_{2l-2}\\ 0 & a_1 & a_3 & \cdots & a_{2l-3}\\ \vdots & \vdots & \vdots & & \vdots\\ 0 & 0 & 0 & \cdots & a_l\end{pmatrix} \qquad l = 1,\ldots,n$$

are positive. It is easy to see that the first two terms in (6) require $H(t) > 0$, which is simply the condition for an (ever) expanding background. Then if there is a constant $H_0 > 0$, such that for all t, $H(t) > H_0$, the conditions in the potential, for its first and second derivatives ($\mathcal{V}'$ and $\mathcal{V}''$) can be inferred —see [1] for details.

Acknowledgments

JAVK was supported by an EPSRC Advanced Research Fellowship and by a project research grant from the Leverhulme Trust (F/07 476/AI). AA and FM are supported by projects FCT/PTDC/MAT/108921/2008 and CERN/FP/116377/2010 and by CMAT, Univ. Minho, through FCT plurianual funding. AA thanks FCT grant SFRH/BD/48658/2008. FM thanks FCT grant SFRH/BSAB/967/2010.

References

[1] Alho A, Mena F C, Valiente-Kroon J, *The Einstein-Friedrich-non-linear scalar field system and the stability of scalar field cosmologies*, arxiv.org/abs/1006.3778
[2] Friedrich H 1998 *Physical Review D* **57** 2317-2322
[3] Reula O A 1999 *Physical Review D* **60** 083507
[4] Kreiss H O and Lorenz J 1998 *Acta Numerica* **7** 203-285
[5] Rahman Q I and Schmeisser G 2002 *Analytic Theory of Polynomials: Critical Points, Zeros and Extremal Properties* London Mathematical Society Monographs (Clarendon Press)

Spanish Relativity Meeting (ERE 2010): Gravity as a Crossroad in Physics IOP Publishing
Journal of Physics: Conference Series **314** (2011) 012100 doi:10.1088/1742-6596/314/1/012100

Cylindrically and toroidally symmetric solutions with a cosmological constant

Jiří Podolský[1] **and Jerry Griffiths**[2]

[1] Institute of Theoretical Physics, Charles University in Prague, V Holešovičkách 2, 180 00 Prague 8, Czech Republic
[2] Department of Mathematical Sciences, Loughborough University, Loughborough, LE11 3TU, UK

E-mail: podolsky@mbox.troja.mff.cuni.cz

Abstract. Cylindrical-like coordinates for constant-curvature 3-spaces are introduced and discussed. This helps to clarify the geometrical properties, the coordinate ranges and the meaning of free parameters in the static vacuum solution of Linet and Tian. In particular, when the cosmological constant is positive, the spacetimes have toroidal symmetry. One of the two curvature singularities can be removed by matching the Linet–Tian vacuum solution across a toroidal surface to a corresponding region of the dust-filled Einstein static universe. Some other properties and limiting cases of these space-times are also described, together with their generalisation to higher dimensions.

1. Introduction

The class of static cylindrically symmetric vacuum solutions, which was found by Levi-Civita in 1919 [1], can be written in the form

$$\mathrm{d}s^2 = -\rho^{4\sigma/\Sigma}\mathrm{d}t^2 + \rho^{-4\sigma(1-2\sigma)/\Sigma}\mathrm{d}z^2 + C^2\rho^{2(1-2\sigma)/\Sigma}\mathrm{d}\phi^2 + \mathrm{d}\rho^2, \tag{1}$$

where $\Sigma = 1 - 2\sigma + 4\sigma^2$. The parameter $\sigma \in (0, \frac{1}{4})$ may be interpreted as the mass per unit length of the source located along the axis $\rho = 0$, while C is the conicity parameter (see [2] for more details). When $\sigma = 0$, Minkowski space in cylindrical coordinates is recovered.

In 1986, a generalisation of (1) to include a non-zero cosmological constant Λ was obtained by Linet [3] and Tian [4] (see also [5]) in the form

$$\mathrm{d}s^2 = Q^{2/3}\Big(-P^{-2(1-8\sigma+4\sigma^2)/3\Sigma}\,\mathrm{d}t^2 + P^{-2(1+4\sigma-8\sigma^2)/3\Sigma}\,\mathrm{d}z^2 + C^2P^{4(1-2\sigma-2\sigma^2)/3\Sigma}\,\mathrm{d}\phi^2\Big) + \mathrm{d}\rho^2, \tag{2}$$

where ρ is a proper radial distance from the axis and

$$Q(\rho) = \frac{1}{\sqrt{3\Lambda}}\sin\left(\sqrt{3\Lambda}\,\rho\right), \qquad P(\rho) = \frac{2}{\sqrt{3\Lambda}}\tan\left(\frac{\sqrt{3\Lambda}}{2}\,\rho\right). \tag{3}$$

Both $\Lambda > 0$ and $\Lambda < 0$ are admitted (in the latter case the trigonometric functions are replaced by hyperbolic ones and Λ by $|\Lambda|$). The metric (2), (3) locally approaches the Levi-Civita solution (1) either as $\Lambda \to 0$ or near the axis as $\rho \to 0$ because $Q \approx \rho$ and $P \approx \rho$ in these limits.

2. Constant-curvature 3-spaces in cylindrical coordinates

In order to understand the global geometrical properties of the Linet–Tian solution, it seems to be important first to rewrite the *maximally symmetric* 3-spaces in cylindrical-like coordinates. Such constant-curvature spaces are usually written in the spherical form

$$\mathrm{d}s^2 = R^2\Big(\frac{\mathrm{d}r^2}{1-k\,r^2} + r^2(\mathrm{d}\theta^2+\sin^2\theta\,\mathrm{d}\phi^2)\Big), \tag{4}$$

where $k = 0, +1, -1$ corresponds to the geometries of E^3, S^3, H^3, respectively (as is well-known from the FLRW cosmology). Performing the transformation

$$\hat\rho = r\,\sin\theta, \qquad \hat z,\ \tan\hat z,\ \tanh\hat z = \frac{r\,\cos\theta}{\sqrt{1-k\,r^2}}, \tag{5}$$

the metric (4) becomes

$$\mathrm{d}s^2 = R^2\Big((1-k\,\hat\rho^2)^{-1}\mathrm{d}\hat\rho^2 + (1-k\,\hat\rho^2)\,\mathrm{d}\hat z^2 + \hat\rho^2\,\mathrm{d}\phi^2\Big). \tag{6}$$

In the case when $k = 0$, this is the flat space E^3 in cylindrical coordinates. Since $\hat z \in (-\infty, +\infty)$, $\phi \in [0, 2\pi)$, the surfaces $\hat\rho =$const. are obviously *cylinders* with topology $R^1 \times S^1$.

In the case when $k = +1$, the metric of the three-sphere S^3 in cylindrical-like coordinates is

$$\mathrm{d}s^2 = R^2\left(\frac{\mathrm{d}\hat\rho^2}{1-\hat\rho^2} + (1-\hat\rho^2)\,\mathrm{d}\psi^2 + \hat\rho^2\,\mathrm{d}\phi^2\right), \tag{7}$$

where $\psi = \hat z - \frac{\pi}{2}$. Since the 3-space is bounded in this case, both ψ and ϕ are *periodic* coordinates with $\psi, \phi \in [0, 2\pi)$, $\hat\rho \in [0, 1]$, and the surfaces $\hat\rho =$ const. are *tori* with topology $S^1 \times S^1$. There are thus *two* nonintersecting circular axes at $\hat\rho = 0$ and $\hat\rho = 1$ around the closed space S^3. This can be explicitly seen from the parametrisation of S^3 as the 3-surface ${x_1}^2 + {x_2}^2 + {x_3}^2 + {x_4}^2 = R^2$ in a flat space $\mathrm{d}s^2 = \mathrm{d}{x_1}^2 + \mathrm{d}{x_2}^2 + \mathrm{d}{x_3}^2 + \mathrm{d}{x_4}^2$. The metric of (7) is obtained by

$$\begin{aligned} x_1 &= R\sqrt{1-\hat\rho^2}\,\cos\psi, & x_3 &= R\,\hat\rho\,\cos\phi, & \text{i.e.,}\quad x_1^2+x_2^2 &= R^2(1-\hat\rho^2), & x_3^2+x_4^2 &= R^2\hat\rho^2,\\ x_2 &= R\sqrt{1-\hat\rho^2}\,\sin\psi, & x_4 &= R\,\hat\rho\,\sin\phi, & \frac{x_2}{x_1} &= \tan\psi, & \frac{x_4}{x_3} &= \tan\phi, \end{aligned} \tag{8}$$

see figure 1.

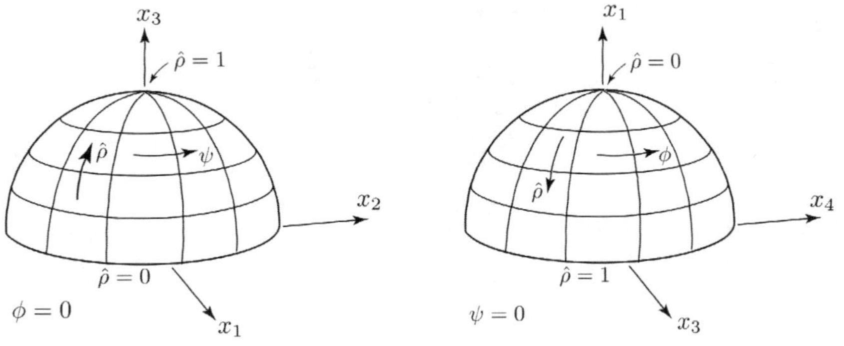

Figure 1. Two sections through the three-sphere S^3, namely $\phi = 0$ (left) and $\psi = 0$ (right). Here ψ and ϕ are "complementary" angular coordinates in different directions. Thin tori $\hat\rho =$const. around the axes $\hat\rho = 0$ and $\hat\rho = 1$ do not intersect.

Journal of Physics: Conference Series **314** (2011) 012100 doi:10.1088/1742-6596/314/1/012100

3. Interpreting the Linet–Tian solution with $\Lambda > 0$

In view of the above geometry, it seems necessary to relabel z in (2) as the angular coordinate ψ and to introduce the related conicity parameter B. The Linet–Tian vacuum metric becomes

$$\mathrm{d}s^2 = Q^{2/3}\Big(-P^{-2(1-8\sigma+4\sigma^2)/3\Sigma}\,\mathrm{d}t^2 + B^2P^{-2(1+4\sigma-8\sigma^2)/3\Sigma}\,\mathrm{d}\psi^2 + C^2P^{4(1-2\sigma-2\sigma^2)/3\Sigma}\,\mathrm{d}\phi^2\Big) + \mathrm{d}\rho^2, \tag{9}$$

where $Q(\rho), P(\rho)$ are given by (3) and $\phi, \psi \in [0, 2\pi)$. Apart from the cosmological constant Λ, the metric has three parameters: B, C and σ. There are curvature singularities along the "axes" $\rho = 0$ and $\rho = \pi/\sqrt{3\Lambda}$, with the deficit angles $2\pi(1 - C)$ and $2\pi(1 - B)$ in the weak-field limit.

Either of these two singularities can be removed and specific values of B and C can be established, for example, by *matching* this vacuum solution across a surfaces on which ρ is constant ($\rho = \rho_1$) to a corresponding *toroidal region of the Einstein static universe.* This well-known homogeneous and isotropic dust-filled universe has the spatial geometry of S^3 with radius $R = 1/\sqrt{\Lambda}$ and thus, using the toroidal coordinates of (7), can be written in the form

$$\mathrm{d}s^2 = -A_1^2\mathrm{d}t^2 + \frac{B_1^2}{\Lambda}\cos^2\Big(\sqrt{\Lambda}\,(\rho-\rho_0)\Big)\mathrm{d}\psi^2 + \frac{C_1^2}{\Lambda}\sin^2\Big(\sqrt{\Lambda}\,(\rho-\rho_0)\Big)\mathrm{d}\phi^2 + \mathrm{d}\rho^2. \tag{10}$$

For convenience, we applied a simple transformation $\hat{\rho} \equiv \sin\Big(\sqrt{\Lambda}\,(\rho-\rho_0)\Big)$ and introduced the free constants A_1, B_1, C_1 and ρ_0 (see [5]).

Usual matching conditions, namely that the metrics (9) and (10) and their first derivatives are continuous across the surface $\rho = \rho_1$, can indeed be consistently satisfied when

$$\cos\Big(\sqrt{3\Lambda}\,\rho_1\Big) = \frac{1-8\sigma+4\sigma^2}{1-2\sigma+4\sigma^2}, \qquad \tan^2\Big(\sqrt{\Lambda}\,(\rho_1-\rho_0)\Big) = \frac{4\sigma(1-\sigma)}{1-4\sigma}, \tag{11}$$

$$A_1 = Q(\rho_1)^{1/3}\,P(\rho_1)^{-(1-8\sigma+4\sigma^2)/3\Sigma}, \tag{12}$$

$$B_1 = B\sqrt{\Lambda}\,Q(\rho_1)^{1/3}\,P(\rho_1)^{-(1+4\sigma-8\sigma^2)/3\Sigma}\Big/\cos\Big(\sqrt{\Lambda}(\rho_1-\rho_0)\Big), \tag{13}$$

$$C_1 = C\sqrt{\Lambda}\,Q(\rho_1)^{1/3}\,P(\rho_1)^{2(1-2\sigma-2\sigma^2)/3\Sigma}\Big/\sin\Big(\sqrt{\Lambda}(\rho_1-\rho_0)\Big), \tag{14}$$

which uniquely determine ρ_1, ρ_0 in terms of $\sigma \in [0, \frac{1}{4}]$ and relate A_1, B_1, C_1 to B, C.

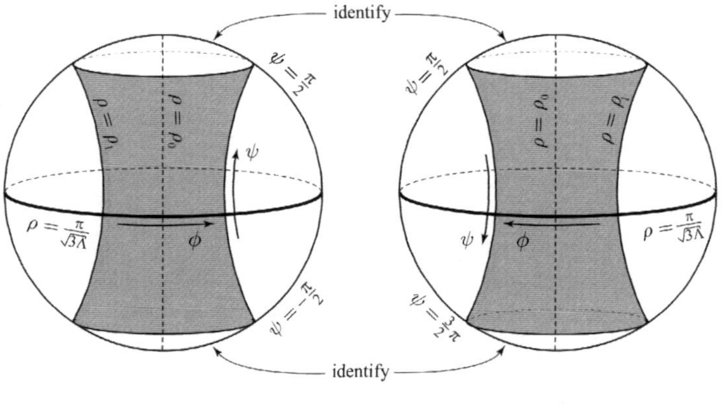

Figure 2. Toroidal region of the Einstein static universe (grey) serves as the dust matter source of the Linet–Tian vacuum solution.

In the resulting composite spacetime, the curvature singularity at $\rho = 0$ is removed. As shown in figure 2, it is replaced by the *toroidal region* $\rho \in [\rho_0, \rho_1)$ which is a part of uniform Einstein static space filled with dust — *the matter source* (this is regular at ρ_0 when $C_1 = 1$). In the *external region* $\rho \in [\rho_1, \pi/\sqrt{3\Lambda})$ there is the *Linet–Tian static vacuum solution.* However, there still remains the curvature singularity at $\rho = \pi/\sqrt{3\Lambda}$ (which could alternatively be removed by a complementary toroidal dust matter source, keeping the singularity at $\rho = 0$).

It is now straightforward to calculate the total mass of such a toroidal dust source of density $\mu = \frac{\Lambda}{4\pi}$, yielding $\int_{\rho_0}^{\rho_1}\int_0^{2\pi}\int_0^{2\pi} \mu\,\sqrt{g_3}\,\mathrm{d}\rho\,\mathrm{d}\psi\,\mathrm{d}\phi = \frac{2\pi B_1 C_1}{\sqrt{\Lambda}}\,\frac{\sigma(1-\sigma)}{(1-4\sigma^2)}$. The *mass per unit length* of the toroid is thus $\sigma(1-\sigma)/(1-4\sigma^2)$, i.e., it is determined just by the parameter σ. This demonstrates that σ in the Linet–Tian class of solutions with $\Lambda \neq 0$ retains its physical meaning known from the Levi-Civita spacetime (for the discussion of possible shell sources see [6]).

Interestingly, the "no source" limit $\sigma = 0$ *is not the (anti-)de Sitter space*, as one would naturally expect! For $\sigma = 0$, the Linet–Tian metric reduces to

$$\mathrm{d}s^2 = p^2(-\mathrm{d}t^2 + B^2\,\mathrm{d}\psi^2) + \frac{4C^2}{3\Lambda}\frac{(1-p^3)}{p}\,\mathrm{d}\phi^2 + \frac{3}{\Lambda}\frac{p}{(1-p^3)}\,\mathrm{d}p^2, \tag{15}$$

where the coordinate $p \equiv \cos^{2/3}\left(\frac{\sqrt{3\Lambda}}{2}\,\rho\right)$ was introduced. This spacetime of algebraic type D belongs to the Plebański–Demiański family and also to the Kundt family of solutions. In fact, it is a generalization of the BIII metric [3,7]. Its geometrical properties still need to be investigated.

4. Extension to higher dimensions

The Linet–Tian class of solutions described above can be extended to any higher D-dimensions. Such toroidally symmetric static vacuum metrics read

$$\mathrm{d}s^2 = R(\rho)^\alpha\Big(-S(\rho)^{2p_0}\,\mathrm{d}t^2 + \sum_{i=1}^{D-2} C_i^2\,S(\rho)^{2p_i}\,\mathrm{d}\phi_i^2\Big) + \mathrm{d}\rho^2, \tag{16}$$

where

$$R(\rho) = \cos(\beta\rho), \qquad S(\rho) = \tan(\beta\rho), \qquad \alpha = \frac{4}{D-1}, \qquad \beta = \sqrt{\frac{(D-1)\Lambda}{2(D-2)}}, \tag{17}$$

$\phi_i \in [0, 2\pi)$, C_i are the corresponding conicity parameters, and the constants p_i satisfy

$$\sum_{i=0}^{D-2} p_i = 1, \qquad \sum_{i=0}^{D-2} p_i^2 = 1. \tag{18}$$

For $D = 4$ this reduces to the Linet–Tian solution (2), (3) with the identification of parameters

$$p_0 = \frac{2\sigma}{\Sigma}, \qquad p_1 = \frac{2\sigma(2\sigma-1)}{\Sigma}, \qquad p_2 = \frac{1-2\sigma}{\Sigma}. \tag{19}$$

The $\Lambda < 0$ counterpart of the metric (16)–(18) has been recently presented in [8,9].

Acknowledgments

We acknowledge financial support from the grants GAČR 202/08/0187, GAČR 202/09/0772, and the Project No. LC06014 of the Czech Ministry of Education.

[1] Levi-Civita T 1919 *Rend. Acc. Lincei* **28** 101
[2] Griffiths J B and Podolský J 2009 *Exact space-times in Einstein's general relativity* (Cambridge: Cambridge University Press)
[3] Linet B 1986 *J. Math. Phys.* **27** 1817
[4] Tian Q 1986 *Phys. Rev.* D **33** 3549
[5] Griffiths J B and Podolský J 2010 *Phys. Rev.* D **81** 064015
[6] Žofka M and Bičák J 2008 *Class. Quantum Grav.* **25** 015011
[7] Bonnor W B 2008 *Class. Quantum Grav.* **25** 225005
[8] Sarıoğlu Ö and Tekin B 2009 *Phys. Rev.* D **79** 087502
[9] Sarıoğlu Ö and Tekin B 2009 *Class. Quantum Grav.* **26** 048001

Action principle for the Z4 and BSSN numerical relativity formalisms

C. Bona and C. Bona-Casas

Institute for Applied Computation with Community Code (IAC^3)
Departament de Fisica, Universitat de les Illes Balears, 07122 Palma de Mallorca, Spain.

E-mail: `cbona@uib.es`

Abstract. A Lagrangian density is provided, that allows to derive the Z4 evolution system from a Palatini-type variation principle. The proposed action includes the supplementary vector Z_μ. The constraint $Z_\mu = 0$ can be imposed just on the initial data, in order to recover true Einstein's solutions. This opens the door to analogous results for other numerical-relativity formalisms, like BSSN, that can be derived from Z4 by a symmetry-breaking procedure.

The action principle for General Relativity was stated by Hilbert since the very beginning of the theory in the Lagrangian form, although the Hamiltonian formulation had to wait for decades [1, 2]. The reason for this long delay is probably related to the complexity of the Cauchy problem for Einstein's equations, which becomes manifest in the 3+1 (space plus time) decomposition [2]. The coordinate gauge freedom produces a mismatch between the number of dynamical fields and that of true evolution equations: four of the field equations are indeed (energy-momentum) constraints. This rich structure opens the door to many different approaches.

Using this freedom, by the end of the past century, some hyperbolic extensions of Einstein's equations were developed with a view on numerical relativity applications [3, 4, 5, 6]. This emergent field is now more mature: there are two main formalisms currently used in numerical simulations. One is BSSN [7, 8], working at the 3+1 level, and the other is the class of generalized harmonic formalisms [9, 10, 11], working at the four-dimensional level. A unifying framework is provided by the Z4 formalism [12], which allows to recover the generalized harmonic one by relating the additional vector field Z_μ with the harmonic 'gauge sources' [9]. On the other hand, it allows to recover (a specific version of) BSSN by a symmetry-breaking process in the transition from the four-dimensional to the three-dimensional formulations [13, 14].

There is a growing interest in incorporating the new hyperbolic formulations into the Lagrangian/Hamiltonian framework. An example is the usage of the 'densitized lapse' [3] as a canonical variable, leading to a modification in the standard form of the canonical evolution equations [15]. Reciprocally, there are very recent attempts of modifying the ADM action [2] in order to incorporate coordinate conditions of the type used in numerical relativity [16, 17], with a view on using symplectic integrators for the time evolution, which could ensure constraint preservation in numerical simulations [18]. On a different context, a well posed evolution formalism developed from a Lagrangian formulation could be a good starting point for Quantum Gravity applications.

In this paper we derive the Z4 formalism from an action principle by introducing a Lagrangian density which generalizes the Einstein-Hilbert one. This is a crucial step towards the Lagrangian formulation of other numerical-relativity formalisms. We actually consider here the BSSN case, following the symmetry-breaking mechanism described in refs. [13, 14]

1. Generalizing the Einstein-Hilbert action principle

Let us consider the generic action

$$S = \int d^4x \, \mathcal{L} \tag{1}$$

with a Lagrangian density which generalizes the Einstein-Hilbert one by including an extra four-vector Z_μ, namely

$$\mathcal{L} = \sqrt{g}\; g^{\mu\nu} \left[\, R_{\mu\nu} + 2\, \nabla_\mu Z_\nu \,\right] \tag{2}$$

(we restrict ourselves to the vacuum case), with the Ricci tensor written in terms of the connection coefficients

$$R_{\mu\nu} = \partial_\rho \Gamma^\rho_{\mu\nu} - \partial_{(\mu} \Gamma^\rho_{\;\nu)\rho} + \Gamma^\rho_{\rho\sigma}\, \Gamma^\sigma_{\mu\nu} - \Gamma^\rho_{\sigma\mu}\, \Gamma^\sigma_{\rho\nu} \,, \tag{3}$$

(round brackets denote symmetrization).

Now let us follow the well-known Palatini approach, by considering independent variations of the metric density $h^{\mu\nu} = \sqrt{g}\; g^{\mu\nu}$, the connection coefficients $\Gamma^\rho_{\mu\nu}$ and the vector Z_μ. From the $h^{\mu\nu}$ variations we get directly the Z4 field equations [12]

$$R_{\mu\nu} + \nabla_\mu Z_\nu + \nabla_\nu Z_\mu = 0 \,, \tag{4}$$

which are currently used in many numerical-relativity developments [14].

From the $\Gamma^\rho_{\mu\nu}$ and the Z_μ variations we get a coupled set of equations which can be decomposed in a covariant way into the tensor equation

$$\nabla_\rho \, g^{\mu\nu} = 0 \,, \tag{5}$$

which fixes the connection coefficients in terms of the metric, and the vector condition

$$Z_\mu = 0. \tag{6}$$

Let us note here the different role of the conditions (5) and (6). As there are much more independent connection coefficients than evolution equations in (4), we will consider condition (5) as a constraint enforcing the metric connection 'a posteriori', that is after the variation process. In this way, we will ensure that equations (4) are identical to the original Z4 equations, rather than some affine generalization. For this reason, we will assume a metric connection everywhere in what follows.

The case of condition (6) is different, as the Z4 equations (4) actually provide evolution equations for every component of Z_μ. Then, (6) is a standard primary constraint and we have a choice among different strategies for dealing with it. If we enforce (6) into the Z4 field equations (4), we get nothing but Einstein's equations. This is not surprising because our Lagrangian obviously reduces to the Einstein-Hilbert one when Z_μ vanishes. The problem is that the plain Einstein field equations do not lead directly to a well-posed initial data problem. This is why the original harmonic formulation [19, 20, 21] was used instead in the context of the Cauchy problem [22]. For the same reason, other formulations (BSSN [7, 8], generalized harmonic [9, 10, 11], Z4 [12, 13]) are currently considered in numerical relativity.

2. Recovering the Z4 formulation

We can alternatively follow a different strategy. Instead of enforcing (6), we can deal with this condition as an algebraic restriction to be imposed just on the initial data, that is

$$Z_\mu \,|_{t=0} = 0 \tag{7}$$

In this way, we can keep the Z4 field equations system, which is known to be strongly hyperbolic when supplemented by gauge conditions, like '1+log' or 'freezing shift', suitable for numerical evolution [13, 14]. The consistency of this 'relaxed' approach requires that the constraint (6) should be actually preserved by the Z4 field equations (4). In this way, the solutions obtained from initial data verifying (6) will actually minimize the proposed action (1).

Allowing for the conservation of the Einstein tensor, which is granted after the metric connection enforcement, we derive from (4) the second-order equation, linear-homogeneous in Z

$$\nabla_\nu \left[\, \nabla^\mu Z^\nu + \nabla^\nu Z^\mu - (\nabla_\rho Z^\rho)\, g^{\mu\nu} \,\right] = 0\,. \tag{8}$$

It follows that the necessary and sufficient condition for the preservation of the constraint (6) is to impose also its first time-derivative conditions in the initial data, that is

$$(\partial_0\, Z_\mu)\,|_{t=0} = 0\,. \tag{9}$$

Note that, allowing for (7) and the Z4 field equations, the secondary constraints (9) amount to the standard energy and momentum constraints, which are then to be imposed on the initial data in addition to (7).

This 'relaxed' treatment of the constraints (6) may look unnatural. But it is just the reflection of a common practice numerical relativity ('free evolution' approach), where four of the ten field equations (the energy-momentum constraints) are not enforced during the evolution, being imposed just in the initial data instead. The introduction of the extra four-vector in the Z4 formalism actually provides a simpler implementation of the same idea.

3. Recovering the Z3-BSSN formulation

Note that in all our developments we have preserved general covariance. Our action integral (1) is a true scalar and, in spite of other alternatives, we have avoided the addition of total divergences which could have simplified our developments to some extent, at the price of adding boundary terms. This means that we keep at this point the full coordinate-gauge freedom.

However, as it is well known, the BSSN formulation is not general covariant. It contains just three additional 'contracted-gamma' quantities, associated to the momentum constraint. But the 3+1 splitting between energy and momentum requires a specific choice for the time coordinate, which breaks general covariance. In refs. [13, 14] a symmetry breaking mechanism is proposed, which allows to recover BSSN from the 'Z3 system' [23], which is obtained in turn from the Z4 one by enforcing the energy constraint in the form

$$Z^0 = 0\,. \tag{10}$$

We can proceed now like in the preceding section, by the following steps:

- Enforcing both the metric connection condition (5) and (10) in the Z4 field equations.
- Splitting the ten resulting equations into the (energy-related) secondary constraint

$$\mathcal{E} \equiv \partial_t Z^0 = 0 \tag{11}$$

plus nine evolution equations for the six space metric components and the space vector Z_i.

Note that this splitting is not unique, as any multiple of $\mathcal{E}$ (times the metric) can be added to the evolution equations, resulting into a modified evolution system, with a different principal part. In this way we obtain the one-parameter family of (generalized) Z3 evolution systems [13, 14]. For a specific value of the parameter, one can recover a flavor of the BSSN formalism (Z3-BSSN) by means of a conformal decomposition of the spatial tensors and the redefinition of space vector Z_i in terms of the equivalent 'contracted-Gamma' quantities, namely

$$\tilde{\Gamma}_i = -\tilde{\gamma}_{ik}\ \tilde{\gamma}^{jk}_{\ ,j} + 2\ Z_i \tag{12}$$

where $\tilde{\gamma}_{ij}$ is the conformal space metric. The space vector Z_i can be interpreted in this context as the difference between the (contracted) conformal metric connection and the BSSN contracted-Gamma quantities. This provides actually time a three-covariant reformulation of the original BSSN formalism.

Acknowledgments

This work has been jointly supported by European Union FEDER funds and by the Spanish Ministry of Science and Education (projects FPA2007-60220, CSD2007-00042 and ECI2007-29029-E). C. Bona-Casas acknowledges the support of the Spanish Ministry of Science, under the FPU/2006-02226 fellowship.

References

[1] P.A.M. Dirac, Proc. Roy. Soc. **A246**, 333 (1958).
[2] R. Arnowitt, S. Deser and C.W. Misner, in *Gravitation: an introduction to current research*, ed. L. Witten (Wiley: New York 1962). gr-qc/0405109.
[3] Y. Choquet-Bruhat and T. Ruggeri, Comm. Math. Phys. **89**, 269 (1983).
[4] C. Bona and J. Massó, Phys. Rev. Lett. **68** 1097 (1992)
[5] C. Bona, J. Massó, E. Seidel and J. Stela, Phys. Rev. Lett. **75** 600 (1995).
[6] A. Abrahams, A. Anderson, Y. Choquet-Bruhat and J. W. York, Phys. Rev. Lett. **75**, 3377 (1995).
[7] M. Shibata and T. Nakamura, Phys. Rev. D**52** 5428 (1995).
[8] T. W. Baumgarte and S. L. Shapiro, Phys. Rev. D**59** 024007 (1998).
[9] H. Friedrich, Comm. Math. Phys. **100**, 525(1985).
[10] F. Pretorius, Phys. Rev. Lett. **95** 121101 (2005).
[11] L. Lindblom et al, Class. Quantum Grav. **23**, S447 (2006).
[12] C. Bona, T. Ledvinka, C. Palenzuela and M. Žáček, Phys. Rev. D **67**, 104005 (2003).
[13] C. Bona, T. Ledvinka, C. Palenzuela and M. Žáček, Phys. Rev. D **69**, 064036 (2004).
[14] C. Bona, C. Palenzuela and C. Bona-Casas *Elements of Numerical Relativity and Relativistic Hydrodynamics*, Lect. Notes Phys. **783** (Springer, Berlin–Heidelberg–New York 2009).
[15] A. Anderson and J.W. York, Phys. Rev. Lett. **81**, 1154-1157 (1998).
[16] J. D. Brown in, *Quantum Mechanics of Fundamental Systems: The Quest for Beauty and Simplicity*, ed. by M. Henneaux and J. Zanelli (Springer, Berlin–Heidelberg–New York 2009). ArXiv:0803.0334 [gr-qc].
[17] D. Hilditch and R. Richter, arXiv:1002.4119v1 [gr-qc] (2010).
[18] C. Di Bartolo, R. Gambini and Jorge Pullin, J. Math. Phys. **46**, 032501 (2005).
[19] T. De Donder, *La Gravifique Einstenienne* Gauthier-Villars, Paris (1921). *The Mathematical Theory of Relativity*, (Massachusetts Institute of Technology, 1927).
[20] K. Lanczos, Ann. Phys. **13**, 621 (1922). Z. Phys. **23**, 537 (1923).
[21] Fock, V.A., *The Theory of Space, Time and Gravitation*, Pergamon, London (1959).
[22] Y. Choquet (Fourès)-Bruhat, Acta Matematica **88**, 141 (1952). 'Cauchy problem' in *Gravitation: An introduction to Current Research*, ed. by L. Witten, Wiley, (New York 1962).
[23] C. Bona, T. Ledvinka and C. Palenzuela, Phys. Rev. D **66**, 084013 (2002).

Spanish Relativity Meeting (ERE 2010): Gravity as a Crossroad in Physics IOP Publishing
Journal of Physics: Conference Series **314** (2011) 012102 doi:10.1088/1742-6596/314/1/012102

Hyperbolicity of Hamiltonian formulations in General Relativity

Ronny Richter
Mathematisches Institut, Auf der Morgenstelle 10, 72076 Tübingen, Germany

David Hilditch
Theoretical Physics Institute, University of Jena, 07743 Jena, Germany

Abstract. Well-posedness of the initial (boundary) value problem is an essential property, both of meaningful physical models and of numerical applications. To prove well-posedness of wave-type equations their level of hyperbolicity is an essential ingredient. We develop helpful tools and classify a large class of Hamiltonian versions of Einstein's equations with live gauge conditions with respect to their hyperbolicity. Finally we find a symmetric hyperbolic Hamiltonian formulation that allows for gauge conditions which are similar to the puncture gauge.

1. Motivation

Hamiltonian formulations play a crucial role in many areas of theoretical physics. Their key properties are the exact conservation of an energy and the symplecticity of the time evolution map. Certain properties of those systems can be obtained in numerical evolutions as well, using either symplectic [1] or energy-preserving integrators [2].

Hamiltonian formulations for General Relativity (GR) were first considered in the 1950's [3, 4]. Today the ADM formulation [5] is the most popular one. Yet, in numerical relativity, where one is interested in an initial value problem (IVP), Hamiltonian formulations are rarely used as the basis of evolution schemes. One can identify a reason for this in the mathematical properties of the ADM system and the numerical schemes used.

The ADM system is composed of six evolution equations and four constraints. Furthermore it possesses a gauge symmetry. It can be shown that the constraint evolution system is closed, i.e. if the constraints are satisfied in the initial hypersurface then they are satisfied for all times.

In numerical applications one usually prefers to solve for the constraints only in the initial data, because the alternative of solving the constraints in every timestep is computationally expensive. However, the IVP for the dynamical subsystem of the ADM equations is ill-posed. It is therefore not possible to build numerical schemes based on the dynamical ADM equations.

A widely used method to avoid this problem and to deal with well-posed IVPs is to fix the gauge and to modify the ADM equations appropriately. The dynamical ADM equations are supplemented by evolution equations for the lapse and shift functions and multiples of the constraints are added to the dynamical ADM equations, such that the IVP for the full system is well-posed (the solution depends continuously on the given initial data). Here we are interested in the question how this technique can be applied while keeping the Hamiltonian structure.

2. Hyperbolicity

In order to construct systems with well-posed IVPs one relies on notions of hyperbolicity. It can be shown that strongly (symmetric) hyperbolic systems possess well-posed initial (boundary) value problems (of course one needs appropriate boundary conditions in the case of IBVPs) [6, 7]. The advantage of this approach is that the proof of hyperbolicity involves essentially linear algebra. Strong and symmetric hyperbolicity are defined as follows [8].

A first order in time, second order in space system of the form

$$\partial_t v = A_1^i \partial_i v + A_1 v + A_2 w + a, \tag{1a}$$

$$\partial_t w = B_1^{ij} \partial_i \partial_j v + B_1^i \partial_i v + B_1 v + B_2^i \partial_i w + B_2 w + b. \tag{1b}$$

is called strongly hyperbolic if the principal symbol

$$P^s = \begin{pmatrix} A_1^i s_i & A_2 \\ B_1^{ij} s_i s_j & B_2^i s_i \end{pmatrix}, \tag{2}$$

has a complete set of eigenvectors (with real eigenvalues) that depend continuously on the spatial vector s_i.

The system (1) is called symmetric hyperbolic if there exists a conserved positive definite energy, $E = \int \epsilon dx$ [8] of the form

$$\epsilon = u_i^\dagger H^{ij}(v) u_j = \begin{pmatrix} \partial_i v \\ w \end{pmatrix}^\dagger \begin{pmatrix} H_{11}^{ij} & H_{12}^i \\ H_{12}^{j\dagger} & H_{22} \end{pmatrix} \begin{pmatrix} \partial_j v \\ w \end{pmatrix}. \tag{3}$$

We denote H^{ij} a *symmetrizer*. Furthermore if there is a matrix $\bar{H}^{ij}$ such that $u_i^\dagger \bar{H}^{ij} u_j$ is a conserved quantity, but not necessarily positive definite then we call $\bar{H}^{ij}$ a *candidate symmetrizer.*

3. Hamiltonian formulations

A Hamiltonian formulation is given by the specification of a Hamiltonian $\mathcal{H}(q,p)$ for the system, constructed from *canonical positions* and *momenta* $(q;p)$. Here the Hamiltonian can be expressed as the integral over space of a Hamiltonian density

$$\mathcal{H} = \int_{\Sigma_t} d^3x\, \mathcal{H}_D(q, \partial_i q, p, \partial_i p), \tag{4}$$

where the Hamiltonian density is a local function of the canonical variables and their spatial derivatives. One obtains the following *canonical equations of motion*

$$\partial_t q = \frac{\partial \mathcal{H}_D}{\partial p} - \partial_i \frac{\partial \mathcal{H}_D}{\partial(\partial_i p)}, \tag{5a}$$

$$\partial_t p = -\frac{\partial \mathcal{H}_D}{\partial q} + \partial_i \frac{\partial \mathcal{H}_D}{\partial(\partial_i q)}. \tag{5b}$$

A system that can be written in the form (5) has *Hamiltonian structure.*

In order to keep the Hamiltonian structure while modifying the evolution system a good strategy is to modify the Hamiltonian density appropriately and derive the equations of motion afterwards. The key ideas in this approach were presented in [9].

One starts from the ADM Hamiltonian density

$$\mathcal{H}_{\mathrm{ADM}} = \int (-\alpha C + 2\beta^i C_i)\sqrt{\gamma} \mathrm{d}^3 x,, \tag{6}$$

where α is the lapse function, β^i is the shift, γ is the determinant of the 3-metric and C, C_i are scalar and vector constraints respectively. The constraints are

$$C = \pi^{ij}\pi_{ij} - \frac{1}{2}\pi^i{}_i\pi^j{}_j - hR\,, \qquad C_i = -2h_{ij}D_k\pi^{jk}, \tag{7}$$

with the 3-metric γ_{ij} and its canonically conjugate momentum π^{ij}. The dynamical degrees of freedom are (γ_{ij}, π^{ij}).

Then in order to provide evolution equations for lapse and shift one needs to include them in the set of dynamical degrees of freedom. This also means that they need conjugate variables, which we denote σ and ρ_i respectively. One finds that these momenta vanish for solutions of GR, i.e. $\sigma = 0$ and $\rho_i = 0$.

After modifying the phase space we must also work with a different Hamiltonian. One adds to the ADM Hamiltonian terms that allow for appropriate gauge conditions, but do not change the physics. We use

$$\mathcal{H} = \mathcal{H}_{\mathrm{ADM}} + \mathcal{H}_{GHG} + \int d^3x \left(\Lambda\sigma + \Omega^i\rho_i\right), \tag{8}$$

where $\mathcal{H}_{\mathrm{ADM}} + \mathcal{H}_{GHG}$ is the Hamiltonian for Brown's generalized harmonic formulation [9]. The terms that appear in the principal part are

$$\begin{aligned}\mathcal{H}_{GHG} &= \beta^i\sigma D_i\alpha + \beta^i\rho_j\partial_i\beta^j + \alpha^2\rho_i\Gamma^i_{jk}\gamma^{jk} - \alpha\rho^i D_i\alpha \\ &+ \frac{1}{8\sqrt{\gamma}}\big(- 4\alpha^2\gamma_{ij}\pi^{ij}\sigma + \alpha^3\sigma^2 - 4\alpha^3\rho_i\rho_j\gamma^{ij}\big).\end{aligned} \tag{9}$$

The terms in Λ and Ω^i that affect the principal part of the equations of motion are linear in the canonical momenta and in the first spatial derivatives of the positions. With the restrictions described in [10] we come to the following form

$$\Lambda = -C_1\alpha^2\gamma^{-1/2}\gamma_{ij}\pi^{ij} + C_4\alpha^3\gamma^{-1/2}\sigma + C_7\left(\alpha D_i\beta^i - \frac{1}{2}\alpha\gamma^{jk}\beta^i\partial_i\gamma_{jk}\right) \tag{10}$$

and

$$\Omega^i = C_2\alpha^2\Gamma^i_{jk}\gamma^{jk} + C_3\alpha^2\Gamma^k_{kj}\gamma^{ji} - C_5\alpha\gamma^{ij}D_j\alpha - C_6\gamma^{-1/2}\alpha^3\rho_j\gamma^{ij}. \tag{11}$$

With this choice the principal symbol depends on the shift in a trivial way, which simplifies the hyperbolicity analysis.

The consequence of using this approach is that one cannot choose equations of motion for lapse and shift and constraint additions independently. In fact, if the gauge conditions are fixed then constraint additions are fixed as well.

4. Crucial Techniques in the hyperbolicity analysis

Given the Hamiltonian (8) with parameters C_i it is straightforward to compute the corresponding equations of motion and their principal symbol (2).

As we describe in [10], since the principal symbol depends only trivially on the shift, it has essentially the following structure

$$P^s = \begin{pmatrix} 0 & X \\ Y & 0 \end{pmatrix}. \tag{12}$$

This matrix decomposes further into scalar, vector and trace-free tensor blocks which simplifies the analysis further. The most complicated part in the proof is then the scalar block. The corresponding sub blocks of X and Y both have upper block triangular form:

$$X, Y = \begin{pmatrix} A_{X,Y} & B_{X,Y} \\ 0 & C_{X,Y} \end{pmatrix}, \tag{13}$$

with 2×2 matrices A, B and C. This structure can be used to reduce the analysis completely to 2×2 matrices. Finally one can identify all three families of strongly hyperbolic formulations [10].

Concerning symmetric hyperbolicity we consider only the strongly hyperbolic formulations, because symmetric implies strong hyperbolicity. The first step is to construct candidate symmetrizers. This can be achieved by making an ansatz for the candidate and solving linear equations for the parameters in that ansatz.

Having a set of candidate symmetrizers one can ask for positive definite matrices in that set. This is the most complicated step in the analysis and we are not able to completely classify the formulations with respect to their symmetric hyperbolicity.

We can however identify strongly hyperbolic formulations that are not symmetric hyperbolic and have also found 2-parameter families of symmetric hyperbolic formulations.

Thus far we have treated the formulation parameters C_i as constants. But one can view them as functions of $(\gamma_{ij}, \alpha, \beta^i)$ as well, which effects neither the physics nor the level of hyperbolicity, but does change the equations of motion. One can use this freedom to derive a formulation that includes a gauge condition very similar to the popular puncture gauge. The equations of motion for lapse and shift are

$$\begin{aligned} \partial_t \alpha &= \beta^i \partial_i \alpha - \mu_L \alpha^2 K + \frac{1}{4\sqrt{\gamma}} \mu_L \alpha^3 \sigma, \\ \partial_t \beta^i &= \beta^j \partial_j \beta^i + \mu_S \gamma^{1/3} \Gamma^i_{jk} \gamma^{jk} + \frac{1}{3} (\mu_S \gamma^{1/3} - \alpha^2) \Gamma^k_{kj} \gamma^{ij} - \mu_S \gamma^{-1/6} \alpha \rho^i - \alpha D^i \alpha. \end{aligned} \tag{14}$$

which is (near the puncture $(\alpha \to 0)$ and if the constraints $\sigma = 0$ and $\rho_i = 0$ are satisfied) very close to the puncture gauge condition

$$\partial_t \alpha = \beta^i \partial_i \alpha - \mu_L \alpha^2 K, \qquad \partial_t \beta^i = \beta^j \partial_j \beta^i + \mu_S \tilde{\Gamma}^i - \eta \beta^i, \tag{15}$$

where

$$\tilde{\Gamma}^i = \gamma^{1/3} {\Gamma^i}_{jk} \gamma^{jk} + \frac{1}{3} \gamma^{1/3} \gamma^{ji} {\Gamma^k}_{kj}. \tag{16}$$

Acknowledgments

This work was supported by DFG grant SFB/Transregio 7 "Gravitational Wave Astronomy".

[1] E. Hairer, C. Lubich, and G. Wanner. Springer Series in Computational Mathematics 31. Berlin: Springer., 2006.
[2] E. Hairer. *J. Numer. Anal. Ind. Appl. Math.* **5** (2010) 73-84
[3] F.A.E. Pirani, A. Schild, R. Skinner. *Phys. Rev.* **87** (1952) 452
[4] P. Dirac. *Proc. Roy. Soc. Lond.*, A 246:333, 1958.
[5] R. Arnowitt, S. Deser, C.W. Misner. in L. Witten (Ed.), Gravitation: An Introduction to Current Research. *Wiley, New York* (1962) 227 [arXiv:gr-qc/0405109]
[6] B. Gustaffson, H. Kreiss, J. Oliger. *Wiley, New York* 1995
[7] G. Nagy, O.E. Ortiz O.A. Reula. *Phys. Rev.* **D70** (2004) 044012 [arXiv:gr-qc/0402123]
[8] C. Gundlach, J.M. Martín-García. *Class. Quant. Grav.* **23** (2006) S387-S404 [arXiv:gr-qc/0506037]
[9] J.D. Brown arXiv:gr-qc/0803.0334 (2008)
[10] D. Hilditch, R. Richter. arXiv:gr-qc/1002.4119 (2010)

Simulations of black holes in compactified spacetimes

Miguel Zilhão[1], Vitor Cardoso[2,3], Leonardo Gualtieri[4], Carlos Herdeiro[1,5], Andrea Nerozzi[2], Ulrich Sperhake[2,6,7] and Helvi Witek[2]

[1] Centro de Física do Porto, Departamento de Física e Astronomia, Faculdade de Ciências da Universidade do Porto, Rua do Campo Alegre, 4169-007 Porto, Portugal
[2] Centro Multidisciplinar de Astrofísica, Deptartamento de Física, Instituto Superior Técnico, Universidade Técnica de Lisboa, Av. Rovisco Pais 1, 1049-001 Lisboa, Portugal
[3] Department of Physics and Astronomy, The University of Mississippi, University, MS 38677-1848, USA
[4] Dipartimento di Fisica, Università di Roma "Sapienza" & Sezione INFN Roma1, P.A. Moro 5, 00185, Roma, Italy
[5] Departamento de Física da Universidade de Aveiro, Campus de Santiago, 3810-183 Aveiro, Portugal
[6] Institut de Ciències de l'Espai (CSIC-IEEC), Facultat de Ciències, Campus UAB, E-08193 Bellaterra, Spain
[7] California Institute of Technology, Pasadena, CA 91125, USA

E-mail: `mzilhao@fc.up.pt`

Abstract. From the gauge/gravity duality to braneworld scenarios, black holes in compactified spacetimes play an important role in fundamental physics. Our current understanding of black hole solutions and their dynamics in such spacetimes is rather poor because analytical tools are capable of handling a limited class of idealized scenarios, only. Breakthroughs in numerical relativity in recent years, however, have opened up the study of such spacetimes to a computational treatment which facilitates accurate studies of a wider class of configurations. We here report on recent efforts of our group to perform numerical simulations of black holes in cylindrical spacetimes.

1. Introduction

Since the 2005 breakthroughs [1–3], the field of numerical relativity has seen remarkable progress. Major achievements range from the discovery of new phenomena—such as black holes (BHs) *kicks* [4–6] and zoom-whirl behaviour [7]—to simulations of high energy collisions of BHs [8–10].

A very interesting research direction, which has recently seen some pioneering works, concerns extending this field to the study of higher-dimensional systems [11–18].

In this work, we wish to study how the compactness of extra dimensions changes the dynamics of such higher-dimensional gravity scenarios. There is considerable literature on Kaluza-Klein BHs and BHs on cylinders [19–22], but, to the best of our knowledge, the full non-linear dynamics of BHs on such spacetimes remains unexplored.

2. Setup

For a five dimensional cylindrical Minkowski spacetime, $\mathbb{M}^{1,3}\times S^1$, the metric can be written as

$$ds^2=\underbrace{-dt^2+dx^2+dy^2+y^2d\phi^2}_{\mathbb{M}^{1,3}}+\underbrace{dz^2}_{S^1}\ . \tag{1}$$

The S^1 direction is parameterised by z, which takes values in the interval $[-L,L]$, with the two endpoints identified and $L\in\mathbb{R}^+$. The coordinate ϕ also parameterises a circle, which is part of $\mathbb{M}^{1,3}$.

Following [15] we take our five dimensional metric ansatz to be

$$ds^2=g_{\mu\nu}dx^\mu dx^\nu+\lambda(x^\mu)d\phi^2\ , \tag{2}$$

where $x^\mu=(t,x,y,z)$. We perform a dimensional reduction by isometry on ∂_ϕ and end up with a four dimensional model of gravity coupled to a scalar field. Performing the standard 3 + 1 decomposition, the evolution equations of the resulting system are

$$\begin{aligned}
(\partial_t-\mathcal{L}_\beta)\,\gamma_{ij}&=-2\alpha K_{ij}\\
(\partial_t-\mathcal{L}_\beta)\,K_{ij}&=-D_i\partial_j\alpha+\alpha\left({}^{(3)}R_{ij}+KK_{ij}-2K_{ik}K^k{}_j\right)\\
&\quad-\alpha\frac{1}{2\lambda}\left(D_i\partial_j\lambda-2K_{ij}K_\lambda-\frac{1}{2\lambda}\partial_i\lambda\partial_j\lambda\right)\\
(\partial_t-\mathcal{L}_\beta)\,\lambda&=-2\alpha K_\lambda\\
\frac{1}{\alpha}(\partial_t-\mathcal{L}_\beta)\,K_\lambda&=-\frac{1}{2\alpha}\partial^i\lambda\partial_i\alpha+KK_\lambda-\frac{1}{\lambda}K_\lambda^2+\frac{1}{4\lambda}\partial^i\lambda\partial_i\lambda-\frac{1}{2}D^k\partial_k\lambda
\end{aligned} \tag{3}$$

For further details and numerical implementation see [15, 23]. We here use periodic boundary conditions along the z direction and Sommerfeld radiative boundary conditions along x and y.

3. Initial data

Following [15, 24], the four-dimensional Brill-Lindquist initial data appropriate to describe non-spinning, non-rotating BHs momentarily at rest, take the form

$$\gamma_{ij}dx^idx^j=\psi^2[dx^2+dy^2+dz^2]\ ,\qquad \lambda=y^2\psi^2\ ,\qquad K_{ij}=0=K_\lambda\ .$$

In a spacetime with standard topology (wherein z parameterises a line), the initial data for two BHs with horizon radius $r_S^{1,2}$ and punctures placed at $(x,y,z)=(0,0,\pm a)$, takes the form

$$\psi=1+\frac{(r_S^1)^2}{4[x^2+y^2+(z-a)^2]}+\frac{(r_S^2)^2}{4[x^2+y^2+(z+a)^2]}\ . \tag{4}$$

3.1. Head on collision of BHs

The appropriate initial data to describe a BH in S^1 can be viewed as having an infinite array of BHs, all with the same mass, separated by coordinate distance $\Delta z=2L$. Since the superposition of various BHs in a line is described by adding up the corresponding initial data, for the infinite array of two BHs in the circle located at $z=\pm a$ $(0<a<L)$ with horizon radii r_S^i, $i=1,2$ (or, equivalently, for two BHs in S^1) the initial data is given by

$$\begin{aligned}
\psi&=1+\sum_{n=-\infty}^{+\infty}\frac{(r_S^1)^2}{4[x^2+y^2+(z-a-2Ln)^2]}+\sum_{n=-\infty}^{+\infty}\frac{(r_S^2)^2}{4[x^2+y^2+(z+a-2Ln)^2]}\\
&=1+\frac{\pi(r_S^1)^2}{8L\rho}\frac{\sinh\frac{\pi\rho}{L}}{\cosh\frac{\pi\rho}{L}-\cos\frac{\pi(z-a)}{L}}+\frac{\pi(r_S^2)^2}{8L\rho}\frac{\sinh\frac{\pi\rho}{L}}{\cosh\frac{\pi\rho}{L}-\cos\frac{\pi(z+a)}{L}}\ .
\end{aligned} \tag{5}$$

where $\rho^2\equiv x^2+y^2$ and in the last equality we have used the result in [21].

Journal of Physics: Conference Series **314** (2011) 012103 doi:10.1088/1742-6596/314/1/012103

4. Results

We will now show some results obtained for a head-on collision (from rest) of BHs with an initial separation of 10.37 r_S, i.e., $a = 5.185\ r_S$. The z ($z \in [-L, L]$) coordinate has been compactified with $L/r_S = 64,\ 32,\ 16$. For comparison purposes, we have also performed a simulation with "standard" outgoing boundary conditions ($L \to \infty$), which will be here referred to as "outgoing".

All results will be presented in units of the Schwarzschild radius $r_S = r_{S,1} + r_{S,2}$.

4.1. Hamiltonian constraint

Figure 1 shows the Hamiltonian constraint along the x and z axis, respectively, for several time steps for the $L = 32$ case. As we can see, the constraint is indeed being satisfied with high accuracy.

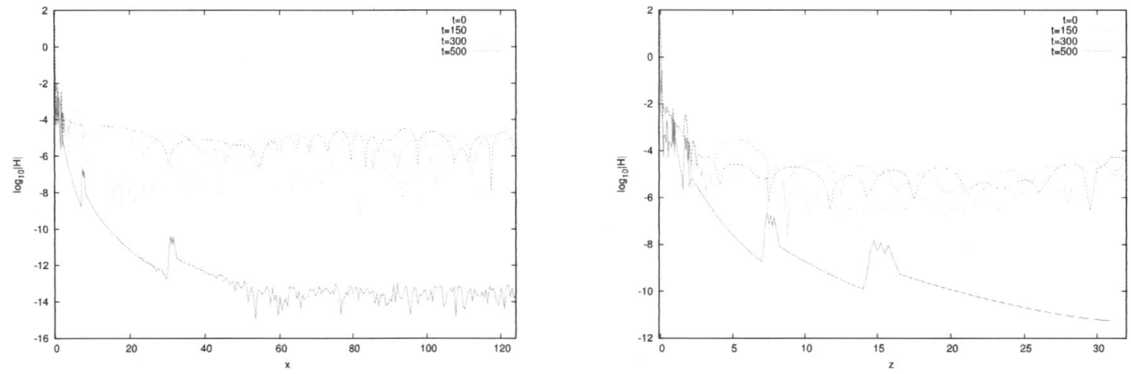

Figure 1. Hamiltonian constraint along the x and z axis, for the $L = 32$ case.

4.2. Collision time

Next, we studied the changes in collision time for different compactification radii. Whereas we don't observe any (noticable) difference for $L = 64$ and $L = 32$ as compared to the outgoing case, the case $L = 16$ shows already a noticable difference. The puncture trajectories for these cases are plotted in Figure 2.

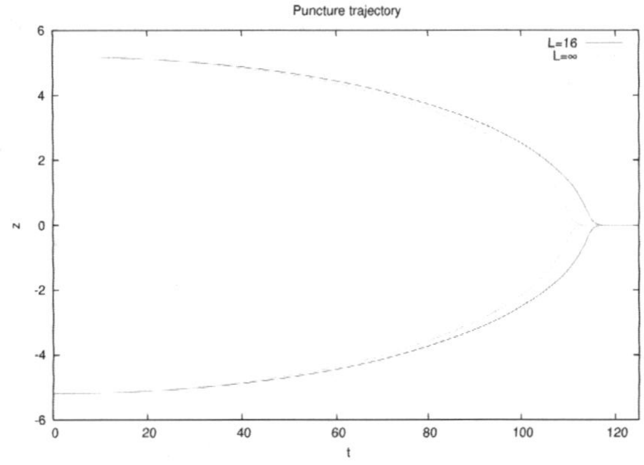

Figure 2. Puncture z coordinate as function of time for the outgoing and $L = 16$ cases.

Recall that one can think of a BH in a cylindrical space as an array of infinite BHs. Therefore, for a head-on collision on such a cylindrical space, each BH will also feel the gravitational pull of all the other BHs. Naïvely, one thus expects that for this cylindrical case it will take longer for the BHs to collide, which is what we observe in Figure 2.

5. Final remarks

Using the formalism introduced in [15], we were able to reduce the head-on collision of (non-spinning) BHs on cylindrical spacetimes (in any dimension) to an effective $3 + 1$ system with a scalar field, and used this procedure to successfully evolve a head-on collision of two BHs on a five-dimensional cylindrical spacetime.

Further issues that we wish to investigate include monitoring the deformation of the BHs' apparent horizon and computing the energy radiated (along the lines of [17]). We further plan to perform simulations with smaller compactification radii and the equivalent six-dimensional system.

Acknowledgments

M.Z. and H.W. are funded by FCT through grants SFRH/BD/43558/2008 and SFRH/BD/46061/2008 U.S. acknowledges support from the Ramón y Cajal Programme of the Ministry of Education and Science of Spain, NSF grants PHY-0601459, PHY-0652995 and the Fairchild Foundation to Caltech. This work was supported by the *DyBHo–256667* ERC Starting Grant and by FCT - Portugal through projects PTDC/FIS/098025/2008, PTDC/FIS/098032/2008, PTDC/CTE-AST/098034/2008, CERN/FP/109306/2009, CERN/FP/109290/2009 as well as NSF grant PHY-0900735. This research was supported by an allocation through the TeraGrid Advanced Support Program under grant PHY-090003 and an allocation by the Centro de Supercomputación de Galicia (CESGA) under project ICTS-2009-40. Computations were performed on the TeraGrid cluster NICS Kraken, the Milipeia cluster in Coimbra, Magerit in Madrid, the Barcelona Supercomputing Centre—Centro Nacional de Supercomputación, Finis Terrae and LRZ in Munich.

References

[1] Pretorius F 2005 *Phys. Rev. Lett.* **95** 121101
[2] Campanelli M, Lousto C O, Marronetti P and Zlochower Y 2006 *Phys. Rev. Lett.* **96** 111101
[3] Baker J G, Centrella J, Choi D I, Koppitz M and van Meter J 2006 *Phys. Rev. Lett.* **96** 111102
[4] Gonzalez J A, Sperhake U, Bruegmann B, Hannam M and Husa S 2007 *Phys. Rev. Lett.* **98** 091101
[5] González J A, Hannam M D, Sperhake U, Brügmann B and Husa S 2007 *Phys. Rev. Lett.* **98** 231101
[6] Campanelli M, Lousto C O, Zlochower Y and Merritt D 2007 *Phys. Rev. Lett.* **98** 231102
[7] Pretorius F and Khurana D 2007 *Class. Quant. Grav.* **24** S83–S108
[8] Sperhake U, Cardoso V, Pretorius F, Berti E and Gonzalez J A 2008 *Phys. Rev. Lett.* **101** 161101
[9] Sperhake U *et al.* 2009 *Phys. Rev. Lett.* **103** 131102
[10] Shibata M, Okawa H and Yamamoto T 2008 *Phys. Rev.* **D78** 101501
[11] Choptuik M, Lehner L, Olabarrieta I, Petryk R, Pretorius F and Villegas H 2003 *Phys. Rev.* **D68** 044001
[12] Sorkin E and Choptuik M W 2009 (*Preprint* `0908.2500`)
[13] Sorkin E 2009 (*Preprint* `0911.2011`)
[14] Yoshino H and Shibata M 2009 *Phys. Rev.* **D80** 084025
[15] Zilhao M *et al.* 2010 *Phys. Rev.* **D81** 084052
[16] Witek H, Cardoso V, Herdeiro C, Nerozzi A, Sperhake U *et al.* 2010 *Phys.Rev.* **D82** 104037
[17] Witek H *et al.* 2010 *Phys. Rev.* **D82** 104014
[18] Witek H, Cardoso V, Gualtieri L, Herdeiro C, Sperhake U *et al.* 2010
[19] Korotkin D and Nicolai H 1994 (*Preprint* `gr-qc/9403029`)
[20] Frolov A V and Frolov V P 2003 *Phys.Rev.* **D67** 124025
[21] Myers R C 1987 *Phys. Rev.* **D35** 455
[22] Harmark T and Obers N A 2005 (*Preprint* `hep-th/0503020`)
[23] Sperhake U 2007 *Phys. Rev.* **D76** 104015
[24] Yoshino H, Shiromizu T and Shibata M 2005 *Phys. Rev.* **D72** 084020

Journal of Physics: Conference Series **314** (2011) 012104 doi:10.1088/1742-6596/314/1/012104

Numerical Relativity in D dimensional space-times: Collisions of unequal mass black holes

Helvi Witek[1], Vitor Cardoso[1,2], Leonardo Gualtieri[3], Carlos Herdeiro[4,5], Ulrich Sperhake[1,6,7], Miguel Zilhão[5]

[1] CENTRA, Departamento de Física, Instituto Superior Técnico, Universidade Técnica de Lisboa - UTL, Av. Rovisco Pais 1, 1049 Lisboa, Portugal.
[2] Department of Physics and Astronomy, The University of Mississippi, University, MS 38677, USA.
[3] Dipartimento di Fisica, Università di Roma "Sapienza" & Sezione INFN Roma1, P.A. Moro 5, 00185, Roma, Italy.
[4] Departamento de Física da Universidade de Aveiro, Campus de Santiago, 3810-183 Aveiro, Portugal.
[5] Centro de Física do Porto — CFP, Departamento de Física e Astronomia, Faculdade de Ciências da Universidade do Porto — FCUP, Rua do Campo Alegre, 4169-007 Porto, Portugal.
[6] Institut de Ciències de l'Espai (CSIC-IEEC), Facultat de Ciències, Campus UAB, E-08193 Bellaterra, Spain.
[7] California Institute of Technology, Pasadena, CA 91109, USA.

E-mail: `helvi.witek@ist.utl.pt`

Abstract. We present unequal mass head-on collisions of black holes in $D = 5$ dimensional space-times. We have simulated BH systems with mass ratios $q = 1, 1/2, 1/3, 1/4$. We extract the total energy radiated throughout the collision and compute the linear momentum flux and the recoil velocity of the final black hole. The numerical results show very good agreement with point particle calculations when extrapolated to this limit.

1. Introduction

Black holes (BH) in four or higher dimensional generic space-times play a key role in fundamental physics; reaching from astrophysics to high energy physics. In high energy collisions of particles, the center of mass energy is beyond the Planck scale and gravity becomes dominant. Then, due to Thorne's hoop conjecture [1] and "no-hair" theorem type arguments, the internal structure of these particles are not important to understand these processes. Therefore, the trans-Planckian scattering of particles is well described by scattering of highly boosted BHs [2]. This becomes especially interesting in the context of TeV gravity scenarios, i.e., theories of gravity with extra dimensions, which have been proposed in order to solve the so-called hierarchy problem [3]. The fact that the effective Planck scale decreases to the order of TeV offers the exciting possibility that BHs could be produced, e.g., at the Large Hadron Collider (LHC) [4]. The LHC thus might provide an opportunity to experimentally test predictions by such higher dimensional theories of gravity. In order to compute the amount of energy that is radiated in form of gravitational radiation and to determine the cross-section rate it is mandatory to evolve BH collisions in full general relativity. Due to the complex structure of Einstein's equations this goal can only

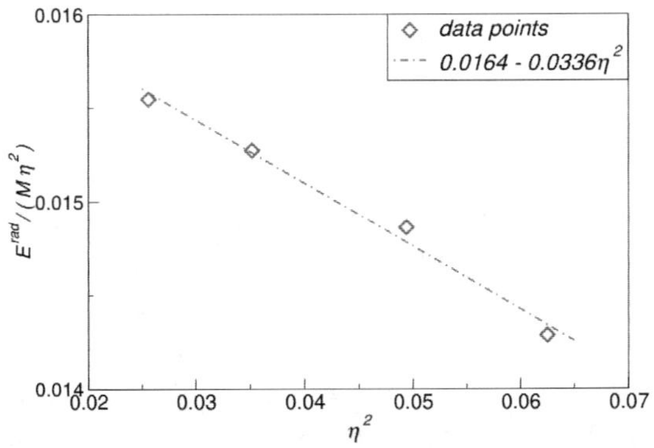

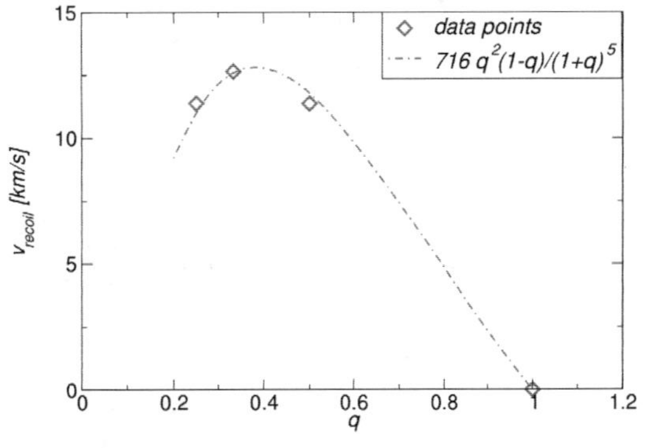

Figure 1. (Color online) Total integrated energy as function of $\eta^2 = [q/(q+1)^2]^2$ (left) and recoil velocity as function of the mass ratio q (right) for a head-on collision in $D = 5$ (red symbols). The curves are the corresponding fitting functions.

be achieved by using numerical methods. Recently, succesful numerical methods have been developed to evolve higher dimensional BH space-times [5, 6, 7].

Here, we intend to extend our previous work [6, 7] to study collisions of unequal mass BH binaries colliding head-on in $D = 5$ space-time dimensions. We calculate the energy and linear momentum emitted throughout this process. We will see that our numerical results, extrapolated to the point particle (PP) limit, are in very good agreement with recently presented PP calculations [8].

2. Numerical setup and results

We have performed numerical simulations of unequal mass BH binaries using the LEAN code [9] extended along the lines of [6]. The LEAN code is based on the `Cactus` computational toolkit [10] and the `Carpet` mesh refinement package [11]. In order to evolve higher dimensional, sufficiently symmetric space-times we performe a dimensional reduction on a $SO(D-4)$ sphere [6]. Thus, the D dimensional vacuum Einstein's equations are cast into an effectively $3+1$ system of gravity which is coupled to a scalar field. The system of BHs, initially at rest, is set up using Brill-Lindquist type initial data in the form presented in Eq. (2.15) in [7]. We evolve this system using the BSSN [12] formulation, together with the moving puncture approach [13]. We extract information about gravitational radiation using the Kodama-Ishibashi (KI) formalism [14]. The radiated energy and energy flux are calculated from the KI gauge-invariant wave function $\Phi^{l}_{,t}$ by Eqs. (2.56), (2.57) in [7]. The momentum flux is given by an infinite series coupling the different multipoles, but the following lowest-order expansion is sufficient for our accuracy purposes [15]

$$\frac{dP}{dt} = \frac{\Phi^{l=3}_{,t}}{4\pi}\left(5\Phi^{l=2}_{,t} + 21\Phi^{l=4}_{,t}\right) . \qquad (1)$$

We have evolved head-on collsions of BHs, initially at rest, with mass ratios $q \equiv M_1/M_2 = r_{S,1}^{D-3}/r_{S,2}^{D-3} = 1,\ 1/2,\ 1/3,\ 1/4$, where M_i is the mass of the i-th BH, in $D = 5$ dimensions. The proper initial separation has been set to $L/r_S = 6.33$. For the details of the initial setup of the simulations in we refer the reader to Table 1 in [15]. The mass parameters of the smaller BH are given by $r_{S,1}^{D-3}/r_S^{D-3} = 0.5, 0.33, 0.25, 0.2$. The values of the second BH are adapted accordingly.

By integrating the energy flux computed from the KI master function, we obtain the total radiated energy. In Table 1 we summarize the radiated energy in units of total mass M. The

q	$E/M(\%)$	$E_{l=2}(\%)$	$E_{l=3}(\%)$	$E_{l=4}(\%)$
1	0.089	99.9	0.0	0.1
1/2	0.073	97.7	2.2	0.1
1/3	0.054	94.8	4.8	0.4
1/4	0.040	92.4	7.0	0.6

Table 1. We show the total radiated energy E/M and the fraction of energy E^l excited in the l-th mode as compared to the total radiated energy, measured from the energy flux at r_{ex} for unequal mass, head-on collisions of BHs

maximum amount of energy is emitted in the equal mass case $E^{\rm rad}/M = 0.089\%$ ($D = 5$) as found previously in Ref. [7]. As reported in [7, 15], the error in the radiated energy is estimated to be about 5 %. We also show the fraction of energy excited in the higher multipoles in Table 1. As expected from calculations in the PP limit, higher multipoles are enhanced as the mass ratio decreases. Post-Newtonian calculations, extended to general dimension, suggest a dependence of the total radiated energy $E^{\rm rad}/M \propto \eta^2$, where $\eta = q/(1+q)^2$ is the dimensionless reduced mass [16]. In the left panel of Fig. 1 we show the ratio $E^{\rm rad}/(M\eta^2)$ which can be seen to depend very weakly on η^2. We make this argument more precise by fitting our numerical results to $E^{\rm rad}/M\eta^2 = A_0 + A_1\eta^2$. We find

$$\frac{E^{\rm rad}}{M\eta^2} = 0.0164 - 0.0336\eta^2\,. \tag{2}$$

Linearized, PP calculations presented in [8] show that in the limit of zero mass ratio one obtains

$$\frac{E^{\rm rad}_{\rm PP}}{M\eta^2} = 0.0165\,, \tag{3}$$

which agrees with the extrapolation of our numerical results within less than 1%. We observe, that the contributions of higher multipoles for $q < 1$ become stronger with increasing dimensionality of the space-time. These results are consistent with the observation in [17] that higher multipoles contribute more to the radiation than in $D = 4$, where for instance the $l = 3$ mode contributed roughly 10% of the total energy in the PP limit. Linearized, point-particle calculations show that the trend is consistent and continues in higher-D [8].

In unequal-mass collisions, the asymmetric emission of radiation along the collision axis causes a net momentum to be carried by gravitational waves. As such, the final BH will "recoil", according to $v_{\rm r} = \left|\int_{-\infty}^{\infty} dt \frac{dP}{dt}\right|$. The recoil velocities for different mass-ratios are shown in the right panel of Fig. 1. We estimate the errors in the recoil velocity to be ≈ 5 %. The general functional dependence of the momentum on the mass ratio is given by $v_{\rm r} = A_0 \frac{q^2(1-q)}{(1+q)^5}$ [16]. By fitting this function to our numerical data, we obtain $A_0 = 716$ km/s. In the PP limit these results are, again, in good agreement with PP calculations [8]. We note that momentum emission is given by a non-trivial interference between different multipoles, so this is a non-trivial agreement.

3. Conclusions and Outlook

We have simulated head-on collision of BHs, initially at rest, with mass ratios $q = 1, 1/2, 1/3, 1/4$ in $D = 5$ space-time dimensions. We have computed the total energy radiated throughout the collision, as well as the fraction of energy excited in the different multipoles. We observe that this fraction is enhanced as the mass ratio decreases. We have extrapolated the numerical results for the total radiated energy to the PP limit and find agreement within 1% with PP calculations [8]. Furthermore, we have computed the recoil velocity of the final BH. We find a maximum of $v_{\rm r,max} = 12.8$km/s ($D = 5$) at $q \simeq 0.38$. Again, we extrapolate our results to the PP limit and find agreement within less than 10% with PP calculations [8, 15].

Acknowledgments

We would like to thank E. Berti and A. Nerozzi for useful discussions. M.Z. and H.W. are funded by FCT through grants SFRH/BD/43558/2008 and SFRH/BD/46061/2008 U.S. acknowledges support from the Ramón y Cajal Programme of the Ministry of Education and Science of Spain, NSF grants PHY-0601459, PHY-0652995 and the Fairchild Foundation to Caltech. This work was supported by the *DyBHo–256667* ERC Starting Grant and by FCT - Portugal through projects PTDC/FIS/098025/2008, PTDC/FIS/098032/2008, PTDC/CTE-AST/098034/2008, CERN/FP/109306/2009, CERN/FP/109290/2009 as well as NSF grant PHY-0900735. This research was supported by an allocation through the TeraGrid Advanced Support Program under grant PHY-090003 and an allocation by the Centro de Supercomputación de Galicia (CESGA) under project ICTS-2009-40. Computations were performed on the TeraGrid cluster NICS Kraken, the Milipeia cluster in Coimbra, Magerit in Madrid, the Barcelona Supercomputing Centre—Centro Nacional de Supercomputación, Finis Terrae and LRZ in Munich.

References

[1] K. S. Thorne, In *J R Klauder, Magic Without Magic*, San Francisco 1972, 231-258.
[2] M. W. Choptuik, F. Pretorius, Phys. Rev. Lett. **104** (2010) 111101.
[3] I. Antoniadis, Phys. Lett. **B246** (1990) 377-384. N. Arkani-Hamed, S. Dimopoulos, G. R. Dvali, Phys. Lett. **B429** (1998) 263-272. L. Randall, R. Sundrum, Phys. Rev. Lett. **83** (1999) 3370-3373.
[4] T. Banks, W. Fischler, [hep-th/9906038]. S. B. Giddings, S. D. Thomas, Phys. Rev. **D65** (2002) 056010. S. Dimopoulos, G. L. Landsberg, Phys. Rev. Lett. **87** (2001) 161602. E. -J. Ahn, M. Cavaglia, A. V. Olinto, Phys. Lett. **B551** (2003) 1-6. M. Cavaglia, Int. J. Mod. Phys. **A18** (2003) 1843-1882. P. Kanti, Int. J. Mod. Phys. **A19** (2004) 4899-4951.
[5] H. Yoshino, M. Shibata, Phys. Rev. **D80** (2009) 084025. E. Sorkin, M. W. Choptuik, Gen. Rel. Grav. **42** (2010) 1239-1286. K. A. Dennison, J. P. Wendell, T. W. Baumgarte *et al.*, [arXiv:1010.5723 [gr-qc]].
[6] M. Zilhao, H. Witek, U. Sperhake *et al.*, Phys. Rev. **D81** (2010) 084052.
[7] H. Witek, M. Zilhao, L. Gualtieri *et al.*, Phys. Rev. **D82** (2010) 104014.
[8] E. Berti, V. Cardoso, B. Kipapa, [arXiv:1010.3874 [gr-qc]].
[9] U. Sperhake, Phys. Rev. **D76** (2007) 104015. U. Sperhake, E. Berti, V. Cardoso *et al.*, Phys. Rev. **D78** (2008) 064069.
[10] Cactus Computational Toolkit, http://www.cactuscode.org/.
[11] E. Schnetter, S. H. Hawley, I. Hawke, Class. Quant. Grav. **21** (2004) 1465-1488. http://www.carpetcode.org/.
[12] M. Shibata, T. Nakamura, Phys. Rev. **D52** (1995) 5428-5444. T. W. Baumgarte, S. L. Shapiro, Phys. Rev. **D59** (1999) 024007.
[13] J. G. Baker, J. Centrella, D. -I. Choi *et al.*, Phys. Rev. Lett. **96** (2006) 111102. M. Campanelli, C. O. Lousto, P. Marronetti *et al.*, Phys. Rev. Lett. **96** (2006) 111101.
[14] H. Kodama, A. Ishibashi, O. Seto, Phys. Rev. **D62** (2000) 064022. H. Kodama, A. Ishibashi, Prog. Theor. Phys. **110** (2003) 701-722.
[15] H. Witek, V. Cardoso, L. Gualtieri *et al.*, [arXiv:1011.0742 [gr-qc]].
[16] M. Lemos, MSc thesis (unpublished, 2010), http://blackholes.ist.utl.pt/?page=Publications.
[17] M. Davis, R. Ruffini, W. H. Press *et al.*, Phys. Rev. Lett. **27** (1971) 1466-1469.

Journal of Physics: Conference Series **314** (2011) 012105 doi:10.1088/1742-6596/314/1/012105

Emission and null coordinates: geometrical properties and physical construction

Bartolomé Coll, Joan J. Ferrando and Juan A. Morales-Lladosa

Departament d'Astronomia i Astrofísica, Universitat de València, 46100 Burjassot, València, Spain.

E-mail: `bartolome.coll@uv.es; joan.ferrando@uv.es; antonio.morales@uv.es`

Abstract. A Relativistic Positioning System is defined by four clocks (emitters) broadcasting their proper time. Then, every event reached by the signals is naturally labeled by these four times which are the emission coordinates of this event. The coordinate hypersurfaces of the emission coordinates are the future light cones based on the emitter trajectories. For this reason the emission coordinates have been also named null coordinates or light coordinates. Nevertheless, other coordinate systems used in different relativistic contexts have the own right to be named null or light coordinates. Here we analyze when one can say that a coordinate is a null coordinate and when one can say that a coordinate system is null. Moreover, we examine the physical construction and the geometrical properties of several "null coordinate systems": the emission and the reception coordinates, the radar coordinates, and the Bondi-Sachs coordinates, among others.

1. Introduction

The relativistic positioning systems were introduced by B. Coll ten years ago in the Spanish Relativity Meeting that took place in Valladolid [1]. In a relativistic positioning system we have four emitters. Then, the four families of future light cones based on the four emitters constitute coordinate hypersurfaces of a coordinate system: the *emission coordinates*. The emission coordinates of an event are the four proper time signals $\{\tau^A\}$ received by any observer at the event.

The basic properties of the emission coordinates have been analyzed by B. Coll and collaborators in several papers [2, 3, 4]. Let us remember that a coordinate system defines and is defined by their associated coordinate hypersurfaces. For the emission coordinates their coordinate hypersurfaces are future light cones and, consequently, the coordinate covectors are null and future-directed. This means that the contravariant metric has zeros in the diagonal ($g^{\alpha\alpha} = 0$) and negative extra-diagonal terms ($g^{\alpha\beta} < 0, \ \alpha \neq \beta$).

These coordinates associated with a relativistic positioning system have been considered by several authors [1, 3, 4, 5, 6, 7, 8, 9, 10] which have given them different names: null coordinates, emission coordinates, GPS coordinates, GNSS coordinates. The first name, null coordinates, has been recommended enough because this name is a reference to their geometrical properties. Nevertheless, we can find in different relativistic contexts other coordinate systems which have similar geometrical properties, a fact which justifies that all of them could be called null coordinates. For this reason we prefer to call *emission coordinates* the coordinates defined by a relativistic positioning system.

Here, we analyze the physical and geometrical construction of several coordinate systems with "null properties". We begin with two conceptual topics: firstly we explain that the physical construction of a coordinate system is linked with the causal character of the coordinate lines, surfaces and hypersurfaces and, secondly, we show that the association of a causal character to a coordinate is not generically coherent. Consequently, it must be explained in what sense we can say that a coordinate is a null coordinate. Later we present a short report on several "null coordinate systems", and we comment on their physical feasibility and on the framework where they has been or could be used.

2. Coordinate systems: causal character and physical construction

In Relativity, lines, surfaces and hypersurfaces of the space-time may be spacelike, lightlike or timelike. These characters, of clear physical meaning, are usually referred to as causal characters. The causal character of lines, surfaces or hypersurfaces is strongly related to the different physical ingredients with which we can physically construct them in the physical space-time (dust, clocks, rods, strings, light beams, light flashes, etc).

By taking three, two or one coordinate as constants, coordinate systems locally define families of coordinate lines, coordinate surfaces or coordinates hypersurfaces, namely they define four three-parametric congruences of lines, six two-parametric families of surfaces and four one-parametric families of hypersurfaces. The causal character of these fourteen geometric elements are generically independent. The ordered set of these causal characters is called the *causal class* of the coordinate system. It is already a well known result that the number of different causal classes of coordinate systems in Relativity is 199 [11] while in Newtonian physics there exists only four causal classes [12].

3. Coordinate parameters and gradient coordinates

When the coordinate system is already known, say $\{x^\alpha\}$, the coordinate geometric elements are: the four one-parameter families of coordinate hypersurfaces $\{x^\alpha = \text{constant}\}$, the six two-parameter families of coordinate surfaces $\{x^\alpha = \text{constant}, x^\beta = \text{constant}\}$, and the four three-parameter families of coordinate lines $\{x^\alpha = \text{constant}, x^\beta = \text{constant}, x^\gamma = \text{constant}\}$ for superscripts α, β, γ all different.

In fact, in any space-time, every coordinate x^α plays two extreme roles: that of a (coordinate) hypersurface for every constant value, of gradient dx^α, and that of a (coordinate) line when the other coordinates remain constant, of tangent vector ∂_α. This simple fact shows that, in spite of our deep-seated custom of associating to a coordinate a causal character, saying that *it is* timelike, lightlike or spacelike, *this appellation is not generically coherent.* Causal characters are generically associated with directions or sets of directions of geometric objects, but not with space-time variables or parameters associated to them. In the case of a coordinate x^α, this generic incoherence appears because its two natural variations in the coordinate system, dx^α and ∂_α, have *generically different* causal characters. *Only* when both causal characters coincide, it is conceptually clear to extend to x^α itself the appellation of the common causal character of its two mentioned variations.

Consequently, we shall say of a coordinate x^α that it is a timelike, lightlike or spacelike *gradient coordinate* (respectively, *coordinate parameter*) when the causal characters of its variation dx^α (respectively, ∂_α) be timelike, lightlike or spacelike.

In addition, of a coordinate t which is a timelike coordinate parameter and a timelike (resp. spacelike) gradient coordinate, we shall say also that it defines a *spacelike* (resp. *timelike*) *synchronization* (the coordinate hypersurfaces t = constant being the synchronous event loci of the coordinate lines t = variable).

Every inertial time is a timelike coordinate parameter that defines a spacelike synchronization. However, a remarkable example where the notion of "timelike coordinate" becomes incoherent is the *local Solar time* (the time shown by a sundial). The local Solar time is the oldest timelike coordinate parameter known by humanity, but it is a spacelike gradient coordinate and, consequently, it not defines a spacelike but a timelike synchronization (see [12] for more details).

4. On the "null coordinate systems"

After the above considerations the appellation "null coordinate system" must be used carefully. One can use this appellation because one or several coordinate lines are null (null coordinate parameters), or because one or several coordinate hypersurfaces are null (null gradient coordinates). Or even because one or several coordinate surfaces are null. A quick glance to the table of the 199 causal classes of frames [11] (see also the recent paper [13]) shows that there are 175 classes with some of these null qualities.

It is worth remarking that no coordinate system exists with the four coordinate being both, null coordinate parameters and null gradient coordinates. At most two coordinates can keep this double null quality. This leads to a causal class which is a special case of the Bondi-Sachs coordinates. Nevertheless, there is a class with four null coordinate parameters, and another class with four null gradient coordinates. In what follows we comment on the interest and physical construction of these "null coordinate systems".

Null coordinate parameters ($g_{\alpha\alpha} = 0$)

When the four coordinates are null coordinate parameters, the four congruences of coordinate lines are null. Then, necessarily, they are spacelike gradient coordinates, that is, the four families of coordinate hypersurfaces are timelike. Moreover, the six families of coordinate surfaces are also timelike.

B. Coll was soon interested in this kind of coordinates and proposed their physical construction by means of light beams [14].

Null gradient coordinates ($g^{\alpha\alpha} = 0$)

When the four coordinates are null gradient coordinates, the four families of coordinate hypersurfaces are null. Then, necessarily, they are spacelike coordinate parameters, that is, the four congruences of coordinate lines are spacelike. Moreover, the six families of coordinate surfaces are also spacelike.

Emission-reception coordinates

Emission-reception coordinates are null gradient coordinates which can be physically built by emission and/or reception of times: an event receives the time broadcast by some emitters and the event broadcasts a signal that reaches some receivers. Then, the coordinate hypersurfaces are future and/or past light cones.

Radar coordinates

In two dimensions, and based in the Poincaré-Einstein protocol of synchronization, radar coordinates $\{t_e, t_r\}$ of an event are defined as follows: a signal is broadcast by an observer at the emission time t_e, and this signal is received by the same observer at the reception time t_r after been echoed by the event. Thus, radar coordinates are emission-reception coordinates defined by a sole emitter-receiver.

In a four-dimensional space-time, Ehlers-Pirani-Schild [15] radar coordinates are emission-reception coordinate defined by two emitter-receivers.

Emission coordinates

Emission coordinates of an event are defined by the four times that the clocks of four *emitters* (satellites) watched when they broadcast signals received at this event. The four families of coordinate hypersurfaces are future light cones with vertex on the emitter world lines.

In his first presentation B. Coll [1] named these coordinates "Coordinate Systems of GPS type", but after he and his collaborators opt for the use of "emission coordinates". The usefulness of these coordinates in building a fully relativistic theory for the Global Navigation Satellite Systems demands to improve their current geometrical and physical comprehension [2, 3, 4, 5, 6, 9, 10].

Reception coordinates
Reception coordinates of an event are defined by the four times that watch the clocks of four *receivers* (satellites) when they receive a signal broadcast from this event. Now, the four families of coordinate hypersurfaces are past light cones with vertex on the receiver world lines.

The reception coordinates will have an important role in a future theory of relativistic stereometry. B. Coll remarked this fact two years ago, in the GraviMAS-FEST [16].

Bondi-Sachs coordinates ($g^{00} = g_{11} = 0$)
The generalized Bondi-Sachs coordinates were introduced in order to study gravitational radiation [17]. Usually in their definition one also imposes the condition $g_{22}g_{33} - g_{23}^2 > 0$. Nevertheless, this condition is a consequence of $g^{00} = g_{11} = 0$. According with [11, 13], the generalized Bondi-Sachs coordinates belong to 13 different causal classes.

In these Bondi-Sachs causal classes there are a null gradient coordinate u and a null coordinate parameter r. The orthogonal lines to the coordinate hypersurfaces $u = constant$ are the null coordinate lines $r = variable$, that is, the (null) gradient du and the vector ∂_r are collinear.

A way to provide Bondi-Sachs coordinates is to consider the future (past) light cones $u = constant$ associated with an emitter (receiver) and to choose a parameter r on the generatrix of the cones.

Acknowledgments
This work has been supported by the Spanish Ministerio de Ciencia e Innovación, MICINN-FEDER project FIS2009-07705.

References
[1] Coll B 2001 *Reference Frames and Gravitomagnetism Proceedings of the XXIIIth Spanish Relativity Meeting ERE-2000* ed J. F. Pascual-Sánchez, L. Floría, A. San Miguel and F. Vicente (Singapore: World Scientific)
[2] Coll B, Ferrando J J and Morales J A 2006 *Phys. Rev.* D **73** 084017; 2006 *Phys. Rev.* D **74**; 104003; 2010 *Phys. Rev.* D **74** 084038
[3] Coll B and Pozo J M 2006 *Class. Quantum. Grav.* **23** 7395
[4] Coll B, Ferrando J J and Morales-Lladosa J A 2010 *Class. Quantum. Grav.* **27** 7395
[5] Bahder T B 2001, Am. J. Phys. **69**, 315
[6] Rovelli C 2002 *Phys. Rev. D* **65**, 044017
[7] Coll B 2006 *A Century of Relativity Physics Proc. of the XXVIII Spanish Relativity Meeting ERE-2005* ed L. Mornas and J. Díaz Alonso (New York: AIP)
[8] Lachièze-Rey M 2006 *Class. Quantum Grav.* **23**, 3531
[9] Bini D, Geralico A, Ruggiero M L and Tartaglia A 2008 *Class. Quantum. Grav.* **25** 205011
[10] Delva P, U. Kostić U and Čedež A 2010 *Adv. Space Res.* doi: 10.1016/j.ars.2010.07.007
[11] Coll B and Morales J A 1992 *Int. J. Theor. Phys.* **31** 1045
[12] Coll B, Ferrando J J and Morales-Lladosa J A 2009 *Found, Phys.* **39** 1280
[13] Coll B, Ferrando J J and Morales-Lladosa J A 2006 *Phys. Rev.* D **73** 084017
[14] Coll B 1985 *Trobades Científiques de la Mediterrània: Actes E.R.E-85*, edited by A. Molina (Servei de Publicacions de l'ETSEIB, Barcelona). See also http//www.coll.cc/
[15] Ehlers J, Pirani F A E and Schild A 1972 The geometry of free fall and light propagation. In Studies in Relativity (Papers in Honour of J. L. Synge), ed. L. O'Raifeartaigh, (Oxford: Clarendon Press)
[16] Coll B 2008 Epistemic Relativity *GraviMAS_FEST workshop in honour of Lluís Mas* (Mallorca: UIB) http://www.uib.es/depart/dfs/GRG/GraviMAS_FEST/
[17] Bondi H, Van Der Burg M G J and Metzner A W K 1962 *Proc. R. Soc.* **269** 2152; Sachs R K 1962 *Proc. R. Soc.* **270** 103126; Fletcher S J and Lun A W C 2003 *Class. Quantum. Grav.* **20** 4153

Journal of Physics: Conference Series **314** (2011) 012106 doi:10.1088/1742-6596/314/1/012106

From emission to inertial coordinates: an analytical approach

Bartolomé Coll, Joan J. Ferrando and Juan A. Morales-Lladosa

Departament d'Astronomia i Astrofísica, Universitat de València, 46100 Burjassot, València, Spain

E-mail: bartolome.coll@uv.es, joan.ferrando@uv.es, antonio.morales@uv.es

Abstract. In a previous work [*Class. Quantum Grav.* **27** (2010) 065013] relativistic positioning systems in Minkowski space-time have been studied, and the transformation from emission to inertial coordinates have been obtained for an arbitrary configuration of the emitters. The formula giving this transformation applies in all the emission coordinate region and involves the orientation of the positioning system (the Jacobian sign of the map which gives the emission coordinates of an event). Nevertheless, there exists an inherent limitation on the applicability of this formula: only the users in the central region of a positioning system can obtain the orientation from the sole emission data. Here an observational method to determine the orientation of a relativistic positioning system is presented. In this procedure, a certain additional information allows any user to obtain its inertial coordinates, irrespectively of its location in the emission region of the positioning system.

1. The coordinate transformation from emission to inertial coordinates

A relativistic positioning system consists of four emitters $\gamma_A(\tau^A)$ $(A = 1, 2, 3, 4)$ broadcasting their respective proper times τ^A by means of electromagnetic signals. Every event reached by the signals is labelled by the four times $\{\tau^A\}$: the *emission coordinates* of this event. With an eye on the development of a relativistic theory for Global Navigation Satellite Systems (GNSS), we consider here relativistic positioning systems in Minkowski space-time. The space-time region covered by the signals is called the *emission region*, $\mathcal{R}$, of the relativistic positioning system. Then, if $x \in \mathcal{R}$, a user at x receives the four broadcast times $\{\tau^A\}$, and then

$$m_A \equiv x - \gamma_A(\tau^A), \quad (A = 1, ..., 4), \tag{1}$$

denote the future oriented light-like vectors that represent the trajectories followed by the electromagnetic signals from the emitters $\gamma_A(\tau^A)$ to the reception event $x \in \mathcal{R}$.

The emission data $\{\tau^A\}$ received at x are the emission coordinates of the event $x \in \mathcal{R}$ and were broadcast when the emitters were at the events $\{\gamma_A(\tau^A)\}$: the *configuration of the emitters* for the event x. Generically, these four events determine an hyperplane (the configuration hyperplane) whose orthogonal direction is that of the vector χ.[1]

[1] Here, we always consider that the four events $\{\gamma_A(\tau^A)\}$ are neither coplanar nor collinear (i. e, that the emitter configuration is regular, $\chi \neq 0$). The study of non-regular (or degenerate) configurations is considered elsewhere.

The space-time region where the gradients $d\tau^A$ are well defined and linearly independent is called the *emission coordinate region*, $\mathcal{C}$, of the positioning system. Indeed, the emitter world-lines are excluded from $\mathcal{C}$ because $d\tau^A$ are not defined at the emission events (cone emission vertices). For all $x \in \mathcal{C}$, the transformation from emission to inertial coordinates is locally well defined.

From now on, let us suppose that the world-lines $\gamma_A(\tau^A)$ of the emitters in an inertial system $\{x^\alpha\}$ are known. In [1], we have obtained the coordinate transformation $x = \kappa(\tau^A)$ from an emission coordinate system $\{\tau^A\}$ to an inertial one $\{x^\alpha\}$, which is covariantly expressed by the following formula:

$$x = \gamma_4 + y_* - \frac{y_*^2\,\chi}{y_* \cdot \chi + \hat{\epsilon}\sqrt{(y_* \cdot \chi)^2 - y_*^2 \chi^2}} \tag{2}$$

where $\gamma_4(\tau^4)$ has been chosen as the reference emitter.

Quantities y_* and χ (the configuration vector) are both computable from the emission data $\{\tau^A\}$ by applying the following expressions:

$$y_* = \frac{1}{\xi \cdot \chi}\, i(\xi) H, \qquad H = *(\Omega_1\, e_2 \wedge e_3 + \Omega_2\, e_3 \wedge e_1 + \Omega_3\, e_1 \wedge e_2) \tag{3}$$

$$\Omega_a = \frac{1}{2}(e_a)^2, \qquad e_a = \gamma_a(\tau^a) - \gamma_4(\tau^4), \qquad \chi = *(e_1 \wedge e_2 \wedge e_3) \tag{4}$$

with $a = 1, 2, 3$, and where ξ is any vector transversal to the configuration, $\xi \cdot \chi \neq 0$; $i()$ and $\wedge$ are, respectively, the interior and the wedge products, and '$*$' stands for the Hodge dual operator.[2]

Quantity $\hat{\epsilon}$ (the *orientation* of the positioning system with respect to x) is defined by

$$\hat{\epsilon} \equiv sgn * (m_1 \wedge m_2 \wedge m_3 \wedge m_4). \tag{5}$$

Note that to obtain x from (2) one needs to determine the orientation $\hat{\epsilon}$, which involves, by substituting (1) in (5), the unknown x. Therefore, in order to show that Eq. (2) does not chase its own tail, one must be able to determine the orientation $\hat{\epsilon}$ at x by using a procedure that does not involve the previous knowledge of x. Next, we study the region where $\hat{\epsilon}$ is computable by the positioning data (the *central region* of the positioning system) which does not cover the whole emission coordinate region $\mathcal{C}$.

2. Emission configuration regions and orientation of a positioning system

Let us consider the map Θ that to every event $x \in \mathcal{R}$ associates its emission coordinates $\{\tau^A\}$, that is $(\tau^A) = \Theta(x)$, and let us denote by $j_\Theta(x)$ its Jacobian determinant,

$$j_\Theta(x) = *(d\tau^1 \wedge d\tau^2 \wedge d\tau^3 \wedge d\tau^4). \tag{6}$$

For each value of A, $d\tau^A$ and m^A define the same future oriented null direction and then, from (5) and (6), the orientation $\hat{\epsilon}$ of the positioning system is nothing but the sign of this Jacobian determinant, $\hat{\epsilon} = sgn\, j_\Theta(x)$. The hypersurface $\mathcal{J}$ where this Jacobian vanishes, $\mathcal{J} = \{x \mid j_\Theta(x) = 0\}$, has been studied by Coll and Pozo [2, 3] who stated that $\mathcal{J}$ is constituted by those events for which any user in them can see the four emitters on a circle in his celestial sphere.

The vector χ provides the causal character of the configuration hyperplane. Taking the metric signature equal to 2ϵ, $\epsilon = \pm 1$, the configuration hyperplane is space-like, light-like or time-like

[2] Using index notation, one write, for instance, $[i(\xi)H]_\nu = \xi^\mu H_{\mu\nu}$, $(e_1 \wedge e_2)^{\mu\nu} = e_1^\mu e_2^\nu - e_2^\mu e_1^\nu$, $[*(e_1 \wedge e_2)]_{\alpha\beta} = \frac{1}{2}\eta_{\alpha\beta\mu\nu}(e_1 \wedge e_2)^{\mu\nu} = \eta_{\alpha\beta\mu\nu} e_1^\mu e_2^\nu$ and $\chi_\alpha = \eta_{\alpha\mu\nu\rho} e_1^\mu e_2^\nu e_3^\rho$, where η is the metric volume element.

if $\epsilon\chi^2$ is, respectively, negative, zero or positive. Thus, $\mathcal{C}$ is the union of three disjoint regions ($\mathcal{C} = \mathcal{C}_s \cup \mathcal{C}_\ell \cup \mathcal{C}_t$): the space-like $\mathcal{C}_s$, the null $\mathcal{C}_\ell$ and the time-like $\mathcal{C}_t$ configuration region, which are respectively defined by the condition $\epsilon\chi^2 < 0$, $\chi^2 = 0$ and $\epsilon\chi^2 > 0$. At every event x in $\mathcal{C}_s$ (respectively, in $\mathcal{C}_l$ or in $\mathcal{C}_t$) a user receives the signals from four emission events that generate a space-like (respectively, null or time-like) hyperplane.

From the emission data $\{\tau^A\}$, the user is able to determine the sign of $\epsilon\chi^2$ and then, from the above definitions, he knows in what of the configuration region, $\mathcal{C}_s$, $\mathcal{C}_\ell$ or $\mathcal{C}_t$ of the positioning system he is. Clearly, one has that for a future pointing time-like vector u, $u \cdot \chi \neq 0$ at any event $x \in \mathcal{C}_s \cup \mathcal{C}_\ell$ because χ is not space-like in this region. On the other hand, from (1) and (4) one obtains:

$$*(m_1 \wedge m_2 \wedge m_3 \wedge m_4) = - * (e_1 \wedge e_2 \wedge e_3 \wedge m_4) = i(m_4)\chi\,, \tag{7}$$

and then, the sign of $u \cdot \chi$ is the same for any u and for all $x \in \mathcal{C}_s \cup \mathcal{C}_\ell$, and this sign is precisely the orientation of the positioning system in both, the space-like and the null configuration regions. This region $\mathcal{C}^C \equiv \mathcal{C}_s \cup \mathcal{C}_\ell$, where the orientation is constant and obtainable from the sole emission data, is the *central region* of the positioning system. Then, one has:

$$\forall x \in \mathcal{C}^C, \qquad \hat{\epsilon} = sgn\,(u \cdot \chi) \tag{8}$$

for any future pointing time-like vector u.

Therefore, from (8) any user in the central region is able to compute the orientation of the positioning system in this region, and then, from (2), (3) and (4), he can obtain his position x in the inertial system from the sole emission data $\{\tau^A\}$, by substituting $\hat{\epsilon} = sgn\,(u \cdot \chi)$ in (2).

As a consequence of (8), the hypersurface $\mathcal{J}$ of vanishing Jacobian is included in the time-like configuration region $\mathcal{C}_t$. In [1], we showed that $\mathcal{C}$ is not a coordinate domain but the union of two disjoint coordinate domains ($\mathcal{C} = \mathcal{C}^F \cup \mathcal{C}^B$): the front $\mathcal{C}^F$, and the back $\mathcal{C}^B$ emission coordinate domains. Further, $\mathcal{C}^F$ is a simply-connected domain but $\mathcal{C}^B$ is not. In fact, in the symmetric stationary case where the four emitters define a (regular) tetrahedron, $\mathcal{C}^B$ is the union of four connected components. Then, the common boundary of both domains is the four-leaf hypersurface $\mathcal{J}$. The orientation $\hat{\epsilon}$ of the positioning system only changes across $\mathcal{J}$: it takes different constant value throughout each coordinate domain.

Suppose that a user is in $\mathcal{C}_t$. For a given set of emission data, there are two events (with distinct inertial coordinates) in $\mathcal{C}_t$ which receive the same set of data. The matter is then: from Eq. (2), how can the user determine its truly localization in the inertial system? Can the users in $\mathcal{C}_t$ know the orientation of the positioning system?

3. Observational procedure to obtain the orientation of a positioning system

What we shall see now is that *all* the users of the coordinate region $\mathcal{C}$, be them in the central region or not, have an *observational* method to determine the orientation $\hat{\epsilon}$.

Consider an arbitrary user, of unit velocity u, at the event x of $\mathcal{C}$; in his relative three-dimensional space π_u, the apparent directions of propagation of the signals m_A are given by the unit vectors $\overline{m}_A$, such that $m_A = -\epsilon(u \cdot m_A)(u + \overline{m}_A)$, and by direct substitution in the definition (5) of $\hat{\epsilon}$ one has

$$\hat{\epsilon} = sgn * \{u \wedge [(\overline{m}_1 \wedge \overline{m}_2 + \overline{m}_2 \wedge \overline{m}_3 + \overline{m}_3 \wedge \overline{m}_1) \wedge \overline{m}_4 - \overline{m}_1 \wedge \overline{m}_2 \wedge \overline{m}_3]\}\,. \tag{9}$$

And because, for any three-form F in π_u, one has $*(u \wedge F) = i(u) * F = *_u F$, where $*_u$ is the Hodge dual operator on π_u, it follows

$$\hat{\epsilon} = sgn\Big(i(\overline{m}_4) *_u (\overline{m}_1 \wedge \overline{m}_2 + \overline{m}_2 \wedge \overline{m}_3 + \overline{m}_3 \wedge \overline{m}_1) - *_u (\overline{m}_1 \wedge \overline{m}_2 \wedge \overline{m}_3)\Big). \tag{10}$$

But, the $\overline{m}_a$ forming a basis of π_u, their dual $\bar{L}^a$, $\bar{L}^a(\overline{m}_b) = \delta^a_b$, are

$$\bar{L}^a = \frac{\epsilon^{abc} *_u (\overline{m}_b \wedge \overline{m}_c)}{2 *_u (\overline{m}_1 \wedge \overline{m}_2 \wedge \overline{m}_3)}. \tag{11}$$

So that

$$\hat{\epsilon} = sgn\Big(*_u \; (\overline{m}_1 \wedge \overline{m}_2 \wedge \overline{m}_3)\Big[i(\overline{m}_4)(\bar{L}^1 + \bar{L}^2 + \bar{L}^3) - 1\Big]\Big). \tag{12}$$

That the square bracket expression vanishes when and only when the Jacobian $j_\Theta(x)$ vanishes is known since [3]. It allows to state that, in the emission region $\mathcal{R}$, the events of the emission coordinate region $\mathcal{C}$ are all those for which the four emitters are not on a circle of the celestial sphere of the users at these events. Now, if the satellites $1, 2, 3$ are increasingly clock-wise on the circle defined by them on the celestial sphere, the sign of $*_u \, (\overline{m}_1 \wedge \overline{m}_2 \wedge \overline{m}_3)$ is positive. Then, the bracket in (12) is positive or negative if the satellite 4 is interior or exterior respectively to this circle.

Thus, if in addition to the received data $\{\tau^A\}$ the user can *observe* the emitters, by applying the above rule he can obtain the orientation $\hat{\epsilon}$ and consequently, from (2), he can determine his position in the inertial coordinate system.

Some final considerations are in order. We have remarked that, in general, the four times that a user receives are insufficient to determine his location in an inertial system (for details, see [1]). This lack of uniqueness was pointed out by Schmidt [6] in connection with the location problem in GNSS, and it was studied by Abel and Chaffee [4, 5]. In this communication, we have outlined a method to obtain the orientation of a relativistic positioning system. Applying (2) and this method one can calculate, in any case, the user space-time location in inertial coordinates. A numerical analysis of the quantities appearing in (2) is not hard to implement and it has been recently accomplished [7].

Relativistic positioning concepts have been recently implemented in an algorithm to obtain the Schwarzschild coordinates of a user from his emission coordinates, and viceversa [8]. In building current models, a fundamental assumption consists in picking out an approximate zero order numerical solution. When gravitational fields are not taken into consideration, expression (2) is the exact covariant solution to the location problem. Thus, in a perturbational scheme based on a Minkowskian background, Eq. (2) provides the exact non-perturbed leading solution: it is the better zero order solution to start with, no matter how the accuracy in the approximation methods may be.

Acknowledgments

This work has been supported by the Spanish Ministerio de Ciencia e Innovación, MICINN-FEDER project FIS2009-07705.

References

[1] Coll B, Ferrando J J and Morales-Lladosa J A 2010 *Class. Quantum Grav.* **27** 065013
[2] Coll B and Pozo J M 2006 *Class. Quantum. Grav.* **23** 7395
[3] Pozo J M 2005 Constructions in 4D: Algebraic properties and special observers. In Proc. Int. School on *Relativistic Coordinates, Reference and Positioning Systems.* B. Coll and J. J. Martín, (eds.) Salamanca, Spain, Jaunary 2005. Universidad de Salamanca (in progress).
[4] Abel J S and Chaffee J W 1991 *IEEE Transactions on Aerospace and Electronic Systems* **27** 952
[5] Chaffee J W and Abel J S 1994 *IEEE Transactions on Aerospace and Electronic Systems* **30** 1021
[6] Schmidt R O 1972 *IEEE Transactions Aerospace and Electronic Systems* **8** 821
[7] Puchades N and Sáez D 2010 (in these proceedings)
[8] Delva P, Kostić U and Čadež A 2010 *J. Adv. Space Res.* doi: 10.1016/j.ars.2010.07.007 (1005.0477 [gr-qc])

Spanish Relativity Meeting (ERE 2010): Gravity as a Crossroad in Physics IOP Publishing
Journal of Physics: Conference Series **314** (2011) 012107 doi:10.1088/1742-6596/314/1/012107

From emission to inertial coordinates: a numerical approach

Neus Puchades and Diego Sáez

Departamento de Astronomía y Astrofísica, Universidad de Valencia, 46100 Burjassot, Valencia, Spain

E-mail: `neus.puchades@uv.es, diego.saez@uv.es`

Abstract. We numerically implement the coordinate transformation between emission and inertial coordinates recently derived –in Minkowski space-time– by Coll et al. [1]. In order to carry out this transformation, we must know both proper times from four satellites and the inertial coordinates of these satellites when they emitted. It allows us to determine the position and the time coordinate of the receiver. Since we need to know the satellite positions at any time, we have simulated satellite constellations similar to those of GPS and Galileo global positioning system. All the satellite positions are calculated, at any time, with respect to an inertial frame with the origin at the Earth center. The orientation criterion described by the same authors [2] has been also implemented. The code has been appropiately tested and, then it has been applied to perform a preliminary study about the structure of the emission coordinate region and the grid region. The errors due to uncertainties in satellite positions have been also estimated.

1. Introduction and generalities

Nowadays Global Navigation Satellite Systems are: GPS in USA, Glonass in Russia, and Galileo (under construction) in the EU. They provide the spatial coordinates and the universal time of any event on the Earth. Position coordinates are calculated thanks to information received from satellites into orbit around the Earth. The GPS constellation of 24 satellites is arranged in six different orbital planes (four satellites per plane), each of them inclined 55 degrees with respect to the equator. To obtain exactly two orbits per day, the satellites are placed at an altitude of 20200km. The nominal error in the resulting receiver positions is approximately 15 m, but their RMS can reach 100 m. The Galileo constellation is composed by 27 satellites (plus three spares), located in three equally spaced orbital planes (9 equally spaced satellites in each plane). These planes are inclined 56 degrees and the altitude of the circular orbits is 23222 km; thus, the orbital period is about 14h. All these positioning systems are based on Newtonian physics, although relativistic post-Newtonian corrections may be performed if necessary.

Relativistic positioning systems are based on relativistic principles from the beginning. The proper times of four satellites (emission coordinates) are sent by means of electromagnetic signals to the receiver, whose inertial coordinates can be found –from the emission ones– by using fully relativistic equations. The general transformation from emission to inertial coordinates –in Minkowski space-time– was given in [1]; so any receiver able to detect the signals from the satellites may use this transformation to get its own space and time coordinates in special relativity. Of course, the formalism must be extended to general relativity to include gravitational fields, accelerated frames, and so on. Realistic 4D implementations of the transformation presented by Coll et al. in [1] require numerical calculations. We have designed and tested a numerical code to do these calculations. It is applied in next sections. The mentioned transformation uses the position of the four satellites when they emitted their respective proper times. We have simulated the motion of GPS and Galileo satellites in an inertial frame with origin at the Earth center; thus, the satellite positions may be calculated at any given time.

In paper [1], a system having a quadratic and three linear equations (hereafter called the main system) was solved to obtain the coordinate transformation between emission and inertial coordinates. This system can have up to two solutions. There are two kinds of solutions, one of them corresponds to *emission signals* travelling from the satellite to the receiver (hereafter emission solutions), whereas the second kind describes *reception signals* propagating in the opposite direction (reception solutions). We are only interested in emission solutions.

Let us now introduce some concepts and notations (extracted from [1] and [2]), which are necessary to understand the following sections. The parameter $\hat{\epsilon}$ appears in the final expression of the solution for the main system and it can take on the values +1 or -1. Vector χ is orthogonal to the hyperplane defined by the relative positions of the satellite pairs. It is called *configuration vector* and χ^2 stands for its square in Minkowski space-time. For $\chi^2 > 0$, the system has two possible solutions, which correspond to two *emission signals*, i.e., two true physical observers at different positions. There are then criteria to distinguish between both solutions. The criterion proposed in [2] is as follows: consider a conical surface with the receiver at the vertex which contains three of the four satellites and then, take $\hat{\epsilon} = 1$ or $\hat{\epsilon} = -1$ depending upon whether the fourth satellite is inside or outside the cone, respectively. Finally, in the case $\chi^2 \leq 0$, the system of equations has an unique emission solution.

2. Grid Structure

Given an arbitrary grid point $(\tau^1,\tau^2,\tau^3,\tau^4)$, the question is: would these proper times be received in some point of Minkowski space? In other words, are there emission solutions of the main system? If affirmative, the points in Minkowski space (receivers) would belong to the so-called emission region and the chosen grid point to the co-region. Our aim in this section is to do a preliminary study about the grid and co-region structure. In order to do this study, we select a point in the grid where the main system has a known solution and, then, we choose quadruplets of proper times (grid points) covering a straight line, L, which contains the selected point. Finally, we study the existence of solutions, for the L-points, by using the coordinate transformation given in [1].

Results are shown in Fig. 1, where the chosen line is represented. The values of the λ parameter are given in the horizontal axis ($\lambda = 0$ for the selected point). There are emission (reception) solutions in the segment marked with stars (triangles), whereas there is no solution in the zone with circles. The segment with stars belongs to the co-region. This result suggests a connected co-region surrounded by grid points where either there are reception solutions or there are no solutions at all (see [3] for a grid theoretical analysis). Many lines containing the selected point must be studied to cover the co-region, but a general study is out of the scope of this paper.

3. Emission region structure

The emission region is the zone of Minkowski space-time where four proper times (emission coordinates) may be received from the satellites. In order to perform a preliminary study of the emission region we proceed as follows: (1) a point on the Earth surface –where the main system has a known solution– is selected, (2) a detector with uniform rectilinear motion leaving from the selected point and meeting one of the four satellites after one hour is considered, and (3) forty points on the detector line world, separated by the same time interval, are considered to find forty associated points in the co-region. These points are numerically calculated taking into account that the distance between associated points is null. This condition leads to a system of algebraic equations which may be solved by using the Newton-Raphson method; thus, we numerically go from inertial to emission coordinates (inverse of the tranformation given in [1]). For each of the forty co-region points, we can calculate the χ^2 value; thus, we know the number of emission solutions at these points. If only one of these solutions exists, it may be verified (as a test) that the resulting point in Minkowski space is just that point initially associated to the co-region point. If there are two emission solutions, we have a second point in the emission region receiving the same proper times from the four satellites. The criterion proposed in [2] to identify the two solutions (see above) has been succesfully used in all cases.

Results are shown in Fig. 2. The lowest empty circle represents the chosen point on Earth. The remaining empty circles are positions of the reveiver for which there are no a second point in the emission region receiving the same proper times; however, when the receiver approaches the crossing satellite, there appear two points of the emission region receiving the same proper times. One of these points (on the receiver line world) is represented by a filled circle and the associated one by a filled square. We have verified that χ^2 changes its sign, from negative to positive, along the receiver world line and, accordingly, the number of solutions changes from one to two. Other motions of the receiver should be studied to perform a general systematic study of the emission region. It well be done elsewhere.

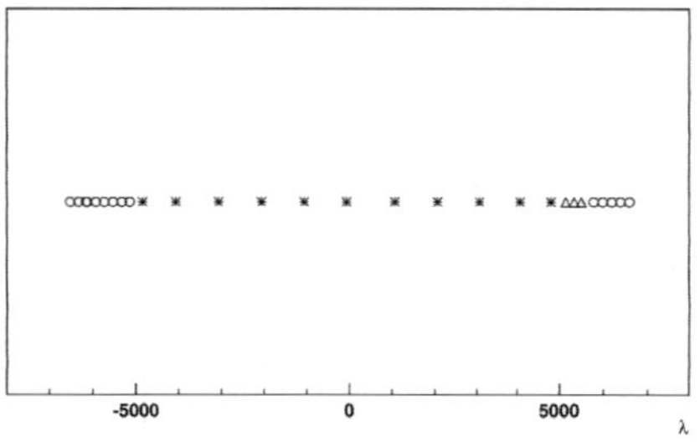

Figure 1. Co-region structure. Emission solutions at different points of a certain straight line in the grid

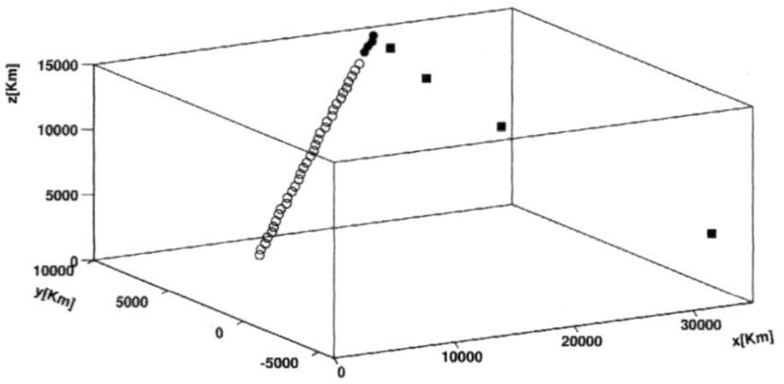

Figure 2. Emission region structure. Single and double emission solutions.

4. Uncertainties in satellite positions and positioning errors

Here we deal with the estimation of positioning errors due to uncertainties in satellite positions. We can choose many groups of four satellites among those of the GPS and Galileo constellations. Moreover, for each group, we can proceed as in previous section; namely, we may find co-region points corresponding to various receiver positions approaching one of the satellites. We first assume that there are no uncertainties in satellite positions and, from each of the above grid points (τ^1,τ^2,τ^3,τ^4), we get the *exact receiver position* by using the transformation of [1]. Then, uncertainties in satellite positions may be simulated as follows: the position of each satellite is randomly placed inside a sphere with radius R (e.g., 1 m) centered on the *exact satellite position.* N different positions of each satellite (close to the exact one) are thus generated. For each of them and the same grid point (proper times are not changed), we may find a *perturbed receiver position* (inertial coordinates). Hereafter, the distance between the exact and perturbed receiver locations is denoted by r. As a result, a distribution of N values of r are obtained in each case. The mean and RMS values of this distribution may be easily calculated. These quantities characterize the positioning errors. In Figs. 3 and 4 we present two distinct cases. Each of them corresponds to four satellites and an exact receiver position. In order to describe these cases, we can consider a cone, with its vertex at the receiver, whose surface contains satellites 1, 2 and 3. Satellite 4 will not be, in general, on the surface. Then, the angle between the axis of the imaginary cone and the direction of the satellite 1 (4) is denoted θ_1 (θ_4). In the Figures of this section one easily see that: (i) the angles θ_1 and θ_4 are rather different in Fig. 3 and rather similar in Fig. 4, and (ii) the positioning errors of Fig. 4 are one order of magnitude greater than those of Fig. 3 (see means and RMS values of r). These facts may be easily understood taking into account that the Jacobian of the transformation from emission to inertial coordinates vanishes for $\theta_4 = \theta_1$; namely, for four satellites observed from the receiver along directions contained in the above cone [4, 5]. Hence, for $\theta_4 \simeq \theta_1$, the Jacobian takes on small problematic values and the uncertainties in satellite positions produce large positioning errors. These situations should be prevented by choosing, for every receiver position, four available satellites with $\theta_4 \neq \theta_1$. More study on this subject is being developed.

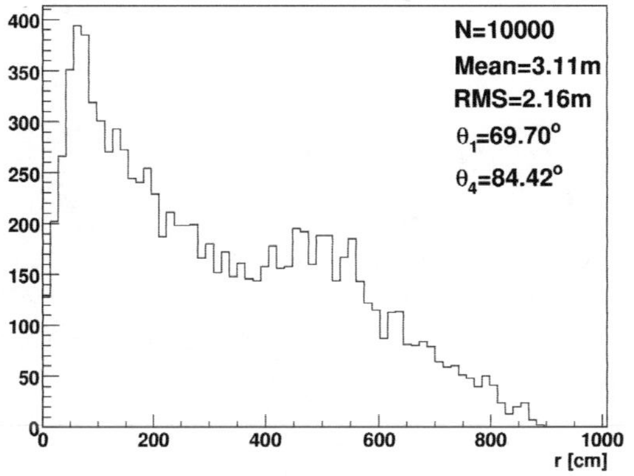

Figure 3. Histogram of the distribution of r deviations for $\theta_1 \neq \theta_4$

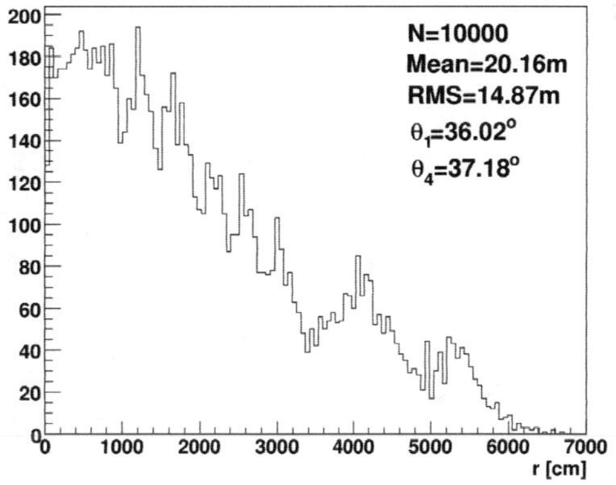

Figure 4. The same histogram as in Fig. 3 for $\theta_1 \simeq \theta_4$

Acknowledgments

This paper has been supported by the Spanish Ministerio de Ciencia e Innovación, MICINN-FEDER project FIS2009-07705. We thanks B. Coll, J.J. Ferrando and J.A. Morales-Lladosa for useful comments.

References

[1] Coll B, Ferrando J J and Morales-Lladosa J A 2010 *Class. Quantum. Grav.* **27** 0605013
[2] Coll B, Ferrando J J and Morales-Lladosa J A 2010 From emission to inertial coordinates: An analytical approach *Gravity as a Crossroad in Physics Proc. of the 33th Spanish Relativity Meeting ERE-2010* Granada, September 2010
[3] Pozo J M 2005 Constructions in 3D. In Proc. Int. School on *Relativistic Coordinates, Reference and Positioning Systems* ed B. Coll and J. J. Martín Salamanca, Spain, Jaunary 2005. Universidad de Salamanca (in progress).
[4] Coll B and Pozo J M 2006 *Class. Quantum. Grav.* **23** 7395
[5] Pozo J M 2005 Constructions in 4D: Algebraic properties and special observers. In Proc. Int. School on *Relativistic Coordinates, Reference and Positioning Systems* ed B. Coll and J. J. Martín Salamanca, Spain, Jaunary 2005. Universidad de Salamanca (in progress).

Journal of Physics: Conference Series **314** (2011) 012108 doi:10.1088/1742-6596/314/1/012108

Studying the microlenses mass function from statistical analysis of the caustic concentration.

T Mediavilla[1], O Ariza[2] , E Mediavilla[3], J A Muñoz[4].

[1,2]Departamento de Estadística e Investigación Operativa, Universidad de Cádiz, Avda de Ramón Puyol, s/n 11202 Algeciras.
[3]Instituto de Astrofísica de Canarias, Avda Vía Láctea s/n, La Laguna.
[4]Departamento de Astrofísica y Astronomía, Universidad de Valencia, Burjassot, Valencia.

E-mail: teresa.mediavilla@ca.uca.es, octavio.ariza@uca.es , emg@iac.es,

Abstract. The statistical distribution of caustic crossings by the images of a lensed quasar depends on the properties of the distribution of microlenses in the lens galaxy. We use a procedure based in Inverse Polygon Mapping to easily identify the critical and caustic curves generated by a distribution of stars in the lens galaxy. We analyze the statistical distributions of the number of caustic crossings by a pixel size source for several projected mass densities and different mass distributions. We compare the results of simulations with theoretical binomial distributions. Finally we apply this method to the study of the stellar mass distribution in the lens galaxy of QSO 2237+0305.

1. Introduction.

Fluctuations induced by microlensing in the observed light curves of quasars contain information about the lensing objects (masses, density), about the unresolved source structure (size of the continuum or broad line regions of the quasar, brightness profile of the quasar) and about the lens system (transversal velocity). So from a comparison between observed and simulated quasar microlensing one can obtain information about the physical parameters of interest. These conclusions can only be done in a statistical sense [1].

Thus, simulating magnification maps for some values of the physical parameters of interest we can try to calculate the probability that microlensing magnification takes a value determined from observations. This statistical analysis faces a problem: experimental errors and sources of variability other than microlensing can significantly affect the data and results.

To solve this problem we propose to simplify it reducing the microlensing effect to a series of discrete events, the caustic crossings. If the source size is small enough (X-ray emitting source) each caustic crossing will appear as a single event. In addition, caustic crossings are events of high magnification (little affected by measurement errors), and they are difficult to mistake with other kind of variability.

2. Statistical analysis of the caustics concentration based on caustic crossing counts.

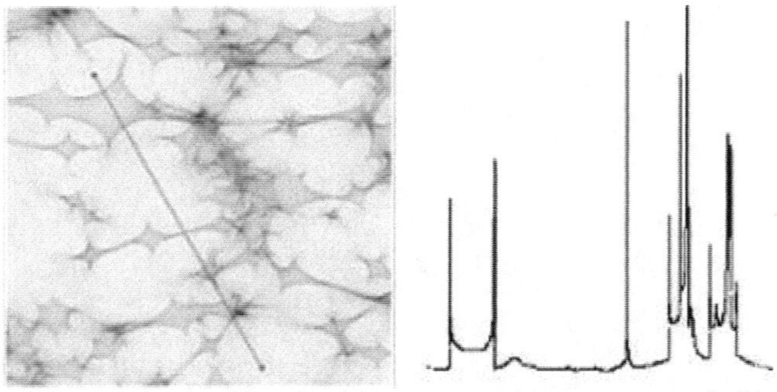

Figure 1. Magnification map crossed by a source (left). Light curve of a one pixel size source (right).

In figure 1 we can see a magnification map. The straight line corresponds to the path of a pixel size source. In the same figure we have represented the source's light curve. Whenever the source crosses a caustic a large increase in the magnification that only can be associated with the phenomenon of caustic crossing is produced. That is, for a source of this size the light curve would provide direct information of an observable very interesting from a statistical point of view: the number of caustic crossings. One source of this size corresponds, in practice, to the X-ray emission.

The distribution of caustics in the magnification maps depends on the characteristics of the distribution of stars and dark matter in the lensing galaxy. In this work, we study how does the existence of a range of masses in the distribution of stars affect to the caustics concentration.

The statistical analysis of the caustics spatial distribution is based on the following steps: (i) simulate magnification maps for different densities of matter and different mass distribution, (ii) identify the caustic curves, (iii) count the number of caustics detected in a one-dimensional window of certain size in pixels for each axis, (iv) estimate the probability of detecting a caustic in a pixel for each axis and (v) compare the experimental distributions obtained in simulations with the theoretical binomial distribution.

We have used the method of Inverse Polygon Mapping [2] to carry out steps (i) and (ii).

3. Application to QSO 2237+0305

We have simulated magnification and caustic maps for the four images of QSO 2237+0305. We have considered two extreme cases in the stars mass distribution: stars distributed in a range of masses (from 0.01 to 1 solar masses with an m^{-1} density law; hypothesis I) and simple distribution of identical stars of one solar mass (hypothesis II).

We use the same simulation for images A and B because their convergence and shear are similar.

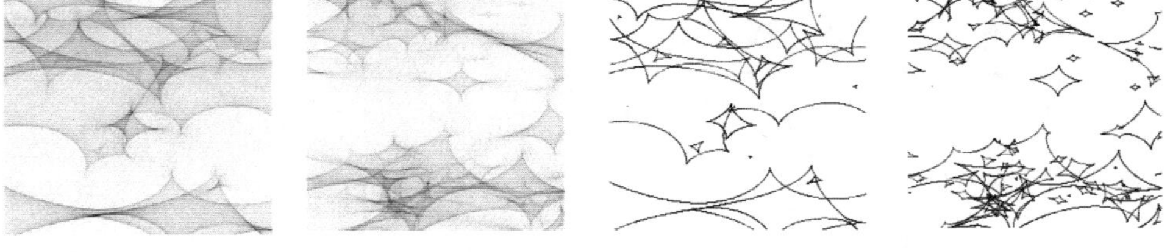

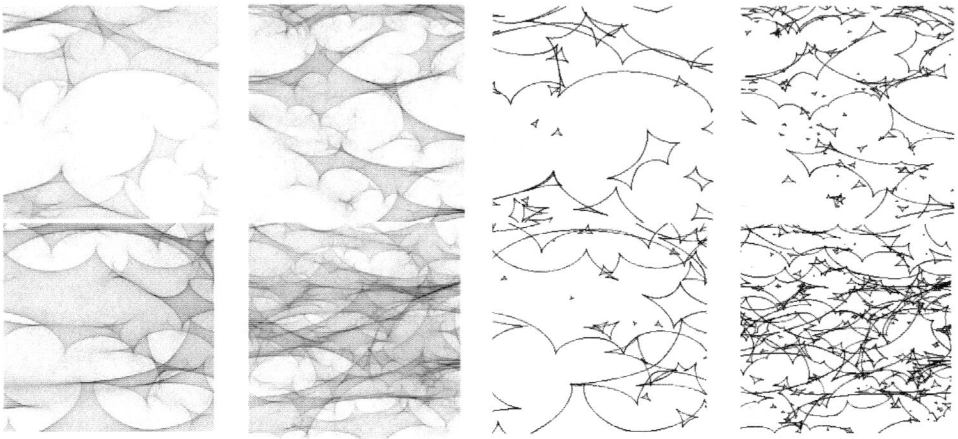

Figure 2. Magnification maps (first and second columns) and caustics maps (third and fourth columns). From top to down images A and B, C and D. Columns first and third: hypothesis II, columns second and fourth: hypothesis I.

Looking at Figure 2 we can see that for the four images the number of caustics is larger when the microlenses are distributed in a range of masses.

The probabilities of detecting a caustic in a pixel for each axis are shown in table 1:

Table 1. Probability of detecting a caustic in a pixel

	P_X	P_Y
A and B (hypothesis II)	0.00664	0.01204
A and B (hypothesis I)	0.01088	0.01745
C (hypothesis II)	0.00515	0.00895
C (hypothesis I)	0.00844	0.01440
D (hypothesis II)	0.00688	0.01100
D (hypothesis I)	0.02126	0.03135

We observe that the probability is higher when the stars are distributed in a range of masses than in the single mass case. The highest probability corresponds to image D when the stars are distributed in a range of masses.

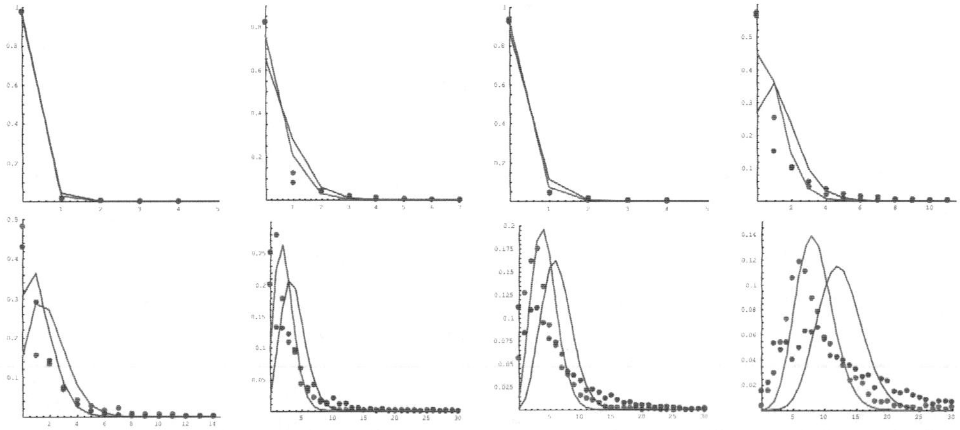

Figure 3. Image D, comparisons between experimental (dotted line) and theoretical

(continuous line) probability distributions in windows of (from left to right) 4, 40, 200 and 400 pixels. Top panels correspond to hypothesis II and bottom panels to hypothesis I. X axis in red, Y axis in Blue.

We have counted the number of caustics detected in one-dimensional windows of 4, 40, 200 and 400 pixels for each axis (1 pixel corresponds to 0.012 Einstein radii for a solar mass star) and we have calculated experimental and theoretical binomial probability distributions. Comparing the different distributions (Figure 3) we have derived the following results for image D:

- Differences between probability distributions are very significant. For example considering a 400 pixel size window for the X axis, the peak and centroid of the number count distribution correspond to 6 or 7 detections when the stars are distributed in a range of masses but only to one detection in the case of 1 solar mass stars.
- From an experimental point of view, a single measure of the number of caustics detected in a window of 400 pixels may be sufficient to distinguish between hypotheses I and II.
- To constrain the stellar mass distribution, the transversal velocity of the quasar (used to define the observing window in suitable Einstein radii units) is needed. Reciprocally, for a given distribution of stellar masses, the transversal velocity may be estimated from the caustic crossing distribution.

4. Conclusions.

Our study reveals that caustic crossing statistics can be significantly affected by the microlenses mass function. For the QSO 2237+0305D image, the detection of even a relatively small number of events (about ten) may allow to constrain the star mass distribution in the lens galaxy.

References

[1] Wambsganss J 1998 *Gravitational lensing in Astronomy Liv. Rev. in Relativity*
[2] Mediavilla E, Muñoz J, Lopez P, Mediavilla T, Abajas C, Gonzalez- Morcillo C and Gil-Merino 2006 *ApJ* **635** 942-953

Journal of Physics: Conference Series **314** (2011) 012109 doi:10.1088/1742-6596/314/1/012109

Charged Rotating Black Holes in Higher Dimensions

Masoud Allahverdizadeh[1], **Jutta Kunz**[1], **Francisco Navarro-Lérida**[2]
[1]Institut für Physik, Universität Oldenburg, Postfach 2503, D-26111 Oldenburg, Germany.
[2]Departamento de Física Atómica, Molecular y Nuclear, Ciencias Físicas, Universidad Complutense de Madrid, E-28040 Madrid, Spain.

E-mail: masoud_alahverdi@yahoo.com, kunz@theorie.physik.uni-oldenburg.de, fnavarro@fis.ucm.es

Abstract. In recent years higher-dimensional black holes have attracted much interest because of various developments in gravity and high energy physics. But whereas higher-dimensional charged static (Tangherlini) and uncharged rotating (Myers-Perry) black holes were found long ago, black hole solutions of Einstein-Maxwell theory, are not yet known in closed form in more than 4 dimensions, when both electric charge and rotation are present. Here we therefore study these solutions and those of Einstein-Maxwell-dilaton theory, by using numerical and perturbative methods, and by exploiting the existence of spacetime symmetries. The properties of these black holes reveal new interesting features, not seen in $D = 4$. For instance, unlike the $D = 4$ Kerr-Newman solution, they possess a non-constant gyromagnetic factor.

1. Introduction

In Einstein-Maxwell (EM) theory in $3 + 1$ dimensions the unique family of stationary asymptotically flat black holes with non-degenerate horizon comprises the rotating Kerr-Newman (KN) and Kerr black holes and the static Reissner-Nordström (RN) and Schwarzschild black holes. EM black holes are uniquely determined by their mass, their electric and magnetic charge, and their angular momentum (see [1] and references therein).

The generalization of black hole solutions to higher dimensions was pioneered by Tangherlini [2] for static vacuum black holes, and by Myers and Perry (MP) [3] for stationary vacuum black holes. Whereas Myers and Perry [3] also obtained charged static black holes in higher dimensional EM theory, higher dimensional charged rotating black holes have not yet been obtained in closed form in pure EM theory, although such black holes are known for some low energy effective actions related to string theory (including additional fields, though) [4, 5, 6, 7].

Higher dimensional black holes received much interest in recent years also with the advent of brane-world theories, raising the possibility of direct observation in future high energy colliders [8], which makes the theoretical study of these higher dimensional objects particularly appealing.

However, finding black hole solutions in higher dimensions is a difficult task due to the size and complexity of the equations. Moreover, contrary to what happens in $D = 4$ the topology of the horizon is not unique in higher dimensions (for instance, ringlike configurations are allowed [9]), which makes the problem even harder. For these reasons the number of solutions known analytically in closed form is very limited. This forces us to use alternative techniques to study these black holes. Here we will briefly analyze such black holes by two methods: the numerical method and the perturbative method.

2. Einstein-Maxwell-Dilaton theory

We will concentrate on Einstein-Maxwell-Dilaton (EMD) theory in D dimensions, as an example of how these methods may be applied. The classical EMD action reads

$$S = \int d^D x \sqrt{-g} \left(R - \frac{1}{2}\partial_\mu \Phi \partial^\mu \Phi - \frac{1}{4} e^{-2h\Phi} F_{\mu\nu} F^{\mu\nu} \right), \tag{1}$$

where R is the scalar curvature, Φ is the dilaton field, $F_{\mu\nu} = \partial_\mu A_\nu - \partial_\nu A_\mu$ is the electromagnetic field tensor, and A_μ is the electromagnetic potential. h is the dilaton coupling constant.

The equations of motion can be obtained by varying the action with respect to the metric $g_{\mu\nu}$, the dilaton field Φ and the gauge potential A_μ, yielding

$$G_{\mu\nu} = \frac{1}{2}\left[\partial_\mu \Phi \partial_\nu \Phi - \frac{1}{2} g_{\mu\nu} \partial_\rho \Phi \partial^\rho \Phi + e^{-2h\Phi}\left(F_{\mu\rho} F_\nu{}^\rho - \frac{1}{4} g_{\mu\nu} F_{\rho\sigma} F^{\rho\sigma}\right)\right], \tag{2}$$

$$\nabla^2 \Phi = -\frac{h}{2} e^{-2h\Phi} F_{\mu\nu} F^{\mu\nu}, \quad \nabla_\mu \left(e^{-2h\Phi} F^{\mu\nu}\right) = 0 . \tag{3}$$

For $h = 0$ the theory reduces to pure EM theory.

These equations admit an analytical solution when the dilaton coupling constant h takes the Kaluza-Klein (KK) value h_{KK} [10, 11]

$$h_{\rm KK} = \frac{D-1}{\sqrt{2(D-1)(D-2)}} . \tag{4}$$

In order to generate charged rotating solutions for generic values of h we will assume that the topology of the horizon is spherical. We will further assume that the black holes are stationary and axisymmetric. Stationary axisymmetric black holes in D dimensions possess $N = [(D-1)/2]$ independent angular momenta J_i associated with N orthogonal planes of rotation [3] ($[(D-1)/2]$ denotes the integer part of $(D-1)/2$, corresponding to the rank of the rotation group $SO(D-1)$). These black hole solutions then fall into two classes: even-D and odd-D-solutions.

When all N angular momenta have equal-magnitude, the symmetry of these black holes is strongly enhanced. In fact, in odd dimensions, the symmetry then increases from $R \times U(1)^N$ to $R \times U(N)$, thus changing the problem from cohomogeneity-N to cohomogeneity-1. As a consequence the original system of partial differential equations in N variables then reduces to a system of ordinary differential equations (ODE's).

3. Numerical and perturbative methods

In what follows we will focus on the case where D is odd, all angular momenta have the same magnitude, and the topology of the horizon is that of a sphere. That case is tractable numerically and perturbatively. Under those assumptions the metric, the gauge potential and the dilaton field can be parametrized with their angular dependence explicitly known

$$\begin{aligned} ds^2 = {} & \frac{m}{f}\left[dr^2 + r^2 \sum_{i=1}^{N-1}\left(\prod_{j=0}^{i-1}\cos^2\theta_j\right) d\theta_i^2\right] + \frac{n}{f} r^2 \sum_{k=1}^{N}\left(\prod_{l=0}^{k-1}\cos^2\theta_l\right)\sin^2\theta_k \left(\varepsilon_k d\varphi_k - \frac{\omega}{r} dt\right)^2 \\ & + \frac{m-n}{f} r^2 \left\{ \sum_{k=1}^{N}\left(\prod_{l=0}^{k-1}\cos^2\theta_l\right)\sin^2\theta_k d\varphi_k^2 - \left[\sum_{k=1}^{N}\left(\prod_{l=0}^{k-1}\cos^2\theta_l\right)\sin^2\theta_k \varepsilon_k d\varphi_k\right]^2\right\} - f dt^2 , \end{aligned} \tag{5}$$

$$A_\mu dx^\mu = a_0 dt + a_\varphi \sum_{k=1}^{N}\left(\prod_{l=0}^{k-1}\cos^2\theta_l\right)\sin^2\theta_k \varepsilon_k d\varphi_k , \quad \Phi = \Phi(r) , \tag{6}$$

where $\theta_0 \equiv 0$, $\theta_i \in [0, \pi/2]$ for $i = 1, \ldots, N-1$, $\theta_N \equiv \pi/2$, $\varphi_k \in [0, 2\pi]$ for $k = 1, \ldots, N$, and $\varepsilon_k = \pm 1$ denotes the sense of rotation in the k-th orthogonal plane of rotation. The spacetime possesses $N+1$ commuting Killing vectors, $\xi \equiv \partial_t$, and $\eta_{(k)} \equiv \partial_{\varphi_k}$, for $k = 1, \ldots, N$. This parametrization employs isotropic coordinates and is appropriate for numerical purposes [12, 13]. However, for perturbative purposes Boyer-Lindquist coordinates simplify the algebra [14, 15].

In order that the solutions correspond to black holes they must possess an event horizon located at $r = r_{\rm H}$, characterized by the condition $f(r_{\rm H}) = 0$ [16]. Such an event horizon is in fact a Killing horizon since the Killing vector

$$\chi = \xi + \Omega \sum_{k=1}^{N} \varepsilon_k \eta_{(k)} \ , \tag{7}$$

is null at and orthorgonal to the horizon, Ω representing the constant horizon angular velocity.

The system of ODE's resulting when Eqs. (5-6) are substituted in Eqs. (2-3) must be supplemented with the appropriate set of boundary conditions. This is obtained by imposing that the horizon is regular and the spacetime is asymptotically flat [11, 12, 13]. Then the system of ODE's is solved by employing a numerical solver (e.g., COLSYS), providing accurate results. The numerical parameters are $\{h, r_{\rm H}, \Omega, Q\}$, with Q being the electric charge. Varying these parameters we obtain families of solutions, with their corresponding values of the physical properties like the mass M, the angular momentum J, etc. An important information the numerical method provides is the domain of existence of black holes in higher dimensions. That domain consists of the region in the parameter space bounded by the extremal solutions. In Figure 1 (left) we exhibit the domain of existence of EMD black holes in terms of the scaled electric charge Q/M and angular momentum $J/M^{(D-2)/(D-3)}$. Clearly the domain depends both on the dilaton coupling constant h and the dimension D.

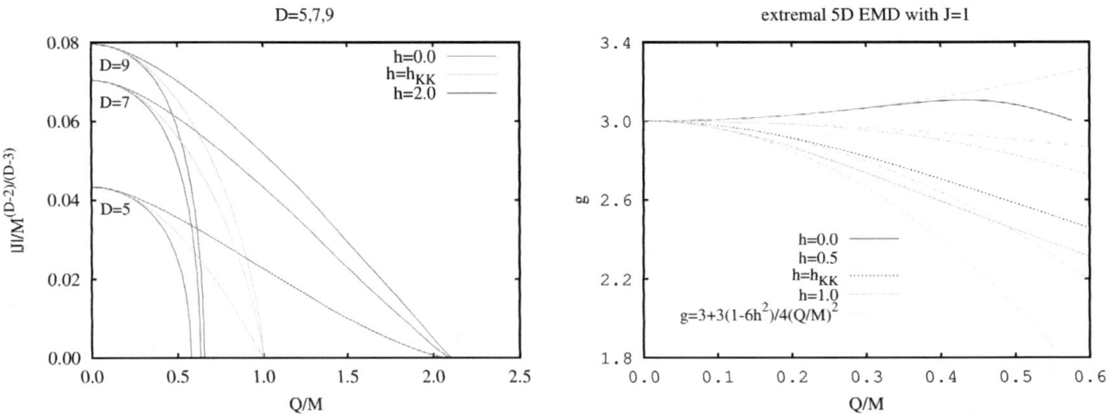

Figure 1. Left: Domain of existence of EMD black holes with equal-magnitude angular momenta in $D = 5$, 7 and 9 dimensions. Right: Comparison of the gyromagnetic factor g of $D = 5$ extremal EMD black holes obtained with numerical and perturbative (cyan) methods.

The accuracy of the numerical method may be tested by means of the analytical KK solutions, as well as by exact relations (like the Smarr mass formula) these black holes satisfy [15].

As a complementary method one may employ perturbations to generate black hole solutions in higher dimensions [14, 15, 17]. The advantage of this method is that it produces analytical formulae, which the numerical method does not. However, the range of validity of these formulae is limited by the accuracy of the order of the perturbations.

A good perturbative parameter to produce charged rotating black hole solutions in higher dimensions is the electric charge. One starts from the MP solution [3] and expands the metric, the gauge potential and the dilaton in series expansions in the electric charge around the MP solution (see [15] for details). Those series are substituted in Eqs. (2-3) and solved order by order in the electric charge. For lower orders the equations are easy to solve but as the order of the perturbations increases the equations become more and more difficult to solve.

An important point when applying the perturbative method is the way the integration constants are fixed. One has to impose regularity of the solutions at the horizon together with the condition of asymptotic flatness. The Smarr mass formula is a useful means to check the accuracy of the solutions.

When the perturbative method is applied to extremal solutions the formulae simplify very much. For instance, one obtains the following formula of the gyromagnetic factor g to second order

$$\frac{g}{D-2} = 1 + \frac{1}{4}\left[1 - \frac{2(D-1)(D-2)}{(D-3)^2}h^2\right]\left(\frac{Q}{M}\right)^2 + o\left(\left(\frac{Q}{M}\right)^2\right) , \tag{8}$$

showing that the gyromagnetic factor is not constant for pure EM solutions ($h = 0$).

4. Comparison of the two methods

Although both methods have a different range of applicability there is a region of the parameter space where both are valid, namely, when the electric charge is small. In that region one can check the consistency of the methods and their level of agreement [15]. This is presented in Figure 1 (right) where the perturbative formula Eq. (8) for extremal solutions is compared to the corresponding non-perturbative numerical results for $D = 5$. We observe that for $|Q/M| \leq 0.2$ both approches are in good agreement. In fact, the perturbative method may be modified to analyze also slowly rotating solutions, using as the unperturbated initial solution the Tangherlini solution and using the angular momentum as the perturbative parameter.

To summarize, these two methods are complementary and provide us with a powerful tool to analyze black holes in higher dimensions when analytical solutions are not available.

References

[1] M. Heusler 1996 *Black Hole Uniqueness Theorems* (Cambridge: Cambrigde University Press).
[2] F. R. Tangherlini 1963 *Nuovo Cimento* **77** 636.
[3] R. C. Myers, and M. J. Perry 1986 *Ann. Phys. (N.Y.)* **172** 304.
[4] J. C. Breckenridge, D. A. Lowe, R. C. Myers, A. W. Peet, A. Strominger and C. Vafa 1996 *Phys. Lett.* B **381** 423.
[5] M. Cvetič and Donam Youm 1996 *Nucl.Phys.* B **476** 118.
[6] D. Klemm and W. A. Sabra (2001) *Phys. Lett.* B **503** 147.
[7] M. Cvetič, H. Lü and C. N. Pope 2004 *Phys. Lett.* B **598** 273.
[8] S. Dimopoulos and G. Landsberg 2001 *Phys. Rev. Lett.* **87** 161602.
[9] R. Emparan and H. S. Reall 2002 *Phys. Rev. Lett.* **88** 101101.
[10] P. M. Llatas 1997 *Phys. Lett.* B **397** 63.
[11] J. Kunz, D. Maison, F. Navarro-Lérida and J. Viebahn 2006 *Phys. Lett.* B **639** 95.
[12] J. Kunz, F. Navarro-Lérida and A. K. Petersen 2005 *Phys. Lett.* B **614** 104.
[13] J. Kunz, F. Navarro-Lérida and J. Viebahn 2006 *Phys. Lett.* B **639** 362.
[14] M. Allahverdizadeh, J. Kunz and F. Navarro-Lérida 2010 *Phys. Rev.* D **82** 024030 .
[15] M. Allahverdizadeh, J. Kunz and F. Navarro-Lérida 2010 *Phys. Rev.* D **82** 064034.
[16] B. Kleihaus, J. Kunz and F. Navarro-Lérida 2002 *Phys. Rev.* D **66** 104001.
[17] F. Navarro-Lérida 2010 *Gen. Rel. Grav.* **42** 2891.

Journal of Physics: Conference Series **314** (2011) 012110 doi:10.1088/1742-6596/314/1/012110

Symmetric solutions of Einstein's equations in higher dimensions

M. Jakimowicz and J. Tafel

Institute of Theoretical Physics, Department of Physics
University of Warsaw, Poland

E-mail: tafel@fuw.edu.pl

Abstract. We present a method of solving the Einstein equations under assumption of $SO(n+1)$ symmetry acting on n-dimensional spheres.

Following [1, 2] we consider (N+n+1)-dimensional metrics of the form

$$g = g_{ab}dx^a dx^b - e^{2f} d\Omega_n^2 \ ,$$

where $d\Omega_n^2$ is the standard metric of a n-dimensional sphere S_n and g_{ab} and f are functions of coordinates x^a, $a = 0, 1, \cdots, N$. For $N > 1$ it is convenient to make the transformation

$$\tilde{g}_{ab} = e^{\frac{2n}{N-1}f} g_{ab} \ , \quad \phi = \sqrt{\tfrac{n(n+N-1)}{(N-1)}} f \ ,$$

which converts the vacuum Einstein equations for g into the Einstein equations for $\tilde{g}$ with matter represented by the scalar field ϕ

$$\tilde{R}_{ab} - \tfrac{1}{2}\tilde{g}_{ab} = \phi_{;a}\phi_{;b} + (V(\phi) - \tfrac{1}{2}\phi^{;c}\phi_{;c})\,\tilde{g}_{ab} \ , \tag{1}$$

$$\phi^{;a}_{\ ;a} = -V_{,\phi} \ , \quad V = -\tfrac{1}{2}n\,(n-1)\,e^{-2\sqrt{\frac{n+N-1}{n(N-1)}}\phi} \ . \tag{2}$$

In order to find solutions to (1) and (2) we assume that

- the normal vector field of surfaces ϕ=const is spacelike and geodesic,
- surfaces ϕ=const are Einstein spaces.

Then

$$\tilde{g} = \tilde{g}_{ij}dx^i dx^j - ds^2 \ , \ \tilde{g}_{ij} = \beta^{\frac{2}{N}}(\gamma e^{P\tau})_{ij} \ , \quad i = 0, \ldots, N-1 \ ,$$

where γ_{ij} and $P^i_{\ j}$ are s-independent matrix fields such that

$$\gamma_{ij} = \gamma_{ji} \ , \ (\gamma P)_{ij} = (\gamma P)_{ji} \ , \ P^i_{\ i} = 0 \ , \ P^i_{\ j}P^j_{\ i} = 2c \tag{3}$$

$$P^k_{\ i;k} = 0 \ , \ R^i_{\ j}(\gamma e^{P\tau}) = \lambda\delta^i_{\ j} \ , \quad \lambda = \text{const}, \quad c = \text{const} \tag{4}$$

and τ, β and ϕ are functions of s, subject to the differential equations

$$\dot{\tau} = 2\beta^{-1} \; , \quad (\beta\dot{\phi})^{\cdot} = -\beta V_{,\phi} \; , \tag{5}$$

$$(1 - \frac{1}{N})\frac{\dot{\beta}^2}{\beta^2} - \frac{2c}{\beta^2} - \dot{\phi}^2 + 2V = -N\lambda\beta^{-\frac{2}{N}} \; . \tag{6}$$

Summarizing, if equations (3)-(6) are satisfied then

$$g = e^{-2\phi\sqrt{\frac{n}{(N-1)(n+N-1)}}} \left(\beta^{\frac{2}{N}}(\gamma e^{P\tau})_{ij}dx^i dx^j - ds^2\right) - e^{2\phi\sqrt{\frac{N-1}{n(n+N-1)}}} d\Omega_n^2 \tag{7}$$

is $(N+n+1)$–dimensional vacuum metric.

Simple examples of metrics obtained in this way are presented in [1]. They include pp-waves of type N

$$g = du\left(dv - r^{1-n}h(u)du\right) - dr^2 - r^2 d\Omega_n^2$$

and the Kundt type metrics with nonconstant scalar invariants

$$g = du\left(\frac{4r^2 dv}{(n+1)(1-uv)^2} - r^{1-n}h(u)du\right) - dr^2 - \frac{n-1}{n+1}r^2 d\Omega_n^2 \; .$$

More advanced examples can be found by virtue of the observation that equations (5)-(6) depend on tensors γ and P via constant c only. Thus, some solutions of (5)-(6) can be extracted from known metrics and then implemented to other forms of γ and P. For instance, in this way one can generate new metrics related to the Gross-Perry solutions [3, 4, 5]. After slight generalization one obtains in this case the following functions β, ϕ, τ and s in terms of a new coordinate r [2]

$$\beta = (r^2 - r_0^2)s_{,r} \; , \qquad e^{\sqrt{\frac{n+N-1}{n(N-1)}}\phi} = (n-1)|rs_{,r}| \; , \qquad \tau = r_0^{-1}\ln\left|\frac{r+r_0}{r-r_0}\right| \tag{8}$$

$$s_{,r} = \pm\frac{1}{n-1}|r|^{-l-1}|r-r_0|^{l-p}|r+r_0|^{l+p} \; . \tag{9}$$

Here p and $r_0 > 0$ are free parameters and

$$l = \frac{n+N-1}{(n-1)(N-1)} \; , \qquad c = 2r_0^2\left[\frac{n}{n-1} - p^2\frac{(n-1)(N-1)^2}{N(n+N-1)}\right] \; .$$

For $N = 2$ and constant matrices γ and P such that $c > 0$ functions (8)-(9) yield (n+3)-dimensional metrics of the form

$$g = \left|\frac{r-r_0}{r+r_0}\right|^{p'-q} dt^2 - \left|\frac{r-r_0}{r+r_0}\right|^{p'+q} dy^2 - \frac{|r+r_0|^{\frac{2p'+2}{n-1}}}{|r|^{\frac{2n}{n-1}}|r-r_0|^{\frac{2p'-2}{n-1}}}\left(\frac{dr^2}{(n-1)^2} + r^2 d\Omega_n^2\right) \tag{10}$$

with parameters p' and q constrained by

$$(n+1)p'^2 + (n-1)q^2 = 2n \; .$$

For $n = 2$ these metrics coincide with the Gross-Perry metrics. For arbitrary n metrics (10) have the following properties:

- vectors ∂_t and ∂_y are additional Killing fields,
- in the limit $r \to \infty$ metrics tend to the flat metric

$$dt^2 - dy^2 - dr'^2 - r'^2 d\Omega_n^2 \,, \quad r' = r^{\frac{1}{n-1}} \,,$$

- they belong to algebraic type I [6, 7],
- they have a coordinate singularity at $r = 0$ and essential singularities at $r = \pm r_0$,
- in the Kaluza-Klein approach they split into asymptotically flat metric on y =const and asymptotically constant scalar field.

If $N = 2$, γ and P are constant matrices and $c < 0$, instead of (10) one obtains stationary metrics of the form

$$g = \left|\frac{r-r_0}{r+r_0}\right|^{p'} \left[\cos\left(q\ln\left|\frac{r+r_0}{r-r_0}\right|\right)(dt^2 - dy^2) + 2\sin\left(q\ln\left|\frac{r+r_0}{r-r_0}\right|\right)dtdy\right]$$
$$- \frac{|r+r_0|^{\frac{2p'+2}{n-1}}}{|r|^{\frac{2n}{n-1}}|r-r_0|^{\frac{2p'-2}{n-1}}}\left(\frac{dr^2}{(n-1)^2} + r^2 d\Omega_n^2\right) \,, \tag{11}$$

where

$$(n+1)p'^2 - (n-1)q^2 = 2n.$$

They have similar properties to (10). In particular, they become flat in the limit of infinite r.

If $N = 2$, $c = 0$ and matrices γ and P are not constants one obtains metrics which belong to the generalized Kundt class

$$g = \left|\frac{r-r_0}{r+r_0}\right|^{\pm\sqrt{\frac{2n}{n+1}}} \left(dudv + \ln\left|\frac{r+r_0}{r-r_0}\right| h(u)du^2\right)$$
$$- \frac{|r+r_0|^{\frac{2}{n-1}\left(\pm\sqrt{\frac{2n}{n+1}}+1\right)}}{|r|^{\frac{2n}{n-1}}|r-r_0|^{\frac{2}{n-1}\left(\pm\sqrt{\frac{2n}{n+1}}-1\right)}}\left(\frac{dr^2}{(n-1)^2} + r^2 d\Omega_n^2\right) \,. \tag{12}$$

They are of algebraic type II_i for $h \neq 0$ and type D for $h = 0$. In the limit $r \to \infty$ they tend to the flat metric. They have coordinate singularity at $r = 0$ and essential singularities at $r = \pm r_0$.

Further metrics can be generated via the presented method combined with symmetry transformations [8]. In this way one can obtain e.g. Kaluza-Klein monopoles and dyons.

Acknowledgments

This work is partially supported by the grant N N202 104838 of Polish Ministerstwo Nauki i Szkolnictwa Wyższego.

[1] M. Jakimowicz and J. Tafel 2008, SO(n+1) symmetric solutions of the Einstein equations in higher dimensions, *Class. Quantum Grav.* **25**, 175002
[2] M. Jakimowicz and J. Tafel 2009, Generalization of the Gross-Perry metrics, *Int. J. Theor. Phys.* **48**, 2876
[3] Gross D.J. and Perry M.J. 1983, Magnetic monopoles in Kaluza-Klein theories, *Nucl. Phys.* **B 226**, 29
[4] Sorkin R. 1983, Kaluza-Klein monopole, *Phys. Rev. Lett.* **51**, 87
[5] Davidson A. and Owen D. 1985, Black holes as windows to extra dimensions *Phys. Lett.*, **B 155**, 247
[6] Coley A., Milson R., Pravda V., Pravdova A. 2004, Classification of the Weyl Tensor in Higher Dimensions *Class. Quantum Grav.* **21**, L35
[7] Coley A. 2008, Classification of the Weyl Tensor in Higher Dimensions and Aplications, *Class. Quantum Grav.* **25**, 033001
[8] Giusto S. and Saxena A. 2007, Stationary axisymmetric solutions of five-dimensional gravity, *Class. Quantum Grav.* **24**, 4269

Journal of Physics: Conference Series **314** (2011) 012111 doi:10.1088/1742-6596/314/1/012111

Higher Dimensional Kerr–Schild Spacetimes with (A)dS Background

Tomáš Málek[1], **Vojtěch Pravda**[2]

[1] Institute of Theoretical Physics, Faculty of Mathematics and Physics, Charles University, V Holešovičkách 2, 180 00 Prague 8, Czech Republic

E-mail: malek@math.cas.cz

[2] Institute of Mathematics, Academy of Sciences of the Czech Republic, Žitná 25, 115 67 Prague 1, Czech Republic

E-mail: pravda@math.cas.cz

Abstract. Geometric properties of higher dimensional Kerr–Schild (KS) spacetimes with (A)dS background are summarised. For Einstein spaces algebraic types of the Weyl tensor compatible with the KS ansatz are established, optical properties of the KS vector are given. Important examples of exact solutions belonging to this class, such as Kerr–de Sitter metrics in arbitrary dimension are also briefly discussed.

1. Introduction

Four-dimensional Kerr–Schild ansatz with Minkowski background was first used in [2]. It led to the discovery of the Kerr rotating black hole. Moreover, many other physically important exact vacuum solutions belong to this class, for instance Kerr–Newmann black hole and Weyl type N Kundt spacetimes including *pp*-waves. The generalization of KS ansatz to higher dimensions led to the discovery of Myers–Perry solutions [3]. General properties of higher dimensional Ricci-flat KS metric were studied recently in [4].

Rotating black hole solutions with a cosmological constant in four, five and higher dimensions [5–7] can be cast to a generalized Kerr–Schild (GKS) form, where the flat background metric is replaced by a de Sitter or an anti-de Sitter spacetime. We will study general properties of GKS spacetimes and generalize the results for the Ricci-flat case [4] to the Einstein spaces.

Generalized Kerr–Schild ansatz is a metric of the form

$$g_{ab} = \bar{g}_{ab} - 2\mathcal{H} k_a k_b \,, \tag{1}$$

where the background metric $\bar{g}_{ab}$ is (A)dS and $\mathcal{H}$ is a scalar function. The KS vector $\boldsymbol{k}$ is null vector with respect to the full metric g_{ab}. Therefore $\boldsymbol{k}$ is also null vector with respect to the background metric $\bar{g}_{ab}$ and the inverse metric has the simple form $g^{ab} = \bar{g}^{ab} + 2\mathcal{H} k^a k^b$. In the following we consider the background metric in the canonical form $\bar{g}_{ab} = \Omega\left(-\mathrm{d}t + \mathrm{d}x_1 + \ldots + \mathrm{d}x_{n-1}\right)$. The conformal factor Ω in the de Sitter and anti-de Sitter background metric is given by

$$\Omega_{dS} = \frac{(n-2)(n-1)}{2\Lambda t^2}\,, \qquad \Omega_{AdS} = -\frac{(n-2)(n-1)}{2\Lambda x_1^2}\,. \tag{2}$$

2. General properties

Let us start with the simplest Ricci frame component of the highest boost weight $R_{00} = R_{ab}k^a k^b$ since many terms in this contraction vanish. Indeed, we obtain

$$R_{00} = 2\mathcal{H} k_{a;b} k^b k^a{}_{;c} k^c \,. \tag{3}$$

From Einstein's equations it immediately follows

Proposition 1 *The null vector $\boldsymbol{k}$ in the generalized Kerr–Schild metric* (1) *is geodetic if and only if the component of the energy-momentum tensor $T_{00} = T_{ab}k^a k^b$ vanishes.*

Note that T_{00} vanishes not only in Einstein spaces but also in spacetimes with aligned matter content, such as aligned Maxwell field $F_{ab}k^a \propto k_b$ or aligned pure radiation $T_{ab} \propto k_a k_b$.

Various quantities constructed from the full GKS metric can be easily used for derivation of the corresponding quantities constructed from the background metric by setting $\mathcal{H}$ to zero. This allows us to compare geodeticity of KS congruence in both metrics

$$k_{a;b}k^b = k_{a\bar{;}b}k^b \,, \quad k^a{}_{;b}k^b = k^a{}_{\bar{;}b}k^b \,, \tag{4}$$

where $k_{a\bar{;}b}$ denotes a covariant derivative with respect to the background metric. Thus the KS vector $\boldsymbol{k}$ is geodetic in the full metric if and only if it is geodetic in the background metric.

From now on we assume $\boldsymbol{k}$ being geodetic, which significantly simplifies our calculations, e.g. boost weight 1 and 0 components of the Ricci tensor vanish.

One may introduce a null frame in the full metric consisting of two null vectors $\boldsymbol{\ell}$, $\boldsymbol{n}$ and $n-2$ spacelike vectors $\boldsymbol{m}^{(i)}$ such that $g_{ab} = 2\ell_{(a}n_{b)} + \delta_{ij} m_a^{(i)} m_b^{(j)}$. Similarly, we can define a null frame $\boldsymbol{\ell}$, $\tilde{\boldsymbol{n}}$, $\boldsymbol{m}^{(i)}$ in the background metric putting $\tilde{n}_a = n_a + \mathcal{H}k_a$. If we identify the frame vector $\boldsymbol{\ell}$ with the KS vector $\boldsymbol{k}$ optical matrices $\tilde{L}_{ij}$ and L_{ij} in the full and background spacetimes obey

$$L_{ij} \equiv k_{a;b} m^{(i)a} m^{(j)b} = k_{a\bar{;}b} m^{(i)a} m^{(j)b} \equiv \tilde{L}_{ij} \,. \tag{5}$$

In other words the KS vector $\boldsymbol{k}$ has the same optical properties in both metrics.

It can be shown [1] that for geodetic KS vector $\boldsymbol{k}$ all positive boost weight frame components of the Weyl tensor vanish

Proposition 2 *Generalized Kerr–Schild spacetime with geodetic vector $\boldsymbol{k}$ is algebraically special with $\boldsymbol{k}$ being the multiple WAND.*

From now on, let us consider only Einstein spacetimes. We employ Einstein's field equations in order to analyze constraints imposed on the parameters of the GKS ansatz. One of the equations [1] can be written as

$$(n-2)\theta(D\log\mathcal{H}) = \sigma^2 + \omega^2 - (n-2)(n-3)\theta^2 \,. \tag{6}$$

The function $\mathcal{H}$ appears in (6) only if $\theta \neq 0$ and, consequently, we will solve the non-expanding case $\theta = 0$ and expanding case $\theta \neq 0$ separately.

3. Non-expanding solutions

Vanishing expansion $\theta = 0$ in (6) implies that the KS congruence $\boldsymbol{k}$ is non-shearing, $\sigma = 0$, and non-twisting, $\omega = 0$, and thus the optical matrix vanishes, $L_{ij} = 0$. It can be shown [1] that consequently all boost weight 0 and -1 components of the Weyl tensor vanish. We can conclude

Proposition 3 *Einstein generalized Kerr–Schild spacetimes with non-expanding KS congruence $\boldsymbol{k}$ are of type N with $\boldsymbol{k}$ being the multiple WAND. Twist and shear of the KS congruence $\boldsymbol{k}$ necessarily vanish and these solutions thus belong to the class of Einstein type N Kundt spacetimes.*

Note that in the Ricci-flat case [4] the class of non-expanding KS spacetimes is in fact equivalent with the class of type N Kundt spacetimes.

4. Expanding solutions

In the expanding case Einstein's equations lead to the "optical constraint"

$$L_{ik}L_{jk} = \frac{L_{lk}L_{lk}}{(n-2)\theta} S_{ij} \tag{7}$$

and it follows that L is a normal matrix, $LL^T - L^TL = 0$. Therefore, in an appropriate frame the optical matrix L_{ij} has a block-diagonal form consisting of 2×2 blocks with vanishing symmetric trace-free part (i.e. there is no shear in the corresponding two dimensional spacelike planes).

The block-diagonal form of L_{ij} allows us to solve Sachs equation ($\mathrm{D}L_{ij} = -L_{ik}L_{kj}$) and thus explicitly determine r-dependence (with r being an affine parameter along $\boldsymbol{k}$) of the optical matrix L_{ij} which consists of p blocks $\mathcal{L}_{(\mu)}$ and a diagonal matrix $\tilde{\mathcal{L}}$

$$\begin{aligned}
&\mathcal{L}_{(\mu)} = \begin{pmatrix} s_{(2\mu)} & A_{2\mu,2\mu+1} \\ -A_{2\mu,2\mu+1} & s_{(2\mu)} \end{pmatrix}, \quad s_{(2\mu)} = \frac{r}{r^2 + (a^0_{(2\mu)})^2}, \quad A_{2\mu,2\mu+1} = \frac{a^0_{(2\mu)}}{r^2 + (a^0_{(2\mu)})^2}, \\
&\tilde{\mathcal{L}} = r^{-1}\mathrm{diag}(\underbrace{1,\ldots,1}_{(m-2p)}, \underbrace{0,\ldots,0}_{(n-2-m)}),
\end{aligned} \tag{8}$$

with $0 \le 2p \le m \le n-2$ and m denoting the rank of L_{ij}. Using this result and integrating equation (6) we get

$$\mathcal{H} = \frac{\mathcal{H}_0}{r^{m-2p-1}} \prod_{\mu=1}^{p} \frac{1}{r^2 + \left(a^0_{(2\mu)}\right)^2}. \tag{9}$$

The KS function $\mathcal{H}$ may introduce a singularity to the full GKS metric whenever $\mathcal{H}$ diverges. Obviously, $\mathcal{H}$ blows up at $r=0$ when $2p \neq m$ or $2p \neq m-1$ and in the special case $2p = m$ (m even) or $2p = m-1$ (m odd) if some of $a^0_{(2\mu)}$ vanish at some points of spacetime. In both cases the Kretschmann scalar shows presence of a curvature singularity.

It can be also shown [1] that expanding Einstein GKS spacetimes are not compatible with Weyl types III and N and thus

Proposition 4 *Einstein generalized Kerr–Schild spacetimes with expanding KS congruence* $\boldsymbol{k}$ *are of Weyl types II or D or conformally flat.*

4.1. Example: 5D Kerr–(A)dS

Higher dimensional Kerr–(A)dS metric is written in the GKS form in [7]. In five dimensions, the background metric $\bar{g}$, KS vector $\boldsymbol{k}$ and KS function $\mathcal{H}$ read

$$\begin{aligned}
\bar{g} = &-\frac{(1-\lambda r^2)\Delta}{(1+\lambda a^2)(1+\lambda b^2)}\mathrm{d}t^2 + \frac{r^2\rho^2}{(1-\lambda r^2)(r^2+a^2)(r^2+b^2)}\mathrm{d}r^2 + \frac{\rho^2}{\Delta}\mathrm{d}\theta^2 \\
&+ \frac{(r^2+a^2)\sin^2\theta}{1+\lambda a^2}\mathrm{d}\phi^2 + \frac{(r^2+b^2)\cos^2\theta}{1+\lambda b^2}\mathrm{d}\psi^2, \qquad \mathcal{H} = -\frac{M}{\rho^2}, \\
k = &\frac{\Delta}{(1+\lambda a^2)(1+\lambda b^2)}\mathrm{d}t + \frac{r^2\rho^2}{(1-\lambda r^2)(r^2+a^2)(r^2+b^2)}\mathrm{d}r - \frac{a\sin^2\theta}{1+\lambda a^2}\mathrm{d}\phi - \frac{b\cos^2\theta}{1+\lambda b^2}\mathrm{d}\psi,
\end{aligned} \tag{10}$$

where $\rho^2 = r^2 + a^2\cos^2\theta + b^2\sin^2\theta$ and $\Delta = 1 + \lambda a^2\cos^2\theta + \lambda b^2\sin^2\theta$. Using the null frame given in [1] we can express L_{ij} in the block-diagonal form with $n=5$, $m=3$, $p=1$. The optical matrix and function $\mathcal{H}$ of 5D Kerr–(A)dS and GKS spacetime are compared in the following table

5D Kerr-de Sitter	GKS ($n=5,\ m=3,\ p=1$)
$L=\begin{pmatrix} \frac{r}{\rho^2} & \frac{\sqrt{\rho^2-r^2}}{\rho^2} & 0 \\ -\frac{\sqrt{\rho^2-r^2}}{\rho^2} & \frac{r}{\rho^2} & 0 \\ 0 & 0 & \frac{1}{r} \end{pmatrix}$	$L=\begin{pmatrix} s_{(2)} & A_{2,3} & 0 \\ -A_{2,3} & s_{(2)} & 0 \\ 0 & 0 & \frac{1}{r} \end{pmatrix}$
$\mathcal{H}=-M\frac{1}{r^2+a^2cos^2\theta+b^2\sin^2\theta}$	$\mathcal{H}=\mathcal{H}_0\frac{1}{r^2+(a^0_{(2)})^2}$
$s_{(2)}=\frac{r}{r^2+a^2\cos^2\theta+b^2\sin^2\theta}$	$s_{(2)}=\frac{r}{r^2+(a^0_{(2)})^2}$
$A_{2,3}=\frac{\sqrt{a^2\cos^2\theta+b^2\sin^2\theta}}{r^2+a^2\cos^2\theta+b^2\sin^2\theta}$	$A_{2,3}=\frac{a^0_{(2)}}{r^2+(a^0_{(2)})^2}$

The presented quantities of five-dimensional Kerr–(A)dS are in accordance with our general results for GKS spacetimes with identification $\mathcal{H}_0=-M$ and $a^0_{(2)}=a^2\cos^2\theta+b^2\sin^2\theta$.

Let us briefly discuss presence of a curvature singularity. In the fist case $a\neq 0$, $b\neq 0$ although $2p=m-1$ corresponds to the special case, $a^0_{(2)}$ is sum of squares and hence there is no singularity. If we set one of the spins to zero, e.g. $b=0$, then $a^0_{(2)}$ vanishes for $\theta=\frac{\pi}{2}$ and there is a ring shaped singularity known from the four-dimensional Kerr solution.

Putting $a=b=0$ leads to (Anti-)de Sitter–Schwarzschild–Tangherlini limit where now $p=0$ and neither $2p=m-1$ nor $2p=m$ and therefore a singularity is present at $r=0$.

Acknowledgments

Some calculations in this paper were performed and many others checked using computer algebra software Cadabra [10, 11]. This work has been supported by research plan No AV0Z10190503. V.P. is supported by research grant GAČR P203/10/0749, T.M. is supported by the project SVV 261301 of the Charles University in Prague.

References

[1] T. Málek and V. Pravda. Kerr-Schild spacetimes with (A)dS background. *Class. Quantum Grav.*, 28:125011, 2011.

[2] R. P. Kerr and A. Schild. Some algebraically degenerate solutions of Einstein's gravitational field equations. *Proc. Symp. Appl. Math.*, 17:199, 1965.

[3] R. C. Myers and M. J. Perry. Black holes in higher dimensional space-times. *Ann. Phys. (N.Y.)*, 172:304, 1986.

[4] M. Ortaggio, V. Pravda, and A. Pravdová. Higher dimensional Kerr-Schild spacetimes. *Class. Quantum Grav.*, 26:025008, 2009.

[5] B. Carter. Hamilton-Jacobi and Schrodinger separable solutions of Einstein's equations. *Commun. Math. Phys.*, 10:280, 1968.

[6] S. W. Hawking, C. J. Hunter and M. M. Taylor-Robinson. Rotation and the AdS-CFT correspondence. *Phys. Rev.*, D59:064005, 1999.

[7] G. W. Gibbons, H. Lu, Don N. Page, and C. N. Pope. The general Kerr-de Sitter metrics in all dimensions. *J. Geom. Phys.*, 53:49–73, 2005.

[8] V. Pravda, A. Pravdová, A. Coley, and R. Milson. Bianchi identities in higher dimensions. *Class. Quantum Grav.*, 21:2873–2897, 2004. See also V. Pravda, A. Pravdová, A. Coley and R. Milson *Class. Quantum Grav.* **24** (2007) 1691 (corrigendum).

[9] M. Ortaggio, V. Pravda, and A. Pravdová. Ricci identities in higher dimensions. *Class. Quantum Grav.*, 24:1657–1664, 2007.

[10] K. Peeters. Introducing Cadabra: A symbolic computer algebra system for field theory problems. arXiv:hep-th/0701238, 2007.

[11] K. Peeters. A field-theory motivated approach to symbolic computer algebra. *Comput. Phys. Commun.*, 176:550, 2007.

Spanish Relativity Meeting (ERE 2010): Gravity as a Crossroad in Physics IOP Publishing
Journal of Physics: Conference Series **314** (2011) 012112 doi:10.1088/1742-6596/314/1/012112

Thermal equilibrium of black di-rings

Hideo Iguchi and Takashi Mishima

Laboratory of Physics, College of Science and Technology, Nihon University, Narashinodai, Funabashi, Chiba 274-8501, Japan

E-mail: `iguchi.h@phys.ge.cst.nihon-u.ac.jp`, `tmishima@phys.ge.cst.nihon-u.ac.jp`

Abstract. We show some physical properties of the systems of regular black diring(the regular gravitational systems with two S^1-rotational black rings arranged in concentric way in ve-dimensional asymptotically at spacetimes). Especially the existence of isothermal systems of black diring are shown, in which both isothermality and isorotation between the inner black ring and the outer black ring are realized. We also give some properties of the thermodynamic black di-ring including discussion about thermodynamic stabilities of the system.

1. Introduction

Since the groundbreaking discovery of S^1-rotating black rings by Emparan and Reall [1], several non-trivial black hole systems have been obtained and used to clarify peculiar features of the higher dimensional gravity. Especially in ve dimensions, solitonic solution-generation methods invented in the legendary era from the 1970s to the 1980s have been proved to be greatly powerful and applied to systematic reconstruction of known solutions including the Myers-Perry black holes [2], the S^1-rotating black rings [3, 4] and also to generation of new solutions: S^2-rotating black rings [5, 6, 7] and doubly rotating black rings [8], for example. As further trial, hunting new solutions have been attempted ambitiously to obtain novel black holes that have never been expected in four dimensions, like black hole systems with multi-horizons [9, 10, 11, 12, 13] or a topologically non-trivial horizon [14, 15].

The solutions of black di-ring among them correspond to concentric con gurations composed of two independently S^1-rotating black rings. The authors rst discovered the regular di-rings by using the solitonic method similar to the Backlund transformation [10]. Successively, Evslin and Krishnan constructed another di-ring solution-set [11]. They used the inverse scattering method that was modi ed by Pomeransky to treat the higher dimensional case (hereafter abbreviated to PISM) [2]. However, because of the complexity of their expressions further investigation of the physics of the di-ring systems remained to be done until quite recently.

In the recent work we con rmed the equivalence of these two di-ring solution sets, and tried to do some investigation of the physics of di-ring systems [16]. So in this place we shall show some properties of regular black di-rings, especially consider the thermodynamic regular black di-ring systems (the states in which both isothermality and isorotation between the inner black ring and the outer black ring are realized). For one of the main purposes to study the higher dimensional gravity is to clarify the phase structure of thermal states of higher dimensional black objects and classify the properties into the universal ones independent of the dimensional number of spacetimes and the peculiar ones depending on the dimensional number. To accomplish this task the study of thermodynamic multihorizon systems is clearly necessary because the multihorizon systems appear naturally in higher dimensions. The thermodynamic black di-ring system may serve as the starting point for the study of thermodynamic multihorizon systems.

For the systems of the black di-rings, we consider ve dimensional spacetimes with three commuting Killing vector elds:a time-like Killing vector eld and two axial-Killing vector elds. We assume further that one of the axial-Killing vector elds is orthogonal to the other. So the line-elements adopted here can be reduced to

$$ds^2 = G_{tt}(dt)^2 + 2G_{t\psi}dtd\psi + G_{\psi\psi}(d\psi)^2 + G \quad (d \quad)^2 + e^{2\nu} \quad d\rho^2 + dz^2 \quad , \tag{1}$$

where the metric coe cients are the functions of (ρ, z) and $\det G = \quad \rho^2$ is imposed. Based on this metric form, we can construct regular di-ring systems using both the solitonic methods mentioned above. It is noteworthy that the original representation of di-ring solutions which was rst found by us can be directly obtained by PISM [16].

Here we omit the explanation of the solution-generating methods, the expressions of the metric form and physical quantities of the di-ring solutions and also the detail of how to con rm the equivalence of the two di erent solution-sets of the black di-rings (one was generated by the authors and the other was by Evslin and Krishnan). In section 2 we show the existence of regular thermodynamic black di-ring systems,and also comment some properties of the thermodynamic black di-rings. In section 3 we give the discussion about thermodynamic stability (instability) of the systems.

2. Existence of regular thermodynamic black di-rings

In this place we show the existence of thermodynamic black di-rings (i.e., systems of regular black di-ring in thermodynamic equilibrium) and their properties. Similar results are also found in [17]. First we must impose the following balance conditions between gravitational attractions and centrifugal forces on the di-ring solution to eliminate conical singularities on the axes:

$$_{\mathrm{L}} = 2 \quad , \qquad _{\mathrm{R}} = 2 \quad . \tag{2}$$

Here $_{\mathrm{L}}$ and $_{\mathrm{R}}$ are the periodic angles to keep regularity on the outer and inner -axes respectively as depicted in Fig. 1.

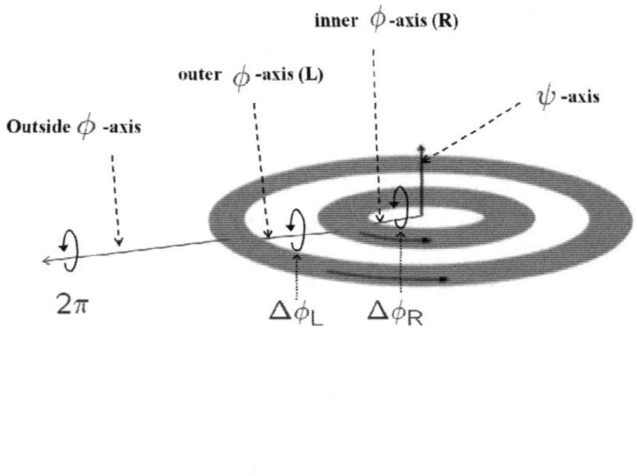

Figure 1. Schematic gure of di-ring. The di-ring has two rotational axes: -axis and ψ-axis.

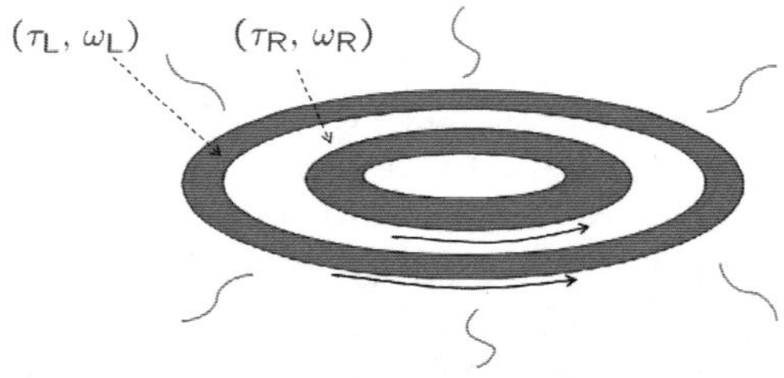

Figure 2. Schematic gure of thermodynamic di-ring. The outer ring(L) and inner ring(R) have the same horizon temperature and angular velocity.

To ensure that a system of regular black di-ring becomes a thermodynamic equilibrium state, the temperatures ($_{\rm L}$, $_{\rm R}$) and angular velocities ($\omega_{\rm L}$, $\omega_{\rm R}$) of the outer (L) and inner (R) black rings of the system must satisfy the following condition,

$$_{\rm L} = \ _{\rm R}\,, \quad \omega_{\rm L} = \omega_{\rm R}\,. \tag{3}$$

It should be noticed that in contrast with the case of black Saturn, to construct a thermodynamic black di-ring with two black rings is not so trivial as pointed out in [18]. Actually black rings with the same temperature and angular velocity have the same shape and size [19], so that we must arrange two identical rings into a thermodynamic di-ring keeping the conditions (2) and (3). That is, it seems rather natural that thermodynamic di-rings do not appear in the phase space. If there remains any possibility of thermodynamic di-rings, some non-linearity may play a key role.

Fortunately we can nd appropriate four independent moduli-parameters, and essentially solve the above four equations of the conditions (2) and (3) in an analytic way (see [16] for more details). So we can easily con rm the existence of regular thermodynamic black di-rings. In Fig. 3 and Fig. 4 the states of thermodynamic black di-ring appear as a continuous thick curve with a cusp. Other thermodynamic black objects are also shown for comparison.

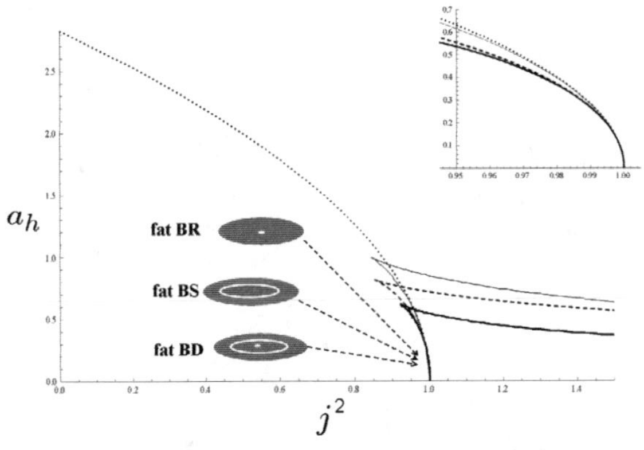

Figure 3. Total area a_h vs squared angular momentum j^2. Dotted, thin, dashed, and thick curves correspond to MP black hole (MP-BH), black ring (BR), black Saturn (BS) and black di-ring (BD) respectively.

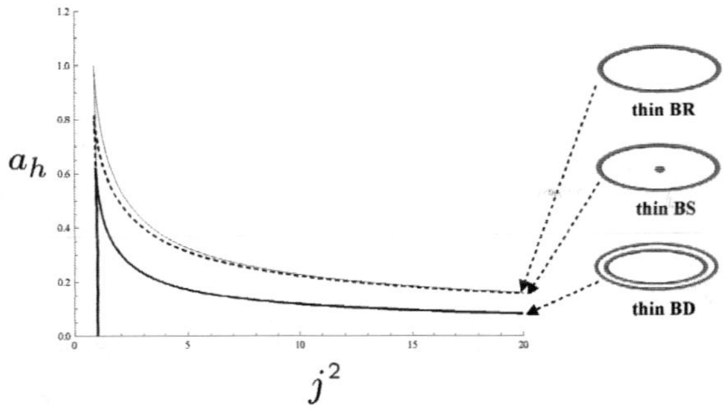

Figure 4. Total area a_h vs squared angular momentum j^2. The behavior of the black objects in the thin branch is shown.

The behavior of thermodynamic di-ring is similar to other thermodynamic objects. The phase of the di-ring has a 'fat ring' branch and a 'thin ring' branch. The state of the di-ring is less dominant than others with respect to entropy (i.e., area a_h) in both branches. In the fat branch, the curves of black ring, black Saturn and black di-ring approach that of MP black hole respectively as depicted in Fig. 3. More detailed behavior of the four black objects is shown in the upper right diagram magni ed near the extremal point. From the graphs in Fig. 4, in the thin branch the curve of black Saturn immediately asymptotes to that of black ring while the black di-ring acts independently.

3. Thermal stability (instability) of thermodynamic black di-rings

Next, as one of important properties of thermodynamic di-rings we discuss whether thermodynamic local stability is realized or not. We follow the approach that was introduced by Evslin and Krishnan to treat the case of black Saturn [20]. They found the existence of meta-stable states of black Saturn. To do this, under the condition of xed mass and angular momentum we shall search for local maxima of the corresponding entropy function, which is a function of appropriate moduli-parameters. If we nd that some maximal eigenvalues of the Hessian of the entropy function becomes negative, we can say that meta-stability occurs at this point. The results of the survey for the black di-ring and black Saturn are shown in Fig. 5 and Fig. 6 respectively. Any stability does not appear in the di-ring case, while the window of meta-stability opens in the black Saturn case as pointed in [20].

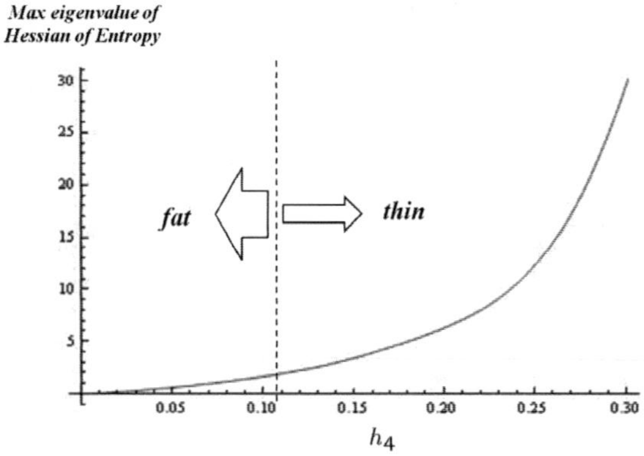

Figure 5. Max eigenvalue of Hessian of entropy of black di-ring vs one of moduli-parameters (h_4)

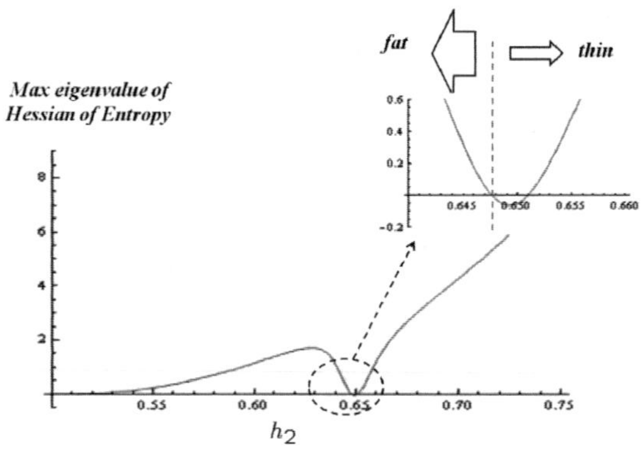

Figure 6. Max eigenvalue of Hessian of entropy of black Saturn vs one of moduli-parameters (h_2)

Acknowledgments

This work is partially supported by Grant-in-Aid for Young Scientists (B) (No. 20740143) from Japanese Ministry of Education, Science, Sports, and Culture.

References

[1] Emparan R and Reall H S 2002 *Phys. Rev. Lett.* **88** 101101
[2] Pomeransky A A 2006 *Phys. Rev.* **D73** 044004
[3] Iguchi H and Mishima T 2006 *Phys. Rev.* **D73** 121501(R)
[4] Tomizawa S and Nozawa M 2006 *Phys. Rev.* **D73** 124034
[5] Mishima T and Iguchi H 2006 *Phys. Rev.* **D73** 044030
[6] Iguchi H and Mishima T 2006 *Phys. Rev.* **D74** 024029
[7] Tomizawa S, Morisawa Y and Yasui Y 2006 *Phys. Rev.* **D73** 064009
[8] Pmeransky A A and Sen'kov R A 2006 *Preprint* hep-th/0612005
[9] Elvang H and Figueras J 2007 *JHEP* **05** 050
[10] Iguchi H and Mishima T 2007 *Phys. Rev.* **D75** 064018
[11] Evslin J and Krishnan C 2009 *Class. Quantum Grav.* **26** 125018
[12] Izumi K 2008 *Prog. theor. Phys.* **119** 757
[13] Elvang H and Rodrigues M J 2008 *JHEP* **04** 045
[14] Evslin J 2008 *JHEP* **09** 004
[15] Chen Y and Teo E 2008 *Phys. Rev.* **D78** 064062
[16] Iguchi H and Mishima T 2010 *Phys. Rev.* **D82** 084009
[17] Emparan R and Figueras J 2010 *JHEP* **11** 022
[18] Elvang H, Emparan R and Figueras J 2007 *JHEP* **05** 056
[19] Elvang H, Emparan R and Virmani J 2006 *JHEP* **12** 074
[20] Evslin J and Krishnan C 2008 *JHEP* **09** 003

Journal of Physics: Conference Series **314** (2011) 012113 doi:10.1088/1742-6596/314/1/012113

Stable Bound Orbits around Black Rings

Takahisa Igata, Hideki Ishihara and Yohsuke Takamori

Department of Mathematics and Physics,
Graduate School of Science, Osaka City University,
Osaka 558-8585, Japan

E-mail: igata@sci.osaka-cu.ac.jp

Abstract. We study stable bound orbits of a free particle around a black ring. Unlike the higher-dimensional black hole case, we find that there exist stable bound orbits in toroidal spiral shape near the ring axis and stable circular orbits on the axis. In addition, radii of stable bound orbits can be infinitely large if the ring thickness is less than a critical value.

1. Introduction

The issue with regard to higher dimensional black objects is presently under active investigation by many researchers. In particular, higher-dimensional black object solutions of the vacuum Einstein theory and various supergravity theories play crucial roles in modern physics [1]. In higher dimensions, there is a wide variety of black hole solutions in contrast to four dimensions. In the framework of the vacuum Einstein gravity, in addition to a generalization of the Kerr metric derived by Myers and Perry in [2], we have the metrics of the black object solutions with various topology of event horizon [3]. Indeed, Emparan and Reall found the five-dimensional black ring solution which has horizon topology $S^2 \times S^1$ in [4]. The existence of the black ring solution shows multiplicity of higher-dimensional black objects. Namely, the black hole uniqueness theorem does not hold in higher-dimensional spacetime as the same form in the case of four-dimensional black hole.

Among many ways of investigating black objects, the study of a free particle motion is a basic approach to understanding physical features of black objects. In recent years, free particle motion in the black ring geometry is studied by several authors [5]. In particular, Hoskisson has investigated extensively geodesics in the black ring geometry in [6]. He has shown that the separation of variable of the Hamilton-Jacobi equation for geodesics does not occur in the ring coordinates.

In this report, we then will focus on existence of stable bound orbits around a black ring. It is worth pointing out that there is no stable bound circular orbit in the higher-dimensional Schwarzschild solution. In contrast, we show that there exist stable bound orbits far from a black ring. The difference of geometry between a black hole and a black ring is clearly distinguishable by geodesic motions.

2. Black Ring Geometry

We begin by giving the geometry of the black ring. The metric is written in the form

$$\begin{aligned} ds^2 &= -\frac{F(y)}{F(x)}\left(dt - CR\frac{1+y}{F(y)}d\psi\right)^2 \\ &+\frac{R^2}{(x-y)^2}F(x)\left(-\frac{G(y)}{F(y)}d\psi^2 - \frac{dy^2}{G(y)} + \frac{dx^2}{G(x)} + \frac{G(x)}{F(x)}d\phi^2\right), \end{aligned} \tag{1}$$

with the range of the ring coordinates $-1 < x < 1$, $-\infty < y < -1$, where

$$F(\xi) = 1 + \lambda\xi, \quad G(\xi) = (1-\xi^2)(1+\nu\xi), \quad C = \sqrt{\lambda(\lambda-\nu)\frac{1+\lambda}{1-\lambda}}, \tag{2}$$

and R, ν, and λ are the parameters of the solution. The three parameters all have a physical interpretation: R, ν, and λ denote the ring radius, the thickness of the ring, the rotation velocity of the ring, respectively, where the parameter range is

$$R > 0, \quad 0 < \lambda \leq \nu < 1. \tag{3}$$

The parameter λ is determined by the regularity conditions at the two rotation axes of the ring in terms of ν as follows:

$$\lambda = \frac{2\nu}{1+\nu^2}. \tag{4}$$

The black ring metrics are stationary and axisymmetric, with Killing vectors ∂_t, ∂_ϕ, and ∂_ψ. The fixed point of ψ-rotation generated by ∂_ψ, which we refer to as the ring axis, exists at $y = -1$, and the fixed point of ϕ-rotation generated by ∂_ϕ, which we refer to as the equatorial plane, exists at $x = \pm 1$. Furthermore, this metric appears to have a singularity at $y = -1/\nu, -1/\lambda, -\infty$. Evaluation of curvature invariants such as $R_{abcd}R^{abcd}$ shows that the singularity at $y = -\infty$ is a true, curvature singularity. The singularities in the metric components at $y = -1/\nu$ and $y = -1/\lambda$ are coordinate singularities, corresponding to the event horizon and the ergo surface with $S^2 \times S^1$ topology, respectively.

3. Stable Bound Orbits in the Black ring geometry

We turn to a discussion of geodesic motion in the black ring geometry. Let p_μ be the 5-momentum of a particle with rest mass m, we define the Hamiltonian of geodesic motion as

$$H = \frac{1}{2}\left[\frac{(x-y)^2}{R^2}\frac{G(x)}{F(x)}p_x^2 - \frac{(x-y)^2}{R^2}\frac{G(y)}{F(x)}p_y^2 + E^2\left(U_{\text{eff}} + \frac{m^2}{E^2}\right)\right], \tag{5}$$

where

$$U_{\text{eff}} = -\frac{F(x)}{F(y)} - \frac{C^2(x-y)^2(y+1)^2}{G(y)F(x)F(y)} + \frac{(x-y)^2}{R^2G(x)}\ell_\phi^2 - \frac{F(y)(x-y)^2}{R^2G(y)F(x)}\ell_\psi^2 + 2\frac{C(x-y)^2(y+1)}{RG(y)F(x)}\ell_\psi \tag{6}$$

is the effective potential of the geodesic motion, and $E = -p_t$, $\ell_\phi = p_\phi/E$, and $\ell_\psi = p_\psi/E$, respectively. The three components, p_t, p_ϕ, and p_ψ, yield constants of motion since the black ring metric admits three commutable Killing vectors.

In what follows, in order to give intuitive picture of particle motion, we use ζ-ρ coordinates which are defined as

$$\zeta = R\frac{\sqrt{y^2-1}}{x-y}, \quad \rho = R\frac{1-x^2}{x-y}. \tag{7}$$

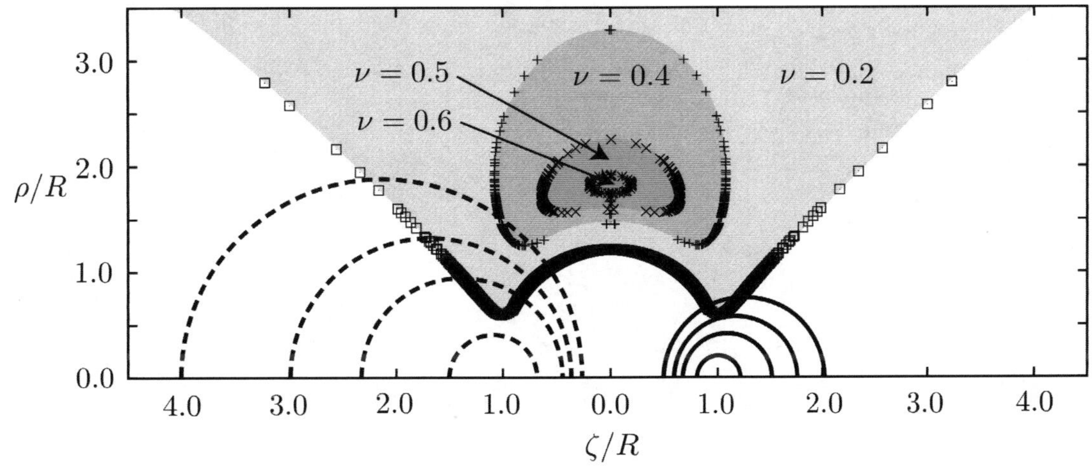

Figure 1. The shaded region shows the domain of stable bound orbits in the black ring spacetime for each ν. The dashed and solid semicircles show the ergo surfaces and the event horizon, respectively, where $\nu = 0.2, 0.4, 0.5$, and 0.6 from the small one to the large one.

In the flat limit, the black ring metric reduces to the flat metric in the form $ds^2 = -dt^2 + d\zeta^2 + \zeta^2 d\psi^2 + d\rho^2 + \rho^2 d\phi^2$.

Let us consider stationary particle motion determined by

$$U_{\text{eff}}(\zeta_{\text{st}}, \rho_{\text{st}}) + \frac{m^2}{E^2} = 0, \quad \frac{\partial U_{\text{eff}}}{\partial \zeta}(\zeta_{\text{st}}, \rho_{\text{st}}) = \frac{\partial U_{\text{eff}}}{\partial \rho}(\zeta_{\text{st}}, \rho_{\text{st}}) = 0, \tag{8}$$

i.e., $(\zeta_{\text{st}}, \rho_{\text{st}})$ are positions of the extrema of U_{eff}. Restricting attention to stable bound orbits, we impose in addition to equation (8) the requirement that

$$\det\mathcal{H}(\zeta_{\text{s}}, \rho_{\text{s}}) > 0, \quad \frac{\partial^2 U_{\text{eff}}}{\partial \rho^2}(\zeta_{\text{s}}, \rho_{\text{s}}) > 0, \tag{9}$$

where

$$\mathcal{H}(\zeta, \rho) = \begin{pmatrix} \frac{\partial^2 U_{\text{eff}}}{\partial \rho^2}(\zeta, \rho) & \frac{\partial^2 U_{\text{eff}}}{\partial \rho \partial \zeta}(\zeta, \rho) \\ \frac{\partial^2 U_{\text{eff}}}{\partial \zeta \partial \rho}(\zeta, \rho) & \frac{\partial^2 U_{\text{eff}}}{\partial \zeta^2}(\zeta, \rho) \end{pmatrix}, \tag{10}$$

so that $(\zeta_{\text{s}}, \rho_{\text{s}})$ denote positions of local minima of U_{eff}.

In Figure 1, the domains of $(\zeta_{\text{s}}, \rho_{\text{s}})$ are plotted by solving the conditions (8) and (9) numerically. The figure shows that there are stable bound orbits on and near the ring axis of the black ring geometry while there is no stable bound orbit on the equatorial plane. The orbits on the time slice of the Killing time have toroidal spiral shape near the ring axis and circular shape on the ring axis, which is generated by the two axial Killing vectors.

We discuss the asymptotic form of U_{eff} near the ring axis to confirm the existence of the stable bound orbit near infinity in analytical way. If $\ell_\psi = 0$, the form of U_{eff} expanded in ρ near asymptotic infinity on the ring axis is

$$U_\infty \equiv U_{\text{eff}}(\zeta = 0, \rho) \simeq -1 - \frac{4\nu R^2 - (1-\nu)\ell_\phi^2}{(1-\nu)^2}\frac{1}{\rho^2} + \frac{2\nu R^2[2(R^2+\zeta^2) - \ell_\phi^2]}{(1-\nu)^2}\frac{1}{\rho^4}. \tag{11}$$

By solving the equation $\partial U_\infty/\partial\rho = 0$ and imposing the two inequalities $4\nu R^2 - (1-\nu)\ell_\phi^2 > 0$ and $2R^2 - \ell_\phi^2 > 0$, we obtain the local minimum point on the ring axis in the form

$$\rho_{\rm s} = \sqrt{\frac{4\nu R^2(2R^2-\ell_\phi^2)}{4\nu R^2-(1-\nu)\ell_\phi^2}}, \tag{12}$$

where conditions (9) hold

$$\det\mathcal{H}(\zeta_{\rm s}=0,\rho_{\rm s}) = \frac{128\nu^2R^4(2R^2-\ell_\phi^2)}{(1-\nu)^4\rho_{\rm s}^{10}} > 0, \tag{13}$$

$$\frac{\partial^2 U_\infty}{\partial\rho^2}(\zeta_{\rm s}=0,\rho_{\rm s}) = \frac{[4\nu R^2-(1-\nu)\ell_\phi^2]^3}{4\nu^2R^4(1-\nu)^2(2R^2-\ell_\phi^2)^2} > 0. \tag{14}$$

If ℓ_ϕ^2 approaches to $4R^2\nu/(1-\nu)$ under the above conditions, the radius of the stable circular orbit ρ_s approaches infinity only in the cases $0<\nu<\nu_\infty := 1/3$. We note that the angular momentum $L_\phi = \ell_\phi E$ for the stable bound circular orbit is finite even if its radius is infinite.

4. Summary

In this short paper, we have discussed stable bound orbits of a free particle in the black ring geometry. By using the way of an effective potential of a free particle motion, we show the existence of stable bound orbits in the black ring spacetime which is a characteristic property of the black ring geometry unlike the black hole case. Furthermore, in the case of $0<\nu<\nu_\infty = 1/3$, we find that the radii of the stable circular orbits can be infinity. The basic mechanism for the appearance of stable bound orbits is different from the case of a four-dimensional black hole. They occur by a balance of Newton's gravitational term and subleading higher multipole term. Detailed discussion is found in our paper [7]. In addition, interesting issue of dynamical particle motion in the black ring geometry is discussed in [8].

References

[1] R. Emparan and H. S. Reall, Living Rev. Relativity 11, (2008), 6. URL (cited on 3 Jun. 2010): http://www.livingreviews.org/lrr-2008-6 .
[2] R. C. Myers and M. J. Perry, Annals Phys. **172**, 304 (1986)
[3] T. Mishima and H. Iguchi, Phys. Rev. D **73**, 044030 (2006)
H. Elvang and P. Figueras, JHEP **0705**, 050 (2007)
H. Iguchi and T. Mishima, Phys. Rev. D **75**, 064018 (2007) [Erratum-ibid. D **78**, 069903 (2008)]
J. Evslin and C. Krishnan, Class. Quant. Grav. **26**, 125018 (2009)
K. Izumi, Prog. Theor. Phys. **119**, 757 (2008)
H. Elvang and M. J. Rodriguez, JHEP **0804**, 045 (2008)
[4] R. Emparan and H. S. Reall, Phys. Rev. Lett. **88**, 101101 (2002)
[5] M. Nozawa and K. i. Maeda, Phys. Rev. D **71**, 084028 (2005)
H. Elvang, R. Emparan and A. Virmani, JHEP **0612**, 074 (2006)
A. Sahay and G. Sengupta, JHEP **0706**, 006 (2007)
[6] J. Hoskisson, Phys. Rev. D **78**, 064039 (2008)
[7] T. Igata, H. Ishihara and Y. Takamori, Phys. Rev. D **82**, 101501 (2010)
[8] T. Igata, H. Ishihara and Y. Takamori, in preparation

Journal of Physics: Conference Series **314** (2011) 012114 doi:10.1088/1742-6596/314/1/012114

The regular cosmic string in Born-Infeld gravity

Rafael Ferraro [1,2] **and Franco Fiorini** [1]

[1] Instituto de Astronomía y Física del Espacio,
Casilla de Correo 67, Sucursal 28, 1428 Buenos Aires, Argentina.
[2] Departamento de Física, Facultad de Ciencias Exactas y Naturales,
Universidad de Buenos Aires, 1428 Buenos Aires, Argentina.

E-mail: ferraro@iafe.uba.ar,franco@iafe.uba.ar

Abstract. It is shown that Born-Infeld gravity –a high energy deformation of Einstein gravity– removes the singularities of a cosmic string. The respective vacuum solution results to be free of conical singularity and closed timelike curves. The space ends at a minimal circle where the curvature invariants vanish; but this circle cannot be reached in a finite proper time.

We shall investigate high energy deformations of General Relativity (GR) based on the following three guiding principles:

a) The theory must reduce to General Relativity in the low field limit.

b) The spacetime dynamics must be described by second order equations.

c) It must provide a proper treatment in order to avoid singularities.

Many deformations can accomplish the condition (a), but just a few of them satisfy (b).[1] Lovelock's extension of GR does it [4], but it only departs from GR for dimensions bigger than four. In turn, a few candidates were proposed following the guidelines (a) and (c) (see e.g. the non-exhaustive list [5]-[10]). Remarkably, there is no conceptual framework accommodating the three requirements. To face the item (c) we will consider one of the prototypes of a GR singular structure: the circular symmetric vacuum solution in (2+1) dimensions [11, 12], or its (3+1)-dimensional analogue, the cosmic string [13]

$$ds^2 = d(t + 4J\,\theta)^2 - d\rho^2 - (1 - 4\mu)^2\,\rho^2\,d\theta^2 - dz^2. \tag{1}$$

In (2+1) dimensions (z is absent), this metric solves the Einstein equations for $T^{00} = \mu\,\delta(x, y)$ and $T^{0i} = (J/2)\,\epsilon^{ij}\,\partial_j\delta(x, y)$ (Cartesian coordinates x, y are defined as usual, and we use hereafter $G = 1$). So the solution looks as a particle of mass μ and spin J at the origin [12]. However, no gravitational field surrounds the particle since the metric is manifestly flat, as is typical of vacuum solutions to Einstein equations without cosmological constant in (2+1)-dimensions. Actually the particle shows itself by means of topological properties: i) the space displays a conical singularity, because the mass μ amounts a deficit angle $8\pi\mu$; ii) the Minkowskian time

[1] For instance, the so called "$f(R)$" theories [1], lead to fourth order equations. They can be reformulated as scalar-tensor theories to give second order equations; however this procedure results in violations of the equivalence principle [2, 3]. Moreover, vacuum singular GR solutions remain as solutions of smooth $f(R)$ theories, so threatening the requirement (c).

$t' \equiv t + 4J\,\theta$ suffers a jump $8\pi\, J$ when the circle is completed. Added to this singular structure, the *cosmon* –as named in Ref. [14]– also exhibits violation of causality because closed timelike curves (CTC) occur. In fact, the interval (1) on the closed curves of constant (t, ρ, z) is

$$ds^2 = \left(\frac{16J^2}{M^2} - \rho^2\right) M^2\, d\theta^2\,, \qquad M \equiv 1 - 4\mu, \tag{2}$$

which means that curves with radio $\rho < 4J/M$ are CTC. This unpleasant feature was avoided in Ref. [14] by resorting to proper boundary conditions. However, we will show that the entire singular structure of cosmons is prevented by a high energy deformation of GR developed in Refs. [15]-[17].

As the starting point for describing the gravitational field, we will propose a *determinantal* Born-Infeld (BI) Lagrangian density [17]

$$\mathcal{L}_{\mathbf{BIG}} = -\lambda/(16\,\pi)\,\left(\sqrt{|g_{\mu\nu} - 2\lambda^{-1}F_{\mu\nu}|} - \sqrt{|g_{\mu\nu}|}\right), \tag{3}$$

where $|\ |$ stands for the absolute value of the determinant. The structure (3) resembles the one used by Born and Infeld in their non linear electrodynamics [18]. In our case, the field tensor $F_{\mu\nu}$ will depend on the dynamical variable describing the gravitational field. The constant λ in Eq. (3) has the units of $F_{\mu\nu}$ and controls the low energy limit alluded in the requirement (a). In fact, since

$$\sqrt{|I - 2\lambda^{-1}F|} = 1 - \lambda^{-1}Tr(F) + \mathcal{O}(\lambda^{-2}), \tag{4}$$

then the Lagrangian (3) goes to

$$\mathcal{L}_{\mathbf{G}} \;=\; 1/(16\,\pi)\,\sqrt{|g_{\mu\nu}|}\;Tr(F), \tag{5}$$

when $F_{\mu\nu}$ takes values much smaller than λ. So, Born-Infeld gravity will reduce to GR if $Tr(F) = -R$. On the other hand, the requirement (b) implies that the quantities involved in the Lagrangian (3) should depend just on the dynamical variables and their first derivatives. This condition could seem to be out of tune with GR, since the curvature scalar depends on the second derivatives of its dynamical field (the metric). However, there is a beautiful way of circumventing this blind alley, by moving out towards an equivalent formulation of GR which depends just on first derivatives of its dynamical variables. Indeed, Einstein theory can be rephrased in a spacetime possessing absolute parallelism [19] where the *vielbein* $\{e^a(x)\}$ –a set of 1-forms– plays the role of the gravitational potentials, while the metric $\mathrm{g}(x) = \eta_{ab}\; e^a(x) \otimes e^b(x)$ is a subsidiary field. The absolute parallelism is ruled by the Weitzenböck connection $\Gamma^{\lambda}_{\mu\nu} \equiv e_a{}^{\lambda}\, e^a{}_{\mu,\,\nu}$, where $\{e_a{}^{\lambda}(x)\}$ is the vector basis dual to the *vielbein*. This connection is metric compatible and curvatureless: Weitzenböck spacetime is flat, but it possesses torsion $T^{\rho}{}_{\mu\nu} = e^{\rho}{}_a\,(e^a{}_{\nu,\,\mu} - e^a{}_{\mu,\,\nu})$ –i.e., $T^a = de^a$–, which is the agent where the gravitational degrees of freedom are encoded. The equivalence between both pictures reveals itself when one discovers that the Levi-Civita curvature scalar R can be rewritten as

$$4\,R[e^a] \;=\; 2\,W_1 \;-\; W_2 \;-\; 4\,W_3 \;+\; 8\,\mathrm{e}^{-1}\,(\mathrm{e}\,T^{\mu}{}_{\mu}{}^{\rho})_{,\rho}\;, \tag{6}$$

where $\mathrm{e} = \sqrt{|\mathrm{g}|}$ is the determinant of the matrix $e^a{}_{\mu}$, and the invariants W_i are

$$W_1 = T^{\mu\nu}{}_{\rho}\,T^{\rho}{}_{\mu\nu}\,,\quad W_2 = T_{\rho}{}^{\mu\nu}\,T^{\rho}{}_{\mu\nu}\,,\quad W_3 = T^{\rho\nu}{}_{\rho}\,T^{\mu}{}_{\mu\nu}\,. \tag{7}$$

Remarkably, the advantage of the teleparallel picture of gravity is that the total divergence term in Eq. (6) can be ruled out to work just with a first order Lagrangian $\mathcal{L}(e^a, \partial e^a)$,

Journal of Physics: Conference Series **314** (2011) 012114 doi:10.1088/1742-6596/314/1/012114

which will preserve the dynamical content of Einstein theory anyway. This distinctive feature makes Weitzenböck spacetime a privileged geometric structure to formulate modified theories of gravitation, since it guarantees that any modified Lagrangian in this language will assure second order field equations. This mechanism is essential to satisfy the requirement (b), otherwise fourth order differential equations would be obtained. Summarizing, we are building a theory of gravity where the fundamental piece is the set of 2-forms $T^a = de^a$ (notice the similarity with Yang-Mills theories), so guaranteeing the requirement (b). To accomplish the requirement (a) we should demand that

$$4\,Tr(F) \;=\; -2\,W_1 + W_2 + 4\,W_3 \;\equiv\; 4\,S_\rho{}^{\mu\nu}\,T^\rho{}_{\mu\nu}, \tag{8}$$

where

$$S_\rho{}^{\mu\nu} = -\frac{1}{4}\left(T^{\mu\nu}{}_\rho - T^{\nu\mu}{}_\rho - T_\rho{}^{\mu\nu}\right) + \frac{1}{2}\left(\delta^\mu_\rho\, T^{\theta\nu}{}_\theta - \delta^\nu_\rho\, T^{\theta\mu}{}_\theta\right). \tag{9}$$

The more general tensor $F_{\mu\nu}$ fulfilling the condition (8) reads $F_{\mu\nu} = \alpha\, S_{\mu\lambda\rho}\, T_\nu{}^{\lambda\rho} + \beta\, S_{\lambda\mu\rho}\, T^\lambda{}_\nu{}^\rho$, $\alpha + \beta = 1$. For simplicity we will set $\beta = 0$; thus, in four dimensions the theory reduces to

$$\mathcal{L}_{\mathbf{BIG}}[e^a] = -\lambda/(16\,\pi)\left(\sqrt{|g_{\mu\nu} - 2\lambda^{-1} S_{\mu\lambda\rho}\, T_\nu{}^{\lambda\rho}|} - \sqrt{|g_{\mu\nu}|}\right). \tag{10}$$

Let us now analyze the requirement (c). We will show that BI gravity (10) prevents the singular structure of the GR cosmon metric (1). First of all, we remark that the cosmon was endowed with spin J by considering the coordinate $t = t' - 4J\theta$ as the continuous time in the place of the Minkowskian time t'; in (2+1) dimensions, this procedure led to a non-zero T^{0i} at the origin. The mass μ came from a similar procedure. In BI gravity, instead, the equations determine not the metric but the vielbein; they contain more integration constants than GR does. In particular, J and $M \equiv 1 - 4\mu$ will appear as labels for the family of cylindrically symmetric vielbeins solving the dynamical equations. We found that the solution of the motion equations is given by the vielbein [17]

$$e^0 \,=\, d(t + 4J\theta)\,, \qquad e^1 \,=\, Y(\rho)\, d\rho\,, \qquad e^2 \,=\, \rho\, M\, d\theta\,, \qquad e^3 \,=\, dz, \tag{11}$$

where the function $Y(\rho)$ solves the cubic equation

$$Y^2(\rho) - Y^3(\rho) = -\frac{16\,J^2}{\lambda M^2}\left(\rho^2 - \frac{16\,J^2}{M^2}\right)^{-2}. \tag{12}$$

The vielbein (11) implies the metric

$$ds^2 = d(t + 4J\,\theta)^2 - Y^2(\rho)d\rho^2 - M^2\,\rho^2\,d\theta^2 - dz^2, \tag{13}$$

then the function $Y(\rho)$ constitutes the sole difference between GR and BI gravity. GR is recovered when $\lambda \to \infty$, as expected from the low energy limit (5). In fact, Eq. (12) says that $Y \to 1$ and the metric (13) acquires the flat form of Eq. (1). For finite values of λ and far away from the string, the low field regime is also restituted. Nevertheless, the BI cosmic string is a geometry with a J-depending non-constant scalar curvature $R = -2Y'/(\rho Y^3)$ depicted in Figure 1 for several values of λ.

According to Eq. (12), Y goes to infinity as ρ approaches the value $\rho_o = 4J/M$. At this minimal circle, R and other curvature invariants such as $R^{\mu\nu}R_{\mu\nu}$ and $R^\lambda{}_{\mu\nu\rho}\,R_\lambda{}^{\mu\nu\rho}$ become zero. Thus, the spacetime is flat not only far from the string (GR region) but at ρ_o as well. It is easy to verify that the proper time to reach ρ_o diverges [17], which means that BI gravity avoids the conical singularity. Moreover, the CTC's are prevented too. Even though the interval on the

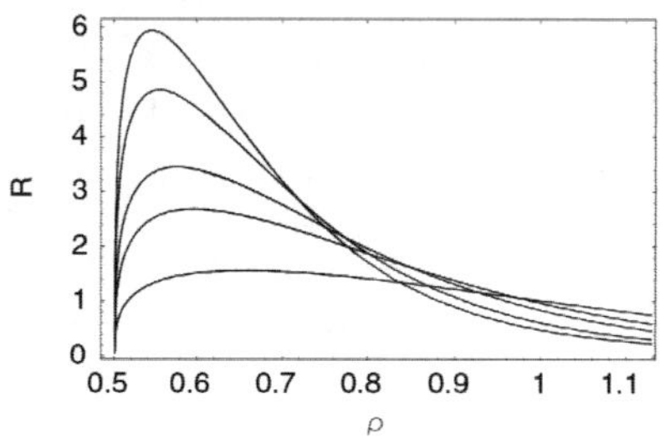

Figure 1. Scalar curvature $R(\rho)$ for $J/M = 1/8$ in units of length (so, R vanishes at $\rho_o = 1/2$). Following the maximum of the curves from bottom to top, it is $\lambda = 1, 3, 5, 10, 15$ in units of inverse square length.

closed curves of constant (t, ρ, z) does not differ from the one of Eq. (2) –since ρ is constant, then $Y(\rho)$ does not play any role–, however CTC's are forbidden because the lower bound of the radial coordinate ρ implies that the interval (2) is negative definite. In this way, the same mechanism responsible for the taming of the conical singularity in BI gravity also provides a natural chronological protection.

It is worth mentioning that the coordinate change $d\xi = Y(\rho)\, d\rho$ in the vielbein (11) allows to regard the curved geometry (13) as a space of a variable deficit angle ranging from $8\pi\mu$ at spatial infinity, to 2π at ρ_0. This feature might have important observational implications on the lensing effect. As another remarkable physical consequence, BI gravity seems to forbid the possibility of packing energy in arbitrarily small regions. Differing from GR, any junction of this vacuum solution with an inner solution has to be made at a radius bigger than $\rho_o = 4J/M$. This property is another manifestation of the Born-Infeld regularization program. As a final remark, we emphasize that the BI determinantal gravity (10) departs from a mere "$f(T)$" theory even at the lowest order in λ^{-1} (T stands for the invariant $S_\rho{}^{\mu\nu} T^\rho{}_{\mu\nu}$). This feature is essential for smoothing the cosmic string singularities, since the solution has $T = 0$ [16].

Acknowledgments
This work was supported by CONICET and Universidad de Buenos Aires.

References
[1] Capozziello S, De Laurentis M and Faraoni V 2009 A bird's eye view of $f(R)$-gravity *Preprint* gr-qc/0909.4672
[2] Brans C H 2005 The root of scalar-tensor theory: an approximate history *Preprint* gr-qc/0506063
[3] Olmo G J 2007 *Phys. Rev. Lett.* **98** 061101
[4] Lovelock J 1971 *J. Math. Phys.* **12** 498
[5] Deser S and Gibbons G W 1998 *Class. Quantum Grav.* **15** 35
[6] Feingenbaum J A 1998 *Phys. Rev.* D **58** 124023
[7] Comelli D and Dolgov A 2004 *JHEP* **0411** 062
[8] Güllü I, Şişman T Ç and Tekin B 2010 *Phys. Rev.* D **81** 104018
[9] Bañados M and Ferreira P 2010 *Phys. Rev. Lett.* **105** 011101
[10] Nieto J A 2004 *Phys. Rev.* D **70** 044042
[11] Staruszkiewicz A 1963 *Acta. Phys. Polon.* **24** 734
[12] Deser S, Jackiw R and 't Hooft G 1984 *Ann. Phys.* **152** 220
[13] Gott III J R 1985 *Ap. J.* **288** 422
[14] Deser S, Jackiw R and 't Hooft G 1992 *Phys. Rev. Lett.* **68** 267
[15] Ferraro R and Fiorini F 2007 *Phys. Rev.* D **75** 084031
[16] Ferraro R and Fiorini F 2008 *Phys. Rev.* D **78** 124019
[17] Ferraro R and Fiorini F 2010 *Phys. Lett.* B **692** 206
[18] Born M and Infeld L 1934 *Proc. R. Soc.* A **144** 425
[19] Hehl F W, McCrea J D, Mielke E W and Ne'eman Y 1995 *Phys. Rep.* **258** 1

Spanish Relativity Meeting (ERE 2010): Gravity as a Crossroad in Physics IOP Publishing
Journal of Physics: Conference Series **314** (2011) 012115 doi:10.1088/1742-6596/314/1/012115

Nonlinear gravitons from the initial value constraints of GR in Ashtekar variables

Eyo Eyo Ita III
Department of Applied Mathematics and Theoretical Physics
Centre for Mathematical Sciences, University of Cambridge, Wilberforce Road
Cambridge CB3 0WA, United Kingdom

E-mail: eei20@cam.ac.uk

Abstract. In this paper we provide a possible realization of Penrose's idea of nonlinear gravitons by constructing a solution to the initial value constraints in Ashtekar variables. The solution inputs are a spatial SU(2) connection and two free functions of position, and can be constructed as a formal operatorial expansion in powers of the cosmological constant about spacetimes of Petrov Type O. We first present the linear case, and then provide a simple nonlinear example to first order using a spatially homogeneous connection.

1. Introduction: The linear graviton

In [1] Roger Penrose takes issue with the standard view of the graviton as a weak-field perturbation of a background spacetime. He proposes the idea that each graviton should carry its measure of curvature, corresponding to a solution of the full nonlinear Einstein equations. Motivated by Penrose's idea, in this paper we will provide a possible realization of the concept of nonlinear gravitons using the initial value constraints of general relativity (GR). The initial value constraints of GR with cosmological constant Λ in the Ashtekar variables [2] are the Gauss' law, diffeomorphism and Hamiltonian constraints (G_a, H_i, H) given respectively by

$$D_i\widetilde{\sigma}^i_a = \partial_i\widetilde{\sigma}^i_a + f_{abc}A^b_i\widetilde{\sigma}^i_a = 0; \quad \epsilon_{ijk}\widetilde{\sigma}^j_aB^k_a = 0; \quad \epsilon_{ijk}\epsilon^{abc}\widetilde{\sigma}^i_a\widetilde{\sigma}^j_b\Big(\frac{\Lambda}{3}\widetilde{\sigma}^k_c + B^k_c\Big) = 0, \tag{1}$$

where A^a_i is a left-handed $SU(2)$ gauge connection and $\widetilde{\sigma}^i_a$ is a densitized triad.[1] We have also defined $B^i_a = \frac{1}{2}\epsilon^{ijk}(\partial_jA^a_k - \partial_kA^a_j + f^{abc}A^b_jA^c_k)$ as the magnetic field of A^a_i, and $D_iv_a = \partial_iv_a + f_{abc}A^b_iv_c$ is the $SU(2)_-$ covariant derivative with structure constants $f_{abc} = \epsilon_{abc}$. Define the Ansatz

$$\widetilde{\sigma}^i_a = -\Big(\frac{3}{\Lambda}\delta_{ae} + \epsilon_{ae}\Big)B^i_e, \quad (\text{det}B) \neq 0, \tag{2}$$

where $\epsilon_{ae} \in SU(2) \otimes SU(2)$. Substitution of (2) into (1) yields

$$B^i_eD_i\{\epsilon_{ae}\} = 0; \quad \epsilon_{dae}\epsilon_{ae} = 0; \quad \text{tr}\epsilon + \frac{\Lambda}{3}Var\epsilon + \frac{\Lambda^2}{3}\text{det}\epsilon = 0, \tag{3}$$

[1] Symbols from the beginning part of the Latin alphabet $a, b, c, \ldots$ denote internal indices, and from the middle $i, j, k, \ldots$ denote spatial indices.

where we have defined $VarM = (\text{tr}M)^2 - \text{tr}M^2$. To obtain the third equation of (3) we have cancelled off a factor of $\text{det}B \neq 0$ as well as any numerical pre-factors. Defining vector fields $\mathbf{v}_e = B^i_e \partial_i$ and the magnetic 'helicity density matrix' $C_{be} = A^b_i B^i_e$, then the Gauss' law constraint G_a can be written as

$$B^i_e D_i \epsilon_{ae} = \mathbf{v}_e\{\epsilon_{ae}\} + (f_{abf}\delta_{ge} + f_{ebg}\delta_{af})C_{be}\Psi_{fg} \equiv \mathbf{w}_e\{\epsilon_{ae}\} = 0. \tag{4}$$

Note for $\epsilon_{ae} = 0$ that (3) is identically satisfied, which corresponds to a particular class of spacetimes of Petrov Type O since all three eigenvalues of ϵ_{ae} are equal. Let us first consider the case of a linear gravitational wave propagating on these background spacetimes. Choose $|\epsilon_{ae}| << \frac{1}{\Lambda}$ as a perturbation and choose a connection $A^a_i = \delta^a_i \alpha + a^a_i$, where $|a^a_i| << \alpha$. Then we substitute this Ansatz into (3) and expand to linear order in ϵ_{ae} and a^a_i. The magnetic field to linearised order is given by $B^i_e = \epsilon^{ijk}\partial_j a^e_k - \alpha a^i_e + \delta^i_e(\alpha^2 + \alpha \text{tr}a)$. Since (3) is already of linearised order in ϵ_{ae}, then we need only expand B^i_e and C_{ae} to zeroth order in a^a_i, since all terms of the form $\epsilon\epsilon$, ϵa and aa are of second order. So we only need $\mathbf{v}_e = B^i_e \partial_i \sim \alpha^2 \partial_e$ and $C_{ae} \sim \delta_{ae}\alpha^3$. Note that using $C_{be} \propto \delta_{be}$ causes the terms in brackets in (4) to drop out due to antisymmetry of the structure constants f_{abc}. Then the linearized counterparts to (3) to zeroth order in Λ reduce to

$$\partial_e \epsilon_{ae} = 0; \quad \epsilon_{dae}\epsilon_{ae} = 0; \quad \text{tr}\epsilon = 0. \tag{5}$$

Equation (5) states that the perturbation ϵ_{ae} is transverse, traceless and symmetric, namely that it corresponds to a linearized gravitational wave.

2. Generalization to the full nonlinear case

Having shown that the initial value constraints can produce linearized gravitons, we will now progress to the full nonlinear case. The full Hamiltonian constraint can be written in the form

$$\text{tr}\epsilon + \Lambda I^{aebf}\epsilon_{ae}\epsilon_{bf} + \Lambda^2 E^{aebfcg}\epsilon_{ae}\epsilon_{bf}\epsilon_{cf} = 0, \tag{6}$$

where we have defined $I^{aebf} = \frac{1}{3}(\delta^a_b\delta^e_f - \delta^a_f\delta^b_e)$ and $E^{aebfcg} = \frac{1}{3}\epsilon^{abc}\epsilon^{efg}$. Defining e^f_{ae} and E^f_{ae} as an orthogonal basis of diagonal and off-diagonal symmetric matrix elements, let us write ϵ_{ae} in the matrix form

$$\epsilon_{ae} = e^f_{ae}\varphi_f + E^f_{ae}\Psi_f = \begin{pmatrix} \varphi_1 & \Psi_3 & \Psi_2 \\ \Psi_3 & \varphi_2 & \Psi_1 \\ \Psi_2 & \Psi_1 & \varphi_3 \end{pmatrix}.$$

Then defining $C_{[ae]} \equiv C_{ae} - C_{ea}$, the Gauss' law constraint (4) is given by $e^f_{ae}\mathbf{w}_e\{\varphi_f\} + E^f_{ae}\mathbf{w}_e\{\Psi_f\} = 0$ with matrix form

$$\begin{pmatrix} \mathbf{v}_1 - C_{[23]} & -C_{32} & C_{23} \\ C_{31} & \mathbf{v}_2 - C_{[31]} & -C_{13} \\ -C_{21} & C_{12} & \mathbf{v}_3 - C_{[12]} \end{pmatrix} \begin{pmatrix} \varphi_1 \\ \varphi_2 \\ \varphi_3 \end{pmatrix}$$

$$+ \begin{pmatrix} C_{22} - C_{33} & \mathbf{v}_3 - C_{12} + 2C_{21} & \mathbf{v}_2 - 2C_{31} + C_{13} \\ \mathbf{v}_3 - 2C_{12} + C_{21} & C_{33} - C_{11} & \mathbf{v}_1 - C_{23} + 2C_{32} \\ \mathbf{v}_2 - C_{31} + 2C_{13} & \mathbf{v}_1 - 2C_{23} + C_{32} & C_{11} - C_{22} \end{pmatrix} \begin{pmatrix} \Psi_1 \\ \Psi_2 \\ \Psi_3 \end{pmatrix} = 0.$$

Assuming invertibility of the matrix of differential operators $E^f_{ae}\mathbf{w}_e$, then we can define $\hat{J}^g_f = -(E^f_{af'}\mathbf{w}_{f'})^{-1}(e^g_{ag'}\mathbf{w}_{g'})$ as a propagator from the diagonal to the off-diagonal elements of ϵ_{ae}. This enables one formally to rewrite the Gauss' law constraint as an embedding map

$$\Psi_f = \hat{J}^g_f\varphi_g \longrightarrow \epsilon_{ae} = (e^g_{ae} + E^f_{ae}\hat{J}^g_f)\varphi_f \equiv \hat{T}^f_{ae}\varphi_f = \hat{T}^f_{ae}[A]\varphi_f. \tag{7}$$

Then defining $Q^{fg} \equiv I^{aebf}\hat{T}^f_{ae}\hat{T}^g_{bf}$ and $Q^{fgh} \equiv E^{aebfcg}\hat{T}^{f'}_{ae}\hat{T}^{g'}_{bf}\hat{T}^{h'}_{cg}$ the Hamiltonian constraint on solutions to the Gauss' law constraint becomes

$$\varphi_3 = -\varphi_1 - \varphi_2 - \Lambda Q^{f'g'}\varphi_{f'}\varphi_{g'} - \Lambda^2 Q^{f'g'h'}\varphi_{f'}\varphi_{g'}\varphi_{h'} \equiv \Lambda Q[\vec{\varphi}; A]. \tag{8}$$

This can be solved by fixed point iteration by defining a sequence $(\varphi_3)_{(n)}$ where $(\varphi_3)_{(0)} = -\varphi_1 - \varphi_2$ and the recursion relation $(\varphi_3)_{(n+1)} = Q[\varphi_1, \varphi_2, (\varphi_3)_{(n)}; A]$. Then the full solution, if convergent, is given by $\varphi_3 = \lim_{n\to\infty}(\varphi_3)_{(n)}$. So the prescription for writing down a solution starts by making a choice of connection A^a_i. This defines the Gauss' law propagator $\hat{J}^g_f = \hat{J}^g_f[A]$ and the integrodifferential matrix operator for the embedding map $\hat{T}^a_{ef} = \hat{T}^a_{ef}[A]$. Then one chooses two free functions φ_1 and φ_2, and constructs the solution

$$\epsilon_{ae} = \hat{T}^1_{ae}\varphi_1 + \hat{T}^2_{ae}\varphi_2 + \hat{T}^3_{ae}\varphi_3[\varphi_1, \varphi_2; A] = \epsilon_{ae}[\varphi_1, \varphi_2; A]. \tag{9}$$

In (9) φ_1 and φ_2 constitute the two physical degrees of freedom, and the connection A^a_i forms an input into $\hat{T}^f_{ae}$ and into φ_3 through the latter.

3. Example: spatially homogeneous connection

We will demonstrate using a spatially homogeneous connection A^a_i, with $\varphi_f = n_f e^{\vec{k}\cdot\vec{r}}$ and $\Psi_f = m_f e^{\vec{k}\cdot\vec{r}}$. For simplicity, we have chosen m_f, n_f and k_f as numerical constants. The following objects follow from A^a_i

$$B^i_a = (\det A)(A^{-1})^i_a;\quad C_{ae} = (\det A)\delta_{ae};\quad \mathbf{v}_a = (\det A)(A^{-1})^i_a\partial_i, \tag{10}$$

and the vector fields $\mathbf{v}_a$ have the following action

$$\mathbf{v}_a\{\varphi_f\} = (\det A)e_a\varphi_f;\quad \mathbf{v}_a\{\Psi_f\} = (\det A)e_a\Psi_f, \tag{11}$$

where we have defined $e_a \equiv (A^{-1})^i_a k_i$. After dividing by $(\det A) \neq 0$, then the Gauss' law constraint propagator $\hat{J}^g_f$ consists of c-numbers, given by

$$\hat{J}^g_f = \frac{1}{2}\begin{pmatrix} (e_1)^2(e_2e_3)^{-1} & -e_2(e_1)^{-1} & -e_3(e_2)^{-1} \\ -e_1(e_3)^{-1} & (e_2)^2(e_3e_1)^{-1} & -e_3(e_1)^{-1} \\ -e_1(e_2)^{-1} & -e_2(e_1)^{-1} & (e_3)^2(e_1e_2)^{-1} \end{pmatrix}.$$

In this case the action of the Gauss' law propagator $\hat{J}^g_f$ is algebraic, due to the functional form of ϵ_{ae}. So (4) is the same as

$$m_1 = \frac{1}{2}\Big[\Big(\frac{(e_1)^2}{e_2e_3}\Big)n_1 - \Big(\frac{e_2}{e_3}\Big)n_2 - \Big(\frac{e_3}{e_2}\Big)n_3\Big];$$
$$m_2 = \frac{1}{2}\Big[-\Big(\frac{e_1}{e_3}\Big)n_1 + \Big(\frac{(e_2)^2}{e_3e_1}\Big)n_2 - \Big(\frac{e_3}{e_1}\Big)n_3\Big];$$
$$m_3 = \frac{1}{2}\Big[-\Big(\frac{e_1}{e_2}\Big)n_1 - \Big(\frac{e_2}{e_1}\Big)n_2 + \Big(\frac{(e_3)^2}{e_1e_2}\Big)n_3\Big]. \tag{12}$$

Using (12), after some algebra we have the following relations

$$Var\epsilon = 2(n_1n_2 + n_2n_3 + n_3n_1 - (m_1)^2 - (m_2)^2 - (m_3)^2)$$
$$= |e|^2\Big[\frac{n_1n_2}{(e_3)^2} + \frac{n_2n_3}{(e_1)^2} + \frac{n_3n_1}{(e_2)^2} - \frac{1}{2}\Big(\frac{e_1}{e_2e_3}\Big)^2(n_1)^2 - \frac{1}{2}\Big(\frac{e_2}{e_3e_1}\Big)^2(n_2)^2 - \frac{1}{2}\Big(\frac{e_3}{e_1e_2}\Big)^2(n_3)^2\Big];$$
$$\det\epsilon = n_1n_2n_3 + 2m_1m_2m_3 - n_1(m_1)^2 - n_2(m_2)^2 - n_3(m_3)^3 = 0 \tag{13}$$

where we have defined $|e|^2 = (e_1)^2 + (e_2)^2 + (e_3)^2$. Note that the determinant of ϵ_{ae} vanishes. Substitution of (13) into (6) yields

$$n_3 = -n_1 - n_2 - \Lambda Q[\vec{n}; \vec{e}]; \quad Q = \frac{1}{3}\sum_{f=1}^{3} I^{fgh}\Big[\Big(\frac{|e|}{e_f}\Big)^2 n_g n_h - \frac{1}{2}\Big(\frac{|e|e_f n_f}{e_g e_h}\Big)^2\Big], \tag{14}$$

where $I^{fgh} = 1$ for even and for odd permutations of $f \neq g \neq h$, and zero otherwise. Equation (14) is a quadratic algebraic equation, which can be solved in closed form for n_3 as a function of n_1 and n_2. But the whole point is to use the fixed point iteration procedure, which applies in the general case where A^a_i is not constant. To first order in Λ this is given by

$$(n_3)_{(1)} = -(n_1 + n_2) + \frac{\Lambda|e|}{3e_1e_2e_3}\Big[-\frac{1}{2}((e_3)^2 + (e_1)^2)n_1n_1$$
$$-\frac{1}{2}((e_2)^2 + (e_3)^2)n_2n_2 + ((e_1e_2)^2 - (e_3)^2|e|^2)n_1n_2\Big]. \tag{15}$$

This can be repeated to any desired order, producing an expansion in powers of Λ whose coefficients depend on (n_1, n_2). The full solution is

$$\widetilde{\sigma}^i_a = -\Big(\frac{\Lambda}{3}\delta_{ae} + (\hat{T}^1_{ae}n_1 + \hat{T}^2_{ae}n_2 + \hat{T}^3_{ae}n_3[n_1, n_2; \vec{e}])e^{\vec{k}\cdot\vec{r}}\Big)(A^{-1})^i_e(\det A). \tag{16}$$

4. Discussion

We have presented a formula for nonlinear gravitons by expansion about Type O spacetimes where $\det B \neq 0$. The solution is fixed by two free functions φ_1 and φ_2 and a connection A^a_i, and can be developed as an formal operatorial expansion by fixed point iteration. We propose this as a possible realization of Penrose's idea [1] at a formal level.

References

[1] Roger Penrose. 'The nonlinear graviton' First Award winning essage of the Gravity research Foundation (1975)'
[2] Ahbay Ashtekar. 'New perspectives in canonical gravity', (Bibliopolis, Napoli, 1988).

Spanish Relativity Meeting (ERE 2010): Gravity as a Crossroad in Physics IOP Publishing
Journal of Physics: Conference Series **314** (2011) 012116 doi:10.1088/1742-6596/314/1/012116

Generating Method Based on Conformal Invariance of the Maxwell Field

J Hruška and M Žofka

Institute of Theoretical Physics, Faculty of Mathematics and Physics, Charles University, V Holešovičkách 2, 180 00 Prague 8 - Holešovice, Czech Republic

E-mail: bukikekh@seznam.cz

Abstract. We use conformal transformation to generate solutions of Einstein-Maxwell equations from other seed electro-vacuum spacetimes. We utilize the fact that the source-free Maxwell field is conformally invariant in four dimensions. Although the transformation is necessarily identical for most seed spacetimes, the *pp*-waves can be non-trivially transformed and may, therefore, play an analogous role as in the Birkhoff theorem.

1. General Background

Finding exact solutions to Einstein equations is very difficult. Therefore, much attention has been paid to methods of generating new solutions from existing ones. One of them is to use a conformal transformation.

Our approach in this article is very simple—we start from a seed metric, which does not even have to represent a physically viable solution to Einstein equations. We define a positive scalar conformal factor varying across the manifold and produce the resulting metric by multiplying the original metric by the conformal factor. We now require the new metric to satisfy Einstein equations. In our case, we actually used a non-vacuum solution to Einstein-Maxwell equations as our seed. This has the advantage that the resulting metric also represents an electrovacuum solution as Maxwell equations are conformally invariant. Therefore, we do not need to deal with the electromagnetic part of Einstein-Maxwell equations. However, it may still happen that we actually do not obtain a new solution but the seed spacetime in a new coordinate system. Therefore, we must carefully check our solutions to make sure they cannot be coordinate-transformed back into the original spacetime.

2. Existing Theorems

There are a number of theorems on conformally related spacetimes. Brinkmann [1] dealt with conformally related Einstein spaces $R_{\mu\nu} = ag_{\mu\nu}$ with a a constant. He showed that any two distinct (properly) conformally related Einstein spaces are either two vacuum *pp*-waves, or Minkowski and (anti) de Sitter. His theorems were later generalized by Daftardar-Gejji [2] who generalized this theorem to the cases where the two Einstein tensors are equal and where they differ by a cosmological constant term. In the former case, both spaces are (not necessarily vacuum) *pp*-waves; in the latter, for perfect fluids with $\mu = 0$, both spaces are Robertson-Walker with equations of state $\mu + 3p = 0$ or $\mu = p$. Van den Bergh [3] established that the only null Einstein-Maxwell fields obtainable by a conformal transformation of a Ricci-flat solution are *pp*-waves. In all cases, the seed metric is Ricci-flat. We are interested in a case where the two spacetimes are solutions of full Einstein-Maxwell equations. Physically, this means that the causal structure of both spacetimes is identical in the corresponding regions. Yet, in general, it turned out that generating new solutions via a conformal transformation does not produce any interesting, non-trivial results. In our paper, we present the first non-trivial explicit example of two, non-vacuum solutions to Einstein equations that are conformally related. Moreover, we show that the two spacetimes are not isometric.

3. The generating method

We exploit the fact that, in a 4-dimensional spacetime, the action of a source-free Maxwell field is conformally invariant and we thus generate solutions of Einstein-Maxwell equations. The invariance tells us that if $F_{\mu\nu}$ is a solution of source-free Maxwell equations on a seed spacetime $(M, g_{\mu\nu})$, then it is also a solution on $(M, \Omega^2 g_{\mu\nu})$. It implies that if we begin with a solution of Einstein-Maxwell equation, then after applying a conformal transformation, it is only Einstein equations we have to worry about, Maxwell equations are satisfied automatically.

3.1. Apparatus

Let us consider a conformal transformation of the seed metric $g_{\mu\nu}$ [1]

$$\widetilde{g}_{\mu\nu} = \Omega^2 g_{\mu\nu} \;, \tag{1}$$

where Ω is a function on M. Then the following equation holds for conformally rescaled Ricci tensor $\widetilde{R}_{\mu\nu}$ computed from the new metric $\widetilde{g}_{\mu\nu}$

$$\widetilde{R}_{\mu\nu} = R_{\mu\nu} - \frac{2}{\Omega}\Omega_{;\mu\nu} + \frac{4}{\Omega^2}\Omega_{,\mu}\Omega_{,\nu} - \frac{1}{\Omega}\Box\Omega g_{\mu\nu} - \frac{1}{\Omega^2}\|d\Omega\|^2 g_{\mu\nu} \;, \tag{2}$$

where all covariant derivatives are taken with respect to the seed metric, $\Box\Omega := g^{\mu\nu}\Omega_{;\mu\nu}$ and $\|d\Omega\|^2 := g^{\mu\nu}\Omega_{,\mu}\Omega_{,\nu}$.

The stress-energy tensor of a source-free Maxwell field has the form

$$T_{\mu\nu} = F_{\mu\rho}F_{\nu}{}^{\rho} - \frac{1}{4}g_{\mu\nu}F_{\rho\sigma}F^{\rho\sigma} \tag{3}$$

and so it transforms as

$$\widetilde{T}_{\mu\nu} = \Omega^{-2}T_{\mu\nu} \;. \tag{4}$$

Suppose we have a triplet $(M, g_{\mu\nu}, F_{\mu\nu})$ which is a solution of Einstein-Maxwell equations. We can now write Einstein equations[2] for $(M, \widetilde{g}_{\mu\nu}, F_{\mu\nu})$:

$$\widetilde{R}_{\mu\nu} = 8\pi\widetilde{T}_{\mu\nu} = 8\pi\Omega^{-2}T_{\mu\nu} = \Omega^{-2}R_{\mu\nu} \;, \tag{5}$$

so the equations we will solve are $\Omega^2\widetilde{R}_{\mu\nu} - R_{\mu\nu} = 0$ or

$$(\Omega^2 - 1)R_{\mu\nu} - 2\Omega\Omega_{;\mu\nu} + 4\Omega_{,\mu}\Omega_{,\nu} - \Omega\Box\Omega g_{\mu\nu} - \|d\Omega\|^2 g_{\mu\nu} = 0 \tag{6}$$

in detail. This system is generally overdetermined, we have 10 equations for a single function Ω, however, in a special case, we might (and will) be able to solve it[3]. The trace of (6) yields a necessary condition $\Box\Omega = 0$, so Ω has to be a harmonic function.

[1] Our signature is (-,+,+,+).
[2] Note that the stress-energy tensor of a Maxwell field is traceless, so the Ricci scalar vanishes.
[3] Actually, there is always the solution $\Omega^2 \equiv 1$, but then the transformation is trivial.

3.2. Application

As could be expected, equation (6) does not allow for a non-trivial solution in most cases. We explicitly checked the Reissner-Nordström, Bonnor-Melvin, Bertotti-Robinson, Tariq-Tupper, and Ozsváth solutions and showed that none of these spacetimes are suitable seeds. However, let us now turn our attention to *pp*-waves, which are spacetimes admitting a covariantly constant null vector field k^μ. The metric can be written in the form

$$ds^2 = -2H(u,\xi,\bar{\xi})du^2 - 2dudv + 2d\xi d\bar{\xi} \ . \tag{7}$$

where $\xi = \frac{1}{\sqrt{2}}(x+iy)$ is a complex coordinate. In these coordinates, the covariantly constant null vector field is simply ∂/∂_v. Not all *pp*-waves carry a non-zero Maxwell field. In order for them to do so, the function H has to satisfy

$$H = f(\xi,u) + \bar{f}(\bar{\xi},u) + 8\pi F(\xi,u)\bar{F}(\bar{\xi},u) \tag{8}$$

and the Maxwell field is then $F_{\mu\nu} = k_{[\mu}F_{,\nu]} + c.c.$ The Ricci tensor of a *pp*-wave is given by $R_{\mu\nu} = 2H_{,\xi\bar{\xi}}k_\mu k_\nu$. Having all the necessary ingredients, we can try to solve (6) for Ω. It turns out that this generally overdetermined system has a nontrivial solution if and only if H can be written as

$$H = \xi\bar{\xi}\Phi^2(u) + (\xi+\bar{\xi})h(u) + g(u) \ , \tag{9}$$

where Φ^2, h, g are arbitrary real functions of u only. Thus, except for the flat case, the spacetimes we are interested in have a non-zero Ricci tensor. There is a transformation of coordinates [4] (p. 383) that preserves the metric in the form (7) which allows us to get rid of the last two terms. A *pp*-wave with such a form of H is called a plane wave and is conformally flat. In fact, McLenaghan *et al.* [5] proved that such a *pp*-wave is the only conformally flat null Einstein-Maxwell field. Then, Ω has to be a function of u only, too, and has to satisfy a second order ODE

$$\Omega\ddot{\Omega} - 2\dot{\Omega}^2 + \Phi^2(1-\Omega^2) = 0 \ , \tag{10}$$

where $\dot{\Omega} := d\Omega/du$. With Ω depending only on u, it can be shown, that the vector field ∂/∂_v is covariantly constant in the new spacetime $(M, \tilde{g}_{\mu\nu})$ as well, so the generated spacetime is again a *pp*-wave, undoubtedly a conformally flat one, in agreement with McLenaghan's results. A question naturally arises, whether the generated wave is generally different from (non-isometric to) the original one or we just generate different coordinate expressions of the same wave. To answer this question, let us examine the equivalence problem for an explicit pair of the seed and the generated wave. A good choice, that will allow us to solve (10) particularly easily is to start with the following *pp*-wave

$$ds^2 = -4\xi\bar{\xi}du^2 - 2dudv + 2d\xi d\bar{\xi} \ , \tag{11}$$

so $H = 2\xi\bar{\xi}$ and $\Phi^2 = 2$. A particular solution of (10) is then $\Omega = \tanh(u)$ and so the generated wave has the following line element

$$\widetilde{ds}^2 = \tanh^2(u)(-4\xi\bar{\xi}du^2 - 2dudv + 2d\xi d\bar{\xi}) \ . \tag{12}$$

In our case, the first covariant derivative of the Ricci tensor is sufficient to prove the non-equivalence. The original spacetime has $R_{\mu\nu} = 4k_\mu k_\nu$ and its covariant derivative vanishes, i.e. $R_{\mu\nu;\rho} = 0$. However, the new spacetime has $\tilde{R}_{\mu\nu} = 4\Omega^{-6}\tilde{k}_\mu\tilde{k}_\nu$ and the covariant derivative no longer vanishes because of the non-zero gradient of the conformal factor Ω, which proves the non-equivalence of ds^2 and $\widetilde{ds}^2$. It follows, that all the *pp*-waves with (9) split into equivalence classes defined by their Maxwell fields.

4. Conclusions and Outlook

To our knowledge, this is the only non-trivial example of applying a conformal transformation to a seed spacetime to produce a non-vacuum solution to Einstein equations. Furthermore, due to the properties of conformal transformations, it is clear that the source remains in the form of an (vacuum) electromagnetic field. The original and resulting spacetimes are explicitly shown not to be identical. Thus, our solutions form the only known pair of conformally related non-vacuum spacetimes. The prospects of using conformal transformations to produce new solutions to Einstein equations may be dim but at least we provided a specific example where this approach works although the resulting solution belongs to the same class as the seed spacetime.

It might be of interest to see whether there may be any other non-trivial example that would not involve *pp*-waves. Note that previous literature always relies exactly on this type of spacetimes. Are there any suitable seeds other than *pp*-waves? Can the method be generalized to be less restrictive?

Acknowledgments

The authors acknowledge funding and support by the following grants of the Czech Grant Agency GAČR-205/09/H033 and GAČR-202/08/0187, of the Charles University Grant Agency GAUK 135409, the project SVV 261301 of the Charles University in Prague, and by the Czech Ministry of Education under the projects MSM0021610860 and LC06014 GAČR-202/09/0772.

References

[1] Brinkmann H W 1925 Einstein spaces which are mapped conformally on each other. *Math. Ann.* **18** 119–45
[2] Daftardar-Gejji V 1998 A generalization of Brinkmann's theorem *Gen. Rel. Grav.* **30** 695–700
[3] Van den Bergh N 1986 Conformally Ricci flat Einstein-Maxwell solutions with a null electromagnetic field. *Gen. Rel. Grav.* **18** 1105–10
[4] Stephani H, Kramer D, MacCallum M, Hoenselaers C and Herlt E 2003 *Exact Solutions of Einstein's Field Equations* 2nd edn (Cambridge University Press)
[5] McLenaghan R G, Tariq N and Tupper B O J 1975 Conformally flat solutions of the Einstein-Maxwell equations for null electromagnetic fields *J. Math. Phys.* **16** 829–31

Journal of Physics: Conference Series **314** (2011) 012117 doi:10.1088/1742-6596/314/1/012117

Unruh Effect Revisited: Poincaré θ-Vacua as Coherent States of Conformal Zero Modes

M. Calixto[1,2], **E. Pérez-Romero**[2] **and V. Aldaya**[2]

[1] Departamento de Matemática Aplicada, Universidad de Granada, Facultad de Ciencias, Campus de Fuentenueva, 18071 Granada, Spain

[2] Instituto de Astrofísica de Andalucía (IAA-CSIC), Apartado Postal 3004, 18080 Granada, Spain

E-mail: calixto@ugr.es

Abstract. We report on a group-theoretical revision of the Unruh effect based on the conformal group $SO(4,2)$, which has been developed by the authors and collaborators. Special Conformal Transformations (SCT) are interpreted as transitions to relativistic uniformly accelerated frames. Poincaré invariant θ-vacua (which turn out to be coherent states of conformal zero modes) are destabilized by SCT and radiate as a black body.

1. Introduction

The Fulling-Davies-Unruh effect [1, 2, 3] has to do with *vacuum radiation* in a non-inertial reference frame and shares some features with the (black-hole) Hawking [4] effect. In simple words, whereas the Poincaré invariant vacuum $|0\rangle$ in QFT looks the same to any inertial observer (i.e., it is stable under Poincaré transformations), it converts into a thermal bath of radiation with temperature

$$T = \frac{\hbar a}{2\pi v k_B} \tag{1}$$

in passing to a uniformly accelerated frame (a denotes the acceleration, v the speed of light and k_B the Boltzmann constant).

This situation is always present when quantizing field theories in curved space as well as in flat space, whenever some kind of global mutilation of the space is involved (viz, existence of horizons). This is the case of the natural quantization in Rindler coordinates [2, 5], which leads to a quantization inequivalent to the normal Minkowski quantization (see next Section), or that of a quantum field in a box, where a dilatation produces a rearrangement of the vacuum [1].

In the reference [6], it was showed that the reason for the Planckian radiation of the Poincaré invariant vacuum under uniform accelerations (that is, the Unruh effect) is more profound and related to the spontaneous breakdown of the conformal symmetry in *quantum* field theory. From this point of view, a Poincaré invariant vacuum will be regarded as a coherent state of *conformal zero modes*, which are undetectable ("dark") by inertial observers but unstable under Special Conformal Transformations (SCT)

$$x^\mu \to x'^\mu = \frac{x^\mu + c^\mu x^2}{1 + 2cx + c^2 x^2}. \tag{2}$$

which can be interpreted as transitions to systems of relativistic, uniformly accelerated observers with acceleration $a = 2c$ (see e.g. Ref. [7, 8, 9]). Indeed, let use take for simplicity an acceleration along the "z" axis, $c^\mu = (0,0,0,\alpha)$, and the temporal path $x^\mu = (t,0,0,0)$. Then the transformation (2) reads:

$$t' = \frac{t}{1-\alpha^2 t^2}, \qquad z' = \frac{\alpha t^2}{1-\alpha^2 t^2}. \tag{3}$$

Writing z' in terms of t' gives the usual formula for the relativistic uniform accelerated (hyperbolic) motion:

$$z' = \frac{1}{a}(\sqrt{1+a^2 t'^2} - 1) \tag{4}$$

with $a = 2\alpha$. Let us say that at least two alternative meanings of SCT have also been proposed [10, 11]. One is related to the Weyl's idea of different lengths in different points of space time [10]: "the rule for measuring distances changes at different positions". Other is Kastrup's interpretation of SCT as geometrical gauge transformations of the Minkowski space [11].

The point of view exposed here is consistent with the idea that quantum vacua are not really empty to every observer. Actually, the quantum vacuum is filled with zero-point fluctuations of quantum fields. The situation is similar to quantum many-body condensed mater systems describing, for example, superfluidity and superconductivity, where the ground state mimics the quantum vacuum in many respects and quasi-particles (particle-like excitations above the ground state) play the role of matter. Moreover, we know that zero-point energy, like other non-zero vacuum expectation values, leads to observable consequences as, for instance, the Casimir effect, and influences the behavior of the Universe at cosmological scales, where the vacuum (dark) energy is expected to contribute to the cosmological constant, which affects the expansion of the universe (see e.g. [12] for a nice review). Indeed, dark energy is the most popular way to explain recent observations that the universe appears to be expanding at an accelerating rate.

In the papers [13] and [14] we construct a conformal-$SO(4,2)$-invariant quantum theory in 3+1 dimensions, giving the Hilbert space and an orthonormal basis for our conformal particle. The construction is based on an holomorphic square-integrable irreducible representation of the conformal group on the eight-dimensional phase space $\mathbb{D}_4 = SO(4,2)/SO(4)\times SO(2)$ inside the complex Minkowski space $\mathbb{C}^4$. In [15], the Poincaré invariance of the ground state is highlighted and the mean energy, partition function and entropy of the accelerated ground state (seen as a statistical ensemble) is explicitly calculated. This leads us to interpret the accelerated ground state as an Einstein Solid, to obtain a deviation from the Unruh's formula (1) and to discuss the existence of a maximal acceleration. Here we shall just deal with the second quantized (many-body) theory, where Poincaré-invariant (degenerated) pseudo-vacua are coherent states of conformal zero modes. Selecting one of this Poincaré-invariant pseudo-vacua spontaneously breaks the conformal invariance and leads to vacuum radiation.

2. Vacuum radiation as a spontaneous breakdown of the conformal symmetry

Let us offer an alternative explanation for the Unruh effect based on symmetry grounds. Actually, in Quantum Field Theory, the vacuum state is expected to be stable under some underlying group of symmetry transformations G (namely, the Poincaré group). Then the action of some spontaneously broken symmetry transformations can destabilize/excitate the vacuum and make it to *radiate*. We argue that this is precisely the case of the Planckian radiation of the Poincaré invariant vacuum under uniform accelerations. Here, the Poincaré invariant vacuum looks the same to every inertial observer but converts itself into a thermal bath of radiation in passing to a uniformly accelerated frame. In fact, we point out that the reason for this radiation is related to the spontaneous breakdown of the conformal symmetry in *quantum* field theory (see [6] and [15] for more details).

Let us denote by φ_n the energy E_n eigenstates. The Fourier coefficients a_n (and a_n^*) of the expansion in energy modes of a state

$$\phi = \sum_n a_n \varphi_n, \tag{5}$$

are promoted to annihilation $\hat{a}_n$ (and creation $\hat{a}_n^\dagger$) operators in second quantization. The fact that the ground state of first quantization, φ_0, is invariant under Poincaré transformations (see [6] and [15]) implies that the annihilation operator of zero-("dark")-energy modes $\hat{a}_0$ commutes with all Poincaré generators. It also commutes with all annihilation operators and creation operators

$$[\hat{a}_0, \hat{a}_n^\dagger] = 0, \; n > 0 \tag{6}$$

of particles with non-zero ("bright") energy. Therefore, by Schur's Lemma, $\hat{a}_0$ must behave as a multiple of the identity in the broken theory, which means that Poincaré "θ-vacua" fulfill

$$\hat{a}_0|\theta\rangle = \theta|\theta\rangle \Rightarrow |\theta\rangle = e^{\theta\hat{a}_0 - \bar{\theta}\hat{a}_0^\dagger}|0\rangle. \tag{7}$$

That is, Poincaré "θ-vacua" are *coherent states of conformal zero modes* (see [16] and [17] for a thorough exposition on coherent states).

In the first-quantized theory, the unitary transformation $\mathcal{U}(c)$ that implements the acceleration (2) does not leave invariant the ground state φ_0 (see [15]) . In fact, a decomposition of the "accelerated ground state" in energy modes is obtained:

$$\varphi_0' = \mathcal{U}(c)\varphi_0 = \sum_{n=0}^{\infty} \phi_n(c)\varphi_n, \tag{8}$$

where $\phi_n(c)$ denotes the probability amplitude of finding φ_0' in φ_n for a given acceleration c. In the second-quantized theory, this acceleration leads to a transformation of annihilation operators

$$\hat{a}_0' = \sum_{n=0}^{\infty} \phi_n(c)\hat{a}_n, \tag{9}$$

so that accelerated Poincaré θ-vacua become:

$$|\theta'\rangle = e^{\theta\hat{a}_0' - \bar{\theta}\hat{a}_0'^\dagger}|0\rangle. \tag{10}$$

The average number of particles with energy E_n in the accelerated vacuum (10) is then given by

$$N_n(c) = \langle\theta'|\hat{a}_n^\dagger\hat{a}_n|\theta'\rangle = |\theta|^2|\phi_n(c)|^2, \tag{11}$$

where $|\theta|^2$ is the total average number of particles in $|\theta\rangle$, and $|\phi_n(c)|^2$ is the probability of finding a particle in the energy state E_n of the accelerated vacuum $|\theta'\rangle$.

In the same way, the probability $P_n(q,c)$ of observing q particles with energy E_n in $|\theta'\rangle$ can be calculated as:

$$P_n(q,c) = |\langle q(n)|\theta'\rangle|^2 = \frac{e^{-|\theta|^2}}{q!}|\theta|^{2q}\,|\phi_n(c)|^{2q} = \frac{e^{-|\theta|^2}}{q!}N_n^q(c). \tag{12}$$

Therefore, the relative probability of observing a state with total energy E in the excited vacuum $|\theta'\rangle$ is:

$$P(E) = \sum_{\substack{q_0,\ldots,q_k:\\ \sum_{n=0}^{k} E_n q_n = E}} \prod_{n=0}^{k} P_n(q_n, c)\,. \tag{13}$$

For the case studied here, this distribution function can be factorized as $P(E) = \Omega(E)e^{-\tau E}$, where $\Omega(E)$ is a relative weight proportional to the number of states with energy E and the factor $e^{-\tau E}$ fits this weight properly to a temperature $T = k_B/\tau$.

One can also compute the mean energy

$$\mathcal{E}(c) = \langle\theta'|\sum_{n=1}^{\infty} E_n \hat{a}_n^\dagger \hat{a}_n|\theta'\rangle = |\theta|^2 \sum_{n=1} |\phi_n(c)|^2 E_n. \tag{14}$$

and see that it is indeed Planckian (see [6] and [15]) .

Acknowledgements

Work partially supported by the Fundación Séneca (08814/PI/08), Spanish MICINN (FIS2008-06078-C03-01) and Junta de Andalucía (FQM219, FQM1951). M. Calixto thanks the "Universidad Politécnica de Cartagena" and C.A.R.M. for the award "Intensificación de la Actividad Investigadora 2009-2010". "Este trabajo es resultado de la ayuda concedida por la Fundación Séneca, en el marco del PCTRM 2007-2010, con financiación del INFO y del FEDER de hasta un 80%".

References

[1] S.A. Fulling, Non-uniqueness of Canonical Field Quantization in Riemannian Space-Time, Phys. Rev. **D7** (1973) 2850-2862
[2] P.C.W. Davies, Scalar production in Schwarzschild and Rindler metrics, J. Phys. **A8** (1975) 609-616
[3] W. G. Unruh, Notes on black-hole evaporation, Phys. Rev. **D14** (1976) 870
[4] S.W. Hawking, Black-hole explosions?, Nature **248** (1974) 30
[5] N. D. Birrell and P. C. W. Davies, Quantum Fields in Curved Space, Cambridge Monographs on Mathematical Physics (1994).
[6] V. Aldaya, M. Calixto and J.M. Cerveró, Vacuum Radiation and Symmetry Breaking in Conformally Invariant Quantum Field Theory, Commun. Math. Phys. **200** (1999) 325-354
[7] E. L. Hill, On accelerated coordinate systems in classical and relativistic mechanics, Phys. Rev. **67** (1945) 358-363.
[8] L. J. Boya and J. M. Cerveró, Contact Transformations and Conformal Group. I. Relativistic Theory, International Journal of Theoretical Physics **12** (1975) 47-54
[9] T. Fulton, F. Rohrlich and L. Witten, Physical consequences to a coordinate transformation to a uniformly accelerationg frame, Nuovo Cimento **26** (1962) 652-671
[10] H. Weyl, "Space, Time and Matter". Dover. NY. First Edition 1922.
[11] H.A. Kastrup, Gauge properties of the Minkowski space, Phys. Rev. **150**, 1183 (1966).
[12] G.E. Volovik: Vacuum energy: myths and reality, Int. J. Mod. Phys. **D15** (2006) 1987-2010
[13] M. Calixto and E. Pérez-Romero, Extended MacMahon-Schwinger's Master Theorem and Conformal Wavelets in Complex Minkowski Space, Applied and Computational Harmonic Analysis (2010), DOI: 10.1016/j.acha.2010.11.004, arXiv:1002.3498, 40 pages
[14] M. Calixto and E. Pérez-Romero, Conformal Spinning Quantum Particles in Complex Minkowski Space as Constrained Nonlinear Sigma Models in U(2,2) and Born's Reciprocity, International Journal of Geometric Methods in Modern Physics **8**, N° 3, 33 pages (2011), to appear, arXiv:1006.5958
[15] M. Calixto, E. Pérez-Romero and V. Aldaya, Unruh Effect and the Spontaneous Breakdown of the Conformal Symmetry, arXiv:1012.4758
[16] A. Perelomov: Generalized Coherent States and Their Apllications, Springer-verlag (1986)
[17] J.R. Klauder and Bo-Sture Skagerstam: Coherent States: Applications in Physics and Mathematical Physics, World Scientific (1985)

Spanish Relativity Meeting (ERE 2010): Gravity as a Crossroad in Physics IOP Publishing
Journal of Physics: Conference Series **314** (2011) 012118 doi:10.1088/1742-6596/314/1/012118

Revisiting a vector-tensor theory of gravitation

R Dale[1,3], D Sáez[2,4]

[1]Departament de Física, Enginyeria de Sistemes i Teoria del Senyal, Universitat d'Alacant, 03690, San Vicent del Raspeig, Alacant, Espanya

[2]Departamento de Astronomía y Astrofísica, Universidad de Valencia, 46100 Burjassot, Valencia, Spain

E-mail: [3]rdale@dfists.ua.es, [4]diego.saez@uv.es

Abstract. A certain vector-tensor theory of gravitation has been recently studied. In this theory, the zero-order energy density of the vector field could play the role of dark energy. In such a case, the question is: could the theory explain current cosmological observations as well as the so-called concordance model? Previous papers on the subject only consider a reduced number of current observations. We consider a wider set of observations including supernovae of type Ia, cosmic microwave background anisotropies, and the power spectrum of the energy density fluctuations. Results imply that, for negligible scalar perturbations of the vector field, the theory does not work.

1. Introduction

A certain vector-tensor (VT) theory of gravitation is revisited with the essential aim of studying its compatibility with a wide set of current observations. It is assumed that the scalar fluctuations of the vector field are negligible (hypothesis to be analyzed elsewhere). Moreover, as it is done in [1,2], we also assume that the energy density of the vector field is the unique dark energy of the universe. This requires a large enough density parameter for the vector field energy and an appropriate equation of state. Predictions of this theory are to be compared either with observational data (χ^2 distribution and associated P-value) or with the predictions of the concordance model, which fit very well most cosmological observations, e.g., high red-shift SNe Ia observations [3-4], WMAP observations of the cosmic microwave background (CMB) anisotropies [5-6], and power spectra estimates based on galaxies surveys. The following parameters are assumed to define the concordance model: $h = 10^{-2} H_0 = 0.704$ (where H_0 is the Hubble constant in units of km s^{-1} Mpc^{-1}), density parameters $\Omega_b = 0.0461, \Omega_{cdm} = 0.2265$, $\Omega_\Lambda = 0.7261$ and $\Omega_K = -0.0012$, for the baryonic, dark matter, vacuum energy and curvature, respectively (so, the matter density is $\Omega_m = \Omega_b + \Omega_{cdm} = 0.2726$), a scalar spectral index $n_s = 0.96$, a reionization optical depth $\tau = 0.087$, a value $\sigma_8 = 0.8065$ normalizing the power spectrum, and an equation of state for the dark energy of the form $p = -\rho$. This model is the best fitting to h-measurements and baryonic acoustic oscillations [6], but it also explains the observations listed above. Quantities G and a are the gravitational constant and the scale factor, respectively. The units have been chosen in such way that the speed of light is $c = 1$.

2. The Vector-Tensor Theory

The theory may be obtained with a variational principle. The action is a particular case (parameters $\omega = 0$, $\varepsilon = 0$, $\tau = -8\pi G$, $\eta = 8\pi G$) of the general action proposed in the '70s by Will & Nordtvedt [7]. For an isotropic homogeneous background, the covariant components of the vector field are $\left(A_0(\tau), 0, 0, 0\right)$ and the field equations of the theory are:

(1)

$$0 = \ddot{A}_0 - 4\frac{\ddot{a}}{a}A_0,$$

$$3\frac{\dot{a}^2}{a^2} = 8\pi G a^2\left(\rho_b + \rho_A + \rho_K\right),\ \rho_A \equiv -\frac{1}{2a^4}\dot{A}_0^2 + \frac{4\dot{a}}{a^5}\dot{A}_0 A_0 - 2\frac{\dot{a}^2}{a^6}A_0^2,$$

$$3\frac{\ddot{a}}{a} = 4\pi G a^2\left[\left(\rho_b + \rho_A + \rho_K\right) - 3\left(p_b + p_A + p_K\right)\right],\ p_A \equiv -\frac{3}{2a^4}\dot{A}_0^2 + 4\frac{\dot{a}}{a^5}\dot{A}_0 A_0 - 2\left(2\frac{\ddot{a}}{a^5} + \frac{\dot{a}^2}{a^6}\right)A_0^2,$$

$$\rho_K = -3p_K \equiv -3K/8\pi G a^2,$$

where dots stand for derivatives with respect to the conformal time, and ρ_b, p_b are the energy density and pressure of the cosmological fluid, respectively.

3. Predictions of the VT theory

We fix $\Omega_K = -0.0012$, as in our version of the concordance model.

3.1. SNe Ia observations.

Once Ω_K has been fixed, the distance modulus redshift relation $\mu = \mu(z)$ only depends on h and Ω_m. The best fit for the VT theory corresponds to $\Omega_m = 0.36$ and $h = 0.73$ with a P-value of 0.82, to be compared with the P-value of 0.91 corresponding to the concordance model.

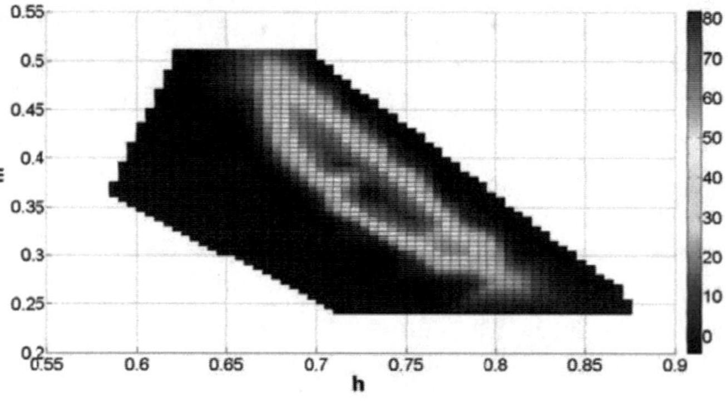

Figure 1. Statistical P-value (in percentage) on the plane $\left(h, \Omega_m\right)$. Points with P<0.05 are statistically ruled out. They correspond to the dark outer region; the admissible values of parameters Ω_m and h belong to the intervals (0.24, 0.5) and (0.65, 0.85), respectively.

3.2. CMB angular power spectrum

For each $\left(h, \Omega_m\right)$ pair compatible with SNe Ia observations, the relation $\Omega_m = \Omega_{cdm} + \Omega_b$ and the observational constraint $\Omega_b h^2 = 0.0213 \pm 0.0010$ (see [8]) almost fix parameters Ω_m and Ω_{cdm}. Other parameters as e.g., σ_8, n_S and τ_o might be fixed by comparing the CMB angular power spectrum from WMAP seven years data and the power spectrum P(k) obtained from galaxy surveys with the corresponding predictions of the VT theory.

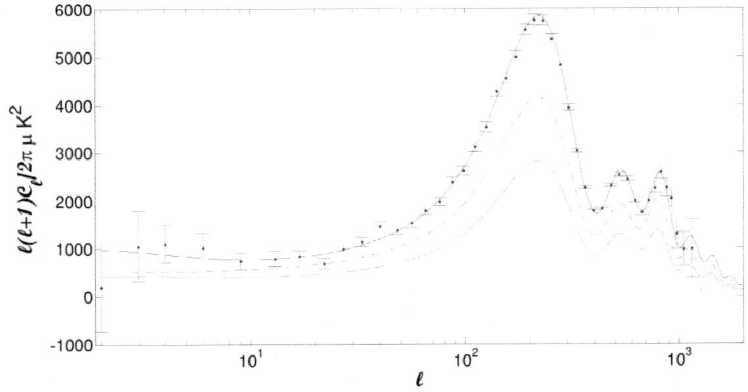

Figure 2. CMB angular power spectrum: (i) for the concordance model (solid line), (ii) for $\Omega_m = 0.25$, $h = 0.82, \Omega_b h^2 = 0.0223$ and the σ_8, n_S, τ_o values of the concordance model (dashed line), and (iii) for the same values of $\Omega_m, h, \Omega_b h^2$ and one of the best possible choices for σ_8, n_S and τ (dotted line).

Even in the best cases (see the dotted line of Fig. 2), it is evident that the predicted spectrum is very different from that of the concordance model (solid line). The same study has been performed for many (h, Ω_m) pairs covering the region of Fig. 1 which is not ruled out by SNe Ia observations. For any pair, the resulting CMB angular power spectrum is worse than for the pair considered in Fig. 2; hence, in the absence of vector field perturbations, the VT theory does not explain WMAP seven years data. On account of these results, and also for the sake of briefness, we do not present the matter power spectrum $P(k)$ and the quantities E_ℓ, B_ℓ, and ET_ℓ, which have been also found by using CAMB [9] and the vector field equation of state numerically obtained from Eqs. (1).

An improved analysis will be presented elsewhere (including a running scalar index and variable Ω_K). In spite of the improvements, the theory does not work for negligible vector field fluctuations. Although these fluctuations require more study, they might be negligible as the tensor modes (gravitational waves); in particular, if such perturbations were generated during an inflationary process driven by some scalar field minimally coupled with the scalar curvature and the vector field A.

4. Acknowledgments.

This work has been supported by the Spanish Ministerio de Ciencia e Innovación, MICINN-FEDER project FIS2009-07705.

References

[1] Beltrán Jiménez J, Maroto A L 2008 *Phys. Rev. D* **78** 063005
[2] Beltrán Jiménez J, Lazcoz R Maroto A L 2009, *Phys. Rev. D* **80** 023004
[3] Astier P, Guy J, Regnault N, et al. 2006, A&A 447,31-48.
[4] Adam G. Riess, Louis-Gregory Strolger, Stefano Casertano, et al. 2007 *APJ* **659** 98-121
[5] Komatsu E, et al 2010 *astro-ph* **1001** 4538
[6] Jarosik N, et al 2010 *astro-ph* **1001** 4744
[7] Will C M 1993 Theory and experiment in gravitational physics (Cambridge University Press, NY)
[8] Pettini M, Berkeley J. Zych, Michael T. Murphy, et al. 2008 *MNRAS* **391**, 1499-1510
[9] Lewis A, Challinor A, Lasenby A 2000 Astrophys. J. **538** 473

Journal of Physics: Conference Series **314** (2011) 012119 doi:10.1088/1742-6596/314/1/012119

High redshift galaxies through gravitational lensing

José A. de Diego[1], Jordi Cepa[2], Mario De Leo[1] & Ángel Bongiovanni[2]

[1] Instituto de Astronomía, Universidad Nacional Autónoma de México, Ciudad Universitaria, 04510 México D,F., MX

[2] Instituto de Astrofísica de Canarias, C/ Vía Láctea, s/n, E38205 - La Laguna (Tenerife), SP

E-mail: jdo@astroscu.unam.mx

Abstract. The technique and some preliminary results about the study of the luminosity function (LF) of Lyman Alpha Emitters (LAEs) through gravitational lensing (GL) is presented. Clusters of galaxies, with suitably modeled gravitational lenses are observed using the tunable filters (TF) of the instrument OSIRIS at the GTC. The combination of these natural gravitational telescopes, and the higher contrast achievable by the TF, will allow the detection of very faint LAEs, not only reaching a step forward in understanding this primordial objects, but also achieving a better determination of the mass distribution on the inner regions of galaxy clusters.

1. Introduction

LAEs and Lyα blobs are distant dust-free galaxies that emit Lyα radiation. Typical LAEs are about 1 kpc in size, and the gas is completely ionized, through photoionization, by the first generation of massive blue stars responsible for the reionization of the universe. The much more extended Lyα blobs have sizes up to 100 kpc, and the ionized gas in these objects is mechanically warmed. LAEs and Lyα blobs are important in cosmology because they trace high redshift galaxies, their dark matter halos and subsequently the evolution of matter distribution in the universe.

2. LAEs detection

Usually, LAEs have been detected in the optical bands ($2 < z < 7$) using deep field narrowband imaging with bandpass around 70 Å to study discrete redshifts (e.g. Subaru Deep Field [1, 2, 3]). The actual record is a $z = 8.6$ galaxy [4]. With OSIRIS and its set of tunable filters we plan to reduce the bandpass to $\sim$ 15 Å and obtain a higher contrast for LAE detections (Fig. 1).

3. Gravitational lensing

We plan to use GL by clusters of galaxies to study faint high redshift sources. Galaxy clusters behave as GL natural telescopes, and they allow the study of different GL regimes, from strong GL forming the impressive arcs around the cluster centers, to the weak GL on the outskirts of the clusters that can be studied through statistical techniques. We expect to detect LAEs at high redshift, reaching luminosities $\sim 10^{42}\,\mathrm{erg\,s^{-1}}$ for highly magnified objects [5]. This will lead to constrict the limits of the LF at high redshift, to study the magnification bias [6, 7], and to improve accurate modeling of the cluster masses. We have already modeled some GL

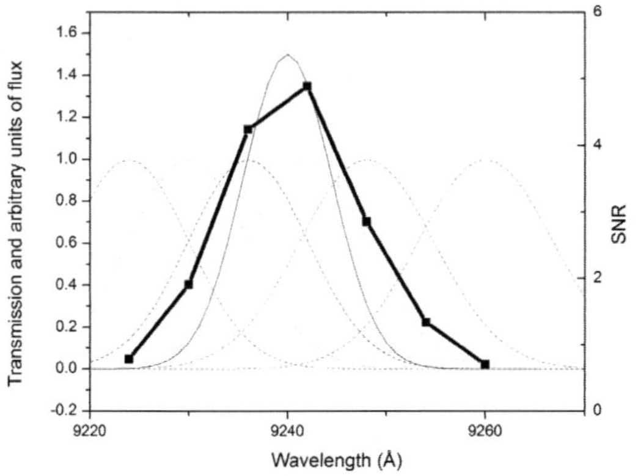

Figure 1. OSIRIS TF with a bandpass of 12 Å (color Gaussians) are shifted half the bandpass for optimum detection. The black Gaussian shows the Lyα emission for a LAE at $z = 6.6$ with a FWHM of ~ 9 Å ($\sim 300\,\mathrm{km\,s^{-1}}$). The thick line with symbols shows the SNR for 200 s exposures with GTC, computed with the OSIRIS calculator developed by J. Ignacio González-Serrano at IFCA.

clusters using archive data from the Keck and Hubble telescopes, such as MS 0440.5+0204, to characterize the critical lines and calculate probabilities of amplification (Figs. 2 and 3).

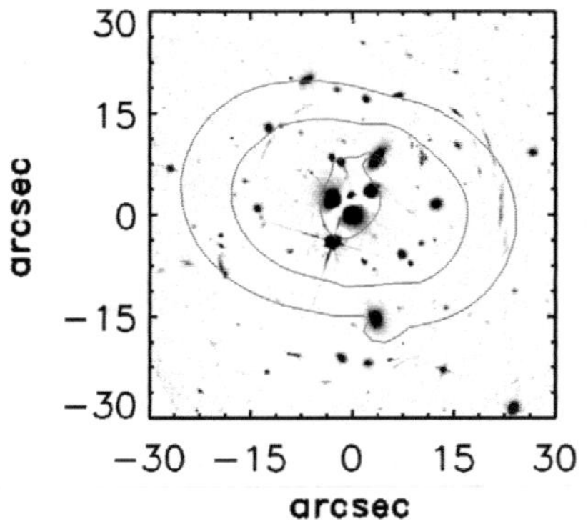

Figure 2. GL model developed for MS 0440.5+0204. (Figure from [8]).

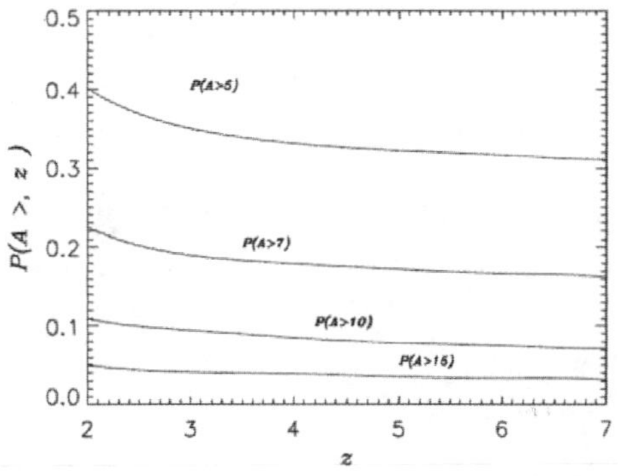

Figure 3. Magnification probability for galaxies in a region around 15″ the caustic lines of MS 0440.5+0204 (Figure from [8]).

4. Methodology

Three spectral regions not contaminated by strong sky lines will be observed: 6750 - 6930; 8100 - 8230; and 9110 - 9250 Å with the red TF. These regions correspond to the Lyα line up between $4.6 < z < 6.6$. From the LF of LAEs [?], we expect to detect about 20 such objects per cluster field with fluxes $\geq 2 \times 10^{-17}\,\mathrm{erg\,s^{-1}\,cm^{-2}}$, to characterize the LF of LAEs, to improve GL models and to study the magnification bias by comparison between blank fields and different lensing regimes available from each $8 \times 8\,\mathrm{arcmin}^2$ OSIRIS field of view. The observations will also be useful to search for SFR of Pop. III dual Lyα -HeII(1640 Å) emitters [10] around $z = 4.6$ (Fig. 4).

5. Available and future data

Observations of MS 0440.5+0204 have been already performed. Data is already reduced and will be analyzed during the next months. Other observations, including 2.5 h of OSIRIS granted time, are in the queue. Figure 5 shows observations of the cluster of galaxies MS 0440.5+0204

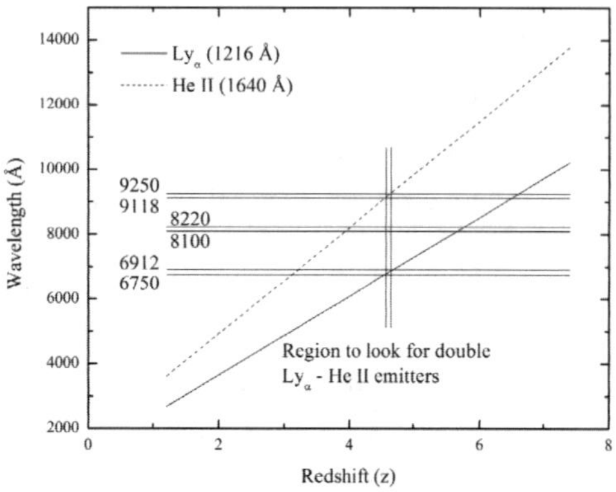

Figure 4. Spectral regions that will be observed. A search for dual Lyα - HeII(1640 Å) emitters at $z \sim 4.6$ will be performed with the OSIRIS data.

obtained during November 2009. Raw data show strong sky features due to the variations of the wavelength across the field of view, an expected characteristics of the OSIRIS TF. Data also suffers from a remarkable dark current stamp which adds an unexpected noise to the observations; fortunately, this dark current problem has been already solved and will not affect future observations.

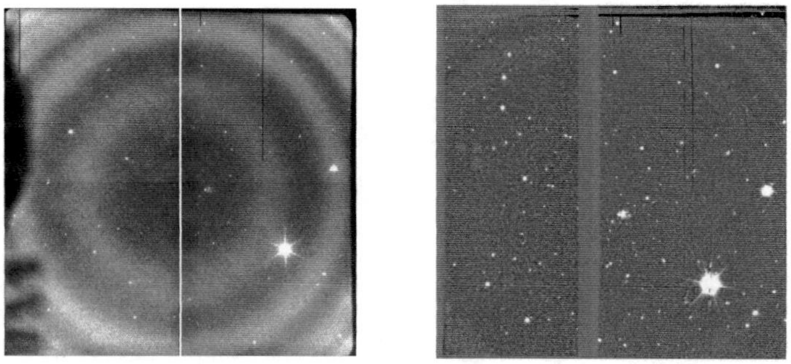

Figure 5. Raw (left) and reduced images of MS 0440.5+0204. observations were performed during November 2009 and February 2010. Wavelengths vary on the OSIRIS field, yielding rings due to sky emission lines. OSIRIS left CCD is vignetted.

This work has been supported by CONACyT grant 50296 and has been partially funded by the Spanish MICINN under the Consolider-Ingenio 2010 Program grant CSD2006-00070: first Science with the GTC (`http://www.iac.es/consolider-ingenio-gtc`).

References

[1] Kashikawa, N., et al., 2006, ApJ 648, 7.
[2] Shimasaku, K., et al., 2006, Publications of the Astronomical Society of Japan 58, 313.
[3] Ouchi, M., et al., 2008, ApJS 176, 301-330.
[4] Lehnert et al. 2010, Nature, 467, 940.
[5] Santos, M.R., et al., 2004, ApJ 606, 683.
[6] Bouwens, R. J. et al., 2009, ApJ, 690, 1764.
[7] Narayan, R. & Bartelmann, M., 1996, astroph, 6001.
[8] Verdugo, T., 2008, PhD Thesis, Instituto de Astronomía, Univ. Nac. Autónoma de México.
[9] Le Delliou, M., et al., 2006, MNRAS 365, 712.
[10] Nagao, T., et al., 2008, ApJ 680, 100.

Journal of Physics: Conference Series **314** (2011) 012120 doi:10.1088/1742-6596/314/1/012120

Uniqueness of static spherically symmetric vacuum solutions in the IR limit of nonrelativistic quantum gravity

Tomohiro Harada, Umpei Miyamoto and Naoki Tsukamoto
Department of Physics, Rikkyo University, Toshima, Tokyo 175-8501, Japan
E-mail: harada@rikkyo.ac.jp

Abstract. We investigate static spherically symmetric vacuum solutions in the IR limit of projectable nonrelativistic quantum gravity, including the renormalisable quantum gravity recently proposed by Hořava, with an emphasis on the uniqueness of the solutions. It is found that the projectability condition plays an important role. Without the cosmological constant, the spacetime is uniquely given by the Schwarzschild solution. With the cosmological constant, the spacetime is uniquely given by the Kottler (Schwarzschild-(anti) de Sitter) solution for the entirely vacuum spacetime. However, in addition to the Kottler solution, the static spherical and hyperbolic universes are uniquely admissible for the locally empty region, for the positive and negative cosmological constants, respectively. This implies that static spherically symmetric entirely vacuum solutions would not admit the freedom to reproduce the observed flat rotation curves of galaxies.

1. Introduction

Hořava [1] proposed a quantum gravity theory, which is supposed to be power-counting renormalisable and ghost free. This theory is called the Hořava-Lifshitz gravity. Subsequently, a variety of vacuum solutions have been discovered. However, it does not seem so clear how unique these solutions are. The uniqueness of the solutions depends on the conditions we impose for the solutions. For simplicity, we now consider static, spherically symmetric vacuum spacetimes with and without the cosmological constant in infrared (IR) limit of the Hořava-Lifshitz gravity. This article is based on Harada, Miyamoto and Tsukamoto [2], and therein the full list of the relevant references is provided.

2. Hořava-Lifshitz gravity

The Hořava-Lifshitz gravity is featured with the following characters. The theory is essentially based on the "3+1" view of the spacetime. There is a preferred spacetime foliation of the spacetime with which the theory is given in a physically meaningful manner. The gravitational fields consist of a three-metric g_{ij}, a shift N_i and a lapse N. The line element in the four dimensional spacetime is given by

$$ds^2 = -Nc^2dt^2 + g_{ij}(dx^i + N^i dt)(dx^j + N^j dt).$$

The Lagrangian is covariant against the foliation preserving diffeomorphisms $\tilde{x}^i = \tilde{x}^i(t, x^j)$ and $\tilde{t} = \tilde{t}(t)$. The kinetic term of the Lagrangian consists of the quadratic form of the first time derivative of g_{ij}. The theory is at the fixed point in the ultraviolet (UV) limit against the anisotropic scaling $\mathbf{x} \to b\mathbf{x}$ and $t \to b^z t$, where the gravitational coupling constant is nondimensional with respect to this scaling. If we additionally impose the detailed balance condition and the relevant deformation to recover the Einstein-Hilbert action with the cosmological constant in the IR limit, the gravitational Lagrangian is fixed to the following form:

$$\begin{aligned} L_g &= \frac{2}{\kappa^2}(K_{ij}K^{ij} - \lambda K^2) + \frac{\kappa^2\mu^2(\Lambda_W R - 3\Lambda_W^2)}{8(1-3\lambda)} + \frac{\kappa^2\mu^2(1-4\lambda)}{32(1-3\lambda)}R^2 \\ &\quad - \frac{\kappa^2}{2w^4}\left(C_{ij} - \frac{\mu w^2}{2}R_{ij}\right)\left(C^{ij} - \frac{\mu w^2}{2}R^{ij}\right), \end{aligned}$$

where κ, λ, μ, Λ_W and w are constants, K_{ij} is the extrinsic curvature and C_{ij} is the Cotton tensor, a symmetric tensor made of the third spatial derivative of g_{ij}. Since the theory is covariant against the foliation-preserving diffeomorphisms, it is natural to restrict the lapse function to N to a function only of t. This is called the projectability condition. By imposing this condition, the constraints constitute a Lie algebra and there appears dark matter as an integration constant [3].

Hereafter we will concentrate on the IR limit of the Hořava-Lifshitz gravity with $\lambda = 1$, where only the terms with the lowest order spatial derivatives in the potential are left. From the comparison with the action of general relativity, we obtain

$$\alpha = \frac{1}{16\pi G c}, \quad \xi = c^2\alpha, \quad \sigma = -2\Lambda c^2\alpha.$$

3. Static spherically symmetric vacuum solution in the IR limit

We assume that the space is static and spherically symmetric, where the line element is given by

$$g_{ij}dx^i dx^j = e^{2\omega(r)}dr^2 + r^2(d\theta^2 + \sin^2\theta d\phi^2).$$

The lapse and shift should be in the form $N = \text{const}$ and $N^i/N = (v(r), 0, 0)$.

Due to the projectability condition, the Hamiltonian constraint from the variation $\delta N(t)$ is a nonlocal equation given by

$$\int dr r^2 e^{\omega}\left[-\alpha(K_{ij}K^{ij} - K^2) + R + \sigma\right] = 0.$$

On the other hand, the momentum constraint from the variation $\delta N^i(t, x^j)$ is local as follows:

$$-2\frac{\omega' v}{r} = 0,$$

where the prime denotes the derivative with respect to r. The evolution equations from the variation $\delta g_{ij}(t, x^k)$ are complicated partial differential equations including the sixth order spatial derivatives in general. However, under the present settings, they become much simpler. The (r, r) and (θ, θ) components of the evolution equations become

$$\alpha\frac{v}{r}\left(2v' + \frac{v}{r} + 4\omega' v\right) + \xi\frac{1 - e^{-2\omega}}{r^2} + \frac{1}{2}\sigma = 0,$$

and

$$\alpha\left[\left(v' + \omega' v + \frac{v}{r}\right)' v + (\omega' v + v')\left(v' + \omega' v + \frac{v}{r}\right) + \left(\frac{v}{r}\right)^2\right] + \xi\frac{\omega'}{r}e^{-2\omega} + \frac{1}{2}\sigma = 0,$$

respectively, and there is no remaining independent equation in the evolution equations.

It is now easy to solve these field equations. The momentum constraint leads to $\omega' = 0$ or $v = 0$.

3.1. Case (a): $\omega' = 0$

Putting $\omega = \omega_0 =$ const, the evolution equation is integrated to give

$$v^2 = -\frac{\xi}{\alpha}(1 - e^{-2\omega_0}) - \frac{1}{6}\frac{\sigma}{\alpha}r^2 + \frac{C}{r},$$

where C is a constant of integration. The right-hand side must be positive. This solution satisfy the Hamiltonian constraint because the integrand of the integral Hamiltonian constraint vanishes identically.

If we make the following foliation-nonpreserving coordinate transformation:

$$Ndt = e^{-\omega_0}dT + \frac{e^{2\omega}v}{c^2 - e^{2\omega}v^2}dr,$$

we obtain the standard form of the Schwarzschild-(A)dS metric, which is not projectable:

$$ds^2 = \left[1 - \frac{1}{3}\Lambda r^2 - \frac{2GM}{c^2 r}\right]dT^2 + \left[1 - \frac{1}{3}\Lambda r^2 - \frac{2GM}{c^2 r}\right]^{-1}dr^2 + r^2 d\Omega^2,$$

where we put $C = 2GM$. Note that the full theory is not covariant against the above coordinate transformation. The IR limit can be justified for sufficiently large r. This is not a solution for $\lambda \neq 1$.

3.2. Case (b): $v = 0$

The evolution equation leads to

$$e^{2\omega} = \left(1 + \frac{\sigma}{2\xi}r^2\right)^{-1}.$$

Although the integrand of the integral Hamiltonian constraint does not vanish for $\sigma \neq 0$, this solution is acceptable as a local solution, if its nonvanishing contribution to the global Hamiltonian constraint can be compensated by that from the nonempty or nonstatic region.

The metric is given by

$$ds^2 = -dt^2 + \frac{1}{1 - \Lambda r^2}dr^2 + r^2(d\theta^2 + \sin^2\theta d\phi^2).$$

This is Minkowski for $\Lambda = 0$, static spherical universe (Einstein's static universe) for $\Lambda > 0$ and static hyperbolic universe for $\Lambda < 0$. The IR limit is justified for sufficiently small $|\Lambda|$. It turns out that this is a solution for any value of λ.

4. Summary

The uniqueness of static, spherically symmetric solutions in the IR limit of Hořava-Lifshitz gravity is investigated. If the Hamiltonian constraint is to be satisfied, the solution must be uniquely Schwarzschild ($\Lambda = 0$) or Schwarzschild-(A)dS ($\Lambda \neq 0$). A local solution may not satisfy the Hamiltonian constraint if we take into account the region outside our observable universe. In spite of this relaxation, there is no additional solution for $\Lambda = 0$, while only a static, homogeneous and isotropic universe is added for $\Lambda \neq 0$. The Schwarzschild ($\Lambda = 0$) and the Schwarzschild-(A)dS ($\Lambda \neq 0$) are subject to $\lambda = 1$, while the static, homogeneous and isotropic universe is a solution for any value of λ. Within this framework, there is no dark matter which explains the observed flat rotation curve of galaxies. See Harada, Miyamoto and Tsukamoto [2] for further details.

References

[1] Hořava P., 2009 Phys. Rev. D**79**, 084008
[2] Harada T., Miyamoto U. and Tsukamoto N., to appear in Int. J. Mod. Phys. D, *Preprint* arXiv:0911.1187
[3] Mukohyama S., 2009 Phys. Rev. D**80**, 064005

Journal of Physics: Conference Series **314** (2011) 012121 doi:10.1088/1742-6596/314/1/012121

Quantum Modifications to Gravity Waves in de Sitter Spacetime

Jen-Tsung Hsiang
Department of Physics, National Dong Hwa University, Hualien, Taiwan, ROC
E-mail: cosmology@gmail.com

L. H. Ford
Institute of Cosmology, Department of Physics and Astronomy, Tufts University, Medford, MA 02155, USA
E-mail: ford@cosmos,phy.tufts.edu

Da-Shin Lee
Department of Physics, National Dong Hwa University, Hualien, Taiwan, ROC
E-mail: dslee@mail.phys.ndhu.edu.tw

Hoi-Lai Yu
Institute of Physics, Academia Sinica, Nankang, Taipei 11529, Taiwan, ROC
E-mail: hlyu@phys.sinica.edu.tw

Abstract. We study the effect of the vacuum fluctuations of a massless conformally invariant field on tensor perturbations of spatially flat de Sitter spacetime. We find that de Sitter spacetime is stable within the regime of the semiclassical theory. Next we investigate the modification of linearized plane gravity waves by the effects of the quantum stress tensor. We find a correction term whose amplitude depends on the initial time η_0. If this initial time is the beginning of inflation, then as long as the energy scale of inflation and the proper frequency of the mode at η_0 are well below the Planck scale, the fractional correction is small. Nonetheless, modes which are transplanckian at the onset of inflation may have a significant correction. The increase in amplitude can be potentially observed through a modification of the power spectrum of tensor perturbations in inflationary cosmology. This enhancement of the power spectrum depends upon the initial time, and is greater for shorter wavelengths.

1. Introduction

Most of the inflationary models assume that the universe has experienced a period of exponential expansion and is approximately described by de Sitter spacetime. Quantum fields in de Sitter spacetime account for the primordial spectrum of scalar and tensor perturbations. In addition, quantum effects may modify the duration of inflation and possibly introduce instabilities [1, 2, 3].

We examine some of the quantum field effects in the semiclassical theory, where gravity couples with the renormalized expectation value of a matter field stress tensor $\langle T_{\mu\nu}\rangle$.

The semiclassical theory has been extensively applied to scalar perturbations of de Sitter spacetime [4], but the topic of this paper focuses on tensor perturbations. We will treat a model in which the matter field is a conformal field, and address two physical questions: the stability of de Sitter spacetime under tensor perturbations, and the effects of one-loop quantum matter field corrections on gravity waves in de Sitter spacetime. We adopt the sign conventions of Ref. [5], and use units in which $\hbar = c = 1$.

2. Weakly perturbed de Sitter spacetime

We are concerned with a portion of global de Sitter spacetime which can be represented as a spatially flat Robertson-Walker universe with the metric

$$ds^2 = a^2(\eta)\,(-d\eta^2 + dx^2 + dy^2 + dz^2)\,, \tag{1}$$

where $a(\eta) = -1/(H\eta)$ and $\eta < 0$ is the conformal time coordinate. We would like to consider tensor perturbations of this geometry, which describe gravitational waves on the de Sitter background. Let the perturbed metric be $g_{\mu\nu} = \gamma_{\mu\nu} + h_{\mu\nu}$, where $\gamma_{\mu\nu}$ is the background metric of Eq. (1), and $h_{\mu\nu}$ is the perturbation. We employ the transverse trace-free gauge defined by $h^{\mu\nu}{}_{;\nu} = 0$, $h = 0$ and $h^{\mu\nu}u_\nu = 0$. Here $u^\nu = \delta^\nu_0$ is the four velocity of a comoving observer, covariant derivatives are taken respect to the fixed de Sitter background, and indices are raised and lowered by the background metric. To first order in the perturbation $h_{\mu\nu}$ [6, 7], we find that the local terms in $\langle T_{\mu\nu}\rangle$ amount to shifts in the cosmological constant and Newton's constant, so we are left with only the effects of the $P_{\mu\nu}$ and nonlocal $Q_{\mu\nu}$ terms on the tensor perturbations, such that in the transverse trace-free gauge the semiclassical Einstein equation reduces to

$$\Box_s h^j_i = -16\pi\ell_p^2\,(P^j_i + Q^j_i)\,, \tag{2}$$

with ℓ_p being the Planck length.

3. Spatially homogeneous solutions

We study the stability of the tensor perturbations of de Sitter spacetime in the presence of the quantum stress tensor of the conformal field. It is sufficient to examine spatially homogeneous solutions of Eq. (2), as these will be the most rapidly growing modes if there is an instability. Let

$$\tilde{h}_{ij} = a^{-2}\,h_{ij} = h^j_i = e^j_i\,(-\eta)^{-b}\,, \tag{3}$$

where e^j_i is a constant polarization tensor and b is a constant. A solution for which $b > 0$ will grow as a power of conformal time as $\eta \to 0^-$. Substituting Eq. (3) into Eq. (2) yields

$$b(3+b) \simeq -\xi\,\left\{6[1+\ln(H\lambda/2)]\,b + [5+\pi^2+11\ln(H\lambda/2)]\,b^2 + O(b^3)\right\}\,, \tag{4}$$

in terms of small parameter $\xi = 64\pi^2\,\ell_p^2\,H^2\alpha$. Within the domain of validity of the semiclassical theory, the only possibility for an unstable solution is one with a small positive value of b. We find that there are no solutions with $b > 0$ so long as $\xi \ll 1$. Hence we conclude that de Sitter spacetime is stable in the semiclasical theory against tensor perturbations.

4. Effects on gravity waves

Now we examine the effect of the quantum stress tensor on gravity waves in de Sitter spacetime. The plane wave solutions of the Lifshitz equation, $\Box_s h^j_i = 0$, are of the form

$$h^\nu_\mu = c_0\,e^\nu_\mu\,(1 + ik\eta)\,\mathrm{e}^{i(\mathbf{k}\cdot\mathbf{x} - k\eta)}\,, \tag{5}$$

where c_0 is a constant and e^ν_μ is the polarization tensor. In the presence of the quantum stress tensor, the modified gravity wave may be expressed as $h^\nu_\mu + h'^\nu_\mu$, and in the limit that $|\eta_0| \gg |\eta|$,

$$h'^\nu_\mu(x) \sim 64\pi^2\, i\, e^\nu_\mu\, c_0 \alpha\, H^2\, k\ell_p^2\, |\eta_0|\, (1 + ik\eta)\, \mathrm{e}^{i(\mathbf{k}\cdot\mathbf{x} - k\eta)}\,, \tag{6}$$

which has the same functional form as h^ν_μ, but is out of phase by $\pi/2$. The modified wave is no longer constant when the mode has a proper wavelength larger than the horizon size, $k|\eta| < 1$. This is in contrast to the unperturbed mode, Eq. (5), whose magnitude is approximately constant when it is outside the horizon. The most striking feature of Eq. (6) is that the correction term due to the quantum stress tensor is proportional to $|\eta_0|$, and hence is larger the earlier the coupling between the quantum stress tensor and the metric perturbation is switched on. We can write the ratio of the magnitude of the correction to that of the original wave as

$$\Gamma = \left| \frac{h'^\nu_\mu}{h^\nu_\mu} \right| = 64\pi^2 \alpha H^2 k \ell_p^2 |\eta_0| = 64\pi^2 \alpha H\, k_P\, \ell_p^2\,. \tag{7}$$

Here $k_P = k/a(\eta_0) = kH|\eta_0|$ is the physical wavenumber of the mode as measured by a comoving observer at time $\eta = \eta_0$. If we require that the curvature of the de Sitter spacetime be well below the Planck scale ($H\ell_p \ll 1$), and that the mode in question be always below the Planck scale ($k_P\ell_p \lll 1$) while it interacts with the quantum stress tensor, then these two conditions together imply that $|h'^\nu_\mu/h^\nu_\mu| < 1$, and hence the quantum correction to the gravity wave is smaller than the original wave. However, if inflation lasts for a sufficiently long time, then modes which are cosmological interest today appear to have been above the Planck scale at the onset of inflation. This is the cosmological version of the transplanckian problem.

The inflationary model predicts a Gaussian and nearly scale invariant spectrum of tensor perturbations [8, 9], which might be found in polarization measurements of the CMB. The tensor perturbations from inflation are less model dependent than the density perturbations, At this time, they leave an imprint on the CMB in the form of a power spectrum of tensor perturbations given by

$$\delta_h^2 \approx \frac{8}{\pi}\, \ell_p^2\, H^2\,. \tag{8}$$

This is an approximately flat spectrum. The effect of the conformal stress tensor is to modify this power spectrum by a factor $|1 - i\Gamma|^2 = 1 + \Gamma^2$.

Let us choose the scale factor to be unity at the end of inflation, E_R be the reheating energy, and S be the factor by which the universe expands from the initial conformal time $\eta = \eta_0$ to the end of inflation. We have

$$\Gamma^2 = 1.34 \times 10^{-78} \left(\frac{10^{25}\,\mathrm{cm}}{\ell_0}\right)^2 \left(\frac{E_R}{10^{15}\,\mathrm{GeV}}\right)^6 S^2\,, \tag{9}$$

if we use the value of $\alpha = 1/(320\pi^3)$ corresponding to the electromagnetic field. Recall that $\ell_0 \simeq 10^{25}$cm corresponds to angular scales of the order of 1° of the present horizon size.

If one has only the minimal inflation needed to solve the horizon and flatness problems, so $S \simeq 10^{23}$, then the effects of the one-loop correction on the tensor perturbation spectrum is negligible. However, larger values of S have the potential to produce significant corrections. For example, $E_R \simeq 10^{15}$GeV and $S \simeq 10^{39}$ would lead to an effect of order unity at 1° scales. One should expect the one-loop approximation to begin to break down, but this can serve as an order of magnitude estimate. In contrast to the nearly flat spectrum, Eq. (8), due to free graviton fluctuations, the one-loop effect is highly tilted toward the blue end of the spectrum. The effect is potentially observable. Thus this possibility does require one to take seriously the contribution of modes which were transplanckian at the beginning of inflation.

[1] L.H. Ford, S.P. Miao, K.W. Ng, R.P. Woodard, and C.H. Wu, Phys. Rev. D **82**, 043501 (2010).
[2] A.A. Starobinsky, JETP Lett. **34**, 438 (1981).
[3] A. Campos and E. Verdaguer, Phys. Rev. D **49**, 1861 (1994).
[4] P.R. Anderson, C. Molina-Paris, and E. Mottola, Phys. Rev. D **80**, 084005 (2009).
[5] C.W. Misner, K.S. Thorne and J.A. Wheeler, *Gravitation*, Freeman (1973).
[6] G.T. Horowitz and R.M. Wald, Phys. Rev. D **25**, 3408 (1982).
[7] J.-T. Hsiang, L.H. Ford, D.-S. Lee and H.-L. Yu, Phys. Rev. D, in print (2011).
[8] L.F. Abbott and M.B. Wise, Nucl. Phys. B **244**, 541 (1984).
[9] A.A. Starobinsky, Sov. Astron. Lett. **11**, 133 (1985).

Journal of Physics: Conference Series **314** (2011) 012122 doi:10.1088/1742-6596/314/1/012122

Some topological properties of the Inverse Lens Mapping

T Mediavilla[1], O Ariza[2], E Mediavilla[3], JA Muñoz[4], A Oscoz[5]

[1,2]Departamento de Estadística e Investigación Operativa, Universidad de Cádiz, Avda de Ramón Puyol, s/n 11202 Algeciras.
[3,5]Instituto de Astrofísica de Canarias, Avda Vía Láctea s/n, La Laguna.
[4]Departamento de Astronomía y Astrofísica, Universidad de Valencia, c/ Dr. Moliner, 50, 46100 Burjassot.

E-mail: teresa.mediavilla@ca.uca.es, octavio.ariza@uca.es , emg@iac.es, jmunoz@uv.es, aoscoz@iac.es

Abstract. Away from critical curves, lens mapping can be seen as a linear invertible transformation of the plane even for regions (cells) of relatively large size. However, close to critical curves the departures from linearity can be very strong. We discuss the topological problems induced by the mapping of regions of the image plane that include critical curves (critical cells).

1. Inverse lens mapping as a linear transformation of the plane.

Lens mapping can be seen locally as a linear invertible transformation of the plane that transforms parallelograms into parallelograms [1]. These transformations are the composition of rotation, dilation (including reflection) and shear (see figures 1,2 and 3). Lens mapping is symmetric too.

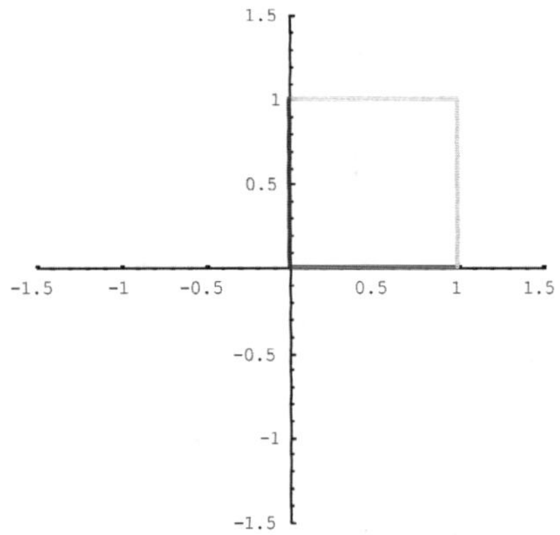

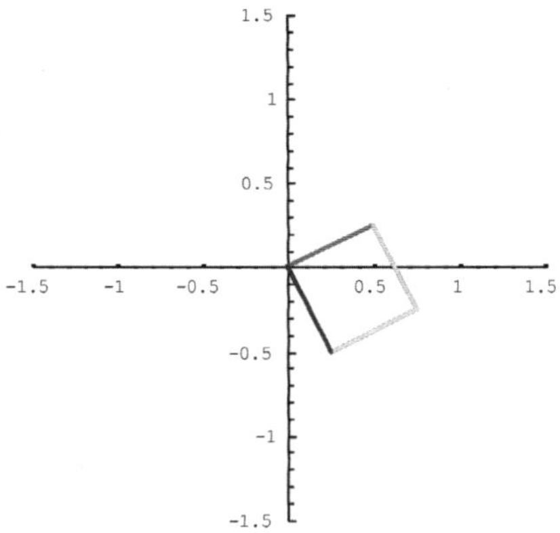

Figure 1. Dilation + rotation.

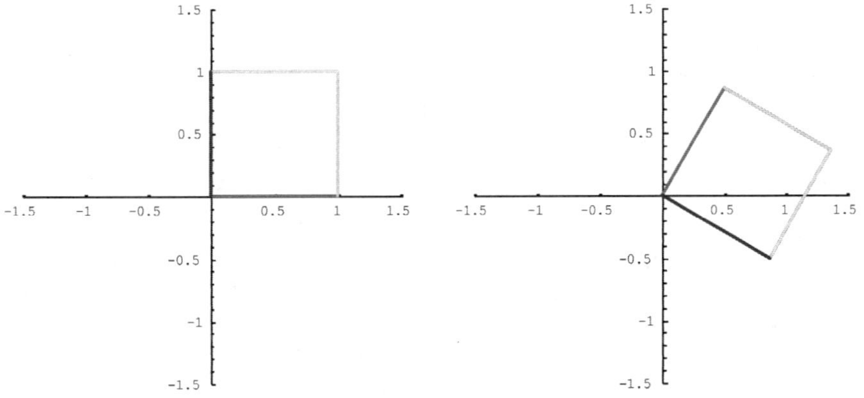

Figure 2. Reflection + rotation.

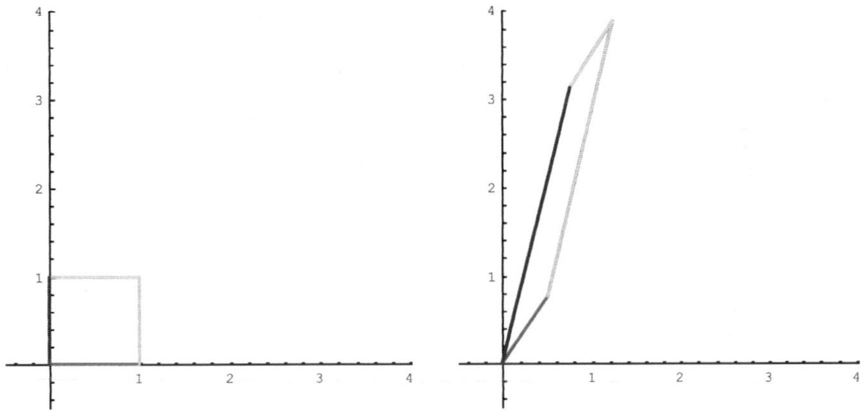

Figure 3. Rotation + shear.

2. Inverse lens mapping close to critical curves.

Close to critical curves the departure from linearity can be strong. Let us consider a small cell located in the neighborhoods of a critical curve. Close to a fold the inverse lens mapping collapse the cell in the direction perpendicular to the caustic (see figure 4). In addition to the collapse the cell is also rotated and suffers a strong shear. Close to a cusp (see figure 5), the inverse lens transformation is clearly non-linear; straight lines blend and parallelism is not preserved. Nevertheless, also in this case the cell collapse in the direction perpendicular to the caustics that join at the cusp.

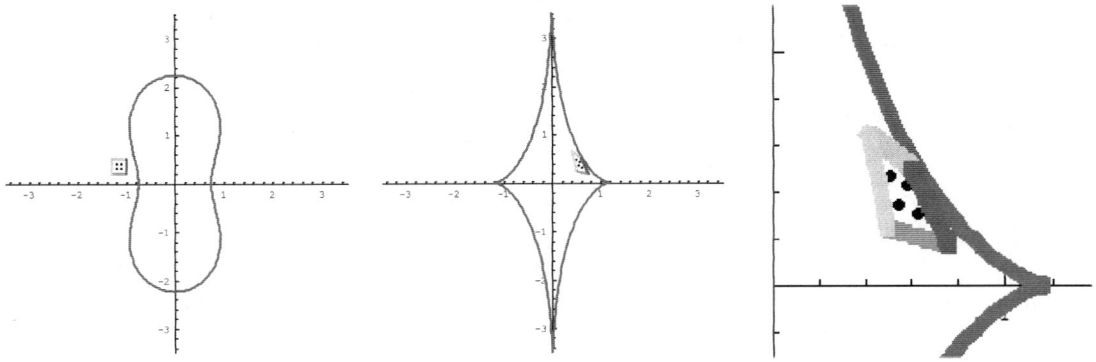

Figure 4. Non-critical cell close to a fold

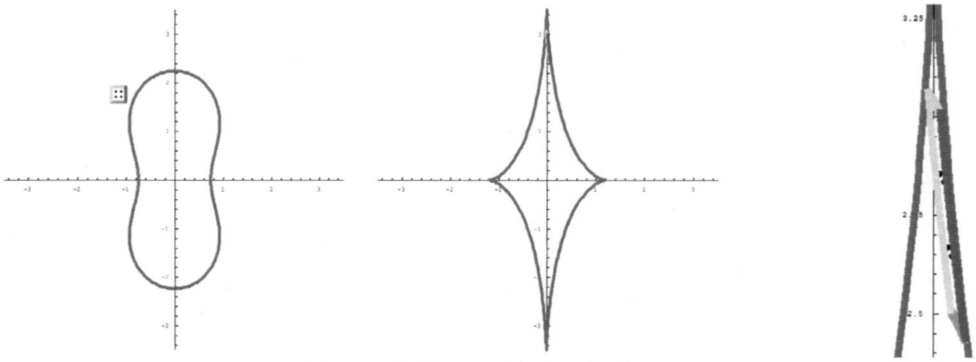

Figure 5. Non-critical cell close to a cusp

3. Inverse lens mapping of critical cells.

If the simply connected region is crossed by a critical curve the transformed region will be not simply connected. This can have very significant consequences. The boundary of the critical cell will be not transformed in the boundary of the transformed region (see figure 6). Another consequence is auto-overlapping (see figure 7). Finally, a someway different example of topological pathology is the transformation of a simply connected region including a point where f is discontinuous (see figure 8).

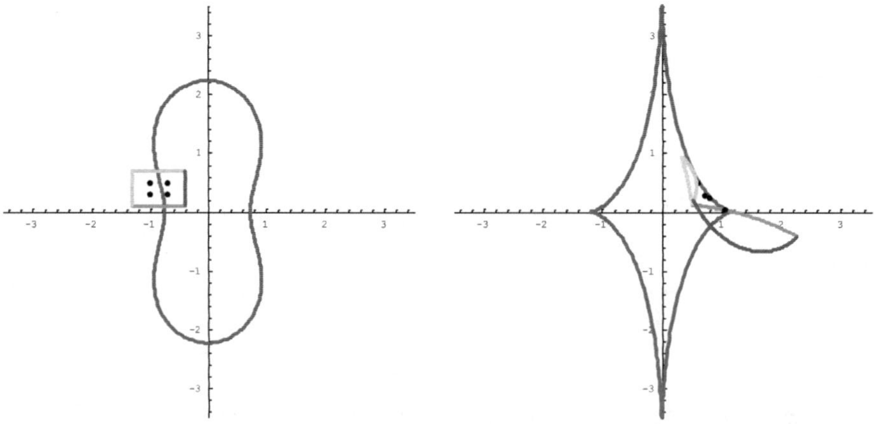

Figure 6. The boundary of the critical curve will be not transformed in the boundary of the transformed region.

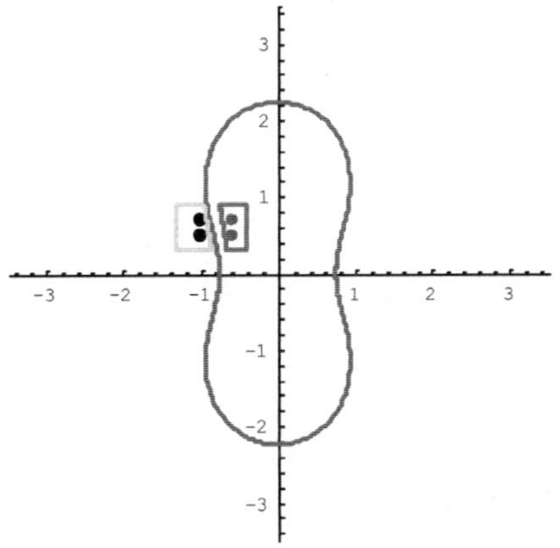

Figure 7. Autooverlaping.

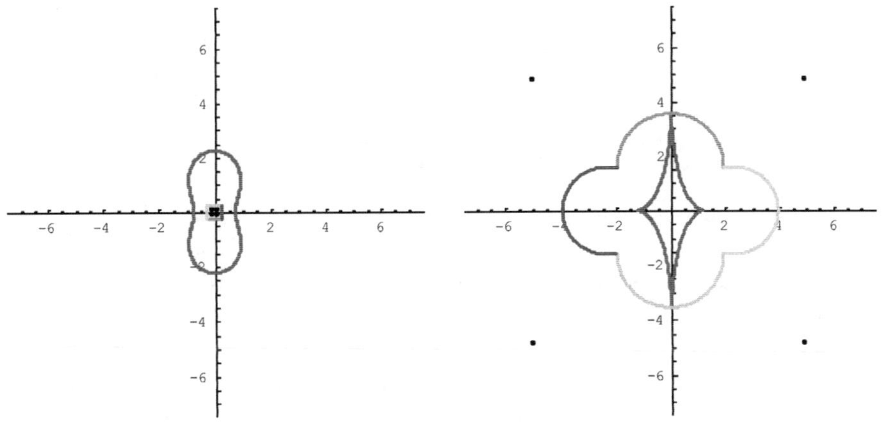

Figure 8. Transformation of a cell including a point where f is discontinuous.

4. Conclusions.

Inverse lens mapping induces strong topological pathologies in the transformation of critical cells. To avoid this problem in the computation of magnification maps, a partition of the image plane in polygonal cells not including critical points (except perhaps at the cells boundaries) should be considered.

References

[1] Mediavilla E, Muñoz J, Lopez P, Mediavilla T, Abajas C, Gonzalez- Morcillo C and Gil-Merino 2006 *ApJ* **635** 942-953

Journal of Physics: Conference Series **314** (2011) 012123 doi:10.1088/1742-6596/314/1/012123

On cosmic censor in high-energy particle collisions

Umpei Miyamoto

Department of Physics, Rikkyo University, Tokyo 171-8501, Japan

E-mail: umpei@rikkyo.ac.jp

Abstract. In the context of large extra-dimension or TeV-scale gravity scenarios, miniature black holes might be produced in collider experiments. In many works the validity of the cosmic censorship hypothesis has been assumed, which means that there is no chance to observe trans-Planckian phenomena in the experiments since such phenomena are veiled behind the horizons. Here, we argue that "visible borders of spacetime" (as effective naked singularities) would be produced, even dominantly over the black holes, in the collider experiments. Such phenomena will provide us an arena of quantum gravity.

The higher-dimensional scenarios with large [1] and warped [2] extra dimensions were proposed to resolve or reformulate the hierarchy between the gravitational and electroweak interactions. In these scenarios the D-dimensional ($D > 4$) Planck energy E_P is related to that in 4-dimension $E_{P(4)}(\sim 10^{19}$ GeV) by $E_{P(4)}^2 \sim L^{D-4}E_P^{D-2}$, where L is the size of an extra dimension. Thus, if L is large enough, the fundamental Planck energy E_P can be as low as a few TeV. For instance, such a small Planck energy is realized if $L \sim 0.1$ cm when $D = 6$, $L \sim 10^{-7}$ cm when $D = 7$. If the Standard Model particles but the gravitons (and possibly other unobserved particles) are confined to our 3-brane these scenarios are consistent with all current observations.

One of the most striking predictions of such scenarios is the productions of a large number of mini black holes in high-energy particle collisions [3]. A simplified picture of the black hole production is put in the following way. The size of a black hole is characterized by Schwarzschild radius R_g, scaling with its energy as $R_g \propto E^{1/(D-3)}$. If colliding particles (partons) have a center-of-mass energy E above a threshold energy of order E_P and an impact parameter less than the Schwarzschild radius, a black hole of energy (mass) E is produced. The black holes so produced will decay thermally via the Hawking radiation and be detected in terrestrial collider experiments such as CERN Large Hadron Collider and in ultrahigh energy cosmic rays. Such possibilities have been extensively studied and known to give rise to rich phenomenology.

Quite recently, the present author and his collaborators pointed out another interesting possibility [4] of the high-energy particle collisions in the TeV-scale gravity. They argued that effective naked singularities called *the visible borders of spacetime* would be generated by high-energy particle collisions. A border of spacetime, proposed in [5] originally, is defined as a domain of spacetime where the curvature becomes trans-Planckian and acts as an border (or boundary) of classical sub-Planckian regimes. That is, the domain $\mathcal{A}$ of spacetime is called the border if the following inequality is satisfied:

$$\inf_{\mathcal{A}} F \geq \alpha E_P^2, \tag{1}$$

where α is a positive constant of order unity and F is given, for instance, by using the scalar polynomials of the Riemann tensor, or components of the Riemann tensor with respect to a parallelly transported orthonormal basis e^a_μ as

$$F := \max\left(|R^a_a|, |R^{ab}R_{ab}|^{1/2}, |R^{abcd}R_{abcd}|^{1/2}, |R_{\mu\nu\rho\sigma}|\right). \tag{2}$$

We denote the union of all borders in spacetime $\mathcal{M}$ by $\mathcal{U}_\mathrm{B}$. We call border $\mathcal{A}$ a visible border if $J^+(\mathcal{A},\mathcal{M}) \cup (\mathcal{M} - \mathcal{U}_\mathrm{B})$ is not empty, where $J^+(\mathcal{A},\mathcal{M})$ is the causal future of $\mathcal{A}$ in $\mathcal{M}$. We can also naturally introduce a globally visible border as an extension of a globally naked singularity [5].

In the rest of this article, we discuss whether the visible borders of spacetime, defined above as effective naked singularities, can be produced in high-energy particle collisions. We will see that the visible borders of spacetime would be generated in the collider experiments such as the Large Hadron Collider.

Let us consider a collision of two elementary particles. For simplicity, suppose that both particles are identical and structureless. The extra dimensions are assumed to be compactified into a length scale considerably larger than the fundamental Planck length E_P^{-1}, whereas the energy of the particles is assumed to be confined within the length scale E_P^{-1} in every direction of the extra dimensions; i.e., within a volume $V_{D-4}E_P^{-(D-4)}$ in the extra dimensions, where V_n is the volume of a unit n-ball $V_n = \pi^{\frac{n}{2}}/\Gamma\left(\frac{n}{2}+1\right)$, where $\Gamma(x)$ is the gamma function. Then, we consider the distribution of two particles in the remaining three-dimensional space.

We denote the center-of-mass (CM) energy of the two particles by E, which is much larger than their rest masses, and assume that the three momenta of these particles are parallel to each other. The minimum uncertainty in the position of each particle measured in the longitudinal direction of their three momenta is approximately E^{-1} due to the Lorentz contraction.

It is assumed that these two particles will form a black hole if they pass through a transverse disk of radius R_g at almost the same time, where R_g is the horizon radius of the Schwarzschild–Tangherlini black hole with gravitational mass E,

$$R_\mathrm{g} := E_P^{-1}\left(\frac{E}{E_P}\right)^{\frac{1}{D-3}}, \tag{3}$$

and where we have defined the fundamental Planck scale E_P so that the reduced Compton wavelength agrees with the gravitational radius if the particle mass is equal to E_P. In this normalization, D-dimensional Einstein's gravitational constant κ ($G_{ab} = \kappa T_{ab}$) is related to E_P in the form $\kappa = \frac{1}{2}(D-2)S_{D-2}E_P^{2-D}$, where $S_n = (n+1)V_{n+1}$ is the area of a unit n-sphere. Based on this picture, the total cross section of black hole formation is given by $\sigma_\mathrm{tot} = \pi R_\mathrm{g}^2$.

We consider the case in which these two particles pass through a transverse disk with a radius of νR_g, where $\nu > 1$. In this case, a black hole does not form. When these two particles pass through a disk of radius νR_g at almost the same time, they can be momentarily confined within a $(D-1)$-volume $V \simeq \pi\nu^2 R_\mathrm{g}^2 \times E^{-1} \times V_{D-4}E_P^{-(D-4)}$, where the second factor E^{-1} is the uncertainty in the longitudinal position of each particle and $V_{D-4}E_P^{-(D-4)}$ is the volume in the extra dimensions. At this moment, the average energy density ρ in this volume is estimated to be

$$\rho \simeq \frac{E}{V} = \frac{E_P^D}{\pi\nu^2 V_{D-4}}\left(\frac{E}{E_P}\right)^{2\left(\frac{D-4}{D-3}\right)}. \tag{4}$$

If $\kappa\rho$ is larger than or equal to αE_P^2, then, through the Einstein equation, the average Ricci tensor satisfies the criterion for a border (1). The average density ρ is a monotonically increasing function of the CM energy E. Thus, a sufficiently large CM energy can produce a visible border,

Table 1. Values of $\alpha\nu_{\rm max}^2$ when $E = 14$ TeV.

E_p	$D=5$	$D=6$	$D=7$	$D=8$	$D=9$	$D=10$	$D=11$
1 TeV	66.0	180	309	437	559	674	783
14 TeV	4.71	5.33	5.89	6.40	6.87	7.31	7.73

since no black hole forms in this situation. The condition for realizing a border, $\kappa\rho \geq \alpha E_P^2$, and Eq. (4) lead to

$$\nu^2 < \nu_{\rm max}^2 := \frac{(D-2)S_{D-2}}{2\pi\alpha V_{D-4}}\left(\frac{E}{E_P}\right)^{2\left(\frac{D-4}{D-3}\right)}. \tag{5}$$

If the upper bound $\nu_{\rm max}^2$ is less than or equal to unity, the border will never be visible. Since $\nu_{\rm max}^2$ will an increasing function of the CM energy, if $E > [2\pi\alpha(D-2)^{-1}S_{D-2}^{-1}V_{D-4}]^{\frac{1}{2}\left(\frac{D-3}{D-4}\right)}E_P$, $\nu_{\rm max}^2$ is larger than unity, then the border will be visible.

The value of $(\nu_{\rm max}^2 - 1)$ expresses the ratio of the production rate of visible borders to that of black holes. The production rate of visible borders increases with increasing CM energy. In the case of the LHC ($E = 14$ TeV), if the fundamental Planck scale E_P is equal to 1 TeV, the production rate of visible borders will be about $70\alpha^{-1}$ times larger than the production rate of black holes for $D = 5$, whereas it will about $800\alpha^{-1}$ times larger than that of black holes for $D = 11$ (see Table 1). Although there is an ambiguous factor α in the criterion for the border (1), the present result, which is based on dimensional analysis, implies that if black holes form at the LHC, more visible borders will be produced than black holes.

Note added

After completing this article, the present author and his collaborators published a paper, in which motivated by the observations in this paper, the particle creation by naked singularities in higher dimensions was investigated [6].

Acknowledgments

The author would like to thank T. Harada, M. Nemoto, K. Nakao, and M. Shimano for collaborations in related works. Works of UM are partially supported by the Grant-in-Aid for Scientific Research Fund of the Ministry of Education, Culture, Sports, Science and Technology, Japan [Young Scientists (B) 22740176], and by Research Center for Measurement in Advanced Science in Rikkyo University.

1. References

[1] N. Arkani-Hamed, S. Dimopoulos and G. R. Dvali, Phys. Lett. B **429**, 263 (1998); I. Antoniadis, N. Arkani-Hamed, S. Dimopoulos and G. R. Dvali, Phys. Lett. B **436**, 257 (1998); N. Arkani-Hamed, S. Dimopoulos and G. R. Dvali, Phys. Rev. D **59**, 086004 (1999).
[2] L. Randall and R. Sundrum, Phys. Rev. Lett. **83**, 3370 (1999); L. Randall and R. Sundrum, Phys. Rev. Lett. **83**, 4690 (1999).
[3] P. C. Argyres, S. Dimopoulos and J. March-Russell, Phys. Lett. B **441**, 96 (1998); T. Banks and W. Fischler, arXiv:hep-th/9906038; R. Emparan, G. T. Horowitz and R. C. Myers, Phys. Rev. Lett. **85**, 499 (2000); S. B. Giddings and S. D. Thomas, physics," Phys. Rev. D **65**, 056010 (2002); S. Dimopoulos and G. L. Landsberg, Phys. Rev. Lett. **87**, 161602 (2001).
[4] K. i. Nakao, T. Harada and U. Miyamoto, Phys. Rev. D **82**, 121501(R) (2010).
[5] T. Harada and K. i. Nakao, Phys. Rev. D **70**, 041501 (2004).
[6] U. Miyamoto, H. Nemoto and M. Shimano, "Particle Creation by Naked Singularities in Higher Dimensions," arXiv:1012.2374 [gr-qc].

Spanish Relativity Meeting (ERE 2010): Gravity as a Crossroad in Physics IOP Publishing
Journal of Physics: Conference Series **314** (2011) 012124 doi:10.1088/1742-6596/314/1/012124

On the stability of black holes at the LHC

Marcos D Maia[1], **Edmundo M Monte**[2]

1 Universidade de Brasilia, Instituto de Fisica, Brasilia, DF, 70910-970, Brazil
2 Universidade Federal da Paraba, Departamento de Fisica, Joao Pessoa, PB, 58059-970, Brazil

E-mail: maia@unb.br

E-mail: edmundo@fisica.ufpb.br

Abstract. The eventual production of mini black holes by proton-proton collisions at the LHC is predicted by theories with large extra dimensions resolvable at the Tev scale of energies. It is expected that these black holes evaporate shortly after its production as a consequence of the Hawking radiation. We show that for theories based on the ADS/CFT correspondence, the produced black holes may have an unstable horizon, which grows proportionally to the square of the distance to the collision point.

1. Introduction

The production of mini black holes at the LHC Tev energy scale is predicted by theories with large extra dimensions and where the gravitational constant is replaced by a Tev scale fundamental constant. In those theories the gauge interactions are confined to four-dimensional space-times embedded in a bulk space defined by the Einstein-Hilbert principle, but gravitation as described by the four-dimensional embedded geometry, can access the extra dimensions at the TeV energy. Thus, the generated black hole will represent a local deformation of Minkowski's space-time resulting from the proton-proton collision. Typical examples are given by the Randall-Sundrum models of brane-world theory defined in the 5-dimensional anti deSitter AdS_5 bulk, using the Israel-Lanczos jump condition at the brane-world, acting as a boundary [1].

The produced black holes are expected to be short lived as consequence of Hawking's radiation. The well known theorem by Hawking of 1975, made use of semi-classical gravitation to prove that virtual particle pairs are formed in the vicinity of a black hole. While one of the particles of the pair falls inside the black hole horizon, the other escapes to infinity producing a thermal radiation of the black hole. Admitting that the radiation is complete the black hole will evaporate away, with the consequence is that the correlation between the spin states of the particles pair is lost, compromising the quantum unitarity [2]. In a recent (2005) revision of his theorem, Hawking made use of the ADS/CFT correspondence within the context of the Horava-Witten $AdS_5 \times S^5$ model of string theory. It was assumed that the black holes in question are charged (like for example, the Reissner-Nordstrom black hole), so that after the evaporation a stable submanifold would remain, in which case some of the information would be stored and could be partially retrieved [3].

In any of the above mentioned models using the same AdS_5 space, the stability of the black hole, seen as a subspace embedded in that bulk must be understood.

2. A black hole from an AdS_5 bulk

Consider the four-dimensional space-time AdS_4, regarded as a hypersurface with negative constant curvature embedded in the five dimensional flat space $M_5(3,2)$, with maximal isometry $SO(3,2)$ [4]. Adding to each of these spaces one extra spatial dimension, we obtain in a very trivial way that the five-dimensional anti de Sitter space AdS_5 is a hypersurface with negative constant curvature embedded in the flat space $M_6(4,2)$, with maximal isometry $SO(4,2)$, which is isomorphic to the conformal group C_0 acting on the Minkowski space-time, so that for each conformal covariant Yang-Mills field in M_4, there corresponds an isometric covariant Yang-Mills field defined in the AdS_5 space.

Since all Yang-Mills fields and their duality properties are consistently defined only in four-dimensional space-times, where 3-forms are isomorphic to 1-forms, the ADS/CFT correspondences must be hold between four-dimensional subspaces of the AdS_5 bulk. Consequently, the isometric invariant Yang-Mills fields must remain confined to a 4-dimensional subspace embedded in the AdS_5 bulk, where the required duality operation is preserved. Furthermore, admitting that the embedding of the 4-dimensional subspace is local and regular, then the inverse function theorem establishes a local 1:1 correspondence between the conformal fields in M_4 and the Yang-Mills fields defined in those four-dimensional subspaces.

The AdS_5 bulk is a solution of the 5-dimensional vacuum Einstein's equations with negative cosmological constant

$$^5\mathcal{R}_{AB} - \frac{1}{2}\,{}^5\mathcal{R}\mathcal{G}_{AB} - \Lambda\mathcal{G}_{AB} = 0, \quad A, B = 1...5 \tag{1}$$

and with constant curvature, characterized by the Riemann tensor

$$^5\mathcal{R}_{ABCD} = \frac{\Lambda}{6}(\mathcal{G}_{AC}\mathcal{G}_{BD} - \mathcal{G}_{AD}\mathcal{G}_{BC}) \tag{2}$$

The embedding is an application $X : V_4 \to AdS_5$, with components $X^A(x_\mu)$, functions of the space-time coordinates x_μ. Together with the unit normal vector η, this defines a Gaussian reference frame on the embedded geometry $\{X^A_{,\mu}, \eta^A\}$. The Riemann tensor of the bulk written in this reference gives the integrability equations for the embedding, the well known Gauss-Codazzi equations [4], which in the case of a constant curvature 5-dimensional bulk can be written as

$$R_{\alpha\beta\gamma\delta} = -(k_{\alpha\gamma}k_{\beta\delta} - k_{\alpha\delta}k_{\beta\gamma}) + \frac{\Lambda}{6}(g_{\alpha\gamma}g_{\beta\delta} - g_{\alpha\delta}g_{\beta\gamma}) \tag{3}$$

$$k_{\alpha[\beta;\gamma]} = 0, \tag{4}$$

The existence of the embedding requires that the metric $g_{\mu\nu}$ and the extrinsic curvature $k_{\mu\nu}$ of V_4 satisfy these equations. Writing (1) in the same Gaussian frame we obtain the gravitational equations of the embedded V_4 [5]

$$R_{\alpha\beta} - \frac{1}{2}Rg_{\alpha\beta} - \Lambda g_{\alpha\beta} + Q_{\alpha\beta} = 0 \tag{5}$$

$$k^\gamma_{\beta;\gamma} - h_{,\beta} = 0 \tag{6}$$

where $h = g^{\mu\nu}k_{\mu\nu}$ is the mean curvature, $K^2 = k_{\mu\nu}k^{\mu\nu}$ is the Gaussian curvature and

$$Q_{\alpha\beta} = (k^\rho{}_\alpha k_{\rho\beta} - hk_{\alpha\beta}) - \frac{1}{2}(K^2 - h^2)\; g_{\alpha\beta}, \quad Q^{\alpha\beta}{}_{;\beta} = 0$$

These equations describe the dynamics of the gravitational field for a space-time embedded in the AdS_5 bulk. Clearly they are more general than the vacuum Einstein's equations in general relativity, because (1) implies that the extrinsic curvature $k_{\mu\nu}$ is also a dynamical variable. When $k_{\mu\nu} = 0$ the usual vacuum Einstein's equations are recovered.

Now, we may discuss the production of black hole at the LHC with the supposition that it happens within the AdS_5 bulk. The experiment is supposed to take place in the Minkowski's space-time M_4 where the protons are defined. However, this M_4 is regarded as a subspace embedded in the AdS_5, so that equations (3) and (4) must apply. The general solution of these equation for a flat space gives $k_{\mu\nu} = \sqrt{\Lambda/6}\ \eta_{\mu\nu}$. This means that although M_4 is flat in the Riemann sense, as an embedded geometry it is also warped like a cylinder, a cone or any ruled surface, where the full translational group of the Poincar symmetry in principle would not apply. However, from (5), it follows that Λ is the cosmological constant, which is too small to mark a significant presence in the local gravitational field of the LHC experiment. Therefore, for any practical purposes we may assume that the experiment starts very approximately as planned, in Minkowski's space-time.

After the collision, the produced black-holes will represent a deformations of the original Minkowski's space-time, transforming into Schwarzschild or Reissner-Nordstrom subspaces embedded in the AdS_5 bulk. This black hole should remain for a very short period, before its eventual evaporation. For a spherically symmetric diagonal metric, the general solution of (4) is of the form $k_{\mu\nu} = \alpha(r)g_{\mu\nu}$, where $\alpha(r) = 1/y(r)$ and where $y(r)$ is the local center of curvature of the embedded geometry, which is a value of the extra coordinate satisfying the condition $\det(g_{\mu\nu} - y(r)k_{\mu\nu}) = 0$ [4]. Replacing this expression in (5), we obtain the Schwarzschild-anti-de Sitter solution

$$ds^2{=}(1{-}\frac{2m}{r}+(3\alpha(r)^2{-}\Lambda)r^2)^{-1}dr^2+r^2d\omega^2-(1{-}\ \frac{2m}{r}+(3\alpha(r)^2{-}\Lambda)r^2)dt^2 \tag{7}$$

Repeating the same for the Reissner-Nordstrom solution, we obtain the Reissner-Nordstrom-anti-de Sitter solution

$$ds^2{=}(1{-}\frac{2m}{r}+\frac{q}{r^2}+(3\alpha(r)^2{-}\Lambda)r^2)^{-1}dr^2+r^2d\omega^2-(1{-}\ \frac{2m}{r}+\frac{q}{r^2}+(3\alpha(r)^2{-}\Lambda)r^2)dt^2 \tag{8}$$

In both cases we obtain a black holes whose horizon grows indefinitely with r^2, thus creating an unstable situation unless we impose the additional condition that $3\alpha(r)^2$ is constant and equal to Λ. However, this implies that the black hole is composed of umbilicus, which is not the case. Even if we neglect the local influence of Λ as before the collision, we cannot neglect $\alpha(r)$ on account of the characteristics of a black hole geometry.

3. Conclusion

We conclude that the exterior gravitational field of a Black hole is not native of an AdS_5 bulk and that the black holes produced by proton-proton collision at the LHC may be unstable. Nonetheless, it is possible that in a higher dimensional bulk $D > 5$, the behavior of the black holes is stable. This follows from the well known example given by the 6-dimensional flat bulk $M_6(4,2)$, whose metric is also invariant under $SO(4,2)$, so that it has the same group of isometries of the AdS_5. Consequently, all arguments of the ADS/CFT correspondence which depend only on the Lie group properties, can be extended without loss of generality to that flat bulk. By the same argument used in the ADS/CFT correspondence, the quantum unitarity of the Yang-Mills fields is maintained in the six-dimensional flat bulk.

References

[1] Randall L and Sundrum R 1999 *Phys. Rev. Lett.* **83** 3370
[2] Hawking S 1975 *Comm. Math. Phys.* **43** 199
[3] Hawking S 2005 *Phys. Rev.* D **72** 084013
[4] Eisenhart L 1966 *Riemannian Geometry* (Princeton: N.J.)
[5] Maia M, Monte E, Maia J and Alacaniz J 2005 *Class. Quant. Grav.* **22** 1623

Journal of Physics: Conference Series **314** (2011) 012125 doi:10.1088/1742-6596/314/1/012125

Observations in a locally inhomogeneous Universe and Dark Energy State Equation

Illa R.L.[1] **and Jordi Cepa**[1,2]

[1] Departamento de Astrofísica, Universidad de La Laguna, C/ Astrofísico Francisco Sánchez, s/n, La Laguna, Tenerife, Spain

[2] Instituto de Astrofísica de Canarias, C/ Vía Láctea, s/n, La Laguna, Tenerife, Spain

E-mail: illa.rivero.losada@gmail.com

Abstract. Determining the redshift of astronomical sources is fundamental for cosmological studies, since it allows deriving distances to the observer. However, the distance-redshift relationship commonly used is obtained from the isotropic, homogeneous Friedmann-Lemaitre-Robertson-Walker (FLRW) metric. By fitting this distance-redshift relations to Supernovae Ia data, used by many authors as standard candles for distance measurements, it was shown that the expansion of the universe is accelerating, since the addittion of equations of state $p < \frac{-1}{3}\rho$ in the cosmological model was required. Whether this dark energy is a cosmological constant or a quintessence is still to be determined, and is the objective of some current and future dark energy studies.

However, the Universe is not homogeneuos. The gravitational lens effect changes the luminosity of distant sources, and it may also change significantly the angular diameter distance, thus modifying the luminosity distance, changing the distance-redshift relation with respect to that of an homogeneous universe. This effect can be considered an additional source of dispersion of the data, affecting the determination of the dark energy equation of state.

We are developing a test for studying departures from the FLRW distance-redshift relation, and its impact on the determination of the dark energy equation of state.

1. Introduction.

The Friedmann-Lemaitre-Robertson-Walker with Λ (ΛFLRW) Cosmology is the current cosmological paradigm. Within it, the goal of determining the large scale geometrical structure of the Universe requires estimating the Hubble parameter H_0, the mass density parameter Ω_m and the cosmological constant Ω_Λ.

However, ΛFLRW Cosmology is homogeneous an isotropic on all scales, so a fourth parameter is needed for taking into account the inhomogeneity of the real Universe. This inhomogeneity affects the model via gravitational lensing.

2. Angular Diameter And Distance Measurement [1]

In a FLRW Cosmology, expressions can be derived for the observed angular diameter Θ and the observed brightness of an object with a known absolute diameter and absolute brightness, as a function of the redshift z. The function $\Theta(z)$ has a remarkable feature, the presence of a minimum when z is approximately equal to 1/2. Moreover, the maximum of the function $f(z) = rH/c\Theta$ (being r the radius of the object) corresponds to a minimun of the angle Θ. Such behavior of Θ is caused by the curvature of space due to the matter filling the Universe, so in a homogeneous Universe, observations of $\Theta(z)$ can provide an independent determination of the average density of matter ρ_0.

The angular diameter Θ subtended by an object AB is determined by the two rays AO and BO emitted by the outer points of the object.

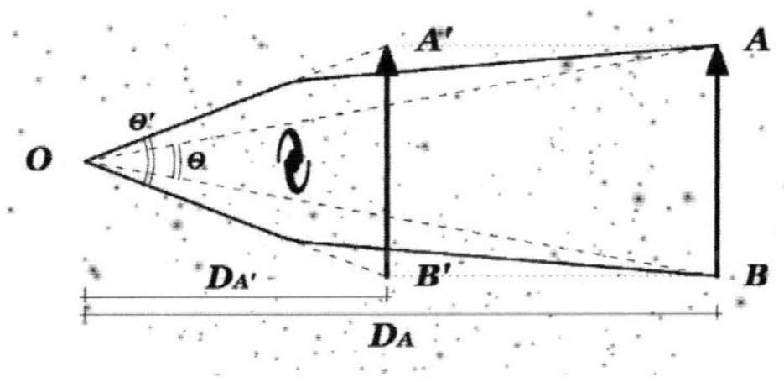

Figure 1. A mass situated between the rays bends the latter in such a way that Θ is increased. This type of bending is the physical reason for the features of the variation of Θ with z. The density of matter within the cone of light rays must be the same as the mean density of matter in the Universe; only in this case will the angle Θ be the same as that obtained in the uniform case.

Inhomogeneous Universe

For aproximating the effect of the non-uniformity of density in the determination of the function $\Theta(z)$ let's assume that we neglect the mass of the intergalactic medium by comparison with the mass of matter in galaxies. In that case, the angular diameter of objects with no galaxies within the cone subtende by them at the observer is $\Theta_1(z) \neq \Theta(z)$ and $\Theta_1(z)$ does not have a minimum.

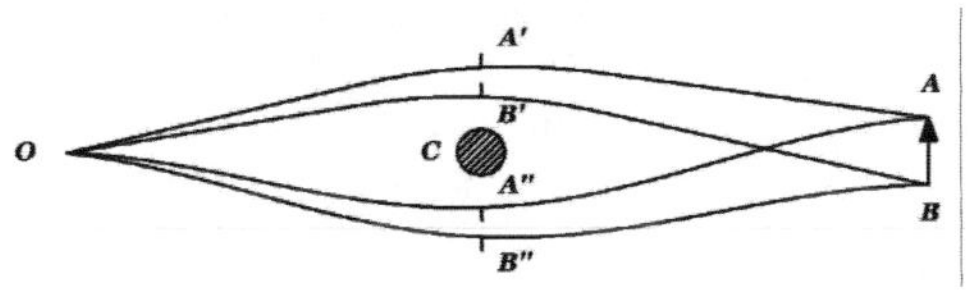

Figure 2. Also, for extended objects, sized AB and that have a galaxy C within their cone, the angular diameter is given by $\Theta_2(z)$ and a ring-shaped blurred image apppears, in the case of perfect aligment, or a more complex pattern if C does not lie on the axis of the cone.

In this case, it is not possible to derive any fixed expression for $\Theta_2(z)$, since Θ_2 depends on the mass of C and the relative distances between AB, O and C.

$\Theta(z)$ represents, in a nonhomoeneous Universe, a weighted mean of $\Theta_1(z)$ and all possible values of $\Theta_2(z)$. The blurring of extended sources in the non-homogeneous case makes it difficult to determine $\Theta(z)$

3. Filling parameter

Then, homogeneous matter inside an observing beam of light, gravitationally focuses the beam in a very different way than a set of clumps of equal total mass. The simplest correction for this gravity-light effect leads to the previously mentioned fourth parameter: ν [2],[3],[4]

$$\frac{\rho_I}{\rho_0} = \frac{\nu(\nu+1)}{6} \; ; \; 0 \leq \nu \leq 2 \tag{1}$$

($\nu = 0$ is for standard 100%-filled-beam FLRW case and $\nu = 2$ is for the empty-beam case).

The matter density of the Universe, ρ_0 is seen as a result of two contributions: homogeneous matter, ρ_H (i.e. intergalactic medium) and inhomogeneous matter (i.e. galactic contribution): $\rho_0 = \rho_H + \rho_I$

In the distance-redshift articles series by Kantowski et al. (2000, 2001, 2003) the procedure required to obtain the distance-redshift relationship for partially filled beam FLRW observations is described. Moreover, the simplified results for the special cases $\nu = 0, 1, 2$ are calculated. They also fit these new analytic Hubble curves to supernovae (SNe) data trying to determine both the mass parameter Ω_m and the beam-filling parameter ν, and the data available comes out to be inadequate for limiting ν.

In our work, we are tackling further studies for estimating realistic values of the filling parameter. In a realistic inhomogeneous Universe ν is a function of the direction and distance the observer is looking at. The further and more homogeneous region we look, the smaller is ν.

Let's assume the simplest case: a circular 2D non-uniform, non-isotropic field of point particles placed in an equispaced grid. We model non-uniformity by removing some of the particles, but maintaining uniformity away from the observer, as shown in the picture. We assume all particles have the same mass m =1 and $R_5 = 1$. In that case, an estimation of densities and filling parameters is straightforward. The total and homogeneous densities are: $\rho_0 = 55.39$ and $\rho_H = 61.43$.
The results obtained agree with the expectation. For the annulus, ν values range from $\nu(R_1) = 1.24$ to $\nu(R_5 - R_4) = 0.15$, approaching zero as space far from the center gains uniformity. For the α_1 case, ν values range from $\nu(R_1) = 1.50$ to $\nu(R_5 - R_0) = 0.70$, therefore non-uniformity is more evident. In the α_2 case, all the values are smaller than 1, since this region is highly uniform.

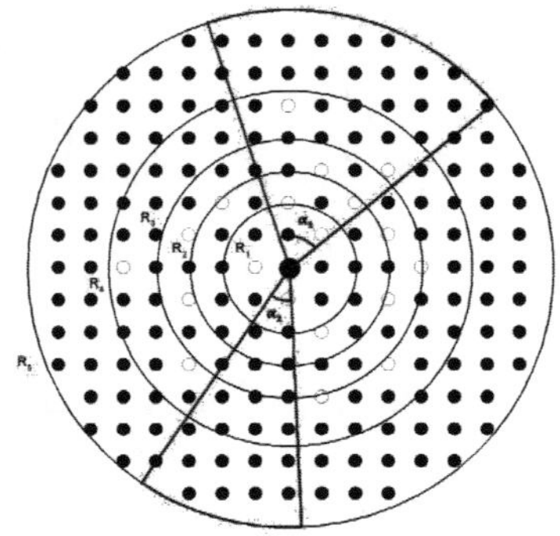

Figure 3. Simplest 2D inhomogeneous model.

In our estimations of ν for more realistic values, our first step is testing average inhomogeneity using the Millenium Simulation database [5] and the second step is using real catalogues, like the Supernova Cosmology Project [6].

One tricky point of the ν estimation is the direction-depending mass distribution calculation. Also, developing a numerical method for measuring distance-redshift relationship is required.

But on sufficiently large scales, the Univers is homogeneous and isotropic, so a estimation of the inhomogeneity length-scale is also important. Sarkar et al. [7] find this length-scale between 60 and 70 $h^{-1}Mpc$, in good agreement with earlier works.

4. Dark Energy equation of State

One of the big unknows of Cosmology is dark energy. Not only its nature is still to be established, but its equation of state is not fixed yet.

The dark energy equation of state relates density with pressure, through the ω parameter: $p = \omega\rho c^2$. A useful method of limiting the ω parameter is fitting observational data of good standard candles to distance-redshift relations. The main problem of this method are uncertainties, given the fact that constraining the dark energy equation of state requires an accurate knowledge of H_0, Ω_{m0}, the curvature and the distances. So, we are going to estimate uncertainties due to real inhomogeneity, compare them with already known uncertainties, and quantify whether dark energy equation of state could be measured with future data.

5. References

[1] Zel'Dovich Y B 1964 *Soviet Astronomy* **8** 13–+
[2] Kantowski R, Kao J K and Thomas R C 2000 *Astrophysical Journal* **545** 549–560
[3] Kantowski R and Thomas R C 2001 *Astrophysical Journal* **561** 491–495 (*Preprint* `arXiv:astro-ph/0011176`)
[4] Kantowski R 2003 *Phys. Rev. D* **68** 123516
[5] Millenium Millenium simulation database `http://www.mpa-garching.mpg.de/millennium/`
[6] Project S C Supernova cosmology project `http://www-supernova.lbl.gov/`
[7] Sarkar P, Yadav J, Pandey B and Bharadwaj S 2009 *Monthly Notices of the Royal Astronomical Society* **399** L128–L131 (*Preprint* `0906.3431`)

Spanish Relativity Meeting (ERE 2010): Gravity as a Crossroad in Physics IOP Publishing
Journal of Physics: Conference Series **314** (2011) 012126 doi:10.1088/1742-6596/314/1/012126

Slowly Rotating Relativistic Stars in TeVeS

Hajime Sotani
Division of Theoretical Astronomy, National Astronomical Observatory of Japan, 2-21-1 Osawa, Mitaka, Tokyo 181-8588, Japan

E-mail: sotani@astro.auth.gr

Abstract. In order to examine the rotational effect around neutron star in tensor-vector-scalar (TeVeS) theory, we consider the slowly rotating relativistic stars with a uniform angular velocity. As a result, we find that similar to the case in general relativity (GR), the angular momentum is proportional to the angular velocity and one could possibly distinguish the gravitational theory in the strong field regime even in the case that the value of the coupling constant K is quite small.

1. Introduction

In the weak gravitational field such as the one in the solar system, there are many experiments and the validity of general relativity (GR) has been shown, while the gravitational theory in the strong field regime is still unconstrained by the observations. However, with the development of technology it will be possible to observe compact objects with high accuracy and those observations could be used as a direct test of the gravitational theory in strong field regime (e.g., [1, 2, 3]). As an alternative gravitational theory, the tensor-vector-scalar (TeVeS) theory has attracted considerable attention, proposed originally by Bekenstein [4] as a covariant theory for modified Newtonian dynamics [5]. In the strong gravitational region of TeVeS, the Schwarzschild solution was found in [6], the Reissner-Nordström solution was found in [7], and the static, spherically symmetric neutron star models were constructed [8]. Additionally, there are some suggestions to distinguish TeVeS from GR by using some observations, i.e., with the redshift of the atomic spectral lines radiated from the surface of neutron star [8], with the Shapiro delays of gravitational waves and photons or neutrinos [9], and with the emitted gravitational waves from neutron stars [10, 11]. In this article, we examine a different way to distinguish TeVeS from GR, i.e., the rotational effect around neutron stars (see [12] for details). For this purpose, we consider slowly rotating neutron stars with a uniform angular velocity.

2. Slowly Rotating Relativistic Stars in TeVeS

We consider slowly rotating relativistic stellar models with a uniform angular velocity $\tilde{\Omega}$, where we assume to keep only the linear effects in the angular velocity. In this case, the star is still spherical, which is constructed according to the recipe derived in [8], because the deformation due to the rotation is of the order $\tilde{\Omega}^2$, and the metric in physical frame is given by

$$d\tilde{s}^2 = -e^{\nu+2\varphi}dt^2 + e^{\zeta-2\varphi}dr^2 + r^2e^{-2\varphi}\left(d\theta^2 + \sin^2\theta d\phi^2\right) - 2\omega r^2 e^{-2\varphi}\sin^2\theta dt d\phi, \qquad (1)$$

where the last term in the right hand side corresponds to the rotational effect and in general $\omega \sim \mathcal{O}(\tilde{\Omega})$ can be expressed as $\omega(r,\theta) = -\omega(r)\partial_\theta P_\ell/\sin\theta$, where $P_\ell = P_\ell(\cos\theta)$ is the Legendre

Published under licence by IOP Publishing Ltd

polynomial of order ℓ. While, the rotational effects in the other components of metric become of the order $\tilde{\Omega}^2$, because those should behave in the same way under a reversal in the direction of rotation as under a reversal in the direction of time. Similarly, the deviations of pressure, density, and scalar field due to the rotation should be of the order $\tilde{\Omega}^2$. Hereafter, in order to distinguish the effects due to the rotation, the deviation from the background properties would be expressed by using variables with δ. Then the fluid velocity in physical frame of the order $\tilde{\Omega}$ can be described as $\delta\tilde{u}^\mu = (0, 0, 0, \tilde{\Omega}\tilde{u}^t)$, where $\tilde{u}^t$ is t-component of four velocity in the case without rotation, i.e., $\tilde{u}^t = e^{-\varphi-\nu/2}$, and non-zero component of the energy-momentum tensor of the order $\tilde{\Omega}$ is only $\delta\tilde{T}_{t\phi}$. By using the normalization condition for the vector field, one can show that $\delta\mathcal{U}^t = \delta\mathcal{U}_t = 0$. So the deviation of vector field from the spherical symmetry can be described as

$$\delta\mathcal{U}^\mu = \left(0, W(r)P_\ell, V(r)\partial_\theta P_\ell, \frac{\mathcal{V}(r)}{\sin\theta}\partial_\theta P_\ell\right). \tag{2}$$

However, the variables of W and V should be of the order $\tilde{\Omega}^2$ because these components are corresponding to the polar parity. The equations expressing the relation between ω and $\mathcal{V}$ can be obtained from the ϕ component of the vector field equation and the (t, ϕ) component of the Einstein equation. Especially, in this article we adopt the case for $\ell = 1$ to reproduce the GR limit of TeVeS. In fact, the other value of ℓ can not be chosen to satisfy the regularity condition at the stellar center and the asymptotic behavior far from the star. Then, one can obtain the two second order differential equations related to ω and $\mathcal{V}$. Integrating these differential equations with the appropriate boundary conditions, the distributions of $\omega(r)$ and $\mathcal{V}(r)$ can be determined.

In order to determine the distributions of $\omega(r)$ and $\mathcal{V}(r)$ with the fixed value of K for the adopted stellar model, we impose two boundary conditions, i.e., the regularity condition at the stellar center and the asymptotic flatness far from the star. In practice, at stellar center we set that $\omega(0) = \omega_0$, $\omega'(0) = 0$, $\mathcal{V}(0) = \mathcal{V}_0$, and $\mathcal{V}'(0) = 0$, where ω_0 and $\mathcal{V}_0$ are some constants. Then we find the correct values of ω_0 and $\mathcal{V}_0$ in such a way that the solutions of $\omega(r)$ and $\mathcal{V}(r)$ should satisfy the asymptotic behavior as mentioned in the previous section, i.e., with the trial values of ω_0 and $\mathcal{V}_0$ the above ordinary differential equations can be integrated outward and we search the correct values by changing the trial values iteratively, where for integration we use the 4th order Runge-Kutta method. In this way, we can determine numerically the distributions of $\omega(r)$ and $\mathcal{V}(r)$ and those behaviors far from the central star can be described as

$$\omega(r) = 2J/r^3 + \mathcal{O}(1/r^4), \quad \text{and} \quad \mathcal{V}(r) = 2\mathcal{V}_c/r^3 + \mathcal{O}(1/r^4). \tag{3}$$

It should be noticed that in GR the distribution of $\omega(r)$ outside the star can be described analytically as $\omega(r) = 2J/r^3$, where the constant J is corresponding to the total angular momentum of the star. While, in the case of TeVeS, we can find the same feature as the case of GR, i.e., the values of $J/\tilde{\Omega}$ and $\mathcal{V}_c/\tilde{\Omega}$ are independent of the value of angular velocity $\tilde{\Omega}$ if only one chooses the non-rotating stellar models.

The left panel of figure 1 shows the distributions of $\omega(r)$ with different values of K for the stellar model with $M_{\rm ADM} = 1.4M_\odot$ and $\tilde{\Omega} = 1$ kHz. At a glance, we can recognize the difference of distributions between in GR and in TeVeS. Especially, with higher value of K those differences become obvious even in the interior of the star. On the other hand, the right panel of figure 1 shows the distributions of $\mathcal{V}(r)$ with different values of K. From this, we can find the important point that as the value of K becomes smaller, the distribution of $\omega(r)$ in TeVeS approaches to that in GR while the induced vector field $\mathcal{V}(r)$ does not vanish. Namely, with smaller value of K the physical metric in TeVeS is almost same as that in GR, but there still exists the additional vector fields induced by the stellar rotation. So if one will consider the stellar oscillations and/or the emitted gravitational waves in TeVeS, this induced vector field could play a role as source term in the linearized equations, and as a result the deviation in frequencies depending on the

Journal of Physics: Conference Series **314** (2011) 012126 doi:10.1088/1742-6596/314/1/012126

gravitational theory can be seen even if the physical metric is not so different from each other. Thus such kinds of observation could tell us the gravitational theory in the strong field regime even if the value of K is quite small.

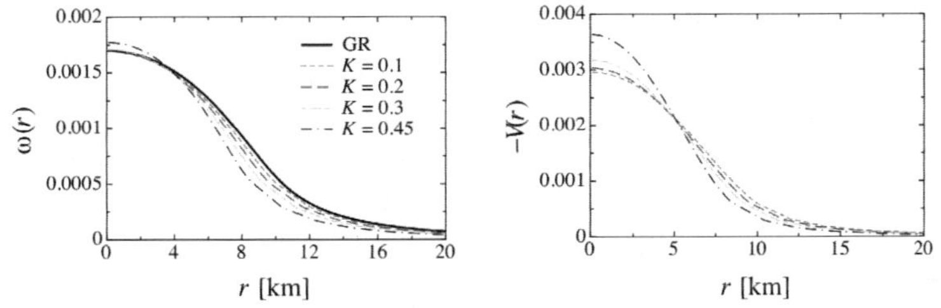

Figure 1. Distributions of $\omega(r)$ and $\mathcal{V}(r)$ with different values of K. In order to compare the results, the solid line shows the distribution in GR.

3. Conclusion

In this article, in order to examine the rotational effect around the neutron star in the tensor-vector-scalar (TeVeS) theory, we consider the slowly rotating relativistic stars with a uniform angular velocity $\tilde{\Omega}$. To deal with this problem in TeVeS, one has to take into account not only the usual frame dragging ω but also the induced vector field $\mathcal{V}$, which is corresponding to the ϕ component. The equations for $\omega(r)$ and $\mathcal{V}(r)$ are derived from the Einstein and vector field equations, and then the distributions of those variables are determined numerically with appropriate boundary conditions.

As a result, we find that, similar to the case in GR, the value of $J/\tilde{\Omega}$ in TeVeS is constant if the stellar mass and value of K are fixed, where J corresponds to the angular momentum. Additionally, with higher value of K, the distribution of $\omega(r)$ deviates obviously from that in GR due to the existence of induced vector field and this deviation can be seen even in the interior region of star. On the other hand with smaller value of K, although ω approaches to that in GR, the induced vector field does not vanish. That is, even if the stellar properties in TeVeS with small value of K, such as the mass and radius, are almost same as those in GR, the induced vector field in TeVeS could still exist. This is a crucial difference depending on the gravitational theory. Thus through the observable phenomena, such as the stellar oscillation and emitted gravitational waves, it is possible to distinguish the gravitational theory in the strong field regime. At the end, recently the toroidal oscillations in the slowly rotating star in TeVeS are calculated. This result also suggests how to distinguish the gravitational theory [13].

References

[1] Psaltis D 2009 *Living Rev. Relativity* **11** lrr-2009-9
[2] Sotani H and Kokkotas K D 2004 *Phys. Rev.* D **70** 084026
[3] Sotani H and Kokkotas K D 2005 *Phys. Rev.* D **71** 124038
[4] Bekenstein J D 2004 *Phys. Rev.* D **70** 083509
[5] Milgrom M 1983 *Astrophys. J.* **270** 365
[6] Giannios D 2005 *Phys. Rev.* D **71** 103511
[7] Sagi E and Bekenstein J D 2008 *Phys. Rev. D* **77** 024010
[8] Lasky P D, Sotani H and Giannios D 2008 *Phys. Rev.* D **78** 104019
[9] Desai S, Kahya E O and Woodard R P 2008 *Phys. Rev.* D **77** 124041
[10] Sotani H 2009a *Phys. Rev.* D **79** 064033
[11] Sotani H 2009b *Phys. Rev.* D **80** 064035
[12] Sotani H 2010a *Phys. Rev.* D **81** 084006
[13] Sotani H 2010b *Preprint* arXiv:1012.2143 [astro-ph,HE]

Spanish Relativity Meeting (ERE 2010): Gravity as a Crossroad in Physics IOP Publishing
Journal of Physics: Conference Series **314** (2011) 012127 doi:10.1088/1742-6596/314/1/012127

An approach to evaporating stars and black holes

F. Fayos and R. Torres
Department of Applied Physics, UPC, Barcelona, Spain.
E-mail: ramon.torres-herrera@upc.edu

Abstract. Many models in which the object under study loses all its mass have appeared in the literature. No doubt, the most eminent examples are the models for evaporating black holes and for evaporating stars. Here we describe a semiclassical study of these evaporating models centered on the evaporating event itself. We pay special attention to the evaporating models as a means of avoiding singularities during the collapse. In case of any pre-existing non-spacelike curvature singularity, we show that these models tend to evaporate it.

1. Introduction

The possibility that an object could radiate away its entire mass was, apparently, first considered by H. Bondi [1] as a means of avoiding the formation of singularities during the collapse of a star. The first model for an evaporating star appeared in an article by Demianski and Lasota [2]. However, it was shown later that this model in fact generates an instantaneous naked curvature singularity at the evaporating event. In this way, it became clear that the evaporating models could also be used to produce counterexamples for the Penrose's *cosmic censorship conjecture.* While many counterexamples of this type appeared later (see, for example, [3] and references therein) the original idea of avoiding the formation of singularities using evaporating models was further pursued by some authors whether in the form of evaporating stars (see [4] and references therein) or in the form of *incipient black holes* (see [5] and references therein). Of course, we cannot forget that at the same time another type of evaporating object, proposed by S. W. Hawking, was being developed: The evaporating black hole.

Our goal here is to study the evaporating models as a whole. In order to get these models we will match a general interior solution with the exterior radiating Vaidya's solution, in such a way that mass will be lost in the form of outgoing incoherent null radiation.

2. The exterior an interior solutions

Locally, Vaidya's metric [6] can be described in radiative coordinates as

$$ds_V^2 = -\bar{\chi} d\bar{u}^2 - 2\, d\bar{u}\, d\bar{R} + \bar{R}^2 \left(d\bar{\theta}^2 + \sin^2\bar{\theta} d\bar{\phi}^2\right) \tag{1}$$

where $\bar{\chi} = 1 - 2\frac{\bar{m}(\bar{u})}{\bar{R}}$ and the *mass function* $\bar{m}$ [7] depends only on $\bar{u}$.

The energy-momentum tensor for the metric (1) is of pure radiation type

$$\bar{T}_{\mu\nu} = -\frac{2}{\bar{R}^2}\frac{d\bar{m}(\bar{u})}{d\bar{u}} l_\mu l_\nu, \qquad l_\mu dx^\mu = -d\bar{u} \qquad l_\mu l^\mu = 0\,. \tag{2}$$

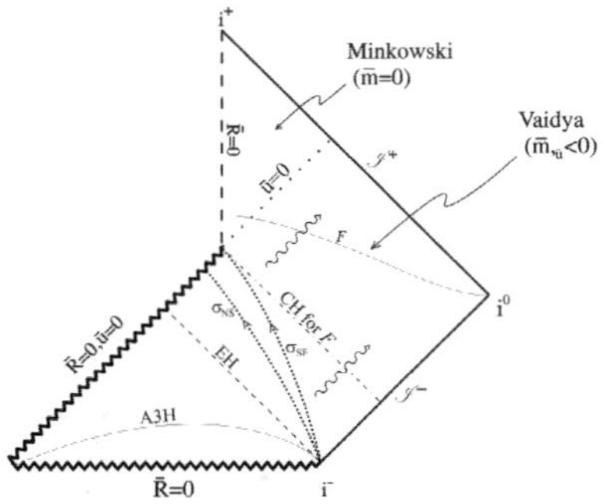

Figure 1. The Penrose diagram for Vaidya's spacetime when $\bar{m}(\bar{u})$ vanishes from some instant $\bar{u} = 0$ on and $-1/16 \leq \Upsilon \leq 0$. Note that a null singularity $\bar{u} = \bar{R} = 0$ appears.

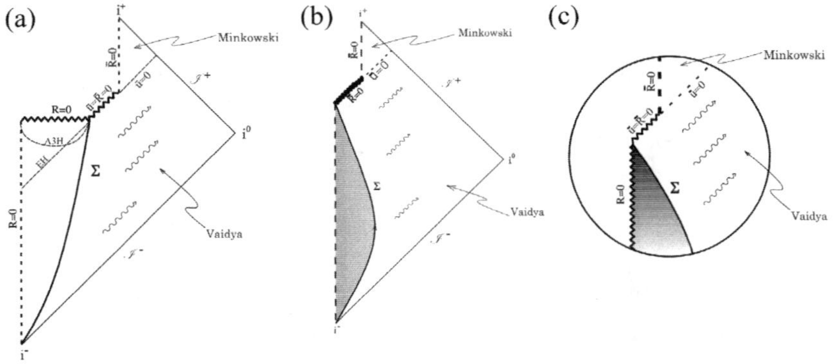

Figure 2. Forbidden possibilities

We arbitrarily set the evaporation event at $\bar{u} = 0$: $\bar{m}(\bar{u} = 0) = 0$. For $\bar{u} \geq 0$ we extend Vaidya's metric with Minkowski's spacetime, i.e. $\bar{m}(\bar{u} \geq 0) = 0$. (See [4]).

It can be shown [3] that the exterior of an evaporating object (see figure 1) must satisfy:

$$-\frac{1}{16} \leq \Upsilon \equiv \lim_{\bar{u}\to 0^-} \frac{\bar{m}(\bar{u})}{\bar{u}} \leq 0. \qquad (3)$$

For the interior we will use a general four-dimensional spherically symmetric space-time $\mathcal{V}$, so that its line-element can be expressed in radiative coordinates $\{x^\mu\} = \{u, R, \theta, \varphi\}$ ($\mu = 0, 1, 2, 3$) as

$$ds^2 = -e^{4\beta}\chi du^2 + 2\varepsilon e^{2\beta}\, du\, dR + R^2\, d\Omega^2, \qquad (4)$$

where $\chi \equiv 1 - 2m/R$, $\varepsilon^2 = 1$, β and m depend on $\{u, R\}$, and $d\Omega^2 \equiv d\theta^2 + \sin^2\theta d\varphi^2$. It is easily checked that $m(u, R)$ is the *mass function* [7].

Our local study is carried out in $\mathcal{U} \equiv \{(u, R)|\ \delta u \leq u \leq 0,\ 0 \leq R \leq \delta R\}$ [3]. According to the values of $m(u, R)$ and $\beta(u, R)$ and their partial derivatives computed at $u = R = 0$ we can classify $\mathcal{U}$ with a fixed causal characterization of $R = 0$ in three groups that have to be studied separately [3]: regular $\mathcal{U}$s, $\mathcal{U}$s with a non-spacelike singularity and $\mathcal{U}$s with a spacelike singularity.

3. Constructing specific evaporating models

If we write down the matching conditions for the interior and the exterior spacetimes and impose that the matched model should collapse and evaporate, we obtain the following results [3]:

Proposition 3.1 *A collapsing star endowed with an interior C^2 mass function cannot develop an exterior light-like singularity at the total evaporation event. (See figure 2).*

Journal of Physics: Conference Series **314** (2011) 012127 doi:10.1088/1742-6596/314/1/012127

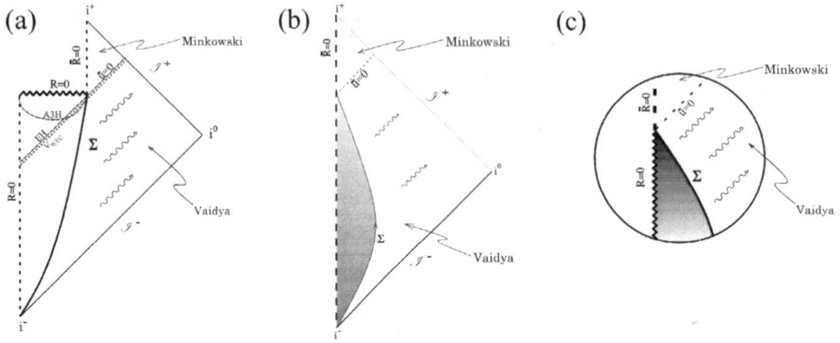

Figure 3. Different allowed evaporating models.

Proposition 3.2 *All conditions for an evaporating collapsing model with a regular interior or with an interior possessing a non-spacelike singularity will be fulfilled if, and only if, the interior satisfies* $m(u<0,R) \overset{\Sigma_{SF}}{>} 0$ *in* $\mathcal{U}-\{u=0,R=0\}$ *and*

$$1 \leq \lim_{\lambda\to 0^-} \frac{m_{,R}}{R\beta_{,R}} < 2 \qquad or \qquad \lim_{\lambda\to 0^-} \frac{2R\beta_{,R}-m_{,R}+2m_{,u}}{2(R\beta_{,R}-m_{,R})} < 0, \tag{5}$$

for $\varepsilon=-1,+1$*, respectively. (See figures 3b and 3c).*

4. Discussion and conclusions

Our aim was to study the evaporating models as a whole. Since the physics of some evaporating models at late stages can be fairly unknown, we did not restrict the energy-momentum tensor in the interior of the model allowing classical or quantum fields.

We have shown (proposition 3.2) that a collapsing stellar model with a regular interior will evaporate if, and only if, the mass function on the matching hypersurface is non-negative and condition (5) is satisfied. In this way, the complete matched model must be singularity free. (See figure 3b). The reader can find a complete family of such models in our article [4]. Next, we have extended our study to stellar models in which an inner non-spacelike singularity (and, therefore, naked) has already formed during the collapse. In this case, we have shown that the evaporation process tends to eliminate any preexisting naked singularity. Figure 3c illustrates this situation around the evaporation event. Finally, we have treated the case in which an inner spacelike singularity preexists. These models violate the weak energy conditions and they all posses an instantaneous naked singularity that does not extend to the exterior of the collapsing object. We have illustrated this type of evaporating black holes in figure 3a.

Acknowledgements

We would like to acknowledge the *Ministerio de Educación y Ciencia* (FIS2007-63034) and the *Generalitat de Catalunya* (grant 2009SGR-00417) for financial support.

References

[1] Bondi H 1964 *Proc. R. Soc. London A* **281** 39
[2] Demianski L and Lasota J P 1968 *Astrop. Lett.* **1** 205
[3] Fayos F and Torres R 2010 *Class. Quantum Grav.* **27** 125011
[4] Fayos F and Torres R 2008 *Class. Quantum Grav.* **25** 175009
[5] Barceló C, Liberati S, Sonego S and Visser M 2008 *Phys. Rev. D* **77** 044032
[6] Vaidya P V 1951 *Proc. Indian Acad. Sci.* **A33** 264
[7] Hernández W C and Misner C W 1966 *Astrophys. J.* **143** 452

AUTHOR INDEX

AUTHOR INDEX

AUTHOR INDEX